INDUSTRIAL AND ENGINEERING APPLICATIONS OF ARTIFICIAL INTELLIGENCE AND EXPERT SYSTEMS

INDUSTRIAL AND ENGINEERING APPLICATIONS OF ARTIFICIAL INTELLIGENCE AND EXPERT SYSTEMS

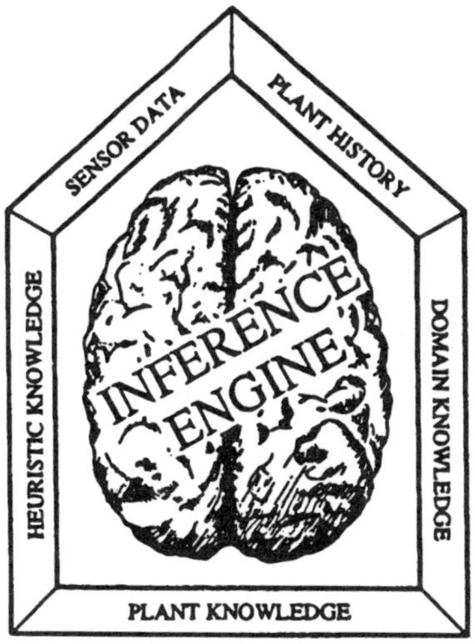

IEA/AIE

95

Proceedings of the Eighth International Conference
Melbourne, Australia, June 6-8, 1995

Edited by

Graham F. Forsyth
Defence Science and Technology Organisation
Melbourne, Australia

Moonis Ali
Southwest Texas State University
San Marcos, Texas, USA

GORDON AND BREACH PUBLISHERS
Australia • Austria • China • France • Germany • India • Japan • Luxembourg • Malaysia •
Netherlands • Russia • Singapore • Switzerland • Thailand • United Kingdom • United States

3 Boulevard Royal
L-2449 Luxembourg

British Library Cataloguing in Publication Data

Industrial Engineering Applications
of Artificial Intelligence and Expert
Systems:Proceedings of the Eighth
International Conference, Melbourne,
Australia, 6-8 June 1995
I. Forsyth, Graham F. II. Ali, Moonis
620.0028563

ISBN 2-88449-198-8

CONTENTS

PREFACE

The editors present the Proceedings of the Eighth International Conference on Industrial and Engineering Applications of Artificial Intelligence and Expert Systems (IEA/AIE - 95) with a great deal of pride and pleasure. This international forum of AI researchers and practitioners attempts to maintain a balance between theory and application. Participants share their experience, expertise, and skills in developing and enhancing intelligent systems technology.

The papers in this volume present a broad spectrum of new ideas, issues, and practicum addressing the needs of real-life applied systems. More than 160 papers from 30 countries were reviewed and just over 100 of these were selected to appear in our conference and this proceedings.

We would like to thank the local Organizing Committee; the collaborating organizations (see full list on page xvii); Cheryl Morriss, who handled the ISAI side of things from Texas; and the conference organizer, AE Conventions. The Organizing Committee has changed personnel over the four years since the decision to bid for the first Southern Hemisphere conference in this series, and we are grateful to John Baxter, Lionel Singer, Mike Croft, Tharam Dillon, Mike Larkin, and Helmi Salem who helped us "get the show off the ground."

The first conference in the Southern Hemisphere will be followed next year by the first Asian conference in the series; 1996 will see us in Fukuoka, Japan.

Finally, we would like to thank all the members of the Program Committee and external referees listed on pages xv and xvi. Without them, no papers would be received and no conference possible.

Graham F. Forsyth, *Program Chair*
Moonis Ali, *General Chair*

COMMITTEE MEMBERS

Organizing Committee

Moonis Ali, Southwest Texas State University, USA	*General Chair*
Graham F. Forsyth, DSTO, Melbourne	*Program Chair*
Elizabeth Hutchinson, AE Conventions	*Organizer*
Grahame Smith, CRC for Intelligent Decision Systems	*Local Arrangements*
Elizabeth Sonenberg, The University of Melbourne	*Local Arrangements*
Rajiv Khosla, LaTrobe University	*Local Arrangements*
Simon Goss, DSTO, Melbourne	*Publicity Chair*
Lyndal Higgins, AE Conventions	*Publicity Co-ordinator*
Jose Alonso, Swinburne University of Technology	*Tutorials Chair*
Alva Purkis, Institution of Engineers, Australia	*Liaison–IEAust*
Bernie Green, Knowledge-Based Systems	*Sponsors and Exhibition Chair*
Meg Brown, AE Conventions	*Sponsorship and Exhibition Co-ordinator*
Chris Marland, Institution of Engineers, Australia	*Proceedings Co-ordinator*
Cheryl Morriss, Southwest Texas State University, USA	*Liaison–ISAI*

Program Committee

Suhayya Abu-Hakima
National Research Council
Canada

Juan J. Alba
Univ. P. Comillas, Spain

Frank Anger
Univ. of West Florida, USA

Paul W.H. Chung
Loughborough Univ. of
Tech., UK

Wai-Keong Foong
Ngee Ann Polytechnic
Singapore

Teruo Fukumura
Chukyo Univ., Japan

David Harris
Univ. of South Australia

Robert Inder
Univ. of Edinburgh, UK

Akio Ito
Communications Research
Lab., Japan

Andrew Jennings
RMIT Univ. of Tech.

Julia Johnson
Univ. of Regina, Canada

K. Kanchanasut
Asian Inst. of Tech., Thailand

Jin H. Kim
KAIST, Korea

Kwong Sak Leung
Chinese Univ. of Hong Kong

Rasiah Loganantharaj
USL, USA

Gillian Lovegrove
Staffordshire Univ., UK

Fumihiro Maruyama
Fujitsu Labs Ltd., Japan

Manton Matthews
Univ. of South Carolina, USA

Laszlo Monostori
Hungarian Acad. of Sci.

Hiroshi Motoda
Hitachi Ltd., Japan

Akitoshi Okumura
IT Labs, NEC, Japan

Paul O'Rorke
Univ. of California, Irvine
USA

Don Potter
Univ. of Georgia, USA

Anil Rewari
Digital Equipment Corp
USA

Rita Rodriguez
Univ. of West Florida
USA

Paul Schutte
NASA Langley Research
Center, USA

Daniel Serain
Digital Equipment Corp
France

Peter Sydenham
Univ. of South Australia

Takushi Tanaka
Fukuoka Inst. of Tech., Japan

Satoshi Tojo
Mitsubishi Research Inst., Japan

Wilson Wen
Telecom Research Labs

Xindong Wu
James Cook Univ.

Xin Yao
Univ. College, ADFA

Wai-Kiang Yeap
Univ. of Otago, New Zealand

All papers received were reviewed by the Program Committee or a panel of external reviewers. In many cases, these external reviews were de-facto members of the Program Committee whose acceptance was not received in time to be placed on the Call for Papers. In other cases, they were last-minute recruits by Program Committee members to assist with the review process. All are thanked sincerely for their efforts.

External Referees

Kai-Hsiung Chang
Auburn Univ., USA

Sung-Bae Cho
Yonsei Univ., Korea

Wei Dai
Telecom Research Labs

H. Ferra
Telecom Australia

Hirotaka Hara
Fujitsu Labs Ltd., Japan

Wong Man Hun
Chinese Univ. of Hong Kong

Hideaki Ito
Chukyo Univ., Japan

Jiansheng Jian
Telecom Research Labs

N. Kasabov
Univ. of Otago, New Zealand

Akira Kawamura
Fujitsu Labs Ltd., Japan

D. Kearney
Univ. of South Australia

Rhee M. Kil
KAIST, Korea

Kawon Kim
KAIST, Korea

Seon Man Kim
KAIST, Korea

Geunbae Lee
Postech, Korea

Seong-Whan Lee
Chungbuk National Univ.

Jongho Nang
Sogang Univ.

Yuiko Ohta
Fujitsu Labs Ltd., Japan

Seishi Okamoto
Fujitsu Labs Ltd., Japan

Cheol Hoon Park
KAIST, Korea

Dong-Jo Park
KAIST, Korea

Adrian Pearce
Curtin Univ.

Miguel A. Sanz-Bobi
Univ. P. Comillas

Ken Satoh
Fujitsu Labs Ltd., Japan

Abdul Sattar
Griffith Univ./TIFR

Peter Sember
Telecom Research Labs

Sandip Sen
Univ. of Tulsa, USA

Ron Sharpe
CSIRO DBCE

Smenby
Univ. of Melbourne

James Soutter
Loughborough Univ., UK

Swee Hor Teh
Utah Valley State College
USA

Man Leung Wong
Chinese Univ. of Hong Kong

Yeung Yam
Chinese Univ. of Hong Kong

Hyunsoo Yoon
CS Dept, KAIST, Korea

Ming Zhao
Telecom Research Labs

Sponsored by

- **INSTITUTION OF ENGINEERS, AUSTRALIA**

Hosted by

- **International Society of Applied Intelligence**

Organized in cooperation with

- American Association for Artificial Intelligence
- Association for Computing Machinery/SIGART
- Canadian Society for Computational Studies of Intelligence
- European Coordinating Committee for Artificial Intelligence
- IEEE Computer Society
- Institute of Measurement and Control
- Institution of Electrical Engineers
- International Neural Network Society
- Japanese Society of Artificial Intelligence
- Australian Computer Society

With support from

- Australian Department of Industry, Science and Technology
- Victorian Department of Business and Employment
- QANTAS, The Australian Airline
- Defence Science and Technology Organisation, Australia

Active Agents

A LINGUISTIC GEOMETRY FOR MULTIAGENT SYSTEMS

Boris Stilman
Department of Computer Science & Engineering
University of Colorado at Denver, Campus Box 109, Denver, CO 80217-3364.
Email: bstilman@cse.cudenver.edu

Abstract: This paper is a new step in the development of the Linguistic Geometry. Linguistic Geometry relies on the formalization of search heuristics of the highly-skilled human experts, which allow for the decomposition of a complex system into a dynamic hierarchy of subsystems, and thus solve intractable problems by reducing the search dramatically. In this paper we report application of the Linguistic Geometry tools to a new example of a solution of simplified 2D optimization problem for the autonomous robotic vehicles in aerospace environment. The novelty of this example with respect to other Linguistic Geometry applications is that in this multiagent problem some agents can move simultaneously.

1. INTRODUCTION

Air Force systems operate in the atmosphere, ionosphere, near-Earth space and through these regions. They include aircraft combat missions, tracking and possible interception of missiles and satellites, surveillance operations, etc. A great number of aerospace problems such as long and short-range mission planning, especially for autonomous navigation, scheduling, aerospace robot control, long-range satellite service, aerospace combat operations control, etc. can be formally represented as reasoning about complex large-scale control systems. The field of efficient aerospace control systems design needs new technology from the science of artificial intelligence (Rodin, 1988; Rodin et al., 1993; Lirov, Rodin et al., 1988). In particular, discrete-event modeling of complex control systems can be implemented as a purely interrogative simulation. These techniques can be based on generating geometrically meaningful states rather than time increments with due respect to the timeliness of actions. As stated in (Lirov, Rodin et al., 1988) this interrogative approach offers much faster execution and clearer simulator definition. For this kind of approach a series of hierarchical dynamic goal-oriented systems should be developed and investigated.

There are many such problems where human expert skills in reasoning about complex systems are incomparably higher than the level of modern computing systems. At the same time there are even more areas, especially in the aerospace problem domain, where advances are required but human problem-solving skills can not be directly applied. For example, there are problems of tactics planning and automatic control of autonomous agents such as aerospace vehicles, space stations and robots with cooperative and opposing interests functioning in a complex, hazardous environment. Reasoning about such complex systems should be done automatically, in a timely manner, and often in a real time. Moreover, there are no highly-skilled human experts in these fields ready to substitute for robots (on a virtual model) or transfer their knowledge to them. There is no grand-master in robot control, although, of course, the knowledge of existing experts in this field should not be neglected – it is even more valuable. Due to the special significance of these problems and the fabulous costs of mistakes, the quality of solutions must be very high and usually subject to continuous improvement.

2. THEORETICAL BACKGROUND

The difficulties we encounter trying to find the optimal operation for real-world complex control systems are well known. While the formalization of the problem, as a rule, is not difficult, an algorithm that finds its solution usually results in the search of many variations. For small-dimensional "toy" problems a solution can be obtained; however, for most real-world problems the dimension increases and the number of variations increases significantly, usually exponentially, as a function of dimension (Garey and Johnson, 1991). Thus, most real-world search problems are not solvable with the help of exact algorithms in a reasonable amount of time. This becomes increasingly critical for the real-time aerospace autonomous and semiautonomous vehicles and robots (Lirov et al., 1988; Strosnider and Paul, 1994).

There have been many attempts to find the optimal (suboptimal) operation for real-world complex systems, in particular, for aerospace applications (Leitmann, 1990; Drabble, 1991; Pigeon et al., 1992). Basically, all the approaches for the limited time search can be broken into four categories: the imprecise computation (Chung et al., 1990), real-time search (e.g., Korf, 1990), approximate processing (Lesser et al., 1988), and anytime algorithms (Dean and Boddy, 1988). According to Strosnider and Paul (1994) the correct pruning in its many manifestations is still the only technique that reduces the worst-case execution time without compromising the goal state. But for real-world applications this reduction is usually insufficient: it does not overcome the combinatorial explosion. Another techniques, such as approximate processing, scoping, and use of domain knowledge, can reduce execution time significantly but they might compromise the goal state.

One of the basic ideas is to decrease the dimension of the real-world system following the approach of a *human expert in the field*, by breaking the system into smaller subsystems. This process of decomposition can be applied recursively until we end up with a collection of basic subproblems that can be treated (in some sense) independently. These ideas have been implemented for many problems with varying degrees of success (see, e.g., Albus, 1991; Knoblock, 1990; Mesarovich et al, 1989; Botvinnik, 1984). Implementations based on the formal theories of linear and nonlinear planning meet hard efficiency problems (McAllester and Rosenblitt, 1991; Chapman, 1987; Nilsson, 1980). An efficient planner requires an intensive use of heuristic knowledge. Moreover it is possible to use both dynamic and static heuristic knowledge in reducing the search variations. The dynamic knowledge can be acquired during the run time and immediately applied for search reduction (Strosnider and Paul, 1994).

In the 1960's, a formal syntactic approach to the investigation of properties of natural language resulted in the fast development of a theory of formal languages by Chomsky (1963), Ginsburg (1966), and others. This development provided an interesting opportunity for dissemination of this approach to different areas. In particular, there came an idea of analogous linguistic representation of images. This idea was successfully developed into syntactic methods of pattern recognition by Fu (1982), Narasimhan (1966), and Pavlidis (1977), and picture description languages by Shaw (1969), Feder (1971).

Searching for adequate mathematical tools formalizing human heuristics of dynamic hierarchies, we have transformed the idea of linguistic representation of complex real-world and artificial images into the idea of similar representation of complex hierarchical systems (Stilman, 1985). However, the appropriate languages should possess more sophisticated attributes than languages usually used for pattern description. The origin of such languages can be traced back to the research on programmed attribute grammars by Knuth (1968), Rozenkrantz (1969).

A mathematical environment (a "glue") for the formal implementation of this approach was developed following the theories of formal problem solving and planning by Nilsson (1980), Fikes and Nilsson (1971), McCarthy (1980), McCarthy and Hayes (1969), and others based on first order predicate calculus.

In the beginning of 80's Botvinnik, Stilman, and others developed one of the most interesting and powerful heuristic hierarchical models. It was successfully applied to scheduling, planning, control, and computer chess. The hierarchical networks were introduced in (Botvinnik, 1984; Stilman, 1977) in the form of ideas, plausible discussions, and program implementations. We consider this model as an ideal case for transferring the developed search heuristics to other domains employing formal linguistic tools.

An application of the developed model to a chess domain was implemented in full as program PIONEER (Botvinnik, 1984). Similar heuristic model was implemented for power equipment maintenance in a number of computer programs being used for maintenance scheduling all over the USSR (Botvinnik et al., 1983; Reznitskiy and Stilman, 1983; Stilman, 1985, 1993a).

3. INTRODUCTION TO Linguistic Geometry

To discover the inner properties of human expert heuristics, which have been successful in a certain class of complex control systems, we develop a formal theory, the so-called *Linguistic Geometry* (Stilman, 1993-94).

The hierarchy of subsystems is represented as a *hierarchy of formal attribute languages* where each "sentence" (a group of "words" or symbols) of the lower level language corresponds to the "word" of the higher level one. Following a linguistic approach each subsystem could be represented as a string of symbols with parameters: $a(x_1)a(x_2)...a(x_n)$, where the values of the parameters incorporate the semantics of the problem domain or lower-level subsystems. The lowest-level language of the hierarchy of languages, the Language of Trajectories (Stilman, 1993a, 1993c), serves as a building block to create the upper-level languages, the Languages of Networks (Stilman, 1993b, 1993c, 1994a, 1994b).

The Language of Trajectories actually is a formalization of the description of the set of various paths between different points of the a complex control system. An element might follow a path to achieve the goal "connected with the ending point of this path." The Language of Networks is a formalization of a set of networks of certain paths unified by the mutual goal. For example, in the chess model such a network represents planning for a local fight, in the robot control model an analogous network of planning paths represents a draft short-range plan for approaching local goal in hazardous environment, i.e., getting over mobile and immobile obstacles. In the scheduling problem it corresponds to the maintenance schedule of a certain power unit including the schedule for the provision of resources required.

The functioning of the system, in a search process, generates a tree of translations of the hierarchy of languages. This tree is represented as a string of the highest level formal language, the Language of Translations (Stilman, 1994b, 1994c).

4. CLASS OF PROBLEMS

A **Complex System** is the following eight-tuple:

$$< X, P, R_p, \{ON\}, v, S_i, S_t, TR>,$$ where

$X=\{x_i\}$ is a finite set of *points*;

$P=\{p_i\}$ is a finite set of *elements*; P is a union of two non-intersecting subsets P_1 and P_2;

$R_p(x, y)$ is a set of binary relations of *reachability* in X (x and y are from X, p from P);

ON(p)=x, where ON is a partial function of *placement* from P into X;

v is a function on P with positive integer values describing the *values* of elements.

The Complex System searches the state space, which should have initial and target states;

S_i and S_t are the descriptions of the *initial* and *target* states in the language of the first order predicate calculus, which matches with each relation a certain Well-Formed Formula (WFF). Thus, each state from S_i or S_t is described by a certain set of WFF of the form $\{ON(p_j) = x_k\}$;

TR is a set of operators, TRANSITION(p, x, y), of transitions of the System from one state to another one. These operators describe the transition in terms of two lists of WFF (to be removed from and added to the description of the state), and of WFF of applicability of the transition. Here,

Remove list: ON(p)=x, ON(q)=y;

Add list: ON(p)=y;

Applicability list: $(ON(p)=x) \wedge R_p(x,y)$,

where p belongs to P_1 and q belongs to P_2 or vice versa. The transitions are carried out with participation of the elements p from P_1 and P_2.

According to the definition of the set P, the elements of the System are divided into two subsets P_1 and P_2. They might be considered as units moving along the reachable points. Element p can move from point x to point y if these points are reachable, i.e., $R_p(x, y)$ holds. The current location of each element is described by the equation ON(p)=x. Thus, the description of each state of the System $\{ON(p_j)=x_k\}$ is the set of descriptions of the locations of the elements. The operator TRANSITION(p, x, y) describes the change of the state of the System caused by the move of the element p from point x to point y. The element q from point y must be withdrawn (eliminated) if p and q do not belong to the same one of the two subsets P_1 and P_2.

The problem of the optimal operation of the System is considered as a search for the optimal sequence of transitions leading from one of the initial states of S_i to a target state S of S_t.

It is easy to show formally that a robotic system can be considered as a Complex System (see below). Many different technical and human society systems (including military battlefield systems, systems of economic competition, positional games) that can be represented as twin sets of movable units (representing two or more opposing sides) and their locations can be considered as Complex Systems.

With such a problem statement for the search of the optimal sequence of transitions leading to the target state, we could use formal methods like those in the problem-solving system STRIPS (Fikes and Nilsson, 1971), nonlinear planner NOAH, or in subsequent planning systems. However, the search would have to be made in a space of a huge dimension (for nontrivial examples). Thus, in practice no solution would be obtained.

We devote ourselves to finding an approximate solution of a reformulated problem.

5. SET OF PATHS: Language of Trajectories

This language is a formal description of the set of lowest-level subsystems, the set of all paths between points of the Complex System. In (Stilman, 1993a) it is shown that geometrical properties of Complex Systems allow us to define a distance function and use it to generate different paths, the so-called trajectories.

A ***trajectory*** for an element p of P with the beginning at x of X and the end at the y of X ($x \neq y$) with a length l is following formal string of symbols $\boldsymbol{a}(x)$ with points of X as parameters:

$$t_0 = \boldsymbol{a}(x)\boldsymbol{a}(x_1)\ldots\boldsymbol{a}(x_l),$$

where $x_l = y$, each successive point x_{i+1} is reachable from the previous point x_i, i.e., $R_p(x_i, x_{i+1})$ holds for i = 0, 1,..., l–1; element p stands at the point x: ON(p)=x. We denote by $t_p(x, y, l)$ the set of all trajectories for element p, beginning at x, end at y, and with length l. $\boldsymbol{P}(t_0)=\{x, x_1, ..., x_l\}$ is the set of parameter values of the trajectory t_0. (To avoid confusion we should emphasize that $\boldsymbol{a}(x)\boldsymbol{a}(x_1)\ldots\boldsymbol{a}(x_l)$ is a formal record and does not mean anything else except what is given above.)

A ***shortest trajectory*** t of $t_p(x, y, l)$ is the trajectory of minimum length for the given beginning x, end y, and element p.

Properties of the Complex System permit us to define (in general form) and study formal grammars for generating the shortest trajectories. A general grammar and its application to generating the shortest trajectories for aerospace robotic vehicles are presented in (Stilman, 1993c).

A ***Language of Trajectories*** $L_t^H(S)$ for the Complex System in a state S is the set of all the trajectories of length less than H. Different properties of this language and generating grammars were investigated in (Stilman, 1993a).

6. NETWORKS OF PATHS: Languages of Trajectory Networks

After defining the Language of Trajectories, we have new tools for the breakdown of our System into subsystems. According to the ideas presented in (Botvinnik, 1984), these subsystems should be various types of trajectory networks, i.e., the sets of interconnected trajectories with one singled out and called the *main trajectory*. An example of such network is shown in Fig. 1.

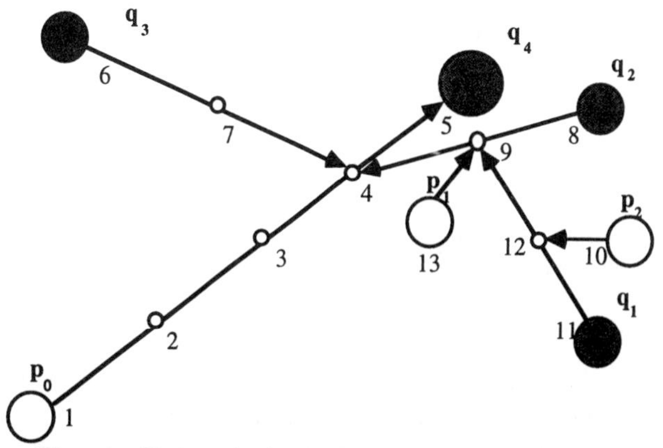

Fig. 1. Network language interpretation.

The basic idea behind these networks is as follows.

Element p_o should move along the main trajectory $a(1)a(2)a(3)a(4)a(5)$ to reach the ending point 5 and remove the target q_4 (an opposing element). Naturally, the opposing elements should try to disturb those motions by controlling the intermediate points of the main trajectory. They should come closer to these points (to the point 4 in Fig. 1) and remove element p_o after its arrival (at point 4). For this purpose, elements q_3 or q_2 should move along the trajectories $a(6)a(7)a(4)$ and $a(8)a(9)a(4)$, respectively, and wait (if necessary) on the next to last point (7 or 9) for the arrival of element p_o at point 4. Similarly, element p_1 of the same side as p_o might try to disturb the motion of q_2 by controlling point 9 along the trajectory $a(13)a(9)$. It makes sense for the opposing side to include the trajectory $a(11)a(12)a(9)$ of element q_1 to prevent this control.

Similar networks are used for the breakdown of complex systems in different areas. Let us consider a linguistic formalization of such networks. The Language of Trajectories describes "one-dimensional" objects by joining symbols into a string employing a reachability relation $R_p(x, y)$. To describe networks, i.e., "multi-dimensional" objects made up of trajectories, we use the relation of *trajectory connection.*

A ***trajectory connection*** of the trajectories t_1 and t_2 is the relation $C(t_1,t_2)$. It holds if the ending link of the trajectory t_1 coincides with an intermediate link of the trajectory t_2; more precisely, t_1 is connected with t_2 if among the parameter values $\mathcal{P}(t_2)=\{y,y_1,\ldots,y_l\}$ of trajectory t_2 there is a value $y_i = x_k$, where $t_1=a(x_o)a(x_1)\ldots a(x_k)$. If t_1 belongs to a set of trajectories with the common end-point, then the entire set is said to be connected with the trajectory t_2.

For example, in Fig. 1 the trajectories $a(6)a(7)a(4)$ and $a(8)a(9)a(4)$ are connected with the main trajectory $a(1)a(2)a(3)a(4)a(5)$ through point 4. Trajectories $a(13)a(9)$ and $a(11)a(12)a(9)$ are connected with $a(8)a(9)a(4)$.

To formalize the trajectory networks, we define and use routine operations on the set of trajectories: $C_A{}^k(t_1,t_2)$, a ***k-th degree of connection***, and $C_A{}^+(t_1,t_2)$, a ***transitive closure***.

Trajectory $a(11)a(12)a(9)$ in Fig. 1 is connected degree 2 with trajectory $a(1)a(2)a(3)a(4)a(5)$, i.e., $C^2(a(11)a(12)a(9), a(1)a(2)a(3)a(4)a(5))$ holds. Trajectory $a(10)a(12)$ in Fig. 1 is in transitive closure to the trajectory $a(1)a(2)a(3)a(4)a(5)$ because $C^3(a(10)a(12), a(1)a(2)a(3)a(4)a(5))$ holds by means of the chain of trajectories $a(11)a(12)a(9)$ and $a(8)a(9)a(4)$.

A ***trajectory network*** **W** relative to trajectory t_o is a finite set of trajectories $t_o,t_1,\ldots,t_k$ from the language $L_t{}^H(S)$ that possesses the following property: for every trajectory t_i from W ($i = 1, 2,\ldots,k$) the relation $C_W{}^+(t_i,t_o)$ holds, i.e., each trajectory of the network W is connected with the trajectory t_o that was singled out by a subset of interconnected trajectories of this network. If the relation $C_W{}^{\mathbf{m}}(t_i, t_o)$ holds, i.e., this is the **m**-th degree of connection, trajectory t_i is called the **m** ***negation trajectory***.

Obviously, the trajectories in Fig. 1 form a trajectory network relative to the main trajectory $a(1)a(2)a(3)a(4)a(5)$. We are now ready to define network languages.

A ***family of trajectory network languages*** $\mathbf{L_C(S)}$ in a state S of the Complex System is the family of languages that contains strings of the form

$$t(t_1, param)t(t_2, param)\ldots t(t_m, param),$$

where *param* in parentheses substitute for the other parameters of a particular language. All the symbols of the string $t_1, t_2,\ldots, t_m$ correspond to trajectories that form a trajectory network W relative to t_1.

Different members of this family correspond to different types of trajectory network languages, which describe particular subsystems for solving search problems. One such language is the language that describes specific networks called Zones. They play the main role in the model considered here (Botvinnik, 1984; Stilman, 1977, 1993b, 1993c, 1994a). A formal definition of this language is essentially constructive and requires showing explicitly a method for generating this language, i.e., a certain formal grammar, which is presented in (Stilman, 1993b, 1993c,1994a). In order to make our points transparent here, we define the Language of Zones informally.

A ***Language of Zones*** is a trajectory network language with strings of the form

$$Z=t(p_o,t_o,\tau_o)\ t(p_1,t_1,\tau_1)\ldots t(p_k,t_k,\tau_k),$$

where $t_o,t_1,\ldots,t_k$ are the trajectories of elements $p_o,p_2,\ldots,p_k$ respectively; $\tau_o,\tau_1,\ldots,\tau_k$ are nonnegative integers that "denote the time allotted for the motion along the trajectories" in a correspondence to the mutual goal of this Zone: to remove the target element – for one side, and to protect it – for the opposing side. Trajectory $t(p_o,t_o,\tau_o)$ is called the *main trajectory* of the Zone. The element q standing on the ending point of the main trajectory is called the *target*. The elements p_o and q belong to the opposing sides.

To make it clearer, let us show the Zone corresponding to the trajectory network in Fig. 1.

$$Z=t(p_o,a(1)a(2)a(3)a(4)a(5),4)\ t(q_3,a(6)a(7)a(4),3)$$
$$t(q_2, a(8)a(9)a(4), 3)t(p_1, a(13)a(9), 1)$$
$$t(q_1, a(11)a(12)a(9), 3)\ t(p_2, a(10)a(12), 1)$$

Assume that the goal of the white side is to remove target q_4, while the goal of the black side is to protect it. According to these goals, element p_o starts the motion to the target, while black starts in its turn to move elements q_2 or q_3 to intercept element p_o. Actually, only those black trajectories are to be included into the Zone where the motion of the element makes sense, i. e., the *length of the trajectory is less than the amount of time (third parameter* τ*) allocated to it.* For example, the motion along the trajectories $a(6)a(7)a(4)$ and $a(8)a(9)a(4)$ makes sense, because they are of length 2 and time allocated equals 3: each of the elements has 3 time intervals to reach point 4 to intercept element p_o assuming one would go along the main trajectory without move omission and all the intercepting elements will move *simultaneously* (if necessary). According to definition of Zone, the trajectories of white elements (except p_o) could only be of the length 1, e.g., $a(13)a(9)$ or $a(10)a(12)$. As element p_1 can intercept the motion of the element q_2 at the point 9, black includes into the Zone the trajectory $a(11)a(12)a(9)$ of the element q_1, which has enough time for motion to prevent this interception. The total amount of time allocated to the whole bunch of black trajectories connected (directly or indirectly) with the given point of the main trajectory is determined by the number of that point. For example, for the point 4, it equals 3 time intervals.

A language $L_Z{}^H(S)$ generated by the certain grammar G_Z (Stilman, 1993b, 1993c,1994a) in a state S of a Complex System is called the ***Language of Zones***.

A practicality of the formal constructions considered in Section 4 as well as the entire hierarchy of languages are demonstrated on the following 2D example of the air combat.

7. AIR COMBAT: PROBLEM STATEMENT

The robotic model can be represented as a Complex System naturally (Fig. 2). The set X represents the operational district, which could be the area of combat operation, broken into smaller square or cubic areas, "points", e.g., in the form of the big square or cubic grid. It could be a space operation, where X represents the set of different orbits, or an air force battlefield, etc. P is the set of robots or autonomous vehicles. It is broken into two subsets P_1 and P_2 with opposing interests; $R_p(x,y)$ represent moving capabilities of different robots for different problem domains: robot p can move from point x to point y if $R_p(x, y)$ holds. Some of the robots can crawl, others can jump or ride, sail and fly, or even move from one orbit to another. Some of them move fast and can reach point y (from x) in "one step", i.e., $R_p(x, y)$ holds, others can do that in *k* steps only, and many of them can not reach this point at all. ON(p)=x, if robot p is at the point x; v(p) is the value of robot p. This value might be determined by the technical parameters of the robot. It might include the immediate value of this robot for the given combat operation; S_i is an arbitrary initial state of operation for analysis, or the starting state; S_t is the set of target states. These might be the states where robots of each side reached specified points. On the other hand, S_t can specify states where opposing robots of the highest value are destroyed. The set of WFF $\{ON(p_j) = x_k\}$ corresponds to the list of robots with their coordinates in each state. TRANSITION(p, x, y) represents the move of the robot p from the location x to location y; if a robot of the opposing side stands on y, a removal occurs, i.e., robot on y is destroyed and removed.

Robots with different moving capabilities are shown in Fig. 2. The operational district X is the table 8 x 8. Robot W-FIGHTER (White Fighter) standing on h8, can move to any next square (shown by arrows). The other robot B-BOMBER (Black Bomber) from h6 can move only straight ahead, one square at a time, e.g., from h6 to h5, from h5 to h4, etc. Robot B-FIGHTER (Black Fighter) standing on a6, can move to any next square similarly to robot W-FIGHTER (shown by arrows). Robot W-BOMBER (White Bomber) standing on c6 is analogous with the robot B-BOMBER; it can move only straight ahead but in reverse direction. Thus, robot W-FIGHTER on h8 can reach any of the points y ∈ {h7, g7, g8} in one step, i.e., $R_{W\text{-}FIGHTER}(h8, y)$ holds, while W-BOMBER can reach only c5 in one step.

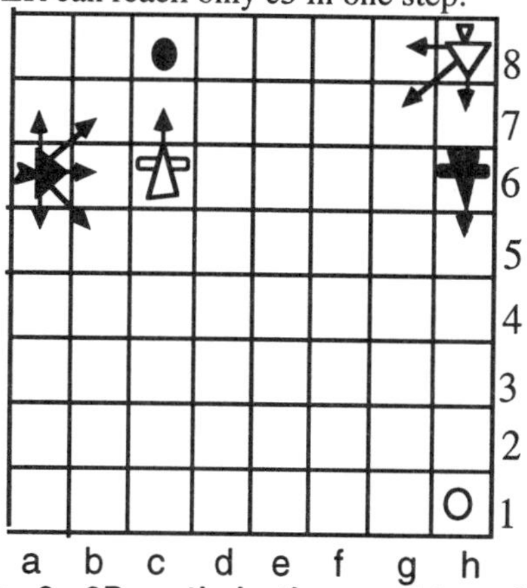

Fig. 2. 2D optimization problem for autonomous robotic vehicles.

Assume that robots W-FIGHTER and W-BOMBER belong to one side, while B-FIGHTER and B-BOMBER belong to the opposing side: W-FIGHTER ∈ P_1, W-BOMBER ∈ P_1, B-FIGHTER ∈ P_2, B-BOMBER ∈ P_2. Also assume that two more robots, W-TARGET and B-TARGET, (unmoving devices or targeted areas) stand on h1 and c8, respectively. W-TARGET belongs to P_1, while B-TARGET ∈ P_2. Each of the BOMBERs can destroy unmoving TARGET ahead of the course; it also has powerful weapons able to destroy opposing FIGHTERs on the next diagonal squares ahead of the course. For example, W-BOMBER from c6 can destroy opposing FIGHTERs on b7 and d7. Each of the FIGHTERs is able to destroy an opposing BOMBER approaching its location, but it also able to protect its friendly BOMBER approaching its prospective location. In the latter case the joint protective power of the combined weapons of the friendly BOMBER and FIGHTER can protect the BOMBER from interception. For example, W-FIGHTER located at d6 can protect W-BOMBER on c6 and c7.

The battlefield considered can be broken into two local operations. The first operation is as follows: robot B-BOMBER should reach point h1 to destroy the W-TARGET, while W-FIGHTER will try to intercept this motion. The second operation is similar: robot W-BOMBER should reach point c8 to destroy the B-TARGET, while B-FIGHTER will try to intercept this motion. After destroying the opposing TARGET the attacking side is considered as a winner of the local operation and the global battle. The only chance for the opposing side to avenge is to hit its TARGET on the next time interval and this way end the battle in a draw. The conditions considered above give us S_t, the description of target states of the Complex System. The description of the initial state S_i is obvious and follows from Fig. 2.

Assume that motions of the opposing sides alternate and each side can participate in both operations ***simultaneously***. It means, for example, that during the current time interval, in case of White turn, both W-BOMBER and W-FIGHTER, one of them, or none can move. Analogous condition holds for Black. There is one exception. If W-FIGHTER hits B-BOMBER while the latter is fully armed, i.e., it is not at its final destination – square h1, W-BOMBER can not move simultaneously during this time interval in order to avoid possible consequences of the B-BOMBER explosion. Similar restriction holds for B-BOMBER: it can not move at the moment when W-BOMBER is destroyed (not at c8).

In this paper we relaxed the most restrictive requirement of participation of the only element in each motion which holds in our earlier papers for the similar problems (Stilman, 1994b, 1994c). This requirement was imposed in the form of the restriction for each side to participate in both local operations simultaneously. It was inherited from the original domain, the game of chess, served as a testbed for the development of the Linguistic Geometry tools (Stilman, 1977; Botvinnik, 1984).

It seems that local operations are independent, because they are located far from each other. Moreover, the operation of B-BOMBER from h6 looks like unconditionally winning operation, and, consequently, the global battle can be easily won by the Black side. The question is: is there a strategy for the White side to make a draw?

Of course, this question can be answered by the direct search employing, for example, minimax algorithm with alpha-beta cut-offs. Theoretical evaluations (Nilsson, 1980) of the lower bounds of complexity of this algorithm for the air combat problem show that, at best, it would result in a search tree of 25 million moves (transitions). In practice, even this number is unreachable. It is very interesting to observe the dramatic reduction of search employing Linguistic Geometry tools.

To demonstrate generation of the Hierarchy of Languages for this problem, we have to generate the Language of Trajectories and the Language of Zones in each state of the search. The details of trajectory generation and generation of different Zones are considered in (Stilman, 1993b, 1993c, 1993d, 1994a).

8. AIR COMBAT: SEARCH GENERATION

Consider how the hierarchy of languages works for the optimal control of the Robotic System introduced above (Fig. 2). We generate the string of the Language of Translations (Stilman, 1994a) representing it as a conventional search tree (Fig. 3) and comment on its generation.

In fact, this tree is close to the search tree of the restricted problem (Stilman, 1994b, 1994c). In our comments on this generation we will emphasize the major steps.

First, the Language of Zones in the start state is generated. The targets for attack are determined within the limit of five steps. It means that horizon H of the language $L_Z(S)$ is equal to 5, i.e., the length of main trajectories of all Zones must not exceed 5 steps. The reasons and the algorithm for picking the right value of the horizon are considered in (Stilman, 1994c). All the Zones generated in the start state within the horizon of 5 are shown in Fig. 4. Zones for FIGHTERs as attacking elements are shown in the left diagram, while Zones for BOMBERs – in the right one. For example, one of the Zones for W-BOMBER, Z_{WB} is as follows:

Z_{WB}=***t***(WB, ***a***(c6)***a***(c7)***a***(c8), 2)***t***(BF,***a***(a6)***a***(b7)***a***(c8), 3)
t(BF,***a***(a6)***a***(b7)***a***(c7), 2)***t***(WB, ***a***(c6)***a***(b7), 1)

The second trajectory of B-FIGHTER ***a***(a6)***a***(b6)***a***(c7) leading to the square c7 is included into different Zone because for each Zone only one trajectory from each bundle of trajectories is taken.

Generation begins with the move 1. c6-c7 in the "white" Zone with the target of the highest value and the shortest main trajectory. The order of consideration of Zones and particular trajectories is determined by the grammar of translations. The computation of move-ordering constraints is the most sophisticated procedure in this grammar. It takes into account different parameters of Zones, trajectories, and the so-called chains of trajectories.

Next move, 1. ... a6-b7, is in the same Zone along the first negation trajectory. The interception continues: 2. c7-c8 b7:c8 (Fig. 5, left). Symbol ":" means the removal of element. Here the grammar cuts this branch with the value of -1 (as a win of the Black side). This value is given by the special procedure of "generalized square rules" built into the grammar. This procedure determined that W-FIGHTER is out of the Zone of B-BOMBER, thus it can not intercept B-BOMBER.

Then, the grammar initiates the backtracking climb. Each backtracking move is followed by the inspection procedure, the analysis of the subtree generated in the process of the earlier search. After climb up to the move 1. ... a6-b7, the tree to be analyzed consists of one branch (of two plies): 2. c7-c8 b7:c8. The inspection procedure determined that the current minimax value (-1) can be "improved" by the improvement of the exchange on c8 (in favor of the White side). This can be achieved by participation of W-FIGHTER from h8, i.e., by generation and inclusion of the new so-called "control" Zones with the main trajectory from h8 to c8. These Zones attack a location where a target is yet to come. Such Zones have been detected (within the horizon 5) in the terminal state after the move 2. ... b7:c8 Fig. 5 (left). Obviously they could not be detected in the initial state of this problem (Fig. 4) because the main element, W-BOMBER, could not "see" the target, B-FIGHTER, within given horizon. However, at the moment of detection it was too late to include them into the search. These Zones have been stored for possible activation at the higher levels of the search tree. The set of different Zones from h8 to c8 (the bundle of Zones) is shown in Fig. 5 (right). The move-ordering procedure picks the subset of Zones with main trajectories passing g7. These trajectories partly coincide with the main trajectory of another Zone attacking the opposing W-BOMBER on h6. The motion along such trajectories allows to "gain time", i.e., to approach two goals simultaneously.

The generation continues with the simultaneous motion of two agents, the double move, W-FIGHTER and W-BOMBER in their respective Zones: 2. h8-g7/c7-c8. The B-FIGHTER intercepts W-BOMBER at c8: 2. ... b7:c8. Now W-FIGHTER is ready to destroy B-BOMBER moving along the attacking trajectory a(7)a(h6): 3. g7:h6.

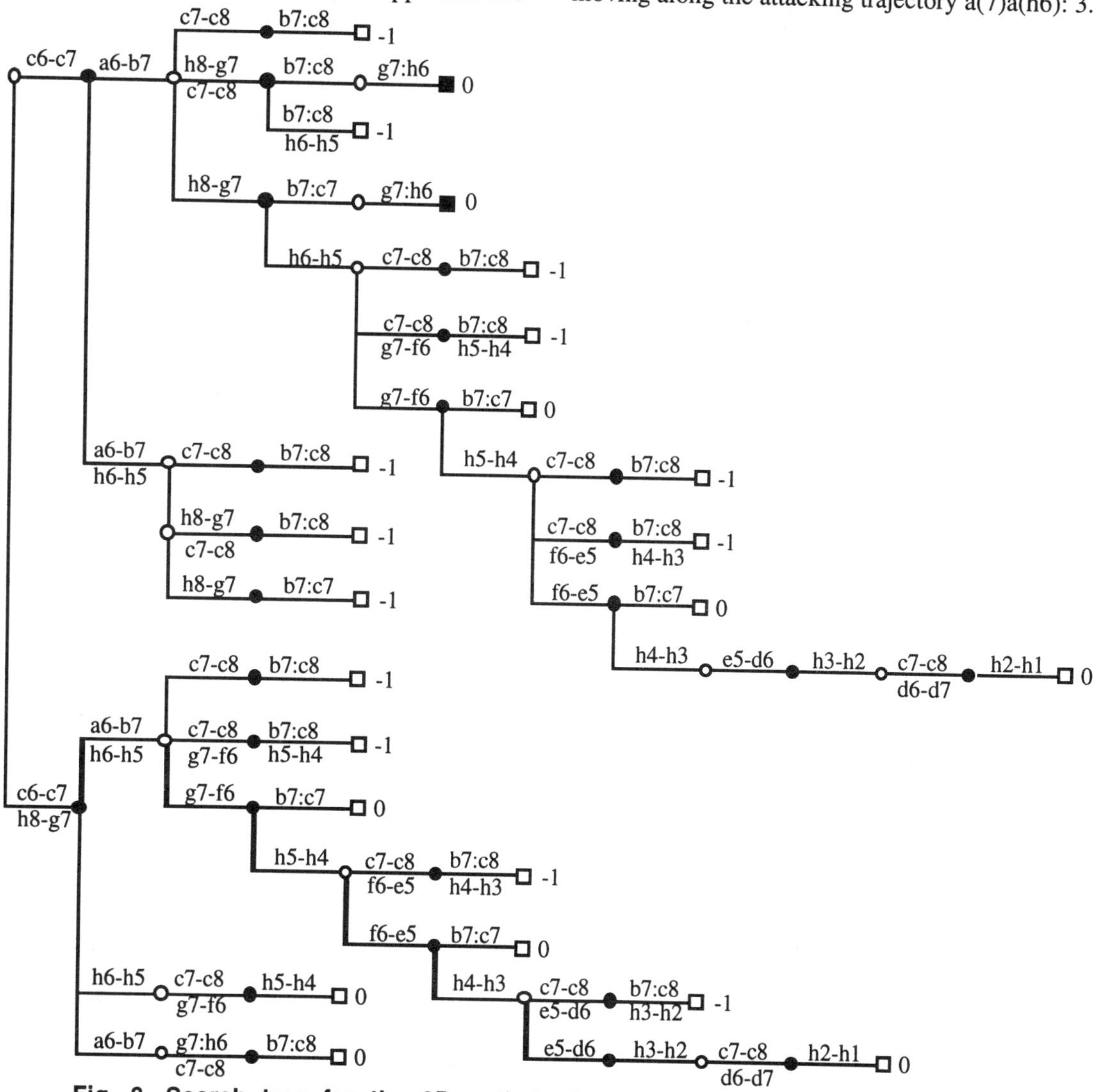

Fig. 3. Search tree for the 2D optimization problem for robotic vehicles.

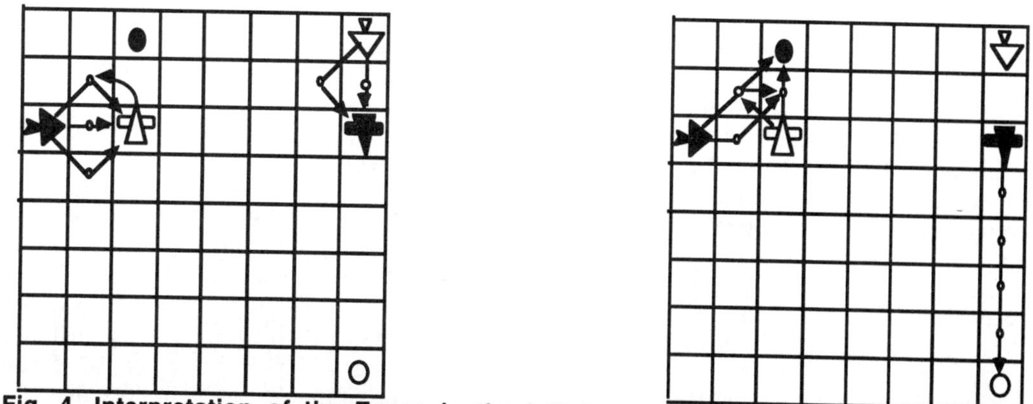

Fig. 4. Interpretation of the Zones in the initial state of the 2D Robot Control Model.

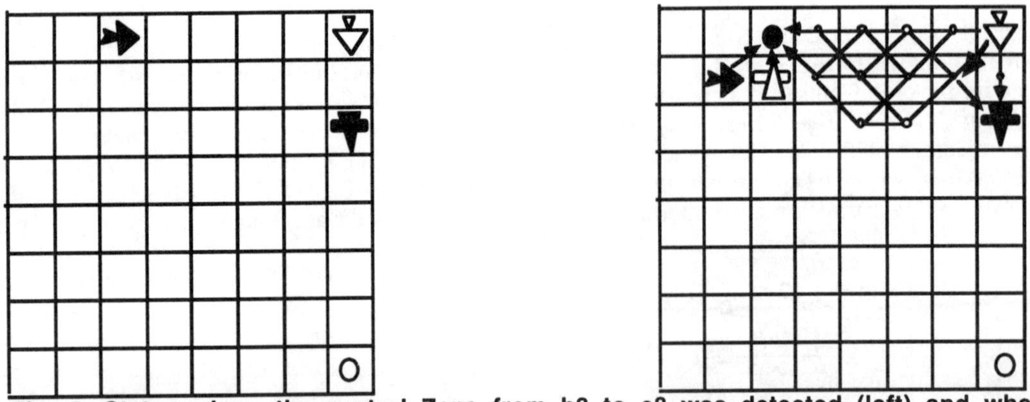

Fig. 5. States where the control Zone from h8 to c8 was detected (left) and where it was included into the search (right)

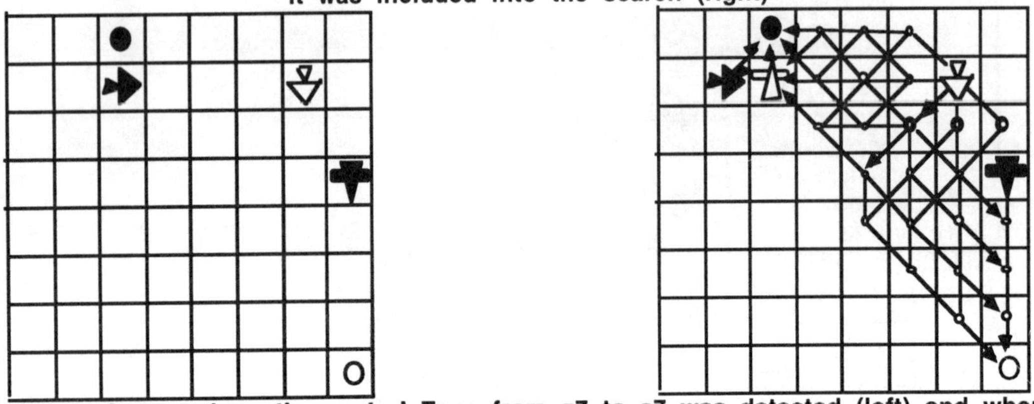

Fig. 6. States where the control Zone from g7 to c7 was detected (left) and where it was included into the search (right).

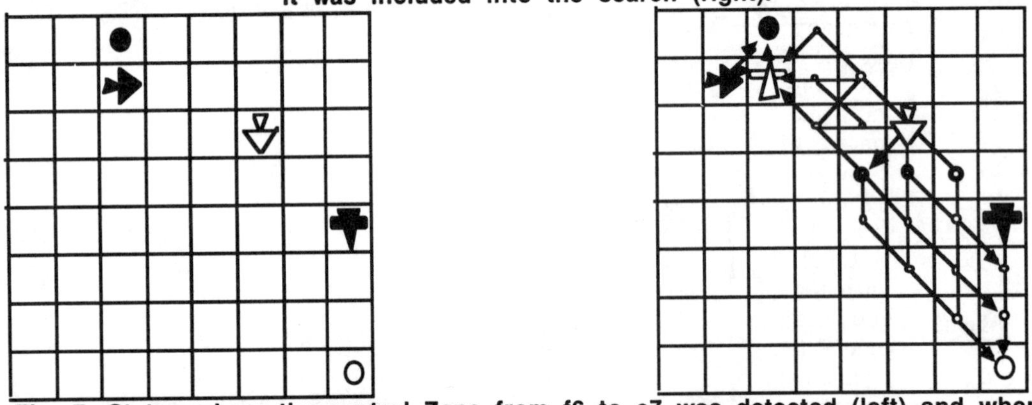

Fig. 7. States where the control Zone from f6 to c7 was detected (left) and where it was included into the search (right).

In this state all the BOMBERs have been destroyed and the grammar evaluates it as a draw (0) and initiates the backtracking climb. Move 2. ... b7:c8 is changed for the double move 2. ... b7:c8/h6-h5. That way the grammar included motion in the Zone of B-BOMBER attacking unmovable target on h1. The "square rule" procedure terminated branch, evaluated it as a win (-1) for the Black side, and initiated backtracking climb. Analogously to the previous case, the inspection procedure determined that the current minimax value (-1) can be improved by the improvement of the exchange on c8. Again, this can be achieved by the inclusion of Zone from h8 to c7. Of course, the best "time-gaining" move in this Zone is 2. h8-g7, and now the grammar includes this move as a single one 2. h8-g7 postponing motion in the Zone of W-BOMBER. In this state (Fig. 6, left) a set of new control Zones of W-FIGHTER from g7 to c7 have been detected and stored as idle to be activated later if necessary. In response, Black hit W-BOMBER at c7: 2. ... b7:c7 along the intercepting trajectory a(b7) a(c7). W-FIGHTER continues its attack of B-BOMBER: 3. g7:h6. This state is evaluated as a draw.

After the cut and climb, the inspection procedure included motion in the Zone of B-BOMBER instead of 2. ... b7:c7. New move 2. ... h6-h5 can not be included as a double move simultaneously with the old one because B-BOMBER can not move at the moment when W-BOMBER is being destroyed not at its final destination (c8). The following motion in 3. c7-c8 b7:c8 resulted in the termination of this branch with the value -1 in favor of Black.

New climb up to the move 2. ... h6-h5 and execution of the inspection procedure result in the inclusion of the groups of new control Zones from g7 to c7 and to c8 in order to improve the exchanges at these locations. Both groups of Zones (to c7 and c8) have been detected earlier in the search tree. The set of Zones with different main trajectories from g7 to c7 and from g7 to c8 is shown in Fig. 6 (right). Besides that, the trajectories from g7 to h4, h3, h2, and h1 are shown in the same Fig. 6. These are "potential" first negation trajectories. It means that beginning with the second symbol ***a***(f6), ***a***(g6) or ***a***(h6) these trajectories become first negation trajectories in the Zone of B-BOMBER on h5. Speaking informally, from squares f6, g6, and h6, Zone gateways, W-FIGHTER can intercept B-BOMBER (in case of White turn). The move-ordering procedure picks the subset of Zones with the main trajectories passing f6. These trajectories partly coincide with the potential first negation trajectories. The motion along such trajectories allows to "gain time", i.e., to approach two goals simultaneously. Thus, the double move 3. c7-c8/g7-f6 is included. After the response 3. ... b7:c8/h5-h4, the branch was terminated with the value -1. The following climb and branching with inclusion of 3. g7-g6 as a single move resulted in 3. ... b7:c7 with terminal value 0. This state is shown in Fig. 7, left. The draw value has been assigned by the "square rule" procedure which detected that W-FIGHTER is in the Zone of B-BOMBER and thus has enough time for interception.

After the climb Black side continued branching 3. ... h5-h4. The following tree generation until 4. f6-e5 is analogous with the previous one after 2. ... h6-h5. Move 4. f6-e5 is selected by the move ordering procedure as the time-gaining move approaching two goals simultaneously, c7 as a goal of the control Zone of W-FIGHTER and one of the gateways (e5, f5, g5) of the Zone of B-BOMBER (Fig. 7, right).

After the change of 4. ... b7:c7 for 4. ... h4-h3, the following branch is pretty straightforward. After 6. c7-c8/d6-d7 h2-h1, it is terminated with the value of 0. The following climb with activation of the inspection procedure in every black node ended at 1. ... a6-b7 which has been changed for the double move 1. ... a6-b7/h6-h5. It seems that this move almost depreciated previous search. The minimax value brought to the top of the tree generated so far is -1. However, the tree generation followed after the change of 1. c6-c7 for the double move 1. c6-c7/h8-g7 showed that previous search was very important. As a result of this search the grammar *learned key networks*, Zones of W-FIGHTER with main trajectories from g8 to c8, from g7 to c7 and c8, and from f6 to c7. These networks have been used successfully in a different context after 1. c6-c7/h8-g7. The optimal branch is shown in Fig. 3 with bold lines.

The total number of moves included into this tree is 63. The maximum depth reached is 12. This means that the branching factor (Nilsson, 1980) of this tree is 1.24, i.e., the search is highly goal-oriented. The unreduced branching factor for this problem is about 17, taking into account the allowance of simultaneous moves. Obviously, 63 is a drastic reduction in comparison with a 17^{12} move trees that would have to be generated by conventional search procedures, or even with the theoretical minimum of the minimax search with alpha-beta cut-offs $(17^{12})^{1/2} = 17^6 \approx$ (25 million).

9. DISCUSSION

Example considered in Sections 7, 8 demonstrate the power of the Linguistic Geometry tools that allowed to transfer heuristics discovered in one problem domain, specifically, in the game of chess, to another domain of simplified aerospace robotic vehicles. It is even more interesting that search reduction achieved in the original domain with one-at-a-time motion of every agent (see, for example, Stilman, 1994b) multiplied tremendously in the new domain with the allowance of synchronous moves. While the total number of moves to be included in to the search tree in order to solve new problem (17^{12}) is by far greater than a million required for the original problem, the actual number of moves generated by the grammar of searches is almost the same (about 60), and the branching factor even decreased from 1.65 to 1.24. Looking at the complexity of the hierarchy of languages which represents each state in the search process, it is very likely that the growth from the single motion case to the synchronous one is limited by multiplication to a constant factor close to one. This means that the complexity of the entire algorithm did not change after allowing synchronous moves, and this allowance is, probably, inherent to the Complex Systems and the entire Hierarchy of Languages.

At the same time the simplified air combat problem considered here is still very close to the original chess domain. It is possible to predict that the power of Linguistic Geometry goes far beyond these limits. The definition of the Complex System (see Section 4) is generic enough to cover a variety of different problem domains. The core component of this definition is the triple X, P, and R_p. Thus, looking at the new problem domain we have to define X, the finite set of points – locations of elements. We do not impose any constraints on this set while the aerospace operational district X considered in this paper as well as the original chess board have different extra features, e.g., 2D space connectivity, which is totally unimportant for these problems. Thus, we can consider X, for example, as a set of orbits where the elements are in constant motion with respect to each other. The set of elements P, e.g., movable units, in our problem is quite small, while their moving capabilities, binary relations of R_p, are non-sophisticated. Indeed, during a one time interval our aircraft can move only to the next area. Even in the game of chess the moving capabilities of different pieces are much more advanced. This is exactly the place for introduction of the variable speed, the gravity impact, the engine impulse duration, etc.

REFERENCES

Albus, J. (1991). Outline for a Theory of Intelligence. *IEEE Trans.on Syst., Man and Cybernetics*(473-509),3.

Boddy, M. and Dean, T. (1989). Solving Time-Dependent Planning Problems, *Proc. of the 11th Int. Joint Conf. on AI*, (979-984).

Botvinnik, M.M. (1984). *Computers in Chess: Solving Inexact Search Problems.* Springer Series in Symbolic Computation, New York: Springer-Verlag.

Botvinnik, M., Petriyev, E., Reznitskiy, A., et al. (1983). Application of New Method for Solving Search Problems For Power Equipment Maintenance Scheduling. *Economics and Mathematical Methods* (1030-1041), 6, (in Russian).

Chapman, D. (1987). Planning for conjunctive goals. *Artificial Intelligence* 32(3).

Chomsky, N. (1963). Formal Properties of Grammars. in *Handbook of Mathematical Psychology*, eds. R.Luce, R.Bush, E. Galanter., 2 (323-418). New York:J. Wiley.

Chung, J., Liu, J. and Lin, K., (1990). Scheduling Periodic Jobs That Allow Imprecise Results. *IEEE Transactions on Computers*, (1156-1174), 39(9).

Drabble, B., (1991) Spacecraft Command and Control Using Artificial Intelligence Techniques, *J. of the British Interplanetary Society*, Vol. 44, (251-254).

Durfee, E.H. and Lesser, V.R. (1987). Incremental Planning to Control a Blackboard-Based Problem Solver. *Proc. of the Workshop on Space Telerobotics*, Vol. 3, NASA Jet Propulsion Lab, July, (91-99).

Feder, J. (1971). Plex languages. *Information Sciences*, 3: (225–241).

Fikes, R.E. and Nilsson, N.J. (1971). STRIPS: A New Approach to the Application of Theorem Proving in Problem Solving. *Artificial Intelligence* 2 (189–208).

Fu, K.S. (1982). *Syntactic Pattern Recognition and Applications*, Prentice Hall, Englewood Cliffs.

Garey, M.R. and D.S.Johnson D.S. (1991). *Computers and Intractability: A Guide to the Theory of NP-Completeness*. W.H. Freeman and Co., San Francisco.

Ginsburg, S. (1966). *The Mathematical Theory of Context-Free Languages*, McGraw Hill, New York.

Knoblock, C.A. (1990). Learning Abstraction Hierarchies for Problem Solving, *Proc. of the 8th AAAI Conf.*, (923-928), Menlo Park, CA.

Knuth, D.E. (1968). Semantics of Context-Free Languages. *Mathematical Systems Theory*,(127–146),2.

Korf, R.E. (1990). Real-Time Heuristic Search, *Artificial Intelligence*, (189-211), 42(2-3).

Leitmann, G., (1990). *Optimization Techniques with Applications to Aerospace Systems*, Academic Press.

Lesser, V.R., Pavlin, J., and Durfee, E. (1988). Approximate Processing in Real-Time Problem Solving, *AI Magazine*, (pp. 49-62), 9(1).

Lirov Y., Rodin, E.Y., McElhaney, B.G., and Wilbur, L.W. (1988). Artificial Intelligence Modeling of Control Systems, *Simulation*, (pp. 12-24), 50(1).

McAllester, D. and Rosenblitt, D. (1991). Systematic Non-Linear Planning, *Proc. of AAAI-91*, (634-639).

McCarthy, J. (1980). Circumscription–A Form of Non-Monotonic Reasoning.*Artificial Intelligence*,(27-39),13.

McCarthy, J. and Hayes, P.J. (1969). Some Philosophical Problems from the Standpoint of Artificial Intelligence. *Machine Intelligence* (463–502), 4.

Mesarovich, M.D. and Takahara Y. (1989). *Abstract Systems Theory*, Berlin: Springer-Verlag.

Narasimhan, R.N. (1966). Syntax–Directed Interpretation of Classes of Pictures. *Comm. of ACM* (166–173), 9.

Nilsson, N.J. (1980). *Principles of Artificial Intelligence*, Palo Alto, CA: Tioga Publ.

Pavlidis, T. (1977). *Structural Pattern Recognition*, New York: Springer-Verlag.

Pigeon, A., Howard, G., and Seaton B. (1992) Operational Aspects of Space craft Autonomy, *J. of the British Interplanetary Society*, Vol. 45, (87-92).

Reznitskiy, A.I. and Stilman, B. (1983). Use of Method PIONEER in Automating the Planning of Maintenance of Power-Generating Equipment. *Automatics and Remote Control* (pp. 147-153), 11, (in Russian).

Rodin E. (1988). Semantic Control Theory, *Applied Mathematical Letters*, (pp. 73-78), 1(1).

Rodin E., Garcia-Ortiz et al. (1993). Application of Semantic Control to a Class of Pursue-Evader Problems, *Int. J. Computers and Mathematics with Applications*, 26(5).

Rozenkrantz, D.J. (1969). Programmed Grammars and Classes of Formal Languages,*J. of ACM* (107–131), 1.

Shaw, A.C. (1969). A Formal Picture Description Scheme as a Basis for Picture Processing System, *Information and Control* (9-52), 19.

Stilman, B. (1977). The Computer Learns: in Levy, D., *1976 US Computer Chess Championship* (83-90). Computer Science Press, Woodland Hills, CA.

Stilman, B. (1985). Hierarchy of Formal Grammars for Solving Search Problems. In *Artificial Intelligence. Results and Prospects, Proceedings of the International Workshop* (pp. 63-72), Moscow, (in Russian).

Stilman, B. (1993a). A Linguistic Approach to Geometric Reasoning," *Int. J. Computers and Mathematics with Applications* (29-57), 26(7).

Stilman, B. (1993b). Network Languages for Complex Systems, *Int. J. Computers and Mathematics with Applications* (51-79), 26(8).

Stilman, B. (1993c). Syntactic Hierarchy for Robotic Systems, *Integrated Computer-Aided Engineering* (57-81), 1(1).

Stilman, B. (1993d). A Formal Language for Hierarchical Systems Control, *Languages of Design* (333-356), 1(4).

Stilman, B. (1994a). Translations of Network Languages. *Int. J. Computers and Mathematics with Applications* (65-98), 27(2).

Stilman, B. (1994b). A Formal Model for Heuristic Search. *Proc. of the 22nd Annual ACM Computer Science Conf.*, (380-389), March 8-10, Phoenix, AZ.

Stilman, B.(1994c). Heuristic Networks for Space Exploration, *Telematics and Informatics, An Int. Journal on Telecommunications & Information Technology*, 11(4), (403-428).

Stilman, B., (1994d). Network Languages for Intelligent Control, *An International Journal: Computers & Mathematics with Applications*., (to appear).

Strosnider, J.K. and Paul, C.J. (1994). A Structured View of Real-Time Problem Solving, *AI Magazine*, (45-66), 15(2).

SCENARIO-BASED MANAGEMENT FOR MULTIAGENT SYSTEMS

Brian J. Garner & Dan Song

School of Computing and Mathematics

Deakin University, Geelong, Australia, 3217

E-mail: {brian, dsong}@deakin.edu.au

ABSTRACT

Scenario-based management offers a unique solution for one particular class of multiagent/distributed problem solving system. The features of this class are as follows:

1. The scenario adopted for the multi-agent system should represent the agreement reached in representing the interests of the participants;
2. Nearly autonomous agents;
3. Real-time instantiation of the compromise scenario; and
4. Correction during a discrete transaction run is not possible.

The problem to be solved is represented by a trading requirement in an open environment, characterized for the purposes of discussion as a "transaction". In a multi-agent system, the transaction (business process) is executed through the cooperation of the multiple agents in the system. The difficulties encountered in this process include firstly;

- While the common scenario guides all agents in the system, each agent is almost autonomous and itself decides which action (or episode in the scenario) should be taken.
- Consulting agents are permitted to co-exist but no one agent has a supervisory or higher level of authority!
- Decentralized control of multi-agent systems implies unique design criteria for the common scenarios/consulting agents.

Secondly; each agent has incomplete knowledge during the execution of one transaction. Each transaction is related to a particular scenario instance, which is derived by the timely instantiation of role players with their required roles.However, how other agents react and the precise status of these agents is unclear! The knowledge possessed by each agent is thus limited to knowledge of the common scenario and to the messages sent by other agents.

Finally, it should be noted that the iterative exchange of control messages to set up a control schema is not feasible during a transaction process. In short, it means that one transaction is executed or terminated immediately following a decision by an agent, and the sequence of this action cannot be changed during the run!

In this paper we illustrate the above difficulties experienced with multi-agent systems and propose a Scenario-based management mechanism that employs actor modelling for problem solving.The contribution of the Scenario-Based Manager is to manage the external/internal integration tasks in the multi-agent system.The novel feature of the design lies in the establishment of an integrated control process for agents of the actor model to be selected in the chosen message transmission environment. The components of each agent in the Scenario-Based management system are the transaction manager, supported by an adaptive mechanism, and the sub-role manager for message generation and planning.

1 INTRODUCTION

It is known (ref. [1]) that the objectives for Open-EDI gateways are external integration (communication with other gateways/agents) and internal integration (automated mapping between internal applications). In order to achieve these objectives, the major tasks at management level include Open-EDI (transaction) operation management and internal subrole management. Because the management services and functions of an Open-EDI gateway can theoretically be supported by Open-EDI scenarios, the advantages of our scenario design for management of multi-agent systems are advanced in this paper. Our studies show that Open-EDI gateway management can be characterized by a multiagent system with a common schema, and problems such as cooperation among autonomous agents with incomplete information can be classified as a subset of the "open system" problem, which is defined in [3]. As Distributed Artificial Intelligence (DAI) provides tools to study such issues, DAI constructs have been re-interpreted in the context of the cited management issues to permit suitable techniques to be designed and experimentally evaluated.

In this paper, therefore, Distributed Artificial Intelligence and its technology are introduced in the context of Scenario-Based Management. Open-EDI gateway management is then characterized using DAI techniques. The fourth section of this paper presents a framework for the management system. The transaction manager, which is the major component of the framework, is introduced in detail in the last section.

2 THE FEATURES OF OPEN-EDI GATEWAY MANAGEMENT

Management of the Open-EDI gateway is one typical application in Scenario-Based Management. Through the analysis of Open-EDI requirements, the features of Scenario-Based Management can be clearly understood.

Open-EDI gateway is a natural and economically attractive method for implementing Electronic Data Interchange (EDI) to realize open trading. Open trading means the trading party opens itself (in some form) to all potential trading partners without negotiating individual (bilateral) agreements with each partner. In order to establish open trading partners hips, common acceptance of a standard (short form) business agreement is required detailing the relevant business activities and the semantics of the electronic trading requirements. On establishment of the common trading policy and environment, Open-EDI gateways can process business transactions automatically on initiation of such transactions by users.

The problem to be solved in an Open-EDI gateway system is called an Open-EDI transaction, which has to be executed and completed through the cooperation of the multiple Open-EDI gateways in the system.

Each gateway represents one trading party who has chosen to be involved in Open-EDI activities. Such trading parties are independent of each other. Subject only to the provisions of the trading party agreement for Open-EDI, each transaction involves largely autonomous agents.

In addition, useful features of the Open-EDI gateway may be deduced from its run-time properties. The open trading agreement defines generic roles instead of specified role players. For a given business activity, it is, therefore, necessary to assign role players to the request roles. Such an instantiation process should be completed through the Open-EI gateway communication procedure at run-time. Each gateway relies solely on the instantiation and business messages sent by other gateways for their knowledge of the system status. The status and action of other agents would not necessarily be known.! Decisions made by each agent is, therefore, based on incomplete knowledge. Moreover, the automated process conducted by the Open-EDI gateway does not allow for correction during run-time of a transaction.

3 SCENARIO-BASED MANAGEMENT AS A MULTIAGENT (MA) SYSTEM

Key features of the management problem identified in the previous section were:

1. agreement by the multiagent system on the appropriate scenario;
2. decisions by nearly autonomous agents based on incomplete knowledge;
3. real time instantiation of the compromise scenario; and
4. no correction during transaction run-time.

Further details of the open trading agreement, defined here as a SCENARIO, in the Open-EDI gateway system are provided in Section 4. Consideration of Distributed Artificial Intelligence (DAI) technologies, specially distributed problem solving (DPS) and multiagent systems (MA) clearly indicated that scenario-based management clearly belongs to the MA class. Such systems provide intelligent behavior in problem solving by co-ordinating the collection (possibly pre-existing) of autonomous "agents" through use of the collective knowledge, goals, skills and plans. Of the various approaches used in DAI, only actor modelling techniques are reviewed here:

Actor problem solving emphasizes an "open system" which can be abstracted as, (1) mutual inconsistency in knowledge or belief; (2) decentralized control; (3) bounded influence or "arms-length relationships" among components; (4) asynchronous operation; and (5) concurrence based on their physical/natural distributed and autonomous properties.

Rather than viewing it as a sequence of choices made by a single decision maker in a web of choice points, a control structure is seen as a pattern of messages passing among a collection of computational agents called actors.

An actor system comprises two parts: the actors, which are the only computational entities; and an operating system, which creates and destroys actors and passes messages among them using the universal communication mechanism called actor transmission.

The definition of each actor also comprises two parts: a script, which defines the behaviors of an actor upon receipt of a message; and a finite set of acquaintances, which are the other actors known to the actor in question.

4 AN FRAMEWORK FOR THE SCENARIO-BASED MANAGEMENT SYSTEM

From the system feature analysis and actor modelling discussion, Scenario-Based Management (SBM) system can be described by MA system, which includes a set of nearly autonomous agents and a set of compromise scenarios to guide the agents' actions. The original concept of scenario was derived from the open trading agreement. In this paper, scenario represents the compromise policy in which every agent in the multiagent system agrees to cooperate. Scenario defines the relationship of the roles involved in stage of cooperation and the range of the roles' activity. Figure 1 shows the structure of a scenario in Conceptual Graph linear form. Based on the structure, one scenario can be set up by defining the roles and episodes. When the multiagent system intends to cooperate to solve problem, the related scenario should be specified by assigning each agent as role player to role in the scenario. The actual cooperation process for the problem will be carried out by executing this instantiated scenario.

This SBM system concentrates on solving integration and management level problems, based on compromise

```
[SCENARIO: x] -> (PART)-->[ROLE: {*a}]

[ROLE: *a]-

          ->(AGNT)-->[ROLE-PLAYER: {*}]

          ->(PART)-->[EPISODE: {*b}]

          ->(ATTR)-->[ROLE-STATE]<--(CURR)

[EPISODE: *b]-

          ->(INIT)-->[TIMER]

          ->(INIT)-->[PARCEL]->(SENR)->[ROLE]

          ->(INIT)-->[ROLE-STATE]

          ->(RSLT)-->[TIMER]

          ->(RSLT)-->[PARCEL]->(REPT)->[ROLE]

          ->(RSLT)-->[ROLE-STATE]
```

FIGURE 1 Scenario structure

scenarios. The agents in SBM system consist of transaction manager, subrole manager, planner, and adapter. (See Figure 2)

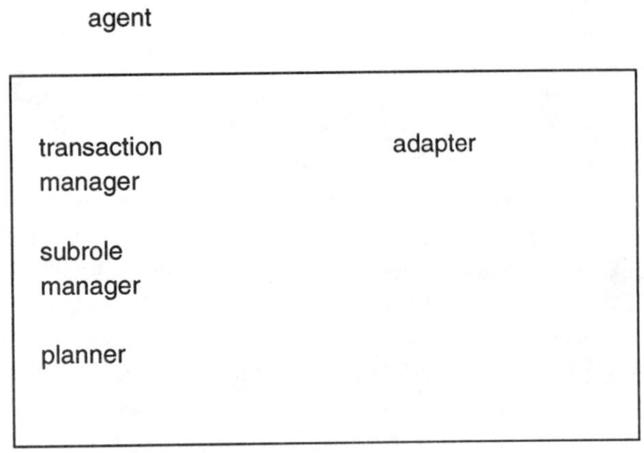

FIGURE 2. The agent in SBM system

Transaction manager is the only actor who is in charge of the external communication function for the agent. For multiagent communication, every agent can be replaced by its transaction manager, and the SBM system shrinks to a simple actor system, which passes messages among transaction managers.

Subrole manager is also an actor for subrole management in which the message generation is the major function to respond to the received messages. Because both transaction manager and subrole manager are actors, the communication between them depends on the messages passed to each other.

The function of the Planner in the agent is to choose the correct scenario for user requirements. The model structure for the planner is quite flexible: actor model is a good option if the planner doesn't have supervisory duty over other managers.

Actually, the "adapter" can be analogous to the assistant managers for transaction management. Transaction manager needs an adaptive mechanism to improve its performance over generations, and the adapter is designed to work in that way.

SBM system provides an operating environment for message passing among agents. As discussed above, as communication among agents is totally through the transaction managers for each agent, the external behavior of the multiagent system is the same as that of the transaction manager model.

5 TRANSACTION MANAGER

In this section, an actor called "transaction manager" is designed for Scenario-Based Management systems. Referring to the four features discussed in previous sections for the multiagent system, the actor structure and message transmission mechanism are designed to satisfy these requirements.

Actor structure is a template for an actor. In Object Oriented Programming language, actor structure can also be referred to as class, and actor as object or instance of a class. Actors are described in two basic parts: acquaintances and behavior. Actor acquaintances are specified in terms of knowledge slots obtained from the actor's structure or class. While actor behavior is specified in terms of procedural code called message-handlers which are attached to the actor's structure.

Actors cooperate via a message passing mechanism; i.e. send a message to an actor, and the actor works by executing the related message handler.

Table 1 depicts the general actor structure for an Open-EDI multiagent system.

TABLE 1. An actor structure

actor structure (class) name
a list of superclasses
shared knowledge slots
local knowledge slots
message handlers

The list of superclasses is the basis for inheritance of the actor structure, i.e., from which the new actor structure inherits knowledge slots and message handlers. In the actor system, we define an actor structure OBJECT as the superclass of itself to prevent endless looping.

Shared knowledge slots are placeholders for knowledge shared by actors. For example, suppose three actors called transaction_manager_1, transaction_manager_2 and transaction_manager_3 to cooperatively work on one Open-EDI transaction. The shared knowledge that they have is one scenario (a class of business rules) related to the transaction.

In contrast, local knowledge slots are placeholders for knowledge owned by individual actors. Consider the same example again! Although the three actors have some shared knowledge to work on the same transaction, each of them represents a nearly autonomous trading party, which has its own knowledge to make decision for itself. This kind of knowledge is local knowledge. For example, the transaction manager keeps some trading history data as its own local knowledge.

In the actor system, we deal with the problem by storing shared knowledge with the actor structure, while local knowledge is retained by the actor, assuming the actors with the same shared knowledge have the same actor structure.

Message handler specifies the behavior of a class of actors in response to a particular message. The basic actions corresponding to the major work for a message are called primary work, such as "send control message SCECOM to another trading party". However, in many cases, the individual actor needs daemons or shadowed handlers to choose a preferred primary action drawing on its local knowledge to responding the receiving message. In this actor system, the concept of "around type" message handler [referring 11] is used, which sets up an environment for the execution of the primary handlers. In the transaction manager model, this kind of message handler includes STARTING, SCECOM, SCEACK and TE.

The actor structure of the transaction manager is shown in Table 2 and Table 3. Table 4 gives an example of how to generate one actor through this actor structure.

In summary, the actor system satisfies the requirements of multiagent systems built from compromise scenarios and nearly autonomous agents. According to the actor structure design and message passing mechanism, assisted by the concepts of local knowledge and daemons, decentralized control and adaptive operation mechanism can be realized for the multiagent systems.

TABLE 2. Transaction manager (defclass part)

```
CLIPS> (ppdefclass
TRANSACTION_MANAGER)
(defclass TRANSACTION_MANAGER
  (is-a USER)
  (slot agent_name
    (initialize-only))
  (slot scenario_id
    (shared))
  (slot transaction_id
    (shared))
  (slot status
    (default idle))
  (slot ori_sender)
  (slot ms_r
    (multiple))
  (slot ms_rp
    (multiple))
  (slot ms_s
    (multiple))
  (slot ds
    (multiple))
  (slot ackis
    (multiple))
  (slot acksender
    (multiple))
  (slot ackcount
    (default 0))
  (message-handler STARTING)
  (message-handler SCECOM)
  (message-handler SCEACK)
  (message-handler TE))

NOTE: There are implicit get- and put- primary
      accessor message-handlers for every slot.

Recognized message-handlers:
init primary in class USER.
delete primary in class USER.
print primary in class USER.
```

TABLE 3. Transaction manager 1

CLIPS> (load "actor_structure") Defining defclass block TRANSACTION_MANAGER CLIPS> (load "message_handlers") Defining defmessage-handler STARTING primary in class TRANSACTION_MANAGER. Defining defmessage-handler TE primary in class TRANSACTION_MANAGER. Defining defmessage-handler SCECOM primary in class TRANSACTION_MANAGER. Defining defmessage-handler SCEACK primary in class TRANSACTION_MANAGER. CLIPS> (make-instance [transaction_manager_1] of TRANSACTION_MANAGER (scenario_id retail-buy) (transaction_id 008) (agent_name transaction_manager_1) (status available)) [transaction_manager_1]	CLIPS> (send [transaction_manager_1] print) transaction_manager_1 of TRANSACTION_MANAGER (agent_name transaction_manager_1) (scenario_id retail-buy) (transaction_id 8) (status available) (ori_sender) (ms_r) (ms_rp) (ms_s) (ds) (ackis) (acksender) (ackcount 0) CLIPS> (instances TRANSACTION_MANAGER inherit) CLASS TRANSACTION_MANAGER: transaction_manager_1 For a total of 1 instance.

CONCLUSION

The concept and features of Scenario-Based Management have been analyzed starting from DAI principles and review of Scenario-Based management. The framework for the SBM system is fundamentally designed by taking advantage of actor modelling. A prototype of one actor model called transaction manager for the multiagent systems has been developed in CLIPS.

REFERENCE

[1] B. Garner & D. Song, (1994). *Scenario-based management for Open-EDI Gateway*, EDI conference Bled., 1994.

[2] H. Durfee, (1991). *The Distributed Artificial Intelligence Melting Pot*, IEEE Transactions on Systems, Man and Cybernetics, Vol. 21, No. 6, Nov/Dec. 1991.

[3] A. H. Bond & L. Gasser, (1988). *An Analysis of Problems and Research in DAI,* Morgan kaufmann, San Mateo, Calif., 1988.

[4] C. Hewitt et al., (1973). *A Universal Modular ACTOR Formalism for Artificial Intelligence*, Proceeding of the International Joint Conference on Artificial Intelligence, Stanford, Aug. 1973.

[5] C. Hewitt, (1977). *Viewing Control Structures as Patterns of Passing Messages*, Artificial Intelligence, Vol 8, 1977.

[6] R. Davis & R. G. Smith, (1983). *Negotiation as a metaphor for Distributed Problem Solving*, Artificial Intelligence, Vol 20, 1983.

[7] Engelmore & Morgan (Eds), (1988). *Blackboard Systems*, 1988.

[8] M. Sloman, (1988). *A Framework for Distributed Systems Management*, distributed Processing, M. H. Barton et al. (Eds), 1988.

[9] G. Uma et al., (1993). *Distributed Intelligent Systems: Issues, Perspectives and Approaches*, Knowledge-Based Systems, Vol 6, No. 2, June 1993.

[10] D. Lukose, (1992). *Goal Interpretation As A Knowledge Acquisition Mechanism*, Ph. D thesis, Deakin University, Australia, Feb. 1992.

[11] (1991). *CLIPS Reference Manual*, Software Technology Branch, Lyndon B. Johnson Space Center, Sept. 1991.

[12] D. Song, (1994). "Open-EDI communication And Its Simulation", Tech report, May 1994.

A NEW SCHEMA FOR THE ENGINEERING DESIGN OF AUTONOMOUS AGENTS IN NON-MANUFACTURING DOMAINS

Robert L. Winter, Jonathan O'Brien and Ron Wakefield
Department of Engineering Construction and Management, School of Civil Engineering,
The University of New South Wales, Sydney 2052 Australia

ABSTRACT

The design of cognitive capabilities for artificial autonomous agents, capable of executing practical tasks in non-manufacturing domains, is the focus of this paper. Practical, as opposed to theoretical tasks, involve the use of valuable resources, and may result in unwanted effects. In such a context, it is proposed that a prerequisite to action is viability, meaning that the utility of an outcome is greater than the expense incurred. As a solution to the problem of designing real-world agents, a general schema for motivation and action, which directs activity to rational, viable outcomes, is presented. A 'viability manager', which embodies this schema, is described. Practical, illustrative application examples taken from construction engineering are presented, indicating the near-term industrial potential of the schema. Features of the schema which may facilitate global integration and multi-agent cooperation are described. Proposed simulation-based testing of agent performance is discussed.

1. INTRODUCTION

Within our society, the non-manufacturing industrial sector is of large scale and economic significance but is not susceptible to current methods of automation. It includes mining, agriculture, forestry, shipbuilding, commercial fishing and construction. In the construction industry [MFU93] there is a strong desire to increase productivity through automation, but there are no appropriate technologies. Current approaches fail in these areas because the nature of the industrial tasks and the general characteristics are very different to those of the orthodox manufacturing sector. These differences are highlighted in Table 1.

From Table 1, it is clear that non-manufacturing industries require some new tools and technology for field task execution. There is some similarity to the military battlefield, where control and coordination cannot be effectively provided from a single source [AES90]. Mobile, intelligent and autonomous (as opposed to merely programmable) systems have the potential to execute field tasks in these domains. With the current trend in engineering systems being towards increasing intelligence and autonomy [LM92, CET93], this paper addresses the design of such systems for direct industrial application.

2. DEVELOPMENT OF AUTONOMOUS TASK EXECUTION AGENTS FOR COMPLEX DOMAINS

Table 1 Comparison of Industrial Domain Characteristics

Manufacturing Domain	Non-manufacturing Domains
Primary Characteristics	
Small, manipulable workpieces able to be passed from operator to operator	Very large, fixed workpieces requiring traversal by operators
High volume of operations and high standardisation	Low-volume, or once-off operations with minimal standardisation
Secondary Characteristics	
Environment may be carefully controlled by enclosing the workspace	Outdoor environment is difficult or impossible to control and is highly variable
Simple operations, requiring low intelligence	Complicated operations, requiring higher intelligence
No spatial or resource conflicts	Spatial and resource conflicts possible
Nature of operations and interactions clearly defined	Operations and interactions need adaptation to prevailing conditions

2.1 System Specification

To be an autonomous and distinct entity, a system must be self-motivated and capable of making its own decisions and judgements, and designed with a particular purpose or function. For simplicity of reference we choose to call a system of this type an **agent** (but also note the concerns [Bro91, Osa93] that others use the term to refer to smaller, software-based computational units).

To operate successfully, an agent needs to:

(i) Have a function or purpose
(ii) Generate, as a result of its function, and in response to the environment, a **goal**
(iii) Be **motivated** to achieve that goal (Its **task** is to move from its current state to the goal state)
(iv) Produce **actions** which **execute** the task (ideally in a cost-effective manner)

The terminology chosen here leads to the description 'task execution agents'. It is assumed here that agents, possessing the above features, are a subset of 'systems'. Tasks are thus a subset of 'operations' and refer specifically to operations performed by agents. For completeness, it is necessary to say that 'autonomy' is used not to mean complete holistic independence, but the ability to act without direct control.

2.2 Precursor To Design - An Analysis Of Domain Requirements

2.2.1 Purposive Or Functional Feature

The operation of a system has a particular sphere of influence, causing changes to the values of ***n*** variables in the environment. The sphere of influence is determined by the designer's ability to perceive influence, not necessarily the total influence of the system.

The user or designer is concerned with a specific effect of the system's operation, which we will call the purpose or **function** of the system. The system is therefore designed and deployed to this end. The function does

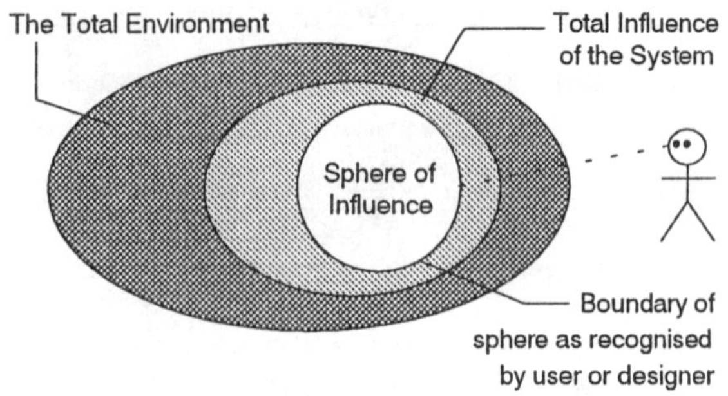

Figure 1 Sphere of Influence as Determined by System Designer

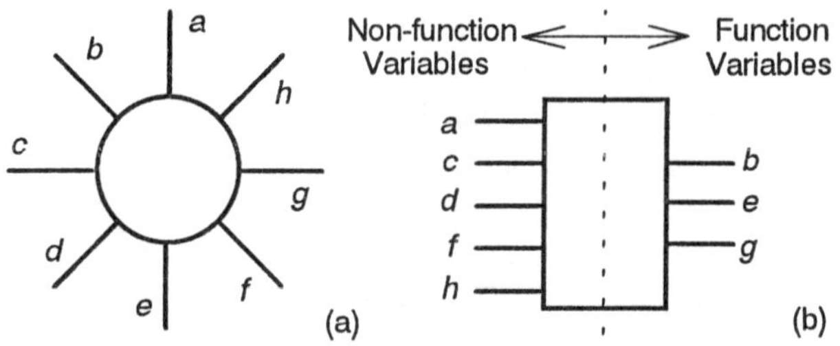

Figure 2 Functional Division of Influence Variables in a System

not detail all of the effects of the system, merely the effect that is desirable in a particular context. This idea is illustrated by an example from Beer [Bee67]: "*A motor-car would normally be regarded as a machine for transporting human beings, and with reason: this was the purpose for which it was built. But if we were studying the wearing qualities of metals, we could with perfect validity call the motor-car a machine for wearing out piston rings. As economists, we might call it a machine for operating garages at a profit.*" Hence, rather than referring to the inputs and outputs of systems, we may refer to functional and non-functional variables. We may define 'function' as that subset of variables, influenced by the system, which is related to the user's or designer's purpose. As an example, the operation of a speedboat may affect the supply of fuel, the stress level of ducks and fish, the amount of CO_2 in the atmosphere and the level of excitement of its passengers. We consider the latter to be its function; changes in other variables are non-functional, or side-effects. Figure 2(a) depicts a generic system with the set of variables within its sphere labelled (a-h). Figure 2(b) depicts a functional division of the **same** system, with these variables separated into 'functional' and 'non-functional' parts.

2.2.2 Viability Feature

Industrial systems exist for teleological reasons. The system and its operations exist because they serve some purpose. The implications are that the user derives some benefit or utility from the operations, or from changes in the function variables. In an industrial context, **changes in the non-functional variables are also important**. Specifically, resources and other inputs and outputs may be costly (i.e. of negative utility) to the user, such that the utility derived from the operation's function is diminished. This may be compared with purely computational agents for which the costs of action are negligible [Sug93]. We define the net utility of an operation, as being:

$$\Delta U = \sum_{i=1}^{n} B_i - \sum_{j=1}^{m} C_j$$

where B_i = the utility benefit derived from the change in function variable i
C_j = the utility cost incurred for side-effect j

We assume that a rational user will only allow the system to operate where the net utility is greater than zero. To describe this, we define a term called **viability**, such that:

If $\Delta U > 0$, operation is viable

If $\Delta U < 0$, operation is not viable

A rational user or designer will try to maximise the value of ΔU, and will only allow operation where it is viable. The respective utility values of B_i and C_j are not necessarily positive and negative. In a particular context, the benefit of a system's function may be negative.

2.3 Engineering Design Of Task Execution Agents

2.3.1 Design for Function

Designing for such autonomous functions as navigation, mobility, vision and manipulation, is an important part of the engineering of these agents, and is considerably more difficult than designing manufacturing systems. We define the design of functional capability to be 'functional design', and an engineering focus to that end to be a 'functional' focus.

2.3.2 Design for Viability

The consideration of viability is an essential feature of the operation of any industrial system, and hence of the actions of agents in such domains. In manufacturing, the fixed nature of the processes involved means that all of the operations and interactions are known at the design stage. Since the operations and interactions are fixed and predictable, utility measures, **which reflect the values of the designer or user**, may be used to calculate the viability of global and local operations. The designer therefore designs the system to be viable, and removes the need for any such assessment to be made during operations.

In complex domains such as construction, however, **each localised action** has an effect within its sphere of influence, which cannot, in detail, be predicted before action commences. Localised actions may therefore be viable, or unviable, from either a local or global position, affecting the outcome of the entire operation. It is therefore necessary to give each agent the ability to assess the viability of all of its actions, so that action depends on both function and viability. We define a design focus directed towards achieving viability to be a 'viability focus'.

3. A PROPOSED VIABILITY SCHEMA FOR INTELLIGENCE DESIGN

The following schema is presented as a possible solution to the problem of incorporating viability considerations into the intelligence of task execution agents. The particular solution is forwarded because of its simplicity, flexibility and generality. The controller or manager which embodies this schema will be called a **viability manager**. Its purpose is to act in parallel with functional intelligence and may be combined with systems of any functional capability. An agent possessing this controller is called a **viable agent**.

3.1 Description Of Viability Manager Features

The viability manager is linked to a functional system which is designed to execute actions which bring about a desired state or goal. In order to evaluate the viability of actions, the manager incorporates an evaluative model and knowledge in the form of a **value set**, which determines the utility of current and potential states as a precursor to action. The value set is divided into a benefit set, which evaluates the utility of function variable states, and a cost set which determines the side-effects and their utility costs. These features are represented in Figure 3.

Information about the actual function state is provided by the functional system and passed to the value set. Potential states are determined by the viability manager's

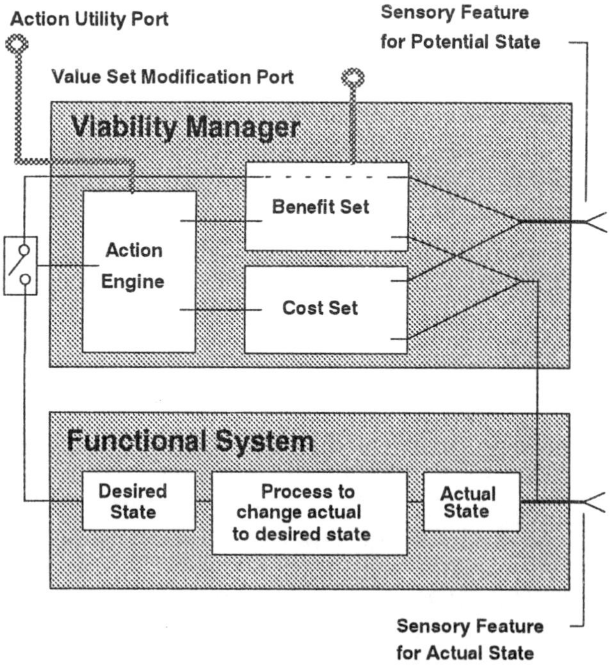

Figure 3 Viability Manager Attached to Functional System

sensory feature and passed to the value set. The benefit set evaluates the utility benefit of a change from the actual function state to the potential function state (ΣB_i). The cost set determines the utility cost of the same change in state (ΣC_j). These two values are passed to the action engine. If the net utility (ΔU) is greater than zero, the engine fires, and passes the potential state information to the functional system, thus forming a new desired state or goal. The functional system is then motivated to reach this goal and executes the necessary tasks. As the environment changes in relation to the agent, the potential function will change and hence the motivation of the agent. The manager contains the preconditions for action, but does not necessarily make plans. 'Reactive' behaviour such as this has the strength of being able to react to unanticipated local events [Hay93]. A steady-state occurs when potential states would not increase the net utility, in which case no action occurs. Action may therefore exhibit latency.

Where a choice between multiple actions is possible, the action engine requires the ability to store and compare the net utility of a number of potential states before firing and passing the best state to the functional system. The functions of the two 'ports' in Figure 3 are described in Section 4.

3.2 Illustrative Examples Of Agent Applications

3.2.1 Road Damage Repair Agent

Consider an artificial agent travelling along a roadway. Its function is to repair potholes or surface damage. It is equipped to sense the depth, d, of surface damage.

Benefit Set: Utility of repairing damage = $10d^2$units (say)

Cost Set: Cost of stopping to repair = 50 + 40d units

The action engine will fire when d > 5mm. The viability manager does not need to solve the quadratic equation in order to initiate action. It is simple in the sense that it merely initiates a fixed action when the environmental conditions are favourable. This is analogous to the implementation of a real-time 'policy'.

3.2.2 Earth Compaction Agent

Consider a soil compacting agent which may incorporate existing, mobile, mechanical compaction equipment. This agent has the capability of continuously changing the function state until it is no longer viable to do so, in a series of incremental actions.

Benefit set: Utility of compaction = 5 D units (say)

Cost Set: Cost of increasing compaction, by D units, = 0.5D - 5M units (D=Density, M=Moisture)

Where action is viable, the compactor will make continual, single passes over a field. Each pass increments the density by D. The cost here relates to fuel costs. Where moisture is present, the energy requirement decreases. Compaction starts at a density of 7, with moisture at zero units.

Table 2 Viability Manager Calculations for an Autonomous Compaction Agent

Actual D	Potential D	Benefit	Cost	Net Utility	Action Status
7	8	5	3.5	1.5	fire
8	9	5	4.0	1.0	fire
9	10	5	4.5	0.5	fire
10	11	5	5.0	0	no fire

If moisture is increased to 0.1 (by rain), the agent will run another cycle.

3.2.3 Concrete Finishing Agent

This agent has a similar function to the compactor in that it will incrementally increase the smoothness of concrete where viable. Here, however, there is a two dimensional slab, and the agent has a **choice** of smoothing any particular section. In this case, the benefit of smoothing decreases as smoothness increases, and thus the agent will first attend to the roughest areas.

4. POTENTIAL FOR INDUSTRIAL INTEGRATION OF VIABLE AGENTS

The viability manager could be used to augment a large number of existing or potential industrial items, ranging from simple to complex equipment and tasks. Its simplicity points to the possibility of implementation in the near term. The following sections illustrate how the above schema may facilitate the industrial implementation of autonomous agents.

4.1 Modification Of Agent Behaviour

By default, the agent described is a closed system in that its actions are a response to the environment and are self-regulated. The attractiveness of autonomy, however, must be balanced by the need to conform to economic, resource, managerial and other dynamic constraints that arise in complexity-rich, non-manufacturing domains. Because of the need to interact and to fulfil global objectives, the agent may become an open system where necessary, by allowing external replacement or modification of the value set through the **value set modification port** shown in Figure 3. This does not alter the function, only the nature of the system's response to the environment. In essence, this means that higher level systems may provide a type of meta-control (this term is taken from [Osa93]) to the agent. The agent is still self-motivated. Modifications may be **long-term changes**, which include the content of the value set or the measured variables, or **short-term changes**, where only the magnitudes of the value function are altered. With short-term modifications, neither the function, nor sphere of influence, are affected.

As an example, the road repair agent may be employed to surface normal roads, or the race track for a grand prix event. To the agent, a road is a road, and according to its function, and its viability manager, will behave in the same manner. Its capabilities do not extend to this consideration. At a higher level, this may be understood, such that the value of filling potholes is five times higher for a race track. To adapt to this, a user needs only to modify the value register by multiplying by five, and the agent will now fill much smaller holes.

If we increase the value of a functional outcome, we simply expand the 'feasible region of action'. This means two things: A previous condition which was not attractive is now attractive; but more importantly, the same outcome now justifies side effects which were previously considered too costly. Thus if the fuel cost of a helicopter flight is fixed, the value of it carrying someone to work may not be justified, but if a sick child is now the passenger, it may then be justified. This may be termed 'contextual rationality'.

4.2 Integration Of Multiple Agents

The authors recognise that the integration of multiple agents still requires many problems to be addressed, including task conflict and interaction, communication and negotiation. However, the schema described is powerful in that the utility measures provide a common 'language' that, if shared by all agents, may facilitate the integration of heterogeneous agents, and also allow the very important capability of being able to replace or upgrade one agent or group of agents without having to alter all others.

In a **hierarchy** of agents, the higher agent is interested in achieving some function. Other agents may perform part of its sub-function. By connecting to the **value set modification port** of any of these agents, the higher agent or user (which may be a person) can, at a high-level, *feed forward* information to control the response of the 'sub-agents'. By connecting to the **action utility port**, which is designed to return the **net utility** of a potential action, the agent may *feed back* information about the viability of the potential actions of each individual agent and thus the combined action. Importantly, with these abilities, the user can **inhibit** action by reducing the magnitude of the value functions, or motivate locally unviable action to occur where the result is globally beneficial. Where agents must **interact** to avoid conflict and to **cooperate**, a form of communication is desired to allow negotiations to take place [Sug93]. By incorporating information from the action utility port, agents may compare, or rank [Kuh93] the benefits of conflicting actions to determine priority, or to coordinate actions to conditions of mutual benefit.

5. TESTING OF VIABILITY CONTROLLER AND SCHEMA

A simulated environment, capable of incorporating the actual physical aspects of task environments, has been developed [WOP94]. The ability to simulate the physics of real environments is important for testing physical, rather than simply computational agents. Logical testing of single agents, subject to meta-control by the user, in this environment, has been commenced by the authors. The pseudo-physical agents are present in the graphical environment which allows the user to alter environmental variables, and to observe the performance of the agents. Testing is to include spatial and temporal variation of environmental variables, such as traction, temperature, wind speed, terrain height, object size and mass, as well as alterations of agent internal value sets. Further testing is planned for multi-agent and hierarchical activity.

6. CONCLUSION

The important differences between manufacturing and non-manufacturing domains have been analysed and it is concluded that conventional automation is inapplicable to the latter areas because of the differences in domain characteristics. In non-manufacturing domains, the nature of operations and their interactions, which take place in variable, spatially-distributed and poorly-controllable environments, prohibit centralised control. Intelligent autonomous task execution agents have been identified as a potential technology for executing tasks in those domains. For the successful design and implementation of industrial systems, issues of both functionality and viability need to be addressed. Localised operations, which together result in a global outcome, must be viable in a global context. The capability to assess the viability of actions needs to be incorporated into the autonomous systems responsible for action. Based on this need, a new, general, flexible agent control schema was presented. Simple domain examples from construction engineering have been used to illustrate the potential for near-term industrial applications, as well as possibilities for the integration of multiple agents. A powerful simulation environment is to be used for the logical testing of agents and their performance in physical, as opposed to computational environments. We consider that this type of approach has significant potential for advancing the application of practical autonomous agents to non-manufacturing domains.

REFERENCES

[AES90] C. Applegate, C. Elsaesser, and J. Sanborn. An Architecture for Adversarial Planning. *IEEE Transactions on Systems, Man and Cybernetics*, 20: 186-194, 1990.

[Bee67] S. Beer. *Cybernetics and Management.* Second Edition. The English Universities Press, London, U.K., 1967.

[Bro91] R. Brooks. Intelligence Without Reason. In *Proceedings of the Twelfth International Joint Conference on Artificial Intelligence (IJCAI)*, Sydney, Australia, August 1991. Morgan Kaufman

[CET93] L. Chaudron, J. Erceau and B. Trousse. Co-operative Decisions and Actions in Multi-Agent Worlds. In *Proceedings of the IEEE International Conference on Systems, Man and Cybernetics*, Le Touquet, France, 1993.

[Hay93] B. Hayes-Roth. Opportunistic Control of Action in Intelligent Agents. *IEEE Transactions on Systems, Man and Cybernetics*, 23: 1575-1587, 1993.

[Kuh93] N. Kuhn. Comparing Rankings of Heterogeneous Agents. In *Proceedings of the 7th Conference on Organizational Computing Systems*, Milpitas, California, USA, 1993.

[LM92] V. Lun and I. M. MacLeod. Strategies for Real-Time Dialogue and Interaction in Multiagent Systems. *IEEE Transactions on Systems, Man and Cybernetics*, 22: 671-680, 1992.

[MFU93] Y. Miyatake, T. Fujimori and T. Ueno. Automation and Robotics in the Construction Industry. In *Robots for Competitive Industries - Proceedings of the International Conference of the Australian Robot Association and the International Federation of Robotics*, Brisbane, Australia, July 1993.

[Osa93] N. Osato. An Action Interpreter of a Robot Control Agent. In *Proceedings of the 1993 IEEE/RSJ International Conference on Intelligent Robots and Systems*, Yokohama, Japan, July 1993.

[Sug93] T. Sugawara. Using Action Benefits and Plan Certainties in Multiagent Problem Solving. In *Proceedings of the 9th Conference on Artificial Intelligence for Applications*, Orlando, Florida, USA, 1993.

[WOP94] R. R. Wakefield, J. B. O'Brien and J. H. Perng. Development of a Real Time, Environmentally Interactive, Robotic Construction Systems Simulator. Submitted for publication in *Robotics in Civil Engineering*, a special issue of *Microcomputers in Civil Engineering.*

Adaptive Architectures

A STOCHASTIC OPTIMIZATION ALGORITHM FOR NEURAL NET LEARNING

Boris Igelnik and Yoh-Han Pao
Department of Electrical Engineering and Applied Physics,
Case Wesren Reserve University,
Cleveland, Ohio 44106, U.S.A.
Email: bxi2.po.cwru.edu

ABSTRACT

A stochastic optimization algorithm is suggested to minimize a multivariate continuous function defined on the closed bounded set. The function is supposed to have the unique global minimum in its domain (but may have any finite number of local minima). The algorithm is based on the limit-integral representation of the coordinates of minimizing point and approximation of the integrals by the Monte Carlo method. This procedure is combined with the iterative decreasing of the volume of search space. An asymptotic analysis suggested the algorithm is efficient for a given class of functions. We present numerical examples which confirm good performance of the algorithm, and discuss possible application of this technique and its possible versions to neural net learning.

INTRODUCTION

This work was motivated by the previous investigations of the authors in neural net computing, namely in the area of one hidden layer feedforward neural nets (FF1) and their applications. One of the authors suggested a specific architecture of the FF1 net, called Functional-Link net [Pao89]. In particular, in one Random Vector version of the Functional-Link net (RVFL) a continuous d -variate function f is expanded in the form

$$\tilde{f}(x) = \sum_{n=1}^{N} a_n g(w_n \cdot x + b_n),$$

$$a_n, b_n \in R, w_n, x \in R^d,$$

where parameters b_n, w_n are chosen randomly, in advance, and only parameters a_n should be learned. The strategy of the Functional-Link net is, therefore, to reduce the problem of learning or optimization to the quadratic one, which allows us to use efficient numerical methods, such as conjugate gradient method, for example. In recent years we received quite enough evidence that this approach combines simple efficient learning and sufficient generalization capability [PT91, PPS92, PPS94, PPE94, PPI94]. We also obtained a mathematical justification for the RVFL approach [IP93, IP94, PI94], based on the central idea of using a limit-integral representation of the function with subsequent evaluation of the integral by the Monte Carlo method. Expansions in terms of randomly placed radial basis functions would constitute another example of that approach.

We see that complexity of learning is the major factor in the overall complexity of constructing and using a neural net. Thus if there is no efficient learning (optimization) procedure, then a trade-off should be made between complexity of learning and generalization capability of the neural net. In other words, the lack of efficient optimization procedure put restrictions on the utilization of the full generalization power of the neural net. That is why we have switched from the search of a good architecture of the neural net to a search of an efficient optimization procedure for *nonquadratic multivariate* functions.

Looking for an efficient optimization procedure, we used the same central idea as in our justification of the Functional-Link net. Namely, if the quantity of interest can be represented by a multiple integral, then the most efficient technique of approximating the integral is use of stochastic or pseudostochastic methods [DR84, Str71]. Fortunately there exists a limit-integral representation for the coordinates of minimizing point if the function to be minimized has unique point of the global minimum in the closed bounded set (without any restrictions on the number of local minima) [Pin68]. At the same time, it is well-known [KW86], that use of the simple Monte-Carlo procedure is often inefficient in the approximation of multiple integrals. One of the ways to use *guided stochastic search* instead of simple Monte Carlo was suggested in [Pin70]. The idea of the method is step by step correction of initial uniform distribution of random points participating in the evaluation of the integral to the final distribution concentrated in the small neiborhood of the point of minimum. This final distribution is obtained in the process of evolving of a simulated Markov chain to its stationary distribution. The idea, good in itself, and actually constituting the mathematical base for simulating annealing [AK89], in practice often leads to inefficient procedure of optimization.

In this paper we suggest a simpler algorithm, also using limit-integral representations [Str71], for minimization of a continuous multivariate function, with possible application in neural net learning. The advantage of our approach is that the complexity of the algorithm is small and predictable. We confirm that by asymptotic analysis and numerical examples. More detailed analysis is left for forthcoming article.

STOCHASTIC OPTIMIZATION ALGORITHMS BASED ON THE LIMIT-INTEGRAL REPRESENTATION

In this section first we give an analysis of basis formulas for the limit-integral representation of the minimizing point coordinates and minimal value of the function. We explain why approximations, derived from these formulas using simple Monte Carlo, are not efficient. Secondly, we describe an algorithm which combines the simple Monte Carlo with succesive decreasing of the volume of the search space. Third, we show that this algorithm is much more efficient than one based on the simple Monte Carlo.

We consider a M-variate continuous function $F(z) = F(z_1,\dots z_M)$ defined on the closed bounded set S. Without loss of generality we assume this set S to be a standard unit cube $I^M = [0,1]^M$. Interpretation of the function in the neural net applications can be system error or error of generalization or total error [Bar93] for FF1, variables $z_1,\dots z_M$ are parameters of the net $a_1,\dots a_N, b_1,\dots b_N, w_{11},\dots w_{1d},\dots w_{N1},\dots w_{Nd}$, $M = N(d+2)$. We suppose that function F has a unique global minimum at point $z_* = (z_{*1},\dots z_{*M})$ on I^M. It was proven in [Pin68] that

$$z_{*m} = \lim_{w\to\infty} \frac{\int_S z_m \exp(-wF(z))dz}{\int_S \exp(-wF(z))dz}, m = 1,\dots M. \quad (1)$$

Nonrigorous derivation of formula (1) is straightforward. The ratio of two integrals in (1) can be represented in the form

$$\frac{\int_S z_m \exp(-w[F(z) - F(z_*)])dz}{\int_S \exp(-w[F(z) - F(z_*)])dz}.$$

If $z \neq z_*$ then $w[F(z) - F(z_*)] \underset{w\to\infty}{\to} \infty$. Therefore only points z which are in infinitely small neiborhood make nonzero contribution in both integrals. For this points we have $z \approx z_*$ and

$$\frac{\int_S z_m \exp(-w[F(z) - F(z_*)])dz}{\int_S \exp(-w[F(z) - F(z_*)])dz} \approx$$

$$\approx z_m \frac{\int_S \exp(-w[F(z) - F(z_*)])dz}{\int_S \exp(-w[F(z) - F(z_*)])dz} = z_m.$$

From formula (1) for sufficiently large w we have

$$z_{*m} \approx \frac{\int_S z_m \exp(-wF(z))dz}{\int_S \exp(-wF(z))dz}, m = 1,\dots M. \quad (2)$$

Using simple Monte Carlo, we obtain from (2)

$$z_{*m} \approx \frac{\sum_{k=1}^{K} z_{km} \exp(-wF(z_k))}{\sum_{k=1}^{K} \exp(-wF(z_k))}, m = 1,\dots M, \quad (3)$$

where $z_k = (z_{k1},\dots z_{kM})$ and all z_{km}, $k = 1,\dots K$, $m = 1,\dots M$ are independent and uniformly distributed in $[0,1]$ random variables. The approximation (3) does not work well in practice and below we give explanation for that.

The total error of approximation (3) can be evaluated as a sum of two errors: error of approximation by formula (2) and error of approximation the integrals by finite sums in formula (3). It is quite evident that error of approximation (2) tends to zero if $w \to \infty$. Indeed the following estimate of this error was obtained in [Pin68]

$$\left| z_{*m} - \frac{\int_S z_m \exp(-wF(z))\,dz}{\int_S \exp(-wF(z))\,dz} \right| <$$

$$< \varepsilon + LV(S - N_\varepsilon)\exp(-w\delta/2)/V(N_\eta), \qquad (4)$$

where ε is arbitrary positive number, δ and η depends on ε, $\delta > 0, \eta > 0$, N_ε, N_η are neighborhoods of the point z_* with radii ε, η correspondingly, $V(B)$ denotes volume of B, and $L = \max_{z \in S} |z_m - z_{*m}|$. For fixed S right side of (4) can be made arbitrarily small by choosing small ε and large w. Note that the same effect can be achieved with smaller w by reducing the volume of S. We use that observation in our algorithm.

The error of approximation of the integral $\int_S \varphi(z)dz$ by the finite sum $\left(\sum_{k=1}^{K} \varphi(z_k)\right)/K$ is proportional to the quantity [KW86]

$$\sigma_\varphi / \sqrt{K},$$

where

$$\sigma_\varphi^2 = \int_S \varphi^2(z)\,dz - \left[\int_S \varphi(z)\,dz\right]^2. \qquad (5)$$

Therefore, this error tends to zero if $K \to \infty$. Thus, the error of approximation of the ratio of two integrals in (1) by the ratio of two finite sums in (3) tends to zero if $K \to \infty$ for a fixed w. At the same time, taking, for example, $\varphi(z) = \exp(-w[F(z) - F(z_*)])$, we obtain (for simplicity of notations temporarily letting $d = 1$)

$$\int_S \exp(-w[F(z) - F(z_*)])dz$$

$$\underset{w\to\infty}{\sim} \int_{N_\varepsilon} \exp(-w[F(z) - F(z_*)])dz$$

$$\underset{w\to\infty}{\sim} \int_{N_\varepsilon} \exp(-w[z - z_*]^2)dz = O(1/\sqrt{w})$$

In the same way we can show that the second integral in (1) also tends to zero if $w \to \infty$. Since both sums in (3) are asymptotically (in K) Gaussians with asymptotically (in w) zero-means, the limiting distribution of their ratio is Cauchy-distribution [Fel66] with a variance depending only on the ratio of the variances of the sums, and, therefore, the variance does not tend to zero with $w \to \infty$. Nethertheless the total error of approximation (3) can be made tending to zero, if we balance w and K so that

$$\operatorname{var}\left\{\frac{1}{K}\sum_{k=1}^{K} \exp(-w[F(z_k) - F(z_*)])\right\} <$$

$$< \frac{1}{t}\int_{I^M} \exp(-w[F(z) - F(z_*)])\,dz,$$

where t is sufficiently large number (in practice 3 or 4 is enough). But in this case we need very large values of K to get reasonable accuracy. That is why simple Monte Carlo is not a good way to use representation (1). Another important conclusion from our analysis is that for each K there exist optimal value of w which brings sum of two errors to the minimum.

In the same manner the following limit-integral representation for the minimal value of the function F in S

$$F(z_*) = \lim_{w\to\infty} \frac{\int_S F(z)\exp(-wF(z))\,dz}{\int_S \exp(-wF(z))\,dz}, \qquad (6)$$

can be derived and analyzied.

Instead of using simulated Markov chain for enhancing simple Monte Carlo [KW86], which results in so much additional complexity, that makes purpose of the method unachievable, we suggest a simple algorithm.

Divide I^M (or in general S) in two sets I^M_{10}, I^M_{11} by the plane $z_1 = 0.5$. If minimal value of the function F in I^M_{10} is less then minimal value of F in I^M_{11}, we decide that

$z_* \in I_{10}^M$, otherwise $z_* \in I_{11}^M$. Those decisions are made on approximations of minimal value of the function by formula (6). Chosen set is then divided by the planes $z_2 = 0.5, \dots z_M = 0.5$. After last division of the search space we come to a new cube I_1^M with an edge equal to 0.5. For a fixed integer R, repeating this procedure $(R-1)$ times, we come to a cube I_R^M with an edge equal $(0.5)^R$. The solution of the problem then calculated by formula (3) with $S = I_R^M$.

Now we describe the asymptotic analysis of the algorithm. This algorithm is based on two simple observations. First, a smaller volume of sample, is needed to determine if the inequality $\min_{z \in I_0^M} F(z) < \min_{z \in I_1^M} F(z)$ is true or not, than to find z_*. Indeed, if

$$\Delta = \left| \min_{z \in I_0^M} F(z) - \min_{z \in I_1^M} F(z) \right|$$

then it is sufficient to have a value of w such that

$$\left| \min_{z \in I_i^M} F(z) - \frac{\int_{I_i^M} F(z)\exp(-wF(z))dz}{\int_{I_i^M} \exp(-wF(z))dz} \right| < \Delta/4, \quad i = 0,1.$$

Then it is sufficient for K to satisfy inequality

$$\mathrm{var}\left[\frac{1}{K} \frac{\sum_{k=1}^{K} F(z_k)\exp(-wF(z_k))}{\sum_{k=1}^{K} \exp(-wF(z_k))} \right] < \Delta/4t$$

This conclusion is true for any set S, subdivided in two halves, if the volume $V(S)$ is large enough. Thus, the probability of a wrong decision can be bounded by a constant $q \ll 1$. Secondly by subdividing the search space in two equal parts we decrease it volume with exponential rate. As our examples have shown one step of the algorithm usually decreases the error about ten times. That is why small errors can be achieved quite fast.

The average value of the edge of the last accepted cube, which contains minimizing point, is bounded by the quantity

$$q + qp/2 + \dots + q[p/2]^{MR} \approx q/2.$$

Therefore the average number of iterations $\bar{I}$ needed to get any accuracy is bounded by the quantity

$$\bar{I} \approx -M \log_2(q/2) \tag{7}$$

That means that average number of iterations increases M almost linearly with M.

The suggested algorithm is supposed to use optimization over parameters w and K on each step. In practice, however, we used optimization over w and K only on the first step. We kept this value of K on the next steps, while multiplying value of w by a number θ, $1 < \theta < 3$ on each next step.

NUMERICAL EXAMPLES

We give two examples for illustration of our algorithm. In first example we considered a function of M variables of the form

$$F(z) = \sum_{m=1}^{M} \left[z_m + \sum_{m'=1}^{M} z_{m'} - 0.33(M+1) \right]^2$$

for $3 \le M \le 10$ and $0 \le z_m \le 1$, $m = 1, \dots M$. The number of steps was $R = 5$. Succesive minimal values of the function into two halves of S are given in the table 1 for last two steps, $M = 8$, $K = 300$.

Table 1 Minimal Values of the Function in the Last 16 Iterations

F1	F2	# of iteration	F1	F2	# of iteration
0.001738	0.000357	25	0.000141	0.000622	33
0.002464	0.000433	26	0.000172	0.001146	34
0.002826	0.000337	27	0.000045	0.001226	35
0.001858	0.000638	28	0.000175	0.000598	36
0.002124	0.001178	29	0.000083	0.000711	37
0.002169	0.000907	30	0.000096	0.000689	38
0.002329	0.000875	31	0.000112	0.000649	39
0.002523	0.000763	32	0.000027	0.000693	40

Exact minimal value of the function $F(z_*) = 0$, while the algorithm provided value $F(z_*) = 0.0000010$. Exact coordinates of the minimizing point are $z_{*m} = 0.33$, $m = 1,\ldots 8$, while the algorithm gave

$$0.329014,\ 0.328569,\ 0.330695,\ 0.330844,$$
$$0.331156,\ 0.329674,\ 0.328828,\ 0.330548.$$

At the same time the simple Monte Carlo with equivalent volume of sample equal 12000 gives only 0.001285 as an estimate of the minimal value of the function. Thus this example illustrates higher efficiency of the algorithm compared with the simple Monte Carlo utilization of the formula (1).

Another example is minimization of the Rozenbrock function [Ban84]

$$F(z) = 100(z_1 - z_0^2)^2 + (1 - z_0)^2.$$

This function is considered to be a difficult one for minimization. Exact minimal value of the function $F(z_*) = 0$ if $z_* = (1,1)$. We used in our algorithm $K = 300$, $R = 5$, $0.2 \le z_i \le 1.2$. Approximations, found by the algorithm, were

$$F(z_*) = 0.000002, \quad z_* = (1.000197, 1.000522).$$

Intermediate results are given in Table 2.

Table 2 Minimal Values of the Function on 10 Iterations

F1	F2	# of iteration
0.153633	0.020327	1
0.041052	0.005024	2
0.007847	0.000521	3
0.001996	0.001161	4
0.000720	0.008333	5
0.000013	0.003095	6
0.000164	0.000926	7
0.000139	0.000211	8
0.000515	0.000043	9
0.000187	0.000039	10

CONCLUSIONS

We suggested an algorithm which can be useful in solving applied optimization problems. An asymptotic analysis of the algorithm and numerical examples confirms efficiency of the algorithm compared with utilization of the simple Monte Carlo. The algorithm is based on the idea of guided stochastic search and simple in implementation. We will consider in future work also other possibilities of guidance in stochastic search based on a limit-integral representation of the minimizing point and the minimal value of the function.

REFERENCES

[AK89] E. Aarts and J. Korst.. *Simulated Annealing and Boltzmann Machine: a Stochastic Approach to Combinatorial Optimization and Neural Computing.* John Wiley, New York, U.S.A., 1989.

[Ban84] B. D. Bunday. *Basic Optimization Methods.* Edward Arnold, UK,. 1984.

[Bar93] A. Barron. Approximation and estimation bounds for artificial neural networks. *IEEE Trans. on Inf. Theory,* 39:930-945, 1993.

[DR84] P. J. Davis and P. Rabinowitz. *Methods of Numerical Integration.* Academic Press, New York, U.S.A., 1984.

[Fel66] W. Feller. *An Introduction to Probability Theory and Its Applications*, volume II. John Wiley, New York, U.S.A., 1966.

[KW86] M. Kalos and P. A. Witlock. *Monte Carlo Methods*, volume I:Basics. John Wiley, New York, U.S.A., 1986.

[IP93] B. Igelnik, Y.-H. Pao. Additional perspectives on feedforward neural nets and the Functional-Link net. In *Proceedings of IJCNN'93-Nagoya,* Nagoya, Japan, October 1993.

[IP94a] B. Igelnik, Y.-H. Pao. Random vector version of the Functional-Link net. In *Proceedings of 28th Annual Conference on Information Sciences and Systems,* Princeton, U.S.A., March 1994.

[IP94b] B. Igelnik, Y.-H. Pao. Stochastic choice of basis functions in adaptive function approximation and the Functional-Link net, Submitted to *IEEE Transactions on Neural Networks*, 1994.

[Pao89] Y.-H. Pao. *Adaptive Pattern Recognition and Neural Networks.* Addison-Wesley, Reading, MA, U.S.A., 1989.

[PI94] Y.-H. Pao, B. Igelnik. Mathematical concepts underlying the Functional-Link approach. In *Proceedings of International Neural Network Society Annual Meeting*, San Diego, CA, U.S.A., June 1994.

[Pin68] M. Pincus. A closed form solution of certain programming problems. *Operation Research*, 16:690-694, 1968.

[Pin70] M. Pincus. A Monte Carlo method for the approximate solution of certain types of constrained optimization problems. *Operation research*, 18: 1225-1228, 1970

[PPE*94] G. H.Park, Y.-H. Pao, K. G. Eyink, S. R. Leclair, and M. S. Soclof. Neural-net based optical ellipsometry for monitoring growth of semiconductor films. In *IFAC Publications, Artificial Intelegence in Real-Time Control*, Valencia, Spain, October 1994.

[PPI*94] G. H. Park, Y.-H. Pao, B.Igelnik, K. G. Eyink, and S. R. Leclair. Neural-net computing for interpretation of semiconductor film optical ellipsometry parameters, Accepted to *IEEE Transactions on Neural Networks,* 1994.

[PPS92] Y.-H. Pao, S. Phillips, and D. J. Sobajic. Neural-net computing and the intelligent control of systems. *International Journal of Control*, 56:263-290, 1992.

[PPS94] Y.-H. Pao, G. H. Park, and D. J. Sobajic. Learning and generalization characteristics of the random vector Functional-Link net. *Neurocomputing*, 6:163-180, 1994.

[PT91] Y.-H. Pao and Y. Takefuji. Functional-Link computing: Theory,system architecture, and functionalities. *Computer magazine*, 3:76-79, 1991.

[Str71] A. H. Straud. *Approximative Integration of Multiple Integrals.* Prentice-Hall, Englewood, NJ, U.S.A., 1971.

SIGNAL TRANSFER AND BASIC LEARNING IN A BIOLOGICALLY-MOTIVATED NEURAL PROCESSOR MODEL

Murray W. Mount and Andrew P. Papliński
Department of Robotics and Digital Technology,
Monash University, Wellington Road, Clayton 3168, Australia.
E-mail: `mwm@rdt.monash.edu.au`

ABSTRACT

A model of a proposed Neural Processor Array which can be used to emulate a variety of biologically motivated neural structures and behaviours is described. The basic building block of the array, the Neural Processor Element, or NPE, provides flexible connectivity with neighboring NPE's, to allow modelling of a range of realistic 3D neural micro-structures. The NPE model is based on a simplified set of neuro-biological mechanisms and can be configured to perform various realistic functions related to modulation, aggregation and generation of neural pulse streams. Short term learning is implemented in a pre-synapse model, while a post-synapse model is used to perform long-term reinforcement style learning. Simulation results illustrating realistic signal processing and basic learning behaviour that can be obtained are presented. Issues such as "spatio-temporal" processing of neural information, implications of modelling short-term and long-term learning and natural ways of combining unsupervised and supervised learning in neural networks are also considered.

INTRODUCTION

Through efficient modelling of basic neuro-biological mechanisms it should be possible to improve the performance of artificial neural systems. Problems involving real-time perception, control and adaptation—traditionally difficult to solve using digital computers based on the Von Neumann model but which are solved very effectively by living creatures—should become more manageable with dedicated, parallel neural processing hardware based on biological principles. Although simplifications and practical assumptions must be made when developing biologically realistic models suitable for hardware implementation, it is possible to devise analog and/or digital VLSI circuits that function in similar ways to well-established neural mechanisms and systems. Such a modelling process should also help to improve theories of neural networks and contribute to the understanding of natural neural phenomena such as learning, memory and intelligence.

The proposed device will employ a reconfigurable 3D array structure which can be used to model realistic and highly interconnected neural systems using only local connections between processors. Configuration of the array to model particular neuronal systems is achieved by setting appropriate switching and continuous valued parameters within each NPE.

After the initial system-level modelling and simulations have been completed, a neural processor microchip will be developed using a hybrid analog-digital approach. Although hardware implementation issues will not be considered in any depth here, some general principles of the design can be stated.

Where possible analog circuits will be used for basic signal processing operations due to their compactness, low energy consumption and suitability for neural computation [Mea91, DFF93, ABP+91]. Pulse-stream techniques, investigated by Murray, Hamilton and others [MTR+91, HMB+92] will be employed in the transmission and modulation of neural output signals. In pulse-generating neurons average output signal strength is encoded by pulse frequency, however precise timing (or phase) information can also be retained. Pulse-encoding of neural states can therefore be of use in various forms of dynamic neural signal processing. All control and interfacing with the microchip will be performed digitally. Due to technical difficulties associated with local storage and update of synaptic weights using analog techniques [MTR+91], some digital circuitry may also be employed for this purpose.

The eventual goal is to build a 3D multi-chip array with suitable parallel input and output ports, under SI-MD control from a host workstation or PC. Using suitable software, the workstation would be responsible for configuration and graphic display of network topology as well as monitoring of transient signals and weight values.

In the next section some typical structures within the cerebral/cerebellar cortex are considered in order to arrive at a sufficiently versatile yet easily implemented interconnection scheme. Following this, the NPE and its sub-systems are outlined. The basic signal transfer functions that can be performed with the neural processor array (NPA) model are then demonstrated. Finally, some preliminary simulation results of the short and long-term learning functions are examined.

PROCESSOR ARRAY STRUCTURE

By investigating some of the basic structures, or micro-circuits of the cerebral/cerebellar cortex, observations can be made about the hardware requirements and an appropriate NPE interconnection scheme for a biologically motivated neural processor array.

The cortex has certain typical characteristics such as a layered organization with specific types of neurons within each layer. One remarkable feature is the way that parallel fibre inputs and the 2D dendritic trees of primary neurons, form along separate orthogonal planes. The parallel fibres therefore only intersect the dendritic tree of a given neuron and form a synaptic connection with it at one position, ensuring that input patterns are widely distributed [Arb89].

In Figure 1, some typical structures found within the cortex are illustrated. These structures are highly simplified and speculative, and have been adjusted to fit within a regular rectangular array. More detailed physiological diagrams of the cortex can be found for example in [Arb89, She94]. See also [Kan92, MR94] for examples of artificial associative memory systems that model structures of the cerebellum.

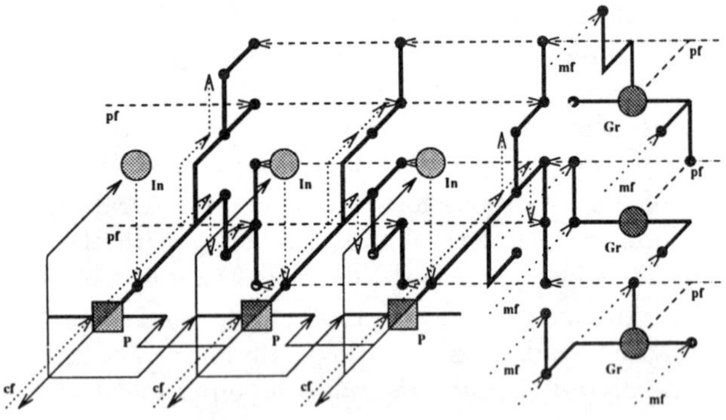

Figure 1: Highly simplified model of cortex.

The approach taken with this project is to provide a framework within which realistic neural structures may be modelled in hardware and their real time functionality explored. For this purpose the exact way in which the various types of cells (primary neurons, granule cells, inhibitory inter-neurons) function does not need to be known, however some general principles are worthwhile considering.

The Primary or Purkinje cells (**P**) receive dispersed sensory inputs from the parallel fibres which intersect their flattened (apical) dendrites and then compete with each others by means of a set of laterally inhibitive interconnections, received on their (basal) dendrites. These networks apparently perform competitive filtering and feature mapping of inputs (along the lines of Kohonen networks [Koh88, Lin86]). Inhibitory inter-neurons (**In**) provide local negative feedback and help to stabilize neuronal systems, although their full purpose is unknown [Arb89]. The parallel fibres (**pf**) are axon collaterals of granule cells (**Gr**), which receive the inputs, from more remote origins via mossy fibres (**mf**). Unlike the highly dispersed parallel fibre inputs, another set of inputs, the climbing fibres (**cf**), form very specific connections with particular **P** cells. These inputs have a pronounced effect on the output state of a given **P** cell, thus providing reinforcement signals which effect the learning behaviour of the cortical **P** cell networks.

By modelling such a scheme, it should be possible to build artificial neural systems that integrate unsupervised (competitive) and supervised (reinforcement or "delta rule") learning in a biologically realistic way, retaining important spatial and temporal information in the input data. In order to properly explore these sort of neural behaviours in an artificial neural system, it is clearly necessary to adopt an NPA interconnection or lattice structure within which structures such as those outlined can be faithfully represented.

What is required could be described as a "generalized cortical micro-circuit", suitably simplified to make it relatively easy to implement in VLSI hardware. Considering the range of neural micro-circuits that need to be modelled, a few observations about NPE connectivity can be made:

1. Using only local inter processor connections it is possible to build up highly inter-connected structures, without the need for multiplexing, by mimicking basic dendritic and axonal branching patterns.

2. A 3D array structure must be adopted in order to model branching neural structures of the cortex properly. Such a structure allows correct topographic mapping and temporal processing of 2D input fields and makes it possible to model the distribution of parallel fibre inputs and 2D dendritic trees along separate orthogonal planes.

3. In order to maximize the range of structures that can be modelled and in the interests of realizing a consistent and easily scalable architecture, the same connectivity should be provided in each of the three axes.

4. As a general rule provision should be made for a greater number of axonal linkages between neighboring NPE's than dendritic ones. (In this case one dendritic and up to 2 axonal links are provided.)

Based on these considerations, an NPA interconnection scheme has been adopted which uses only nearest neighbour axonal and dendritic links to connect NPE's organized in a regular orthogonal manner. The basic scheme is illustrated in Figure 2.

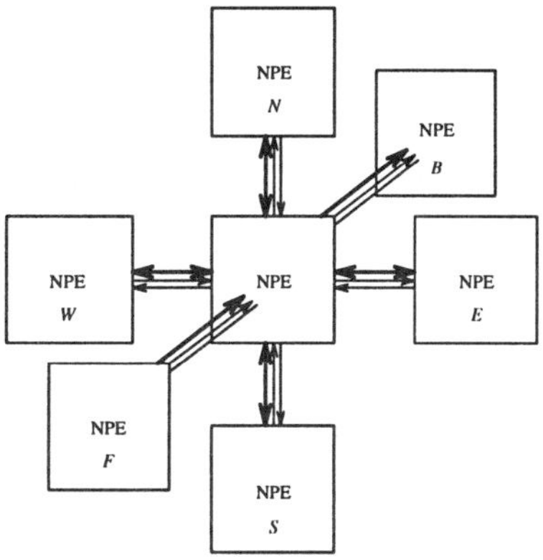

Figure 2: Dendritic and axonal connections in 3D NPA

The bold links represent bi-directional *dendritic connections* which convey transient voltage information between neighboring NPE's forming part of a given model neuron. The thinner links represent uni-directional *axonal connections*, which convey pulse stream information between separate model neurons, possibly via several intervening NPE's. For simplicity, the two axonal links between neighboring NPE's convey information in opposite directions. This assumption hopefully does not place too many restrictions on the type of neural micro-circuits that may be modelled.

In the current simulation model a simpler 2D structure has been used as shown in Figure 3. Each input or output path now represents an axonal connection and one half of the bi-directional dendritic connection.[1]

Using such a 2D NPE, a 1×3 SIMULINK NPA model, illustrated in Figure 4 was produced for use in simulations. This structure can easily be expanded into a full 2D array. The same axonal and dendritic connections shown in Figure 3 are used in this model, however some re-arrangement is necessary because the inputs and outputs must be on opposite sides of the SIMULINK NPE blocks.

[1] In order to implement this bi-directional link in SIMULINK, it was necessary to use 2 actual connections; one for communicating a local membrane voltage, the other for transmitting a positive or negative current.

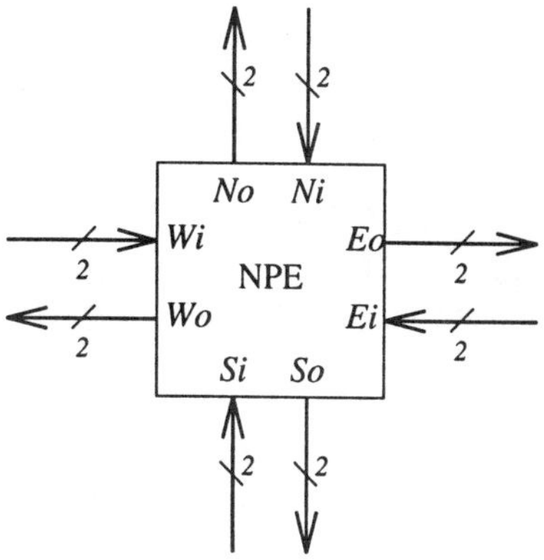

Figure 3: Linking scheme in 2D SIMULINK model

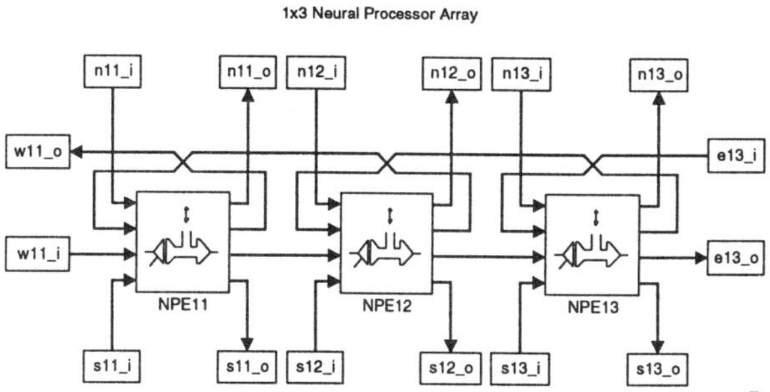

Figure 4: 1x3 NPA model used for simulations.

OVERVIEW OF NEURAL PROCESSOR ELEMENT

The NPE model represents a multi-purpose processor element with programmable function and connectivity. The top-level SIMULINK model of the NPE is shown in Figure 5.

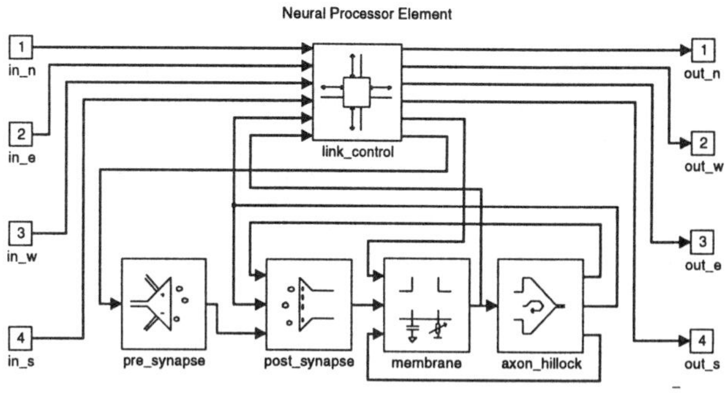

Figure 5: Neural processor element

One of the main principles underlying the NPE design is the integration of synaptic input and neural output functions in a single, homogeneous processor element. Given that real neurons in the cortex can have in the order of 10^4 synapses [Arb89] and the importance of the synapse in learning, it is sensible to include some form of synaptic function in a general-purpose neural processor. Examples of hardware implementations along these lines can be found in [PK91, BSB+91]. Often a non-linear neural output function is implemented in a different type of processor on a separate chip, however this obviously limits the range of neuronal systems that may be modelled. Some examples of implementations that successfully combine input and output functionality in a single generic processor can be found in [LVB93, SMB+92].

The specific functions performed within each subsystem in the NPE model are summarized below.

Link-control: The link control block is responsible for the interconnection of axon and dendrite links between neighboring NPE's. North-south and east-west dendrite links are controlled with two axial conductance parameters and are effectively non-existent if the conductance is set to zero (open circuit)[2]. A 3-bit selector is employed at each axonal output from the link-control block to choose between one of 5 axonal inputs, or not connected.[3] Delay buffer stages, representing the Nodes of Ranvier in biological axons [She94, Mea91], are included at each axonal output.

Pre-synapse: The pre-synapse model implements short-term, *attentive* learning. This can be used to model temporary habituation, or depression of a direct synaptic input and sensitization through the use of an optional second input. The short-term amplitude modulated signal is transferred to the post-synapse model.

Post-synapse: Long term, reinforcement style learning is implemented in the post-synapse model. The learning law can be adjusted to perform a combination of associative (Hebbian), differential Hebbian and "time prediction" learning. This sub-system is intended for use in the modelling of neural networks that can perform both unsupervised and supervised learning and which can re-produce realistic classical conditioning phenomena.

Membrane: The membrane model represents the equivalent cable properties for a segment of a compartmentalized neural membrane [Ral89] and performs spatio-temporal summation (or aggregation) of transient dendritic signals. Conductances due to synaptic inputs, self-excitation of the membrane and passive leakage are implemented using combined controllable Sodium and Potassium ion channel models.[4]

Axon-hillock: Self-excitation of the membrane and pulse generation is controlled by the axon-hillock model. By adjustment of gain and timing parameters, it can be used to generate rapid output spikes, slower dendritic spikes, or a simple linear transfer function.[5]

MODULATION AND TRANSFER OF NEURAL SIGNALS

In this section the basic signal modulation, aggregation and neural signal transfer functions that can be performed with the modelled NPA are presented. Simulations of the basic signal transfer functions were conducted using fixed short and long-term weights, so the pre-synapse and post-synapse models are not considered. The operation of the membrane and axon-hillock models are examined more closely.

The 3-NPE model shown in Figure 4 was used in the following simulations. The connected NPE's represent a simple non-branching model neuron. NPE11 and NPE12 represent segments of dendrite with linear output functions, which receive excitatory synaptic inputs $X1^+$ and $X2^+$ from SIMULINK sources **S11_i** and **S12_i** respectively. NPE13 is configured as the soma/axon-hillock which produces non-linear self-excitation of the membrane and generates output spikes. It also receives inhibitory synaptic input $X3^-$, provided by **S13_i.**

Signal Aggregation in Membrane Model

The local membrane voltage V_m is calculated in the membrane model by summing currents due to the synaptic input I_s, active self-excitation I_a, passive leakage I_p and dendritic flow I_d on to a (fixed) membrane capacitance C_m. These currents are obtained using the combined excitatory and inhibitory conductances, G_{Na} and G_K, due to the modelled Sodium and Potassium ion channels respectively. The grouping of these conductances is described below:

$$\begin{aligned}
\dot{V}_m &= C_m^{-1}(I_s + I_a + I_p + I_d) \\
&= C_m^{-1}(G_s(V_s - V_m) + G_a(V_a - V_m) \\
&\quad + G_p(V_p - V_m) + I_d) \\
&= C_m^{-1}((G_{sNa} + G_{aNa} + G_{pNa})(V_{Na} - V_m) \\
&\quad + (G_{sK} + G_{aK} + G_{pK})(V_K - V_m) + I_d)
\end{aligned}$$

[2] In the 3D version a third up-down axial conductance will also be included.

[3] Selection will be made between 7 inputs, or not connected in the 3D version.

[4] In the hardware implementation, two trans-conductance amplifiers [Mea91] will be used for this purpose.

[5] If a non-linear transfer function is not required at an NPE configured as a synaptic input, this sub-system can still be utilized efficiently as it provides a threshold detection facility that is used by the post-synaptic learning mechanism.

Finally, we have:

$$\dot{V}_m = C_m^{-1}(G_{Na}(V_{Na} - V_m) + G_K(V_K - V_m) + I_d) \quad (1)$$

where $V_{Na} = V_{DD}$, $V_K = V_{SS}$.

By setting the axial conductances between the three connected NPE's (G_{e11} and G_{e12}) relatively high, an approximately *iso-potential* neuron model is realized [BZMC92]. In Figure 6, aggregation of synaptic input pulse streams in the 3-NPE model is illustrated. Between $t = 50$ and $t = 150$ excitation of the membrane due to an input pulse stream, $X1^+$ of relatively low stochastic probability is shown. The membrane voltage becomes saturated at a value of approximately 1.5 volts, which is just below the pulse-generating threshold. Between $t = 200$ and $t = 300$ excitatory input $X1^+$ with a higher probability, raises the membrane potential at NPE13 above the threshold, resulting in pulse generation. Between $t = 350$ and $t = 450$ aggregation and saturation of the membrane due to the inhibitory pulse stream $X3^-$ is shown.

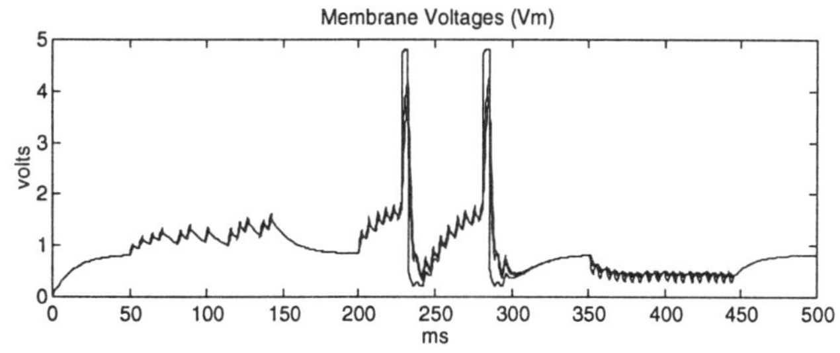

Figure 6: Excitation and inhibition of 3 connected NPE's.

In Figure 7 dispersion and spatio-temporal summation of individual synaptic input pulses in the 3-NPE model is illustrated. G_{e11} and G_{e12} have been lowered to increase the voltage transients along the dendrite.

In the first case (at approximately $t = 50$) an excitatory pulse from $X1^+$ (indicated by solid lines in the *Synaptic Inputs* plot), occurs *after* a pulse from $X2^+$ (dash-dot lines.) The pulses are not temporally correlated at NPE13 and the threshold for pulse generation is not exceeded. In the second case (at about $t = 100$) a pulse from $X1^+$ occurs just *before* a pulse from $X2^+$ This time the two pulses are temporally correlated at NPE13 and V_m3 is raised above the threshold, resulting in the generation of a pulse. The pulse can be seen to progressively disperse at NPE12 and NPE11 as it travels back along the modelled dendrite segment. The third case is similar to the second, except that now an initial, strong inhibitory pulse from $X3^-$ lowers the membrane potential and prevents the generation of the pulse.

This form of spatio-temporal summation of synaptic inputs should facilitate the development of neural systems that can be used for recognition and generation of pattern sequences and for other interesting forms of dynamic neural processing. As the relative timing of neural inputs affects the output state of the neuron, which in turn is utilized in associative learning mechanisms, the spatial and temporal structure of input patterns can effectively become encoded in the synaptic weights. Some general implications of spatio-temporal processing are discussed in [Tho92, BZMC92]. One interesting application that is possible using delayed synaptic inputs, is modelling of neurons that are tuned detect motion in specific shapes, as is done in the visual cortex [Arb89]. A method for using pulse-level neural processing in high-speed competitive filter networks is described in [DFF93]. The role of neural pulse dynamics in *feature linking* or *perceptual binding* and in the synchronization of cortical functions is discussed in [EKS91] and related articles.

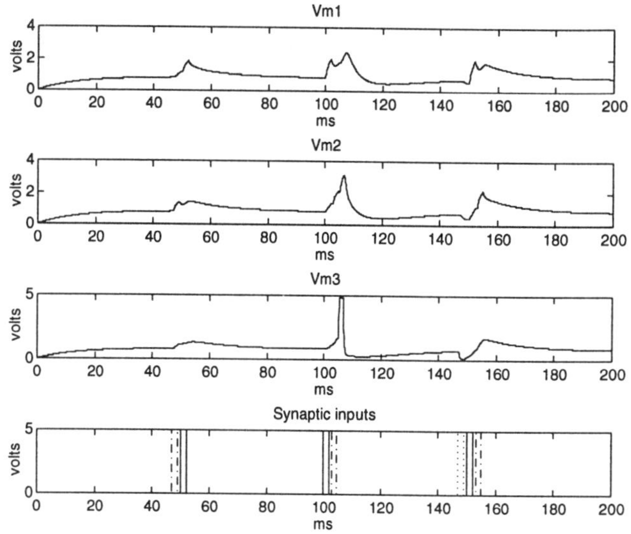

Figure 7: Aggregation of synaptic inputs in 3-NPE model.

Use of Axon-Hillock for Threshold Detection and Pulse Generation

The axon-hillock model produces the active self-excitatory conductances G_{aNa} and G_{aK}, which represent the Sodium and Potassium conductances through "voltage-gated" ion channels described very precisely by the Hodgkin/Huxley equations [HH52]. These equations are rather complex however and describe the characteristic behaviour of these conductances rather than specific mechanisms responsible for them [Mac87].

In general terms, the neural pulse is caused by sudden changes in conductance of the cell membrane to Sodium (Na^+) and then Potassium (K^+) ions. When a certain threshold voltage is initially exceeded, the voltage-gated Sodium-specific channels begin to open, the influx of Na^+ increases and there is a rapid depolarization of the membrane. This sudden voltage increase is the "rising edge" of the neural pulse. Soon after this, the influx of Na^+ falls off and there is a rapid increase in the efflux of K^+, which re-polarizes the membrane and produces the falling edge of the pulse. The K^+ conductance continues for a short time afterwards, giving rise to the *refractory period* during which generation of subsequent pulses is impossible or highly unlikely.

A relatively simple method of producing this behaviour, which was implemented in the axon-hillock model, is as follows: Firstly the difference between the local membrane voltage, V_m and a second signal, representing the greater of a threshold function, Θ_m, or a delayed version of the membrane voltage, V_m', is calculated. This produces suitable "trip-points" for the alternate positive and negative feedback components. This difference, Δ_m, is then passed through a non-linear sigmoidal function which produces an axonal output pulse, Y. This is summarized below:

$$\Delta_m = V_m - \max(V_m', \Theta_m)\ , \quad Y = \tanh(k_b \Delta_m) + 1 \tag{2}$$

where k_b is the threshold sensitivity biasing constant.

G_{aNa} and G_{aK} can then be generated by using Y as input to an active conductance control system, implemented as an "imperfect differentiator". The differential equation describing the combined active conductance G_a is given by:

$$\dot{G}_a = \tau_a^{-1}(\dot{Y} - \frac{1}{k_g} G_a) \tag{3}$$

where k_g is the active conductance gain constant,
τ_a is the time constant for the system.

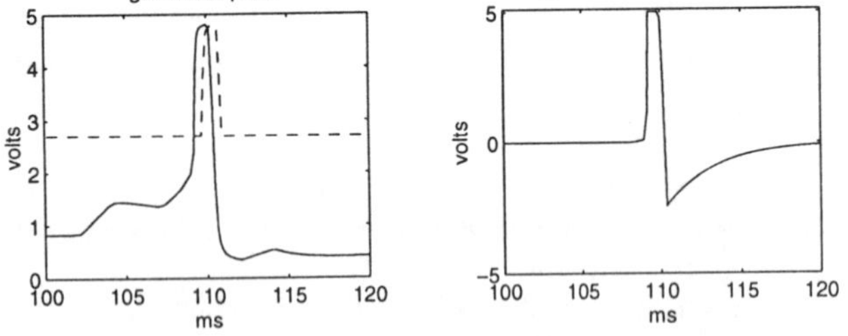

Figure 8: Generation of neural pulse.

Figure 8 shows the pulse generated by this method and the active conductances involved.[6]

By decreasing kg, kb and increasing the delay between V_m and V_m', it is also possible to emulate slow Calcium induced pulses at intermediate points along neural dendrites. These can be used in various non-linear dendritic computations and in producing neural bursting behaviour [BZMC92, Mac87].

Adaptation of the threshold Θ_m can be included using a simple system based on the Hill model [Mac87]:

$$\dot{\Theta}_m = \tau_\Theta^{-1}(aV_m + \Theta_0 - \Theta_m) \tag{4}$$

where Θ_0 is the resting threshold,
a is the accommodation factor.

If the average membrane voltage level is high, then the threshold will increase, preventing the generation of further pulses. This is illustrated in Figure 9.

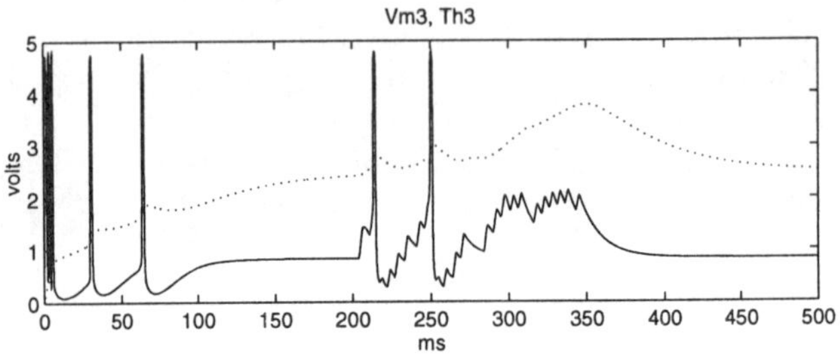

Figure 9: Positive accommodation of pulse-generating threshold.

Following an extended period of strong inhibition, bursts of neural activity can also occur. This effect is known as post inhibitory rebound (PIRB) and is shown in Figure 10. A "spontaneous" pulse can also be seen to occur at approximately $t = 315$, which is the point at which the threshold reaches its lowest value. Similar behaviour is also observed in real neurons.

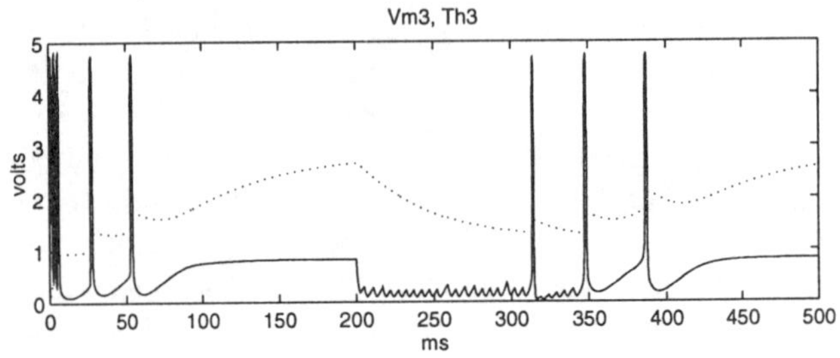

Figure 10: Negative accommodation of pulse-generating threshold.

SYNAPTIC LEARNING MODELS AND SIMULATIONS

Basic short and long-term learning functions are performed in the pre-synapse and post-synapse models respectively. Some preliminary simulation results illustrating these functions are presented below. In the

[6] Note that positive conductance corresponds to G_{aNa} and negative conductance corresponds to G_{aK}.

pre-synapse it is possible to model the effects of habituation and sensitization (or facilitation), while the post-synapse implements reinforcement style learning, based on a combination of Hebbian / differential Hebbian laws and "temporal prediction" of neural activity. Through repeated reinforcement, the short-term weight component stored at the pre-synapse is stored more permanently at the post-synapse model. This provides a means of consolidating Short-Term Memory (STM) into Long-Term Memory (LTM).

Short-Term Learning in Pre-Synaptic Model

The pre-synapse model is based on the model proposed by Kandel et al. to describe short-term attentive learning in *Aplysia* [Kan76]. In a famous series of experiments, repeated "inoffensive stimulus" to the gill results in depression or *habituation* of the gill-siphon reflex. This behaviour is explained by a temporary reduction in the amount of neuro-transmitter released across the synaptic cleft to the post-synaptic membrane, due to depletion of a "ready-releasable" neuro-transmitter pool. A short, "offensive stimulus" was found to rapidly re-sensitize the pre-synapse. The suggested explanation for this behaviour is that the arrival of a separate *facilitating input* causes a temporary change in the conductive properties of the pre-synaptic membrane, due to blocking of Potassium (or Sodium) specific channels. This affects the amount of Calcium influx following direct input, which in turn is required for the release of the neuro-transmitter molecules.

In the SIMULINK pre-synapse model, the effects of habituation and facilitation on a short-term weight, W_s, are implemented with a simple first order system which can receive negative or positive inputs. The weight is modified according to:

$$\dot{W}_s = \tau_s^{-1}(W_{s0} + fX_f - hX_d) \qquad (5)$$

where τ_s is the time constant associated with W_s
W_{s0} is the resting value of W_s
X_d is the direct input
h is the habituation gain
X_f is the facilitating input
f is the facilitation gain

In Figure 11, modification of the short-term pre-synaptic weight is illustrated.[7] Initially W_s (dotted line in Figure 11) is steady at approximately 3.5 volts. A prolonged burst of direct input from X_d occurs between $t = 1200$ and $t = 1500$, which gradually reduces, or habituates W_s. The output pulse train X_p (solid line) can be seen to reduce accordingly. Between $t = 1700$ and $t = 1800$, a facilitatory burst from X_f rapidly raises W_s to approximately 5 volts. Subsequent direct input between $t = 1800$ and $t = 1900$ produces a series of output pulses X_p with increased amplitude.

[7] Note that a small value for the pre-synaptic time constant, τ_s has been used for illustrative purposes.

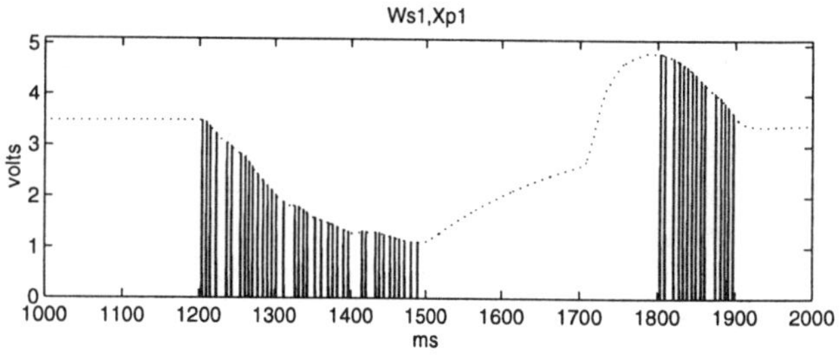

Figure 11: Short term learning effects in pre-synapse.

Using this short-term learning capability, it should be possible to model neuronal systems which can perform attentive pre-processing of input signals. Prolonged ("familiar") direct sensory inputs would produce a lowered neural response, but facilitating inputs, which could be provided by modelled *Central Systems* [She94] could increase "interest" in particular sensory stimuli. In this way a simple method for determining the relative value of input patterns can be implemented. Such discriminating short-term learning behaviour could also be useful in the development of artificial neural systems for autonomous robots for example, which could perform actions within their particular environments based on internal drives.

Long-Term Learning in Post-Synaptic Model

As with the pre-synaptic terminals of biological neurons, the precise mechanisms responsible for the adaptive behaviour observed at the post-synaptic terminals at dendritic spines have not been conclusively determined, although many theories have been suggested.

In a hardware realization of a biologically realistic learning mechanism that can not be altered as easily as in a software implementation, it seems reasonable to incorporate some flexibility into the learning law adopted. Another desirable attribute of the learning law is that it is locally implementable.

The learning law employed in this model was chosen to provide a basis for construction of biologically motivated neural networks as found in the cortex, within which both supervised and unsupervised forms of learning seem to be used and in which spatio-temporal relationships between neural input and output signals are preserved.

The general idea that associations between pre- and post-synaptic activity result in an increase in the efficiency of the synaptic connection, first proposed by Hebb [Heb48], is widely accepted. The original idea of Hebbian learning in usually modified such that the synaptic weight is either increased or decreased depending on the average pre-synaptic activity level, to ensure stability in the learning law. Associative learning laws can be used in *unsupervised*, self-organizing neural networks based on competitive filtering [HPK91, Koh88, Kos92].

Differential Hebbian laws are used in reinforcement style learning [Kos92, Klo82, SB91]. Reinforcement learning is similar to delta-rule learning [W+73], which is used in *supervised* neural networks such as simple perceptrons and back-propagation networks [HPK91]. However they are considered to be more biologically feasible and often more suitable for hardware realization, as they can be implemented locally.

In a biologically motivated neural processor, which allows reasonably accurate representation of neural signal dynamics, it should be possible to encode spatio-temporal relationships between input and output activities in the synaptic weights. Ideally it should also be possible to reproduce various phenomena described in psycho-physical theories and experiments on *classical conditioning*, in which timing plays a critical role.

Examples of associative learning laws which emphasize neural dynamics include the Klopf "Drive-Reinforcement" (DR) model [Klo82] and the "Time-Difference" (TD) model of Sutton & Barto [SB91]. These models can learn to predict the onset and offset times of reinforcement signals, or unconditioned stimuli (US) from correlated synaptic excitation produced by conditioned stimuli (CS).

The post-synapse model of the NPE employs a flexible learning law, which can be used to implement Hebbian, differential Hebbian and temporal prediction type learning by setting appropriate parameters.

In the model, the long-term weight is modified according to:

$$\dot{W}_l = \eta R(V_m) E(X_p) \tag{6}$$

in which

$$\begin{aligned} \dot{E}(X_p) &= \tau_x^{-1} X_p \\ R &= b\dot{P}_m - cP_m + Y_m - V_{m0} \\ \dot{P}_m &= \tau_p^{-1} \Delta_m \end{aligned}$$

Here $E(X_p)$ is the synaptic eligibility function, which represents a recent history of synaptic input activity along the lines of the Klopf DR model [Klo82]. $R(V_m)$ represents the reinforcement signal, Y_m the neural output state, measured locally using the threshold detection system in the axon-hillock model, V_{m0} is the resting membrane potential and P_m is the temporally smoothed prediction signal, which is also obtained from signals produced in the axon-hillock model. The parameters b and c are used to include temporal prediction in the learning law and for determining the degree of *imminence weighting*, as in TD model [SB91].

Results are preliminary and show a form of reinforcement learning using the simple 3-NPE model shown in Figure 4. **S11_i** provides an adaptive excitatory input, $X1^+$ (representing the CS), **S12_i** is used for positive reinforcement signals (or US^+) and negative reinforcement signals (US^-) are provided from **S13_i**.

In Figures 12 and 13 an example of positive reinforcement is shown. Positive reinforcement of the long-term weight can be seen to occur in Figure 12, as synaptic activity (CS) between $t = 100$ and $t = 250$ is temporally correlated with the US^+ between $t = 300$ and $t = 400$.[8] In Figure 13, W_{l11} has increased following a number of trials and has stabilized at a level for which UR^+ is predicted, or in other words, the CS now produces the conditioned response (CR.)[9] As the post-synaptic activity is no longer unexpected, no further weight change takes place. In a similar way the effects of negative reinforcement of the long-term weight can also be shown.

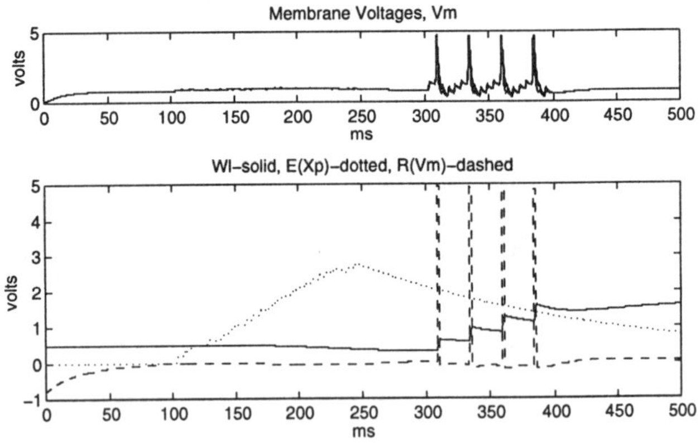

Figure 12: Positive reinforcement of W_{l11} (before)

The hybrid learning law employed in this model is well suited for supervised learning models, as positive and negative reinforcement signals can operate in a similar way to error signals in delta-rule learning. Specifically, if the expected output is high, but the actual, reinforced output low, negative reinforcement results (analogous to a negative error signal or delta.) Conversely, if low output is expected but a high output is enforced, positive reinforcement results (analagous to positive error signal.) Note that the direction of weight change is determined by sign of the reinforcement signal only.

In unsupervised, competitive learning on the other hand, the synaptic weights of the *winning neuron*,

[8] For illustrative purposes, a large value for the learning gain, η, has been used.

[9] To predict the actual onset time of the US would require a sequence of CS inputs leading up to the US, rather than a single input as in this case.

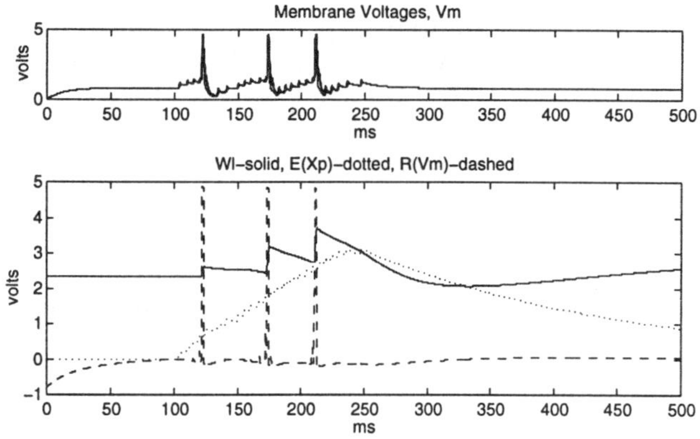

Figure 13: Positive reinforcement of W_{l11} (after)

are normally increased if corresponding inputs are active and decreased if they are inactive. If the active output state is considered to be equivalent to positive reinforcement, it can be seen that the reinforcement signal no longer determines the direction of the weight change. This causes problems if an integrated unsupervised and supervised learning mechanism is required. However, perhaps it is possible to implement unsupervised learning using such a reinforcement learning law if one considers how it might operate in a recurrent competitive filter (or 'Winner Take All') network.

In such a network, initial excitation of the "output layer" of inter-connected neurons, may not result in an *expected* active output. However following the rapid recurrent competition process, an actual high output could result. This result would be unexpected, which would result in positive reinforcement of synaptic weights (they would be increased if they were "eligible".) This is analogous to the "winning" condition. Through successive trials the weights would increase up to the point where the active output was predicted (or expected). In order to reduce weights, we can hypothesize a "losing" condition. In this case there would be an initial expectation of output activity, but following the rapid feedback process, a low output would result. This would produce negative reinforcement which would cause eligible weights to decrease.

CONCLUSIONS

A model of a proposed biologically motivated neural processor array, employing a novel 2D or 3D array structure with sufficient flexibility to allow modelling of a range of realistic neural structures was described. The various communications and signal processing sub-systems that make up the NPE model were outlined and their use in emulating a variety of realistic neural behaviours was discussed.

Simulation results illustrating the various signal transfer functions that can be modelled were presented. These include aggregation of excitatory and inhibitory synaptic input pulse streams on an equivalent neuron membrane model, spatio-temporal summation and correlation of synaptic inputs and generation of realistic neural spikes. Positive and negative accommodation of the pulse generating threshold was also shown.

The use of separate pre-synaptic and post-synaptic models was outlined and the short and long-term learning mechanisms employed in each was described. Preliminary simulations of short-term, attentional learning in the pre-synapse model and long-term associative/reinforcement learning in the post-synapse model were presented.

Future experiments using a simple 3-NPE neuron model, will attempt to reproduce other classical conditioning phenomena such as *second order conditioning*, *blocking* and *over-shadowing* [SB91, VS92]. Following this, a small cortical P-cell network, similar to that illustrated in Figure 1 will be modelled and tested. In particular the integration of supervised and unsupervised learning functions will be examined.

After the system-level simulations have been completed, equivalent analog and digital VLSI circuits will be developed for the neural processing sub-systems of the NPE and the necessary interfacing. SPICE simulations of the electronic models will then be conducted in an effort to obtain similar signal transfer and basic learning results to those presented here.

REFERENCES

[ABP+91] A. G. Andreou, K. A. Boahen, P. O. Poliquen, et al. Current-mode subthreshold MOS circuits for analog VLSI neural systems. *IEEE Transactions on Neural Networks*, 2(2), March 1991.

[Arb89] M. A. Arbib. *The Metaphorical Brain 2*. John Wiley, 1989.

[BSB+91] B. E. Boser, E. Săckinger, J. Bromley, et al. An analog neural network processor with programmable topology. *IEEE Journal of Solid-State Circuits*, 26(12), December 1991.

[BZMC92] T. H. Brown, A. M. Zador, Z. F. Mainen, and B. J. Claiborne. Hebbian computations in hippocampal dendrites and spines. In *Single Neuron Computation*, chapter 4. Academic Press, 1992.

[DFF93] M. DeYong, R. L. Findley, and C. Fields. The design, fabrication, and test of a new

VLSI hybrid analog-digital neural processing element. *IEEE Transactions on Neural Networks*, 3(3), May 1993.

[EKS91] A. K. Engel, P. Kőnig, and W. Singer. Correlated neuronal firing: a clue to the integrative functions of cortex? In J. G. Taylor, E. R. Caianello, et al., editors, *Neural Network Dynamics*. Springer-Verlag, 1991.

[Heb48] D. O. Hebb. *The Organization of Behaviour: A Neuropsychological Theory.* John Wiley, 1948.

[HH52] A. L. Hodkin and A. F. Huxley. A quantitative description of membrane current and its application to conduction and excitation in nerve. *Journal of Physiology*, pages 400–544, 1952.

[HMB+92] A. Hamilton, A. F. Murray, D. J. Baxter, et al. Integrated pulse stream neural networks: Results, issues and pointers. *IEEE Transactions on Neural Networks*, 3(3), May 1992.

[HPK91] J. A. Hertz, R. G. Palmer, and A. S. Krogh. *Introduction to the Theory of Neural Computation.* Addison-Wesley Publishing Company, 1991.

[Kan76] E. R. Kandel. *Cellular Basis of Behaviour.* W. H. Freeman, 1976.

[Kan92] P. Kanerva. Associative-memory models of the cerebellum. In I. Aleksander and J. Taylor, editors, *Artificial Neural Networks, 2*, volume 1. North-Holland, 1992.

[Klo82] A. H. Klopf. *The Hedonistic Neuron: A Theory of Memory, Learning and Intelligence.* Hemisphere, 1982.

[Koh88] T. Kohonen. *Self-Organization and Associative Memory.* Springer-Verlag, 2nd edition, 1988.

[Kos92] B. Kosko. *Neural Networks and Fuzzy Systems.* Prentice-Hall, 1992.

[Lin86] R. Linsker. Self organization in a perceptual network. *Computer*, pages 105–117, March 1986.

[LVB93] C. Lehmann, M. Viredez, and F. Blayo. A generic systolic array building block for neural networks with on-chip learning. *IEEE Transactions on Neural Networks*, 4(3), May 1993.

[Mac87] R. J. MacGregor. *Neural and Brain Modelling.* Academic Press, Inc., 1987.

[Mea91] C. Mead. *Analog VLSI and Neural Systems.* Acadamic Press, 1991.

[MR94] C. F. Miles and C. D. Rogers. The microcircuit associative memory: A biologically motivated memory architecture. *IEEE Transactions on Neural Networks*, 5(3), May 1994.

[MTR+91] A. F. Murray, L. Tarassenko, H. M. Reekie, et al. Pulsed silicon neural networks - following the biological leader. In U. Ramacher and U. Rűckert, editors, *VLSI Design of Neural Networks.* Kluwer Academic Publishers, 1991.

[PK91] S. J. Prange and H. Klar. Architectures for a biology-oriented neuroemulator. In U. Ramacher and U. Rűckert, editors, *VLSI Design of Neural Networks.* Kluwer Academic Publishers, 1991.

[Ral89] W. Rall. Cable theory for dendritic neurons. In C. Koch and I. Segev, editors, *Methods in Neuronal Modelling.* MIT Press, 1989.

[SB91] R. S. Sutton and A. G. Barto. Time derivative models of pavlovian reinforcement. In M. Gabriel and J. W. Moore, editors, *Learning and Computational Neuroscience.* MIT Press, 1991.

[She94] G. Sheperd. *Neurobiology.* Oxford Uni. Press, 3rd edition, 1994.

[SMB+92] J. Van der Spiegel, P. Műller, D. Blackman, et al. An analog neural computer with modular architecture for real-time dynamic computations. *IEEE Journal of Solid-State Circuits*, 27(1), January 1992.

[Tho92] S. J. Thorpe. Spike arrival times: A highly efficient coding scheme for neural networks. In R. Eckmiller et al., editors, *Parallel Processing in Neural Systems and Computers.* North-Holland, 1992.

[VS92] J. Vico and F. Sandoval. A learning by reinforcement model from the neurobiological outlook. In I. Aleksander and J. Taylor, editors, *Artificial Neural Networks, 2*, volume 1. North-Holland, 1992.

[W+73] B. Widrow et al. Punish/reward: Learning with a critic in adaptive threshold systems. *IEEE Transactions on Systems, Man, and Cybernetics*, (3):455–465, 1973.

ORBITS AND QUOTIENT CIRCUITS IN NEURAL NETWORKS.

Eric Fimbel, Fran Brito,
Departamento de Electricidad, UNEXPO, Puerto Ordaz, Venezuela.
E-mail fbrito@conicit.ve

ABSTRACT.

The use of orbits in neural networks to store and retrieve information is discussed, and an important theorem about the possible orbits of a network is given. Instead of reaching an equilibrium, a recurrent neural network may follow a chaotic trajectory or reach an **orbit** (a periodic sequence of states). Orbits can represent information by mean of **frequency coding:** heterogeneous information is stored in separate frequency ranges, and retrieved by mean of filters.

The oscillations in an isolated network result from the features of the node functions and the circuits between nodes. We construct an **equivalence relationship** based upon the distance between nodes and we work in the **quotient network** built with the equivalence classes. We prove that the periods of oscillation of a node is defined by the size of the **quotient circuit** its belongs to, and the **cycles of the quotient node functions**.

I. INTRODUCTION.

The dynamics of Recurrent Neural Networks may be complex. Instead of a steady state, the network often reaches an **orbit**, i.e. repeats a sequence of states periodically. In previous works, we suggested that this periodical behavior can be used to store information [FIM94].

Due to its temporal dimension, an orbit allows the use of **frequency coding**. Heterogeneous information can be stored in separate frequency ranges; the retrieval of information can be done by mean of **filters**. Biological experiments seems to evidence frequency coding in the mammalian visual cortex[BUL92].

This paper gives a short introduction to networks dynamics: **equilibriums, orbits, and chaotic trajectories.** The use of frequency coding and the existence of efficient filters for information retrieval are discussed.

In the second part, we present the **factors of oscillations** in a neural network: extern signals, node functions, synaptic weights and circuits between nodes. The effect of **cycles** in the node futiti ((x such that f(f(..f(x))))=x) is discussed. We show that in the general case, almost every frequency can be generated by simple circuits.

The third part deals with self generated oscillations. We define an **equivalence relationship** based upon the distance between nodes. In the **quotient network** built with the equivalence classes, the size of the **quotient circuits** and the cycles of the **quotient node function** define the possible periods of oscillation in a straightforward manner. We give an algorithm to construct the quotient circuits and we prove a theorem that define the possible periods of oscillation of the nodes. We use this theorem in two important particular cases: binary node functions, and increasing node functions with positive synaptic weights.

1.1 Asymptotic behavior of neural networks with feedback.

From an initial state, a neural network can present three kinds of asymptotic behaviors:

- convergence towards an **equilibrium**, i.e. a steady state;
- convergence towards an **orbit,** i.e. a finite sequence of states that repeats itself periodically. When an orbit is reached, each node may have a different period. The global period of the network is the composition of the node periods. It may be very large, even if the node periods are short. However, the orbit can be **identified by direct observation** of the network;
- **chaotic behavior:** the network never repeats the same state. In order to identify exactly the trajectory, you have to know the the initial state of the network. However, the trajectory of the network in the phase space (the space of possible states) can be **roughly identified** by its **strange attractor** , i.e. the set of its adherence points.

The kind of behavior depends on several elements : the structure of the network, the synaptic weights, the node function and the initial state of the network. If the network is not well trained, very close initial states may lead to opposite behaviors.

1.2 Using equilibriums to represent information.

J. Hopfield suggested that the convergence of a network can be used to solve problems. From an initial state, which represents the input data, the networks converges towards an equilibrium that represents the result of the computation[HOP82]. This technique uses vector coding, i.e. the state of the network is seen as a vector in the phase space representing the stored information.

The equilibriums are defined as the local minimums of some energy function, and convergence can be described as an energy minimization, like in isolated systems in physics [HOP82].

In order to store a new pattern, the learning algorithms change the shape of the energy function by mean of synaptic weights modifications, until an attractor (i.e. a "hole") appears at the desired point.

Equilibriums are fragile. Weight modifications can loose previously learned states. Parasite equilibriums can appear (for instance, the complementary of the learned pattern). Furthermore, in the general case, the number of possible minimums is unknown [HER90].

1.3 Using chaotic trajectories to represent information.

W. Freeman and al. suggest that neural networks are fundamentally chaotic systems. Biological experiments seems to give evidence of chaotic behavior in the mammalian cortico-olfactory system. Their idea is that the strange attractor of the trajectory of the network starting from a given initial state defines the recognized pattern. In [FRE90], they present an application based upon this idea.

Strange attractors seem to be a powerful information representation, but the computational cost may be high, because the network may have to run for a long time in order to "draw" clearly the attractor.

1.4 Using orbits to represent information.

We suggested in previous works that orbits can be used to store and retrieve information by mean of **frequency coding** [FIM94]. Orbits have been used in several works to represent temporal sequences of patterns. When it generates a sequence of patterns, the networks "plays it" periodically. Several orbits learning algorithms are given in [CHU92], [MOZ93], [WAN89].

However, frequency coding, allows the orbits to represent static information as well. When a network is in orbit its state is periodic. Thus, its frequency spectrum is discrete and can be obtained by means of Fourier Series [OPP83]. Separated frequency range can represent different kind of information. An example is given in section 1.5.

When using frequency coding, the retrieval of information can be done by mean of **filters.** The frequency spectrum of a network in orbit is discrete, so the existence of efficient filters, that can separate any frequency from the others, is granted [FIM94],[OPP83].

Frequency coding and filters are general concepts which may be used in many neural networks applications (for instance signal processing, networks with internal state neurons, correlative learning, etc.). Some applications can be found in [FH94],[HOP91].

I.5 An example of frequency coding.

In the following example, the nodes 1,2,3 present a periodical state. In order to simplify the example, we suppose that the node states are sinusoidal functions of time, of frequencies 1,2,4.

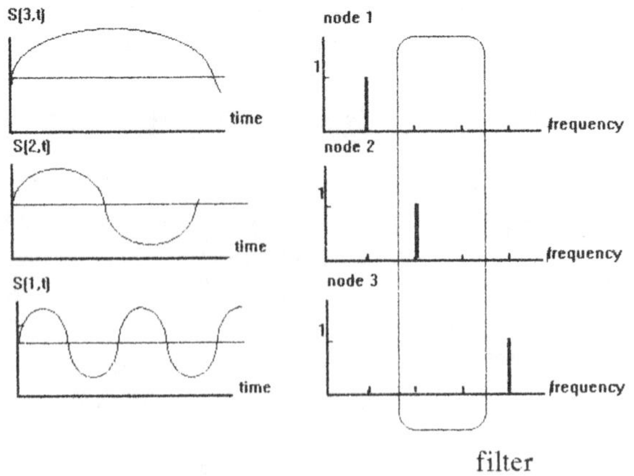

FIGURE 1. Frequency analysis for 3 sinusoidal functions.

In the previous example [Fig.1] , the frequency spectrum of each node is reduced to one single frequency. When we apply a filter to the range of frequencies [2,3], where only node 2 has a component, we get the array of values (0,1,0). Each range of frequencies corresponds to a state vector.

1.6 Frequency coding in the mammalian brain.

Some experiments seem evidence frequency coding in the mammalian visual cortex. When several objects are shown to a rhesus monkey (bars with several colors and orientations), some neurons in areas V3, V4 and neocortex present synchronous pitches [BUL92] . Correlation analysis upon registered neural activities show that each object is responsible of some pitches, and that all the objects "share" most of the neurons in that areas. If we apply a high pass filter to the brain, we will "see" the objects successively. If we apply a low pass filter, we will probably see global image properties (color information, light, orientations, etc.). The kind of information retrieved from the cortex activity changes with the frequency range.

On the other hand, most of the results on neural plasticity (LTP, etc.) has been obtained sending a given frequency to the entry of the neurons (for instance 100Hz, 100Mv, 1") [CHU92][HAM91]. They are commonly interpreted in term of levels ("high input", "stronger response", etc.). It could be interesting to give a new interpretation of these experiments from a frequency coding point of view.

1.7 Four important questions about orbits.

Orbits can obviously store much more information than simple states. Now, four questions arise:

What is the set of possible orbits of a given neural network ? When a network does not receive external signals, the orbits result from self generated oscillations. The theorem proven in section 3 may allow to define the possible oscillatory periods from a previous examination of the network.

What is the computational cost of orbit coding ? It can be observed that an orbit is totally defined by the successive states of the network until all the nodes complete a period. Any later state can be computed from the first ones. Hence, the knowledge of the possible periods of each node gives an upper value for the computational cost.

Which of the orbits can be learned ? The knowledge of a network is embedded in the synaptic weights, and an intrinsic limit can be expected, i.e. not all the orbits can be learned together. More practical results will be available when efficient learning algorithms will be constructed and tested.

- **Which of the orbits are stable to initial conditions and/or noisy entry signals** ? As a general rule, dynamic behaviors are less robust than equilibriums. Hence, orbits may be sensible to initial conditions. However, we believe that the orbits can be stabilized by periodic external signals, which is impossible with equilibriums. For instance, orbits may be used in process control applications, where the networks are processing signals from a plant.

2. GENERATION OF PERIODIC STATES IN A NEURAL NETWORK.

2.1 Orbits, Periods, Definition time.

Def.1 : an **orbit of period T** in a neural network is a sequence of states S(0)..S(T-1), such that if the network is in the state S(i) ($i < T$) at time t, it is in the state S(i+1 modulo T) at time t+1.

This definition is available for **synchronous** networks, in which all the nodes change of state together. It can easily be generalized to **asynchronous networks,** in which each node changes of state at any time [BRI94]. In that case, time measures numbers of events, not real durations.

Def.2.: when the network reaches an orbit of period T, the state of each node is a periodic function. Let **T(i,O)** be the **period of node i in the orbit O,** i.e. the period of the state function of node i, S(i,t) when the network reaches the orbit O.

Def.3.: the **definition time of an orbit O** is the largest of the node periods in the orbit O.

Furthermore, it is the lowest number of initial cycles which have to be observed in order to compute the state of the network at any time.

For instance let us consider a 5 nodes network, with unitary weights and binary node functions:

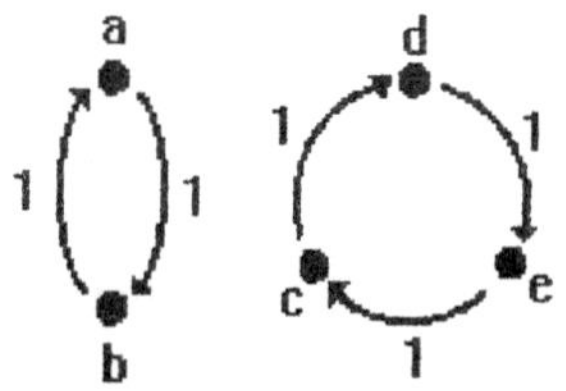

FIGURE 2. A network in orbit.

The network of [Fig.2] has the following features:
- The nodes follow the Pitts Mc Culloch rule:

$$S(i,t+1) = (E(i,t) > 0), \quad \text{with } E(i,t) = \sum_j W ji*S(j,t)$$

where:
- S(i,x) is the state of node i at time x.
- E(i,x) is the entry of node i at time x,
- Wji is the synaptic weight from node i to node j.
- The network is synchronous.

A typical trajectory of the network may be:
S(1,t) = (1,0), period 2.
S(2,t) = (0,1), period 2.
S(3,t) = (1,0,0), period = 3.
S(4,t) = (0,1,0), period = 3.
S(5,t) = (0,0,1), period = 3.

The orbit has a period 6, but the definition time is only 3: if we observe the network during 3 cycles, the rest of the trajectory is fully defined.

2.2 Factors of oscillations.

The oscillations of a network come from several factors:

Extern signals. Their periodicity can induce periodical changes in the state of the nodes. It is said that the network is working in **Forced Mode**.

Circuits between the nodes. They introduce closed loops with delay in the nodes functions, and can create stable periodic modes.

Node function. In dynamic systems theory, a lot of work has been done about cyclic functions (i.e. there exists x such that f(f..(x))=x). Any cyclic functions can obviously create oscillations. Some cyclic functions can generate any kind of period. On the other hand, many acyclic functions oscillate when they are working in closed loop.

Synaptic weights. A change in weight W(i,j) may change the feedback coefficients of some closed loops, hence it may introduce or eliminate oscillations in the state of some nodes

2.3 Periodic extern signals.

When its entries are periodic, **the network works as a filter:** it amplifies some frequencies and eliminates other. Very simple circuits can filter the entry signal. For instance, the network of [Fig.3] only allows the frequencies n*(1/3) of the input signal. If the extern signal has frequency 1/5, the network "stops" it, and the output stays in the "1" state.

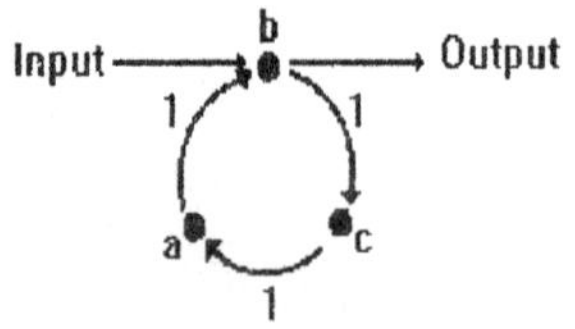

FIGURE 3. A simple filter.

The network of [Fig.3] has the following features:
- The node function is: S(i,t+1) = (E(i,t) > 0),

$$\text{with } E(i,t) = \sum_j W ji*S(j,t),$$

- The entry signal I(t) is added to E(b,t).
- The network is synchronous.
- The output signal is the state of node b, S(b,t).

When the entry has period 5, from an initial state where all the nodes are in the "0" state, we get the following trajectory:
I(t) = (1,0,0,0,0,1,0,0,0,0,1,0,0,
S(b,t)=(0,1,0,0,1,0,1,1,0,1,1,1,1,
S(c,t)=(0,0,1,0,0,1,0,1,1,0,1,1,1,
S(a,t)=(0,0,0,1,0,0,1,0,1,1,0,1,1,

It can be seen that the network is filtering the entries. On the other hand, periodic entries can sometimes force oscillations and lock the network in an orbit that would be impossible otherwise.

2.4 Cyclic node functions.

A simple way to generate oscillations is the use of a cyclic node function and simple circuits which create closed loops. A function is n-cyclic if and only if there exists x such that f(f(...f(x)))))= x
n times.

The simplest circuit which can be combined with a cyclic node function is shown in [Fig.4}.

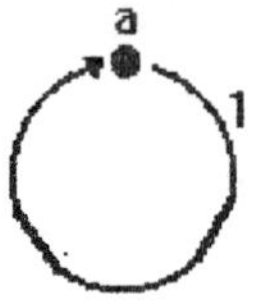

FIGURE 4. The simplest circuit .

The state of the node is given by the recurrent sequence: S(a,0)=So, S(a,t+1)=f(S(a,t)).

Any period can be obtained. For instance, if there exists a cycle of order 3 (F is 3-cyclic), there exist initial conditions that give cycles of any order, and other that lead to a chaotic sequence of states.

On the other hand, if f is 2-cyclic, or more generally 2**n-cyclic, it has no chaotic behavior, and all the cycles are power of 2[DEL92].

Now, if we combine cyclic functions with circuits, the set of possible periods changes. A circuit of size c introduces a delay of c: S(a,t+c)=F(S(a,t)), where F is a composition of node functions. New cycles appear and the situation turns complex. For instance, if we use a 12-cyclic function in a circuit of size 3, we can obtain a period 4, as shown in the network of [Fig.5]

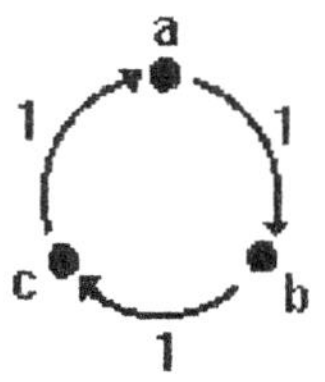

FIGURE 5. generating a period 4 with 3 nodes.

The network of [Fig.5] has the following features:
- The node function is: S(i,t+1)=(E(i,t)+1)modulo 12,

$$\text{with } E(i,t) = \sum_j W ji*S(j,t)$$

- The network is synchronous.

If we start from state (1,5,9), we get:
S(a,t) = (1 , 10, 7 , 4 , 1 , ...
S(b,t) = (5 , 2 , 11 , 8 , 5 , ...
S(c,t) = (9 , 6 , 3 , 0 , 9 , ...

The period in this trajectory is 4. With other initial conditions, the period can be 12.

Obviously, we prefer to use "good" node functions, with only equilibriums (1-cycles), or low order cycles. As a matter of fact, the common node functions are monotone (sigmoide, threshold functions, etc.), and **monotone functions can only have equilibriums.**

On the other hand, logical functions such as XOR and NOT are 2-cyclic and they are proven to be "safe" [DEL92]: they may be used as node functions.

2.5 Negative weights.

The negative synaptic weights can create oscillations. A negative weight W(i,j) can introduce a negative closed loop in node j, i.e. a source of unstability. Furthermore, they can combine with other factors to change existing oscillatory modes. For instance:

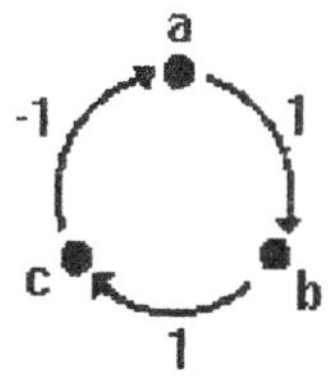

FIGURE 6. Oscillations due to negative weights.

The network of [Fig.6] has the following features:
- The node function is:
S(i,t+1) = if (E(i,t) > 0), 1 else -1

$$E(i,t) = \sum_j W ji*S(j,t),$$

- The network is synchronous.

From the initial state (1,1,1) we get the following trajectory:
S(a,t)= (+1,-1,-1,-1,+1,+1,+1,...
S(b,t)= (+1,+1,-1,-1,-1,+1,+1,....
S(c,t)= (+1,+1,+1,-1,-1,-1,+1,...

The period is 6 due to the negative weight Wca. With positive weights the period can be 1 or 3.

2.6 Circuits between nodes.

The circuits introduce closed loop from the nodes output to their input. With increasing node functions and positive synaptic weights, the circuits allow stable periodic modes. When a node belongs to a circuit of size C, the possible periods are divisors of C. When it belongs to several circuits of size C1,..Cn, the possible periods are divisors of all the Ck. However, when it belongs to interconnected circuits, the definition of possible periods is quite difficult. The reader should try on the following example:

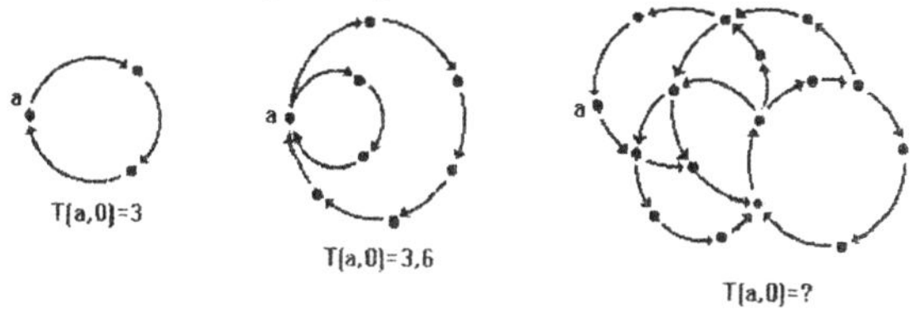

FIGURE 7. Increasingly complex circuits.

Furthermore, negative weights and cyclic node functions can generate counter intuitive results. Hence, the direct study of the circuits is not a good solution in order to define the possible periods.

3 POSSIBLE PERIODS OF SELF GENERATED OSCILLATIONS.

3.1 Simplifying the network.

We will now focus on self generated oscillations, so we suppose the network does not receive extern signals. As it appears in section 2.6, we have to simplify the network. We introduce an equivalence relationship basedpopon the distance between nodes, and we work on the **quotient network**, where all the nodes that are at the same distance of a given point are grouped together. We present an algorithm which constructs the quotient network.

In the quotient network, a node only can belong to one circuit, so the topography is quite simple. Now, the node function is defined as a multidimensional function, since each "node" is now a class of nodes.

Obviously, in a circuit, the state of a node at time t defines the state of the following node at time t+1, etc., and, when the loop closes n times later, it defines its own state a time t+n; we are dealing with recurrent sequences (Xk+n = f(Xk)).

Hence, the size of a quotient circuit (circuit of the quotient network), and the possible cycles of the **self node function** (that defines the state of a class at time t+c from its state a time t), give the possible periods of oscillation in a quite straightforward manner. We give two practical applications of this result: binary node functions, and increasing node functions, with positive weights.

3.2 Notations and definitions.

Let us consider any given network, working in a synchronous mode, with no external entries. The synaptic weight between node i and node j is noted Wij or W(i,j). Wij may be different from Wji. The node equation is:

S(i,t+1) = fi(E(i,t)), with

$$E(i,t) = \sum_j Wji*S(j,t)$$

The node function fi may be different for each node.

Def.4: We say that a **weight Wij exist** if and only if it is positive (Wij > 0).

Def. 5: Node i is previous to node j if and only if Wij is not zero.

Def.6: a **path of size n** from I to J is a sequence of nodes I0..In, such that:

I0=I;
In=J;
W(k,k+1) exist for any k in [0,n[;

Remark: a path of size 0 is a single node.

Def.7: Node i is connected to node j if and only if there is a path of any size from i to j.

Def.8: A **Circuit of size n** is a path of size n-1, I0..In-1 such that:

all the Ik are different;
W(n-1,0) exists.

Remark: there are no circuit of size 0.

3.3 An equivalence relationship between nodes.

Def.9: let **E(i,j)** be the equivalence relationship between the nodes such that:

E(i,i) for any i of the network.
if E(i,j), and Wik, Wjl exist, E(k,l).

Obviously, an equivalence class contains all the nodes which are at the same distance from a given node. For instance:

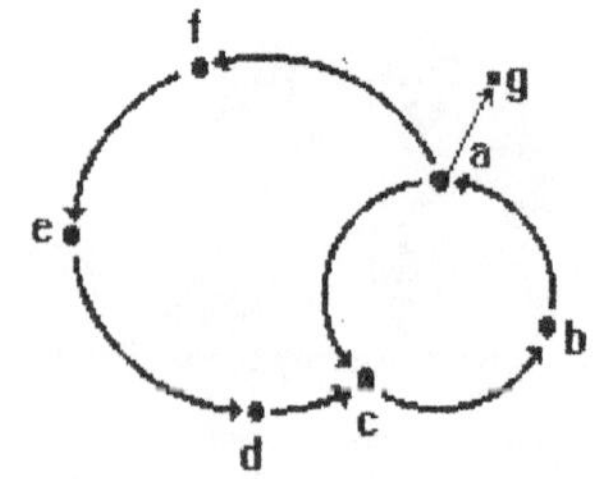

FIGURE 8. An example of equivalence classes.

The network of [Fig.8] contains the following classes:

C0={a,d},C1={b,e},C2={c,f,g}

An algorithm for the construction of the equivalence classes is given in 3.7. Meanwhile, the reader can verify that C1,C2,C3 are the equivalence classes of E(.,.).

3.4 Quotient network.

Let us consider the equivalence classes for the relationship E. We can draw a new network, where the classes are nodes. In this network, the weights are binary: Wcd=1 if and only if Wij exists for some i of C and some j of D.

Def.10: The **quotient network** if defined by:

- the set of equivalence classes for E ;
- weights Wcd between classes C and D such that:
  if Wij exists for some i of C and some j of D Wcd=1,
  else Wcd=0.

All the concepts defined in section 3.2 (existence of Wij, i before j, i connected to j, path of size c, circuit, etc.) are available in the quotient network.

In the quotient network, each node C is a class, and can be considered as an ordered sequence of nodes C = (I1,..Ik). The state of cla C C is a vector which takes values in a k-dimensional space.

Def.11: let C be a class of the quotient network, C=(I1,..Ik). The **state** of C is defined as:

S(C,t)=((S(I1,t),..S(Ik,t)).

We now define the equivalent of the node function for the classes of the quotient network. Due to the variable dimension of the states of the classes, we cannot use weighted sums and we have to give a less specific definition.

Def.12: let C be a class of the quotient network. The **quotient node function** of class C it the function fc such that:

S(C,t+1)=fc(S(C1,t),..S(Cn,t)), where
C1,..Cn are the classes before C.

Each quotient node function fc can be different, due to the high dimension of the arguments S(C1,t),..S(Cn,t). Moreover, fc can present cycles even if the node functions do not. For instance, if class C is {i1,i2}, and S(i2,t+1)=S(i1,t), S(i1,t+1)=S(i2,t), fc permutes the states of nodes i1 and i2, yet fc is 2-cyclic.

3.5 Quotient circuits and their properties.

The **quotient circuits** are the circuits in the quotient network. An algorithm for quotient circuit construction is given in 3.7. We now prove an important property of quotient circuits: in the quotient network, a class can belong at most to one circuit. First of all, let us prove an intermediary result.

Teo.1.: let C0..Cn be a quotient circuit, i a node of C0. For any node j, there exists a path of size k from i to j if and only if j belongs to Cm, where m=k modulo n.

Proof:

- if there exists a path of size 0 from i to j, i.e., if j=i, j belongs to C0.
- Let us suppose the property is true for paths of size <= s.
- Let j be any node such that there exists a path of size s+1 from i to j.
- There exists a node k such that there exists a path of size s from i to k, and Wkj exists [Def.6].
- From the hypothesis, k belongs to Cm, with m=s modulo n.
- From the definition of quotient circuit, j belongs to Cl, with l=m+1 modulo n [Def.8],
- m+1 modulo n = s+1 modulo n, so the property is proven.

Teo.2.: a node can belong at most to the first class of **one** quotient circuit.

Proof:

- Suppose the node i belongs to the first class of 2 quotient circuits C0..Cn and D0..Dp.
- For any k in [0,p], any node of Dk belongs to Cm with m=k modulo n. [Teo.1].
- Equivalence classes are disjoints so for any k in [0,p], Dk=Cm, with m=k modulo n.
- p cannot be greater than n, otherwise some Dk would be identical [Def.8].
- A symmetric argument proves that p=n, and that Dk=Ck for any k in [0,n].

Corollary: given a class C of the quotient network, there is at most **one** circuit that starts with C. It is obvious, because if a node of C belongs to the first class of a quotient circuit, the class C **is** the first class of this circuit.

Teo.3.: in a quotient circuit of size c, the state of any node Ik of Ck at time t+k is defined from the state of C0 at time t, i.e.:

for any j of Ck, k in [1,c[, there exists a function Fj such that:

S(j,t+k) = Fj(S(C0,t))

for any j of C0, there exists a function Fj such that

S(j,t+c) = Fj(S(C0,t)).

The function Fj defined for the nodes of C0 is called the **self node function** of j.

Proof:

- Let (X1,..Xn) be S(C0,t), where Xk = S(Ik,t).
- Let i be a node of C1.
- We know that S(i,t+1)=f(E(i,t)), where

$$E(i,t)=\sum_{j\in C0} Wji * S(j,t).$$

- The property is proven with

$$Fi(X1,..Xn)=F(\sum_h Whi * Xh),$$

- Let us suppose the property is true for any node of Ck, k in [1,c[.
- Let i be a node of Cm with m=k+1 modulo c.
- We know that S(i,t+k+1)=fi(E(i,t+k)), where

$$E(i,t+k)=\sum_{j\in Ck} Wji * S(j,t+k).$$

- By hypothesis, we can write S(i,t+k) as
S(j,t+k)= Fj(X1,..Xn)

- The property is proven with

$$Fi(X1,..Xn)=F(\sum_{j \in Ck} Wji * Fj(X1,..Xn)).$$

Corollary: in a quotient circuit of size c, the state of any class Ck at time t+k is defined from the state of C0 at time t, i.e.:

for any k in[1,c[, S(Ck,t+k) = Fk(S(C0,t))

S(C0,t+c)=F0(S(C0,t))

F0 is called the **quotient self node function** of the class C0.

This results immediately from [Teo.3], using the definitions of the state of a class [Def.11] and the quotient node function [Def.12].

3.6 Construction of quotient circuits.

We present an algorithm which constructs the quotient circuits of a network. It derives from a graph unification algorithm [FIM85]. The first pass constructs the equivalence classes. Let S be a set of classes. S will be transformed into the set of the equivalence classes. At any time, each node i belongs to an unique class of S, Class(i).

```
S := U {k} for any node k of the network.
repeat
  changes := false
  for any Ck in S
    merges := false
    h := nil
    for any i in Ck.
      for any j such that Wij exists
        if h = nil then h := j
        else if (j does not belong to Class(h)) then
          Class(h) := Class(h) U Class(j)
          Class(j) := Empty_set
          merges := true
        endelse
      endfor
    endfor
    if merges then changes := true
  endfor
until changes = false
remove any empty class from S.
```

The second pass of the algorithm constructs the quotient circuit that starts from a given class. QC will be a sequence of classes of S.When the algorithm stops, QC is the desired quotient circuit or an empty sequence of classes, if no circuit starts from C

```
QC := ( )
D = C
j := any element of D
while there exists k such that Wjk exists and
                    Class(k) is not element of QC
  D := Class(k)
  QC := concat (QC ,D)
  j := any element of D

  /* QC was a path, not a circuit */
  if no Wjk can be found then QC := ( )

  /* the circuit QC did not close on C: C is not a
     circuit */
  else if D != C then QC := ( )
endwhile
```

3.7 Quotient circuits, quotient node functions and periods.

Teo.4: if a class C of the quotient network does not belong to any quotient circuit, the state of C reaches an equilibrium.

Proof: if C does not belong to any quotient circuit, then for any node i of C, the set of nodes that are connected to i does not contain any circuit (see section 3.6). The node i is in the n-th layer of a feed forward network. The state of any input node j (such that no Wkj exist) connected to i is constant since it has no entries: S(j,t)=fj(0). Whatever the initial conditions may be, the state of i will be constant for times t higher than the distance from any entry node j to i.

Teo.5: let C be a class of the quotient network. If C is the first class of a quotient circuit, and if C reaches the period T, i.e. S(C,t+T) = S(C,t), then all the nodes of the quotient circuit reach the same period T.

Proof:

- Let i be a node of the k-th class of the circuit, Ck, and Fi its node self function.
- S(C,t) = S(C,t+T) for any t >= t0, implies that Fi(S(C,t)) = Fi(S(C,t+T));
- Hence S(i,t+T+k) = S(i,t+k) for any t>= t0.
- The periodicity of S(i,t) is proven for any t >= t0+k.

Teo.6: let C be a class of the quotient network, such that C is the first class of a quotient circuit of size n. Let Fc be the quotient self node function of C. The nodes of C can reach the period p if and only if Fc has a cycle of order m, and p divides n*m.

Proof:
From left to right. Suppose the nodes of C reach the period p, i.e.there exists t0 such that for any t >= t0, S(C,t+p) = S(C,t)
- Let S be the initial state, i.e. S= S(C,t0).
- Let q be a multiple of both p and n. There exists k and m such that q=k*p=m*n.
- p is a period, hence so is q: S(C,t) = S(C,t+q) for any t >= t0. If we take t=t0, we get:
- S(C,t0) = S(C,t0+q),
- S = S(C,t0+m*n).
- S = Fc(...Fc(S(C,t0))..) [Teo.3].
  m times
- S = Fc(..Fc(S))
  m times.
- This proves that Fc has cycles of order m, such that k*p = m*n.

From right to left. Suppose Fc has a cycle of order m: Fc(..Fc(S)) = S
  m times
- Let Ck be the k-th class of the circuit We start with the initial conditions:

S(C,0) = S.
S(C,1) = S, S(C1,1) = F1(S)
S(C,2) = S. S(C1,2)=F1(S), S(C2,2)=F2(S).
...
S(C,n-1) = S, S(C1,n-1)=F1(S),....S(Cn-1,n-1) = Fn-1(S)

where the Fk are the functions defined in [Teo.3].
- These conditions ensure that for any t in [n,2n[, S(C,t)=Fc(S), and generally, for any t in [n*q,n*(q+1)[, S(C,t)=Fc(..Fc(S))
  q times.
- Let t be any time in [0,n[.
- S(C,t+n*m) = Fc(...Fc(S(C,t)))
  m times.
- From the initial conditions S(C,t) = S, thus:
  S(C,t+n*m) = Fc(..Fc(S))
  m times
- From the hypothesis, Fc(..Fc(S)) = S, thus:
  m times
  S(C,t+n*m) = S , hence
- For any t in [0,n[, S(C,t+n*m) = S(C,t).
- The same argument applies with t in [n*q,n*(q+1)[, using Sq = Fc(.. Fc(S))
  q times
instead of S (we know that Fc(..Fc(Sq)) = Sq).
  m times
- Hence, for any t in [0,n*m[, S(C,t) = S(C,t+n*m), the nodes of c have period n*m.
- Let p be **the** period of the nodes of C (i.e. the smallest period) : we see that p divides n*m.

3.8 Interpretation and important particular cases.

The result of [Teo.6.] is simple to use. For instance, if the quotient circuit starting from C is of size 3 and the quotient self node function Fc has cycles of order 8,4,2,1, the periods of the nodes of C are divisors of 24.

However, we need to **know the function Fc** in order to use the theorem. Fc is defined from the node functions and the weights, and this may lead to a complex expression. The cycles of Fc should be defined without developing the function expression. This is possible in several particular cases: binary node functions, and increasing functions with positive weights, in some configurations of circuits.

Teo.7. If the node functions are binary, and the class C has c elements, the order of any cycle of Fc is lower than 2**c.
Proof: 2**c is the number of possible states of S(C,t), so no cycle can be greater than this number.
Corollary. If the node functions are binary, the class C has c elements, the quotient circuit starting from C is of size n, the period of the nodes of C is lower than n* 2**c.

This result is quite important, and can be extended easily to **finite state node functions**. It can be improved if we only consider the elements of C **that are really in the circuit**. For instance, in the example of [Fig.8], the node G is out f the circuit; hence its state is defined by the other nodes. Moreover, we can look for **the smallest class of the circuit** before to apply [Teo.7], since all the nodes of the circuit have the same period [Teo.5].

Teo.8. If C has a single element, the nodes functions are increasing, the weights Wij are positive, the function Fc can only have cycles of order 1, i.e. x such that x = F(x).
Proof:
- Let i be the single element of C.
- With our hypothesis, Fc is an increasing function of S(i,t) (proved by induction upon the size of quotient circuit).
- Functions of R that have k-cycles, k > 1 are no monotone, so the result is proven.

Corollary. If the class C has a single element, the nodes functions are increasing, the weights Wij are positive, the quotient circuit starting from C is of size n, the node of C and all the quotient circuit can only have periods divisors of n.

This result can be extended to negative weights combined with decreasing node functions. It can be generalized to classes C with more than one element, when only one element of C belongs really to a circuit. Moreover, it can be applied if some class of the circuit has only one element, since all the nodes of the circuit have the same period [Teo.5].

CONCLUSIONS.

Orbits and frequency coding provide an interesting alternative to vector coding. The storage and retrieval of information are proven to be efficient. The computational cost of identifying orbits can be estimated as N times the cost of identifying equilibrium states. The factor N can be estimated thanks to the results given in section 3. Obviously, we expect to get more applications of our general result [Teo.6].

However, we still have to resolve practical and theoretical problems: efficient learning algorithms, better than those used in sequence recognition, stability to noisy signals, storage capacities, etc.

REFERENCES.

[BRI94] Brito F. Fimbel E., **Periodicidades en redes neurales,** Publicación interna UNEXPO, August 94.

[BUL92] Bullier J. Salin P. Girard P. , **Le cerveau en temps réel.** La Recherche , November 92.

[CHU92] Churchland P. Seknowski T., **The computational brain**, MIT Press, 92.

[DEL92] Delahaye J.P. **Information, complexité et hasard,** Hermes, 94.

[FH94] Fimbel E. Hernandez G., **Detección de fuentes de señales desconocidas mediante una red neural,** Panel 94, Monterrey, Mexico, 94.

[FIM85] Fimbel E., **Les Reseaux Miroirs**, These de l Université Paris 6, Publications du CNRS, 85.

[FIM94] Fimbel E. Brito F., **Uso de órbitas para almacenar información en redes neurales,** Iberamia 94, Caracas, Venezuela (presented in IBERAMIA 94, to be published in INTERFAZ, january 95).

[FRE90] Freeman W. Yong Y., **Model of biological pattern recognition with spatially chaotic dynamics.** Neural Networks, Vol 3, Num.2, 90.

[HER90] Herz J. Krogh A. Palmer R., **Introduction to the theory of neural computat**ion, Addison Wesley, 90.

[HOP82] Hopfield J. **Neural networks and physical systems with emergent collective computational abilities**. Proceedings National Academy of Science, Vo. 79, 82.

[HOP91] Hopfield J., **Olfactory computation and object perception**, Proceedings National Academy of Science, Vol. 88, 91.

[HAM91] Hammond C., Tritsch D., **Neurobiologie cellulaire**, Douin Editeur, 90.

[MOZ93] Mozer M., **Neural net architectures for temporal sequence processing,** to be published in predicting the future and understanding the past, Addison Wesley, 93.

[OPP83] Oppenheim A. Willsky A., **Signals and systems**, Prentice Hall, 83.

[WAN89] Wang D. Arbib M., **Complex temporal sequence learning based on short term memory,** proceedings of IEEE, Vol. 78 No. 9, september 89

USING A COMBINATION OF GENETIC ALGORITHM AND HILL CLIMBING TECHNIQUES FOR THE SOLUTION OF A DIFFICULT LETTER RECOGNITION PROBLEM.

John R. Podlena and Tim Hendtlass
Centre for Intelligent Systems,
School of Biophysical Sciences and Electrical Engineering,
Swinburne University of Technology,
P.O. Box 218 Hawthorn 3122.
Email : jrp@bsee.swin.edu.au & tim@bsee.swin.edu.au

ABSTRACT

The combination of the broad problem searching capabilities of a genetic algorithm with the local maxima location capabilities of a hill climbing algorithm can be a powerful technique for solving classification problems. Producing specialist artificial neural networks, each an expert on one category, can be beneficial when solving problems in which the categories are distinct. This paper describes combining genetic algorithms, hill climbing and sets of specialist artificial neural networks to solve a difficult character recognition problem.

INTRODUCTION

The genetic algorithm (GA) is a paradigm which is based on both gene recombination and the Darwinian "survival of the fittest" theory. A group of individuals in some environment have a higher probability to reproduce if their fitness (their ability to thrive in this environment) is high. Offspring are created via a crossover operation (as in gene recombination) and mutation. The algorithm, described by Holland [Hol92], aims to increase the average fitness of the individuals over a number of generations. The notion of combining the broad search capabilities of the genetic algorithm and the hill climbing ability of the artificial neural network has occurred to many researchers [Co92]. This has lead to much work in the areas of weight and topology optimisation individually, but less on the two combined. In the area of weight optimisation, a population of fixed architecture networks is "bred" (using crossover and/or mutation operations) to create offspring whose weights will hopefully be closer to a solution than their parents [Gol89]. An individual's breeding chance is directly related to their "fitness" (e.g. average error over a test set). This method has shown promise when using either crossover or mutation as the main GA operator. One of the problems with such net training is that the GA has trouble completing the solution to the problem and thus the GA is often only applied initially, the best network after a number of generations being left to home in on a solution using its normal learning rule [SJW92].

Research has also been done on the topological design of neural nets employing various adaptations of the GA, most with successful results [Smi91],[CC93]. Some use back propagation networks, applying a genetic algorithm to the number of hidden nodes which subsequently produces the most structurally fit individual for a given task. Often this uses a specific initial architecture, and "grows" a network which has a unique (and not necessarily maximally connected) structure. One method for the addition and removal of these nodes is to map the topology into the chromosome (or genotype) such that it represents both nodes and connections, subsequent crossover causing fluctuations in the overall number of nodes [SJW92]. However, this method does not pass on weight information between successive individuals. Each new individual must be trained from an initial weight state before it can be placed in the new generation, increasing computational time.

The combination of weight training by both GA and normal techniques has also been researched [Co92]. The weights are initially evolved using a GA for a fixed period. The best resulting network completes training by conventional methods. Some studies involving the simultaneous evolution of both the topology of the net and the weights have been reported. These involve the use of weight crossover and have shown restrictions on network complexity [ASP94].

THE ADAPTED GENETIC ALGORITHM

The work described in this paper uses an adaptation of the standard genetic algorithm to develop the most appropriate topology for a given architecture and to find an appropriate learning coefficient for all nodes. Gradient descent and genetic mutation are the only mechanisms used to converge the network weights to a solution. The variation to the standard genetic algorithm described below was designed to overcome the problems reported by other researchers.

The genotype/chromosome structure used consists of a template of initially empty "slots", each slot able to hold a hidden node, complete with weights, learning coefficient and a Boolean "activation" flag. When the system is initialised, the first N nodes are activated for each individual in a given population, and the learning coefficient and weights set to random numbers. This creates a chromosome with some slots filled/activated and others left empty. At the time of breeding these slots are crossed over, the probability for a slot from one parent being placed in the child chromosome being 0.5. This initially causes little change in structure, however when combined with a switch on/off node mutation (with equal mutation probability for on/off) the crossover method enables greater diversity in structure. A simplified example is shown below, displaying crossover at each node point, whether used or "free", and a mutation in the child.

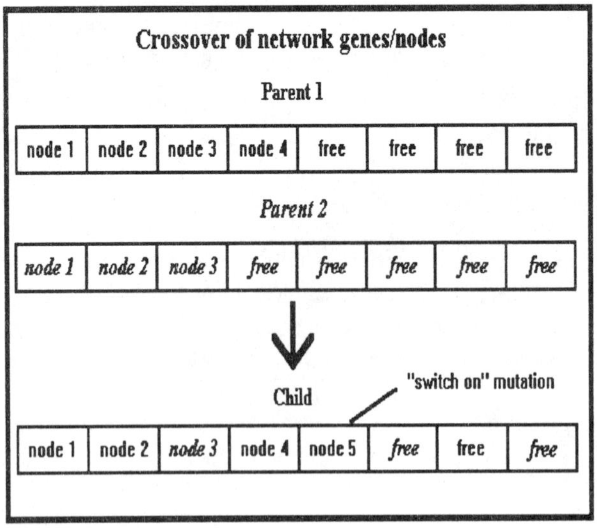

There is a low probability that the on/off state of any node may be toggled during the birth process. When a node is activated there is an equal probability that either the weights previously stored in the switched off node or an initialized set of weights will be used. This operation has biological parallels, some genes being "switched off" in cell development but still being retained in the DNA for possible next generation use. It also has practical value in its ability to retain what may be a successful component of a previously successful individual (this gene retention is called Lamarkian learning [HL93]). It has been seen [ASP94] that one problem with crossover is that parents with differing structures do not necessarily produce offspring with valid structures. By keeping the switch on/off mutation rate low, a population can at any time include a number of individuals of a particular structure, improving the probability of compatible parents breeding valid individuals.

In the adapted algorithm described in this paper, mutation is the only GA operator affecting the weights values of nodes. Nodal weights of all individuals can be mutated if selected via a mutation probability (usually low). Individuals do not lose nodal weight information from one generation to the next through weight crossover methods (correlation between the two parent's hidden unit individual weights values is unlikely [ASP94]). The probability of "stagnation" of the evolution process may be diminished by attempting to ensure a diverse population of individuals. To help achieve this, the probability for mutation can be made inversely proportional to the difference between the current parents.

The algorithm implemented is as follows:

For all individuals:
  initialise i nodes to initial learning coefficient
  initialise all weights to random values
until of best individual's error < minimal value Loop
  train nets in population for a fixed number of iterations
  calculate fitness based on net error
  calculate probability for parenting new generation
  set n worst individuals to zero mating probability
    (n=Ndead)

copy m of the best individuals into new population
(m=Ncopied)
breed individuals to fill next generation pool,
EndLoop

BREEDING STRATEGY

The above algorithm displays two features not yet described. The first is the "culling" of the worst N individuals in the population at breeding time by setting their probability of breeding to zero. This effectively increases the next generation's chances of producing more viable individuals by giving the 'fit' breeding parents larger "slices" of the breeding probability "pie". The second feature implemented is a further deviation from the classic genetic algorithm. Using the standard breeding method, once a parent is selected for breeding a crossover probability (often 0.5) determines whether the individual will be crossed over with another parent, or whether it will be copied intact into the next generation. One problem with such a method is that relatively low fitness individuals can be copied straight into the next generation, while highly fit individuals can be crossed over and thus have the possibility of losing their important genetic material. The benefits of always copying N of the most fit parents directly into the new generation are two fold. Information from the best parents can not be lost via unfortunate crossover combinations, and the best individuals have a chance to make progress based on their local heuristic over a number of generations. To combat the possibility of highly trained parents, which are stuck in local minima, forever being copied directly into the next generation, and to further encourage diversity, each parent may have an associated "age penalty" coefficient which effectively reduces the parents fitness to breed with time.

The strategy described above for parent retention differs from the more general forms of the genetic algorithm. In one method, called the Generational Replacement Genetic Algorithm (GRGA), the new generation replaces all the previous individuals (except those parents not crossed over). Another common method, Steady State Genetic Algorithm (SSGA), replaces only one or two members of the population. Results from the variation of the number of N individuals retained from the previous generation are outlined below.

FOUR BIT PARITY PROBLEM

The performance of the algorithm on the four bit parity problem (a classically difficult problem for the back propagation network) has previously been reported [PH95]. Trials involving changing the amount of least fit individuals removed from the breeding pool (Ndead) showed that setting this to any value greater than zero improved the performance, the exact value used appeared not to be important. Similarly setting Ncopied to a non zero value also assisted, a value of approximately 1/2 the population size appearing optimum for this problem and approximately halved the time required to find the solution. The problem of individuals with equivalent structural functionality but different symetry competing in the breeding process ("competing conventions", [Co92]) has been found to be insignificant in the work described in this paper, and thus prevention has not necessary.

RESULTS FROM A DIFFICULT LETTER RECOGNITION PROBLEM

The algorithm was applied to Frey and Slate's [FS91] letter recognition problem data. The problem set consists of 20,000 unique letter images generated by randomly distorting pixel images of the 26 uppercase letters from 20 different commercial fonts. The fonts used included a full range of character types including Script, Italic, Serif and Gothic. Below is an example of part of the character set and the distortions involved.

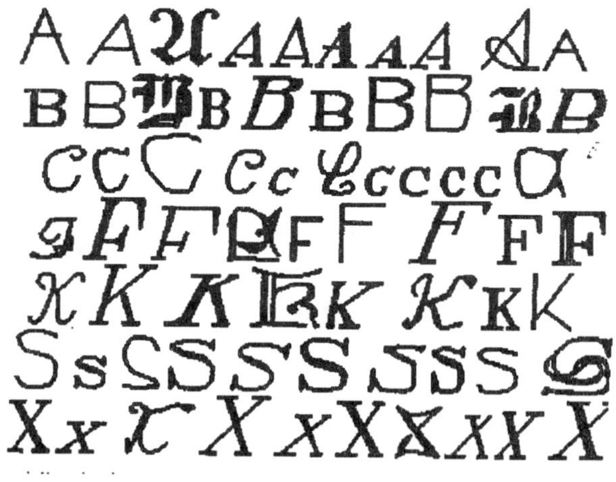

Sixteen numerical attributes are provided to summarise each example (e.g. dimensions of the encapsulating box, number of 'on' pixels etc.). Frey and Slate used "Holland style adaptive classifiers" (a type of machine rule induction) to solve the classification problem, training their system using the first 16,000 examples. Testing was carried out using the remaining 4,000 examples. Our algorithm was applied to multiple sub-populations or "families" of networks, each responsible for creating individuals that recognises one specified letter (for example, the 'A' recognition task is handled by one population subset, while 'B' is handled by another). This "Bottom up" approach was chosen to reduce the complexity of the problem via the familiar "divide and conquer" paradigm [VJ93]. Each family used a training data set which consisted of around 700 examples of the letter assigned to that family and 700 drawn from the pool of all other letters. Each family had a population size set to 40 networks. Each network was trained using a full sequential run through this data set after which the adapted GA described above was applied to the entire family. It was found that after the GA had produced an appropriate network, it was advantageous to "switch off" the GA and let the best network in a family home in on the solution by the normal back propagation training algorithm.

The ultimate individuals produced by the different families had a range of different numbers of hidden nodes and average learning coefficients (figure 1).

Figure 1 indicates the diversity of the solutions obtained by the different families for each of their respective letters. Note the deviation of learning coefficient and number of hidden nodes from their means (0.44 and 22 respectively); most letters have a learning coefficient deviation from the mean that correlates with their hidden node number deviation. It is interesting, but may not be significant, that those letters which do not show this correlation tended to have higher post training error rates than those which did.

After an average of 350 GA epochs, at which stage the GA had usually found a satisfactory network, the GA was "switched off". The best performing network from each family was then selected and subsequently trained for around a further 8,000 cycles using just the back propagation learning rule. The winning net from each of the families was shown each of the 20,000 examples. The category assigned was that of the net with the overall highest output.

Histograms of each network's average output for each letter example revealed interesting interplay between competing letter networks (figure 2). For example, the 'F' network had difficulty in distinguishing between the 'F' character and the 'P' character. The average response to an 'P' input was not far behind the response to an 'F' input. For some individual cases the response of the 'F' network to a 'P' input actually exceeded the output of the 'P' network giving an incorrect resulting classification. Inspection of the histogram of average responses in figure 2 reveals no surprises in that the letters that the 'F' net responses to most are indeed geometrically similar (F,P,T and to a lesser extent Y). The response to geometrically dissimilar letters (for example A) is lower.

Figure 1. Learning Coeff. and # of Hidden Nodes for Best Individual Nets

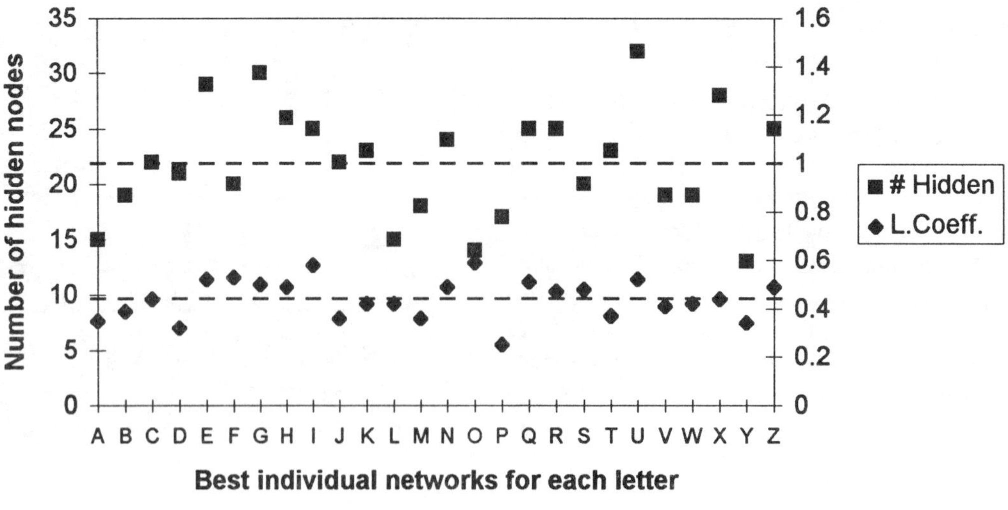

Figure 2. Average outputs from F network for each test letter example

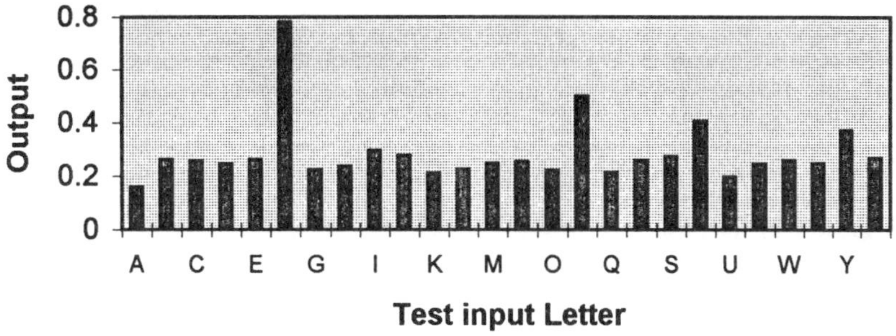

The overall result obtained was 87.5% (17510 correct, 2490 incorrect). With the addition of hand tuned thresholding this could be lifted 88.6% correct responses (17725 correct, 2275 incorrect). These results compare favourably with Frey and Slate's best result of 82.7% which also required hand tuning, and are superior to our attempt to solve the problem on a single large network using commercial software which failed to yield results over 75%.

Another attempt to solve the problem was made by Fogarty [Fog92], who used the "first nearest neighbour" algorithm for classification. This entailed retaining 16,000 examples in memory and using them to classify each member of the testing set. To find which class a testing example belonged to, the Euclidean distance between it and each of the 16,000 examples in memory was calculated using the 16 primitives of both the unknown and of the example being considered. The unknown was assigned the class of the closest example found. This method produced good results (average of 95.7%), however for each test input it needs to make 16,000 distance calculations (using 16 coordinates each) to arrive at each classification. Using only 1,600 examples reduced the average result to 82.57% correct over the 4,000 testing examples. Since our algorithm produces 26 networks with small memory requirements giving almost instant classification it obviously has significant advantages.

CONCLUSION

The implemented algorithm successfully breeds and trains a suitable network configuration for each letter of a difficult letter recognition problem. With low mutation rates, the system changes its learning coefficient and hidden node numbers to explore different areas of the problem space. In such problems where success is highly dependent on the initial conditions and architecture of an individual network, the algorithm shows promise in its ability to change such parameters as it learns. Some of the unique features of the above algorithm are: a) its "slot" mechanism for a dynamic nodal topology, with crossover at each node point, b) Lamarkian probability mutation (the probability of keeping crossed over individual node's weights intact), c) removing the worst "Ndead" individuals from a breeding pool and d) copying "Ncopied" directly into the next generation.

The expansion of this system to a multiple sub-population algorithm, with test problems being broken down into subgroups to be learned by population subsets, was found to have superiority over the use of machine rule induction or large single neural networks for a difficult character recognition problem. The large memory and computational requirements of the nearest neighbour solution for classification on test examples make it less of a practical solution than our method.

Future work in this area may include the storage of multi-generational information for each sub-population, such as previously invalid topology configurations or initial weights, to be used by future generations to avoid going over old paths. The algorithm could be expanded further to include a "distributed" genetic algorithm, which would break each sub-population down further into subsets (e.g. two 'A' families with populations of ≈ 20 each), increasing diversity and therefore the chances of finding a near optimal solution. The interaction between the two population subsets within a family would be restricted to the occasional "migration" of a fit individual from one subset to another (as in Brill, Brown and Martin's work [BBM92]).

REFERENCES

[ASP94] Peter J. Angeline, Gregory M. Sanders, Jordan B. Pollack, An Evolutionary Algorithm that Constructs Recurrent Neural Networks, IEEE Transactions on Neural Networks, vo. 5, no.1, January 1994.

[BBM92] Frank Z. Brill, Donald E. Brown, Worthy N. Martin, Fast Genetic Selection of Features for Neural Network Classifiers, IEEE Transactions on Neural Networks, Vol. 3, No. 2, March 1992.

[CC93] C.H. Chu, C.R. Chow, A Genetic Algorithm Approach to Supervised Learning for Multilayered Networks, World Conference on Neural Networks, vol. 4, pages 744-747, 1993.

[Co92] International Workshop on Combinations of Genetic Algorithms and Neural Networks, COGANN-92, IEEE Computer Society Press, California, 1992.

[Fog92] Terence C. Fogarty, First Nearest Neighbour Classification on Frey and Slate's Letter Recognition Problem, Machine Learning, 9, 387-388, 1992.

[FS91] Peter W. Frey, David J. Slate, Letter Recognition Using Holland-Style Adaptive Classifiers, Machine Learning, 6, 161-182, 1991.

[Gol89] D.E. Goldberg, Genetic Algorithms in Search, Optimisation, and Machine Learning, Addison-Wesley Publishing Company, Reading, Mass. 1989.

[HL93] Ackley D. H., Litman M. L., A Case for Lamarkian Evolution, In C. G. Langton (Ed.), *Artificial Life III*, Reading, MA: Addison-Wesley.

[Hol92] John H. Holland, Adaptation in Natural and Artificial Systems, second edition, MIT Press, 1992.

[PH95] John R. Podlena, Tim Hendtlass, A Modified Genetic Algorithm Applied to Neural Network Design, Proc. The Sixth Australian Conference on Neural Networks, ACNN95, February 1995.

[SJW92] W. Schiffmann, M Joost, R.Werner, Synthesis and Performance Analysis of Multilayer Neural Network Architectures, University of Koblenz, Technical Report 16. 1992.

[Smi91] Guy Smith, Back Propagation With Dynamic Topology and Simple Activation Functions, International Journal of Neural Networks, Vol.2, No2/3/4, June - December 1991.

[VJ93] Christine L. Valenzuela, Antonia J. Jones, Evolutionary Divide and Conquer : A Novel Genetic Approach to the TSP, Evolutionary Computation, Volume 1, Number 4.

Artificial Neural Networks

A NEW METHOD OF TRAINING NEURAL NETWORKS FOR CONTROL

M. Bahrami and K. E. Tait
School of Electrical Engineering,
University of New South Wales,
Sydney NSW 2052, Australia

ABSTRACT

This paper presents a new method of training neural networks of Radial Basis Function (RBF) type which are used for controlling nonlinear plants. This method of designing neuro-controllers does not require identification of the plant or its inverse model. Direct inverse controllers discussed in the literature require the Jacobian of the plant or the sign of the Jacobian which may not be available for an unknown plant. We approximate the output of the plant with the output of its reference model in a model reference adaptive control scheme and use the Jacobian of the reference model instead of the plant. Simulation results show a satisfactory performance.

I. INTRODUCTION

Different types of Artificial Neural Networks (ANNs) have been proposed and successfully used for identification, control and fault detection in different systems. ANNs have certain characteristics that make them particularly useful for these tasks. Among these characteristics are: - inherent parallel and distributed nature that allows a fast response if implemented in hardware,
- learning from examples,
- ability to approximate virtually any mapping function,
- capacity for generalization,
- simple design,
- fault tolerant.

A ANN used as a controller in a Model Reference Adaptive Control (MRAC) scheme is shown in Fig. 1, in which Δ denotes a time delay. The controller in this specific example uses the past two values of the output of the plant and input to the plant. The number of tapped delay lines required for the controller in general depends on the order and relative degree of the plant. In the MRAC scheme, a reference model is used that provides the desired form of the output for each given input.The function of the neuro-controller (NC) in this structure is to provide the $u(k)$ from the inputs that can generate an output close to the desired output $y_m(k)$. If we knew what plant input $u(k)$ can generate the desired output $y_m(k)$, we could train a network to provide that input $u(k)$ for any input. Generally we do not have this information.

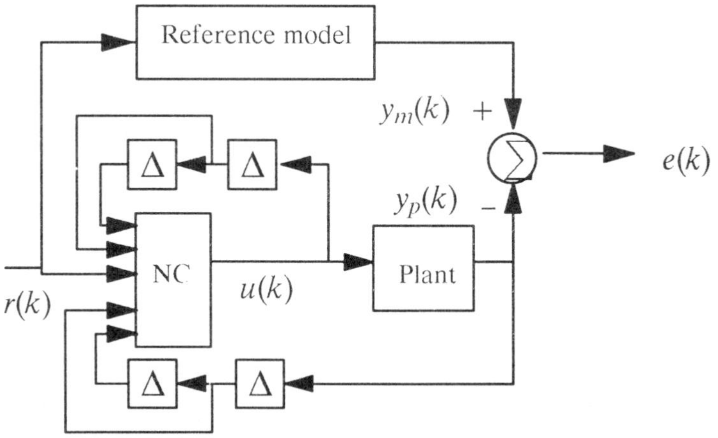

Fig. 1. A neuro-controller in an MRAC scheme.

II. THE NEW METHOD OF TRAINING

For a single-input-singe-output (SISO) plant, we have:

$$\frac{\partial e}{\partial u} = \frac{\partial e}{\partial y_p}\frac{\partial y_p}{\partial u} \qquad (1)$$

where u and y_p are the input and the output of the plant respectively. For an unknown nonlinear plant, the Jacobian of the plant ($\frac{\partial y_p}{\partial u}$) is not known. If the Jacobian of the plant was known, we could back propagate error $e(k) = y_p(k) - y_m(k)$ through the plant and adjust the weights of the neuro-controller accordingly [PSY87]. Saerens and Soquet [SS89] suggested using the sign of the Jacobian, instead of its real value in the training of the neuro-controllers. The sign of the Jacobian of the plant may not be available in general and it may not be constant for the whole range of the input. Here we propose another method which overcomes this limitation.

By approximating the output of the plant y_p with the output of the reference model y_m in an MRAC system, we will have

$$\frac{\partial y_p}{\partial u} \approx \frac{\partial y_m}{\partial r} / \frac{\partial u}{\partial r} \qquad (2)$$

Now, instead of the Jacobian of the plant, we can use the Jacobian of the reference model to calculate the backpropagated error at the output of the controller. Calculation of the $\frac{\partial u}{\partial r}$ may not be easy for a general Multilayer Perceptron (MLP) network, but it is possible for a simplified Radial Basis Function (RBF) network. Each node in the hidden layer of an RBF network is sensitive to a specific region of the input space. These nodes which are called *kernel* nodes compute radially symmetric functions that are maximum when the input is near the centre of a node. Weights from kernel nodes to the output nodes are trained using delta rule.

Kernel functions typically have Gaussian shapes with a fixed or varying widths. This activation function can be approximated by:

$$f(x) = \frac{1}{1 + \beta(x - x_0)^2} \qquad (3)$$

where x_0 indicates the centre of the kernel and β indicates the width of the function. During training we only adjust output weights of the network, the location of the kernel units x_0 and their width β remain fixed.

Let J denote the cost function defined by:

$$J = \sum_{i=1}^{A}\sum_{k=1}^{T}[y_{mi}(k) - y_{pi}(k)]^2 \qquad (4)$$

where y_{mi} and y_{pi} are the outputs of the reference model and the plant for the ith input, A is the number of sample inputs used for training, and T is the time during which we want the output of the plant follow the reference model. For an *on line* or *per sample* training we can use a simplified cost function as

$$J = (y_m - y_p)^2 \qquad (5)$$

For a steepest descend learning, adjustment of the weights should be made in a direction to decrease this cost function, therefore we should have

$$\Delta w_j = -\eta\frac{\partial J}{\partial w_j} \qquad (6)$$

where w is the output weight of the network, η the learning rate and

$$\frac{\partial J}{\partial w_j} = \frac{\partial J_i}{\partial u}\frac{\partial u}{\partial w_j}. \qquad (7)$$

For an RBF network with a single output we have

$$u = \sum_{j=1}^{L} w_j x_j \qquad (8)$$

where x_j is the output of the jth kernel unit and L is the number of kernel units. Considering the simplified activation function given in Eq. (3) we have

$$x_j = \frac{1}{1 + \beta s_j} \qquad (9)$$

where s_j is defined as

$$s_j = \sum_{k=1}^{N}(p_k - R_{jk})^2. \qquad (10)$$

In this equation N is the number of inputs to the RBF network, p_k is the kth input and R_{jk} indicates the centre of the jth kernel unit for the kth input. Combining Eqs. (5) to (10) and using the approximation given in Eq. (2), results in the following equation:

$$\Delta w_j = -\frac{\eta}{\beta}(y_m - y_p)x_j\frac{\frac{\partial y_m}{\partial r}}{\sum_{l=0}^{L} w_l(r - R_{l0})x_l^2} \qquad (11)$$

where $r = p_0$ is the reference input.

The difference between this weight update equation and the conventional ones is in replacing $\frac{\partial y_p}{\partial u}$ by $\frac{\partial y_m}{\partial r} / \frac{\partial u}{\partial r}$. While $\frac{\partial y_p}{\partial u}$ may not be known for a given plant, the partial derivative of the reference model with respect to its input can always be determined.

The denominator in Eq. (11) is related to the derivative of u with respect to r. It is possible for this term to be zero or a very small value. There are several possible ways of handling this situation. One method that was used in our simulations is using a new weight update equation based on a function like $h(.)$ as defined below.

$$\Delta w_j = -\frac{\eta}{\beta}(y_m - y_p)x_j \frac{\frac{\partial y_m}{\partial r}}{h\left(\sum_{l=0}^{L} w_l(r - R_{l0})x_l^2\right)} \tag{12}$$

where

$$h(x) = \begin{cases} x & |x| > \epsilon \\ sign(x)\epsilon & otherwise \end{cases} \tag{13}$$

and ϵ is a small positive number.

III. SIMULATION RESULTS

Different nonlinear plants were used for simulation but for space limitations the results of only one system is discussed. Signals with amplitudes uniformly distributed in the interval [−1, 1] were chosen as the reference input for training. The values chosen for learning rate η and width parameter β in Eq. (12) were 0.2 and 1 respectively. Up to 20,000 iterations were carried out for the neuro–controller training. After completion of training, the performance of the plant with controller was compared with the reference model for different inputs. The plant with trained controller was able to follow the reference model reasonably well which indicates that training has been successful.

For a model II plant [NP90] described by

$$y_p(k) = \frac{y_p(k-1)y_p(k-2)(y_p(k-1) + 2.5)}{1 + y_p^2(k-1) + y_p^2(k-2)} + u(k-1)$$

$$y_m(k) = 0.6y_m(k-1) + 0.2y_m(k-2) + r(k-1)$$

Figs. 2A to 2C show the response of the reference model and plant with controller and without controller for three different inputs.

IV. CONCLUSIONS

The controller discussed here uses an RBF network which has less training time compared with MLP networks at the expense of requiring more kernel units and weights. RBF networks also make the performance of the network mathematically tractable.

Since the adjustments of weights according to Eq. (12) is based on approximating the output of the plant with the output of the reference model, the performance of the training method to a large extent depends on the choice of the reference model. The method of training proposed here does not perform a true gradient descent and does not promise a faster training. Compared with methods proposed by [PSY87] and [SS89], it gives a third possibility for training neuro–controllers which is based on the characteristics of the known reference model and not the possibly unknown plant.

REFERENCES

[PSY87] D. Psaltis, A. Sideris, and A. Yamamura "Neural Controllers" *IEEE International Conference on Neural Networks*, pp. IV 551–558., 1987.

[SS89] M. Saerens and A. Soquet "A Neural Controller" *First IEE Conference on Artificial Neural Networks*, pp. 211–215, 1989.

[NP90] K. S. Narendra and K. Parthasarathy, "Identification and Control of Dynamical Systems Using Neural Networks" *IEEE Transactions on Neural Networks*, **1**, pp. 4–27, 1990

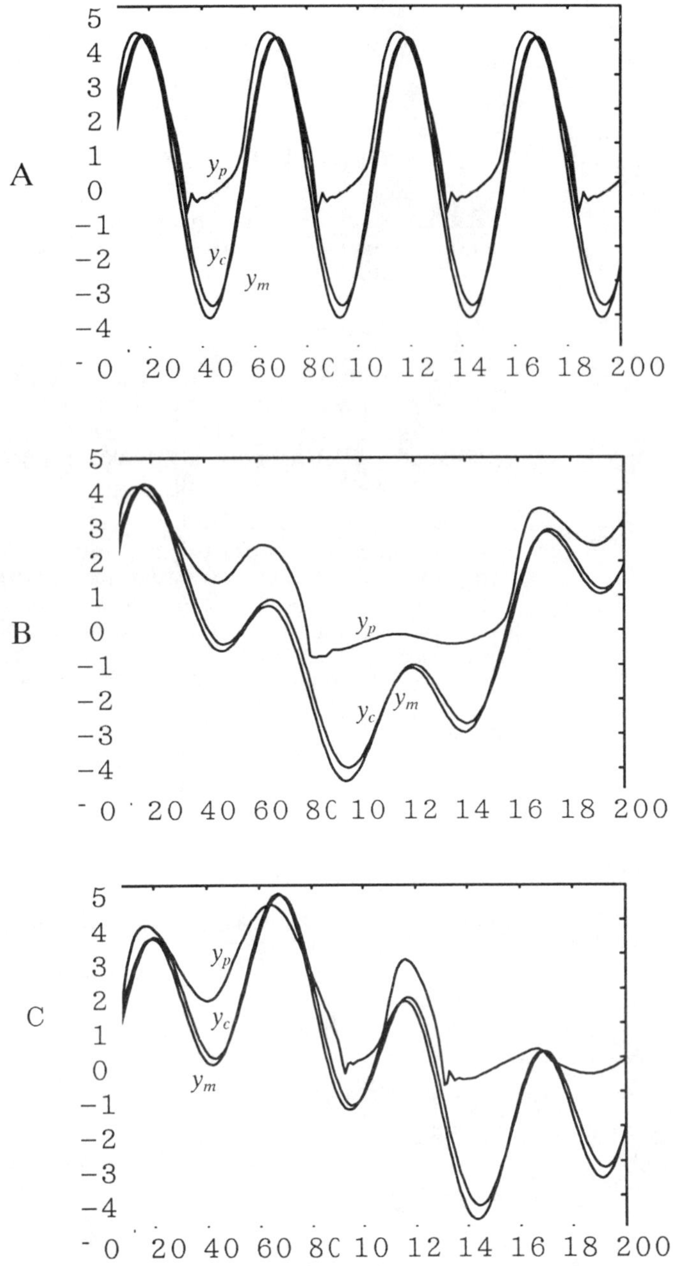

Fig. 2. Output of the plant without control y_p, the reference output y_m, and plant with control for the following inputs A: $sin(2\pi k/50)$, B: $0.4sin(2\pi k/50)+0.6cos(2\pi k/200)\pi$, and C: $0.6sin(2\pi k/50)+0.5sin(2\pi k/200)$.

A PROBABILISTIC NEURAL NETWORK APPROACH TO JET ENGINE FAULT DIAGNOSIS

Richard Eustace & Graeme Merrington
DSTO, Aeronautical and Maritime Research Laboratory
richard.eustace@dsto.defence.gov.au
graeme.merrington@dsto.defence.gov.au

ABSTRACT

In this paper a neural network approach is used to diagnose faults in gas turbine engines. In particular, faults are required to be diagnosed from many engines within a fleet of aircraft. The problem of diagnosing faults across a fleet is difficult because any faults have to be identified using parameters which vary naturally from engine to engine within a healthy fleet. Such variations in no-fault performance arise due to tolerances in manufacture and construction of critical engine components. Any in-field technique which is applied to a number of engines needs to allow for or accommodate this variation in the no-fault performance but still recognise abnormal behaviour. For this study a low by-pass ratio turbofan engine was tested with a number of faults implanted in the variable geometry schedules of the compressor and exhaust nozzle. A probabilistic neural network is employed for diagnosis with encouraging results. The method is shown to be capable of often diagnosing faults even when the parameter changes due to the fault are less than the no-fault engine-to-engine variation

INTRODUCTION

Fault diagnosis in a complex mechanical system, such as a gas turbine engine, often involves comparing measured parameter values from a suspected faulty engine with some suitable no-fault reference values. In the process, the differences between the measured and reference values form residuals which become significant if they exceed predetermined statistical uncertainty limits. Meaningful comparisons can be made by using a mathematical model to generate the reference system states at the same condition at which the tests actually occurred. Assuming the model is accurate, the residuals contain meaningful information which may be used for detection and diagnosis of faults. Methods such as likelihood ratio tests, chi-squared tests or Bayes' decision process can be applied [Wil76].

Recently, knowledge-based methods are being offered as another tool for examining residuals for fault diagnosis [Fra90]. Such techniques include expert systems which make use of heuristic knowledge to provide a qualitative analysis [Fre91]. Expert systems allow for 'rule-of-thumb' procedures based on empirical knowledge gleaned from maintenance personnel to be factored into the overall diagnostic process in the form of 'if-then' statements. These statements can then be chained together to link symptoms with known fault conditions.

Neural networks are also being fostered as a fault diagnostic method, especially for rapid real time diagnosis [DM92]. After the process of residual generation, a neural network uses the residuals to

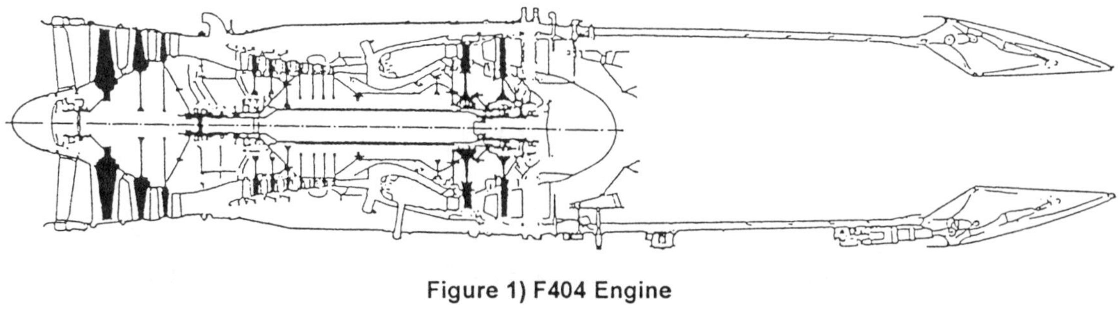

Figure 1) F404 Engine

perform the diagnosis. Networks can also be used to replace the model used for generating the residuals provided sufficient training data exist. In situations where the system being monitored is complex, and an accurate model is not available, a neural network can be used for both generating the residuals and diagnosing faults from them. Applications include the Space Shuttle Main Engine [WFA90, WKA90, LG90] process systems [SKK91] and aircraft engines [DKA89, Gen89].

In this paper, a neural network is applied to the problem of diagnosing faults in gas turbine engines. While recent work [Eus93] has demonstrated the feasibility of using neural networks for diagnosing faults in a single engine, this paper applies neural networks to the task of diagnosing faults in any engine within a fleet. The task of identifying faults from a fleet of engines is more difficult because engine-to-engine variations increase the resultant no-fault scatter from which abnormal behaviour must be recognised.

ENGINE

The engine selected for this investigation was the General Electric F404 low by-pass ratio afterburning turbofan (Figure 1), used by the Royal Australian Air Force (RAAF) in the F/A-18. This engine is modular in construction and designed to be maintained 'on-condition'. It is made up of six main modules; fan, compressor, high pressure turbine, low pressure turbine, and afterburner/final nozzle section. Variable geometry is used in the fan, compressor and final nozzle in order to optimise the performance of the engine over a wide operating range. An effective on-condition maintenance policy requires operational fault conditions to be easily identified by maintenance personnel. Faulty modules can then be replaced with repaired units and the engine returned to service quickly.

For this investigation eight engine parameters were used. These consist of fan spool speed (NL), compressor spool speed (NH), compressor entry total temperature (T25), compressor outlet static pressure (Ps3), fuel flow (WF), low pressure turbine outlet total pressure (PT5), low pressure turbine outlet total temperature (T5), and exhaust nozzle area (A8). The selection of these parameters is based on a statistical correlation study and explained in detail in the following section.

FLEET ENGINE NO-FAULT DATABASE

A large set of no-fault data is required to establish the fleet data base of nominal engine behaviour as a reference. These data were obtained from production acceptance records. Following construction of each F404, engine acceptance testing provided a set of no-fault

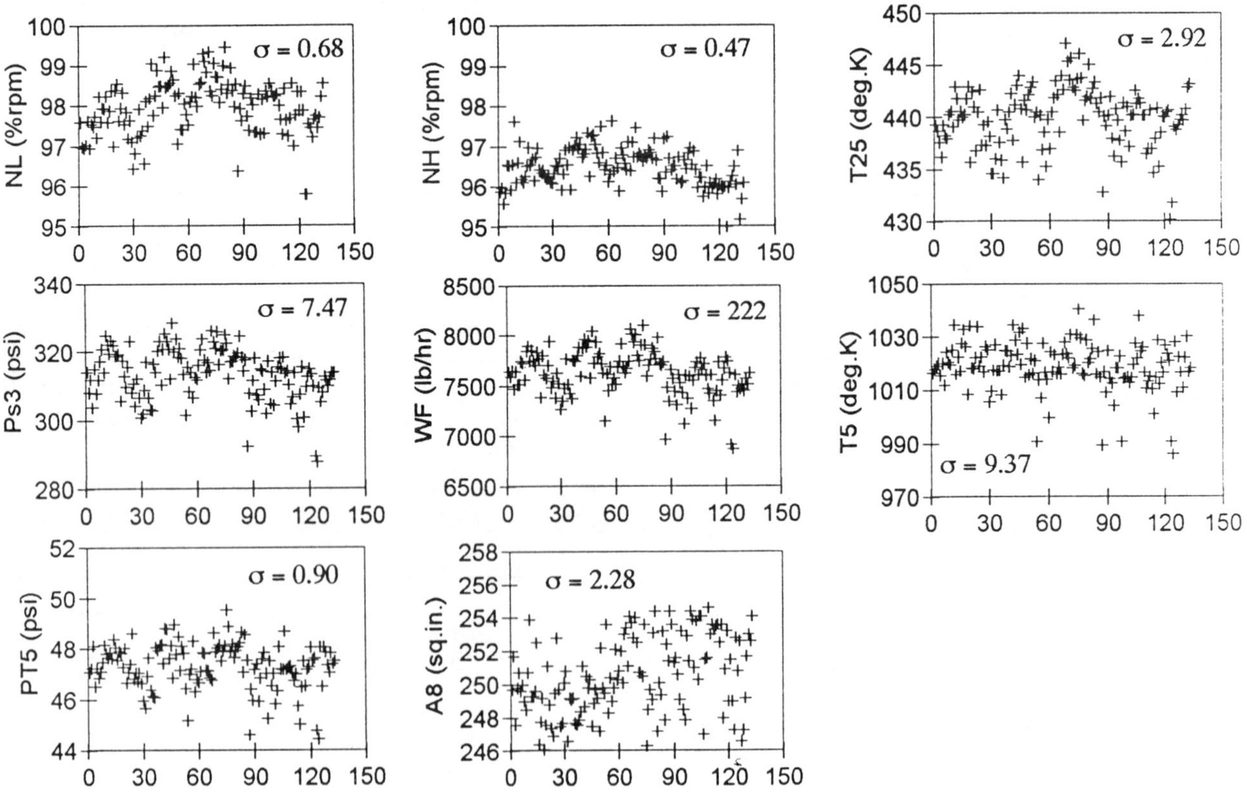

Figure 2) Plots of corrected acceptance test data for 130 no-fault engines

engine performance data at comparable power settings. The data set contained 130 engines and therefore provided a good indication of the no-fault parameter variation within a fleet of new engines (Figure 2). The acceptance testing was performed under Sea-Level-Static (SLS) conditions and data acquired at a power lever setting of approximately 94 degrees were utilised for the present analysis.

However, it is not possible to set the throttle at exactly the same point for each engine, so some variability in the resultant measured parameters can be attributed to this. In order to compare engines with one another, one parameter such as fuel flow (WF), high pressure spool speed (NH), or engine pressure ratio (EPR) needs to be designated as a baseline reference parameter to facilitate the process of monitoring the other parameters for the effect of faults.

A common approach is to chose one of the above parameters for the baseline, and set it equal to a reference value. The other engine parameters are then adjusted for the difference between the engine's actual measured baseline value and the reference value using the method of influence coefficients. The influence coefficients are usually generated from a computer cycle deck of the engine by perturbing the baseline parameter around the point of interest. The effects of these perturbations can then be seen on the other engine parameters, allowing parameter variations from the fleet-average baseline values to be ascertained. This approach relies on having a computer model available which is capable of generating accurate influence coefficients. Unfortunately such a model was not available when this work was performed.

A different approach was therefore adopted. Examination of the data showed that certain parameters,

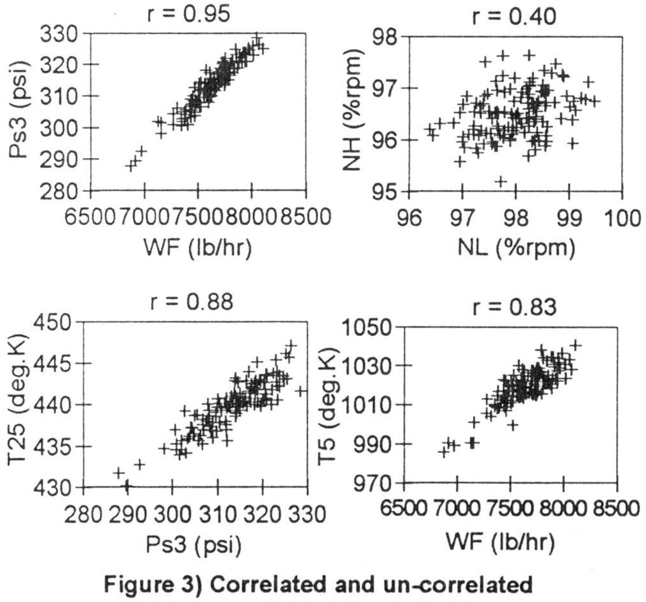

Figure 3) Correlated and un-correlated parameter pairs

such as fuel flow (WF) and Ps3, are strongly related. This relationship is not unexpected, as the fuel flow in many engines, including the F404, is scheduled as a function of compressor outlet pressure over at least part of the operating range. As Figure 3 shows, there is a strong correlation in these parameters which is not masked by the engine-to-engine variation. Other parameter pairs, such as NL and NH, which in any one engine would be expected to be correlated due to the aerodynamic coupling, show no such feature in the fleet engine data (Figure 3)

From an examination of the data, five pairs of parameters were found to be strongly correlated. Furthermore, the average performance of a fleet engine around an operating point could be represented by simple linear models. For each pair, one parameter is designated as being independent, and the other dependent. The five pairs used, with their associated correlation coefficients (r), are:

$$Ps3 = a_1 \times WF + b_1 \qquad r = 0.95 \qquad (1)$$

$$NL = a_2 \times T25 + b_2 \qquad r = 0.88 \qquad (2)$$

$$T2 = a_3 \times Ps3 + b_3 \qquad r = 0.88 \qquad (3)$$

$$PT = a_4 \times WF + b_4 \qquad r = 0.87 \qquad (4)$$

$$T5 = a_5 \times WF + b_5 \qquad r = 0.83 \qquad (5)$$

These five equations are then used to generate residuals of the dependent parameter. Given the actual values of the independent parameters for each engine, the residual is given by the difference between the calculated and actual value of the dependent parameters. For example, the Ps3 residual is obtained by subtracting the calculated value derived from Equation 1 (using the measured fuel flow) from the measured value of Ps3. Similarly, when Equation 2 is used to calculate NL, and this is compared with the measured NL value, the NL residual is now formed at the measured value of T25. Thus this approach does not utilise a constant reference point, but uses a relationship as a measure of baseline performance, and each residual is obtained from the difference between the appropriate baseline parameter value and the actual measured engine value.

Fault-free engines will produce residuals within a defined band, defined by the engine-to-engine build differences. However, if the engine includes a fault which changes the relation between one of the five pairs of parameters, then the residual for this engine will be noticeably different.

FAULT DATA-BASE

The task of identifying faults from a fleet of engines requires a suitable data base of engines with and without faults. While the acceptance testing provides a suitable set of no-fault data, faulty-engine data are much more difficult to obtain. As the purpose of this paper is to determine if neural networks can recognise particular faults in any engine within a fleet of engines, fault data from many engines with and without the same faults are required. Ideally, these data could be obtained by implanting the same fault in a number of engines, or from in-service data. However, since engine fault implant tests are both costly and time consuming, and suitable in-service data are usually not available, an alternative method was sought.

The diagnostic approach adopted in this paper uses information obtained from fault implant tests in a single-engine. The fault deltas derived from these tests were then superimposed on the no-fault acceptance data to form a fleetwise faulty-engine database. The underlying assumption in this step is that the linear superposition principle applies. The justification for this is that the magnitude of the implanted faults were small because the RAAF, the engine's owner, stipulated that each implanted fault had to fall with the normal adjustment range for serviceable engines so that the engine could be made available for normal operation at short notice.

The above superposition approach enables typical fault data for a fleet of engines to be generated cheaply, albeit approximately, and also retain the important underlying statistical information associated with real engine-to-engine variability effects.

FAULT IMPLANT TESTING

Fault data were obtained form a fault implant test program on a single F404 engine using the same test-cell as was employed for the original acceptance testing. Faults implanted included off-nominal adjustments to both the compressor variable geometry (CVG) and the exhaust nozzle final area (A8). Two separate CVG bias faults were implanted, consisting of two degrees closed (cl2), and six degrees closed (cl6). In a similar manner two adjustments were also made to the variable area exhaust nozzle actuator. The first fault condition corresponded to setting the nozzle to its minium physical area position (min) and the second, to increasing the resultant nozzle area to 5% above the nominal setting (+5%). Finally, a combined fault (cbl) was examined which consisted of bleeding a given quantity of air from the compressor customer bleed port in association with the '+5%' nozzle bias fault. In addition to the five fault conditions, the engine was also tested in the no-fault (nf) condition.

Steady-state data obtained at 90 degrees PLA were used for the present analysis because it was the closest available to the no-fault acceptance data (of approximately 94 degrees PLA). Subsequent tests have shown that the fault signatures are similar at these two power settings. The gas path parameters were recorded at a frequency of 32 Hz in the test cell.

Data processing included passing the raw data through a simple median filter to remove unwanted measurement spikes followed by a mean filter to reduce the effects of measurement noise. Standard correction factors were applied to correct the data to standard-day conditions. The value of the engine parameters at the fault test points (cl2 cl6, min, +5%, cbl) were then subtracted from the no-fault (nf) performance to obtain the "test deltas" . The test deltas are shown in Figure 4. Further details of this testing program can be found in [EWM94].

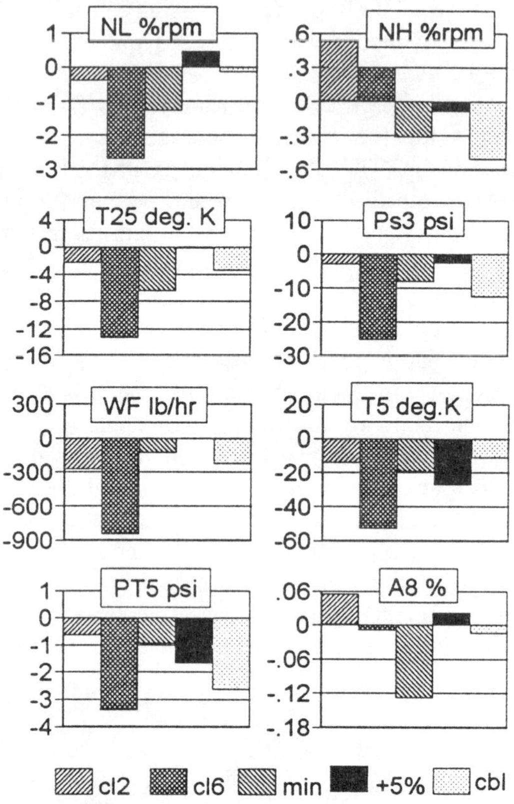

Figure 4) Test deltas for the eight recorded parameters

SUPERPOSITION OF TEST DELTAS

This fault implant superposition process is shown diagrammatically in Figure 5. Initially, a no-fault acceptance engine ('+') could be expected to depart from the average linear fleetwise model by a small amount. This small difference due to the engine-to-engine variation corresponds to the no-fault residual ('*nf. res.*'). To create a 'cl6' fault for this engine, the cl6 test deltas are added to each parameter pairing. For example, with WF selected as the independent parameter and Ps3 the

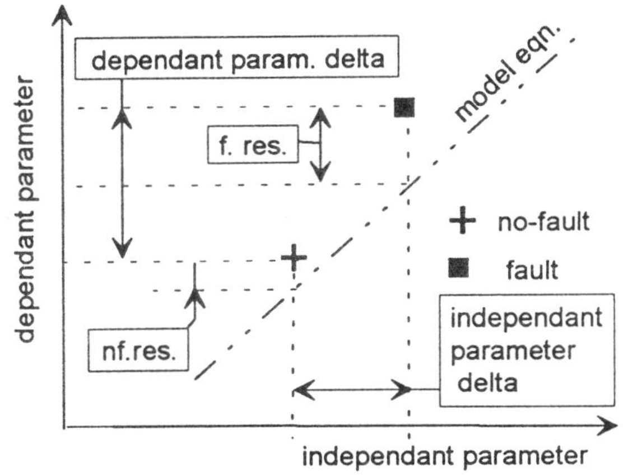

Figure 5) Test delta superposition process

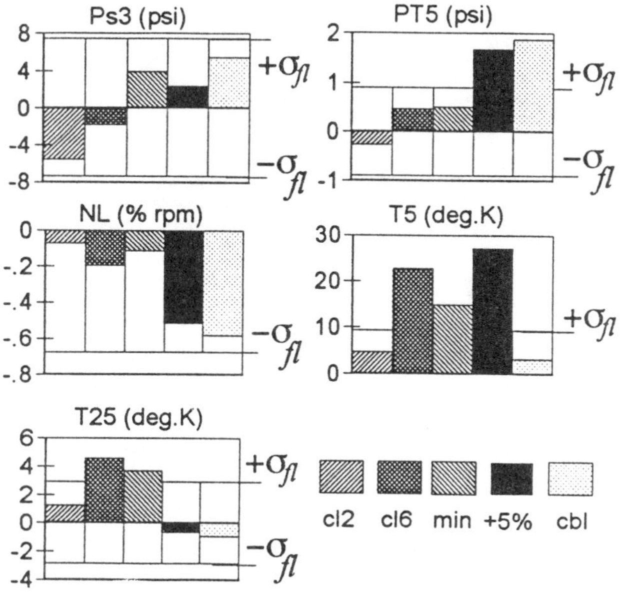

Figure 6) Effect of faults compared with fleet engine standard deviation (σ *fl*)

dependent parameter, the WF and Ps3 test deltas from Figure 4 are added to the no-fault point ('+'), resulting in a new cl6 fault point ('■') for this engine. From the fault point '■' the fault residual (*'f. res.'*) is obtained with respect to the model. Assuming the fault affects the correlation between WF and Ps3, the fault residual will be different to the no-fault value and can be used for the diagnosis. This process can be repeated for all faults across the fleet of engines for the five parameter pairings.

It is important to recognise that while the test deltas contain important fault information, they also contain variations due to small differences in power setting between the fault and no-fault points. However, since the fault residual is measured with respect to the model at the same value of the independent parameter, any component of the test delta resulting from differences in the independent parameter is then effectively removed.

The true effect of the faults can then be easily obtained by using the models developed from the no-fault data. Again, Figure 5 is useful as an illustration. Here, a difference in the independent parameter occurs between the '+' and '■' points. This is partly responsible for the size of the delta in the dependent parameter. Thus, when the fault residual is measured, it is much smaller than the dependent parameter delta. However, the engine already had a no-fault residual. Thus, the real effect of the fault is somewhat less than the value of the fault residual. It is the difference between the fault residual and the pre-existing no-fault residual that is important.

Fault deltas for a typical fleet engine have been isolated by the above procedure and are shown in Figure 6. In this figure the effect of the faults are also compared with the standard deviation (σ_{fl}) applicable to the fleet engines (Figure 2). It is evident that the effects of the fault are often smaller than the standard deviation levels of some of the no-fault engine parameters.

NEURAL NETWORKS

The operation of neural networks is often explained in terms of human qualities such as the ability to 'recognise' or 'generalise' about 'concepts'. While networks do exhibit some characteristic attributes of human behaviour, explaining a mathematical system in terms of such traits often serves to confuse the underlying process.

In essence, neural networks are a diagrammatic representation of a mathematical equation. Values, in this case the parameter residuals, are input to the network and each fault (and the no-fault condition) is represented by an output. The network thus represents a multi-variable equation, with five inputs and six outputs:

$$(\text{nf, cl2, cl6, min, +5\%, cbl}) = f(\text{Ps3, NL, T25, PT5, T5 residuals}) \qquad (6)$$

When the residual values are input, output values are calculated and the fault is recognised when the output assumes a value of one, while all other outputs are negative one. For example, given the residual input:

Ps3 (psi)	-2.26
NL(%rpm)	0.01
T25 (deg.C)	5.26
PT5 (psi)	0.79
T5 (deg.C)	26.0

the output is:

nf	-1
cl2	-1
cl6	1
min	-1
+5%	-1
cbl	-1

indicating the presence of a cl6 fault. Treated as a 'black box', the neural network behaves in a similar fashion to a mathematical process, however the internal operation of the network is the key to achieving the correct outputs for the input.

Structurally, the network is comprised of layers which contain elements called nodes. Each node can be considered as a single term in the equation, where a relatively simple mathematical operation occurs, such as the summing of values, or the application of a function. Nodes are then connected together by links, which transfer the output from one node to another. Connection of the nodes together forms a complete equation from the separate terms. Usually several nodes perform identical operations on different values, and then pass their results via the connections to another set of nodes. As such, each set of nodes is identified as belonging to the same layer.

While neural networks tend to operate in a similar way to a mathematical equation, the unique feature is that the resultant coefficients are determined from a training process. In this process, input samples are presented to the network along with known (correct) output values. The network uses these known solutions to establish the appropriate internal coefficients. The neural network can then be used to closely reconstruct the known solutions for the training data. Ideally, the network will then be able to recognise the same faults patterns in unseen data sets.

As different types of identification problems for neural networks exist, different architectures and rules are used for constructing networks to solve these problems. For example, some tasks may have outputs which are mutually exclusive (is this fruit an apple or an orange?), while others have outputs which may be continuous (is this fruit big or small?). In this application a Probabilistic Neural Network (PNN) is used. Such a network was found to achieve consistent performance after training and is ideal for solving classification type problems using statistics to identify from which class an unknown input most likely belongs.

THE PROBABILISTIC NEURAL NETWORK

The PNN is a neural network implementation of Bayesian theory of conditional probability. Given a vector of residuals to be diagnosed,

$$X? = \left[Ps3_{res}\, NL_{res}\, T25_{res}\, PT5_{res}\, T5_{res}\right] \quad (7)$$

Bayes' decision rule is that (in a simple two-case scenario for example)

$$\begin{aligned} X? &= X_{cl2} \ \ if\ \ f_{cl2}(X?) > f_{nf}(X?) \\ X? &= X_{nf} \ \ if\ \ f_{nf}(X?) > f_{cl2}(X?) \end{aligned} \quad (8)$$

where $f_{cl2i}(X?)$ is the probability density function (PDF) of the 'cl2' category. The estimation of PDFs from observed experiments is known and can be shown [Spe90] to be given by

$$f_{cl2}(X?) = \frac{1}{(2\pi)^{p/2}\,\sigma^p}\frac{1}{m}\sum_{i=1}^{m} \times e^{-\left[\frac{(X?-X_{cl2i})^t(X?-X_{cl2i})}{2\sigma^2}\right]} \quad (9)$$

in certain circumstances, where

i	=	training sample number
m	=	total number of training samples
X_{cl2i}	=	i'th training sample of the *cl2* category
σ	=	smoothing parameter
p	=	dimension of measurement space

Thus the PNN uses the training data to develop probability distribution functions of each class. That is, given *m* samples of 'cl2' and 'nf' engines, the PNN develops a statistical profile of these classes from those examples. When given an unknown input vector, the PNN calculates the values of $f(X?)$ for each class and assigns the unknown input to the class with highest result. In this paper 60 engines were used for training while the remaining 70 (from the complete data set of 130) were used to test the network.

Whilst the calculation of the PDF appears complicated (Eqn. 9), its operation can be understood when explained in conjunction with the structure of the neural network, shown in Figure 7. The first step uses a simplification that

$$e^{\left[\frac{X? \cdot X_{cl2i}-1}{\sigma^2}\right]} = e^{-\left[\frac{(X?-X_{cl2i})^t(X?-X_{cl2i})}{2\sigma^2}\right]} \qquad (10)$$

if $X?$ and X_{cl2i} are normalised to unit length. This occurs in the first layer of the network, where the residuals (in engineering units) are subjected to a polar normalisation algorithm. The algorithm maps the scalar residuals to points on a unit circle and represents the data by a vector of angle θ from the center of the circle to the new point. The algorithm operates in a non-linear way such that residuals near the mean are given a distribution over a wide angular sector, while outlying residuals are mapped to a small sector. This non-linear algorithm is essentially making the assumption that points further from the mean are less likely to occur [Neu90].

After normalisation, the values are distributed to the pattern layer. The pattern layer contains 60 nodes for each output condition, corresponding to the 60 training examples of that condition. Each node in the pattern layer holds the data in vector form of one training example. When an unknown sample is input to the network, it is also normalised to vector form. The dot product on the left-hand side of Eqn. 10 is then calculated between the unknown vector sample and the data in each pattern node [Spe90]. By using vector multiplication rather than scalar, the multiplication result is only large if both vectors originated from data of similar magnitude. Hence each vector multiplication gives an answer which indicates how well the unknown

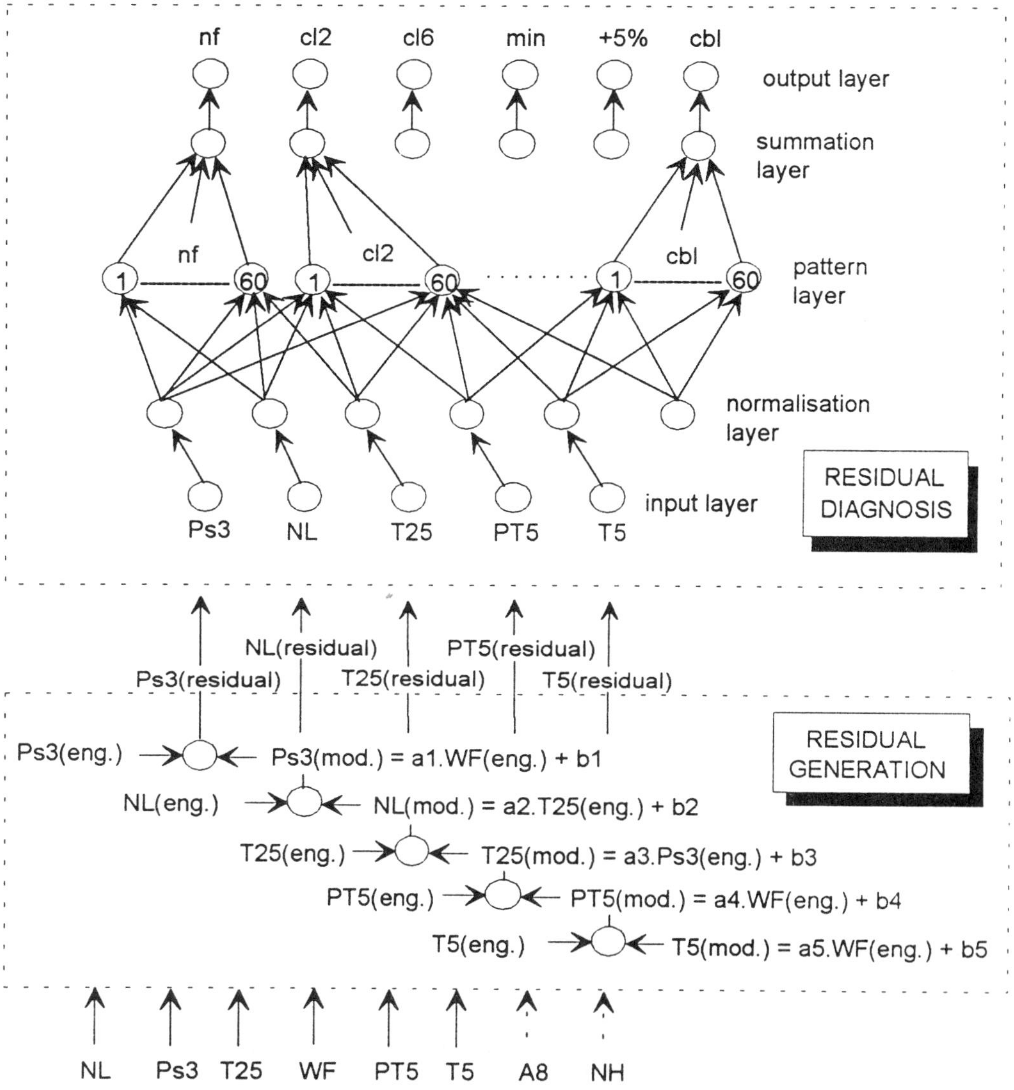

Figure 7) PNN structure and schematic of diagnostic process

data compares with each training example of all six output conditions.

At the summation layer (Figure 7) the individual vector product in each class are added, giving a result which indicates how well the unknown sample belongs to that class. The fit for each class is compared in the output layer and the class with highest probability is chosen. The output of the winning class is thus set to one, with all others set to zero.

Probabilistic neural networks can also include different relative frequencies of each fault, and loss functions associated with giving a wrong answer. For example, if the network was trained with only 30 samples of a fault, it would include a factor that this fault is half as likely to occur than one trained on 60 samples. As the network here is not intended for immediate operation in the field, these features have not been included.

RESULTS

The PNN was trained on 60 examples of each condition and tested on these data to check that these had been correctly learned. The network is run using NeuralWare's graphical software. The results in Table 1 show the network achieves 100% recall accuracy for all faults. However, the real test for any neural network is when trialed on unseen data. Using the remaining engines, 70 new examples of each fault were presented to the network. Table 2 show the diagnosis varied from 97.1 % to 81.4 % with an average accuracy of 87.6%. Confusion has occurred between the 'nf' and 'cl2' conditions, as well as between 'cl6' and 'min' faults. Twelve 'nf' engines have been incorrectly identified as 'cl2', and eleven 'cl2' engines have been miss-identified as 'nf'. Similar errors exists between the 'min' and 'cl6' faults.

class	score /60	score %	engines incorrectly diagnosed were mistaken as: nf	cl2	cl6	min	+5%	cbl
nf	60	100.0	-	0	0	0	0	0
cl2	60	100.0	0	-	0	0	0	0
cl6	60	100.0	0	0	-	0	0	0
min	60	100.0	0	0	0	-	0	0
+5%	60	100.0	0	0	0	0	-	0
cbl	60	100.0	0	0	0	0	0	-

Table 1. Network results when tested on learning data.

In order to find results which are typical and not based on a favourable set of training engines, the process was repeated 50 times. For each case, the 60 learning engines were selected at random and the remainder used for testing. Figure 8 shows the average performance in diagnosing the learned engines was 99.8%, and on the 'unknown' test engines 87.1%.

class	score /70	score %	engines incorrectly diagnosed were mistaken as: nf	cl2	cl6	min	+5%	cbl
nf	57	81.4	-	12	0	1	0	0
cl2	58	82.9	11	-	1	0	0	0
cl6	60	85.7	0	0	-	10	0	0
min	57	81.4	0	0	13	-	0	0
+5%	68	97.1	0	0	0	2	-	0
cbl	68	97.1	1	0	0	0	1	-

Table 2. Network results when tested on 'unknown' data

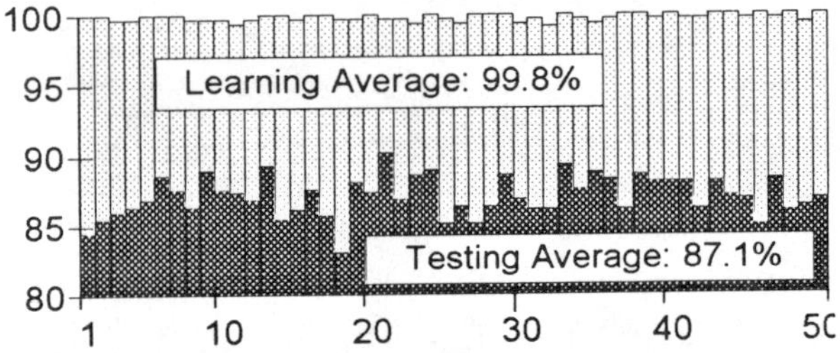

Figure 8) Network results for 50 randomly selected data sets

DISCUSSION

The performance of the neural network is best demonstrated by displaying the variation in the engine data which the neural network must diagnose. For this purpose, 5 engines from the 70 test engines have been chosen, and the faults have been superimposed on them and included with the no-fault data. This produces a simulated set of 30 engines; the original 5 engines in no-fault form, and repeated with each of the five faults added. The residuals of these 30 engines are shown in Figure 9. The figure shows no obvious indication that some of the engines represent faults.

If such data were presented to a human operator, it would not be apparent that the 30 engines presented represent five different categories since the graphs show no clustering in any groups. However, the neural network performs remarkably well. From the 30 engines 27 were correctly identified, an average result of 90%, 3% higher than the networks performance when tested on the larger group. Once again confusion has occurred between the 'nf' and 'cl2' conditions (engines 5,9), as well as between 'cl6' and 'min' (engine 11). This behaviour indicates that the pattern of residuals is similar between the mistaken conditions.

Apart from expanding the size of the training set, which slows learning and testing, two alternative approaches exit which may improve network performance.

The first approach aims at improving performance by enhancing the clarity of the patterns in the input data, as the neural network is essentially performing a pattern recognition role. This requires finding new or improved relationships that are used to generate the residuals. The more accurately the models describe the performance of the average engine and the smaller the no-fault residuals can be made, the more distinct the faults will become. Further examination of the data occurred from both a statistical view point and from knowledge of engine thermodynamics. For example, engines which had a high Ps3 residual were examined to determine if another residual or parameter was also correlated with the Ps3 residual. Alternatively, from a thermodynamic view point, engines with larger exhaust nozzle areas should show a lower exhaust gas temperature. However, despite a thorough search for better algorithms to calculate the residuals, the original five equations still remain the most accurate. As these relationships were formed from six of the eight measured parameters, the potential still exists to further improve diagnostic accuracy if the remaining two parameters, namely A8 and NH, can be used in a meaningful way.

The second method of improving network performance involves assigning an extra output node which may be selected by the network for samples recognised as likely to cause incorrect diagnosis. In this way, only samples the network is confident of are attributed to specific faults and others are indicated as suspect and requiring further investigation. As the results show, confusion often occurs between the 'nf' and 'cl2' engines. If the extra output '?' existed for engines the network confused, then the remaining attributed to either 'cl2' and the 'nf' conditions should all be correctly diagnosed. Thus network behaviour would be improved if engines which are diagnosed to specific conditions are always correct, at the expense of allowing some engines to be undiagnosed as the '?' condition, requiring further attention.

CONCLUSIONS

A Probabilistic Neural Network has been used to diagnose faults in a gas turbine engine by examining patterns in the residuals of measured parameters. The diagnosis task involved recognising unknown turbofan engine data samples, from a fleet of aircraft, as belonging to either the no-fault condition or one of five fault conditions. The problem of identification across a fleet is made more difficult because of the inherent engine-to-engine variability that occurs. However, this fleet variability did not appear to unduly mask the important features of the fault signatures embedded within the data samples. The neural network achieved an average diagnosis accuracy of 87.1%.

The process of residual generation used five simple linear models to represent the average performance of a no-fault engine. The residuals determined using these equations then exposed the fault behaviour in the engine parameters, which were detected by the neural network. The results show that while substantial variation can occur in no-fault data across engines within a fleet, the method can diagnose faults even when some of the parameter shifts due to the faults fall with the normal no-fault engine-to-engine variability. Moreover, such faults are unlikely to be detected and diagnosed by conventional techniques. Although not fully developed for in-field use, the neural network approach shows promise as a diagnostic tool.

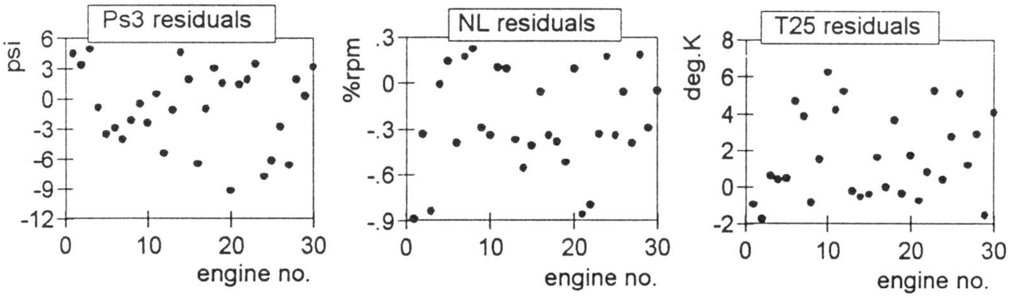

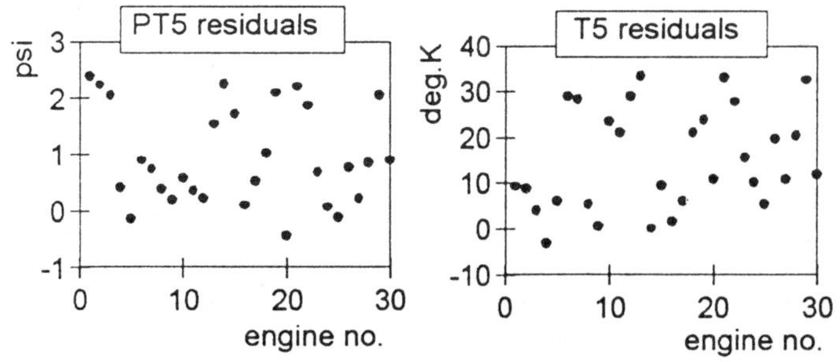

	correct response	network response
nf	4,5,8,9,17	4,8,17
cl2	16,20,24,25,27	16,20,24,25,27,**<u>5,9</u>**
cl6	6,7,10,12,26	6,7,10,12,26,**<u>11</u>**
min	11,18,23,28,30	18,23,28,30
+5%	13,19,21,22,29	13,19,21,22,29
cbl	1,2,3,14,15	1,2,3,14,15

Errors printed in bold and underlined.

Figure 9) Performance of PNN diagnosing 30 simulated engines

REFERENCES

[DKA89] Dietz, W.E., Kiech, E.L, and Ali, M., "*Jet and Rocket Engine Fault Diagnosis in Real Time*",Journal of Neural Network Computing, Summer, 1989, pp. 5-18.

[DM92] Duyar, A., and Merrill, W., 1992, "*Fault Diagnosis for the Space Shuttle Main Engine*", Journal of Guidance, Control, and Dynamics. Vol. 15, No, 2, March-April, pp. 384-389.

[EWM94] Eustace, R.W., Woodyatt, B.A., Merrington, G.L., and Runacres, A., "*Fault Signatures Obtained from Fault Implant Tests on an F404 Engine*", ASME Journal of Engineering for Gas Turbines and Power, Vol. 116, No. 1, January 1994, pp. 178-183.

[Eus93] Eustace, R.W. 1993, "*Neural Network Fault Diagnosis of a Turbofan Engine*", XI International Society of Air Breathing Engines, Tokyo, Japan, September 19-24, 1993, pp. 937-946 .

[Fra90] Frank, P.M., "*Fault Diagnosis in Dynamic Systems Using Analytical and Knowledge-based Redundancy - A Survey and Some New Results*", Automatica, Vol. 26, No. 3, pp. 459-474.

[Fre91] Freyermuth, B., "*Knowledge Based Incipient Fault Diagnosis of Industrial Robots*", Proceedings of the IFAC/IMACS Symposium on Fault Detection, Supervision and Safety for Technical Process, Baden-Baden, September 1991, pp. 31-37.

[Gen89] Gengo, T.J., "*Application of Neural Networks to the F/A-18 Engine Condition Monitoring System*", Thesis, Naval Postgraduate School, Monterey, California, 1989.

[LG90] Luce, H., and Govind, R., "*Neural Network Pattern Recognizer for Detection of Failure Modes in the SSME*", AIAA/SAE/ASME/ASEE 26th Joint Propulsion Conference, July 1990, Orlando, Florida. AIAA Paper 90-1893.

[Neu90] NeuralWare Inc., '*Reference Guide*', 1990

[WFA90] Whitehead, B., Ferber, H., and Ali, M., "*Neural Network Approach to Space Shuttle Main Engine Health Monitoring*", AIAA/SAE/ASME/ASEE 26th Joint Propulsion Conference, July 1990, Orlando, Florida. AIAA Paper 90-2259.

[WKA90] Whitehead, B., Kiech, E., and Ali, M., "*Rocket Engine Diagnostics Using Neural Networks*", AIAA / SAE /ASME / ASEE 26th Joint Propulsion Conference, July 1990, Orlando, Florida. AIAA Paper 90-1892.

[Wil76] Willisky, A.S., "*A Survey of Design Methods for Failure Detection in Dynamic Systems*", Automatica, Vol. 12, pp. 601-611, 1976.

[SKK91] Sorosa, T., Koivo, H., and Koivisto, H., "*Neural Networks in Process Fault Diagnosis*", IEEE Transactions on Systems, Man, and Cybernetics, Vol. 21, No. 4, July/August 1991, pp. 815-825.

[Spe90] Specht, D.F., "*Probabilistic Neural Networks*", Neural Networks, Vol. 3, pp. 109-118, 1990

NEURAL NETWORK REGION-BASED INEXACT RELATIONAL MEDICAL IMAGE MATCHING

Steven M. Vajdic[1], A. Bouzerdoum[2], M. J. Brooks[3], A. R. Downing[4] and H. E. Katz

[1,2]*Department of Electrical and Electronic Engineering*
[3]*Department of Computer Science*
The University of Adelaide, Adelaide, S. Australia 5005
[4]*School of Engineering, Flinders University of SA, Adelaide, S.A. 5001*
[5]*ISCS Inc., 77-14 113th St, Suite 1B Forest Hills, NY 11375, USA*
E-mail: svajdic@eleceng.adelaide.edu.au

ABSTRACT:

A neural network region-based inexact relational image matching algorithm is introduced. It is implemented in the domain of medical imaging and is based on Artificial Neural Network (ANN) and Artificial Intelligence (AI) paradigms —in particular, Unsupervised Competitive Learning (UCL) and tree search. The 2D reference and target images are selected from 3D sets and are segmented into non-touching and non-overlapping regions similar to those of human organs using unsupervised competitive learning. Selected region attributes are calculated. Region matches are obtained using a tree search, and the error is minimized by evaluating a "goodness" of matching function based on similarities of region attributes. Once the matched regions are found and the spline geometric transform is applied to regional centres of gravity, images are ready for fusion into a single 2D/3D image of higher clarity.

KEYWORDS

2D/3D Medical Imaging, Artificial Intelligence, Artificial Neural Networks, Unsupervised Competitive Learning, Multi Modal, Inexact Relational Matching, Image Region Attributes/Relationships, Tree Search, Evaluation of Matching, Image Fusion, 3D Visualisation

INTRODUCTION

Image fusion (composing images of the same objects taken by different means) has played a significant role in disciplines such as medical imaging, pattern recognition, computer vision and remote image sensing. The idea is to combine images of a single object obtained using either different sensors, different frequencies/energies, with perhaps different sections of the object observed at different times, from various positions or viewing angles, under different illumination conditions or polarisations. The resultant single image would be of a better clarity and contain more information than could be obtained from any of the individual "matched" images.

The idea of combining images in medicine becomes very important because of the necessity to improve the diagnosis of diseases and the treatment and care of patients. 2D/3D fusion of medical images acquired by medical scanners e.g. Computerized Tomography (CT), Magnetic Resonance Imaging (MRI), Positron Emission Tomography (PET) and Single Proton Emission Computer Tomography (SPECT) is of special interest for medical experts because of the different image characteristics produced by each device that are useful for combining (Figure 1).

It is known that CT and MRI can provide anatomical information (structure, texture, grey-level shapes) of human organs and PET and SPECT can provide functional information (tissue function, blood vessels, cell function, metabolism). Moreover, CT can provide good bone structure with less distortion but with less tissue texture, as opposed to MRI that can give a high quality structure description but with more distortion due to secondary magnetic field effects formed around the examined object. As a consequence, any one of the following 5 image pairs — (CT,PET), (CT,SPECT), (MRI,PET), (MRI,SPECT), (CT,MRI) is useful for fusion. The first four are classified as structure-function fusion, the last one as structure-structure fusion. The first component in each pair

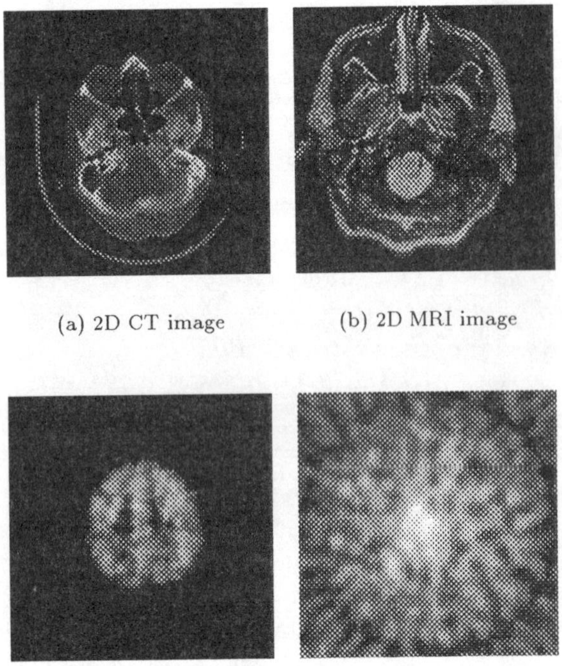

(a) 2D CT image (b) 2D MRI image

(c) 2D PET image (d) 2D SPECT image

Fig. 1: Typical 2D representations (slices, scans) of 3D medical images.

is referred to as a reference image (the one that is not changed), the second component is referred to as a target one (the one that is changed). The mutual characteristic for all four modalities is that they contain a finite set of 2D cross-sections (slices, scans) of a human organ, forming a 3D structure of the organ, with a chosen distance between the 2D slices.

MATCHING/FUSION STRATEGY

Before the images are fused they have to be matched (registered); that is, the target image needs to be modified somehow to match the reference image. All image matching/fusion algorithms developed and classified ([1],[2],[3],[4]) so far generally adopt a four step strategy: (i) extraction of key structures, (ii) matching of key structures, (iii) evaluation of similarities between structures and (iv) transformation of target image.

Classification of matching/fusion algorithms is a very complex task in general since it involves many classification criteria [2] such as image dimensionality (2D, 3D), image properties (attributes and their interrelationships - extrinsic, intrinsic) and image property coupling (interpolation, approximation), domain (global, local) and elasticity (rigid, affine, projective, curved) of transformation, parameter determination (direct, search-oriented) and interaction (interactive, semi-automatic, automatic). A complex classification with all explanations is beyond the scope of this paper. Yet, it would be worth mentioning that different classifications would be possible from the following global set of principles of matching methods: (i) control points, (ii) moments, (iii) knowledge base representation of object characteristics, (iv) similarity optimisation, (v) stochastic sign change, (vi) Fourier analysis, (vii) cross correlation, (viii) eigenvalue decomposition, (ix) warping, (x) 3D surface fitting, (xi) anatomical atlas, (xii) internal landmarks, (xiii) external landmarks, (xiv) principal axes, (xv) inexact relational matching and (xvi) neural network segmentation. Which method is to be chosen depends on the type of images to be matched/fused (e.g. natural scene images, stereo images, Landsat images or medical images). Moreover, a set of possible problems related to the matching task and type of images would be given as: (i) choosing a matching method, (ii) choosing the reference image, (iii) selecting image properties, (iv) imposing constraints, (v) evaluating "goodness of matching", (vi) choosing an appropriate geometric transformation, (vii) coping with distortion, and (viii) choosing a fusion method.

This project is concerned with medical images. Hence, with regard to the suggested four step matching methodology and related problems, the following key steps are chosen: (i) image regions, similar to the anatomical shapes of human organs, and their attributes, are extracted from 2D slices to make less complex the image matching procedure and the finding of the best match; (ii) CT or MRI image modality is selected to be the reference since they provide anatomical information which is closely related to segmented regions; (iii) tree search is employed as the best method for finding the region matches, and similarity between regions (for finding the best match) is evaluated by applying functions that incorporate comparisons of region attributes; (iv) a spline geometric transform is applied to correct the local distortion. Based on these key steps and knowledge base representation (IF-THEN rule-based analysis) 2D and 3D generalised approaches were proposed and discussed ([5],[6]). A generalised 2D/3D model was proposed (Figure 2).

In order to get a single 2D (3D) composite image out of two modality components, two (three) major procedures needed to be performed: matching and fusion (and 3D visualisation). These procedures were broken into steps: (i) selection of two 3D medical image data sets, (ii) resampling of the 3D target modality using bilinear interpolation (getting the same dimension of both 3D sets (number and size of slices)), (iii) selection of 2D slices of interest, (iv) generation of non-touching regions (region segmentation via iterative thresholding and/or using knowledge base representation of the anatomical shapes of human organs), (v) calculation of region attributes and their interrelationships, (vi) application of inexact relational region matching, (vii) application of the geometric transformation (spline transform applied to the region centres of gravity as the best transform for eliminating local distortions [7], (viii) fusion of images [8] - composition of weighted average of corresponding pixels (voxels) of either any chosen pair of 2D slices or complete 3D sets. In the case of 3D fusion a 3D visualisation was also performed. Two distinct, sequential step-designed approaches were proposed: (i) initially supervised segmentation based on iterative thresholding to form the image regions, region-based relational matching, and a tree search for finding the best match, (ii) unsupervised knowledge-based segmentation to form image regions, region-based relational matching, and a tree search for finding the best match. The shapes of closed region boundaries, produced by segmentation, were as similar as possible to those based on the anatomical structure of human organs. Segmentation by iterative thresholding was manually supervised only for the initial trials, either until the shapes of significant parts of organs in all of the matching modalities became similar to those initially suggested, or until the number and sizes of regions became close enough to match. Once the set of threshold values T(m)=t1m, t2m,...,tnm (where m represents modality — CT, MRI, PET, SPECT) was found, it could be used as the set of initial values for any future iterative threshold procedures. That would provide an automatic approach without significant human intervention. Also, separate sets of threshold values would have to be found and stored for different organs or structures such as the head, chest, spine or neck. An automatic knowledge-based unsupervised segmentation procedure was based on prior knowledge about the regions that expectyed to be obtained e.g. in the case of brain images, a definition of knowledge-based shapes of characteristic parts - thalamus, sinus, skull, cerebellum, white matter, gray matter, etc. was demanded. Each small region, obtained using initial low-level segmentation, was checked against the prior knowledge about the chosen anatomical region shapes and merged into larger regions. The prior knowledge was represented in the form of masks consisting of anatomical shapes of specific organs, and was driven by rule-based (IF-THEN) facts that used the image features and guided the region merging process. After segmentation regions of both images were to be matched.

In this paper the emphasis is given to the neural network segmentation in obtaining image regions closely related to the anatomical shapes of human organs, the similarity measure for comparing region attributes/relationships and the tree search methods in order to evaluate the inexact relational matching procedure and find the best match within a set of "good" matches (AI-based tree search methods).

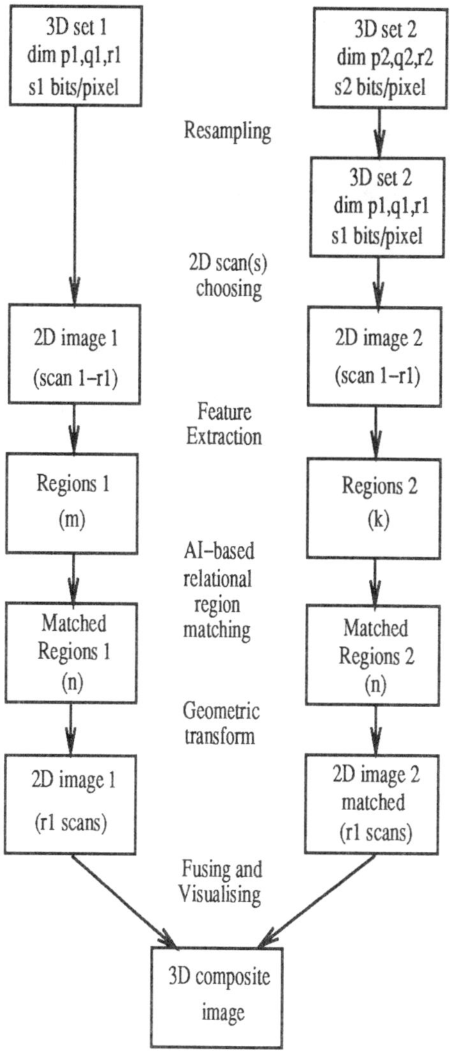

Fig. 2: 2D/3D image fusion block diagram.

SEGMENTATION USING NEURAL NETWORKS

The main task of region segmentation is to yield non-overlapping image regions, each having some homogeneous characteristic. The shapes of the closed boundaries are to be as similar as possible to those of human organs. The greater the number of model regions initially chosen, the more accurate the matching algorithm becomes, but at the cost of increased computation. These opposing aspects have to be taken into account when realising an experimental model.

A sequential step-designed approach is adopted, involving segmentation to form image regions or clusters using an unsupervised competitive learning neural network (UCL), region-based inexact relational matching, and a tree search for finding the best match.

Segmentation based on UCL involves training the neural network to learn the image patterns of anatomical shapes of human organs, then presenting the image to the trained network for segmentation. The training process involves presenting an input vector simultaneously to each of the output neurons. The neurons compare the input vector to their own weight vectors and one of them is declared the winner. The winning neuron then updates its weight vector to become similar to the input vector. This process is repeated until some stopping criterion is reached. The network is trained only on images with undeformed shapes, and shapes normally positioned within human organs and related to anatomical parts of the organs.

The UCL neural network used here has a single processing-layer whose output is described in the matrix-vector form as

$$\mathbf{y} = W^T \mathbf{x} \tag{1}$$

where $\mathbf{x} = [x_1 x_2 ... x_n]^T$ and $\mathbf{y} = [y_1 y_2 ... y_m]^T$ are the input and output vectors respectively, and W is the n x m weight matrix

$$\begin{bmatrix} w_{11} & w_{12} & \cdots & w_{1m} \\ w_{21} & w_{22} & \cdots & w_{2m} \\ \vdots & \vdots & \vdots & \vdots \\ w_{n1} & w_{n2} & \cdots & w_{nm} \end{bmatrix}$$

with w_{ij} being the weight from input i to output j.

Both reference and target images are considered as inputs, each to a slightly different network. Each image pixel is presented to the neuron input node. Typically, the whole image is divided into small neighbourhood regions of 2 or 4 pixels that are presented to the network simultaneously (with 2 or 4 input nodes). The number of outputs is related to the number of image gray level clusters, typically 8-16 (Table 1). All input units are connected to each output unit and compete under a "winner-takes-all" rule. The competitive transfer function returns "0" neuron outputs for all neuron inputs, except for the winner where the output is "1". In the end, an input image is clustered and represented as a region-based output image (Figure 3).

The network configuration is based on three basic procedures: learning anatomical shapes of organs,

Modality	x	y	IN	OUT
CT	512	512	2-4	8-16
MRI	256	256	2-4	8-16
PET	100	100	2-4	8-16
SPECT	128	128	2-4	8-16

Table 1: UCL input-output characteristics

recognition of similar features in the input pair of images and segmentation of those features into regions similar to those in learning phase. It is, also, able to locate the positions of recognised regions from the pair of input images to be fused. Once the regions are segmented and located, all the necessary attributes/relations can be computed for the matching stage. A refining procedure also needs to be applied to extract very small and very big regions from clusters that do not represent parts of human organs since they would lead to an enormous pre-matching calculation and would be most likely rejected by the matching algorithm due to lack of pair-wise matches.

INEXACT RELATIONAL REGION MATCHING

Among many image primitives (points, lines, regions), regions are adopted because of the special segmentation methods applied for their extraction (unsupervised competitive learning). More attributes can be extracted from regions than from points and lines. Nevertheless, by applying Hough or medial axis transformations, points and lines could also be used with the same algorithm proposed here (the network parameters would need to be modified though), but the accuracy of matching may well be below the level that can be reached with regions.

Image region attributes including perimeter, area, shape, mean/max/min radius, centre of gravity, and amplitude, histogram and average gray values are computed. Either the attributes of the regions, or calculable inter-relationships between them, can be used. If the inter-relationships are used, then attributes of each region within either the reference or target images are recalculated in terms of their relationships with other regions in order to provide uniformity. A major experimental search for the best combination of region attributes and/or relationships is underway. The main criterion for a good choice is invariance under linear transformations (translation, rotation, scaling, skewing) and possibly non-linear distortions. The non-linear distortions are to be corrected mainly by applying the spline geometric transformation using the

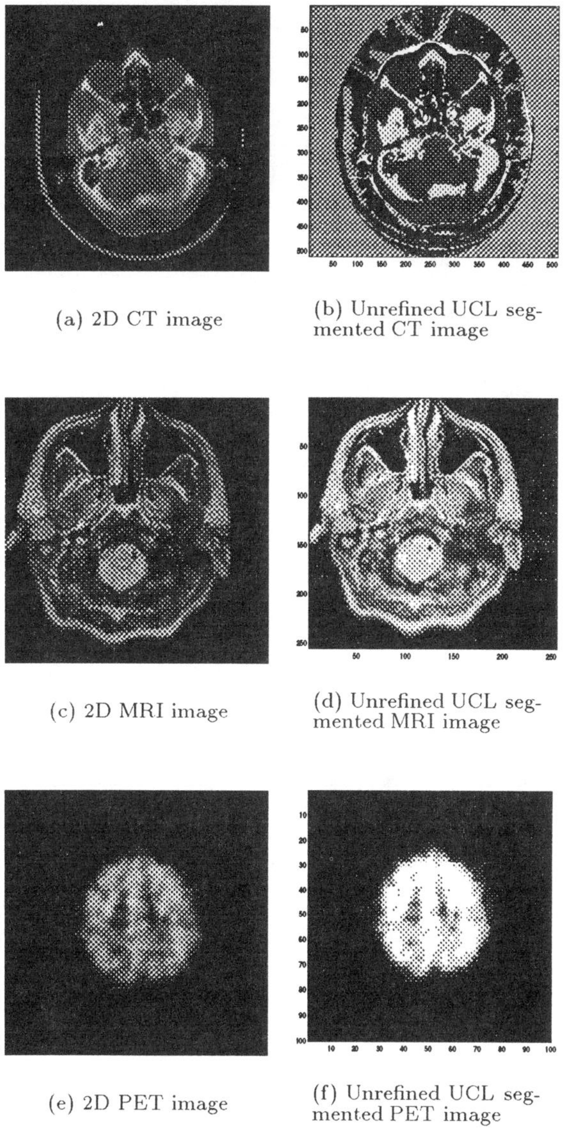

(a) 2D CT image

(b) Unrefined UCL segmented CT image

(c) 2D MRI image

(d) Unrefined UCL segmented MRI image

(e) 2D PET image

(f) Unrefined UCL segmented PET image

Fig. 3: UCL neural network segmentation.

centres of gravity of matched regions after the region matching has been performed.

Let sets $R = \{R_1, R_2, ..., R_m\}$ and $T = \{T_1, T_2, ..., T_n\}$ represent regions as image primitives in the reference and target images, where m and n are the numbers of regions in each (in general, m is not equal to n). Each region is described by a finite number N_p of region properties p_k with values v_k, $k = 1, 2, ..., r$ (the term "property" will stand for attributes and their inter-relationships (usually called relations) - numerical or symbolical) to form hierarchical structural image pattern descriptions - representations. All numerical property values are scaled into the interval $[0, 1]$ using the expression

$$v'_k = (v_k - v_{kmin})/(v_{kmax} - v_{kmin}) \qquad (2)$$

or, in case of Gaussian-distributed properties, normalized using the expression

$$v'_k = (v_k - v_{mean})/v_{st-dev} \qquad (3)$$

where v_{kmin} and v_{kmax} are the minimum and maximum values, and v_{mean} and v_{st-dev} are the mean value and standard deviation of property p_k. Properties are weighted depending on what pair of modalities (CT, MRI, PET, SPECT) is to be fused. The property weights differ from each other, especially when a structure-function fusion takes place, due to different regions being extracted in the segmentation process. For example, properties "area" and "perimeter" (or "relative area" and "relative perimeter" in the case of forming relations) are given a higher weight than shape; the property "mean radius" (or "relative mean radius") is considered more influential than property "distance between region centres of gravity", etc. Unfortunately, there is no mathematical apparatus that can theoretically describe the way of assigning weights to properties — experiments are the only alternative researchers have for the time being. The matching procedure is complete when l regions ($l \leq min(m, n)$) are found in both matched images. Relational region matching is solved by means of tree search methods. The matching procedure is guided by the match evaluation function that compares similarities between weighted region property values in the reference and the target images in order to find the best match. A tree search is applied only to find region matches prior to the application of a geometric transformation.

EVALUATION OF MATCHING

The match evaluation function has to be applied to two data descriptions (here, reference and target image regions as primitives and their properties — attributes and their relationships-relations) in order to guide the search method in finding the best match. The best match between two sets of image region primitives is to be found as the mapping for which the regions show the best similarity in their property values. The major problems of definition of similarity of properties — what property is more important and/or how to combine numerical and symbolical properties — need to be overcome. Thus, the property values have to be weighted and their inter-relationships defined in terms of all the relations over the set of property values, and/or a conditional probability function of properties needs to be employed.

Three types of evaluation function that compare similarities of two descriptions (here, sets of regions) have been proposed so far: a distance measure (cost function) and a merit function, both involving mainly comparison of pairs of primitive property values, and the function that involves more sophisticated parameters, e.g. the number of primitives, the number of properties and the number of successful matchings of each pair of properties. The distance measure (distance function) is to be minimised, the merit function is to be maximised, and the third function could be defined in such a way to allow either minimization or maximization. In the case of the two sets of region primitives defined in Section , the cost function of similarity between regions Ri and Tj is usually defined as the sum of the absolute or square value differences, or even as the square root of the sum of the square value differences of all region properties. These forms can be applied to numerical properties only. In the case of either only symbolical properties or a combination of both symbolical and numerical properties, a conditional probability function can be successfully applied [9] or some penalty for non-similarity needs to be incorporated into an evaluation function expression. The conditional probability (density) function (or conditional information) $P_{p_k}(v_k(R_i)|v_k(T_j))$ of property p_k is defined as the likelihood that the property $p_k(T_j)$ of the primitive T_j in the target image will have the value $v_k(T_j)$ if it is known that the property $p_k(R_i)$ of the corresponding primitive R_i in the reference image has the value $v_k(R_i)$. Then, the contribution to the distance measure between these two primitives from property p_k could be defined as:

$$DM_{p_k}(v_k(R_i)|v_k(T_j)) = -logP_{p_k}(v_k(R_i)|v_k(T_j)) \quad (4)$$

Minimizing the sum of all distance measures will lead to a maximizing of the product of all probabilities and, consequently, a maximizingof the likelihood mapping. The maximum likelihood mapping can be obtained by minimizing the sum of the square root property value differences in the case of Gaussian-distributed properties or by minimizing the sum of the absolute property value differences in the case of Laplacian-distributed properties. In the case of symbolical properties, beside adding a penalty value to the total distance measure if compared property values are not the same, which is rather a rough method since it does not distinguish the level of similarity, the only smooth method involves the above mentioned probabilistic approach.

Boyer and Kak [10] developed a cost function whereas Vosselman [9] developed a merit function based on probabilistic information theory and the formula defined in Equation 4. At issue here is the calculation of property conditional probabilities. All methods (analytical, numerical and hand-training) require either complex functions to be found (analytical and numerical methods) or many matches to be done by hand to guarantee correctness.

Two well-known cost (distance) functions were designed by Sanfeliu and Fu [11] and by Shapiro and Haralick ([12], [13]). The first one defines a distance measure on graphs (primitives are graph nodes, branches are inter-relationships/relations between primitives). The function primarily includes a measure of differences between normalized attribute values, and some heuristics involving structural differences between graphs. The second one is based on comparing weighted primitives, weighted attributes (using attribute value thresholds) and weighted relation vectors/tuples, and adding a normalized distance (between 0 and 1) for each corresponding pair of primitive properties (relation vectors/tuples) that is inexactly matched. Price and Reddy [14] and Faugeras and Price [15] extended this weighted attributes comparison to assigning a strength for attribute contribution in particular matching tasks, e.g. the same attributes may have more or less influence on the evaluation of matching depending on the type of images being used and primitives being extracted. Difficulties here are the need for an experimental (non-theoretical) search for weights and strengths.

Barrow and Popplestone [16] proposed an evaluation function involving more parameters than a comparison of attributes and their relationships:

$$f = 1 - n_s/n_r + 0.5/n_R \quad (5)$$

where n_s is the number of successful matchings of relations (inter-relationships between region attributes), n_r and n_R are respectively the number of relations and the number of regions in the reference image region descriptions. Relations between reference and target images (r_R and r_T) are considered to be matched if their difference is within some threshold. The threshold is chosen experimentally to be $3\sigma_R$, where σ_R is the standard deviation of the relation r_R. The drawback is that neither attribute/relation weights have been introduced nor penalties have been applied for combining numerical and symbolical region properties.

To overcome some of the problems mentioned above the authors propose a set of new evaluation functions:

$$\begin{aligned} f_1 &= W_R \frac{|N_{R_R} - N_{IR}|}{N_{IR}} + \\ &\quad + W_T \frac{|N_{T_R} - N_{IR}|}{N_{IR}} + \end{aligned}$$

$$+W_P\frac{|N_P - N_{SM}|}{N_P} + \sum_{r_1}^{r_2} w_k \quad (6)$$

$$\begin{aligned} f_2 &= \frac{W_R}{N_{IR}}\sqrt{(N_{R_R} - N_{IR})^2} + \\ &+\frac{W_T}{N_{IR}}\sqrt{(N_{T_R} - N_{IR})^2} + \\ &+\frac{W_P}{N_P}\sqrt{(N_P - N_{SM})^2} + \sum_{r_1}^{r_2} w_k \quad (7) \end{aligned}$$

$$\begin{aligned} f_3 &= \frac{W_R}{(N_{IR})^2}(N_{R_R} - N_{IR})^2 + \\ &+\frac{W_T}{(N_{IR})^2}(N_{T_R} - N_{IR})^2 + \\ &+\frac{W_P}{(N_P)^2}(N_P - N_{SM})^2 + \sum_{r_1}^{r_2} w_k \quad (8) \end{aligned}$$

where W_R, W_T and W_P are the weights that define the level of contribution of the number of regions segmented in reference and target images and the number of the properties used; N_{R_R} and N_{T_R} are the numbers of regions in reference and target images; N_{IR} is the number of initially suggested anatomical regions that should be segmented; N_P is the number of properties extracted; and N_{SM} is the number of successful matchings of property values. The factor $\sum_{r_1}^{r_2} w_k$ represents the sum of penalty values for each pair of unmatched properties. The first two factors contribute to the function as the relative difference between the number of extracted and initially suggested regions in both images. All the functions are to be minimized for good matches. All properties are weighted and are compared with regard to the threshold given by $K_k\sigma_{R_Pk}$, $k = 1, ..., r$, where K_k is a constant to be found experimentally, and σ_{R_Pk} is the standard deviation of all region property values of one kind in the reference image. If the difference between two property values is above the threshold, a penalty, equal to the weight of that property, is added to the evaluation function. Properties of regions of both images can either be attributes or relations (all attributes of one kind for all regions are recalculated with regard to each other forming inter-relationships/relations).

TREE SEARCH METHODS

Tree search methods have been well developed in AI. Here, they are used to search the space of similarities between region property values of reference and target images and find the best match based on the proposed evaluation function. In general, the search for the best match is very time consuming since there are many regions and associated properties involved. Search time could be reduced by imposing some constraints. Hard constraints limit the set of all mappings within the search space to those for which the values of some attributes/relations or the values of the evaluation function are inside experimentally predefined thresholds. For example, if the values of very few (even only one) attributes/relations are incompatible, then two primitives being compared cannot be matched even if all the other attributes/relations are compatible. Soft constraints define a relative preference over the space of possible mappings by implementing some heuristics in the evaluation function that direct the search method towards "better mappings" (informed search methods). To minimise the search time for the best match, both hard and soft constraints could be employed. However, it should be mentioned that the hard constraints strategy is based on decreasing the number of primitives that are compared (leaving out all primitive mappings due to incompatibility of some properties) while the soft constraints employ a strategy of finding "the best match" faster over the whole set of primitives and their attributes/relations — employing some heuristics. These heuristics could either be included in an evaluation function, or applied within a tree search, or both.

The search space of the relational matching problem between the two descriptions (reference image region set $R = \{R_1, R_2, ..., R_m\}$ and target image region set $T = \{T_1, T_2, ..., T_n\}$) can be represented in a tree form (Figure 4).

Several forms of the tree search methods (including some heuristics implemented to improve the search efficiency) have been proposed ([10], [12], [13], [17], [18], [19], [16], [20]) so far: backtracking, backtracking with forward checking, backtracking with one-step lookahead, backtracking with two-step lookahead, hill-climbing search, best-first search, parallel algorithms. With the exception of the latter, where several processors are used to speed the search, the other algorithms are based on the standard backtracking method.

The standard backtracking method works as follows: the attributes/relations of the primitives in the reference image (usually called units of the set R) are compared with the corresponding attributes/relations of the primitives in the target image (usually called labels of the set T); the measure of the similarity of comparisons is defined by the match evaluation function; at level (row) 1 (there are m levels, Figure 4), the first unit (R_1) is compared with all the labels ($T_j, j = 1, ..., n$); only those labels that satisfy the criteria set up in the match evaluation function (T_1, T_2, T_j) are assigned to this unit; at each chosen node (T_1, T_2, T_j) the method begins to assign the labels that

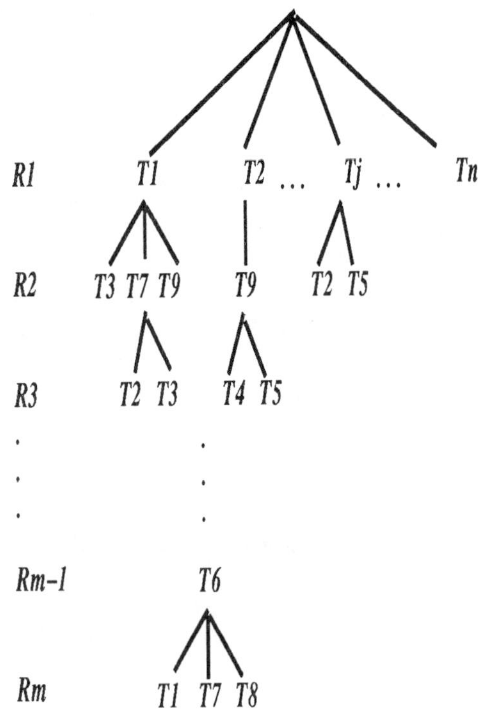

Fig. 4: A generalised top-bottom tree search space representation. Paths from the top node to the bottom nodes form a set of possible mappings — the search space for the best mapping.

match the second unit (R_2 at level 2); this is repeated until the last level (R_m) is assigned to some labels; this node-expanding strategy is repeated until either all nodes have been visited while remembering the best path or an admissible set of matched primitives in both descriptions has been found. If the nodes are expanded by keeping the breadth levels unchanged while increasing the depth levels (going down), the tree search is characterised as "depth-first"; if the nodes are visited by keeping the depth levels unchanged while increasing the breadth levels, the search is called "breadth-first". Moreover, if there is no information about where to look for the solution and/or how to guide the search, the methods are "blind"; if there is knowledge that guides the search for "good nodes", the methods are said to be "informed". Blind methods are inefficient since all nodes must be treated and visited equally. Informed methods allow heuristics to guide the search and improve their efficiency.

With regard to the defined evaluation functions (equations 6, 7, 8) the authors propose the following heuristics (soft constraints) for the informed (guided) backtracking tree search algorithm:

(i) Decrease the time for the calculation of the evaluation matching function by first comparing those properties with high weights:

- after each comparison the partial value of the evaluation function of a currently compared pair of unit-label regions is calculated and compared with the stored minimum value of the function of some previously completed region comparisons;
- the process of evaluation (and, consequently, the process of the comparison of yet non-calculated region properties) is stopped when the current value exeeds the minimum stored value of another pair of compared regions;
- the current unit-label pair is ruled out;
- the track of "the best match" is always kept.

(ii) Decrease the number of nodes needing to be visited by invoking prior knowledge about the region properties of anatomical shapes of human organs:

- change both the order of unit primitives (reference image description) and the order of label primitives (target image description) in such a way that primitives whose attributes/inter-relationships are similar to those initially suggested take higher (or lower) positions in the tree scheme (Figure 4: depth — from 1 to m; breadth — from 1 to n);
- going down and right (or going up and left) will be less likely (or more likely) to yield "better matches";
- the search method starts from level (1,1) (or level (m,n)) and finds a solution faster, and without visiting all nodes.

This procedure works well with ordered unit regions since they are extracted from reference images (anatomical information (CT, MRI)). It works less well with ordered label regions since they are selected in target images (functional information — (PET, SPECT)).

(iii) Interrupt the search after finding a sufficient number of unit-label pairs (for instance 10-15 in the case of human brain images). This heuristic can only be combined with that described in (ii). For this process to succeed, the matches must be ranked so that better ones are searched first.

CONCLUSION

2D/3D medical image fusion approaches based on ANN and AI principles have been considered. It has been intended to convey some ideas that have not previously been applied in 2D/3D medical image fusion. While some non-ANN and non-AI problems have been considered here (geometric transformation for mapping target images onto reference ones, image fusion of compositing pairs of images into single ones), some ANN and AI topics are under investigation: unsupervised competitive learning neural network segmentation and tree search in image region inexact relational matching. Although both relaxation labelling ([20],[21]) and relational matching methods can be used for matching relational image descriptions, the latter has been chosen for this work because it always finds the best mapping between the primitives of image descriptions (in this case, the image regions). The trade-off between the computational cost and accuracy also needs to be taken into account whereby the more image regions and region attributes/relationships adopted, the better the accuracy, but the greater the computation. The work is underway to test the algorithms under a large set of medical sample images of all combinations of structure-structure and structure-function fusions.

ACKNOWLEDGMENTS

The authors are grateful to Prof. Marilyn Noz of the Department of Nuclear Medicine, New York University, for her unreserved support in supplying matched medical image data sets used in this project.

REFERENCES

[1] Brown L. G. (1992), "A survey of image registration techniques", *ACM Computing Surveys*, **24**(4), pp. 325–376.

[2] Van den Elsen P. A., Pol D. E. & Viergever M. A. (1993), "Medical image matching – A review with classification", *IEEE Engineering in Medicine and Biology*, pp. 26–38.

[3] Chen Q. (1993), *Image Registration and its Applications in Medical Imaging*, PhD thesis, Vrije University Brussel, Faculty of Applied Sciences, Brussels.

[4] Chapnick J. V., Noz M. E., Maguire G. Q. Jr., Kramer E. L., Sanger J. J., Birnbaum B. A. & Megibow A. J. (1993), "Techniques for multimodality image registration", *Proceedings of the 1993 IEEE 19th Annual Northeast Bioengineering Conference*, pp. 221–222.

[5] Vajdic S. M., Katz H. E., Downing A. R. & Brooks M. J. (1994), "Artificial Intelligence and registration of functional and structural images: Generalised 2D approaches", *Proceedings of the 1994 IEEE International Conference on Systems, Man and Cybernetics*, Vol. 1, pp. 1030-1035, 1994.

[6] Vajdic S. M., Katz H. E., Downing A. R. & Brooks M. J. (1994), "AI-based relational matching and multi-modal medical image fusion: Generalised 3D approaches", *Proceedings of the 1994 IEEE International Conference on Visual Communications and Image Processing*, Vol. 2308, pp. 1565-1572, 1994.

[7] Goshtasby A. (1988), "Registration of images with geometric distortions", *IEEE Transaction of Geoscience and Remote Sensing*, **26**, pp. 60–64.

[8] Porter T. & Duff T. (1984), "Compositing digital images", *Computer Graphics*, **18**(3), pp. 253–259.

[9] Vosselman G., (1992), *Relational Matching*, Berlin Heidelberg: Springer-Verlag.

[10] Boyer K. L., Kak A. C., (1988), "Structural stereopsis for 3-D vision", *IEEE Transactions on Pattern Analysis and Machine Intelligence*, **10**(2), pp. 144–166.

[11] Sanfeliu A., Fu K. S., (1983), "A distance measure between attributed relational graphs for pattern recognition", *IEEE Transaction on Systems, Man and Cybernetics*, **13**(3), pp. 353–362.

[12] Shapiro L. G., Haralick R. M., (1981), "Structural descriptions and inexact matching", *IEEE Transactions on Pattern Analysis and Machine Intelligence*, **PAMI-3**(5), pp. 505–519.

[13] Shapiro L. G., Haralick R. M., (1987), "Relational matching", *Applied Optics*, **26**, pp. 1845–1851.

[14] Price K., Reddy R., (1979), "Matching segments of images", *IEEE Transactions on Pattern Analysis and Machine Intelligence*, **PAMI-1**(1), pp. 110–118.

[15] Faugeras O. D., Price K. E., "Semantic description of aerial images using stochastic labelling", *IEEE Transactions on Pattern Analysis and Machine Intelligence*, **PAMI-3**(6), pp. 633–642.

[16] Barrow H. G., Popplestone R. J., (1971), "Relational description in picture processing", *Machine Intelligence*, **6**, pp. 377–396.

[17] Ullman J.R. (1976), "An Algorithm for Subgraph Isomorphism", *Journal of the Association of Computing Machinery*, **Vol. 23**, pp. 31–42.

[18] Haralick R.M. & Elliot G.L. (1980), "Increasing Search Efficiency for Constraint Satisfaction Problems", *Artificial Intelligence*, **Vol. 14**, pp. 263–313.

[19] Pearl J. (1984), *Heuristics: Intelligent Search Strategies for Computer Problem Solving*, Addison-Wesley, Reading, MA.

[20] Rosenfeld A., Hummel R. A. & Zucker S. W. (1976), "Scene labeling by relaxation operations", *IEEE Transaction on Systems, Man and Cybernetics*, **SMC-6**, pp. 420–453.

[21] Medioni G. & Nevatia R. (1984), "Matching images using linear features", *IEEE Transaction on Pattern Analysis and Machine Intelligence*, **PAMI-6**(6).

Artificial Intelligence meets Artificial Insemination:
An application of rule extraction from neural networks

Emanoil Pop, Alan B. Tickle, Ross Hayward
Don Simonetta, Joachim Diederich
Neurocomputing Research Centre
School of Computing Science
Queensland University of Technology
Brisbane Q 4001 Australia

ABSTRACT

The concepts learned by neural networks are difficult to understand because they are hidden in the network's architecture and weights. One approach to understanding the classification behavior of a trained neural network is to extract symbolic rules which describe this behavior. We present here two methods (RULE-NEG and DEDEC) which efficiently extract symbolic rules from trained neural networks by exploiting the fact that a neural network can be queried for classifying any pattern from the problem domain. The results of the empirical tests performed using these method have shown that 1) the extracted rules closely reproduce the network from which they were extracted, and 2) the extracted rules are human comprehensible. We apply both rule extraction form neural network techniques to a neural network which aims at predicting the milk production of an offspring of a dairy dam and sire.

INTRODUCTION

The rapid and successful proliferation of applications incorporating Artificial Neural Network (ANN) technology in fields as diverse as commerce, science, industry and medicine, offers a clear testament to the capability of the ANN paradigm. Of the three salient characteristics of ANNs which underpin this success, the first is the comparatively direct and straightforward manner in which Artificial Neural Networks acquire information/`knowledge' about a given problem domain through a training phase. This process is quite distinct from the more complicated knowledge engineering/acquisition processes of symbolic AI systems. The second characteristic is the compact (albeit completely numerical) form in which the acquired information/`knowledge' is stored within the trained ANN and the comparative ease and speed with which this `knowledge' can be accessed and used. The third characteristic is the robustness of an ANN solution in the presence of `noise' in the input data. In addition to these characteristics, one of the most important advantages of trained Artificial Neural Networks is the high degree of accuracy reported when an ANN solution is used to generalise over a set of previously unseen examples from the problem domain.

However the success of the ANN paradigm is at a cost - an inherent inability to explain in a comprehensible form, the process by which a given decision or output generated by an ANN has been reached. For Artificial Neural Networks to gain a even wider degree of user acceptance and to enhance their overall utility as learning and generalisation tools, it is highly desirable if not essential that an `explanation' capability becomes an integral part of the functionality of a trained ANN. Such a requirement is mandatory if, for example, the ANN is to be used in what are termed as `safety critical' applications such as airlines and power stations. In these cases it is imperative that a system user be able to validate the output of the Artificial Neural Network under all possible input conditions. Further the system user should be provided with the capability to determine the set of conditions under which an output unit within an ANN is active and when it is not, thereby providing some degree of transparency of the ANN solution.

THE IMPORTANCE OF RULE-EXTRACTION ALGORITHMS

Since rule extraction from trained Artificial Neural Networks comes at a cost in terms of resources and additional effort, an early imperative in any discussion is to delineate the reasons why rule extraction is an important, if not mandatory, extension of conventional ANN techniques. The merits in including rule extraction techniques as an adjunct to conventional Artificial Neural Network techniques include:

USER EXPLANATION

Within the field of symbolic AI the term `explanation' refers to an explicit structure which can be used internally for reasoning and learning, and externally for the explanation of results to a user. Users of symbolic AI systems benefit from an explicit declarative representation of knowledge about the problem domain, typically in the form of object hierarchies, semantic networks, frames etc. The explanation capability of symbolic AI also includes the intermediate steps of the reasoning process e.g. a trace of rule firings, a proof structure etc., which can be used to answer `How' questions.

Experience has shown that an explanation capability is considered to be one of the most important functions provided by symbolic AI systems. In particular, the salutary lesson from the introduction and operation of Knowledge Based systems is that the ability to generate even limited explanations (in terms of being meaningful and coherent) is absolutely crucial for the user-acceptance of such systems. In contrast to symbolic AI systems, Artificial Neural Networks have no explicit declarative knowledge representation. Therefore they have considerable difficulty in generating the required explanation structures. It is becoming increasingly apparent that the absence of an `explanation' capability in ANN systems limits the realisation of the full potential of such systems and it is this precise deficiency that the rule extraction process seeks to redress.

SAFETY-CRITICAL PROBLEM DOMAINS

While the provision of a `user explanation' capability is one of the key benefits in extracting rules from trained ANNs, it is certainly not the only one. For example within a trained Artificial Neural Network the capability should also exist for the user to determine whether or not the ANN has an optimal structure or size. A concomitant requirement is for ANN solutions to not only be transparent as discussed previously but also for the internal states of the system to be both accessible and able to be interpreted unambiguously. Satisfaction of such requirements would make a significant contribution to the task of identifying and if possible excluding those ANN-based solutions that have the potential to give erroneous results without any accompanying indication as to when and why a result is sub-optimal. Such a capability is mandatory if neural-network based solutions are to be accepted into a broader range of applications areas such as air traffic control, the operation of power plants, medical surgery etc.

SOFTWARE VERIFICATION

A requirement of increasing significance in software-based systems is that of verification of the software itself. While the task of software verification is important it is also acknowledged as being difficult, particularly for large systems. Hence if Artificial Neural Networks are to be integrated within larger software systems which need to be verified, then clearly this requirement must be met by the ANN as well. At their current level of development, rule-extraction algorithms do not allow for the verification of trained Artificial Neural Network i.e. they do not prove that a network behaves according to some specification. However rule extraction algorithms provide a mechanism for either partially or completely `decompiling' a trained Artificial Neural Network. This is seen as a promising vehicle for at least indirectly achieving the required goal by enabling a comparison to be made between the extracted rules and the software specification.

IMPROVING GENERALISATION

Where a limited or unrepresentative data set from the problem domain has been used in the ANN training process, it is difficult to determine when generalisation can fail even with evaluation methods such as cross-validation. By being able to express the knowledge embedded within the trained Artificial Neural Network as a set of symbolic rules, the rule-extraction process may provide an experienced system user with the capability to anticipate or predict a set of circumstances under which generalisation failure can occur. Alternatively the system user may be able to use the extracted rules to identify regions in input space which are not represented sufficiently in the existing ANN training set data and to supplement the data set accordingly.

KNOWLEDGE ACQUISITION

One of the principal reasons for introducing machine learning algorithms over the last decade was to overcome the so-called `knowledge acquisition' problem for symbolic AI systems. The notion of using trained Artificial Neural Networks to assist in the knowledge acquisition task has existed for some time. An extension of these ideas is to use trained Artificial Neural Networks as a vehicle for synthesising the knowledge that is crucial for the success of knowledge-based systems. Alternatively domain knowledge which is acquired by a knowledge engineering process may be used to constrain the size of the space searched during the learning phase and hence contribute to improved learning performance.

A CLASSIFICATION SCHEME FOR RULE EXTRACTION ALGORITHMS

Over time a number of different strategies have been developed to achieve the desired goals of both extracting rules from trained Artificial Neural Networks and the use of ANNs to refine an existing rule set. However apart from a rudimentary classification scheme used by Craven and Shavlik (1994), to date little has appeared in the form of a rigorous and systematic attempt to develop a coherent and consistent taxonomy for classifying these approaches. Given the proliferation of rule extraction/ rule refinement techniques, the need for such a schema is now well established. The method of classification proposed here is in terms of: (a) the expressive power of the extracted rules; (b) the `translucency' of the view taken within the rule extraction technique of the underlying Artificial Neural Network units; (c) the extent to which the underlying ANN incorporates specialised training regimes; (d) the `quality' of the extracted rules; and (e) the algorithmic `complexity' of the rule extraction/rule refinement technique.

The primary dimension in our proposed classification scheme labeled as `the expressive power of the extracted rules' focuses directly on the actual output presented to the end user from the rule extraction/rule refinement process viz. the rules themselves. To date one line of effort in the development of rule extraction techniques has been directed towards presenting the output as a set of rules expressed using conventional (i.e. two-valued Boolean) symbolic logic in the form if...then...else.... A substantial parallel effort has also been directed towards expressing the knowledge embodied in the ANN using concepts drawn from `fuzzy' logic. This allows rules to be expressed in a form which also use an if...then...else.... structure but in lieu of two-valued logic, such rules use the concept of membership functions to deal with what are termed `partial' truths e.g. if x is low and y is high then z is medium; where low, high, and medium are fuzzy sets with corresponding membership functions.

Hence for the purposes of this exercise the two primary categories selected to represent this dimension are that the rules extracted be in the form of: (a) propositional/Boolean (i.e. if...then...else) or (b) non-conventional (i.e. the resulting representation can include probabilistic or `fuzzy logic' rules which may also incorporate an if...then...else structure). While not deemed necessary at this stage, at some future point it may expedient to introduce a third category to cater for extracted rules represented in first-order logic form i.e. rules with quantifiers and variables.

The second proposed dimension to the classification scheme extends the classification schema used by Craven and Shavlik (1994). The `translucency' dimension of classification is designed to reveal the relationship between the extracted rules and the internal architecture of the trained ANN. It comprises two basic categories of rule extraction techniques viz. `decompositional' and `pedagogical' and a third - labeled as `eclectic' - which combines elements of the two basic categories.

The distinguishing characteristic of the `decompositional' approach is that the focus is on extracting rules at the level of individual (hidden and output) units within the trained Artificial Neural Network. Hence the `view' of the underlying trained Artificial Neural Network is one of `transparency'. A basic requirement for rule extraction techniques in this category is that the computed output from each hidden and output unit in the trained Artificial Neural Network must be mapped into a binary (yes/no) outcome which corresponds to the notion of a rule consequent. Hence each hidden or output unit can be interpreted as a `step' function or a Boolean rule which reduces the rule extraction problem to one of determining the situations in which the `rule' is true i.e. a set of incoming links whose summed weights guarantee the unit's bias is exceeded regardless of the activation value present on other incoming links. The rules extracted at the individual unit level are then aggregated to form the composite rule base for the ANN as a whole.

The translucency dimension - `pedagogical' is given to those rule extraction techniques which treat the trained ANN as a `black box' i.e. the view of the underlying trained Artificial Neural Network is `opaque'. The core idea in the `pedagogical' approach is to `view rule extraction as a learning task where the target concept is the function computed by the network and the input features are simply the network's input features' (Craven & Shavlik, 1994). Hence the `pedagogical' techniques aim to extract rules that map inputs directly into outputs. Where such techniques are used in conjunction with a symbolic learning algorithm, the basic motif is to use the trained Artificial Neural Network to generate examples for the learning algorithm.

As indicated above the proposed third category in this classification scheme are composites which incorporate elements of both the `decompositional' and `pedagogical' (or `black-box') rule extraction techniques. This is the `eclectic' group. Membership in this category is assigned to techniques which utilise knowledge about the internal architecture and/or weight vectors in the trained Artificial Neural Network to complement a symbolic learning algorithm.

The next classification dimension used is the extent of the requirement within a given rule extraction technique for a specialised ANN training regime. This is an important consideration because it provides some measure of the `portability' of the rule extraction technique across various ANN architectures. An additional facet of this classification dimension is whether the underlying ANN is modified or left intact by the rule extraction process.

The fourth dimension in our proposed classification scheme is the `quality' of the extracted rules. The specific intent here is to attempt some measure of how well the task of extracting the required `explanation' has been performed. Criteria for evaluating rule `rule-quality' include: (a) accuracy; (b) fidelity; (c) consistency; and (d) comprehensibility.

In this context a rule set is considered to be accurate if it can correctly classify previously unseen examples. Similarly a rule set is considered to display a high level of fidelity if it can mimic the behaviour of the Artificial Neural Network from which it was extracted by capturing all of the information embodied in the ANN. An extracted rule set is deemed to be consistent if, under differing training sessions, the Artificial Neural Network generates rule sets which produce the same classifications of unseen examples. Finally the comprehensibility of a rule set is determined by measuring the size of the rule set (in terms of the number of rules) and the number of antecedents per rule.

The final dimension in the proposed classification scheme is that of the algorithmic `complexity' of a given rule extraction procedure. The inclusion of such a dimension reflects an almost universal requirement for the algorithms underpinning the rule extraction process to be as efficient as possible. In particular a crucial issue in developing a rule extraction process/algorithm is how to constrain the size of the solution space to be searched. As in the case for using rule quality as a classification dimension, the difficulty is that not all authors have reported on the issue, and the discussion by those who have is occasionally superficial.

OVERVIEW

We first describe the architecture of the neural network. This network aims at the prediction the milk production of an offspring given information about the mother and the herd the animals live in. The network itself uses data from one individual bull with a large number of offsprings only (i.e. the network itself represents the sire). We briefly describe the data that has been used for neural network training and continue with a full description of the rule extraction algorithms that are used in this study. Finally, we present the rules that have been extracted and conclude with a discussion of the results.

THE APPLICATION

The Australian Dairy Herd Improvement Scheme (ADHIS) provides Australian Breeding Values (ABVs) for dairy sires used in Australia. The ABVs are provided for three sets of traits (Production, Workability and Type). IN 1992 there were 12000, 1800 and 1500 Holstein-Friesian sires for each group respectively and a subset of 1100 sires which have ABVs for all three groups.

It is certainly difficult to scan the large database for the purpose of selecting the most appropriate bulls to meet criteria specified by one particular farmer. This would correspond to re-ranking all the sires using a specific list of criteria. This problem is a difficult task as the number of sets of possible criteria is limitless.

Non-Australian attempts in the past have used statistical techniques, linear programming as well as neural networks to assist farmers in scanning the volume data for the purpose of selection. We propose to exploit the generalisation capacity of neural networks to service the uniquely Australian requirements.

THE ARCHITECTURE OF THE SYSTEM

The methodology allows a connectionist model to be formed by training using breeding examples. Observable traits of a dam are linked in a non-linear way with observable traits of the resulting offspring for one particular bull.

As an input to the network, the milk production of the mother per year (in litres) is given, as well as the fat and protein production. Furthermore, information about the desired offspring (age) and the herd is provided. We use the following input-output representation for Cascade2 and Backpropagation networks (i.e. fully-connected feedforward networks):

Input	
dam milk	litres per year (or as a Boolean variable (low, medium, high)
dam fat	kg per year (or as a Boolean variable (low, medium, high)
dam protein	kg per year (or as a Boolean variable (low, medium, high)
dam age	in days
daughter age?	1 bit (Boolean- 0 if <= 3 years, 1 otherwise)
same_year?	1 bit (Boolean- 1 if dam & daughter lactate in the same year, 0 otherwise)
dam_1st_lact?	1 bit (Boolean- 1 if dam's first valid recorded lactation and if dam's age <= 3 years, 0 otherwise)
daughter age	in days
Output	
daughter milk	litres per year, or as a Boolean variable

In this application, data from one particular Australian sire is used. The mean milk production of daughters of this bull is 5096 litres per year (SD 1061 litre). The total data set includes 10452 records with data from 648 herds. The largest herd includes 196 dam records, the mean herd size is 16 records for dams only. Neural networks are trained with data from one single herd only.

RULE-EXTRACTION FROM NEURAL NETWORKS

RULE-NEG AND ITS APPLICATION TO HERD-SPECIFIC NETWORKS

Rule-extraction is viewed as a learning task and the target concept is the function computed by the network while the input features are the patterns used in training the network. Because every symbolic rule in propositional calculus can be expressed as a disjunction of conjunction, the method concentrates on extracting conjunctive rules only. It is not a problem if the network has to learn disjunctive concepts, because we use the training patterns and the net for rule extraction. For the network to be able to learn a disjunctive concept, this concept has to be expressed explicitly in different training patterns. What the method extracts in this case is two (or more) conjunctive rules (some of them may have only one antecedent) which have to be "or-ed" to express the disjunctive concept learnt.

A conjunctive rule holds only when all of the antecedents in the rule are true. In other words by changing the truth value of one of the antecedents the outcome of the rule changes. This is the basic idea employed by this method. Given a trained neural network and the patterns used for training the net, by presenting the patterns one by one to the net the rules which are learnt by the net from the patterns are extracted. Some of the patterns might encode two or more conjunctive rules; in those cases the algorithm does not extract any rules from that pattern because negating only one input feature at the time might cause that one of the conjunctions will fail having one of its antecedents negated. But the remaining (satisfied) conjunctions will keep the output class unchanged. However this is not a problem because the conjunctive concepts have to appear separately in different patterns if the network is to be able to learn each of them.

An outline of the RULE-NEG algorithm is given below:

```
initialise the Rule-Holder to empty
for every pattern "p" from the training set
    find the class "C" for "p" by use of the ANN
    /* C = ANN(p) */
    if "p" is not classified by the existing rules
      initialise a new rule "r" for class "C"
      for every input "i" into the network
          make a copy "p`" of "p"
          negate the "i"-th entry in "p`"
          find the class "C`" for "p`" by use of
           the ANN
          /* C` = ANN(p`) */
          if "C" is not equal to "C`"
              Add "i"-th input and its truth value to "r"
      /* end for every input */
      add "r" to Rule-Holder
 /* end for every pattern */
```

As it can be seen from this outline, RULE-NEG will always pick only one conjunctive rule per pattern, but will still be able to extract all the rules learned from the patterns.

We trained a neural network using the Backpropagation algorithm using 134 training patterns. The extracted rules are:

1) IF Daughter Age >= 3 THEN Milk Production is greater than median.

2) IF Daughter Age < 3 THEN Milk Production is less than median.

3) IF Dam milk is NOT high AND Daughter Age <= 2 THEN Milk Production is less than median

4) IF Dam fat is low AND Daughter Age <= 2 THEN Milk Production is less than median

DEDEC AND ITS APPLICATION TO HERD-SPECIFIC NETWORKS

One of the approaches surveyed by Andrews et al. (1995) is that of Craven and Shavlik (1994). This Rule-extraction-as-learning approach (classified as "pedagogical" by Andrews et al,. 1995) is based on viewing the underlying Artificial Neural Network as a "black box". The core idea in their approach is to "view rule extraction as a learning task where the target concept is the function computed by the network and the input features are simply the network's input features".

The reported success of this technique provided the initial impetus for the development of the DEDEC technique. While DEDEC uses the same basic "pedagogical" motif as the Rule-extraction-as-learning, an important difference is that it utilises the knowledge embedded in the weight vectors/ architecture of the trained network to rank the inputs in terms of their relative significance. This additional information is then used to direct the strategy for generating a (minimal) set of cases for the learning phase. The DEDEC technique also utilises a different algorithm for the rule extraction phase (Orlowski, 1993).

SCHEMATIC OUTLINE OF THE DEDEC TECHNIQUE

The steps in the DEDEC algorithm are as follows:

1. Assign each propositional variable from the problem domain (i.e. a rule antecedent) to an input unit for an artificial neural network based on the "Cascade" incremental learning algorithm. (Fahlman & Lebiere, 1990). The network output unit corresponds to the decision outcome of the artificial neural network applied to an input vector from the problem domain - i.e. in essence the decision outcome represents a rule consequent.

2. Train the neural network.

3. Rank the input units in order of the relative share of their contribution to the output prediction using a weight partitioning algorithm based on Garson (1991).

4. Initialise the set of (symbolic) rules as the attribute value(s) for the minimum set of input units required to form one valid case/ example in the problem domain;

> repeat
> select the next input unit from the ranked list (created at Step 3);
>
> use the trained neural network to give the decision/ classification for the set of cases/ examples from the problem domain involving this input unit and all previously selected input

units;

for each case/ example generated at the previous step

(a) update the rules for classifying each existing case/ example in the presence of this new case;
(b) determine the rules for classifying the new case/example in the presence of the existing cases; remove all rules which are not minimal

until stopping criteria met.

The "stopping criteria" used are based on the following considerations:
(a) the significance of the input unit added at the current iteration relative to the unit selected at the previous iteration;
(b) a "patience" criterion indicating no change in the set of extracted rules for a specified number of iterations.

The following rules were extracted by use of DEDEC:

IF age of daughter >= 3 years
OR mother's milk production is high
AND mother's protein content is high
THEN the daughter's milk production is >= the median for the herd.

IF mother's milk production is low
AND age of daughter is =< 1 year
OR mother's milk production is low
AND age of daughter is 2 years OR mother's milk protein is low
AND age of daughter is =< 1 year OR mother's milk protein is low
AND age of daughter is 2 year
THEN the daughter's milk production is < median for the herd

CONCLUSION

Both rule extraction algorithms, RULE-NEG and DEDEC, have produced similar results. Both rule sets are comprehensible and identify the important input features (e.g. the mother's fat and protein production).

A similar result could have been achieved by multiple experimentation as it is practice in Neurocomputing: remove one input from the data set, train a network and compare the result with previous ones. Rule extraction, however, offers efficient (i.e. in the case of RULE-NEG O(n)) algorithms which work on the trained neural network and deliver comprehensible results. Frequent re-training is therefore avoided. Rule extraction algorithms therefore offer a considerable service to the field of Neurocomputing.

The DEDEC algorithm discovered the importance of the inputs "mother fat" and "mother protein" via the ranking of the weight's share score. This can be interpreted as evidence that a hybrid "decompositional"/"learning" algorithm has advantages.

ACKNOWLEDGMENT

This work was supported by an infrastructure grant of the Queensland Department of Primary Industries to the Neurocomputing Research Centre at Queensland University of Technology. Gerard Finn, Robert Young, Hans Moulder, Raymond Lister, Thomas Szabo and Liang-gang Zhang are investigating neural network architectures for the prediction of dairy production. Thanks to Scott Fahlman who made the original CASCADE2 code available to us.

REFERENCES

Andrews, R.; Diederich, J. & Tickle, A.B., A Survey and Critique Of Techniques For Extracting Rules From Trained Artificial Neural Networks. Knowledge-Based Systems, 1995, in press.

Craven, M.W. & Shavlik, J.W., Using Sampling and Queries to Extract Rules from Trained Neural Networks. In: Cohen, W.W.; Hirsh, H. (Eds.): Machine Learning: Proceedings of the Eleventh International Conference. Morgan Kaufmann, San Francisco, CA, 1994.

Fahlman, S.E.; Lebiere, C.: The Cascade-Correlation Learning Architecture. In: Touretzky, D.S. (Ed.): Advances in neural information processing systems. San Mateo, Ca.: Morgan Kaufmann Pub., 2, 1990.

Garson D. G., Interpreting Neural-Network Connection Weights". AI Expert (April), 47-51, 1991.

Orlowski M., On The Integration Of Knowledge Bases. In: Advances in Computers and Information Engineering' G.E Lasker, editor. The International Institute for Advanced Studies in Systems Research and Cybernetics, Canada, 1993.

Tickle, A.B.; Orlowski, M.; Diederich, J., DEDEC: Decision Detection by Rule Extraction from Neural Networks. Submitted, 1994.

Towell, G.G. & Shavlik, J.W. The Extraction of Refined Rules from Knowledge-Based Neural Networks. Machine Learning 13 (1993) 1, 71-101.

NEURAL NETWORK ARCHITECTURES: MEDICAL IMAGE SEGMENTATION APPROACH AS A STEP TOWARDS MEDICAL IMAGE FUSION

Steven M. Vajdic[1], A. Bouzerdoum[2], M. J. Brooks[3], A. R. Downing[4] and H. E. Katz

[1,2]*Department of Electrical and Electronic Engineering*
[3]*Department of Computer Science*
The University of Adelaide, Adelaide, S. Australia 5005
[4]*School of Engineering, Flinders University of SA, Adelaide, S.A. 5001*
[5]*ISCS Inc., 77-14 113th St, Suite 1B Forest Hills, NY 11375, USA*
E-mail: svajdic@eleceng.adelaide.edu.au

ABSTRACT:

An Unsupervised Competitive Learning neural network architecture for medical image segmentation is presented. It is considered as a crucial part of the 2D/3D region-based inexact relational multi-modality medical image matching/fusion algorithm. The 2D reference and target images are selected from 3D sets and are segmented into non-touching and non-overlapping regions similar to those of human organs. The networks are trained on images with undeformed shapes, and shapes normally positioned within human organs and related to anatomical parts of the organs. Both reference and target images are considered as inputs, each to a slightly different network. In the end, input images are clustered and represented as region-based output images ready for further processing towards matching and fusion.

KEYWORDS

2D/3D Medical Imaging, Artificial Neural Networks, Unsupervised Competitive Learning, Multi Modal, Image Segmentation, Regions.

INTRODUCTION

Image fusion (composing images of the same objects taken by different means) has played a significant role in disciplines such as medical imaging, pattern recognition, computer vision and remote image sensing. The idea is to combine images of a single object obtained using either different sensors, different frequencies/energies, with perhaps different sections of the object observed at different times, from various positions or viewing angles, under different illumination conditions or polarisations. The resultant single image would be of a better clarity and contain more information than could be obtained from any of the individual "matched" images.

The idea of combining images in medicine becomes very important because of the necessity to improve the diagnosis of diseases and the treatment and care of patients. 2D/3D fusion of medical images acquired by medical scanners e.g. Computerized Tomography (CT), Magnetic Resonance Imaging (MRI), Positron Emission Tomography (PET) and Single Proton Emission Computer Tomography (SPECT) is of special interest for medical experts because of the different image characteristics produced by each device that are useful for combining.

It is known that CT and MRI can provide anatomical information (structure, texture, grey-level shapes) of human organs and PET and SPECT can provide functional information (tissue function, blood vessels, cell function, metabolism). Moreover, CT can provide good bone structure with less distortion but with less tissue texture, as opposed to MRI that can give a high quality structure description but with more distortion due to secondary magnetic field effects formed around the examined object. As a consequence, any one of the following 5 image pairs — (CT,PET), (CT,SPECT), (MRI,PET), (MRI,SPECT), (CT,MRI) is useful for fusion. The first four are classified as structure-function fusion, the last one as structure-structure fusion. The first component in each pair is referred to as a reference image (the one that is not changed), the second component is referred to as a target one (the one that is changed). The mutual characteristic for all four modalities is that they contain a finite set of 2D cross-sections (slices, scans) of a human organ, forming a 3D structure of the organ, with a chosen distance between the 2D slices.

BACKGROUND

Before the images are fused they have to be matched (registered); that is, the target image needs to be modified somehow to match the reference image. All image matching/fusion algorithms developed and classified ([1],[2],[3],[4]) so far generally adopt a four step strategy: (i) extraction of key structures, (ii) matching of key structures, (iii) evaluation of similarities between structures and (iv) transformation of target image.

Classification of matching/fusion algorithms is a very complex task in general since it involves many classification criteria [2] such as image dimensionality (2D, 3D), image properties (attributes and their interrelationships - extrinsic, intrinsic) and image property coupling (interpolation, approximation), domain (global, local) and elasticity (rigid, affine, projective, curved) of transformation, parameter determination (direct, search-oriented) and interaction (interactive, semi-automatic, automatic). A complex classification with all explanations is beyond the scope of this paper. Yet, it would be worth mentioning that different classifications would be possible from the following global set of principles of matching methods: (i) control points, (ii) moments, (iii) knowledge base representation of object characteristics, (iv) similarity optimisation, (v) stochastic sign change, (vi) Fourier analysis, (vii) cross correlation, (viii) eigenvalue decomposition, (ix) warping, (x) 3D surface fitting, (xi) anatomical atlas, (xii) internal landmarks, (xiii) external landmarks, (xiv) principal axes, (xv) inexact relational matching and (xvi) neural network segmentation. Which method is to be chosen depends on the type of images to be matched/fused (e.g. natural scene images, stereo images, Landsat images or medical images). Moreover, a set of possible problems related to the matching task and type of images would be given as: (i) choosing a matching method, (ii) choosing the reference image, (iii) selecting image properties, (iv) imposing constraints, (v) evaluating "goodness of matching", (vi) choosing an appropriate geometric transformation, (vii) coping with distortion, and (viii) choosing a fusion method.

This project is concerned with medical images. Hence, with regard to the suggested four step matching methodology and related problems, the following key steps are chosen: (i) image regions, similar to the anatomical shapes of human organs, and their attributes, are extracted from 2D slices to make less complex the image matching procedure and the finding of the best match; (ii) CT or MRI image modality is selected to be the reference since they provide anatomical information which is closely related to segmented regions; (iii) tree search is employed as the best method for finding the region matches, and similarity between regions (for finding the best match) is evaluated by applying functions that incorporate comparisons of region attributes; (iv) a spline geometric transform is applied to correct the local distortion. Based on these key steps and knowledge base representation (IF-THEN rule-based analysis) 2D and 3D generalised approaches were proposed and discussed ([5],[6]) and improvements to evaluation of region similarities and searching for good region matches were made ([7], [8]).

Matching and fusion were performed for a 2D image composing, 3D visualisation was added to these two procedures for a 3D image composing. First, 3D sets were resampled to the same dimension and 2D slices/scans of interest were selected. Then, these 2D slices were segmented into regions (iterative thresholding and/or rule-based knowledge representation of the anatomical shapes of human organs). After region properties (attributes and their interrelationships/relations) were extracted and an inexact relational region matching procedure was applied to provide a set of best region matches (by employing an evaluation function to compare similarities of region properties and guide a tree search for finding the best match), a spline geometric transform was introduced as the best method for eliminating local distortions [9]. Fusion, as the final step, was performed by means of composing weighted average pixel (2D) or voxel (3D) intensities [10].

In this paper the emphasis is given to the neural network segmentation in obtaining image regions closely related to the anatomical shapes of human organs. The similarity measure for comparing region attributes/relationships and the tree search methods in order to evaluate the inexact relational matching procedure and find the best match within a set of "good" matches (AI-based tree search methods) are mentioned to make the matching procedure complete.

SEGMENTATION USING NEURAL NETWORKS

The main task of region segmentation is to yield non-overlapping image regions, each having some homogeneous characteristic. The shapes of the closed boundaries are to be as similar as possible to those of human organs. The greater the number of model regions initially chosen, the more accurate the matching algorithm becomes, but at the cost of increased computation. These opposing aspects have to be taken into account when realising an experimental model.

A sequential step-designed approach is adopted, involving segmentation to form image regions or clusters using an unsupervised competitive learning neural network (UCL), region-based inexact relational matching, and a tree search for finding the best match.

Segmentation based on UCL involves training the neural network to learn the image patterns of anatomical shapes of human organs, then presenting the image to the trained network for segmentation. The training process involves presenting an input vector simultaneously to each of the output neurons. The neurons compare the input vector to their own weight vectors and one of them is declared the winner. The winning neuron then updates its weight vector to become similar to the input vector. This process is repeated until some stopping criterion is reached. The network is trained only on images with undeformed shapes, and shapes normally positioned within human organs and related to anatomical parts of the organs.

The UCL neural network used here has a single processing-layer whose output is described in the matrix-vector form as

$$\mathbf{y} = W^T\mathbf{x} \tag{1}$$

where $\mathbf{x} = [x_1 x_2 ... x_n]^T$ and $\mathbf{y} = [y_1 y_2 ... y_m]^T$ are the input and output vectors respectively, and W is the n x m weight matrix

$$\begin{bmatrix} w_{11} & w_{12} & \dots & w_{1m} \\ w_{21} & w_{22} & \dots & w_{2m} \\ \vdots & \vdots & \vdots & \vdots \\ w_{n1} & w_{n2} & \dots & w_{nm} \end{bmatrix}$$

with w_{ij} being the weight from input i to output j.

Both reference and target images are considered as inputs, each to a slightly different network. Each image pixel is presented to the neuron input node. Typically, the whole image is divided into small neighbourhood regions of 2 or 4 pixels that are presented to the network simultaneously (with 2 or 4 input nodes). The number of outputs is related to the number of image gray level clusters, typically 8-16 (Table 1). All input units are connected to each output unit and compete under a "winner-takes-all" rule. The competitive transfer function returns "0" neuron outputs for all neuron inputs, except for the winner where the output is "1". In the end, an input image is clustered and represented as a region-based output image (Figure 1).

The network configuration is based on three basic procedures: learning anatomical shapes of organs, recognition of similar features in the input pair of images and segmentation of those features into regions similar to those in learning phase. It is, also, able to locate the positions of recognised regions from

Modality	x	y	IN	OUT
CT	512	512	2-4	8-16
MRI	256	256	2-4	8-16
PET	100	100	2-4	8-16
SPECT	128	128	2-4	8-16

Table 1: UCL input-output characteristics

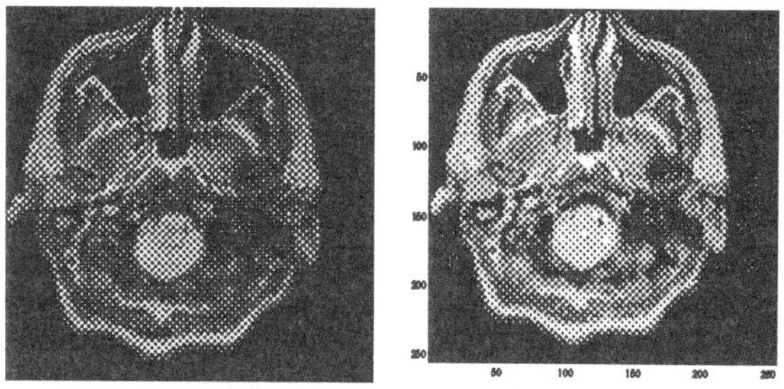

(a) 2D MRI image (b) Unrefined UCL segmented MRI image

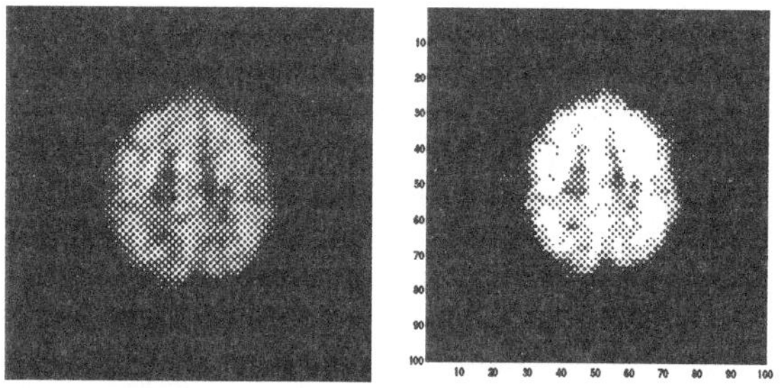

(c) 2D PET image (d) Unrefined UCL segmented PET image

Fig. 1: UCL neural network segmentation.

the pair of input images to be fused. Once the regions are segmented and located, all the necessary attributes/relations can be computed for the matching stage. A refining procedure also needs to be applied to extract very small and very big regions from clusters that do not represent parts of human organs since they would lead to an enormous pre-matching calculation and would be most likely rejected by the matching algorithm due to lack of pair-wise matches.

INEXACT RELATIONAL REGION MATCHING

Among many image primitives (points, lines, regions), regions are adopted because of the special segmentation methods applied for their extraction (unsupervised competitive learning). More attributes can be extracted from regions than from points and lines. Nevertheless, by applying Hough or medial axis transformations, points and lines could also be used with the same algorithm proposed here (the network parameters would need to be modified though), but the accuracy of matching may well be below the level that can be reached with regions.

Image region attributes including perimeter, area, shape, mean/max/min radius, centre of gravity, and amplitude, histogram and average gray values are computed. Either the attributes of the regions, or calculable inter-relationships between them, can be used. If the inter-relationships are used, then attributes of each region within either the reference or target images are recalculated in terms of their relationships with other regions in order to provide uniformity. The attributes/relations are normalised and weighted according to their influence to the match evaluation function. A major experimental search for the best combination of region attributes and/or relationships is underway. The main criterion for a good choice is invariance under linear transformations (translation, rotation, scaling, skewing) and possibly non-linear distortions. The non-linear distortions are to be corrected mainly by applying the spline geometric transformation using the centres of gravity of matched regions after the region matching has been performed.

The matching procedure is complete when the same number of regions is found in both matched images. Region matching is solved by means of tree search methods. The matching procedure is guided by a region attributes/inter-relationships based match evaluation function. A tree search is applied only to find the best region matches prior to the application of a geometric transformation.

EVALUATION OF MATCHING

The match evaluation function has to be applied to two data descriptions (here, reference and target image regions as primitives and their properties — attributes and their relationships-relations) in order to guide the search method in finding the best match. The best match between two sets of image region primitives is to be found as the mapping for which the regions show the best similarity in their property values. The major problems of definition of similarity of properties — what property is more important and/or how to combine numerical and symbolical properties — need to be overcome. Thus, the property values have to be weighted and their inter-relationships defined in terms of all the relations over the set of property values, and/or a conditional probability function of properties needs to be employed.

Three types of evaluation function that compare similarities of two descriptions (here, sets of regions) have been proposed so far: a distance measure (cost function) ([11], [12], [13], [14], [15], [16]) and a merit function [17], both involving mainly comparison of pairs of primitive property values, and the function that involves more sophisticated parameters, e.g. the number of primitives, the number of properties and the number of successful matchings of each pair of properties ([7], [18]). The distance measure (distance function) is to be minimised, the merit function is to be maximised, and the third function could be defined in such a way to allow either minimization or maximization. In the case of the two sets of region primitives (descriptions), the cost function of similarity between regions is usually defined as the sum of the absolute or square value differences, or even as the square root of the sum of the square value differences of all region properties. These forms can be applied to numerical properties only. In the case of either only symbolical properties or a combination of both symbolical and numerical properties, a conditional probability function can be successfully applied [17] or some penalty for non-similarity needs to be incorporated into an evaluation function expression.

The matching procedure used here is guided by a novel match evaluation function [7] that involves the weighted attribute/relation comparison, the weighted penalty for non-similarity, the number of attributes/relations, the number of successful attribute/relation matches, and the weighted relative difference between the segmented and initially suggested anatomical regions of human organs of both the reference and the target images.

TREE SEARCH METHODS

Tree search methods have been well developed in AI. Here, they are used to search the space of similarities between region property values of reference and target images and find the best match based on the proposed evaluation function. In general, the search for the best match is very time consuming since there are many regions and their properties of both images involved. The search time could be reduced by imposing some constraints. Hard constraints limit the set of all mappings within the search space to those for which the values of the evaluation function are inside experimentally predefined thresholds. Soft constraints define a relative preference over the space of possible mappings by implementing some heuristics in the evaluation function that direct the search method towards "better mappings". To minimise the search time for the best match, both hard and soft constraints could be employed. For example, with regard to the defined evaluation functions [7] the authors propose soft constraints that direct the search algorithm to compare the properties with high weights first; after each comparison the partial value of the evaluation function of currently compared regions is calculated and compared with the stored minimum value of the function of some previously completed region comparisons; the process of the calculation of the current function (and, consequently, the process of the comparison of the rest region properties) is stopped when its value exeeds the minimum stored value of another pair of compared regions; the current match is then assigned the tag "not the best one"; the track of "the best match" is always kept.

CONCLUSION

2D/3D medical image fusion approaches based on ANN and AI principles have been considered. It has been intended to convey some ideas that have not previously been applied in 2D/3D medical image fusion. While some non-ANN and non-AI problems have been considered here (geometric transformation for mapping target images onto reference ones, image fusion of compositing pairs of images into single ones), some ANN and AI topics are under investigation: unsupervised competitive learning neural network segmentation and tree search in image region inexact relational matching. Although both relaxation labelling ([19],[20]) and relational matching methods can be used for matching relational image descriptions, the latter has been chosen for this work because it always finds the best mapping between the primitives of image descriptions (in this case, the image regions). The trade-off between the computational cost and accuracy also needs to be taken into account whereby the more image regions and region attributes/relationships adopted, the better the accuracy, but the greater the computation. The work is underway to test the algorithms under a large set of medical sample images of all combinations of structure-structure and structure-function fusions.

ACKNOWLEDGMENTS

The authors are grateful to Prof. Marilyn Noz of the Department of Nuclear Medicine, New York University, for her unreserved support in supplying matched medical image data sets used in this project.

References

[1] Brown L. G. (1992), "A survey of image registration techniques", *ACM Computing Surveys*, **24**(4), pp. 325–376.

[2] Van den Elsen P. A., Pol D. E. & Viergever M. A. (1993), "Medical image matching – A review with classification", *IEEE Engineering in Medicine and Biology*, pp. 26–38.

[3] Chen Q. (1993), *Image Registration and its Applications in Medical Imaging*, PhD thesis, Vrije University Brussel, Faculty of Applied Sciences, Brussels.

[4] Chapnick J. V., Noz M. E., Maguire G. Q. Jr., Kramer E. L., Sanger J. J., Birnbaum B. A. & Megibow A. J. (1993), "Techniques for multimodality image registration", *Proceedings of the 1993 IEEE 19th Annual Northeast Bioengineering Conference*, pp. 221–222.

[5] Vajdic S. M., Katz H. E., Downing A. R. & Brooks M. J. (1994), "Artificial Intelligence and registration of functional and structural images: Generalised 2D approaches", *Proceedings of the 1994 IEEE International Conference on Systems, Man and Cybernetics*, Vol. 1, pp. 1030-1035, 1994.

[6] Vajdic S. M., Katz H. E., Downing A. R. & Brooks M. J. (1994), "AI-based relational matching and multi-modal medical image fusion: Generalised 3D approaches", *Proceedings of the 1994 IEEE International Conference on Visual Communications and Image Processing*, Vol. 2308, pp. 1565-1572, 1994.

[7] Vajdic S. M., Brooks M. J., Downing A. R. & Katz H. E. (1994), "AI and Medical Imagery: Strategy and Evaluation of Inexact Relational Matching", *Proceedings of the Second Australian National Conference on Complex Systems*, pp. 125–132, 1994.

[8] Vajdic S. M., Brooks M. J., Downing A. R. & Katz H. E. (1994), "Artificial Intelligence and Image Matching/Fusion: A Tree Search", *Proceedings of the 1994 International Conference on Artificial Intelligence and Expert Systems Applications*, to appear, December 1994.

[9] Goshtasby A. (1988), "Registration of images with geometric distortions", *IEEE Transaction of Geoscience and Remote Sensing*, **26**, pp. 60–64.

[10] Porter T. & Duff T. (1984), "Compositing digital images", *Computer Graphics*, **18**(3), pp. 253–259.

[11] Boyer K. L., Kak A. C., (1988), "Structural stereopsis for 3-D vision", *IEEE Transactions on Pattern Analysis and Machine Intelligence*, **10**(2), pp. 144–166.

[12] Sanfeliu A., Fu K. S., (1983), "A distance measure between attributed relational graphs for pattern recognition", *IEEE Transaction on Systems, Man and Cybernetics*, **13**(3), pp. 353–362.

[13] Shapiro L. G., Haralick R. M., (1981), "Structural descriptions and inexact matching", *IEEE Transactions on Pattern Analysis and Machine Intelligence*, **PAMI-3**(5), pp. 505–519.

[14] Shapiro L. G., Haralick R. M., (1987), "Relational matching", *Applied Optics*, **26**, pp. 1845–1851.

[15] Price K., Reddy R., (1979), "Matching segments of images", *IEEE Transactions on Pattern Analysis and Machine Intelligence*, **PAMI-1**(1), pp. 110–118.

[16] Faugeras O. D., Price K. E., "Semantic description of aerial images using stochastic labelling", *IEEE Transactions on Pattern Analysis and Machine Intelligence*, **PAMI-3**(6), pp. 633–642.

[17] Vosselman G., (1992), *Relational Matching*, Berlin Heidelberg: Springer-Verlag.

[18] Barrow H. G., Popplestone R. J., (1971), "Relational description in picture processing", *Machine Intelligence*, **6**, pp. 377–396.

[19] Rosenfeld A., Hummel R. A. & Zucker S. W. (1976), "Scene labeling by relaxation operations", *IEEE Transaction on Systems, Man and Cybernetics*, **SMC-6**, pp. 420–453.

[20] Medioni G. & Nevatia R. (1984), "Matching images using linear features", *IEEE Transaction on Pattern Analysis and Machine Intelligence*, **PAMI-6**(6).

"TOP DOWN" PROCEDURE USING NEURAL NETWORKS FOR PRELIMINARY STRUCTURAL DESIGN

Aldo Cauvin
Professor of Structural Engineering
Department of Structural Mechanics
University of Pavia
Via Abbiategrasso 211
27100 Pavia,Italy

ABSTRACT

The framework of an Expert System which can usefully assist the structural engineer in performing the general choices which need be made in the preliminary design of structures is briefly illustrated.
The system has two main characteristics:
-the design procedure is performed at different "levels" beginning with the most general "strategic" choices and ending with the most particular ones.As the design proceeds in this "top down" way boundary conditions deriving from previous choices are added to the very general ones with which the procedure has been started.
-Instead of using conventional "forward" or "backward" inference procedures for performing choices,simple two layered "perceptron" neural network are adopted.
These network are "instructed" using data derived by suitably organized data bases containing informations relative to an adequate number of existing structures.
The istruction procedure is rather mechanical and does not require an high degree of design expertise.
The possible application to design of tall building and bridges is briefly illustrated.
The system is designed to operate in a CAD environment so that results can be presented and manipulated in graphic form.

INTRODUCTION

The applications of Artificial Intelligence to structural design have begun to develop only recently[D90][M84][FM85][F89].

It is now recognised that they can help in all the phases of the design process where non algorithmic mainly qualitative decisions need be made.They are therefore particularly useful in the preliminary phases of design when shapes and materials are chosen,expecially for those structures,such as bridges and tall buildings ,which are both important and thus require a careful preliminary evaluation and selection of types ,and can be classified using limited and well defined structural schemes and components.

The first applications used rule based Expert System shells and procedures previously adopted for different purposes in different fields(medical,geological),mainly diagnostic.These shells utilize deductive inference engines which are not particularly suited for fuzzy logic processes like design,which must lead to non univocal,often ambiguous answers.

These projects ,in analogy with KBES prepared for other fields,and often using "general purpose" shells,use inference mechanisms based on deductive processes(forward and backward chaining),or "abductive" ones[CS94/2][CS95/1].

In the preparation of a system of this kind two main difficulties are met:

- The preparation of a consistent and reasonably objective sets of rules is a difficult task both because expert designers are not willing to reveal their expertise but also because this expertise is necessarily limited , somewhat hydiosincratic , subjective and difficult to formalise.

-Structural design is a complex procedure in which many choices among a great number of types need be made.It is therefore necessary to reduce the design space both grouping,by a suitable classification, the number of structural types among which choices need be made and individuating several hyerarchic levels of design such that decisions performed within the upper levels are to some extent independent from the ones to be made at the lower levels.

How these difficulties were overcome in the resulting expert system,named STRUCNET, is explained in the following paragraphs.

CAN NEURAL NETS BE USED IN STRUCTURAL DESIGN?

As known,over other techniques of Artificial Intelligence Neural Networks possess some advantages,the most important of which is the possibility to "train" them using procedures which do not require the presence of experts capables of codifying their knowledge in sets of rules.Another precious advantage is that a limited number of input errors made during the "instruction" phase does not affect in practice the network performance.

However the use of this technology has been confined to well defined fields,mostly related to pattern mapping,pattern classification,recognition and completition(handwriting recognition is one of the most successful applications)or weather and financial forecasting.

One may wonder whether this approach can be useful in such a different field as structural design.

However the mentioned difficulty in defining a knowledge base organized in rules(and the difficulty to find experts able to do so)was a strong incentive toward investigating the applicability of neural nets to design.

It was a surprise to discover that the problem could be to some extent solved adopting a very simple two layers net,the well known "Perceptron" net.

ADOPTED NEURAL NET ARCHITECTURE

As results from fig.1 ,a knowledge base used for design purposes can be represented by a net of units and connections which can be interpreted as a "neural network" characterized by two layers:

-The units of the first layer represent all the possible design choices

-The units of the second layer represent all the possible performances(expected behaviours) corresponding to all the possible design choices.

Each unit in the first layer is connected with each unit in the second layer,while units in the same layer are not connected among them.

This is a "classical" neural net called "Perceptron",widely used in pattern recognition works[D90].Here however its use is completely different.

The connections between the two layers can be represented by a boolean matrix where 1 represents a connection between a design choice and a performance(and therefore the existence of a rule or "model" which expresses the fact that that performance is one of the consequences of that choice),while 0 denotes no connection(see fig. 2 where the boolean matrix corresponding to the net of fig.1 is represented.

That matrix therefore defines syntetically and univocally the knowledge base.

Now the question is:how can this matrix be built and,if it is the case, modified?

"INSTRUCTION" OF THE NET

The more rational way to build the knowledge base is by careful examination of an adequate number of existing structures.

It is therefore necessary to build a data base containing informations about these structures[CS94/1][CS94/2].By careful examination of these informations the boundary conditions and therefore the required behaviours according to which the structure was designed should be detected.In other words we use in this case a procedure which is inverse to that of design:instead of making a design choice according to given boundary conditions,given a choice we detect the boundary conditions according to which the choice was made.

Once for an adequate number of structures the connections between choices and behaviours have been detected,it becomes possible to "instruct" the net so that the knowledge base can be gradually accumulated in it.For this instruction procedure,once the structural types and possible required behaviours have been individuated and classified,it is not necessary to use a Specialist,the procedure being essentially mechanical.

We could proceed in this way(see fig.3):

-At the beginning of the procedure the boolean matrix $[\beta]$ is empty(every $\beta ij=0$).Its dimensions are given by the number of possible design choices(number of rows) and by the maximum assumed number of expected behaviours.

-Sheets are prepared for every example containing a given structural choice and the connected behaviours.

-Counters are associated to every design choice(first layer) and to each connection.

-Every time a sheet is considered the counters of the corresponding choice i (Ni)and connections(Mij) are incremented by one.

-The ratio Rij=Ni/Mij is computed for every connection.

-Whenever Rij becomes greater than a given threshold T(let's say 0.8 for example) βij is set=1(connection on)

-Whenever Rij<T β ij is kept(or set again)=0(connection off)

-The "instruction" of the net is terminated when an adequate number of examples(more than 100) has been processed in this way.

The meaning of the threshold is obvious:when an adequate number of examples has been considered and more than 80% of a given structural type have been found associated with a given behaviour,it is reasonable to assume that the same behaviour has contributed in an essential way to determine that choice.It is therefore reasonable,as required behaviours must match performances(expected behaviours) to adopt a rule connecting choice and behaviour.

The choice of the threshold must be made carefully,according to the distribution of available data.It is not therefore an "a priori" choice.

We could say that the threshold defines the degree of importance of the rule.

The association found for a single example could be the result of misjudgements by the designer;it is however unlikely that 80% of designers who have chosen a given type have made the same mistake.

The kind of knowledge which can be acquired in this way is therefore based on past experience,on the assumption that most(but not all) considered designers were competent.

It is not in this way that innovative designs can be elaborated;however in structural design past experience has always be,also for safety reasons,the main source of information.

Even in the cases where innovative solutions are required the solutions suggested by these procedures could be useful for reference:the "degree of innovation" could be evaluated,with the connected risks.

Summing it up,the expert system is not the "boss" of the designer.Its only purpose is to trasmit to him in the most rational way the past experience.

USE OF THE "INSTRUCTED" NET

At this point,the "instructed net" can be used to perform design choices:

The layer containing performances is used as input layer for the required behaviours

The matrix $[\beta]$ becomes the matrix of weights.

The units of the output layer process the signals coming from the connections as sums of the wheights of the connections with units representing the required behaviours(fig.4) .The threshold value is represented in this case by the number of required performances.When the threshold is reached in a given unit,it "fires" that is the design choice corresponding to that unit is made.

EXAMPLE OF USE OF NEURAL NETWORK IN STRUCTURAL DESIGN

On fig.5 an example net is given permitting to perform choices of structural shapes of bridges according to given boundary conditions which in this example are limited to span ranges and basic layouts.

This is the second step in the preliminary design process as described in the following paragraphs and in refs[CS94/2] and [CS95/1].

Two boundary conditions are given;as a consequence the threshold is =2. Ony three units reach this threshold:those corresponding to a continuous beam(prestressed concrete or mixed structure) or a frame(prestressed concrete).

In practice the net is more complicated,given to the much greater number of possible boundary conditions which can be encountered in practice.

MODULAR NATURE OF STRUCTURAL DESIGN

The process of preliminary structural design can be divided in the following phases:

-Definition of basic boundary conditions
-Definition of general required behaviours(specifications)
-Definition of basic layout
-Definition of shapes
-Preliminary Dimensioning
-Preliminary quantities and costs evaluation.

The first two phases constitute the formulation of the problem and require the use of deductive processes.

The following three phases constitute the design part of the process and can in turn be divided in a number of "levels" as specified in fig.6

Each level can be processed in an independent way,with its own design space,proceding from top levels down to the lower ones(that is from more general to more specific and detailed designs)provided that specifications directly deriving from choices performed at the preceding levels are added to the general specifications and the specific boundary conditions of the problem at the level under consideration.

The preliminary dimensioning phase which follows does not utilize procedures of Artificial Intelligence but rather statistical data transformed into suitable diagrams,and logical considerations related to the shape and compatibility of the chosen parts of the design,and also "rules of thumb" derived by common design dimensioning practice.

Once the various parts of the design are dimensioned they can be assembled in a "prototype" design which ,although generical,stereotypical and not detailed,should contain enough information for the evaluation process of the chosen structural type which must follow.

At last the quantities evaluation,obtained by using a spreadsheet associated with each part of prototype design which automatically update itself as dimensions are changed,provides an analytical mean of computing expected costs.A more syntetical mean would be provided by manipulation of parametric statistical data concerning costs extracted by a number of similar executed designs.

The choice and dimensioning of structural types and components,as well as their manipulation and assemblage in prototype designs is greatly facilitated if the system is programmed within a CAD system which can provide an enourmous amount of graphic functions and facilities.This is made possible by the use of programmable CAD environments such as AUTOCAD™ and MINICAD™.

Let us at this point consider with some detail the nature and the tasks to be performed at each level,illustrating the process with hypotetical design examples concerning a bridge and a tall building.

Definition of basic layout

The first level of design concerns the determination of basic layout of the structure. In the case of bridges five types of basic layouts have been individuated,in which most of the possible cases can be included(fig.8).

In the example of fig.8,concerning the preliminary design of a cable-stayed bridge the inference procedure leads to the choice of two layouts between which the layout consisting of a large central span with two lateral smaller ones was selected.

In the case of tall buildings the basic layout is the architectural grid.

Definition of shape of the bridge

The complete individuation of the bridge shape is accomplished in the next six levels:

The second level concerns the determination of the general longitudinal shape(structural scheme) ,which is associated closely with the choice of the building material.

To make this choices the specification deriving from the preceding level are added to the basic specifications .

This procedure is repeated in the following levels,every time a given choice increases the number of boundary conditions to be accounted for.

The third level concerns the refinement of the longitudinal shape .In the given example,this refinement consists in the choice of arrangement of the stay cables.

From this stage downwards the choices are being made among well defined components of the design which can be parametrized.

The fourth level concerns the choice of the type of deck cross section

The fifth level concerns the choice of vertical supports(towers,piers and abutments).

The sixth level concerns the choice of foundation types.

At last the seventh level concerns the choice of construction method

In the scheme of fig.8 the path which was followed for the given example,as well as the possible alternatives which could have been reasonably chosen are illustrated.

Definition of shape of the tall building

In the case of the tall building example (fig.9)the complete individuation of the shape is accomplished in the levels two and three

The second level concerns the determination of the general shape(structural scheme)

In the given example a concrete central core is chosen

The third level concerns the refinement of the shape .In the given example,this refinement consists in the addition of steel outrigger beams.

From this stage downwards the choices are being made among well defined components of the design which can be parametrized.

The fourth level concerns the choice of the type of floor and framed structure.In the given example concrete one way slabs and a mixed framed structure is chosen

The fifth level concerns the choice of basement and foundations.

DIMENSIONING AND PROTOTYPE ASSEMBLAGE

At this point the general shape of the bridge and tall building(and of their main components) is clearly defined.

To evaluate the obtained design and,in case,to perform preliminary structural analyses on it,it is necessary to complete the definition of dimensions.

The sizes needed to univocally determine the project can be divided into four categories:

-Sizes which are defined from the beginning(for example the bridge deck width) or as a consequence of the basic layout choice(the length of the spans,the height of piers,the architectural grid of the tall building)

-Sizes which can be defined statistically by comparison with designs of about the same shapes and span length(or building height in the case of the tall building).

-Sizes which can be determined using empirical "rule of thumb" rules in function of the already determined shapes.For example the depth of th web in a beam can be established in function of its height.

-Sizes which are established to insure compatibility between connected parametrized parts of the design.

Using these criteria the main parts of the structure are completely defined and can be assembled in a "prototype" on which the design evaluation can be performed .

MACRO FLOW-CHART OF SYSTEM

The self explanatory macro flow chart of the system is represented on figs. 6 and 7.

It can be divided into three parts according to the functions to be performed:formulation ,design and evaluation.

It is interesting to notice that,at each design level the use of the inference engine is not mandatory,but the design choice can be made directly in those cases where the knowledge base is not considered sufficient,reliable or it lacks completely;it is thus possible to use the procedure even when the collection of knowledge is far from complete.

GENERAL CLASSIFICATION OF STRUCTURES

As already said the first step in the preparation of an expert system for structural design is the definition of structural types among which a selection must be made in the preliminary stages of the design process.This definition is an uneasy task and,given the enormous variety of structural shapes and layouts which can be used in practice,could never be complete and entirely satisfying.

For some categories of structures this task can be made easier by the fact that in most cases the structural types are well defined and limited in number.Such is the case of bridge structures and tall buildings[M84][CS94/1][CS94/2][CS95/2].

All this considered,we think that it would be useful to individuate a broad classification of structural types which can include most of those structures whose design requires a careful attention.These structures,excluding some very specialized kinds such as dams,reservoirs,retaining walls and so on can be classified in three broad categories(see Tab.1):

a)Structures whose most important elements lie essentially in a vertical plane; this is the case of bridges:the main structural elements to be designed are decks(horizontal layout) and piers(vertical layout)

b)Structures whose most important elements lie essentially in an horizontal plane or whose dimensions in the two horizontal directions prevail on the vertical one.This is the case of long span low rise buildings,such as industrial buildings,multistoreys parking lots,assembly and sports halls.The main structural elements in this case are floors ,roofs and theirs supporting beams(all horizontal or sub horizontal elements)

c)Structures whose elements must constitute a system capable of resisting strong horizontal actions.

This system is composed of frames,shear walls,tube elements,braced frames connected by floors and is contained in a 3d space.This is the case of tall buildings and seismic structures.

Of course there are some cases which lie somewhere between situation b and c.

Within each of these broad categories it is possible to individuate a number of basic layouts with which a number of structural shapes can be considered compatible.

According to this classification,structural spans and heights,shapes,loads and other relevant parameters ypermedia data bases organized by "sheets",containing both drawings and numerical data, can be prepared.Each sheet (or group of sheets)represents a design example.

Information contained in a sheet can be used to "instruct" the neural networks of the system.

The organization of this data base is represented on figs. 11 to 13.

CONCLUSIONS

It is interesting to note that a simple two layered,"Perceptron type" neural net can usefully assist in building ,departing from generic technical information concerning realized designs,the needed knowledge base to be used in an expert system devoted to preliminary structural design.

The same network,duly instructed,can also be used as an inference engine to perform the needed choices.

This procedure permits to build in a simple,mainly mechanical way the knowledge base needed in the preparation of an expert system for structural design.

On the other hand he advantages of the chosen"top down refinement plus constraints propagation" approach can be summarized as follows:

-The division of design and therefore of the knowlege base in different stages both simplifies the neural nets architectures and reduces the number of operations to be performed at each stage.

This approach also has drawbacks in the sense that the exploration of different and basically independent design spaces can lead to disregard some connections among the different stages;however the fact that design proceeds hyerarchically"top down" that is,from general to more particular problems permits to conclude that the design spaces at the lower levels are in some way included in those at the upper ones (and therefore their rules do not normally contradict those at the upper levels).Also the procedure is usually an iterative one,in the sense that it is repeated several time to get a number of possible design among which economical comparisons can be made.During this process contradictions can be detected and eliminated.

-The adopted approach does not eliminate artificially the inherent ambiguity of the design process and permits to obtain not a single but a range of possible designs.In this way the creativity of the designer is ,even at this stage, inhibited only to the extent that absurd or irrational choices can be avoided.

.The final decision stems from the comparison of these designs,which can be obtained quickly,with a sufficient degree of detailing to make these comparisons highly meaningful and reasonably objective.

-The inclusion of the process in a CAD environment which can be performed easily ,given the availability of programmable CAD systems,permits to interact graphically with the system in a very intuitive and clear way,avoiding lengthy and ambiguous descriptions of results.

-The modular nature of the process permits to skip some of the stages by directly making the decision at those stages.

In this way the system can be usefully exploited even if the knowledge base is far from complete.

REFERENCES

[C94]Cauvin A,Use of Neural networks for preliminary Structural Design,research Report,Department of Structural Mechanics,University of Pavia,Pavia,Italy,1994

[CPS95]Cauvin A.,Passera R.,Stagnitto G.,Collection of relevant data and definition of a knowledge base for structural design using expert systems.To be published on the Proceedings of IABSE COLLOQUIUM,Bergamo,1995

[CS94/1]Cauvin A,Stagnitto G.,Classification of Structures and organization of hypermedia data bases for Structural Design,First workshop of EG-SEA-AI,EPFL,Lausanne 1994.

[CS94/2]Cauvin A,Stagnitto G.,A "Model Oriented" Expert system for the preliminary design of bridge structures,First workshop of EG-SEA-AI,EPFL,Lausanne 1994.

[CS95/1]Cauvin A.,Stagnitto G.,Framework of a general purpose expert system for preliminary structural design. Proceedings of IABSE COLLOQUIUM,Bergamo,1995 Session 1-Knowledge support for Structural Design and Construction

[CS95/2]Cauvin A,Stagnitto G.,A "Top down" procedure for preliminary design of Tall Building Structures using expert systems. Proceedings of the World Conference on Tall Buildings and Urban Habitat,Amsterdam 1995.

[D90]Dayhoff J.,Neural Network Architectures,Van Nostrand Reinhold,New York,1990

[F89]Fenves S J,Expert Systems,Expectations versus realities,Proceedings of IABSE Colloquium "Expert Systems in Civil Engineering",Bergamo 1989.

[FM85]Fenves S J ,Maher M.L.,Sriram D.,Knowledge based Expert Systems in Civil Engineering",IABSE Periodica,Number 4,1985

[Fa93]Faltings B,Reasoning Strategies for Engineering Problems,Proceedings of IABSE Colloquium "Knowledge Based Systems in Civil Engineering",Bejing 1993.

[M84]Maher M.L.,Fenves S.J.,HI-RISE:an Expert System for preliminary design of High Rise buildings,Technical Report,Dept of Civil Engineering,Carnegie Mellon University,1984

[S92]Steinberg L,Design as Top Down Refinement Plus Constraint Propagation,from Chapter 8 of book:Artificial Intelligence in Engineering Design,Vol.1,Academic Press Inc.,1992

[TS92]Tong C,Sriram,Artificial Intelligence in EngineeringDesign,Vol.1,chapt.1,introduction, Academic Press Inc.,1992

AKNOWLEDGMENTS

This research has been performed with the partial financial support of CNR(Italian Council for research) and MURST(Italian Ministry for University)The Author warmly express his thanks to Doctor s Giuseppe and Diego Stagnitto for their help.

GROUP	MAIN STRUCTURAL ELEMENTS	TYPE OF CONSTRUCTION
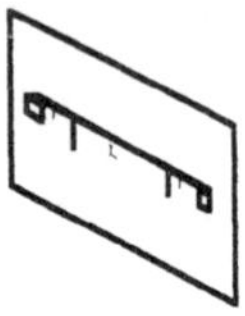 1	DECKS (and their supporting structures such as arches and stays) PIERS	BRIDGES OVERPASSES
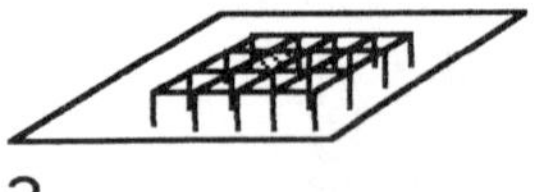 2	FLOORS ROOFS (and their supporting beams)	INDUSTRIAL BUILDINGS ASSEMBLY HALLS SPORTS HALLS MULTISTOREY PARKINGS etc
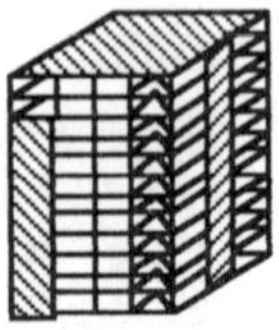 3	ALL THE LATERAL FORCES RESISTING 3D SYSTEM	TALL BUILDINGS SEISMIC STRUCTURES

Tab.1-Groups of structures according to given classification

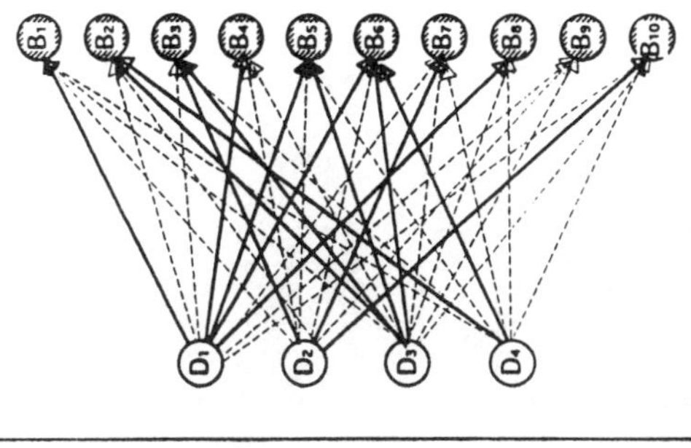

Fig.1-Example neural net

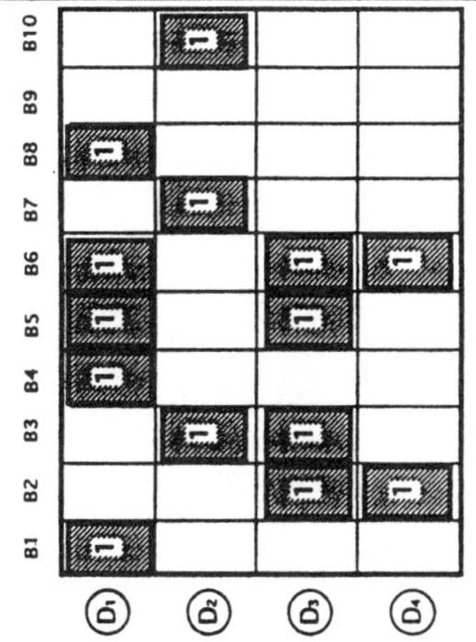

Fig.2-Corresponding boolean matrix [β]

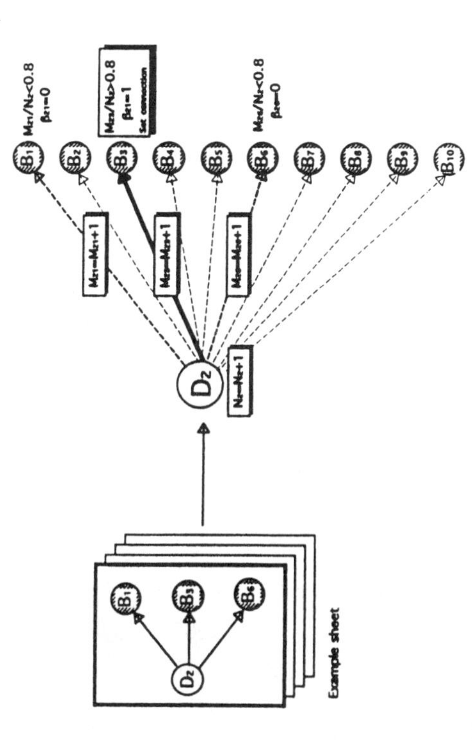

Fig.3-Instruction procedure for Neural Net

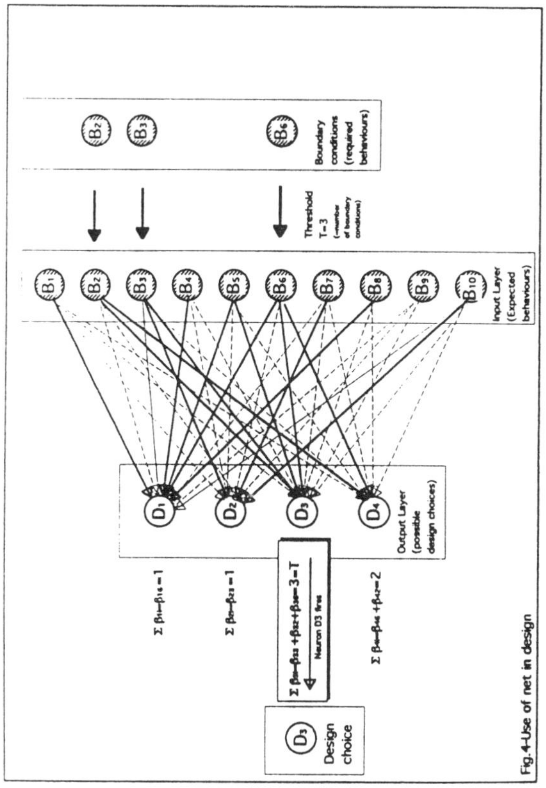

Fig.4-Use of net in design

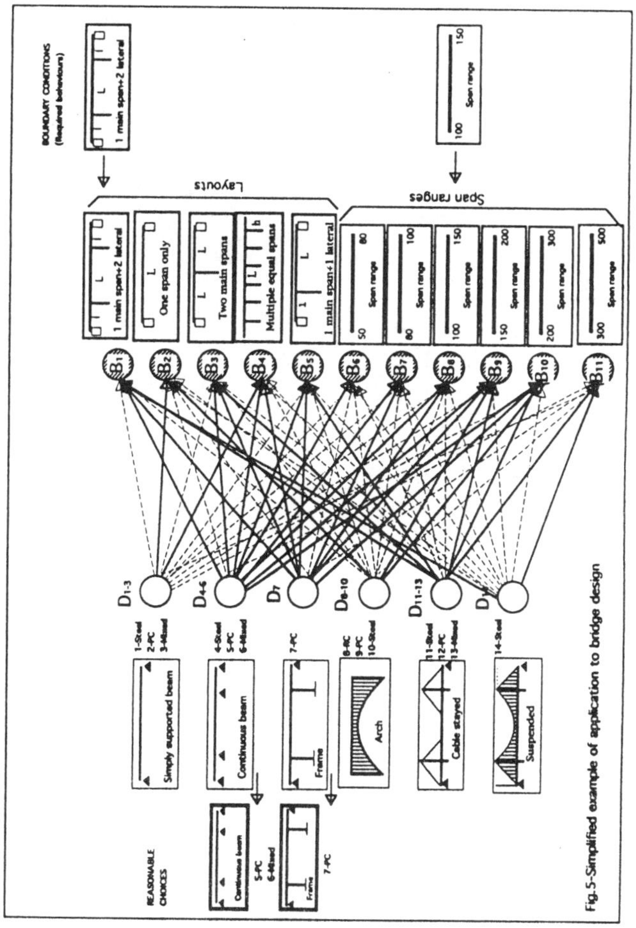

Fig.5-Simplified example of application to bridge design

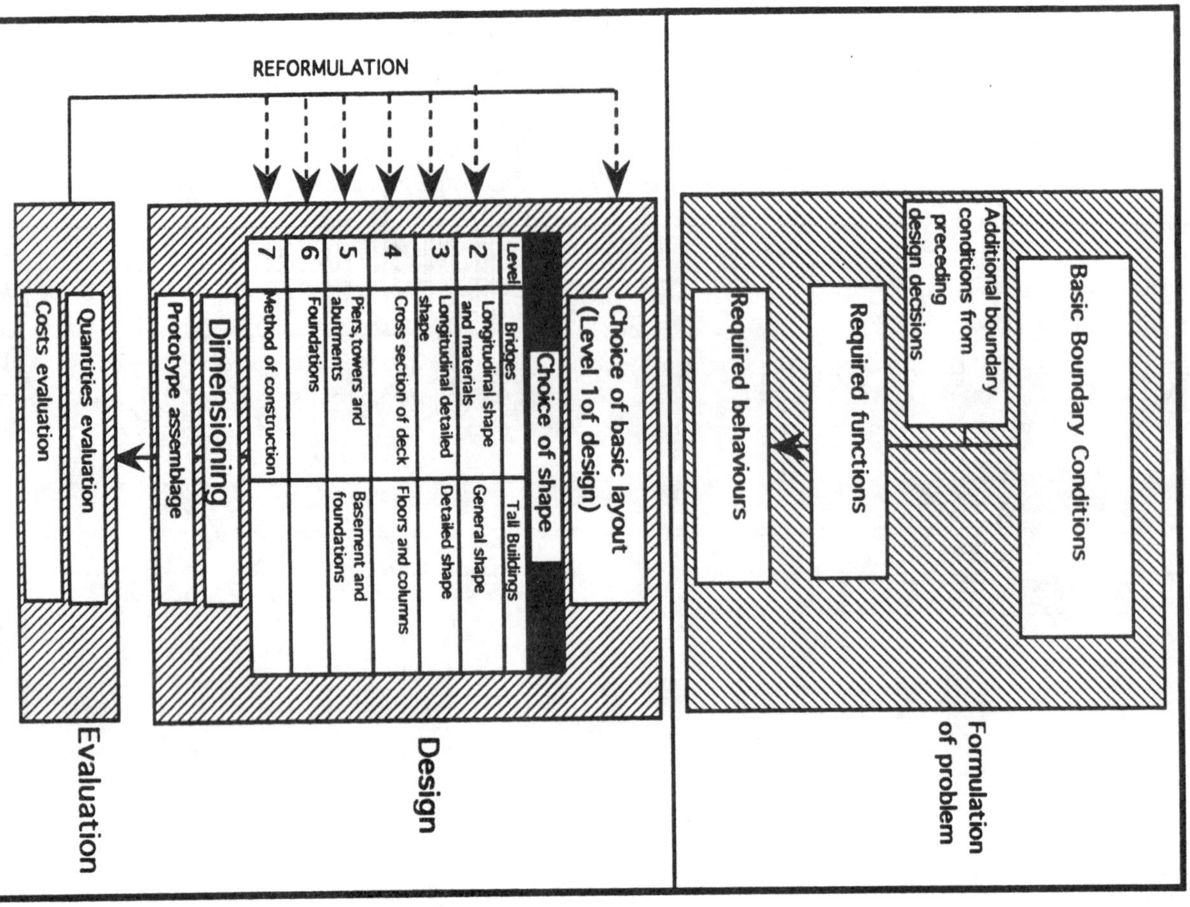

Fig.6-Phases of Structural Design and connected activities

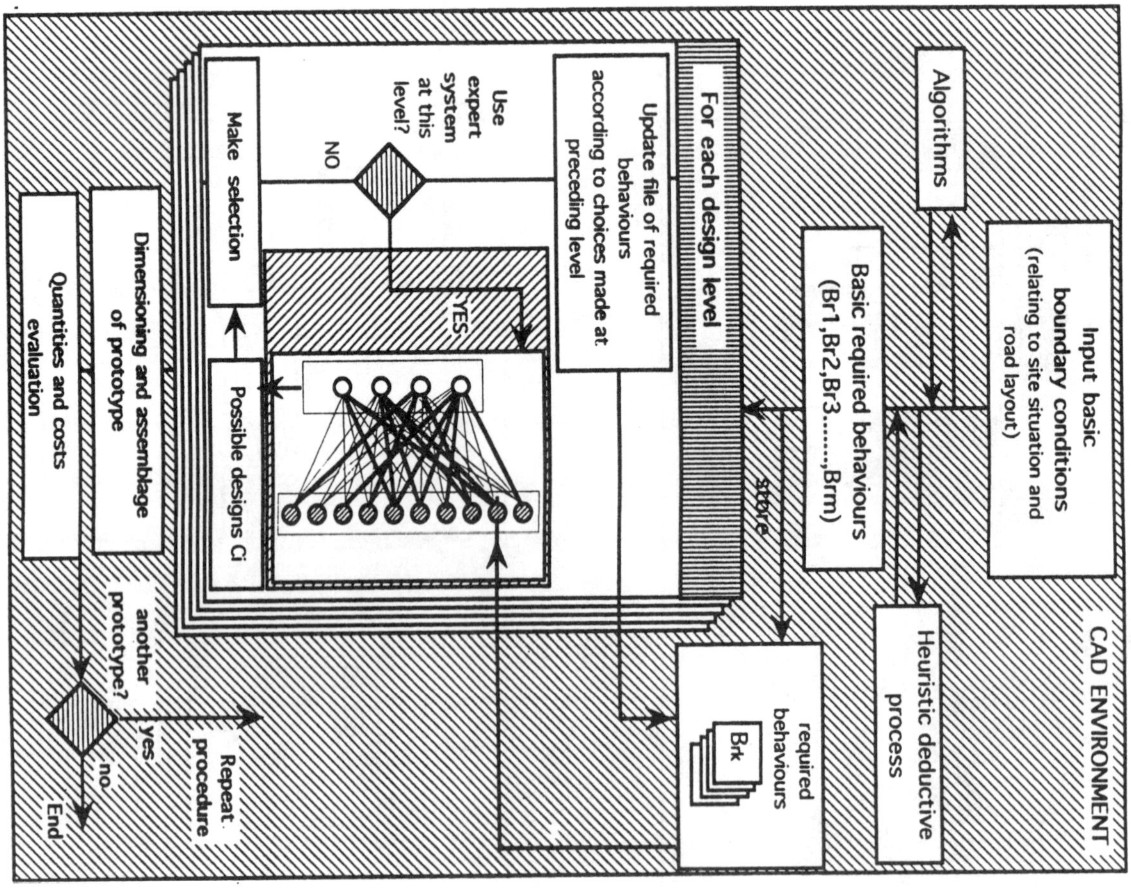

Fig.7-Macro flow-chart of Expert System

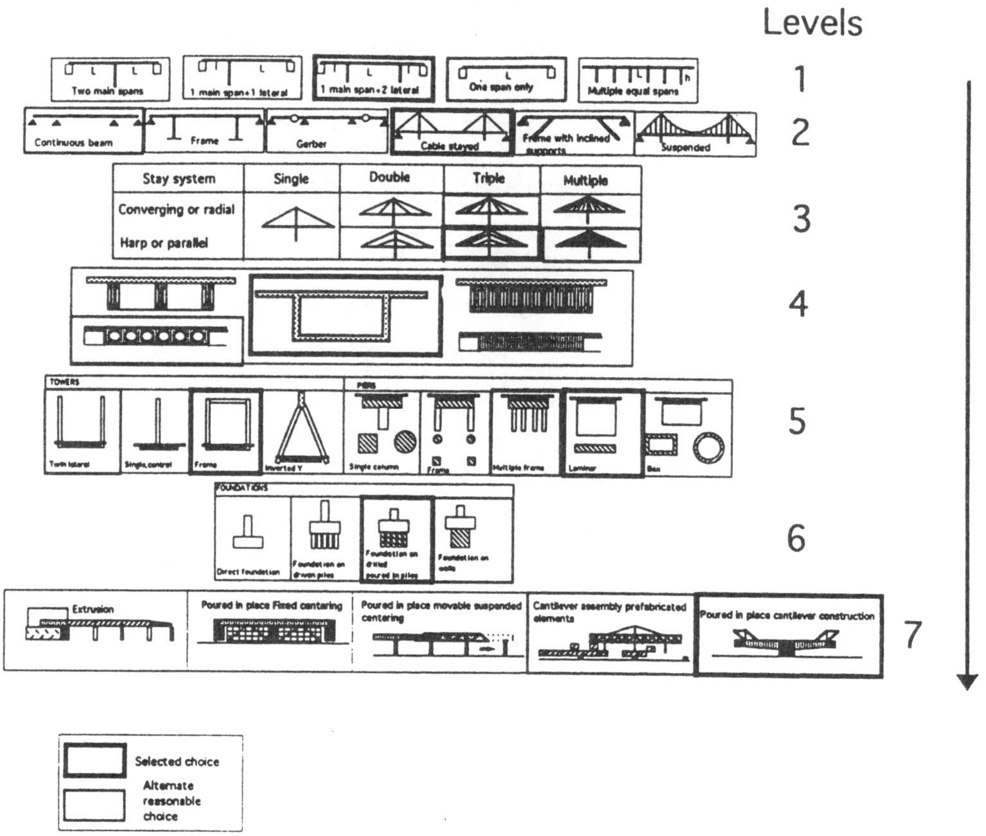

Fig.8-Example of possible choices along chosen path(bridges)

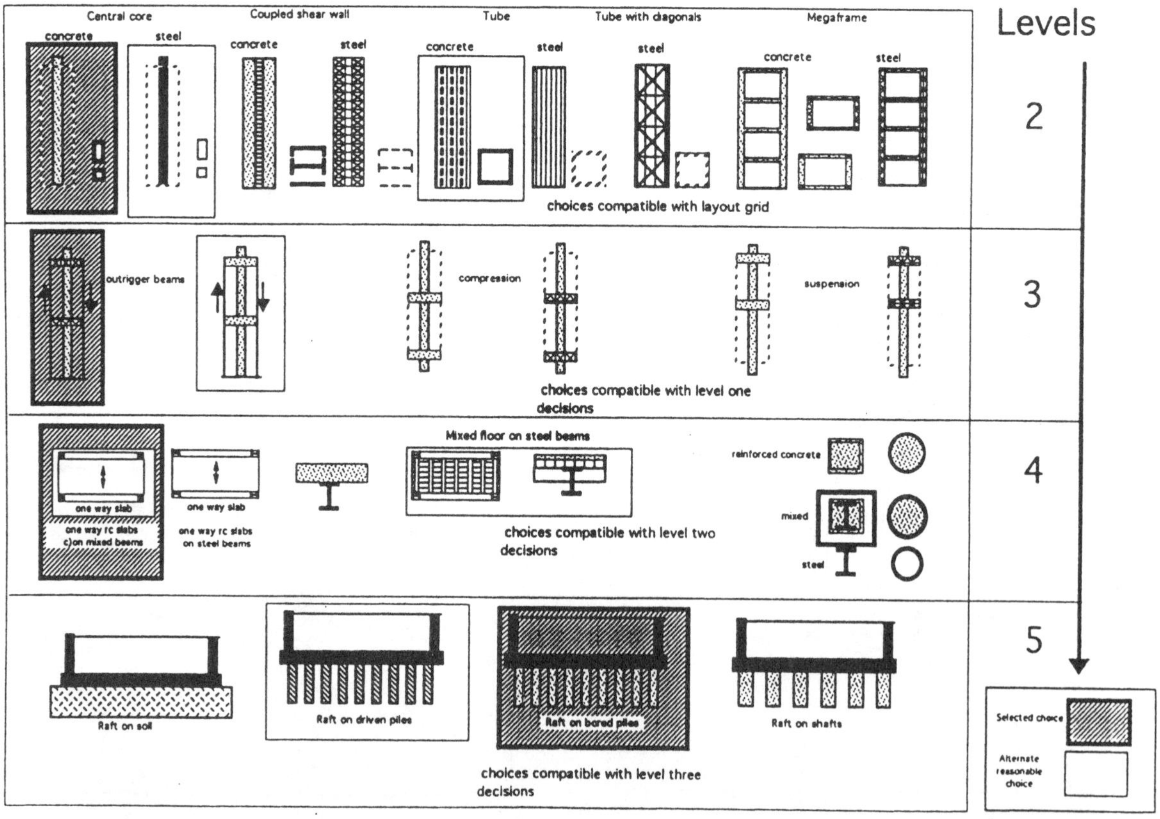

Fig.9-Example of possible choices along chosen path(tall buildings)

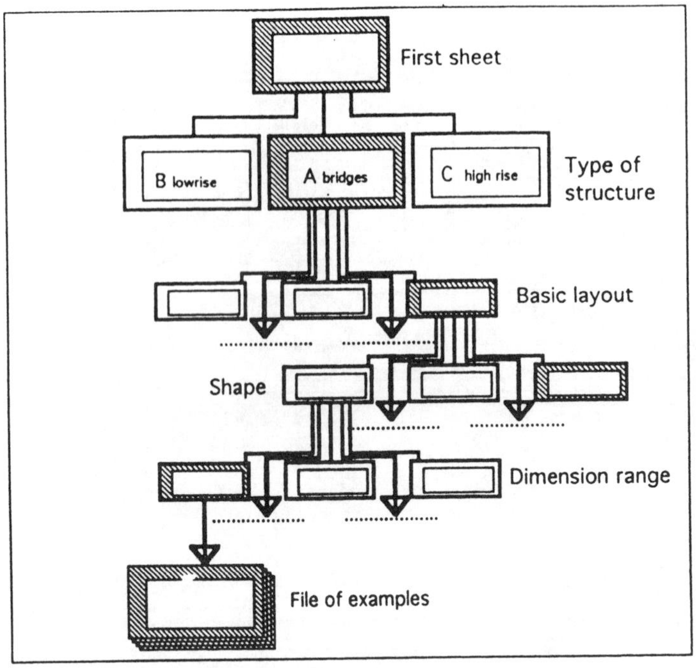

Fig.11-structure of ypermedia data base for bridges
(Class A)

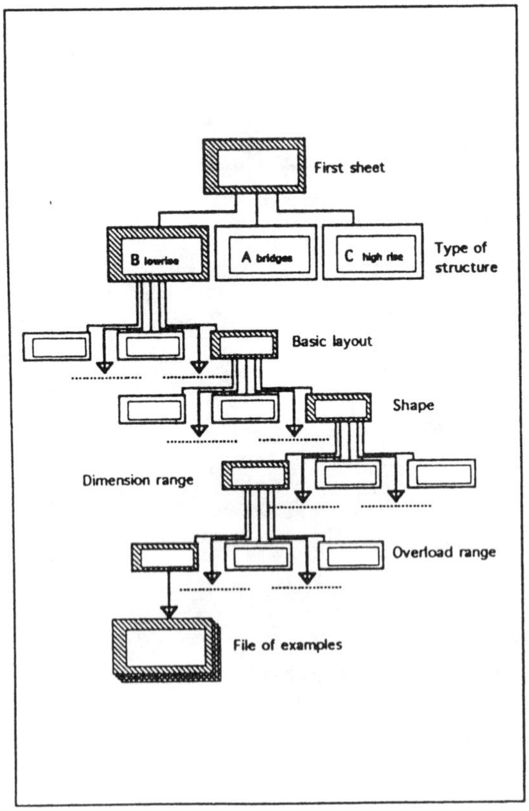

Fig.12-structure of ypermedia data base for low-rise buildings
(class B)

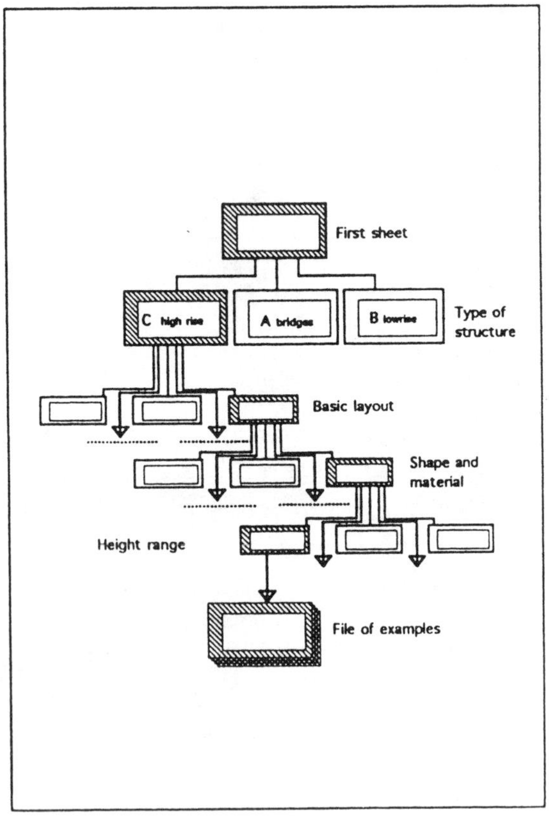

Fig.13-structure of ypermedia data base for tall buildings
(class C)

A METHOD OF SELECTING SIMILAR LEARNING DATA BASED ON CORRELATION COEFFICIENTS IN THE PREDICTION OF TIME SERIES USING NEURAL NETWORKS

Hisashi Shimodaira
Research and Development Department, Nihon MECCS Co., Ltd.
4-13-27 Midori-Cho, Koganei-Shi, Tokyo, Japan, 184

ABSTRACT

This paper explores a method of improving the predictive performance for the multi-layer feedforward neural network in time series predicting. In the similar data selective learning method (SDSL method), learning samples which are similar to the data used for a prediction, are selected using weighted distances between them. For this method, we propose a method of weighting the distance by a function of correlation coefficients for the time series (CSDS method). The results of numerical experiments show that for the case of a time series of which nature was suitable for the ability of the SDSL method, weighting the distance by power functions of correlation coefficients in a proper degree is considerably effective to improve the predictive performance. From these results, it seems that the SDSL method with the CSDS method can be a promising method alternative to the previously proposed other methods.

INTRODUCTION

The neural network is promising as one of the tools to predict the future values of a time series using its past values. Neural networks, such as multi-layer feedforward types and their variants, and recurrent types, are applicable to this problem, whereas the predictive performance and the learning time depends on the type used. For instance, while the predictive accuracy for recurrent types was higher than that for feedforward types, recurrent types took much more learning time than feedforward types [LH91]. Considering the application to the industry in practice, it is desirable to explore a method of obtaining accurate predictions using the feedforward type, because in its nature the structure is simple and the learning time is considerably small. The objective of this paper is to explore such a method.

The predictive performance for feedforward neural networks depends on how to select learning samples from the historical database of a time series. In the method which we refer to as the moving window data learning method (MWDL method) [PHK90], all the data in a moving window which is allocated in the immediate previous range of a prediction point, are used as learning samples. In the method which we refer to as the similar data selective learning method (SDSL method) [PHK92], data which are similar to the data used to make a prediction, are selected from the historical database as learning samples. In the method which we refer to as the whole data learning method (WDL method) [WM90], all the past data which are previous to the starting point of the prediction, are used to train the neural network.

In the SDSL method, the similarity between the data group used to make a prediction and a data group which is a candidate for learning samples, is measured with the distance between them. In this paper, we propose a method of weighting the distance by the function of correlation coefficients for the time series, which we refer to as the correlation coefficient based similar data selection method (CSDS method). In the proposed method, distances in which we attach importance to past data which have more influence on the value to be predicted, are calculated. We present the results of numerical experiments which were performed to investigate the predictive performance for the SDSL method with the CSDS method, the MWDL method, and the WDL method.

OUTLINE OF PREDICTION PROCEDURES

To facilitate the discussion, an outline of the general prediction procedure for a single time series, is described.

In a time series $x_i (i=1, \cdots, t, \cdots, \infty)$, when the values of points previous to t were observed and x_t is to be predicted, the group of d data which are immediate previous to x_t, is used to predict x_t:

$$(x_{t-d}, \ldots, x_{t-2}, x_{t-1}) \longrightarrow X_t \tag{1}$$

where X_t is the predicted value of x_t. The d data are applied to the input nodes and X_t is produced in the output node as the predicted value. To make such predictions, the network needs to be trained by learning samples. At the training stage, using the group of d data which are immediate previous to $x_{t'}$ as a learning sample, $X_{t'}$ is produced by the network:

$$(x_{t'-d}, \ldots, x_{t'-2}, x_{t'-1}) \longrightarrow X_{t'} \tag{2}$$

The weight values of the network are estimated by iteration calculations using the error backpropagation method [RM86] so that the sum of the errors between $x_{t'}$ and $X_{t'}$ concerning n learning samples is minimized. At the prediction stage, using these estimated weight values, X_t is produced.

In the chaos theory, d is referred to as the embedding dimension. For a chaotic time series, it is theoretically proved that the original characteristics of the chaos can be reconstructed from a single time series by using a proper embedding dimension [FS87].

THE SDSL METHOD

In this method, learning samples are selected from the range for selecting learning samples in the historical database in the order that the similarity between the data used to make a prediction and the candidate for learning samples, is greater. In order to quantify the degree of the similarity, both the data groups are regarded as two points in the d-dimensional Euclidean space and the distance between them is calculated. It is assumed that the similarity is great, as the distance is smaller; and n learning samples are selected in the increasing order of the distance.

In the CSDS method, we propose to use the following measurements of the distance. The one is the weighted Manhattan distance defined by

$$D = \sum_{i=1}^{d} |\rho_i|^m |x_{t-i} - x_{t'-i}| \tag{3}$$

The other is the weighted Euclidean distance defined by

$$D = \sqrt{\sum_{i=1}^{d} |\rho_i|^m (x_{t-i} - x_{t'-i})^2} \tag{4}$$

where ρ_i ($0 \leqq \rho_i \leqq 1$) is the autocorrelation coefficient between x_t and x_{t-i}. In the above equations, the mth power of ρ_i is used to adjust the influence of its magnitude on the distance calculation. The optimum value of m is a part of the investigation.

Eqns (3) and (4) are basically the weighted distances traditionally used to identify clusters of data in a multidimensional space. We propose to employ the function of autocorrelation coefficient as the weight in these distances. The autocorrelation coefficient measures the degree of the dependence between x_t and x_{t-i}; and the great influence of x_{t-i} is exerted on x_t, as the values is greater. Therefore, the coordinate is scaled by the great factor, as the autocorrelation coefficient is greater. With these equations, distances in which we attach importance to past data which have more influence on the value to be predicted, are calculated.

Let us define the first point of the range for selecting learning samples as x_s. Strictly, ρ_i needs to be calculated by the data sets (x_s, x_{s+i}), (x_{s+1}, x_{s+i+1}), $\cdots$, (x_{t-i-1}, x_{t-1}), (x_{t-i}, x_t). However, because x_t is unknown, the approximate value of ρ_i is calculated by the data sets except (x_{t-i}, x_t).

In eqns (3) and (4), it is assumed implicitly that the correlation coefficients between x_t and x_{t-i}, and $x_{t'}$ and $x_{t'-i}$ are the same. If this assumption holds, we can fix x_s and enlarge the range for selecting learning samples, as observed values are added to the database, to utilize the past data effectively. If the correlation coefficient for t' which is apart from t, differs from that for t, we need to determine the size of the range for selecting learning samples so that the assumption holds and move it (sift x_s) forward according to a point to be predicted.

While in the Manhattan distance, the differences of coordinates themselves are considered, in the Euclidean distance, the squares of them are considered. Thus, the latter is more affected by the larger values of differences of coordinates than the former. The effect of the definition of distances used on the predictive performance, is investigated by the numerical experiments.

The method above mentioned can be applicable to the case where explanatory variables as well as response variables are considered as the input. In this case, each variable needs to be standardized by, for instance, the following equation to eliminate the influence of the unit in the distance calculation.

$$x_{is} = (x_{ir} - \bar{m})/\sigma \tag{5}$$

where x_{ir} is an observed value, x_{is} is the standardized value, $\bar{m}$ is the mean value, and σ is the standard deviation. The terms associated with the explanatory variable, are added to eqns (3) and (4), where ρ_i is a correlation coefficient between the response and the explanatory variable.

THE MWDL AND THE WDL METHOD

In the MWDL method, all the data in the moving

window which is allocated in the immediate previous range of a prediction point, are grouped to form the learning samples according to eqn (2). In this paper, we define the size of the moving window by the number of the learning samples (n).

In the WDL method, the weights of the network are calculated using the learning samples which are formed by grouping all the data previous to the starting point for the prediction according to eqn (2). The predictions are made with these weights using the observed values as data for the prediction.

OVERVIEW OF NUMERICAL EXPERIMENTS

In the following sections, the results of the numerical experiments which were performed to investigate the predictive performance for the SDSL, the MWDL, and the WDL method using three kinds of time series, are presented.

The neural network used consists of one input layer with d nodes, one hidden layer with h nodes, and one output layer with one node. The predictive performance depends on the range from which learning samples are selected, the number of data d, and the number of learning samples n. An optimum value of the node number (h) of the hidden layer depends on d and n as well as the characteristics of the time series. d, n, and h are not fixed; but determined as a part of the investigation in the numerical experiments.

As the activation function for the hidden and the output layers, the logistic function ($g(x)=1/(1+e^{-x})$) is used. The input data are scaled so that the values are within the range [0.2, 0.8]. For the SDSL and the MWDL method, the magnitude of the input data may vary according to each prediction point. This scaling allows the magnitude to be in the same range. We used 0.1 as the learning rate for the backpropagation method and 0.9 as the coefficient for the inertia term. The network was trained until the half of the sum of the errors in the output node becomes less than 0.001.

For the SDSL method, the first point of the range for selecting learning samples was allocated to the first point of the time series data (namely s=1), and from the point next to the half of the time series data, the learning and the prediction were performed in turn. The effect of m in eqns (3) and (4) was investigated by taking 0, 1, 2, 3, or 4 as its value. The optimum combination of the values of the parameters (d, n, and h) was sought by performing simulations in which these values were changed little by little. For the MWDL method, from the point next to the half of the time series data, the learning and the prediction were performed in turn. The optimum combination of the values of the parameters (d, n, and h) was also sought. For the WDL method, the first half of the time series data was used to train the network and the predictions were performed using the last half. The optimum combination of the values of the parameters (d and h) was also sought.

To measure the predictive performance, we use correlation coefficients between the actual and the predicted values (CRC), mean values ($mean$) and standard deviations (σ) of the absolute errors (ABE), and mean values and standard deviations of the relative errors (RLE). The correlation coefficient between the actual and the predicted values shows whether the trends of variations of both values are similar to each other. The absolute error is defined by

$$ABE = |X_i - x_i| \times 100/x_{max} \qquad (6)$$

where x_{max} is the maximum value of the actual values in the range where the predictions were performed. With the ABE, we can see the degree of the magnitude of an absolute error compared with x_{max}. The relative error is defined by

$$RLE = |X_i - x_i| \times 100/x_i \qquad (7)$$

It should be noted that the RLE can become considerably large, because the numerator and the denominator can be in the same degree in magnitude in the case where the values to be predicted is very small.

To investigate the nature of the time series in advance, the autocorrelation coefficients were calculated using all the data. The autocorrelation coefficient tends to reflect the essential smoothness of the time series [S88]. Very smooth series exhibit autocorrelation coefficients that stay large even when the time lag (τ) is large, whereas very choppy series tend to have autocorrelation coefficients that are nearly zero for large τ's.

PREDICTIONS ABOUT A TIME SERIES OF MEASLES CASES

We performed numerical simulations on the time series of monthly measles cases in New York city between 1928 and 1963 (432 points) [YL73]. These data are used as a standard for testing new techniques [SM90, WM90]. Figure 1 shows the plot of the case number versus time (month). Figure 2 shows the correlogram which is the plot of the autocorrelation coefficient versus the time lag (τ). We can see from Figure 2 that the time series has a rather choppy nature, because the autocorrelation coefficient shows periodic peaks, whereas the peak values are not so large. It is pointed out that this time series is chaotic and the long-term prediction is impossible [SM90, WM90].

For the SDSL method, because according to preliminary tests, the predictive performance for the case where x_s was shifted was better than that for the case where x_s was fixed, we used the former thereafter. Table 1 shows the optimum parameter

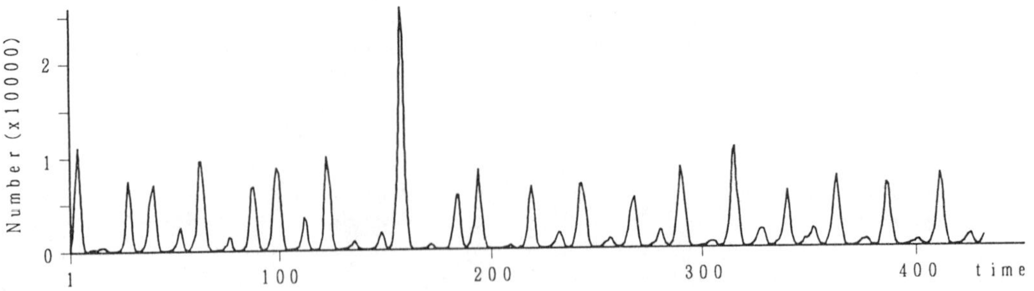

Figure 1 Time Series for Numbers of Measles Cases

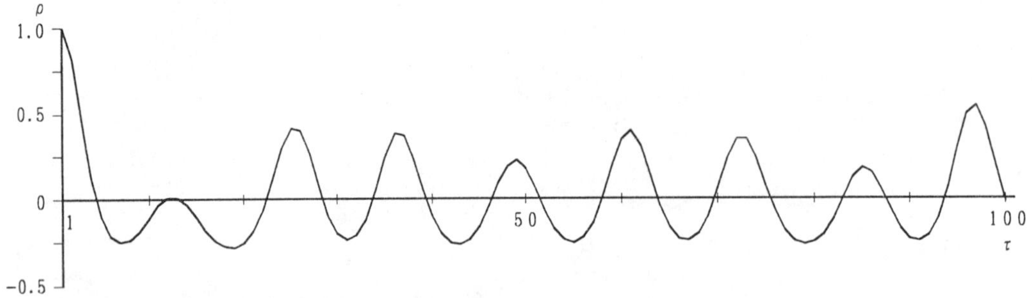

Figure 2 Correlogram for Time Series of Measles Cases

Table 1 Optimum Parameter Values for Time Series of Measles Cases

Method	Network	*n*
SDSL	8-4-1	16
MWDL	30-3-1	50
WDL	8-2-1	(209)

Table 2 Prediction Results for Numbers of Measles Cases with SDSL Method

DST	*COR*	*m*	*CRC*	*ABE* mean	*ABE* σ	*RLE* mean	*RLE* σ
1	0	—	0.953	3.31	5.36	73.5	162.
	1	1	0.955	3.19	5.20	54.1	70.3
		2	0.961	3.03	4.72	44.7	38.0
		3	0.959	3.12	4.73	51.8	69.7
		4	0.952	3.62	4.98	79.6	210.
2	0	—	0.955	3.29	5.11	76.1	156.
	1	1	0.952	3.34	5.34	60.6	83.6
		2	0.956	3.37	4.97	65.8	121.
		3	0.960	3.10	4.78	49.2	58.7
		4	0.957	3.25	4.91	58.8	105.

Table 3 Learning Data Selection Methods vs. Prediction Results for Numbers of Measles Cases

Method	*CRC*	*ABE* mean	*ABE* σ	*RLE* mean	*RLE* σ
SDSL	0.961	3.03	4.72	44.7	38.0
MWDL	0.933	4.23	5.77	97.2	215.
WDL	0.925	5.08	5.60	123.	144.

values by which the best prediction performance was obtained, where, for example, 8-4-1 denotes the network with 8 input nodes, 4 hidden nodes, and one output node. Table 2 shows the results for the SDSL method in which these parameter values were used. In the table, *DST*=1 denotes the Manhattan distance; and *DST*=2, the Euclidean distance. *COR*=0 denotes that the weighting is not used in eqns (3) and (4), and *COR*=1 denotes that the weighting is used.

These results show that the predictive performance for the SDSL method was considerably improved by weighting the distance using autocorrelation coefficients in a proper degree. The optimum value of *m* for the Manhattan distance is 2, and that for the Euclidean distance is 3. Although the predictive performance does not differ so much by the kind of the distance used, that is the best in the case of the Manhattan distance with m=2. Compared with the case where the weighting was not used, this case offers a 0.8% increase in the correlation coefficient between the actual and the predicted values; an 8.5% and an 11.9% reduction in the mean value and the standard deviation of the absolute errors, respectively; and a 39.2% and a 76.5% reduction in the mean value and the standard deviation of the relative errors, respectively.

Table 3 shows the results for the three methods in which the optimum values of the parameters were used (for the SDSL method, DST=1, COR=1, and m=2). The best performance is obtained by the SDSL method, and followed by the MWDL method and the WDL method. Compared with the MWDL method, the SDSL method offers a 3% increase in the correlation coefficient between the actual and the predicted values; a 28.4% and an 18.2% reduction in the mean value and the standard deviation of the absolute errors, respectively; and a 54.0% and an 82.3% reduction in the mean value and the standard deviation of the relative errors, respectively.

PREDICTIONS ABOUT AN ARTIFICIALLY GENERATED TIME SERIES

We performed numerical simulations on the time series (500 points) which was generated by the following equations [WM90].

$$x_{t+1} = 3.8x_t(1 - x_t) \quad \text{for even t} \tag{8}$$

$$x_{t+1} = (1 - x_t) \quad \text{for odd t} \tag{9}$$

Eqn (8) represents the logistic map which is one of the deterministic chaoses. Eqn (9) represents a periodic function which generates identical values at every other points. 0.1 was taken as the initial value for even t; and 0.3, as that for odd t. Figure 3 is the plot of the value of the series versus t. Figure 4 shows the correlogram which is the plot of the autocorrelation coefficient, which was calculated using all the data, versus the time lag (τ). The autocorrelation coefficient shows periodic peaks, whereas the peak values are not so large. It is conjectured from the generation process that the nature of this series is chaotic and periodic.

For the SDSL method, x_s is fixed, because the magnitude of the autocorrelation coefficients does not change so much by the part where they are calculated. Table 4 shows the optimum parameter values by which the best predictive performance was obtained. Table 5 shows the results for the SDSL method in which these parameter values were used. These results also show that the predictive performance for the SDSL method was considerably improved by weighting the distance using autocorrelation coefficients in a proper degree. The optimum values of m are the same as those for the case of

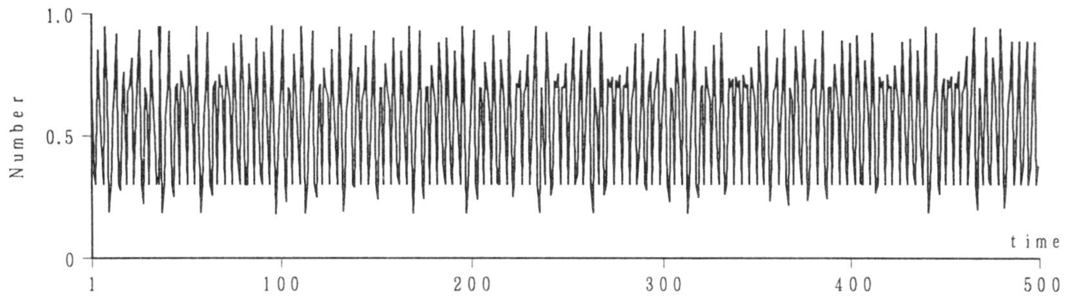

Figure 3 Chaotic and Periodic Time Series

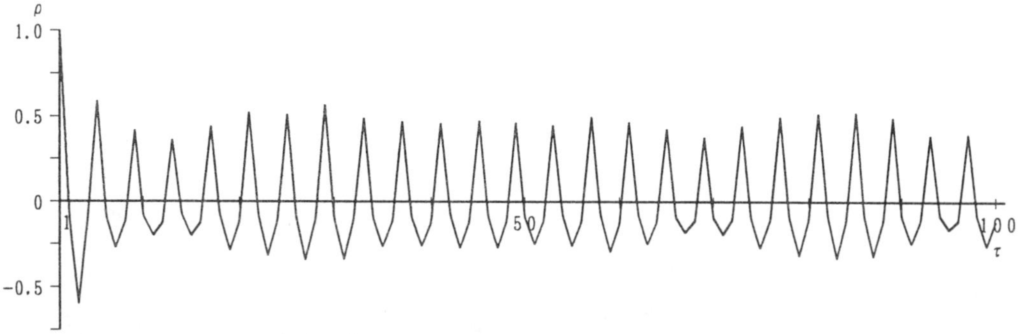

Figure 4 Correlogram for Chaotic and Periodic Time Series

Table 4 Optimum Parameter Values for Chaotic and Periodic Time Series

Method	Network	n
SDSL	5-2-1	2
MWDL	40-9-1	120
WDL	20-9-1	(231)

Table 6 Learning Data Selection Methods vs. Prediction Results for Chaotic and Periodic Time Series

Method	*CRC*	*ABE* *mean*	*ABE* *σ*	*RLE* *mean*	*RLE* *σ*
SDSL	0.997	1.30	1.17	2.50	2.75
MWDL	0.966	4.01	4.87	7.82	12.8
WDL	0.975	3.62	4.07	7.16	10.1

Table 5 Prediction Results for Chaotic and Periodic Time Series with SDSL Method

DST	*COR*	*m*	*CRC*	*ABE* *mean*	*ABE* *σ*	*RLE* *mean*	*RLE* *σ*
1	0	—	0.995	1.38	1.96	2.62	4.03
	1	1	0.995	1.51	1.89	2.84	3.96
		2	0.997	1.36	1.21	2.51	2.43
		3	0.995	1.72	1.68	3.04	2.91
		4	0.995	1.84	1.58	3.28	2.89
2	0	—	0.996	1.40	1.78	2.73	3.97
	1	1	0.995	1.47	1.94	2.82	4.12
		2	0.996	1.43	1.59	2.80	3.56
		3	0.997	1.30	1.17	2.50	2.75
		4	0.997	1.51	1.35	2.80	2.80

the measles cases. The best predictive performance is obtained in the case of the Euclidean distance with m=3. Compared with the case where the weighting was not used, this case offers a slight increase in the correlation coefficient between the actual and the predicted values; a 7.1% and a 34.3% reduction in the mean value and the standard deviation of the absolute errors, respectively; and an 8.4% and a 30.7% reduction in the mean value and the standard deviation of the relative errors, respectively.

Table 6 shows the results for the three methods in which the optimum values of the parameters were used (for the SDSL method, the case of *DST*=2, *COR*=1, and m=3). The best performance is obtained by the SDSL method, and followed by the WDL method and the MWDL method. Compared with the MWDL method, the SDSL method offers a 3.2% increase in the correlation coefficient between the actual and the predicted values; a 67.6% and a 76.0% reduction in the mean value and the standard deviation of the absolute errors, respectively; and a 68.0% and a 78.5% reduction in the mean value and the standard deviation of the relative errors.

PREDICTIONS ABOUT A TIME SERIES OF CHICKENPOX CASES

We performed numerical simulations on the time series of monthly chickenpox cases in New York city between 1928 and 1972 (534 points) [YL73]. Figure 5 shows the plot of the case number versus time (month). Figure 6 shows the correlogram which is the plot of the autocorrelation coefficient, which was calculated using all the data, versus the time lag (τ). The autocorrelation coefficient shows periodic peaks with a considerably large value, and the peak values are almost the same throughout all τ's. Thus, it is conjectured that this time series is smooth and has a strong periodic nature.

For the SDSL method, because according to preliminary tests, the predictive performance for the case where x_s was shifted was better than that for the case where x_s was fixed, we used the former thereafter. Table 7 shows the optimum parameter values by which the best predictive performance was obtained. Table 8 shows the results for the SDSL method in which these parameter values were used. In this case, the predictive performance was not improved by weighting the distance using autocorrelation coefficients, but rather degraded slightly. The best predictive performance is obtained in the case of the Euclidean distance without using weighting. Table 9 shows the results for the three methods in which the optimum values of the parameters were used (for the SDSL method, the case of *DST*=2, and *COR*=0). The predictive performance for the MWDL and the WDL method is almost the same, and slightly superior to that for the SDSL method.

SUMMARIES OF THE RESULTS OF THE NUMERICAL EXPERIMENTS

The results of the numerical experiments shows that the predictive performance for the three methods depends on the nature of the objective time

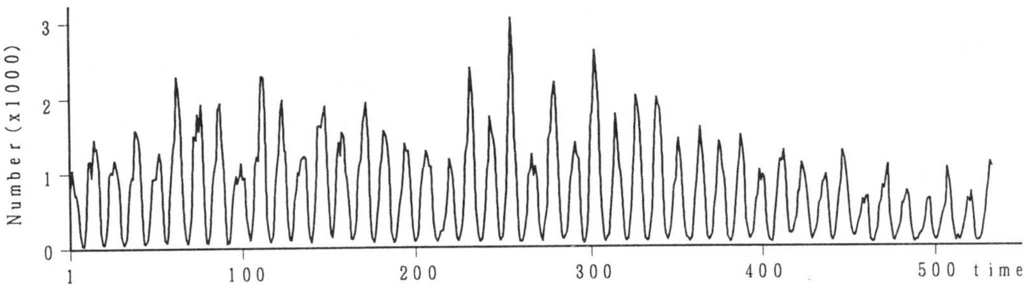

Figure 5 Time Series for Numbers of Chickenpox Cases

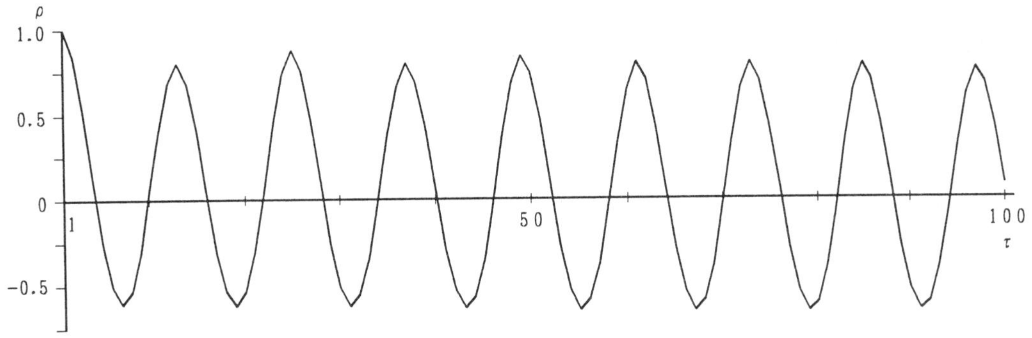

Figure 6 Correlogram for Time Series of Chickenpox Cases

Table 7 Optimum Parameter Values for Time Series of Chickenpox Cases

Method	Network	*n*
SDSL	32-3-1	35
MWDL	30-3-1	150
WDL	31-4-1	(237)

Table 8 Prediction Results for Numbers of Chickenpox Cases with SDSL Method

DST	*COR*	*m*	*CRC*	*ABE* *mean*	*ABE* *σ*	*RLE* *mean*	*RLE* *σ*
1	0	—	0.956	4.62	3.71	43.4	51.2
	1	1	0.959	4.59	3.48	45.1	58.7
		2	0.958	4.72	3.60	44.9	50.8
		3	0.958	4.69	3.58	44.7	49.6
		4	0.958	4.80	3.58	45.1	50.3
2	0	—	0.959	4.49	3.71	41.8	52.0
	1	1	0.959	4.63	3.55	46.4	59.0
		2	0.957	4.75	3.65	48.5	60.6
		3	0.956	4.71	3.73	47.5	62.9
		4	0.953	4.94	3.74	51.8	72.1

Table 9 Learning Data Selection Methods vs. Prediction Results for Numbers of Chickenpox Cases

Method	*CRC*	*ABE* *mean*	*ABE* *σ*	*RLE* *meam*	*RLE* *σ*
SDSL	0.959	4.49	3.71	41.8	52.0
MWDL	0.966	4.03	3.51	31.8	37.0
WDL	0.964	4.19	3.72	27.4	29.1

series. For the time series of measles cases of which nature is chaotic and rather choppy, the performance for the SDSL method with the CSDS method was considerably better than that for the other methods. For the artificially generated time series of which nature is chaotic and periodic, the performance for the SDSL method with the CSDS method was remarkably better than that for the other methods. In addition, it should be noted that compared with the other methods, the SDSL method with the CSDS method needed a very small network size and very small learning samples to obtain the best performance. It should be also noted that it showed very good performance for the case of the time series which was generated by the two mechanisms. For the time series of chickenpox cases which has a smooth and strong periodic nature, the performance for the SDSL method was slightly inferior to that for the other methods.

For the case of the time series of which nature are choppy and chaotic, that predictive performance for the SDSL method is better than that for the MWDL method, seems to mean that to learn the nature and structure of the series, data near the prediction point are not necessarily sufficient; and similar data which are selected from the past broad range, are required.

For the case of the time series of which nature was suitable for the ability of the SDSL method, weighting the distance by correlation coefficients in a proper degree using the CSDS method was considerably effective to improve the predictive performance. The predictive performance does not differ so much by the kind of the distance used. The proper kind of distance to be used seems to depend on the nature of the time series. The optimum degree in which autocorrelation coefficients are used, is 2 for the Manhattan distance; and 3 for the Euclidean distance.

RELATED WORKS AND FEATURES OF THE PROPOSED METHOD

FS87 proposed a method of predicting chaotic time series. In the method, after embedding the time series, nearest neighbors of the data used for a prediction are selected using a metric of distances. However, the details of the metric of distances were not described. To make a prediction, a linear polynomial fitted to the neighbors, was used.

PHK92 proposed a method for short-term load forecasting in which learning samples were selected using weighted distances. In the distance calculation, derivatives of the output variable with respect to the input variables were used as the weights. For the ordinary feedforward neural networks, the predictive performance for the case where the weighting was used, was rather inferior to that for the case where the weighting was not used. For the network in which the input is linked to the output by both nonlinear and linear terms, the former was better than the latter; and the former offers a 5.4% and a 17.1% reduction in the mean value and the standard deviation of the relative errors, respectively, over the latter. However, the derivative of the output variable with respect to the input variable is a value which relates the variations of the variables, and does not necessarily represent the dependence of the output variable on the input variable.

The originality of the CSDS method proposed in this paper consists in that the correlation coefficient which has the theoretically obvious foundation as a value representing the dependence between two points in a time series, is employed as the weight for the distance calculation. It is devised so that using the correlation coefficient in its *m*th power function allows its magnitude to adjust the degree of weighting. Moreover, this paper offers new information on the predictive performance for the SDSL, the MWDL, and the WDL method; and their advantages and disadvantages. As far as we know, concrete comparisons among the performance for these methods were not performed in the previous works.

CONCLUSIONS

The results of the numerical experiments show that the predictive performance for the three methods depends on the nature of the objective time series. Although we can not draw a complete conclusion from only these results, the SDSL method seems to be suitable for the prediction about a time series of which nature is choppy and/or chaotic. For the case of a time series of which nature is suitable for the ability of the SDSL method, weighting the distance by correlation coefficients in a proper degree using the CSDS method is considerably effective to improve the predictive performance. From these results, it seems that the SDSL method with the CSDS method can be a promising method alternative to the MWDL and the WDL method. We hope to apply the proposed method to other time series in which explanatory variables are available and to verify its effectiveness.

REFERENCES

[FS87] J.D.Farmer and J.J.Sidorowich. Predicting chaotic time series. *Physical Review Letters*, 59(8):845-848, 1987.

[KR90] L.Kaufman and P.J.Rousseeuw. *Finding groups in data* (p.13). John Wiley & Sons, Inc., New York, 1990.

[LH91] J.Lambert and R.Hecht-Nielsen. Application of feedforwad and recurrent neural networks to chemical plant predictive modeling. In *Proceedings of International Joint Conference on Neural Networks*, I:373-378, 1991.

[PHK90] T.M.Peng, N.F.Hubele, and G.G.Karady. Conceptual approach to the application of neural network for short-term load forecasting. In *Proceedings of the 1990 IEEE International Symposium on Circuit and Systems*, New Orleans, Louisiana, May 1-3, 1990, 2942-2945.

[PHK92] T.M.Peng, N.F.Hubele, and G.G.Karady. Advancement in the application of neural networks for short-term load forecasting. *IEEE Transactions on Power Systems*, 7(1):250

-257, 1992.

[RM86] D. E. Rumelhart and J. L. McClelland. *Parallel distributed processing*, 1 (Chapter 8), MIT Press, Cambridge, MA, 1986.

[SM90] G. Sugihara and R. M. May. Nonlinear forecasting as a way of distinguishing chaos from measurement error in time series, *Nature*, 344·19:734-741, 1990.

[S88] R. H. Shumway. *Applied statistical time series analysis* (p. 17). Prentice Hall, Englewood Cliffs, NJ, 1988.

[YL73] J. A. Yorke and W. P. London. Recurrent outbreaks of measles, chickenpox and mumps. *American Journal of Epidemiology*, 98(6):469-482, 1973.

[WM90] D. M. Wolpert and R. C. Miall. Detecting chaos with neural networks. In *Proceedings of*

DEALING WITH UNCERTAINTY IN FINE MOTION: A NEURAL APPROACH

Enrique Cervera, Angel P. del Pobil,

Computer Science Department
Jaume I University
Campus Penyeta Roja
E-12071 Castelló, Spain
{ecervera, pobil}@inf.uji.es

Edward Marta, Miguel A. Serna

Robotics & AI Laboratory
C.E.I.T & University of Navarra
Manuel de Lardizábal, 13-15
E-20009 San Sebastián, Spain
{emarta, maserna}@ceit.es

ABSTRACT

A new approach based on the use of neural networks to deal with uncertainty in fine motion is presented. The six force and torque signals from the wrist sensor of a robot arm are fed into the network. It is a self-organizing map which learns without need of a teacher the different contact states. The method is intended to work properly in complex real-world situations involving fine motion and grasping, for which a geometric analytical model may not be feasible, or too difficult. It is used for assembly and assembly-like tasks, particularly complex insertion operations in a flexible manufacturing system. Real examples are presented.

1. INTRODUCTION

The research field of robotic assembly and task planning plays an important role in the automation and flexibility of manufacturing systems [Kak 90]. Starting from high-level descriptions of a task, the system should be capable of generating a plan to perform the task. In this plan the low-level operations relative to motion and perception must be clearly specified.

A key problem in task planning is the unavoidable presence of uncertainty. For a given design, once the assembly parts and operations are selected, the resources needed for each task have been planned, and the order of the operations are established, a sequence of subtasks has to be determined for each operation. Each particular subtask requires, in its turn, a lower level plan that may involve gross motion, fine motion, or grasping actions. For robots working under real-world conditions, all these motions are affected by uncertainty.

In contrast with gross motion planning [del Pobil & Serna 95], fine motion planning deals with small clearances and contact, for which the existence of uncertainty may render useless a synthesized plan. The use of sensors -mainly force and torque sensors- is necessary to get information about the actual situation of the process. A similar problem arises in motions involving grasps.

This paper addresses the problem of uncertainty in robot assembly and assembly-like tasks involving fine motion and grasping operations. An approach based on unsupervised neural networks and force/torque sensing is proposed. The technique is oriented toward applications in real-world manufacturing environments, for which a geometric analytical model may not be feasible. In these cases the relations between the six sensor signals and the contact states are very complex. The network will learn to distinguish the different contact states, without the need of a teacher, and will be used to monitor the execution of the plans. In addition, due to the short time required for the learning process and its simplicity, it allows for a great flexibility.

2. UNCERTAINTY IN FINE MOTION AND GRASPING

Robot tasks involving contact between parts and, particularly insertion and extraction, are very common in assembly and manufacturing in general. A robot arm must grasp a part, carry it to its destination and insert it adequately. This is a very error-sensitive task, since, because of uncertainty, the part may be badly inserted or extracted, or even damaged. Errors are typically caused by a deficient geometric model of the environment and uncertainty in the initial pose (sensor) and control. In order to be able to cope with this kind of errors, the development of sensors and their integration into robots is of fundamental importance in the application of robotics to assembly [Sanderson & Perry 83]. Force and torque sensors must be used to monitor the contact state between the arm and its environment, and try to determine if the assembly task is proceeding correctly.

Much work has been recently devoted to fine motion planning with uncertainty. Model uncertainty is been dealt with in [Gottschlich & Kak 91], [Donald 90], [Smith & Cheeseman 86]. Lozano-Pérez, Mason and Taylor [84] first proposed a representation for initial pose (sensor) and control uncertainty and the concept of *preimage backchaining*; compliant motion strategies have been used in [Dakin & Popplestone 91] and [Peshkin 90]; Donald [88] proposes a geometric approach to error detection and recovery with uncertainty; another approach is the use of *backprojections* [Erdmann 86]; Donald [89] discusses error detection and recovery at length. Geometric approaches may become complex for non-trivial cases specially in three dimensions [Canny & Reif 87]. Natarajan [86] considers the complexity of generating strongly guaranteed geometric motion strategies.

The application of our method will take place in the context of an actual flexible manufacturing system. In this system, there is a machining center which works with several types of tools. These tools are fed into the unit by a robot arm (ABB IRB 2000), which also picks the tools from a robot vehicle (Fig. 1). The tools are carried from this vehicle to the machining center and vice-versa (Fig. 2). The robot arm is very dependant on the spatial positions of the tools. A small displacement of the vehicle can led to a defficient grasp or ungrasp operation of the tool, which could even be dropped by the arm.

Figure 1. The robot arm picks or places a tool in the vehicle

Obviously, the greater source of error is the control of the vehicle. An additional amount of uncertainty is acrued in every displacement. Although the positioning uncertainty is being improved with the use of ultrasonic sensing, errors of a few milimeters may be expected after several trips. The position uncertainty of the arm can also lead to an incorrect insertion or extraction of the tool in the machining center.

Figure 2. The robot arm is about to extract a tool from the machining center.

On top of the vehicle a loading/unloading unit for pallets is placed. The tool pallet slides into this unit and can be carried by the vehicle. The pallet can hold six tools, for each one there is a clamping device with two flexible claws. The two parallel jaws of the robot arm move a special gripper with the shape of the tools.

A wrist force/torque sensor is attached to the robot arm. The force and torque signals should help us detect the uncertainty which leads to incorrect insertions or extractions. Moreover, they should be detected while it is still possible to recover from an error. Our system should learn from examples of the task, and be able to distinguish the contact states between the arm and its environment in order to detect good and bad insertions at an early enough stage of the process.

It is worth noting that the complexity of these actual operations and similar manufacturing tasks is considerably greater than the usual test-cases in geometric approaches. First, it is a 3D problem with a non-trivial geometry due to the shape of the tool and the insertion place. Second, due to the existence of two flexible claws in the clamping device, the arm must exert a force that makes them yield for the tool to be properly inserted or extracted.

3. KOHONEN'S SELF-ORGANIZING MAPS

In our problem, we have complex relations between force and torque magnitudes which correspond to the different contact states. A formal analysis of these relations using other techniques will not generally be possible in real-world manufacturing situations. Instead, in this approach these relations will be learned by a neural network.

The unsupervised learning scheme has been selected for this particular problem because in a real application there is no a-priori knowledge of all the possible contact states of the system. The network must discover for itself patterns, features, regularities, correlations or categories in the input data and code for them in the output [Hertz, Krogh & Palmer 91]. Among all the unsupervised schemes, a network based on Kohonen's algorithm [Kohonen 90] has been chosen, because it shows interesting topological properties and it is usually very efficient. The neural networks which use this algorithm are called Self-Organizing Maps (SOM), and are one of the most widely used artificial neural networks based on unsupervised learning (see [Rappa & Depackere 92] for a survey).

SOMs are usually represented as a two dimensional neural network sheet, whose units -neural cells or neurons- are fully connected to the inputs and become tuned to different input signal patterns. Those units which are best tuned to a given signal pattern become active and a response is concentrated in the area with the most active units in the sheet. The important difference from other neural network models is the phenomenon occurring in the SOM that after learning, the responses are topologically arranged in the map. That is, two similar input patterns which are near each other in the signal space, are correspondingly located near each other in the map sheet. The ordering takes place automatically without external supervision based only on the internal relations in the structure of input signals themselves, and on the coordination of the unit activities through the lateral connections between the units. The responses to different input patterns are organized in an ordered fashion, similar input patterns producing similar responses.

4. THE SOM AND THE ORGANIZATION OF CONTACT STATES

A crucial issue is to visualize the complex relations among the variables in an adequate and clear way. The SOM defines a mapping between the multidimensional space of the input variables and the bidimensional space of the cells. This mapping generally preserves the topological relationships among input vectors.

Our approach is based on labelling the cells after the learning process. For that purpose, we introduce some known contact states into the network. A label is assigned to the cluster of cells that are activated for those particular states. The process is repeated for all available contact states. Sometimes, some cells are not properly activated for any state, this kind of cells are usually located near the borders between states.

Once the map has been labeled, an unkown sensor signal can be monitored. When it is introduced into the network it will activate some particular cells. The label of these cells will provide us with the contact state that corresponds to the given input signal.

In a general case, we have six input signals: the three spatial components of the force and torque vectors acting on the robot wrist. The network is a two-dimensional lattice of cells with rectangular shape. Every cell is input with the six signals. Different cells will learn different values of signals and, after training, will become more active when presented with those, or similar, signals. Thus the network will be organized in well defined regions of cells, or clusters, and we expect that these clusters will be related to the contact states of the system. If we identify the state associated to each cluster, we will be able to detect uncertainty in an assembly insertion/extraction task by observing the network response to the on-line measured force signals, i.e., which of the map clusters is more active, and thus, which type of contact state is occurring.

The reponse of neuron i for a particular input vector is given by:

$$o_i = \exp(-k\, |\mathbf{w}_i - \mathbf{i}_i|),$$

where:

o_i is the ouput, a real value that is represented in the figures with gray levels, with white for 1 (maximum activation), and black for 0 (minimum activation)

$\mathbf{w}_i$ is the weight vector

$\mathbf{i}_i$ is the input vector

k is a coupling parameter between weights and inputs. The greater k is, the closer the input must be to the weight vector to produce a response.

The same input vector is presented to all the neurons. The computation of the neuron responses results in characteristic gray-level patterns. In this patterns, the state associated with the given input can be readily identified as the brighter region on the net.

We can also visualize the region activated by a whole sequence of inputs, instead of just a single input. To do so, we just add the activation for each instant, obtaining the resulting pattern for the sequence. These regions can be reduced to individual cells if only the one giving the maximum response is considered. This neuron is usually known as the winner. Thus, we can visualize the sequence on the map as the trace of the winner cells for each input.

5. APPLICATION TO UNCERTAINTY IN ROBOT FINE MOTION

We want to apply Kohonen's unsupervised neural networks to the situation described in section 2. The neural network will be fed with the six signals of a real force/torque sensor attached to the wrist of the robot arm. We will limit the analysis to the fine motion involved in the tasks of inserting the tool in the pallet on the robot vehicle or the machining center. A similar treatment could be done for the task of extracting the tool.

5.1 Insertion in the machining center with positive uncertainty

In the first experiment, the tool is inserted in the tool-exchange device at the machining center shown in figure 2. The uncertainty has been simulated by introducing small position errors in the arm. The network was trained with samples of 100 sensor signals for each insertion operation. Four situations were considered: offsets of 0 mm (correct insertion), +1, +2 and +3 mm (incorrect insertions) on OY axis for the wrist. In figure 3 these signals are shown. The differences are clearly apparent from the sensor signals, but an interpretation of these signals is very difficult.

After training the network with data from a set of examples, we have to label the regions on the map. In this case, we assign a label to each different error. In figure 4 the activation patterns for complete insertion sequences are shown. Each pattern correspond to a different error. In each pattern, the brighter cells are the ones that were more activated in some moment of the process. On the figures, trajectories are also shown, which correspond to the neuron with maximum response at each step of the process.

Figure 4(a) shows the cell activation for the correct insertion with no error. The most active cells are located in the lower part of the map. As a greater error is introduced, we can see in the other three figures 4(b-d) how active cells are located in the upper zones on the map, in such a way that the greater the error, the upper the most active cells are. This continuity is a consequence of the topological properties of the SOM that preserves neighborhood relations: neighbor cells tend to respond jointly to similar inputs.

Uncertainty in fine motion planning for the insertion task can be dealt with by means of this network. The resulting map can be partitioned into four regions, according to the activation borders (Fig. 5). Each region is labeled with the process corresponding to no error, or +1, +2, or +3 mm errors. Now, we can consider a new unknown motion for which a positive uncertainty in OY axis for the actual goal position may be expected in the range [0 mm, +3 mm]. The network activation pattern will permit not only to detect an incorrect insertion, but also to identify approximately the magnitude of the error. Indeed, the further the activation pattern from the lower region D0 is and towards the upper region, the greater the positive error will be. A correct insertion will stay between the borders of region D0. Borders should not be seen as precise, a certain overlap exists between regions.

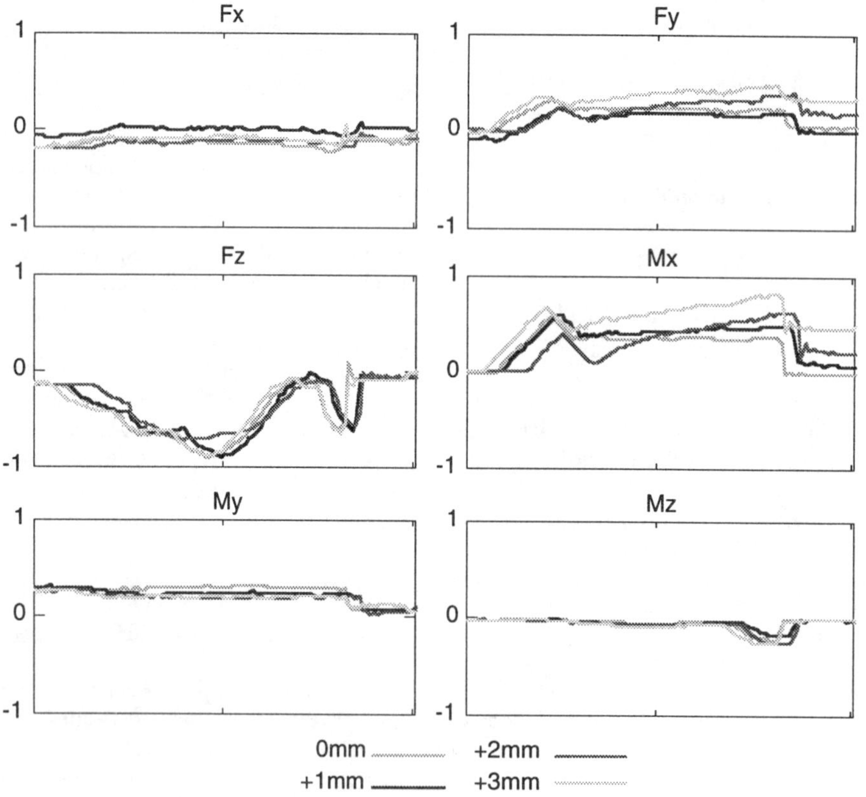

Figure 3. Six sensor signals for the insertion task at the machining center with different errors (offsets 0, 1, 2 and 3 mm).

The existence of two additional regions B and E0 can be accounted for as follows. B corresponds to a common starting zone, since the sensor signals for the insertion tasks have been taken a few instants before actual contact. Similarly, there is a final ending zone E0 because the ungrasp action is included in the monitored process. E0 could be extended to include the ending areas for cases D1, D2 and D3.

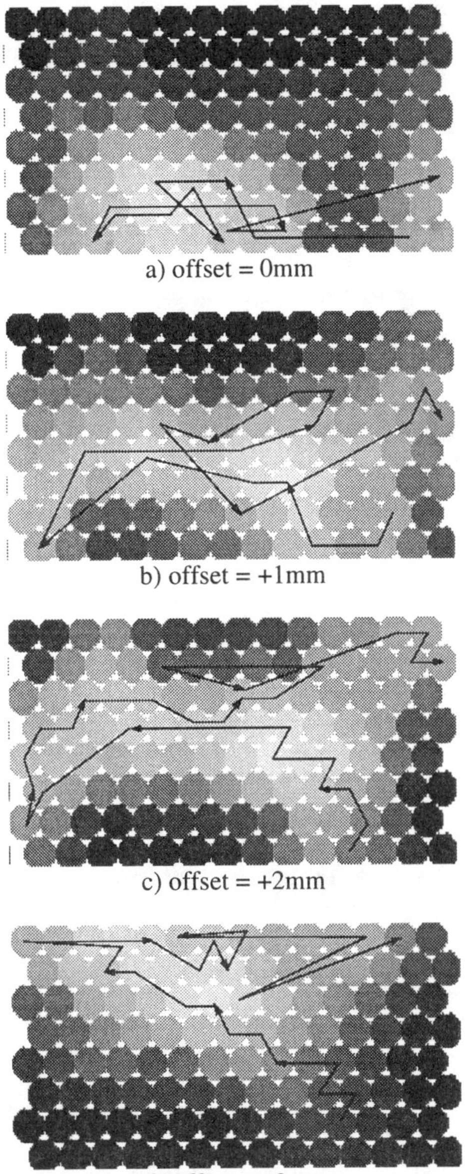

a) offset = 0mm

b) offset = +1mm

c) offset = +2mm

d) offset = +3mm

Figure 4. Activation patterns for the insertion task in the machining center. Traces of winner cells are also shown.

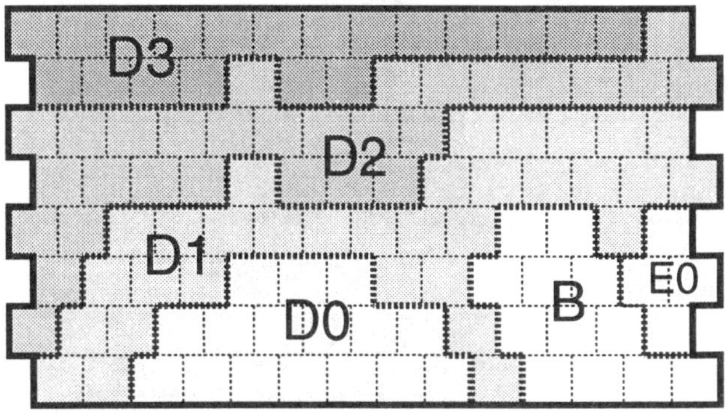


```
Beginning zone B (for all offsets)
Working Zones:
D0: offset = 0 mm     D1: offset = 1 mm
D2: offset = 2 mm     D3: offset = 4 mm
Isolated ending zone  E0: offset = 0 mm
```

Figure 5. Network regions for the insertion task in the machining center.

5.2 Insertion in the tool pallet with positive and negative errors

To perform the second experiment, the tool-pallet has been removed from the robot vehicle and placed in a fixed position. The number of different possible situations that the network must recognize has been increased by considering uncertainty in the range [-4 mm, +4 mm] on OY axis (in the direction along tool centers on the pallet). To train the network, we have used a set of sensor signals corresponding to insertion tasks with offsets of -4, -3, -2, -1, 0, +1, +3 and +4 mm. For each task, a sample of 55 signals was used. Once the network has been trained, the activation pattern for the complete sequence in each case is obtained. The result can be seen in figure 6, it is more complex than in the previous case, since more states have been introduced. The correct insertion is located approximately in the middle of the map, the processes with positive errors move towards the left side while the negative errors give rise to regions that move towards the right side and lower right corner. The common beginning and ending zones exist because these phases of the process are not dependant on the offset, since there is no contact between the gripper and the structure.

Figure 7 shows the partitioned map with labeled regions. We have used a bigger network than in the previous example, because the problem is more complex as more states have been introduced. The topology of the map is similar to the previous case, with two regions for the beginning and ending states, and the rest of regions varying continuously with the error. The regions on the upper right corner may disappear with additional training, although in complex situations it might not always be possible to map the six-dimensional input space onto the two-dimensional output space in a perfectly continuous way. Errors for a new situation will be detected as explained for the previous case. The further the resulting

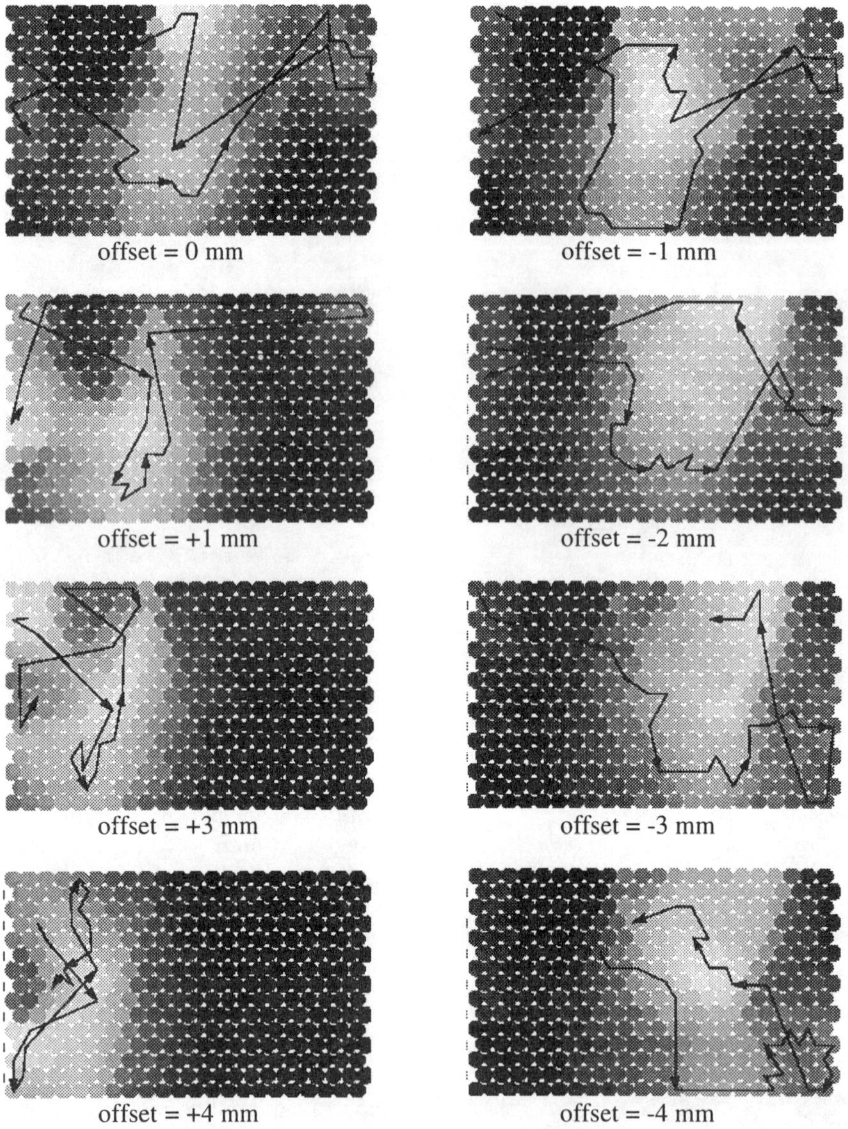

Figure 6. Activation patterns for the insertion task in the tool pallet. Traces of winner cells are also shown.

activation pattern is from region N0 the greater the error (positive or negative) will be. In figure 7 the different regions have been classified into safe or dangerous using grey levels.

6. CONCLUSION

An approach for dealing with uncertainty in robot fine motion based on unsupervised neural networks has been presented. The six force and torque signals from the wrist sensor of a robot arm are fed into the network. The method has proved to work properly in complex real-world situations for which a geometric analytical model may not be feasible, or too difficult.

The training of the network is very straightforward, because there is no teacher. We have shown how the network must be labelled in order to be used to monitor the process. Despite the simplicity of the network, the self-organizing map manages to store complex information about the states of the task, and allows to detect error conditions easily. Due to the short time required for the learning process and its simplicity, it allows for a great flexibility and new tasks can be readily learned without a long process of analysis.

Future directions include expanding the training procedure to allow the network to learn on-line when the task is running, and do the labelling process automatically. We are also dealing with temporal sequences of signals, so we could use the temporal

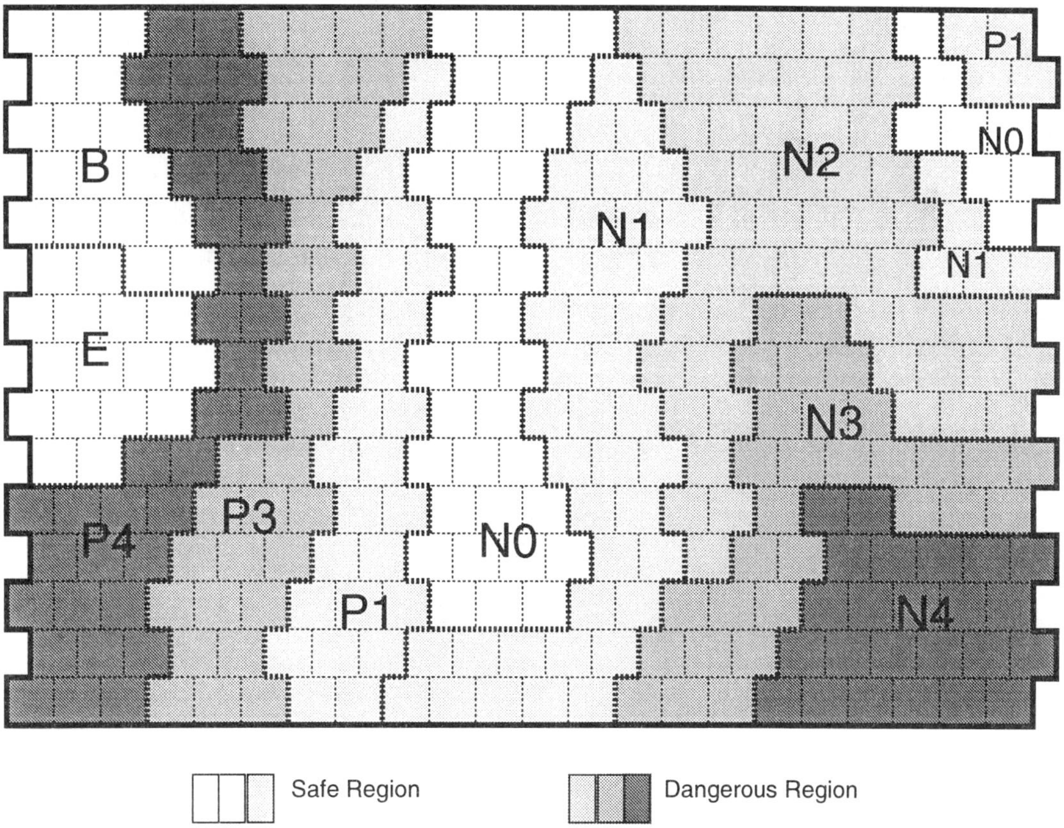

Figure 7. Network regions for the insertion task in the tool pallet.

information associated to the signals, e.g., by adding more inputs with the values of the signals in past instants. An extension to the system will be the addition of recovery actions associated to cells of the network labelled with error states. We are currently working in that direction.

ACKNOWLEDGEMENTS

This work has been funded by the CICYT under project TAP92-0391, and by a grant of the FPI Program of the Spanish Department of Science and Education. We wish to thank Prof. Kohonen, Prof. Oja and Dr. Kangas for making possible a stay of one of the authors at the Laboratory of Computer and Information Science at Helsinki University of Technology, for granting access to the the laboratoty facilities for the preparation of this paper, and for helpful discussions.

REFERENCES

[Canny & Reif 87] Canny, J., Reif, J., *New Lower Bound Techniques for Robot Motion Planning Problems*, Proc. of the 28th IEEE Symp. on Foundations of Computer Science, 49-60, Los Angeles, 1987.

[Dakin & Popplestone 91] Dakin, G.A., Popplestone, R.J., *Augmenting a nominal assembly motion plan with a compliant behavior*, Proc. National Conf. on Artificial Intelligence, 1991.

[del Pobil & Serna 95] del Pobil, A.P., Serna, M.A., *Spatial Representation and Motion Planning*, Springer-Verlag, (to appear in 1995).

[Donald 88] Donald, B.R., *A geometric approach to error detection and recovery for robot motion planning with uncertainty*, Artificial Intelligence, 37:223-271, 1988.

[Donald 89] Donald, B.R., *Error Detection and Recovery in Robotics*, Springer-verlag, 1989.

[Donald 90] Donald, B., *Planning multi-step error detection and recovery strategies*, International Journal of Robotics Research, 9(1):3-60, 1990.

[Erdmann 86] Erdmann, M., *Using Backprojections for fine motion planning with uncertatinty*, /nternational Journal of Robotics Research, 5(1):19-45, 1986.

[Gottschlich & Kak 91] Gottschlich, S., Kak, A.C., *Dealing with uncertainties in cad-based assembly motion planning*, Proc. National Conf. on Artificial Intelligence, 1991.

[Hertz, Krogh & Palmer 91] Hertz, J.A., Krogh, A. and Palmer, R.G., *Introduction to the Theory of Neural Computation*. Addison-Wesley, 1991.

[Kak 90] Kak, A.C. (ed.), *Special issue on robotic assembly and task planning, AI Magazine*, 11(1), 1990.

[Kohonen 90] Kohonen, T., *The self-organizing map*. Proceedings of the IEEE, 78:1464-1480, 1990.

[Lozano-Pérez, Mason & Taylor 84] Lozano-Pérez, T., Mason, M.T., Taylor, R.H., *Automatic Synthesis of Fine-Motion Strategies for Robots*, Intl. J. Robotics Research, Vol. **3**, No. 1, 1984, pp. 3-24.

[Mason 81] Mason, M.T., *Compliance and force control for computer controlled manipulators*, IEEE Trans. on Systems, Man and Cybernetics, 11:418-432, 1981.

[Natarajan 86] Natarajan, B.K., *The Complexity of Fine Motion Planning*, Tech. Rep. TR-86-734, Dept. of Computer Science, Cornell University.

[Peshkin 90] Peshkin, M.A., *Programmed compliance for error corrective assembly*, IEEE Trans. on Robotics and Automation, 6:473-482, 1990.

[Rappa & Depackere 92] Rappa, M.A., Depackere, K., *Neural Network community survey*. T. R. E52-538, Massachusetts Institute of Technology, Cambridge, May 1992.

[Sanderson & Perry 83] Sanderson, A. C. and Perry, G. (1983). *Sensor-Based Robotic Assembly Systems: Research and Applications in Electronic Manufacturing*, Proc of the IEEE, 71, 7, pp. 856-871.

[Smith & CheeseMan 86] Smith, R.C., Cheeseman, P., *On the representation and estimation of spatial uncertainty*, International Journal of Robotics Research, 5:56-68, 1986.

APPENDIX

Some technical data about the experiments are included here.

For the experiment described in section 5.2, training was done in two phases. The first is the ordering phase during which the reference vectors of the map units are ordered. During the second phase the values of the reference vectors are fine-tuned.

Data for experiment of section 5.2:

Network size: 22 x 15 units

Initial values: random [-1, 1]

Training parameters:

	Iterations	*Learning rate*	*Neighbd. radius*	*CPU time*
1st phase	4000	0.2	15	6 sec
2nd phase	40000	0.04	5	52 sec

The neighbourhood radius decreases to one during training, while the learning rate decreases to zero.

CPU is a MIPS IP22 processor runing at 150MHz on a Silicon Graphics Indigo 2 workstation.

Time is proportional to the number of iterations and network size.

Memory space required is proportional to the number of units and dimension of the input (6 in our case).

Training sets for experiments of sections 5.1 and 5.2:

	Samples	***Processes***	***Offsets***
1st case	2425	24	0,1,2,3
2nd case	580	11	-4,-4,-3,-2, -1,0,1,3,3,4,4

All the simulations were done using the public-domain software package SOM_PAK (Self-Organizing Map Program Package) developed in Laboratory of Computer and Information Science at Helsinki University of Technology (Finland).

Estimating blue-green algae counts with neurocomputing techniques

Joachim Diederich, Don Simonetta and Gerhard Paass
Neurocomputing Research Centre
Queensland University of Technology
GPO Box 2434
Brisbane Q 4001 Australia
e-mail: J.Diederich@qut.edu.au

ABSTRACT

Blue-green algae (BGA; cyanobacteria) are naturally occurring organisms. Some of the blue-green algae are toxic and pose potential health problems for humans and animals. The objective here is to estimate the amount of BGA (measured in organism per milliliter) in one storage area (Lake Baroon, Sunshine Coast Hinterland, Queensland) given the presence of nutrients in the water (in particular nitrate and phosphorus), the turbidity of the water and the rainfall of the last 14 days. The estimation is done by artificial neural networks (Cascade 2 networks and multi-layer perceptrons). The network accepts as input measurements of important parameters (nutrients, rainfall etc.) and produces an estimate of the BGA count as output. Our current results show that we can estimate the number of blue-green algae with an average error of 2390 organisms/ml. This result makes neurocomputing techniques a strong candidate for BGA estimation and prediction systems.

INTRODUCTION

Blue-green algae (cyanobacteria) are naturally occurring organisms (this section is closely based on: Queensland Water Quality Task Force (1992): Freshwater Algal Blooms in Queensland, ii-iii; 1). They are one of a number of groups generally described as phytoplankton; microscopic organisms which are present in all surface waters. They have a similar appearance to algae but are closely allied to bacteria. Because of their blue-green color they are known scientifically as cyanobacteria. Blue-green algae are very much a part of the Australian water environment but are not unique to Australia; they occur worldwide.

An algal bloom occurs when favorable conditions exist for the rapid multiplication of cells and the organisms are present in such large numbers that a surface scum or discolored water results. Such blooms can result in water quality problems, particularly with taste and odor. Some of the blue-green algae are toxic and pose potential health problems for humans and animals. They can cause death in animals and may have significant effects on human health if heavily infested water is used for recreational activities or consumption.

The onset of a bloom can result from various combinations of contributing factors: high concentration of nutrients; high light penetration; low turbidity; calm water conditions; and generally, but not always, warm water temperatures.

Blue-green algal blooms have been recorded in Australia since the 1830's (South Australia). The first recorded stock deaths were at Lake Alexandrina in South Australia in 1878. Since then, stock deaths associated with blue-green algae have been widely reported in Australia and throughout the world. On the human health side, no deaths in Australia have been attributed to blue-green algae.

While blue-green algal blooms are natural phenomena in Australia, the frequency and intensity of their occurrence can be influenced by human activities such as disposal of nutrients into watercourses from sewage treatment plants; nutrient rich urban runoff; intensive agricultural industries such as cattle feedlots and piggeries; and extensive agricultural practices resulting in nutrients sourced from fertilizers and stock wastes.

Periodic blue-green algal blooms have occurred in Queensland, particularly in the storages of South East Queensland, but have never really posed a serious problem until the summer of 1991-92, when simultaneous blooms in a number of water storages resulted in the closure of some for many months. The closure of storages denied the community access for recreational use and in some cases, notably Warwick and Palm Island, restricted the availability of water for domestic water supply.

These closures, together with the 1000 km bloom along the Darling River in November 1991, caused public concern about the vulnerability of Queensland's waterways and storages to algae and the need for nutrient management.

MAJOR INFLUENCES TO THE GROWTH OF BLUE-GREEN ALGAE

The BGA genera of interest are: anabaena, microcystis, nodularia [A, page 29]. Some BGA can get into depth with optimal light and nutrients [C, page 5] The algae may position themselves close to the thermocline thus getting access to nutrient rich water below this location [C5]. If there is too little phosphorus (P) or nitrate (N) the BGA sink further to hypolinnion which is rich in nutrients before moving up again [C5]. Therefore, they may simultaneously have optimal nutrients and light [C5].

Some BGA may produce akinetes (~spores) which are packed with food and are resistant to temperature and may exist in sediments for years. Then they may seed the water again [C5]. Some BGA (anabaena, aphanizomenon, not microcystis) can fix nitrogen from the atmosphere [A2] [C6]. If P and trace metals are available, they have an advantage over other planktonic algae [C5].

Furthermore, some BGA are capable of producing toxins [C5]. The toxicity of a bloom cannot be determined by any visual feature (color, cell morphology) [C6]. Some toxic species of the genera in Queensland are: Anabaena, Microcystis, Cylindrospermopsis, Oscillatoria, Nodularia [C6].

An algal bloom has occurred if the chlorophyll levels reach 20 mug/l [A1]. Possible effects in humans are: liver damage, gastro-enteritis [C10], neurotoxin, tumor promotion, skin allergies [C11]. However, there is probably a finite dose below which there is no toxicity [A22]. The provisional safety level are 2000-4000 microcystis cells/ml [A22] and 10000-20000 aeruginosa cells/ml [A23]. The alert level are: 2000 cells/ml [A24] toxic level: 15000 cells/ml [A24].

Favorable conditions for a BGA bloom are [C6]: high nutrients; high light penetration; low turbidity; calm water conditions; long periods of stable weather; and generally, but not always, warm water temperatures [C6]. The species composition is influenced by seed source, turbulence and water mixing.

The nitrate/phosphate ration (N:P ratio) seems to be of great importance, as well as the bio-availability of P. High P and low N/P ratio favors some species of cyanobacteria [Aiv]. At low N:P ratio, N is limiting, and at high ratios, P is limiting [A3]. A ratio of 10:1 is considered as balanced [A3]. The reduction of N is insufficient if P is not reduced too [A3]. In general aquatic flora contain N and P in the atomic ratio of 16:1 [C6]. Other nutrients (carbon, silica, iron, sulfur, molybdenum) may also play a significant role.

THE OBJECTIVE

The objective here is to estimate the amount of BGA (measured in organism per milliliter) in one storage area (Lake Baroon near Caloundra, Queensland) given the presence of nutrients in the water (in particular nitrate and phosphorus), the turbidity of the water and the rainfall of the last 14 days. In some experiments, the pH-level and the rainfall over the past six weeks is considered. The estimation is done by an artificial neural network. The network accepts as input measurements of important parameters (nutrients, rainfall etc.) and produces an estimate of the BGA count as output.

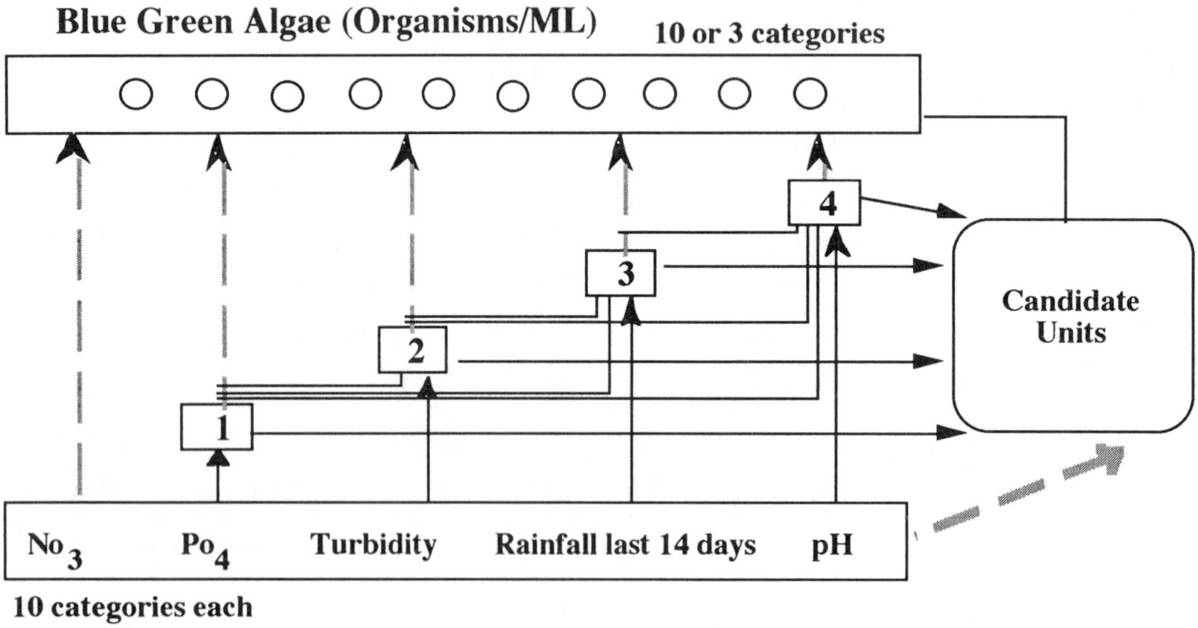

Figure 1: The Cascade 2 architecture for the estimation of BGA counts.

The remainder of this paper is organized as follows. First we give an introduction to the special-type artificial neural network architecture that is used (Cascade 2). Next, we describe the network that is used in preliminary experiments, and finally, we present first results.

CASCADE CORRELATION AND THE CASCADE 2 ALGORITHM

Incremental learning is currently an important issue in neural network learning. Learning algorithms such as cascade correlation [F], weight decay [I] and recruitment learning [E] all change network structures and units during learning (see Alpaydin (1991; [D]) for a summary of incremental learning architectures). Results from computational learning theory [H] indicate that the general learning problem for neural networks (the "loading" and generalization problem) is NP-complete and that learning algorithms must be capable of changing network structures during the learning process.

The cascade correlation architecture realises an incremental learning technique and was first introduced by Fahlman and Lebiere in 1990. This algorithm is for building a network dynamically. It adds units to a hidden layer between the input and output layers inducing the network to increase its accuracy in responding to given input patterns.

The algorithm is a three step process. Initially, patterns are presented to the input layer. The difference in the expected output and the actual output from the network is then used to train the weights on the connections into the units in the output layer. The next step is to train a unit (candidate unit) for insertion into the hidden layer of the network. The hidden layer begins with no units and as candidate units are trained they are added to this layer. The candidate at this stage has connections from the input units and all the units that may be in the hidden layer. The weights on these connections are trained so that the correlation between the output of the candidate unit and the difference seen in the output layer in maximised.

Finally the candidate is inserted into the network. The weights on the connections to the new unit from the input and hidden units remain constant for the life of the network. Insertion is done by allowing the units in the output layer to receive a connection each from the newly trained unit. These three steps are repeated until the network can give an acceptably accurate response.

A slight variation on the above algorithm is Cascade 2. It differs from Cascade Correlation by replacing the correlation between the output of the candidate unit and the error, with a set of candidates each trying to minimise the error at the output unit. The candidates, as before, have connections from units in the input layer and hidden layer. They now also have pseudo connections to each of the units in the output layer. After the first step of training the weights into the output layer, the total sum square error is retained for use in training the candidates.

The candidates are given an initial score of the retained value at the beginning of every pass through the training patterns. As training proceeds the weights on the connections into and out of the candidates are updated. Candidates scores are also lowered by an amount that indicates how little the particular unit will minimise the error at the output. The candidate with the highest score after a maximum number of passes through the patterns is then implanted into the hidden layer as in the first algorithm.

The C implementation of the Cascade 2 algorithm has provided success in building networks for many situations where a prediction must be obtained from limited data sets. Due to these successes, there is a keen interest in further increasing the efficiency of the Cascade 2 algorithm. For example; by implementing strategies to reduce the number of connections to the hidden layer; or using a more robust learning algorithm to adjust the weights on connections from the one currently implemented. To achieve this a more flexible program than the C version was required.

The ICSIM neural network simulator (Gomes, 1993, [G]) from the International Computer Science Institute (ICSI) in Berkeley provides a powerful environment for use in the construction of different types of networks. The simulator is written in Sather, an object oriented language, and provides a rich base of object classes to build networks. Many networks are already implemented in ICSIM, including the Cascade Correlation network and many features are provided that could be used to enhance the Cascade 2 algorithm. To this end, we have implemented the Cascade 2 algorithm in Sather using classes already written from ICSIM. This allows the program to be altered simply but still provide the same degree of control over variables as the C implementation.

THE NETWORK ARCHITECTURES

Two different Cascade 2 networks and one multi-layer perceptron are combined in this application. The first Cascade 2 network is used for a classification task, the second Cascade 2 system is a function approximation network, and a multi-layer perceptron is used to predict the alert levels only.

The function approximation Cascade 2 network has five inputs and one output. The input and output measurements are divided by 1000 and are directly used to train the network. The Cascade 2 network allocates hidden units as necessary.

The multi-layer perceptron is a classification network using sparse binary coding and has three output units. These categories correspond to the alert level identified by the Queensland Water Quality Task Force. The first output unit is activated by the network if the BGA count is <2000, the second output unit is activated if the BGA count is 2000-15000, and third output unit is activated by the network if the BGA count is >15000. Figure 1 shows the Cascade 2 architecture.

In the second Cascade 2 network, sparse binary coding instead of real values are used. That is, input values are translated into one of 10 categories and presented to the network. A NO_3 value of 0.5, for example, would be presented to the network as a binary pattern: 0 0 0 0 0 0 0 0 1 0. The task of estimating the BGA count is therefore turned into a classification problem which can be learned by a neural network.

Currently, NO_3, PO_4, Turbidity, Rainfall (last 14 days) and pH are presented to the network as part of the input and the network produces an estimate of the BGA count as part of the output.

THE DATA SET

For this preliminary study, a limited data set from Lake Baroon is used. The approach omits a number of data points from the analysis and applies the learning procedure to the rest. Subsequently, the error of the network for the omitted data is determined. To get comprehensive results this procedure is repeated by omitting different sets of observations. This procedure is a strong test of the generalization capabilities of the network.

The data was gathered for purposes other than statistical or neurocomputing analysis, and for a reliable neurocomputing application, the amount of data is not sufficient. We deal with this problem in the following way: (1) Only a few input parameters are used in order to avoid over-parameterisation and to allow good generalization, (2) for evaluation purposes, cross-validation (leave-one-out) is used (this is a strong test of the generalization capability of the network), and (3) we developed a scheme for the utilization of all data which we now discuss.

Because the data set has been created for different purposes, the measurement of the nutrients, for instance, has not necessarily been made on the same day as the BGA count (it might be a few days earlier or later). We account for this by a scheme which allows us to *weigh data points*. If the measurement of the nutrients has been made on the same day as the measurement of the BGA count, the corresponding data pattern becomes full weight (this is achieved by simply repeating the pattern in the data set). If the measurement of the nutrients has been made a day earlier or later as the measurement of the BGA count, the corresponding data pattern gets a lower weight (and so forth). This way, a total data set of 2591 patterns has been created.

RESULTS

A statistical analysis of the data revealed a correlation between Nitrate and Phosphate (.598) and between Turbidity and Rainfall over the last 14 days (.815). There was no correlation between Nitrate and the BGA count (-.082) and Phosphate and BGA count (-.082). All Cascade 2 networks used a non-linearity, even when sufficient learning time for a linear mapping was available. The function approximation network, however, has only one hidden unit. The non-linearity is weak as an analysis of the weights of the hidden units revealed. Rule-extraction from the Cascade 2 network resulted in one rule only: High turbidity and low rainfall contributes to a relatively high BGA count.

While the amount of data that we are using at this point in time is still not sufficient, the early results are encouraging. Our current results show that we can estimate the number of blue-green algae in 90% of all test cases with an average error of less than 2250 organisms/ml. This result has been achieved with the Cascade 2 classification network. The average error for the Cascade 2 function approximation network (all test cases) is 2390 organisms/ml.

A percentile comparison between the Cascade 2 network predictions and the real values (test cases) showed a tendency to overestimate if the BGA count is low, and to underestimate if the BGA count is high. This is a result which has been observed in other Cascade 2 applications as well.

If a network with only 3 output units is used (corresponding to the alert levels) and the values of the previous measurement are provided as an additional input, in 11 tests with cross-validation (leave-one-out) method no error was found. That is, in all test cases the correct alert level was produced. This cross-validation test was made with a restricted data set without repetition of data. In another test with the same network and 2490 training patterns and 101 test patterns, the generalization performance was perfect as well (no bit error).

ACKNOWLEDGMENT

This work was supported by an infrastructure grant of the Queensland Department of Primary Industries to the Neurocomputing Research Centre at Queensland University of Technology. We would like to thank Gerard Finn, Elizabeth Gurrie as well as Graham Baxter (Department of Water Resources, Brisbane City Council) for comments on earlier versions of this paper. Thanks to Scott Fahlman who made the Cascade 2 code avaliable to us and Ross Hayward who implemented Sather Cascade2.

REFERENCES

[A] Murray-Darling Basin Commission (1993): Algal Management Strategy. April 1993, ISBN 1 875209 11 5.

[B] Lake Baroon Blue-Green Algae Information Summary.

[C] Queensland Water Quality Task Force (1992): Freshwater Algal Blooms in Queensland. Volume 1, Oct. 1992. GPO Box 2454, Brisbane Q 4001, Australia.

[D] Alpaydin, E.: GAL: Networks that grow when they learn and shrink when they forget. Berkeley, Ca.: International Computer Science Institute, März 1991 = ICSI-TR-91-032.

[E] Diederich, J.: Instruction and High-Level Learning in Connectionist Networks. Connection Science 1 (1989) 2, 161-180.

[F] Fahlman, S.E.; Lebiere, C.: The Cascade-Correlation Learning Architecture. In: Touretzky, D.S. (Ed.): Advances in neural information processing systems. San Mateo, Ca.: Morgan Kaufmann Pub., 2, 1990.

[G] Gomes, B.: ICSIM. Master Thesis. CS Division, University of California,
Berkeley, January 1993.

[H] Lin, J.-H.; Vitter, J.C.: Complexity Results on Learning by Neural Nets. Machine Learning 6 (1991) 211-230.

[I] Weigend, A.S.; Huberman, B.A.; Rumelhart, D.E.: Predicting the Future: A Connectionist Approach. Palo Alto, Ca.: Xerox Palo Alto Research Center, System Sciences Laboratory, April 1990 = Xerox Parc TR SSL-90-20.

APPENDIX: THE BLUE-GREEN ALGAE ESTIMATION SYSTEM

NEUROCOMPUTING RESEARCH CENTRE
QUEENSLAND UNIVERSITY OF TECHNOLOGY

QUEENSLAND DEPARTMENT OF PRIMARY INDUSTRIES
BLUE-GREEN ALGAE ESTIMATION SYSTEM

Version 0.1

You have the option of using either Function Approximation or

Classification Networks. An Alert Level Network will also be run.

Function Approximation(F), Classification(C) or Both(B) Networks? : b
Enter Nitrate Level : 1.1
Enter Phosphate Level : .617
Enter Turbidity Level : 39.5
Enter amount of Rainfall in last 14 days : 587.5
Enter pH Level : 6.8

Running through network ...

Function Approximation Network:

Estimated number of Blue Green Algae (in organisms/ml):
1670
Blue-Green Algae Alert Level : Low

Classification Network:

Estimated number of Blue Green Algae (in organisms/ml):
Output Category is 1
Category is from 78 to 1596.2
Middle of Output Category is 837.1
Blue-Green Algae Alert Level : Low

Alert Level Network:

Estimated Blue Green Algae Alert Level:

LOW

NEURAL NETWORK CONTROL - A CASE STUDY

S.-G. Cong, D. G. Holmes and W. A. Brown
Dept. of Electrical & Computer Systems Engineering, Monash University
Clayton, Vic 3168, Melbourne, Australia
Email: scong@basil.eng.monash.edu.au

ABSTRACT

This paper describes the development of a neural network controller (neurocontroller for short) for a servo system. In the application of neural networks to control, some issues have to be dealt with, such as structure, training methods and the means of implementation. Other concerns include training speed and convergence. In the example described in this paper, these issues are addressed. The whole system consists of a digital position controller, an analog speed controller, and a brushless DC motor to drive a robot arm. A three layer feedforward neural network is applied to perform position control which takes only position errors as inputs. Training is carried out off-line using the impulse response of the plant. The experimental results confirm the superior performance of neurocontrollers with regard to dynamic response, the ability to adapt to parameters variations and to deal with the uncertainty of plant models.

1. INTRODUCTION

The application of neural network control has grown quickly in recent years. The impetus for using neural networks in control arose from the desire for neural networks, with inherent self-learning, nonlinear mapping, and fault-tolerant capability, to deal with increasingly complex control environments where traditional control methodologies have experienced difficulties [Tul91].

The applications of neural networks in control can be found in many areas, such as drive systems [Bor93], robotics [Poo92], process control [Thi92] and others e.g. [Lig92][Fuk92]. In these examples, neural network control has proved to be superior than traditional schemes.

Although these applications confirm the power of neural network control, there are still difficulties in applying this new technique in a more systematic way. The major causes are the lack of efficient training schemes and the lack of knowledge for the best use of neurocontrollers for a particular process. The difficulty is more significant when the model of the process is inadequate or even unknown. In such circumstances, it is suggested that experimental data should be used as training information [Psa88]. However it is necessary that experimental data containing adequate information about the process can be obtained without causing severe disturbance to the operation of the plant.

The authors have investigated these issues and proposed a set of training algorithms based on the impulse response and step response [Cong92][Cong94]. To make training successful, other relevant issues such as the optimal structure of neurocontrollers, fast training schemes, the influence of feedback on the training, have also been explored and corresponding solutions are proposed [Cong94]. A SIMULINK based program has been developed to serve the training requirements.

To verify the feasibility of these proposed algorithms and approaches, a neural network controller has been implemented for a servo system which is used to drive a robot arm. This paper describes the whole process of developing this system, including structure design, training, and implementation. The experimental results demonstrate strongly the superior performance of the neurocontroller and hence confirm the feasibility of the proposed schemes.

The contents are organized as follows: Section 2 describes some issues related to neurocontrollers; Section 3 deals with design of the servo system and training of neural network controllers; experimental results are given in Section 4 and finally the conclusions are provided in Section 5.

2. ISSUES ABOUT NEUROCONTROLLERS

To apply neural networks for control purposes, three fundamental issues have to be considered:

- Determine a *controller structure* including the topology of a neural network;
- Acquire plant knowledge as *training information*;
- Using a predetermined training scheme, optimize the weights of the neural network in terms of the plant information and control objective, to achieve the final control goal. This process is known as *training*, or from the neural network's point of view, *learning*.

a) *Training Information*

The more information is known about the plant, the better a neurocontroller learns. If the dynamic model of the plant to be controlled is available, or persistent excitation is permitted for the plant, it is easier to get training information and off-line training will usually suffice for optimizing the behavior of the neurocontroller.

In contrast, if the knowledge about the plant is incomplete and/or persistent excitation is not allowed, the performance of a neurocontroller may not be acceptable and the credibility of training is more of a question.

In practice it is rare that very precise or very little knowledge is available for a plant. Furthermore, many systems cannot sustain persistent excitation. As the consequence, there needs to be a compromise between the demands for training information and the restrictions of the physical environment. As it is not always possible to get a complete description for the plant dynamics, the use of simple experiments to attain a relatively precise knowledge of the plant is sensible and desirable.

b) *The Structure of Neurocontrollers*

The widespread application of PID controllers indicates that an industrial controller should be simple, easy to use and require minimum information. Such requirements are also true for neurocontrollers.

Although the merits of a neurocontroller can justify its relative complexity compared to a PID controller, it should still be as simple as possible from both training and implementation perspectives. This means that a simple neural network topology requiring minimum feedback information is to be preferred, provided that excellent control performance can be maintained.

c) *Training Schemes*

A training scheme is usually developed for a predetermined neurocontroller structure, taking into account available training information and an appropriate optimization algorithm.

The well known backpropagation approach deals with only static feed forward information and uses a gradient descent method. The input and output pairs for a neural network are constants and the input is independent of the architecture of the neural network. But applying a neural network in a control system is much more complicated because of the presence of dynamics and feedback.

An ideal training algorithm should be simple and fast, converge asymptotically and work for all applications, no matter how complex the dynamic systems are. This is of special importance for on-line training. Unfortunately, there is no such algorithm available so far, and furthermore, it has been found that some algorithms may work in certain applications, but may behave poorly for the others.

d) *Tools for Performing Training*

In general, off-line training should be carried out before a neurocontroller is applied to its physical environment, due to the concern of stability. Therefore, a tool to manage the training process is essential in the application of neurocontrollers.

The requirements for such a tool are rather complicated. It should allow change of neural network model and controller structure. It should provide a number of functions e.g. optimization algorithms, to carry out training tasks. In addition, it must be able to perform dynamic system evaluation, either continuous or discrete, as the performance of a neurocontroller should be assessed in terms of its dynamic behavior rather than static. Unfortunately, such a tool is commercially unavailable at present.

The extensive investigations about these issues and corresponding solutions were presented in [Cong94]. The major points of this work are:

- use the step or impulse response as training information when precise knowledge about the plant is unavailable;
- use PID signals as controller inputs; for digital implementations, only using error signals.
- a fast and stable training algorithm was developed which enables excellent training results;
- a extensive simulation package for off-line training of neurocontrollers was developed.

In order to verify these concepts, a neurocontroller was implemented to control a real-time servo system. In the following sections, a complete procedure of developing this system is presented.

3. A DIGITAL SERVO SYSTEM USING NEURAL NETWORKS

a) *The System Structure*

Fig. 1 shows the configuration of the system which is mounted to drive a robot arm through a gearbox with 1:200 ratio.. The system includes:

- a digital Position Controller which contains three MiniMicro boards: a Main Processor board, a Signal Processor board, and an Interface board;
- a speed control device: SC401 Analog Speed Controller, and a tachometer;
- a brushless DC motor;
- an optical encoder for position measurement; and
- an IBM compatible PC (not shown in the figure).

The digital position controller consists of an 8-bit master processor which is a Hitachi HD64180, a slave Signal Processor TMS34010 which is capable of executing 32-bit calculations. The master processor carries out housekeeping duty while the slave DSP is used to accomplish complicated calculations required by neural networks.

b) Topology of Neurocontroller

A three layer neural network with one input layer, one hidden layer and one output layer, was used for the system.

In the system the inputs to the neural network have been chosen as the current error between the desired position and the actual position $e(k)$, one step delay error $e(k\text{-}1)$, and two step delay of error $e(k\text{-}2)$. These signals provide the dynamic information about the

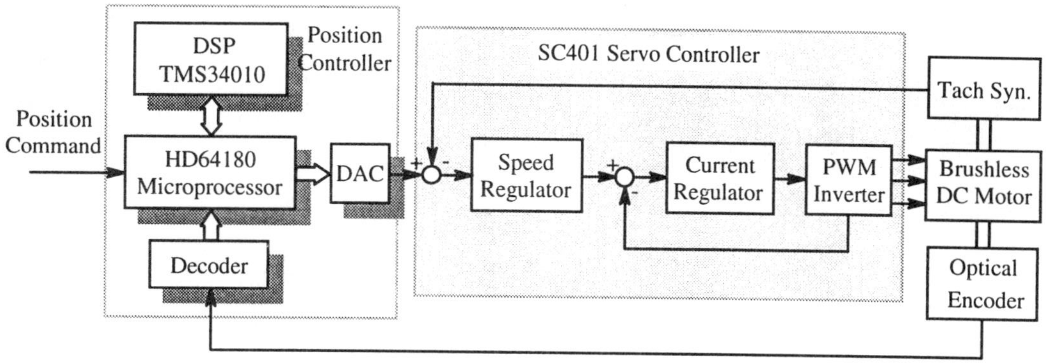

Fig. 1 The configuration of brushless DC motor position control system

system to the neural network. It will be noted that there is no integral term in the controller inputs.

As the output of the position controller is speed demand, so only one output neuron is required. The activation function for the output neuron is linear.

The determination of the number of hidden neurons for a particular plant is a complicated issue. It is widely believed that the performance of neurocontrollers may be improved by including more hidden neurons. However less neurons will reduce the time for training and the cost in implementation. Furthermore, if a neural network is going to be realized by software instead of hardware, less neurones require less calculation and hence a higher sampling rate can be used. This is important in real-time digital control.

It has been found through extensive simulation study that six hidden neurons are sufficient for this system. Meanwhile, hypertangent function is used as activation function for hidden layer neurons.

So the neural network used is a 3-6-1 one with the error $e(k)$ also fedforward (to bypass the hidden layer).

c) Training of the Neurocontroller

The parameters of neurocontrollers cannot be designed manually, they can only be determined through a training process. Therefore, training is a key issue for use of neurocontrollers.

Training was carried out off-line with a neuro-controller simulator which is developed within the MATLAB package, with supporting SIMULINK blocks to facilitate the construction of control systems (see Fig. 2) and a number of C-based subroutines to speed the process.

At the start of training, the neuron weights (i.e. the parameters of neural networks) are initialized with random numbers ranging between -1 and 1. Conventionally, a typical neural network training process requires several hundred or even thousands of iterations to converge, but with the algorithm developed in [Cong94], this process can be reduced to 30 ~ 200 iterations which varies depending on the initial weights. The training process is illustrated in Fig. 3 which shows the plant step response improving after each iteration. However training may fail if care is not taken, as demonstrated in Fig. 4. The tactics associated with the training of neurocontrollers are discussed extensively in [Cong94].

Eventually, the off-line training converges to a minimum as illustrated by Fig. 5, which corresponds to the optimal performance of controlled system. However, this good performance is only for reference input signal. To test if the trained neurocontroller is able to work well for unlearned inputs, i.e. if it can generalize properly, its performance was evaluated by applying various input signals to the system. The respective responses (not shown here) confirms that the trained neurocontroller is able to track various command signals very well.

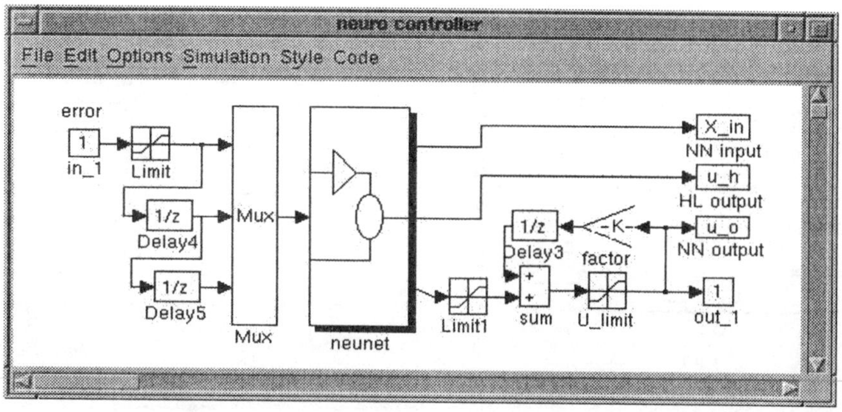

Figure 2 Construction of a neural network controller by means of SIMULINK

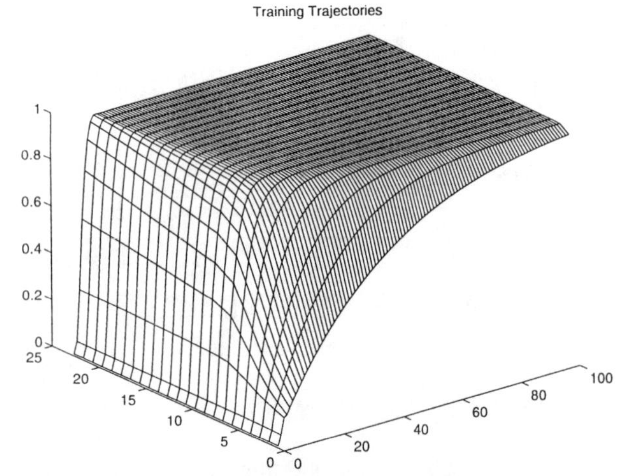

Fig. 3 Training trajectory

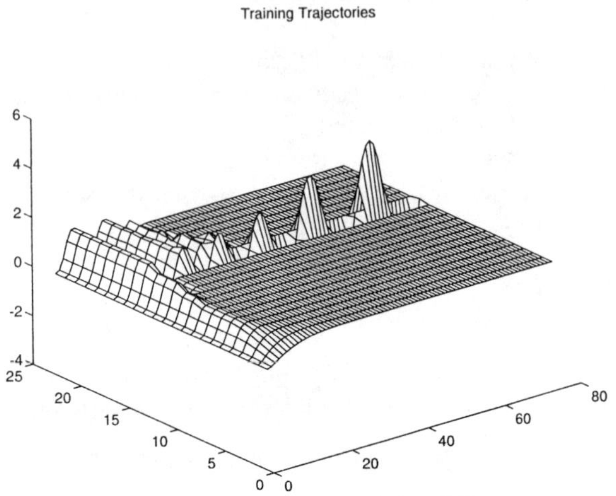

Fig. 4 A example of training failure

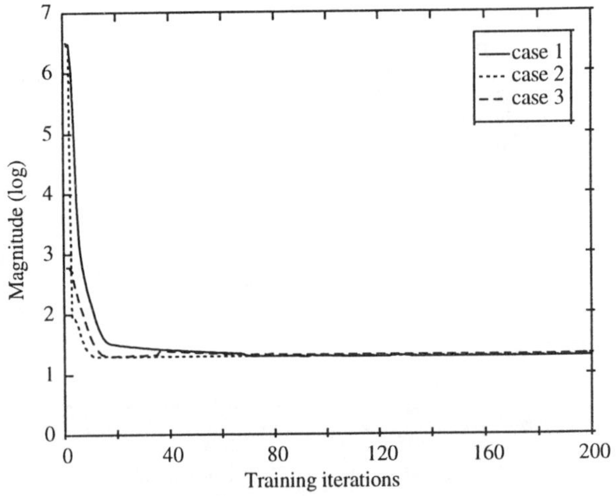

Fig. 5 Training process (log scale)

d) Generalization of Trained Neurocontroller

The trained neurocontroller can produce an excellent dynamic response to the reference input. However, whether it can still work well with unlearned signals remains a question. Various reference signals were tested and the results (not shown here) confirm the ability of generalization by trained neurocontrollers.

e) The Effect of Initial Weights

Initial weights play an important part in the success of training neurocontrollers. Different initial weights may result in different optimal weights. Wherever initial weights are close to optimal weights, the training time can be short, and vice versa. Fig. 6 gives two Hinton diagrams which show the difference of initial weights with the corresponding optimal weights. In the figure, each weight is represented by a square whose area

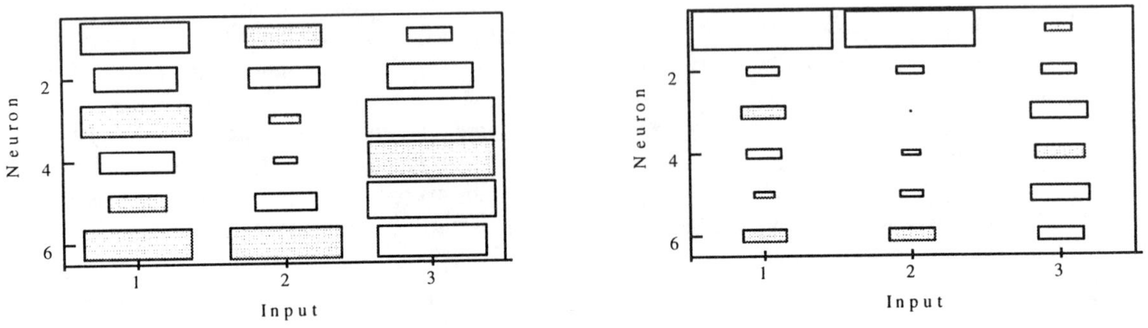

(a) Left: initial weights; right: optimal weights

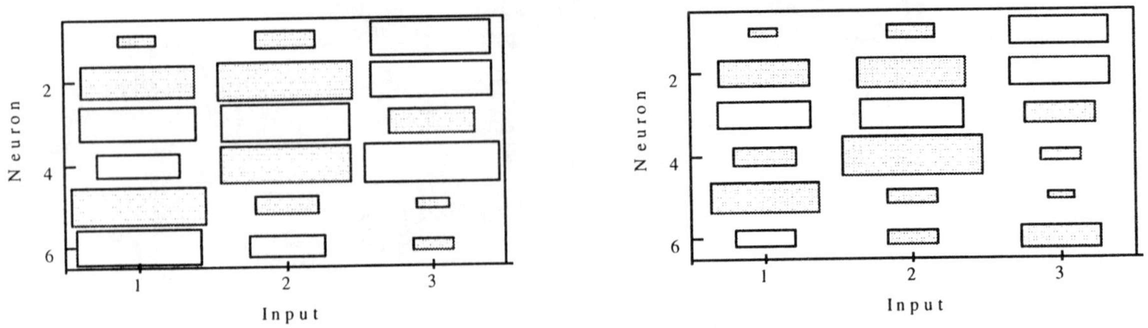

(b) Left: initial weights; right: optimal weights

Fig. 6 Weights change during training

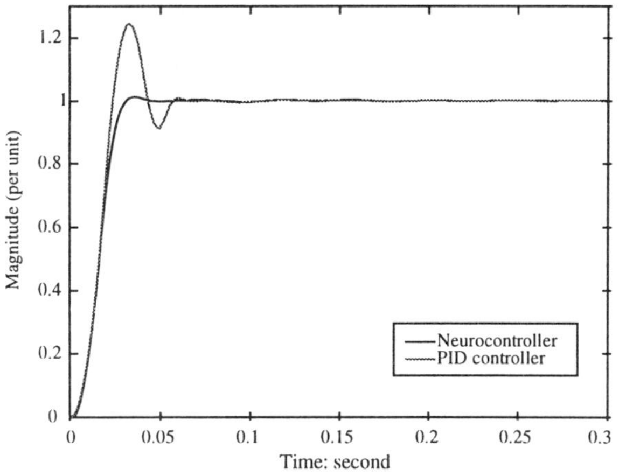

Fig. 7 The step response of both controllers

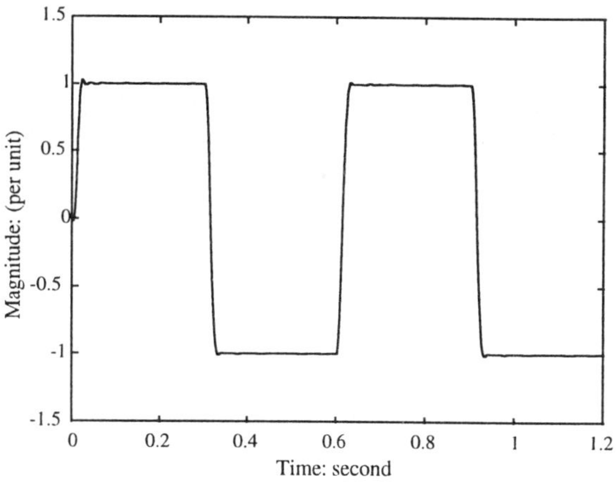

Fig. 8 Tracking square waveform

corresponds to the magnitude of that weight. The shaded squares are positive numbers. Fig. 6(a) shows that the final optimal weights are very different from that of initial ones, while in Fig. 6(b) the optimal weights are rather close to the initial values.

4. EXPERIMENTAL RESULTS

The experiments were carried out by supplying a step change command to the system and observe the corresponding per unit change of motor position. Fig. 7 gives the response of the system with a PID controller and a neural network controller respectively. In the figure it can be seen that the response of the system controlled by an optimally tuned PID controller has large over-shoot and takes long time to settle to steady state, while the response with the neural network controller is nearly optimal and much better than the PID controller in all cases.

Since the neurocontroller is trained to follow a positive rise step input, its response to a negative step must be considered. Fig. 8 shows the system response to a square wave input. This figure confirms that the system is able to generalize to unlearned input signals. Finally, the influence of sampling period is illustrated in Fig. 9, where it can be seen that a decreased sampling period can significantly improve the system performance. This is why a slave DSP is used to carry out the

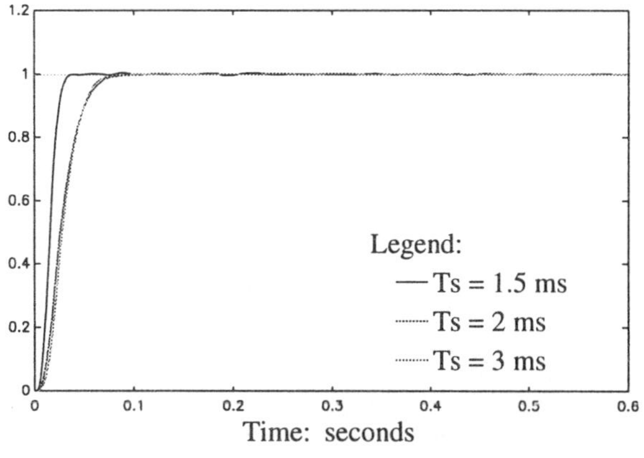

Fig. 9 The effect of sampling period

complicated calculations required by neural networks.

5. CONCLUSION

In this paper, a comprehensive procedure is presented for a neural network controlled digital servo system. Some important aspects regarding to the training neurocontrollers are discussed. The neural network controller is trained in terms of the impulse response of the plant and is realized using one micro-processor and one signal processor in a master-slave fashion. Experimental results show the excellent response of the system in all aspects.

REFERENCES

[Bor93] L. E. Borges da Silva, G. Lambert-Torres, E.C Saturno, A. P. Alves da Silva, and D. Olivier, *Neural Net Adaptive Scheme for DC Motor Drives*, Proc. of IEEE IAS Annual Meeting, 1993, Vol. 2, pp. 2043-2048.

[Cong92] Cong Shi-guo and D. G. Holmes, "A method for Training In-line Neural Network Controller for Linear Plant Systems", *Proceedings of Australasia Universities Power and Control Engineering Conference*, Brisbane, Oct. 1992, pp. 8-13.

[Cong94] Cong Shiguo, *Investigation and Implementation of Neural network Controlled systems*, PhD thesis (to be submitted), Monash University, 1994.

[Fuk92] T. Fukuda, and T. Shibata, "Theory and Applications of Neural Networks for Industrial Control Systems", *IEEE Trans. Industrial Electronics*, Vol. 39, No. 6, pp. 472-489, Dec. 1992.

[Lig92] G. Lightbody, Q. H. Wu and G. W. Irwin, "Control applications for feedforward networks", *Neural Networks for Control & Systems*, Warwick edit., 1992, pp. 51-71.

[Poo92] A.N. Poo, M.H. Jr. Ang, C.L. Teo, and Qing Li, "Performance of a neuro-model-based robot controller: Adaptability and noise rejection",

Intelligent Systems Engineering, Vol. 1, No. 1, Fall 1992, pp. 50-62.

[Psa88] D. Psaltis, A. Sideris, and A.A. Yamamura, "A Multilayered Neural Network Controller", *IEEE Control Systems Magazine*, April 1988, pp. 17-21.

[Thi92] J. Thibault, B. P. A. Grandjean, "Neural networks in process control - a survey", *IFAC Symposia Series*, No. 8, 1992, pp. 251-260.

[Tul91] E. Tulunay, "Introduction to Neural Networks and their Application to Process Control", *NEURAL NETWORKS: Advances and Applications*, North-Holland, pp. 241-273, 1991.

Case-Based Reasoning

CASE-BASED REASONING FOR THE STRUCTURAL DESIGN OF BUILDINGS

Dong Mei Zhang and Mary Lou Maher
Key Centre of Design Computing
Department of Architectural and Design Science
University of Sydney NSW 2006 Australia

ABSTRACT

Case-based reasoning provides a knowledge-based approach to solving problems that uses past experience directly, rather than requiring the formalisation of generalised knowledge. The use of case-based reasoning in a complex domain such as the structural design of buildings raises a number of issues related to the representation of design cases and the adaptation of cases to produce new design solutions. We present the alternatives and advantages of several representation paradigms and memory organisation techniques. We then outline an approach to design case adaptation that uses a propose-verify-modify cycle. Through an example we show how these techniques can come together to provide a case-based reasoning tool to assist in the design of structural systems.

INTRODUCTION

Case-Based Reasoning (CBR) provides a process model for applying prior experience to new problems. It involves retrieving relevant previous cases and adapting the solution from a previous case if necessary to solve new problems. CBR as a design model is appealing intuitively because much of design knowledge comes through the experience of multiple, individual design situations. For many domains where design knowledge is difficult to acquire and may not be objectively applicable, the case-based paradigm provides a model for the acquisition, organisation and reuse of specific design knowledge. Therefore, there is a large and growing interest in the use of CBR approach for building knowledge-based systems that aid in the process of design.

Structural design, as a specific design domain can be viewed as a sequence of three stages: preliminary design, analysis and detailed design. The preliminary structural design of buildings involves the selection of a feasible structural configuration satisfying a few key constraints. Preliminary design is chosen as as a focus because this phase of design is not well supported by design methods and procedures. During preliminary design the form of design solution is identified. Domain knowledge provides design principles and performance theoretical guidance. Structural design episodes serve as resources in the structural design process because they record examples of solutions.

The applications of case-based reasoning to design can be classified as design aiding systems and design automation systems. Some system have been built only to assist human designers. These systems assist designers by finding appropriate design cases to a given problem and by allowing designers to browse and view those relevant cases. The reasoning processes in these systems are left to human designers. For example, ARCHIE (Goel and Kolodner, 1991) and its successor ARCHIE-II (Domeshek and Kolodner, 1992), and CASECAD (Maher and Balachandran, 1994) are tools for aiding designers doing conceptual design in architectural and structural design. Some other case-based design models have been developed to solve problems automatically. For example, DDIS (Wang and Howard, 1991), CADRE (Hua and Falting, 1993), CADET (Sycara and Navinchandra, 1989), and CADSYN (Maher and Zhang, 1993). We adopt an approach to case-based design that includes some automated reasoning but focuses on a tool to aid rather than fully automate design.

A case-based system for design problem solving is often developed as a hybrid system which combines CBR with other reasoning methods and domain knowledge. DDIS is an integrated system for structural engineering design in which CBR and rule-based reasoning are combined. Generalized knowledge is represented as rules. CADRE combines constraint solving with CBR for topological and geometric architectural design. Generalised grammars and rules on topological description are used to support the adaptation. CADET, a mechanical device design system uses CBR and qualitative reasoning to perform design synthesis. A design case is a graph-based behaviour model about a particular device. In this approach, case-based reasoning is viewed as analogical reasoning of selecting and applying various device models rather than specific episodes. CADSYN integrates a general decomposition approach with case-based reasoning. Adaptation of design cases is modelled as a constraint satisfaction process, in which generalized constraints knowledge and design models are applied.

Hybrid case-based design systems are developed for the following reasons. First, certain generalized domain knowledge compensates incompleteness of specific design knowledge in design cases. Because of incomplete design knowledge in design drawings, documents, and interviews with designers, it is not realistic to acquire rich design cases covering a wide range of knowledge such as design rationale, design history, reasoning steps, and specific design knowledge from real design projects. Second, the cognitive activity of designers requires the combination of generalized and specific design knowledge. Facing a new design problem, experienced designers do not design strictly based on abstract design principles, nor do they exhaustively search a space of previous design cases, testing whether each exactly matches the current design criteria. Third, in some situations, a certain amount of design knowledge as general formalisms is already available prior to the development of case-based designs. Case-based systems should make use of what ever the knowledge is available.

Generally, the development of case-based reasoning for design raises two major issues: representation of design cases and adaptation of a design case. Specifically, a case-based design system raises the issue of integrating different types of knowledge and methods of reasoning within the framework of case-based reasoning. The types of available design knowledge and reasoning methods vary from context to context, thus, leading to alternatives for design case representation, case memory organization and reasoning process combination. This paper discusses two fundamental issues: representing design cases and adapting structural design cases. A case-based structural design model which uses specific building cases and generalised design models is presented. At the end, a summary is drawn.

DESIGN CASE MEMORY

The representation of design case memory is the essence of a case-based design system. The details of design case memory indicates what the case memory can be used for, how cases can be retrieved, and whether there is enough information for case adaptation. When considering the representation of case memory for design, three major issues are raised:

- identification of the contents of a design case;
- representation scheme for design cases; and
- organization of design cases and generalized knowledge.

We present three alternative schemes for identifying the content and representation of design cases: a case/subcase hierarchy; function-behaviour-structure; and the use of text and graphics.

The organisation of case memory is concerned with the efficient retrieval of cases. The issue of case memory organisation is closely linked with case indexing schemes, retrieval strategies and update of case memory. A good case memory organisation might facilitate the efficiency and accuracy of retrieval. We present three approaches to the organization of case memory: a flat structure, a partitioned structure and a tree structure.

Cases as a Hierarchy of Subcases

A design description usually is a composition of multiple component systems, e.g., in structural design the components may be the lateral-system and the gravity-system. Using a hierarchy of multiple subcases is one approach to represent design information. Each design case can be described as a set of components that are linked to their parent by part-of relations. Each component may itself be a structured system with components in turn. Component systems in this form of representation are viewed as subcases. Subcases contain pointers that can be used to reconstruct the entire case. For each subcase, specific design information can be described by a set of attributes. In such a way, each subcase can be represented in terms of attribute-value pairs. The attributes define the vocabulary for describing the previous designs and the values identify the information specific to one design case. Thus, design decision making can be viewed as determining the feasible values for the relevant design attributes in a particular design system.

An example of a hierarchical representation of part of a structural design of a building is illustrated in Figure 1. In this example, "case1/2D-lateral-x-system", "case1/gravity-system", and "case1/lateral-x-beam" are three subcases in the hierarchy. Each subcase is described by attribute-value pairs. As shown in Figure 1, a gravity-system can be described by attributes "floor-type", "support", "action" etc.

A hierarchy of subcases makes it easier to handle complex design information. Furthermore, it provides access to parts of design cases to solve parts of new problems. Central to this representation scheme is the determination of a hierarchy paradigm. In many domains such as building design, the hierarchy of subcases is often defined based on design theory and principles which imply a decomposition ontology from a more general task to more specific subtasks. For the design tasks which can not be decomposed into loosely coupled subtasks, a hierarchy of subcases is not a suitable representation scheme. On the other hard, using the hierarchy of subcases often implies a fixed form for representing specific design information, which may be restrictive.

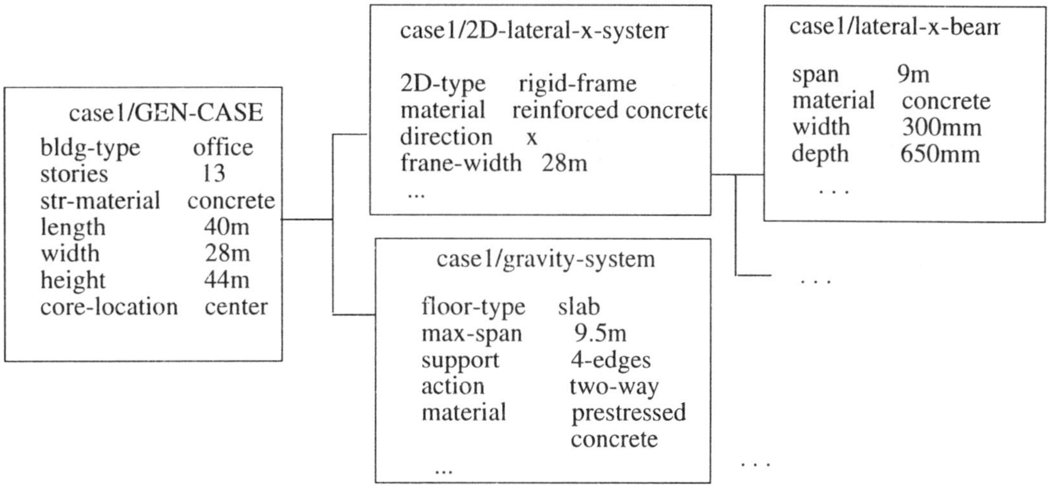

Figure 1. Part of a building design case (case1) using hierarchy of subcases.

Cases as Function-Behaviour-Structure (FBS)

Using a set of attribute-value pairs to represent design subcases does not indicate the different roles that design attributes in a design system play. Using the prototype scheme (Gero, 1990), the design information can be classified into three categories: Function, Behaviour and Structure. Design attributes in a design case thus can describe structural, functional or behavioural aspects of a case.

The function attributes are those referring to the purpose of the design case. The purpose of the design makes explicit what the design is to be used for, e.g., the purpose of a beam is to carry load over a distance; the purpose of a cup is to hold liquid. The function of an object is related to the perception of the object's use, rather than inherently related to its structure. The behaviour attributes describe the response of the design case to its use. They describes how the design case works, for example, the beam carries load through bending stress; the cup holds liquid by providing an enclosed volume. Behaviour attributes provide a means of testing the performance of a design to determine whether it achieves the intended function. The structure attributes describe the physical characteristics of the object. Structure attributes define geometry, material, colour, etc. Since the explicit definition of a design is an abstraction of the actual object, the structure attributes used to describe a specific design are typically related to its behaviour, so the performance of the design can tested, and to its construction or manufacture, so the design can be realised. Structure attributes that are not relevant to the design may not be specified in design case memory.

Including these three characteristics of a design case satisfies the need to include information that can be indices for a new design problem since a new design can be described in terms of its required function, behaviour, and possibly some required structural features. Figure 2 depicts a partial description of the FBS attributes of the structural system of a building.

FUNCTION ATTRIBUTE	building-type
BEHAVIOUR ATTRIBUTE	cost-of-construction
STRUCTURE ATTRIBUTE	building-stories material building-height building-width

Figure 2. Example of FBS attributes in building design.

Using FBS attributes to represent the contents of design cases makes explicate the roles the design information plays in the design process. In the retrieval of design cases, this scheme of representation provides the flexibility of recalling design cases based on required function, behaviour or structure. Two aspects need to be considered in choosing FBS representation scheme. The first problem is repetition in the function, behaviour, and structure attributes and their values across many design cases. Another one is the articulation of three categories of information. In some design domains, the function, behaviour and structure attributes of a design case are difficult to articulate.

Cases as Text and Graphics

The above two representation schemes use attribute-value pairs to encode specific design knowledge. In some designs the use of attribute-value pairs is too restrictive to represent the designers' intentions in a previous design. In fact, real world design information actually appears in the forms of text and graphics.

BUILDING-1
Beam Design Description:
The structural system uses wide and shallow beams since the floor-to-floor height is critical. The beam shells were designed with all positive beam reinforcement and cast in place.

Figure 3. A text description of a beam design in a building.

In some situations it may be appropriate to represent a design case as a text description. The text description can convey anecdotal information about the design or describe the structure of the design in a more holistic way. The use of text as a computational model of the previous design requires the identification of keywords, or indices, to the story for retrieval. The text description does not necessarily provide a structure from which the case-based design system can reason about the relevance of a story to a new design problem or about how to adapt the text description to fit the needs of the new design problem. An example of a text representation to describe the beams in a building is shown in Figure 3. The information conveyed in this representation is substantially different to the attribute-value representation. Even though the text could be converted to an attribute-value representation, some of the meaning would be lost. We suggest that text is used in addition to attribute-value pairs.

Graphics or design drawings provide another approach to represent a design case in order to provide more information. Graphics have been an unique and most effective form of communication in design. The graphical representation of design allows the designer to understand and to evaluate the design easily and quickly regardless of the complexity involved. Sometimes, using existing computer-based drawings for case-based reasoning results in less effort to create case memory. Using graphics or drawings as contents of design cases requires that graphics or drawings are computationally stored in a design case base. An example of a design drawing to describe a design solution is shown in Figure 4. Using graphics to represent design information leads to two main concerns: indexing of graphics and reasoning about graphics. The ability of computer systems to reason about graphics is still fairly primitive. Therefore, we suggest that attribute-value pairs provide the indices to the drawings and are the basis for reasoning about the drawings. The use of graphics is to assist the human designer in comprehending the information in a design case.

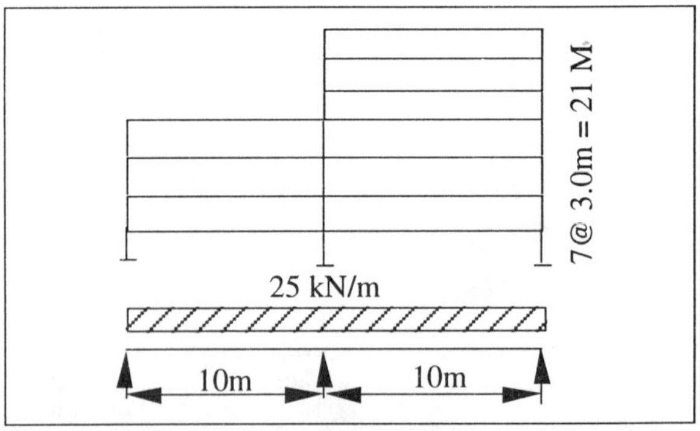

Figure 4. A graphical representation of a frame system.

Organising Cases in a Flat Structure

The simplest approach to design case memory organisation is to store cases in a flat structure. A design case memory organisation can be regarded as a flat structure if design cases are put in a sequential data structure such as a list. Computationally, this flat structure can be a simple list, an array, or a file. Moreover, in a flat structure, the name of each design case is stored in a flat data structure as indices. The flat organization structure is suitable for a case memory with a small number of cases since the search time is not excessive. The advantage to a flat structure is that all cases can be considered as candidates for adaptation without bias. The disadvantage is that as the number of cases grows, the search time increases and retrieval becomes very inefficient.

Organising Cases in a Partitioned Structure

Another approach for organizing case memory is to partition design cases into several groups. A set of critical design features are selected to describe the major characteristics of the design cases in a group. In this approach, the selected features and their values serve as indices. Figure 5 gives an example of a partitioned structure of case memory. Design cases are organized by a partitioned structure based on three features, building-height, critical-height-ratio, core-location. In retrieval, the values of attributes "building-height", "core-location" and "critical-height-ratio" in a given problem are used in searching for relevant cases.

This method for the organization of case memory is more efficient than a flat structure when case memory is large, however, it has the disadvantage of localising the search for a relevant case on the basis of a few key features. This approach also has problems when case memory becomes very large since the number of groups is fixed, the number of cases in a group can become too large for efficient retrieval.

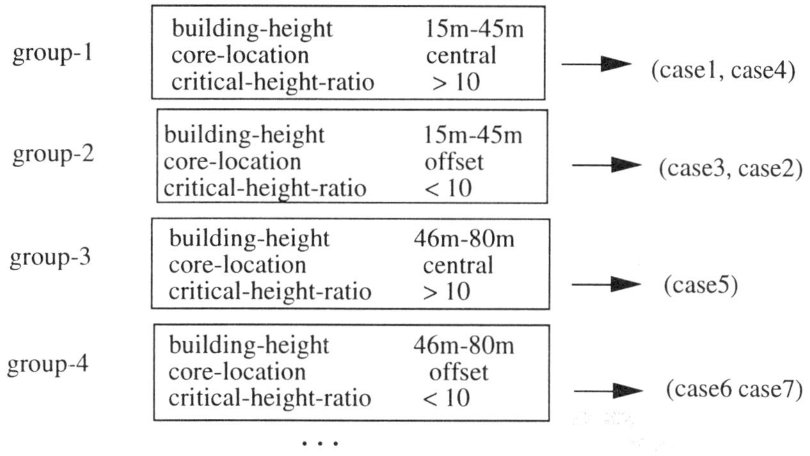

Figure 5. The groups of design cases based on a partitioned structure.

Organising Cases as a Tree Structure

Another approach for organizing a large case memory is based on the idea of tree structure. The tree structure of case memory organization can be built by using a clustering method or Dynamical Memory concept (Shank, 1982). Using clustering methods, design cases are clustered according to design features. Features shared by many cases are identified. These features provide a basis for partitioning cases. This is similar to the partitioned memory organisation except that the features are represented in a tree and a path in a tree leads to a group of cases. The advantage of clustering-based tree are that it supports an incremental retrieval of design cases and the number of design cases actually considered is reduced to a subset of all cases. However, this tree structure implies that the hierarchy of nodes corresponds to the importance of features. If unimportant features are high in the tree, there is a chance that cases that share important features with the new problem will not be retrieved. Updating case memory with such a tree structure is more complex since the tree structure could be altered during update.

An example of tree organization based on clustering is illustrated in Figure 6. Cases are first clustered according to the bldg-type as office or hotel, then according to the ranges of stories.

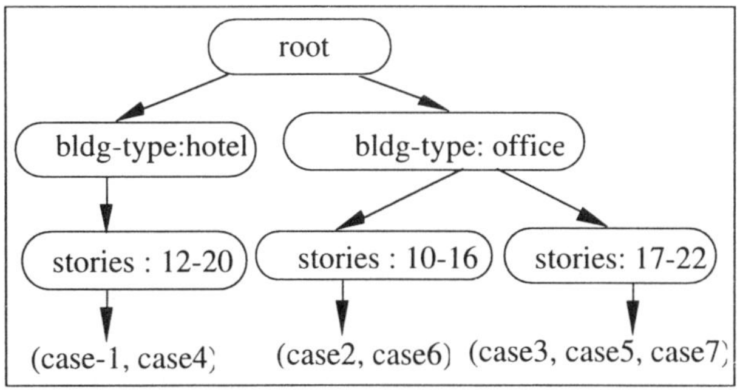

Figure 6. A tree organization based on clustering.

In the dynamic memory organization, design features and their values that are used to described the contexts of design cases are used as indices in a tree. The attributes and their values in the design cases are organized in a two level tree. First level contains the names of the attributes. For each attribute node, its possible values form the second level. Each value of an attribute points a set of design cases which have such an attribute-value pair. Given a new design problem in terms of a set of attribute-value pairs, each attribute-value pair is used as an index to find relevant cases. This structure is dynamic (Shank, 1982), in the sense that the memory can be reorganized without altering memory structure when new cases are incorporated, or cases are removed. Dynamic memory supports the retrieval of all relevant cases in case memory to the given problem because every design case which has one or more attribute-value pairs matched with the problem can be found.

ADAPTING DESIGN CASES

Adapting design knowledge from past experience to the solution of a new problem or to the process of constructing a solution to the new problem is problem solving process itself. In theory, transformational analogy and derivational analogy (Carbonell 1983, 1986) provide two basic transfer inferences to adapt designs. But how to practically apply the ideas suggested in these methods varies with each design domain. Moreover, the complexity of structural design problem solving has been consistently underestimated in the past literature with a tendency for it to be dismissed as merely requiring the application of simple algorithms and heuristics.

Following the generate and test principle, the adaptation of design cases consists of a generator, which defines a design solution; and an evaluator, which verifies solutions. In fact, in developing the method for carrying out the adaptation of design cases, it is necessary to consider domain-specific knowledge and underlying

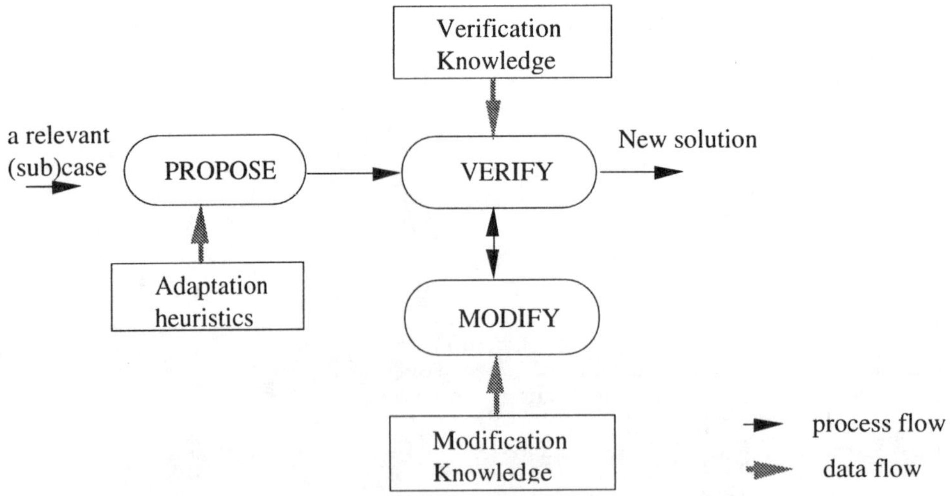

Figure 7. A general paradigm for adapting previous designs.

problem solving approaches. A general paradigm for adapting structural design cases is presented in this section, which assumes that case retrieval provides a description of a specific design solution that is acceptably close to the final solution, and adapts those aspects of the proposed design solution that are inconsistent.

Propose-Verify-Modify

The general adaptation of design cases is described as a transformational process of changing previous design descriptions along with the necessary verification and modification of the adapted design into a new design description. Propose, verify and modify can be viewed as three elementary tasks The adaptation process involves efficiently proposing plausible solutions; verifying the validity of those solutions; and modifying invalid design parts as necessary as illustrated in Figure 7.

In the propose operation, the design problem specifications of a previous case are compared with that of a new design problem. A plausible design to the new problem is proposed by structurally considering the previous design solution as a new one, during which the discrepancies of the design specifications between new and old problems are resolved. This phase of adaptation utilizes an empirical methodology to structurally or syntactically adapt the previous design case. Specifically, this phase of adaptation is concerned with changing the design case to establish the consistency of old and new design contexts. A set of operations can be applied to previous designs, i.e., adding new design specifications; substituting design features with those in the new design context; and removing those features of the description unrelated to the new design context. The construction of a proposed design based on consistent mapping from old to new contexts introduces inconsistencies within the adapted design, so the potential solution must be checked for validity.

Verifying the proposed design solution ensures the solution is valid by testing if it conforms to general design principles and if its actual functions and behaviours meet expected ones. The knowledge used to justify and verify design decisions is design task specific. When an inconsistency is found in a design solution adapted from previous design cases, the adaptation process must determine what parts contribute to the inconsistency, then take actions to repair the infeasible part of the adapted design. The verification and modification of the adapted design are often combined as a analytical process. This analytical process uses domain specific knowledge and specific reasoning models during the assessment of the proposed design's feasibility and the repair of any invalid design elements. The strategies for the evaluation and modification of a proposed design case are generally influenced by the representation of domain knowledge and reasoning processes. Techniques for analytical adaptation often rely on underlying problem solving processes, e.g., constraint satisfaction, qualitative reasoning, model-based reasoning, heuristic reasoning etc.

Assumed General Knowledge for Adaptation

Previous designs can not be reused without making some changes. Reasoning about those changes requires domain specific knowledge to guide design decision making during adaptation. In addition, adaptation is a type of plausible reasoning and the result of adaptation is often plausible but inconsistent. Adaptation of structural design cases necessarily uses knowledge to revise the adapted design and maintain the feasibility of the adapted design.

In general, knowledge about adaptation represents causal relations or reasons for decision making. This knowledge defines the capability of case-based structural design systems to generate design solutions. According to the role domain knowledge plays in the adaptation of design cases, domain knowledge for adaptation can be characterized as verification knowledge and modification knowledge. During the adaptation of design cases, verification knowledge is used to revise the case for consistency with the new problem and compare the adapted design against design criteria. This type of knowledge can be represented as general rules or constraints. Modification knowledge guides the refinement of the adapted design by modifying infeasible design decisions. Considering the decision making in terms of attributes and their values, the representation of modification needs to provide design attributes and the range of its alternatives value in order to support the repair of invalid decisions.

The knowledge for adaptation can be represented through formalisms as generalized design knowledge, or embodied in individual design cases as part of the content of a design case. If the knowledge for adaptation is kept within each case, design cases may be very large, or there may be a representation of the same general knowledge in many cases. Generally, the reasoning methods used for adaptation determines the representation of adaptation knowledge.

EXAMPLE OF CASE-BASED STRUCTURAL DESIGN

A case-based structural design system has been developed based on the case representation schemes, organization structures and adaptation strategies that are described previously. This section presents examples to illustrate the retrieval and adaptation of design cases in our model.

New Design Problem

The problem description is specified as a set of attribute-values pairs that states the design requirements. An example of specifications for a new problem is given in Figure 8. The new design task is a 12-story hotel with certain requirements specified, e.g., the location of core, design load and geometric grid information etc.

bldg-type	hotel
cost	12 million
stories	12
length	36m
width	23m
height	41.5m
typical-fl-ht	3.4m
core-location	center
plan-shape	rectangular
wind-load	20psf
...	
grid-lable	(ab c d e f)
grid-offset-x	(8m 6m 8m 6m 8m)
. . .	

Figure 8. Specified requirements for a hotel design.

Case Retrieval

The retrieval process in our model searches the design case memory for relevant design cases to the given problem. A design case is represented as a hierarchy of subcases in terms of attribute-values. Each case is comprised of a supercase and a set of subcases. The supercase provides a general description of the structural design for a building. Each subcase represents a description for a structural system, e.g., gravity system in a particular design context. Design cases in our model are organized as a tree structure using the concept of dynamic memory. All attribute-value pairs used to describe the context of design cases and subcases are used as indices in organizing design cases. The organization of design (sub)cases is a two-level tree, where the first level is attributes' names and the second level is the relevant values for each attribute. Design (sub)cases are pointed by the value of attributes.

As illustrated in Figure 9, attributes used to describe the design case and subcases form the first level, e.g, "bldg-type" and "core-location" for the overall cases, "floor-type", and "action" for subcases about gravity-system. Each value of attributes point to a set of cases containing subcases that have such attribute-value. For example, "case1" and "case3" are two office buildings, "case1/gravity-system" uses slab as floor-type.

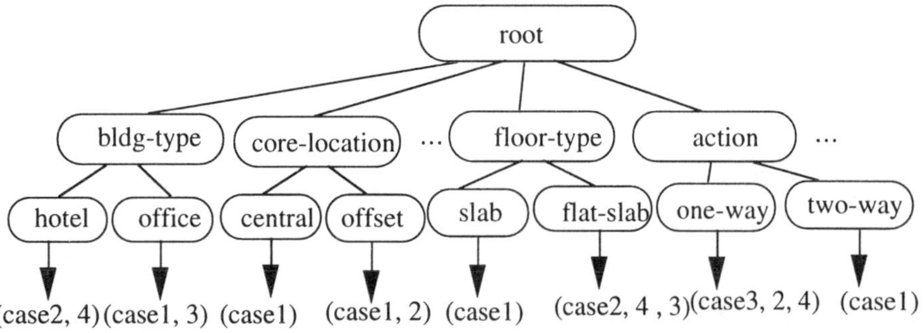

Figure 9. The tree structure of case memory organization.

```
bldg-type        hotel
cost             12 million
stories          12
length           36m
width            23m
height           41.5m
typical-fl-ht    3.4m
core-location    center
plan-shape       rectangular
wind-load        20psf
...
grid-lable       (ab c d e f)
grid-offset-x    (8m 6m 8m 6m 8m)
. . .
```

```
Gravity-system

max-aspect-ratio   1.56
typical-span       8m
floor-type         slab
support            4-edges
action             two-way
thickness          180mm
material           concrete
...
```

. . .

Figure 10. Proposed solution to hotel design problem in Figure 8.

Given the new design problem as shown in Figure 8, relevant design cases to it are retrieved based on the organization in Figure 9. In retrieval of design cases, each attribute-value pair in the new problem description is used as an index to search for design cases. For example, "case1"and "case3" are retrieved using attribute "bldg-type" with value "hotel", "case1" is retrieved based on a central core-location. Retrieved design cases for the problem in Figure 8 include "case1" and "case3" according to the organization of case memory. Then "case1" is selected as the most relevant design case to the problem since it has the most features in common with the problem. The partial description of "case1" is illustrated in Figure 1.

Case Adaptation

The adaptation of the structural design case in our model is performed by a structural substitution of attribute-value pairs for the propose operation and a constraint checking approach for the verification and modification operations.

At the beginning of the adaptation process, a potential design solution is generated by applying adaptation rules favouring the new design requirements. Figure 10 shows the proposed solution using the case in Figure 1 for the new design task in Figure 8. This potential solution is produced by eliminating the differences in problem specifications between the selected case and the new problem. In the potential solution of the gravity system, the typical-span is adapted using the new design context, whereas floor-type and support etc stay the same. Some inconsistencies are caused by this substitution process.

The proposed solution is verified by checking design constraints. The violated constraints identify the attributes that should be modified. The left-hand part of Figure 11 shows examples of design constraints. CONSTRAINT-1 specifies that the slab is normally not compatible as floor type in such a building as a hotel or apartment. CONSTRAINT-2 and CONSTRAINT-3 indicate whether the flat-plate and flat-slab work as one-way or two-way given their typical-span. According to the design constraints, the attribute-value pairs with bold headings in Figure 10 are found invalid and need to be modified.

```
CONSTRAINT-1
  (bldg-type) in (hotel apartment)
  (gravity-system/floor-type) = slab

CONSTRAINT-2
  (gravity-system/floor-type) in (flat-plate flat-slab)
  (gravity-system/action) = one-way
  (gravity-system/typical-span) < 10m

CONSTRAINT-3
  (gravity-system/floor-type) in (flat-plate flat-slab)
  (gravity-system/action) = two-way
  (gravity-system/typical-span) > 9m
```

```
system GRAVITY-SYSTEM

  ...
  floor-type   (flat-plate flat-slab slab)
  action       (one-way two-way)
  support      (cols-only 2-edges 4-edges)
  ...

end
```

Figure 11. Design constraints and the description of the "gravity-system" model.

To modify the invalid attribute-value pairs, the generalized design models are used as modification knowledge. The generalized model gives the design attributes and their alternative values. As illustrated in the right-hand part of Figure 10, the generalized "gravity-system" model provide alternative values for attributes "floor-type", "support" and "action". To modify the invalid design decisions in gravity-system, the values of "floor-type" and "action" are replaced by "flat-slab" and "one-way". The revised solution is then verified again. This process proceeds until all constraints are satisfied and a feasible design solution is found. The completely consistent gravity-system solution to the new problem is eventually produced as illustrated in Figure 12.

Gravity-system	
max-aspect-ratio	1.56
typical-span	8m
floor-type	flat-slab
support	4-edges
action	one-way
thickness	180mm
material	concrete
...	

Figure 12. The final solution to gravity-system.

CONCLUSION

In developing a case-based design system, the representation and adaptation of design cases are two fundamental issues to be addressed. Three representation schemes have been discussed to represent specific design information. They are characterized by the approaches to encode design knowledge and the roles they play in case-based design systems. A hierarchy of subcases is a scheme to represent design knowledge as multiple levels of abstraction, which provides a way to make use of an entire design case and/or specific part of a design case. The FBS scheme makes explicate the roles design features play in the design process. When new designs are described in terms of required function, behaviour and possibly some required structural features, this scheme can supports flexible retrieval of design cases based on functional, behavioural or structural features. In using a hierarchy of subcases and/or FBS as representation schemes, design information in real world projects needs to be formalized into attribute-value pairs. Sometimes, direct formulation of the case data from design projects may be difficult due to the complexity of design process. However, documents in terms of text and drawings are readily used as a powerful medium to handle the complexity of design information. For example, drawings used during conceptual design express an alternative design solution through sketches of form, interaction, and intended behaviour. Since it is difficult to reason about text and drawings, the text and drawing are used as a scheme to represent design cases in case-based design aiding systems rather than design automation systems.

To organize design cases, three structures have been discussed. In general, the size of case memory plays an important role in choosing an appropriate approach to organize design cases. A flat structure is often used in a relatively small case memory. A partitioned structure of case memory supports selective retrieval. A tree structure provides a more efficient search of relevant design cases. More specifically, the determination of an organization of case memory is associated with the accuracy of similarities between the retrieved cases and new problems. For example, some potentially similar design cases may not be retrieved in a tree structure built by clustering methods.

The adaptation process is central to a case-based design automation system. In theory, adaptation of cases performs changing old cases to fit new situations by proposing a new solution. In adapting design cases, the nature of design imposes new requirements. Since proposing a solution by adapting the design context of cases to fit a new context leads to inconsistencies within the proposed design solution, there is the need for verification of the proposed design and modification of the proposed design if required. Consequently, it is essential to apply underlying reasoning methods and domain knowledge in adaptation of design cases. In different case-based design systems, the types of available design knowledge and reasoning methods vary from context to context. A general adaptation paradigm has been described to characterize the adaptation generally. The propose, verify, and modify are characterized as three tasks in adaptation.

REFERENCES

[Car83] Carbonell, J.G. Learning by analogy: formulating and generalizing plans from past experience, Machine Learning Volume 1: An Artificial Intelligence Approach, Tioga, Palo AltO, Calif. pp.137-161, 1983

[Car86] Carbonell, J.G. Derivational analogy: a theory of reconstructive problem solving and expertise acquisition, Machine Learning Volume 2: An Artificial Intelligence Approach. Morgan Kaufmann, San Mateo, CA. pp.371-391. 1986.

[DK92] Domeshek, E.A. and Kolodner, J.L.. A case-based design aid for architecture, Artificial Intelligence in Design'92. J.S. Gero (ed.) Kluwer Academic. pp.497-517, 1992

[Ger90] Gero, J.S. Design Prototypes: A Knowledge Representation Schema for Design. AI Magazine, 26-36, Winter, 1990

[GK91] Goel, A. and Kolodner, J.L. Toward a case-based tool for aiding conceptual design problem solving, in Proceedings of the DARPA Workshop on Case-Based Reasoning, Washington, pp.109-120, 1991.

[HF93] Hua, K. And Faltings, B. Exploring case-based building design CADRE, AI Engineering Design Analysis and Manufacturing Journal, 7(2): 135-143, 1993.

[MZ93] M.L. Maher and D.M. Zhang. CADSYN: A Case-Based Design process Model. AI Engineering Design Analysis and Manufacturing Journal, Vol 7, No. 2. 97-110, 1993.

[MB94] Maher, M.L., and Balachandran, M.B. Flexible retrieval strategies for case-based design, in J.S. Gero (ed.) Artificial Intelligence in Design'94, Kluwer Academic Press, 1994.

[Sha82] Shank, R. Dynamic Memory: A theory of Learning in Computers and People. Cambridge University Press, 1982.

[SN89] Sycara, K.P. and Navinchandra, D. Integrated case-based reasoning and qualitative reasoning in engineering design, Artificial Intelligence in Design, J. S. Gero (ed.), Springer-Verlag. pp.231-250, 1989.

[WH91] Wang, J. and Howard, H. C. A design-dependent approach to integrated structural design, Artificial Intelligence in Design'91. J.S. Gero (ed.), Butterworth-Heinemann, Oxford, pp.151-170, 1991.

USE OF NEURAL NETWORKS FOR CASE-RETRIEVAL IN A SYSTEM FOR FASHION SHOE DESIGN

Julie Main, Tharam S. Dillon and Rajiv Khosla

Dept of Computer Science and Computer Engineering, La Trobe University,
Bundoora, Victoria, Australia, 3083.
also Expert and Intelligent Systems Laboratory, Applied Computing Research Institute
Email: main@latcs1.lat.oz.au

ABSTRACT

Case-based reasoning is a powerful mechanism for developing machines that can learn from, and adapt past experiences. This paper looks at how Case-Based reasoning can be applied to a shoe-designing system and also examines the possibility of using neural networks as a method of retrieval in such case-based systems. It looks at the representation of the various forms of data encountered and the representation of it to enable it's use in the case base and the neural networks. Also, the methods, issues and decisions involved in implementing the system.

INTRODUCTION

Fashions change from season to season, and even mid-season. However fashion trends have a cyclic behaviour. A fashion 'look' will often go 'out' and a number of years later make another appearance. Another factor in women's fashion is the diversity. Even within a single trend, there are a multitude of variations. People usually want something that is 'in', but a little different, or a little better than the other shoes on the market. A company that can create new designs to satisfy this demand increases its chance of success by giving itself an edge in the marketplace.

There are hundreds of designs produced each year and a designer forgets all but a few of the main ones in a relatively short span of time. Thus, when a fashion trend makes a second (or third) appearance, the previous designs have been forgotten and the new shoes are re-designed from scratch. This forgetting and starting over procedure is a problem that if overcome could lead to a faster design stage and also to the input of a wider variety of ideas than the designer has access to at any given time. Therefore, if a computerised system could be implemented that would retain a memory of previous designs, could retrieve appropriate ones and then modify them to current needs, much of the design work could be taken out of shoe designing. This would decrease the response time to markets. One group of techniques that offers promise here is case-based reasoning.

Case-based reasoning is a subset of the field of artificial intelligence that deals with storing past experiences or cases and retrieving relevant ones to aid in solving a current problem. Case-based reasoning consists of: storing and indexing cases so that the appropriate ones are retrieved; retrieval of the appropriate case; and then adapting the retrieved case to help solve the current problem case.

Problems remain with the case-based approach both in deriving efficient characteristics for indexing and with the retrieval methods used for recovering the past case(s). There are several main types of retrieval.

The simplest way of retrieving cases is using a nearest neighbour search. A nearest neighbour search looks for those cases stored in memory that have the greatest number of characteristics that are the same as the current problem. The weighting of characteristics used has to be determined by the implementer. It is frequently difficult to decide which characteristics are more important than the others, in that they contribute more to the successful outcome of a particular case.

Inductive approaches to case-retrieval use statistical algorithms to determine which characteristics of cases are most important in the retrieval of those cases. Inductive approaches are often better at achieving correct weightings than the weightings used in nearest neighbour search, but inductive indexing methods have their own drawbacks. The first of these problems is that a large number of cases are required for deriving satisfactory weightings for the attributes of the cases. The other problem is that not all cases are determined by the same order of importance of their characteristics. For example, one case may be primarily determined by three particular characteristics and those characteristics may have little or no importance in determining the outcome of another case.

Knowledge based approaches to the retrieval of cases use domain knowledge to determine the characteristics that are important for the retrieval of a particular case, for each individual case. In other words, a case is indexed in this type of approach by the features that are important in determining that case's success or failure. This is a good method for case indexation as it is efficient in retrieving the correct case. In using this type of indexation, the cases can be stored in a hierarchical fashion, thus limiting the cases to be searched. Unfortunately, to implement this type of indexing system, a large amount of domain knowledge must be known, acquired and applied to the cases.

It can be seen that none of the above three mechanisms seems entirely appropriate for a shoe design system which contains a large number of cases, where factors in different cases are more important than those factors in some other cases and where formal domain knowledge is very limited. Therefore to build the proposed system, a different method of correctly and efficiently retrieving the cases was tried. One approach that was believed could overcome the expected problems and was worth exploring was the use of neural networks as a method of indexing and retrieving cases.

PROBLEM STATEMENT

The problem that the project aimed to solve was the creation of a system which will aid a footwear designer in creating a conceptual level design of a new shoe. In making the description of this problem more precise, one first needs to characterise the nature of the input data, the nature of the outputs from the system and the nature of the knowledge.

In planning this shoe designing system, numerous types of data were encountered that had to be represented. What follows is a brief summary of each input type, how it was represented and an example of each type that is relevant to the footwear domain.

Boolean - Boolean data, or the type of variables that can be true or false, are needed to model questions about a shoe design that can be answered with a yes or no. For example, we can ask whether a shoe has an open toe. It either does or it does not.

Continuous - In this system, an example of a continuous variable is 'percentage perforation'. The percentage of the shoe that is open or 'cut-out'. Such a value can be 9%, 10% or 11%. It can be 10.5%, 9.11% or 6.76%. Hence, the input for this type of data to the neural network will be a decimal fraction between 0 and 1.

Multi-valued - Multi-valued data can have several different yet deterministic values. In the shoe domain, an example of such a multivalued variable is the attribute 'fastening', which describes the method by which the shoe stays attached to the foot. This can occur in a number of different ways: laces, buckle, zipper, elastic or none and hence 'fastening' can take one of these five values. It is interesting to note at this point that while 'fastening' is a multi-valued variable it may be used to define other (namely boolean) types. In the class hierarchy devised for the system, one of the distinguishing features at the top level is whether the shoe or boot has laces. If the value of 'fastening' is laces then this new attribute 'laces?' is true, and if the attribute 'fastening' is any of the other four values 'laces?' will be false. It can be seen from this that multivalued variables can be modelled as a collection of boolean variables.

Fuzzy - Perhaps the most interesting type of data encountered is fuzzy data. It is best explained by an example. The height of a shoe can be low (for example, a type of sandal called a mule), normal (for a shoe), and high (for a boot). We have to classify the data into three ranges of values to correspond to low, medium and high as already described. A shoe with an inside-back height of 0.5" is going to be low, a shoe of inside-back height of 1.5" is medium (a shoe), and one with a value of 2" or above is high (a boot). However, what is a shoe with a height of 1.9"? It is a little higher than a traditional shoe, but not really a proper boot either. It could in fact be classified as either. Therefore, it was decided to represent attributes such as shoe height as fuzzy variables and this is discussed further in a following section.

The output from this system should be a set of previously made designs that most closely approximate the desired design. These designs will be used in a future system, to develop adaptations that provide the desired design.

There are two very distinct components of knowledge available, both of which have to be used together to determine the required solution. These are:

i. The large collection of previously existing designs. To create a case base for use in this system, Main Jenkin Pty. Ltd, a Melbourne based footwear manufacturer, provided a hard copy of its designs from the previous few years. These designs needed to be sifted through to find the relevant designs that fit into the women's 'fast' fashion category.
ii. The expertise of the designers.

ASPECTS OF SOLUTION METHODOLOGY

There were a number of different aspects of the solution methodology that needed to be considered at this point:

i. *Representation of Cases*
   The cases obtained needed to be converted to computer readable data. This involved manual input of facts into the case base and also scanning in pictures of the designs and converting them to the appropriate format.

ii. *Choosing sub-classes and classes*
   One of the more difficult tasks of this project was grouping the cases obtained into their respective classes and subclasses. There are no hard and fast rules in the industry for this sort of categorisation so this stage involved extensive consultation with shoe designers and experts.

iii. *Choosing distinguishing characteristics of cases*
   Once the classes and subclasses of cases were determined it had to be decided which features of those cases led to their categorisation in a particular class. Again this stage required extensive consultation with industry experts.

iv. *Implementing the neural networks*
   To use artificial neural networks to retrieve cases, the neural networks first had to be created and then trained. Training the networks involved converting the data obtained into normalised inputs for input into the nets, and then running the networks through a training phase.

v. *Retrieval of appropriate cases*
   After the cases had been entered into the case base and had been appropriately indexed and classified, a method needed to be determined for using the networks for retrieval of the appropriate case. This could have been done using a system of rules, methods or daemons, whichever was found to be more efficient in the shell chosen for implementation.

vi. *Interaction with user*
   An appropriate interface with the user needed to be provided so an input specification can be entered into the case-based system and results provided.

vii. *Case adaptation*
   When a retrieved case does not meet all of the requirements specified by the user, or a user wishes to alter the case, a mechanism needs to be provided for case adaptation. Whether this takes the role of converting the retrieved case or merely providing suggestions, modifications will be determined taking into account the resources available in the creation of the system and the determined importance of this stage of the reasoning process. This stage is not addressed in this paper.

SOLUTION STRATEGY

Once it was decided that neural networks would be used to retrieve the cases in this system, more detail on the type and number of networks in the implementation was necessary.

Firstly the number of neural networks in this system had to be determined. With over 400 cases collected and grouped into classes and sub-classes, it was necessary that more than one network be used. It is not feasible that a single neural network be used to classify the number of cases obtained into so many subclasses in the one hit. Therefore it seems reasonable to have a neural network at each classification level. This means that one neural network is used to classify all the cases into the highest level classes. For each of those high level classes, a further neural network will be used to divide the cases in that class into its sub-classes. This is repeated for each sub-class until the lowest level sub-classes are reached.

The domain that is represented in this system determines that the classification of cases at each level is non-linear. This means that for each classification a single-layer network is not sufficient. A multi-layered network is required. Therefore, for this system three layered networks, ie. with one hidden layer, are used at each level to determine the classification. The use of a three layered network allows the networks to build their own internal representation of the inputs and determine the best weightings. As it is known how each particular case is classified into its classes and sub-classes, supervised learning is taking place. In other words because the input features and the output goal are provided, a teacher can be said to be training the network, hence 'supervised' as opposed to 'unsupervised' learning is taking place.

It seems an obvious step from here to define the learning algorithm that is used to determine changes in weights during training. Taking into account the use of multilayer networks and supervised learning, the use of the back-propagation algorithm for learning was an obvious choice.

FUZZY REPRESENTATION OF SOME DATA VALUES

As mentioned in the example of classifying a shoe as high, medium or low, some shoes would not fit in the crisp separations of categories. In such a case as described previously, that is a shoe with a height of 1.9", it would be nice to say that it is partly a shoe and partly a boot. This can be represented for input to the neural network by using 3 attributes. 'low', 'medium' and 'high'. Each shoe would have a value for each of these three variables. A low shoe (or mule) would have attributes with these values:

low:1 medium:0 high:0

A boot or shoe with a large back-height would have the following:

low:0 medium:0 high:1

A traditional shoe (of 1.5") would have the following values

low:0 medium:1 high:0

This is very straight forward for the instances that fall exactly into one of the main categories. However, for those cases that do not fit so nicely (for example the 1.9" case), we needed to define a function that determines the value for in-between cases.

This function can take many forms, but as an initial function we used straight line functions. The actual function to determine the values associated with the height of the shoe is shown in Figure 1.

In this case the values for the shoe with height 1.9" can be seen to be:

low: 0 medium: 0.2 high: 0.8

Fuzzy data can be represented using two or more functions. They do not have to be straight or even continuous functions. In this project, only two and three continuous functions were used to represent the values for a fuzzy variable.

USING OBJECTS TO REPRESENT CASES

The next question that arose is how would the cases be represented for storage. As the shoes are first to be divided into classes and subclasses and each shoe is an instance of a class, an object-oriented representation is

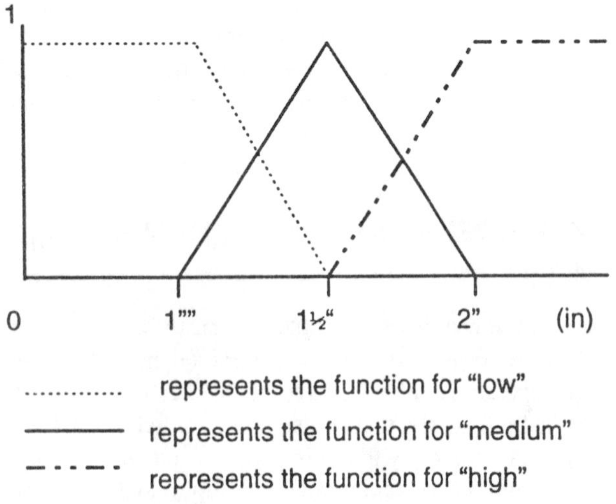

FIGURE 1: Fuzzy representation of the height of a shoe.

the obvious solution. Particular shoe designs can be represented as objects, being instances of particular classes. Groupings such as 'sandals' or 'mules' can be represented as subclasses and every subclass or individual object is a division of the superclass of shoes. Conceptual modelling of this domain as outlined in Dillon & Tan [DT93] was carried out.

Individual shoes could then be attached statically to the lowest level classes in the hierarchy to be easily retrieved when a neural network has determined the lowest sub-class to which the current input can be associated. The number of objects that can be attached to a sub-class is large, but it needs to be limited to a small number so that only the most relevant cases are retrieved. Each sub-class has its own number depending on how many cases are associated with it. This may range from as little as 1 to as many as 20.

META-CONTROL STRATEGY

To control the reasoning process a reasoning or meta-control strategy needed to be decided upon. Rules were a possibility and so were methods, though it was believed that the most appropriate strategy for this system is to use daemons attached to slots within the objects. Thus, not classical objects but objectfs as described by Dillon and Tan [DT93] should be used.

Rules combined with calls to the neural networks would be able to control the reasoning process, however they are deficient in their lack of a structuring facility. They make modification and addition to the system difficult as it involves comprehensive restructuring of the rules.

Using methods associated with the objects is also a possibility and this overcomes the rule's modularity problem. Methods ensure encapsulation and aid in the ease in which the system may be adapted and maintained.

Daemons were chosen however as not only are they encapsulated within the objects, the meta-control strategy is more flexible, and control is passed down the hierarchy as the system narrows down the possible classification through the class - sub-class hierarchy. NEXPERT OBJECT, the shell chosen for implementation has the facility to easily incorporate these into the reasoning process.

To actually use these slots and daemons for the retrieval process, the following occurs: Firstly, a slot in the highest level class ('shoe') is changed, and this sets off the daemon attached to the if-changed part of the slot. The daemon that is set off calls the highest level neural network (external to the object). This network returns a value that determines which sub-class the current specification is best classified in. On receipt of this information, the daemon changes a slot in the appropriate sub-class and the process repeats again. The slot that is changed in the sub-class likewise sets off a daemon which calls a lower level network. This continues until the lowest level sub-classes are reached.

Once the cases have been classified as members of particular low-level sub-classes, simple rules are used to display the relevant cases (or objects) attached to that sub-class. If the low level category still contains a reasonably large number of cases, but not enough for a further classification, simple rules could be used to retrieve only the best cases. In such a situation where there are a number of cases that cannot be divided into smaller groups, the cases tend to be very similar, with very little variation between them.

NEURAL NETWORK ISSUES

Structuring the features of classes

The determination of the features that are important in the classification of cases into each class was a major knowledge engineering task. For example to determine the features that should be used as inputs into the highest level neural network we have to look at the classes into which the super-class is to be divided, and how the cases in each of those sub-classes differ from each other.

Distinguishing these features is not easy. If it were, we could easily construct rules to do the classification. Determining the correct features involved looking at the cases, seeing the major differences between them, and using these features to attempt to successfully train a network. When a network learnt the input patterns to give a low sum of errors, then the features we had determined were correct. If there were patterns that this network could not classify, we needed to add new features that would distinguish between these cases. For example, sometimes the inputs to a neural network were the same for cases in more than one output class. When this occurred obviously there was a difference between these cases that the attributes defined had not captured.

It was still helpful when the networks did not train as the cases that were not learnt led us to distinguishing important features that were previously undetected. Once the additional features had been determined, the input patterns for the neural network needed to be modified to include the additional attributes and the networks retrained. This training and testing cycle continued until all the networks had low error rates.

Hierarchy of classes

Creating the hierarchy of classes was another significant task. Each class (except the lowest level) had to be divided into component sub-classes, and at the lowest level, instances (particular designs) had to be attached to the appropriate sub-classes.

In each layer of classification, the division is not clear cut. For each class there is more than one way of sub-dividing the class. The desired classification was the one in which the cases that are retrieved most closely match the problem specification and those cases in the class will be useful in the new shoe design. Hence, when choosing the sub-classes of each class, it was important to choose the classifications that groups the cases that are most similar conceptually and at a design level.

It is never possible in a domain such as footwear to arrive at a definitive solution. However it is possible that in the majority of cases the class hierarchy will group the cases usefully. Therefore design of the class structure for the system was carried out in a way that divided cases into groups that are similar conceptually and from a design perspective.

COLLECTION OF DATA

The first step in implementing a case-base reasoning footwear-design system involved collecting previous designs from Jenkin. Jenkin keeps the information they use in the production of shoes on paper 'specification sheets'. Collecting these involved finding and copying the relevant sheets. Over 400 different previous designs were obtained using this method. While specification sheets do not contain all the necessary information to implement the case-based reasoner, they contain enough information including a picture of the design, to deduce the data that is not specifically mentioned on the sheet. A sample specification is shown in figure 2.

USING NEXPERT OBJECT

Nexpert Object was chosen as the tool in which to construct the system. Nexpert has a number of features that particularly leant itself to this system. The classes and objects can be transferred directly into it, and inheritance of attributes and slots is automatic. Slots and daemons can also be directly transferred into Nexpert as it has provision for if-changed meta-slots and the firing of daemons. Nexpert can also make external calls to the neural networks outside the system and can be fully integrated with C programs.

Therefore, the next thing that was done was to transfer all the data obtained into Nexpert Object. The classes and objects were easily transferred and the attributes of the instance objects could be defined and initialised using the 'initvalue' command in the order-of-sources part of the meta-slots.

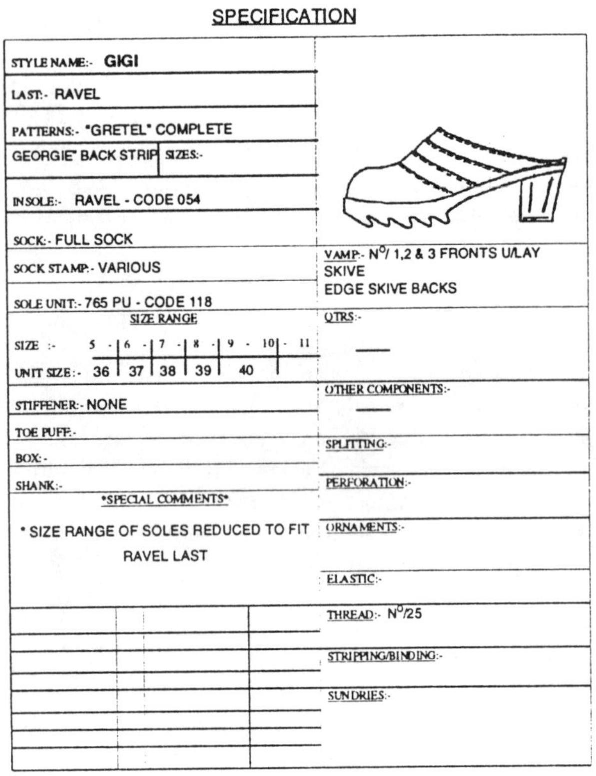
SPECIFICATION

STYLE NAME:- GIGI
LAST:- RAVEL
PATTERNS:- "GRETEL" COMPLETE
GEORGIE" BACK STRIP SIZES:-
INSOLE:- RAVEL - CODE 054
SOCK:- FULL SOCK
SOCK STAMP:- VARIOUS
SOLE UNIT:- 765 PU - CODE 118
SIZE RANGE
SIZE :- 5 - 6 - 7 - 8 - 9 - 10 - 11
UNIT SIZE:- 36 37 38 39 40
STIFFENER:- NONE
TOE PUFF:-
BOX:-
SHANK:-
SPECIAL COMMENTS
* SIZE RANGE OF SOLES REDUCED TO FIT RAVEL LAST

VAMP:- N°/ 1,2 & 3 FRONTS U/LAY SKIVE
EDGE SKIVE BACKS
QTRS:- ——
OTHER COMPONENTS:- ——
SPLITTING:-
PERFORATION:-
ORNAMENTS:-
ELASTIC:-
THREAD:- N°/25
STRIPPING/BINDING:-
SUNDRIES:-

FIGURE 2 A Specification Sheet

The meta-control strategy that we decided on was also able to be transferred into Nexpert. As Nexpert has the facility to implement slots and daemons, implementing the control flow was simply a matter of placing calls in the if-changed slots. More difficult however was using these daemons to call the neural networks.

The neural network code had to incorporate Nexpert routine handlers and a number of other functions tailored to Nexpert.

The daemons that are fired by changes to the value of a slot, call functions written in the C part of the implementation. These C functions do three main things:

They access the appropriate slots of the current case to find the user's specifications. The values of these attributes are read and then converted to normalised inputs for use in the neural networks.

The next action these functions perform is to call the neural networks which are also contained in the C part of the implementation. This involves retrieving the network's specification and the weights that were determined during training. The input data is given to the networks, which return a set of output values. The highest of these values is taken to be the class into which the current case should be classified.

The final action performed by the C functions which were called by the daemons, is to change a slot in the appropriate subclass; that class which was determined by the neural network. By changing the value of the slot in the sub-class, another daemon is triggered and the classification process propagated to the lower levels.

The inclusion of Nexpert Handlers is simply a matter of defining (in the appropriate syntax) which 'Execute' calls in Nexpert trigger which C functions.

The daemons other than those that called the external routines were easier to implement. They simply involved the invocation of rules, or commands to show pictures stored in bitmap files.

Pattern matching rules were used to implement the displaying of data. Once a low level category is reached, the daemon triggers a pattern matching rule that simply takes all the instances of that subclass and changes the value of a slot in those cases. The slots changed in those cases likewise trigger daemons, but these ones simply use a 'Show' command to display a bitmap image stored in a file. A picture is displayed for each case.

RESULTS

Learning in the Neural nets

Learning in the neural networks was rapid and did not take as long as expected. In almost all of the networks the initial training learnt most, but not all of the input patterns. The highest level neural network trained satisfactorily the first time, and the lower level networks learnt approximately 80 to 90 per cent of their input patterns on the first training. Looking at those patterns that did not train the first time enabled us to find additional characteristics that were needed to differentiate between sub-classes. When the input files were changed to incorporate the additional attributes, re-training invariably was satisfactory, with the total sum of errors for each below 0.02.

Structure of model

The hierarchical structure that was designed and then incorporated into Nexpert enabled satisfactory classification. When a low level sub-class was reached and the cases in this subclass displayed, the cases were relevant to the input specification. Each of the designs retrieved, whilst not necessarily having the exact same specifications, could be modified with small alterations to create a new design that exactly matched the specification.

The number of levels that were used and thus the number of cases in the bottom most sub-classes was also good, as the number of cases retrieved for a particular set of input characteristics was enough to give a variety of solutions while still being a manageable number. All the cases retrieved, while varying, were relevant

The properties that were used to distinguish differences between the categories of shoes were also satisfactory. Firstly, as far as the user is concerned, the number of properties that had to be defined was not large, and the values of the properties were relevant to the shoe domain, not requiring any conversion for the computer to accept them. For example, the values for toe shape were 'rounded', 'pointed', 'chiselled' and 'bump'. These are both the terms used by the designers and also what is input into the system making the system accessible for designers.

The other way that the properties were successful was in the ways that they helped the system reach the proper sub-class during the reasoning process. The properties, in combination with the weights placed upon them by the neural networks, were sufficient for deciding an input specification's classification.

Use of fuzzy variables in determining case classification

To decide whether the use of fuzzy variables was successful in this system, it is possible to look at the cases retrieved concentrating on those attributes that were defined as fuzzy. For example, if we look at the shoe height attribute, it can be seen that all the cases retrieved for a particular set of input characteristics matched the specification in regard to the height of the shoe. In fact, the cases retrieved tended to match the current case most closely in regard to the fuzzy variables used in the classification.

Retrieval of correct cases

As has been briefly mentioned above, the system built retrieves cases that are very similar to the input specifications. The correct cases for a particular input specification can be defined as those cases that most closely match the current case and that can be adapted to exactly meet the user's requirements. While retrieval is the aspect of case based reasoning that is addressed by the system, and case adaptation is not undertaken, it can be seen that the cases that are retrieved have enough in common with the input case so as to make future case adaptation possible.

In the testing, it was found that the cases retrieved did match the current case the closest, in at least 95 per cent of the tests. The cases in other sub-classes not retrieved were not as similar to the current case. In the other 5 per cent, the cases retrieved were still very close and almost the best cases.

When a case was determined to be incorrectly classified, the cases retrieved were not wrong, however another category could be considered to be better. The reason for the incorrect retrieval of these cases was that the combination of characteristics in those cases was very unusual and the networks were obviously not trained to deal with the combination of characteristics found in them.

The classification of cases, however, was excellent and the problem above only minor as it can be remedied by adding more cases when training the networks.

Summary of Results

It has been seen that the system that was implemented was successful for a number of reasons. As a case based reasoner, the retrieval of relevant cases was good. This implies that using neural networks as a mechanism for retrieval is a good and viable option in the development of case-based reasoners. The choice of classes and properties on which to base retrieval was also successful as cases similar to the input specifications were retrieved.

Thus it is seen that the creation of a system to retrieve past shoe designs to meet certain criteria has been successful due to the methods used in the indexing and retrieval process.

CONCLUSION

A system was designed that retrieved past designs that were close to the user's requirements. The designs retrieved by the system do approximate the user's specification. This system was achieved by exploring the possibility of using hierarchical neural networks for retrieval. Neural networks were found to be viable as the retrieval method for this system.

Additionally, neural networks were found to have some advantages over other forms of case retrieval methods. Neural nets eliminated the necessity of deciding the weightings that specific attributes should have. In a traditional system one needs to define those weightings or use other algorithms to work them out. Other algorithms generally require more specific data or a very large number of cases. Neural nets also allow for imprecision and uncertainty, meaning that retrieval can still take place when attribute values are not known.

An additional issue addressed by this system was the use of fuzzy functions to determine some of the inputs into the neural networks. Modelling some of the attributes in this way was successful as the cases retrieved tended to match most closely on the fuzzy attributes.

Hence we can say that the solution given to the problem being addressed is a successful one and the theoretical aspects it explores and implements, that is, the use of neural networks and fuzzy values, were likewise successful.

REFERENCES

[ASH90] Ashley, K.D. *Modeling Legal Argument: Reasoning with Cases and Hypotheticals.* Massachusetts, MIT Press, 1990.

[BJ90] Beale, R. and Jackson, T. *Neural Computing: An Introduction.* Hilger, Bristol, 1990

[DT93] Dillon, T.S. and Tan, P.L. *Object-Oriented Conceptual Modeling.* Australia, Prentice Hall, 1993.

[Ham86] Hammond, K.J. 'CHEF: A Model of Case-Based Planning'. *Proceedings of the Fifth National Conference on Artificial Intelligence* (AAAI-86). 1986:267-271.

[HH92] Hennessey, D. and Hinkle, D. 'Applying Case-Based Reasoning to Autoclave Loading'. *IEEE Expert.* 7(5):21-26, October 1992.

[Kol93] Kolodner, J. *Case Based Reasoning.* California, Morgan Kaufmann, 1993.

[RS89] Riesbeck, C. K. and Schank, R.C. *Inside Case-based Reasoning.* New Jersey, Lawrence Erlbaum Associates, 1989.

[Sim92] Simoudis, E. 'Using Case-Based Retrieval for Customer Technical Support'. *IEEE Expert.* 7(5):7-12, October 1992.

Footwear Design:

[Mil76] Miller, R.G. (Ed.) *Manual of Shoemaking.* Bristol, U.K., Clarks Ltd. Printing Department, Second Edition, 1976.

[Pat83] Patrick, H.J. *Modern Pattern Cutting and Design.* Northants, England, U.K., Mobbs Miller, Fourth Edition, 1983.

Computer Aided Manufacturing

FACILITATING KNOWLEDGE ACQUISITION FROM INDUSTRIAL PROCESS DATA BY AUTOMATING FEATURE SELECTION

David J. Urpani and James A. Sykes
School of Information Systems, Swinburne University of Technology
Hawthorn, Vic 3122, Australia.
Email: dju@stan.xx.swin.edu.au

ABSTRACT

In complex and poorly defined domains, such as those found in some industrial processes, knowledge acquisition is currently a challenging problem holding back the development of decision support for process supervisors. We present here a method for facilitating knowledge acquisition by identification of sub-groups of process variables which discriminate between different process conditions. The method uses a combination of a genetic algorithm working in tandem with a nearest neighbour classifier. Identification of the more pertinent variables also leads to an improvement in classification accuracy. In terms of the knowledge acquisition process this suggests that rules induced from such a reduced variable set will be of higher quality, less complex and have a greater generalising power.

1. INTRODUCTION

The need to be globally competitive is an operating reality for most manufacturing concerns. Of the three technologies available to industry: process technology, control technology and knowledge technology, only the latter has the qualities of being unique and containable within an organisation [CD88]. This in-house knowledge technology therefore, can deliver a lasting market advantage .

Deming [Dem86] has pointed out the need for manufacturing organisations to continuously strive to improve the efficiency of their processes and the quality of their products. The increasing use of automation and information systems technologies, in conjunction with the development of new sensors and instruments has provided organisations with new opportunities for achieving this. Efficiencies and quality can be improved if organisations are able to learn, on a continuous basis, from the vast amounts of data that these new technologies generate.

The volume of data generated by complex industrial processes such as those found in aluminium smelters, petrochemical refineries and nuclear power stations is enormous. It is not unusual to find processes generating several thousand variables. If any benefits are to be gained from this resource it is essential to convert the raw sensor data (usually available as non-symbolic, numerical data) into information which can be used by process supervisors to improve their decision making.

In the 1980's factory automation boom, one of the biggest oversights was not recognising the vital role played by the human operator [Swe91]. Instead the emphasis seemed to be more on replacing the human input. Human operators have an important, integrative role to play in making sense of the diverse, often contradictory aspects of the plant's operations. A system which complements the human operator is more likely to succeed in achieving process improvements. The development of expert systems which provide decision support, termed here intelligent decision support systems (IDSS), is an acknowledgment that these systems are to work in tandem with the human operator and offer advice [Hol89] , in this case for process improvements.

2. THE PROBLEM

In the complex and poorly defined environments found in some industrial processes, the knowledge acquisition bottleneck is the main barrier to the development of decision support systems [JA91]. A lot of the data is inaccurate, noisy and redundant. Furthermore process experts are often unable to reach a consensus on the optimal way of managing the process and on which data should be used to interpret different process conditions.

Unfortunately, at present, there is very little theory available to help select the optimal (according to some agreed error estimation technique) set of sensor data

[Fu82, Fuk82] with respect to a given classification problem. In real life most features are selected solely on heuristic or experimental evidence [Pau82].

The ability to distinguish between data that is necessary for discriminating between the various process conditions and data that is superfluous and redundant is crucial to the success of IDSS development in this area. Superfluous data reduces classification accuracy and together with redundant data increases the overheads required in computing such classifications. Furthermore reducing the number of features used results in inducing simpler rules, more easily understood by human operators. Feature selection can also improve process knowledge by highlighting those variables that play a determining role in deciding the different process conditions. A method for achieving feature selection in the ill-defined and noisy environments that prevail in many industrial processes would therefore be of great help to IDSS developers.

This paper reports on work being carried out with the aim of automating this knowledge acquisition phase. This requires a methodology for the selection of relevant features and the elimination of redundant ones. The approach taken here is to view the problem as essentially one of pattern recognition, where labelled training samples are to be classified. The aim of feature selection in this context is to transform the data vector space to a new coordinate system eliminating those measurements that contain no useful information with respect to the problem under investigation.

Our work shows that genetic algorithm (GA) search in tandem with the k-nearest neighbour (kNN) algorithm can indeed be used to automate identification of variables that play an important role in determining the state of a particular process from those variables which are redundant and do not convey any new information. We also show that in most cases this reduction in dimensionality improves our ability to discriminate between different process conditions.

3. A BRIEF LITERATURE REVIEW

The selection of features is still an unsolved problem and an active area of research [Aha90; AD91, KR92; Sch93]. Selecting features requires closer examination of two issues. The first is the criterion to be adopted when selecting between different feature sets and the second is the search procedure used to examine the space of all possible combinations of features.

3.1 The Search Criterion

3.1.1 The Parametric Approach

The criterion most frequently adopted in practice is the error criterion. In statistical literature this is also referred to as the error probability. The optimal error criterion is Bayes error probability, in the sense that it minimises the probability of misclassification. This however requires complete knowledge of the probability distributions and class conditional density estimates [Rip87], a requirement which cannot possibly be met in industrial process (and other real world) applications . Approximations to Bayes error usually involve relaxing one or more of the assumptions on which it is based resulting in a sub-optimal feature selection criterion [Dev82].

The most common non-Bayesian approach is to assume multivariate Gaussian densities [JC82]. This is due to the relative simplicity of the resulting mathematical analysis and to the fact that many real world classifications problems can be approximately modelled as multivariate Gaussian distributions. It is further assumed that classes have equal covariance matrices and known mean vectors. As one by one even these assumptions are abandoned for other methods with less stringent a priori knowledge requirements, estimates of the error criterion shift from parametric to non-parametric.

3.1.2 The Non-Parametric Approach

The attraction of the non-parametric approach when using industrial sensor data is that no prior assumptions need be made on the structure of the data. Changes in process operating conditions might make any such assumptions invalid. This lack of structure is not necessarily at the cost of classification performance. Several authors have reported that it is not unusual for non-parametric decision rules to perform better than their more informed (parametric) counterparts [WK91; JC82].

One of the most popular, non-parametric error estimates that does not require any knowledge of the form of the underlying densities is the k nearest neighbour (kNN) method. The kNN method assigns an unclassified sample vector by majority vote to the class of the k nearest of a set of correctly labelled samples or prototypes. For an infinite number of training samples it can be shown [CH67; Dud73] that the kNN is bounded from above by twice the Bayes error rate. Also as k (the number of nearest neighbours) tends to infinity kNN is identical to the optimal Bayes rule. These rules are, in general, applicable irrespective of the choice of distance measure. A commonly adopted distance measure is the Euclidean distance. This is the square root of the difference between the values of each feature, squared and then summed over all features.

The behaviour of the kNN classification rule under more realistic conditions of finite training sample size exhibits less optimal performance. Bailey et al [BJ78] observed that for a fixed number of training samples per class m, and for a given dimensionality N, classification improves as k increases, reaching a maximum before starting to deteriorate. This is often referred to as "peaking performance". Empirical results, reported by Loftsgaarden et al [LQ65] imply that a good choice for k is one that is proportional to the square root of m.

In practice classification performance improves with increasing dimensionality up to a certain point and then deteriorates as more features are added. Jain et al [JC82] refer to this peak in performance as the "curse of the finite sample size". It seems to point to the existence of an optimal measurement complexity when the size of the training set is finite. Simply put, even if the measurements (features) are correct for the classification at hand an increase in their number requires a corresponding increase in the size of the training sample. If they are redundant or noisy then the only way to improve performance is to eliminate such features. As a general rule, for a given classification performance on a training data set a simpler classifier (having a reduced feature set) will have better prediction accuracy on an unseen test data set than a more complex classifier [WK91].

3.2 The Search Procedure

As mentioned in the introduction to this section the second issue in feature selection is the search procedure. This involves examining all the possible combinations of features for optimal classification based on a predetermined criterion. In the approach we describe in this paper the criterion is the percentage of correct classifications on a training sample.

3.2.1 Common Search Strategies

Probably the simplest (and weakest) search strategy is the one that assumes no interdependency between the candidate features and then chooses only those features that pass a simple test of statistical significance. A better approach is to choose the first single best feature then sequentially the next best feature accumulatively. Another similar procedure is to progressively eliminate the worst features from a full set of features. A disadvantage of these strategies is that latter choices are influenced by previous ones as no backtracking is allowed.

If the latter two strategies are combined the search process will now allow a limited form of backtracking. The branch-and-bound method works on a search tree considering one node at a time. This method is efficient when using a monotonic search criterion as a subtree will no be examined if computation at its root reveals it is sub-optimal. However misclassification error criteria do not normally satisfy the monotonicity condition.

A problem with 'incremental' search approaches is that the best n feature subset may not necessarily be made up of the n best individual features, or even the best sub-group of features [Cov74, CC77]. This is true even when the features are shown to be conditionally independent [Tou88]. Essentially this implies that the total number of combinations of n features for a complete, exhaustive search is 2^n which in most cases of practical significance is too vast a search space to be computationally feasible.

3.2.2 Genetic Algorithm Approach

A guided, random search technique that could be useful for this type of problem is genetic search as carried out by genetic algorithms (GA). GAs are computational models that approximate Darwinian evolutionary theories and work on a 'population' of strings using operators such as random mutation and crossover to generate new specimens from the current population. Mutation involves randomly changing the value of one or more bits making up a string. Crossover refers to a mating operation by two parent strings to generate new offspring. Evaluation of each new specimen is according to the fitness function defined as part of the search algorithm. This cyclic procedure is repeated as many times as necessary depending on the required residual error or until a predetermined number of generations have been evaluated. GAs are particularly well suited for the solution of complex optimisation problems.

The search characteristic of GAs which renders them particularly well suited for the feature selection problem is their ability to consider any combination of attributes without any sequential order. Therefore no assumptions on the independence of class conditional probability distributions need to be made. Another valuable characteristic of GAs with respect to this problem is their ability to perform well in noisy environments. Some GA practitioners [FG88] have gone as far as saying that due to their ability not to be misled by noise, GAs may be 'the search technique of choice for machine learning systems in complex environments'.

Guo et al [GU92] used GAs to select features as a means of preprocessing the data before presentation to a neural net. An n-dimensional problem is represented by n binary digits where the 1's and 0's signify the use and elimination of a particular feature respectively. The fitness function adopted in this case is the number of misclassifications. Classification can be carried out using any suitable algorithm. Some workers have used the kNN, [BBM92, KD91] while others have used a modified version of the backpropagation neural network [RMP88]. In both the work of Guo et al and Brill et al feature selection was carried out with the sole aim of improving neural network classification performance.

4. METHODOLOGY & DATA

The program developed for the work reported here, Classer, uses an integer valued GA coupled with a kNN classifier to derive a new set of binary feature weights. The GA (see description below) has been empirically developed to carry out successful optimisation in the feature selection space.

Classer accepts a set of labelled training examples and modifies the weighting (0 or 1) of each feature to improve its classification rate. Classification of a new sample is assigned on the basis of the class of the nearest neighbour as measured by the weighted Euclidean distance between the two. The number of misclassifications of a labelled training set is used to provide a measure of the error estimate. This is referred to as the approximate classification rate (ACR). An unbiased estimate of the error rate is also obtained by using a test sample set (unseen by the algorithm during training). This is the true classification rate (TCR).

In experiments (not reported here) different values of k, the number of nearest neighbours, were tried. No appreciable improvement in the classification rates were observed for increasing values of k and therefore one nearest neighbour was chosen for the experiments reported in this paper.

4.1 The Genetic Algorithm

Development of the GA used in these trials was empirically based. Classer uses a steady state population of binary strings (the feature vector weights). An elite population of the best 50 candidates is maintained. Parent selection is essentially random with some bias towards the best or the most recent. In this case therefore, selection is extinctive, meaning that some weak performing strings may have zero probability of reproducing. Extinctive selection combined with a high mutation rate as reported in the work of Back [Bac92] is known to result in an aggressive search.

One point crossover operators are used throughout. Mutation takes the form of either a random flipping of one bit (from 1 to 0 and vice-versa) or alternatively the random creation of a new string. DeJong [DeJ85] reports that this naive evolution approach has been found to give very good results on a variety of optimisation problems. In this approach mutation is perceived to play more than just a background role and is in fact viewed as another search operator.

The fitness function adopted in this work is dominated by the number of correct classifications of the 1-NN classifier. However for the same classification rate a string with a larger number of zeros is awarded a higher fitness value.

In the work reported here the GA has been viewed simply as a tool for achieving successful search over the feature space and not much time was spent on discovering more efficient structures for this problem. This empirical work has suggested several avenues for the improvement of GAs particularly suited for the feature selection problem, which will be reported elsewhere.

4.2 The Data

Six runs of the GA were performed for each experiment. The data used for these experiments are from three domains: a benchmark data set, data from a simulated chemical process and real world industrial data. All data submitted to the algorithm were standardised with a mean of zero and a standard deviation of one.

4.2.1 The Iris Data

This is a well-known pattern recognition data set originally created by R.A.Fisher [Fis36]. The data set is made up of 150 samples with four features each. There are three classes, 50 samples each, referring to a type of iris plant: Versicolour, Virginicia and Setosa.

4.2.2 The Batch Reactor Simulator Data

To demonstrate the application of GA/kNN to feature selection in a simulated dynamic process environment an exothermic reaction occurring in a batch reactor was modelled. This is a common piece of equipment in the chemical and biochemical industries. An irreversible, first order reaction takes place in the jacketed tank as shown in Fig.1.

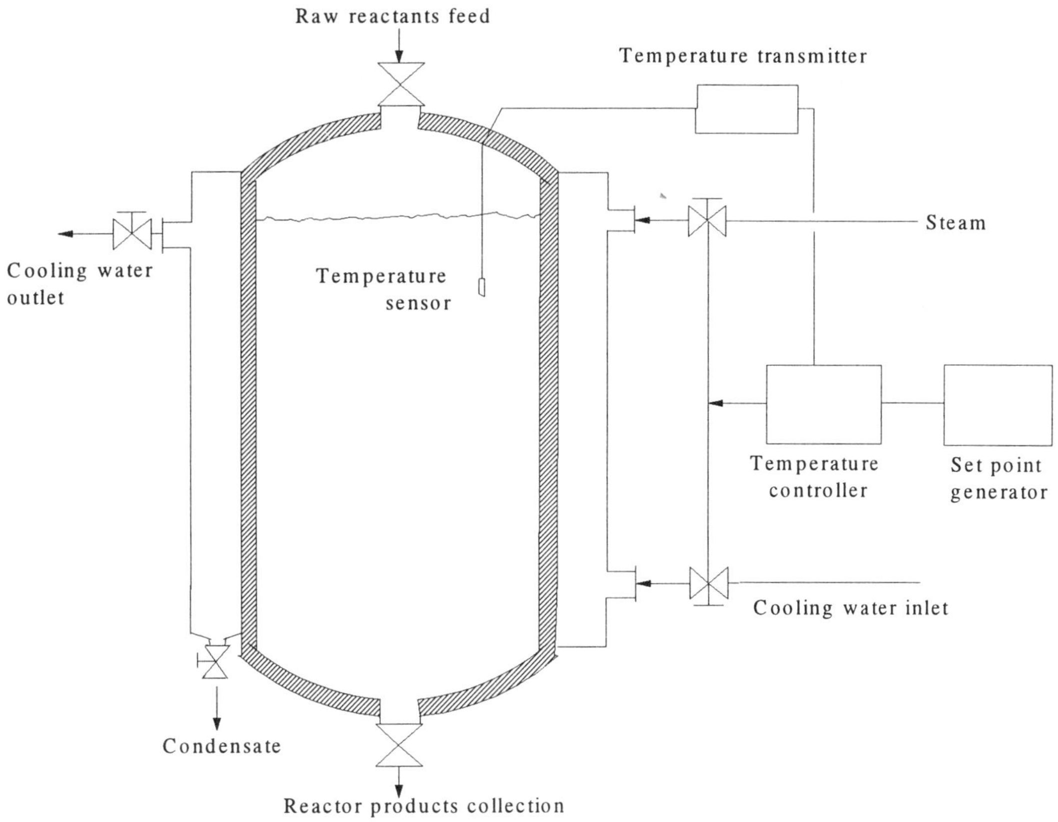

FIGURE 1 Batch Reactor Process Schematic

After raw material is fed into the tank, steam is allowed to flow through the jacket to bring the reaction mass up to the required temperature. Once the reaction starts, cooling water is used to control the exothermic heat so that the process follows a predetermined temperature profile. This is achieved by means of a time varying set point signal sent to a temperature controller. Deviation from this profile results in the creation of an undesirable by-product or in the non-conversion of the raw material.

Altogether thirteen process and control variables are logged. These include water and steam valve positions, reactant temperature, vessel wall and jacket temperatures, water flow and volume of water in jacket, steam pressure and chemical conversion rates.

Seven representative fault conditions have been simulated: 1) Low reactant concentration at the reactor inlet, 2) High temperature at reactant inlet, 3) Low temperature at reactant inlet, 4) Drop in Water Head Pressure 5) Drop in Steam Supply Pressure, 6) Temperature Set Point too high, 7) Temperature sensor transmitter malfunctioning. The simulator uses interval halving with adaptive stepsize control to solve the differential equations governing heat flow and chemical conversion.

The training data used in these simulations contained 300 training vectors (patterns) and 150 test patterns. Two classes are represented (50% each class), one with a particular fault condition present and the other with no fault present.

4.2.3 The Aluminium Smelter Data

An example of a complex industrial process for which no quantitative model exists is aluminium smelting. Aluminium is produced by the electrolytic reduction of aluminium oxide in a reduction cell or pot. Carbon anodes are suspended in a solution of molten cryolite into which aluminium oxide is fed. A very large current at low voltage traverses the pot from anode to cathode decomposing the aluminium oxide. The aluminium pad which settles to the bottom of the pot is then siphoned off.

The data used in this study was obtained from the Portland Aluminium Smelter. At this smelter pot operation is monitored every twelve hours by a state vector made up of 19 process variables. These include a combination of sensor and manually acquired data:

temperature, voltage, metal and liquid levels in the pot, chemical concentrations in the pot, feed rates, tapping distances and others. Some of the measurements are very noisy.

Metal pad fluctuations, floating debris, accumulation of deposits at the bottom of the pot and other phenomena contribute to make pot operation inefficient by shorting the flow of current. This results in the so called 'noisy' pot. In this experiment data vectors are separated into two classes: noisy and not-noisy. There are 230 training samples and 114 test samples.

5. RESULTS

5.1 The Iris Data

A manual, exhaustive best feature search of the Iris data, using the 1-NN classification algorithm on the full 150 sample data set, showing the correct classification rate is shown in Fig. 2. This search shows that the best feature set is one which makes use of feature three and four, the petal length and petal width, respectively.

Over six separate runs, Classer was each time able to select a reduced feature set in which the first two features are eliminated. As shown above this is the optimum solution for this data set.

5.2 The Batch Reactor Simulator Data

For the Batch Simulator data, classification results using the reduced feature set obtained by Classer are compared with the classification results obtained using a unity vector. This is a vector in which all the features present in the training set are used for the classification task. The results are shown in Fig. 3. For each of the seven faults there was a reduction in the number of process variables necessary for discriminating between the presence or otherwise of the fault. As an example, Fault 4 (Drop in Water Head Pressure) could be identified by using just three out of the thirteen variables originally presented. Similarly, Fault 1 (Low Reactant Concentration) needed only five out of the original eleven variables presented. It is importantto note that in both cases this reduction in dimensionality was not accompanied by a loss in classification accuracy. In fact, on average, a small improvement in the approximate and true classification rate is evident.

For most fault conditions Classer found the same (reduced) feature set. The exceptions are Fault 5 and 7, where in both cases one weight vector was different from the other five.

In the first case the same ACR was obtained for the different vectors indicating that the alternative feature sets found compensate perfectly and can be replaced by each other. This is an example of the multimodal search

FIGURE 2 Iris Data Results: Classification rate for all feature combinations

WEIGHT VECTOR	CLASSIFICATION %
0 0 0 0	0.00
0 1 0 0	48.00
1 0 0 0	58.67
1 1 0 0	71.33
0 0 1 0	88.00
0 0 0 1	88.00
0 1 1 0	88.67
1 0 1 0	90.67
1 1 1 0	90.67
0 1 0 1	92.00
1 0 0 1	92.67
1 1 0 1	94.00
1 0 1 1	95.33
1 1 1 1	95.33
0 1 1 1	95.33
0 0 1 1	96.00

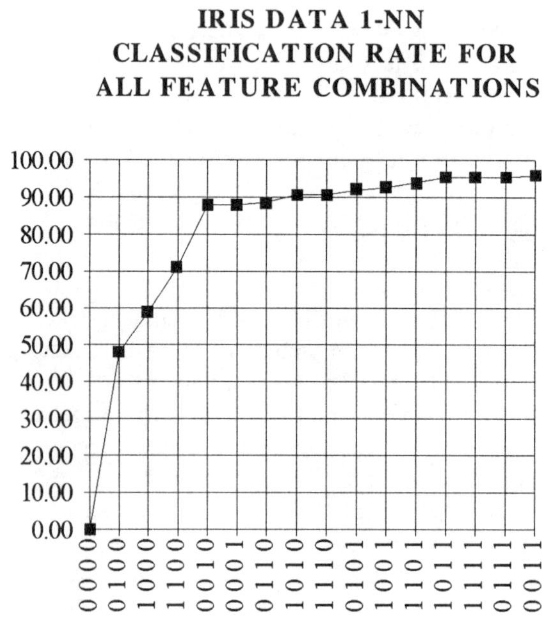

FIGURE 3 Batch Simulator Results : All results are average of six runs 1000 Iterations, 300 Training Samples and 150 Test Samples, 1 - NN

FAULT	BEST WEIGHT VECTOR	ACR (U) %	ACR %	% CHANGE ACR	TCR (U) %	TCR %	% CHANGE TCR	NO. FEATURES ELIMINATED
1	0 0 0 1 0 1 0 0 0 0 0 1	96.33	99.00	2.77	92.00	92.67	0.73	6 out of 11
2	1 0 0 0 0 0 0 0 0 0 0 1	95.33	99.33	4.20	85.67	92.67	8.17	10 out of 12
3	1 0 0 0 0 0 0 0 0 0 1 1	91.00	99.67	9.53	86.67	96.00	10.76	9 out of 11
4	0 0 0 1 0 1 0 0 0 0 0 0 1	95.33	96.00	0.70	76.67	77.33	0.86	10 out of 13
5*	0 0 0 0 0 1 1 0 0 0 0 0 1	99.00	99.67	0.68	99.33	99.33	0.00	10 out of 13
6	0 0 0 0 0 1 0 0 0 0 0 0 1	99.33	99.67	0.34	98.67	98.67	0.00	8 out of 13
7**	0 1 1 0 0 0 0 1 1 0 0 0 1	83.00	89.22	7.49	83.33	88.22	5.87	8 out of 13

ACR(U	Apparent Classification Rate (i.e. on Training Samples) using Unit Weight Vector
ACR	Apparent Classification Rate (i.e. on Training Samples) using Classer evolved Weight Vector
TCR(U	True Classification Rate (i.e. on Test Samples) using Unit Weight Vector
TCR	True Classification Rate (i.e. on Test Samples) using Classer evolved Weight Vector

* One weight vector was different from the rest : 1 0 0 0 0 0 0 0 0 0 0 1 1

** One weight vector was different from the rest: 0 0 0 0 1 0 0 1 0 0 1 1 1

space which the GA has to navigate. In such circumstances, given the same classification rate, selection of the alternative (reduced) feature set would have to depend on other domain heuristics, such as the expense of acquiring one set of sensor data over another.

In the second case the ACR was marginally below the other vectors (89.0% as compared to 89.33%). This is a clear indication that, on this occasion, Classer has become trapped on a local optimum, albeit one not much worse than the better alternative. Of course GA search does not guarantee that the global optimum will be found. In circumstances such as these the system developer would normally choose the better alternative (feature set which provides the best classification performance). However, if for some reason process experts again deem a given sensor to be preferable over another, then the close classification rates leave the choice open to these sort of domain heuristics.

5.3 The Aluminium Smelter Data

In the Aluminium Smelter data comparison is again made with a unity weight vector. The results are shown in Fig.4. In this case also there was a reduction in the number of process variables required for classification. For the majority of cases (four out of six) five features were eliminated out of nineteen. Also, as in the previous data set, this reduction was accompanied by an average improvement in ACR of 7.62% and a TCR average of 2.58%. Four out of the six weight vectors were identical with the two dissimilar vectors having a lower ACR (86.96% and 85.65% as compared to 87.39%).

6. DISCUSSION

The existence of more than one reduced feature set shows that although there seem to exist features that are always eliminated there are also features that compensate for each others presence or absence. This demonstrates the multimodal nature of the feature search space.

In the feature selection problem the location of a global optimum for non-trivial data may not be possible within the limits imposed by time and resources. There is also the possibility that more than one global optimum exists.

Such a situation has advantages in a decision support scenario. Once the 'obviously' redundant features have been removed, selection of the final feature set from a group of candidate reduced feature sets needs new domain knowledge input. The user can base final selection on acquisition cost, reliability, ease of monitoring or physical interpretation of the different process variables.

It may also be useful to induce rules from, say, two separate (or partially separate) but equally accurate sets of features. This provides a measure of inbuilt redundancy in the rule set should a sensor become faulty. There is also an added measure of reliability in that agreement of two rules (using different sensor data) with respect to classifying a particular process condition would increase operator confidence in its validity. In the. case of two or more rules providing conflicting diagnosis the operator is given an indication that either one or more sensors are faulty or alternatively that rule reliability needs to be improved

FIGURE 4 Aluminium Smelter Results
1000 Iterations, 230 Training Samples and 114 Test Samples, 1 - NN

RUN	BEST WEIGHT VECTOR	ACR (U) %	ACR %	% CHANGE ACR	TCR (U) %	TCR %	% CHANGE TCR	NO. FEATURES ELIMINATED
1	1 1 1 1 1 1 1 1 1 0 1 0 0 0 0 1 1 1 1	80.87	87.39	8.06	78.95	79.82	1.10	5 out of 19
2	1 1 1 1 1 1 1 1 1 0 1 0 0 0 0 1 1 1 1	80.87	87.39	8.06	78.95	79.82	1.10	5 out of 19
3	1 1 1 1 1 1 1 1 1 0 1 0 0 0 0 1 1 1 1	80.87	87.39	8.06	78.95	79.82	1.10	5 out of 19
4	1 1 1 1 1 1 1 1 1 0 1 0 0 0 0 1 1 1 1	80.87	87.39	8.06	78.95	79.82	1.10	5 out of 19
5	1 1 1 1 1 1 1 0 0 0 1 1 0 0 0 1 1 1 1	80.87	86.96	7.53	78.95	80.70	2.22	6 out of 19
6	1 1 1 1 0 0 0 0 1 1 0 0 0 0 0 0 0 0 0	80.87	85.65	5.91	78.95	85.96	8.88	13 out of 19
Averages			87.03	7.62	78.95	80.99	2.58	

ACR (U) Apparent Classification Rate (i.e. on Training Samples) using Unit Weight Vector
ACR Apparent Classification Rate (i.e. on Training Samples) using Classer evolved Weight Vector
TCR (U) True Classification Rate (i.e. on Test Samples) using Unit Weight Vector
TCR True Classification Rate (i.e. on Test Samples) using Classer evolved Weight Vector

7. CONCLUSIONS

The work reported here has shown that a genetic algorithm incorporating 1-NN classification results within the fitness function can successfully identify a sub-set of features from a bigger set submitted to it without any reduction in classification accuracy even in situations involving noisy industrial data. This is good news for expert system developers who want to better understand the roles played by different process variables under different operating conditions. The identification of only the features which matter will result in the creation of simpler more compact rules for inclusion in the knowledge base.

A bonus is that, in the cases shown in this work, the reduction in the feature set size is nearly always accompanied by an improvement in the classification rate. True classification error rates also reveal that there is a parallel improvement in the generalising power. The implications on the quality of the rules generated are that these will have greater classification accuracy and higher generalisation capabilities coupled with a reduction in complexity.

The successful demonstration of feature selection on noisy, industrial data shows that the method adopted has the potential to be successfully applied in the development of IDSS for industrial process operator support.

8. FURTHER WORK

In future work we will examine in greater detail the GA operators, population management and structures which will make genetic search more suited for searches in the feature space. The hope is that searches will be faster and will have a greater probability of finding a high quality solution. The identification of a more sensitive objective function other than just the number of misclassifications is a problem currently being studied. An empirical comparison with other search techniques is envisaged.

Further work on Classer will include the ability to synthesize new features (especially temporal ones such as first and second order deriviatives) before carrying out feature selection.

An important component of this work is to understand the nature and implications of the bias introduced when inducing rules from data which has been preprocessed by a feature selection algorithm (such as Classer). A study of the quality and complexity of rules induced from a decision tree before and after feature selection by Classer, with and without pruning is therefore planned.

9. ACKNOWLEDGMENTS

The first author would like to gratefully acknowledge the financial support provided by a Portland Aluminium Pty. Ltd. scholarship. Access to Portland Aluminium's process experts and data is also acknowledged.

10. REFERENCES

[AD91] H.Almuallin and T.Dietterich, Learning with many irrelevant features. In *Proceedings of the Ninth National Conference on Artificial Intelligence*,San Jose, CA, 547-552, 1991

[Aha90] D.Aha, A Study of Instance-Based Algorithms for Supervised Learning: Mathematical, Empirical and Psychological Evaluations, *PhD Thesis*, Dept of Information and Computer

Science, University of California, Irvine, 1990.

[Bac92] T.Back, Self-Adaptation in Genetic Algorithms. In *Proceedings. of the First European Conference on Artificial Life*, Varela, F.J. & Bourgine, P. (Eds.), MIT Press, Cambridge, MA, 263-271, 1992.

[BBM92] F.Z. Brill, D.E. Brown and W.N. Martin, Fast Genetic Selection of Features for Neural Network Classifiers, *IEEE Trans. Neural Networks*, 3, No.2:324-328, 1992.

[BJ78] T.Bailey and. A.K. Jain, A Note on Distance-weighted K-Nearest Neighbour Rules, *IEEE Trans. Sys., Man, Cybernet.*, 8: 311-313, 1978.

[CC77] T.M. Cover and J.M. Campenhout, On the possible Orderings in the Measurement Selection Problem, *IEEE Trans. Sys., Man, Cybernet.*, 7, No.9: 657-661, 1977.

[CD88] C.Cahill and A.Demers, Using knowledge technology to gain a competitive advantage in manufacturing through predictive control. *Intelligent Manufacturing*, .337-355. Benjamin/Cummings Pub. Co. Inc., Michael Oliff, Ed, 1988.

[CH67] T.M. Cover and P.E. Hart, Nearest Neighbour Pattern Classification, *IEEE Trans. Inform. Theory*, 13: 221-227, 1967.

[Cov74] T.M. Cover, The Best Two Independent Measurements are not the Two Best, *IEEE Trans. Sys., Man, Cybernet.*, 116-117, 1974.

[DeJ85] K. DeJong, Genetic algorithms: A 10 year perspective, In *Proc. First International Conf. on Genetic Algorithms*, Grefenstette, J.J. (Ed.), Lawrence Erlbaum Assoc. 167-177.

[Dem86] W.Deming, Out of the Crisis. MIT, Centre for Advanced Eng. Study, 1986.

[Dev82] P.A. Devijver, Statistical pattern recongnition. In *Applications of Pattern Recognition*, Fu, K.S. (Ed.), CRC Press, Boca Raton, Florida, 15-36, 1982.

[Dud73] R.O. Duda, Pattern Classification and Scene Analysis, Wiley, New York, 1973.

[FG88] J.M. Fitzpatrick and J.J. Grefenstette, Genetic Algorithms in noisy environments, In *Machine Learning* 3, Kluwer Academic Pub., 101-120, 1988.

[Fis36] R.A. Fisher, The use of multiple measurements in taxonomic problems, In *Annual Eugenics*, 7: Part II:179-188, 1936.

[Fu82] K.S. Fu, Applications of Pattern Recognition, CRC Press, Boca Raton, Florida. 1-13.

[Fuk82] K. Fukunaga, Intrinsic dimensionality extraction. In *Classification, Pattern Recognition and Reduction of Dimensionality*, Krishnaiah P.R., Kanal L.N. (Eds.), North-Holland Pub., 347-360, 1982.

[GU92] Z. Guo and R.E Uhrig, Using genetic algorithms to select inputs for neural nets. In *Proceedings of the International Workshop on Combinations of Genetic Algorithms and Neural Networks*, Baltimore, Maryland. IEEE Computer Soc. Press, CA, 223-234, 1992.

[Hol89] S., Holtzman, Intelligent Decision Systems, Addison-Wesley, 1989.

[JA91] G. Johannsen and J.L. Alty, Knowledge engineering for industrial expert systems, *Automatica*, 27, No.1: 97-114, 1991.

[JC82] A.K. Jain and B. Chandrasekaran, Dimensionality and sample size considerations in pattern recognition practice, In *Classification, Pattern Recognition and Reduction of Dimensionality*, Krishnaiah, P.R. & Kanal, L.N. (Eds.). North-Holland Pub, 835-855, 1982.

[KD91] J.D. Kelly and L.Davis, A Hybrid Genetic Algorithm for Classification. In *Proc. of the 12th IJCAI*, Sydney, 1991.

[KR92] K.Kira and L.Rendell, A practical approach to feature selection. In *Proceedings of the Ninth International Conference on Machine Learning*, Aberdeen, Scotland. Morgan Kaufmann. 249-256, 1992.

[LQ65] D.O. Loftsgaarden and C.P. Quesenberry, A Non-parametric Estimate of a Multivariate Density Function, *Ann. Math. Statist.*, 36:1049-1051, 1965.

[Pau82] L.F. Pau, Application of pattern recognition to failure detection and analyis in non-digital systems. In *Applications of Pattern Recognition*, Fu, K.S. (Ed.), CRC Press, Boca Raton, Florida: 155-168, 1982

[Rip87] B.D. Ripley, An introduction to statistical pattern recognition. In *Interactions in artificial intelligence and statistical methods*, Phelps, B. (Ed.). Technical Press: 176-187, 1987.

[RMP88] D., Rumelhart, J.L. McClelland and PDP Resarch Group, Parallel Distributed Procesing, Volume 1: Foundation, Volume 2: Psychological and Biological Models, Cambridge, MA, MIT Press, 1988.

[Sch93] J. Schlimmer, Efficiently inducing determinations: A complete and systematic search algorithm that uses optimal pruning. In *Proceedings of the Tenth International Conference on Machine Learning*, Morgan Kaufmann: 284-290, 1993.

[Swe91] L.M. Sweet, People and technology in the globally competitive factory - keynote speech to the 1991 Amer. Contr. Conf. IEEE Control Sys. 4-8, 1991.

[Tou88] G.T. Toussaint, Note on Optimal Selection of Independent Binary-valued Features for Pattern Recognition, In *IEEE Trans. Inform. Theory* (Corresp.), 17:618, 1988.

[WK91] S.M. Weiss and C.A. Kulikowski, Computer Systems that Learn, Morgan Kaufmann Pub., San Mateo, CA: 51-79, 1991.

FUZZY LOGIC APPLICATIONS IN IRON AND STEEL MAKING

S.S.Shivathaya*, X.D.Fang*, J.G.Williams**

* Department of Mechanical Engineering, University of Wollongong, NSW 2500, Australia

** BHP Slab and Plate Product Division, Port Kembla, NSW, Australia

Email: g9276610@uow.edu.au

ABSTRACT

Iron and steel making is an area characterised by the inadequacy of control theories and operation research methods, difficulty in modelling, and dynamically changing operation conditions. This makes it desirable to apply the fuzzy techniques to systematise the processes, reduce the dependence on experts, save the time taken to perform the knowledge-intensive complex tasks and more importantly to efficiently manage the uncertainty which suits particularly fuzzy logic methods. This paper reviews various fuzzy techniques in iron and steel making, highlighting the salient features of the successful systems and the benefits derived from these systems. The trend of developing hybrid systems combining the fuzzy and operation research (OR) technologies is discussed with the aim of providing a guide to the main fuzzy techniques used. The ongoing research at BHP Steel in the fuzzy techniques is presented. Finally the bottleneck areas in fuzzy system development and the future research needed is briefly discussed.

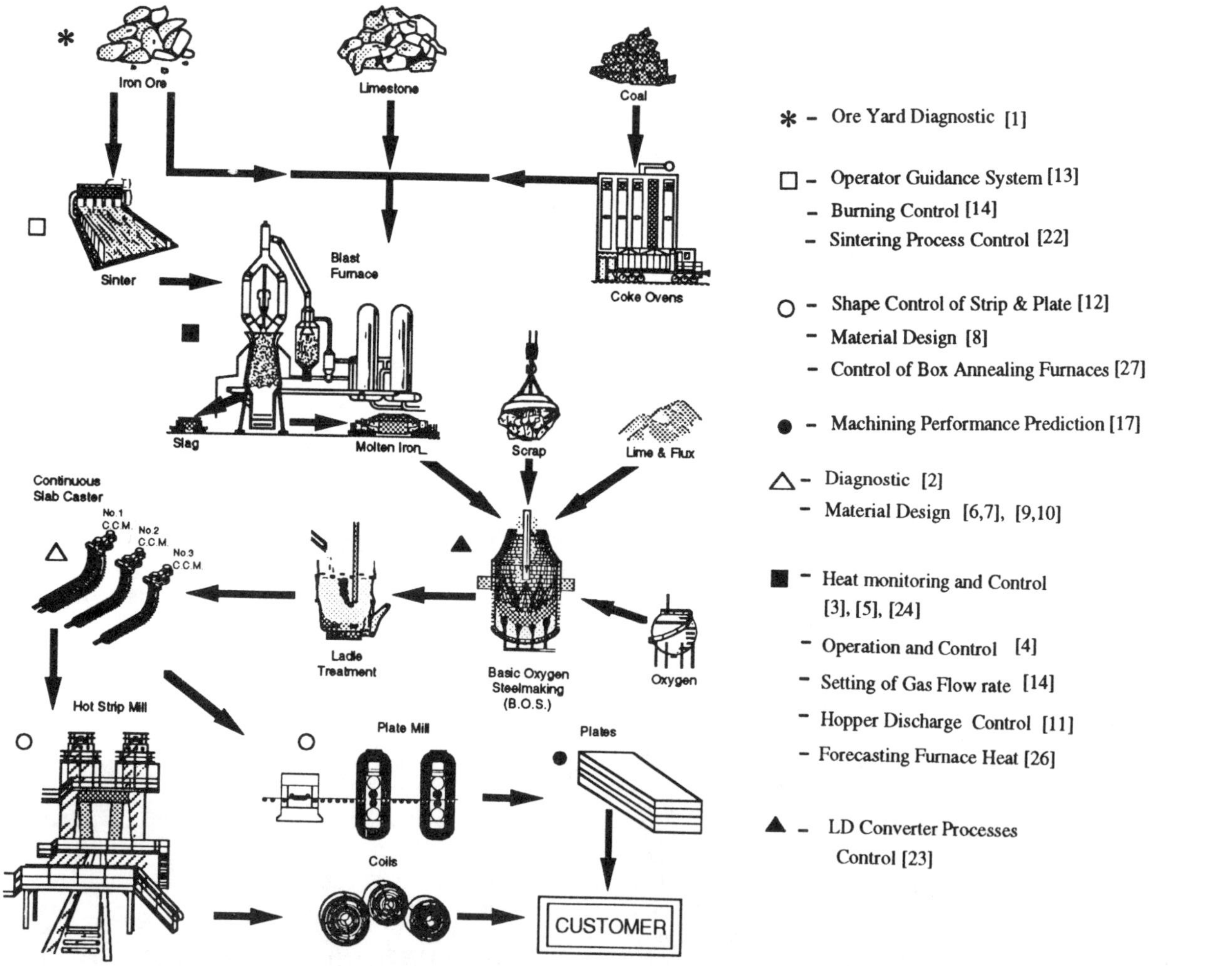

FIGURE 1 Fuzzy Systems in Iron and Steel Making

1.0 INTRODUCTION

The steel industry is facing the need for increased automation to improve product quality and productivity. The steel makers also face the challenge from non ferrous materials which is expected to intensify further in the future. The operating philosophy of steel makers is also switching from mass production at low unit cost to small volume production at low unit cost, and from higher quality at higher cost to higher quality at reasonable cost. There are several important areas in steel making such as control of blast furnace, scheduling of steel making processes and material design, where the dependence on the experts is very much. There are some fields in steel making that have been not yet systematised because of inadequacy of the control theories and operation research methods available today to solve the problems, due to difficulties in modelling, and due to dynamically changing operation conditions. These problems are normally solved by the experts who use the vast knowledge they have amassed over a long period of time, in addition to heuristic and intuitive knowledge which they possess. In order to reduce the dependence on experts and enable reduction in time taken to perform these knowledge intensive complex tasks, it becomes desirable to apply the fuzzy techniques to steel making practices.

Fuzzy set theory, a theory of graded concepts is a technique which deals with complex phenomenon which cannot be analysed by classical methods based on probability theory and bivalent logic. Real-world systems are very often uncertain or vague. Fuzzy logic has been successfully applied to several areas in steel making in which uncertainty or vagueness is inherent and fuzzy expert systems developed which deal with diagnostic problems [TKT+91, BKS92]; blast furnace control problems [TAT+89, OMK+92, TTH+87]; design problems [VSW+89, VWS89, WIO+92, WSW93, WSW+92]; and control problems [TKT+91, ITU+92, HT91, LMT+90, MTO+91, IHI+88, KYY+90, NSA90, NAM89, Yon81] as depicted in figure 1. The phenomenon of uncertainty can be modelled adequately than is being done so far with the use of fuzzy set theory in order to capture human thinking and perception through artificial intelligence (AI). Fuzzy logic uses linguistic variables to describe the fuzziness which makes the communication between the system and human beings easy. Another advantage of using fuzzy logic is that it manages the uncertainties involved in a complex process and can be applied to ambiguity expressions to prevent excessive increase in the number of rules.

A comprehensive review of the application of Artificial Intelligence and Expert Systems in Steel Industry has already been undertaken [Car93]. The purpose of this paper is to review several existing applications of fuzzy logic to iron and steel making area and to highlight the benefits derived from the application of fuzzy logic. This is also an attempt to indicate to the researchers of the work being done in the area of fuzzy logic applied to iron and steel making and to point to the future direction of research. In this paper the fuzzy applications in steel making are classified into appropriate categories and the salient feature of each application is reviewed. The ongoing research at BHP Steel in the field of fuzzy logic are then briefly discussed. At the end the future trends in the application of fuzzy logic and the areas which need further attention are discussed.

2.0 BACKGROUND OF FUZZY SET THEORY

The theory of fuzzy logic was introduced by Professor Lotfi A. Zadeh of the University of Berkeley in 1965 [Zad65]. Conventional computer logic is incapable of manipulating data representing subjective or vague human ideas, such as "an attractive person" or "pretty hot". Computer logic previously envisioned reality only in such simple terms, as on or off, yes or no, and black or white. Fuzzy logic was designed to allow computers to determine valid distinctions among data with shades of gray, working similarly in essence to the processes which occur in human reasoning. Accordingly, fuzzy technologies are designed to incorporate fuzzy theories into modern control and data processing, to create more user friendly systems and products.

In recent years considerable attention has been given to use fuzzy set theory to quantify the qualitative and approximate evaluations and judgments of human beings. A successful application of using such an idea is the quantitative assessment of total machinability for arbitrary selection of operation parameters including steel materials [FJ90, FJ94].

Fuzziness in expert systems could be represented mainly by two methods- by introduction of a certainty factor (c.f.) into the rules, or through fuzzy theory using membership functions. The central concept of Zadeh's fuzzy-set theory is the membership functions which represents numerically the degree to which an element belongs to a set. This function takes on values between 0 and 1. The membership function is assessed subjectively in any instance, small values representing low degree of membership and high values representing high degree of membership.

A brief summary of the basic properties of fuzzy sets such as the membership function, goals, constraints, decisions, and α- Cuts described in [Zad65, BZ70, DP80, Zim91] is given below in order to define the concepts and terms which will be used in this paper.

Fuzzy Set. If $X=\{x\}$ denotes a collection of objects (points) denoted generically by x, then a fuzzy set A in X is a set of ordered pairs

$$A=\{(x,\mu_A(x))\},\ x\in X \qquad (1)$$

where $\mu_A(x)$ is termed as the grade of membership of x in A and μ_A:X$\rightarrow$M is a function from X to a space M called the membership space.

Intersection. The intersection of A and B is denoted by $A\cap B$ and the membership function of $A\cap B$ is given by

$$\mu_{A\cap B}(x)=Min(\mu_A(x),\mu_B(x)),\ x\in X \qquad (2)$$

where

$$Min(a,b)=a \quad if\ a\leq b$$
$$and \quad Min(a,b)=b \quad if\ a>b$$

Union. The union of A and B denoted as $A\cup B$, and the membership function of $A\cup B$ is given by

$$\mu_{A\cup B}(x)=Max(\mu_A(x),\mu_B(x)),\ x\in X \qquad (3)$$

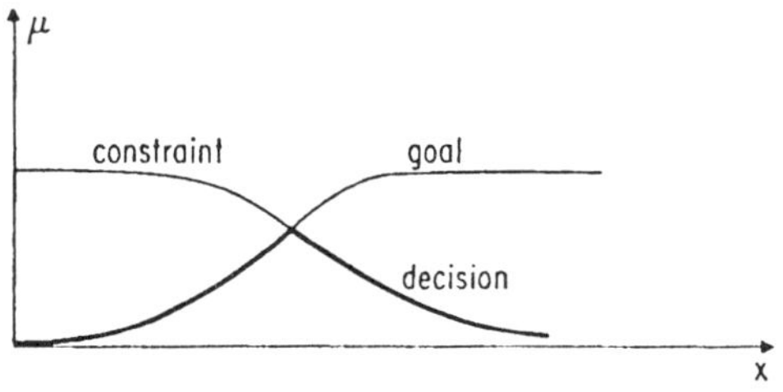

FIGURE 2 Relation between Goal, Constraint and Decision [Zad65]

where

$$Max(a,b) = a \quad if\ a \geq b$$
$$and \quad Max(a,b) = b \quad if\ a < b$$

Decision. Intuitively, a decision is basically a choice or a set of choices drawn from the available alternatives. If a fuzzy goal G and a fuzzy constraint C is given in a space of alternatives X, then, G and C combine to form a decision, D, which is a fuzzy set resulting from intersection of G and C. Symbolically

$$D=G \cap C \tag{4}$$

More generally, if n goals $G_1, G_2, \ldots G_n$ and m constraints $C_1, C_2, \ldots C_n$ are given, then the resultant decision is the intersection of the goals $G_1, G_2, \ldots G_n$ and the given constraints $C_1, C_2, \ldots C_n$, that is

$$D= G_1 \cap G_2 \cap \ldots \cap G_n \cap C_1 \cap C_2 \cap \ldots \cap C_m \tag{5}$$

The relation between Goal, Constraint, and Decision is depicted in figure 2. In defining a fuzzy decision D as the intersection or, more generally, as the confluence of the goals and constraints, it is assumed that all of the goals and constraints that enter into D are, of equal importance. However, if some of the goals and perhaps some of the constraints are of greater importance than other, then D might be expressed as a convex combination of the goals and constraints, with the weighting coefficients reflecting the relative importance of the constituent terms. More explicitly, $\mu_D(x)$ may be expressed as

$$\mu_D(x) = \sum_{i=1}^{n} \alpha_i(x)\mu_{G_i}(x) + \sum_{j=1}^{m} \beta_j(x)\mu_{C_j}(x) \tag{6}$$

where α_i and β_j are membership functions such that

$$\sum_{i=1}^{n} \alpha_i(x) + \sum_{j=1}^{m} \beta_j(x) \equiv 1 \tag{7}$$

Subject to this constraint, then, the values of $\alpha_i(x)$ *and* $\beta_j(x)$ can be chosen in such a way as to reflect the relative importance of

$$G_1, G_2, \ldots, G_n \text{ and } C_1, C_2, \ldots, C_m.$$

α-Cuts. To exhibit an element $x \in X$ that typically belongs to a fuzzy set A, *its* membership value may be demanded to be greater than some threshold $\alpha \in [0,1]$. The ordinary set of such elements is the α-cut A_α of A given by

$$A_\alpha = \{x \in X, \mu_A(x) \geq \alpha\} \tag{8}$$

3.0 APPLICATION OF FUZZY LOGIC IN STEEL MAKING

The applications of fuzzy logic in iron and steel making could be classified into the following three main categories- Diagnostic, Control and Design. The design category

Table 1 Classification of Fuzzy Systems in Steel Making

Developed By	**Application**	**Type**	**References**
Kawasaki Steel	Belt Conveyor Control	Diagnostic	[1]
Rautarauukki Oy	Continuous Steel Casting Diagnostic	Diagnostic	[2]
Nippon Steel	LD Converter Process Control	Control*	[16]
Nippon Steel	Control of Box Annealing Furnaces	Control	[19]
Nippon Steel	Shape control for cold strip mill	Control	[12]
Nippon Kokan	Sintering Process Control	Control	[15]
Nippon Kokan	Blast Furnace Control	Control*	[3], [5], [17]
Kawasaki Steel	Control of Hopper Discharge Rate	Control	[11]
Kawasaki Steel	Burning Control of Sintering Machine	Control	[14]
Kawasaki Steel	Hot Stove Combustion Control	Control	[14]
Kawasaki Steel	Ore Yard Operation Control	Control	[1]
BHP Steel	Sinter Plant Operator Guidance System	Control	[13]
Kobe Steel	Forecasting Furnace Heat	Control	[18]
Sumitomo Metals	Blast Furnace Control	Control*	[4]
Bethlehem Steel	Optimal Metallurgical Grade Design	Design	[6-7],[9-10]
Nippon Steel	Material Design	Design*	[8]

* Expert System Building Tool or Expert System Shells used.

Table 2 Knowledge Base Representation*

Input Casting Variables							Diagnostic Results	
liquid steel superheat ladle treatment	heat content of ladle	mass of skull in ladle	holding time of ladle before casting	ladle stirring time	ladle position in the casting sequence	ladle casting time	problems at the beginning of casting	problems at the end of casting
H	FH	L	FL	FL	FL	FL	No	No
H	FH	FL	L	L	FL	L	No	No
H	FH	FL	L	FL	L	FL	No	No
H	FH	FL	L	FL	L	L	No	No
H	FH	FL	L	FL	FL	L	No	No
H	FH	FL	L	FL	FL	FL	No	No
H	FH	FL	FL	FL	FL	L	No	No
FL	L	L	L	L	L	L	Yes	No
FL	L	L	FL	L	FL	L	Yes	No
FH	FH	L	FL	FL	FL	L	Yes	No
H	FH	L	L	L	L	L	Yes	No
H	FH	FL	FL	FL	L	L	Yes	No
L	L	L	L	FL	FL	FL	No	Yes
FL	L	L	L	L	L	FL	No	Yes
FL	L	L	FL	L	FL	FL	No	Yes

* Reorganised based on [BKS92]   L - Low; FL - Fairly Low; FH - Fairly High and H - High.

includes the planning and scheduling applications also. Diagnosis problems can be further divided into 3 main categories: equipment diagnosis, operational diagnosis and product quality diagnosis. Table 1 depicts the above categories of fuzzy systems developed for iron and steel making applications.

3.1 DIAGNOSTIC APPLICATIONS

Diagnostic problems are easier to handle than planning problems from the view point of the amount of computation involved and hence a large proportion of fuzzy applications developed are of diagnostic type.

Fuzzy expert system for continuous steel casting diagnosis based on an artificial neural network called productive neural network is developed and implemented at Rautaruukki Oy, Raahe works in Finland [BKS92]. Productive neural networks are a kind of neural networks developed specifically for fuzzy logic based reasoning. Heuristic knowledge acquired from the operational experience of engineers and operators partly in the form of a decision tree has qualitative data for the effect of some central variables on the outcome of the casting process. This was represented in a tabular form and a part of it is shown in table 2.

It is claimed that this approach may save a lot of time in developing the expert system. Further reasoning is almost instantaneous and no expensive expert system development tools are required. The work demonstrated that productive networks can be used for convenient storage and retrieval of the knowledge for use as a fuzzy expert system. Decision making in continuous steel casting to decide whether to accept or reject the liquid content of a steel ladle (heat) was tackled by using productive neural networks.

A system was developed at Mizushima works of Kawasaki Steel for the operation of the complicated connecting parts of the belt conveyor system [TKT+91]. The system for conveyor control is forward looking production system using IF-THEN rules, and it uses fuzzy certainty factor (c.f.) to express human priority judgments and success probability rates. The system has over 100 rules including system diagnostic rules and operation guidance rules. It is reported that the belt conveyor control system has resulted in automatic operation control of complicated belt conveyor connections.

3.2 CONTROL APPLICATIONS

Control problems are becoming important applications of fuzzy logic in steel making. These problems find applications mainly in the blast furnace and sintering machine which are characterised by complex reactions and thus difficult to be expressed by numerical models.

Nagoya works of Nippon Steel has developed a shape control system for cold strip mill [HT91]. The actual shape of strip being rolled is detected by a shape meter and is feed back controlled by work roll bending, screw down levelling and coolant zone control. The shape control system is characterised by the application of fuzzy set theory in improving control accuracy. The manipulated variables are determined on the basis of fuzzy set theory. The shape deviation for the example of work roll bending is evaluated in detail by a membership function on a seven-point scale from PB (positive big) to NB (negative big) as illustrated in figure 3. The centre of gravity of the entire shaded area is calculated to determine

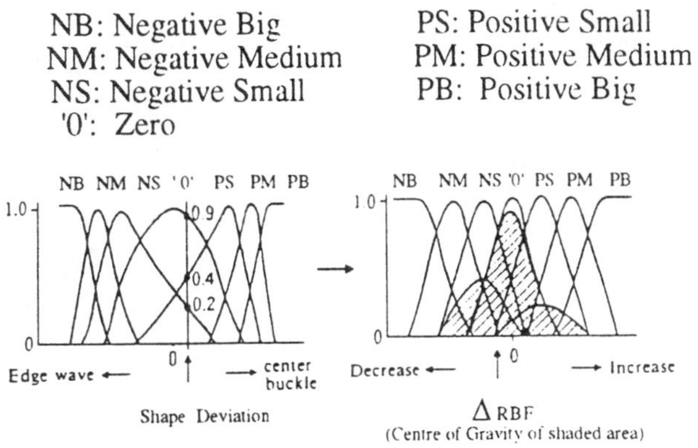

FIGURE 3 Shape Deviation Evaluation by Membership Function [MTO+91]

the required change Δ RBF in work roll bending force. It is claimed that this system has greatly contributed to the improvement of productivity of product quality.

The expert system developed by NKK for blast furnace control consists of " an abnormal condition forecasting system" which forecasts abnormal in-furnace conditions such as burden slip or channelling and "a furnace heat monitoring and control system" which controls the furnace heat represented by molten iron temperature [TAT+89, TTH+87, NSA90], To deal with the fuzziness present in the large amount of sensor data required to judge the thermal condition, certainty factors (c.f. values) are given to fact data and 3-dimensional fuzzy membership functions are applied. The self-learning of membership function is illustrated in figure 4. If the hot metal temperature to be controlled deviates considerably from the target value, the membership function is re-established. The aim of introducing fuzzy theory in the process to determine a fuzziness is to prevent expansion of system rules and simplify knowledge representation. The system is in practical use and is reported to be operating favourably, after having undergone modifications as necessary and setting several other new rules.

Control of hopper discharge rates of blast furnace at Kawasaki Steel is done through a system utilising fuzzy logic [ITU+92]. Fuzzy control is applied to unloading from the ore bin and the control rule used by the skilled operator is applied to adjust the gate opening by fuzzy inference and feed-back. The fuzzy variable involved in this are the gate opening, the actual discharge rate and the predicted rate deviation. This grain control method uses fuzzy logic to achieve real-time and high speed operation by PLC and the logic type IF-THEN control algorithms are used in parallel. It is claimed that this system has resulted in faster response in unloading control from the ore bin, than was possible with the manual method and the segregation of material brands during raw material discharging is eliminated.

In the sintering plant at Mizushima works of Kawasaki Steel a fuzzy control system for uniform burning control was developed [MTO+91]. Fuzzy control was applied to the control of charged density (cut volume), where it is used to operate five-split sub-gates based on the information obtained by waste gas thermometers installed at four points in the width direction along five lines in the longitudinal direction. Inference is conducted based on some 28 fuzzy production rules with temperature deviation as the IF part and cut volume at each point in the width direction as the THEN part. It is reported that burning rate uniformity in the pallet width direction was improved considerably .

Fuzzy control was implemented at blast furnace of Kawasaki Steel to the setting and calculation of the gas flow rate and calorie levels during combustion based on information on the residual stove heat value and brick temperature distribution [MTO+91]. Fuzzy inference is conducted based on 36 rules. It is claimed that automatic setting of the gas volume has resulted in an increase of about 3% in thermal efficiency.

Sumitomo Metals had developed a hybrid expert system similar to fuzzy control system for blast furnace operation incorporating the numerical model and the knowledge base approach utilising experience rules [OMK+92] as depicted in figure 5. This system combined the advantage of model controllability at the stationary state with the advantage of experience rule stability at the non-stationary state. It is aimed at controlling the blast furnace stably for a long time by use of a general situation judgment rule and by checking model appropriateness. The application possibility of the blast furnace condition model is judged and operation action taken mainly from experience rules whenever necessary. It is claimed that the scattering in the hot metal temperature and the silicon percentage in hot metal is remarkably lowered in comparison to the conventional manual control.

Remote automatic control system for sizing plant to control the crusher clearance and apron feeder were developed using fuzzy control with the aim of labour

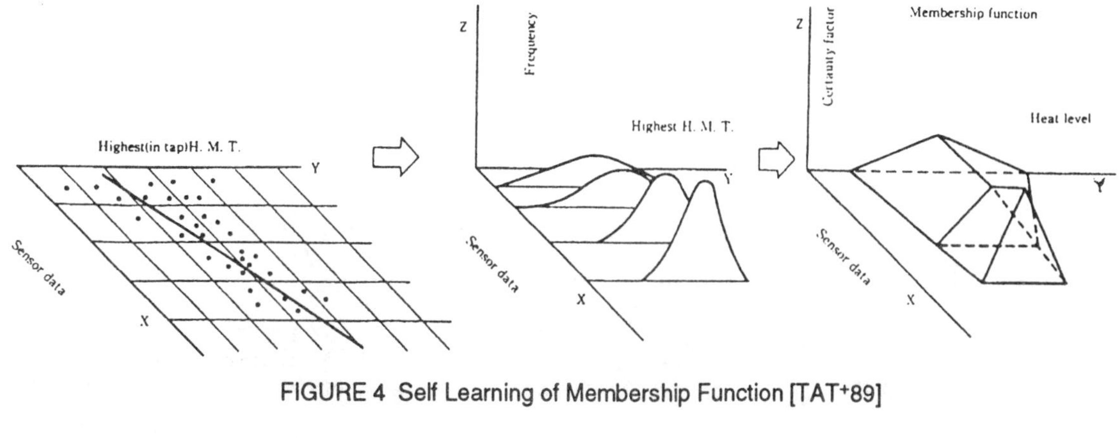

FIGURE 4 Self Learning of Membership Function [TAT+89]

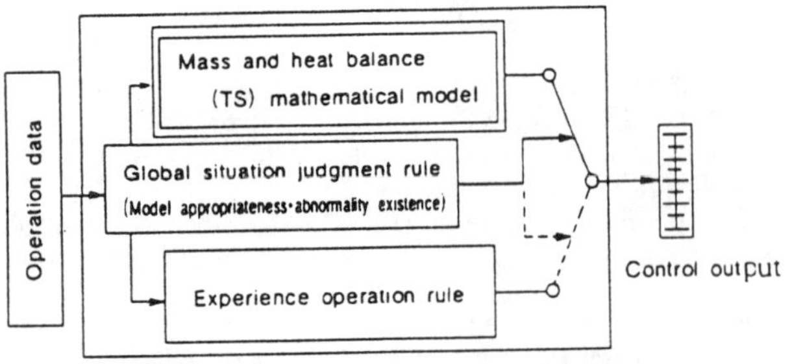

FIGURE 5 Hybrid Control System [OMK+92]

saving and energy saving at Kawasaki Steel, ore yard operation [TKT+91]. Development of fuzzy control system involved the development of a crushing-state estimate model using the electric current value and electric power value, preparation of a control logic based on operation know-how, testing of the control logic and parameter tuning. The control logic incorporates fuzzy theory, and is composed of an energy saving logic and quality control logic. It is claimed that this system consistently maintains the size distribution of crushed products within specified range.

Fuzzy logic has been successfully applied to the control of iron ore sintering process at Nippon Kokan Company (NKK) in Japan [IHI+88]. It is reported that the deviation of return fine hopper level decreased from 12 to 4% and the amount of return fine ore decreased by about 2 Kg per ton, resulting in considerable reduction in product cost.

Nippon Steel has developed an expert system (LD-ES) to control the LD converter processes [KYY+90]. The main objectives of the system included improved blowing accuracy, labour savings and increased flexibility. The system has two modules- static control and dynamic control. Static control utilises rule based reasoning based on the empirical knowledge of the operators to correct errors in the control variables. Dynamic control is realised through fuzzy reasoning where membership functions are used to define the reliability of condition and conclusion parts. An example of membership function used for the estimation of phosphorous concentration is shown in figure 6. As a large number of process information X_i are involved for estimating the phosphorous concentration, the final phosphorous concentration is predicted by calculating the weighted mean of all the rules. Fuzzy reasoning is also used to estimate slag formation. The fuzzy reasoning resulted in the reduction of number of rules and hence the reasoning speed. It is reported that the LD-ES performed better than a skilled operator in controlling the LD converter processes.

Another fuzzy expert system has been developed to estimate the furnace heat from hot metal temperature [NAM89]. The system has been implemented at No. 3 blast furnace at Kobe Steel, Japan and it is reported that it has resulted in successful forecasting from 60% to 86%.

The control of box annealing furnaces has been achieved at Nippon Steel by applying fuzzy logic [Yon81]. In box annealing the coil weight and size differ from charge to charge and so the heat transfer characteristics also vary. Due to this the traditional PID controllers are not very effective. A fuzzy expert system having some 22 rules combined with simulation techniques was used for the control of furnaces.

3.3 DESIGN APPLICATIONS

The design problems are usually ill structured, difficult to systematise and a large number of rules are involved and hence the design problems have been given priority in fuzzy applications to steel making. The current trend in the growth of AI/Fuzzy applications as shown in figure 7, is in design and planning systems [YI92]. In the field of material design, many complex factors interact and the design standards also frequently change due to introduction of new materials.

A fuzzy set based system is developed at Bethlehem Steel Corporation, USA, for optimally assigning metallurgical grades [VSW+89, VWS89, WSW93, WSW+92]. In this system a list of all applicable grades is prepared by utilising the metallurgical expertise and statistically derived predictive equations along with the characteristics and past performance of every grade and various standards. This list is then ranked according to their membership functions based on the mechanical properties, chemistry requirements, and desirability of producing the grades. The weighted sum membership functions are computed by summing up the composite membership function components which are weighted, based on the equation

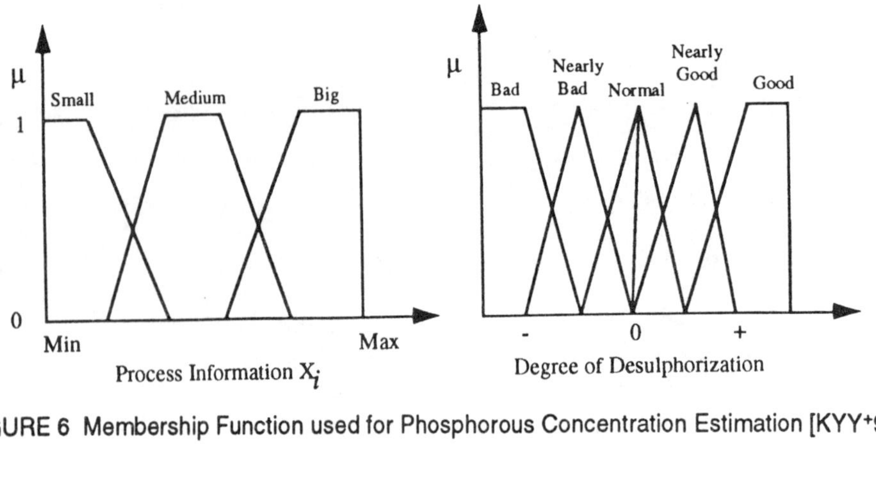

FIGURE 6 Membership Function used for Phosphorous Concentration Estimation [KYY+90]

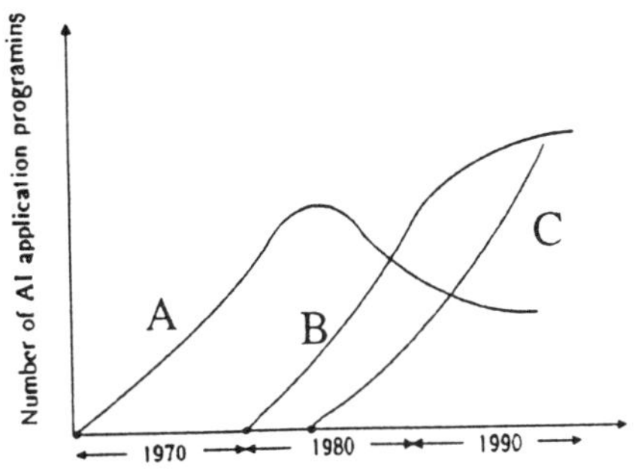

A- Diagnosis Type; B- Planning Type; C- Finance and Economy Type

FIGURE 7 Trend in AI/Fuzzy Applications [YI92]

$$f_{A(i)}(G_j)=\begin{cases}\beta_1 f_{K(i)}(G_j)+\beta_2 f_{L(i)}(G_j)+\beta_3 f_p(G_j) & \text{if } \min\{f_{K(i)}(G_j), f_{L(i)}(G_j)\}\neq 0\\ 0 & \text{if } \min\{f_{K(i)}(G_j), f_{L(i)}(G_j)\}=0\end{cases} \quad (9)$$

where L refers to the Chemistry, K the mechanical property, and P is a fuzzy sub set of G indicating the plant's desirability of producing grade G_j. $\beta 1 + \beta 2 + \beta 3 = 1$ and $\beta 1 > \beta 2 > \beta 3$

β's are determined based on metallurgical considerations.

Optimisation to reduce the number of grades melted and to maximise the likelihood of meeting all customer specifications without difficulty is then done using several alpha-cut offs together with the defined weighted sum membership functions.

To design steel grades which are not standard, a design type expert system was developed and implemented at Nippon Steel's head office and Sakai works [WIO+92]. The core of the system is the reasoning function that performs the manufacturability study. The system gives a concrete form to the expertise and thinking capability of the expert engineers by making the best use of various reasoning techniques like case-based reasoning, neural network reasoning, fuzzy reasoning and hypothetical reasoning. The system was built using the expert system building tool ART™

The fuzzy reasoning technique applied to the expert system proved effective in raising the reasoning accuracy and simplifying the system configuration without difficulty. Fuzzy reasoning was applied to judge whether the specified toughness can be guaranteed as explained in figure 8. This expert system was put into operation in May 1990 and it is claimed that it has shortened the product manufacturability study task time to one-sixth. The higher-order reasoning techniques adopted in the expert system result in reduction in the knowledge acquisition load, system simplification, and enhancement of the problem solving efficiency.

4.0 RESEARCH IN FUZZY LOGIC AT BHP STEEL

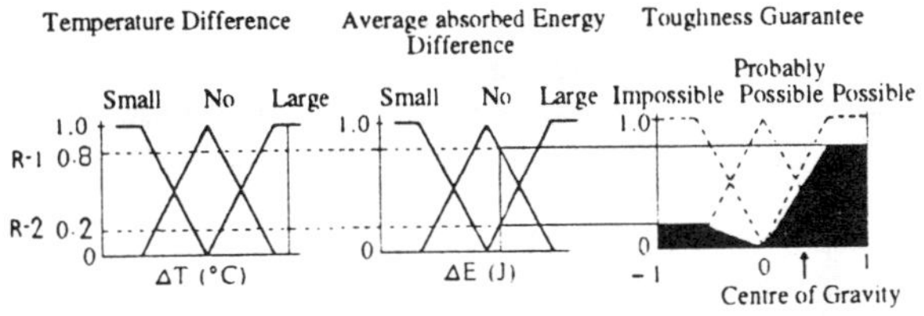

FIGURE 8 Fuzzy Reasoning for Toughness Prediction [WIO+92]

4.1 SINTER PLANT OPERATOR GUIDANCE SYSTEM (OGS)

A prototype sinter plant OGS which utilises fuzzy logic to determine the strength of the recommended parameter is developed at BHP Steel [LMT+90]. Sample data set was taken from the Slabs and Plates Product Division's sinter plant data acquisition computer and transferred to the SHERPA (prototyping tool) work station. The pre processing and operator display modules were used to process and display the data to the expert's satisfaction. Previously developed knowledge base was entered into TABLEAUX (knowledge acquisition tool) for analysis and modification in conjunction with the expert.

Fuzzy logic is used to determine the strength of each parameter, which is then displayed graphically on a gauge to the left of the parameter's time series trace. The strength value is determined by locating the data point on a linear membership function. The upper and lower limits for the membership functions are determined by the experts. The determination of fuzzy strength for a single parameter is depicted in figure 9. If more than one rule contributes towards the truth of a particular hypothesis, the highest strength rule arbitrarily has its strength assigned to that hypothesis. The recommendations and its relative strength are then displayed graphically on the action gauges. Fine tuning can be achieved through the adjustment of the shape of membership functions or the methods by which they are combined.

4.2 HYBRID MATERIAL DESIGN SYSTEM

Another research currently being undertaken at BHP Steel is in material design. The main objective of this research is to determine the steel making aim chemistry utilising iterative and knowledge based approach. A list of alternate steel making aim chemistries is obtained and then

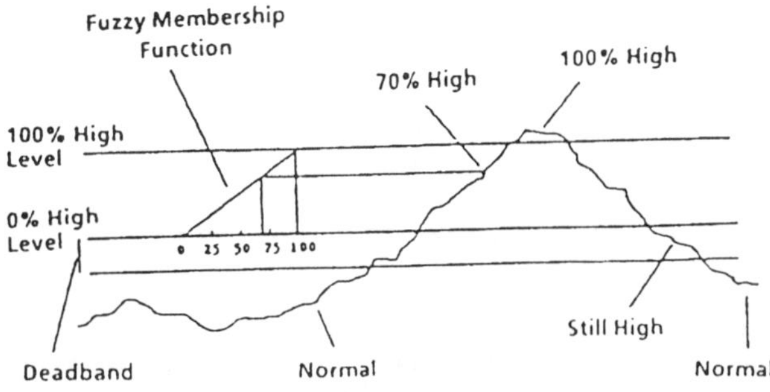

FIGURE 9 Fuzzy Strength Determination [LMT+90]

fuzzy logic is applied to rank these aim chemistries based on the likelihood of the grades meeting the mechanical properties, chemical composition, cost and complexity of producing the grades mainly based on the capacity of bottleneck production unit. This is done by defining fuzzy membership functions describing the ability of the grades in meeting the mechanical properties, chemical composition, cost and the complexity. The membership functions are essentially the probability density functions assuming that all the properties under consideration are normally distributed. The system utilises various knowledge bases along with various data bases and statistically derived empirical relations to determine the appropriate metallurgical grades satisfying the customer requirements.

5.0 CONCLUSION AND FUTURE PROSPECTS

Iron and steel making is an area where the application of fuzzy techniques systematises the processes, reduces the dependence on experts, reduces the time taken to perform the knowledge intensive complex tasks and efficiently manages the uncertainty or fuzziness and ambiguity. In spite of the diversity and complexity of iron and steel making processes, an attempt is made to identify a common core of techniques from fuzzy systems that may be applied to solve the problems in the bottleneck areas in iron and steel making. A review of various fuzzy techniques used in iron and steel making is presented stressing the higher order reasoning techniques and the salient features of the successful systems. The later half of 1980's has been a period in which more and more problems in the steel making field were tackled successfully through the use of fuzzy logic applications especially the hybrid type of fuzzy systems using AI and OR technologies. This recent trend of developing hybrid systems by fusing AI and OR seems to be very promising in future, with OR applied to quantitative optimisation computation and AI applied to qualitative inference. It can also be said that the new AI technologies such as case-based reasoning, fuzzy reasoning and neural networks can be integrated to solve more challenging practical problems faced by the steel industry in future. The trend of manpower getting more costly and computer becoming cheaper will result in further interest in the applications of fuzzy systems in future.

The number of fuzzy logic applications in iron and steel making is increasing rapidly due to its good controllability in systems where it is difficult to apply OR methods, traditional PID control and modern control theory based on process models. The contents of fuzzy systems is easily understood by operators as the empirical knowledge of operators is incorporated in the system in the form of rules. However to increase the scope of applications of fuzzy systems in iron and steel making, in order to increase the productivity of development of fuzzy systems, and also to reduce the number of fuzzy systems which fail in meeting their targets, the following areas need to be given more attention and researched further: Knowledge structuring and multistage fuzzy inference processes, automatic tuning of membership function using neural network or other means and development of tools for building efficient fuzzy systems and tools for supporting the knowledge acquisition and knowledge representation.

ACKNOWLEDGMENTS

This research work was funded by the Australian Research Council. The authors acknowledge the support of the Commonwealth.

REFERENCES

[BKS92]] Bulsari, A.B., Kraslawski, A., Saxen, H., 1992, Continuous steel casting diagnostic fuzzy expert system in an artificial neural network, *Scandinavian Journal of Metallurgy*, Vol. 21, 146-150.

[BZ70] Bellman, R.E., Zadeh, L.A., 1970, Decision-making in a fuzzy environment, *Management Science*, Vol. 17, No. 4, (Dec), B141-B164.

[Car93] Carayannis, G., 1993, Artificial Intelligence and Expert Systems in the Steel Industry, *JOM*, (Oct), 43-51.

[DP80] Dubois, D., Prade, H., 1980, Fuzzy Sets, *Fuzzy Sets and systems: Theory and Applications*, Academic Press Inc., 9-35.

[FJ90] Fang, X.D., Jawahir, I.S., 1990, An Expert System Based on a Fuzzy Mathematical Model for Chip Breakability Assessments in Automated Machining, *ASME Proc. Manufacturing International (USA)*, Vol. 4, 31-37.

[FJ94] Fang, X.D., Jawahir, I.S., 1994, Predicting Total Machining Performance in Finish Turning Using Integrated Fuzzy-Set Models of the Machinability Parameters, *International Journal of Production Research*, Vol. 32, No. 4, 833-849.

[HT91] Hasegawa, A., Taki, F., 1991, Development of fuzzy set theory-based shape control for cold strip mill, *Nippon Steel Technical Report*, No. 49, (April), 59-62.

[IHI+88] Iwamoto, M., Hashimoto, K., Inaba, M., Kobayashi, T., Hattori, M and Komatsu, O., 1988, Application of fuzzy control for iron ore sintering process, *Transactions ISIJ*, Vol. 28, No. 5, (May), 341-345.

[ITU+92] Iida, O., Taniyoshi, S., Uetani, T., Sawada, T., Hashimoto, M., Onoda, D., 1992, Application of AI technique to blast furnace operation, *Kawasaki Steel Technical Report*, No. 26, (June), 30-37.

[KYY+90] Kanemoto, M., Yamane, H., Yoshidha, T. and Tottori, H., 1990, An Application of Expert System to LD Converter Processes, *ISIJ International,* Vol. 30, No. 2, (Feb), 128-135.

[LMT+90] Lee, L.G.L., Mcnamara, A.R., Teh, K.C., Lie, H.M., Orenstein, B.J., Brown, D.J.H., 1990, Rapid prototyping tools for real-time expert systems in the steel industry, *ISIJ International,* Vol. 30, No. 2, 90-97.

[MTO+91] Maki, Y., Takashima, N., Obata, H., Iida, O., Nakashima, K., Sawada, T., 1991, Application of fuzzy theory to iron making process control, *Kawasaki Steel Technical Report,* No. 24, (April), 8-15.

[NAM89] Nagai, N., Arai, A. and Matsuda, K., 1989, Application of Expert System to Forecast Heat Level at Kobe No. 3 Blast Furnace, *CAMP-ISIJ,* No. 2, 10-13.

[NSA90] Niwa, Y., Sumigama, M. and Aoki, T., 1990, Application of Self-Learning Function to an Expert System for Blast Furnace Heat Control, *ISIJ International,* Vol. 30, No. 2, (Feb), 111-117.

[OMK+92] Otsuka, K., Matoba, Y., Kojima, M., Yoshida, M., Matsuoka, S., 1992, Expert system for blast furnace operation, *The Sumitomo Search,* No. 50, (July), 43-50.

[TAT+89] Takekoshi, A., Aoki, T., Takahashi, I., Tomishima, H., Yoshino, M., 1989, Application of knowledge engineering for iron and steel making works, *NKK Technical Review,* No. 56, 1-9.

[TKT+91] Taniyoshi, S., Kaneda, T., Tamura, T., Onoda, D., Miyazaki, Y., Ikeda, T., 1991, Computer control system for ore yard operation at Mizushima works, *Kawasaki Steel Technical Report,* No. 24, (April), 1-7.

[TTH+87] Tsunozaki, Y., Takekoshi, A., Hashimoto, K., Aoki, T., Wakimoto, K., Sakurai, M., 1987, An expert system for blast furnace control at Fukuyama works, *Nippon Kokan Technical Report,* No. 51, 1-10.

[VSW+89] Vasko, F.J., Stott, K.L., Wolf, F.E., Woodyatt, L.R., 1989, A fuzzy approach to optimal metallurgical grade assignment, *Applications of Fuzzy Set Methodologies in Industrial Engineering,* (ELSEVIER) 285-298.

[VWS89] Vasko, F.J., Wolf, F.E., Stott, K.L., 1989, A set covering approach to metallurgical grade assignment, *European Journal of Operational Research,* Vol. 38, 27-34.

[WIO+92] Watanabe, M., Iwata, Y., Obama, N., Shirahata, K., Suematsu, K., Meiga, T., Yamane, H., 1992, Development of shape steel quality design expert system, *Nippon Steel Technical Report,* No. 53, (April), 29-33.

[WSW+92] Woodyatt, L.R., Stott, K.L., Wolf, F.E., Vasko, F.J., 1992, Using Fuzzy Sets to Assign Metallurgical Grades to Steel, *JOM,* (Feb), 28-31.

[WSW93] Woodyatt, L.R., Stott, K.L., Wolf, F.E., 1993, An application combining set covering and fuzzy sets to optimally assign metallurgical grades to customer orders, *Fuzzy Sets and Systems,* No. 53, 15-25.

[YI92] Yamasaki, J., Iwamura, T., 1992, Artificial Intelligence applications at Kawasaki Steel, *Kawasaki Steel Technical Report,* No. 26, (June), 171-177.

[Yon81] Yonekura, M., 1981, The Application of Fuzzy Set Theory to the Temperature Control of Box Annealing Furnaces using Simulation Techniques, *Proc. of the 8th Triennial World Congress of IFAC* (Oxford U.K., Pergamon Press), 773-778.

[Zad65] Zadeh, L.A., 1965, Fuzzy Sets, *Information and Control,* No. 8, 338-353.

[Zim91] Zimmermann, H.J., 1991, *Fuzzy Set Theory and its Applications,* Kluwer Academic Publishers, 11-21.

EXPERT SYSTEM DETERMINATION OF FEASIBLE ASSEMBLY OPERATIONS

Maria Joao Abrantes and Simon D. Hill
Robotics and Digital Technology,
Monash University,
Australia
Email: mja@rdt.monash.edu.au

ABSTRACT

This paper describes the encapsulation of assembly knowledge in a rule-based system which allows efficient generation of feasible assembly sequences for mechanical products. The set of feasible sequences is generated from the topological and mechanical descriptions of the connections between component parts of the product. The rule-based system interprets the connection information, enhances it and evaluates the feasibility of assembly operations under the control of procedural code. The method presented here has two advantages. One is that it does not need to interrogate the designer for supplementary data as the sequence-generation evolves. The second is that it is suitable to be used in an assembly-planning concurrent-engineering tool, early in a product development cycle, because it does not require any geometric information about the product.

1 INTRODUCTION

It is now widely recognised that product development requires a concurrent engineering approach for effective cost, quality and delivery times. Concurrent Engineering creates an environment in which all the disciplines that contribute to the product development cycle (design, manufacturing, quality assurance, marketing, etc.) perform their functions cooperatively, allowing ease of manufacture and assembly, quality control, maintenance etc., to be considered right from the conceptual design phase of the product.

It is at the conceptual design phase that the design team has the greatest opportunity to explore alternatives and identify the best product and manufacturing process. At this stage commitments to hardware and constraints due to the interactions between the various specialities involved in the product's life cycle are not yet tight. But, because of the large number of options available and because the decisions taken at this stage will seriously constrain the options available at later phases of the product development cycle, these are the most difficult and influential decisions in the whole process.

For this reason it is fundamental to provide tools to the concurrent engineering team that facilitate effective communication and identification of downstream implications of the influential early decisions. There are in fact several types of tool currently available to concurrent engineering teams, which range from management based tools to Computer-Aided Design (CAD), Quality Function Deployment (QFD), Design for Assembly and Manufacture (DFA/M), etc. However, there is an important activity which still lacks an efficient computer-aided tool for use early in the product development cycle. This activity is the planning of the product-assembly sequence. DFA methodologies and related software tools successfully assist the designer in reducing the cost of a product by improving its assemblability, but the planning of the sequence is not part of the DFA procedure.

Assembly sequence planning has a central role in product and assembly system design. The choice of assembly sequence can impact directly on the efficiency and cost effectiveness of the assembly process. The sequence in which parts are assembled influences, for example, the number of changes in assembly orientation and in tools (e.g. robot grippers), the number and complexity of fixtures, and the complexity of the assembly operations. In addition, the assembly sequence can affect, and be affected by, several factors such as the

*This work was supported by a grant from the Faculty of Computing Research Fund, and by funding from the Department of Robotics and Digital Technology.

choice and layout of equipment, product testing and repair strategies, and the assembly method; manual or automated. Different sequences can lead to totally different assembly times and costs. Without considering the complete set of feasible assembly sequences it is possible that a poor choice is made.

The set of alternative assembly sequences of a product is primarily determined by the design of the product. Small changes in the design may change the set of sequence options, and rule out a sequence that could lead to a more efficient assembly process. Furthermore, for a typical product there may be thousands of possible assembly sequences. Hence, a tool that can be used very early in the design stage of the product, when CAD drawings may not yet be available is fundamental to deal efficiently with the combinatorial complexity of the problem and allow an analysis of the impact of design changes on the set of valid sequences of assembly.

2 REVERSE ASSEMBLY APPROACH

The system detailed in this paper is capable of generating the complete set of feasible assembly sequences from the model of a product. It combines algorithmic and rule-based solution methods, which have been implemented on a Unix workstation in C++ and CLIPS [NAS93] expert-system code respectively.

A *Reverse Assembly Approach* is used in the generation of the assembly sequences. This method, often called in assembly planning literature *Disassembly Approach* [Del92, HL91, MS91], consists of generating all possible decompositions of the initial assembly into pairs of subassemblies, and then applying the same procedure to each of the subassemblies obtained, until the new subassemblies are only single parts. The term subassembly has the meaning of a non-empty subset of the set of parts that constitute an assembly that either has only one part, or is such that every part has at least one connection with another part in the subset.

Each decomposition corresponds to an assembly operation that joins two subassemblies to form a larger subassembly. However, not all decompositions generated will correspond to feasible assembly operations. Some operations cannot be performed because of geometric, mechanical or stability constraints.

To obtain the complete set of feasible decompositions, a generate and test approach is used. The decompositions are generated by the procedural module of the system in the form of an AND/OR graph, where the nodes correspond to unique subassemblies and the hyperarcs to the assembly operations. Each decomposition generated is immediately evaluated for feasibility by the expert system module (if its feasibility cannot be inferred from a previous evaluation of another operation) and pruned from the graph if infeasible. The set of feasible assembly sequences is then directly extracted from the AND/OR graph.

The decomposition approach to the problem and the use of the AND/OR graph as an explicit representation of all assembly sequences of a product was proposed by De Mello [Mel89]. The methods being developed evolved from the work of De Mello and Sanderson [MS90, MS91]. However, their assembly sequence planner is intended to be used as part of an integrated control system for the assembly process, i.e., much later in the product development cycle, and consequently these researchers propose the use of algorithmic methods to the evaluation of the geometric feasibility of assembly operations, from CAD representations of the product.

Other researchers, such as, De Fazio and Whitney [BAL91] and Delchambre [Del92] derive precedence constraints between connections from which the assembly sequences are then generated. Delchambre's precedence constraints are derived from a CAD representation of the product, whilst De Fazio and Whitney's are obtained through a question-and-answer procedure.

The method used in this work to decompose the initial assembly is based on graph theoretic principles and was described in detail in [AH93, HAL94]. The current implementation of the system includes a more efficient method of generating the subsequent subassemblies decompositions. These are directly derived from the sets of connections "broken" by each decomposition of the initial assembly by extracting from each set the subset of connections present in the respective subassembly.

3 INITIAL MODEL OF THE PRODUCT

The initial model comprises relational, topological and mechanical information, which combine to describe the relative position of parts and eventual fasteners in the complete assembly, the types of connections established between them, and the directions in which these connections are established. Additionally, it may also contain user defined assembly constraints.

This model contains solely the relevant information that cannot be obtained by other means, and hence must be provided by the user. The system was designed so that it generates itself a higher level description of the product with all necessary and sufficient data for the generation of feasible assembly sequences. It automatically infers pertinent data from the initial model and the static assembly knowledge encoded in the knowledge-base of the system (refer to Sections 4.1 and 4.2).

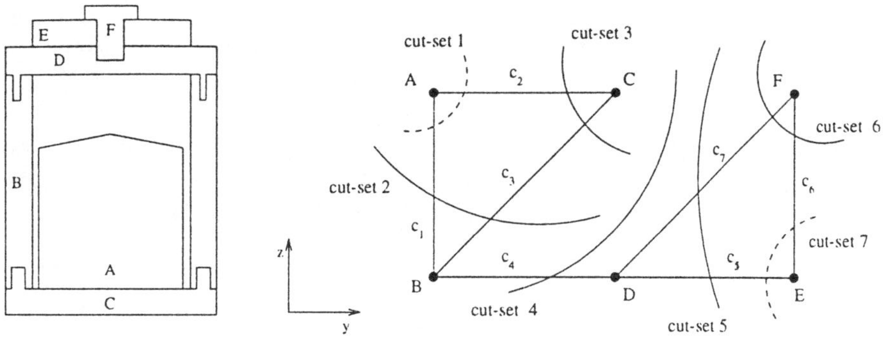

(a) (b)

$$\begin{array}{c} A \\ B \\ C \\ D \\ E \\ F \end{array} \begin{bmatrix} -1 & 1 & 0 & 0 & 0 & 0 & 0 \\ 1 & 0 & 1 & -1 & 0 & 0 & 0 \\ 0 & -1 & -1 & 0 & 0 & 0 & 0 \\ 0 & 0 & 0 & 1 & -1 & 0 & -1 \\ 0 & 0 & 0 & 0 & 1 & -1 & 0 \\ 0 & 0 & 0 & 0 & 0 & 1 & 1 \end{bmatrix}$$

{z inside, z planar, z pin, z pin, z planar, z pin, z pin}
{x 1 0 x x x x}

(c)

Figure 1 Cross-section of didactic assembly (a), graph of connections (b), and assembly model (c)

Relational information is used for the generation of all decompositions of the product into its subassemblies, whilst topological, mechanical and constraint information is used for checking the feasibility of each candidate assembly operation.

The cross-section of a didactic assembly and its model are shown in Figure 1. This assembly is formed by a housing, part B, closed at both ends by parts C and D, with *pin type* connections. Part A has a planar connection with part C and is enclosed by parts B, C and D, whilst part D has a planar connection with part E and a pin connection with part F, which also has a pin connection with E.

The graph of connections, which was first introduced by Bourjault [Bou84] for the purpose of assembly sequence planning, and has since been used by several other researchers [BAL91, Del92, Lee92, MS91] is encoded as an incidence matrix as shown in Figure 1 (c). Each column of the matrix corresponds to a connection and each row to a component part of the product.

The "ones" in each column of the matrix identify the connections between pairs of parts, whilst the signs encode topological information. The latter data describes the relative position of each part with respect to a reference coordinate frame. This data is combined with the mechanical information contained in the *Connection Information Vector*, shown in Figure 1 (c) below the incidence matrix, to define the type and direction of each connection.

For the example in Figure 1 the first column of the matrix combined with the first element of the *Connection Information Vector*, "z inside" defines part A to be inside part B, and the connection is established in the z-axis positive or negative direction. The second column and element define parts A and C to be joined in the z-axis direction and that C is closer to the origin of the coordinate frame than part A, i.e., "A is above C".

The initial model may also contain user defined technological and strategic assembly constraints. These constraints are not directly determined by the product's design, and therefore their inclusion in the model is optional. Technological constraints derive from the knowledge of particular characteristics of the assembly system, process or of component parts of the product. Examples of this type of constraints are the choice of a base part, the maximum number of fixtures, operations executed in parallel, or subassembly reorientations. A typical strategic constraint is the imposition of building specific subassemblies, when different product models use common subassemblies.

An example of a user-defined constraint on the assembly of Figure 1 is the following connection precedence constraint: part A must be assembled to part C after the connection between B and C has been established. This constraint is encoded in the line below the connection information vector.

The data contained in the initial product model is translated by the C++ module into a suitable format and loaded in the data-base of the expert system. The model of the assembly in Figure 1 is shown below in CLIPS *facts* format:

(connection A B z inside)
(connection C A z planar)
(connection C B z pin)
(connection B D z pin)
(connection D E z planar)
(connection E F z pin)
(connection D F z pin)

The constraint information is not included in the CLIPS model, because this type of constraint in handled by the C++ module during the generation of subassemblies.

4 ASSEMBLY KNOWLEDGE REPRESENTATION

Assembly knowledge is represented in the knowledge-base of the expert system to enable it to make inferences in the same way a design, process or manufacturing engineer would infer the degree of difficulty or the infeasibility of an assembly operation from a sketch of the product.

This inferential reasoning uses only knowledge of the connections types and topology of the assembly. The actual dimensions of each part are not required in the knowledge-base. It then follows that assembly sequences can be determined from a sketch of the product rather than requiring fully-dimensioned engineering drawings.

Originally, the facts in the knowledge-base are the initial assembly model described in the previous section. The implicit knowledge the engineer uses in analysing an assembly operation is explicitly encoded in the knowledge-base of the system in the form of production rules. This knowledge includes, for example, the realisation of the physical impossibility of performing an operation due to the obstruction of the insertion path of a part or the impossibility of accessing a fastener, as well as, the underlying knowledge that leads to these conclusions. The latter knowledge, which allows the automatic enhancement of the product's model (i.e., the addition of new facts to the expert system's data-base), includes the understanding of eventual interferences between parts that may not be directly connected in the operation, the set of connections that are established by the operation their characteristics and possible establishment directions.

4.1 Derivation of Fundamental Contacts

The first step in the automatic enhancement of the initial model is the transformation of the connection information into fundamental contacts and respective feasible and infeasible establishment directions.

Several types of connections found in current products have been analysed. Figure 2 shows some of those connections and their respective potential assembly directions.

For example, in the assembly of Figure 1, the connection fact (connection D F z pin) yields three new contact facts (contact D F z pin), (contact D F not x), and (contact D F not y). From the same example, connection fact (connection D E z planar) originates five new contact facts (contact D E z planar), (contact D E x planar), (contact E D x planar), (contact D E y planar), and (contact E D y planar).

The current implementation of the system includes four fundamental contacts: planar, pin, inside, and threaded contacts.

4.2 Generation of Virtual Contacts

The second enhancement of the initial model is provided by the *Part Virtual Contact Generation Rules*. When, for example, a part is *inside* another part, there may exist a third part in contact with the latter which obstructs the assembly path of the former, without however, having any direct or real contact with it. The *Part Virtual Contact Generation Rules* detect this type of "interference" and generate virtual contacts by propagating other existing contacts.

Figure 1 shows that although A and D are not connected, if part D is joined to BC, part A cannot be assembled because Part D obstructs its assembly path. The *Part Virtual Contact Generation Rules* detect this situation and generate a planar contact between A and D by propagating the pin contact between B and D:

(contact A B z inside) and (contact B D z pin) => (contact A D z planar).

These rules simplify the user's job of modelling the assembly. The user needs only to specify "real" connections between parts in the assembly, which obviously only exist among adjacent parts, and are easy to identify.

Virtual contacts are initially generated from fundamental contacts, since these are the only facts in the knowledge-base of the expert-system. However, the generation becomes recursive in the sense that virtual contacts may lead to the generation of more virtual contacts. This is shown in the example of Figure 3. When modelling this assembly, the user specifies three connections. These three connections are processed by the *Fundamental Contact* rules to generate eleven fundamental-contact facts. Three of which are:

1. (contact D C z inside+)
2. (contact C B z inside+)
3. (contact B A z planar)

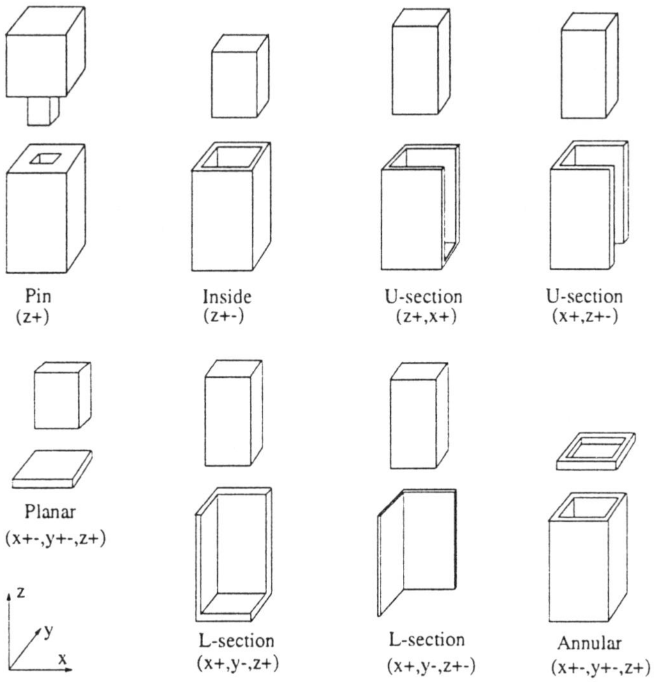

Figure 2 Connection Types

The *Part Virtual Contact Generation Rules* add to these fundamental-contact facts the following three new contact facts:

4. (contact C A z planar)
5. (contact D A z planar)
6. (contact D B z inside+)

The planar contact between C and A (fact 4) was generated by "propagating" the planar contact of B with A (fact 3), because C is "inside+ B" (fact 2):

(contact C B z inside+) and (contact B A z planar) => (contact C A z planar).

Similarly, fact 6 is generated from facts 1 and 2:

(contact D C z inside+) and (contact C B z inside+) => (contact D B z inside+).

But fact 5 is generated from fact 1 and from the virtual contact in fact 4:

(contact D C z inside+) and (contact C A z planar) => (contact D A z planar).

4.3 Generation of Contacts Between Subassemblies

The principle behind the generation of contacts between two subassemblies is that the set of contacts between two subassemblies is the union of all contacts (verified in the fully assembled product) between pairs of parts with one part in each subassembly. Hence, *if in the assembled product there is a contact C between a pair of parts, and the parts belong to two subassemblies to be assembled together, then those subassemblies have the same contact C between them.*

Every time a new pair of subassemblies is generated by decomposition of a larger subassembly, the C++ module stores the contents of the subassemblies in a CLIPS *fact* format. It then calls the expert system, which generates the set of *contact facts* that define all the contacts between the two subassemblies.

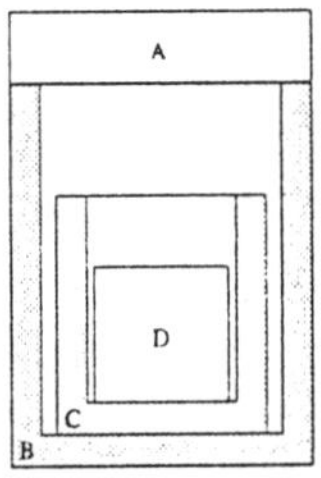

Figure 3 Interference between parts without any real contact

4.4 Evaluation of Assembly Operation Feasibility

An assembly operation is feasible if [Mel89]:

1. There is a collision-free path to bring the two subassemblies into contact.
2. It is possible to establish the attachments that act on the contacts between the two subassemblies (e.g.: the contacts can be accessed by the fasteners).
3. The two subassemblies that are joined and the resulting subassembly are stable.

A set of rules, the *Assembly Operation Feasibility Rules*, have been developed to evaluate the feasibility of each candidate assembly operation. These rules are activated after the generation of all virtual-contact facts and after the generation of the subassembly-contact facts for the pair of subassemblies involved in the operation.

Some rules account for situations that make an assembly operation infeasible, such as, the obstruction of subassembly and fastener assembly paths or the obstruction of fastener access caused by another part, subassembly, or fixture. Others, account for the situations that make an operation feasible as, for example, when the operation involves the establishment of only one connection or all connections are compatible and established in a common direction.

These rules are based on the specific characteristics of each fundamental contact type, on the feasible and infeasible assembly direction information, and on eventual user specified constraints.

4.5 Accesses to CLIPS Knowledge-base

To improve the system's efficiency, the rules are partitioned into three distinct knowledge-bases. One of them contains the *Fundamental Contact* and *Virtual Contact* rules and is accessed by the C++ program only once. The other two knowledge-bases contain, respectively, the *Subassembly Contact Generation* and the *Assembly Operation Feasibility* rules. Furthermore, the accesses to the last two knowledge-bases have been reduced to a minimum, without precluding any sequence from being generated, by concurrently applying three measures.

Two of these measures, the *generation of the AND/OR graph of unique subassemblies* and the *immediate pruning of any infeasible operation and associated subassemblies generated*, have already been mentioned in Section 2. This approach guarantees that the second and third knowledge-bases, mentioned above, are never accessed for any subassembly associated with an infeasible operation nor accessed more than once for each subassembly generated in the process. Since each unique subassembly has a unique set of decompositions, the feasibility of each corresponding set of assembly operations is never evaluated more than once.

The third measure consists of *inferring the feasibility (or infeasibility) of an operation based on a previous evaluation of another operation* by the application of rules encoded in the procedural module of the system.

These rules evolve from rules used by De Fazio [BAL91] and Delchambre [Del92], and originally derived from Bourjault's work [Bou84]. As stated in [BAL91] these rules are:

The Subset Rule:
If two subassemblies can mate, then removing any part, itself not associated with the mating liaison(s), from either subassembly, will not change this situation.

The Superset Rule:
If two parts or subassemblies cannot mate due to interference in the approach path, then adding a part to either set will not change this situation.

By concurrently applying the two measures mentioned, these rules can be expanded to include a much larger number of assembly operations. The Subset Rule is expanded to include operations where the parts removed are associated with the mating connections and a second Subset Rule is added.

<u>In the generation process of the AND/OR graph of feasible assembly operations:</u>

Operation Feasibility Inference Rule:
If the assembly operation joining subassemblies A_1 and A_2, respectively constituted by the sets of parts S_{A_1} and S_{A_2}, is found feasible, then it is also feasible to join any other pair of subassemblies B_1 and B_2 generated in the process and respectively constituted by the sets of parts $S_{B_1} \subseteq S_{A_1}$ and $S_{B_2} \subseteq S_{A_2}$.

Operation Infeasibility Inference Rule 1:
If the assembly operation joining subassemblies A_1 and A_2, respectively constituted by the sets of parts S_{A_1} and S_{A_2}, and establishing the set of connections C, is found infeasible, then it is also infeasible to join any other pair of subassemblies B_1 and B_2 generated in the process and respectively constituted by the sets of parts $S_{B_1} \subseteq S_{A_1}$ and $S_{B_2} \subseteq S_{A_2}$, and whose assembly operation establishes the same set of connections C, provided that the infeasibility of joining A_1 and A_2 is not due to the obstruction of fastener access by any part(s) removed.

Operation Infeasibility Inference Rule 2:
If the assembly operation joining subassemblies A_1 and A_2, respectively constituted by the sets of parts S_{A_1} and

S_{A_2} is infeasible, then it is also infeasible to join any other pair of subassemblies B_1 and B_2 respectively constituted by the sets of parts $S_{B_1} \supseteq S_{A_1}$ and $S_{B_2} \supseteq S_{A_2}$.

The application of the above measures and the breadth-first generation of the AND/OR graph of feasible assembly operations of the product in Figure 1 reduced the number of operations generated and tested from 175 to 8 operations tested and a total of 38 operations generated, i.e., 8 accesses to the third knowledge-base and 38 to the second one. The operations checked are the seven operations generated in the first level of the AND/OR graph and the operation joining A to BC which is found feasible when level two of the graph is generated.

5 CONCLUSION

Computerised methods to generate a complete set of feasible assembly sequences from a relational model of a product have been developed. The software system is written in both C++, and CLIPS expert-system code. The procedural part generates all possible decompositions of the product and subassemblies in the form of an AND/OR graph, using graph-theory principles along with information provided by the expert-system. It also extracts the feasible assembly sequences from the AND/OR graph. The expert system completes the initial assembly model and ascertains the physical feasibility of each candidate assembly operation.

The use of the expert-system (in contrast to geometric algorithms) to interpret the relationships between the parts in the assembly and to check assembly operation feasibility, allows the generation and evaluation of assembly sequences earlier in the design stage of a product when CAD drawings may not yet be available. This approach makes the method developed suitable to be used in a concurrent engineering software tool and it can also provide an effective filter of feasible assembly sequences to reduce the number of operations that are presented to a path planner for collision free trajectory checking.

REFERENCES

[AH93] M. J. Abrantes and S. D. Hill, *Expert-System Generation of Assembly Sequences*, in Proceedings of the International Conference on Assembly Automation, Adelaide, Australia, pp. 69-76, November 1993.

[BAL91] D. Baldwin, et al, *An Integrated Computer Aid for Generating and Evaluating Assembly Sequences for Mechanical products,* IEEE Trans. on Robotics and Automation, Vol. 7, No. 1, February 1991.

[Bou84] A. Bourjault, *Contribution a une Approche Methodologique de L'Assemblage Automatise: Elaboration Automatique des Sequences Operatoires*, These d'Etat, Universite de Franche-Comte, 1984.

[Del92] A. Delchambre, *Computer-Aided Assembly Planning*, Chapman & Hall, London, 1992.

[HAL94] S. D. Hill, M. J. Abrantes, and G. S. Lowe, *Determination of Feasible Assembly Sequences using an Expert-System*, in Proceedings of the First Vietnam Conference on Automation, Hanoi, April 1994.

[HL91] Y. F. Huang and C. S. Lee, *A Framework of Knowledge-Based Assembly Planning*, in International Conference on Robotics and Automation, Sacramento, California, pp. 599-604, April 1991.

[Lee92] S. Lee, *Backward Assembly Planning with Assembly Cost Analysis*, in International Conference on Robotics and Automation, Nice, France, pp. 2382-2391, May 1992.

[Mel89] L. Homem de Mello, *Task Sequencing Planning for Robotic Assembly*, PhD. thesis, Carnegie Mellon University, Pittsburgh, PA, 1989.

[MS90] L. Homem de Mello, and A. Sanderson, *AND/OR Graph Representation of Assembly Plans*, IEEE Transactions on Robotics and Automation, Vol. 6, No. 2, April 1990, pp. 188-199.

[MS91] L. Homem de Mello, and A. Sanderson, *A Correct and Complete Algorithm for the Generation of Mechanical Assembly Sequences*, IEEE Trans. on Robotics and Automation, Vol. 7, No. 2, April 1991.

[NAS93] Lyndon B. Johnson Space Center, Huston, TX, *CLIPS User's Guide - Volume One, Rules*, 1993.

A COMPARISON OF FUZZY AND COMPUTED-TORQUE CONTROL OF A ROBOT ARM

Tim McFarland and Simon D. Hill

Department of Robotics and Digital Technology,
Monash University, Wellington Road, Clayton 3168, Australia.
E-mail: ttm@rdt.monash.edu.au

ABSTRACT

This paper reports on experiments which compare the performance of a computed-torque controller and a fuzzy controller in guiding a one-link robot arm over a predetermined fifth-order polynomial trajectory. The design techniques described for each controller provide comparative evidence that the fuzzy controller is considerably simpler to design and implement. It is shown that the computed-torque controller delivered superior performance in a dynamically simple environment, but the fuzzy controller excelled where an accurate dynamic model of the system was not readily derived.

INTRODUCTION

The growing acceptance of robots in industry has led to the desire to apply them to ever more complex tasks. Consequently, a need for more capable non-linear control strategies has arisen. Traditionally, control was achieved using model-based methods such as computed-torque control [Sch90]. These methods rely on detailed knowledge of the dynamic characteristics of the robot expressed in an accurate mathematical model. However, as robots and their tasks become more complex, the effectiveness of model-based control methods begins to decline. Firstly, the often highly non-linear dynamic model of the robot becomes more difficult to establish with sufficient accuracy. Particularly, effects such as joint friction and link flexibility can make accurate models difficult to derive. Secondly, strategies such as computed-torque control, which require that the robot's dynamic model be recalculated during every sample period, can place exceptional demands on computing resources, especially where rapid sampling is required.

In recent years, the use of fuzzy logic to control robot motion has been described by several authors [Lee90, Ber92]. This method requires no direct knowledge of the dynamic characteristics of the robot, thus avoiding the complications inherent in model-based control. Instead, it utilizes a set of rules reflecting expert knowledge of how to respond to particular system states while performing a given task. In robot control, the system state is often a set of joint-position and joint-velocity errors realized while following a desired trajectory.

This paper describes the design techniques and subsequent experimental verification of those techniques carried out on a single-link arm in the authors' laboratory. These activities examined the design complexity and servo performance, measured by position and velocity errors, of both computed-torque and fuzzy control methods. This work is a prelude to performing a fuller comparison on the four-link AdeptOne robot.

COMPUTED-TORQUE CONTROLLER DESIGN

Computed-torque control makes use of an open-loop dynamic model of the arm to cancel the nonlinear effects of manipulator inertia, centrifugal and Coriolis forces and gravity, producing a linear closed loop system which can be given a desired dynamic performance by manipulating the gains of the position and velocity errors which drive the controller [LWP80].

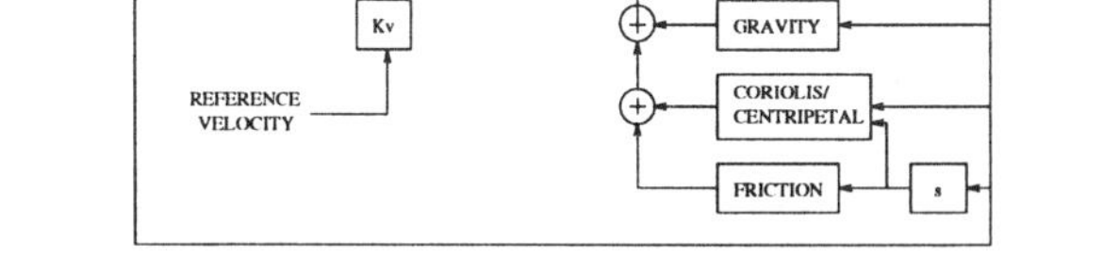

Figure 1: Computed-Torque Controller

Derivation of Dynamic Model of Arm

The primary task in designing the computed-torque controller, and the most time-consuming, was the derivation of an accurate dynamic model of the arm and motor along with estimation of the relevant system parameters. The required model is one which relates motor input voltage V to arm angular position θ, angular velocity $\dot{\theta}$ and angular acceleration $\ddot{\theta}$. The model of the arm without the motor was first derived symbolically for horizontal motion by Lagrange's method [Sch90]. Letting I represent the arm's moment of inertia about the joint axis and τ represent the torque applied to the arm:

$$\tau = I\ddot{\theta} + \nu\dot{\theta} + k\text{sign}(\dot{\theta}) \tag{1}$$

where ν is the coefficient of dynamic friction and k is the coefficient of static friction. The mechanical time constant of the motor was assumed to be small compared to that of the arm, so the motor's torque-speed characteristic was used to relate the input voltage to the torque produced at the motor shaft [Ele75].

The motor voltage constant K_e was determined from a known applied voltage V and the resulting no-load shaft speed ω_{nl}. The motor's maximum no-load speed ω_{nl}^{max} could then be found given the maximum input voltage V^{max}:

$$\omega_{nl}^{max} = \frac{V^{max}}{K_e} \quad \text{where } K_e = \frac{V}{\omega_{nl}} \tag{2}$$

The motor's speed regulation constant R_m was then determined as described in [Ele75] and the torque generated by the motor T_g was written as a function of the motor speed ω:

$$T_g = \frac{\omega_{nl}^{max} - \omega}{R_m} \tag{3}$$

The efficiencies of the motor gearing and belt were assumed to be 0.94 and 0.97 respectively [BAI78], giving an overall efficiency μ of 0.9118 and the gear ratio γ between the arm and motor was measured as 196.5:1. The model which relates motor input voltage to arm motion was thus found to be:

$$V = \frac{K_e R_m}{\gamma\mu}[I\ddot{\theta} + \nu\dot{\theta} + k\text{sign}(\dot{\theta})] + K_e\gamma\dot{\theta} \tag{4}$$

This suggests a controller with the control law

$$V = \frac{K_e R_m}{\gamma\mu}[I\ddot{\theta}^* + \nu\dot{\theta} + k\text{sign}(\dot{\theta})] + K_e\gamma\dot{\theta} \tag{5}$$

where

$$\ddot{\theta}^* = \ddot{\theta}^d + K_v(\dot{\theta}^d - \dot{\theta}) + K_p(\theta^d - \theta) \tag{6}$$

and K_p and K_v are the controller's position-error gain and velocity-error gain respectively.

The analysis for the arm mounted in the vertical plane was similar to the above. The model for the arm alone contains a term to account for the effects of gravity, so that Equation 1 becomes

$$\tau = (I + md^2)\ddot{\theta} + mgd\sin\theta + \nu\dot{\theta} + k\text{sign}(\dot{\theta}) \tag{7}$$

Table 1: Estimates of inertial parameters

Parameter		Horizontal	Vertical
Moment of Inertia	I	0.0043	0.0097
Coeff. of dynamic fric.	ν	0.0076	0.0061
Coeff. of static fric.	k	0.0119	0.0054
Gravity term	mgd	N/A	0.0102

where m is the mass of the arm and d is the distance between the spindle and the centre of mass of the arm. The analysis then proceeded as above, yielding a control law similar in form to Equation 5:

$$V = \frac{K_e R_m}{\gamma\mu}M + K_e\gamma\dot{\theta} \tag{8}$$

where

$$M = (I + md^2)\ddot{\theta}^* + mgd\sin\theta + \nu\dot{\theta} + k\text{sign}(\dot{\theta}) \tag{9}$$

Estimation of Arm Parameters

The parameters to be estimated were the arm inertia, gravity and friction terms in Equations 1 and 7. The arm was run over a known fifth-order polynomial path for ten seconds and readings of arm position in encoder counts and the corresponding motor voltage were recorded at a rate of 100Hz. The arm velocity and acceleration were estimated at each sample time and were filtered to reduce the resulting noise levels. The motor input torque at each sample time was estimated using the motor's torque-speed characteristic as described previously. The arm model was then expressed in terms of a vector of motor input torques T, a matrix of arm velocity and acceleration variables Φ and a vector of arm parameters P. QR factorization was used to estimate the solution, the parameter vector P, to the overdetermined system 10.

$$\Phi P = T \quad \text{where } P = \begin{bmatrix} I \\ \nu \\ k \end{bmatrix} \tag{10}$$

The resulting parameter estimates are given in Table 1.

Determination of Error Gains

The position and velocity error gains were kept as low as possible to reduce instability problems caused by unmodelled dynamics and sampling delay [AAH88], while also being chosen to produce a critically damped response [Kuo87].

FUZZY CONTROLLER DESIGN

The design of the fuzzy controller was a rather more subjective process than the design of the computed-torque controller as it was based largely on empirical knowledge of the system, not on an established mathematical model. The controller requires 4 components: a fuzzification strategy, a knowledge base (the rule set), an inference engine and a defuzzification strategy [Cox94].

Fuzzification

The control variables were chosen to be the arm's measured position error and estimated velocity error. The universe of discourse of each error variable was defined by 5 sets as shown in Figures 2 and 3.

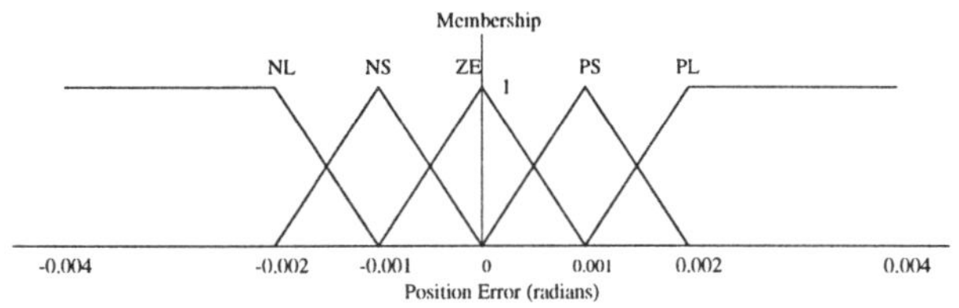

Figure 2: Arm Position Error expressed as a Fuzzy Variable

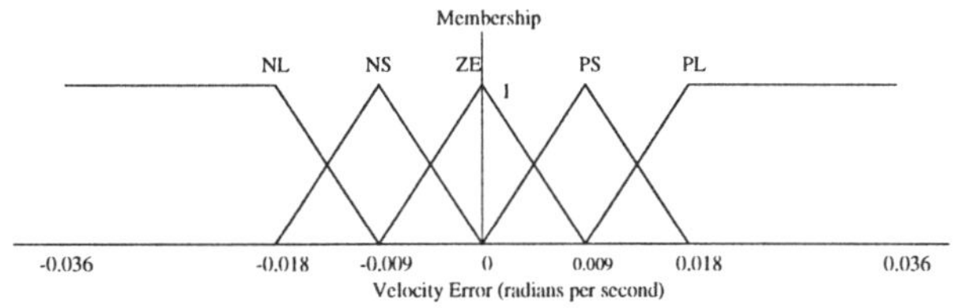

Figure 3: Arm Velocity Error expressed as a Fuzzy Variable

A triangular profile was used for the sets in accordance with current control convention as it increases the simplicity and speed of calculation of membership grade. In order to define the domains of these sets, a performance objective had to be established. Given the purpose of comparing the two controllers, the performance objective of the fuzzy controller was chosen to match the performance obtained with the computed-torque controller. Early trials showed the mean position error achieved by the computed-torque controller to be 0.001 radians and the mean velocity error to be 0.009 radians/second. Accordingly, for the fuzzy controller a position error was considered to be *zero* (ZE) if its magnitude was less than 0.001 radians, *small* (NS or PS) if it was between 0 and 0.002 radians and *large* (NL or PL) if it was greater than 0.001 radians, and similar for the velocity error.

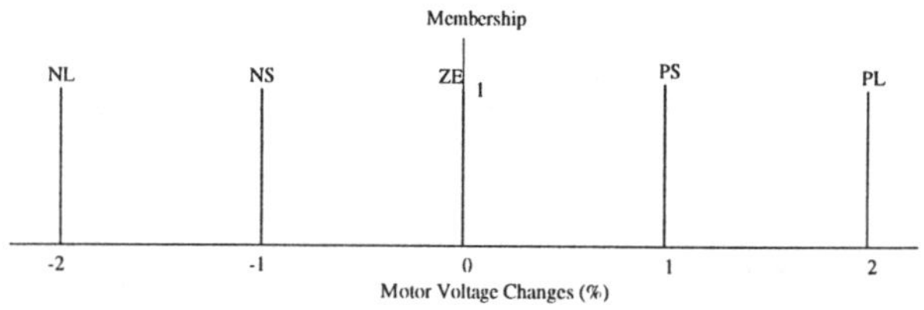

Figure 4: Motor Voltage Increments expressed as Fuzzy Singletons

The output variable was chosen to be a change in the motor voltage(Figure 4). It was expressed using the singleton geometry representation to increase the speed of defuzzification.

Knowledge Base

Derivation of the fuzzy control rules required specifying both the number of rules and the form and content of each rule.

As it was not possible to identify any combination of error and change in error which could never be realized, it was decided to create a rule for every combination. Thus with two control variables each described by five fuzzy sets, an exhaustive set of 25 rules was used in the controller.

MacVicar-Whelan [MW76] described an algorithm for forming control rules for a fuzzy controller which uses a single variable and its time derivative in rules of the form

IF (variable condition) AND (derivative condition) THEN (output condition)

In this case, where the variable is position error of the arm e, its derivative is $\dot{e}$ and the controller output is a change in motor voltage δV, the algorithm is essentially

1. If e and $\dot{e}$ are both zero, do not change the voltage ($\delta V = 0$).

2. If e is non-zero but will go to zero within an acceptable time, do not change the voltage ($\delta V = 0$).

3. If e will not go to zero within an acceptable time, set δV such that the direction of the change acts to reduce e and the magnitude of the change is proportional to the magnitudes of e and $\dot{e}$.

Applying these rules to the fuzzy variables e and $\dot{e}$ gives the fuzzy associative memory [Kos91] shown in Figure 5, with antecedent error conditions on the labelled axes and consequent motor voltage increments in the body of the figure.

VELOCITY ERROR \ POSITION ERROR	NL	NS	ZE	PS	PL
NL	NL	NL	NL	NS	PS
NS	NL	NS	NS	ZE	PS
ZE	NL	NS	ZE	PS	PL
PS	NS	ZE	PS	PS	PL
PL	NS	PS	PL	PL	PL

Figure 5: Fuzzy Associative Memory used in this Controller

Inference Engine

Inference is the process of deriving a fuzzy output (consequent) from a fuzzified input (antecedent). In this case it corresponds to mapping a point in the system's state space, defined by position error and velocity error, to a singleton point on the universe of discourse of the consequent variable. Inference is performed using Zadeh's AND operator within each rule, which specifies that the minimum membership grade from the position error and velocity error be applied to the consequent. Zadeh's OR operator is then used to apply the maximum resulting membership grade for each consequent point to the corresponding point in the final fuzzy region [Cox94, YZ92].

Defuzzification

Defuzzification is a mapping from the output fuzzy region to a crisp controller output value. In these experiments defuzzification was performed by finding the centre of gravity of the fuzzy region resulting from firing each rule. With n fuzzy singletons in the consequent region, each positioned at domain value d_i with membership μ_i the centre of gravity was found by evaluating

$$COG = \frac{\sum_{i=1}^{n} d_i \mu_i}{\sum_{i=1}^{n} \mu_i} \tag{11}$$

RESULTS

A computed-torque controller and then a fuzzy logic controller were designed and implemented in software, then applied to the control of a one-link robot arm as described below. Both controllers were written in C++. The arm was run over a fifth-order polynomial path while the arm's position and motor command data were sampled at 100Hz.

Hardware Used

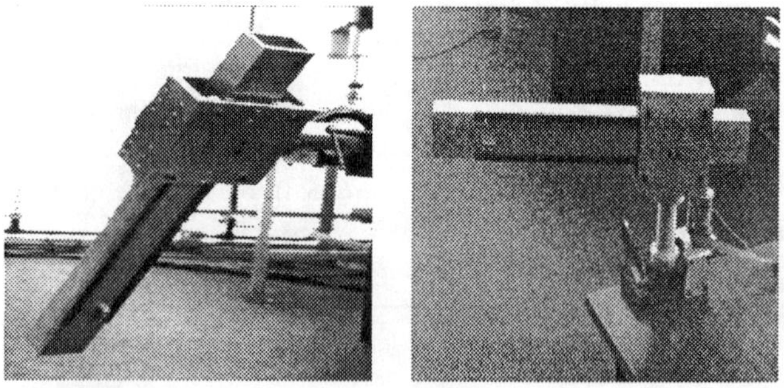

Figure 6: Arm Used in these Experiments

The arm being controlled is shown in Figure 6. It consists of a single link with movable centre of mass mounted on a revolute joint. Power is by a geared DC motor through a toothed belt to the arm spindle. An optical encoder is attached to the motor shaft. The motor is supplied with power through a pulse-width modulated amplifier which is controlled by a Hewlett-Packard HCTL1100 General Purpose Motion Control IC mounted in a PC. The inbuilt control functions of the chip have not been used in the experiments, its only functions being to set the PWM duty cycle and read encoder counts from the motor as directed by the software controllers.

Input Trajectory

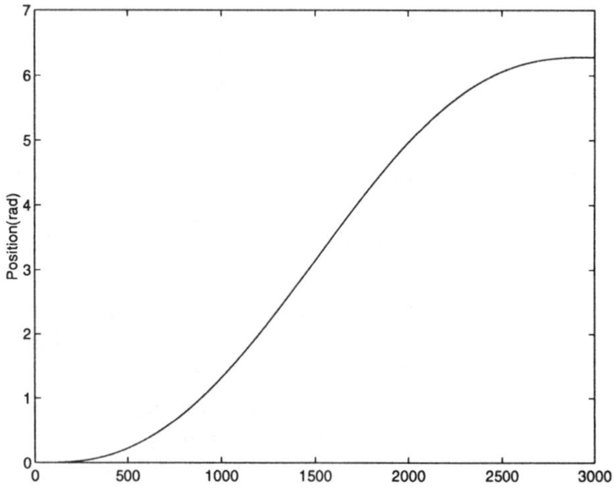

Figure 7: Input Arm Trajectory

The arm's trajectory is shown in Figure 7. In the first set of experiments this path was followed in the horizontal plane over 30 seconds with the arm centred on its spindle, giving a linear system. In the second set the path was followed in the vertical plane, also over 30 seconds, with the centre of mass of the arm offset from the spindle, giving a system with significant nonlinearities. The centre of gravity of the arm was initially vertically below the joint. In the third set of experiments the path was followed in the horizontal plane again, but this time over 10 seconds.

Comparison of Servo Performance

Table 2 gives a summary of the servo performance of each controller in the form of RMS position-error and velocity-error values realized over each trajectory. Detailed results for each trajectory appear in Figures 8 to 11. From these results a number of observations can be made.

The fuzzy controller exhibited better performance with respect to position errors than velocity errors relative to the computed-torque controller. This was likely because the universe of discourse of the fuzzy variable "velocity error" was too large and so the velocity error gain was too small. Early trials showed that increasing controller gain by changing the universe of discourse of the input or output variables improved tracking accuracy up to a point beyond which performance deteriorated due to chattering about the reference input. Further tuning of the universe of discourse of the velocity error variable may alleviate this problem.

However, when reading the RMS error values given in Table 2 it is important to realize that while the fuzzy controller errors and the computed-torque velocity errors appear to be randomly distributed about their means, the computed-torque position error seems to be proportional to the load on the motor which suggests a deficiency in the dynamic model of the arm. This is especially obvious in Figure 9 where the position errors are small when the arm is near vertical at the ends and midpoint of the path, but increase as the arm's centre of gravity becomes horizontally displaced from the spindle during the upward and downward arcs. Indeed, in the dynamically simple "horizontal, 30s" trajectory, where the arm model is likely to be the most accurate, the performance of the computed-torque controller was equal to or better than the fuzzy controller with respect to position and velocity errors, but in the more complex trajectories, where unmodelled effects were more significant, the performance of the computed-torque controller suffered accordingly. This indicates that if the deficiencies in the arm model could be rectified the computed-torque controller would give superior performance in all respects.

Taking the horizontal, 30s path as a reference, the position error performance of the computed-torque controller deteriorated when the same path was traversed in 10s, where unmodelled inertial effects are more significant, whereas the fuzzy controller's performance did not. Similarly when moving to the highly non-linear vertical trajectory, the position errors realized by both controllers increased significantly, but the computed-torque controller more so than the fuzzy controller even though the fuzzy controller was identical for both the linear horizontal and non-linear vertical cases. This lends support to the frequently reported ability of the fuzzy controller to adapt more effectively to poorly defined systems than can a model based controller.

Figure 11 shows that both controllers placed similar demands on the motor but that the fuzzy controller caused slightly more chattering of the motor about its mean setting. This is most likely related to the poorer velocity error performance of the fuzzy controller, as the motor voltage is directly proportional to the arm velocity.

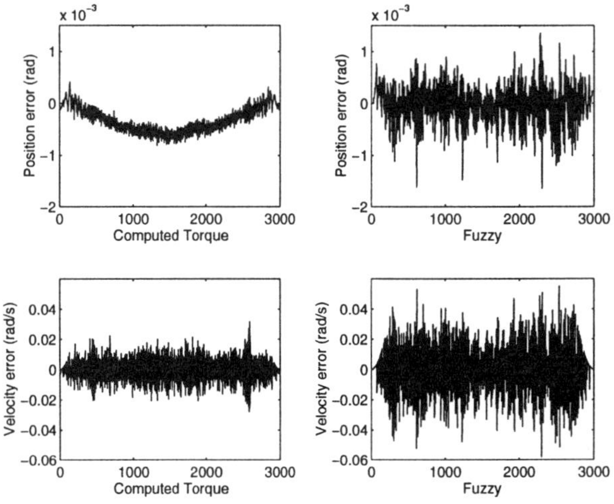

Figure 8: Horizontal Plane Motion, 30 Second Trajectory

CONCLUSION

The results for the low-speed horizontal trajectory, dynamically the simplest of the three, indicate that where an accurate dynamic model of the system can be derived, it may be expected to provide better control than a fuzzy controller. However, where the system becomes more complex, such as when following the vertical path or when moving more rapidly, unmodelled dynamics can quickly erode the performance of a model-based controller. In these situations a fuzzy controller can deliver relatively better performance as it does not depend on having detailed knowledge of a complicated system.

However, it should be noted that the performance of a fuzzy controller depends heavily on the skill of the

Table 2: RMS Errors Achieved Over Each Trajectory

trajectory	computed-torque		fuzzy logic	
	position error	velocity error	position error	velocity error
horizontal, 30s	0.0004	0.0071	0.0004	0.0164
vertical, 30s	0.0035	0.0110	0.0020	0.0451
horizontal, 10s	0.0012	0.0068	0.0003	0.0130

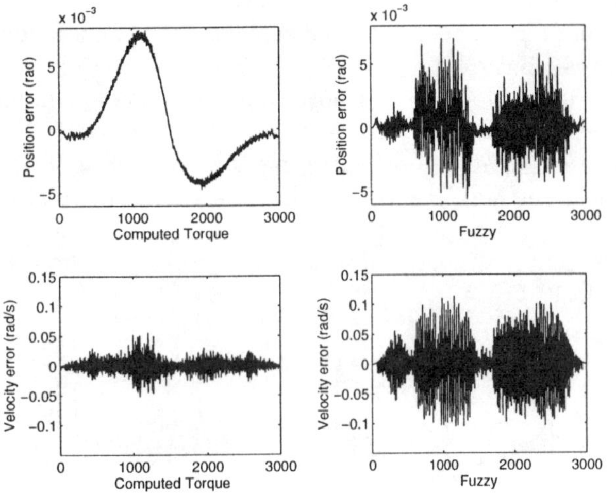

Figure 9: Vertical Plane Motion, 30 Second Trajectory

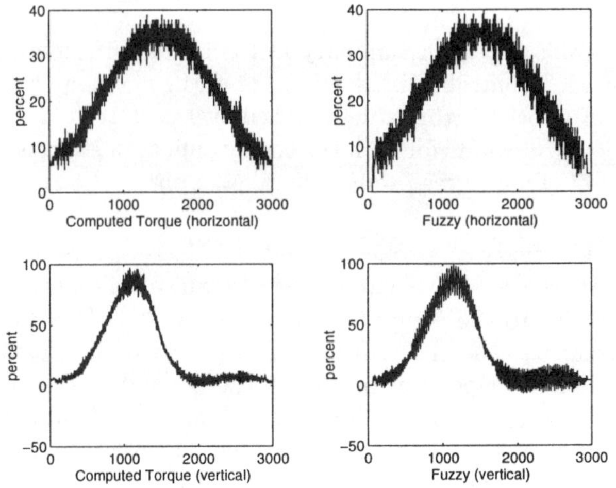

Figure 11: Motor Voltage Increments Applied in the Horizontal and Vertical Planes

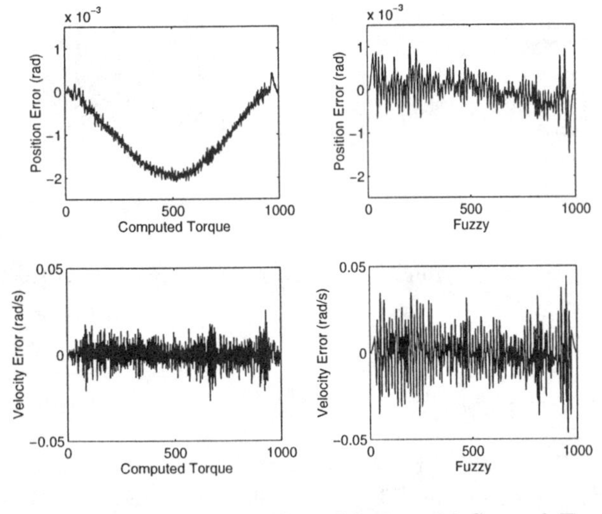

Figure 10: Horizontal Plane Motion, 10 Second Trajectory

designer, much more so than model-based controllers, as virtually all the design parameters are determined from heuristic knowledge of the system and its performance objectives, rather than from established mathematical principles. These parameters include the domains and shapes of fuzzy sets describing control variables, the form and number of rules in the rule base, the choice of a consequent for a given set of antecedents and the method of defuzzification. This means that when interpreting the results given here, it should be remembered that a change in any or all of the above parameters may affect the performance of the fuzzy controller. Some authors have tackled this problem by introducing a self-tuning capability to the fuzzy controller, such that the universe of discourse of one or more variables or the fuzzy control rules can be modified according to prevailing error conditions, for example[BK89].

It should be noted that the design and implementation time for the fuzzy controller was around one third that of the computed-torque controller. This was primarily due to not having to model the dynamic behaviour of the arm and motor, and to the ease of forming the required empirical control rules. This time saving represents a major advantage of the fuzzy controller, provided acceptable performance can be obtained.

References

[AAH88] Chae H. An, Christopher G. Atkeson, and John M. Hollerbach. Model-based control of a direct drive arm. In *Proceedings of the 1988 IEEE International Conference on Robotics and Automation*, pages 1374–1379, April 1988. Philadelphia, PA.

[BAI78] Theodore Baumeister, Eugene A. Avallone, and Theodore Baumeister III, editors. *Marks' Standard Handbook for Mechanical Engineers.* McGraw-Hill, 8th edition, 1978.

[Ber92] Hamid R. Berenji. *Fuzzy Logic Controllers.* Kluwer Academic Publishers, Boston, MA, 1992.

[BK89] C. Batur and V. Kasparian. Application of a self-tuner using fuzzy control techniques. In *Proceedings of IEA/AIE-89*, pages 235–244, 1989.

[Cox94] Earl Cox. *The Fuzzy Systems Handbook.* AP Professional, New York, 1994.

[Ele75] Electro-Craft Corporation, Hopkins, MN. *DC Motors, Speed Controls, Servo Systems*, 3rd edition, 1975.

[Kos91] Bart Kosko. *Neural networks and fuzzy systems.* Prentice-Hall, Englewood Cliffs, NJ, 1991.

[Kuo87] Benjamin C. Kuo. *Automatic Control Systems.* Prentice-Hall, Englewood Cliffs, NJ, 5th edition, 1987.

[Lee90] Chuen Chien Lee. Fuzzy logic in control systems: Fuzzy logic controller — parts I & II. *IEEE Transactions on Systems Man and Cybernetics*, 20(2):404–435, March/April 1990.

[LWP80] John Y.S. Luh, Michael W. Walker, and Richard P. C. Paul. Resolved-acceleration control of mechanical manipulators. *IEEE Transactions on Automatic Control*, AC-25(3):468–474, June 1980.

[MW76] P.J. MacVicar-Whelan. Fuzzy sets for man-machine interaction. *International Journal of Man-Machine Studies*, 8:687–697, 1976.

[Sch90] R. J. Schilling. *Fundamentals of Robotics: Analysis and Control.* Prentice-Hall, Englewood Cliffs, NJ, 1990.

[YZ92] Ronald Yager and Lofti Zadeh. *Introduction to Fuzzy Logic Applications in Fuzzy Systems.* Kluwer Academic Publishers, 1992.

PROCESS PARAMETER OPTIMIZATION FOR COMPUTER AIDED MANUFACTURING

Robert G. Fenton and Robert F. Moss
Department of Mechanical Engineering, University of Toronto,
Toronto, Ont. Canada, M5S 1A4

ABSRACT

The objective of this paper is to describe an algorithm developed to optimize machine tool and cutting-tool material selection for constrained multi-pass machining operations. The purpose of the work is to provide a process planning aid to improve production system efficiency.

MACHINING ECONOMICS

In this work the selection of the optimum machine tool - cutting tool - workpiece system elements and process variable settings is described, to minimize the production cost or production time. In most instances the cost dominates process planning decisions. However, if a deadline is to be met, or the process under analysis is a bottleneck, then production time is likely to have the greatest significance.

Most research to date on this topic has focused on the optimum selection of process variable settings through the use of predictive models [G50], [B58], [B62], [OH64]. Relatively little attention has been given to the selection of machine tool and tool materials [FG93], [S82], with industry still relying heavily on sub-optimal machining reference books and databases [MDH72]. Thus, the motivation for this paper is to examine this aspect of the machining economics problem from a practical, computer-oriented perspective.

This will be accomplished by building on the traditional machining economics theory on the work of [FG93]. Subsequent elaborations will focus on the most commonly used metal cutting operation - cylindrical turning.

The production time, T_p needed to complete the machining of one part is a function of the actual machining time, the prorated tool-replacement time and the idle time. For cylindrical turning operations it can be expresses as

$$T_p = f\,(T_m\,,\ T_{cp},\ T_n) \qquad (1)$$

The unit production cost for cylindrical turning, C_T is function of machining time, unit machining cost and tool cost.

$$C_T = g\,(T_p,\ X,\ C_t\,) \qquad (2)$$

The tool life can be considered as a statistical quantity. Based on the Taylor tool-life model, the unit production time and production cost can be expressed in terms of the cutting velocity, V and the feed-rate, f as

$$T_p = f\,(V,\ f) \qquad (3)$$

and

$$C_T = g\,(V,\ f) \qquad (4)$$

To arrive at a realistic solution of the machining economics problem the following constraints should be taken into account [AB69]: maximum feed-rate; maximum machine tool spindle speed; maximum machine tool power; machine tool feed and speed steps; and maximum allowable cutting force.

All of these items clearly depend on the machine tool design. In addition, another important constraint to be considered is the required surface finish.

Central to the findings of [FG93] is the concept of examining the machining economics problem in a coordinate system defined by unit production time and unit production cost. Both criteria can be expressed as a function of the cutting speed, V, and tool life, T.

Assigning unit production time to the y-axis and unit production cost to the x-axis defines the criteria space. Equations (3) and (4) then specify a nose-shaped tool utilization curve. Several such curves are shown in Figure 1. The areas within the curves illustrate feasible regions for specific levels of tool life utilization. Unique sets of process conditions (feed, cutting speed, depth of cut, etc.) produce discrete T_p and C_T pairs that can be represented in criteria space. If a pair falls on the curve for $k = .6$, the set defines a feasible set of operating conditions that will result in a tool life utilization of 60%. If the point falls inside the curve, the tool utilization is less than 60%. Conditions that result in T_p and C_T pairs lying outside the boundary curve are infeasible, since such operating conditions will cause the expiration of more than 100% of the tool life.

Clearly, the value of k must be in the range $0 < k \leq 1$. A tool life corresponding to $k > 1$ (i.e. greater than 100%) will be considered infeasible.

In most instances two additional factors must be taken into account to arrive at the optimal selection of feasible T_p and C_T values. Firstly, it is desirable to replace tools after an integer number of parts have been machined. Secondly, many lathes operate at a discrete number of spindle speeds.

To accommodate both of these points, it is necessary to view parts per tool edge and spindle speed curves in criteria space. This is accomplished by the use of transforms. The first transformation requires that the T_p and C_T become functions of the integer number of parts produced per tool edge, P, and the spindle speed, N, is defined by the relation

$$P = \text{INTEGER} (T / T_m) \tag{5}$$

where T is the tool life. A similar procedure results in an expression for the N curves.

Figure 2 shows various P and N curves, both are parallel straight lines. By superimposing the N lines on a tool utilization curve for a particular value of k, the optimal T_p and C_T values can be easily found for the available spindle speeds. These occur near or at the point where the N lines, closest to the ideal minimum T_p and C_T values, intersect the tool life utilization curve , near the tip of the nose of the curve.

Further, P lines can also be included in this optimization routine to indicate when the machine tool operator should replace the cutting tool to achieve the desired tool life utilization. The intersection point of the appropriate P line and the optimal available N line, when lying within the feasible region, designates the number of parts that may be machined per tool edge for a particular k value.

Figure 3 displays the above procedure graphically. The tool utilization curve is intersected by spindle speed lines closest to the minimum ideal T_p and C_T values. In this case, the spindle speed associated with the optimal unit production cost is 1120 RPM. Similarly, for best unit production time the corresponding speed is 1400 RPM. Feasible P-N intersections show that the integer number of parts per tool edge for minimum C_T, is 89; for minimum T_p, it is 34.

For a situation in which several candidate tools are available for the operation, this procedure must be repeated for each tool-workpiece combination. The resulting optimal points for both production criteria can then be compared to identify which tool results in the minimum overall values of T_p and C_T see Figure 4 . This represents the optimal tool material choice.

It is noteworthy at this point to state that this procedure identifies the optimal tool material from a set of candidate tools, however, it does not optimize the process variables, nor does it take into consideration any constraints other than that of the spindle speed steps.

MULTIPLE PASS OPERATIONS WITH CONSTRAINTS

Expanding the single pass method to a multi-pass machining situation with constraints requires some modifications be made to the developments presented in the previous section. Still, the constraints applied and the various underlying assumptions need to be stated in order to provide a clear picture of the strengths and limitations of the proposed mathematical model.

The model allows for m identical roughing passes and one finishing pass to be considered. The objective is to determine the best tool material and number of passes for roughing, as well as the best tool material for finishing, and the corresponding optimal speeds and parts produced per tool-edge, for both production criteria. Furthermore, this can occur in a scenario where several machine tools are available to perform the machining operation.

Process variables such as roughing feed, finishing feed and depth of finishing cut are inputs. It is assumed that surface finish requirements are taken into account before the optimization procedure begins. Machine tool constraints include the maximum desired feed and power levels as well as the discrete spindle speed steps.

An interactive computer program was developed for roughing and finishing cuts using different sets of candidate tool materials for both operations. The program is capable of selecting the most suitable machine tool and cutting- tool materials.

The purpose of the program is to identify the optimum machine tool, tool material, number of roughing cuts, spindle speeds, and parts produced per tool edge for both production criteria.

To achieve the objectives, a complex numerical algorithm, based on the relations presented earlier, is used. It consists of the following major parts: process parameter and constraint definition; optimization; and graphical representation.

The first step involves inputting the required data for all subsequent calculations. Data required for each machine tool includes the spindle speeds, the maximum power available, the maximum available feed rates, and the charge rate. For the purpose of enforcing the power constraint, empirical coefficients are required for each tool material - workpiece combination.

Cutting tool information consists of the cost per edge, tool utilization factor closest to the tip of the "nose-shaped" curve.

Once the optimum available points are located, it is necessary to evaluate the corresponding number of parts produced per edge at those spindle speeds and process conditions. To do this, the *P-N* intersection points are determined and then checked to see if they lie within the feasible region defined by the tool life utilization curve. The largest *P* value inside the feasible region is taken to be the *P* value associated with the optimal production criteria, see Figure 7.

Often, due to power constraints, one roughing pass is not feasible. To compensate for this, up to three identical roughing passes can be investigated. This is accomplished by dividing the total depth of roughing cut into two or three equal-depth passes, which has a scaling effect on the utilization curve, see Figure 8.

Upon ascertaining the optimal production criteria values and corresponding *P* and *N* values for each machine tool-cutting tool-workpiece combination, the program then sorts the information to find the best roughing tool material, the number of roughing cuts, and the corresponding T_p, C_T, *P* and *N* values, for each machine tool.

Another sort is then done to compare which machine tool under consideration should be chosen to perform the operation. This is a function of the spindle speed step, maximum desired feed and maximum power constraints stemming from individual machine tool capacity.

Once this is accomplished a similar procedure is undertaken to determine the best finishing tool material on the already-specified machine tool.

Then the program simply displays colour-coded optimal results for both roughing and finishing operations, with the best overall values highlighted.

EXAMPLE

Consider a rough turning operation performed on a quenched and tempered SAE 4130 workpiece with a Brinell hardness of 180.

Three machine tools will be considered as summarized as followed:

Machine Tool 1:
R_c = $.583 /min
Power = 10 hp
Spindle Speeds: 56, 71, 90, 112, 180, 229, 280, 355, 450, 560, 710, 900, 1120, 1400

Machine Tool 2:
R_c = $.50 /min
Power = 6 hp
Spindle Speeds:60,85,115,165,240,340,465, 560, 815, 1160, 1600

Machine Tool 3:

R_c = \$.40 /min
Power = 3 hp
Spindle Speeds:75,120,200,300,500,800,1250, 2000

The tool materials considered for this example are WIDIA carbide insert grades TTR, TK15, TN25 and TN35 [KWC]. The desired tool utilization is 100% .

The machining requirements and process parameters are as follows:

D_{outer} = 7.0 in. L = 2.0 in.

t_{nr1} = 4.5 min t_1 = 4 min

D_{final} = 6.8 in. f_r = .01 in.

t_{nr2} = .5 min

Finally, the empirical coefficients for the power constraint relationship are C =150,000 , a=.9 and b = 1 [10].

The computed results are listed in Table 1.

LIMITATIONS OF THE MULTIPLE PASS OPTIMIZATION ALGORITHM.

For the multi-pass procedure described in this work both feed and depth of cut are inputs. An approach which determines optimum process variable settings for each roughing pass and the finishing pass, in addition to the optimal machine tool and tool material, is beyond the scope of this paper. If this tact were pursued, it would require the tool utilization curve and P and N lines to be continually re-evaluated during the search for optimal machining system elements and conditions.

For this reason, the only compensation made when power constraints restrict the number of available spindle speeds is the possibility of removing the desired amount of material in two or three identical roughing cuts. The depth of cut per pass in this case simply represents the desired amount of material removed divided by the number of roughing cuts.

In essence, the opportunity to execute multiple, identical roughing cuts, rather than a single one, is a form of constraint [Y83] and it represents a effort to surmount machine tool power restrictions.

CONCLUSIONS

The algorithm and computer program presented in this paper automated the earlier developed [FG93] tool material selection procedure and successfully expanded it to multi-pass machining situations in which several candidate machine tools are available for the operation. The investigations revealed that: one tool material in a candidate set tends to dominate the others, provided that no power constraint is imposed; at spindle speeds outside the dominant tool's optimal range, other tool materials may be superior; and the minimum unit cost criterion is heavily dependent on charge rate level, so much so that two identical roughing cuts on a lower cost machine tool may be cheaper than one on a more expensive model.

REFERENCES

[FG93] Fenton, R.G., and Gagnon, M.F.J., 1993, Computer-Aided Tool Material Selection for Metal-Cutting Operations, Annals of the CIRP, Vol. 42, pp. 565-568.

[G50] Gilbert, W.W, 1950, Economics of Machining, Machining Theory and Practice, American Society of Metals, pp. 465-485.

[B58] Brewer, R.C., 1958, On the Economics of the Basic Turning Operation, ASME, J. of Engineering for Industry, Series B, Vol. 81,No. 3, pp.1479-1489.

[B62] Brown, R.H., 1962, On the Economical Selection of Machining Rates, Int. Journal of Production Research, Vol. 1, No.2, pp. 1-22.

[OH64] Okushima, K., and Hitomi, K., 1964, A Study of Economical Machining: An Analysis of the Optimum Cutting Conditions, ASME , J. of Engineering for Industry, Vol. 3, No.1, pp. 73-78

[S82] Shaw, M.C., 1982, Optimization of the Metal Cutting Processes, On the Art of Cutting Metals - 75 Years Later, ASME , PED-Vol. 7, Phoenix, AZ.

[MDH72] Machining Data Handbook, 2nd Ed., 1972, MetCut Research Associates Inc., Cincinnati, OH.

[AB69] Armarego, E.G.A., and Brown, R.H.,1969, The Machining of Metals, Prentice-Hall, NJ.

[KWC] Krupp Widia Corp., Recommended Data for Turning Ferrous Metals,Publ. No. W5.3-10.3 e 786.

[Y83] Yellowley, I., 1983, A Fundamental Examination of the Economics of the Two-pass Turning Operation, Int. J. of Production Research, Vol.21, No.5, pp.617-626.

TABLE 1

Best machine tool	3	Best machine tool	1
Workpiece material	SAE 4130 BHN180	Workpiece material	SAE 4130 BHN180
Length of cut	2.0 in.	Length of cut	2.0 in.
Utilization factor	1	Utilization factor	1
ROUGHING		ROUGHING	
Best tool material	TN25	Best tool material	TK15
Cos, $/part	**2.406**	Production time, min/part	**5.424**
Production time, min/part	6.006	Cost, $/part	3.245
Depth of cut, in	0.1	Depth of cut, in	0.1
Feed, in/rev	0.01	Feed, in/rev	0.01
Spindle speed, rpm	200	Spindle speed, rpm	710
Parts machined/tool edge	664	Parts machined/tool edge	28

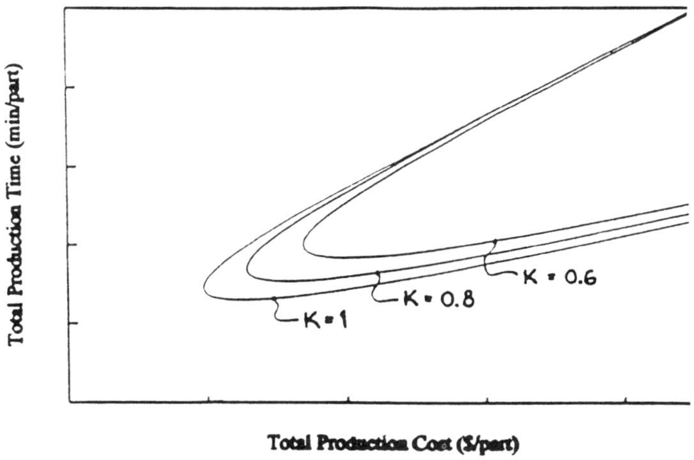

Figure 1 - Tool Utilization Curves in Criteria Space

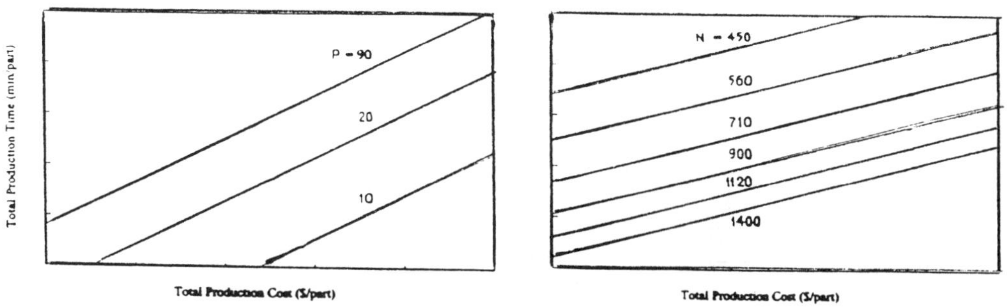

Figure 2 - P and N Curves in Criteria Space

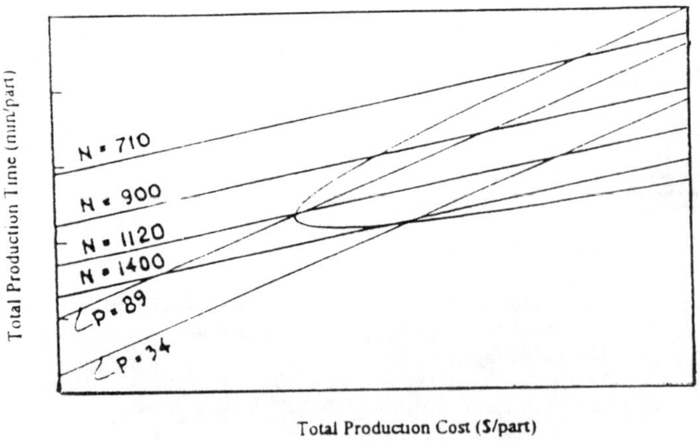

Figure 3 - Tool Material Optimization

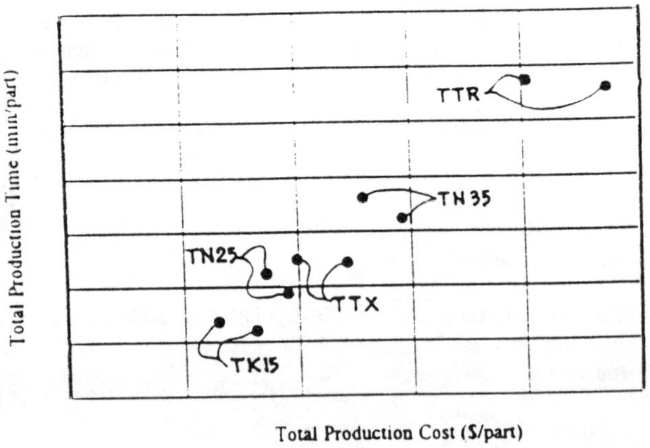

Figure 4 - Comparison of Several Candidate Tool Materials

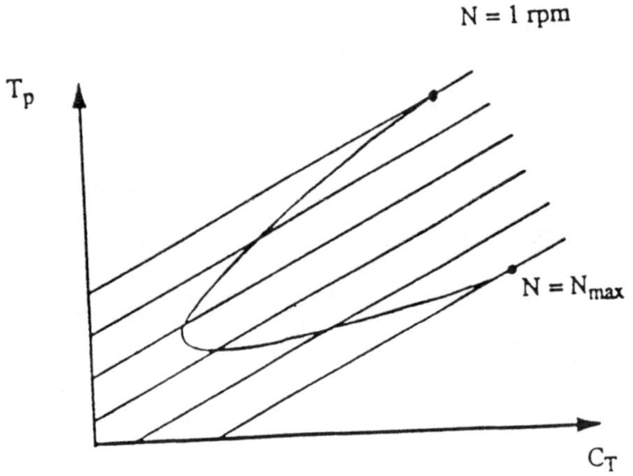

Figure 5 - Plotting the Utilization Curve

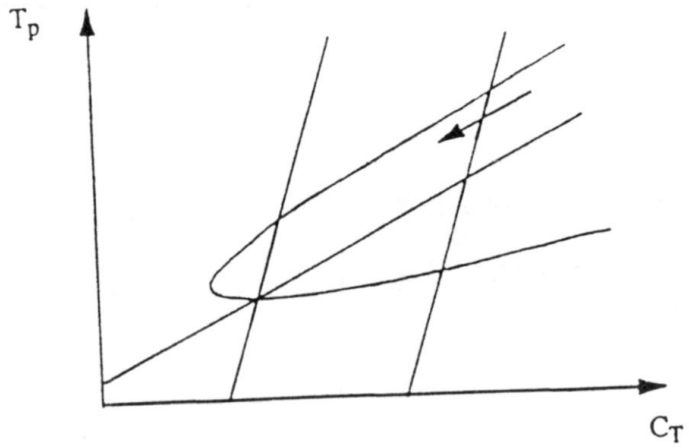

Figure 6 - P- Evaluation for an Optimal Spindle Speed

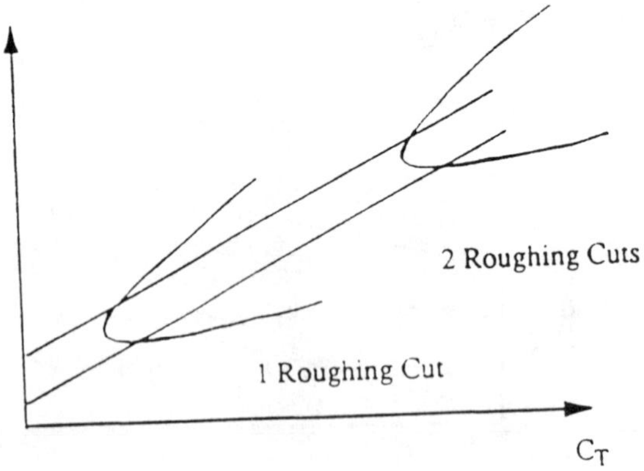

Figure 7 - Scaling Effect of Identical Multiple Roughing Cuts

EXPLAINING CONTROL LOGIC TO PROCESS OPERATORS

Pertti Huuskonen, Kari Kaarela, Jyrki Okkonen and Anne Väisänen
VTT Electronics, P.O.Box 1100, FIN-90571 Oulu, Finland
Email pertti.huuskonen@vtt.fi

ABSTRACT

Industrial plant staff sometimes have problems with control logic. Its behaviour and designers' intentions are sometimes obscure, calling for reverse engineering. We outline ways to improve presentations of the logic, emphasising the need for greater understanding. Knowledge-based support systems are a way to promote this understanding. We propose three kinds of explanations about control logic: to clarify the intentions behind designs, to demonstrate the behaviour of logic, and to allow for what-if analysis. The ideas are demonstrated with the help of a research prototype. Explanation search has potential for combinatorial explosion in the case of large logic networks. Fortunately, human users can guide the search, which reduces the complexity. Modelling the diagnostic knowledge would cause extra work to designers, and therefore design tools should support modelling and knowledge transfer to control systems. The techniques have to be embedded into control systems to be accepted.

1. INTRODUCTION

Much of the automation found in a typical plant consists of control logic. In some problem cases the operators have to reverse engineer the behaviour of the logic, if the control systems are not able to give sufficient information about it. The operators may have to keep track of states of numerous signals and perform boolean logic manipulations. This is a hard task and could be aided with suitable support systems. In this chapter we point out some problems that operators have with control logic, and in subsequent chapters we propose a number of improvements.

1.1 Logic in Process Control

Control logic is needed to automate discrete processes, typically batch operations and operational sequences. Most continuous processes need some supporting discrete automation as well. There are four main classes of logic used at a plant: *(1) Controls*. Logic is used to drive a device from one state to another, avoiding undesirable states. *(2) Sequences*. Many operations, especially in batch processes and manufacturing, need to be performed in a certain order and are encoded in sequences. They are a class of finite state machines, each state consisting of a set of operations that affect the process. The transitions between states are triggered when certain process variables reach given limits, or after a certain time. *(3) Interlocks*. Devices are often protected with interlocks, i.e. logic that allows their operation only when given conditions are met. Furthermore, interlocks are frequently required to ensure safety and environmental protection. *(4) Alarms*. Interlocks guard the process under normal conditions, but if there is a significant disturbance in the process, an alarm (or a warning) is given to signal that human intervention is necessary to bring the process back to a suitable state. Logic is utilized both in recognising situations that require an alarm (detection logic), and to suppress unnecessary alarms (suppression logic). Critical alarms usually lead to automatic actions.

Combinatorial logic functions in automation usually consist of simple AND, OR and NOT operations. Various kinds of delays are also heavily used to include the time dimension into the logic. Monostable flip-flops and delayed transitions are common varieties. Other common functions include set-reset gates, voting (n of m) gates, and counters. Sequences contain a variety of other functions.

1.2 Operator and Logic

Automation logic contains various kinds of information, most of which is not meant to be seen by the operators[1]. Normally control systems should take care of the logic functions and the operator sees only the end result. This is true for the majority of functions. However, some less common alarms can be rather confusing, as they can occur in seemingly unconnected situations. Complex processes exhibit complex behaviour, and apparently unrelated events may cause an alarm that can be hard to understand. The same is true for interlocks. Under normal conditions

1 Yet in problem cases they have to interpret this information! This is one of the ironies of automation [Bainbridge 83].

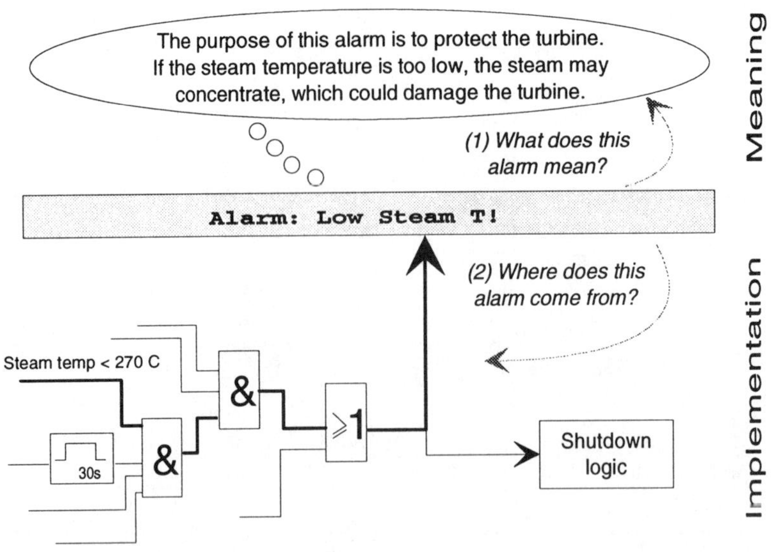

Figure 1: Meaning of an alarm vs. its implementation

they work as expected, but in certain typical yet unplanned cases (such as equipment failure) they may cause rather confusing side effects.

At a power plant we studied, more than 95% of alarms and interlocks arc considered normal and well known by the operators. The remaining few per cent cause major trouble. Sometimes the operators (possibly aided by other staff) have to find out the reason for a strange alarm by interpreting logic diagrams together with other documentation. This is a knowledge-intensive task, requiring both rules of thumb and more involved reasoning. The needed information is rather specialised (meant for control system designers) and spread all over the plant's documentation and information systems.

Finding out the reason for a strange alarm may take considerable time and mental effort. Demanding tasks like this can make the work more interesting and be educative, but in critical situations there is not enough time to spend on such detective work. It is especially in rapidly changing situations (e.g. disturbances) that most strange alarms and interlocks are found. When there is not enough time to fully consider the problem, decisions may be made on the basis of vague assumptions, or sometimes even wild guesses. Afterwards it is hard to reconstruct the state of the process at the time of the alarm to understand what actually happened. The process variables related to the case may have changed completely. Systems that could help in understanding the logic would be needed.

1.3 Implementation vs. Meaning

Let us look more closely what happens when operators encounter an unfamiliar alarm message. Figure 1 describes a possible case. The alarm message shows up on the automation system's screen. The operators acknowledge the message, though it does not immediately make sense. Trying to understand the significance of the message, they first identify the subprocess to which the alarm is connected. Then they seek the documentation of this subprocess to find out the designer's aim for the alarm. The text (if available) only seems to offer general guidelines, and therefore the operators resort to a separate logic diagram to find out how the alarm was generated. Part of the diagram is shown in the figure.

From the diagram, the operators (possibly helped with other personnel) search through the circuit from the concerned output to each of the inputs affecting that output. While traversing the network, they ignore parts of the diagram as irrelevant by ruling out passive gates and non-related signals. Finally, they come to the hypothesis that a certain input was the primary cause for the alarm. The path that was traversed is highlighted in the figure.

This simplified case exposes a number of problems with current automation and documentation practices:

1. Note that the operators are actually trying to find the meaning (or purpose) of the alarm (1), i.e. why this alarm was given. However, current automation systems do not provide much support in understanding the intended function of the alarm, or what devices or process parts it serves to protect. The alarm messages are at a very low level: they merely comment on the state of a process variable, e.g. that a process variable has exceeded a predefined alarm limit (e.g. T123 > 300°C). Such messages are not always understandable by themselves, but the operators have to infer both their meaning and cause. To accomplish this, they only have descriptions of the implementation (2) to turn to.

2. The operator must find relevant documents by manual search. The information is spread out in many

places, and often the documents are not sufficiently cross-referenced. Some modern automation systems may offer on-line documentation, but rarely in a context sensitive fashion, which would be necessary to truly support quick and easy access to the relevant information.

3. Operational procedures detail what should be done in an alarm case. They are increasingly made available in the control systems. However, the designers may not have been able to predict all the possible scenarios that cause an alarm, and the instructions may even be misleading. Predesigned procedures may only apply in limited situations.

4. The logic diagrams are not always available in the automation system. In such cases, the operator has to infer the states of the signals in the diagram. Analysis of a logic circuit requires keeping track of dozens of entities (gates, signals, states) at one time and can be a hard mental task.

5. There are too many non-related details. The diagrams are relatively flat, non-structured, and the operator must build a mental hierarchy of the circuit to be able to analyze it. All the details are visible all the time; there's no context sensitivity. Note that this violates the principles of information hiding: show only those details that are necessary to understand the message, but no more. Currently the operator has to filter out the unnecessary details.

1.4 Reusing the logic

Industrial designs are frequently reused. Solutions in existing plants are regularly taken as a basis when designing new ones. Moreover, much of the maintenance work at a plant consists of modifying existing designs.

Changes to designs are inevitable. The preconditions that led to the design may change several times during a plant's lifetime. Since it can be hard to predict the behaviour of a process, it can be be impossible to foresee all possible cases and constraints that a design should cover. Moreover, even plain errors, present in most designs, may appear only later. Changes are especially frequent in relation to control logic.

It is the maintenance staff of a plant that is usually responsible for carrying out the modifications at the plant. To do this, they must be able to understand the original design to a large extent. According to Fischer, maintenance can be divided to two clearly distinct phases: *understanding the existing solution* and *making changes* [Fischer 91]. This implies that one has to understand the purpose, intended function, and the implementation of a design to be able to alter it.

Unfortunately the maintainers often have difficulties in understanding the intentions behind the designs. Usually the available documentation describes more the implementation of the logic than its objectives. Another important prerequisite for making changes is to anticipate the effects of the change. This is of uttermost importance when modifying logic that could affect the safety of the plant.

The maintainers spend a considerable factor of their time reverse engineering existing designs. Clearly the maintainers could benefit from design rationale (design issues, alternative solutions, criteria behind choices, etc.) to recover those objectives [Lee 91, Kaarela 94]. This, in turn, would help them to better plan the changes.

The designers often face the same problems as the maintainers. When designing new plants based on existing ones, or when modifying existing designs, they have to recover part of the reasoning that led to the original design. This is even harder for novice designers, who can not rely on their experience to understand the design. They spend much of their time reverse engineering the solutions of other designers.

It would be clearly beneficial to all parties that are concerned with logic — operators, maintainers, designers — to be able to capture knowledge about logic circuit designs and to be able to explain it. The rest of this paper describes how this problem can be alleviated.

2. IMPROVING CONTROL LOGIC

Current control systems offer little high level information concerning logic functions. To recover this information, the user is engaged in non-trivial reverse engineering. In this chapter we propose some techniques to help this task. Better presentations of the logic can help to some extent, but in our view most flexible help is given by knowledge-based support systems that can explain the logic when needed. This improves not only presentations but also understanding.

2.1 Improving Presentations

We see a number of possible improvements to the current logic-based automation. An obvious improvement would be to use more meaningful alarm texts. In addition to announcing a change in a process variable, an alarm text should carry information about the purpose of the alarm. In many cases, this is not possible with predefined text, because a change in the process may be due to a number of reasons. If the text is too specific, it will suit only a subset of these reasons and be extremely confusing in others.

A better way – but more expensive – is to use more specific logic conditions. Instead of one general alarm, several ones would be realised, each one with a specific text for a specific case. This, of course, increases the amount of logic considerably. Besides, it is unlikely that all possible process state combinations could be predicted, especially with complex continuous processes. It would then be impossible to design a specific solution for each case.

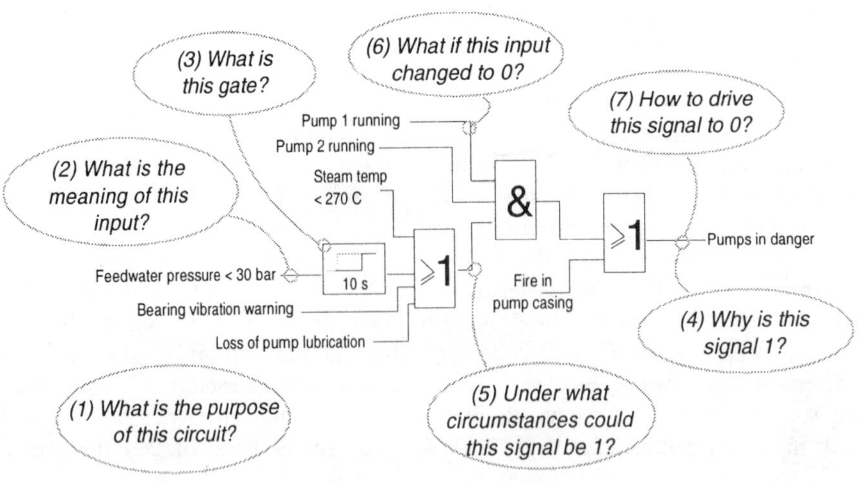

Figure 2: Possible questions concerning a logic network

A third way is to use dynamic, knowledge-based help to assist the operator. This is the approach we have taken. The help system acts in a context-dependent manner, offering information that suits the situation at hand. The operator can take a detailed look or a more general summary as desired.

At the very least, logic diagrams should be displayed on operator's screen when needed, with live data. This is rarely done because of the cost. The logic diagrams, although designed using CAD systems, often have to be hand-crafted to the displays with limited sizes. It is likely that raw CAD diagrams would not give much help anyway. Their level of detail is usually too high to be directly useful to the operator.

A more informative display would be structured, with large diagrams broken down into smaller parts. Each of the parts should be understandable in itself, and when it is combined with other parts, it would be abstracted to a less detailed representation. This would create a hierarchical structure to the logic, which should help the operator to better understand it. Note that this is the usual practice for mimic displays, but less common for logic, since the operators are not expected to know about it. Again, the cost is prohibitive if the displays need to be hand-crafted.

2.2 Improving Understanding

The operator's work is only partially enhanced by making the logic diagrams accessible in the automation system. This still leaves their interpretation to the operator, although making the logic states visible helps in this task. What is really needed is help in analysing the logic and explaining its intended function.

If the automation logic is modelled in data structures, it can be interpreted on-line. When the operator is uncertain about a circuit, the system can examine it and explain it in a more understandable form. This goes beyond mere displays, offering tools for analysis and understanding. We consider explanations of logic circuits to appear in three main categories:

A. *Explaining the meaning.* Educative help on the intended purpose of the design (e.g. an interlock).
B. *Explaining the behaviour.* Detailed demonstration of why the network reached a certain state.
C. *Speculative help.* Finding possible ways to reach a given state, "what-if"-analysis.

These three categories correspond closely to the three kinds of explanation defined by Chandrasekaran: (1) Clarification of domain concepts and underlying deep principles, (2) Justifications on the reasoning performed by an expert system, and (3) Strategic explanations about the general problem solving methods and approaches used in a system [Chandrasekaran 89].

2.3 Explanation Categories

Figure 2 outlines possible ways in which logic circuits used in process automation could be explained to the user. Note that although interaction in this example is in natural language, it is not necessarily the approach taken in a plant environment. Interaction possibilities for support systems are discussed further in [Feiner 90, Huuskonen 92].

Category A allows the designer's intentions for a specific logic circuit (e.g. an interlock) or its elements to be communicated to the user: the meaning, purpose or criteria behind a design choice, i.e. design rationale. This type of information is usually only implicitly stated in today's documentation. For instance, question (1) could be answered: "This circuit protects the feedwater pumps by turning them off when operating conditions are not met." This answer could then be carried further by listing these conditions if the user so desires. For question (2), the answer can be expressed informally with verbal texts attached to each signal: "This signal, when active, indicates that feedwater pressure is lower than normal, which could signify a fault.". Less well known logic elements can be explained, as with question (3): "This gate delays the signal by 10 seconds. If the input signal is a stable 1 during this time, the output also goes to 1. The output goes to 0 immediately when the input goes to 0."

Category B helps the user to follow the operation of the logic. Normally, this task is left to the user, requiring keeping track of states of numerous signals and performing boolean logic manipulations. This mechanistic work should be helped by the automation system. Question (4) could be answered: "Pumps are in danger because pump 1 is running, pump 2 is running, and feedwater pressure after the pumps is below 30 bar". It would also be possible to let the user to step through the circuit one gate at a time, following the propagation of the logic states.

Category C allows the user to try out various possibilities, which could help in selecting a good strategy for recovering from a problem situation. This would be necessary, for instance, to reconfigure the process after a malfunction. In case of question (5), the user wants to find out the various ways that a signal could be driven to a certain state. The system can answer by listing the possibilities: "This signal would be 1 if pump lubrication is faulty, if there's too much vibration in pump bearings, if feedwater pressure is below 30 bar for 10 seconds, or steam temperature is below 270°C". Simulation would be needed to answer question (6): "The pumps would be safe". After this answer, question of type (4) would be useful to trace the behaviour. Question (7) is actually a variation of question (5) used for a specific case: "In this case, these conditions would have to be filled: either pump 1 is stopped, pump 2 is stopped, or feedwater pressure is above 30 bar".

3. RESULTS

We have developed a research prototype called LEPO (acronym for Logic Explanation for Process Operators) to demonstrate our ideas about logic explanations. The prototype simulates and explains simple automation logic behaviour. It contains sufficient knowledge to answer many of the proposed question types and has been used mainly to facilitate discussion about the possibilities of logic explanation. We now describe the prototype, raising issues concerning its efficiency and embedding. The relation of logic explanation to functional modelling and diagnosis are discussed as well.

3.1 Prototype

Figure 3 describes the architecture of LEPO. The prototype consists of three main modules: Simulator, Explainer and User Interface. The Simulator simulates the

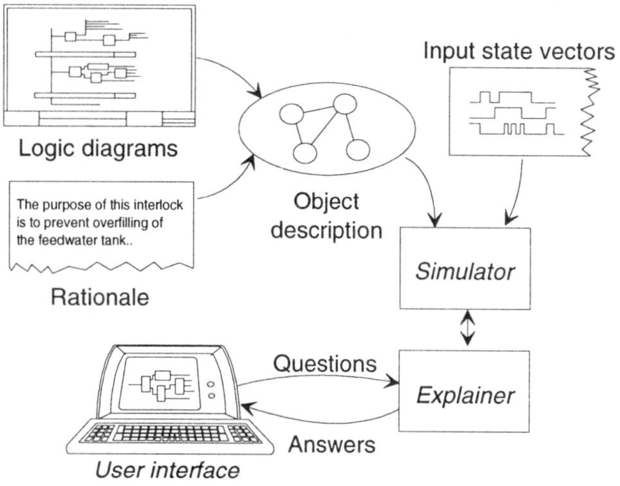

Figure 3: The architecture of LEPO

time-dependent behaviour of simple networks, limited to a few specific types of logic gates. The Explainer interprets the simulation results and explains them to the user. The User Interface interacts with the user, displaying explanations and receiving questions. These questions are transferred to the Explainer, that then starts the necessary simulations to obtain the answers.

Simulation would be unnecessary if all the logic states in a network could be obtained from outside. This may not be possible, because automation systems usually only export a few logic variables to the outside world. Category A needs no state information, but relies heavily on the documentation produced in the design phase. To obtain category B explanations, some simulation is needed to find the missing states. In the case of category C, the speculated events have never taken place in the real world and therefore can only be simulated.

The logic network is described to the prototype as object descriptions. The rationale texts are stored in the objects' properties. States are represented incrementally: only changes after the previous step have to be stored. This is advantageous for networks with little activity. When there are frequent changes, other representations should be considered. In addition to binary states, the simulator can also handle unknown states, i.e. when there is no information about the logic state of a signal at a certain time. Simulation is based on a simple propagation algorithm, where the network is depth-first scanned backwards from an output until given inputs are found. At each gate, logic states are propagated forward.

The current version of the prototype was coded in a generic manner with Xlisp, a public domain Lisp interpreter. The user interface was implemented in Toolbook under Microsoft Windows running on an personal computer. In the future, a C++ version of the prototype using X Windows is foreseen that would be directly embeddable to automation systems.

The current user interface is functional but far from ideal. The user sees a Toolbook page with the logic network diagram. With a set of buttons, one can show or hide additional items: network node labels, the logic timing diagram, and the explanation output window. When the user points to a node label, a pop-up menu appears, offering the available questions that concern this node. Once the user selects a question, simulation and explanation are started. The answers are shown in windows as text. Figure 4 shows an example screenshot of the prototype.

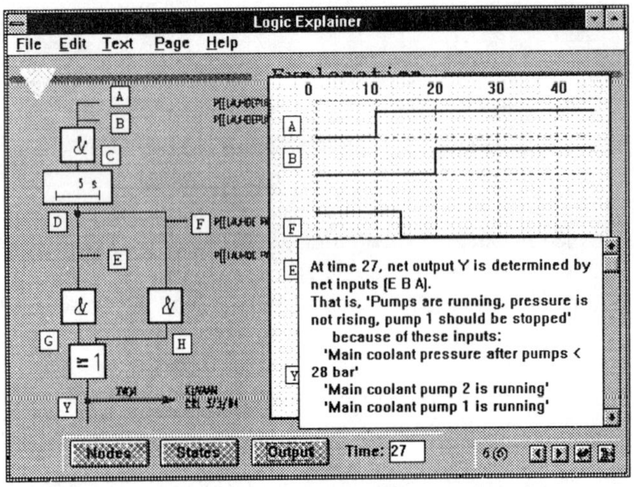

Figure 4: The LEPO research prototype

The prototype currently offers some explanations of categories A and B. Category C is currently offered only implicitly. The possible questions are *"What is the meaning of this signal?"*, *"What is the state of this signal at time <time> ?"*, *"Why is the state of this signal <state> at time <time> ?"* and *"What inputs caused the state of this signal at <time>?"*. There are both recursive and non-recursive forms of the question. The recursive form searches through the network to all the inputs that affected the given signal, while the non-recursive form only searches one gate level and leaves further search to the user.

3.2 Knowledge sources

LEPO needs to know the topology of the logic circuit, the logic states of the nodes at various points of time, and the explanatory knowledge. The topology is available in CAD design, from where it can be interpreted automatically (especially with the emergence of such standards as IEC 1131-3, see [IEC 93]). The diagrams include labels for the input and output nodes that contain data about the circuit's connections to the process. Data about the nodes' logic states is necessarily incomplete. The automation system may be able to produce a history of the states of the input and output nodes, but the intermediate state data is missing, and has to be reconstructed via simulation.

With current design tools, LEPO would increase the designer's workload. After designing a circuit and drafting a diagram, the designer would have to enter the explanations manually. This order is unnatural: first the purpose is translated into a design, then the purpose is recovered. Design tools would have to be enhanced to capture the explanatory information already at the design phase, where it is immediately available [Kaarela 93a, 94]. Many designs are reused in other projects, and they have similar explanations for the design choices. If the explanations were stored with the design, they could be reused as well. The explanations would also make the designs more understandable at the reuse time. This would no doubt make it more attractive to invest into devising meaningful explanations.

Fewer handcrafted explanations are needed if they can be derived from models. In our view, means-end modelling, as pioneered by Rasmussen, is particularly useful for this aim. Multilevel Flow Modelling (MFM) that describes a plant as a man-made artifact at several levels of abstraction, was first described in 1982 by Lind. In 1986, Sembugamoorthy and Chandrasekaran presented their work on functional representation, including a representation of behaviour as a causal sequence of states. This work was further developed by Keuneke. [Chandrasekaran 93, Kaarela 94, Keuneke 91, Lind 82, Lind 90, Rasmussen 86, Sembugamoorthy 86].

The meaning or purpose of any part in a system can be explained via its function: what function was it designed to serve in the system. Figure 5 shows a simple example, where the physical implementation of a logic function (e.g. Interlock 34) is tied to the more abstract function/purpose (Prevent a pump from starting without fluids) that it was designed to fulfill. These abstract functions are in turn tied to the overall goals of the plant (economy). Questions about the purpose of the logic can be answered by following the links.

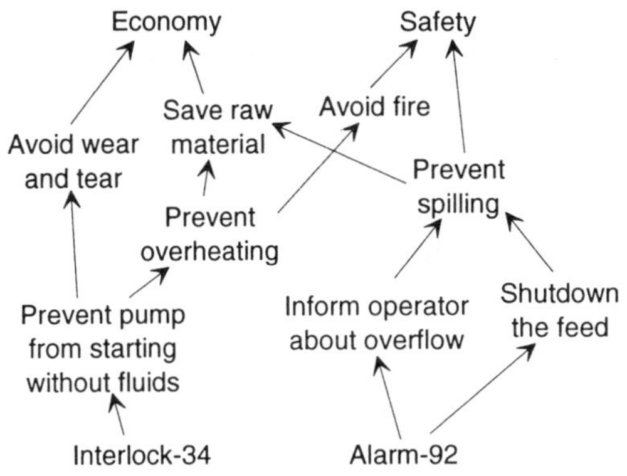

Figure 5: The purpose of the logic can be explained through the functions it performs.

3.3 Embedding

One of the primary requirements of an explanatory system of the kind of LEPO is that it should be integrated in the automation system. According to the experiences of our industrial partners, standalone systems — no matter how useful — will not be used at the plant in the long run. The automation configuration systems should be extended to support explanation creation. Only then would the necessary information be available for explanation. The developed research prototype uses Toolbook as an interface and Lisp as a representation and inference language. Clearly the prototype is not meant to be embedded into the automation system, but to introduce the principles that can be used there.

Figure 6 outlines one possible architecture for embedding LEPO. The automation system's console functions as the human-computer interface, displaying text and graphics output and receiving user's queries. LEPO resides in one of the process stations of the distributed control system, interpreting the queries, retrieving relevant data from the databases and sending the results to the console. The database contains the graphical documentation about the circuits to be displayed, as well as the information for reasoning and explaining. The state data is obtained from the process databases.

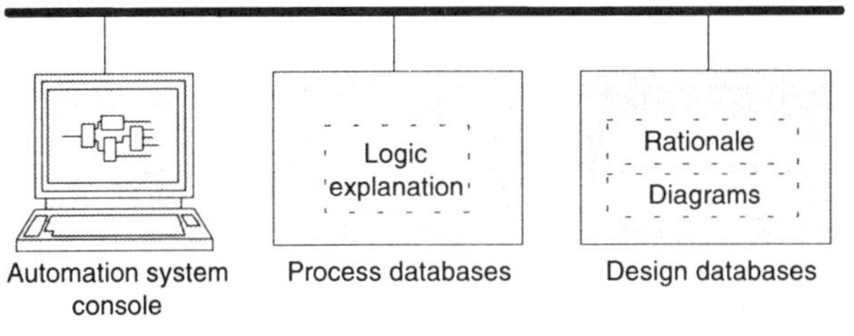

Figure 6: A possible architecture for embedding LEPO

3.4 Efficiency

The topology of the logic network determines the complexity of the search. Logic circuits in automation systems are typically trees, with several inputs being abstracted to fewer outputs. For explanation, the trees are traversed top-down, from the output towards the inputs. This search has potential for combinatorial explosion with large circuits. It may, for a single time step, require an exhaustive search through the 2^n possibilities for an n-input network, which is only practical and representable for small n. Fortunately the search tree is pruned by passive gates (that is, ones that do not affect the node's state).

Further pruning can be achieved through heuristics. A simple way is to consider those inputs first where the signal has changed at the time of interest. Modelling the inputs, sorting them according to their relative importance or to their source types, helps further. Adopting conventions from digital circuit design, the network can be divided into regions of interest that are then simulated separately. These topics are still under study.

The time dimension complicates the matter: delays and state machines require retrieval of prior state data. The feedback used in some designs causes loops that are not handled by the current version. Sequential logic also requires further work. Currently LEPO can only explain some time dependencies, but a clear state machine representation is missing in the present version.

The conclusion is that logic explanation can be computationally intractable. Fortunately, LEPO can act in a consultative mode. The user can interactively guide the search to interesting parts of the logic, alleviating the problem. Some of the complexity can be avoided by limiting search to one level at a time and letting the user choose the direction of further search. It is evident, however, that such explanations would prove insufficient in autonomous applications. Fortunately it appears that many of the logic networks used in industry are sufficiently small not to cause major computational problems.

3.5 Explanation as Diagnosis

The large body of work on diagnosis and automated reasoning on discrete devices [Barrow 84, Davis 84, Davis 88, deKleer 84, Genesereth 84, Hamscher 91, Malhotra 92] is clearly relevant to our work. Logic explanation is, in essence, a form of diagnosis in which the aim is not to locate faults in the logic network but to explain its behaviour. Of particular interest regarding our domain is to find "important" events at the network inputs. These tasks, diagnosis and explanations, call for similar techniques. We have adopted several of the techniques from diagnosis research, including representations and search algorithms.

We share Chandrasekaran's views on modelling explanations [Chandrasekaran 89], although our emphasis is in explaining the domain, not the reasoning of a knowledge based system. The explanatory techniques of LEPO are still primitive. They should be augmented with better text and graphics generation, presentation planning, and user modelling [Feiner 90, Korteniemi 91, McKeown 85, Paris 87, Swartout 91]. Suitable interaction techniques have been discussed for instance in [Huuskonen 92, Moore 90].

It is encouraging that systems similar to LEPO are starting to appear in research laboratories and industrial systems [Hayes-Roth 94, Zinser 94]. Often the text and displays for these systems are handcrafted (as is the case for the current version of LEPO), which makes them impractical for large scale use. There is a tendency towards more dynamic, model-based explanations, that would ease the construction task. For instance, the GDA package (G2 Diagnostic Assistant) offers some built-in capabilities of this kind [Finch 91], although it concentrates on category B explanations only. The explanations and simulations are immediately available on screen once the circuit topology has been drafted.

4. CONCLUSION

We studied the role of control logic in industrial plants and pointed out how it affects the work of plant staff. The behaviour of logic was seen problematic to them. We outlined ways to improve presentations of the logic, emphasising the need for greater understanding. We proposed three kinds of explanations to clarify the behaviour and the intentions behind logic designs, which helps to better understand the process. The ideas were demonstrated in a research prototype.

Logic explanation is expected to be useful for disturbance management, helping operators to more quickly understand the behaviour of logic in abnormal situations. At the same time it also serves monitoring and sequential control activities, allowing better understanding of sensor data. This provides a better view to the process, which in turn helps the operator take proper control actions. This should in turn lead to increased safety and quality. Maintenance and redesign tasks make heavy use of knowledge about design choices. The proposed types of explanations would no doubt allow for a better understanding of the design, which would help making changes at the plant.

Comments from industry have been so far very promising. In a case we studied, the plant staff clearly expressed their need for higher level information concerning control logic. They felt that the kind of help given by LEPO could increase their understanding of operational situations and clarify the intentions behind the designs. Control system manufacturers and application developers have emphasised that explanations would have to be embedded inside control systems and the required knowledge should be derived from design phases. Unfortunately, required changes to the design practices were seen hard to implement, as both individuals and organisations resist change. Computational efficiency was not felt to be an issue, although it would have to be considered for more realistic cases.

In the future we plan to extend logic explanation to cope with sequential logic. The means-end approach for explaining purposes and design rationale needs more work before practical implementations. Prototypes will be embedded in control systems, and where possible, use existing design data. Standards will have to be considered. Finally, to evaluate how operators benefit from logic explanation, field studies have to be conducted and analysed.

ACKNOWLEDGEMENTS

This research has been funded mainly by the Technology Development Centre (TEKES), Finland. Financial support has also been provided by the Technical Research Centre of Finland (VTT), the University of Oulu, Imatran Voima Oy, Valmet Automation Inc., and Tampella Power Inc. We would like to thank our industrial partners for their support and feedback and especially Hannu Paunonen, Jarmo Ervast, and Esa Salonen of Valmet Automation for their constructive comments.

REFERENCES

[Bainbridge 83] Bainbridge, L.: "Ironies of Automation", Automatica, Vol. 19, No. 6, pp. 767 - 773, 1983.

[Barrow 84] Barrow, H.G.: "VERIFY: A Program for Proving Correctness of Digital Hardware Designs", Artificial Intelligence 24, 1984, pp. 437-491.

[Chandrasekaran 89] Chandrasekaran, B. et al: "Explaining Control Strategies in Problem Solving", IEEE Expert, Spring 1989, pp. 9-24.

[Chandrasekaran 93] Chandrasekaran, B. et al., "Functional Representation as Design Rationale", IEEE Computer, Vol., 26 No. 1, 1993, pp. 48-56.

[Davis 84] Davis, R: "Diagnostic Reasoning Based on Structure and Behaviour", Artificial Intelligence 24, pp. 347-410, 1984.

[Davis 88] Davis, R., Hamscher, W.: "Model-based Reasoning: Troubleshooting", in H.E.Shrobe (ed): Exploring Artificial Intelligence: Survey Talks from the National Conf. on Artificial Intelligence, pp. 297-346, Morgan Kaufmann, San Mateo, California, 1988.

[deKleer 84] Kleer, J.D.: "How Circuits Work", Artificial Intelligence 24, 1984, pp. 205-280.

[Feiner 90] Feiner, S., McKeown, K.: "Coordinating Text and Graphics in Explanation Generation", Proc. AAAI-90, 8th Nat. Conf. on Artificial Intelligence, July-August 1990, pp. 442-449.

[Finch 91] Finch, F.E., Stanley, G.M., Fraleigh, S.P.: "Using the G2 Diagnostic Assistant for Real-time Fault Diagnosis", European Conference on Industrial Applications of Knowledge-Based Diagnosis, Segrate (Milan), Italy, October 17-18, 1991.

[Fischer 91] Fischer, G. et al., "Making Argumentation Serve Design", Human-Computer Interaction, Vol. 6, 1991, pp. 393-419.

[Genesereth 84] Genesereth, M.R.: "The Use of Design Descriptions in Automated Diagnosis", Artificial Intelligence 24, 1984, pp. 411-436.

[Hamscher 91] Hamscher, W.C.: "Modeling digital circuits for troubleshooting", Artificial Intelligence 51(1-3), 1991, pp. 223-271.

[Hayes-Roth 94] Hayes-Roth, F., Jacobstein, N.: "The State of Knowledge-Based Systems", Communications of the ACM, Vol.37, No3, March 1994, pp. 27 - 39.

[Huuskonen 92] Huuskonen, P., Korteniemi, A.: "Explanation Based on Contexts", Proc. 8th IEEE Conference on Artifical Intelligence for Applications, Monterey, California, March 2-6, 1992, pp. 179 - 185.

[IEC 93] IEC 1131-3, "Programmable controllers - Programming languages", International Electrotechnical Commission, 1993, 207 p.

[Kaarela 93a] Kaarela, K., Huuskonen, P., Leiviskä, K., "The Role of Design Knowledge in Industrial Plant Projects", Proceedings of the International Conference on Cognitive and Computer Sciences for Organizations, Montreal, Canada, May 4-7, 1993, pp. 173-183.

[Kaarela 93b] Kaarela, K., Huuskonen, P., Jaako, J.: "Providing Plant Design Knowledge to the Operators", Proc. Fifth Intl. Conf. on Human-Computer Interaction, August 8-13, 1993, Orlando, FL, USA, Vol. 1, pp. 546-551.

[Kaarela 94] Kaarela, K., Oksanen, J., "Structuring and recording plant design knowledge", Proceedings of the IFIP international conference on Feature Modelling and Recognition in Advanced CAD/CAM Systems, Valenciennes, France, May 24-26, 1994, Vol. 2, pp. 853-866.

[Keuneke 91] Keuneke, A.M., "Device Representation. The Significance of Functional Knowledge". IEEE Expert, Vol. 6, No. 2, 1991, pp. 22-25.

[Korteniemi 91] Korteniemi, A., Huuskonen, P.: "An Expert System with Explanations Tailored to Each User", 3rd Intl. Symposium on Expert Systems Application to Power Systems, Tokyo, Japan, April 1-5, 1991, pp. 381-384.

[Lee 91] Lee, J. et al., "What's in Design Rationale?", Human-Computer Interaction, Vol. 6, 1991, pp. 251-280.

[Lind 82] Lind, M., "Multilevel Flow Modelling of Process Plant for Diagnosis and Control", Risø-M-2357, Risø National Laboratory, Denmark, 1982.

[Lind 90] Lind, M., "Representing Goals and Functions of Complex Systems - An Introduction to Multilevel Flow Modelling", Report No. 90-D-381, Institute of Automatic Control Systems, Technical University of Denmark, 1990.

[Malhotra 92] Malhotra, P., Seviora, R.E.: "Object Oriented Framework for Generating Machine Understanding of Digital System Design", Proc. IEA/AIE Conf. Industrial and Engineering Applications of Artificial Intelligence and Expert Systems, Paderborn, Germany, 1992, pp. 690-700.

[McKeown 85] McKeown, K.R., Wish, M., Matthews, K.: "Tailoring Explanations for the User", Proc. IJCAI-85, Los Angeles, California, August 1985, pp. 794-798.

[Moore 90] Moore, J.D., Swartout, W.R.: "Pointing: A Way Toward Explanatory Dialogue", Proc. 8th Nat. Conf. Artificial Intelligence (AAAI-90), July 29-August 3, 1990, pp. 457-464.

[Paris 87] Paris, C.L.: "Combining Discourse Strategies to Generate Descriptions to Users along a Naive/Expert Spectrum", Proc. 10th IJCAI, Morgan Kaufmann, Los Altos, Calif. 1987, s. 626-632.

[Rasmussen 86] Rasmussen, J.: "Information Processing and Human-Machine Interaction", North-Holland, Amsterdam, 1986, 215 pages.

[Sembugamoorthy 86] Sembugamoorthy, V. et al., "Functional Representation of Devices and Compilation of Diagnostic Problem-Solving Systems", in Experience, Memory, and Reasoning, J.L. Kolodner and C.K. Riesbeck, eds., Lawrence Erlbaum Associates, Hillsdale, New Jersey, 1986, pp. 47-73.

[Swartout 91] Swartout, W.R., Paris, C.L., Moore, J.: "Design for Explainable Expert Systems", IEEE Expert, Vol. 6, No. 3, 1991, pp. 58-64.

[Zinser 94] Zinser, K., Frischenschlager, F.: "Multimedia's push into power", IEEE Spectrum, July 1994, pp. 44-48.

TPMS - A SYSTEM FOR SEQUENCING OPERATIONS IN PROCESS PLANNING

João Rocha, Carlos Ramos
Departamento de Engenharia Informática, Instituto Superior de Engenharia do Porto
Rua S. Tomé, 4200 PORTO, PORTUGAL
Email: csr@garfield.fe.up.pt

ABSTRACT

This paper deals with a system named TPMS[1] (Task Planning for Manufacturing Systems). The final goal of this system is to automatically generate programs or sets of commands for robots, AGV's, numerical control machines and other components of the system without the need of spending much time programming the Flexible Manufacturing System components. We will describe how to automatically generate a high level plan for the task.

The sequence of operations to be handled by the Manufacturing System depends on the constraints for the task. Three types of constraints are defined and explained: processing constraints; feasibility constraints and geometric constraints. The symbolic plan for the task is represented in the form of a precedence graph.

The main contributions are the following ones: to put together symbolic planning and execution planning; it is being developed to work for complete Manufacturing Systems (and not for controlling only one robot); different industrial operations (assembly, welding, drilling,...) can be considered at the same time.

1. INTRODUCTION

For the past years, we have observed fundamental changes on markets for manufacturing products. Today, Manufacturing Systems markets are more global, complex and competitive than before.

An important challenge is the reduction of the products life-cycles and the reduction of the portion dimension. This implies to be in a need of frequently redesigning products and reprogramming Manufacturing Systems. However, Manufacturing Systems reprogramming is a complex task involving a considerable set-up time. Most of this time is spent in defining how the new product will be manufactured and in programming the different machines of the Manufacturing System (e.g. CNC, robots, AGV's, conveyors, ...).

The increase of system's flexibility is very important to achieve efficiency and productivity improvements. Now the emphasis is on versatility and intelligence leading to ideas such as Intelligent Design and Automatic Planning and Programming.

This paper deals with Task and Execution Planning for Manufacturing Systems. Task and Execution Planning involve two phases. In the first one a symbolic plan for the Manufacturing Task is generated. In the second one the symbolic operations of the high level plan are converted in sets of programs, instructions and commands to control the Manufacturing System's equipment. For this purpose we are developing a system named TPMS [RR94a] [RR94b] [RR94c], (Task Planning for Manufacturing Systems) that will be presented in this paper.

2. STATE OF THE ART AND BACKGROUND

Some systems have been developed to deal with Automatic Task Planning and Execution. Most of them only take into account high level symbolic planning without machine or robot programming.

Bourjault [Bou84] has described an interactive system for assembly sequence generation of a product. It is created a liaison graph representing parts connections. However this system has not inference capabilities concerning geometric knowledge about the assembly. This knowledge is furnished by the user who answers questions about the sequence of operations.

De Fazio [DFW87] has improved this methodology reducing the number of questions to the user. In this

[1] This work is partially supported by FLAD ("Fundação Luso-Americana para o Desenvolvimento") under project 443/93 - SFF

system the user writes logic formulas representing the sequence of operations. The user needs to be familiar with this syntax and, sometimes, it is difficult to map from assemblies to the formulas. It is also created a liaison graph. Consider the example assembly shown in figure 1. Each node represents the set of all components, assembled or not and each path, an assembly operation.

Homem de Mello [HM89] uses AND/OR graphs to represent assemblies. With this kind of graphs, the number of nodes is drastically reduced, namely for tight coupled assemblies. It is possible to generate all feasible assemblies from the AND/OR graph. However, the combinational explosion problem may happen here. Some feasibility tests (geometric feasibility, mechanical feasibility and stability feasibility) are applied to test if a specific assembly sequence is really feasible. In figure 2 the same assembly sequence than figure 1 is shown, using an AND/OR graph. In this case, each node represents an assembly, sub-assembly or individual component and each path an assembly operation.

Wilson [Wil92] has improved the efficiency of plan generation from AND/OR graphs using some kind of geometric reasoning. Now, only feasible sequences of operations are generated.

Sukhan Lee [Lee92] has developed COPLANNER, a distributed and co-operative approach to Planning with the Blackboard paradigm. Concerning Planning, this system uses the Backward Assembly Planning (BAP). Sukhan Lee illustrates some examples in which a plan cannot be achieved by the inversion of the plan generated backwards. However, BAP is able to detect these problems, achieving feasible solutions.

The above described five systems did not consider the existence of a manufacturing system for performing the assembly task.

The system from Huang and Lee [HL91] [LeeC92] accepts as input a product description from a CAD system and generates an assembly plan, subject to some resource restrictions inherent to a manufacturing cell.

There is a "world" database representing the components, relations among them, industrial operations, machines and tools from the manufacturing cell. A "world" simulated model for the various stages that exist during the assembly is generated. There is a knowledge acquisition mechanism with three modules: precedence between operations acquisition module, fixing operations specification's module and tools selection module. The plan generator makes inferences on a set of production rules like "IF <pre-condition> THEN <actions>". Actions correspond to a Delete List and a Junction List. The plan is obtained by inversion of the dismount sequence.

All the systems above described did not consider the Execution Planning, i.e., how to program machines and/or robots to perform the generated high level symbolic plan. However some systems dealt with this question in the area of Robotics. The group of Lozano-Perez [LJMD89] has implemented HANDEY, a system for performing grasping and regrasping operations for pick and place problems. However, this system has some limitations concerning the reasoning about action. Hutchinson [Hut90] has presented another interesting approach to this problem in SPAR system. SPAR uses a three-level hierarchy with operational, geometric and uncertainty-reduction goals. Unfortunately, collision avoidance is not considered in this work.

TPMS takes advantage of CIARC (Cooperative Intelligent Assembly Robotic System). CIARC has been implemented for Task Planning and Execution of Manipulation and Assembly Tasks using one robot arm programmed by a textual language [RO92a] [RO92b] [RO93]. The conversion of symbolic plan operations is situation dependent and some aspects, such as collision avoidance and intelligent execution, were considered in CIARC system.

While including the ideas of Intelligence and Flexibility, CIARC was developed for Assembly and Manipulation Robotics, controlling the operation of only one robot.

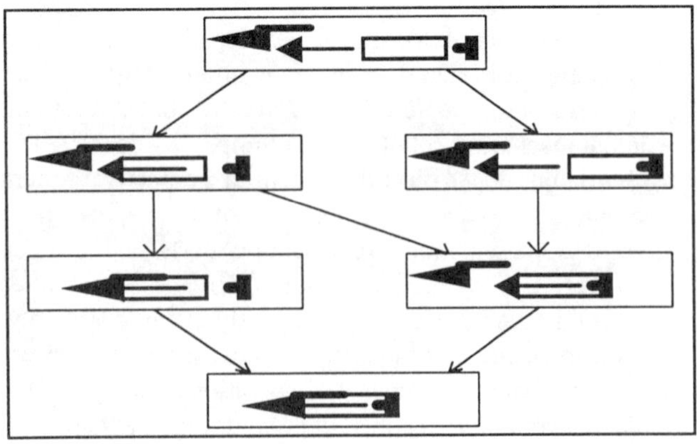

FIGURE 1 - An example using a liaison graph

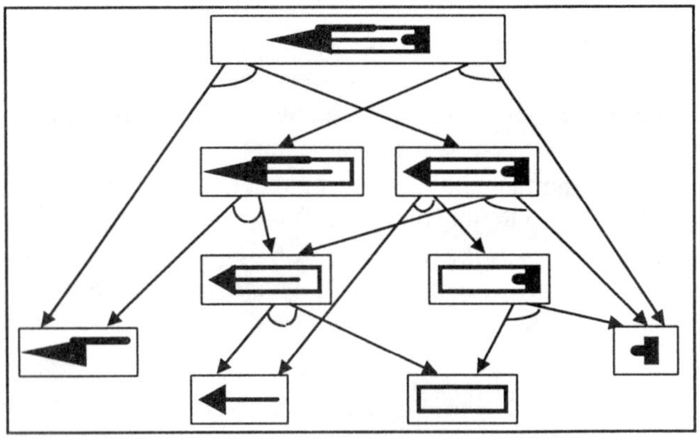

FIGURE 2 - The same example using an AND/OR graph

TPMS will be the extension of CIARC to complete Manufacturing Systems. The final goal is to automatically generate programs or sets of commands for robots, AGV's, numerical control machines and other components of the system without the need of spending much time programming the several Manufacturing System components.

In the current manufacturing cells the operator is the one who defines the sequence of operations to be carried out and the way these ones will be performed by the different components of the system (robots, numerical control machines, AGV's, ...). The purpose of the research work addressed in this paper is to develop a software to replace (or help) the manufacturing cell programmer, i.e., to automatically obtain a plan to the industrial task and to manage the control code (program, instructions) of each machine and robot.

This is an important contribution to reduce the time spent for task activation, since it helps the programmers to develop programs for the manufacturing of new products. This work points out in the direction of the new trends of Manufacturing Systems: intelligence, agility and flexibility [SJ93] [Kus90].

Readers interested in a classification of Assembly Task Planning may consult reference [GRL94].

3. HIGH LEVEL TASK PLANNING

3.1. INTRODUCTION

Task and Execution Planning involve two phases: first a high-level plan composed of symbolic operations is automatically generated and represented (e.g. put A on B, weld C on D, insert E in F); then the programs for controlling the several machines of the Manufacturing System are generated (e.g., a program for controlling a robot to insert an object into a hole of another object).

The high-level planning makes possible to generate a plan considering all processing, symbolic and geometric constraints. It is also possible to generate not only one, but several plans that could be dynamically activated according to the current situation.

Let us consider some basic definitions about "High Level Symbolic Planning". Planning is a research area in which one tries to answer the following question: "How to solve a problem", or in other words "How to achieve a goal or a set of goals". To solve this problem we need to go from an initial state to a desired final state.

Operators are the elements that can be used to change the state of the "world". Operators can be seen as general procedures in which the variables have not values. For example, we can consider the following operators:

- pickup(Ob)
- insert(Ob1,Ob2)
- put_on(Ob,LOb)
- drill(Ob)
- weld(LOb)
- cut(Ob1, Edge1)
- solder(Ob1, Ob2)
- lay(Ob1)

Operations are instantiations of operators; an operation corresponds to the instantiation of all variables in the operator arguments. If our basic objects or parts are in the set [a, b, c, d, e, f] then the following operations are valid:

- pickup(a)
- insert(a,b)
- put_on(c,[a,d])
- drill(e)
- weld([e,a])
- cut(a, 1)
- solder (a, b)
- lay(a)

Once an operation is applied the state of the parts′ world changes (e.g. figure 3).

Some operators are able to transform parts or change their shapes (e.g. drilling or milling) and other are able to join parts (e.g. welding). This means that after an operation we may have a transformation of the parts handled by the operation. To represent an operation and the new part that results from the operation we will use the following syntax:

operation => result

Let us see some examples using this format:

drill(a) => a_d	means that part a was drilled giving origin to a new part named a_d
weld([a,b]) => ab	means that parts a and b are welded giving origin to a new part named ab

The general goal we want to achieve is composed by a set of subgoals. We will use operators to achieve these subgoals. The sequence of operations for the task is the symbolic plan for the task.

However the subgoals cannot always be achieved in any order. Often the accomplishment of some goals is restricted by constraints.

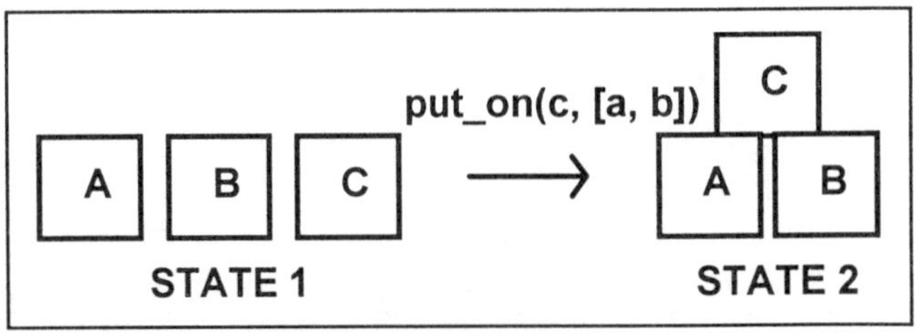

FIGURE 3 - Transition from state 1 to state 2, applying operator put_on

For Manufacturing Tasks the main constraints are the process constraints, feasibility constraints and geometric constraints. The constraints imply precedence relationships to guarantee that operations will be executed in the correct order. The syntax we will use will be the following one:

operation1 ***before*** operation2

As examples of precedence relationships between operations we can have:

- drill(a) before weld([a,b])
- cut(W2, 1) before solder (Rl2, W2)

It is important to acquire the knowledge about these constraints. The knowledge about the feasibility and geometrical constraints can be automatically acquired with some geometric reasoning capabilities [RO92a] [RO92b] [RO93].

As an example of feasibility constraints' detection we can consider the case in which two subgoals must be achieved: one to drill object b and another to insert object a in the hole of b. The second one can be accomplished only after the accomplishment of the first one.

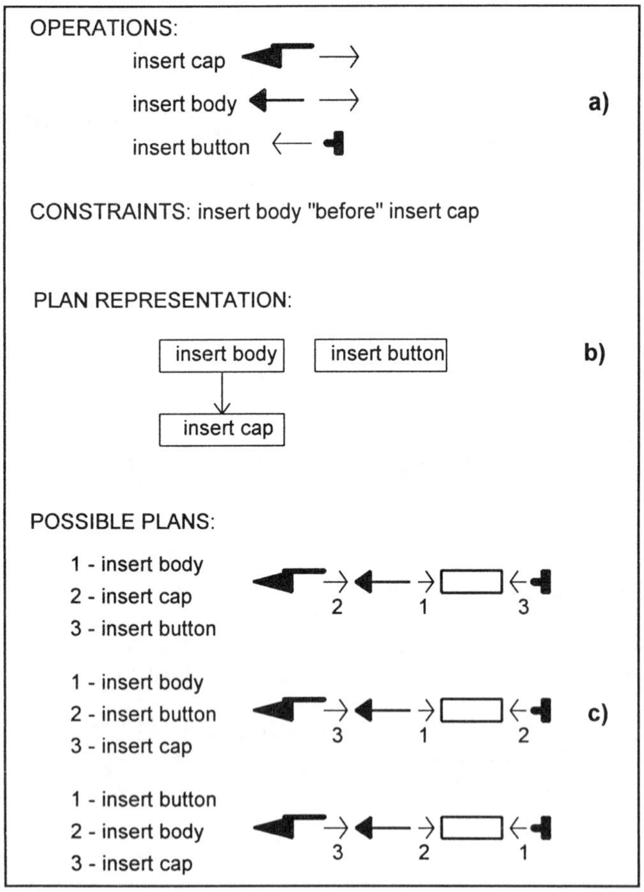

FIGURE 4 - Operations, graph and possible plans

Feasibility constraints are detected by the verification of a set of feasibility constraint rules. An example of a rule of this type may be the following one:

For insert(Ob, Ob1)
- Ob must be available
- Ob1 must be available
- the hole of Ob1 must be performed.

Geometric Constraints are obtained by the analysis of the geometry of the operation. Collisions are a complex problem to deal with in Manufacturing Tasks. The geometric constraints related with the collision analysis can be automatically obtained as it happens in the CIARC system where two kinds of constraints related with collisions are automatically acquired: constraints to avoid collisions between a robot and the objects and constraints to avoid collisions between objects being handled and other objects [RO92a] [RO92b] [RO93].

Figure 4 illustrates this methodology for the same example assembly shown in figure 1. Each node represents an operation and each path a constraint between operations. The operations are shown in figure 4(a), the graph (plan representation) in figure 4(b), while all possible plans are shown in figure 4(c).

One important aspect of figure 4(b) is that the graph in not necessarily conexed, i.e., it is possible to have two or more graphs for represent one plan. The plan generation can be, therefore, simpler than only with one graph.

The graph of figure 4(b) is much simpler than those of figure 1 and figure 2, although it also allows the generation of all feasible plans.

A more complex example is presented in next section.

3.2. AN EXAMPLE

Let us consider an example for assembling a wire harness system based on the example referred in [AR94] (figure 5). There are 16 parts: 4 wires (*W1*, *W2*, *W3*, *W4*) with different lengths; 3 ring lugs (*Rl1*, *Rl2*, *Rl3*), 3 lugs (*L1*, *L2*, *L3*), 2 terminals (*T1*, *T2*), 1 snap-in connector (*SI*) and 3 straps (*S1*, *S2*, *S3*).

The 4 wires must be cut and stripped by a cutting machine and their ends must be terminated with the ring lugs, lugs or terminals by a soldering machine. Then the snap-in connector must be inserted. The harness is then assembled on a harness jig or board. Once all the wires have been correctly laid, the harness can be bundled by a strapping machine.

By default, we do not know the order by which the parts will be processed and assembled. The sequence of operations will be obtained by the feasibility and geometric constraints between parts and tools. There are also some processing constraints.

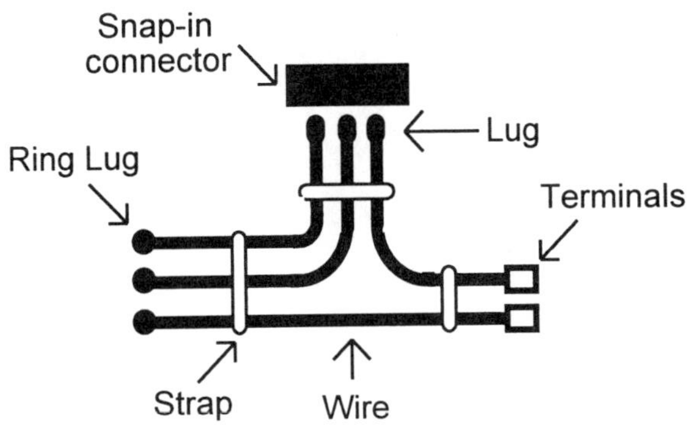

FIGURE 5 - An example

The equipment which composes the manufacturing cell for this example is as follows:

- 1 strapping machine
- 1 cutting machine
- 1 soldering machine
- 1 robot

The operations to be considered are the following:

- cut *W1* edge *1* → *cut(W1, 1)*
- cut *W1* edge *2* → *cut(W1, 2)*
- cut *W2* edge *1* → *cut(W2, 1)*
- cut *W2* edge *2* → *cut(W2, 2)*
- cut *W3* edge *1* → *cut(W3, 1)*
- cut *W3* edge *2* → *cut(W3, 2)*
- cut *W4* edge *1*→ *cut(W4, 1)*
- cut *W4* edge *2*→ *cut(W4, 2)*
- solder *Rl1* and *W1* → *solder* (*Rl1*, *W1*)
- solder *Rl2* and *W2* → *solder* (*Rl2*, *W2*)
- solder *Rl3* and *W3* → *solder* (*Rl3*, *W3*)
- solder *L1* and *W1* → *solder* (L*1*, *W1*)
- solder *L2* and *W2* → *solder* (*L2*, *W2*)
- solder *L3* and *W4* → *solder* (*L3*, *W4*)
- solder *T1* and *W3* → *solder* (*T1*, *W3*)
- solder *T2* and *W4* → *solder* (*T2*, *W4*)
- insert *SI* into [*W1*, *W2*, *W4*] →
  insert (*SI*, [*W1*, *W2*, *W4*])
- laid *W1* → *laid (W1)*
- laid *W2* → *laid (W2)*
- laid *W3* → *laid (W3)*
- laid *W4* → *laid (W4)*
- strap *S1* to [*W1*, *W2*, *W3*] →
  strap (*S1*, [*W1*, *W2*, *W3*])
- strap *S2* to [*W1*, *W2*, *W4*] →
  strap (*S2*, [*W1*, *W2*, *W4*])
- strap *S3* to [*W3*, *W4*] →
  strap (*S3*, [*W3*, *W4*])

However we do not know in which order they will be carried out. To know this, we will generate all feasibility, geometric and processing constraints.

They are the following ones:

<u>processing constraints:</u>

- for this class of assemblies we must insert the snap-in connector before bundle. This means:
  insert(SI) before *strap*(S1)
  insert(SI) before *strap*(S2)

<u>feasibility constraints:</u>

- if we must solder some components to the wires, then the wires must have been cut. This means:
  cut(W1, 1) before *solder* (*Rl1*, *W1*)
  cut(W1, 2*)* before *solder* (*L1*, *W1*)
  cut(W2, 1) before *solder* (*Rl2*, *W2*)
  cut(W2, 2*)* before *solder* (*L2*, *W2*)
  cut(W3, 1) before *solder* (*Rl3*, *W3*)
  cut(W3, 2*)* before *solder* (*T1*, *W3*)
  cut(W4, 1) before *solder* (*L3*, *W4*)
  cut(W4, 2*)* before *solder* (*T2*, *W4*)
- if we want to insert the snap-in connector into the lugs, the lugs must have been soldered. This means:
  solder (*L1*, *W1*) before *insert(SI)*
  solder (*L2*, *W2*) before *insert(SI)*
  solder (*L3*, *W4*) before *insert(SI)*

<u>geometric constraints:</u>

- concerning laying
  if we consider strapping before laying we will see that the correct layout of the cable cannot be achieved. This means:
  - *lay(W1)* before *strap(S1)*
  - *lay(W2)* before *strap(S1)*
  - *lay(W3)* before *strap(S1)*
  - *lay(W1)* before *strap(S2)*
  - *lay(W2)* before *strap(S2)*
  - *lay(W4)* before *strap(S2)*
  - *lay(W3)* before *strap(S3)*
  - *lay(W4)* before *strap(S3)*

The precedence graph that is automatically generated by our planning methodology is shown in figure 6 (next page).

Notice that the parallel branches of the precedence graph are non-preemptible sets of operations. This happens because some operations are related with the same resource (one cutting or soldering machine). With two cutting machines we could have interleaving between some operations sets (e.g. cut edge one of wire one in one cutting machine, cut edge one of wire two in other cutting machine; solder ring lug one into wire one, ...). It is then important to notice that the layout of the Manufacturing System must be take in account when we want to generate the execution plan.

The plan generation is achieved considering all the precedence relations (obtained from constraints), which can be represented by a graph.

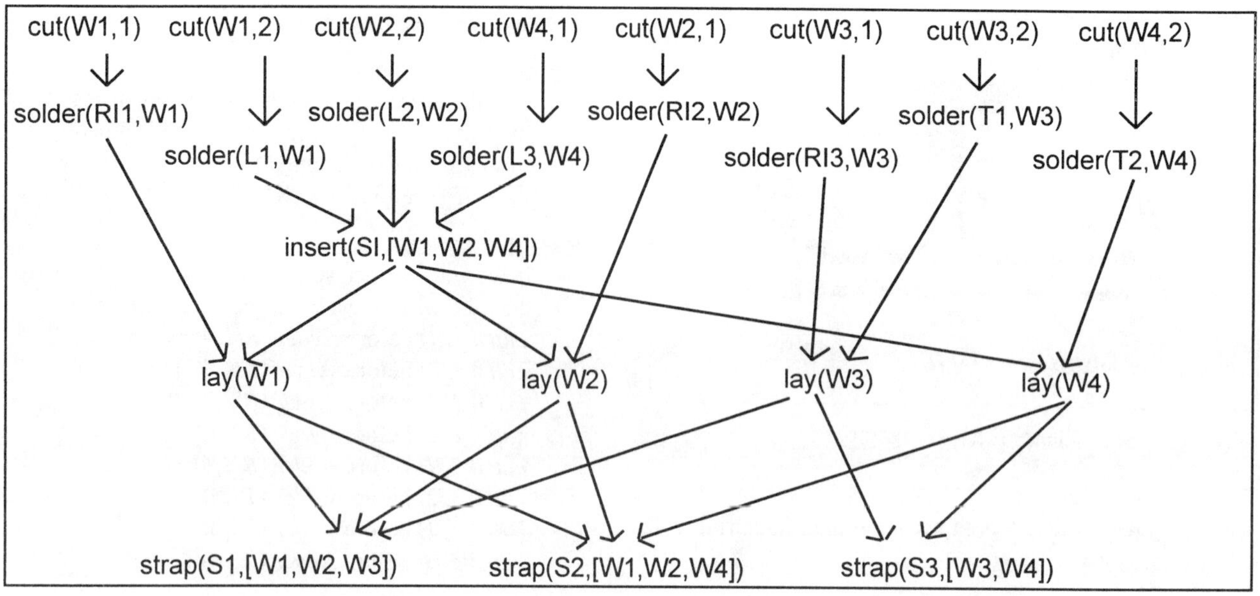

FIGURE 6 - Precedence graph generated by our planning methodology

The algorithm for generating our plan is as simple as follows:

```
{
  repeat until number of nodes = 0
    {
      choose one operation with no precedences
      arriving to the corresponding node          N
      remove N and all precedences starting from N
    }
}
```

If the generation of all possible plans is desired, we can use a similar algorithm with backtracking (in this case, nodes and precedences are marked in spite of being removed). However one must notice that this generation of all plans can be a much time-consuming phase for complex precedence graphs.

To solve this problem a new algorithm was implemented using a methodology similar to A*, considering, for each operation, the time needed for concluding that operation as well as the time for changing from one operation to another (e.g., change the tool, wait a certain time before a specific operation due to the involved process and so on). Then it will be possible to select the best plan considering the operation's time.

The cost for each node (corresponding to an operation) is obtained by adding the cost of the two components described above with the cost of the parent node. The cost of first node (node 0) is 0.

It is also necessary to estimate the merits of each node we generate in order to search more promising paths first. We do that by considering the sum of times for all changing tools' operations needed to obtain a plan. Since this is an underestimation of the total time needed to execute the plan, the algorithm is guaranteed to find an optimal (as defined before) plan as determined by the time.

The algorithm is as follows:

```
{
  first node is number 0 (considering all operations
        and constraints)
  node 0 is a terminal node
  repeat
  {
      expand node i
      choose next node for expansion
  }
  until there is no non-terminal node with lesser
        cost than terminal node
}
```

For expanding a node:

```
{
  repeat
  {
      select an operation i with no precedences
          arriving to the corresponding node N
      remove operation i from operations list
      remove all precedences starting from N
          from precedences list
      generate a new node with new operation
          and precedences lists
      mark new node as a terminal node
  }
  until there is no operations with no precedences
          arriving to the corresponding node N
}
```

For choosing a node

```
{
    find all terminal nodes
    calculate cost of all terminal nodes
    choose node with lesser cost
}
```

In this algorithm, the configuration of the Manufacturing System is also considered, since the sequence of operations is important due to the time needed for changing from one operation to another.

In the example of figure 6 we must notice the 8 first parallel branches of the graph. This situation can easily generate a problem of combinational explosion, if we want to obtain all feasible plans.

Using the methodology described in the algorithm above it is possible to generate an optimal plan (e.g. figure 7), since all times involved in the operations and tools' changing are considered. In the example, as the number of different operations is small, but the number of total operations is much bigger, the main problem is not the operations time, but the tools' changing time.

This is why the best plan is the one that minimizes the number of tools' changing operations.

```
cut(w1, 1)
cut(w1, 2)
cut(w2, 1)
cut(w2, 2)
cut(w3, 1)
cut(w3, 2)
cut(w4, 1)
cut(w4, 1)
solder(rl1, w1)
solder(l1, w1)
solder(l2, w2)
solder(l3, w4)
solder(rl2, w2)
solder(rl3, w3)
solder(t1, w3)
solder(t2, w4)
insert(si, [w1, w2, w4])
lay(w1)
lay(w2)
lay(w3)
lay(w4)
strap(s1, [w1, w2, w3])
strap(s2, [w1, w2, w4])
strap(s3, [w3, w4])
```

FIGURE 7 - An optimal plan

4. CONCLUSIONS AND FUTURE WORK

This paper has described and illustrated TPMS (Task Planning for Manufacturing Systems). Some definitions related with symbolic task planning have been presented and precedence relationships are obtained from feasibility constraints, geometric constraints and process constraints.

The main contributions of TPMS are the following ones:

- to put together symbolic planning and execution planning (as in CIARC)
- to be prepared to work for Manufacturing Systems (and not for controlling only one robot).
- to be prepared to deal with different industrial operations and not only for assembly

The future trends of TPMS will lead to some new performances, namely:

- the integration with a scheduling policy
- considering the precedence graph, suggest possible configurations of the manufacturing system
- the improvement of the automatic conversion of the symbolic plan into programs to control machines, robots and other equipment
- the interaction with an Intelligent Design System.

Concerning the integration with scheduling, the main idea is to generate a structure for representing symbolic plans and programs for controlling the FMS equipment in such a way that this structure could be dynamically handled. This means that changing the scheduling will correspond to select new symbolic plans (or parts of plans) and to adopt new programs to control the equipment of the FMS.

REFERENCES

[RR94a] Rocha, J. and Ramos, C - Task and Execution Planning for Flexible Manufacturing Systems, Proc. of the Second International Conference on Integrated Logistics and Concurrent Engineering - ILCE'94, pp. 131-139, Montpellier, France, 1994

[RR94b] Rocha, J. and Ramos, C - Generating and Converting Plans for Intelligent Manufacturing Systems, Proc. of IEEE International Symposium on Industrial Electronics - ISIE'94, pp. 304-309, Santiago, Chile, 1994

[RR94c] Rocha, J. and Ramos, C - Task Planning for Flexible and Agile Manufacturing Systems, Proc. of IEEE International Conference on Intelligent Robots and Systems - IROS'94, pp. 105-112, Munich Germany, 1994

[Bou84] Bourjault, A. - Contribution a une approche méthodologique de l'assemblage automatisé: Elaboration automatique des séquences opératoires, Thèse d'État, Univ. de Franche-Comté, Besançon, France, 1984

[DFW87] De Fazio T. and Whitney, D. - Simplified generation of all mechanical assembly sequences, IEEE Journal of Robotics and Automation, vol. 3, n. 6, pp. 640-658, December 1987

[HM89] Homem de Mello, L. - Task sequence planning for robotic assembly, PhD Dissertation, Carnegie Mellon University, 1989

[Wil92] Wilson, R. - On geometric assembly planning, PhD Dissertation, Stanford University, Rep. n. STAN-CS-92-1416, 1992

[Lee92] Lee, S. - Backward Assembly Planning with assembly cost analysis, Proc. of IEEE International Conference on Robotics and Automation, pp. 2382-2391, 1992

[HL91] Huang, Y. and Lee, C. - A framework of knowledge-based assembly planning, Proc. of IEEE International Conference on Robotics and Automation, pp. 599-604, 1991

[LeeC92] Lee, C. - Proc. of the first workshop on assembly planning: theory and implementation Workshop of IEEE International Conference on Robotics and Automation 1992

[LPJM89] Lozano-Pérez, T.; Jones, J.; Mazer, E.; O'Donnel, P. - Task level planning of pick-and-place robot motions, IEEE Computer, vol. 22, n. 3, pp. 21-29, March 1989

[HK90] Hutchinson, S. and Kak, A. - SPAR: A planner that satisfies operational and geometric goals in uncertain environments, AI Magazine, Spring 1990, pp. 30-61, 1990

[RO92a] Ramos, C. and Oliveira, E. - Closing the loop of Task Planning, Action and Sensing, Proc. of IEEE/RSJ International Conference on Intelligent Robots and Systems, pp. 909-916, Raleigh, NC, USA, 1992

[RO92b] Ramos, C. and Oliveira, E. - Sensor-based Reactive Planning and Execution for Robotics Assembly Tasks, Proc. of IEEE International Conference on Systems Engineering, pp. 135-138, Kobe, Japan, 1992

[RO93] Ramos, C. and Oliveira, E. - CIARC: A Multi-Agent Community for Intelligent Assembly Robotics Systems with Real-Time Capabilities, Proc. of IEEE International Conference on Systems, Man and Cybernetics, Le Touquet, France, 1993

[SJ93] Strobel, R. and Johnson, A. - Pocket pagers in lots of one, IEEE Spectrum - Manufacturing à la Carte, pp. 29-32, September 1993

[Kus90] Kusiak, A. - Intelligent Manufacturing Systems, Prentice-Hall International Series in Industrial and Systems Engineering, 1990

[GRL94] Gottschlich, S; Ramos, C.; Lyons, D. - Assembly and Task Planning - a Taxonomy, IEEE R&A Magazine, n. 3, pp. 4-12, September 94

[AR94] Aguire, E and Raucent, B - Performances of Wire Harness Assembly Systems, Proc. of IEEE International Symposium on Industrial Electronics - ISIE'94, pp. 292-297, Santiago, Chile, 1994

PLANNING ARCHITECTURES FOR OPERATING PROCEDURE SYNTHESIS

James K. Soutter and Paul W. H. Chung
Department of Chemical Engineering, Loughborough University of Technology, Loughborough, Leicestershire, LE11 3TU, United Kingdom.
Email: J.K.Soutter1@lut.ac.uk

ABSTRACT:

Operating Procedure Synthesis (OPS) is the study of the automatic synthesis of operating procedures for process plants. Although there has been twenty years of research in OPS, it is difficulty to see how any current system will lead to the creation of a general and practical OPS tool. Based on an understanding of the limitations of earlier work, we have developed a new OPS architecture which avoids many of these limitations. Through our prototype system, CEP (Chemical Engineering Planner), we show that our new architecture is able to solve commercially useful problems that were impractical using earlier OPS systems.

1 INTRODUCTION

Operating Procedure Synthesis (OPS) is the study of the automatic synthesis of operating procedures for process plants. As a field of research, OPS is very similar to AI Planning. Both AI Planning and OPS study the predictive synthesis of plans to carry out complex tasks. However, OPS tasks are often more complex than the toy planning tasks that appear in the planning literature. For example, the concept of safety in OPS can not easily be represented by most AI Planning systems.

AI Planning techniques have not previously been used in OPS because it was thought that these techniques could not be extended to solve OPS problems [RR74, LS88a]. Instead, OPS systems have been created around purpose built planning architectures.

In this paper, we re-examine the idea of using AI Planning techniques for OPS through the development of an AI Planning based OPS system, known as CEP (Chemical Engineering Planner). We demonstrate that a planner like CEP can solve useful OPS tasks that are impractical for earlier systems. We also discuss how CEP can be extended to solve more technically complex problems.

The paper is organised as follows. In the first section we look at the motivations for OPS research. We go on to present a critical review of previous OPS work. Given this background, we look at the development of the CEP architecture in terms of the problems it solves and the problems it has yet to overcome. We demonstrate CEP by providing a problem that is easy to solve using our approach but very difficult to solve using earlier systems. The paper then ends with a short conclusion.

2 MOTIVATIONS

Current OPS systems create operating procedures using a knowledge base of the process plant or component or system to be operated. Part of this knowledge base describes the actions that can be performed on the process plant. Opening a valve is an example of an action. Part of this knowledge base describes situations to avoid. For example, the planner should not allow hydrogen and oxygen to mix. Part of the knowledge base describes the relation between components in the plants. For example, pipe-1 is connected to valve-1. The benefits of OPS come from the reuse of the information within this knowledge base.

One such benefit is a saving of time. This will be especially noticeable and important after plant modification. After the modification of a plant already described in the knowledge base, very little effort will be required to create the procedures for the modified plant. However, accidents have been known to occur when the operating procedures for the plant are not updated after plant modification. By making it easier to update the procedures for the modified plant, OPS may help prevent similar accidents

An OPS system can also be used as a tool to verify the operability of a plant during design. Currently, operating procedures are generated very late in the design process. At this late stage, a plant may be found to be inoperable, resulting in expensive modifications to the design. If operability were considered throughout the design process, some design errors could be removed earlier, thus reducing development costs.

Finally, OPS may help in the creation of error free operating procedures. An OPS system can be viewed as a formal method for creating procedures. An error in an

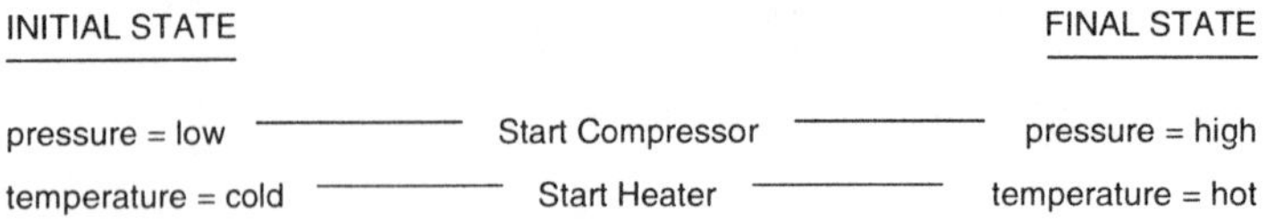

Figure 1: Action selection in [FP87]

Figure 2: Some goals can not be solved by a single action.

automatically generated procedure is logically derived from the knowledge base used to generate that procedure. If a knowledge rule is used many times then an error in the rule will manifest its self in many different ways. This may make the error relatively easy to spot.

3 THE HISTORY OF OPS RESEARCH

Previous research into the creation of a general purpose OPS system can be broken down into two paradigms. Both these paradigms have major limitations and neither is clearly superior to the other. Our objective has been to create a new method for the synthesis of operating procedures which will support the creation of a practical system.

3.1 Forward Chaining Systems

The first attempt to create an OPS system was made by Rivas and Rudd [see RR74]. This original system was designed to solve valve sequencing problems[1]. However, the algorithm was later adapted in [FP87] to solve more general tasks.

Rivas and Rudd's system is based on a forward chaining search engine. That is, procedures are created by applying actions successively to the current state of the plant until one of a set of goal states are achieved. This search strategy is inherently time consuming because, at worst, the planner must consider all possible operating procedures before finding the one procedure which has the desired effects. Heuristic search is used to improve the performance of the system. The search heuristic forbids actions which can not be evaluated, a priori, to contribute to the achievement of a goal. This heuristic is inadmissible (prevents the consideration of some valid solutions) because some goals can only be achieved by the combination of two or more actions. For example, in order to create a flow of chemical to a particular vessel, an OPS system may require opening two or more valves in sequence. Rivas and Rudd's OPS system is also limited because it does not backtrack, and so must be guided to make the correct choice first time.

The work by [FP87] can be seen to continue Rivas and Rudd's paradigm by creating a forward chaining OPS system utilising a similar goal selection heuristic. The major differences between Fusillo and Powers work, and the earlier work, is the range of problems which can be expressed. Rather than solving only valve sequencing tasks, Fusillo and Powers system will solve any problem in which each goals of the procedure is a single action away from the starting state. For example, if a particular vessel is cold and at low pressure and if an operating procedure requires the vessel to be hot and pressurised, Fusillo and Powers' system will propose starting up a heater and a compressor (see figure 1). However, Fusillo and Powers system uses Rivas and Rudd's inadmissible search heuristic and as a result, can not solve goals requiring more than one action. Figure 2 shows a problem that the system can not solve.

[TH*89] improve on the problem solving ability of [FP87] by using both backward and forward chaining search. The backward chaining engine decides how to achieve each goal in the operating procedure. The forward chaining engine then merges these solutions together and checks the side effects of the final procedure. Using this approach, this system can solve problems which require more than one action to solve each goal, such as the example shown in figure 2

The authors have a limited understanding of the work of [TH*89] because most of the earlier and later work is written in Japanese. However, we believe that the approach will have problems because of interaction between sub-plans. The backward chaining engine in the system will produce a sub-plan for each goal of the operating procedure. Merging sub-plans together is a non-trivial task. In the worst case, two sub-plans may be mutually exclusive. Even when sub-plans can be integrated, new actions may have to be added during this integration. An example is provided in figure 3. It is not clear how the forward chaining search engine in the system can deal with this complexity.

[1] A valve sequencing task involves the creation of a flow of chemical from one point in a process plant to another. These tasks are more difficult when they involve constraints on the mixing of chemicals because often these constraints can only be satisfied by purged (flushed out) components by a second flow of chemical.

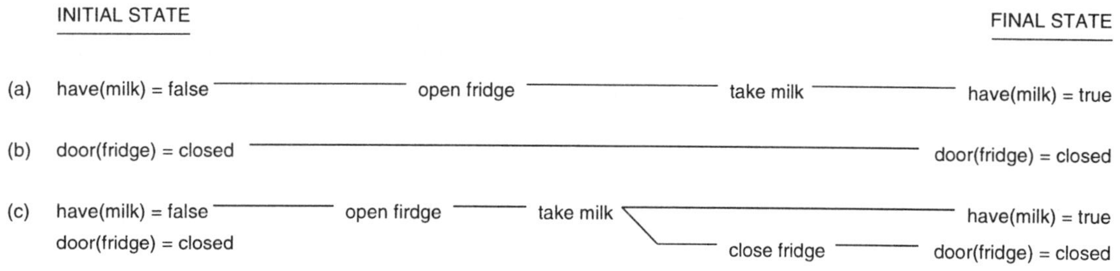

Figure 3: Two subgoals may interact.

3.2 State Graph Planning

A very different approach to OPS is state graph planning. The state graph paradigm was proposed independently in [IKPR80] and [KUO81]. In this approach, problems are represented as a graph of the possible states of the planning domain (process plant). Directed arcs in the graph representing the use of an action to move from one state to another. Using this representation, an OPS planner must find the path through the graph from the start node to the goal node. All the modelling problems and the safety issues have been solved during the creation of the graph.

State graph planning has many advantages. A state graph based OPS system can provide valid solutions to almost all OPS problems. State graph planning does not require complex techniques to deal with safety or valve sequencing operations. Finally, planning time using this approach is bounded by a polynomial function on the number of arcs in the state graph. Most other planners have an exponential time complexity.

The limitation of state graph planning is the difficulty involved in creating a state graph. The size of a state graph is related exponentially to the number of variable in the planning domain. For instance, in the example OPS problem for the first OPS system [RR74], there are 17 valves. If we assume that these valves may only be open or closed, there are 2^{17} or 131,072 states in the state graph for the plant. Even though some of these states will be practically impossible and can be removed, extra variables will be needed to model the contents of the pipes these valves connect. As a result, a practical model of the plant would be expected to contain at least a hundred thousand states.

State graphs are also difficult to modify. For instance, if a new valve is added to the 17 valve plant then the size of the state graph may double. The difficulty involved in modifying a state graph means that a state-graph based OPS system is not ideal for verifying the operability of a process plant during design or for helping to regenerate operating procedures after plant modification.

4 THE ARCHITECTURE OF CEP

An OPS system is made up of many components or agents, as shown in figure 4.

The core component of an OPS system is a planner. The planner is responsible for finding a sequence of actions to achieve the goal of the current OPS problem. Surrounding the planner are agents to monitor the operating procedure being created. The majority of these agents prevent the planner creating undesirable procedures. For example, one agent watches out for conflict between actions in the plan, and so will resolve cases where one sub-plan needs a valve to be open and another sub-plan needs the valve to be shut. Other agents advise the planner on how to achieve complex goals. Still other agents guide the planner in decision making.

The design of the planner in an OPS system will affect the design of the other agents in the system. Hence a practical OPS system must find a balance between the needs of all these agents.

In this section, we describe the design of our prototype OPS system, know as CEP (Chemical Engineering Planner). CEP is significant because it provides a new trade off between the design requirements of its components. In the next few subsections, we consider the implementation of the components which make up CEP.

4.1 The Planning Engine

The core of an OPS system is a planner, the tool

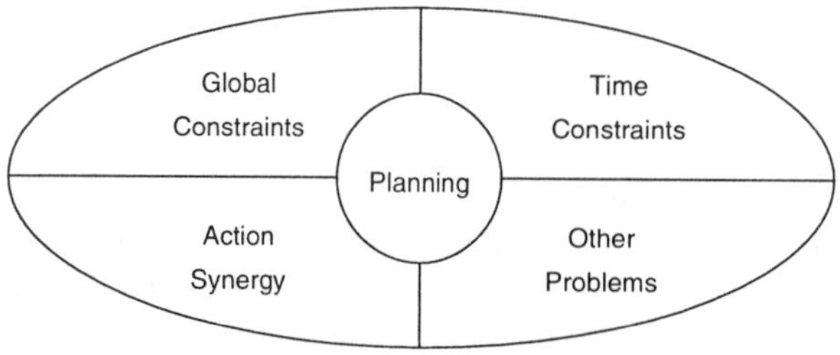

Figure 4: The problem solving components of an OPS system.

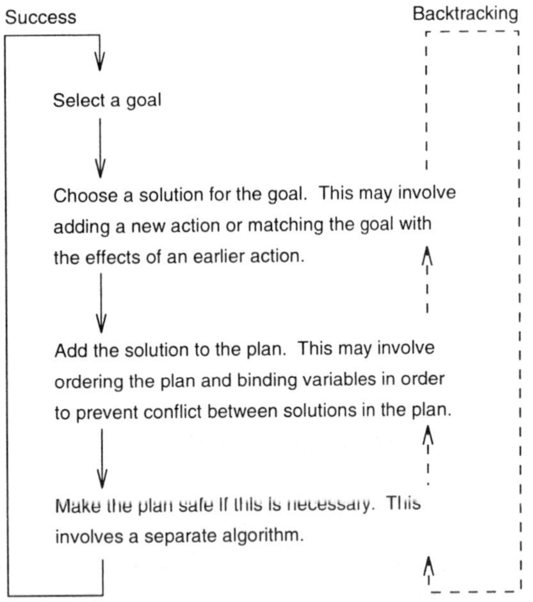

Figure 5: The planning cycle for CEP.

which finds a sequence of operations to achieve the objective of the operating procedure. For example, in state graph planning, the planner is the search engine which finds a path through the graph from the current state to a goal state.

An OPS system is more than a planner. The system will also need integrated tool to solve problems like valve modelling and safety management. Some of these tools have been studied by the AI Planning community although we would say that they are distinct from the planner.

Ideally a planning engine should use a backward chaining search so that heuristics like means-ends analysis can be used. AI Planning research has shown that very efficient backward chaining planners can be created [see CT91].

The tools surrounding the planner in an OPS system seem to require a forward chaining approach. For example, the natural way to verify the safety of an operating procedure is to simulate the effects of the procedure step by step. If procedures are created in order, from the first action to the last action, then this natural safety checking algorithm can be used. However, if the OPS system uses a backward chaining search, and especially if the planner uses a least commitment approach, then safety checking become more complex. Other related problems, like action synergy, are also made harder to solve by the use of backward chaining.

Most current OPS systems are based on forward chaining search strategy. However, these forward chaining planners are limited because of problems with planning time. For example, [RR74] and [FP87] have needed to use inadmissible search heuristics to improve the performance of their systems. Hybrid systems, utilising distinct backward and forward chaining stages [see TH*89] are also limited because the forward chaining engine will still have to plan to correct and integrate the output from the backward chaining search.

We take the view that the planner is the most complex part of an OPS system. As a result, CEP is base on backward chaining search in order to take advantage of the extensive body of AI Planning research. In the AI Planning terminology, CEP could be described as a partial-order, constraint posting, STRIPS (STandford Research Institute Problem Solver) based planner [see FN71, Sac75, Ste81].

The planning cycle for CEP is shown in figure 5. Rather than generating and ordering actions separately, as in [FP87] and [TH*89], CEP orders actions as they are generated. This means that the system is much better at resolving conflict between the solutions to its goals. For example, CEP can easily solve the problem shown in figure 3. As you will recall, this type of problem was not solvable using the earlier forward chaining systems.

4.2 Global Constraints:

In many toy planning domains, a procedure is acceptable if it is correct. In other words, if a plan achieves a particular set of goals then that is all that matters. In OPS, an operating procedure must be correct and safe. For example, if an operating procedure achieves its goal but also creates an explosive mixture then that procedure is not acceptable.

CEP assumes that an operating procedure is safe if it does not create an unsafe situation. For example, mixing hydrogen and oxygen is prohibited because this will cause a situation in which some pipe or vessel contains an explosive mixture of chemicals.

A global constraint[2] is simply a description of a set of unsafe states or situations. In CEP, a *global constraint* is represented by a list of conditions which, if all these conditions are true in a particular state, will characterise that state as unsafe. For example, figure 6 shows CEP's representation of the global constraint “do not mix chlorine and methane unless the system temperature is high” used in [FP87]. A planning domain may contain many such constraints.

Planning with global constraints is NP-Hard by reduction to the satisfiability problem [see Sou93]. An OPS system must contain some representation of safety and certainly this must include the prevention of unsafe states. Hence we expect all practical OPS systems to be NP-Hard. The state graph planning algorithm is not NP-Hard only because the NP-Hard step is the users task of creating a state graph for a particular problem.

To ensure that a procedure does not violate a set of global constraints, CEP checks the procedure after each plan modification. If the procedure starts from a safe state and each plan modification does not possibly cause

[2]Global constraints were first used in [FP87]. These constraints were not needed in the earlier state graph planning work. In state graph planning, safety constraints are represented implicitly within the graph.

```
restriction
{
    system ?s;
    temperature ?t;

    ?t != high;

    prevent
     temperature of ?s is ?t;
     contains(?s, chlorine) is true;
     contains(?s, methane) is true;
    end
}
```

Figure 6: The representation of a global constraint in CEP.

a global constraint violation then the finished plan must also be safe. A similar algorithms was used in [FP87] and [RR74] but our work is complicated by CEPs use of least commitment search. Checking a plan incrementally is more efficient than simply checking the safety of the finished procedures because unsafe procedures are discarded as early as possible.

When a possible global constraint violation is found, the system must decide whether the violation can be prevented. Some global constraint violations can only be prevented by adding new actions to the plan. For example, when planning with the global constraint shown in figure 6, the only acceptable way to add methane to a cold system containing chlorine is to heat the system first.

When Fusillo and Powers system finds a global constraint violation, it backtracks. This is insufficient because action selection heuristics prevent the system from later adding the actions needed to resolve some violations. CEP explicitly considers adding actions to resolve global constraint violations. CEP will also try to resolve a violation by reordering the plan or by binding plan variables. As a result, CEP is the first OPS system to work with global constraints in a complete way.

4.3 Action Synergy

STRIPS based planners contain a relatively simple language for modelling the world. Essentially, an action in STRIPS has two components, a list of *preconditions* and a list of *effects*. It is assumed that if all the preconditions of an action are true in the state immediately before the action is executed, then all the effects of the action will be true immediately after the action is executed. It is further assume that all conditions true immediately before and action and not explicitly negated by the effects list of the action will be true immediately after the action. In other words, the effects or a STRIPS action must be no more and no less than the declared effects of the action. This assumption causes problems when attempting with some real world tasks.

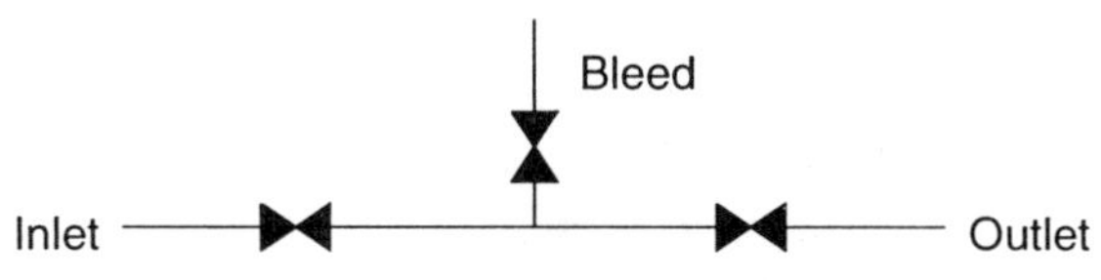

Figure 7: A block and bleed valve arrangement.

In the real world, a group of actions may exhibit an effect which is not shown by any individual action. This is action synergy. For example, consider the task of fixing a chain of Christmas Tree lights, where the bulbs are wired in series. If two bulbs in the chain have broken, then each of the two bulbs must be replaced before the chain can be said to have been repaired.

Action synergy is particularly a problem for OPS when sequencing valve operations. Consider the plant shown in figure 7. Opening any one valve (represented by butterfly symbol) will allow only a limited movement of chemical. However, if both the inlet and outlet valves are open, a flow of chemical will be created.

From the examples so far, it is apparent that an understanding of action synergy is required to achieve some goals. What we have not shown is that action synergy can also cause a plan to be unsafe. Consider again the problem of creating a flow of chemical from inlet to outlet in the plant shown in figure 7. If the bleed valve is open when the flow of chemical is created then a flow from the inlet to the bleed valve will also be created. This bleed pipe is not designed for this kind of use and so it would have to be closed before the creation of the inlet-outlet flow.

In the following section, we look at methods for planning with action synergy.

4.3.1 The Inference Engine

CEP contains a backward chaining inference engine to help solve goals requiring an understanding of action synergy. The inference engine is used to rewrite a high level goal into a list of more detailed goals. For example, consider the problem of creating a flow of chemical in the plant shown in figure 7. The high level goal is *flow(inlet, outlet)*. This can be rewritten by the inference engine as *open(inlet, middle)* and *flow(middle, outlet)*. The CEP code for this rewrite operator is shown in figure 8. After rewriting *flow(middle, outlet)* in a similar way, the goals needed to create the flow will be created.

It is important to note the distinction between planning and inference. The distinction is that planning actions happens in time while inference steps happen instantaneously. If a planning action was used to deduce *flow(inlet, outlet)* in the example above then *close(inlet, middle)* could be achieve between the point at which *flow(inlet, outlet)* was achieved and the point at which it was required. Of course the close operation would negate the flow condition, although the planner would not realise this. By using a deductive step, the goal is rewritten in place and so can not be clobbered as easily.

The use of inference in planning is not a new idea. Summary actions [Sac75] provide a more powerful representation of the same thing. We have chosen to use distinct inference and planning operators simply because we find this representation to be cleaner.

4.3.2 Valve Operations

Valve sequencing tasks require the use of valve operations to move chemicals around a process plant. There are two difficulties with these tasks. Firstly, inference is needed to find a sequence of valve operations that will create the desired chemical flow. Secondly, the OPS system must prevent undesirable streams and mixtures of chemical from forming. This second task is more difficult because a least commitment planner, like CEP, can work on a large number of plans at the same time and it is too time consuming to check all the effects that can be inferred for each of these possible plans.

One solution to this second problem is to make use of the valve sequencing algorithm system proposed in [LS90]. This algorithm has been optimised but it is basically a forward chaining search engine. At each stage in the search, the planner creates a complete description of the process plant and then decides how the new flow of chemical will disturb the plant. If the new flow will cause a mixture of undesirable chemicals, the planner breaks the problem into two subproblem. The first is to remove one of the undesirable chemicals and the second is to move from the end state for the first subproblem to the original goal state.

This approach is not ideal for CEP because it requires a complete knowledge about the state of the plant when the valve operation is to take place. It is very difficult for a backward chaining planner to provide complete state information. As a result, if this algorithm is used, all the valve sequencing tasks would have to be added after the body of the plan has been created. This would cause problems if a valve sequencing task was found to be over constrained because it would be difficult to communicate this failure back to the planner so that the planner could repair the plan quickly.

```
operator FindFlow
{
    pipe ?a;  pipe ?b;
    pipe ?c;

    expand
      * flow(?a, ?b)
    using
      open(?a, ?c)
      flow(?c, ?b)
    end
}
```

Figure 8: A CEP inference operator.

Another possible solution, which we have also not implemented, is to simplify the modelling of valve operations. In most operating procedures, chemicals are mixed only in special reaction vessels. Hence, two streams of chemical should not be directed down the same section of piping. Using this knowledge, we can remodel the effect of opening a valve as a process of adding a new section of pipe to the flow of a particular chemical. By closing all the valves surrounding the new pipe before connecting it to the existing flow, we can ensure that the flow will not expand into unchecked areas of the plant. In other words, we only have to worry about the mixing of chemical in the new pipe with the chemicals already in the flow. Using this new representation, opening a valve only has a simple effect given that a flow of chemical can be modelled as a simple condition.

5 AN EXAMPLE PROBLEM

In this section we look at how CEP can be used to generate an operating procedure to shutdown a compressor. The model operating procedure for this problem is taken from [Sut92], and is shown in figure 11. Our aim is only to have CEP generate the ordered sequence of instructions given in the Detailed Procedure section of the model solution. We are largely unconcerned about the formatting of the procedure that CEP creates.

The first section of the domain file for the compressor problem is a declaration of the objects in the domain. In CEP, objects are represented as instances in a frame hierarchy (see figure 9). Frames in this hierarchy represent sets of objects and should be viewed

```
frame(compressor).
instance(V-206 is a compressor,
        [ PressureGuage is PI-2065,
          Switch is SW-2066,
          Meter is M-2066 ]).
```

Figure 9: The Definition of an Object.

```
operator CheckIdle
{
    compressor ?c;

    achieve
      * checkIdle of ?c is true;
    using
      state of ?c is off;
    end

    print
      'Confirm the current through meter '
      (?m) [ Meter of ?c is ?m; ] ' is zero';
}
```

Figure 10: A STRIPS Style Operator.

TITLE	How to Shut Down C-206 The Off-Gas Recycle Compressor	MODULE NO. REVISION NO. DATE	ABC-200-246 2 04/08/1991

CHECKLIST

Close XV-2067.

Wait until V-206 pressure (PI-2065) is less than 35 psig (normally 2 minutes).

Stop compressor (Switch-2066).

Confirm current meter 2066 is zero.

Close discharge valve XV-2068.

Open XV-2069 until V-206 (PI-2065) pressure is zero.

Close XV-2069.

DETAILED PROCEDURE

200.24.13 Isolate the Suction Line

> C A U T I O N: The Isolation valve, XV-2067, may be partialy plugged with corosion products. Ensure that the valve is completely closed before stopping the compressor.

(1) Close valve XV-2067.

(2) Wait until the pressure is V-206 (local pressure gauge PI-2065) is less than 35 psi (this will usually take about 2 minutes). (PI-2065 is shown in Module ABC-200-126).

200.24.14 Stop the Compressor

(1) Stop the compressor by pressing switch SW-2066 on the outside control panel.

(2) Confirm that the current through meter 2066 is zero.

(3) Close the discharge valve, XV-2068 (Shown in Module ABC-200-121).

200.24.15 Vent the Compressor Casing to the Flare

(1) Open bleed valve XV-2069 until the pressure in V-206 is zero (indicated by PI-2065).

(2) Close Bleed Valve XV-2069.

Figure 11: The Compressor Problem.

as defining properties in the domain. For example, natural gas may belong to the set of flammable, the set of gases and the set of chemicals.

Slots in the object definitions are interpreted as facts which are expected to remain true throughout the plan. In the example shown in figure 9, *PressureGage of V-206 is PI-2065* is expected to remain true.

The second section of the domain file lists the inference operations in the domain. In this problem we use inference to break down high level requirements into the low level goals that the system can solve directly. For example, we have chosen to specify the goal of this procedure as "shutdown of V-206 is true" and to use inference operators to rewrite this goal as shown in figure 12.

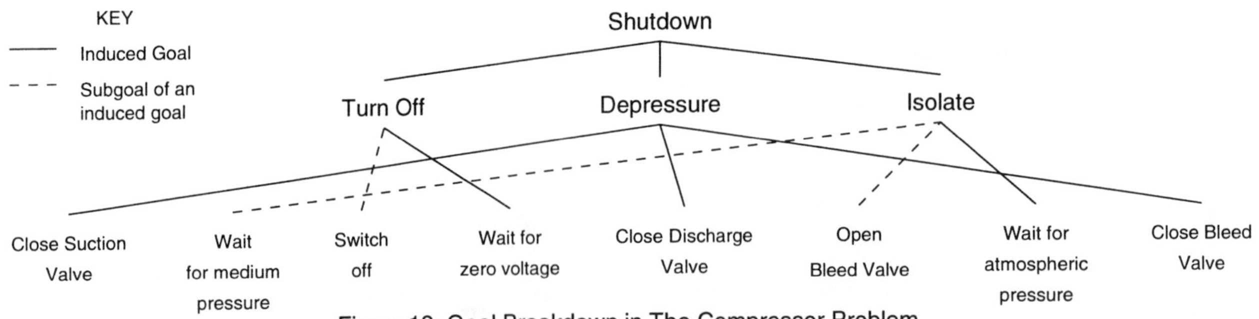

Figure 12: Goal Breakdown in The Compressor Problem.

The second section of the domain file also lists the planning operators in the domain. An example planning operator is given in figure 10. Planning operators in CEP are similar to STRIPS operators. The difference is that CEP uses a single effect list, the achieve list, rather than using STRIPS style add and delete lists. This change comes from the different representation of conditions in the two planners. STRIPS represents conditions as facts which can either be made true (added) or made false (deleted). CEP represents conditions as functional literals of the form *f(x)* is *y* such that *f(x)* is *?a* and *f(x)* is *?b* implies *?a = ?b* [see Ver83]. Hence, operators in CEP change the value of functions rather than asserting and retracting new conditions.

The final section in the domain file specifies the problem to be solved in terms of a start and end state. The goal state for the compress problem is simply "shutdown of V-206 is true". The start state for the compressor problem is a simple list of seven conditions.

Once CEP has been given this full problem description then it can start work. Computation for the compressor problem takes 0.15 seconds. The resulting output resembles the Detailed Procedure in figure 11 except for CEP's omission of the caution and the section headings.

The compressor problem was the first practical tasks that CEP was able to solve. After CEP could solve the compressor task, we extended the system to use global constraints. This changed the order in which the goals in some problems were solved. We were surprised to find that, when solving the compressor problem again, CEP now had no preference for the order in which the bleed valve was open and the discharge valve was closed. However, opening the bleed valve before closing the discharge valve is undesirable because it might cause back flow. We had previously forgotten to add a rule to prevent this back flow and the problem was easily fixed by adding a global constraint. This is a good example of how CEP's repeated use of rules in a knowledge base provides good opportunity to spot and corrected errors.

6 CONCLUSION

OPS was first studied in 1974 by Rivas and Rudd. At the time, computers were relatively slow and expensive and so the computational complexity of OPS made it impractical.

In 1987, there was a new burst of OPS research with the realisation that computational power was now much cheaper and more widely available. However, this burst seems to be floundering because of an inability to overcome the limitations of the original systems.

We have developed a new OPS architecture based on AI Planning. We have shown that this architecture can easily solve problems that previous system find difficult or impossible to solve. We believe that our OPS system is more efficient and better understood that the previous OPS systems, because our system is based on the large body of AI Planning research rather than on new planning techniques. We also believe that our OPS architecture is extendible to new problem areas.

REFERENCE

[AHT90] J. Allen, J. Hendler and A. Tate (Eds.). *Readings in Planning*. Morgan Kaufmann, San Mateo, California, 1990.

[CT91] K. Currie and A. Tate. O-Plan: The Open Planning Architecture. *Artificial Intelligence*, **52**, 49–86, 1991.

[FN71] R. E. Fikes and N. J. Nilsson. STRIPS: A New Approach to the Application of Theorem Proving to Problem Solving. *Artificial Intelligence*, **2**, 189–208, 1971. Also in [AHT90].

[FP87] R. H. Fusillo and G. J. Powers. A Synthesis Method for Chemical Plant Operating Procedures. *Computers chem. Engng.*, **11**(4), 369–382, 1987.

[IKPR80] V. A. Ivanov, V. V. Kafarov, V. L. Perov and A. A. Reznichenko. On Algorithmization of the Start-Up of Chemical Productions. *Computer Cybernetics*, **18**, 104–110, 1980.

[KUO81] A. Kinoshita, T. Umeda and E. O'Shima. An Algorithm for Synthesis of Operational Sequences of Chemical Plants. In *14th*

European Symposium on Computerized Control and Operation of Chemical Plants, pp. 363–368, Laxenburg near Vienna, 1981.

[LS88a] R. Lakshmanan and G. Stephanopoulos. Synthesis of Operating Procedures for Complete Chemical Plants—I. Hierarchical Structured Modelling for Non-Linear Planning. *Comput. chem. Engng.*, **12**(9/10), 985–1002, 1988.

[LS90] R. Lakshmanan and G. Stephanopoulos. Synthesis of Operating Procedures for Complete Chemical Plants—III. Planning in the Presence of Qualitative, Mixing Constraints. *Comput. chem. Engng.*, **14**(3), 301–317, 1990.

[RR74] J. R. Rivas and D. F. Rudd. Synthesis of Failure-Safe Operations. *AIChE Journal*, **20**(2), 320–325, 1974.

[Sac75] E. D. Sacerdoti. The Non-Linear Nature of Plans. In *Proc. IJCAI*, pp. 206–214, 1975. Also in [AHT90].

[Sou93] J. K. Soutter. Knowledge Based Planning Architectures for Operating Procedure Synthesis, a first year report. Department of Chemical Engineering, Loughborough University of Technology, 1993.

[Ste81] M. Stefik. Planning with Constraints–MOLGEN: Part 1. *Artificial Intelligence*, **16**, 111–140, 1981. Also in [AHT90].

[Sut92] I. S. Sutton. Writing Operating Procedures. In *International Conference on Hazard Identification Risk Analysis*, Rugby, UK, 1992. IChemE publication 1293.

[TH*89] S. Tomita, K. S. Hwang, E. O'Shima and C. McGreavy. Automatic Synthesizer of Operating Procedures for Chemical Plants by use of Fragmentary Knowledge. *Journal of Chemical Engineering of Japan,* **22**(4), 364–372, 1989.

[Ver83] S. Vere. Planning in Time: Windows and Durations for Activities and Goals. *Pattern Analysis and Machine Intelligence*, **5**, 246–267. Also in [AHT90].

Constraint

A REFINEMENT APPROACH TO SEARCH AND CONSTRAINT SATISFACTION PROBLEM

Bing Liu

Department of Information Systems and Computer Science
National University of Singapore
Lower Kent Ridge Road
Singapore, 0511
Republic of Singapore

E-mail: liub@iscs.nus.sg

ABSTRACT

A typical practical search problem has more than one objective to optimize. Some of these objectives may even conflict with one another. In a conventional approach, these objectives are combined to form a cost function, and a search technique or algorithm is then used to optimize the cost function. However, it is well known that such a combined cost function is very difficult to form. Even if it is formed, it is difficult to assess how reliable it is in reflecting the real requirements of the problem. In this paper, this traditional approach to search is questioned. A new heuristic technique is proposed that tries to meet different objectives in different search processes, rather than using a single cost function to consider them in a single search. Experimental results have shown this technique can substantially improve the final solution.

1. INTRODUCTION

Most realistic search problems are also optimization problems. Achieving a good solution, which may not be the optimal solution, is essential for almost every application. Typical formal approach to solving such an optimization problem is to form a cost function and to optimize the cost function. A typical heuristic method or knowledge-based method uses an evaluation function to evaluate each alternative state in the search process and then extends the state that is more likely to achieve a good solution. In other words, such an evaluation function is used to focus the attention of the search and to prune off those parts of the search space that are unlikely to contain a good solution. Pruning off search space is necessary because most practical search problems are NP-hard. Exhaustive search to find the best solution will be impossible. More advanced approaches in heuristic search also allow modification of evaluation functions during the course of search. This paper questions these approaches. In particular, we highlight the following two main problems:

- Reliability of cost functions or evaluation functions.

  In almost any realistic search problem, we have more than one objective. They typically interact with each other and even conflict with each other. To express them with a single cost function or an evaluation function may not be reliable as it is very difficult to see how they interact with each other and how their interactions affect the final result. Even if we manage to find the best solution to a problem regarding the cost function, it may not be the true best solution in the real world. Many researchers have realized this problem [e.g., Mah88, Mer84, Mor86, Smi86].

- Effectiveness of a single search process with a cost function or an evaluation function in solving a problem with multiple objectives.

  Most existing techniques for search use a single search process combined with sophisticated cost (or evaluation) functions to achieve a good solution. We believe this is ineffective in meeting all the objectives, especially those secondary objectives. The reason is that the primary objectives in a problem always dominate the cost function. The secondary objectives are not fairly considered. We believe more can be done to meet these secondary objectives. The example in Section 4 will illustrate the point.

In this paper, we propose a refinement approach to search, which concentrates on meeting different objectives in different search processes. Hence, the problem solving in this approach involves more than one search. Each new process progressively refines and improves the previous results. The key assumption of this method is that not every choice in the search space has the same effect on the final solution. Typically, choices in certain parts of the search space have a great impact on the final result while choices in the other parts have a less impact and thus can be manipulated with little effect on the structure of the final solution. We argue that this assumption is realistic in many practical search problems, and we believe this refinement technique is generic for improving search results.

The technique has a significant effect on how the cost functions or evaluation functions are formulated and how each objective in a problem is considered in the problem solving process. There will no longer be a single cost function that covers all objectives. Instead, each objective is considered semi-independently. Since the technique is based on multiple search processes, each process may also use a different search technique for meeting different objectives. Test results in a scheduling domain have shown that this technique can produce substantially better results than those produced with a traditional approach.

The paper is organized as follows: In the next section, we briefly review previous techniques for optimization. Section 3 presents the refinement approach. Here, we will first introduce the concept of constraint satisfaction range, and then discuss how constraint satisfaction ranges can be used to improve a solution. Constraint satisfaction range is actually the key to the refinement approach. Section 4 gives an example in the job shop scheduling domain. Some testing results are provided. These results are also compared with those from a system that uses a traditional approach. Section 5 discusses the related work, and Section 6 concludes the paper.

2. PAST WORK

A typical optimization problem in Operations Research (OR) can be formulated as follows:

Optimize:	<a cost function>
Subject to:	<constraint 1>
	<constraint 2>
	...
	<constraint n>

The cost function is a combination of the objectives of a problem written in a mathematical form. The constraints are the restrictions that must be satisfied in the problem solving process. After the problem is formulated, an optimization technique is used to optimize the cost function. The reliability of the cost function in relation to the actual problem is not a direct concern of OR. Instead, it is more concerned with the optimization algorithms.

Depth-first search, breadth-first search, best-first search, branch and bound, etc., are some of the search techniques that can be used for optimization. They are independent of how a problem is formulated, and may be used for solving many types of problems. Apart from these general techniques, there are also many specific algorithms for solving specific optimization problems.

Due to the NP-hard nature of the most practical search problems, it is almost impossible for a pure optimization algorithm to find the best solution. In real problem solving, many optimization problems are dealt with various heuristic search methods to improve search efficiency. Heuristic search improves the search mainly by ordering the choices so that the most promising ones are explored first. Various measurements are used to determine a reasonable ordering. There are many heuristic search methods, e.g., beam search and A* [Dec88] search, etc. These methods mainly differ in the type of measurements they use and how the search is conducted. Although the measurement that each technique uses may be different, they have similar effects. As normal cost functions, their effectiveness and reliability are also in question.

Another branch of the work on search is in the knowledge-based systems. Here, expert knowledge is used to guide the search as well as to prune the search space so that only a small space is searched. In these systems, evaluation functions as well as expert knowledge are used to measure the goodness of a state. The evaluation functions and the expert knowledge are similar to the measurements used in a normal heuristic search.

Many people have questioned the effectiveness of evaluation functions and cost functions. A number of suggestions to improve them have also been made. For example, evaluation functions should be modified during the course of the search according to the problem solving need [Mah88, Mer84, Mor86]. This helps in the situations where objectives do change from time to time, and we have the knowledge to change them. Not much work has been done in improving the results after a search process has been performed, and possibly using more than one search process to achieve a better solution. This paper suggests such a method. This method may be used in many domains. Its implementation is, however, domain dependent.

3. MEETING DIFFERENT OBJECTIVES WITH DIFFERENT SEARCH PROCESSES

This section presents the refinement approach. But before that, we first discuss various constraints, objectives and their relationships, and then introduce the concept of constraint satisfaction range, which is the key to the refinement approach.

3.1. Objectives and constraints in a search optimization problem

A typical optimization problem consists of a number of objectives to be optimized and a set of constraints to be satisfied. Constraints can be classified into two categories, i.e.,

- Hard constraints, and
- Soft constraints.

Hard constraints are those causal constraints that have to be satisfied in the search process if the result produced is to be legal. Soft constraints are those whose satisfaction can be relaxed during the search, but we wish to minimize their relaxation.

The objectives, in fact, are soft constraints. For example, meeting due date is one of the objectives of scheduling. It can be seen as a soft constraint as follows: If some due date cannot be met in scheduling, it has to be relaxed. In this situation, we would like to minimize the delay. Or, in other words, such a relaxation carries a cost and we wish to minimize the cost.

On surface, many objectives may not be easily translated into soft constraints, but they normally can be. For example, minimizing space usage S can be interpreted as a relaxible (or soft) constraint $S = 0$. Anything greater than 0 for the space usage S means a relaxation of the constraint and carries a cost. In fact, the cost function or evaluation function of a normal optimization problem is a combination of these soft constraints.

In the following discussion, constraints (soft and hard) will be used, which should be taken to represent both normal constraints and objectives. Meeting objectives will be seen as optimizing the satisfaction of their corresponding soft constraints.

3.2 Constraint satisfaction range

Constraint satisfaction range is the range within which a constraint remains satisfied. In most constraint satisfaction situations more than one position in the search space can satisfy a constraint, or typically, there is a range in which a constraint can remain satisfied. Within this range, choices can be manipulated in order to satisfy other constraints or improve their satisfaction, while also maintaining the existing constraint satisfaction. In other words, there may be some point in this range that the constraints that cannot be satisfied originally may also be satisfied. If a solution could be manipulated to reach such a situation, the final result produced will certainly be better. We call this the positive way of satisfying constraints.

For example, in the case of the due date for a job in scheduling, if it is day 5, anything between day 0 and day 5 satisfies the due date constraint. This gives us some space to manipulate. See Figure 1 below. It shows the schedules of a job, which has three operations. We can see clearly the constraint satisfaction ranges as there are spaces in between operations and in between operation 3 (Op3) which is the last operation and the due date. If there are other constraints to be satisfied, we can manipulate within these ranges to satisfy them while also maintaining the due date constraint satisfaction.

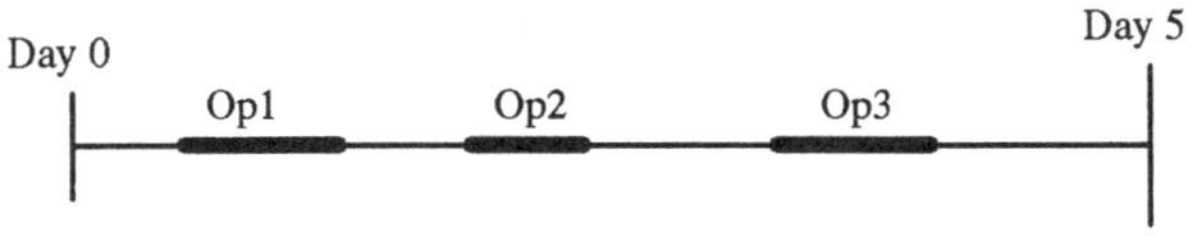

Figure 1. An schedule example to show the constraint satisfaction range

3.3 Refinement approach

In a search process, at each state a decision has to be made to choose the right move from a set of alternatives to forward the search. Obviously, the criterion for making such a decision will be the satisfaction of constraints. Hence, the most important thing to do at each choice point is to assess the constraint satisfaction situation on each alternative and to select the most favorable one according to the evaluation results.

However, constraints typically interact with and/or conflict one another [Liu93, Smi86]. For example, in scheduling, in order to meet due dates, it is good to schedule jobs earlier. However, this is likely to cause long work-in-process time. To reflect these conflicting concerns, compromise decisions often have to be made (soft constraints can be relaxed or compromised but not hard constraints). In order to achieve an effective compromise, reliable information about the satisfaction of the constraints involved and their interactions is essential. But, it is very difficult to assess these at a local search state as many of the constraint problems only manifest themselves after some or all of the search has been done. The refinement technique helps in these situations.

The refinement approach, as mentioned before, is characterized by satisfying different soft constraints with different problem solving processes. Indeed, problem solving in this method involves multiple search processes.

The essence of the approach is to find constraint satisfaction ranges and to use them in decision making for achieving better constraint satisfaction. But it is very hard to know what the satisfaction range is for a constraint when the system is actually trying to satisfy it since the system does not know what will happen in the future. In the refinement approach, the system tries to satisfy a subset of constraints in each search process while only considering the others opportunistically. After this subset of constraints (old constraints) has been considered, it will be easy to see what their satisfaction ranges are and how other constraints (new constraints) can be better satisfied within these ranges. Or, in other words, it will be easy to see how the new constraints may interact with the old ones and how these interactions may be used for better constraint satisfaction. Basically, the constraint satisfaction range serves as a basis for assessing future constraint interactions between the new constraints and the old constraints. Later processes analyze the results of the earlier searches to find out these constraint satisfaction ranges and to manipulate in them to satisfy the new constraints and also to maintain the satisfaction results of the old ones.

An important policy of this approach is that in considering the new constraints, the system assumes that previous searches have done their best to satisfy those old constraints and the new processes should not try to improve their satisfaction but only to maintain their satisfaction situations. The reason for this is that if the old constraints are allowed to be modified in the new processes, their old satisfaction ranges will become unreliable for satisfying the new constraints.

This heuristic technique depends on two important assumptions:

Assumption 1: In practice, feasible results are sufficient.

Obviously, the refinement technique cannot be used to find the best solution (assuming such a solution can be precisely defined). However, this assumption is realistic and is acceptable with almost any practical search application. The reason is that practical search problems are mostly NP-hard problems and it is very difficult to find the best solution. Worse still, it is very difficult to even define the optimal solution and the cost function.

Assumption 2: Not every choice in the search space affects the final solution a great deal. Only choices in certain parts of the search space significantly influence the outcome of the search.

This assumption ensures that the manipulation within the satisfaction range of a constraint to improve the satisfaction of new constraints is possible. Without this assumption, i.e., change of any choice will have a great impact on the search, then the old constraint's satisfaction cannot be maintained after the change. This makes the approach pointless.

This assumption is also realistic in practice because in most search problems it is always some outstanding issues that cause bad results (or there are certain subspaces in the search space that are critical to the final solution). For example, in planning, scheduling and resource allocation problems, it is always the limited resource capacities that restrict the satisfaction of various constraints. Without resource requirements or if resource capacities are infinite, constraint satisfaction will not be a problem. However, in most situations, not all the resources cause problems all the time. In certain periods of time, some resources will not influence the problem solving much as their capacities are well above the required capacities, while others will influence the problem solving a great deal as their capacities cannot meet the demand so that the satisfaction of some constraints will be hampered. Those time periods where resource capacities are well above the required capacities will not affect scheduling a great deal, while plans or schedules on those congested resources or bottlenecks are critical to the final solution.

3.4. How to use the approach

There can be at least two ways to use this approach:

- The first one is to find a new solution with a completely new search by employing the old solution as the guideline. The new process will preserve the satisfaction of the old constraints in the old result, and also improve the satisfaction of the new constraints. This method may significantly improve the final result because such a new search process may change the global picture of the solution. The example presented in the next section uses this strategy.
- The other one is to modify the old solution in order to improve it without using a completely new search process. This is normally called fine-tuning. This process will not destroy the previous solution. To do this, the system needs to find out which parts of the solution may be modified such that the new constraints could be satisfied without violating the old constraints. This is quite similar to our human experience in refining a solution by fine-tuning the result. However, such an improvement is quite limited because only local modifications are possible.

4. AN EXAMPLE FROM JOB SHOP SCHEDULING DOMAIN

In this section, we discuss a scheduling system RESS-II that uses the refinement approach. Its results are compared with those of an earlier version of the system,

RESS-I. RESS-I employs the conventional approach of using a single evaluation function with one scheduling process.

4.1. Scheduling problem

Generally, a scheduling environment can be described as: Suppose that it is necessary to execute q tasks, T_1, T_2, T_3, ... T_q, and each task T_i consists of a number of subtasks, t_1, t_2, ... t_n, where the subtasks have to be performed in a specified sequence and where each subtask requires a resource, to perform these subtasks we have k resources; each resource can only be used to perform one subtask at a time, but it can execute any number of subtasks in succession.

In a job shop, these tasks are the orders (i.e., customer orders for products); the subtasks are the operations in each order and resources are machines. Job shop scheduling is about prescribing when and where each operation is to be performed such that the production is carried out in a timely and cost effective manner. It is a difficult task. Much of the difficulty comes from the need attend to a large and diverse set of constraints and objectives [Liu93, Smi86].

The basic shop resources are machines. The basic shop constraints are:

(1) Objectives (soft constraints)

a. Global preference (soft) constraints

These constraints are the primary concerns of the factory. They are called global preference constraints because on one hand they are not necessarily satisfied in the scheduling process, and on the other, their satisfaction can be assessed only when all the scheduling has been done. They include:

meeting delivery due dates, and
minimizing work-in-process time.

b. Local preference (soft) constraints

These constraints are secondary objectives. They express the preferred choices among alternatives at each scheduling decision-making state. Their satisfaction will improve a schedule. This class of constraints includes:

minimizing set-up time, and
meeting machine preferences of operations

(2). Precedence (hard) constraint

This constraint constrains the ordering of operations to be performed, e.g., a precedence constraint specifies an operation must be performed before or after other operations.

(3). Resource availability (hard) constraint

This constraint says that a machine must be available before an operation is performed.

4.2 Using the refinement approach in scheduling

We have used the refinement approach in our scheduling system RESS-II. This system has, in total, 4 stages of scheduling. Each stage is a complete scheduling process. Later stages improve the results from the earlier stages. The first stage builds a capacity plan to show the possible resource problems or the conflict situation between the due date constraint and the resource availability constraint. The second stage uses this capacity plan to find out the satisfaction range of the resource availability constraint and to try to maximize the satisfaction of the due date constraint within this range. Its results also show the problems between the due date constraint and the work-in-process constraint. The third stage tries to reduce the work-in-process time by manipulating the schedule in the satisfaction range of the due date constraint obtained from the scheduling results of the second stage. The fourth stage improves the schedule produced from the third stage in order to minimize set-up time and to meet machine preferences of operations. A brief description of each stage of the scheduling is given below. For more details on RESS-II, please refer to [Liu93].

4.2.1 Building capacity plan

The capacity plan is built by assuming that every order will be finished on time, hence constructing a rough schedule based on machines without considering their finite capacity. The resulting rough schedule shows the capacities needed in different periods of time on different machines. This plan provides information on how the due date constraint and the resource availability constraint interact with each other, and consequently on the satisfaction ranges of the resource availability constraint in the different time periods.

4.2.2 Meeting due dates

This stage tries to satisfy the due date constraint of each order by using a capacity analyzer. This capacity analyzer analyzes the capacity plan to obtain the resource availability situation of the factory. This information serves as a guideline for decision making to satisfy the due date constraint. The resulting schedule also shows the satisfaction range of the due date constraint. The next step uses this result to reduce the work-in-process time.

4.2.3 Minimizing work-in-process time

After the second stage of scheduling is finished, the system has sufficient information to reduce the work-in-process time. The interaction between the due date constraint and the work-in-process constraint has been exposed in the result from this schedule. The objective of this stage is to take this schedule as guidelines to generate a schedule that will reduce the work-in-process time and also maintain the due date constraint's satisfaction from the second stage.

The rationale behind this is that typically at certain machines in certain periods of time (especially at bottlenecks), the waiting list is very long so that some of the operations have to wait for a long time before they are performed. This means that the satisfaction range of the due date constraint could be quite large due to these long waitings. It is also these long waiting times that cause long work-in-process times. Based on the information from the schedule at the second stage, if the orders that have long waiting operations can be started later, the work-in-process time will be reduced. Since the schedules on these bottleneck machines will be least affected and since this stage observes the due date constraint satisfaction range, its satisfaction from the last stage will still be maintained.

4.2.4 Minimizing set-up time and meeting machine preferences

This last stage improves the schedule constructed by trying to satisfy as many local preference constraints as possible. The reason for not considering the local preference constraints earlier is that they are heavily dependent on the local scheduling situation and their early consideration may hinder the satisfaction of those global objective constraints (due dates and work-in-process). After all the global objective constraints have been considered, minor adjustments to the schedule produced within the satisfaction ranges of the global constraints will improve the satisfaction of local preference constraints while also leaving the satisfaction of those more important constraints unchanged.

4.3. Experimental results

A number of experiments have been conducted with the scheduling system. The results from different stages of RESS-II are compared to see how the refinement approach influences the scheduling results. Along with this, we also compare it with an earlier version of the system (RESS-I) that combine all the objectives in a single evaluation function. As there is no actual factory case that is available to be used in this research, all the test data are generated. However, the generation was based on a general factory model [Hal80].

We have used three shop environments to test our system. Different shops consist of different sets of machines, different sets of tools and different sets of process routings (which describe how many kinds of product can be produced in the shop and what route each product should follow in manufacturing). Within each environment 12 different tests have been carried out, and each test schedules a set of orders. So, altogether, 36 tests have been performed in three environments. This section will show the results from the experiments in one shop environment.

The test results show that the average tardiness does not change much with different systems (RESS-I and RESS-II) and with different stages of RESS-II. The reason for this is that although RESS-II has more stages of scheduling than RESS-I, only its second stage (which is almost the same as the detailed scheduling in RESS-I) tries to satisfy the due date constraints and the other stages maintain this satisfaction result. The figure about tardiness is not shown here as there is little change. Below, we give three figures. They represent comparison about work-in-progress time, the number of set-ups saved and machine preferences met, respectively.

Figure 2 shows the average work-in-process time per order over different tests. We can see that the work-in-process time has been dramatically reduced by RESS-II. The result from the second stage of RESS-II is nearly the same as the result from RESS-I. The explanation for this is that although RESS-I considers work-in-process time in its evaluation function, it does not have a clear picture about its interaction with due dates and consequently its decision making concentrates on the most important constraint, i.e., the due date.

Similarly, Figure 3 and Figure 4 illustrate that the number of set-ups saved and machine preferences met have also been improved substantially by RESS-II over RESS-I.

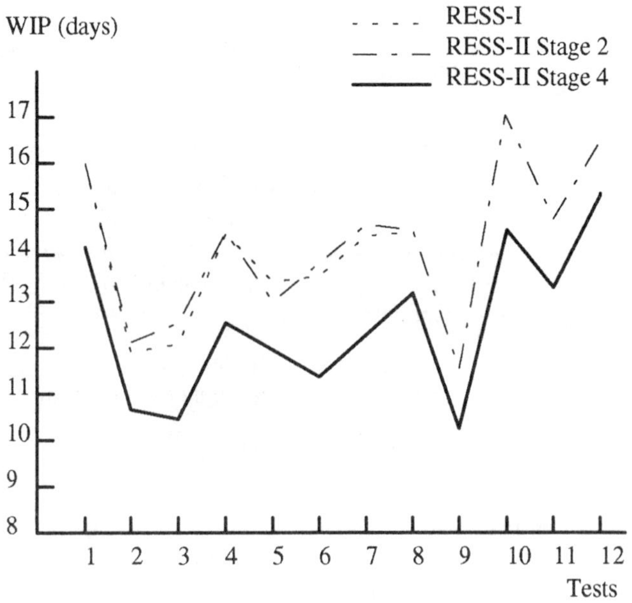

Figure 2. Average work-in-process (WIP) time per order over different tests

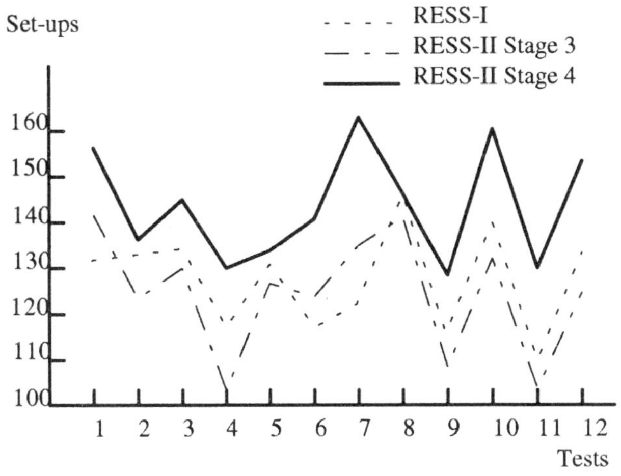

Figure 3. Number of set-ups saved over different tests

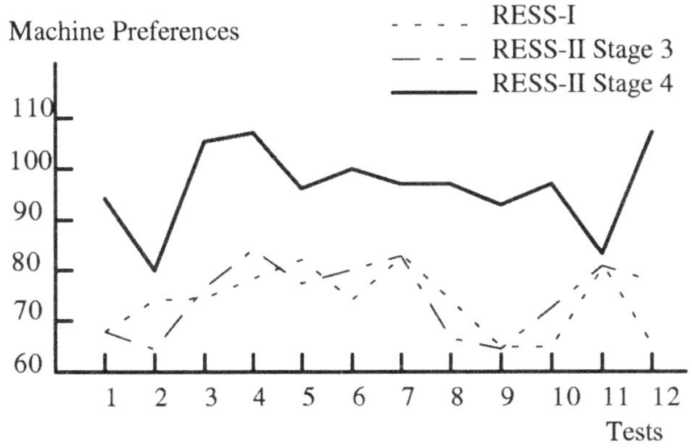

Figure 4. Number of machine preferences met at different states over different tests

5. RELATED WORK

To the best of our knowledge, not much work has been done in using multiple search processes to improve constraint satisfaction and to meet search objectives. Most systems use only one search process with some evaluation function to produce a result. They assume that through this evaluation a good compromise will be reached among the competing objectives. However, if a system does not know what the constraint satisfaction situation is and where the conflicts among constraints may occur, it will have no ground on which to arrive at a good compromise decision. The compromises produced will not be reliable. We have shown this can be improved with our refinement approach.

Some practical systems [e.g., Cho93] have used local refinement approach (or fine-tuning) to improve the solution (without using a new search process). Local manipulation can be quite useful in many situations. We human beings do this quite often. However, unlike the global manipulation, local approach has only limited effects on the final result.

The closest approach to our global refinement is the capacity analysis using in many scheduling systems [e.g., Sad93, Smi86]. The capacity analysis basically shows where the resource contentions are likely to be. This information is used in the search process to guide the decision making. However, this is often not sufficient because the capacity analysis can only show one type of problem, i.e., resource bottlenecks. It cannot show the interactions of other constraints. In fact, our scheduling system also has a capacity analyzer, and we have illustrated that the result produced with capacity analysis can still be further improved.

Our approach is clearly different from the technique of modifying evaluation functions [Mah88, Mer84] in the course of the search. We believe the refinement approach can also be used to improve this type of search.

6. CONCLUSION

In this paper, a refinement approach to search is presented for solving search problems involving multiple objectives. We do not claim that this technique is universally applicable. In fact, it will not work with the problems in which any modification to choice making could completely change the picture of the final result, as this technique depends on the later stages to preserve the satisfaction results of the earlier ones while trying to satisfy the new constraints. In these problems, the results of the previous stages are impossible to preserve. Indeed, the refinement approach depends on two key assumptions discussed in Section 3. However, we argue that these assumptions are reasonable for many problem domains, such as planning, scheduling and resource allocation. In these domains, it is always those congested resources that affect greatly the final results. Planning or scheduling on the other resources are not that critical. This gives us freedom to satisfy more constraints with minimal effects on the structure of the solution.

We believe the refinement approach is a good heuristic technique for solving practical planning, scheduling, and resource allocation problems. So far only the scheduling domain has been used in our testing. In our future work, we will also try it in other domains.

REFERENCE

[Cho93] K.P. Chow andd C. K, Hui, "Knowledge-Baseed System for Rostering," Expert Systems with Applications: An International Journal, Vol 6. No. 3, pp.361-377, 1993.

[Dec88] R. Dechter and J. Pearl, "The Optimality of A*," in Search in Artificial Intelligence, L. Kanal and V. Kumar (Eds), New York, Springer-Verlag, pp. 166-199, 1988.

[Hal80] Halevi, G. The Role of Computers in Manufacturing Processes, John Wiley & Sons, Inc. 1980,

[Kor88] R. Korf, "Optimal Path Finding Algorithm," in Search in Artificial Intelligence, L. Kanal and V. Kumar (Eds), New York, Springer-Verlag, pp. 223-267, 1988.

[Liu93] B. Liu, "Knowledge-based Factory Scheduling: Resource Allocation and Constraint Satisfaction," International Journal of Expert Systems with Applications, Vol.6, No. 3, pp349-359, 1993.

[Mah88] A. Mahanti and R. Ray, "Network Search Algorithms with Modified Heuristics," in Search in Artificial Intelligence, L. Kanal and V. Kumar (Eds), New York, Springer-Verlag, pp. 200-222, 1988

[Mer84] L. Mero, "A Heuristic Search Algorithm with Modified Estimate," Artificial Intelligence, Vol 23, pp.13-27, 1984.

[Mor86] T.E. Morton, MRP-STAR: PATRIARCH's Planning Module, Graduate School of Industrial Adminsistration, Carnegie-Mellon University, Technical Report, 1986.

[Nad88] B. Nadel, "Tree Search and Arc Consistency in Constraint Satisfaction Algorithms", in Search in Artificial Intelligence, L. Kanal and V. Kumar (Eds), New York, Springer-Verlag, pp. 287-342, 1988.

[Sad90] N. Sadeh and M. Fox, "Variable and Value Ordering Heuristics for Activity-Based Job-Shop Scheduling". Proceedings of the Fourth International Conference on Expert Systems in Production and Operations Management, 1990.

[Sad93] N. Sadeh, "MICRO-ROSS: A Micro-Opportunistic Factory Scheduler,", Expert Systems with Applications: An International Journal, Vol. 6. No. 3, pp.377-393, 1993.

[Smi86] Stephen F. Smith, M. Fox & Peng. Si, Ow, Constructing and Maintaining Detailed Production Plans: Investigations into the Development of Knowledge-Based Factory Scheduling Systems. Al Magazine, pp.46-61, 1986.

A TEMPORAL CONSTRAINT SATISFACTION TECHNIQUE FOR NONLINEAR PLANNING

Chung-Ming Cheng and Cheng-Seen Ho
Department of Electronic Engineering
National Taiwan Institute of Technology
43 Keelung Road, Section 4
Taipei, TAIWAN 106

ABSTRACT

A constraint satisfaction technique centered in the interval-represented temporal knowledge is proposed to support nonlinear planning. Given rough user-specified temporal information for a task, this approach employs a temporal propagation procedure to properly refine the temporal information of each action. The information is then used to help order the actions. This technique is demonstrated in a conflict resolution procedure, which integrates the technique in the resource-level conflict detection and resolution scheme to support nonlinear planning. The major advantage of this approach is that the integration of temporal constraints and resource usage help dramatically reduce the search space of a nonlinear planner. Using **IQTCN** as a representation and reasoning structure for the temporal knowledge in the technique makes the reasoning on temporal constraints effective and efficient, which further improves the performance of a nonlinear planner.

1. INTRODUCTION

Nonlinear Planning was originally proposed as a least commitment strategy for operator ordering [Sac77]. It generates parallel subplans for each goal state and then detects and resolves the conflicts among the subplans. A conflict occurs when the precondition of an operator is destroyed by the postcondition of another one. Several well-known systems have been reported to employ specific techniques trying to solve the conflicting problem, e.g. **NOAH** [Sac77] using TOME, **NONLIN** [Tat77] using GOST, **SIPE** [Wil84], and **TWEAK** [Cha87] using the necessary truth criterion. These systems, however, have invested too much in detecting conflicts. The resource-level conflict detection and resolution scheme, proposed by Lee and Chung, detects the conflicts when any two parallel resource-usage branches compete or share the same resources [LC89]. It, thus, avoids many unnecessary searches.

This paper goes one step further by studying how temporal knowledge can be integrated with the resource-level information to help perform nonlinear planning. We will propose an interval-based quantitative temporal constraint network (**IQTCN**) to represent the temporal knowledge in a nonlinear planner. It stores the temporal constraints about the actions and their relationships. It provides two reasoning mechanisms, including temporal propagation and actions ordering. Given rough initial temporal information about a task, the temporal propagation mechanism can refine the information. The actions ordering mechanism then uses the information to order the actions. We will demonstrate the technique in a conflict resolution procedure, which integrates the technique into the resource-level conflict detection and resolution scheme to support nonlinear planning.

Section 2 will introduce our architecture for a nonlinear planner supported by the technique as the conflict detection and resolution scheme. Section 3 describes the kernel of the technique, i.e., the **IQTCN** and its related reasoning mechanisms. The complete conflict resolution procedure is then described and exemplified in Section 4.

2. OUR ARCHITECTURE FOR NONLINEAR PLANNING

A nonlinear planner usually generates a partial-ordered plan via backward planning, during which a cycle of *linking*, *expansion*, and *conflict detection/resolution* is repeated until all the involved subgoals are solved [LC89]. Fig. 1 shows our conceptual architecture for a nonlinear planner. A task in terms of its desired goals is fed to the goals manager (GM). GM

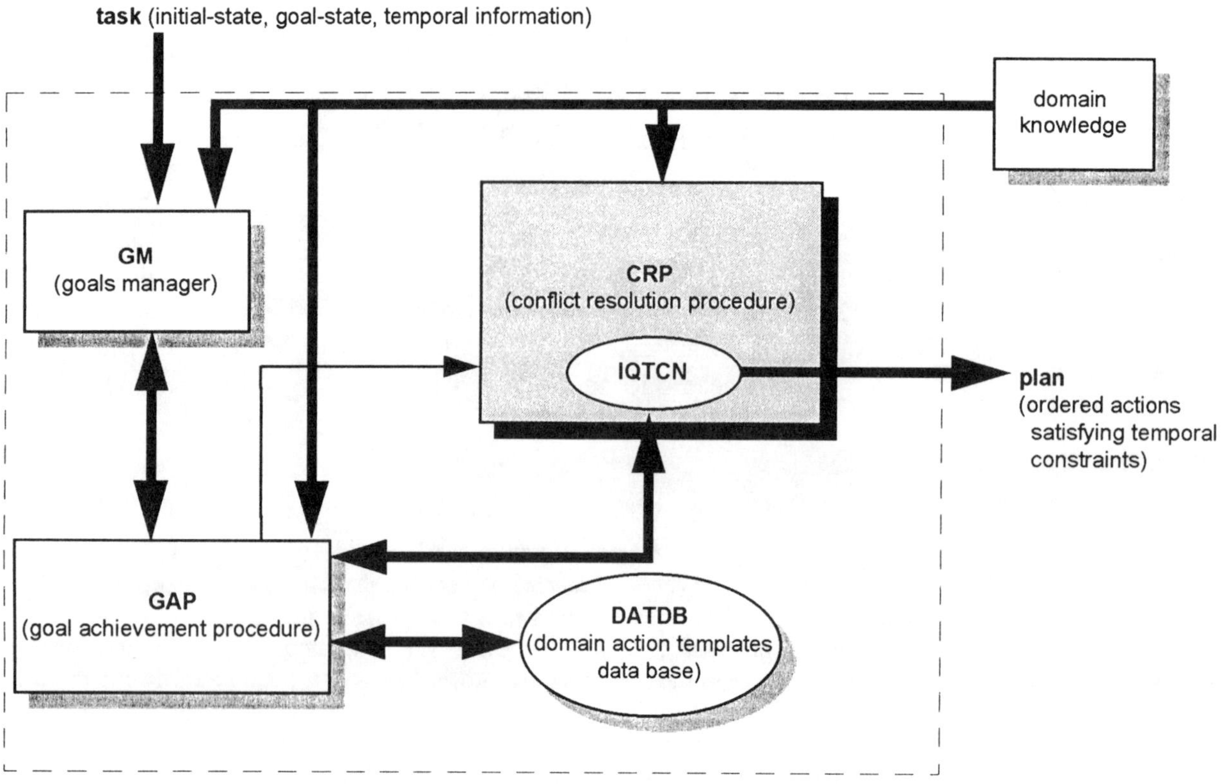

FIGURE 1 Conceptual architecture of a nonlinear planner

manipulates the goals by dispatching them to the goal achievement procedure (GAP) according to the domain knowledge. GAP tries to achieve the designated goal from GM. If the designated goal can't be solved under the current situation, it will be returned to GM. If some actions can be proposed to solve the goal, their preconditions will be checked against the current situation. They will be sent back to GM as new goals, if they can't be solved. Conflict resolution procedure (CRP) is invoked whenever a new action is proposed by GAP. It is responsible for ordering the proposed action according to the relevant temporal constraints into a network, represented by **IQTCN**. When all the goals in GM are solved, planning is terminated. A final **IQTCN** is returned as a plan, which contains ordered actions with temporal constraints for the user-specified task.

CRP employs resource-level conflict detection and resolution scheme, which uses a table called Resource-Usage Flow Table (RUFT) to record the usage of each individual resource during planning. There are two kinds of resources, i.e., active resources and passive resources, in the table. The passive resources are resources to be manipulated by the active resources. For example, robots are active resources and blocks are passive resources in the blocks world. An action may involve several passive and active resources. Each field in the RUFT contains a RUF-structure (a partially ordered network) for a resource. When a new action is added to **IQTCN**, the corresponding resources' RUF-structures need to be updated. Each updated RUF-structure will be checked to see whether it needs to be serialized with any other parallel branches through the causal relationship. If so, a conflict is detected and resolved by merging the two resource-usage branches into a larger one. The RUFT will be properly updated accordingly.

3. DETAILS OF IQTCN

3.1 BASIC CONCEPTS

First, an **IQTCN** represents the temporal relationships among the actions in terms of nodes. There are three fields in a node to store the triggering time τ_{tr}, execution time τ_e, and deadline time τ_{dl} of an action [HH94]. Each temporal constraint is represented by an ordered pair. For example, "Triggering a_j at $\tau_{trj}=(\xi_{trj1},\xi_{trj2})$" means that a_j must be triggered in the interval between

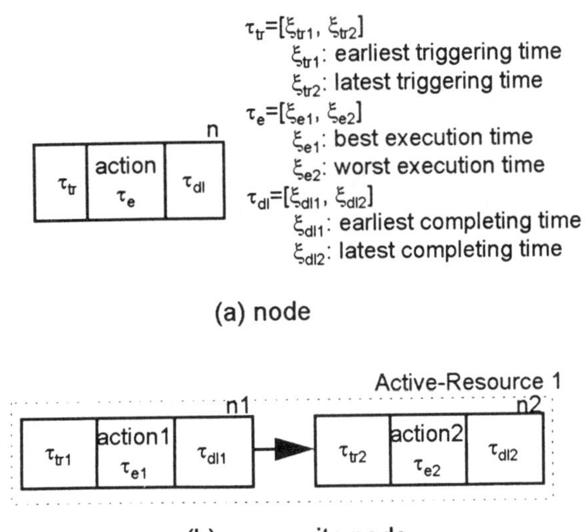

FIGURE 2 Notations used in IQTCN

ξ_{trj1} and ξ_{trj2}. "Completing a_j at $\tau_{dli}=(\xi_{dlj1},\xi_{dlj2})$" means that a_j must be completed in the interval (ξ_{dlj1},ξ_{dlj2}). Finally, "execution time for a_j is $\tau_{ej}=(\xi_{ej1},\xi_{ej2})$" meaning it may take at least ξ_{ej1} and at most ξ_{ej2} to execute a_j. In other words, ξ_{ej1} is the best and ξ_{ej2} is the worst execution time of a_j. Fig. 2(a) details the node structure. For explanation convenience, a number is labeled on the upper right of each node. If node n1 precedes node n2, an arrow will point from n1 to n2. If no relationship is known, there are no arcs between them.

If two actions are in the RUF-structure of an active resource, and no other actions, which involve the same active resource, can be inserted between them, then we can form a *composite node* containing these two nodes as shown in Fig. 2(b). This is actually an undecomposable subplan for this particular active resource, and can be treated as a single macro operator during subsequent planning. This reduces the planning space.

One more basic concept. If a new node n1 is proposed to satisfy the preconditions of node n2, its triggering and deadline time must be constrained by τ_{tr2}, the triggering time of n2. This involve two possible cases. Fig. 3(a) shows the one where n1 is completed during τ_{tr2}. This case brings to the surface some "legitimate" constraints on the temporal intervals. For instance, Fig. 3(a) says $\tau_{dl1} \leftarrow \tau_{tr2}$, meaning τ_{dl1} must be modified to be the same as τ_{tr2} under the circumstance. Fig. 3(b) shows the other case where n1 completes before τ_{tr2} and the associated temporal constraints.

3.2 TEMPORAL PROPAGATION

Now let's explain the first reasoning mechanism, temporal propagation, on an **IQTCN**. In an **IQTCN**, if n1 precedes n2, n1 must be completed before n2 is triggered. That is, the temporal relationship between τ_{dl1} and τ_{tr2} must be one of {eq,b,m,o,s,fi}. (We use Allen's temporal logic [All83] to describe the relationships among temporal intervals.) If their relationship is not one of the cases, there exits a temporal conflict between τ_{dl1} and τ_{tr2}. To resolve the conflict, τ_{dl1} and τ_{tr2} need to be adjusted. Table 1 summarizes the rules for temporal conflict resolution.

If the temporal information of a node is refined, this should be propagated to the other nodes of the **IQTCN**. Algorithm **Temporal-Propagation** does the job.

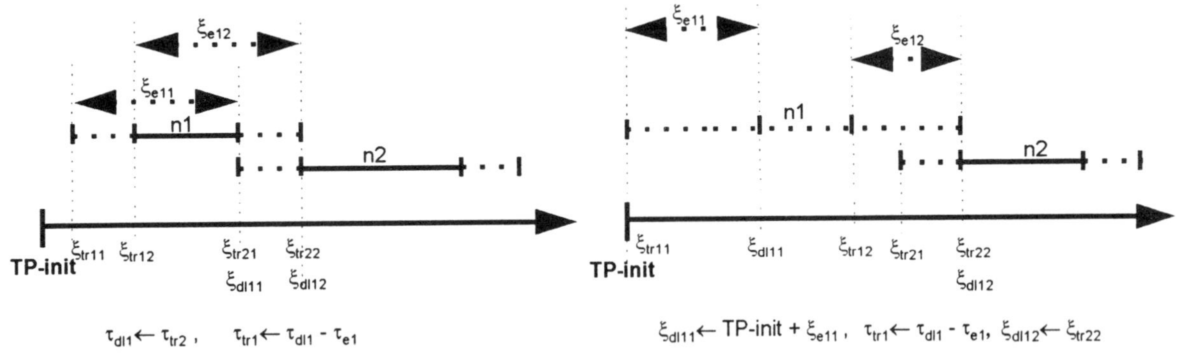

(a) n1 completed during the triggering time of n2

(b) n1 completed before n2

FIGURE 3 Temporal constraints for a newly proposed node

Algorithm **Temporal-Propagation**
Input: node **N**, **IQTCN**
Output: **IQTCN**

1. *Push*(**Stack**, **N**)
2. while **Stack** not empty do
3. begin
4. $\mathbf{N_{now}} \leftarrow$ *Pop*(**Stack**)
5.   for each $\mathbf{N_f}$ in father($\mathbf{N_{now}}$) do
6.   begin
7.   if $\mathbf{N_f}$ = **START** then go to 12
8.   *Temporal-Conflict-Resolution-Rules*($\mathbf{N_f}$,$\mathbf{N_{now}}$)
9.   if 'error' then return "temporal error"
10.   if '$\mathbf{N_f}$ change' then *Push*(**Stack**, $\mathbf{N_f}$)
11.   if '$\mathbf{N_{now}}$ change' then *Push*(**Stack**, $\mathbf{N_{now}}$); go to 21
12.   end /* end of for */
13.   for each $\mathbf{N_c}$ in child($\mathbf{N_{now}}$) do
14.   begin
15.   if $\mathbf{N_c}$=**FINISH** then go to 20
16.   *Temporal-Conflict-Resolution-Rules*($\mathbf{N_{now}}$,$\mathbf{N_c}$)
17.   if 'error' then return "temporal error"
18.   if '$\mathbf{N_c}$ change' then *Push*(**Stack**, $\mathbf{N_c}$)
19.   if '$\mathbf{N_{now}}$ change' then *Push*($\mathbf{N_c}$, $\mathbf{N_{now}}$); go to 21
20.   end /* end of for */
21. end /* end of while */

Procedure *Temporal-Conflict-Resolution-Rules* at statements 8 and 16 uses Table 1 to resolve temporal conflicts.

Table 1 Rule set for temporal conflict resolution

	relation between τ_{dl1} and τ_{tr2}	resolution
n1 precedes n2	bi, mi	error
	eq, b, m, o, s, fi	no conflict
	d	$\xi_{tr21} \leftarrow \xi_{dl11}$
	f	$\xi_{tr21} \leftarrow \xi_{dl11}$
	di	$\xi_{dl12} \leftarrow \xi_{tr22}$
	si	$\xi_{dl12} \leftarrow \xi_{tr22}$
	oi	$\xi_{tr21} \leftarrow \xi_{dl11}$ $\xi_{dl12} \leftarrow \xi_{tr22}$
n1 precedes n2 and n1 and n2 to be composed	b, m, bi, mi	error
	eq	no conflict
	o	$\xi_{dl11} \leftarrow \xi_{tr21}$ $\xi_{tr22} \leftarrow \xi_{dl12}$
	s	$\xi_{tr22} \leftarrow \xi_{dl12}$
	d	$\xi_{tr21} \leftarrow \xi_{dl11}$ $\xi_{tr22} \leftarrow \xi_{dl12}$
	f	$\xi_{tr21} \leftarrow \xi_{dl11}$
	oi	$\xi_{tr21} \leftarrow \xi_{dl11}$ $\xi_{dl12} \leftarrow \xi_{tr22}$
	di	$\xi_{dl11} \leftarrow \xi_{tr21}$ $\xi_{dl12} \leftarrow \xi_{tr22}$
	si	$\xi_{dl12} \leftarrow \xi_{tr22}$
	fi	$\xi_{dl11} \leftarrow \xi_{tr21}$

3.3 ACTIONS ORDERING BY TEMPORAL CONSTRAINTS

The resource-level conflict detection and resolution scheme may suffer from a condition, where two resource usage branches for a passive resource can not be ordered due to temporarily insufficient resource usage constraints. This may cause some subsequent backtracking operations. If each action is equipped with proper temporal constraints, we can solve the problem at an earilicar stage and save the backtracking process. This is because, for an active resource, the usage flow from the initial to the goal state must be serial. Thus, via the comparison of the completion times of the actions, we can order their relationships. The following **Ordering-Actions-by-Temporal-Constraints** algorithm does the job. As a matter of fact, this algorithm can be used in another case, where there exist no other constraints to order the actions of a plan except temporal information.

Algorithm **Ordering-Actions-by-Temporal-Constraints**
Input: new node N, **IQTCN**
Output: **IQTCN**

1. **NAR** ← *Active-Resource*(**N**)
2. (N_f, N_b) ← *Composite-Node*(**N**)
3. *Push*(**Stack**, child(**START**))
4. while **Stack** not empty do
5. begin
6. **U** ← *Pop*(**Stack**)
7. if Visited(**U**)=true or **U** ∈ *descendant*(**N**)
8. then go to 21
9. if *Active-Resource*(**U**) ≠ **NAR** or **U** ∈ *ancestor*(**N**)
10. then go to 19
11. (U_f, U_b) ← *Composite-Node*(**U**)
12. if $\tau_{dlN} < \tau_{dlU}$ then
13. father(U_f)= N_b ∪ father(U_f) - *ancestor*(N_b);
14. child(N_b)= U_f ∪ child(N_b) - *descendant*(U_f);
15. go to 20
16. if $\tau_{dlU} < \tau_{dlN}$ then
17. father(N_f)= U_b ∪ father(N_f) - *ancestor*(U_b);
18. child(U_b)= N_f ∪ child(U_b) - *descendant*(N_f);
19. *Push*(**Stack**, child(**U**))
20. Visited(U) ← true
21. end /* end of while */

Note that for any node N, *Active-Resource*(N) returns the active resources involved in node N. *Composite-node*(N) returns the first and last nodes of the composite node that includes node N.

4. THE CONFLICT RESOLUTION PROCEDURE (CRP)

Now we can describe the conflict resolution procedure that integrates the technique into the conflict detection and resolution scheme. When a new action is added to an **IQTCN**, the following basic steps are conducted to detect and resolve possible conflicts.

Step 1: Modify the **RUFT** to account for the resource usage of the newly added node.

Step 2: Determine the temporal constraints (triggering and deadline time) for the new node, according to Fig. 3.

Step 3: According to the new status of resource usage in **RUFT**, order the nodes in **IQTCN**. Execute **Temporal-Propagation** algorithm in the **IQTCN**.

Step 4: Use **Ordering-Actions- by-Temporal-Constraints** algorithm to order under-constrained nodes in **IQTCN**.

Step 5. If we have re-ordered the nodes, re-do **Temporal-Propagation** algorithm.

Let's take the blocks world as an example to briefly explain how our system (Fig. 1) works. Fig. 4 shows the example. There are two kinds of blocks and two robots R1 and R2. R1 is responsible for picking up the small blocks, while R2 is for picking up the large ones. User-specified temporal information includes several deadlines for the subgoals and a deadline for the whole task. We assume that the execution time of every domain action is known, e.g. stack(a,b) takes τ_e=[15,18]. Now, the system can be used to perform planning on the example. Fig. 5 shows the final IQTCN and the associated RUFT.

5. CONCLUSIONS

We have proposed a new technique that integrates temporal constraints and resource usage to support conflict resolution in nonlinear planning. It provides an interval-based quantitative temporal constraint network (**IQTCN**) to represent and reason about the temporal knowledge. Its related reasoning mechanisms include temporal propagation and actions ordering. By means of temporal propagation, the temporal information of each action can be made more precise. The refined temporal information can be used to order the actions in a plan.

The major advantage of this approach is that the integration of temporal constraints and resource usage

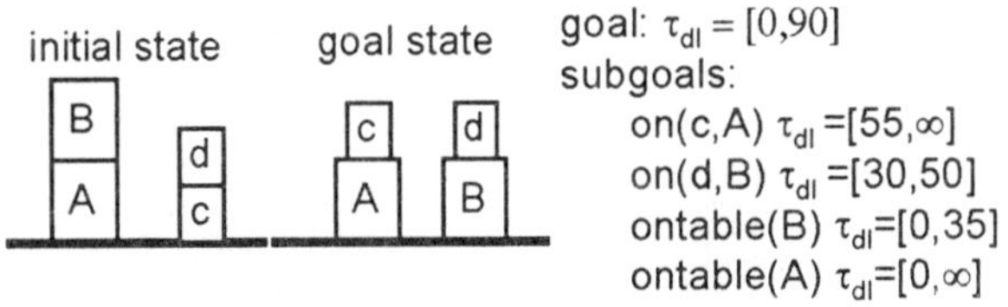

FIGURE 4 A blocks world example

help dramatically reduce the search space of a nonlinear planner. Using **IQTCN** as a representation and reasoning structure for the temporal knowledge in the technique makes the reasoning on temporal constraints effective and efficient, which further improves the performance of a nonlinear planner.

REFERENCES

[All83] J. F. Allen, "Maintaining Knowledge about Temporal Intervals," *Communications of ACM*, vol. 26, no. 11, pp. 832-843, 1983.

[Cha87] D. Chapman, "Planning for Conjunctive Goals," *Artificial Intelligence*, vol. 32, pp. 333-377, 1987.

[HH94] C. P. Hwang and C. S. Ho, RTPN: A Knowledge Representation and Verification Tool for Real-Time Systems, Tech. Rept. NTIT-ET-TR94004, Department of Electronic Engineering, National Taiwan Institute of Technology, Taipei, Taiwan, 1994.

[LC89] S. Lee and K. Chung, "A Parallel Architecture for AI Nonlinear Planning," *IEEE International Workshop on Tools for Artificial Intelligence*, pp. 51-58, 1989.

[Sac77] E. D. Sacerdoti, A Structure for Plans and Behavior. American Elsevier, 1977.

[Tat77] A. Tate, "Generating Project Networks," *Proceedings of the Internation Joint Conference on Artificial Intelligence*, pp. 888-893, 1977.

[Wil84] D. E. Wilkins, "Domain-independent Planning: Representation and Plan Generation," *Artificial Intelligence*, vol. 22, pp. 269-301, 1984.

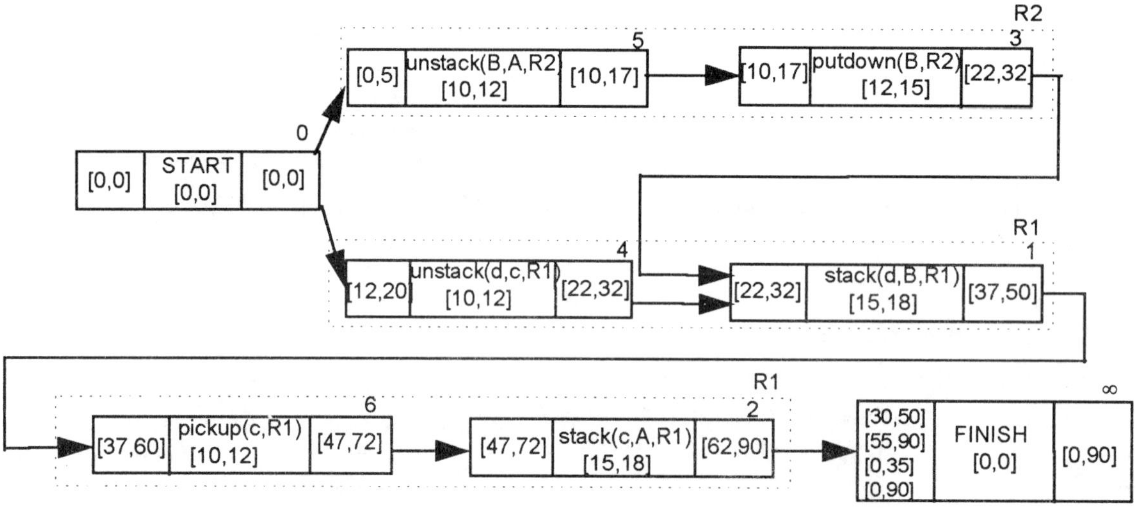

(a) Interval-based quantitive temporal constraint network (IQTCN)

Resource type	Resource Name	Partial Ordered Resource-Usage Network
Passive	A	5-2
	B	5-3-1
	c	4-6-2
	d	4-1
Active	R1	4-1 6-2
	R2	5-3

(b) Resource-Usage Flow Table (RUFT)

FIGURE 5 Final IQTCN and RUFT for the example

ARC-CONSISTENCY FOR DYNAMIC CONSTRAINT SATISFACTION PROBLEMS OVER REAL INTERVALS

Duong Tuan Anh and Kanchana Kanchanasut
Computer Science Program, School of Advanced Technologies, Asian Institute of Technology
GPO Box 2754 Bangkok 10501, Thailand
Email: {anh@emailhost,kk@cs}.ait.ac.th

ABSTRACT

This paper presents a method to deal with constraint deletion in Constraint Satisfaction Problems (CSPs) over real intervals in which we have to redo only the local propagation in the part of the constraint network dependent on the deleted constraint while parts independent of the deletion are not reconsidered. From this result we can propose an efficient algorithm to maintain arc-consistency in Dynamic CSPs over real intervals.

1. INTRODUCTION

Arc-consistency is the most commonly used technique for solving constraint satisfaction problems (CSPs) over finite domains. One of the typical arc-consistency algorithm for finite domains is AC-3 (see [Mac77]). Some authors, such as Davis [Dav87], Older & Vellino [OlV90], Hyvonen [Hyv89], and Sidebottom & Havens [SiH92] developed arc-consistency algorithms for CSPs where the domains of the variables are real intervals. These algorithms make use of Interval Arithmetic [Moo66] to compute the refined domains. However, all the techniques already developed for solving CSPs over real domains deal only with the problems where there is one fixed set of constraints (also called Static CSPs).

Many real life problems involve reasoning in dynamic environments. For example, in a design process, the designer may add more constraints to specify further the problem, or relax some constraints when there is no solution. Every time a constraint has been added or removed we need to check if there still exists solutions in the CSP. We can modify the traditional arc-consistency algorithm for static CSPs to become the algorithm to handle the addition of a single constraint. But the main difficulty in developing arc-consistency algorithm for dynamic CSPs is the problem involving the deletion of a constraint: how to avoid resolving the remaining system from scratch when a constraint is removed from the CSP.

In this paper we propose a method to deal with constraint deletion in which a reexecution from scratch can be avoided and describe an incremental algorithm to maintain arc-consistency in a dynamic CSP over real intervals.

The paper is organized as follows. Section 2 reviews some basic definitions and previous results on CSPs. Section 3 presents the algorithm achieving arc-consistency for Dynamic CSPs over real intervals. Related research in arc-consistency algorithm for Dynamic CSPs over finite domains and real intervals is briefly outlined in section 4. Section 5 contains a summary and some final remarks.

2. DEFINITIONS

In this section we recall some basic definitions and previous results on arc-consistency for real intervals.

2.1 Real Intervals

A *real interval* is specified by a *lower* and an *upper bound*. A bound consists of a real numeral and a bracket symbol. A square bracket indicates that the bound is closed and a round bracket indicates the bound is open. For instance, [0,1] denotes the set $\{ x \mid 0 \leq x \leq 1 \}$ and [0,1) denotes $\{ x \mid 0 \leq x < 1 \}$.

2.2 Constraint Satisfaction Problems over Real Intervals

A *static* CSP involves a set of *variables* V and a set of *constraints* S. Each variable X in V takes value in its respective real *domain* D_x. A constraint specifies which values from the domains of its variables are compatible. Here constraints can be in form of equations and inequalities.

A solution of a CSP will be the assignment of the form: $x_1 \leftarrow I_1$, $x_2 \leftarrow I_2$,.. $x_i \leftarrow I_i$,...where each I_i is a real interval. That means we deal with interval constraint satisfaction (ICS) in which variable domains are real intervals.

Example 2.1. Suppose we have the CSP over real intervals $D_X = [1,10]$, $D_Y = [3,8]$, $D_Z = [2,7]$, $D_U = [0,15]$ and $D_V = [-20,20]$ and the constraints

$$x + y = z \text{ (C1)} \quad y \leq x \text{ (C2)} \quad u = 2*v \text{ (C3)}$$

Let v(C) denote the set of variables in constraint C. If $v(C) = \{X_1,...,X_k\}$, we write $C(X_1,...,X_k)$ to make explicit the variables of C. The constraint network of a CSP is described here as a *directed hypergraph* where variables are associated with nodes and each constraint C is associated with a set of *directed hyperarcs* of the form (X,C) for each variable $X \in v(C)$. In this paper we deal with hypergraphs and hyperarcs which are a natural generalization of graphs and arcs respectively. Hyperarcs may connect one, two or more than two nodes. Given a directed hyperarc (X,C) of a constraint hypergraph, X is called the *target* and the rest of the variables in v(C) are called *sources*, and the arrow of the hyperarc is from the target and to all the sources. For example, given the constraint C: X+ Y = Z, the hyperarcs (X,C), (Y,C) and (Z,C) respectively represent the *solution functions*: X = Z-Y, Y = Z-X and Z = X+Y. Given a CSP, an arc consistency algorithm deletes inconsistent values from target variable domains.

Consider a set of hyperarcs describing the constraint system S and containing the hyperarc (T,C), we denote DH(T,C,S) *the set of all hyperarcs in S dependent on the hyperarc* (T,C) (i.e. $DH(T,C,S) = \{(T',C') \in S \mid T \in v(C')\backslash\{T'\} \text{ and } C \neq C'\}$). In a local propagation, if the hyperarc (T,C) has changed the domain of T, all the hyperarcs in DH(T,C,S) will be activated to further the propagation.

Computing Projections

A useful function for describing arc consistency algorithms over real domains is *projection*, denote Π, which maps a constraint C and a variable $X \in v(C)$ to a subset of D_X, the domain of X. $\Pi_X(C)$ is the set of values for X which is consistent with the constraint C and with all the domains of the other variables.

Given a constraint C and a related variable X, computing the projection of C on X is very important in the arc-consistency algorithm over real intervals. This computation can be done using *Interval Arithmetic* with some generalizations(c.f. [SiH92]).

Assume that from the constraint C we can derive $X_i = f_E(X_1,...,X_{i-1},X_{i+1},...,X_k)$ describing how to compute the target X according to the sources $X_1,...,X_{i-1},X_{i+1},...,X_k$. Let $D_1,...,D_{i-1}$, D_{i+1}, ...,D_k be the current real domains of $X_1,...,X_{i-1},X_{i+1},...,X_k$. Using Interval Arithmetic, we can compute the *interval function*

$$F_E = \{ f_E(a_1,...,a_{i-1},a_{i+1},...,a_k) \mid \exists(a_1,...,a_{i-1},a_{i+1},...,a_k) \in D_1 \times ... \times D_{i-1} \times D_{i+1} \times ... \times D_k)\}$$

If D_X is the current real domain of X, then the projection

$$\Pi_X(C) = [\min F_E, \max F_E] \cap D_X.$$

Example 2.2: Given C: z = x+y and $D_X = [1,10]$, $D_Y = [3,8]$, $D_Z = [2,7]$.

We derive y = z-x. And applying the formula [a,b]-[c,d]=[a-d,b-c] from Interval Arithmetic, we get

$$\Pi_y(C)= ([2,7]-[1,10]) \cap [3,8] = [-8,6] \cap [3,8] = [3,6]$$

To reduce the complexity of computing projections, there should be a restriction on the form of constraints. Each equality contains at most one function symbol and each inequality contains no function symbols. That is, constraints are of the form 'A1 = A2', 'A1 <= A2', 'A1 + A2 = A3','A1.A2 = A3', 'A1=sin(A2)', etc where A1, A2, and A3 are either real variables or real constants.

The basic approach to solve CSPs over real intervals is to combine local arc-consistency algorithm descending from AC-3 (see [Mac77]) and interval arithmetic (see [Moo66]). Benhamou and Older [BeO93] present the fundamental properties of domain narrowing function and using narrowing function to describe an arc-consistency algorithm for solving CSPs over real intervals.

Arc-consistency algorithms computes locally consistent solution for a constraint network by local propagation. One important property of local propagation is that the order of activating constraints is unimportant.

Property 1 (Gusgen and Hertzberg, 1988). *In local propagation the order of activating constraints is unimportant as long as fairness is guaranteed: Every constraint of a network must be activated at least once; and if a constraint C has changed any variable domain then all of C's neighbor constraints must be activated thereafter.*

For full description of this property, see [GuH88].

2.3 Dynamic Constraint Satisfaction Problems

The characteristics of dynamic constraint satisfaction problem (DCSP) is that the constraint network changes during the reasoning. We can consider a DCSP as a sequence of static CSPs each resulting from an addition or a retraction of a constraint to/from in the preceding static CSP imposed by "the outside world". This change can be a *restriction* (a new constraint is imposed on a subset of variables) or a *relaxation* (a constraint which was present in the CSP is removed because it is no longer interesting or because the current CSP has no solution).

A *state* δ of a dynamic constraint system S at a given moment is the set of the domains each of which is associated to each variable in S at that moment.

Dynamic CSPs have non-monotonic behaviors since adding a new constraint or deleting an old constraint can produce a quite different solution to the new CSP.

3. AN ARC-CONSISTENCY TECHNIQUE FOR DYNAMIC CSPS OVER REAL INTERVALS

In this section, we propose DACR, a new incremental arc-consistency algorithm for Dynamic CSPs over real intervals. The algorithm consists of two procedures: one for adding a new constraint to and another for deleting an old constraint from a constraint system.

3.1 Addition of a new constraint

The procedure for addition of a new constraint C to the network S is given in Fig. 1. It is a modification of Davis's arc-consistency algorithm for static CSPs (c.f. [Dav87]). In the arc-consistency algorithm for static CSPs, first we put all the hyperarcs of all the constraints in the system (in any order) into the set Q. But in our procedure for constraint addition, first we put the hyperarcs of the new constraint into the set Q instead. That means the domains of C's variables will be refined by the procedure REVISE which implements the fundamental operation of the arc-consistency algorithm as introduced by Mackworth [Mac77]. The changes in the domains of these variables will invoke some other constraints to be considered for interval propagation. This propagation goes on until the system becomes *stable*, i.e., there is no more domain to be narrowed.

In most cases, when a constraint is added into a constraint network, its effect to the network is local. That means the effects of the new constraint will only propagate through a small part of the network. This observation is important for a technique to deal with the constraint deletion problem.

The algorithm given in Fig. 1 has the same structure of the arc-consistency algorithm given in [Dav87]. Line 5 and 6 of procedure *Revise* specify that D_T is updated only if *Revise* succeeds in refining it. Line 10 of procedure *Add* initializes Q to the set of hyperarcs of the new constraint. If Revise(T,C) refines D_T in line 16, then Q is updated in line 17 by adding DH(T,C,S), the set of hyperarcs in S dependent on the hyperarc (T,C), into Q so that these hyperarcs could be further revised. These are of the form (T',C') with $T \in v(C')\setminus\{T'\}$ and $C \neq C'$. This is because T is a source variable of C' so the arc-consistency of some values in $D_{T'}$ may have depended on values deleted from D_T.

The main difference between the algorithm given in Fig. 1 and the algorithm given in [Dav87] is that in this paper hyperarcs are used while in the Davis's paper constraints themselves are used. Hyperarc used in this paper is a generalization of the concept *solution function* in [Hyv89] which is used in the cases that constraints are equations.

3.2 Deletion of a Constraint

Now we explain the technique to deal with constraint deletion. In most cases it will not require the recomputation of the entire constraint system.

Definition: A constraint c is *immediately related* to c' written c $\sim$ c', iff c and c' share at least one variable, i.e. $v(c) \cap v(c') \neq \emptyset$.

Given a constraint c, it is possible to compute N(c), the set of all the constraints immediately related to c and $N^*(c)$, the set of all constraints *related to* c (*related to* is transitive closure of $\sim$).

Proposition 1: *When a constraint c is added to a constraint system S, only the domains of the variables in the subnetwork made from $N^*(c)$ can be changed and the domains of the variables in the part $S\setminus N^*(c)$ of the network remain the same.*

Proposition 2: *When a constraint c is deleted from a constraint system S, only the domains of the variables in the subnetwork made from $N^*(c)\setminus\{c\}$ can be changed and the domains of the variables in the part $S\setminus N^*(c)$ of the network remain the same.*

The constraint deletion process can be divided into two basic steps:

1. Reset all variables belonging to the subnetwork $N^*(c)$ of the deleted constraint c to their initial domains
2. Recompute new domains for all variables Y s.t.

```
procedure Revise(T,C)
begin
  if Π_T(C) = ∅ then halt /* the original constraints were inconsistent */
  else
      CHANGED := Π_T(C) ⊂ D_T;
  if CHANGED then D_T := Π_T(C);
  return CHANGED
end

procedure Add(c,S)
begin
  Q:= A;  /* A is set of hyperarcs of the added constraint c */
  while Q not empty do
    begin
      select and delete any hyperarc(T,C) from Q;
      if Revise(T,C) then
        Q := Q ∪ DH(T,C,S)
    end
end.
```

FIGURE 1: The procedure for addition of a constraint

(Y,c') ∈ DH(X,c,S); the changes in the domains of these variables will activate the local propagation through the subnetwork $N^*(c)\setminus\{c\}$.

Removing a constraint c from the system S is done by using the procedure *Delete* given in the Fig. 2.

Lines 3-6 in procedure *Delete* reset the domains of all variables belonging to the subnetwork $N^*(c)$. Lines 7-9 initialize Q to the set of all hyperarcs immediately related to the variables of the deleted constraint. Lines 11-15 propagate the changed domains through the subnetwork. The propagation process in the procedure *Delete* can be seen as a flow goes through only the subnetwork which is related to the deleted constraint.

Constraint Deletion has a special feature. Whereas a restriction of a new constraint narrows the domains of a subset of variables in the network through propagation process, a constraint deletion is an **"resetting"** propagation which makes the related domains larger. The enlarging propagation is performed through the network as long as there are still domains that should be enlarged.

Correctness of the Procedure *Delete*

We outline the key ideas for a complete proof of the correctness of the algorithm for constraint deletion.

After deleting c from the set of constraints S, now the system can be seen as consisting of two parts: $S\setminus N^*(c)$ and $N^*(c)\setminus\{c\}$. To resolve the constraint system $S\setminus\{c\}$, we should reset all the variables in $S\setminus\{c\}$ to their respective initial domains and then apply local propagation through the whole network. In the local propagation we put all the hyperarcs of $S\setminus N^*(c)$ and then all the hyperarcs of $N^*(c)\setminus\{c\}$ into the set Q since the order of activating constraints does not affect the result (Property 1). Doing so, the propagation works through the two parts separately. However, due to Proposition 2, the resetting and local propagation on the subnetwork $S\setminus N^*(c)$ would bring the domains of all variables in $S\setminus N^*(c)$ to the same domains as they have in the state before deleting c. Since this state is already known, there is no need to do any computations for the part $S\setminus N^*(c)$. So we only have to reset the variables in $N^*(c)\setminus\{c\}$ to their initial intervals and perform local propagation through this part as the procedure *Delete* does.

3.3 An Example

Consider the CSP given in Example 2.1 in which the constraints are added into the system in the order (C1, C2 ,C3) and then C1 is removed from the system. The algorithm DACR will perform the sequence of computations given in Table 1 to maintain the arc-consistency of the system and provide the state of the system after each computation as in Table 1.

So after deleting z = x+y, the state of the constraint system becomes D_X =[3,10], D_Y =[3,8], D_Z =[2,7], D_U =[0,15] and D_V =[0,7.5].

Now let resolve the constraint system as if we had only {C2,C3} from the beginning, the sequence of computation

```
procedure Delete(c,S)

begin
  Let N*(c) be the set of all constraints related to c
/* Resetting all variables in N*(c)  */
  forall c'∈ N*(c) do
    forall X ∈ v(c') do
      D_X := D_X^I  /* reset D_X to its initial interval */
/* Recomputing   */
  Q := ∅
  forall X ∈ v(c) do
    Q:= Q ∪ DH(X,c,S\{c});
   while Q not empty do
     begin
       select and delete hyperarc(Y,c') from Q
       if Revise(Y,c') then
          Q:= Q ∪ DH(Y,c',S\{c})
      end
 end.
```

FIGURE 2. The procedure for deleting a constraint

will be given as in Table 2. Again we obtain the same result as in Table 1. with D_X =[3,10], D_Y =[3,8], D_Z =[2,7], D_U =[0,15] and D_V =[0,7.5]. However, the computation work in Table 2 is nearly double of the work in Table 1 at the constraint deletion since the constraint u=2v still had been reconsidered.

The computation overhead of deleting a constraint depends on the question of how large the subnetwork related to the deleted constraint is. If the part consists of only few constraints, the computation overhead of resetting and recomputing is small. (Note that arc-consistency is achievable in time almost linear in the number of constraints as Mackworth and Freuder shows in [MaF85]). If there is no constraint in the subnetwork related to the deleted constraint, no computation is needed. In the worst case, if every constraints in the network depends on the deleted constraint, the resetting and propagation will go through the whole network and the overhead will be equal to that of recomputation from the beginning using all the remaining constraints. In real life problems, a constraint always has local effects in a small part of the constraint network. In comparison to addition of a constraint, deletion of a constraint c requires an extra work for resetting all variables in the subnetwork $N^*(c)$. This extra work depends on how large the undirected graph representing the relation "*immediately related to*" among constraints in the set $N^*(c)$ is. The resetting step can be implemented by a depth-first search through this undirected graph.

3.4 Deletion of a Group of Constraints

In some situations we need to delete a group of constraints at the same time from a Dynamic CSP. One way to deal with this problem is to apply the procedure Delete(c,S) repeatedly by removing one constraint in the group at a time. But doing that we still have to consider some constraints which in fact will be deleted later. This causes some redundancy of work. We can save a lot of resetting and propagation by using the procedure Delete_Group, given in Fig. 3. which is an extension of the procedure Delete(c,S). This procedure handles the deletion of a group of constraints G from the constraint system S.

Assume that c1,c2 are the two constraints in G which are *related to* each other. In that case, $N^*(c1) = N^*(c2)$ and if we use the procedure Delete_Group to deal with deleting G, all the variables in $N^*(c1)$ (also $N^*(c2)$) can be resetted to their initial domains only one time and all the hyperacrs involving with $N^*(c1)$ (also $N^*(c2)$) can be put into the set Q only once. Whereas using procedure Delete(c,S) repeately, all the variables in $N^*(c1)$ (also $N^*(c2)$) can be resetted to their initial domains twice and all the hyperarcs involving with $N^*(c1)$ (also $N^*(c2)$) can be put into the set Q twice: once for deleting c1 and once for deleting c2. That means in this case the overhead of resetting and propagation in using procedure Delete repeatedly is almost double of the overhead in using procedure Delete_Group. This observation shows that we can save a lot of work by using the procedure Delete_Group instead of using the procedure Delete

Table 1 The sequence of computations performed by the algorithm DACR

Next Hyperarc	New value	Q
Add z = x+y		z=x+y, y=z-x, x=z-y
z = x+y	([1,10]+[3,8])∩[2,7]=[4,7]	y = z-x, x = z-y
y = z-x	([4,7]-[1,10])∩[3,8]=[3,6]	x = z-x
x= z-y	([4,7]-[3,6])∩[1,10]=[1,4]	
Add y ≤ x		y ≤ x, x ≥ y
y ≤ x	D_Y ≤ [1,4] ⇒ D_Y =[3,4]	x ≥ y, z = x+y
x ≥ y	D_X ≥ [3,4] ⇒ D_X =[3,4]	z = x+y
z=x+y	([3,4]+[3,4])∩[4,7]=[6,7]	
Add u= 2v		u = 2v, v = u/2
u = 2v	(2*[20,20])∩[0,15]=[0,15]	v = u/2
v = u/2	(0.5*[0,15])∩[20,20]=[0,7.5]	
Delete z =x+y	D_X=[1,10], D_Y=[3,8], D_Z=[2,7]	y ≤ x, x ≥ y
y ≤ x	D_Y ≤ [1,10] ⇒ D_Y = [3,8]	x ≥ y
x ≥ y	D_X ≥ [3,8] ⇒ D_X = [3,10]	

Table 2 The computations for the constraint system consisting of {C2,C3}

Next Hyperarc	New value	Q
Add y ≤ x		y ≤ x, x ≥ y
y ≤ x	D_Y ≤ [1,10] ⇒ D_Y =[3,8]	x ≥ y
x ≥ y	D_X ≥ [3,8] ⇒ D_X = [3,10]	
Add u= 2v		u = 2v, v = u/2
u = 2v	(2*[-20,20])∩[0,15]=[0,15]	v = u/2
v = u/2	(0.5*[0,15])∩[-20,20]=[0,7.5]	

repeatedly when deleting a group of constraints.

3.5 Constraint Deletion in Dynamic CSPs vs. Backtracking in CLP

Notice that constraint deletion in a Dynamic CSP over real intervals and constraint removal due to backtracking in a CLP language for real intervals are different in nature. Their differences can be listed as follows.

(1) The former occurs randomly whenever the user want to relax a constraint while the latter occurs whenever a goal failure or a constraint inconsistency happens in a derivation and the system has to go back to the most recent choice point.

(2) When some constraints are removed due to the chronological backtracking in CLP, the CLP system goes back to the status of the constraint system at the point where the constraints were introduced, i.e. choice point. From the choice point, the computation can continue with the addition of some new constraints arising from an alternative choice of the rule in question. So in CLP, the status of all variable domains must be *kept on trail* for

```
procedure Delete_Group(G,S)
begin
  NN := ∅;
  forall c ∈ G do
    NN := NN ∪ N*(c);
/* Resetting      */
    forall c' ∈ NN do
      forall X ∈ v(c') do
         D_X := D_X^I   /* reset D_X to its initial interval */
/* Recomputing    */
  Q := ∅;
  forall c ∈ G do
    forall X ∈ v(c) do
      Q:= Q ∪ DH(X,c,S\G);
  while Q not empty do
    begin
      select and delete hyperarc(Y,c') from Q
      if Revise(Y,c') then Q:= Q ∪ DH(Y,c',S\G)
    end
end.
```

FIGURE 3. The procedure for delete a group of constraints

each choice point so that the variable status at the choice point can be recovered in case of backtrack later (see [AgB91]). In the technique we propose here for constraint deletion we do not involve with choice points, so we do not have to do any trailing to record the status of variable domains at each choice point.

In short, constraint removal due to backtracking in a CLP requires more extra work for recomputation and trailing mechanism while the technique proposed here for constraint deletion in a Dynamic CSP requires less work for recomputation and no extra work for trailing.

4. RELATED WORK

Motivated by some techniques from Truth Maintenance Systems, Bessiere [Bes91] proposed an arc-consistency algorithm for dynamic CSPs over finite domains. During adding a constraint this algorithm keeps track of "justifications" for every value removed from a domain and "killers" for each tuple deleted so that during a relaxation of the constraint the algorithm resorts to the help of "justifications" and "killers" to put the previously deleted values back to their domains correctly. For a CSP over finite domains, we can keep track of each value whenever it is removed from a certain domain in an arc-consistency algorithm. However, for a CSP over real intervals we can not do the same bookkeeping since subintervals are removed from a certain domain and their justifications can not be determined easily. Therefore, Bessiere's method is not applicable to Dynamic CSPs over real intervals.

Hua et al. [HFH90] suggested a Dynamic CSP algorithm over real domains. This algorithm requires a partial ordering on set of variables in the constraint system and the constraints have to meet some pre-defined condition. The constraint network in their paper must have a special property that interval propagation can flow only in one direction.

In a short paper, Codognet and Diaz [CoD93] presented an algorithm to deal with incremental deletion of constraints. Their algorithm is for constraints over finite domains and the constraints are of basic form *X in r* where *X* is a domain variable and *r* is a range. Their method avoids the recomputation from scratch by redoing only the subcomputations dependent on the deleted constraint. Our work can handle constraints on real intervals and our constraints are of general forms.

5. CONCLUSIONS

We have defined what we call Dynamic CSPs over real intervals. We have noticed that each constraint in the constraint network always has local effects and developed a method to deal with constraint deletion without resolving the system from scratch. We have achieved the objective of minimal recomputation as parts of the network which are independent of the deletion are not reconsidered. Then we have provided DACR an efficient algorithm incrementally achieving arc-consistency in Dynamic CSPs over real intervals.

The algorithm DACR has been implemented and tested with a number of Dynamic CSPs over real intervals. It can also be applied in an incremental solver for constraint hierarchies as defined by Borning and Maher [BoM88] but over real intervals instead of over exact values. Our further research will focus on this application.

REFERENCES

[AgB91] A. Aggoun and N. Beldiceanu. *Overview of the CHIP Compiler System*. The 8th Int. Conf. Logic Programming, Paris, France, MIT Press, 1991.

[BeO93] F. Benhamou and W. J. Older. *Applying Interval Arithmetic to Real, Integer and Boolean Constraints*, Technical Report, Bell Northern Research, Computing Research Laboratory, Ottawa, Ontario, Canada, 1993.

[Bes91] C. Bessiere. *Arc-Consistency in Dynamic Constraint Satisfaction Problems*. Proc. AAAI'91, pp 221-226.

[BoM88] A. Borning and M. Maher. *Constraint Hierarchies and Logic Programming*. Tech. Report 88-11-10, Comp. Sc. Dept., Univ. of Washington, Nov. 1988.

[CoD94] F. Codognet and D. Diaz. *Defeasible Constraint Solving over Finite Domains*. Technical Report, INRIA, France, 1994.

[Dav87] E. Davis. *Constraint Propagation with Interval Labels*. Artificial Intelligence, Vol 32, 1987, pp 281-331.

[GuH88] H. W. Gusgen and J. Hertzbert. *Some Fundamental Properties of Local Constraint Propagation*. Artificial Intelligence, Vol. 36, 1988, pp 237-247.

[HFH90] K. Hua, B. Faltings and D. Haroud. *Dynamic Constraint Satisfaction in a Bridge Design System*. Int. Workshop on Expert Systems in Engineering: Principle and Applications, Vienna, Austria, Sep. 1990.

[Hyv89] E. Hyvonen. *Constraint Reasoning based on Interval Arithmetic*. Proc. IJCAI'89, pp 1193-1198.

[Hyv92] E. Hyvonen. *Constraint Reasoning based on Interval Arithmetic: The Tolerance Propagation Approach*. Artificial Intelligence, Vol. 58, 1992, pp 71-112.

[Mac77] A. K. Mackworth. *Consistency in Networks of Relations*. Artificial Intelligence, Vol. 8, 1977 pp 99-118.

[MaF85] A. K. Mackworth and E. C. Freuder. *The Complexity of Some Polynomial Network Consistency Algorithms for Constraint Satisfaction Problems*, Artificial Intelligence, Vol. 25, 1985, pp 65-74.

[Moo66] R. E. MOORE. *Interval Analysis*, Englewood Cliffs, New Jersey, Prentice-Hall, 1966.

[OlV90] W. Older and A. Vellino. *Extending Prolog with Constraint Arithmetic on Real Intervals*. Proc. of Canadian Conference on Electrical and Computer Engineering, 1990.

[SiH92] G. Sidebottom and W. S. Haven. *Hierarchical Arc Consistency for Disjoint Real Intervals in Constraint Logic Programming*. Proc. of Post-conference Workshop on Constraint Logic Programming Systems: Design and Applications, Washington D.C., USA, Nov 1992.

SPATIAL RELATIONS AND CONSTRAINT SYSTEMS: A SHORT VIEW ON COMPLEXITY

Manfred Kopisch
University of Hamburg, Department of Computer Science
Vogt-Kölln-Str. 30; D - 22527 Hamburg
Email: kopisch@informatik.uni-hamburg.de

ABSTRACT

Several configuration problems can not be solved without consideration of spatial relations between its objects. Constraint systems are a good choice for processing these spatial relations. They offer the advantage of processing spatial and non-spatial relations in a single system under consideration of interdependencies. But constraint systems have their limits too. In this paper we want to have a look at questions about complexity and representation possibilities of constraint systems.[1]

INTRODUCTION

Some configuration tasks need spatial knowledge in order to find a correct solution. We are only interested in one part of this spatial knowledge: spatial relations between objects. Two questions are important to us:

- Which relations have to be expressible by a system?
- How efficient can a system deal with these relations?

We will introduce two domains that need such spatial knowledge in section 2: configuration of a kitchen and cabin layout of a passenger aircraft. Examples from both domains will give us an overview about needed spatial relations. We describe the representation and processing of these relations with the help of a constraint system in section 3. The limits that appear especially in time consumption are shown in section 4. We introduce our proposal for a better time behavior of the system in section 5. In the conclusions we give a valuation of the usefulness of the introduced method.

1. This work is supported by the German Ministry for Research and Technology (BMFT) under grant ITW9101A6 (project PROKON). Responsibility for this paper lies with the author.

SPATIAL RELATIONS IN CONFIGURATION PROBLEMS

A configuration is built from a set of basic parts. To find a correct solution for a configuration problem we have to consider given constraints. These constraints may be spatial constraints. We will introduce two domains that show which spatial constraints have to be considered for a correct solution.

Configuration of a kitchen

For the configuration of a kitchen we have to place several items like an oven, a fridge, a dish washer, etc. on a given area. Constraints like "The fridge must not be placed beside the oven." or "A cupboard must not be placed in front of a window." have to be considered in order to find a correct solution.

This domain is known to everybody and thus gives good examples for necessary spatial relations. A typical kitchen has a rectangular form with rectangular objects in it that are placed parallel to the walls.

Until now we introduced the location relations "beside" and "in front of". The relation "beside" is in fact an adjacency relation where the distance between the objects has to be zero. The "in front of" holds also for distances greater than zero. Thus we have to think first about the exact meaning of a relation before we are able to express it with pre-defined relations.

The kitchen domain is used by several authors (e.g. [BF90] and [Sch93]) in order to show the capabilities of their systems for the generation of layouts. Both use a similar repertoire of spatial relations, which will be introduced in section 2.

Before, we want to have a look on a second domain that uses similar spatial relations.

Configuration of the passenger cabin of an aircraft

A cabin layout defines the passenger cabin of an aircraft by the number, the placement and the type of cabin interior components. Cabin interior components are objects like galleys, lavatories, seats, cabin attendant seats, etc. All cabin interior components together define the capacity, the comfort, the possible service, etc. of the passenger cabin.

In order to produce a cabin layout we have to place cabin interior components on a restricted area under consideration of certification rules and technical constraints in a way that fulfils the customer's requirements for the number of seats, the seat pitch, etc. in an optimal way [Kop91], [Kop91a].

Spatial relations are found in all types of restrictions. There is for example a certification rule that demands 22 inch of free space between the front of a seat and a wall. The number of galleys behind door 4 is restricted to a maximum of four by technical restrictions. The seat pitch is a customer's requirement.

This domain can be described by rectangles that a parallel to the co-ordinate system. The spatial relations between the objects are similar to the relations in the kitchen domain. Location, adjacency and distance relations are also found. In addition to that we find overlapping and containing relations and we differentiate the relations into object-centered (in front of a seat) and global relations (behind door 4). [Kop93] gives an overview on the relations that are needed in this domain.

In the next section we cover the most important spatial relations in both domains.

Spatial Relations in Domains with Rectangles that are parallel to the co-ordinate system

Both domains have in common that rectangles have to be placed on a given rectangular area. Therefore we find similar spatial relations in both domains. Theses relations are ordered be several authors ([BF91], [Kop93] and [Sch93]) in a similar way. All types of relations are found in four types for each of the possible directions.

- object-centered relations
  - location relations
    (in-front-of, behind, left-of, right-of)
  - distance relations (e. g. front-distance)
  - adjacency relations (e. g. left-adjacent)
  - overlapping relations (e. g. overlaps-right)
  - orientation relations (e. g. parallel-to)
- global relations
  - location relations
    (east-oft, west-oft, north-of, south-of)
  - distance relations
  - adjacency relations
  - overlapping relations
  - containing relations (contains, is-contained-in)

Figure 1 shows some spatial relations. B lies east-of A and in-front-of A. B and E are parallel to each other. The back-distance between B and D equals the right-distance between A and D.

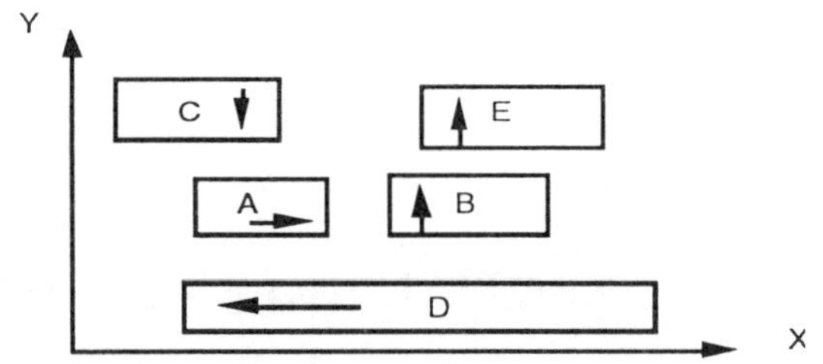

FIGURE 1 Spatial relations

The different authors introduce several other relations (e.g. beside or does-not-overlap). But these relations can easily be built by using the existing ones. For example beside := left-of or right-of.

Thus there is a broad consensus about the necessary relations. In the next section we describe the representation and processing of spatial relations with a set of pre-defined relations and a constraint system.

PROCESSING OF SPATIAL RELATIONS WITH A CONSTRAINT SYSTEM

In configuration problems spatial relations between objects restrict their placement and dimensioning. Many systems which deal with such spatial restrictions use constraint systems for their processing (e.g. [Her90], [BF90], [Güs91] and [Kop93]).

Using the three-stage constraint integration of PLAKON [CGS91] as an example we will show how the representation and processing of spatial relations with a constraint system works.

Spatial relations like 'front-distance', 'left-of', 'inside', etc. are defined as domain-independent **constraint relations**. These pre-defined constraint relations are used for the definition of **conceptual constraints** such as: "There have to be 36 inches of free space in front of a galley." These conceptual constraints are instantiated during the configuration process whenever an affected object is instantiated. All **instantiated constraints** form a constraint net. This constraint net adapts all connected parameters if one parameter is changed. The constraint net automatically considers all connections.

The following example shows the use of domain-independent constraint relations for the representation of spatial relations in the cabin layout domain.

```
(def-conceptual-constraint
   :name free-space-in-front-of-galleys
   :variable-pattern-pairs ( (?k :name galley)
                             (?e :name cabin-interior-item)
                                 :parameters ((front-neigh-
                                                  bour ?k))))
   :local-variables         (distance)
   :constraint-calls        ( (front-distance ?k ?e distance)
                              (greater-equal 36 distance)))
```

This conceptual constraint is instantiated during the configuration process whenever there exists a cabin interior item which is 'front-neighbor' of an instantiated galley. Every object has four slots which contain the neighbor(s) to each side.

The following constraint ensures the non-overlapping of two objects.

```
(def-conceptual-constraint
   :name no-overlap-allowed
   :variable-pattern-pairs ( (?e1 :name cabin-interior-item)
                             (?e2:name cabin-interior-item))
   :constraint-calls       ( (does-not-overlap ?e1 ?e2)))
```

This declarative form of the knowledge base ensures the easy maintenance of the knowledge. But constraint systems have their draw backs too. Time limits their usefulness in large knowledge bases.

LIMITS OF THE USE OF CONSTRAINT SYSTEMS FOR SPATIAL PROBLEMS

Constraint satisfaction problems (CSP) are np-complete. Thus it is unlikely that general CSPs can be solved in polynomial time. Therefore constraint systems are used for the deletion of these values in given sets that would cause a local inconsistency. Algorithms for different types of consistency - node, arc, or path consistency (for a description see [Mac77]) can be solved in polynomial time [MF85].

But, what does polynomial time mean in praxis? Let's take AC-3, a well known algorithm for path consistency [MF77]. It has a worst case time complexity of $O(d^3e)$, where d is the number of elements in the largest domain and e the number of constraints with a worst case order of n^2 (n = number of variables). Other algorithms, e.g. AC-4 [MH86], behave better in the worst case but may be worse than AC-3 on average problems [Wal93]. Therefore we have to look not only on the algorithms but also on the domains in order to get valuable results.

Let's take the best known algorithmus AC-6 [CB93] with a worst case time complexity of $O(d^2e)$ and have a look on the cabin layout domain for a valuation. In this domain we have to place nearly 200 objects. Many of these objects are dimensonable. Therefore we need simple constraints like (x-min obj) + (x-extent obj) = (x-max obj). The domain for the position of a passenger seat contains nearly 6000 values (the passenger cabin is 2000 inches long and has seat rails on both sides and in the middle (s. figure 2). Each seat rail allows an attachment in 1-inch distances). The objects have 11 spatial variables (length, width, front, rotation, orientation, x-min, x-extent, x-max, y-min, y-extent, y-max). Thus the worst case for securing local consistency (not for finding the solution!) computes like this: $e = 0.5 * 200 * 11 * (200 * 11 - 1) = 2418900$;
$d = 6000 \Rightarrow 6000^2 * 2418900 \approx 8.7E13$.

By using some properties of spatial constraints it is possible to save some constraint checks by inferring knowledge about their necessity. For example we can use symmetry or the fact that the constraints are not directed. [Fre94l gives some general proposals which lead to AC-7. These proposals can also be used for spatial constraints. But one has to weigh between the costs for the inferences and the costs for the constraint checks.

We propose for domains like the ones we introduced in section 2 an additional method that uses spatial nearness. Thus it is possible to instantiate far less constraints than without this method.

USING NEARNESS FOR IMPROVEMENT

Remember the example that prohibits the overlapping of objects. The formulation of the conceptual constraint in section 3 leads (under the assumption of 200 objects) to the instantiation of 0.5 * 2200 * 2199 = 2418900 constraints. By introducing an additional structure where every object is put into a matrix we can decrease the number of constraints enormously.

The cells of this matrix have not to be squares. By checking the knowledge base one knows how many objects a configuration may contain. Additional knowledge about the length and width of the free space is used to compute an array with cells that contain whenever possible only one object. For example one cell in the cabin layout domain represents a rectangle with a length of 40 inches

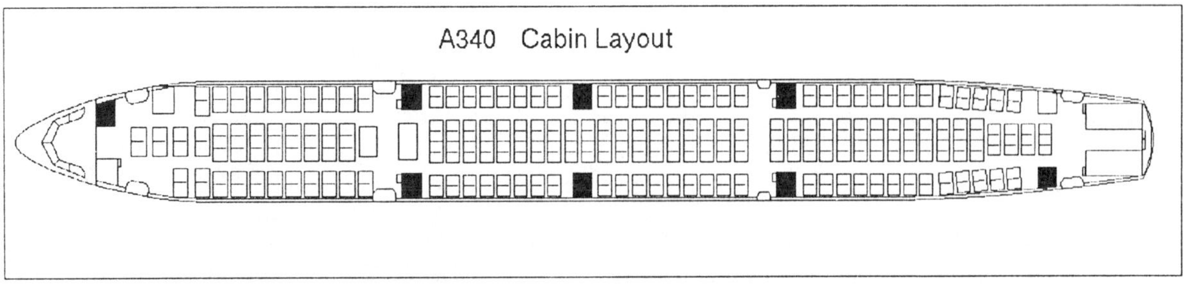

FIGURE 1 Cabin Layout of an A340.

and a width of 70 inches. This structure does not need to instantiate constraints between each object but only for objects that share a cell of the array. In our example we have ca. 4 objects in a cell - depending on the region of the passenger cabin and the possibilities for the placement of objects in it. Thus only ca. 189200 = 200 * (0.5 * 44 * 43) constraints have to be instantiated. For other constraints, e.g. the prohibition of nearness, we find similar advantages. For example if it is not allowed to place lavatories beside galleys, we do not have to instantiate a constraint for each lavatory when we instantiate a galley. We only have to instantiate constraints for each lavatory in neighbor cells of the galley which may be none at all.

By the use of nearness, by restricting the number of objects from all to objects in a well defined near distance we can diminish the number of constraints in the constraint net to under 10%.

CONCLUSIONS

Using two domains as examples we listed a repertoire of spatial relations that are needed for the description of spatial relations in domains with rectangles that are parallel to the axes of the co-ordinate system. The pre-definition of these relations for the use in a constraint system enables a knowledge engineer to represent spatial relations in a very easy way. But this is only one part of the job - representing the knowledge. The other part - processing of the knowledge - is the harder one. Problems that do not only process eight queens need an amount of time that limits the usefulness of constraint systems in praxis. We proposed a method especially for spatial constraints that reduces considerably the number of constraints. This is done by using two steps: In a first step an array is scanned for objects that may be candidates. Only these candidates are instantiated and processed in the constraint system. This two step method from rough to fine allows the handling of such spatial problems like the ones we introduced in section 2. The price of the additional structure and its management - especially for the filling and deleting of cells is outdone by the reducing of instantiated constraints. But the progress is a small one. A reduction of the number of constraints by a factor of 10 means that we are able to handle 200 objects in the passenger cabin instead of 20 in the kitchen. But spatial relations are nevertheless a time consuming problem. A knowledge engineer who uses a constraint system should always be aware of its limits and try to reduce the number of constraints.

REFERENCES

[BF90] C. A. Baykan and M. S. Fox. Constraint Satisfaction Techniques for Spatial Planning. In P. ten Hagen and P. Veerkamp, editors, *Intelligent CAD Systems 3: Practical Experience and Evaluation*, 1990, eurographics Seminar Series, Springer

[BF91] C. A. Baykan and M. S. Fox. WRIGHT: A Constraint Based Spatial Layout System. In C. Tong and D. Sriram, editors, *AI in Engineering Design*

[CB93] M. Cordier and C. Bessiere. Arc consistency and arc consistency again, *AAAI-93*, 108-113

[CGS91] R. Cunis, A. Günter and H. Strecker, editors, *Das PLAKON-Buch.* Informatik-Fachberichte Nr.266, 1991, Springer

[Fre94] E. C. Freuder. Using Metalevel Constraint Knowledge to Reduce Constraint Checking. In: *Workshop ECAI 94*, 1994

[GDG+91] A. Günter, H. Dörner, H. Gläser, B. Neumann, C. Posthoff and H. J. Sebastian. *Das Projekt PROKON -- Problemspezifische Werkzeuge für wissensbasierte Konfigurierung* PROKON-Bericht Nr. 1, 1991

[Güs91] H. W. Güsgen. Feasible Reasoning about Space.In *Proc. 5. Workshop Planen und Konfigurieren*, 1-15, Hamburg, 1991

[Her90] D. Hernàndez. Using Comparative Relations to Represent Spatial Knowledge. In W. Hoeppner, editor, *Workshop Räumliche Alltagsumgebungen des Menschen*, 69-80, Universität Koblenz-Lindau, 1990

[Kop91] M. Kopisch. Konfigurierung der Passagier–kabine eines Verkehrsflugzeuges. In *Proc. 5. Workshop Planen und Konfiguieren, Hamburg*, 176-179, Hamburg, 1991

[Kop91a] M. Kopisch. *Expertensystemgestützte Konfigurierung der Passagierkabine eines Verkehrsflugzeuges*, Diplomarbeit, Universität Hamburg, Fachbereich Informatik, 1991

[KG92] M. Kopisch and A. Günter. Configuration of a Passenger Aircraft Cabin based on Conceptual Hierarchy, Constraints and flexible Control. In F. Belli and F. J. Radermacher, editors, *Industrial and Engineering Applications of Artificial Intelligence and Expert Systems Proc. of 5th International Conference IEA/ AIE-92*, 421-430, Springer, 1992

[Kop93] M. Kopisch. Spatial Relations for Configuration Tasks in General and for the Cabin Layout of the Airbus A340 in Particular. In P. W. H. Chung, G. Lovegrove and M. Ali, editors, *Industrial and Engineering Applications of Artificial Intelligence and Expert Systems*
Proc. of 6th International Conference IEA/ AIE-93, 450-457, 1993, Gordon and Breach

[Mac77] A. K. Mackworth. Consistency in Networks of Relations
In *Artificial Intelligence 8*, 95 - 118, 1977

[MF85] A. K. Mackworth and E. C. Freuder. The Complexity of Some Polynomial Network Consistency Algorithms for Constraint Satisfaction Problems
In *Artificial Intelligence 25*, 65 - 74, 1985

[MH86] R. Mohr and T. Henderson. Arc and path consistency revisited
In *Artificial Intelligence 28*, 225-233, 1986

[Sch93] S. Schmitgen. *Räumliche Fragestellungen bei der Konfigurierung*
Dissertation, Universität - GH - Paderborn, Fachbereich Mathematik/Informatik, 1993

[Wal93] R. Wallace. Why AC-3 is almost always better than AC-4 for establishing arc consistency in CSPs. In *Proc. IJCAI-93*, 239-245

Diagnosis

Enabling Technology for Diagnostic Applications

R. Khosla T. Dillon

Expert and Intelligent Systems Laboratory
Department of Computer Science and Computer Engineering
La Trobe University
Bundoora, Victoria-3083, Australia

ABSTRACT

In this paper we describe a distributed object-oriented integrated architecture of symbolic (e.g. expert systems) and connectionist (neural networks) systems. We address issues like information processing, learning and knowledge representation associated with symbolic-connectionist integration. We show how the object-oriented paradigm facilitates the development of intelligent distributed symbolic-connectionist objects. The integrated architecture has been applied in a real time alarm processing system with encouraging performance results.

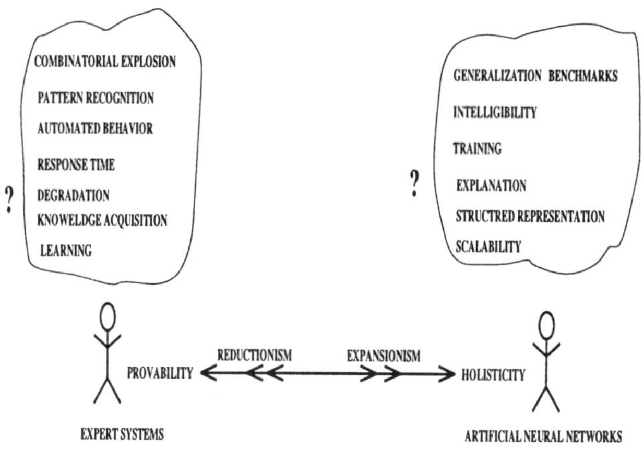

Fig.1: Need for an Integrated Architecture

Introduction

Alarm Processing is a problem which has been bothering the minds of power system researchers since the 1977 New York Blackout. Various problems associated with real-time alarm processing include power system size, enormity of alarms during a system emergency, response time constraints, temporal reasoning, incomplete information and incorrect information[5, 6]. The existing expert systems [12, 13, 19, 20] approaches have been useful in addressing various issues related to alarm processing. However, as stated by [4, 5, 6] in their surveys on practical use of Expert Systems(ES) in power systems some of the problems like response time, incomplete information and temporal reasoning remain only partially addressed or solved. The existing artificial neural network(ANN) approaches[1, 18] on alarm processing have been on small scale systems [11]. They do not adequately address the problems with respect to the size of power systems and generalization problems with neural networks.

In fact the problems expressed above are a manifestation of more general problems (shown in Fig. 1) associated with the expert system and neural network paradigms. These problems have created a need for integration of the two paradigms in order to develop more powerful problem solving strategies especially when dealing with real world problems of sufficient complexity.

In this paper we address the above problems by building a distributed, object-oriented symbolic-connectionist computational architecture for complex, knowledge intensive, time critical diagnostic domains. We validate the architecture by applying it in a real-time alarm processing system for a regional power system control centre.

In order to achieve these two goals, the paper is divided into the follow sections.

Section 2 briefly surveys the previous work in integration of the two paradigms. Section 3 describes the integrated information processing architecture, namely Generic Neuro-Expert System (GENUES) Architecture. Section 4 briefly covers the learning knowledge and strategy of the GENUES architecture. Section 5 describes the distributed object-oriented integrated architecture for symbolic-connectionist computation. Section 6 describes the implementation

Previous Work

The integration of expert systems and artificial neural networks has been broadly attempted in two directions. One is in developing unifying architectures like connectionist expert systems where rules are represented in a connectionist framework([7, 8], and others). Cognitively most of these architectures do not reflect human ways of reasoning[7, 2, 3]. The other direction(which we relate to) is in integrating ANNs and ESs while maintaining their identity as two separate paradigms which have two separate roles to play. In this direction there are some applications in place which attempt to integrate ANNs and ESs ([9, 10], and others).

However, most of these architectures are adhoc in nature and do not adequately address the questions like i) What is the task structure level architecture for applications with symbolic-connectionist constituents ii.) What generic strategy should be adopted to combine the two together on a cognitive and practical basis? In other words what information processing architecture needs to be adopted for applications involving symbolic-connectionist methods? This becomes important when attempting to solve complex real world problems; iii) What overall computational architecture needs to be adopted for capturing task structure level architecture as well as the methods employed in it? iv) What is the knowledge content employed by symbolic and connectionist methods? For example, What learning knowledge and strategy needs to be adopted for training neural networks in such integrated architecture; v.) What architecture should be adopted to reduce the cognitive barriers between human and computer and facilitate comprehensibility, intelligibility and practical use of such integrated architectures?; v.) What architecture/model should be adopted to facilitate scalability, cost effectiveness of the systems so realized? and a few others[14]. In this paper only questions 'ii', 'iii', and 'iv' are addressed.

Generic Neuro-Expert System(GENUES) Architecture

The integrated information processing architecture, namely the Generic Neuro-Expert System (GENUES) architecture shown in Fig.2 has been built from various perspectives including, philosophical, neurobiological, cognitive science, physical systems and practical perspectives [14] as well as pragmatic constraints like cognitive compatibility, forms of knowledge, learning, notion of real time, reasoning with incomplete and incorrect information, temporal reasoning, diagnostic reliability, problem complexity, generalization, knowledge sharing and reuse.

It essentially consists of five phases namely, Decomposition Phase, Control Phase, Decision Phase, Preprocessing Phase and Postprocessing Phase. In the decomposition phase the goal is to decompose a global concept into a set of abstract heterogeneous classes. Artificial neural networks are chosen as medium for accomplishing this goal. This is so because from a cognitive view point, humans are very quick and accurate in elementary classification in various domains. For example, it would take an operator in a control centre no time in elementarily classifying an alarm as a circuit-breaker alarm or a communication alarm. In other words, elementary classification generally forms a part of our unconscious or automated behavior which is best reflected by neural networks. In the control phase the goal is to determine which decision-level classes within each heterogeneous class need to be activated. Decision level classes are those classes inference on which is of importance to the system's user. This goal is accomplished by using artificial neural networks or symbolic methods like knowledge based systems. The division of responsibility between use of neural networks or symbolic methods is based on class complexity and the desired classification granularity by the problem solver and other pragmatic constraints mentioned above. Control phase is also used for resolving the conflicts between decision level classes in the decision phase. Thus the extent of communication level between the control phase and decision phase is higher than the communication between decomposition phase and the control phase.

In the decision phase specific classifications within each decision level class or instances of a decision level class are determined. This is for resolving conflicts between different decision level classes in

It is in the decision phase that human experts engage in generalization. For example, in a power system control center, operator is accurate and quick in detecting whether a set of alarms are circuit breaker(CB) alarms or communication alarms, or whether they are 220kv CB alarms or 66kv CB alarms. However, the operator generalizes on whether these set of alarms depict a bus fault or multiple line fault. That is, if a bus fault occurs as a result of eight CB alarms in a bus section, the operator is likely to generalize the same fault for six CB alarms in real time. Thus ANNs which have generalization capabilities can be used in this phase to compute component faults. Preprocessing involves processing prior to decomposition phase and processing within each phase. Thus it is categorized as a) Global Preprocessing, and b) Local Preprocessing

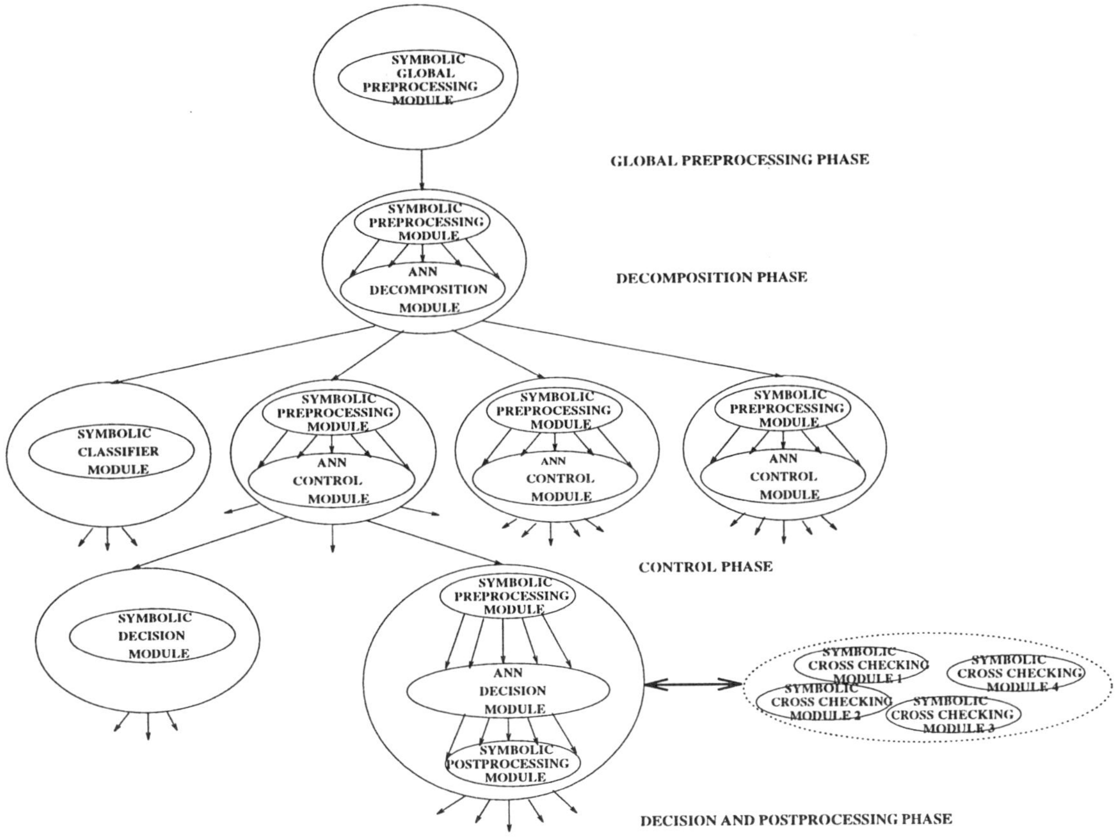

Fig. 2: Generic Neuro-Expert System Architecture

Global Preprocessing is done at the top most level using symbolic heuristics in case of alarm processing for filtering out noise which is peculiar to the domain as a whole or common to many classes. Local preprocessing is done in the decomposition, control and decision phases using symbolic preprocessing modules for checking the relevance and validity of data and preprocessing of symbolic input data, etc.

Logic and provability which are the hallmarks of our conscious behavior[2, 3] explains our underlying unconscious behavior. Thus in the Postprocessing phase symbolic modules are used to explain and validate decisions made by the ANN/s in the decision phase. This is done by using symbolic-cross checking modules. These symbolic cross-checking modules can be fault propagation model/s or other application dependent cross-checking module. More details on the GENUES architecture can be found in [15].

In order to facilitate the use of the GENUES architecture on complex data intensive problems one needs to address questions like how is one going to train the neural networks in different phases of the GENUES architecture?

Learning

The GENUES architecture provides a *priori* structure in which learning can take place. This a *priori* structure which has been built from philosophical, neurobiological, cognitive and practical viewpoints facilitates concept learning. It facilitates learning at different levels of abstraction (i.e. decomposition, control and decision) which reduces the complexity of learning large amounts of data. Further, by learning at different levels of abstraction enables making provision for redundancies (for noise) as well as removing redundancies in the data and eliminating irrelevant features.

The a *priori* structure of the GENUES architecture is used for designing learning knowledge and learning strategy needed for training neural networks. Learning knowledge (i.e the decomposition, control, and decision knowledge with respect connectionist methods) determines the type of input features and output classes which need to be selected or extracted in each phase (i.e. decomposition, control and decision phases respectively) of the GENUES architecture. Learning strategy on the other hand determines the method of extraction/selection of input features and output classes in different phases. It also includes the stopping criteria in each phase. Its

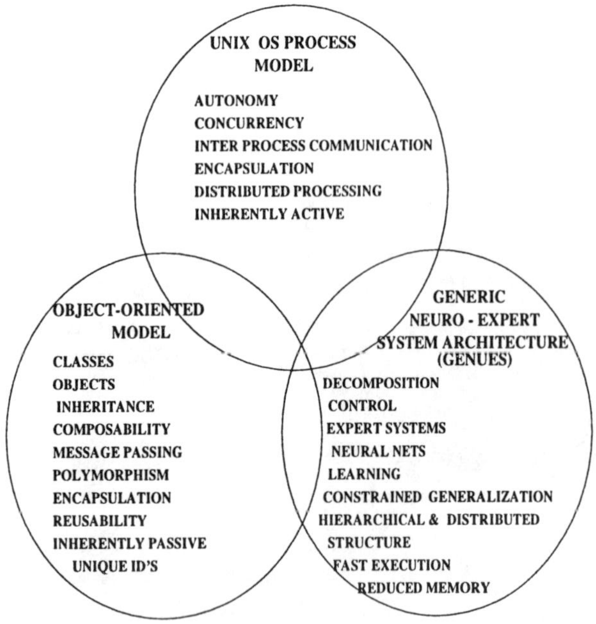

Fig.3: The Integrated Architecture

application is shown in the implementation section.

Integrated Architecture for Symbolic Connectionist Computation

In order to develop real world applications, we need a computational architecture in which we can realize all the five phases of the the GENUES architecture and the concepts built in them. The computational architecture has been built by integrating the GENUES architecture with an object-oriented model and, a UOSP model as shown in Fig.3. Fig. 3 shows the important features of each of the three sub-models of the integrated architecture. Fig.4 shows the the distributed object-oriented symbolic-connectionist computational architecture which is realized through this integration. Some of the features of the computational architecture are described below.

Generic Software Structure Classes

The object-oriented model helps to derive the notion of an ANN class, an ES class, and a Unix Operating System Process (UOSP) class. This facilitates the development of generic data structures and methods(polymorphism), a multiple inheritance structure and, formation of heterogeneous objects which help to realize the concept of abstraction espoused in the GENUES architecture. In addition the UOSP model provides for parallel distributed processing and competing objects for reliable inference.

Three Layers of Knowledge

The integrated computational architecture in Fig. 4 represents three forms of knowledge. The conceptual knowledge is represented by objects and symbolic heuristics (symbolic attributes and methods). The intuitive or sub-conceptual knowledge is represented by neural networks (sub-symbolic attributes and methods). The logical knowledge is represented by the symbolic cross-checking objects (e.g. fault-propagation model of a particular component).

Global and Local Level Parallelism

In the integrated computational architecture parallelism is realized at two levels, i.e the global level and the local level. At the global level the decomposition, control and decision objects of the integrated architecture are processed in a parallel and distributed manner. Within each object (i.e. ES-ANN object), parallelism at the local level is realized through use of neural networks.

Two Level Hierarchical Communication Model

Fig. 4 represents a hierarchical communication model. In this model the communication takes place at the process object level and the domain object level. The inter-process communication facilitates dynamic inheritance of methods between objects belonging to different class structures. However, if required lateral communication between heterogeneous objects at the same level can also be realized.

Real Time System Constraints

The five phases of the GENUES architecture guarantee a macro level execution precedence for an application. The one step reasoning property of the ANNs in the heterogeneous (ES-ANN) objects at different levels of abstraction has a flattening effect on the system hierarchy and reduction in execution time. This property of artificial neural networks helps to overcome the problems associated with expert systems of an unpredictable solution hierarchy during run time. All these properties combined with object-oriented properties of modularity, encapsulation and uniform communication interface increases the predictability level of execution precedence and execution time in complex real-time domains.

Implementation

The distributed object-oriented integrated architecture has been used by us for developing a real

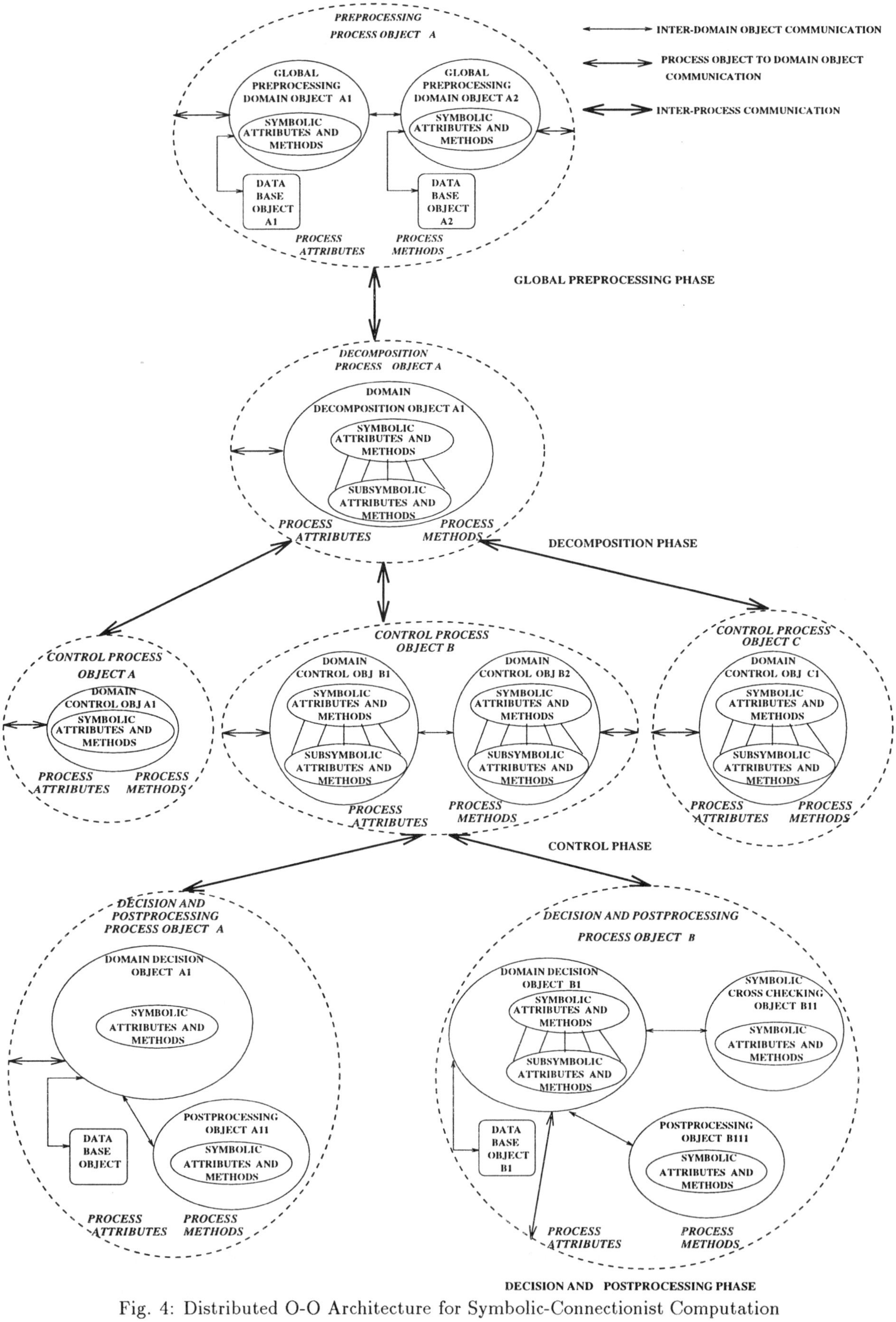

Fig. 4: Distributed O-O Architecture for Symbolic-Connectionist Computation

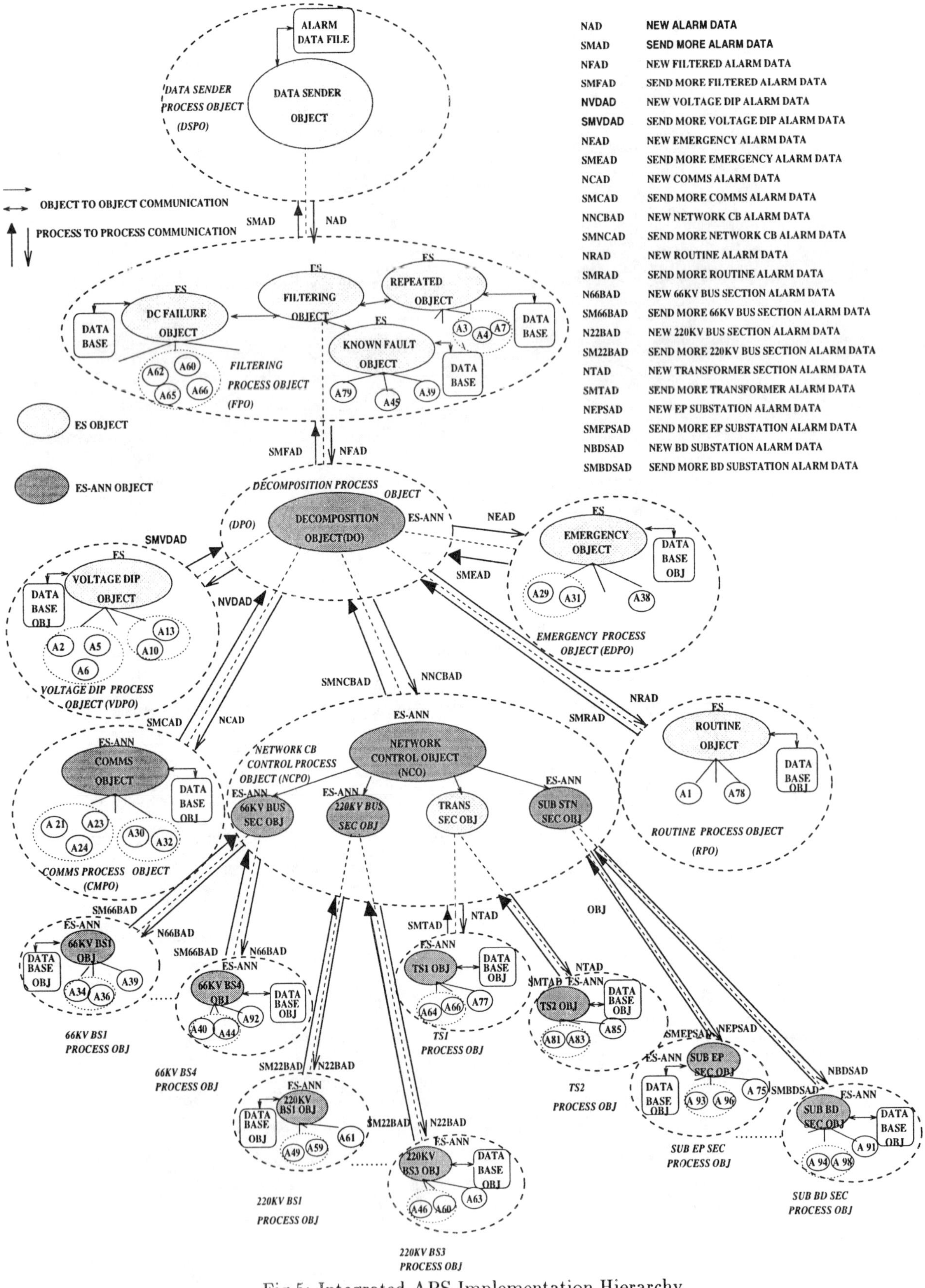

Fig.5: Integrated APS Implementation Hierarchy

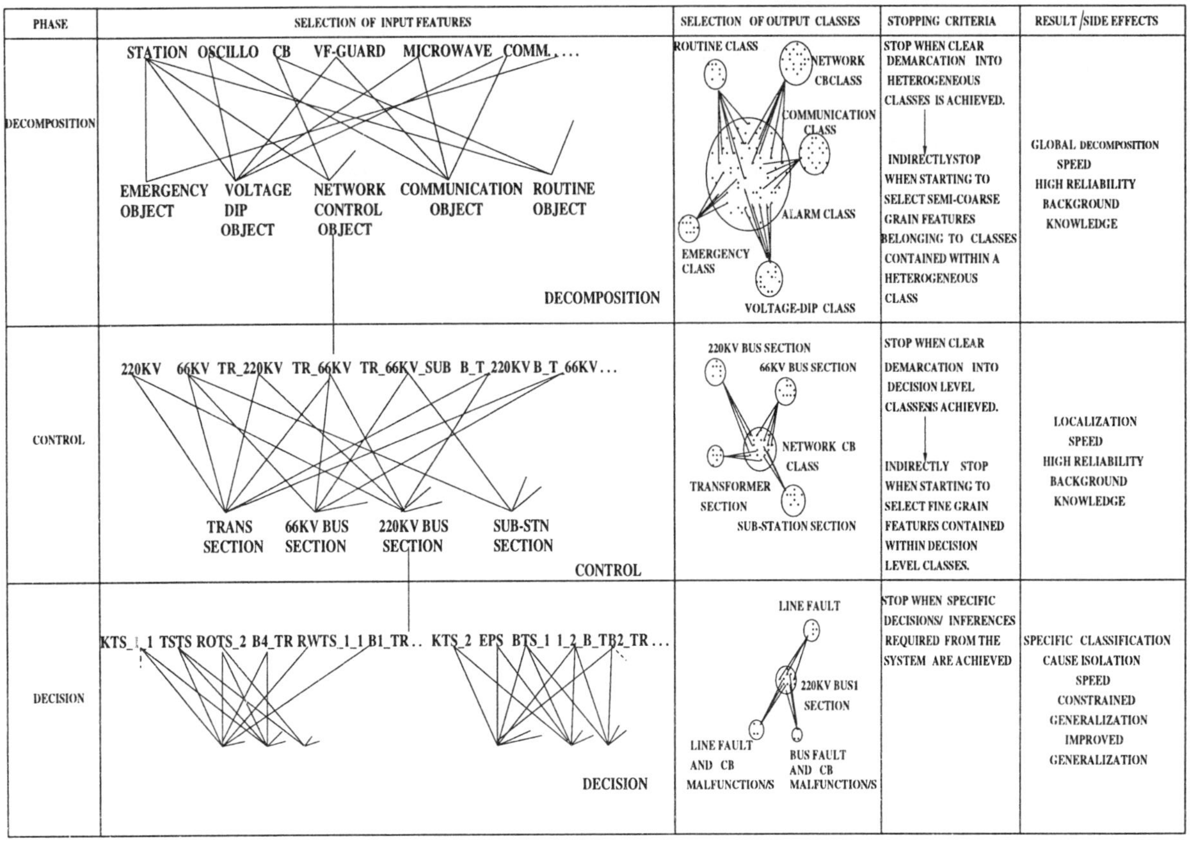

Fig.6: Learning Knowledge and Strategy of Integrated APS

time alarm processing system for the 220kv automated Thomastown Terminal Power Station(TTS) and its 16, 66kv connected substations. The alarms are received by the regional control centre (under jurisdiction of State Electricity Commission, Victoria) from the automated TTS and its connected substations.

The objective of the Alarm Processing System(APS) is to assist the operator in a) Isolating the event, b) Isolating the cause of the event, and c) achieving a response time range of 4 secs (for simple faults) to 30 secs (for severe multiple faults).

The network topology, real time alarm data and operator's problem solving solving behavior have been taken into account for developing APS. APS has been analyzed, designed [16] and implemented in an object-oriented continuous, parallel and distributed manner. In Fig 6 BS1, BS2, etc. represent individual bus section objects. TS1, TS2, are transformer section objects and EP, BD, etc are substation section objects.

Nexpert[21] object along with external C routines have been used to implement APS. APS uses alarm data used by control room staff for real time inferencing. ES objects in the implementation hierarchy (Fig.5) are shaded in light grey where as ES-ANN objects are shaded in dark grey. In Fig.5, the light dashed lines circles represent instances of each object. Light dashed line circles with solid line circles inside represent composite instances. "A78" represents an alarm with id 78.

All the bus section objects shown in Fig.5 compete with each other in order to infer single/multiple faults on a particular alarm snapshot. There are 60 rules and 28 ANNs in the APS at the moment. The ANNs are trained using combination of Kohonen nets and back propagation algorithm with discrete inputs and outputs. The learning knowledge and strategy used for training a subset of ANNs in different phases is shown in Fig.6.

Coping with Size of TTS

There are 65 Circuit Breakers (CB) associated with TTS and its associated 16 substations. There are 190 input features and 110 output classes associated with TTS at the moment. The use of the integrated computational architecture has facilitated effective handling of problem complexity associated with power systems. The TTS network topology has been sectionalised into 66kv and 220kv bus section objects, substation section objects and transformer section objects. This task sectionalisation has en-

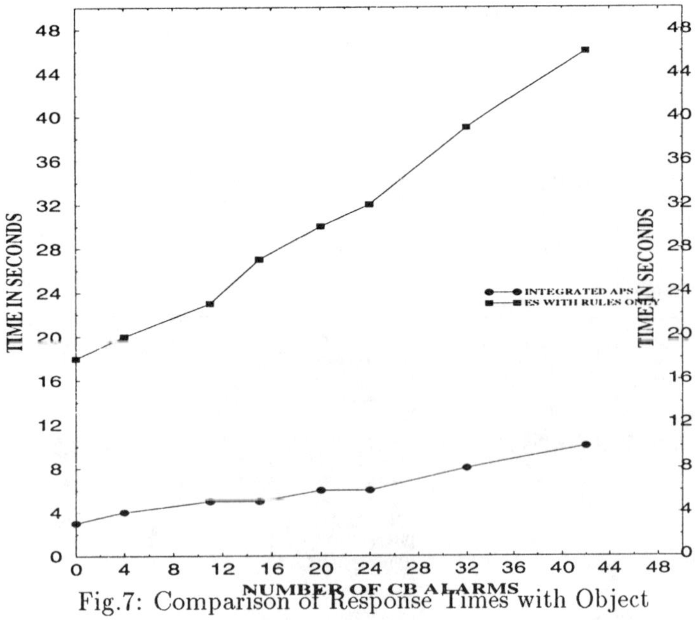

Fig.7: Comparison of Response Times with Object Loading Time

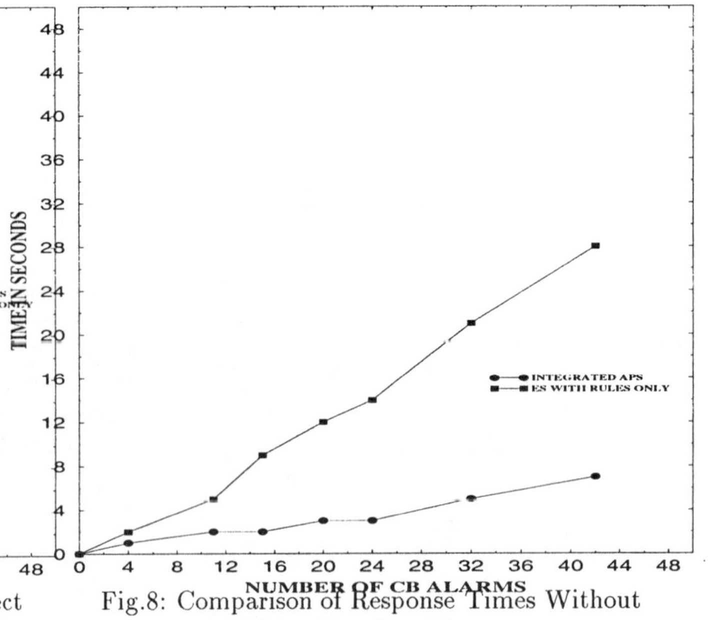

Fig.8: Comparison of Response Times Without Object Loading Time

abled us to train all the ANNs in APS with 100% generalization and obviated the need for using any symbolic fault propagation cross-checking models in the present version.

Temporal Reasoning

The status of the power system constantly evolves over time. The messages coming from different plants have different transmission times. The meaning of a message depends on what happened before and, in certain cases, on what happened next.

All inferenced alarms are stored in a temporal history data base by different domain objects, thus making a provision for re-inferencing and retracting previous conclusions. In case of multiple network circuit breaker alarms where ANNs are used, we only store/retrieve the previous alarm snapshot input/output activation/s and time of the last alarm id in the previous alarm snapshot. This allows for very quick re-inference because of the massive parallelism inherent in ANNs and consequent minimal search time. The present and previous alarm snapshot activations are used for re-inferencing. The previous and present alarm data has a time bound validity. The concept of time bound inferential validity is also used in the APS. The re-inferencing time is largely dependent on the maximum time difference between reading of two alarm snapshots from the alarm data file. The capacity of the APS to cope with temporal reasoning is significantly better than the existing expert system based alarm processing systems which can take up to a few minutes for re-inferencing because of large search time.

Response Time

A comparison of the response times (on a Sun 4/280 workstation) with a expert system with objects and rules (executed in a parallel and distributed manner) only and integrated APS is shown in Fig. 7 and 8. The comparison is based on 42 multiple Circuit Breaker (CB) alarms representing multiple faults. As can be seen the response time of the integrated APS is significantly better than the expert system with object and rules only. In general the integrated APS is able to infer 180 alarms in 58 secs on a Sun 4/280 workstation. Some of the inferences on 180 alarms are shown in Fig.9 The rightmost large window in Fig.9 shows the creation and exit times of different APS objects which are executed in parallel.

Conclusion

In this paper we have developed a distributed object-oriented architecture with symbolic-connectinist objects for complex knowledge intensive domains. We have proposed an integration of the two paradigms keeping in view the cognitive and practical viewpoints. From a computational point of view we have proposed a set of generic software classes which can be used for symbolic-connectionist applications. Through the integrated architecture we have defined a set of conceptual constructs (e.g. preprocessing, control, decision, postprocessing) which can be used by a system developer for building complex symbolic-connectionist systems. We have also outlined the learning knowledge for training neural networks in the integrated architecture. Thus the

Fig.9: Inferences of Different Alarms

integrated architecture can be seen as an enabling architecture for building symbolic-connectionist applications especially in the diagnostic domain.

We have applied the integrated architecture in a real time alarm processing system and shown encouraging performance results. Thus the difference in the symbolic approach and the integrated approach lies not only at the cognitive level but also at the performance level.

The development and application of the integrated architecture has also shown how the object-oriented paradigm can be used as a as a unifying computational paradigm for integration of AI problem solving architecture, software technology and database technology.

References

[1] E. H. P. Chan,"Application of neural-network computing in intelligent alarm processing," in*Proc. Power Industry Computer Applications Conf.*, 1989, pp. 246-251.

[2] Chandrasekaran, B., "What kind of information processing is intelligence," *Foundations of Artificial Intelligence*, Cambridge University Press, 1990, pp14-45.

[3] Fodor, J. A,. and Pylyshyn, Z. W., "Connectionism and Cognitive Architec ture: A Critical Analysis," *Cognition*, 1988, vol.28, pp3-71.

[4] CIGRE International Task Force WG38-06-02, Convenor - T. Dillon, "Survey on Expert Systems in Alarm Handling", *Electra* , No. 139, 1991, pp. 133-147.

[5] Liu, C. C., "Knowledge Based Systems in Electric Power Industry", *Pro ceedings of the IEEE*, 1992,Vol. 80, No. 5, pp. 659-662.

[6] C.C.Liu, "Practical Use of Expert Systems in Power Systems," Invited paper in *Proc. 4th Symp. on Expert System Applications in Power Systems*, 1993.

[7] Gallant, S., "Connectionist Expert Systems", In *Communications of the ACM*, 1988, February pp. 152-169.

[8] Shastri, L. and Ajjanagadde, V., 1989, "A Connectionist System for Rule Based Re asoning with Multi-Place Predicates and Variables ", *Technical Report MS-CI S-8906*, Philadelphia,

PA: University of Pennysylvania Computer and Information Science Department.

[9] Becraft, R., Lee, P. L. and Newell, R. B., 1991, "Integration of Neural Networks and Expert Systems for Process Fault Diagnosis", in *International Joint Conference on Artificial Intelligence*, pp. 832-37.

[10] Handelman, D.A., Lane, S.H., and Gelfand, J.J., 1990, "Integrating Neural Networks and Knowledge-Based Systems for Intelligent Robotic Control", In IEEE Control Systems Magazine, pp. 77-86.

[11] D. S. Kirschen and B. F. Wollenberg, "Intelligent Alarm Processing in Power Systems," in *Proc. of the IEEE*, 1992, pp. 663-672.

[12] F. Eickhoff, E. Handschin and W. Hoffmann, "Knowledge Based Alarm Handling and Fault location in Distributed Networks," in *Proc. of the Conference on Power Industry Computer Applications*, 1991, pp495-498.

[13] A. G. Jongepier, H. E. Dijk and L. V. D. Sluis, "Neural Networks Applied to Alarm Processing," *Proc. 3rd Symp. Expert System Applications in Power Systems*, 1991, pp. 615-620.

[14] R. Khosla and T. Dillon, 'Perspectives for Integration of Artificial Neural Networks and Expert Systems', in *IEEE Australia-New Zealand Conference on Information Systems*, Perth, pp. 624-628, Nov. 1993

[15] R. Khosla and T. Dillon, 'Learning Knowledge and Strategy of a Generic Neuro-Exp ert System Model', in *AAAI International Symposium on Integrating Knowledge and Neural Heuristics*, Florida, USA, May 1994.

[16] R. Khosla and T. Dillon, 'An Object-Oriented Analysis and Design of a Alarm Processing System using a Symbolic-Connectionist Model', *Third IEEE International Symposium on Electricity Distribution and Energy Management*, Singapore, Oct. 1993, pp. 675-80.

[17] Smolensky, P., "Connectionism and Foundations of AI", in *The Foundati ons of AI: A Sourcebook*, Cambridge, UK: Cambridge University Press, 1990, pp. 306-326 .

[18] K. Kim and J. Park, "Application of Hierarchical Neural Network to Fault Diagnosis of Power System", *Proc. 3rd Symp. Expert System Applications to Power Systems*, 1991, pp. 323-327.

[19] M. Munneke and T. S. Dillon, "An Object-Oriented Approach to Alarm Handling", Expert Systems Applications in Power Systems, Prentice Hall, London, 1991, pp343-382.

[20] D. Tesch et al., "A knowledge-based alarm processor for an energy management system," *IEEE Trans. Power Syst.*, vol.5, 1990, pp. 268-275.

[21] Nexpert Object manuals, Neuron Data Inc., Palo Alto,California, 1989.

[22] Zeidenberg, M., Neural Networks in Artificial Intelligence, 1990. Ellis Horwood Limited, West Sussex, England.

A CONCEPTUAL MODEL FOR MONITORING AND FAULT IDENTIFICATION IN INDUSTRIAL INSTRUMENTS USING DATA FROM A DISTRIBUTED CONTROL COMPUTER

Philip A. Collier

Department of Computer Science, University of Tasmania, GPO Box 252C, Hobart, 7001, Australia

ABSTRACT

I present an outline theory for monitoring and fault identification in industrial instruments (MFIII) associated with a hydrometallurgical process. This is implemented in a prototype expert system for Pasminco Metals-EZ at Hobart. The theory is presented here in the form of a conceptual model expressed using the KADS framework in a form that is reusable. I conclude with a discussion of the suitability of KADS for this enterprise.

INTRODUCTION

Any system for automatically monitoring or diagnosing faults in an industrial process relies on measurements of properties of the process. (This contrasts with fault diagnosis systems that interact with operators, who may make direct observations of the process and report these on demand to the system.) One of the first tasks of such a system is to validate the measurements.

Measurements from an industrial process will come from instruments of various sorts. Each type of instrument will have various quirks possibly with signatures that can be used to assist with detailed fault diagnosis of the instrument. Some instruments are relatively complex devices that have non-trivial internal behaviour that may also be subject to deeper analysis.

In this paper I will be concerned with a generic theory for monitoring and fault identification in industrial instruments (MFIII). The main issues that I will consider are:

- Instrument readings that are not mutually consistent with other nearby readings.
- Instruments that fail low or high, with a sharp change from their previous good values.
- Instruments that give an indication of having a bad value, because they are outside their range of calibration.

The challenge in considering instrument problems is to sort out the genuine instrument failures from temporary aberrations that are caused by other effects in the process. In the case of a temporary problem, a reasonable value for the measurement may need to be substituted. Once instrument values are validated then the process of monitoring and diagnosing the underlying process can proceed with more confidence.

The style of generic theory will be model-based to a appropriate degree. It trades the sophisticated form of semi-quantitative simulation used in [DK91], for re-useability encouraged by KADS.

BACKGROUND

The work in this paper is based on a prototype expert system developed for the Strong Acid Leach (SAL) section of the Pasminco Metals–EZ (PMEZ) Zinc Refinery in Hobart, Tasmania. The system identifies several classes of faults similar to those outlined by [BL93]: "operator error, malfunction or failure of process equipment, or undesired changes in the conditions of the process environment".

The purpose of the SAL plant is to dissolve zinc and iron from a slurry, leaving a silver/lead residue which is a valuable by-product of the zinc process.

Process control at PMEZ is by a Taylor MOD300 Distributed Control System (DCS). Data from the distributed control system is accessed using an RS232 serial link.

The prototype of SAL Operator's Assistant (SALOA) runs on a PC. It is implemented in NEXPERT OBJECT 2.0. Each minute a data acquisition program runs, updates a data base of historic values, and produces files containing raw data and derived features. At regular intervals, SALOA produces a report of the new problems, or a report of all outstanding problems. SALOA has run in prototype form for several months.

There are five classes of problems reported by SALOA: *Phantom problems* include certain bad signals and process documentation for inputs and outputs in *tracking* mode. *Controller problems* are indicated by deviation of measured variable from set point or tanks overflowing or empty. *Instrument failures* is the work reported here. *Process upsets* includes checks that agitators and rakes are running, manual settings on valves are appropriate, and acid tank alarms. *Process failures* check for failures in motors, agitators, rakes, etc.

One design parameter of SALOA is to avoid any direct input from plant operators. The operators are already using a sophisticated computing environment, and an extra screen would probably not have been accepted. The main effect of this decision is that all monitoring and diagnosis had to be carried out solely with the data recorded routinely by the DCS. Messages from SALOA are routed back to the DCS for display on the standard operator's console.

The significance of this framework is that SALOA has to infer for itself the accuracy or otherwise of instrument values, prior to using these values for other forms of monitoring and diagnosis.

The theory for diagnosing instrument problems is expressed here in a style similar to KADS, Knowledge Analysis and Documentation System [SWB93], [GB92]. This is one of the best developed amongst several different modelling formalisms, see for example [Deb89], [FB93], [Bar93].

A major contribution from the KADS methodology is that knowledge should be represented in layers. The layers encourage a principled representation of knowledge, where the domain itself is modelled directly. This discourages ad hoc production rules that are encouraged by many expert system shells. Long-term maintenance of the knowledge is facilitated by a principled knowledge representation.

The layers in KADS can be classified as follows:

- *Domain knowledge* This is where the concepts, structures, sets, relations and their properties are described. These model the underlying domain and its processes. The relations are often concerned with (sets of) expressions. These can be used to model causality and other regular processes in the domain.
- *Inference knowledge* This comprises *knowledge sources* and associated *meta-classes*. Knowledge sources are tightly constrained to belong to one of a few different types, although this is not followed strictly in papers showing outlines of KADS applied to different domains. Knowledge sources have one or more inputs and one output. Each meta-class is associated with something in the domain layer that plays some role for the knowledge source. For example a concept can act as input for one knowledge source and the resulting hypothesis for another.
- *Task knowledge* This is the outline algorithm for the use of the knowledge. It is a description of the order in which knowledge sources shall be used, together with other bookkeeping functions.
- *Strategic knowledge* This is an opportunity to specify how and when task knowledge can be used. In terminology of MYCIN, this is where the meta-rules are specified.

These layers are illustrated by example in the next

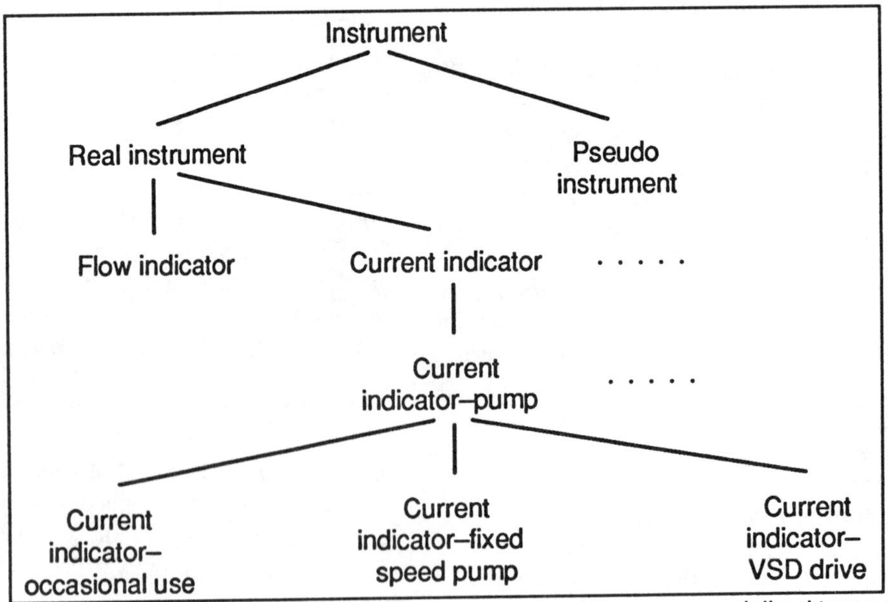

Figure 1 The *IS-A* relation between the generic concept *Instrument* and the more specialised types of instrument.

Table 1 A domain schema for MFIII

Primitive	Name	Description
Concept	Instrument	A physical instrument that measures a property of the process.
Property	Instrument:property-value	A property and its current value for an instrument.
Relation between concepts	Instrument IS-A Instrument	Sub-type hierarchy of different types of instruments.
Relation between expressions	Set-of Instrument:property-value INCONSISTENT-WITH Instrument:property-value	A set of instrument property values that are not consistent under normal operating conditions.
Relation between expressions	Set-of Instrument:property-value MAY-CAUSE Instrument:property-value	A set of instrument property values that may cause another instrument property value.

Table 2 Properties associated with an *Instrument.* †=not used in the MFIII.

Name	Type	Description
Alarm_HI	Number	†Value of high alarm
Alarm_HIHI	Number	†Value of high-high alarm
Alarm_LO	Number	†Value of low alarm
Alarm_LOLO	Number	†Value of low-low alarm
Full_range	Number	MV_high – MV_low, ie the magnitude of the range of values for the Measured_variable
Measured_average	Number	Mean of the Measured_variable over the last 5 minutes
Measured_slope	Number	Slope of the Measured_variable over the last 5 minutes
Measured_variable	Number	Current measurement from the instrument
MV_close_to_zero	Boolean	Measured_variable is within 1% of MV_low and not increasing.
MV_high	Number	Maximum value of Measured_variable for the instrument
MV_low	Number	Minimum value of Measured_variable for the instrument
Failed	Boolean	The instrument has failed.

section.

The KADS framework shares its roots in entity-relation constructs and in object-oriented design methodologies, for example [RBP*91]. From an OOD point of view the domain layer is similar to an object model (plus relations), the task knowledge is similar to a functional model.

KADS ANALYSIS FOR MFIII

The domain schema for MFIII in table 1 identifies the

Table 3 Additional properties associated with a *Real Instrument*

Name	Type	Description
Bad_signal	Boolean	Indication for the instrument about whether the signal is outside the normal operating range. If it is bad, then no Measured_variable is supplied.
Rapid_change	Boolean	Has the signal changed rapidly, that is by some percentage over some period?
Steady_measurement	Boolean	Has the Measured_variable been identical for a time period?

Table 4 Additional properties associated with an *Current indicator—pump.* FSP=fixed speed pump, VSD=variable speed drive

Name	Type	Description
Flow_measured_slope	Number	(FSP, VSD) Slope of the Measured_variable for the flow over the last 5 minutes
Speed_signal	Number	(VSD) The speed signal sent to the VSD drive.

Pump:Status = RUN
Alternate_pump:Status = STOP
Flow:Measured_variable < 0.05*Flow:Full_range + Flow:MV_low
Pump:Speed_signal > 0.0
INCONSISTENT-WITH
Current indicator—VSD drive:MV_close_to_zero

Figure 2 A part of the relation *INCONSISTENT-WITH*. This relation needs to be constructed specially for every application of the theory of MFIII, as it depends on identifying redundancy in instrument values at various points in the process. In this case, we are concerned with a *Current indicator—VSD drive* in the context of duplicate pumps. Care is needed at times when flow lines can suffer from syphoning.

fundamental generic concept *Instrument* and how it is related to different types of instrument.

As usual, properties of *Instrument* are inherited down to the specific types of instrument. The properties of an *Instrument* are shown in table 2. Additional properties of a *Real Instrument* are shown in table 3, and those for different types of *Current Indicator—pump* are shown in table 4.

The final two components of the domain schema are relations between expressions. These represent some of the relevant underlying processes in the domain. The relation *MAY-CAUSE* is a coarse-grained form of model-based reasoning about causality. The essence of these two relations is shown in figures 2 and 3. Several of the expressions in these figures refer to aspects of the domain that are not modelled in this paper, for example *Pump:Status*.

In summary, the domain knowledge is entirely concerned with instruments, not with how problem solving will be achieved.

The inference knowledge is a tightly specified part of the KADS methodology. This facilitates the existence of

Effect of recent change in Status of a command to a pump:
Pump:Time_since_STA_change < 5.0
MAY-CAUSE
Current indicator—pump:Rapid_change

Low level in a tank may cause a rapid change in the *Current indicator—pump* of the pump(s) that pump out of the tank:
Level:Measured_variable < Level:Alarm_LO
MAY-CAUSE
Current indicator—pump:Rapid_change

A change in flow set point may cause a rapid change in the *Current indicator—pump* of the pump(s) that pump out of the tank, and of the Flow from the tank:
ABS(Flow:Set_point_slope) > threshold
MAY-CAUSE
Current indicator—pump:Rapid_change

ABS(Flow:Set_point_slope) > threshold
MAY-CAUSE
Flow:Rapid_change

An instrument showing a high average value may slip out of range with a Bad value:
Real_instrument:Measured_average + 0.05*Real_instrument:Full_range > Real_instrument:MV_high
MAY-CAUSE
Real_instrument:Bad_signal

An instrument with a rapid change may overshoot the range with a Bad value:
Real_instrument:Rapid_change
Real_instrument:Measured_slope _ 0.0
MAY-CAUSE
Real_instrument:Bad_signal

Figure 3 A part of the relation *MAY-CAUSE*. Several of these issues are of general applicability, providing that the appropriate flows are connected with the pumps that drive the flows etc.

generic *interpretation models* for various types of tasks. As such they constitute a specialised shell for problem solving in that task, with only the domain knowledge to be filled in. Interpretation models consist of *inference structures* and *task knowledge*.

For MFIII I have chosen not to use one of the generic interpretation models. The model for monitoring [SWB93, p38] is a reasonably close fit, but the specialised model presented in figure 4 accurately describes a framework to fit this domain.

The knowledge sources in figure 4 are defined in figure 5. Some of these definitions refer to relations that were foreshadowed earlier, others are defined only informally here.

Most of the meta-classes in figure 4 will be connected with *Instrument:property-values* as defined in table 1. In this model we assume that all information about an instrument, including complaints are stored as properties of an Instrument.

Task knowledge is the outline algorithm for the use of the inference knowledge. The task knowledge for MFIII is shown in figure 6. There are two ways that an instrument problem can be detected. One way is that features are detected by *compute-1* that need investigation. These are *Rapid_change* or *Bad_signal*. Investigation involves computing possible acceptable values for those discrepancies, and seeing whether the value can be explained. If the value can be explained then it may be necessary to *assign-value* a good value to take the place of a bad value.

A direct detection of a faulty instrument is possible using *compute-3*. This looks for mutually inconsistent instrument readings.

There is no strategic knowledge provided MFII. This is in common with many KADS models.

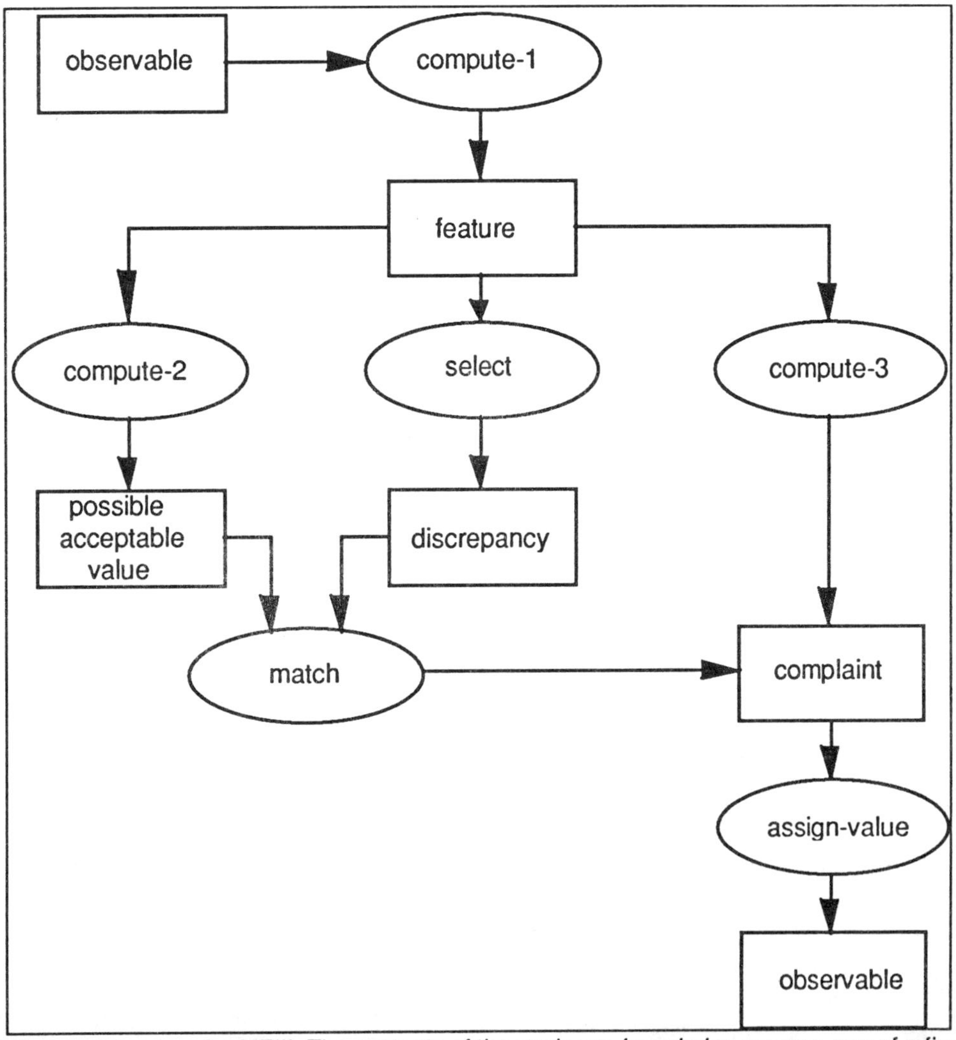

Figure 4 An inference structure for MFIII. The contents of the ovals are *knowledge sources*, one of a fixed number of allowed in KADS. The rectangles represent *meta-classes*. The meaning of these is elaborated in the text.

compute-1(observable → feature)
This uses domain knowledge that is not included in the domain model in this paper. Most observables, which are instrument measurements, status indications etc, are transmitted directly as features. Additional features such as *Measured_average, Measured_slope, MV_close_to_zero, Rapid_change, Steady_measurement* are computed in this phase. The *Measured_average* and *Measured_slope* are computed over the last five minutes. *MV_close_to_zero* is true if the *Measured_variable* is within 1% of the *Full_range* of *MV_low* and not rising significantly. *Rapid_change* is parameterised on each instrument type. It is *true* if there is a change of *x*% over t_1 minutes. *Steady_measurement* is true if the instrument value, which is quantised, is identical over t_2 minutes. Computed features are assigned to appropriate properties.

select(feature → discrepancy)
Various features are potentially discrepancies that may indicate an instrument is not working. This knowledge source picks up any Instruments having Bad_signal property or Rapid_change with value true.

compute-2(feature → possible-acceptable-value)
This knowledge source is driven by the Instrument that needs to know a possible_acceptable_value. For this instrument the MAY-CAUSE relation is used to try to find a possible acceptable value for the Instrument.

compute-3(feature → complaint)
This uses the INCONSISTENT-WITH relation. If all the expressions in the relation are satisfied by the current property values, then the Instrument mentioned in the expression in the second argument is faulty. This is a new complaint.

match(possible-acceptable-value + discrepancy → complaint)
If a possible acceptable value matches the discrepancy , then this indicates that the discrepancy can be explained by some legitimate process. In this case an empty set is returned as result, otherwise the discrepancy is a complaint and is returned as result.

assign-value(complaint → observable)
If the complaint is explained by some legitimate process, then it may be necessary to update some of the observed values to make them legitimate. For example if an instrument has been showing a high value and then changes to be BAD, the following assignments are required:
Real_instrument:Measured_variable = Real_instrument:MV_high
Real_instrument:Bad_signal = FALSE

Figure 5 The knowledge sources for MFIII.

DISCUSSION OF KADS

The KADS philosophy of separating knowledge into layers provides sound guidance for the formation of a clear conceptual model. The domain layer provides a firm foundation stone for the building of a model of expertise. Using an *IS-A* hierarchy with properties and inheritance encourages the domain layer to be implemented in an object-oriented language like C++ or an object-oriented shell like NEXPERT. Care in this enterprise should provide great flexibility for later building of the inference knowledge and task knowledge.

The use of relations to model various processes in the domain layer is both innovative and restrictive. The independence from later use of the knowledge contributes to the clear conceptual model. However relations are restricted in their expressiveness. For example, only two 'arguments' are allowed, and they are often typed weakly as (set of) expression. This means that I cannot formally specify that certain expressions will refer to certain types of objects. Also the last tuple in figure 3 should really have three arguments, the third being the assignment operations to be performed to recover from the problem. At present these are linked only informally in the description of *assign-value(complaint → observable)*.

One of the most difficult features of KADS to grasp from reading the literature is the issue of the primitives used for knowledge sources and their meaning. This seems to be acknowledged by the KADS group itself, see [SWB93, p94]. The primitives appear to be important as they are used extensively in the library of interpretation models. However, the impact of these is not entirely clear. There are several examples of work by the KADS group where different primitives are introduced.

The knowledge source presented in figure 4 differs from many in the KADS literature mainly because it lacks a knowledge source to specify normal values for

```
task check-instrument-readings
    goal:
    validate the instrument readings,
        report faults where appropriate and attempt to correct other bad instrument measurements.
    input:
    synchronous snapshot of instrument values and data base of recent values
    output:
    set of valid complaints that correspond to faulty instruments
    task-structure:
    check-instrument-readings(observable → observable) =
        compute-1(observable → feature)
        valid-complaints = {}
        select(feature → discrepancy)
        FOR EACH discrepancy ∈ discrepancies
            compute-2(feature → possible-acceptable-value)
            match(possible-acceptable-value + discrepancy → complaint)
            valid-complaints = valid-complaints ∪ complaint
            assign-value(complaint → observable)
        END
        valid-complaints = valid-complaints ∪ compute-3(feature → complaint)
```

Figure 6 The task knowledge for MFIII.

the situation under consideration so that they may be compared with the actual values. This method requires a (set of) powerful methods for model-based reasoning, which may be possible with more sophisticated interpretation of instrument measurements [For87], [WD92], [CL93].

The inference structure for MFIII takes an alternative position. It has limited knowledge about how certain features of instrument values may propagate to generate other possible features in the process. (For example if a tank is nearly empty then the flow out may drop sharply.) If the possible features turn out to be present, then they are explained by the antecedent feature. In general there are probably several equally valid inference structures for a domain, and this needs to be recognised in KADS.

The domain of MFIII seems to fit well with the KADS style of conceptual modelling. A project where the underlying domain involves people and their actions seems to be much less easy to model in this way. Here we have certain people and their actions associated with certain outcomes. The relations in the KADS domain knowledge are much more elusive in this situation, as there is no clear notion of causality.

CONCLUSION AND DISCUSSION

This paper provides a generic theory for MFIII. As such it provides a strong foundation for any system where such reasoning is needed. It takes the KADS methodology a step further than just providing an interpretation model.

The theory here still leaves a few open problems. It does not describe the order in which instruments should be considered. Under the reasonable assumption that only one instrument fails at a time this is probably not serious, as otherwise the failure of one instrument could be seen to legitimise the failure of another. Since the whole problem tackled is under-constrained the theory here can be summarised as a combination of relative simplicity with non-trivial abilities.

ACKNOWLEDGEMENTS

This work has been supported by a CSIRO Collaborative Programme in Information Technology and an Australian Research Fellowship (Industry).

REFERENCES

[Bar93] Barbuceanu, M., Models: toward integrated knoweldge modelling environments *Knowledge Acquisition*, **5**, 245–304 (1993)

[BL93] Becraft W. R. and P. L. Lee, An Integrated Neural Network/Expert System Approach for Fault Diagnosis, *Computers chem. Engng.*, **17**, 1001–1014 (1993)

[CS90] Cheung, J. T.-Y. and G. Stephanopoulos, Representation of process trends—Part I. A formal representation framework, *Computers chem. Engng.*, **14**, 495–510 (1990)

[Deb89] Debenham, J. K., *Knowledge Systems Design* Prentice-Hall, New York (1989)

[DK91] Dvorak, D. and B. Kuipers, Process Monitoring and Diagnosis: A Model-Based Approach, *IEEE Expert*, **6**, 3, 67–74 (1991)

[GB92] Gaines, B. R. and J. H. Boose, *The KADS approach to knoweldge engineering*, Special issue of *Knowledge Acquisition*, **4**, 1 (1992)

[For87] Forbus, K. Interpreting Observations of Physical Systems, *IEEE Trans. on Systems, Man and Cybernetics*, **SMC-17**, 3, 350–359 (1987).

[FB93] Ford, K. M. and J. M. Bradshaw, *Knowledge Acquisition as Modelling*, Wiley, New York (1993)

[RBP*91] Rumbaugh J., M. Blaha, W. Premerlani, F. Eddy and W. Lorensen, *Object-oriented modelling and design*, Prentice-Hall Inc., New Jersey (1991).

[SWB93] Schreiber, G., B. Wielinga and J. Breuker, *KADS: A Principed Approach to Knoweldge-Based System Development*, Academic Press, London (1993)

[WD92] Whiteley, J. R. and J. F. Davis, Knowledge-Based Interpretation of Sensor Patterns, *Computers chem. Engng.*, **16**, 329–346 (1992)

EVOLUTION OF A MULTI-PARADIGM BLACKBOARD BASED NETWORK DIAGNOSTIC PLATFORM

B. Pagurek, J. Bryant, N. Dawes
Dept. of Systems and Computer Engineering
Carleton University, Ottawa, Canada K1S5B6
Phone: (613) 788-5724 Email: pagurek@sce.carleton.ca

ABSTRACT

This paper describes the evolution of a dual role agent architecture for a network diagnosic platform. Based on a blackboard model, the platform employs agents capable of cooperative diagnostic reasoning as well as management protocol communication. Several of the current diagnostic agents are also briefly discussed.

INTRODUCTION

In a previous paper we reported on a multiparadigm diagnostic system for communication networks that was based on a blackboard architecture[PDK92][Ign92]. Success with the blackboard architecture in an environment where the network was simulated or previously measured data was transported off line and fed to the blackboard, led us to consider it as the underlying platform architecture as we moved into the real world of intricate communication protocols. In bringing our blackboard and diagnostic tools to the real world we were faced with several problems:

1/ The necessity of accommodating multiple diagnostic tools in a cooperative environment
2/ The necessity of introducing a flexible environment in which new diagnostic tools could be introduced, explored, and tested without disrupting operational network management. The complexity of introducing and testing new network management applications appears to be one of the greatest inhibitions to implementing them on line.
3/ The necessity of working within the network management protocols which determine both the type of messages transmitted and, more importantly, the information base used.

Network management protocols like SNMP[Ros94] (The Simple Network Management Protocol) are actually very complex and cumbersome to use. In their basic form they are limited to using a predefined knowledge base called a MIB which is in effect a schema for a distributed data base. In this environment a client - server relationship is typically used in which a manger process at the network operation centre is a client requesting information from network management agents (servers) which are distributed throughout the network and reside on critical network devices. One of the main difficulties in introducing new diagnostic tools is that perhaps new measurements need to be introduced and at least new intermediate results need to be computed and transmitted that were not taken into account when the system was designed and built. As a result the network management world has developed several methods for extending such MIBs dynamically through the use of what is termed subagent technology [WCC[+]94] [RWW94].

Another difficulty is that in complex networks, there is often not a pure client - server relationship because distributed and hierarchical structures are used where mid-level managers act sometimes as manager and sometimes as agent. In exploring this world we recognised that the Knowledge System (KS) structure we employed for reasoning agents, with dynamic registration and asynchronous operation, could also be used to develop subagents to extend SNMP agents when new information was required. Moreover the blackboard architecture could be used to fully integrate diagnostic reasoning tools with management protocol tools in a way that facilitated dynamic introduction of new reasoning agents and new knowledge bases without interrupting the regular operation of the network management process. The end result desired therefore is dual role agents which can serve as extensible network management agents and diagnostic reasoning agents

This paper contains four sections. The first summarizes the previous system and its primary reasoning and diagnostic agents. The second describes new agents which have been developed and tested. The third describes the motivation, requirements, and architecture for the next generation of Pegasus systems. The last section summarizes where we are and our short term and medium range plans.

1. THE ORIGINAL PEGASUS BLACKBOARD ARCHITECTURE AND SUBSYSTEMS

The Pegasus network diagnostic system is organized around a fairly basic blackboard architecture used to provide centralized communication, a global database, and control for multiple cooperating knowledge systems. It was written in the 'C' language using the Unix facilities of a SUN workstation. As reported earlier the blackboard was divided into several read/write protected areas such as events, problems, hypotheses, knowledge source activation records (KSARs) etc. A Knowledge Source (KS) could be written in any language as long as it could call C interface functions which provide controlled access to the blackboard data structures. Memory management was achieved by assigning a separate semaphore for each area on the blackboard. Current KSs include C and Pascal processes but also included one Prolog KS and a forward chaining rule system written in Cosmic Clips. Several KSs are described briefly in sections 1.1 to 1.4 below.

1.1 THE WAN SIMULATOR

The WAN simulator is a full scale high speed discrete event simulator and is capable of simulating a wide variety of WANs of realistic international scale. For the sample network it generates and handles packets in Internet Class and ATM Protocols. Data collection is done with SNMP and PING as in the sample network but OSI management standards can be accommodated. All status changes are reported to the blackboard through a program interface. The simulator is used to allow for the dynamic introduction of device malfunction and repair through the blackboard interface as well as for the testing of device functionality. Simulation device data and statistics are available as in the real world. A second major use of the simulator is as a network model for model based reasoning.

1.2 THE PROBABILISTIC REASONER

The belief propagation reasoner called BANES [DAP95] collects evidence of network faults from the status messages emitted at regular intervals from the data collectors. Every three minutes BANES issues a report on its current beliefs including an almost certainly broken list and an intermittently failing list. This information is posted to the blackboard as diagnostic hypothesis and can be used as additional information by the other KSs. BANES has proved to be enormously successful as a passive monitor and hypothesis generator. Its main limitations are that it is passive, acting only on status messages and not additional tests, and that it was not designed to respond to end to end faults as would typically be reported in a user's complaint phone call.

1.3 THE RULE-BASED EXPERT NETWORK DIAGNOSTICIAN

The rule-based expert network diagnostic KS, called END, is a classical expert system in that it implements human expert reasoning acquired from true expert network operators in the form of rules. Implemented in the shell Cosmic Clips, it provides a forward chaining inference tool. The expert system is meant to be an active KS. For example it can respond to a user complaint, order tests and respond to the results. The expert system implemented several expert diagnostic scenarios exactly as the expert performed them. END responds to telephone complaints for problems that result from higher level protocol errors that do not cause low level status error messages and are thus hidden from BANES.

1.4 THE NEURAL NETWORK ANALYZER

One of the goals of the target system is that it should be able to recognize certain problems early and to react before the problem reaches a critical stage. This is a difficult task because the environment is very complex and there is a great deal of data including status messages, error rates, and performance data to absorb. To be proactive the system must be able to recognize subtle changes in network performance. The neural network analyzer (LEAPR) was our first attempt at using machine learning to provide intelligent monitoring in the network domain. The basic idea is to train the system what the operator considers to be anomalous or normal behaviour without having to explain why he thinks so, and to use the neural network to recognize such patterns in the future.

2.0 NEW KNOWLEDGE SOURCES

The results from the previously reported system, Pegasus

I, were sufficient to encourage further development resulting in several new experimental KSs being developed for abnormality detection in communications systems, particularly faults associated with TCP/IP network routers.

2.1 CBR

A Case Based Reasoning prototype represented previous anomalous situations in terms of patterns of status messages and traffic statistics. For example for a router case traffic features included mean interface traffic volume out, traffic trends, and relative contribution to zonal or neighbouring traffic. The type of faults we were interested in detecting included faults like data storms, router table corruption and large file transfer anomalies. Diagnosis was attempted by comparing the current situation to recorded past generalized cases. A significant effort was applied to feature analysis and case representation. The method used an interesting graph representation of sequences of status messages which permitted convenient representation and efficient partial match retrieval of cases. The prototype as implemented didn't do case adaptation so it is perhaps better termed a memory based reasoner. The retrieval accuracy in retrieving best cases was about 80% for the class of problems considered. There were however problems of scalability for the large network of interest and case library acquisition was difficult. This led to research into automated case and symptom discovery as discussed below for the ADDS process.

2.2 ML

A Machine learning tool that employed the CN2 rule induction learning algorithm was developed. to further explore automated rule acquisition for rule based diagnosis. Network data was divided into a training sets for rule learning and a test set. Again much effort was spent on feature selection for a class of problems that were not the result of simple break faults and didn't result in explicit status messages but were the result of more subtle problems like router table corruption. The purpose of these methods was to come up with a more proactive diagnostic process that could detect anomalous situations before complete failure occurred. Application of CN2 also involved the development of a pruning technique used to avoid the overfitting problem due to noise such as overlapping classes in the training data and to learn better which features discriminated better between faulty and healthy routers.

2.3 ADDS

The above two new methods worked reasonably well on a class of overlapping problems but none could be classified as an outstanding success in its class as was BANES for break faults, In the process of investigating feature selection the best ideas from the previous approaches were incorporated into a new umbrella diagnostic process called ADDS for Automatic Discovery Of Diagnostic Symptoms. ADDS [DAW94] which essentially subsumes the two previous methods is based on a unified theory for abnormality discovery, machine learning, and case classification for diagnosis. It is far quicker and much more accurate than the methods it replaces. ADDS will continuously monitor network devices, detect faults, and alert operators. It will decide if new abnormalities are occurring, learn them and ask the operator if these are real problems to be typed or if things it should ignore in the future. In this way its integration of abnormality detection, learning of diagnostic features, and generalization to cases used for diagnosis can be exploited in real time diagnostic domains.

In operational use all router communication lines (thousands of them) are continuously monitored. For each line data is aggregated into events consisting of measurements of eight variables including a time stamp, input data rate, input error rate etc. Events are recorded every 5 minutes or so depending on the importance of the device being monitored. A case is defined as a sequence of events of arbitrary length. ADDS is capable of handling hundreds of kilobytes of monitored data per second. Labelled training data will be used first for case descriptions if it has been previously acquired but this is optional since ADDS is capable of unsupervised learning of abnormalities.

The ADDS process discovers the probably best symptoms to discriminate between faulty cases and healthy ones by exploring a variety of functions of measured attributes to come up with derived diagnostic metrics to be used in real time. It automates the discovery process to a large extent by applying a new formulation of sequential Bayesian decisions to attribute combinations. ADDS has been successfully tested on a variety of simulated communications faults, on some real traffic data, and on a previously unsolved medical diagnosis problem. At the time of writing a series of field trials on real communications networks is planned for the next six months. These networks will include heterogeneous TCP/IP and ATM systems. Fault diagnosis from the application layer down to the physical layer will be attempted.

In fact some preliminary results from these field trials are available and can be summarised as follows. On the experimental systems network, ADDS has been provided with sets of monitoring information generated both when conditions were normal and when problems were introduced. ADDS has found the general diagnostic solution to all the problems provided, such that normal and problem conditions were completely separated using derrived metrics found by ADDS This solution has been recoded and will be installed and tested over the next two months. This method also recognises faults and detects abnormal behaviour in complex state table systems.

Data from the operational network has been exhaustively analysed. Particular solutions were readily found. The common features of these were analysed and a possible general solution to recognising very many abnormal and problem patterns in traffic measurements has been defined. This will be further analysed and then tested against new samples from the operational network before installation is attempted.

2.4 OPERATIONAL REQUIREMENTS FOR ADDS

The ADDS system is the newest development in our investigation of diagnostic algorithms and has operational requirements typical of other such systems. for example:

- It is to be deployed on two very different networks, one a large mature operational IP/TCP network with an established SNMP management system and the other an experimental ATM network with only a few device proprietary monitoring tools available. In the first case any disruption of the network management and operation is not tolerable. In the second case the management framework will be subject to continuous development.
- ADDS will need traffic data from the net and the raw data will need to be saved briefly on the blackboard for analysis of several possible tools.
- The diagnostic system will need to coexist with other diagnostic tools some of which will have overlapping data requirements, and some of which are not completely defined yet.
- ADDS will report alarms and hypotheses to the blackboard along with other tools. Some higher level coordination will be required both in determining the sampling of network data and in the correlation of diagnostic hypotheses.

3.0 PEGASUS II: MOTIVATION, REQUIREMENTS, AND ARCHITECTURE

Several major conclusions about the development process arose from our previous work. The first is that in arriving at one or more diagnostic tools in a real world environment, there often arise several partially successful overlapping tools to be compared and results correlated as well as several versions of one tool to be tested. Secondly, evaluating experimental tools on the actual network was hard because installation was very difficult and possibly disruptive. It was for these reasons that we originally developed the simulator for trials. However as results became more promising, we had to seriously consider how our diagnostic algorithms could be implemented in the real world. Finally, current communications networks can produce an enormous amount of data both raw and aggregated into status messages. A multiparadigm diagnostic system must be capable of acquiring such diverse types of data and make it available to a variety of diagnostic tools. Some of the tools of course wouldn't even exist at system inauguration and in fact some of the measurement variables might also have to be introduced a posteriori in response to the needs of new tools. In other words the system needs to be extensible in terms of both the diagnostic algorithms employed and the types of management information employed. This led us to PEGASUS II.

3.1 THE SNMP ENVIRONMENT

SNMP is the de facto standard protocol for enterprise network management and in reality it is hardly simple. The basic architecture is a client server relationship in which the manager is the client and the servers are remote agents distributed throughout the network on critical devices and who respond to requests from the manager - possibly more than one. Agents maintain a local database of device statistics such as error and traffic rates for each of the ports etc. The database description or schema is called the MIB or Management Information Base. The term MIB is often used to represent the data instances themselves rather than the schema. In the MIB, management object types and instances are identified by unique OIDs or Object Identifiers. Retrieval from remote agents is accomplished by SNMP GET and SET operations in which packets encoded with management variables are sent. The entire management system constitutes a distributed database and has all the complexities associated with such databases.

Large amounts of data can be involved and management traffic can easily swamp the network if care is not taken. SNMP uses inband signalling. That is, it uses the network itself for transmission of management information. This is complicated by the fact that management is needed most when the network is at its worst and is failing. Typically provision is made for out of band access to nodes using telnet and modems.

Protocol oriented management systems are very complex and cumbersome to modify, This makes it difficult to add management information to the MIBs after construction. Such addition is however necessary to accommodate new devices, subnetworks, and services as well as new diagnostic algorithms. At the present time a tremendous amount of effort is being addressed to the extensibility issue - that is the providing architectures to allow the addition of new MIBs and management information. The main direction of this effort is in the development of subagent technology for extensible agent construction. The main idea is that to add new management information one can develop a subagent to support it and provide the required database data. Subagents can be attached to master agent in one of several ways, either as a separate process communicating with the master agent using some form of IPC or by linking the subagent code in with the master agent code. Fig.1 represents such relationships.

When a master agent receives a request for information from a manager it demultiplexes the request by deciding which subagent(s) supplies that information and sends the request to the appropriate agent. When the subagent(s) responds the master then returns an aggregated SNMP response to the requesting manager. To facilitate demultiplexing the master needs a dispatch table of OIDs and what subagent can supply the required OID instances. Such a directory is built when a subagent registers with the master agent. This can be done dynamically during execution.

In principle a manager should not have to know how an agent handles requests or that subagents exist. Internal communication between subagent and agent can be accomplished in many ways but the data content of the communication and the registration process are usually specified in a subagent protocol such as the DPI protocol (Distributed Protocol Interface)[WCC+94]. Again, new subagents can be separate processes communicating with the master using IPC or can be dynamically linked into the master agent process. The attachment process will depend on whether the subagent is to be local or is to reside on a remote machine but the main point is that subagents should be able to register dynamically with a running master. It is interesring to note the similarity of this model to that of a knowledge system registering with a blackboard as a reasoning agent.

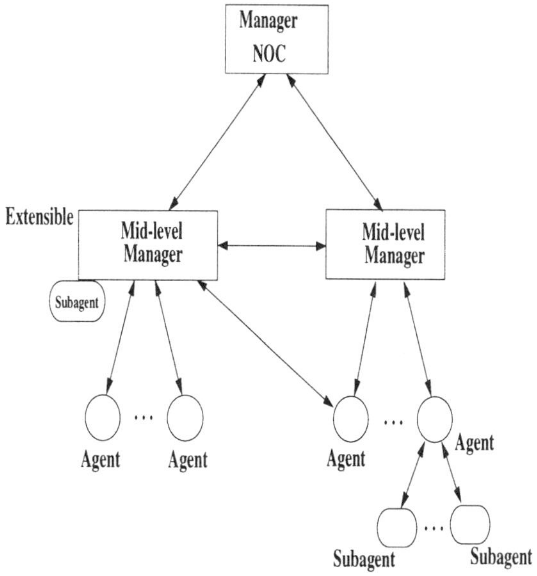

Figure 1 Possible Management Relationships

3.2 MID-LEVEL MANAGEMENT

Another major thrust in network management is that of distributed management, usually hierarchical, but not limited to this relationship. The main reason for this approach is the reduction of WAN management traffic and the introduction of management intelligence out into the network. Fig. 1 also shows the position of the mid-level manager in the hierarchical setup. By performing statistical aggregation and other value added duties the MLM need only pass on higher level information to the NOC. For example in a wide area network which includes multiple geographically distributed local area networks the MLM can be involved in local diagnosis of LAN segments and communication with the NOC or other MLM for correlation purposes, In order for new diagnostic algorithms to be introduced and tested there may be new management communication required. In other words there is the need for extensible mid-level managers to facilitate dynamic addition and experimentation.

We have identified therefore two separate needs, the protocol need for management communication and extensibility through subagent technology and the diagnostic algorithm needs for cooperating distributed diagnostic agents which come and go. In order to make such a complex system as simple as possible It was our goal to unify the handling of the two tasks as much as possible, sharing technique where ever possible. Our successful experience with the robust blackboard type of platform with its KS/Agents communicating through shared memory led us to explore this approach further and led to an implementation of Pegasus II.

3.3 EVOLUTION OF THE PEGASUS IMPLEMENTATION

The original Pegasus blackboard, designed as it was to suit the requirements of a particular collection of knowledge systems, is not entirely suited to the implementation of extensible mid-level managers and agents. In view of this, a new and more general version (Pegasus II) has been developed and is currently being brought into use.

One fundamental difference between the old and new blackboards is that, while Pegasus provides a fixed number of blackboard areas, with each having a specific role (e.g. there is an "events" area, a "hypotheses" area, and so on), Pegasus II allows for any number of general purpose areas. Moreover areas are not static (as in the original backboard) but can instead be created and destroyed as required. At the time a blackboard is brought into existence (by invoking a supervisory process which continues to exist for the life of the blackboard) it contains no areas at all. User processes may, however, freely create new areas by sending appropriate requests to the supervisory process, which then looks after the actual mechanics of area creation. Similarily, once the need for an area has passed, a user process may request that it be destroyed.

Another key difference lies in the nature of communications. Pegasus conforms to the classical concept of a blackboard in providing what might be termed "one-to-many" or "broadcast" communications. Whenever anything is posted to a blackboard area, it is, in effect, received by all user processes which have registered an interest in that area. This is ideal in situations in which a broadcast capability is in fact required, but less than desirable in cases in which the sender knows, a priori, that a particular posting will only be of interest to some particular user. In the case of an extensible agent, for example, the master agent always knows which subagent any given request is to be sent to, and replies generated by subagents are always directed to the master agent. If a blackboard is to be an attractive basis for the implementation of subagents, it must thus make some provision for "point to point" communications, and Pegasus II does so. Indeed, at its present stage of development only such communications are supported, making Pegasus II, for the moment, more of a message exchange than a blackboard. The general framework also allows for more traditional blackboard communications, however, and it is our intention to support them in due course, so making Pegasus II into a completely general communications facility.

At present Pegasus II supports two kinds of blackboard areas: "message" areas and "data" area. Message areas are simply a wrapped version of Sun's IPC message queue facility and are intended to provide a basic inter-process communications facility within the blackboard framework (as opposed to forcing processes with simple message exchange requirements to either i) use overly powerful blackboard constructs or ii) communicate outside the blackboard environment). Data areas are implemented using shared memory areas and are perhaps best thought of as a form of shared heap. A process which wishes to post information to a data area does so by obtaining space in it and building a structure containing the desired information. Except in that the memory involved is shared, and that space is obtained by calling a function in the Pegasus II API, the process is analogous to building a structure using memory obtained by calling "malloc". Once a structure is complete, control of it is passed to one of the area's other users, which is then free to use and/or modify it as it sees fit before either releasing the space involved or passing control on to another user. Note that, while it is presently only possible to send a structure to a particular user, one might also allow a structure to be send "to all". In this case each recipient would be allowed only "read only" access, no "retransmission" of a received structure would be permitted, and the space occupied by a structure would be physically released only upon all of the recipents having logically released it. Another possibility being considered is a form of "serial" broadcast in which a structure is passed from user to user until all have seen it, and yet another is the invention of a third type of area specifically designed for broadcast communications.

User processes interact with the Pegasus II blackboard (or, more correctly, with a Pegasus II blackboard, since many of them may exist at once) through an API consisting of a collection of C functions. Before doing anything else, a process must register with the blackboard as a whole by calling "BB_check_in". Once checked in, a process may create new areas and/or gain access to existing ones by making appropriate function calls. Though a detailed description of the API obviously cannot be given here, it is perhaps worth noting that, in registering an interest in a data area, a user process can specify how it is to be notified of the arrival of structures. The options are i) no notification (presumably the process will poll the area from time to time), ii) signal a specified (SUN IPC) semaphore, and (iii) send a message identifying the data area to a specified message area. The last of these options is particularily useful in that, by waiting upon the specified message area, a process may await either the arrival of data in any one of possibly numerous data areas or the arrival of messages from other processes.

At the time of writing, Pegusus II is in the process of development and, in addition to providing for broadcast communications, work is also underway upon rendering it compatible with multi-threaded programs (in particular with Solaris Threads).

3.4 THE OVERALL ARCHITECTURE

Figure 2 illustrates the proposed mid level manager architecture. A number of diagnostic processes (D_1 through D_n) require data from the network in order to carry out their diagnoses. This data is obtained by sending requests (e.g. "please sample such-and-such variables every so many seconds") to a data collector process, which carries out the required sampling and makes the results available to the diagnostic processes. Both the requests to the data collector and the results to its sampling flow through the blackboard. In the latter case a simple area to which diagnostic processes can post requests for the data collector is all that is needed, but in the first case the possibilities are rather more complex, and, while one might want to post all data obtained to a single area, one might prefer to have separate data areas for each of the diagnostic processes. Note that the use of a single data collector process allows data requirements to be consolidated, thus reducing sampling to the absolute minimum required. This is important because it is to be expected that, in general, the data requirements of multiple diagnostic processes will overlap considerably.

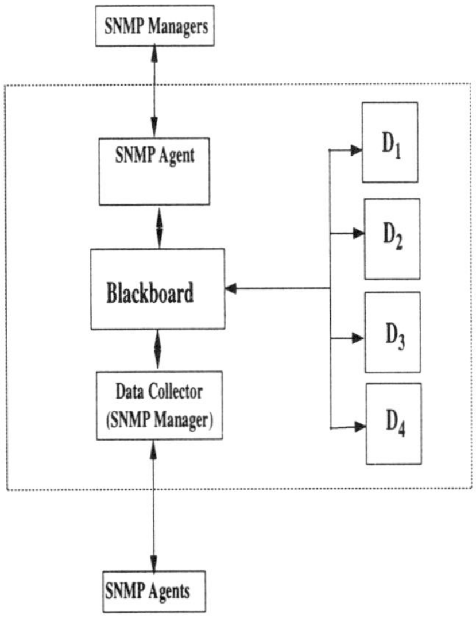

Figure 2 Mid Level Manager Architecture

The model does not require any particular relationship between the various diagnostic processes, and in fact there need not be any relationship at all. It may be that the diagnostic processes all operate quite independently of each other, and that none are even aware of the existence of others. While no relationship is required, any conceivable relationship can, however, be readily achieved. Thus, for example, one might have a master diagnostic process which gathers no data at all but instead simply serves to integrate the hypotheses produced by the other diagnostic processes. The blackboard not only permits such inter-diagnostic communication, but is in fact ideally suited to it. Blackboards were, after all, originally developed to serve exactly this type of need.

Conclusions reached by the diagnostic processes must, of course be somehow communicated to higher-level managers. There are essentially two ways of doing this. One is to have the mid-level manager send unsolicitated SNMP INFORM notification's[Ros94] containing diagnostic results to the managers, and the other is to have managers extract information from the mid-level manager by issuing SNMP GET's. Note that, in practice, the two ways are in fact complementary, and that, after a problem has been brought to a managers attention by means of an INFORM, that manager will typically want to use GET's to obtain further information. Diagnostic processes which wish to cause an INFORM to be sent do so by sending a request (via the blackboard) to the SNMP Agent architecture component, which then looks after actually transmitting the INFORM. Allowing managers to obtain information from diagnostic processes is rather more complex, though those familiar with the "extensible agent" problem will recognize the situation at once. Basically each diagnostic process which wishes to make information available supports a collection of variables known as a MIB and registers the existence of these variables with the SNMP Agent. Upon receiving a GET request which pertains to variables supported by diagnostic processes, the SNMP Agent interacts with the processes involved to obtain the corresponding values, and then uses them in generating a response. In the terminology of extensible agents, the diagnostic processes are acting as "subagents". Note that, in addition to allowing managers to extract information, a diagnostic process' MIB can also be used as a means of controlling the process' operation (by having the manager use SNMP SET's to modify the values of control variables).

A number of ways of implementing extensible agents have been proposed but are somewhat similar to blackboards in that they involve dynamic registration and deregistration. Moreover may take the form of protocols which might be used in conjunction with any number of transport mechanisms, including simple blackboard areas. Indeed one recent proposal[Whi94] (MCSA) involves a

"Message Exchange" whose functionality is not all that far removed from the simplified blackboard presented earlier. The main point is that a blackboard is a very reasonable proposition for subagent implementation, and its use it has great advantage that diagnostic processes need only deal with a single interface.

4.0 SUMMARY AND CONCLUSIONS

The original Pegasus system (the blackboard, simulator and KSs) was robust, efficient, and successful but the KSs developed were not deployed operationally except for live trials. Although the trials for BANES were extremely successful, operational requirements made it difficult for companies to add Pegasus or its subsystems to their existing network operations systems. We have invented new diagnostic agents, particularly ADDS, which address many problems suggested to us by operations managers. We want to deploy ADDS, BANES, and other agents, first testing them on operational networks and then letting the operators use them operationally. To accomplish this we need to have a comprehensive architecture which minimally interferes with operational systems on installation and as those operational systems change and grow. The Pegasus II architecture will hopefully satisfy these requirements. At the time of writing a prototype message area and a prototype data area of Pegasus II have been set up and the basic communication mechanisms established. The rest will be completed over the next several months. ADDS will be tested on two operational networks over the next six months.

REFERENCES

[PDK92] Pagurek B., Dawes N., and Kaye R., "A MultipleParadigm Diagnostic System for Wide Area Communication Networks" IEA/AIE-92, 5th International Conference on Industrial and Engineering Applications of AI and Expert Systems, Paderborn, June 1992.

[Ign92] Iqneibi S. M., "A Blackboard Architecture to Support Network Fault Diagnosis" M.Eng Thesis, Dept. of Systems and Computer Engineering, Carleton University, May 26, 1992. Published as Departmental Report SCE-92-11.

[DAP95] Dawes, N.W., Altoft, J., Pagurek, B., 'Network diagnosis by reasoning in uncertain nested evidence spaces. To appear in IEEE Transactions on Communications, Feb. 1995.

[Daw94] Dawes N.W, "Automated Discovery of Diagnostic Symptoms :ADDS" Report SCE-94-28 Carleton University, Dept. of Systems and Computer Engineering, October 1994.

[Ros94] Rose M., "The Simple Book: An Introduction to Internet Management" 2nd Edition, Prentice Hall Publishing Co. 1994.

[WCC+94] Wijnen B., Carpenter G., Curran K., Sehgal A., and Waters G., "Simple Network Management Protocol Distributed Protocol Interface Version 2.0", Internet RFC 1592, March 1994.

[RWW94] Romanov A., White W.H., and Wilson P., "The Open SNMP Agent" v 0.2 Paul Freeman Associates, March 28 1994. Available via ftp from ftp.uu.net : tmp/openagent.txt.

[Whi94] White B. et al "Open Agent Architecture, Multiple Component SNMP Agent (MCSA)" October 1994. Available by anonymous ftp from sunsite.unc.edu: /pub/micro/pc-stuff/ms-windows/winsnmp/winsnmp_agent/mcsa_v 08.txt

A conceptual clustering technique and its application to computer workload characterisation

K. I. Shihab

Department of Applied Computing and Mathematics, University of Tasmania at Launceston, PO Box 1214, Launceston, Tasmania, 7250.

&

J. A. Campbell

Department of Computer Science, University College London, Gower Street, London WC1E 6BT, UK.

ABSTRACT

This work is a part of a project that aims to design and build a workload-modelling system to be used for computer performance evaluation and capacity planning. It describes a conceptual clustering method for constructing a reliable classification of data. The underlying scheme is based on qualitative and quantitative measures of similarity in order to create classes. The method takes into consideration the objectives of the classification study that has not dealt with by previous work. It is computationally efficient and has high classification accuracy. The SOM (self-organising map) method is used to show the clustering capability of this technique. The simulation results of the application of these methods to computer workload characterisation indicate that the conceptual clustering has more potential for producing meaningful categories of the workload components than the traditional clustering methods.

1. Introduction

In order to evaluate the performance of a computer or a communication system, it is necessary to model the existing workload on these systems. Workload modelling mainly requires a clustering of the workload components that has been achieved by using numerical clustering techniques. However, these techniques have been under sever criticism because they are only based on similarity measures of the quantitative data.

A concept-based clustering approach provides an accurate and efficient way to develop a workload modelling system for analysing and constructing a representative workload model of a given workload. The produced models can be used by computer and communication system performance analysts and planners in academic and industrial settings to project performance for evolving computer systems, and to tune and compare such systems.

Also, the workload modelling system can be used as a bridging program between a software or accounting monitor and a widely used queueing network package (a performance modelling package).

2. Conceptual clustering techniques

Traditional clustering techniques based on numerical taxonomy do not seem to be adequate and they are recently experiencing severe criticism [SM86, Fis87, HB89]. The major criticism appears to be that the existing techniques arrange objects into classes in such a way that components in the same cluster reveal a similar numerical measure without regard to the contribution of these components for determining a meaningful concept. Consequently, the problem of the cluster interpretation is simply left to the data analysts [SM86]. This limitation can be overcome if a clustering technique takes into account beforehand both the concept and the numerical measure, rather than solely the numerical measure.

A conceptual clustering technique accepts a set of components and produces clusters (groups) of components based on quantitative and qualitative similarity measures. A description should be assigned to each cluster produced.

Researchers have considered conceptual clustering techniques as powerful and important of summarising and explaining data. It was first suggested by Stepp and Michaliski in 1980 [SM86]. They implemented their method in a program called CLUSTER/2 and presented comparison studies with 18 numerical clustering methods. It was revealed that the clustering obtained from CLUSTER/2 tend to match solutions most satisfactory for people. Chen and Fu [CF85] used conceptual clustering to organise expert system knowledge. Fisher [Fis87] described his incremental conceptual clustering method, which implemented by COBWEB program, as a mean of improving performance. He claimed that his program has the ability to deal with constraints imposed by real world environments. Hanson and Bauer [HB89] developed a system called WITT. They divided the clustering problem into two problems: the clustering and the characterisation. A similar approach to a numerical clustering technique was used for identifying a set of object classes. The characterisation problem involved determining useful concepts for each object class that is produced by the first method.

3. Computer Performance Evaluation

Satisfactory computer services depend greatly on the choice of configurations and the amount of capacity available on the computer systems. Performance evaluation of computer and communication systems helps not only in determining how well they are performing and whether any improvements need to be made, but also in understanding their behaviour in order to plan and to design the systems of the future. As the hardware cost of these systems is decreasing, their complexity and the demands being placed upon them are increasing dramatically. Therefore evaluating their performance is vitally important to manufacturers and users to assist them in the optimisation and planning of existing systems, and guide them in the design of new generations. Manual approaches based on intuition and experimentations with existing similar systems are unreliable and require expensive, exclusive use of resources. Hence performance evaluation based on formal models is necessary for efficient and reliable design and optimisation [HP93].

Although many computer-modelling packages have been used for performance projection and capacity planning, they require construction of baseline models as their inputs [LZG84]. These baselines are extracted from representative workload models produced by workload characterisation of existing computer systems. Without a reliable predictive model, the results from these performance modelling packages will be flawed. Therefore, it is imperative to study the input side of computer systems and to model the behaviour of users and their programs. An accurate predictive workload model is also indispensable in the construction of benchmark workload models and for workload forecasting [CH90, Ter92].

Building effective workload models require the experience and judgement of experts in performance evaluation and modelling beyond that normally available to an individual or a group [Shi94]. At present, workload models are developed on a case-by-case basis by experts who employ techniques such as numerical clustering and data analysis. Workload modelling deserves a more thorough investigation than it has received and AI (artificial intelligence) offers powerful tools to be exploited for automatic construction of workload models. Experimental studies conducted by many researchers have shown that conceptual clustering algorithms and artificial neural networks tend to be more effective than numerical clustering methods [SM86]. This is mainly because a real workload consists not only of quantitative data (e.g. elapsed time, start time, processor time, page faults, number of disk I/O blocks transferred) but also qualitative data (e.g. process type, username, ID, group, location or source).

4. Workload Characterisation and Parameterisation

This stage involves the collection, classification and implementation of a suitable set of representative examples for different types of workload.

Data have been collected from different sites running a Unix system. Our first intention is to find the most important traffic measures and their roles in the selection of a subset of workload components, and in the classification of these components. Also we are investigating the effect of the groups to which the users belonged and of the source of the traffic in order to assign an appropriate weight for each group and for each source. A piece of software has been designed and implemented to extract the data from the daily job accounting records, based on our characterisation scheme. The classification principles are:

* Classes should consist of components that reveal similar performance measures. (For example, I/O bound components should not be grouped in a single class that also contains CPU-bound components).

* Classes must distinguish workload components of special interest. (For example, if we are only interested in the response time of the interactive components, then we must have one class representing the interactive components and one or more classes representing the rest).

* Classes must distinguish different workload types (transaction, terminal and batch).

* Classes must not be limited to a small number. (For example we may have several classes representing batch services: One class for short jobs, one class for medium size jobs, and one for all others).

* Classes must distinguish different user groups and locations.

5. Methodology

The objective, therefore, is to produce a grouping of workload components based on conceptual clustering techniques. The results are compared with the results of the self-organising map (SOM).

Two conceptual clustering methods will be used and run in parallel. The weighted inter-cluster criterion is used to select the better clustering result. These methods are outlined as follows:

Given:

1. A selected sample of data extracted from a daily accounting routine. (The quantitative data should then be scaled. The standard measure or Z score was used [CH90].)

2. The classification principles that are mentioned in the progress report (see Workload Characterisation and Parameterisation).

3. A distance measure (inter-cluster criterion) which is given as follows: Let Z1 and Z2 be any two components (vectors) such that,

$$Z_1 = \{z_{11}, z_{12}, z_{13}, \ldots, z_{1n}\};$$
$$Z_2 = \{z_{21}, z_{22}, z_{23}, \ldots, z_{2n}\}$$

$$D_{ij} = \sqrt{\frac{\sum_{k=1}^{n} w_k (z_{ik} - z_{jk})^2}{\sum_{k=1}^{n} w_k}}$$

$$Q_{ij} = \sum_{k=1}^{n} \frac{W_k \left| \left(Z_i - \{z_j\}^{j \neq k} \right) \cap Z_j \right|}{|Z_i \cup Z_j|}$$

$$T_{ij} = D_{ij} + Q_{ij}$$

D_{ij} = the weight Euclidean distance between component *i* and component *j*

Q_{ij} = the weight qualitative distance between component *i* and component *j*

T_{ij} = the total distance measure between component i and component *j*

W_k = the weights that are assigned to workload parameters. Initially, these values are based on the objectives of the study and assigned to each parameter to reflect its relative importance, and then they will be changed during clustering according to the following rule of thumb.

$$W_k = W_k(1 - W_0); \qquad W_0 = \frac{1}{N+1}$$

where N is the number of cycles.

Goal:

Distinguish clusters $C_1, C_2, \ldots, C_n$, using Method 1 and Method 2 described below, and then select the better result according to the weighted inter-cluster criterion (ie. when inter-cluster similarity tends to be minimised and intra-cluster similarity tends to be maximised).

Method 1

This method is a hierarchical and incremental method. It is inspired by WITT's clustering algorithm [HB89].

1. Randomly divide the sample into two subsets, namely S_1^n and S_2^m; where n and m are the dimensions (i.e. the number of components) of S_1 and S_2 respectively.

2. Use S_1^n and start with n clusters, each containing a single component.

3. Apply the distance measure and the classification criteria to join the two closest clusters. This will continue until:

   a. a sufficient number of clusters has been formed (this depends on the complexity of the real workload and the objectives of the evaluation study), and

   b. either

      i. a sudden large change in the linkage distance has been reached, or

      ii. the values of W_k have increased too much.

4. If no clustering is achieved, change the weights W_k and repeat from step 2.

5. Derive concepts for the clusters. The components that are near the centroid of a cluster will lead to an appropriate concept for that cluster. (For example if requests arrive at constant rate, then the concept is *transaction type* [LZG84]). If a concept can not be derived for a cluster, then the components of this cluster should be added to the subset S_2 of unclassified components.

6. If two clusters belong to the same concepts, they may be merged according to the objectives and the complexity of the study.

7. Try to place the components of S_2 into the resulting concepts according to the above similarity measure and the description of each concept. Creation of a new concept occurs when the intra-cluster similarity resulting from placing a new component exceeds a specified threshold. The new component will be placed in a new cluster and a concept should be generated for that cluster. The fraction of components that can not be assigned to the existing concepts can be used as another measure of the clustering quality. If this fraction is small then stop. Otherwise save this state, reset weights W_k and repeat from step 1 until you reach a consistent state or a pre-defined number of cycles.

Method 2

Method 2 is inspired by the ideas of Stepp and Michaliski [SM86].

1. Select k disjoint components from the sample based on the given dissimilarity measure and classification criteria.

2. Derive m concepts ($m \leq k$) from the produced clusters. This step is the same as step 5 of Method 1. Merging of concepts may occur here as in step 6 of Method 1.

3. Try to place the rest of the components into the derived concepts (i.e. the m concepts are used as sieves for filtering the rest of the components).

4. Create new classes for those components that do not fit into any concept: A component does not fit into a class if the inter-cluster measure (dissimilarity distance) between the newly created cluster (which would result from placing a new component into the original cluster) and the original cluster exceeds a pre-specified distance (see step 5 of Method 1). This step proceeds as follows:

Let m_i = the number of derived concepts in step i; where $m_0 = m$.

n_i = the number of components that do not fit into m_i.

c_i = the concept number i.

If n_i/m_i is small, then repeat steps 1 - 4 on these components until you reach a consistent state (i.e. when $|n_i/m_i - n_{i+1}/m_{i+1}| \leq \varepsilon$, for small ε), else change the weights w_k and repeat steps 1 - 4 on the whole sample.

5. Manipulate the components that fit into or appear in more than one concept as follows:

a. Remove these components and place them on an exceptions list. Place the concepts that have their components removed on a test list.

b. Evaluate $D_j = Min\{d_1, d_2, \ldots, d_p\}$; where d_i for $i = 1 \ldots p$ is the dissimilarity distance measure between the newly formed cluster, which resulted from placing the component j into c_i ($i = 1, \ldots, p$), and the original one for each.

c. Use the minimum distance measure d_i to allocate the component j into the concept c_i.

6. An artificial neural network (ANN) approach to data classification

Neural networks are being used in data classification, and satisfactory results have recently emerged. Therefore, one of the aims of this project is to develop an unsupervised pattern classifier that will be able to respond to patterns which have some similarities to already learned patterns [Koh92]. Kohonen's Self-Organising Map (SOM) has been obtained, by file transfer via AARNet. The intention is to extend this unsupervised neural network and use it as a test case for our system.

The Kohonen's self-organising map (SOM) uses an unsupervised approach to learning. It defines a mapping from the input data space (input layer) R^n onto a regular two-dimensional array (output layer) of nodes. Each of these nodes is a composite profile or typical representative of all cases that fit that profile. The problem lies in which node should represent a case (input vector), the solution is to let the nodes compete for the right to represent a class of input vector. The winner will adjust its weights so that it becomes more like the input vector. Therefore, the winner is closest to the input vector. The neighbours of the winning node also adjust their weights to become closer to the input. This means that neighbouring SOM nodes should be similar or should represent similar SOM input vectors.

7. Simulation Results

The experimental data were obtained from daily reports that were collected on Sparc machines at the University of Tasmania, running under the Solaris 2.3 operating system. The reports, which relate to approximately 50 commands, form the input to the conceptual clustering and the SOM. They contain a command label and nine dimensions of data, such as the CPU time consumed, the memory space required, the number of disk I/O blocks transferred.

Before raw data were used with both systems, they were converted into a suitable format. Furthermore, the data were scaled by using Z score. Several utility programs specially written for the current study were used to carry out the conversion and the scaling activities. Table 1 shows a summary of the experimental results obtained.

The experimental results that obtained from applying several numerical clustering are not reported here since the produced classes are by no means acceptable. This is mainly because of the treatment of the qualitative data that are replaced by binary variables.

Table 1: The experimental results for the eight classes that were identified. The percentages represent the clustering accuracy relative to a manual classification.

	Clusters							
	1	2	3	4	5	6	7	8
Manual Classification	18 entries	2 entries	9 entries	2 entries	1 entries	14 entries	1 entries	3 entries
Conceptual Clustering	83.33%	100%	67%	100%	100%	100%	100%	66.67%
SOM	89.89%	50%	44.44%	50%	100%	78.57%	100%	0%

8. Conclusion

This article has described effective methods for designing and building a software tool for construction of reliable workload models of given real workloads. The methods are based on conceptual clustering and neural techniques. The main advantage of these methods is their capability of producing clusters of data based on quantitative and qualitative similarity measures. Using this approach, the produced workload models will have a general applicability to system performance evaluation and capacity planning.

References

[CF85] Cheng Y & Fu K (1985) "Conceptual Clustering in Knowledge Organisation", IEEE Transaction on Pattern Analysis and Machine Intelligence, 7, 592-598.

[CH90] Cady J and B Howarth (1990) "Computer Systems Performance Measurement and Capacity Planning", Prentice-Hall, Inc.

[CS85] Campbell JA & KI Shihab (1985) "Applications of Artificial Intelligence in Computer Performance Modelling and Simulation", Proc. International AMSE Conference, Modelling and Simulation, pp 11-25.

[Fis87] Fisher D (1987) "Knowledge Acquisition Via Incremental Conceptual Clustering", in Machine Learning, 2.

[HB89] Hanson S & M Bauer (1989) "Conceptual Clustering, Categorisation, and Polymorphy", in Machine Learning, vol. 3, pp 343-372.

[HP93] Harrison PG & NM Patel (1993), "Performance Modelling of Communication Networks and Comuter Architectures", Addison-Wesley.

Analysis Using Queueing Network Models", Prentice-Hall, Inc.

[Shi89] Shihab KI (1989) "ESYMA: An Expert System for Analysis of Computer Performance", CMGF International Conference Modelling and Measurement, pp 112-127, Nice, France.

[Shi94] Shihab, K (1994) "Learning and Reasoning by Analogy and their Applications to Computer Performance Evaluation", International Conference on Data and Knowledge Systems for Manufacturing and Engineering, Chinese University of Hong Kong, May, 1994, pp 651 - 657.

[SM86] Stepp RE and RS Michalski (1986), "Conceptual Clustering: Inventing goal-directed classifications of structured objects", in RS Michalski at al. (eds.), Machine Learning: An artificial intelligence approach, vol. 2, Morgan Kaufman.

[Ter92] Terplan K (1992), "Communication Networks Management", Prentice-Hall, Inc .

MODELING AND PREDICTION OF DYNAMIC BEHAVIOR FOR MODEL-BASED DIAGNOSIS

Sabine Kockskämper
University of Hamburg
Laboratory for Artificial Intelligence
Vogt-Kölln-Str. 30, 22527 Hamburg, Germany
kockskaemper@informatik.uni-hamburg.de

ABSTRACT

The majority of machines and artifacts in our world exhibit dynamic behavior, e.g. their output does not only depend on their current input but also on previous values. Representing time is therefore crucial for many tasks such as the monitoring and diagnosis of technical systems. This paper is concerned with behavior prediction for dynamic systems, i.e. the derivation of new facts by applying general knowledge about the behavior of a technical device to facts from a temporal data base. We describe properties of the modeling formalism and the analyzing method for dynamic behavior. We also sketch the steps necessary to integrate our predictor into a model-based (consistency-based) framework.

INTRODUCTION

Dealing with time-varying behavior is crucial for many tasks such as the monitoring and diagnosis of dynamic systems. But so far, most of the existing diagnosis applications are based on an important simplifying assumption: they neglect the explicit representation of time, i.e. they restrict themselves to static systems.

In this paper, we propose a framework for representing and analyzing time-varying behavior for the purpose of consistency-based diagnosis, a subclass of model-based diagnosis [dKW87]. In model-based systems, structure and behavior of technical devices is usually described and analyzed by means of constraints and constraint propagation. But when considering dynamic aspects (e.g. devices with state, feedback), traditional constraint systems are not appropriate. Episode propagation has been proposed as a way to integrate temporal reasoning into a constraint framework [Wi86], [De88], [Guc91] [Nö91]. One drawback of these approaches is that they support either qualitative or quantitative temporal relations but do not integrate both[1]. Qualitative simulation, another means to dynamic behavior analysis, is not well suited for model-based diagnosis because it represents system behavior as sequences of global system states which imposes a total ordering on the involved events. This often leads to irrelevant behavior distinctions which blows up the search space.

In the last years, the original consistency-based approach has been extended in several directions. For instance in [DS92], [BD93] a framework is provided to reason with multiple models. This approach also allows - at least in principle - to reason on different levels of temporal granularity which is an important means for efficient troubleshooting. In some situations a temporally coarse description suffices, whereas other situations require a more detailed specification of the device or component under consideration. [FL91], [NG93,94] propose extensions of the consistency-based approach in order to deal with temporal misbehavior. But so far, work has mainly focussed on the development of the diagnostic framework[2]. With our system we want to provide appropriate modeling and analyzing mechanisms in order to be able to exploit such a powerful framework.

In order to meet these new requirements on behavior modeling and prediction, we introduce a modeling formalism which uses explicit quantitative and qualitative temporal relations to describe complex temporal behavior and which is able to exploit temporal abstractions in order to represent behavior on various levels of granularity. We also present a predictive engine (model analyzer) for such models which is based on episode propagation.

The remainder of this paper is organized as follows. The following section introduces two examples of dynamic systems. After a brief overview of the underlying diagnostic framework, the following sections present the modeling language and the prediction procedure. Both are illustrated by the above mentioned examples. Afterwards, we sketch the steps necessary to integrate

1. or only to some extent

2. Besides the above mentioned temporal constraint propagators, two exceptions, i.e. systems which represent and analyze dynamic behavior, are [CPDT92], [NG93,94], see Section 7.

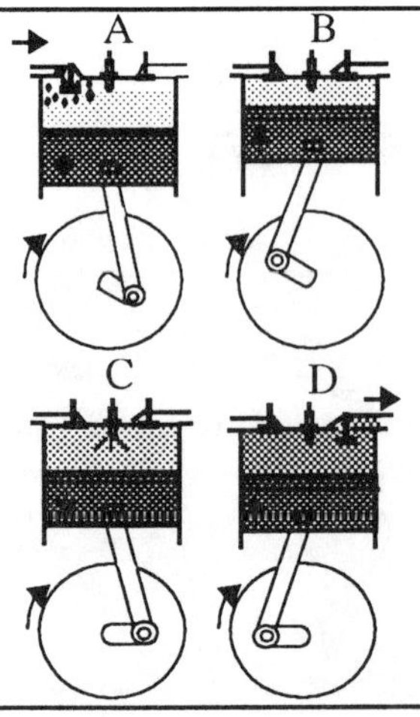

The Gas Engine. The gas engine is a machine for converting heat into mechanical energy. The operation of a four-stroke cycle engine is represented in Figure 1. Diagram A shows the intake stroke. At the beginning both valves (the inlet and the outlet valve) are closed. While the piston is moving downward, the inlet valve opens, and a mixture of air and gasoline vapor is being taken in. When the piston reaches the lowest point, the inlet valve is already closed. The piston now moves upward, compressing the mixture of air and gasoline vapor, as shown in B.

When the compression is complete, an electric spark ignites the gas. The heat of combustion of the vapor greatly increases the pressure, driving the piston downward (C). This is the power stroke. It is during this movement that the fuel delivers energy. After passing the lowest point, the piston moves again upward, the outlet valve opens, and the spent gases are discharged, as shown in D. The outlet valve closes shortly before the piston reaches the uppermost point. The fourth or exhaust stroke completes the cycle and a new cycle begins with the next intake stroke, as in A (after [CGK$^+$59], description slightly modified).

FIGURE 1: Four-stroke cycle of a gas engine

our predictor into consistency-based framework, in particular we discuss how it is combined with a truth maintenance system and which problems could arise. A discussion of our approach and related work will conclude the paper.

TWO EXAMPLES

In this section, we introduce two example systems where dynamic aspects play an essential role. The gas engine (exactly: one of its cylinders) has been chosen because it is a cleverly thought-out system of events which fit in exactly with each other, i.e. its correct functioning highly depends on the temporal order in which events occur[1]. The second system, a JK-flipflop, has been selected because logical circuits have become the "standard" demonstration domain for temporal constraint propagators, e.g. [De88], [Guc89]. In digital logic reasoning about time is essential, for instance, when hazards have to be identified. Considering only the boolean algebra does not suffice if one has to reason about phenomena that occur between clock pushes, e.g. set-up or hold times. Fig.2 c) shows a detailed behavior description which is more appropriate for such situations[2].

Taking a closer look at the first example, two things can be recognized. Firstly, we notice that the description of this system is purely *qualitative*. Several kinds of *temporal and value abstractions* are used to explain the functioning of the device, e.g. the piston is `moving`, the pressure is `increasing` etc. Secondly, it is remarkable that some of the *causal dependencies are replaced by temporal ones*. This is often found in descriptions of systems that involve complex behavior. For instance "event1 causes an event2", is translated to "event2 cannot start before event1" or "event2 starts within event1". This temporal interpretation of causality can be used to replace, for instance, a complicated mechanical interaction by inducing a temporal order on the things that should happen. An example: the fact that the inlet valve opens while the piston is moving downward (during the intake stroke) is caused by a mechanical linkage. When the connecting rod reaches a certain position, the valve is

1. this is an example for a (temporally) coarse behavior description

2. this description will serve as the example in the following

a)

J(t)	*K(t)*	*Q(t+1)*	*¬Q(t+1)*
low	*low*	*Q(t)*	*¬Q(t)*
low	*high*	*low*	*high*
high	*low*	*high*	*low*
high	*high*	*¬Q(t)*	*Q(t)*

b)

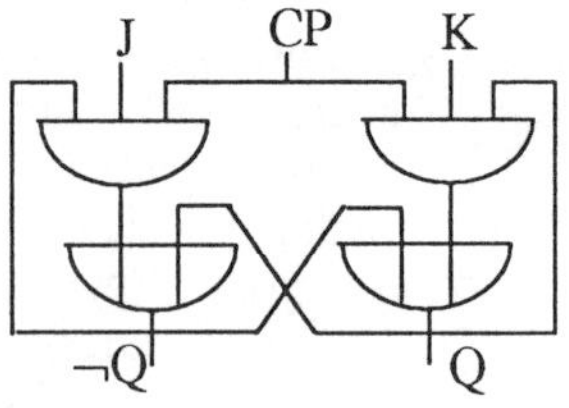

c) Detailed specification corresponding to the third row:

If a clock pulse occurs and J has been high from at least 30 ns before the clock pulse (CP) to at least 5 ns after the clock pulse and K is low in the same interval then Q is high after a delay time of 30 ns and ¬Q is low after 20 ns.

FIGURE 2: a) Discrete specification of the JK-flipflop, b) Block diagram of the JK-flipflop opened. But this linkage is neither mentioned in the text nor in the figure. Instead a temporal dependency of the expected events is given.

The level of behavior description of the second example (Fig 2.c) is quite different from that of the first one. Here we need the specification of *exact* times, i.e. the *duration* of events and the *temporal distance* between two events play an important role.

We will use both examples in order to illustrate the expressiveness our modeling formalism and the inference mechanism.

SKETCHING THE DIAGNOSTIC FRAMEWORK

In this section, we give a brief introduction to the diagnostic framework in order to give the reader an idea how our modeling and analysis methods can be used in a model-based system. We also point out which requirements the model analyzer has to met.

We follow a framework provided by [SD92], [BD93] with extensions towards temporal diagnosis by [FL91], [NG93,94].

As already mentioned, model-based diagnosis starts from an explicit model of structural and functional relations (the system description *SD*) of the device under consideration. Diagnosis is performed by tracing discrepancies between predicted and observed behavior (*OBS*) back to the components which cause the malfunctioning, i.e. which are responsible for the conflict. For each component a set of behavioral modes can be modeled. Because we have to cope with time-varying behavior, we associate a time interval with each mode, i.e.

$mode_i(c, [int.B\ int.E])$

denotes that a component c is in the mode $mode_i$ during the interval *[int.B int.E]*[1].

Hence, a temporal diagnosis is defined as an assignment of sequences of behavioral modes Π_t to the involved components which is consistent with the observations and the system description, and which covers a specified part of the time line such that

$SD \cup OBS \cup \Pi_t$ *is satisfiable.*

As already mentioned, switching between alternative models for a component is also supported. For this reason a set of working hypothesis is maintained characterizing, for instance, the model granularity which is currently used. Thus, a model of a component is represented as follows:

whyp $\wedge$ *mode(c, [int.B int.E])* $\rightarrow$ *<model-constraints>*,

The flipflop example illustrates how such model chunks can look like. Because we have not introduced our modeling language so far, we use a verbal description for the model constraints. On the coarse level time is abstracted away, whereas the detailed view considers set-up and hold times[2].

1. .B denotes the start point of the interval, .E the end point respectively

coarse-view $\wedge$ *ok(flipflop, [int.B int.E])*
$\rightarrow$ *<third row of table in Fig. 2a)>*
detailed-view $\wedge$ *ok(flipflop, [int.B int.E])*
$\rightarrow$ *<specification of Fig. 2c)>*

A diagnosis is then only valid with respect to the current set of working hypothesis, i.e.

$SD \cup OBS \cup Whyp \cup \Pi_t$ *has to be satisfiable.*

From the presented diagnostic framework, the requirements which have to be met by an underlying predictive engine can be derived: it has to detect discrepancies, i.e. to check the consistency between the expected and the observed behavior and it has record dependencies in order to identify the behavioral modes/ working hypothesis which are inconsistent with the observations.

INTEGRATING TEMPORAL INFORMATION INTO SD

Temporal Abstractions

As already mentioned, using abstractions is an important means to model complex temporal behavior. The use of temporal abstractions was also motivated by [Ha91]. He argued that abstractions account for models appropriate for troubleshooting. Hence, it is not always necessary to consider details of behavior, i.e. often a coarse approximation of the actual behavior suffices to detect inconsistencies of given observations or measurements. The necessary model granularity depends on the relevant faults and the observables.

[St93] pointed out that the term abstraction subsumes several kinds of representation transformations: we can change the structure of time, dropping or switching metrics, changing time ontology etc. He gave an example for an abstraction which is very common in technical domains and which is also used in our approach: parameters are sampled at equidistant time points. Thus, time is considered as a linearly ordered set of finite time instants $TI \subset \mathbf{R}$ (other time instances are ignored). Further abstractions have to be performed e.g. when using time intervals instead of time points or when using "abstract" terms like something is moving, changing etc. Terms of this kind are used in the gas engine example. Other examples for abstractions can be found in [Ha91], [St93], [KNS93].

In our approach, value and temporal abstractions are performed by the evaluation of qualitative *predicates*[3]. A predicate can be considered as an abstraction operator, it checks e.g. the value of a quantity being a certain

2. This is a simple example, in more sophisticated applications more than two levels could be appropriate.

3. For measured or observed variables, evaluation of appropriate predicates and determination of episodes is performed by a measurement interpretation system (see [KNS93] for details).

range, the derivation of a quantity being in a certain direction, whether something is moving, turning around, etc. An *episode* is a qualitative property assertion holding over a period of time. A *maximal episode* represents the maximal temporal extent over which a predicate holds.

Episode: `(J-high (10 11] [30 31))`

"The value of J is high within the specified time interval. (10 11] denotes the interval for the beginning of the episode, [30 31) for the end resp. e.g. 10 is the latest time point at which the predicate evaluates to false, 11 the first time at which it holds"

A *history* is a set of maximal episodes which are ordered according to their occurrence time.

Temporal Relations

Obviously three kinds of temporal relations are necessary to describe the (dynamic) behavior of our example systems[1]:

- the relative order of episodes
- the duration of an episode
- the temporal distance between different episodes

In order to be able to represent these kinds of temporal knowledge, we provide *unary relations* to restrict single timemarks by assigning a range of permitted values to them. *Order relations* (<, ≤, =, ≥, >) are used to express binary order relationships between timemarks [VKvB90]. A fragment of Allen's interval language[2] is included to impose *qualitative temporal relations* on episodes. These qualitative relations are automatically translated into conjunctions of end point relations of episodes (using the above mentioned set of order relations). Furthermore, we allow *bounded differences* [DM87], i.e. the difference between two timemarks can be bounded from above or below.

All temporal relations (except the unary ones) are represented internally by linear inequalities of the form $x + c \leq y$. By using algebraic relationships between timemarks as the basis for temporal constraints, it is possible to effectively combine qualitative and quantitative constraints in one theory.

We assume that a maximal time resolution of one basic time quantum Δt is necessary to describe the temporal structure of incoming data. For further processing, the time point relations <, >, =, and ≥ are translated to ≤-relations by adding the time quantum Δt, = is expressed by two ≤-relations [KNS94].

The temporal relations are be modeled by relational graphs (time nets), where the nodes represent timemarks and the directed edges are labeled with the binary relations between timemarks. Each node itself is labeled with a minimum and a maximum value corresponding to unary relations. In an empty net minimum and maximum of a timemark are typically unrestricted (this is denoted by $-\infty$ and ∞) For an example, see Figure 5.

Putting Episodes and Temporal Relations Together: Model Constraints and Mode Transition Graphs

In this section we describe how episodes and temporal relations can be used to describe the temporal behavior of a device.

As already mentioned, model constraints are used to specify the behavior of components. A *model constraint* consists of a name, episode binding specifications, temporal relations, and a specification of the possible inference directions (*behavioral rules*)[3]. An *episode binding specification* consists of a local name (constraint scope, see below), a predicate(type), arguments and timemarks which indicate the beginning and the end of the episode[4]. During rule invocation, predicate and arguments of an episode binding specification are matched against the temporal data base in order to find concrete episodes. Starting and end interval of a concrete episode are bound to the timemarks of the episode binding specification.

The following examples show an episode specification (Fig.3) and the model constraint (Fig.4) for the flipflop example. Figure 5 shows the (automatically generated) time net which is the internal representation of the temporal relations of the model constraint.

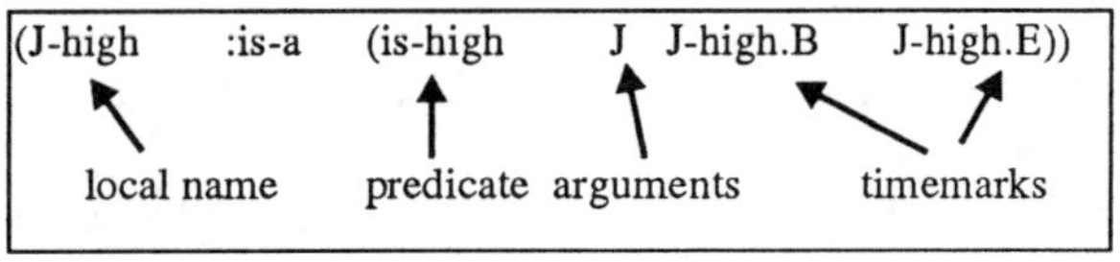

FIGURE 3: Example of an episode binding specification

It has to be mentioned that a behavioral rule is only a part of the model, therefore the declaration of the input, output, and state variables is not included in the rule, e.g. the variables J, K, CP, Q and Not-Q correspond to those of Fig.2b).

As can be seen, the direction slot (Fig. 4) contains the behavioral rules that realize the model constraint. It states in which directions conclusion can be drawn. If all episodes of the antecedent of a behavioral rule are observed or derived (by other model constraints) and if their temporal order and duration satisfy the temporal constraints, then the episodes specified in the "consequence" part of the rule are deduced.

As already mentioned in Section 3, a component can be in one of several behavior modes. To express the possible transitions between modes, we adopt the notion of a *mode transition graph* as it is used in [Nejdl,-Gamper94]. Like in model constraints, the temporal rela-

1. this does not only hold for the examples, but also for a wide class of dynamic systems

2. the 'convex' relations [Nö91]

3. A model constraint is realized as a set of rules, in the following. We will call this set of rules *the behavioral rules*

4. <episode-name>.B denotes the starting point of an episode (named by episode-name), <episode-name>.E the end point, respectively.

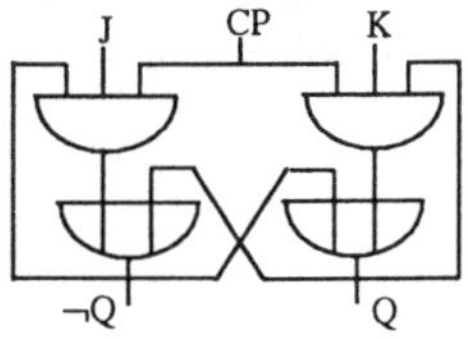


```
Name:          row-3-of-JK-flipflop
Episode Binding Spec.:
( (CP-low       :is-a  (is-low   CP     CP-low.B      CP-low.E))
  (CP-high      :is-a  (is-high  CP     CP-high.B     CP-high.E))
  (J-high       :is-a  (is-high  J      J-high.B      J-high.E))
  (K-low        :is-a  (is-low   K      K-low.B       K-low.E))
  (Q-high       :is-a  (is-high  Q      Q-high.B      Q-high.E))
  (Not-Q-lo w   :is-a  (is-low   Not-Q  Not-Q-low.    Not-Q-low.E) )
Temp. Relations:
( (CP-low  (meets)   CP-hig h)             "The episodes CP-low and CP-high meet each other"
  (subint:  J-high intersects   K-low      subint. B           subint.E)
                                           " J-high and K-low must have a common subinterval"
  (subint.B + 30 ≤ CP-high.B)              "The common subinterval of J-high and K-low has to start at
  (subint.E ≥ CP-high.B + 5)               least 30 ns before CP-high and has to last at least until 5 ns after the
                                           beginning of CP-high."
  (CP-high.B + 30 = Q-high.B)              "The episode Q-high starts 30 ns after the beginning of CP-high"
  (CP-high:B + 20 = Not-Q-low.B))          " The episode Not-Q-low starts 20 ns after the beginning of CP-high"
Directions (behavioral rules):
( (forward:     CP-low, CP-high, J-high, K-low             → Q-high, Not-Q-low)
  (backward-1:  CP-low, CP-high, K-low, Q-high, Not-Q-low  → J-high)
  (backward-2:  CP-low, CP-high, J-high, Q-high, Not-Q-low → K-low) )
```

FIGURE 4: Model constraint for the JK-flipflop, "intersect" is an abbreviation for the Allen relation {equal, overlaps, during, starts, finishes, overlapped-by, contains, started-by, finished-by}. The interval denoted by subint is called an implicit (see [Kockskämper et al.94]) because it is a locally defined interval.

tions expressing possible mode transitions are internally represented as a time net[1] (mode transition graph). Figure 6 shows the mode transition graph for a cylinder.

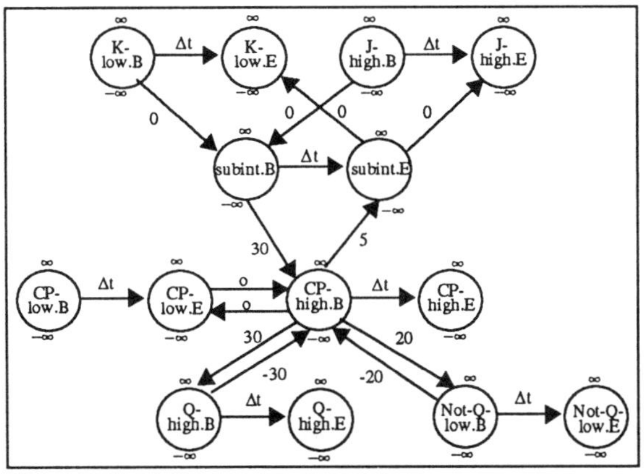

FIGURE 5: Time net of the model constraint of the JK-flipflop

BEHAVIOR PREDICTION

Predicting the behavior of a system means deriving new facts (new episodes) from already known ones by applying general knowledge about its behavior (e.g. given in form of model constraints) to facts from the temporal data base.

The time net of a model constraint specifies the conditions which must be met to be able to infer the consequence of one of the behavioral rules of a model constraint. New episodes can be derived by checking the temporal data base for episodes that match the episode binding specifications of the antecedent of a behavioral rule. Rule invocation takes place for each combination of episodes that matches the episode binding specification. After finding appropriate episodes, they have to be checked for temporal consistency, i.e. whether they satisfy the temporal constraints of the antecedent part. This is performed by binding concrete times (of the episodes) to the timemarks in the time net of the behavioral rule. This instantiation process[2] can be carried out effectively as windows of expectations are kept for each timemark which indicate the earliest and latest possible time for a certain timemark based on the restrictions of the time net. The window of expectation is given by the current min/max-values of a timemark. If a timemark can be instantiated within this window, then it is consistent with the specified constraints of the time net.

Ideally, instantiation of a timemark amounts to the reduction of its range of possible values to one concrete value. In general, however, a range of uncertainty may be remaining.

Two situations can occur during an instantiation attempt:

- An instantiation attempt of a timemark leads to a conflict, i.e. the value is outside the window of

1. a mode could also be a failure mode, for the sake of simplicity failure modes are not shown in the example.

2. Binding concrete episode values to timemarks is carried out for the episodes specified in the antecedent of a behavioral rule. The values of the timemarks of the episodes of the consequence are derived due to these instantiations.

```
Name:          intake-stroke-constraint
Episode Binding Spec.:
(  (down-piston   :is-a  (dec-height   piston      down-piston.B   down-piston.E))
   (iv-open       :is-a  (is-open      in-valve    iv-open.B       iv-close.E))
   (iv-close d    :is-a  (is-closed    in-valve i  iv-closed.B     iv-closed.E)) )
Temp. Relations:
(  (iv-open       (during)        down-piston)
   (iv-open       (meets)         iv-closed ))
Directions:
   ((forward:     down-piston     →   iv-open, ov-closed, iv-closed))
```

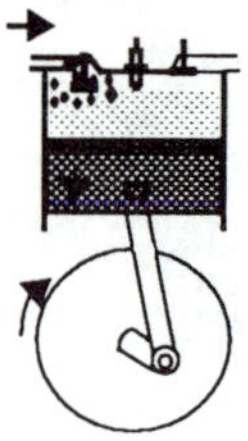

```
Name:          air-gas-vapor-constraint
Episode Binding Spec.:
(  (iv-open       :is-a  (is-open      in-valve    iv-open.B       iv-open-.E))
   (ov-close d    :is-a  (is-closed    out-valve   ov-open.B       ov-open-.E))
   (air-gas       :is-a  (air-gas      in          air-gas.B       air-gas.E)) )
Temp. Relations:
(  (iv-open       (starts-within)  ov-closed)
   (air-gas       (equal)          iv-open) )
Directions:
(  (forward:      iv-open          →  air-gas)
   (backward -1   air-gas          →  iv-open) )

Name:          compression-stroke-constraint
Episode Bin.ing Spec.:
(  (up-piston     :is-a  (dec-height   piston      down-piston:B   down-piston.E))
   (iv-close d    :is-a  (is-closed    in-valve    iv-closed.B     iv-closed.E)) )
Temp. Relations::
(  (up-pisto n    (during)         iv-closed))
Directions:
(  (forward:      up-piston        →  iv-closed) )
```

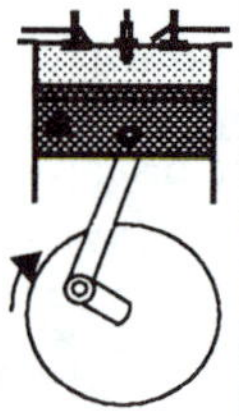

```
Name:          compressed-air-gas-vapor-constraint
Episode Binding Spec.:
(  (air-gas          :is-a  (air-gas         in   air-gas.B     air-gas.E))
   (compr-air-gas    :is-a  (compr-air-gas   in   air-gas.B     air-gas.E)) )
Temp. Relations::
(  (compr-air-gas    (after)        air-gas ))
Directions:
(  (forward:      air-gas          →  compressed-air-gas)
   (backward-1:   compr-air-gas    →  air-gas) )
```

FIGURE 7: Some model constraints for the gas engine example

expectation for an episode (exactly: for one of its timemarks), i.e.
in case of a new min-value:
$[\min(tm_i)\ \max(tm_i)] \cap [\text{new-min}\ \infty] = \emptyset$
in case of a new max-value:
$[\min(tm_i)\ \max(tm_i)] \cap [-\infty\ \text{new-max}] = \emptyset$

- A relevant episode starts or ends within the window(s) of expectation, i.e. the temporal constraints of the corresponding time net are met, i.e.
in case of a new min-value:
$[\min(tm_i)\ \max(tm_i)] \cap [\text{new-min}\ \infty] \neq \emptyset$
in case of a new max-value:
$[\min(tm_i)\ \max(tm_i)] \cap [-\infty\ \text{new-max}] \neq \emptyset$
Then a new (restricted) window of expectation is computed,

Propagation. If the instantiation of a timemark constrains its window of expectation, the new information has to be propagated to the other timemarks in the net by applying the specified relations to the new values in order to update these windows of expectation. A new minimum value has to be propagated along the directed edges. A new maximum value has to be propagated against the specified direction. After propagation the time net is globally consistent. Conflicts can only arise when instantiating timemarks. The propagation of a minimum (maximum) value is performed in O(n) where n is the number of nodes in a time net. Because propagation may occur for each timemark in the antecedent (suppose there are m), instantiation of one behavioral rule requires

Name: `mode transition`

Episode Binding Spec.:

```
( (intake       :is-a  (intake       mode   intake.B       intake.E))
  (compression  :is-a  (compression  mode   compression.B  compression.E))
  (power        :is-a  (power        mode   power.B        power.E))
  (exhaust      :is-a  (exhaust      mode   exhaust.B      exhaust.E)))
```

Temp. Relations::

```
( (intake       (meets)        compression)
  (compression  (meets)        power)
  (power        (meets)        exhaust)
  (exhausts     (loop-meets)   intake))
```

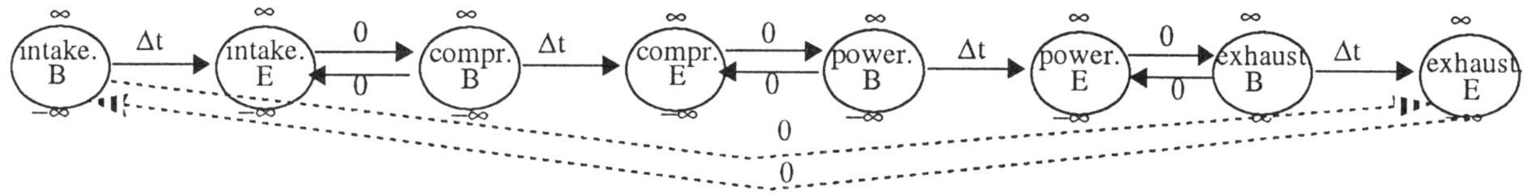

FIGURE 6: Mode transition specification and the respective transition graph for the cylinder (loop-meets characterizes the possible recurrence of the mode sequence

$O(n^2)$ propagation runs.

If a new episode can be deduced, it has to be added to the appropriate history. In this case we have to take care for the maximality of the episodes contained in the history. Otherwise unnecessary inference steps have to be performed. Furthermore, for each episode a list of justifications has to be maintained, this will be one of the topics of the next section.

PREDICTOR DATA BASE

Maintaining dependencies

A predictor data base - usually realized as an ATMS (assumption-based truth maintenance system [dKl86]) - is necessary to store predictions and inconsistencies together with the underlying assumptions.

In order to trace discrepancies back to the underlying assumption (to find faulty components), for each episode a list of those justifications is maintained that establish a certain episode. A justification consists of the model constraint/rule that establish the episode for a certain time interval, the underlying mode assumptions and working hypotheses.

The predictor also requires truth maintenance support in order to maintain alternative predictions. Consider for example an AND-Gate. Assume the output and one of the inputs of the gate is observed to be *low*. Then we can infer the second input to be *high* or *low*. Thus, we have two alternatives which have to be pursued in future inference steps unless there is another way to determine a unique value for the input.

Reasoning in multiple contexts[1] can lead to substantial computational overhead, although a variety of focussing techniques has been developed so far [Dr90]. Orthogonal to different contexts, we have also to consider different time positions at which propositions hold. [DF94] proposes the introduction of temporal labels into the ATMS in correspondence to the logical ones in order to be able to update them along the dependencies if the temporal extension of a value is extended. This works for the "intersection" constraint [KNS94] as well as for simple delays, but for other temporal constraints further work has to be done.

Value contradictions

Consistency checking is a crucial task in a consistency-based approach to diagnosis. Inconsistencies arise if contradictory episodes with a temporal overlap occur. Assume, for instance, the following episodes have been derived

ep1: (J-low (80 81] [160 161))

ep2: (J-high (140 141] [200 201))

Both episodes claim different values for J in a common subinterval. This can be detected by installing an appropriate nogood in the ATMS (stating that J-high and J-low are contradictory).

Maintaining dependencies and detecting value contradictions are requirements by consistency-based diagnosis. In principle, both tasks can be performed by an ATMS. But work remains to be done towards efficient dependency management for time-dependent data as already mentioned in the previous section.

CONCLUSIONS

Modeling and analyzing the dynamic behavior of technical devices are important steps towards a broader applicability of model-based diagnosis. We proposed an approach which uses temporal abstractions, qualitative and quantitative temporal relations to model the behav-

1. a context is formed, for instance, by behavioral mode assignments to the involved components

ior of dynamic systems. We have also shown how the behavior of such a system can be predicted by means of episode propagation and how this fits into a consistency-based framework.

Episode propagation has been first proposed by Williams [Wi86] in his TCP system. The original approach has been extended in several directions. While in TCP episodes are processed according to their derivation order, the propagators EP [De88], EEP [Guc91] and our predictor process them independent of their occurring time. In contrast to EP and TCP our behavioral rules and also those presented in [Guc91] can refer to intervals, not only to time points.

Our work differs from Guckenbiehls in the following points: In EEP, the intersection of intervals is automatically computed by an elegant method called persistency resolution. We decide to specify intersection of episodes explicitly in the model constraints (for this reason we offer the notion of implicit intervals) in order to gain the possibility to refer explicitly to end points of episodes. EEP supersedes our system in one direction: it provides means to express arbitrary constraints on parameter values. Our modeling language allows constraints on parameter values (represented as predicates) only to be modeled by enumerating all possible combinations. But in contrast to our system, EEP gives only limited support to express qualitative relations.

The modeling formalism used in [NG93,94] is very similar to ours. A network of qualitative relations is used to express complex temporal behavior. In opposite to our approach quantitative constraints[1] are handled separately. They provide no means to refer to the intersection interval of two episodes as it is necessary for instance in the description of the JK flipflop. Besides the consistency-based diagnosis, abductive diagnosis is supported.

The presented approach is concerned with behavior modeling and analysis for the purpose of model-based diagnosis. But there exist other tasks to be performed during diagnosis which are also affected by time-varying behavior. Measurement probing, for instance, has to be performed in order to discriminate candidates. A good selection method is essential for an efficient localizing of faults. For dynamic models, standard probe selection has to be extended. Solutions to this problem have been proposed by [GS-R89], [Nö91], [GN94].

ACKNOWLEDEMENTS

I would like to thank Bernd Neumann, Michael Trowe, Ulli Vogel and my colleagues in the BEHAVIOR project for many discussions and helpful comments. I also appreciate the help of Ralf Möller and Andreas Günter who commented on earlier drafts of this paper. This work has been supported by the Ministry of Research and Technology of the Federal Republic of Germany as part of the joint project BEHAVIOR.

1. unfortunately, it does not become clear what kinds of quantitative constraints they allow

REFERENCES

[BD93] C. Böttcher, O. Dressler: Holding the diagnostic trackhound in Leash, in: Proc. IJCAI'93,

[CGK+59] E. J. Cable, R. W. Getchell, W. H. Kadesch, W.J. Poppy, H.E. Crull: The Physical Sciences, Prentice Hall,1959

[CPDT92] L. Console, L. Portinale, D. T. Dupre, P. Torasso: Diagnostic reasoning across different time points, in: Proc. ECAI-92

[DM87] T. L. Dean, D. V. McDermott:Temporal Data Base Management, Artificial Intelligence 32(1),87

[dKl 86] J. de Kleer: An Assumption-Based TMS, Artificial Intelligence 28 (2), 1986

[dKlW87] J. de Kleer, B.C. Williams: Diagnosing Multiple Faults, Artificial Intelligence 32, 1987

[De88] R. Decker: Modeling the Temporal Behavior of Technical Systems, in: Proc. GWAI, 1988

[Dr90] O, Dressler: Annahmenbasierte Begründungsverwaltung für modellbasierte Diagnose, PhD-thesis, Hamburg, 1990

[DF94] O. Dressler, H. Freitag: Prediction Sharing Across Time and Contexts, Proc. AAAI, Seattle 1994

[DS92] O. Dressler, P. Struß: Back to the Defaults: Characterizing and Computing Diagnosis as Incoherent Assumption Sets, in: Proc. ECAI´92, Wien 1992

[FL91] G.Friedrich, F. Lackinger: Diagnosing Temporal Misbehavior, in: Proc. IJCAII-91, Sidney, 1991

[Guc89] Th. Guckenbiehl: Ein Ansatz zur Erweiterung der Episodenpropagierung, in Proc. ÖGAI-Jahrestagung, Springer, 1989

[Guc91] Th. Guckenbiehl: Formalizing and Using Persistency; in: Proc. IJCAI-91, Sidney, 1991

[GS-R89] Th. Guckenbiehl, G. Schäfer-Richter: SIDIA: Extending Prediction Based Diagnosis to Dynamic Models; in: G. Gottlob, W. Nejdl (eds.): Expert Systems in Engineering, Springer, 1990

[Ha92] W. Hamscher:Modeling Digital Circuits for Troubleshooting, in: Readings in Model-Based Diagnosis, 1992

[KNS.93] S. Kockskämper, B. Neumann, M. Schick: Using Event Models to Cope with Time-dependent Phenomena in Dynamical Systems, behavior memo 04-93, University of Hamburg, 1993

[KNS94] S. Kockskämper, B. Neumann, M. Schick: Extending Process Monitoring by Event Recognition, in Proc. Intelligent Systems Engineering ISE'94

[NG93] W. Nejdl, J. Gamper: Model-Based Diagnosis with Qualitative Temporal Uncertainty, in: Proc.

of the 10th Conference on Uncertainty in Artificial Intelligence, 1993

[NG94] W. Nejdl, J. Gamper: Harnessing the Power of Temporal Abstractions in Modelbased Diagnosis of Dynamic Systems, in: Proc. ECAI'94

[GN94] J. Gamper, W. Nejdl: Proposing Measurements in Dynamic Systems, report RWTH Aachen, May 1994

[Nö91] K. Nökel: Temporally Distributed Symptoms in Technical Diagnosis, Lecture Notes in Artificial Intelligence, Springer, 1991

[St93] P. Struß: On Temporal Abstractions in Qualitative Physica, in: Working Papers QR'93

[VKvB90] M. Vilain, H. Kautz, P. van Beek: Constraint propagation Algorithms for Temporal Reasoning - A revised Report, in: D. S. Weld, J. de Kleer (eds.): Readings in Qualitative Reasoning about Physical Systems, Morgan Kaufmann, 1990

[Wi86] B.Williams: Doing Time: Putting Qualitative Reasoning on Firmer Ground, in: Proc. AAAI'86

DEVELOPMENT AND INTEGRATION OF AI APPLICATIONS IN CONTROL CENTERS : THE SPARSE EXPERIENCE

Zita A. Vale
A. Machado e Moura

University of Porto
Faculty of Engineering/DEEC
Rua dos Bragas - 4099 Porto Codex
Portugal
Fax: 351-2-2000808
Email: zav@fe.up.pt

M. Fernanda Fernandes
Couto Rosado
Albino Marques
Electricity of Portugal
NationalTransmission Network
Rua Cidade de Goa, 4 - 2685 Sacavém
Portugal
Fax: 351-1-9411318

ABSTRACT

Modern Power System Control Centers must operate and control medium and large electrical networks that usually cover a large geographic area. The accomplishement of these tasks has been made possible due to important developments either in computer and transmission of information techniques.

Nowadays, complex Power Systems require a very accurate and efficient control what makes Control Centers especially important. These centers are equipped with SCADA (Supervisory Control and Data Acquisition) systems allowing to acquire information about the Power System and its transmission to Control Centers in real-time.

Due to the symbolic nature of the reasoning involved in some control tasks, Artificial Intelligence techniques appeared as a promising solution. Some years ago, some of the most important electrical utilities around the world began to develop knowledge-based applications to incorporate in their Control Centers. These applications deal with the information available in the Control Center and help operators in decision making.

On the other hand, the duration of this kind of project is still rather long and the total number of deployed systems all over the world is still very low.

This paper deals with what we consider the most important points for the success of Artificial Intelligence applications in Control Centers.

INTRODUCTION

Modern Power System Control Centers (CC) must operate and control medium and large electrical networks that usually cover a large geographic area. The accomplishement of these tasks has been made possible due to important developments either in computer and transmission of information techniques.

Nowadays, complex Power Systems require a very accurate and efficient control what makes Control Centers especially important. These centers are equipped with SCADA (Supervisory Control and Data Acquisition) systems allowing to acquire information about the Power System and its transmission to Control Centers in real-time.

Modern Control Centers use several computer applications that include load flow, state estimation, contingency analysis and short-circuit analysis among many others. However, human operators still play the most important role in supervising the Power System and in decision making.

These tasks can be very complex, especially during emergency situations when several hundreds of messages may be received during one minute and decisions must be taken under great stress and, sometimes, in the absence of the most experienced operators. This led to the necessity of developing a new generation of alarm processors able to provide more intelligent information.

Due to the symbolic nature of the reasoning involved in alarm processing, Artificial Intelligence (AI) techniques appeared as a promising solution. Some years ago, some of the most important electrical utilities around the world began to develop knowledge-based applications to incorporate in their Control Centers [Wol86, Tal86, Ame86, Kir89, Kir92, Cig93]. These applications deal with the information available in the Control Center and help operators in decision making.

In this paper, we present the experience of developing an Intelligent Alarm Processor for portuguese Substations Control Centers, the difficulties encountered and the results achieved. This application - SPARSE - performs not only an intelligent alarm processing of the messages but also gives advice to operators namely regarding the correct actions concerning service restoration.

This paper deals with the most important issues arosen during the development of SPARSE and also during its integration in Control Centers. These issues include the complex temporal reasoning required, user interface features and interfacing with existing software. These aspects are discussed, the solutions that have been achieved are presented and an example concerning a real situation is shown.

INTEGRATION ISSUES

Artificial Intelligence applications for Control Centers are very demanding because they must deal with enormous volumes of information in real-time and be integrated with the existing software.

This means that the current state of the art in what concerns Artificial Intelligence applications is not usually sufficient for AI applications for Power System Control Center. Due to this reason, these projects require an important effort being very time consuming usually requiring several years to be accomplished.

On the other hand, the integration of these applications in Control Centers is a key issue for the success of these projects. This integration must consider the existing software and hardware and guarantee that AI applications do not interfere with control functions.

In what concerns the integration, one of the first decisions to be taken is to choose in what machine the AI application will be installed. There are mainly two different approaches: installing the AI application in the machine that performs control functions and installing it in another machine, connected with the main machine in the Control Center through a Local Area Network (LAN).

The first solution is very often desirable because most of Control Center machines are already heavily loaded whereas the second allows a minimum interference of the AI application in normal Control Center functions.

If the AI application is installed in another machine, there are two possibilities for interfacing it with the Control Center operators: to use the display of the machine where the AI application runs or to use the console used for the remaining Control Center applications.

This decision usually depends on the level of satisfaction towards the user interface that is used for Control Center functions. If this is considered of good quality, it is desirable to use it and to integrate the results of AI applications in this interface.

On the other hand, if the user interface used in Control Center functions presents some problems it is usually better to use the display of the machine where the AI application runs to communicate with operators. As this display is likely to be a full-graphics one, in Control Centers where semi-graphic displays are used for Control Center functions, the use of the AI application may be considered the first experience in using full-graphics displays. This may be a very important experience not only for developing a good user interface for the AI application but also to settle down ideas about the requirements to be specified when a new software/hardware is to be acquired.

Another importante issue regarding the integration of AI applications in Control Centers is the hardware and software used in Control Centers. The present trend towards more open and distributed systems will ease this integration process. Distributed systems allow the machine where the AI application runs to be integrated in the Control Center hardware. However, the integration in terms of software seems to be more critical. In fact, the software used usually makes it very difficult to interface the AI application with the already existing software.

SPARSE OVERVIEW

The architecture of our system [Val93a, Val93b, Val93c, Val93e, Val94a] includes a pre-processing module, a Knowledge Base, an Inference Engine and an Explanation Module. The software organization is presented in figure 1.

Our system is mainly written in Prolog, includes a pre-processing module written in C language and runs on a Decstation 5000/240, under Ultrix.

Integration in Control Centers

Figure 2 shows the hardware architecture that supports the integration of SPARSE in Control Centers (CC).

CC machines are microVAX II that are installed since 1989, being presently quite overloaded and not well adapted to Artificial Intelligence applications. Due to this, it has been decided to install SPARSE in another machine: a DECstation 5000/240 with ULTRIX environment. This guarantees a minimum interference in control functions, either during installation and test stages and during normal operation.

This machine is connected through a Local Area Network Ethernet of duplicate configuration with the two microVAX II machines in the CC. One of these machines is always on-line, supporting SCADA and control functions, while the other is in hot standby.

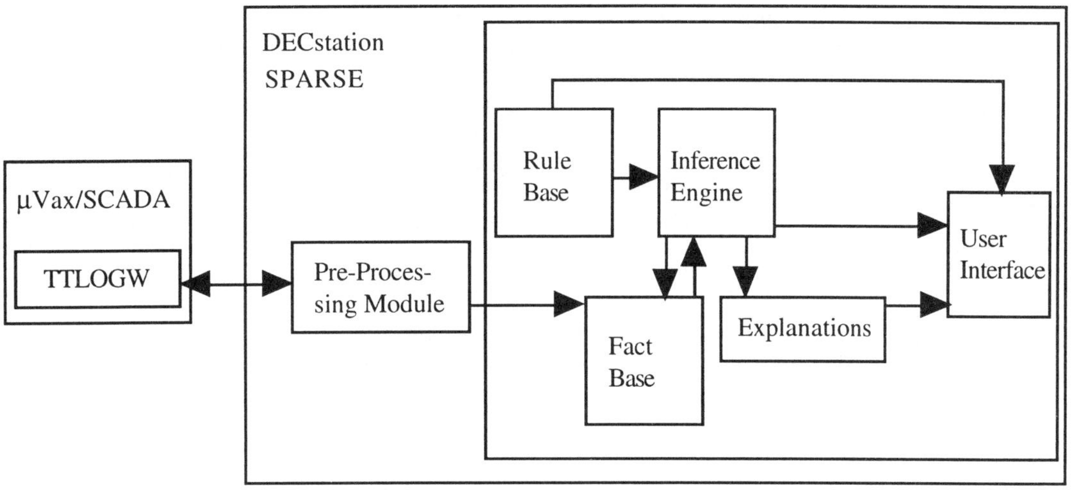

Figure 1 - Software organization

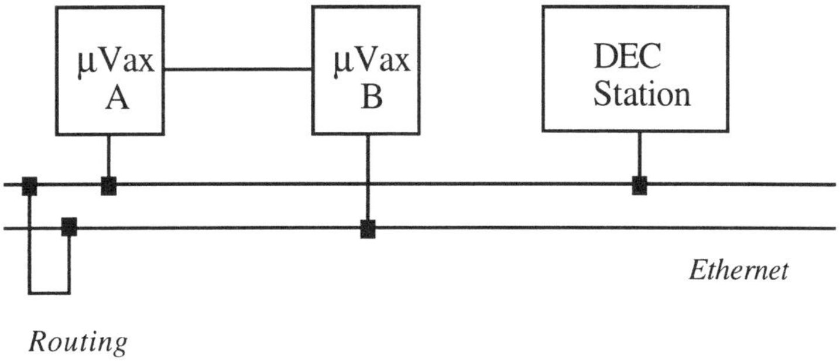

Figure 2 - Hardware configuration

In order to send to SPARSE the required real-time information, a software application - TTLOGW [Ros93] - has been developed and installed in the microVAX II machines. TTLOGW acquires the information of a time-tigged message list named TTLOG. This application is developed in FORTRAN and includes the following features:

- as SPARSE is intended to perform incident analysis, the only information sent to it under normal conditions concerns state changes (e.g. changes in the state of breakers and disconnectors and changes in the mode of operation of substations)
- under incident conditions, all the information arrived to the CC is sent to SPARSE
- when TTLOGW is initialized or the machines boot, TTLOGW sends to SPARSE a "state protocol"; this protocol includes all the states relevant to the network considered and the corresponding messages are time-tagged with the system present time
- in case of communications or RTU faults, a message "RTU off-line" is generated and time-tagged with the system present time; when the fault is cleared, a message "RTU on-line" is generated and the state protocol of the corresponding substation is sent to SPARSE time-tagged with the system present time
- when there are several RTU faults within a small period of time, the state protocol is not generated
- when there are repetitive messages, TTLOGW detects this situation and stops sending them to SPARSE and sends a message reporting this fact; when the repetition disappears, TTLOGW sends a new time-tagged message with the present system time
- all the messages generated by TTLOGW, that are not copies of TTLOG messages, are time-tagged with the present system time.

In what concerns the use of the information of TTLOG, one must remember that although TTLOG messages are time-tagged in RTU, these messages arrive at the CC with a delay with respect to the current time and not necessarily in chronological order.

In order to obtain a correct set of TTLOG messages in real time, the following issues must be considered:

- TTLOG information is acquired by SCADA system and is stored in a database that is based on a circular list containing a record for each stored message
- as new TTLOG messages arrive, they are stored in the following records in the database till the last record (1200) is occupied; when this occurs the next message to arrive occupies the record corresponding to the most ancient information
- due to the mechanism of storing TTLOG information, TTLOG messages are stored in the record in a non-sequential way requiring two pointers to indicate the record that contains the previous and the next TTLOG messages in terms of chronological order; these pointers are updated when a new TTLOG message is stored in the database
- in order to guarantee that the acquired information is correct, these pointers cannot be modified, i.e., SCADA process should not interrupt the access of TTLOGW to the database; this can be accomplished in two different ways:
  - assigning high priority to TTLOGW
  - optimizing TTLOGW making its access to the database very fast;

  in our case, we optimized TTLOGW code and assigned it with priority 10, do not disturbing SCADA which priority level is 12
- as SPARSE should not disturb the SCADA system, the process TTLOGW is normally sleeping and acquires TTLOG information periodically with a frequency that can be modified on-line and that is presently set to 30 seconds; when the beginning of an incident is detected, TTLOGW is waked up by SCADA
- this procedure does not guarantee that the information received by SPARSE is chronologically ordered; to minimize the effort of reordering the information, TTLOGW considers another time parameter that is defined as the time difference between the most recent TTLOG message and the last one that is acquired by TTLOGW and sent to SPARSE; this parameter may also be modified on-line
- in order to minimize the loss of information, each acquisition of information begins a specified interval of time before the last acquired message; this parameter may also be modified on-line.

This short description of the software application that allows to interface the knowledge-based application with the SCADA system shows that this is a major issue in developing knowledge-based applications for Control Centers. In fact, the development of TTLOGW has been quite time consuming and required a high level of expertise in what concerns the SCADA system. One important factor is that this kind of application should be developed in parallel with the knowledge-based application and not only when the knowledge-based application is almost completely developed. In fact, this interface with the SCADA system may impose several and severe constraints that must be considered during the development of the knowledge-based application.

On the other hand, the development of TTLOGW has resulted in the building of a new database that stores the information generated by TTLOGW. This information includes already some filtering (e.g. in the case of repetitive messages) and the new messages generated by TTLOGW. Due to the better quality of this information, presently CC operators usually consult the images based on the information generated by TTLOGW.

Knowledge-Based System and Temporal Reasoning

The development of SPARSE required several kinds of knowledge [Val93e]:

- knowledge about Power System elements and topology (e.g. power plants, substations and lines)
- knowledge about the alarms that are generated at each moment (alarm lists)
- knowledge about the information transmission system (e.g. how data is transmitted and what is transmitted)
- knowledge concerning alarm interpretation.

The first two kinds of knowledge are included in the Fact Base, under the form of Prolog facts. The remaining knowledge is embedded in the Rule Base. A major issue in the knowledge included in the Rule Base concerns temporal aspects. In fact, temporal aspects are the key for alarm interpretation in what concerns the events occurred and also the transmission of information to Control Centers. According to this, SPARSE rules include the required features in terms of temporal questions consideration [Val93e].

The Left Hand Side (LHS) of the rules may be a disjunction of conjunctions of several conditions. The Right Hand Side (RHS) contains the actions/conclusions to be taken if the LHS is verified. Let us consider the following rule:

```
rule xx: 'EXAMPLE' :
[
[C1 and C2 and C3]
or
[C4 and C5]
]
==>
[A1,A2].
```

Conditions considered in the LHS (C1 to C5 in this example) may be one of the following types:

L1) a fact which truth must be proved
L2) a temporal condition.

The actions/conclusions to be taken (A1 to A2 in this example) may be of one of the following types:

R1) generation of facts (conclusions to be asserted to the knowledge base)
R2) killing of facts (conclusions to be deleted from the knowledge base)
R3) interaction with the user interface.

Let us now consider a simplified version of a rule concerning the closure of a line, under fault, by the operator:

```
rule d8: 'THREE PHASE TRIPPING WITH
RECLOSURE'
[[
message(Date1,Time1,[Plant1,Panel1,
[Plant2,NL]],'>>>> TRIP ORDER', 'BEGIN') at T1
and
message(Date2,Time2,[Plant1,Panel1,
[Plant2,NL,'BREAKER']],'BREAKER','OPEN')
at T2
and
panel_in_service(Plant1,Panel1)
and
condition(mod_dif_time_less_or_equal (T2,T1,15))
and
message(Date3,Time3,[Plant1,Panel1,
[Plant2,NL,'BREAKER']],'BREAKER','CLOSED')
at T3
and
condition(dif_time_less_or_equal(T3, T2, 50))
]]
==>
[kill_fact(breaker_closed(_,_,_,Plant1, Panel1,
Plant 2,NL),_,T2),
generate_fact(breaker_open(Date2,Time2,trip,
Plant1, Panel1,Plant2,NL),T2),
kill_fact(breaker_open(_,_,_,Plant1,Panel1,Plant2, NL),
_, T3),
generate_fact(breaker_closed(Date3,Time3,
tp_reclosure,Plant1,Panel1,Plant2,NL),T3)].
```

Temporal issues are also of major importance for the inference mechanism. The Inference Engine of our system [Val93e] has been completely developed in Prolog especially for this application. It allows an efficient processing of alarm messages in real time, treating more than 1200 alarm messages per minute.

The main requirements that led us to develop our own Inference Engine were the following:

- the amount of the involved information is very large
- the processing must be done in real time, according to the arrived information
- the analysis involves complex temporal reasoning
- the power system is a dynamically changing environment requiring non-monotonic reasoning.

The inference Engine of our system uses a forward chaining strategy of reasoning that triggers the appropriate rules when a new fact, external (alarm) or internal (conclusion) arrives, in order to derive new conclusions.

The Inference Engine uses metaknowledge in order to select the appropriate rules and guide the reasoning process, improving the system efficiency (figure 3).

For each arriving fact (alarm or conclusion) the Rule Selector selects the rules to be triggered. The Rule Selector generates facts as the following:

```
trigger(NF,NR,T)
```

that means that rule number NR should be triggered at instant T, due to the arrived fact number NF.

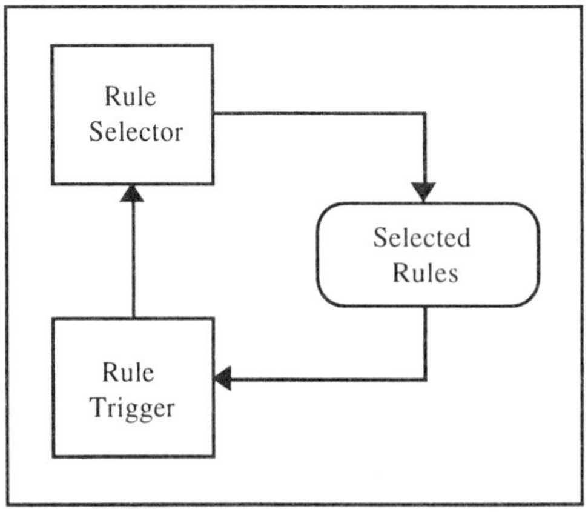

Figure 3 - Inference mechanism

The metaknowledge used by this module specifies, in each case, the time that the system should wait before triggering the selected rules. This allows the introduction of a delay in the triggering of the rules. This delay allows to:

- wait, for some time, for the arriving of facts that are conditions of the rule to be triggered; in this case, there are two possible situations:
  - the delay corresponds to a tolerance allowed in the arriving of the messages (to consider delays in the transmission of the information)
  - the delay corresponds to the time required for the arrival of other messages involved in the same rule (e.g. the maximum time interval between a "trip order" message and a "breaker open" message)
- wait, for a certain period of time, for the arrival of facts that are, in the negative form, conditions of the rules to be triggered (if one of the extremities of a line opens, due to tripping, and the other extremity does not open within a defined time interval, the operator is informed about the tripping in the other end).

The inference engine of an Intelligent Alarm Processor must deal with complex temporal reasoning, mainly due to the following reasons:

- the sequence of events is especially important for the interpretation of alarm lists
- the meaning of a message depends on what happened before and, in certain cases, on what may happen next.

The Rule Trigger module selects, for the present moment, among the Selected Rules, what is the next one to be triggered. For each triggered rule, the veracity of the conditions is verified. If this veracity is proved, the rule is successful and the actions/conclusions of its RHS are taken.

When all the rules selected to be triggered until the present moment have been considered, the Rule Selector considers the facts that have arrived to the system and that have not yet been considered.

As Power Systems are dynamically changing environments, the corresponding information also changes dynamically. This requires dynamically changing knowledge bases for Intelligent Alarm Processors and inference engines supporting non-monotonic reasoning. In fact, alarm messages reflect changes in the Power System and the arrival of a message may turn false conclusions previously achieved. The inference engine must be able to test the veracity of previous conclusions according to the most recent information.

In our system, the facts that are found out to be false are not simply erased from the Knowledge Base. They are recorded as "old facts" allowing its later use for explanation facilities. The following fact:

old_fact(504,urg_al('SFA','102'),'23609096','23609100').

means that a fact reporting an urgent alarm in panel 102 of Ferreira do Alentejo substation (SFA) became true at instant 23609096 and became false at instant 23609100.

The User Interface

User Interfaces are a key issue in real-time applications and especially in critical applications such as a knowledge-based system that assists Control Center operators under incident and emergency situations [Val93d, Val93f, Val94b].

In the case of portuguese CC, whose SCADA machines are equipped with semi-graphic displays, the introduction of a workstation equipped with a full-graphics displays opened a wide range of new possibilities.

On one hand, the use of a multi-window system with zooming and panning facilities allows to present more information, of general and detailed nature, in a more flexible way. On the other hand, we must keep in mind who are the end-users of the knowledge-based application and develop the user interface according to their specific needs.

In fact, the use of dragging and resizing possibilities proved to be not well accepted by operators. In emergency situations, operators should have as few tasks as possible and the most relevant and urgent information should be presented without requiring much effort from Control Center operators.

A multi-window system with window manipulation controlled by the application seems to be the better solution in emergency situations, when it is important to "oblige" operators to see the most urgent information.

On the other hand, under normal conditions, the user interface should be more flexible and let operators decide about a wide range of options. This raises the question of the need of more intelligent user interfaces that are able to adapt themselves to the present situation in each instant. For instance, the information presented over the image of the network during an incident should provide operators with a dynamic view of the evolution of the situation.

Explanation Module

Another important issue for the practical use of knowledge-based systems is a good explanation module. In SPARSE, we have designed an explanation module that furnishes operators easy to read explanation sentences when they select, using the mouse, one or more conclusions generated by the knowledge-based system.

On the other hand, SPARSE includes a tutor module that allows its off-line use for trainning purposes. This module uses SPARSE inference engine in order to analyse lists of alarms that may concern past incidents or be built for a special trainning program [Val94c].

Example

Let us consider an example based on a real incident occurred on the 20th of April of 1993 in the portuguese transmission network:

18:51:59.774	SRR	201	VALDIGEM I
OTHER ALARMS		BEGIN	
18:51:59.804	SRR	201	VALDIGEM I
>>>> TRIP ORDER		BEGIN	
18:51:59.834	SRR	201	VALDIGEM I
BREAKER		MOVING	
18:51:59.894	SRR	201	VALDIGEM I
>>>> TRIP ORDER		END	
18:51:59.901	SVG	203	RECAREI I
OTHER ALARMS		BEGIN	
18:52:00.081	SVG	203	RECAREI I
>>>> TRIP ORDER		BEGIN	
18:52:00.131	SVG	203	RECAREI I
BREAKER		MOVING	
18:52:00.284	SRR	201	VALDIGEM I
BREAKER		CLOSED	
18:52:00.734	SRR	201	VALDIGEM I
OTHER ALARMS		END	
18:52:00.801	SVG	203	RECAREI I
>>>> TRIP ORDER		END	
18:52:01.731	SVG	203	RECAREI I
BREAKER	CLOSED		
18:52:01.801	SVG	203	RECAREI I
OTHER ALARMS		END	
18:52:01.874	SRR	201	VALDIGEM I
URGENT ALARM		BEGIN	
18:52:01.904	SRR	201	VALDIGEM I
>>>> TRIP ORDER		BEGIN	
18:52:01.944	SRR	201	VALDIGEM I
BREAKER		MOVING	
18:52:02.033	SRR	201	VALDIGEM I
BREAKER		OPEN	
18:52:02.394	SVG	203	RECAREI I
URGENT ALARM		BEGIN	
18:52:02.941	SVG	203	RECAREI I
>>>> TRIP ORDER		BEGIN	
18:52:03.001	SRR	201	VALDIGEM I
>>>> TRIP ORDER		END	
18:52:03.099	SRR	201	VALDIGEM I
URGENT ALARM		END	
18:52:03.041	SVG	203	RECAREI I
BREAKER		MOVING	
18:52:03.056	SVG	203	RECAREI I
BREAKER		OPEN	
18:52:03.246	SVG	203	RECAREI I
>>>> TRIP ORDER		END	
18:52:05.767	SVG	OPA	AUT OPER
AUTOMATIC OPERATOR		BEGIN	
18:55:08.547	SVG	203	RECAREI I
BREAKER		MOVING	
18:55:08.555	SVG	203	RECAREI I
BREAKER		CLOSED	
18:55:08.767	SVG	OPA	AUT OPER
AUTOMATIC OPERATOR		END	

For this set of messages, SPARSE presents the following conclusions:

18:51:59.804			BEGINNING OF INCIDENT
18:52:03.056			Line SVG-SRR 1 - one phase tripping with reclosure followed by definite tripping in two ends
18:52:03.056	SVG	203	>> Wait service restoration by OPA (3 minutes)
18:55:08.555	SVG	203	Breaker closed by OPA
18:55:08.555	SRR	201	>> Close breaker

SPARSE strongly reduces the number of messages presented to operators. On the other hand, it does not only present a synthesis of the incident but also gives advices about the actions that operators should take. These advices take into account the usual order of service restoration and also the installed automatic operators and their characteristics [Abr93].

CONCLUSIONS

This paper dealt with the most important aspects in the development and integration of an Intelligent Alarm Processor. In what concerns the development, temporal reasoning requirements led to the need of developing a special purpose inference engine making inadequate the use of a commercial shell. The user interface and the explanation module are also important issues that may condition the success of this kind of application.

The integration of knowledge-based applications in Control Centers is especially important because it must attend to severe constraints imposed by the already existing software and hardware. These constraints must be considered during the phase of knowledge-based application development. In fact, the required software for interfacing the new application with the already existing software should be developed in parallel with the knowledge-based application.

Knowledge-based applications in general, and Intelligent Alarm Processors in particular, are presently considered important software applications for Control Centers. In terms of future, the integration of this kind of applications requires a deep analysis of integration effort, not only for already existing Control Centers but also for future Control Centers. In fact, the integration of new applications, that may be knowledge-based or not, is an important issue that is presently rather complicate but can become easier with the trend of using more open and distributed systems in Control Centers.

REFERENCES

[Abr93] A. P. Abreu, M.F. Fernandes, "Automation in Portuguese Transmission Substations: a Pragmatic Approach", IEEE Transactions on Power Delivery, Vol. 8, no. 3, pp. 1088-1094, July 1993

[Ame86] Herman Amelink, Anthony M. Forte and Robert P. Guberman, "Dispatcher Alarm and Message Processing", IEEE Transactions on Power Systems, vol. 1, n. 3, pp. 188-194, August 1986

[Cig93] CIGRE - Working Group 38.06.03, "Expert Systems: Development Experience and User Requirements", Electra, n. 146, pp. 29-67, February 1993

[Kir89] Daniel S. Kirschen, Bruce F. Wollenberg, Guillermo D. Irisarri, Jeffery J. Bann, Bradley N. Miller, "Controlling Power Systems During Emergencies: The Role of Expert Systems", IEEE Computer Appplications in Power, pp. 41-45, April 1989

[Kir92] Daniel S. Kirschen, Bruce F. Wollenberg, "Intelligent Alarm Processing in Power Systems", Proceedings of the IEEE, vol. 80, n. 5, pp. 663-672, May 1992

[Ros93] Couto Rosado, "Process TTLOGW", EDP Report RESP/SCDS 20/93, July 1993

[Tal86] S. N. Talukdar, E. Cardozo, T. Perry, "The Operator's Assistant - An Expandable Program for Power System Trouble Analysis", IEEE Transactions on Power Systems, vol.1, n.3, pp. 182-187, August 1986

[Val93a] Zita A. Vale, A. Machado e Moura, "An Intelligent Alarm Procesor for Portuguese Control Centers", 4th International Symposium on Expert Systems Application to Power Systems, Melbourne, Australia, pp. 193-195, 4 - 8 January, 1993

[Val93b] Zita A. Vale, A. Machado e Moura, "An Expert System with Temporal Reasoning for Alarm Processing in Power System Control Centers", IEEE Transactions on Power Systems, vol. 8, no. 3, pp. 1307-1314, August 1993

[Val93c] Zita A. Vale, A. Machado e Moura, Albino Marques, "Artificial Intelligence Applications in a Regional Substations Operational Center", 11th Power Systems Computation Conference (PSCC), Avignon, France, vol. 2, pp. 1163-1169, 30 August - 3 September, 1993

[Val93d] A. Almeida do Vale, M. J. Dias Gonçalves, Zita A. Vale, "Improving Man-Machine Interaction in Control Centers: The Importance of a Full-Graphics Interface", 11th Power Systems Computation Conference (PSCC), Avignon, France, vol. 1, pp. 273-279, 30 August - 3 September, 1993

[Val93e] Zita A. Vale, A. Machado e Moura, "Intelligent Alarm Processing in Control Centers", Athens Power Tech - Joint International Power Conference, Athens, Greece, vol. 2, pp. 926-930, 5 - 8 September, 1993

[Val93f] A. Almeida do Vale, M. J. Dias Gonçalves, Zita A. Vale, "User Interface for Power System Control Centers: A New Approach Using Full-Graphics Displays", Athens Power Tech - Joint International Power Conference, Athens, Greece, vol. 1, pp. 379-383, 5 - 8 September, 1993

[Val94a] Zita A. Vale, A. Machado e Moura, M. Fernanda Fernandes, Albino Marques, "SPARSE - An Expert System for Alarm Processing and Operator Assistance in Substations Control Centers", 2nd International Conference on the Practical Application of

Prolog, London, United Kingdom , pp. 549-568, 26 - 29 April, 1994

[Val94b] Zita A. Vale, M. J. Dias Gonçalves, A. Almeida do Vale, M. Fernanda Fernandes, Albino Marques, "Man-Machine Interaction in Power System Control Centers", 1994 IEEE International Conference on Systems, Man and Cybernetics, San Anbtonio, USA, pp. 1092-1097, 2 - 5 October, 1994

[Val94c] Zita A. Vale, "Intelligent Tutoring Systems for Power System Control Centers", 1994 IEEE International Conference on Systems, Man and Cybernetics, San Anbtonio, USA, pp. 1832-1837, 2 - 5 October, 1994

[Wol86] Bruce F. Wollenberg, "Feasibility Study for an Energy Management System Intelligent Alarm Processor", IEEE Transactions on Power Systems, vol.1, n.2, pp. 241-247, May 1986

A HYBRID DIAGNOSIS SYSTEM FOR DIGITAL CIRCUITS

Benjamin Han, Hsin-Tai Yang and Shie-Jue Lee
Department of Electrical Engineering[1]
National Sun Yat-Sen University, Kaohsiung, Taiwan 80424
E-mail: {ben | styang | leesj}@water.ee.nsysu.edu.tw

ABSTRACT

Given a description of some system and an observation of the system's behavior, if the observation conflicts with the expected one, we can deduce all possible sets of diagnoses, i.e., possibly faulty system components, from first principles [Rei87]. To further refine the sets of diagnoses, we have to make necessary system measurements [Hou94]. However, the efficiency of such kind of diagnosis systems greatly depends on the ordering of the measurements taken, hence some mechanism for determining the optimal ordering is required. In this paper, we suggest the possibility of incorporating heuristic information about the system being diagnosed into a diagnosis system in order to help determine the optimal ordering of the measurements. To this end, we describe a diagnosis system for digital circuits (DOC). DOC incorporates heuristic information about the system being diagnosed by adopting *genetic algorithms* [Gold89] for generating measurement requests (MR), therefore the system can be categorized as a "***hybrid diagnosis system*** (HDS)".

1 INTRODUCTION

Under the assumption that the design of a system is consistent, the diagnostic problem arises if the observation of the system conflicts with the way the system is meant to behave. A diagnosis of such a malfunctioning system should point out those possibly defective components which introduce the fault.

Many researchers have developed theories and systems to tackle the diagnostic problem, and the approaches can be roughly categorized into two kinds. In the first kind of the approaches, the diagnosis system tries to codify the statistical intuitions and heuristic rules for future operation. When a conflict occurs between the supposed and the actual observation, the diagnosis system tries to locate the abnormal components by searching its knowledge base and trying to apply all pertinent heuristic rules. A notable system based on such approach is MYCIN [BS84]. Such a system, which is based on "shallow knowledge", cannot be guaranteed to be sound *and* complete in *all* cases.

The method of the other kind for determining diagnoses *deduces* all possible diagnoses from the system description and the observation. In particular, no heuristic information about the system being diagnosed is required in this kind of approach, hence the name "diagnosis from *first principles*" is given. Also, since the method derives all possible diagnoses from the design and the structure of the system, it is sometimes categorized as the "deep knowledge" method. Reiter [Rei87] has built and formalized the major theorems of diagnosis from first principles upon the work of de Kleer [Kleer76] and Genesereth [Ge84]. Other notable examples are [Dav84, KW87, HD87, RNW83, RNW85]. Note that this kind of approach is both sound and complete, but may spend more time in exhausting all possibilities.

When locating a diagnosis from first principles, what we get might be a set of possible diagnoses. Upon this point, one is confronted with the problem of discriminating among the competing diagnoses. The normal solution is to make system measurements in order to obtain additional observations. In [Hou94], Hou provides formally an incremental method for computing new diagnoses given a new measurement, based on the previous diagnoses predicting the opposite. However, the efficiency of this approach depends greatly on the ordering of the measurements taken, and no definite rules can be followed to obtain the optimal ordering. To adequately solve this problem, we suggest the possibility of incorporating heuristic information about the system being diagnosed into the diagnosis system. Such a system, which can be categorized as a ***hybrid diagnosis system*** (HDS), is both sound and complete, and to some extent, efficient.

In this paper, we describe a hybrid diagnosis system for digital circuits (DOC). DOC receives a circuit de-

---

[1]*Tel:* 886-7-5316171 ext 4036, *Fax:* 886-7-5615137.

scription and an observation expressed by DOC Language (DOCL), which is a superset of the propositional calculus, and generates the initial set of diagnoses by a propositional calculus prover (PC prover). The method of HS-dag pruning, which is a revised algorithm of Reiter [Rei87], is proposed by Greiner, Smith and Wilkerson [GSW89] and is used to generate the initial diagnoses. Upon this point, DOC will automatically generate the most promising measurement request (MR), to which we must answer with the result of the measurement. Then, DOC will update the set of diagnoses based on the measurement, generate another MR, and so on. By *most promising MR*, we mean that the corresponding measurement can help us to *refine* the set of diagnoses most *effectively*. In DOC, we adopt *genetic algorithm* [Gold89] to generate the most promising MR based on the past experience, therefore improve the efficiency of the system.

In order to make this paper more self-contained, Section 2 and 3 present a concise review of Reiter's and Hou's theories. Section 4 reveals the general picture of DOC, and Section 5 to 9 describe each part of DOC in more detail. In addition, Section 10 offers a circuit example which is processed by DOC. Finally, we summarize the characteristics of DOC in Section 11.

2 THEORY OF DIAGNOSIS FROM FIRST PRINCIPLES

This section provides a brief review of Reiter's theory of diagnosis from first principles. To make the theory domain-independent, Reiter defines a *system* as a triple (SD,COMPONENTS,OBS), where SD is the system description, COMPONENTS is a finite set of constants which stand for the system components, and OBS is the system observation [Rei87, Def. 2.1 and 2.3]. Both SD and OBS are expressed in first-order sentences. A predicate AB is inserted in each sentence which defines the function of a particular component, e.g., ANDG(x)$\wedge$ $\neg$AB(x) $\supset$ out(x) = and(in1(x), in2(x)) states if x is an AND gate in digital circuits and it is not *abnormal*, then it performs function *and* on its two inputs. Thus the addition of predicate AB enables us to define a *diagnosis* for (SD,COMPONENTS,OBS) as a minimal set $\Delta\in$ COMPONENTS such that SD $\cup$ OBS $\cup$ {$\neg$AB(c) | $c\in$COMPONENTS-Δ} is consistent [Rei87, Def. 2.4 and Prop. 3.4].

A conflict set for (SD,COMPONENTS,OBS) is defined as a set $\{c_1,...,c_k\} \subseteq$ COMPONENTS such that SD $\cup$ OBS $\cup$ {$\neg$AB(c_1), ..., $\neg$AB(c_k) } is *inconsistent* [Rei87, Def. 4.1]. Moreover, a *minimal* conflict set is defined as a conflict set such that no proper subset of it is a conflict set.

Suppose C is a collection of sets, a hitting set H for C is defined as a set $H \subseteq \bigcup_{S\in C} S$ such that $\forall S \in C$, $H \cap S \neq \{\}$ [Rei87, Def. 4.3]. Quite like the definition of a minimal conflict set, a *minimal* hitting set is defined as a hitting set such that no proper subset of it is a hitting set.

By the definition of conflict sets Reiter shows that a diagnosis Δ for (SD,COMPONENTS, OBS) is a minimal set such that COMPONENTS-Δ is *not* a conflict set for (SD, COMPONENTS,OBS) [Rei87, Prop. 4.2].

With this consequence and the definition of hitting sets, Reiter shows that $\Delta\in$COMPONENTS is a diagnosis for (SD,COMPONENTS,OBS) *iff* Δ is a *minimal hitting set* for the collection of *minimal conflict sets* for (SD, COMPONENTS,OBS) [Rei87, Theorem 4.4 and Corollary 4.5]. The theorem is intuitively simple for that if Δ does not intersect with some conflict set, say c, then COMPONENTS-Δ must contain c, which implies that Δ cannot be a diagnosis.

Now finding all possible diagnoses is straightforward: find all minimal conflict sets first, then find all minimal hitting sets for the collection of minimal conflict sets. However, the task of finding all minimal conflict sets is very time-consuming, for it requires a call to a theorem prover to find each minimal conflict set. To improve the efficiency, Reiter proposes an algorithm, which does not require all minimal conflict sets *explicitly*, to find all possible diagnoses. The algorithm manipulates on *HS-tree* (hitting-set tree) [Rei87, Def. 4.6], whose efficiency can be further improved by two methods: the first method is to reuse the node labels previously generated by a theorem prover if possible, and the other method is to prune the edge that is guaranteed to be useless.

After getting the set of possible diagnoses, one has to obtain further information about the system by performing measurements to get a more refined set of diagnoses. To do this, we first define the prediction Π of a diagnosis Δ for (SD,COMPONENTS,OBS) as a first-order sentence such that [Rei87, Def. 5.1] SD $\cup$ OBS $\cup$ {AB(c) | $c\in\Delta$} $\cup$ {$\neg$AB(c) | $c \in$ COMPONENTS-Δ}. Moreover, diagnoses are either preserved under confirming measurements or are rejected under disconfirming measurements [Rei87, Prop. 5.5]; if it is rejected, then it is a *strict subset* of some diagnosis for (SD, COMPONENTS, OBS $\cup$ {Π}) [Rei87, Theorem 5.7]. This suggests an open question about what the relation between the diagnosis and the measurement is, to which Hou gives an answer in a formal way.

3 THEORY OF MEASUREMENT IN DIAGNOSIS FROM FIRST PRINCIPLES

In this section we provide a concise review of Hou's theory of measurements in diagnosis from first principles. Firstly, a conflict set resulting from a measurement Π is defined as a conflict set for (SD, COMPONENTS, OBS $\cup$ {Π}) but not for (SD,COMPONENTS,OBS) [Hou94,

Def. 3.2]. If Δ is a diagnosis for (SD, COMPONENTS, OBS) which predicts ¬Π, then according to the definition above, COMPONENTS-Δ is a conflict set for (SD,COMPONENTS,OBS∪{Π}) resulting from Π [Hou94, Prop. 3.3]. This suggests us to find all minimal conflict sets for (SD,COMPONENTS,OBS∪{Π}) resulting from Π by first computing all COMPONENTS-Δ sets where Δ are diagnoses for (SD,COMPONENTS,OBS) predicting ¬Π. Furthermore, it turns out that every minimal conflict set for (SD,COMPONENTS,OBS∪{Π}) is either a minimal conflict set for (SD, COMPONENTS, OBS) or a minimal conflict set for (SD, COMPONENTS, OBS∪{Π}) resulting from Π [Hou94, Theorem 3.7]. Hence to compute all diagnoses for (SD,COMPONENTS, OBS∪{Π}), one needs to consider the collection of minimal conflict sets for (SD, COMPONENTS, OBS) and *only* the set of minimal conflict sets for (SD, COMPONENTS,OBS∪{Π}) resulting from Π [Hou94, Corollary 3.8].

In [Hou94, Theorem 3.13], Hou describes an incremental method for computing all diagnoses for (SD, COMPONENTS, OBS∪{Π}). Moreover, Hou devises the notion of homogeneous diagnoses for enhancing conflict recognition [Hou94, Theorem 4.5], therefore improving the efficiency.

However, one underlying problem remains untouched in Hou's paper — how is the measurement chosen? Enumeration is the simplest approach, but it could seriously degrade the performance. As will be indicated in Section 8, DOC adopts genetic algorithm [Gold89] to decide the priority of various measurements, hence some heuristic information is embedded to improve the efficiency. Therefore, DOC can be categorized as an HDS.

4 A GENERAL PICTURE OF DOC

DOC is a diagnosis system for digital circuits based on the work of Reiter and Hou. Although the system produces all possible diagnoses by a sound and complete theorem prover, still some heuristic information about the system being diagnosed is incorporated into DOC to improve the efficiency.

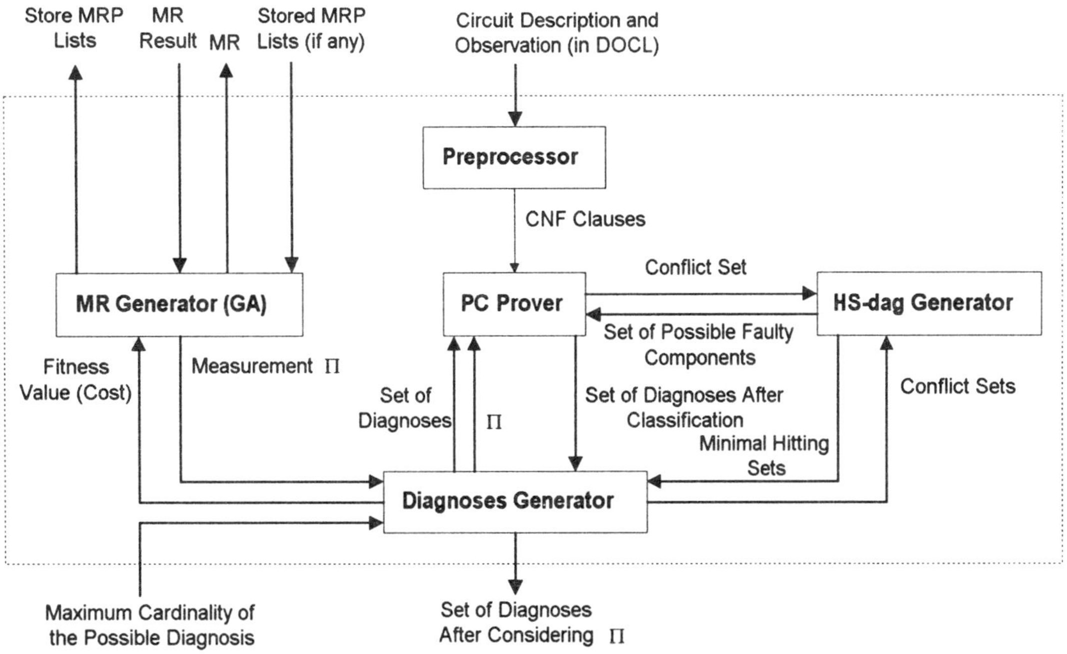

Fig. 1. The Block Diagram of DOC

As shown in Fig.1, DOC is primarily composed of five parts: the *preprocessor*, the *PC prover*, the *HS-dag generator*, the *MR generator* and the *diagnoses generator*. The system description and the observation expressed by DOC Language (DOCL) are fed into the preprocessor, which translates the DOCL clauses into their equivalent CNF clauses. Also, the preprocessor prepares the first generation of MR priority lists (MRP lists) by either load it from disk or generate it randomly. The generation of MRP lists will be used by the genetic algorithm (GA) in order to produce the more promising next generation.

The HS-dag generator calls the PC prover when it needs a new conflict set for labeling the HS-dag node.

After all necessary nodes of the HS-dag are explored and labeled, the HS-dag generator gives out a set of possible diagnoses as its output.

At this point, the MR generator automatically generates an MR from the set of possible diagnoses in accordance with an MRP list. The MR with a higher priority will be fired earlier in the hope that the MR with higher priority will help to refine the set of possible diagnoses at a higher rate. To respond with the request, the user shall perform a physical test on the requested component of the circuit and feed back the result of the measurement.

The diagnoses generator in DOC receives the result of the measurement and the set of possible diagnoses, operates in accordance with the algorithm proposed by Hou [Hou94], and gives all new diagnoses as its output. During the process the cost of refining the set of diagnoses is recorded and is transformed[2] to the fitness value of the corresponding MRP list. Hence after all MRP lists receiving their fitness value, the generation is processed by GA in the MR generator to generate the next better generation. The output of the diagnoses generator can be fed back into the MR generator and the diagnoses generator itself to further refine the set of diagnoses. The whole process will continue until one of four stopping criteria holds [Hou94, Section 6.2].

5 THE PREPROCESSOR AND DOCL

The goal for the preprocessor is to provide a more convenient interface for users. Instead of writing CNF clauses directly, users are allowed to specify the circuit description and the observation in simple DOCL clauses.

Table 1. DOCL predicates and their equivalent CNF clauses of the propositional calculus

DOCL predicates[3]	CNF clauses of the propositional calculus
and (a, b)	$a \wedge b$
or (a, b)	$a \vee b$
xor (a, b)	$(a \vee b) \wedge (\neg a \vee \neg b)$
not (a)	$\neg a$
nand (a, b)	$\neg a \vee \neg b$
nor (a, b)	$\neg a \wedge \neg b$
xnor (a, b)	$(\neg a \vee b) \wedge (a \vee \neg b)$
implication (a, b)	$\neg a \vee b$
equ (a, b)[4]	$(\neg a \vee b) \wedge (a \vee \neg b)$

DOCL is a superset of the propositional calculus and is defined in terms of the set of equivalent *conjunctive normal form* (CNF) clauses as shown in Table 1. Note that by implicitly mapping 0/1 signal of digital circuits into TRUE/FALSE constant of the propositional calculus, we embed *Huntington's postulates* [Hays93] for Boolean algebra into DOCL clauses. The advantage of embedding Huntington's postulates into DOCL clauses is to avoid explicitly listing the postulates, therefore reduce the number of clauses fed into the PC prover. However, when a more general system for diagnosis is proposed, one has to discriminate the domain of the propositional calculus and the domain of the application, hence the postulates must be listed explicitly.

To illustrate how to write the description in DOCL, consider the simple example bellow:

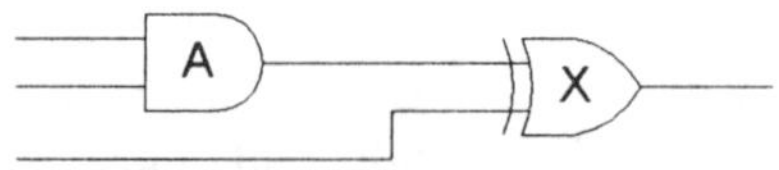

Fig. 2. A Simple Circuit

Table 2. (A) The description of Fig. 2 in DOCL

```
// Definition of gates
ANDG(x) := out(x) = and(in1(x), in2(x));
XORG(x) := out(x) = xor(in1(x), in2(x));

// Declaration of components
ANDG(A);
XORG(X);

// Connections
out(A)=in1(X)
```

Table 2. (B) The CNF Equivalent of Table 2A

$AB(A) \vee \neg out(A) \vee in1(A)$
$AB(A) \vee \neg out(A) \vee in2(A)$
$AB(A) \vee out(A) \vee \neg in1(A) \vee \neg in2(A)$

$AB(X) \vee \neg out(X) \vee in1(X) \vee in2(X)$
$AB(X) \vee \neg out(X) \vee \neg in1(X) \vee \neg in2(X)$
$AB(X) \vee out(X) \vee in1(X) \vee \neg in2(X)$
$AB(X) \vee out(X) \vee \neg in1(X) \vee in2(X)$

$\neg out(A) \vee in1(X)$
$out(A) \vee \neg in1(X)$

Table 2 (A) shows the description of the circuit shown in Fig. 2 in DOCL and Table 2 (B) shows its equivalent in CNF. Note DOCL is case-sensitive. Also note the use of

2The transformation of the cost into the fitness value is defined in a unique way such that the lower the cost, the higher the fitness value.

3We assume that all predicates except for not, which accepts only one variable, accept only two variables. It is trivial to compose a predicate accepting more then two variables in terms of the ones listed in Table 1.

4It is equivalent to the expression "$a = b$", which is the operator version of the *equ* predicate.

';' as the delimiter and the use of '//' to initiate a comment. ':=' should be interpreted as "is defined as" and is different from '='.

6 THE PROPOSITIONAL CALCULUS (PC) PROVER

From Reiter's theorem, diagnoses can be determined by calculating the minimal hitting sets for the collection of conflict sets. In order to test whether a set of components is a conflict set, we need a theorem prover to solve the satisfiability problem. However, Reiter's work makes no claim to the preferred theorem prover. In [BB92], an SAT competition is held and the Davis-Putnam-based [DP60] program is considered the quickest one for deciding the satisfiability of propositional formulas in CNF. This encourages us to choose the Davis-Putnam procedure as the underlying algorithm of our PC prover. In short, Davis-Putnam procedure consists of the following rules as described in [CL73]. Let S be a set of propositional clauses;

(1) *Tautology Rule*: Delete all the clauses in S that are tautologies. If S becomes an empty set, then S is valid. Otherwise S is unsatisfiable iff the remaining set S' is.

(2) *One-literal Rule*: Suppose L is a unit clause in S. If $\neg L$ is also a unit clause, then S is unsatisfiable. Otherwise, delete all the clauses that contain L. If the remaining set S' is empty, S is satisfiable. We obtain a set S'' by delete $\neg L$ from S', S is unsatisfiable iff S'' is.

(3) *Pure-literal Rule*: A literal L is pure if $\neg L$ does not appear in S. If L is pure, then we can obtain a set S' by deleting all the clauses containing L. If S' is empty, S is satisfiable. Otherwise, S is unsatisfiable iff S' is.

(4) *Splitting Rule*: If the set S can be put into the form $(A_1 \vee L) \wedge \ldots \wedge (A_m \vee L) \wedge (B_1 \vee \neg L) \wedge \ldots \wedge (B_n \vee \neg L) \wedge R$, where A_i, B_i and R do not contain L and $\neg L$, then obtain the sets $S_1 = A_1 \wedge \ldots \wedge A_m \wedge R$ and $S_2 = B_1 \wedge \ldots \wedge B_n \wedge R$. S is unsatisfiable iff both S_1 and S_2 are unsatisfiable.

As a new measurement is added, the PC prover is used not only to test a conflict set, but also to derive all minimal conflict sets from a conflict set. However, the satisfiability problem is *NP-complete* [GJ79]. Hence the efficiency of the PC prover has a great influence on that of the whole diagnosis system. Zhang and Stickel [ZS94] propose a technique for implementing Davis-Putnam algorithm by tries in order to reduce the complexity of the unit propagation (i.e., one-literal rule) to sublinear, therefore greatly improves the efficiency. Moreover, if the number of failures of each component in the past has been recorded, the PC prover can be designed to exploit such statistical information for better efficiency.

Since the smaller the conflict set is, the better efficiency we have. Our prover should have some additional mechanism to reduce the size of the generated conflict sets. We take advantage of dependency rules [Lee90] to get the conflict sets as small as possible.

7 HS-DAG GENERATOR

Reiter [Rei87] proposes an algorithm for computing all possible diagnoses of (SD,COMPONENTS,OBS). However, it turns out that the technique for closing nodes conflicts with the tree pruning method in some cases, therefore causes loss of some diagnoses. To avoid the possible incompleteness, Greiner, Smith and Wilkerson [GSW89] revise Reiter's algorithm to exploit the HS-dag instead of HS-tree. The algorithm adopted in the HS-dag generator of DOC is essentially the same as their proposal.

Note that finding all of the minimal hitting sets with cardinality less then k from a collection of subsets of some set is known as an *NP-Complete* problem [GJ79]. The three techniques used in construction of HS-dag can only confine the size of the dag to some extent, but they cannot reduce the time complexity order.

8 MEASUREMENT REQUEST (MR) GENERATOR

In diagnosis from first principles, we need a measurement to update the initial set of diagnoses. The choice of the measurement affects the cost which the diagnosis system must pay for the updating task. To illustrate this, recall that in [Hou94], one must obtain all minimal conflict sets for (SD,COMPONENTS,OBS $\cup$ $\{\Pi\}$) resulting from Π (i.e., *new* minimal conflict sets) before one can compute all diagnoses for (SD, COMPONENTS, OBS $\cup$ $\{\Pi\}$). To do this, one has to derive all new minimal conflict sets from the sets COMPONENTS-Δ_j, where Δ_j is a diagnosis for (SD,COMPONENTS,OBS) which predicts $\neg\Pi$. If we choose a particular Π such that the number of Δ_j is rather small, then the number of COMPONENTS-Δ_j is small and the computation cost[5] to generate CS-tree (conflict set tree) is small. However, less information is added by such a Π because the size of the set of *new* minimal conflict sets is also small. Hence more updating *turns* are required to refine the set of diagnoses. On the contrary, choosing a particular Π such that the number of Δ_j is rather large results in larger cost

[5]Here we define the cost as the number of times the PC prover is called.

in each turn but less turns to refine the diagnoses (both situations are shown in Fig. 3). Therefore, no simple and definite rule can tell us how to choose a better Π to lower the total cost required by the updating process. Note the problem addressed here is a typical optimization problem, in which the search space is too large to be thoroughly explored

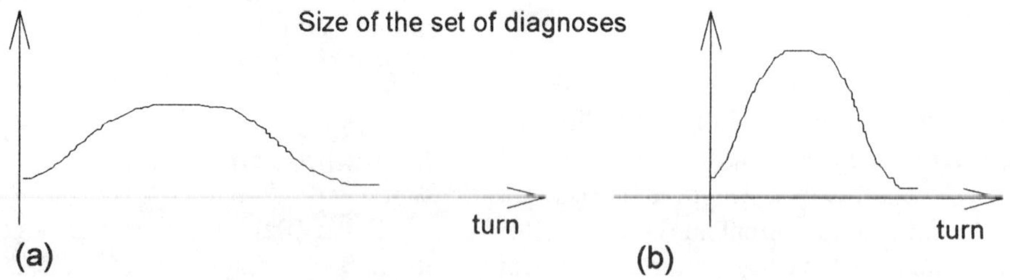

Fig. 3. (a) Smaller number of Δ_j; (b) Larger number of Δ_j.

Various approaches have been proposed to solve the optimization problem. Among them we choose genetic algorithms [Gold89] as the underlying mechanism to generate a promising MR. The reasons are listed below:

(1) Genetic algorithms do not require the fitness function to be well-defined, or even explicitly stated.
(2) Since genetic algorithms operate on a set of strings instead of only one, and use stochastic operators instead of deterministic ones, it can be very robust and, to some extent, efficient.
(3) Genetic algorithms operate on strings, which is a natural representation of our problem here.
(4) Genetic algorithms implicitly find similarities among high-performance strings.
(5) The operators of genetic algorithms are simplicity itself.

For each possible MR, we give it a number as its priority, and form all MRs into an MRP (MR Priority) list. To generate an MR, the MR generator simply picks the one with the highest priority. After the updating task, the corresponding MRP list receives the cost from the diagnoses generator and transforms it into its fitness value. When all MRP lists of the initial generation receive their fitness values, the MR generator first transforms all MRP lists into the corresponding binary strings, then it reproduces their offspring in accordance with their survival probabilities, crossovers the offspring and mutates them with a mutation probability, in order to generate the next generation of binary strings, which in turn are transformed back to the corresponding MRP lists. Furthermore, the set of MRP lists can be saved on disk and loaded when the same circuit is diagnosed. It is a rather intuitive analogy with a doctor: a doctor learns a lot from past experience about the same syndrome, and applies the experience in diagnosing the patient with the similar syndrome.

9 DIAGNOSES GENERATOR

Given the set of diagnoses for (SD, COMPONENTS, OBS) and the measurement Π, the diagnoses generator (DG) is responsible for producing a refined set of diagnoses considering Π. By Hou's theorem, DG has to call the PC prover to classify the diagnoses into those predicting Π and those predicting ¬Π; let us denote the former by Θ and the latter by Ψ. Since all elements of Θ must intersect with the minimal conflict sets for (SD, COMPONENTS,OBS∪{Π}) resulting from Π (i.e., *new* minimal conflict sets), Θ is preserved. On the other hand, each element of Ψ is replaced by the union of the minimal hitting set for the new minimal conflicting sets that do not intersect with it and itself. Therefore DG has to call the HS-dag generator to generate the proper minimal hitting set.

Since DG requires new minimal hitting sets for computing new diagnoses, the derivation of such sets from new conflict sets is necessary. However, such derivation is very time-consuming. To improve the efficiency, Hou proposes the notion of homogeneous diagnoses. This approach is incorporated in DG.

However, in order to detect homogeneous diagnoses initially, one must find out the initial equivalence relation between system components from the sets of minimal conflict sets. This implies that the PC prover has to find out the exact *minimal* conflict sets initially, hence the overall efficiency of the diagnosis system is incurred. To work around this dilemma, DOC obtains the initial equivalence relation from the initial minimal hitting sets and the initial conflict sets instead of the initial minimal conflict sets.

10 EXAMPLE

This section provides an example to illustrate how DOC analyzes a circuit, generates the initial set of diagnoses, fires promising MRs and refines the set of diagnoses. The example circuit is shown in Fig. 4 with the observation labeled on all inputs and the output. The shaded gates (A1, A3 and A4) are assumed to be abnormal and all intermediate signals are labeled in the figure.

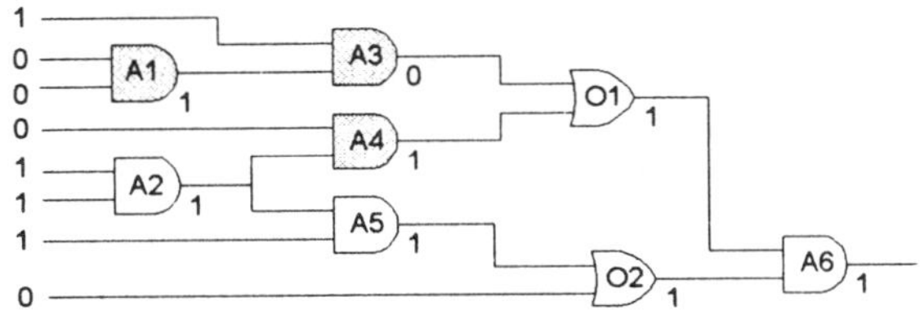

Fig. 4. Example circuit

After converting the system description in DOCL into the propositional logic CNF clauses, the HS-dag generator produces the initial set of 5 diagnoses: {A1}, {A3}, {A4}, {A6} and {O1}. The possible MRs here are the output of A1, ..., A5 and O1, O2, totally 7 possibilities. Hence there are 7! = 5,040 different MRP lists! As pointed out in Section 8, it is the responsibility of the MR generator to find out the most promising MRP list in such a huge search space. We shall follow the execution process of DOC to see how GA tries to do this.

The MRP list is defined as a list $\{p_1, p_2, ..., p_7\}$ where p_i is the priority of the ith element of the list $\{A_1, A_3, A_4, A_2, A_5, O_1, O_2\}$ and $p_i \in \{1,2,\ldots,7\}$. For each MRP list, its permutation number is defined as

(1) The permutation number of {1, 2, 3, 4, 5, 6, 7} is 0.

(2) If $\{a_1, a_2, ..., a_7\}$ is an MRP list, the next MRP list $\{b_1, b_2, ..., b_7\}$ must be such that
   (a) $a_i = b_i$, $1 \le i \le 6$, and $a_m < b_m$, for the largest possible m.
   (b) b_m is the smallest element among $a_{m+1}, a_{m+2}, \ldots, a_7$ that is larger than a_m.
   (c) $b_{m+1} < b_{m+2} < \ldots < b_7$.

The strings manipulated by GA are the binary strings of the permutation numbers; this is shown in Table 3.

Table 3. The initial generation of MRP lists and their permutation numbers

	MRP list	Permutation Number	Binary String
MRP1	{1, 2, 5, 4, 3, 6, 7}	54	0000000110110
MRP2	{5, 1, 6, 3, 2, 4, 7}	2958	0101110001110
MRP3	{7, 2, 1, 5, 6, 3, 4}	4456	1000100101000
MRP4	{6, 3, 5, 4, 7, 2, 1}	3923	0111101010011
MRP5	{7, 4, 5, 1, 2, 6, 3}	4753	1001010010001

Suppose for some reasons, we are quite sure that the true diagnosis consists of three components. For each MRP list, DOC adds the measurement in the order of increasing priority to generate the next set of diagnoses. The process continues until a unique diagnosis with three components is generated. We list the results produced by DOC in Table 4.

Table 4 (a).

MRP1	out(A1)=1	out(A3)=0	out(A5)=1	out(A2)=1	out(A4)=1
Diag. predicting ¬Π	{A3}, {A4}, {A6}, {O1}	{A1}	None.	None.	{A1, A3, A6}, {A1, A3, O1}
New min. CS (cost)	{A1} (95)	{A3} , {A4, A6, O1} (99)	None.	None.	{A4} (23)
New Diag.	{A1}	{A1, A3, A4}, {A1, A3, A6}, {A1, A3, O1}	{A1, A3, A4}, {A1, A3, A6}, {A1, A3, O1}	{A1, A3, A4}, {A1, A3, A6}, {A1, A3, O1}	{A1, A3, A4}
Total cost	95	99	0	0	23

Table 4 (b).

MRP2	out(A3)=0	out(A5)=1	out(A2)=1	out(O1)=1	out(A1)=1	out(A4)=1
Diag. predicting ¬Π	{A3}, {A4}	None.	None.	{A6}	{A4}, {O1}	{A1, A3, O1}
New min. CS (cost)	{A4, A6, O1} (63)	None.	None.	{A4, O1} (63)	{A3}, {A1} (111)	Ignored.
New Diag.	{A4}, {A6}, {O1}	{A4}, {A6}, {O1}	{A4}, {A6}, {O1}	{A4}, {O1}	{A1, A3, A4}, {A1, A3, O1}	{A1, A3, A4}
Total cost	63	0	0	63	111	0

Table 4 (c).

MRP3	out(A4)=1	out(A3)=0	out(O1)=1	out(O2)=1	out(A2)=1	out(A5)=1	out(A1)=1
Diag. predicting ¬Π	{A1}, {A3}, {A6}, {O1}	None.	None.	None.	None.	None.	{A4}
New min. CS (cost)	{A4} (95)	None.	None.	None.	None.	None.	{A1}, {A3} (111)
New Diag.	{A4}	{A4}	{A4}	{A4}	{A4}	{A4}	{A1, A3, A4}
Total cost	95	0	0	0	0	0	111

Table 4 (d).

MRP4	out(O2)=1	out(O1)=1	out(A3)=0	out(A2)=1	out(A4)=1	out(A1)=1
Diag. predicting ¬Π	None.	{A6}	{A1}, {A3}	None.	{O1}	{A4}
New min. CS (cost)	None.	{A1, A3, A4, O1} (23)	{A4, O1} (63)	None.	{A4} (95)	{A1}, {A3} (111)
New Diag.	{A1}, {A3}, {A4}, {A6}, {O1}	{A1}, {A3}, {A4}, {O1}	{A4}, {O1}	{A4}, {O1}	{A4}	{A1, A3, A4}
Total cost	0	23	63	0	95	111

Table 4 (e).

MRP5	out(A2)=1	out(A5)=1	out(O2)=1	out(A3)=0	out(A4)=1	out(O1)=1	out(A1)=1
Diag. predicting ¬Π	None.	None.	None.	{A1}, {A3}	{A6}, {O1}	None.	{A4}
New min. CS (cost)	None.	None.	None.	{A4, A6, O1} (39)	{A4} (95)	None.	{A1}, {A3} (111)
New Diag.	{A1}, {A3}, {A4}, {A6}, {O1}	{A1}, {A3}, {A4}, {A6}, {O1}	{A1}, {A3}, {A4}, {A6}, {O1}	{A4}, {A6}, {O1}	{A4}	{A4}	{A1, A3, A4}
Total cost	0	0	0	39	95	0	111

To calculate the cost of each MRP list, we sum up all costs listed in Table 4, together with 20 for each measurement required; e.g., the cost of MRP1 is 92 + 99 + 23 + 20 * 5 = 317. The reason of adding 20 for each additional measurement taken is that we want to minimize the number of measurements required to refine the set of diagnoses. We list the cost and the fitness value of each MRP list, together with the number of its offspring in Table 5.

Table 5. Cost, fitness value and the number of offspring of each MRP list

	MRP1	MRP2	MRP3	MRP4	MRP5
Cost	317	357	346	412	480
Fitness value	480	357	412	346	317
Probability	0.2510	0.1867	0.2155	0.1810	0.1658
# of offspring	2	1	2	0	0

Average Cost=382.4

Table 6. The second generation of MRP lists and their permutation numbers

	MRP list	Permutation Number	Binary String
MRP1	{1, 2, 5, 4, 3, 6, 7}	54	0000000110110
MRP2	{1, 2, 6, 7, 4, 3, 5}	94	0000100101000
MRP3	{6, 5, 3, 7, 4, 2, 1}	4150	1000100101000
MRP4	{5, 1, 6, 3, 4, 2, 7}	2984	0101110101000
MRP5	{7, 1, 3, 6, 5, 2, 4}	4366	1000100001110

Next, the MR generator initiates all operators of GA, i.e., reproduction, crossover and mutation) to produce the next generation of MRP lists (shown in Table 6). Then the process of locating the diagnosis is repeated to obtain the cost of each MRP list of the second generation. The result is listed in Table 7, and the cost of each MRP list is listed in Table 8.

Table 7 (a).

MRP1	out(A1)=1	out(A3)=0	out(A5)=1	out(A2)=1	out(A4)=1
Diag. predicting ¬Π	{A3}, {A4}, {A6}, {O1}	{A1}	None.	None.	{A1, A3, A6}, {A1, A3, O1}
New min. CS (cost)	{A1} (95)	{A3} , {A4, A6, O1} (99)	None.	None.	{A4} (23)
New Diag.	{A1}	{A1, A3, A4}, {A1, A3, A6}, {A1, A3, O1}	{A1, A3, A4}, {A1, A3, A6}, {A1, A3, O1}	{A1, A3, A4}, {A1, A3, A6}, {A1, A3, O1}	{A1, A3, A4}
Total cost	95	99	0	0	23

Table 7 (b).

MRP2	out(A1)=1	out(A3)=0	out(O1)=1
Diag. predicting ¬Π	{A3},{A4},{A6},{O1}	{A1}	{A1, A3, A6}, {A1, A3, O1}
New min. CS (cost)	{A1} (95)	{A3} {A4, A6} (95)	None.
New Diag.	{A1}	{A1, A3, A4}, {A1, A3, A6}, {A1, A3, O1}	{A1,A3,A4}
Total cost	95	99	0

Table 7 (c).

MRP3	out(O2)=1	out(O1)=1	out(A4)=1	out(A5)=1	out(A3)=0	out(A1)=1
Diag. predicting ¬Π	None.	{A6}	{A1}, {A3}, {O1}	None.	None.	{A4}
New min. CS (cost)	None.	{A1,A3,A4,O1}(39)	{A4}(95)	None.	None.	{A1},{A3}(111)
New Diag.	{A1}, {A3}, {A4}, A6},{O1}	{A1}, {A3}, {A4}, {O1}	{A4}	{A4}	{A4}	{A1,A3,A4}
Total cost	0	39	95	0	0	111

Table 7 (d).

MRP4	out(A3)=0	out(O1)=1	out(A2)=1	out(A5)=1	out(A1)=1	out(A4)=1
Diag. predicting ¬Π	{A1},{A3}	{A6}	None.	None.	{A4},{O1}	{A1,A3,O1}
New min. CS (cost)	{A4,A6,O1}(39)	{A4, O1} (95)	None.	None.	{A1},{A3}(111)	
New Diag.	{A4}, {A6}, {O1}	{A4}, {O1}	{A4}, {O1}	{A4}, {O1}	{A1,A3,A4}, {A1,A3,O1}	{A1, A3, A4}
Total cost	39	95	0	0	111	0

Table 7 (e).

MRP5	out(A3)=0	out(O1)=1	out(A4)=1	out(A5)=1	out(O2)=1	out(A2)=1	out(A1)=1
Diag. predicting ¬Π	{A1},{A3}	{A6}	{O1}	None.	None.	None.	{A4}
New min. CS (cost)	{A4,A6,O1}(39)	{A4},{O1}(95)	{A4}(95}	None.	None.	None.	{A1}, {A3} (111)
New Diag.	{A4}, {A6}, {O1}	{A4}, {O1}	{A4}	{A4}	{A4}	{A4}	{A1, A3, A4}
Total cost	39	95	95	0	0	0	111

Table 8. The cost of each MRP list of the second generation

	MRP1	MRP2	MRP3	MRP4	MRP5
Cost	317	254	365	365	495

Average Cost=359.2

Note by comparing the average cost in Table 5 with that in Table 7 (382.4 vs. 359.2), we find that DOC does benefit from the past experience of diagnosing the same circuit.

11 SUMMARY

DOC is a diagnosis system based on Reiter's and Hou's theory of diagnosis from first principles. The major characteristic of DOC is that it incorporates heuristic information about the system being diagnosed, therefore it can be categorized as one of "hybrid diagnosis systems (HDS)". DOC is guaranteed to be both sound and complete, and to some extent, efficient. By incorporating *genetic algorithms*, DOC benefits from the past experience it *learns*.

The other characteristics of DOC include the use of a Davis-Putnam-based PC prover, which exploits tries data structure to find conflict sets, and the manipulation on HS-dag instead of HS-tree to generate minimal hitting sets. We also devise a method to detect the initial equivalence relation between system components without explicitly using the initial minimal conflict sets. Moreover, the PC prover of DOC takes advantage of the recorded failure frequency from past experience to further improve the efficiency.

REFERENCES

[BB92] M. Buro and H. K. Büning, Report on a SAT Competition, Universität Paderborn, 1992.

[BS84] B. G. Buchanan and E. H. Shortliffe (Eds.), Rule-based Expert Systems: The MYCIN Experiments of the stanford Heuristic Programming Project, Addison-Wesley, Reading, MA, 1984.

[CL73] C. L. Chang and R. C. T. Lee, Symbolic Logic and Mechanical Theorem Proving, Academic Press, New York, 1973.

[Dav84] R. Davis, Diagnostic Reasoning Based on Structure and Behavior, *Artif. Intell.* **24** (1984) 347-410.

[DP60] M. Davis and H. Putnam, A Computing Procedure for Quantification theory, J. *ACM*, vol. **7**, no. **3** (1960), 201-215.

[Kleer76] J. de Kleer, Local Methods for Localizing Faults in Electronic Circuits, *MIT AI Memo* **394**, Cambridge, MA, 1976.

[KW87] J. de Kleer and B. C. Williams, Diagnosing Mutiple Faults, *Artif. Intell.* **32** (1987) 97-130.

[Ge84] M. R. Genesereth, The Use of Design Descriptions in Automated Diagnosis, *Artif. Intell.* **24** (1984) 411-436.

[GJ79] M. R. Garey and D. S. Johnson, Computers and Intractability: A Guide to the Theory of NP-Completeness, Bell Laboratories, N.J., 1979.

[Gold89] D. E. Goldberg, Genetic Algorithms in Search, Optimization & Machine Learning, Addison-Wesley, 1989.

[GSW89] R. Greiner, B. A. Smith and R. W. Wilkerson, A Correction to the Algorithm in Reiter's Theory of Diagnosis, *Artif. Intell.* **41** (1989/90) 79-88.

[Hays93] J. P. Hayes, Introduction to Digital Logic Design, Addison-Wesley, 1993.

[HD87] W. Hamscher and R. Davis, Issues in Model Based Troubleshooting, *MIT AI Memo* **893**, Cambridge, MA, 1987.

[Hou94] A. Hou, A Theory of Measurement in Diagnosis from First Principles, *Artif. Intell.* **65** (1994) 281-328.

[Lee90] S. J. Lee, CLIN: An Automated Reasoning System Using Clause Linking. PhD thesis, University of North Carolina at Chapel Hill, 1990.

[Rei87] R. Reiter, A Theory of Diagnosis from First Principles, *Artif. Intell.* **32** (1987) 57-95.

[RNW83] J. A. Reggia, D. S. Nau and Y. Wang, Diagnostic Expert System Based on a Set Covering Model, *International Journal of Man-Machine Studies* **19** (1983) 437-460.

[RNW85] J. A. Reggia, D. S. Nau and Y. Wang, A Formal Model of Diagnostic Inference I. Problem Formulation and Decomposition, *Information Science* **37** (1985) 227-256.

[ZS94] H. Zhang and M. Stickel, Implementing the Davis-Putnum Algorithm by Tries, typescript, University of Iowa, 1994.

OPERATOR SUPPORT SYSTEM ON ALARM CONDITION FOR COMBINED CYCLE POWER PLANT

Satoshi Tanaka, Masao Okamachi, Ryouji Ito, Yoshiyuki Kita and Masumi Nomura
Power Plant Engineering Department Takasago Machinery Works
Mitsubishi Heavy Industries Ltd.
2-1-1 Shinhama, Arai, Takasago, Hyogo, Japan, 676

ABSTRACT

We developed the Operator Support System which enables inexperienced operators as well as experienced operators to cope with abnormal conditions of Combined Cycle Power Plant. The features of this system are the Estimate of Emergency Level for Operation and the Prediction of Subsequent Abnormality, adding to the Diagnosis of Cause and the Operation Guidance.
Moreover in this system, Diagnosis of Cause was improved by using our original method and support screens can be displayed for educational means in normal condition as well.

INTRODUCTION

In recent years, Combined Cycle System, which consists of Gas Turbines and Steam Turbines, is being increasingly applied to Thermal Power Plants in all over the world, because of its high performance. Moreover, the trouble and manual operation have been decreased by the improvement of plant reliability and the development of a fully automatic control system. Therefore, in normal operation, even operators without much experience can operate the power plant skillfully enough.

However, if adequate operation is not taken when an abnormal condition occurs in the plant, the machines may be damaged and the plant may trip. For the purpose of improving the plant operational reliability, we developed Operator Support System, referring various Support Systems. [KMT87, MHIT91, SAOM88]

DESIGN CONCEPTS

When a plant alarm occurs, Operator would take action as shown in Fig.1. In this series of actions, the following issues are different between experienced operators and inexperienced operators.

- Estimate of emergency level for operation.
- Diagnosis of cause.
- Prediction of abnormality which will probably happen subsequently.
- Planning of countermeasure operation.

Therefore, it is necessary for Operator Support System to advise above - mentioned subjects in order to compensate this difference.

The diagnostic results should be summarized on one CRT screen as shown in Fig.2 in order not to misunderstand and misoperate. Even in no alarm condition, Operator can use these displays for studying purposes.

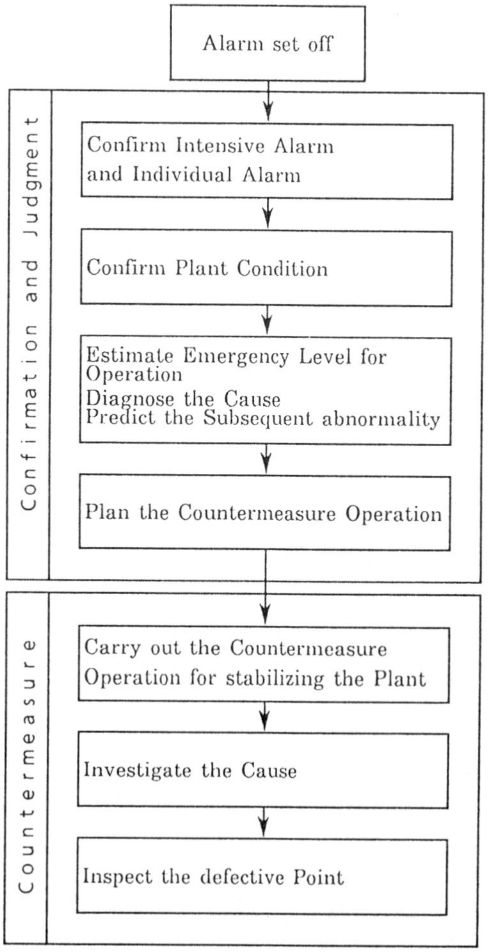

Fig.1 Analysis of Operator's Action in Alarm Condition

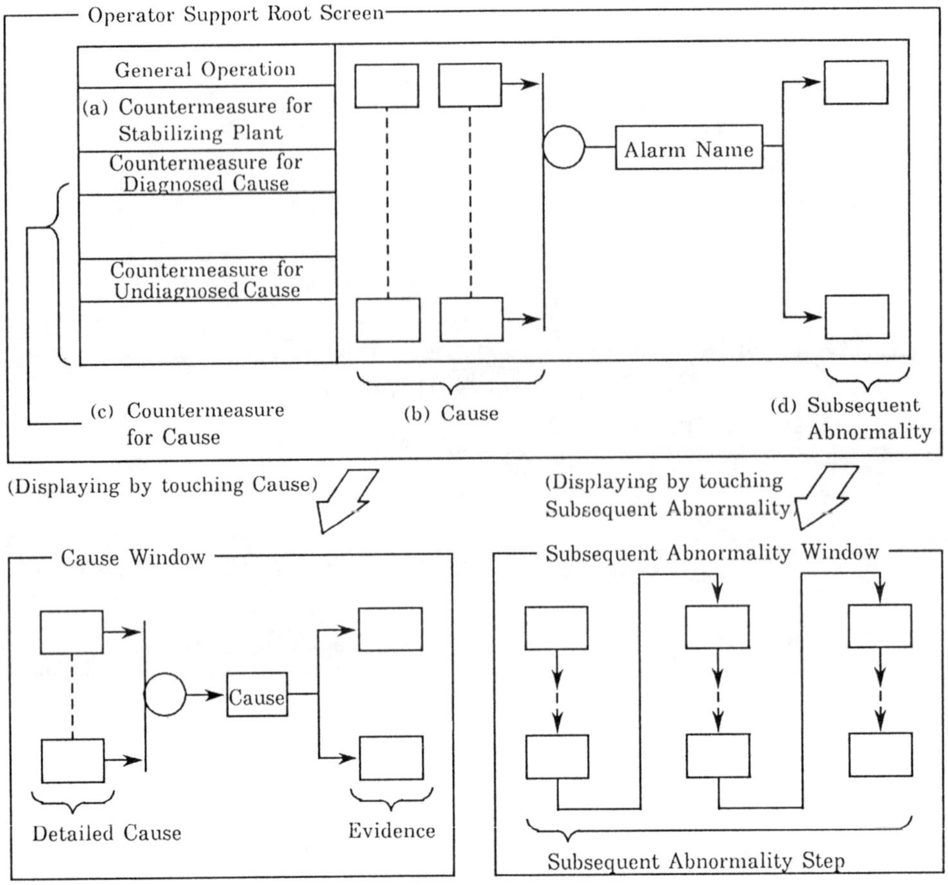

Fig.2 Operator Support Screen in Alarm Condition

FUNCTION

Operator Support System contains Diagnosis Flow Diagram for each plant alarm as shown in Fig.3, as Knowledge Database and diagnoses the following items by using these Diagnosis Flow Diagrams.

(1) Estimate of Emergency Level for Operation Supposing that the incline of the process continues for a few minutes, Operator Support System concludes that it is an emergency case if the process exceeds the plant trip limit in a few minutes (typically 3 minutes).

(2) Diagnosis of Cause
All the alarms of a Combined Cycle Power Plant are described by Fault - Tree Diagram as shown in Fig.3 (a). Operator Support System calculates each estimate value by using our original method as shown in Fig.3 (a) and calculates the certain factor of each cause by using the calculated estimate value. Then Operator Support System displays the cause with the highest certain factor as the cause of this alarm.

(3) Prediction of Subsequent Abnormality
Operator Support System predicts the abnormalities which are expected to occur subsequently, based on the cause of this alarm, by using Subsequent Abnormality Prediction Flow Diagram as shown in Fig.3 (b).
Operator Support System informs operator of the stage of subsequent abnormalities which have already occurred, so that operator might understand how long the plant progresses for the subsequent abnormality.

(4) Planning of Countermeasure Operation
Operator Support System suggests the countermeasure operation, based on the cause of the alarm, by using Countermeasure Operation Planning Flow Diagram as shown in Fig.3 (c).
Moreover, Operator Support System calculates the margin of the suggested operation for checking whether this operation causes other abnormalities for the plant or not.

(5) Education
By touching cause and subsequent abnormality, Operator can study guidance of general operation, countermeasure for cause detailed cause, evidence and subsequent abnormality step, even in no alarm condition.

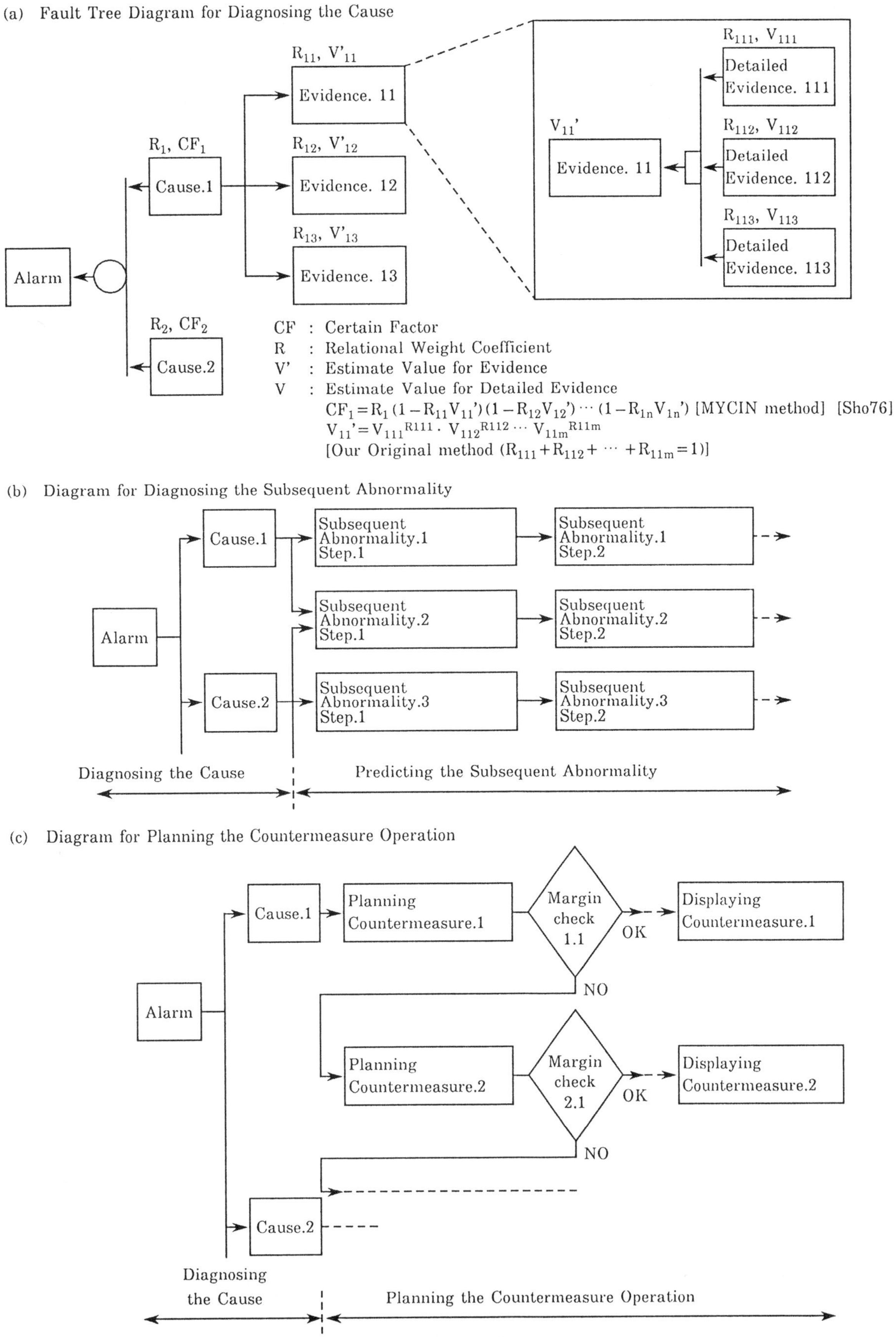

Fig.3 Diagnostic Flow Diagram

APPLICATION

Operator Support System can be installed easily in the plant control system as shown in Fig.4. Plant data which are the base of Support information are transmitted from all of plant control systems and it is not necessary to install special sensors for this Operator Support System. The outputs of the system are displayed on CRT for plant operation. Therefore, seeing the contents of Support, Operator can excute the adequate countermeasure for the alarm on CRT. UNIX based Engineering Workstations are applied for the system. All programs are made with C language.

SUPPORT EXAMPLE

To clarify the system functions, we explain the system behavior with an example.

(Gas Turbine Bearing Metal Temperature High Alarm)

(1) Diagnosis of Cause

Fault - Tree Diagram for "Gas Turbine Bearing Metal Temperature High Alarm" is shown in Fig.5 (a).

In Gas Turbine Lubricating System, in the case that all bearing oil temperatures and lubricating oil temperature are high and cooling water pressure is low but cooling water temperature is normal, Operator Support System diagnoses "Cooling Water Flow Low" as the cause of this alarm.

In detailed cause of "Cooling Water Flow Low" there are "Cooling Water Pump Trip", "Cooling Water Leak", and "Cooling Water Differential Pressure Control Valve Abnormal," and so on.

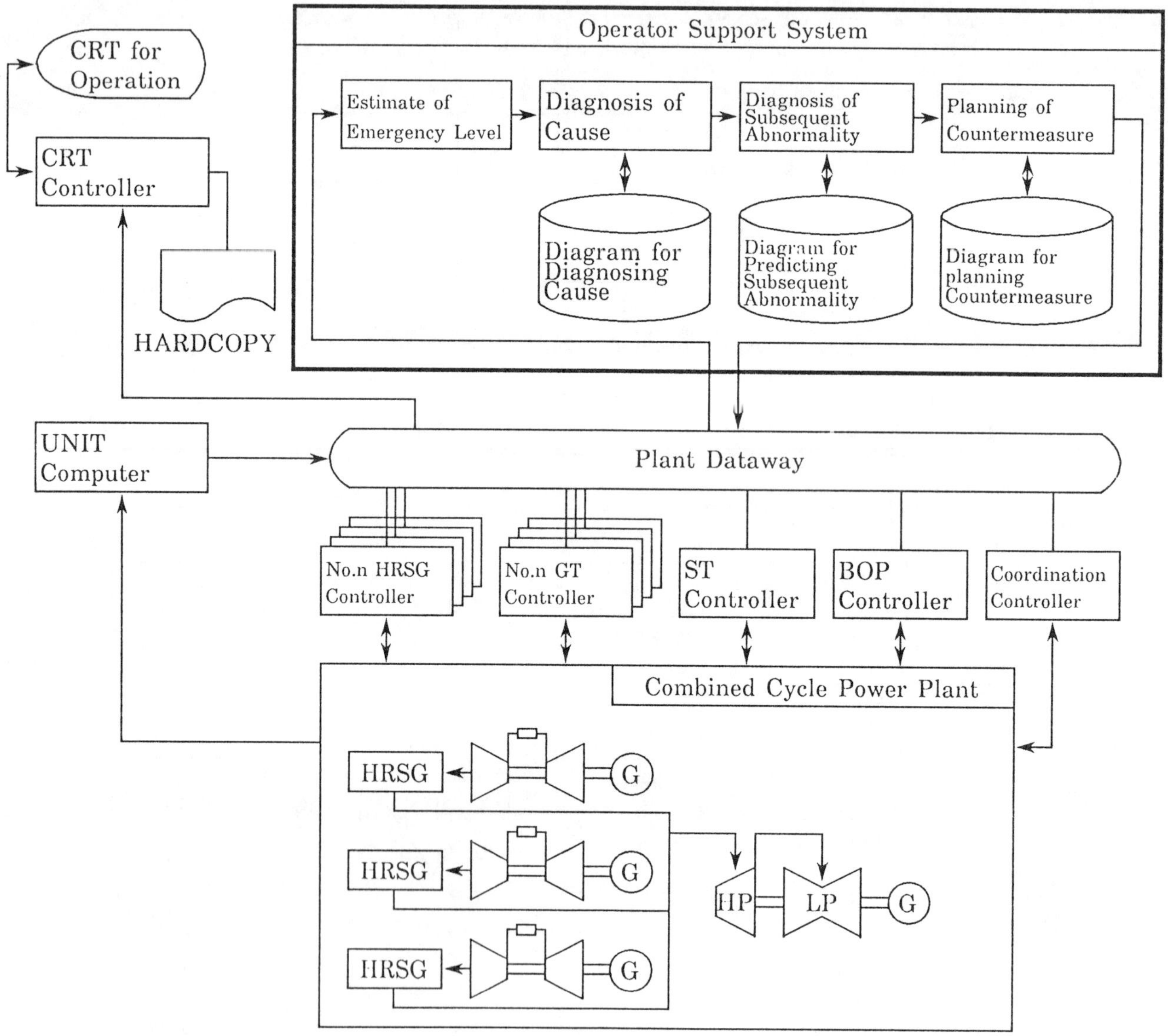

Fig.4 Operator Support System and Plant Control System

(2) Estimate of Emergency Level for Operation
In the case that Gas Turbine bearing metal temperature rises rapidly, Operator Support System estimates that Operator has to take the countermeasure urgently, because there is not enough time for a plant trip.
In the case that Gas Turbine bearing metal temperature rises gradually, Operator Support System estimates that Operator should monitor the process variables and has to take the countermeasure after the temperature rises still more, because the plant trip may not occur in a few minutes.
According to the above - mentioned estimation of emergency level, Operator Support System changes the operation guidance message, as shown in the following.
(a) Emergency Case (In the case that bearing metal temperature rises rapidly)
"Stop Gas Turbine manually immediately."
(b) No Emergency Case (In the case that bearing metal temperature rises gradually)
"① Monitor Gas Turbine bearing metal temperatures, bearing oil temperatures and bearing vibrations.
② If the alarm is not cleared or occurs again, stop Gas Turbine manually."

(3) Prediction of Subsequent Abnormality
In the case that "Gas Turbine Bearing Metal Temperature High Alarm" is caused by "Cooling Water Flow Low", there is the fear that lubricating oil stickiness becomes low and Gas Turbine vibration will become higher.
If Gas Turbine bearing metal temperature rises still more, Operator has to stop Gas Turbine manually for protection.
These subsequent abnormalities are shown in the diagram of Fig.5 (b).
Therefore, when this alarm occurs Operator Support System displays these subsequent abnormalities and requests that Operator make preparation for these subsequent abnormalities.

The example of Operator Support System screen is shown in Fig.6.

(a) Gas Turbine Bearing Metal Temperature High Alarm Fault-Tree Diagram

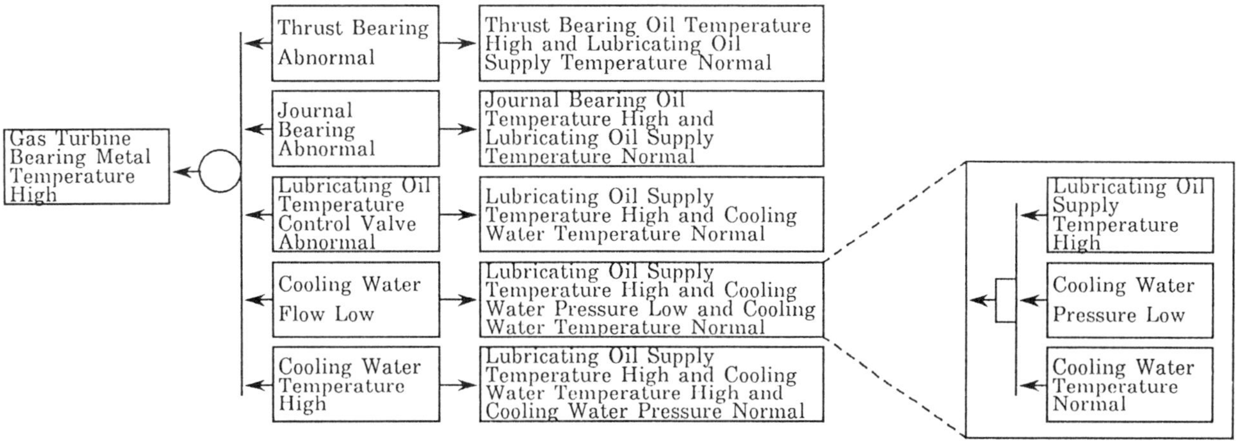

(b) Gas Turbine Bearing Metal Temperature High Alarm Subsequent Abnormality

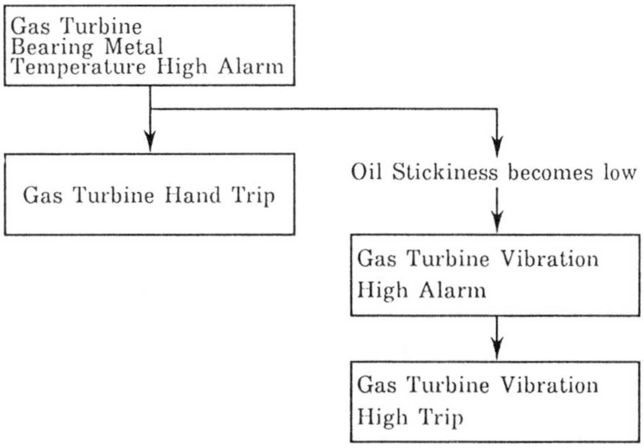

Fig.5 Diagnostic Flow Diagram of Gas Turbine Bearing Temperature High Alarm

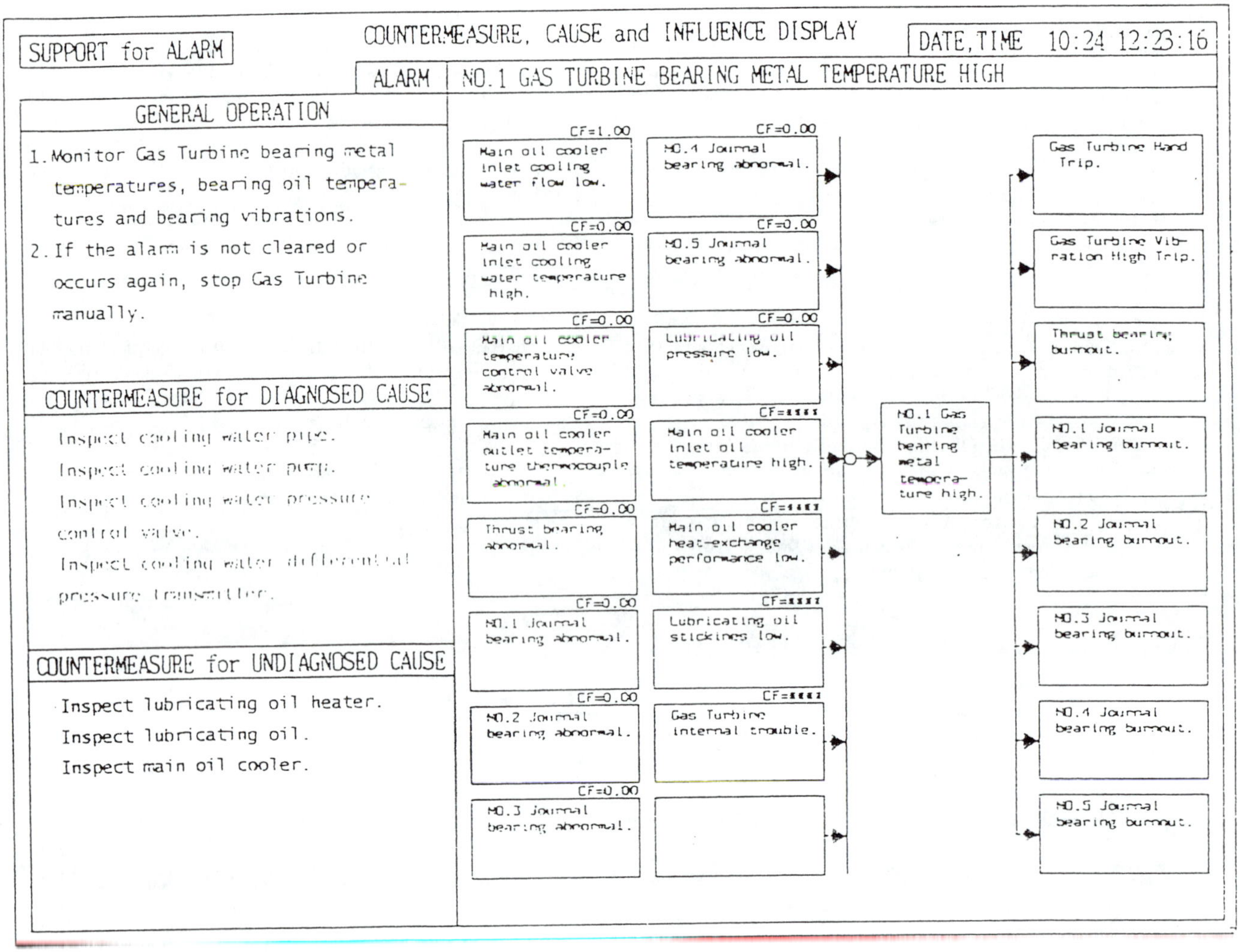

"CF=****" means that the certain factor can not be calculated from physically measured values.

Fig.6 Example of Operator Support System Screen

CONCLUSION

In this paper, we introduced Operator Support System, which is applied to Combined Cycle Power Plants. When the alarm occurs in Combined Cycle Power Plant, Operator Support System provides Operator with Emergency Level Estimation, Diagnosed Cause, Subsequent Abnormality and adequate Countermeasure Operation.

By taking the countermeasure according to the information of Operator Support System, less experienced operators can cope with abnormal conditions in the same way as an experienced operator.

In the future, we intend to expand Operation Support Item and develop Support System which covers overall operation and maintenance working of power plant.

REFERENCES

[KMT87]
Y. Kitamura, M. Mizuno, T. Kojima, An Expert System for Supporting Coal Fired Plant Operation, EPRI Seminar on Expert System Application on Power Plants (May 1987)

[MHIT91]
H. Maruyama, S. Hosaka, S. Ishikawa, M. Tani
Study on Fast Predictor in Anomalous Conditions for Operational Guide, A191 Frontiers in Innovative Computing for the Nuclear Industry (Sep, 1991)

[SAOM88]
T. Sato, H. Abe, M. Okamachi, R. Murata.
Development of Diagnostic Expert System for PWR Feedwater System (IAEA - CN - 49 / 22P)

[Sho76]
Shortliffe. E. H. 1976 Computer - based medical consultation, MYCIN, New York, American Elsevier

A LEARNING APPRENTICE ARCHITECTURE FOR INDUSTRIAL CONTROL AND DECISION SUPPORT

Mohammad Jamil Sawar
Department of Computer Science, Military College of Signals
National University of Sciences and Technology
Rawalpindi, Pakistan.

ABSTRACT

In this paper we present a Learning Apprentice Architecture incorporating a decision support system for industrial plant control and diagnostics. The Learning Apprentice acquires control and diagnostic knowledge by following the case studies and exemplars presented to the decision support system. It employs three machine learning techniques to learn: Exemplar Based Learning is used to acquire new control and diagnostic knowledge, Explanation Based Learning and Similarity Based Learning are used to specialize known cases or discriminate against other cases.

INTRODUCTION

Decision support systems are typically based on knowledge acquired from domain experts, traditionally using techniques of question answering, form filling, and knowledge editing. Though advances in such techniques have eased the process of knowledge acquisition, this process still remains a bottleneck in developing new decision support systems.

This research argues that systems can automatically acquire new knowledge by solving problems and letting the expert provide advice and criticism. Thus, an automated knowledge acquisition system can be constructed by integrating machine learning techniques and decision support systems technology. A learning apprentice architecture for industrial plant control and diagnostics is proposed which actively acquires new diagnostic knowledge from routine control scenarios and the expert's criticism of its diagnostic and control actions.

Learning Apprentice Systems (LAS) provide a very flexible way to acquire new knowledge and overcome the problem of knowledge acquisition bottle-neck faced by many expert systems. One of the major advantages of LAS is in helping the system to gain autonomy, since greater autonomy will be required in advanced stages of the system than in the early stages [Por90].

A very comprehensive definition of LAS is given by Mitchell: "We define learning apprentice systems as the class of interactive knowledge based consultants that directly assimilate new knowledge by observing and analysing the problem solving steps contributed by their users through their normal use of the system" [Mit90]. Another definition of LAS is provided by Segre: "A learning apprentice is a system, usually embedded in the software tools used by a human expert, that gradually increases its own problem-solving abilities by unobtrusively monitoring and analyzing the performance of the expert system" [Seg88]. Yet another definition of LAS is provided by Wilkins: "Apprenticeship learning is a form of learning by watching, in which learning occurs as a by-product of building explanations of human problem solving actions" [Wil88].

Our own definition of LAS is somewhat restricted and employs an active methodology [Saw90] in which the learning apprentice system monitors the control actions and diagnoses generated by the decision support system, and whenever there is a discrepancy between the expert's actions and the decision support system's actions, the learning apprentice system tries to learn a new model of control or diagnosis candidate, or modify its already existing candidate.

We first outline the general architecture of the underlying decision support system, its knowledge base and temporal modeler, then we go on to describe the learning apprentice architecture in detail.

DECISION SUPPORT SYSTEM

The decision support system interactively/directly receives real time data from the user or the plant . From this data, the system presents an ordered list of very-likely, likely, possible and not-likely decisions or diagnoses and can answer questions on how the recommendation was reached, what control or diagnostic action would be most suitable, and what further investigative action could be taken to focus on a particular decision. The data is time tagged and inserted in the object-oriented *current model* (see Fig. 1). The data is also maintained in a history file, which can be used to infer data trends and other time dependent analysis.

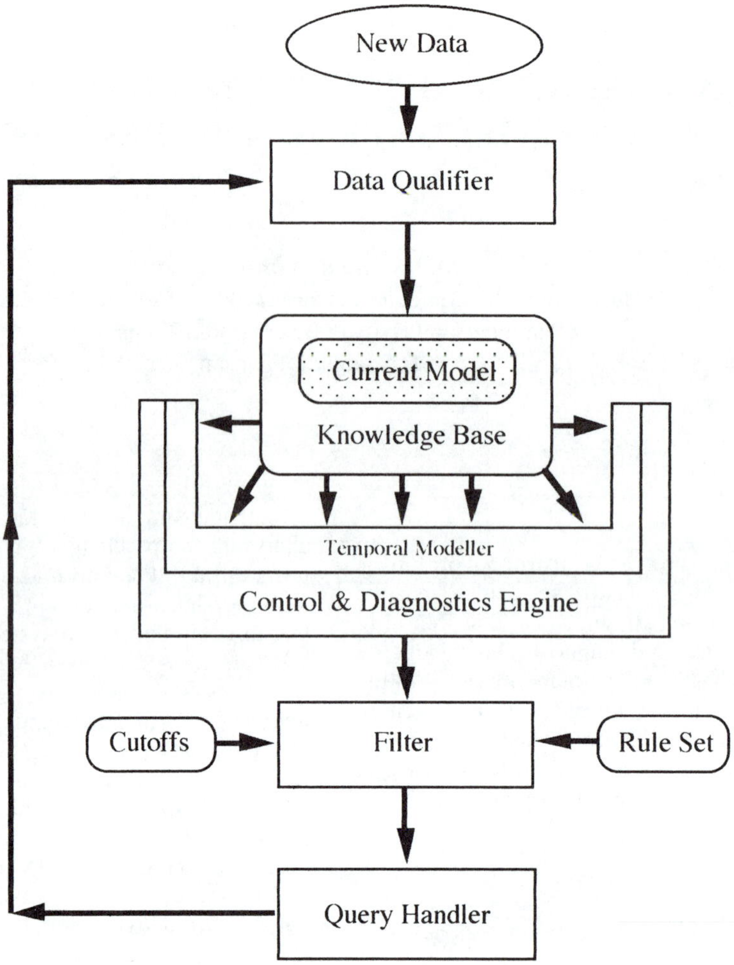

Fig. 1. Architecture of the decision support system

Knowledge Base

The knowledge base is a hierarchical structure of interconnected data objects, with frames representing each object. The data objects are divided into three categories:

- Domain based knowledge (structure of plant, control strategy, hazards, etc.)
- Decision or Diagnosis candidates (here after termed as **candidates**)
- Control or diagnostic Actions

The data objects in these categories are interconnected by semantic links subdivided into: (a) Support links, connecting data objects in domain based knowledge to candidates. (b) Action links, connecting candidates to actions. (c) Expectation links, connecting actions to data objects and candidates.

The candidates possess the actual decisional or diagnostic information (i.e. support links and action links), and are mainly acquired by the learning apprentice. They are composed of two parts: (i) The specific model of the candidate used by the decision support system. (ii) A collection of case studies which were used to learn the specific candidate model.

The support links in a specific candidate model are grouped, by their support level, into: discriminatory slot, strong-support slot, supporting slot and exclusionary slot. Where, for example, the support links in discriminatory slot have a higher weight for likelihood than support links in strong-support slots.

An individual support link is represented as a data-object value pair (also termed as feature), like *temperature very-high* or *pressure very-low*. The advantage of these data-object value pairs is that the decision support system can easily maintain dependency links between the data-object given in a support link and the candidate containing the support link (see section on Control and Diagnostic Engine).

Temporal Modeler

Every data object added to the current model is time stamped to take account of its recency at the time of decision, and to identify the trend of the data. For this the decision support system has an internal clock and a

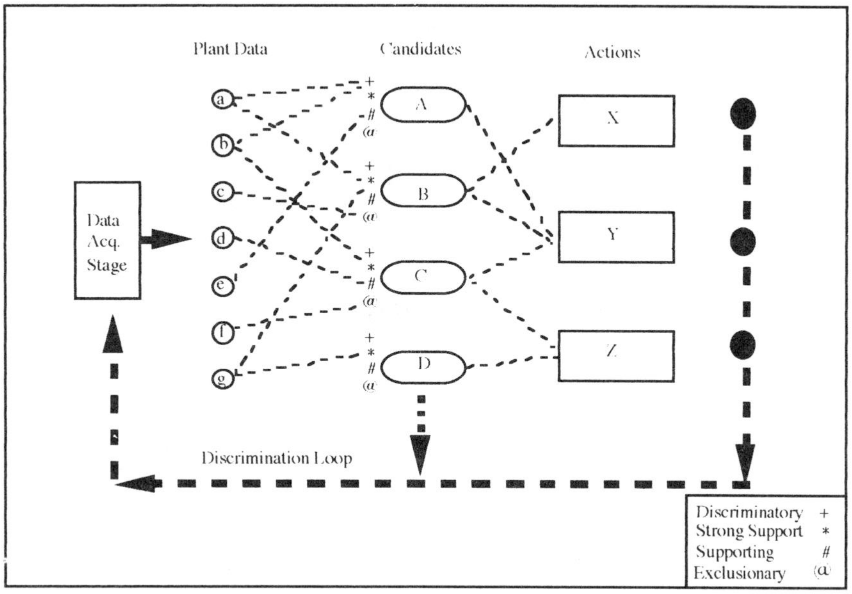

Fig. 2. Decision Process

dynamic temporal modeler which manages the following tasks:

- Trend Modelling
- Time Tagging

When new data values are added, they are compared with the earlier values and with the trend information stored in the data object. If the new value fits the trend criteria a trend qualifier is added to the data object's value. For example *Increasing* data-name or *Decreasing* data-name.

The dynamic role of temporal modeler shows up in time tagging, when the control and diagnostic engine requests the current values of the data objects entered. The temporal modeler evaluates the time tag by comparing the period of time passed since the data was observed with the shelf life[1] of the object. If the time period is:

- Within shelf life: return current-value and no time tag (e.g. *Feed low*)
- Within seven days: return current-value with "Recently" time-tag (e.g. *Recently Feed low*)
- Within two months: return current-value with "Sometime-ago" time-tag (e.g. *Sometime-ago Feed low*)
- Above two months: return current-value with "Long-ago" time-tag (e.g. *long-agoFeed low*)

Control and Diagnostic Engine

Once sufficient data has been added to the current model, the system can automatically suggest a certain control decision, or the user can request the system to display the current applicable decision. This may be one very likely candidate or spread to a list of other likely to possible candidates. These decisions are reached by an incremental decision process which evaluates the likelihood of individual candidates as the new data is acquired or added. The incremental process is based on a constraint propagation mechanism [Gui89]. A simple graphical representation of the decision process is given in Fig. 2, where all the data objects entered into the system through the data acquisition stage are shown on the left, with their links in place. The diagnosis candidates and their discriminatory, strong support, supporting and exclusionary slots are shown, with their dependency links attached to data objects on one side and actions on the other side.

[1] Time period allowed for the data value to remain current.

LEARNING APPRENTICE

The learning apprentice acquires control and diagnostic knowledge through the normal use of the decision support system [Saw91, Saw92a, Saw92b, Saw93]. It is capable of refining the already existing candidates, or of acquiring new ones. The learning process starts when the expert rejects a decision presented by the decision support system, by confirming another candidate. The rejection can be due to three reasons: (i) The system has no model for the expert's confirmed decision. (ii) The system has a wrong model for the expert's confirmed decision. (iii) The system has a correct model of the confirmed decision, but is listing other likely or very-likely candidates which should not be so likely. Fig. 3 shows the architecture of the learning apprentice system.

The learning apprentice employs an extended Exemplar Based Learning module [Por90] to control the learning process and to drive two other modules based on Explanation Based Learning (EBL) and Similarity Based Learning (SBL). This is in contrast to other learning systems which follow the EBL driven SBL or SBL driven EBL approaches [Mit86, DeJ86, Paz88, Saw90].

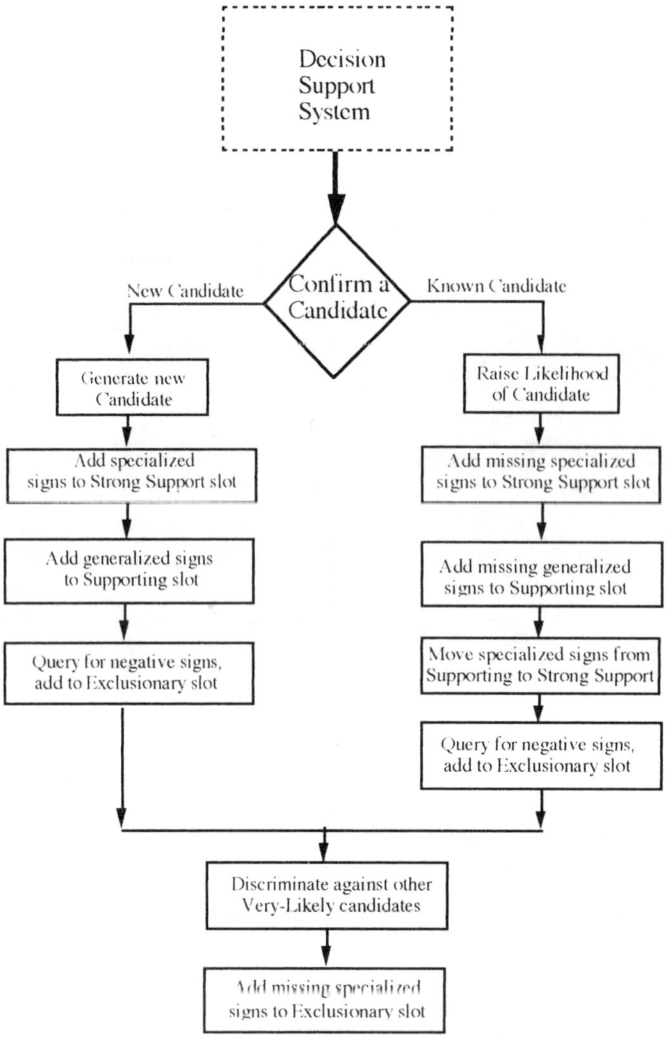

Fig. 3. Learning Apprentice System

The advantage of having an Exemplar Based Learning module on top of EBL and SBL modules is two folds. Firstly the learning apprentice gets a chance to analyze the control scenarios presented and convert it into an exemplar (see below). Secondly it records the scenarios with the relevant diagnosis candidate for future refinement.

Whenever a new candidate is to be learnt, the learning apprentice creates a new model for the candidate and attaches the current scenario to it. To create a new candidate model, the learning apprentice tries to specialize the scenario to produce an exemplar by removing most[2] of the data objects with normal or inferred values. After this specialization, it performs a generalization process where some general data objects which belong to the hierarchical structure of the current set of data objects are also added to the exemplar. Once this process is complete the Exemplar Based Learning module stores this exemplar as a new model for the candidate.

Though the major bookkeeping and storing of the cases with their relative candidates is performed by the Exemplar Based Learning module, when the system has to modify an already known candidate it uses Discriminatory Explanation-based Learning (DEBL) [Saw93] to generalize the known model with the new data. To modify an already known candidate, e.g. by raising or lowering its likelihood for the current set of data, the learning apprentice system first tries to find all the data objects which would have supported that diagnosis candidate. Then it tries to compare this set with the specialized and generalized exemplar (derived as mentioned above from the study) to determine the data objects which have to be added to the candidate's support slots.

After applying one of the above strategies, the learning apprentice uses Discriminatory Similarity-based Learning (DSBL) [Saw93] to discriminate the confirmed candidate against other very-likely or likely candidates returned by the decision support system. DSBL uses the current

2 Most of the normal or inferred data objects are removed, not all, since some normal or inferred data objects can act as exclusionary signs for a candidate.

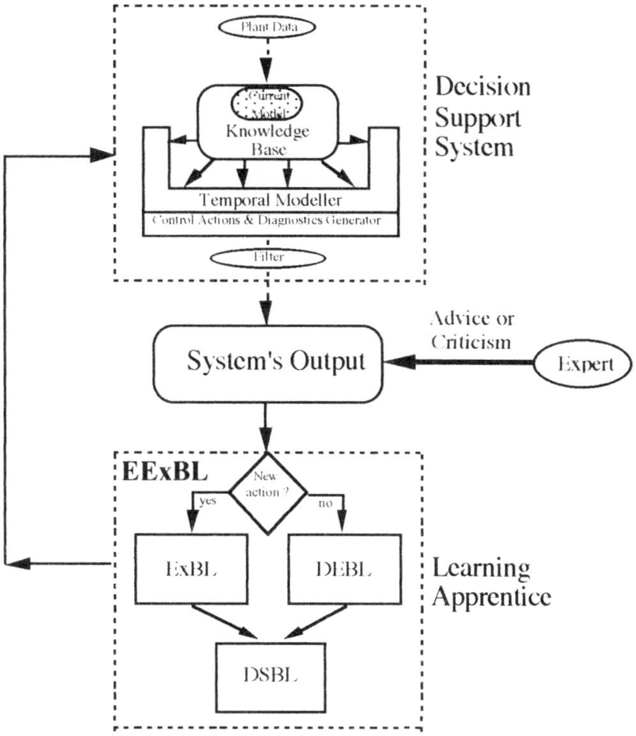

Fig. 4 The learning apprentice architecture (an abstract view)

exemplar as a negative example for the selected candidates and tries to reduce their likelihood by adding missing data objects into their exclusionary slots.

CONCLUSIONS

The learning apprentice architecture described in this paper follows the working definition of *human apprentices*[3]. It has a decision support system which embodies the domain specific knowledge and carries out the control and diagnostic tasks. In the process of carrying out these tasks the system may come across scenarios which it diagnoses incorrectly or is unable to control. In such circumstances the system relies on the learning apprentice to learn from the expert's advice or criticism, which point it to the correct control or diagnosis candidate and provide any missing information.

This general architecture is shown in Fig. 4, which presents an abstract view of the system being developed in this research. It divides the system into two conceptual blocks: (i) Decision support system, and (ii) Learning apprentice. These two blocks are connected by an intermediate stage which presents the system's recommended candidates to the expert and lets the expert provide any advice or criticism. The three learning strategies employed by the systems are briefly summarised below:

Exemplar-based Learning (ExBL): is used to learn a new candidate from the exemplar (derived from control scenarios), when the confirmed candidate is not known by the system.

Discriminatory Explanation-based Learning (DEBL): is used to refine the confirmed candidate and learn discriminatory knowledge from the exemplar (derived from scenario), when the confirmed candidate is known by the system.

Discriminatory Similarity-based Learning (DSBL): is used to learn discriminatory knowledge for other candidates with similar likelihood (i.e. which are similar to the confirmed candidate), by using the exemplar (derived from scenarios) as a negative example for them.

---

3 *Human Apprentices*, who learn from their experience in factories, garages, hospitals, etc. and from the advice and criticism received from their seniors. Thus, in their routine tasks controling and diagnosing various complications, the *human apprentices* gain experience. When they come across some difficult situation which they are unable to diagnose they are advised by their seniors, and when they diagnose incorrectly they are criticised by their seniors. This input from the seniors (or in technical terms *experts*) allows them to learn new diagnostic skills and new diagnosis candidates which they did not know already.

REFERENCES

[DeJ86] G. DeJong and R. Mooney "Explanation-Based Learning: An Alternative View" *Machine Learning*, vol. 1, no. 2, pp. 145-176, 1986.

[Gui89] G. Guise *KR: Constraint-Based Knowledge Representation*. Technical Report CMU-CS-89-142, Carnegie Mellon University, April 1989.

[Mit86] T. M. Mitchell, R. M. Keller, and S. T. Kedar-Cabelli "Explanation-Based Generalization: A Unifying View" *Machine Learning*, Vol. 1, no. 1, pp 47-80, 1986.

[Mit90] T. M. Mitchell, S. Mahadevan, L. I. Steinberg "LEAP: A Learning Apprentice for VLSI Design", in *Machine Learning an Artificial Intelligence approach, Volume III*, Ed. Y. Kodratoff, R. Michalski, Morgan Kaufman, 1990.

[Paz88] M. J. Pazzani "Integrated Learning with Incorrect and Incomplete Theories" *in Proceedings of Fifth International Conference on Machine Learning*, pp. 291-297, June 1988.

[Por90] B. W. Porter "Concept Learning and Heuristic Classification in Weak Theory Domain", Artificial Intelligence, Vol. 45, No 1 & 2, 1990.

[Saw90] M. J. Sawar and R. C. Thomas "Learning Apprentice System for Turbine Modelling" in *Proceedings of IEA/AIE-90*, Charleston, U.S.A. July 1990.

[Saw91] M. J. Sawar, T. G. Brennan, A. J. Cole and J. Stewart "POEMS (PostOperative Expert Medical System)" in *Proceedings of IJCAI-91 One Day Workshop : "Representing Knowledge in Medical Decision Support Systems"*, 1991.

[Saw92a] M. J. Sawar, T. G. Brennan, A. J. Cole and J. Stewart "An Expert System for PostOperative Care (POEMS)", in *Proceedings of MEDINFO-92*, Geneva, Switzerland, Sep. 1992.

[Saw92b] M. J. Sawar, A. J. Cole, and B. M. Smith "A Learning Apprentice for Decision Support Systems", accepted for publication in Proceedings of EXPERSYS-92, Paris, France, October 1992. Also available as Research Report 92.19, School of Computer Studies, University of Leeds, August 1992.

[Saw93] M. J. Sawar "A Learning Apprentice for Medical Decision Support Systems", Ph.D. Thesis, 1992, School of Computer Studies, University of Leeds, U.K.

[Seg88] A. M. Segre, "Machine Learning of Robot Assembly Plans", Kluwer Academic Publishers, 1988.

[Wil88] D. C. Wilkins, W. J. Clancey, B. G. Buchanan "Knowledge Base Refinement by Monitoring Abstract Control Knowledge", *Knowledge Acquisition for Knowledge Based Systems Volume I*, Ed. B. Gaines, 183-195, Academic press, 1988.

DIAGNOSIS OF AUTOMOBILE FAILURES USING FUZZY LOGIC

Brennan T. Hamilton[1] and Yi Lu[2]
[1]Diagnostic Systems Department
Powertrain Division
Ford Motor Company
Dearborn, MI 48128
[2]Department of Electrical and Computer Engineering
The University of Michigan - Dearborn
Dearborn, MI 48128-1491
E-mail: lu@umdsun2.umd.umich.edu
Telephone: (313) 593-5420
Telex: (313) 593-5028

ABSTRACT

This paper describes a project that applies fuzzy logic to the field of automobile fault diagnosis. We present a fuzzy logic based fault diagnostic system for testing Lambse fails in an Electronic Engine Controller(EEC) in a vehicle. The system is designed to be a component of an End-of-Line test system at automobile assembly plant. The system has been tested on log files recorded from assembly plant test sites and has proven to outperform the existing semi-manual testing procedure.

1. INTRODUCTION

As automotive electronic control systems have become more advanced and sophisticated in recent years, malfunction phenomena have also become increasingly more complicated. It has been well recognized in the automotive industry that effective vehicle diagnostic systems will play a key role in the competitive market of the 1990's.

The Ford Motor company is in the process of launching an End-of-Line test system at every North American assembly plant. In order to accomplish this task, a system is designed to collect and analyze Electronic Engine Controller(EEC) data while the vehicle is dynamically tested. Operators drive the vehicles through a preset profile and the vehicles are either passed or failed according to the data collected during the tests. The decision is made based on two information sources, the EEC on-board tests and off-board test performed by the vehicle test system on EEC generated data. This paper addresses the second problem.

The existing techniques used in assembly plants for the off-board test are mainly on a trial and error basis and largely dependent on individual engineers' experience. The mechanics in the assembly plants have varying degrees of vehicle diagnostic expertise. They are in general very weak in fundamentally EEC operations, signals, and etc. Therefore, there is a strong need for developing an intelligent system to automatically diagnose EEC failures. The nature of the application requires such an intelligent system to provide prompt and reliable diagnosis. The work presented in this paper is the first step towards this goal.

Many of the past applications involving diagnosis have been crisp rule-based systems in which productions rules are used to provide a mapping between the possible causes and the inputs of the systems and the possible faults. This type of technique is not effective to diagnose faults occurring in today's vehicles. The electronic control components are more like a blackbox. It is difficult for mechanics to diagnose faults accurately unless they are thoroughly familiar with the system specifications and functions. As a result, it is extremely difficult to develop reliable crisp expert rules that can fully answer all the questions in terms of faults.

The theory of fuzzy is aimed at the development of a set of concepts and techniques for dealing with sources of uncertainty or imprecision which are nonstatistical in nature. It has found many applications in control theory when gradual adjustments are necessary. One of the most successful stories is its application in household appliances in Japan. In the USA, more applications of fuzzy logic has its application widely spreaded in automotive, air and spacecraft and even the stock exchange. The application of expert systems and fuzzy logic to automobiles has been explored by several authors. Filljov, Marinov, and Ovcharov[FiO88] have discussed the application of an engine diagnostic expert. Tennant[Ten90] has explored where the opportunities and problems lie with onboard knowledge systems in vehicles. Xiaojun, Shuzi, Anfa, and Hanmin[ZYZ88] have described a knowledge based system for automobile engines which employs a hierarchical and modular architecture. This paper presents a project that uses fuzzy logic to solve the problem of Lambse failures in the End-of-Line test in automobiles.

2. PROBLEM DESCRIPTION

The modern automobile is a complex operating system. Hundreds of signal inputs are processed per second to determine the proper air/fuel ratio, optimal transmission shift points, and etc. One Electronic Engine Controller (EEC) parameter that can give insight into the overall condition of the engine system is the desired air/fuel flow 'Lambda' (λ). Based on the stoichiometry, i.e. 14.64 lb. of air to 1 lb. of fuel is the chemically correct ratio for complete combustion, the desired air/fuel flow λ can be expressed as the air/fuel ratio divided by 14.64, i.e.

$$\lambda = \frac{air}{14.64 fuel}.$$

In the ideal case, λ equals to 1 which is the stoichiometry(See Figure 1). λ is then converted to a fuel pulse width for the injectors, based on the following five parameters:

- engine RPM(Revolution per Minute),
- number of cylinders,
- injector flow characteristics,
- number of injectors, and
- desired injection frequency.

During Closed Loop mode of operation(see Figure 1), the computer ramps the desired lambda value (LAMBSE) in a limited cycle around stoichiometry; and the LAMBSE value is clipped at the rich and lean end. Using the oxygen sensor (HO2S), the EEC increases or decreases LAMBSE as a calculated rate of change. When the HO2S sensor indicates that the quality of the combustion was too lean (an excess amount of oxygen) the computer changes the LAMBSE value to force the air/fuel mixture to go rich. When the HO2S sensor indicates the quality of the combustion was too rich, the computer forces the LAMBSE value lean. In this fashion, the average air/fuel mixture will be stoichiometry as shown in Figure 1 in dashed line. A vehicles can have either one HO2S (mono) or two HO2S's (stereo). Stereo HO2S systems control fuel separately from cylinder bank to cylinder bank e.g., LAMBSE1 is for cylinder bank #1 and LAMBSE2 is for cylinder bank #2. The system we study in this project has two HO2S's.

There are many factors that can influence the air/fuel ratio in the engine. If there is a vacuum leak, more air is entering the combustion process than is calculated to be flowing past the air meter, causing the combustion to burn lean. In this case the value of LAMBSE will be rich ($\lambda < 1$). If the air flow sensor is reading more air than is actually entering the combustion process, less air is entering the combustion process than is calculated from the air flow sensor; causing the combustion to burn rich. In this case, the value of LAMBSE will be lean ($\lambda > 1$). Similarly, any engine component that affects air, fuel, or spark will have an effect on the calculated value of LAMBSE. Therefore, the Lambse signal is one of the signals used in **off-board calculations.** The LAMBSE signal is compared with a calibratable max and min and an error code is set if the signal is out-of-range for a given number of records. Once an error code is set, the vehicle must be repaired and retested until it passes.

Diagnosing Lambse failures is very difficult because Lambse is a symptom, not a cause. Common Lambse failures are generated by engine vacuum leaks, as discussed above. However the Lambse codes may also be set by HO2S failures, fuel systems problems, intake manifold problems, ignition problems, etc. In a stereo HO2S vehicle, the vehicle computer receives two channel data, i.e., HO2S-bank 1 and HO2S-bank 2, and the vehicle computer can adjust the fuel ratios differently from bank to bank. A LAMBSE failure on one side and not the other in a stereo HO2S vehicle may be caused by the following sources:

- A manifold leak,
- injector failure,
- cam mistimed (retarded),
- rocker misadjusted,
- exhaust leak,
- misseated spark plug, etc.

A LAMBSE failure on both sides of an engine may be caused by the following sources:

- vacuum leak,
- faulty Manifold Air Flow Sensor,
- a stuck-open Exhaust Gas Recirculating Valve,
- low fuel pressure, and
- the Canister Purge Valve stuck-open, etc.

At this step of research we focus on the vacuum leak problem.

3. A FUZZY LOGIC BASED SYSTEM FOR LAMBSE FAILURE DIAGNOSIS

The vacuum leak detection algorithm is developed to operate in the assembly plant environment. The algorithm was developed based on the three primary constraints, speed, computer platform and available knowledge data. Because the average test time of a vehicle is approximately 1.5 minutes, the algorithm we are developing should not add more than a few seconds to the overall test time. The application environment requires the algorithm to run on the DOS based PC's. The algorithm we are developing uses only those parameters that are currently collected.

The model for the Vacuum Leak Detection System is

based on fuzzy logic. In fuzzy logic, beliefs in the validity of the data are represented by membership functions for various data variables. As we discussed in the last section, the vacuum leak is one of the five sources that can cause both parameters Lambse_1 and Lambse_2 to failure. Based on the engineering knowledge, we believe three more parameters are needed to verify a vacuum leak, they are throttle position, idle speed DC, and mass air flow. The amount of air entering a gasoline engine is controlled by a throttle plate that opens and closes. A throttle position sensor sense the angle that the throttle is open and transforms it into a voltage that is read by the vehicle controller. If there is a vacuum leak, the throttle position should be low. The idle speed duty cycle refers to the pulse train that modulates the solenoid. The mass air flow tells us the mass of air entering the engine. Figure 2 shows the LAMBSE fail diagnose system.

The inference engine contains the membership functions(MBF) for all five variables. The membership functions were developed by using the existing parameter calibrations which have proven over time to be affective in detecting vehicles with vacuum leaks. Three membership functions, LOW, MEDIUM, and HIGH, were developed for each of the five variables. For each variable x ∈ {throttle position, Lambse_1, Lambse_2, Idle speed DC, Mass air flow}, the three membership functions for x are:

$$MBF\_LOW_x(t) = \begin{cases} 1 & X\_absolute\_\min \le t < x\_\min \\ \dfrac{2t - x\_\min - x\_\max}{x\_\min - x\_\max} & x\_\min \le t < \dfrac{x\_\max + x\_\min}{2} \\ 0 & \dfrac{x\_\max + x\_\min}{2} \le t \le x\_absolute\_\max \end{cases},$$

$$MBF\_MED_x(t) = \begin{cases} 0 & x\_absolute\_\min \le t < x\_\min \\ \dfrac{2(t - x\_\min)}{x\_\max - x\_\min} & x\_\min \le t < \dfrac{x\_\min + x\_\max}{2} \\ \dfrac{2(x\_\min - t)}{x\_\max - x\_\min} & \dfrac{x\_\min + x\_\max}{2} \le t < x\_\max \\ 0 & x\_\max \le t < x\_absolute\_\max \end{cases},$$

$$MBF\_HIGH_x(t) = \begin{cases} 0 & X\_absolute\_\min \le t < \dfrac{x\_\max + x\_\min}{2} \\ \dfrac{2t - x\_\min - x\_\max}{x\_\max - x\_\min} & \dfrac{x\_\max + x\_\min}{2} \le t < X\_MAX \\ 1 & \dfrac{x\_\max + x\_\min}{2} \le t \le x\_absolute\_\max \end{cases}$$

where the x_max and x_min are the calibratable limits for variable x, and x_absolute_min and x_absolute_max are the lowest and highest possible values for x respectively. Figure 3. (a) illustrates the derivation of the three membership functions for variable Lambse_1. The start of the maximum point for the low lambse_1 belief was set to Lambse_1_min, and the lowest Lambse value was set to the lowest possible value, x_absolute_min. The maximum point for the $MBF\_MED_{lambse\_1}$ is at (x_max + x_min)/2. The start of the maximum point for the high belief was set to Lambse_1_max, and the highest Lambse value was set to the highest possible calculated Lambse value, Lambse_1_absolute_max. Figure 3 (a) through (e) illustrate the membership functions for all five variables. Figure 3 (f) illustrates the membership function for vacuum leak. The fuzzy rules used in the Inference Engine in the system model are as follows:

IF Lambse_1 IS LOW AND
Lambse_2 IS LOW AND
Mass_Air_Flow IS MED AND
Idle_Speed_DC IS LOW AND
Throttle Position IS NOT HIGH THEN
Vacuum Leak = HIGH

IF Lambse_1 IS LOW AND
Lambse_2 IS NOT LOW THEN
Vacuum_Leak = LOW

IF Lambse_2 IS LOW AND
Lambse_1 IS NOT LOW THEN
Vacuum_Leak = LOW

The fuzzy rules reflect the engineering diagnostic knowledge. The inference steps for vacuum leak is as follows. The values of five variables are extracted from a log data file and are sent to the Inference Engine. The Inference Engine then calculates the belief for LOW, MED and HIGH for each variable. Based on these belief values, the fuzzy rules are searched and the matched rules are fired. The vacuume leak belief are obtained from the fuzzy rules. We use the mass centroid method for defuzzification. The vacuum leak belief value is used to make a repair recommendation to engineers, for example, check for vacuum leak. The following is an example of fuzzy reasoning about vacuum leak.

Input data extracted from a log file:

Idle_speed_DC = 20%
Mass_Air_Flow = 0.75 volzDC
Lambse_1 = 0.76
Lambse_2 = 0.79
Throttle Position = 190 counts

MBF values for individual variables:

$MBF\_LOW_{idle\_speed\_DC} = 1.0$
$MBF\_MED_{idle\_speed\_DC} = 0$
$MBF\_HIGH_{idle\_speed\_DC} = 0$

$MBF\_LOW_{Mass\text{-}Air\text{-}Flow} = 0$
$MBF\_MED_{Mass\text{-}Air\text{-}Flow} = 1.0$
$MBF\_HIGH_{Mass\text{-}Air\text{-}Flow} = 0$

$MBF\_LOW_{Lambse\text{-}1} = 1.0$
$MBF\_MED_{Lambse\text{-}1} = 0$
$MBF\_HIGH_{Lambse\text{-}1} = 0$

$MBF\_LOW_{Lambse\text{-}2} = 1.0$
$MBF\_MED_{Lambse\text{-}2} = 0$
$MBF\_HIGH_{Lambse\text{-}2} = 0$

$MBF\_LOW_{Throttle\text{-}position} = 0.18$
$MBF\_MED_{Throttle\text{-}position} = 0.81$
$MBF\_HIGH_{Throttle\text{-}position} = 0$

RULE 1 fired:

$MBF\_LOW_{idle\_speed\_DC} = 1.0$ AND
$MBF\_LOW_{Lambse\text{-}1} = 1.0$ AND
$MBF\_LOW_{Lambse\text{-}2} = 1.0$ AND
$MBF\_MED_{Mass\text{-}Air\text{-}Flow} = 1.0$ AND
Not $MBF\_HIGH_{Throttle\text{-}position} = 1.0$

$\Longrightarrow$
$MBF\_HIGH_{Vacuum\text{-}Leak} = 1.0$

Defuzzification:

the vacuum leak possibility is 83%

Recommendation:

check vacuum leak

4. EXPERIMENTS AND RESULTS

The algorithm was implemented under Microsoft C 7.0 and runs under DOS 5.0 on 80386 or 80486 based processors. Currently, the data for vehicle End-of-Line tests are collected and transmitted hourly from test site to Ford Motor Company in Dearborn, Michigan. "Failed" vehicle log files and "Passed" vehicle log files are separated into different directories under a Julian data structure. In a Julian date structure, PC files are stored in directories corresponding to the date they were created. For example, January 1 is Julian date 1 and December 31 is Julian date 365, and all files created on November 9 (Julian date 313) were stored in C:\DATA\313 . The experiments are confined to determining whether or not LAMBSE failures are related to vacuum leaks on 4.6L 1994 Thunderbirds.

For the purposes of our experiment, approximately 1,500 vehicle log files were downloaded to examine for vacuum leaks. Of those 1,500 files, 25 had failed for "Lambse Low". Our system positively identified four out of the 25 failures as valid vacuum leaks, which are consistent with the mechanics comments included in the vehicle log file. Additionally, our algorithm identified three more vehicles as candidates for vacuum leaks. The experiment results indicate that the vacuum leak diagnose system performs very well in comparison with engineering experts. The system was shown to be able to detect vacuum leaks and had the added benefit of identifying vehicles that were incorrectly failed for low LAMBSE readings. It is because of the success of the fuzzy logic in the detection of vacuum leaks, we are currently investigating fuzzy logic in the detection of automatic transmission shift anomalies.

5. ACKNOWLEDGEMENT

This work was support in part by a grant from the Center for Engineering Education and Practice, School of Engineering, University of Michigan-Dearborn.

6. REFERENCES

[FiO88] Filljov, M. Marinov, S. Ovcharov, "*Engine Diagnostic Expert System*", 18th International Symposium on Automotive Technology & Automation, 30th May - 3rd June 1988.

[Kau75] A. Kaufmann, "*Introduction to the Theory of Fuzzy Subsets, Vol . 1*," New York, Academic Press, Inc., 1975.

[Ten90] Harry Tennant, "*Onboard Knowledge Systems in Vehicles*", Vehicle Electronics in the 90's: Proceedings of the International Congress on Transportation Electronics, October 1990.

[YOT87] R.R. Yager, S. Ovchinnikov, R. M. Tong, and H. T. Ngyen, "*Fuzzy Sets And Applications*," New York, John Wiley & Sins, 1987

[Zad69] Lotfi Zadeh, "*Toward a Theory of Fuzzy Systems,*" Technical Report, Electronics Research Laboratory, The University of California, Berkeley, California, ERL-69-2, 1969

[Zim91] H. J. Zimmermann, "*Fuzzy Set Theory and Its Applications*," Massachusetts, Kluwer Academic Publishers, 1991.

[ZYZ88] Zheng Xiaojun, Yang Shuzi, Zhou Anfa, Shi Hanmin, "*A Knowledge-Based Diagnosis System for Automobile Engines*", The International Journal of Advanced Manufacturing Technology, 1988.

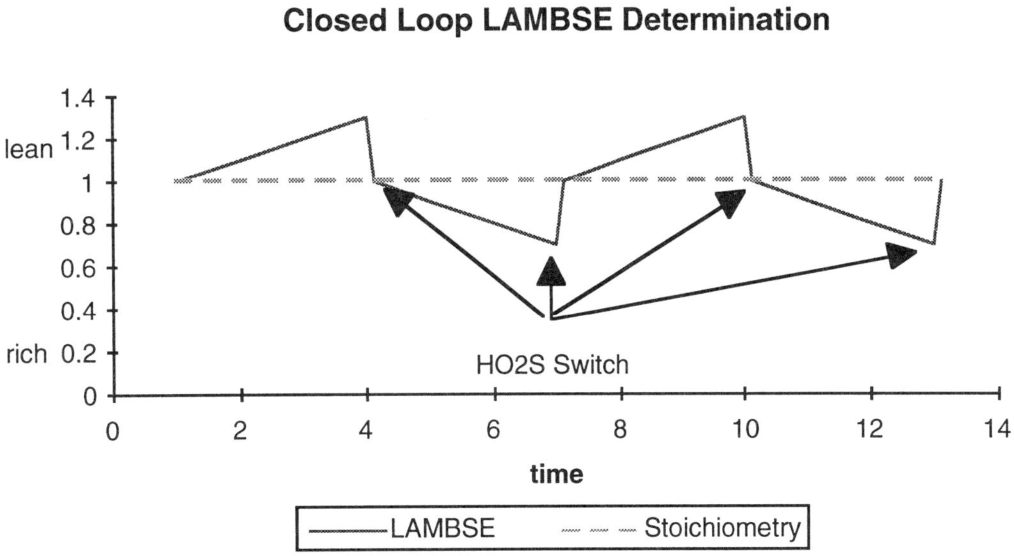

Figure 1. Closed Loop Lambse Determination

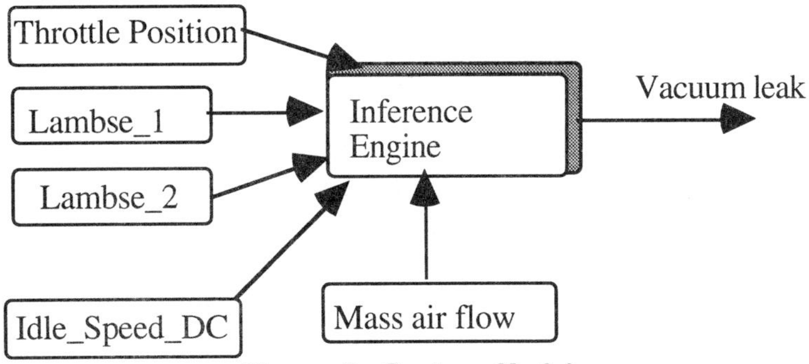

Figure 2. System Model.

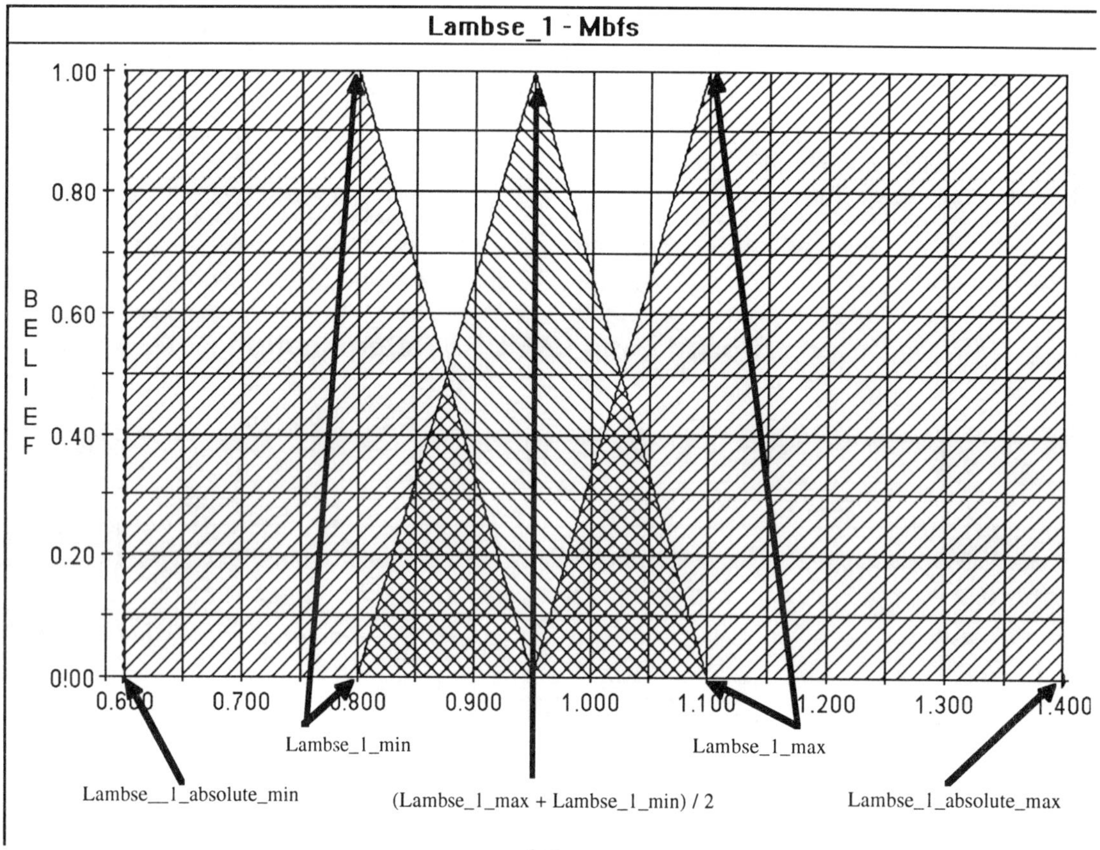
Lambse_1 - Mbfs
1.00
0.80
0.60
0.40
0.20
0!00
BELIEF
0.600
0.700
0.800
0.900
1.000
1.100
1.200
1.300
1.400
Lambse_1_min
Lambse_1_max
Lambse__1_absolute_min
(Lambse_1_max + Lambse_1_min) / 2
Lambse_1_absolute_max

(a)

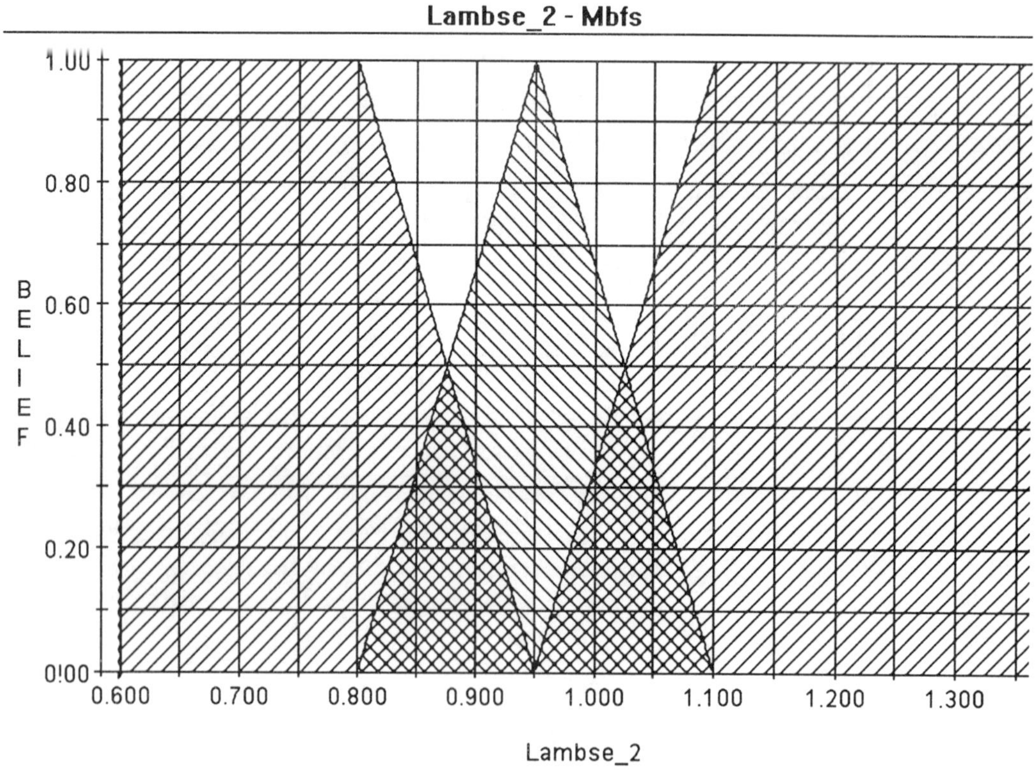
Lambse_2 - Mbfs
1.00
0.80
0.60
0.40
0.20
0!00
BELIEF
0.600
0.700
0.800
0.900
1.000
1.100
1.200
1.300
Lambse_2

(b)

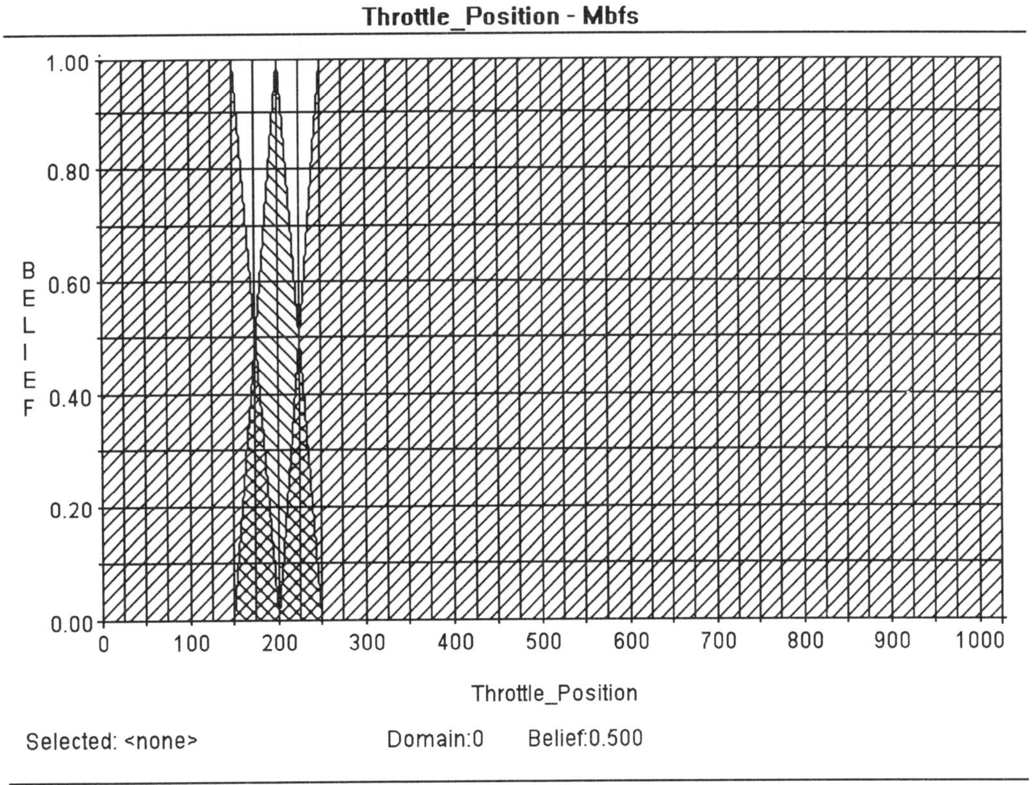
Throttle_Position - Mbfs
1.00
0.80
0.60
0.40
0.20
0.00
BELIEF
0
100
200
300
400
500
600
700
800
900
1000
Throttle_Position
Selected: <none>
Domain:0
Belief:0.500

(c)

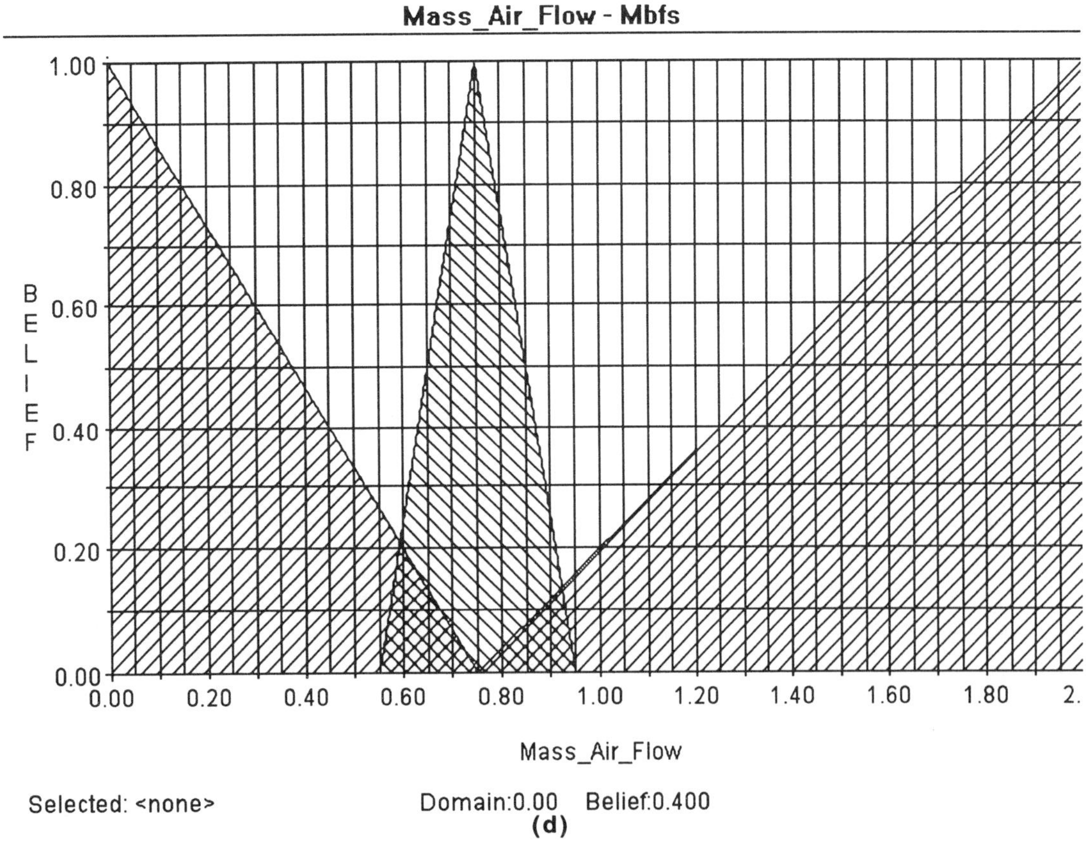
Mass_Air_Flow - Mbfs
1.00
0.80
0.60
0.40
0.20
0.00
BELIEF
0.00
0.20
0.40
0.60
0.80
1.00
1.20
1.40
1.60
1.80
2.
Mass_Air_Flow
Selected: <none>
Domain:0.00
Belief:0.400

(d)

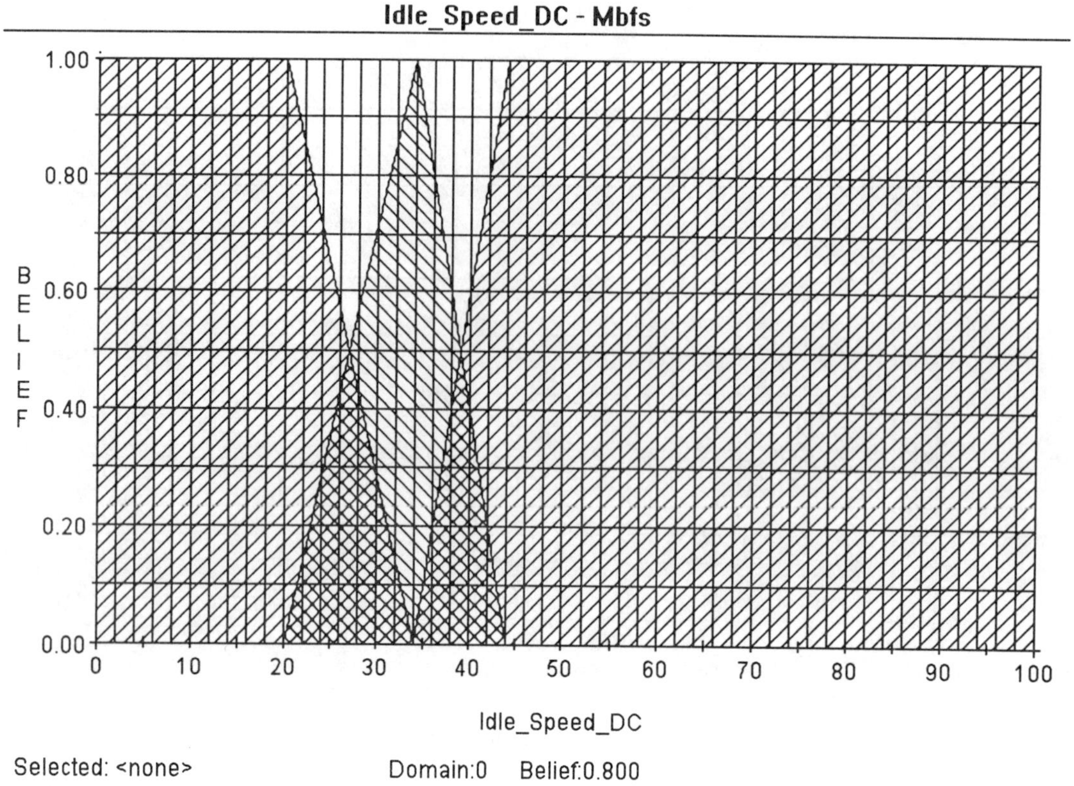

(e)

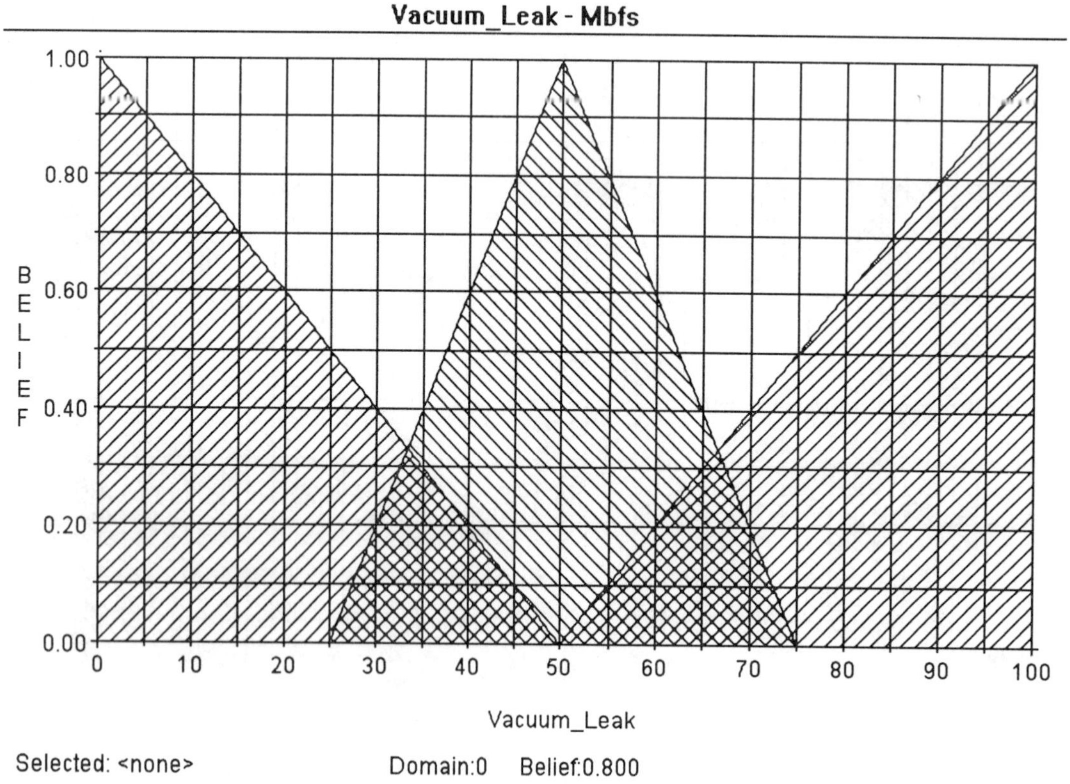

(f)

Figure 3 Fuzzy membership functions for the diagnostic parameters and vacuum leak.

DESIGN AND IMPLEMENTATION OF REAL-TIME EXPERT SYSTEM FOR TROUBLESHOOTING IN INTERNATIONAL TELEPHONE NETWORKS

Kazunori Matsumoto, Kazuo Hashimoto and Sadao Obana

R&D Laboratories, Kokusai Denshin Denwa Co., Ltd., (KDD)

2-1-15 Ohara Kamifukuoka-shi, Saitama 356, Japan
Email: matsu@kddnews.nes.lab.kdd.co.jp

ABSTRACT

This paper discusses the key points of successful design and implementation of an expert system for troubleshooting in international telephone networks, that is, (1) case-based optimization of mapping from quantitative to qualitative values, (2) software architecture for real-time diagnosis. The implemented system is evaluated through practical use with regard to diagnosis capabilities, which are trouble detection and cause identification, and response time. The system has achieved 79% of the human capability of cause identification, and sufficient ability for practical use in other points.

1 INTRODUCTION

The goal of network management is to maximize the utilization of total network resources. However, as there exist a number of constraints in obtaining information about the network resources of foreign networks, it is often the case that network operators in international telephone networks are forced to react to trouble with insufficient information. It is heuristic know-how accumulated through experience that can compensate for lack of information. This is the major reason for introducing expert systems (ESs) .

Many attempts to develop ESs have been made for network management [1]-[4]. However, since these ESs assume that sufficient network resource information be accessible, they are not necessarily suitable for a domain of troubleshooting of international telephone networks, in which sufficient information is not given.

We have developed an ES suitable for this domain. For the development, there are two major technical issues into account to be taken.

First is the qualitative treatment of quantitative values. Network operators identify phenomena in the networks in terms of qualitative values of network management parameters (NMPs) such as SZ[1] and ANS[2]. For instance, the trend of a NMP of the traffic to any foreign network is represented by a set of symbols, such as 'sharp increase', 'slight increase', etc. As this is a convenient way to represent and understand, the heuristic knowledge in the ES is qualitatively represented. On the other hand, the observed values of NMPs are given in quantitative terms, and mapping from quantitative to qualitative values depends on a network operator, the foreign network, circuits-group[3], and etc. As qualitative

[1]SZ (Seizure): Number of calls succeeding in connecting to foreign switching equipment through a specified circuits-group per unit time

[2]ANS (Answer): Number of calls which are connected successfully through a specified circuits-group per unit time

[3]circuits-group: All the communication circuits in a physical link

knowledge assumes that mapping is appropriately carried out, the mapping should be designed to maximize diagnostic capability.

Second is software architecture for real-time diagnosis. The system should repeatedly undertake the surveillance of networks and the selection of a diagnostic objective. The interval of a repeated cycle should be shorter than the given limit.

This paper discusses the above two major technical issues and describes fundamental design and implementation of the system. The paper, firstly, summarizes troubleshooting in international telephone networks and clarifies the scope of the system. Second, the paper describes a qualitative treatment of quantitative values. A technique of case-based optimization of the mapping from quantitative to qualitative values is introduced as a solution. Then, the paper discusses the software architecture for real-time diagnosis. Finally, the implemented system is evaluated through practical use with regard to the capability of trouble detection and cause identification, and response time.

2 TROUBLESHOOTING FOR INTERNATIONAL TELEPHONE NETWORKS

ITU-T (International Telecommunication Union, Telecommunication standardization sector) has defined a set of network management parameters (NMPs) to be monitored, such as SZ and ANS, and recommended a series of operations for troubleshooting at each local network management center [5]. Figure 1 illustrates a schematic view of network operations for troubleshooting. Each operation in the figure is as follows.

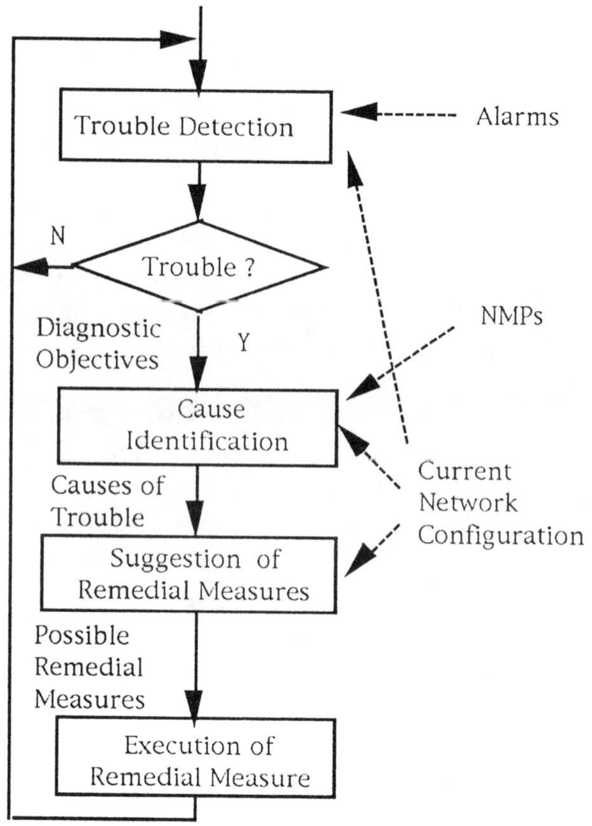

Figure 1: Troubleshooting in international telephone networks

Trouble Detection: to detect deterioration of success connection rate by monitoring alarms and considering the configuration of foreign networks where alarms are issued. The criterion of detection depends on the characteristics of the foreign networks under consideration, and the period of alarm continuation. When several alarms are issued at the same time, network operators infer whether the alarms are due to a single fault or to multiple faults distributed over different foreign networks.

Cause Identification: to identify causes of the detected trouble. Network operators collect time-series data for NMPs related to the trouble, and determine causes by trend analysis of NMPs and information about current network configuration.

Suggestion of Remedial Measures: to find possible remedial measures to deal with the deterioration such as "setting up re-routing paths" based on the identified causes and the current network configuration.

Execution of Remedial Measures: to adopt a final remedial measure from the possible alternatives, including negotiating with a foreign network management office.

The scope of the ES discussed in this paper includes operations of 'Trouble Detection,' 'Cause

Identification,' and 'Suggestion of Remedial Measures.' The last operation of the troubleshooting, 'Execution of Remedial Measure', is excluded, since it includes a process of negotiation with a foreign network management office.

3 FUNDAMENTAL DESIGN

A qualitative treatment of quantitative values and software architecture for real-time diagnosis are important in the fundamental design of the ES.

3.1 Qualitative treatment of quantitative values

3.1.1 Quantitative and qualitative representation

Network operators identify phenomena in the networks in terms of qualitative values. For instance, the trend of SZ of any circuits-group is represented as 'sharp increase', 'slight increase', 'constant', 'slight decrease', 'sharp decrease'. The most recent value of SZ is represented as 'very large', 'large', 'medium', 'small', 'very small'. As it is easy to understand and manage network operators' knowledge with qualitative values, diagnostic knowledge of the ES is represented in terms of qualitative values.

Network management parameters (NMPs) are, however, given as quantitative values from network resources. It is necessary to map observed quantitative values to qualitative ones.

Followings are examples of a quantitative representation and qualitative ones.

(1) rate of change in SZ is 5.21 per minute. (quantitative)
(2) trend of SZ is sharp increase. (qualitative)
(3) trend of SZ is slight increase. (qualitative)

One, who assigns a quantitative range $[5.0, \infty]$ of SZ as an interpretation of 'sharp increase', interprets (1) as (2). The other, who assigns $[4.5, 5.5]$ as 'slight increase', interprets (1) as (3). Thus, the interpretation of a qualitative value depends on a network operator. Furthermore, the interpretation also depends on foreign network, circuits-group, and etc. Therefore, it is difficult for human operators to determine a proper mapping function.

We discuss realization and optimization of a mapping function in **3.1.2** and **3.1.3**, respectively.

3.1.2 Mapping function from quantitative to qualitative values

Network operators are usually concerned with long-term trends rather than short-time trends of time-series data of NMPs. As shown in Fig. 2,

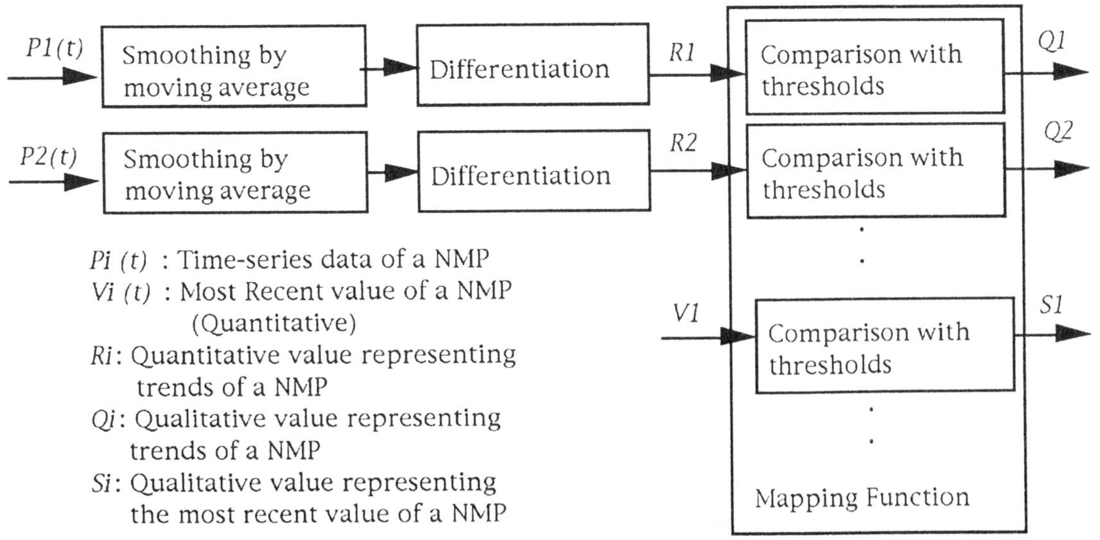

Figure 2: Mapping from quantitative to qualitative values

therefore, a quantitative value R_i representing a long-term trend of each NMP is obtained after smoothing and differentiation. Quantitative values are converted to qualitative values by a mapping function. Thresholds in the mapping function should be properly predefined.

A mapping function is generally defined as

$$M : R_1 \times R_2 \times \cdots \times R_n \rightarrow Q_1 \times \cdots \times Q_m.$$

We construct the mapping function with a set of mappings $\{M_1 : R_1 \rightarrow Q_1, \cdots, M_n : R_n \rightarrow Q_n\}$.

This construction of a mapping function needs less parameters to be adjusted than the general construction, and is known to be effective in many domains [6].

3.1.3 Case-based optimization of a mapping function

Let $v_1, v_2, \cdots, v_n$ be quantitative values which are input to a mapping function. The range of v_i is divided into l_i regions by an ordered set of thresholds $\theta_i^1, \theta_i^2, \cdots, \theta_i^{l_i-1}$. Then the mapping function is defined as a vector of thresholds, i.e.

$$\vec{\theta} = (\theta_1^1, \theta_1^2, \cdots, \theta_1^{l_1-1}, \theta_2^1, \theta_2^2, \cdots).$$

Training data consists of cases of network trouble. Attributes of a case are $v_1, v_2, \cdots, v_n$. For each case, values of attributes and a real cause are given. With a mapping function $\vec{\theta}$, every case is represented by a sequence of qualitative values.

The evaluation function $E(\vec{\theta})$ is defined as below:

$$E(\vec{\theta}) = E_1(\vec{\theta}) + wE_2(\vec{\theta}),$$
$$E_1(\vec{\theta}) = m_{err}(\vec{\theta})/m,$$
$$E_2(\vec{\theta}) = H(\vec{\theta})/(m\ log\ \omega\ \Pi_{i=1}^n l_i),$$

where w is weight, m_{err} is the number of wrong diagnoses of cases, m is the number of cases, $H(\vec{\theta})$ is the number of bits required to encode all cases represented by qualitative values, and ω is the number of types of causes.

$E_1(\vec{\theta})$ represents the rate of wrong diagnosis, and its range is $0 \leq E_1(\vec{\theta}) \leq 1$. $E_2(\vec{\theta})$ is normalized as $0 \leq E_2(\vec{\theta}) \leq 1$. Calculation of $H(\vec{\theta})$ is based on a method in [7].

The optimization problem is to find $\vec{\theta'}$ such that $E(\vec{\theta'}) = min\ E(\vec{\theta})$. $E_1(\vec{\theta})$ decreases as m_{err} decreases; if there is no wrong diagnosis, then $E_1(\vec{\theta})$ is 0. $E_2(\vec{\theta})$ decreases as the length of a code to encode the training data decreases. It is known as the Minimal Description Length (MDL) Principle [8]-[10] that the best classification derivable from training data minimize the length of a code to encode training data. We apply Simulated Annealing [11] to solve this optimization problem [12],[13].

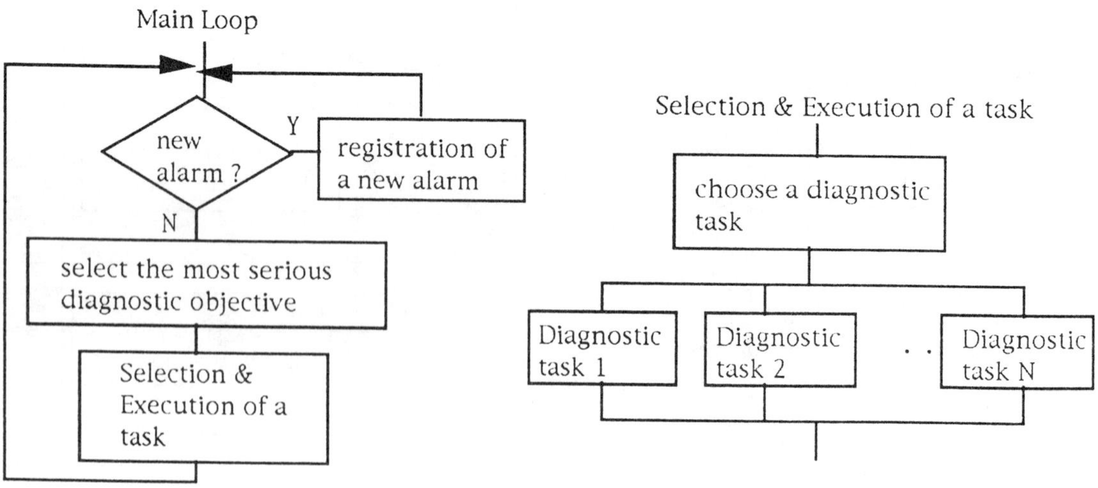

Figure 3: Mechanism of dispatcher

3.2 Software architecture for real-time diagnosis

The ES should be able to operate even when receiving a large number of alarms, and also to react to the most serious trouble first. To assure real-time response, the following conditions must be satisfied:

(1) When a new alarm is received by the ES, the alarm is recognized as a working memory element within the given limit (3 seconds).
(2) If the recognized alarm is more important than the diagnosis currently being performed, the ES starts a new diagnosis for it.

For knowledge base systems where reasoning tasks run concurrently, the strategy to control inference for real-time response is discussed in papers [4], [14]-[16]. It is the major requirement that ES should monitor changes in states of network resources as frequently as possible to improve the real-time response. To satisfy this requirement, it is commonly agreed that each diagnostic task should be simple enough to be completed in a short time.

By following the same approach, we decompose a task into smaller tasks until each task is completed within the given limit. We design a dispatcher illustrated in Fig. 3. Registration of an alarm as a working memory element, and selection of diagnostic objectives in the order of seriousness, are tasks for inference control. They are given a higher priority than diagnostic tasks.

4 IMPLEMENTATION

This section describes the implementation of the system based on the fundamental design described in **3**.

4.1 Implementation environment

The real-time ES shown in Fig. 4 runs under ULTRIX 4.3 on DECStation 5000/25 (29MIPS, 40MB). Knowledge Base (KB) is written in OPS83 and C, and the other processes are written in C. The ES works in cooperation with Network Monitoring Equipment (NME) which has the following functions:

Alarm Issuing: to issue alarms by observing network management parameters (NMPs) every 5 minutes.

Server of NMP: to return stored NMPs in response to a retrieval request by the ES.

Server of Network Configuration: to provide information about a current network configuration, such as routing information, periodically.

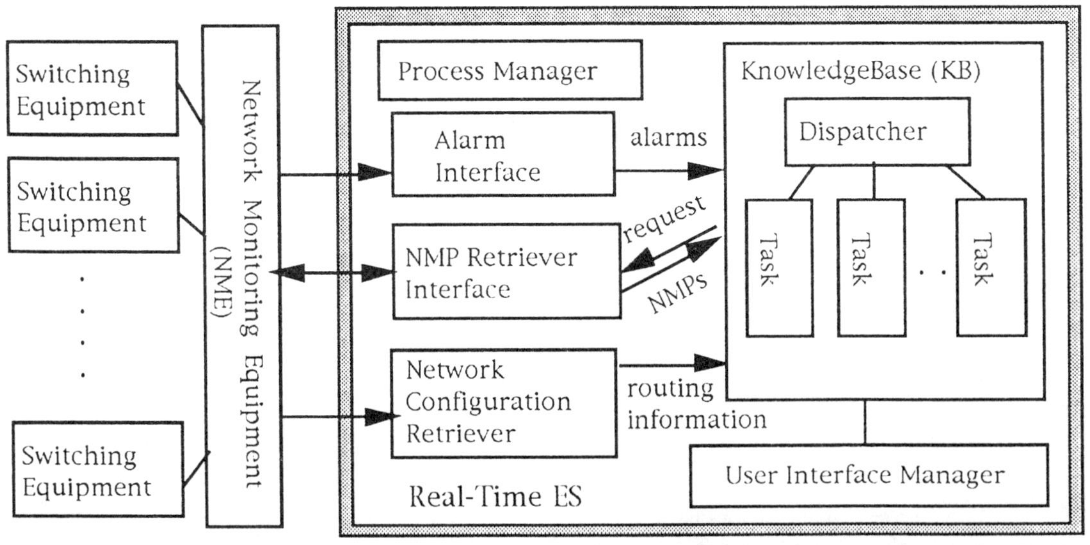

Figure 4: Configuration of real-time ES for troubleshooting in international telephone networks

4.2 Software configuration

The ES consists of the following processes:

Process Manager: monitors the execution of all the processes of the ES. It restarts each process when a fatal fault in the ES occurs.

Alarm Interface: queues alarms from NME.

NMP Retriever Interface: performs queueing to transmit retrieval requests from KB to NME, and transmits the results from NME to KB.

Network Configuration Retriever: retrieves information about a current network configuration.

User Interface Manager: displays results of diagnosis such as possible remedial measures. Also has functions to set parameters such as a priority of a diagnostic objective for each foreign network, etc.

Knowledge Base (KB): executes operations of 'Trouble Detection,' 'Cause Identification,' and 'Suggestion of Remedial Measures' described in **2**. Diagnostic knowledge except inference control is represented in seven decision tables in Table 1. About 1050 rules are generated from these decision tables. About 450 meta rules are prepared for inference control. These meta rules choose a diagnostic task to be executed, and relax conditions when probable causes of trouble cannot be identified. This KB can identify 11 types of causes as shown in Table 2.

5 EVALUATION AND DISCUSSION

For the evaluation of the system, we use actual network data. The data consists of network management parameters (NMPs) and alarms during the period from Jan. 16 to 22, 1994. The period includes the day (Jan. 17) of the earthquake in Los Angeles in the U.S. The influence of congestion caused by the earthquake is noticed by a number of alarms, and the deterioration of service quality was observed in related circuits-groups.

Table 1: Decision tables in Knowledge Base

No	Function of a decision table
1	Deciding a target diagnosis from distributed alarms over different networks
2	Deciding a target diagnosis from increases and decreases of alarms
3	Proposing candidates of causes from routing information
4	Proposing candidates of causes from distributed alarms over different networks
5	Proposing candidates of causes from NMPs related to foreign networks of a diagnostic objective
6	Proposing candidates of causes from NMPs related to circuits-groups of a diagnostic objective
7	Proposing possible remedial measures from a cause and current network configuration related to a diagnostic objective

Table 2: Types of causes in networks

No	Causes
1	Faults in overseas switching equipment
2	Faults in local switching equipment
3	Faults in circuits-groups connected directly to deteriorated areas
4	Faults in unobservable circuits-groups used for transit
5	Faults in circuits-groups in local networks
6	Congestion in overseas networks in deteriorated areas
7	Congestion in overseas networks used for transit
8	Increase of traffic demand for deteriorated areas
9	Increase of transit calls for deteriorated areas
10	Decrease of connection rate due to overseas holidays
11	Increase of non-valid calls for deteriorated areas

5.1 Capability of trouble detection and cause identification

The system extracts 116 cases as trouble from 14,171 alarms. The detection is examined by human network operators, and it is proved that the system succeeds in extracting all the trouble without missing. The rate of information compression is 122.1 (=14,171/116) on average. In 116 cases of detection, 105 cases are recognized as actual trouble, 11 cases are not. The capability of trouble detection is 90.5% (=105/116).

The capability of cause identification is evaluated by comparing with the capability of human network operators. The evaluation is measured in terms of the rate **I** that the real cause of a trouble is included in the final list of candidates, and the number of candidates **C** per trouble. **I** becomes 100% if the real cause is in candidates for each detected trouble, and **C** becomes 1.0 if the system proposes only one cause for each detected trouble.

The system achieves **I** = 79%, **C** = 1.5, which means that the system achieves 79% of the human capability of cause identification. Before we apply the case-based optimization described in **3.1.3**, **I** was less than 48% and **C** was over 2.0, in spite of much human effort to refine KB. This shows that the case-based optimization drastically improves the capability of cause identification.

5.2 Response time

To evaluate the real-time response of the main loop illustrated in Fig. 2, we measured the selection interval of a diagnostic objective. The interval is 220 milliseconds on average, and 2.1 seconds in the worst case. This result means that the system can reconfirm every interval shorter than 2.1 seconds that the most appropriate diagnostic objective is selected. At the peak of alarm occurrence, new alarms are issued at the rate of 3 seconds per alarm. The dispatcher of the ES satisfies the requirement of real-time for practical use.

As the diagnosis of one trouble is completed by execution of a set of diagnostic tasks, the total diagnosis time for one trouble is given as a summation of processing time of related tasks. The response time of diagnosis is 6.0 second per trouble on average. The response time of diagnosis is sufficiently short without distributing tasks over multiple CPUs.

6 CONCLUSION

This paper discussed the fundamental design and implementation of a real-time ES for troubleshooting in international telephone networks, and discussed key technical issues, that is, (1) the qualitative treatment of quantitative values, and (2) the software architecture for real-time diagnosis.

In order to design an optimal mapping from quantitative to qualitative values, we introduced the case-based optimization, which determines the most appropriate set of thresholds in a mapping function from the training data, that is, the history of diagnosis conducted by human network operators. This approach proved to be very effective in improving the capability of cause identification measured with **I** from 48% to 79%.

For software architecture for real-time diagnosis, we developed a dispatcher to control the execution of diagnosis, under which one diagnostic task is decomposed into smaller tasks recursively until each task is completed within the given limit. Although such a decomposition is too ad hoc, it is a practical solution at present.

The implemented system, NEMS (Network Management Support System), has been in practical use since May, 1993. It is probably the first practical real-time ES used in international telephone network operation.

ACKNOWLEDGEMENTS

The authors wish to acknowledge Dr. Y. Urano, Director of KDD R&D Laboratories, and Mr. K. Maya, Deputy Director, for their continuous guidance and kind suggestions. We also wish to thank Mr. S. Kamimura, Manager of the Network Management Center and Mr. Y. Ishibashi, Administrative Officer, for their help in developing and evaluating the ES knowledge base and in practical use of the system.

REFERENCES

[1] E. Ericson, L.Ericson, and D. Minoli (editors): "Expert Systems Application in integrated Network Management," Archtech House, Inc. 1989.

[2] J. Goldman, J. Miller: "Real Time Autonomous Expert Systems in Network Management," Proc. of International Symposium on Integrated Network Management, pp.599-624, 1989.

[3] M. Frontini, J. Griffin, and S. Towers: "A Knowledge-Based System for Fault Localization in Wide Area Networks," Proc. of International Symposium on Integrated Network Management, pp.519-530, 1991.

[4] C. Joseph, A. Sherzer, K. Muralidhar: "Knowledge Based Fault Management for OSI Networks," Proc. of International Conference on Industrial & Engineering Applications of Artificial Intelligence & Expert Systems, Vol. 1 pp.61-69, 1990.

[5] ITU Recommendations E.401-E.427, BLUE BOOK, Volume II, Fascicule II.3 1989.

[6] J. deKleer: "How Circuits Work," Artificial Intelligence Vol. 24, pp. 205-280, 1984.

[7] J. deKleer and B. C. Williams: "Diagnosing multiple faults," Artificial Intelligence Vol. 32, pp. 97-130, 1987.

[8] J. Rissanen: "Modeling by shortest data description," Automatica, Vol. 14, pp. 465-471, 1978

[9] J. Rissanen: "A universal prior for integers and estimation by minimum description length, " Ann. of Statis., Vol. 11, pp. 416-432, 1983

[10] J. Rissanen: "Complexity of strings in the class of Markov sources," Vol. 32 (4), pp. 526-531, 1986

[11] S. Kirkpatrick, C. D. Gelatt and M. P. Vecci: "Optimization by simulated annealing," Science, Vol.220, No.4958, pp.671-683,1983.

[12] Matsumoto, Hashimoto, Shimizu, Obana, and Asami: "A Method for Applying Qualitative Knowledge to Real-Time Network Management," Technical Report of the Institute of Electronics Information Engineers of Japan, AI93-37, pp.9-14, 1993. (in Japanese)

[13] Matsumoto, Hashimoto, Shimizu, Obana: "Evaluation of Optimization method for Mapping Function from Qualitatives to Quantitatives with Simulated Annealing," Autumn Conference of Information Processing Society of Japan, Vol. 2 pp.19-20, 1994. (in Japanese)

[14] M. Sutter and P. Zeldin: "Designing Expert Systems for Real-Time Diagnosis of Self-Correcting Networks," IEEE network Magazine. Vol.5, No.4 pp.43-51, 1988.

[15] J. L. Seamons: "Network Traffic Management in Telecom Australia," Telecommunication Journal of Australia, 38 PP.15-24, 1988.

[16] J. Schroder and W. Schodl: "A Modular Knowledge Base for Local Area Network Diagnosis, " Proc. of International Symposium on Integrated Network Management, pp.493-503, 1991.

APPLICATION OF A RADIAL BASIS NEURAL NETWORK IN THE FAULT DETECTION AND DIAGNOSIS OF THE TEMPERATURE SENSORS OF A HEAT EXCHANGER

Daniela Andreia Hanomolo, Universitatea "Politehnica" Bucuresti, Facultatea de Automatica si Calculatoare, Splaiul Independentei 313, sector 6, Bucuresti, Romania, e-mail: d0096@mvs4381.cc.pub.ro, fax:+40-1-3122106
Michel Kinnaert, Université Libre de Bruxelles, Faculté des Sciences Apliquées, Service d'Automatique, Av. R.D. Roosevelt 50, 1050 Bruxelles, Belgique

ABSTRACT

Most of the methods employed for the fault detection and diagnosis (FDD) are based on the use of linear models. Neural Networks (NN) have proved successful for non-linear systems because of their non-linear representation capabilities. NN have been mostly used in FDD based on classification. When present in schemes based on modelling and estimation, only certain types of faults are considered. In this paper we propose a fault detection and diagnosis scheme that could be regarded in some sense as a generalisation to nonlinear systems of the parity relations approach for sensor fault detection. A radial basis function network is used for the modelling of a nonlinear system and for the detection and isolation of abrupt as well as of incipient faults on the temperature sensors of a heat exchanger. The performance of two types of NN based on two different learning algorithms is studied and the influence of initial conditions is evidentiated.

I. INTRODUCTION

Fault detection and isolation are fundamental issues for complex control systems from the viewpoint of plant safety and reduced operation costs. The most advanced methods developed for fault detection and diagnosis are those based on pattern recognition and those based on estimation and modelling. Initially neural networks (NN) have been employed in pattern recognition techniques (the system operation is classified in normal and faulty situations). The performance of different types of feed-forward NN was analysed [Sorsa91]. Faults during process transients were detected by classifying time-series into fault classes [Venkatasubramanian91].

In what concerns the methods based on modelling and estimation the classical techniques require linear models for successful operation. They can be extended to nonlinear systems but they require some knowledge about the type of non linearity. On the other hand, Artificial Neural Networks represent a technology for processing signals without requiring any particular knowledge about the studied process. Radial basis function (RBF) networks have been proposed [Sorsa93] for the nonlinear system identification and the Bayes theorem was used for the fault detection and isolation [Sorsa93]. A fault detection and diagnosis (FDD) scheme based on physical system parameters estimation was developed for a hydraulic test rig [Daley94] and a dynamic neural structure was proposed for the FDD of a turbo charger [Ayoubi94].

This paper presents a new fault detection scheme employing ANN which can be seen in some sense as a generalisation of parity relations to nonlinear systems (faults that influence the input or the output of the process will result in changes of residuals). The method exploits the nonlinear functions approximation capability of a radial basis function neural network and its application to the fault detection and isolation of the temperature sensors of a heat exchanger is carried out for two types of learning algorithms: the unsupervised selection of centres and the supervised selection of all the network parameters by means of a generalised gradient descent algorithm.

II DESCRIPTION OF THE PROCESS AND OF THE AVAILABLE DATA

The experimental study is carried out on a pilot heat exchanger consisting of three subsystems: the air circuit, the water circuit, the exchanger itself (fig.1) which performs the heat exchange between two fluids: air and water [Simonart93].

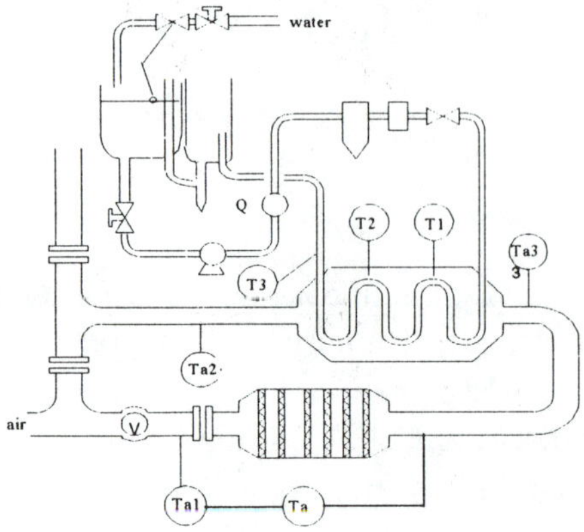

Fig. 1 Scheme of the heat exchanger

The two inputs to the process are the heating power, p (0-10kW) and the water flow, Q (0-180l/h) (fig.2). During the measurements the sampling period was chosen equal to 2 seconds. PI controllers were used for the control of the heating power and the water flow so that the two inputs are constant between two sampling instants. As the measured values (the air and water temperatures) are strongly affected by noise an analog anti-aliasing filter was used.

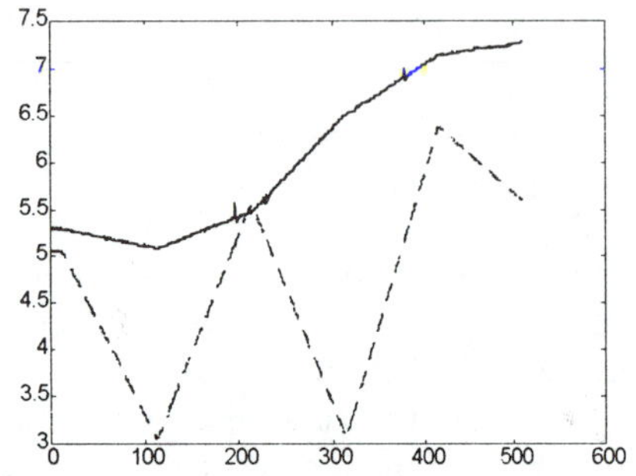

Fig.2 Evolution of the heating power ('--') and water flow ('-')

The following measures were available (fig.2):
- the heating power (p);
- the water flow (Q);
- the air temperature: Ta;
- the water temperature at the first third of the exchanger, T1;
- the water temperature at the second third of the exchanger, T2;
- the water temperature at the exit of the exchanger, T3.
- the air temperatures: Ta1, Ta2, Ta3;

The evacuation valve opened 100% and the recycling valve was completely closed.

The measured temperatures are in the range: 0- 100° and they are represented as numbers between 0 and 10, as well as p and Q.

III. NEURAL NETWORK STRUCTURE FOR THE APPLICATION

In [Chen89, Chen90] it is shown that under some mild assumptions, any non-linear, discrete and time invariant system can be represented by a NARX model:

$$y(t) = f[(y(t-1),\dots,y(t-n),u(t),\dots,u(t-m)] + e(t) \quad (1)$$

where y is the output of the system, u, the input and e, the noise. As any continuous function can be arbitrarily well approximated by polynomial models, relation (1) can be written as a *linear regression model*:

$$y(t) = \sum_{i=1}^{M} p_i(t)\theta_i + \xi(t) \quad t=1,..,N \quad (2)$$

where $p_i(t)$ are called regressors and they depend on y(t-1),..., y(t-n), u(t),...,u(t-m), θ_i represent the weights or the parameters and $\xi(t)$ is a modelling error. Thus, the model becomes linear in the regressors p_i its non linearity being preserved in the structure itself of these regressors. In the general case when nothing is known, it will be demonstrated experimentally that the use of radial basis functions as regressors is efficient. The employed NN is a radial basis function one with the gaussian function as activation function for the hidden neurons (fig. 3).

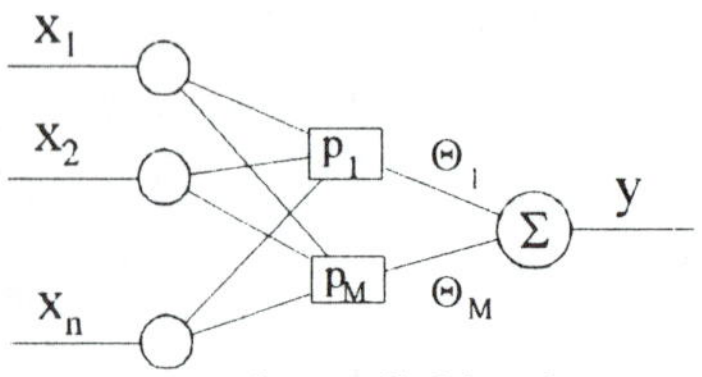

Fig. 3 The structure of a radial basis neural network

In fig. 3 p_i are the gaussian functions (outputs of the hidden nodes):

$$p_j = \exp(-\frac{1}{2\sigma_j^2}\sum_{k=1}^{n}(x_k - c_{kj})^2) \quad j=1..M \quad (3)$$

and y is the output of the network:

$$y = \sum_{i=1}^{M}\theta_i p_i \quad (4)$$

with M the number of hidden nodes, n the number of inputs, $c_{k,j}$ the k-th component of the jth node, σ_j the width of the jth node and θ_i the value of the ith weight.

IV. LEARNING ALGORITHMS

Two learning algorithms are used: the self-organised selection of centres and the supervised selection of all network parameters.

A. SELF-ORGANISED SELECTION OF CENTRES

The network undergoes a hybrid learning process: self-organised for the centres and supervised for the weights [Duda73]. An unsupervised clustering technique was used to put in practice the learning procedure and it consists in finding natural groupings in the set of data using as criterion function for clustering the sum-of-squared-error criterion:

$$J_e = \min J_i$$

where:

$$J_i = \sum_{X_i} \|x - m_i\|^2 \tag{5}$$

with: $m_i = \frac{1}{n_i} \sum_{X_i} x_i$

In these expression, X_i denotes the i-th class of elements from the set of data. Each class contains n_i samples. Moreover, $\sum_{X_i}$ indicates a summation over all the elements x_i of X_i, so that m_i is the mean value of the samples of cluster X_i. The final values of m_i are the centres values c_i.

For the weights selection a least mean squares technique was used, precisely the modified Gram-Schmidt algorithm which consists in finding the parameter estimate $\tilde{\Theta}$ which minimises $\|Y - P\Theta\|$ where Y is the vector of outputs from the training set, P the regression matrix and Θ the weights matrix.

B. SUPERVISED SELECTION OF ALL PARAMETERS

A gradient descent algorithm is applied for all parameters [Renders93] in order to minimise the criterion:

$$J = \frac{1}{2}(y - y_d)^2$$

with y the output of the network and y_d the desired (measured) output. The parameters are adapted according to relations (6) where the notations are those from relations (3) and (4).

$$\begin{aligned} \theta_j(t+1) &= \theta_j(t) - \eta_\theta \frac{\partial J}{\partial \theta_j(t)} \\ c_{j,k}(t+1) &= c_{j,k}(t) - \eta_c \frac{\partial J}{\partial c_{j,k}(t)} \\ \sigma_j(t+1) &= \sigma_j(t) - \eta_\sigma \frac{\partial J}{\partial \sigma_j(t)} \end{aligned} \tag{6}$$

$\eta_\theta, \eta_c, \eta_\sigma$ represent the learning rates. In order to increase the convergence speed the *projection algorithm* [Astrom89, Renders93] was used which consists in adapting the learning rates:

$$\eta = \frac{\gamma}{\alpha + \sum_{j=1}^{M} p_j^2} \tag{7}$$

where $0<\gamma<2$ and $\alpha \geq 0$. This algorithm was applied on one side only for the weights learning rate and on the other side for all parameters and the results were compared.

Two types of NN were studied each of them based on one of the training algorithms described.

V. IDENTIFICATION OF THE SYSTEM

The identification procedure is performed in the training phase of the network. The aim is to get a maximum absolute error (8) as small as possible in order to be able to detect faults of magnitudes as small as possible:

$$\varepsilon = \max_i |y_{di} - y_i| \qquad i = 1,..,M \tag{8}$$

where M is the dimension of the training set, y_{di} the i-th measured output from the training set and y_i the i-th network output.

The identification procedure has three steps:

1. *structural identification*: determine the measures on which the output will depend;
2. *parametric estimation*: determine the numerical values of the model coefficients: the centres, the variances and the connection weights;
3. *validation*: check the obtained model.

Several models were built for the estimation of T1. For all the experiments 250 examples (taken randomly from the whole set of data) were used for training and 260 were used for the validation.

For each model a modelling error ε_{mi} was computed by evaluating (8) with the measurements y_{di} taken from the validation set. Some experiments were made for the choice of the best NN structure.

For the same initial conditions the number of centres was modified in order to determine the optimal number of centres. For the obtained number of centres the initial conditions were changed in order to choose the most convenient values (from the maximum absolute error point of view) for the NN parameters.

A. RESULTS FOR THE FIRST TYPE OF NN

The implementation of the training algorithm is rather simple and the time needed for learning is very small, but there is the problem of selecting the starting point.

Two models were built for T1:

- T1(t)= f_1(p(t), Q(t), T2(t), T2(t-1))
- T1(t)= f_2(p(t), Q(t), T3(t-1), T3(t-2)).

During the experiments it was noticed that J_e didn't decrease with the increase of the number of centres. This was expected and can be explained by the fact that the clustering procedure can lead to local minima. In table 1 the results of ten experiments -carried out for the first model- are presented in which the initial conditions vary. This clearly shows that for a fixed model structure a series of trials is needed to determine initial conditions leading to a satisfactory performance.

For both models a structure with 51 centres was chosen and the maximum absolute modelling errors corresponding to the best achieved performance are: 0.52° for the first model, 0.55° for the second one.

Table 1 The dependence of J_e and ε_m on the initial conditions when T1(t)=f_1( ) (validation)

Trial	Criterion value (J_e)	Maximum estimation error (°C)
1	0.8747	**0.52**
2	1.03	0.606
3	1.004	0.659
4	0.911	0.662
5	1.076	0.678
6	1.08	0.694
7	1.065	0.741
8	1.078	0.923
9	0.965	1.425
10	1.08	1.545

B. RESULTS FOR THE SECOND TYPE OF NN

The learning procedure was applied in pattern mode: the parameters updating was performed after the presentation of each training example.

Apart from ε defined in (8) another performance criterion was used for the validation of the models: the mean square error:

$$\varepsilon_{sqr} = \sqrt{\frac{1}{M}\sum_{i=1}^{M}(y_{di} - y_i)^2} \qquad (9)$$

Two models were built for the estimation of T1:

-T1(t) = f_1(p(t), Q(t), T2(t), T2(t-1))

-T1(t) = f_2(p(t), Q(t), Ta2(t), Ta2(t-1)).

This time (table 2) the behaviour of the NN is more expected than the first one in the sense that the mean square error decreases with the number of centres. For every experiment the initial conditions are different (random values between 1 and 10). The projection algorithm was applied for the weights learning rate (in the first four lines of tables 2 and 3, with: $\eta_c = \eta_v = 0.1$, $\gamma = 1.5$ and α=0.5) and also for all the parameters (in the last line of tables 2 and 3, with: $\gamma = 0.5$ and α=1.5). The optimal number of centres was chosen 30 for both models and the best achieved modelling errors are 0.29 for the first model and 0.43 for the second.

From table 2 we can notice as inconvenient the slowness of the algorithm (the big number of epochs).

Table 2 Results for T1(t)=f_1( )

Number of centres	Number of epochs	Mean square error
25	1372	0.054
30	1384	0.0127
35	948	0.00984
40	1296	0.00781
30	1285	0.0086

Table 3 reflects the high power of generalisation of this type of network as the modelling errors in the training and validation phases are very close to each other.

Table 3 Results for T1(t)=f_1( )

Number of centres	ε_m (training phase)	ε_m (validation)
25	0.4162	0.417
30	**0.29**	**0.29**
35	0.428	0.428
40	0.4875	0.488
30	0.395	0.396

Different values for the parameters of the projection algorithm were chosen and, as it can be noticed from table 4 (with results for the first model), the significant difference is in the number of epochs, the modelling error varying in a slight manner.

Table 4 Maximum estimation error dependence on the parameters of the projection algorithm

γ and α	Number of epochs	ε_m
$\gamma = 1.5$ $\alpha = 0.5$	1384	0.29
$\gamma = 1$ $\alpha = 0$	2000	0.282

VI. FAULT DETECTION AND DIAGNOSIS

We shall design a residual generator for detection and isolation of the faults on the sensors measuring T1, T2 and T3 for the first type of NN and T1, T2 and Ta2 for the second type, assuming that only one fault is present at any moment.

We'll use each of the models identified previously, namely:

1) $T1(t)=f_1(p(t), Q(t), T2(t), T2(t-1))$ and $T1(t)=f_2(p(t), Q(t), T3(t-1), T3(t-2))$ for the first type of NN and

2) $T1(t)=f_1(p(t), Q(t), T2(t), T2(t-1))$ and $T1(t)=f_2(p(t), Q(t), Ta2(t), Ta2(t-1))$ for the second one

in order to build a residual. Remember that a residual is a signal whose nominal value is zero (or close to zero) in the absence of failure. Moreover, upon occurrence of specific faults, its value should become significantly different from zero. Note that the parameters of the NN characterised by $f_1(\ )$ and $f_2(\ )$ are fixed to their previously identified values from now on.

At each sampling time t, the residuals $\tilde{\varepsilon}_1(t)$ and $\tilde{\varepsilon}_2(t)$ are given by (for each type of NN):

$$\tilde{\varepsilon}_1(t) = T1(t) - \overline{T}1_1(t) = T1(t) - f_1(\ )$$

$$\tilde{\varepsilon}_2(t) = T1(t) - \overline{T}1_2(t) = T1(t) - f_2(\ )$$

where $\overline{T}1_i(t)$ denotes the estimate of T1(t) provided by the i-th model. Normally the value of $\tilde{\varepsilon}_i(t)$ should be of the order of the modelling error of the i-th model, ε_{mi} in the absence of failure.

For each residual $\tilde{\varepsilon}_i$ we can define a detection threshold:

$$\varepsilon_{fi} > \varepsilon_{mi} \qquad (10)$$

above which the discrepancy between T1 and $T1_i$ is too big to be justified by modelling errors. If $\tilde{\varepsilon}_i > \varepsilon_{fi}$ a sensor failure occurred. From the definition of $\tilde{\varepsilon}_1(t)$ and $\tilde{\varepsilon}_2(t)$ it is obvious that the decision logic defined in table 5 can be used to determine which sensor is faulty for the first type of NN. The same logic is used for the second type of NN, with the only difference that we replace T3 with Ta2.

Table 5 Fault detection and isolation logic for the first model

Residual	$\varepsilon_i > \varepsilon_{fi}$		
1	Y	Y	N
2	Y	N	Y
Decision	T1 faulty	T2 faulty	T3 faulty

Although the principle of the proposed FDD system is explained for a set of three sensors, it can be generalised in a straightforward way to an arbitrary number of sensors.

In the experiments reported here, abrupt faults of 1° between iterations 100 and 130 and incipient faults of 0.1°/iter, between iterations 100 and 120 were simulated. The aims were to study the fault detection and isolation performance, namely, the promptness of a fault detection and the sensitivity to incipient faults.

A. RESULTS FOR THE FIRST TYPE OF NN

For both residuals fault thresholds of 0.6° were chosen in order to satisfy relation (10).

The experiments proved that the network can satisfactorily detect the abrupt and incipient faults occurring at the sensor which measures the temperature corresponding to the network output (T1 here).

In what concerns faults occurring at the sensors measuring temperatures that are inputs to the network (T2 and T3 here), a certain delay can be noticed in the reaction of the residuals (table 6 and 7). In the case of incipient faults we can also notice that the residuals react only during a few iterations. This could lead to the situation of a not detected fault!

Table 6 Instants between which the residuals react to abrupt faults

Residual	Abrupt fault at T1	Abrupt fault at T2	Abrupt fault at T3
1	100-130	116-130	-
2	100-130	-	117-130

Table 7 Instants between which the residuals react to incipient faults

Residual	Incipient fault at T1	Incipient fault at T2	Incipient fault at T3
1	105-120	116-120	-
2	105-120	-	117-120

In fig. 4 and fig. 5 it is illustrated the situation with an abrupt fault of 1^0 occurring at the sensor which measures

T1. Both residuals (of both neural networks) increase over the imposed limit indicating the presence of a fault (following the logic of table 5) between iterations 100 and 130.

If the fault occurred at the sensors measuring T2 or T3, only the residual corresponding to the network which has as input one of these temperatures would increase over the fault threshold the other one remaining in the limit of the modelling error.

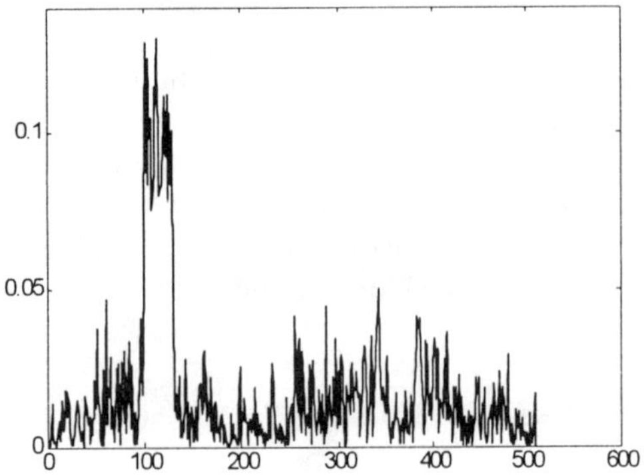

Fig. 4 The residual for $T1=f_1(\,)$ (abrupt fault of 1°C at the sensor for T1)

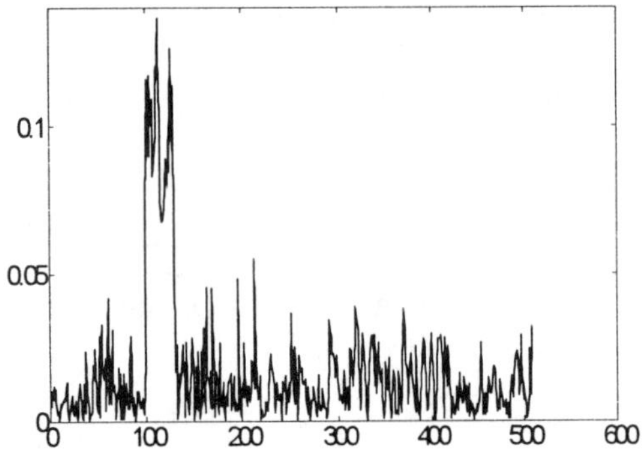

Fig. 5 The residual for $T1=f_2(\,)$ (abrupt fault of 1°C at the sensor for T1)

B. RESULTS FOR THE SECOND TYPE OF NN

For the second type of NN, as the modelling errors are smaller, smaller detection thresholds have been chosen: 0.4 for the first model, 0.5° for the second.

The second learning algorithm provides more sensitive to both abrupt and incipient faults. Tables 8 and 9 reflect the more rapid reaction of the residuals to faults occurring at the sensors whose measurements are inputs to the NN, compared to the results obtained with the first type of NN (compare with tables 6, respectively 7).

Table 8 Instants during which the residuals react to abrupt faults

Residual	Abrupt fault at T1	Abrupt fault at T2	Abrupt fault at Ta2
1	100-130	101-130	-
2	100-130	-	101-107

Table 9 Instants during which the residuals react to incipient faults

Residual	Incipient fault at T1	Incipient fault at T2	Incipient fault at Ta2
1	104-120	104-121	-
2	105-120	-	113-121

CONCLUSIONS

A FDD scheme based on the residual analysis was proposed and two types of neural networks were built according to two different learning algorithms: a self-organised selection of centres and a supervised selection of all parameters. A system for the detection and isolation of both abrupt and incipient faults was implemented in Turbo Pascal 6.

From the identification point of view we can notice that the first algorithm is very fast (it takes only a couple of seconds). This allows one to carry out many experiments for different initial values of the parameters. On the other hand, the second one provides smaller modelling errors even though it needs almost always more than 1000 iterations. The first type of NN is much more sensitive to initial conditions and it can be trapped in local minima. As the initial conditions depend on the training set, representative data should be chosen. The second type of NN provides a very good generalisation for the considered data set: the estimation errors for the training and validation phases are very close to each other. For the first type, great differences occur between the two values.

In the experiments reported here, both NN can perform a good fault detection and isolation; they react very quickly to faults occurring at sensors whose measurements are reconstructed by the NN. The second one behaves more satisfactorily when faults occur at sensors corresponding to inputs to the NN: its sensitivity is better for both abrupt and incipient faults. The same NN, because of its better estimation performance can detect smaller magnitude faults, smaller fault threshold being possible.

Studying the residuals of more NNs (that approximate the same measure) in the same time we can also say which of the sensors is faulty thus, a good fault isolation being possible.

Compared to conventional methods based on modelling and estimation, this method has first of all the advantage that it doesn't need a deep knowledge of the process. Even methods based on the parameter estimation implemented with NN can necessitate some knowledge of the process [Daley94]. For our case only input and output data are needed. Secondly it allows the fault detection and isolation for a non-linear case. Usually, the conventional methods give good results for domains around different operating points. In what regards the use of NN for the implementation of modelling and estimation techniques, usually it can be noticed a limitation to only a certain type of fault [Sorsa93] while this proposed method can cope with any kind of abrupt or incipient fault.

To be able to draw more general conclusions on the feasibility of FDD schemes based on the proposed approach, the neural networks performance should be studied for different data sets.

REFERENCES

[Astrom89] K. J. Astrom, B. Wittenmark, *"Adaptive Control"*, Addison-Wesley Publishing Company, 1989

[Ayoubi94] M. Ayoubi, *"Fault Diagnosis with dynamic neural structure and application to a turbo charger"*, IFAC Symposium on fault detection, supervision and safety for technical processes, SAFEPROCESS '94, Espoo, Finland, pages 618-623, June 13-16, 1994

[Chen89] S. Chen, S.A. Billings, W. Luo, *"Orthogonal least squares methods and their application to non-linear system identificatio"*, International Journal of Control, pages 1873-1896, volume 50, no. 5, 1989

[Chen90] S. Chen, S.A. Billings, C.F.N. Cowan, P.M. Grant, *"Practical identification of NARMAX models using radial basis functions"*, International Journal of Control, pages 1327-1350, volume 52, no. 6, 1990

[Daley94] S. Daley, H. Wang, *"A neural network approach to the detection and diagnosis of faults in a hydraulic test rig"*, IFAC Symposium on fault detection, supervision and safety for technical processes, SAFEPROCESS '94, Espoo, Finland , pages 637-641, June 13-16 1994

[Duda73] R.O. Duda, P.E. Hart, *"Pattern classification and scene analysis"*, John Wiley & Sons Inc, 1973

[Haykin94] S. Haykin, *"Neural Networks, A comprehensive foundation"*, MacMillan College Publishing Company, New York, Maxwell MacMillan Canada, Maxwell Macmillan International, New York, 1994

[Patton89] R.J. Patton, P.M. Franck, R.N. Clark, *"Fault diagnosis in dynamic systems"*, Prentice hall, 1989

[Renders94] Jean-Michel Renders, *"Metaphores biologiques appliquées à la commande de processus"*, Thèse de doctorat, 1994

[Simonart93] X. Simonart, *"Etude et mise en oeuvre de méthodes de detection de pannes de capteurs sur l'échangeur de chaleur du laboratoire"*, Master thesis, Universite Libre de Bruxelles, 1993

[Sorsa91] T. Sorsa, H. Koivisto, H.N. Koivo, *"Neural networks in process fault diagnosis"*, IEEE Trans. Systems, Man and Cybernetics, volume 21, pages 815-825, 1991

[Sorsa93] T. Sorsa, J. Suontausta, H. N. Koivo, *"Dynamic fault diagnosis using radial basis function networks"*, TOOLDIAG '93, International conference on fault diagnosis, Toulouse, France, pages 160-168, April 5-7, 1993

[Venkatasubramanian91] V. Venkatasubramanian, R. Vaidyanathan, *"Diagnosing noisy process data using neural networks"*, Preprints of IFAC/IMACS Symposium SAFEPROCESS '91, Baden-Baden, Germany, volume 2, pages 375-380, 10-13 September 1991

Evolutionary Programming

ADAPTIVE HARDWARE OPTIMIZATION BASED ON GENETIC ALGORITHMS

Mehrdad Salami and Greg Cain
Department of Electrical Engineering,
Victoria University of Technology,
PO Box 14428 MMC, Melbourne, VIC 3000 Australia.
Email: salami@cabsav.vut.edu.au , gregc@cabsav.vut.edu.au

ABSTRACT

We have investigated the possibility of constructing a digital logic device which directly executes a family of search and optimisation procedures known as genetic algorithms. These algorithms are significant to engineering as a means of providing additional adaptive capability to known and existing control mechanisms. This paper discusses the underlying mechanisms of genetic algorithms and a model of a computing architecture which implements these features. The model has been developed and simulated using the hardware descriptive language VHDL.

I - INTRODUCTION

In recent times some new classes of probabilistic search algorithms have been developed to handle difficult or intractable problems. Principal among these developments are Genetic Algorithms [Holland92, Goldberg89] and the method of Simulated Annealing [Kirkpatrick83]. The search method used in Genetic Algorithms (GAs) mimics some of the mechanisms and principles of natural evolution in biological systems. They have exhibited an almost unique ability to solve difficult problems in discrete configuration spaces where solutions are found as unconnected points in an N-dimensional space rather than as points on differentiable surfaces. Examples include adaptive game-playing, biological cell simulation, machine learning, pattern recognition, VLSI microchip layout and job scheduling. These discrete configuration problems usually have an enormous set of candidate solutions which expands in a non-polynomial fashion with the problem dimension (N). Such problems are thus known as *NP Hard* and algorithms which seek exact solutions are generally only used when N is small. GAs handle large configuration spaces by sampling and processing randomly selected points in the space. Figure 1 shows the range and performance of GAs relative to conventional search algorithms [Goldberg89].

Many optimisation algorithms are designed for solving problems represented by continuous functions and use methods such as hill-climbing or gradient descent which are sensitive to the topology of the surface on which the solutions lie. These methods are easily deceived by anomalies in the surface and often become trapped at a local optimum point [Ackley89]. GAs cannot compete with the speed of gradient descent methods but can be applied successfully in many situations where conventional methods fail. They can be used for solving an equation and for constructing models to emulate observed behaviour. They can also be applied in situations where there is no model but where a fitness value can be determined from system responses.

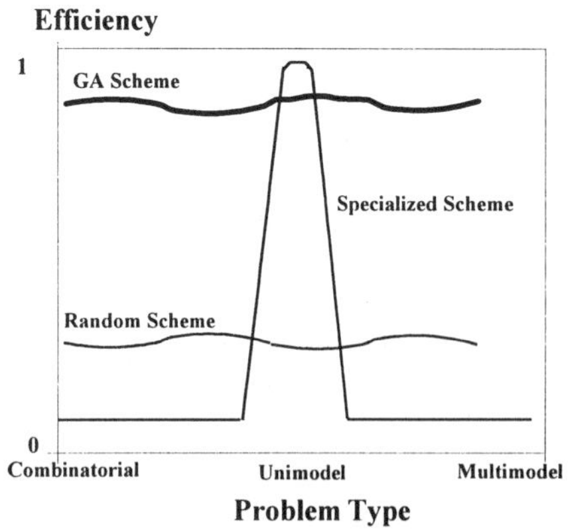

Figure 1 . Efficiency versus Problem Type

In summary GAs are often the best choice for finding solutions in the following situations.

- When there is no known algorithm for a solution or when an initial search is necessary to construct an algorithmic model.
- When the number of search parameters is large.
- When the problem is NP complete or the solution space has a very large number of discrete solutions.

In searching for a solution a GA represents each attempt as a bit string known as a *genetic string* (or chromosome). In many problems where the GA is required to find an optimal path through a network or maze, a solution can be regarded as a sequence of conditions and actions. Selecting a suitable bit string representation of these *condition-action* sequences is often the most difficult aspect of applying a genetic algorithm method. To provide some direction to the evolutionary process a genetic algorithm must include an *objective function* which measures the performance or evolutionary *fitness* of a genetic string. Rather than attempting to optimise a single string variable, a GA maintains a *population* of strings. The GA includes mechanisms to create new member strings using bit level operations on existing member strings. A GA can now be seen as a series of steps. Starting from a population of random strings we attempt to improve the overall fitness of the population by a continual cycle of selection, manipulation and assessment. This cycle is moderated by probabilistic processes.

The manipulation of the existing members of the population is carried out using the following string operators:

1 - Reproduction
2 - Crossover
3 - Mutation

Reproduction is a process in which individual strings are copied according to their fitness values. The fittest members of the existing population are duplicated and will form the basis of the new population. Note however that weak members of this population are not automatically replaced by stronger candidates which may emerge. A string with a higher fitness value has a higher probability of being copied into the new population. After reproduction, pairs of strings are selected for crossover. Each string in a pair is cut at a random position and the corresponding section (or gene) is exchanged to create a new pair of strings. The final step in forming a new generation is mutation in which one or more bits in a member string are flipped. A small percentage of the members in the new population are selected to be mutated. These operations introduce new and often fitter strings into the population. The majority of the weak strings which are also introduced by these operations tend to be replaced by the next generation.

II - ADAPTIVE GENETIC ALGORITHMS OPTIMIZATION

As a simple application for genetic algorithms consider the optimization task shown in Figure 2. The load RL changes in time due to unknown external factors. The Optimization Part (OP) is required to adjust a resistor RS to equalize voltage on RL and RS. In theory this task is achieved by measuring RL or current (I) through the resistors and then adjusting RS. We can control this adaptive task with a genetic algorithm without any knowledge of the current/resistance relationship inside the GA.

The principal advantage of using GAs in adaptive optimization are algorithmic robustness and ease of implementation. The main disadvantages are search speed and incompatibility with some aspects of dynamic behaviour.

Because GAs are very robust, they can be applied to different problems without regard to the type of problem. If the problem is considered as a black box then GAs will present a string or number to the black box and receive the result as a fitness value. Generally GAs don't require any other information about the black box except fitness value however additional information could be structured into the GA to solve the problem more easily.

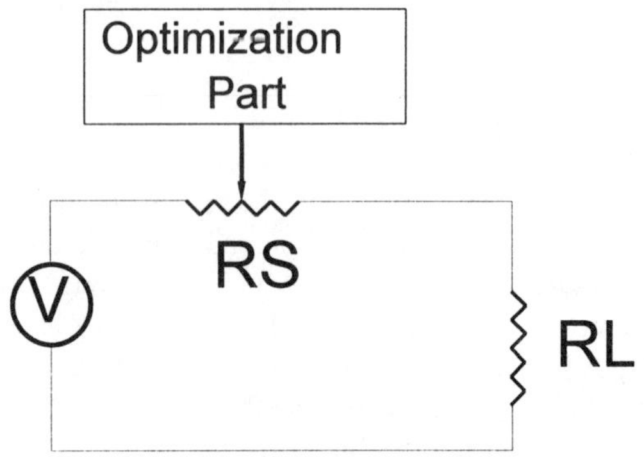

Figure 2 . A simple optimization task

It is easy to implement a genetic algorithm. Basically there are three operators for implementing GAs. The first is the selection algorithm of which there are different kinds such as roulette wheel, ranked and scaled selection. All of these selection algorithms are easy to implement in software and even easy to implement in hardware. The next operator is crossover. Single point, two point and uniform crossover are simple and easy to implement in software and hardware. The last operator is mutation which is very simple. Thus all of the essential elements of all genetic algorithms can be implemented easily.

The first disadvantage of genetic algorithm is the speed of searching. The GA works on a population of solutions and tries to optimize the population. The GA must manipulate and evaluate all members of the population before an improvement in performance can be seen. Indeed GAs are very slow compared with deterministic search methods, but taking into consideration the simplicity and extent of the search space which can be covered, they are quite efficient in comparison with other methods.

The speed problem can be addressed with specialized hardware for both the evaluation of the genetic algorithm and evaluation of the fitness function. The latter case must be handled by hardware specific to the problem at hand. Generic GA processing hardware can readily implemented in digital hardware as all of the internal operation are quite straightforward.

The second disadvantage of GAs is incompatibility with some aspect of dynamic systems. The operation of a GA is unsuited to controlling many dynamic systems where the member strings can radically alter the state of the system. The GA always tries to search the whole problem space to find the best solution. This kind of search is not applicable to dynamic systems. If some random value is applied as a control signal to a dynamic system, it might hang the system or put the system in a defective state.

Therefore the controlled system must include a facility to discriminate and eliminate signals outside of preset limits and then return a fitness value of zero for these cases. It is not always possible to modify an existing control system to be compatible with a GA controller. GAs are best suited to secondary refinements of a dynamic trajectory or to detector tuning tasks such as focussing a lens or tuning a filter.

For dynamic systems we need dynamic GAs which change their internal parameters to match with the system. This will be explained later in the second example section.

Figure 3 shows a similar system which can be implemented in hardware to control a dynamic system. The GA optimization part consists of two modules: Memory Module and the main part Genetic Algorithm Processor (GAP). The dynamic system receives all signals from the control section and returns the fitness value to the GAP. The GAP stores its population in memory and applies the command signals to the control section and receives the fitness value. In this process the GAP tries to improve the whole population in memory.

In the following sections the Genetic Algorithm Processor will be explained and in an example it will be shown how the GAP can be applied to a static or dynamic system to optimize the performance of the system.

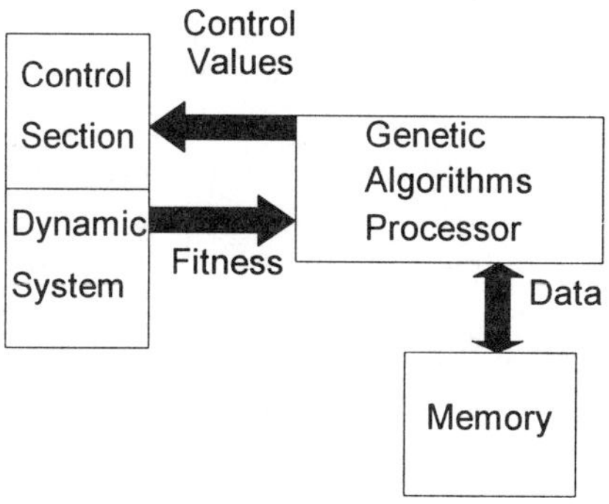

Figure 3 . A simple model of the dynamic system optimization with GA

III - THE GA PROCESSOR MODEL

A genetic algorithm processor can be constructed to directly execute the operation of a genetic algorithm [Salami94]. It is shown in the Figure. 4. Such a processor can be used in situations where high throughput is required and where the logic of the genetic algorithm is expressible in simple units which can be synthesised in hardware. This is generally the case as genetic algorithms are inherently simple and contain only a few logic operations.

The GAP consists of six functional blocks whose main purpose is to increase the fitness values of a population of genetic strings held in *Memory Module*. These blocks are the *Memory Interface Module, Read Module, Selection Module, Crossover Module, Mutation Module,* and *Fitness Module*. The GA processor must also be connected to the problem under investigation through an interface which would receive the genetic string as one or a sequence of digital input words of fixed length and would respond by returning an output word, the fitness value or performance measure.

This external subsystem shown here as the *Control Module* is required to complete the processing loop.

The GAP modules operate concurrently and together form a coarse-grained pipeline. All modules are modelled in VHDL and are independent of the operating environment and implementation technology (e.g. Xilinx FPGAs or fabricated chips) except for the Memory Interface Module. The functionality of this module varies according to the physical memory attached to it and the desired interface between the GAP and Control Module.

Each of the GAP modules in Figure 4 is described in more detail as follows.

A. Memory Module

The memory serves to store both genetic strings and fitness values. In realistic applications the fitness value would normally be an integer of 8...16 bits but the genetic strings would normally be much longer (16...128 bits) depending on the type of problem for which the machine is to be used. In our model these values are defined in a data file. The model must be recompiled whenever any of these values is changed. At first the memory *is* filled with random strings with equal fitness. In operation the Selection Module takes two strings from memory for processing. Following processing, successful strings and their associated fitness values will be rewritten to memory by the Fitness Module.

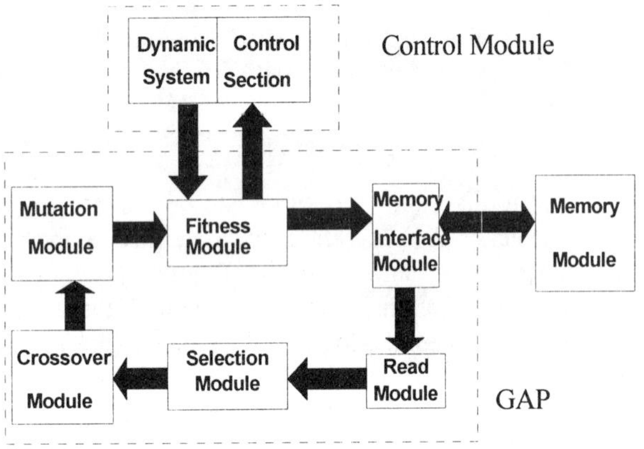

Figure 4 . Implementation of the GAP in separate modules

The GAP section creates the necessary control signals for reading and writing in an actual memory chip. It creates read/write, output enable and chip select signals for communicating with a memory chip. The address bus is not very critical because the average population number is about 100. 128 locations (seven lines) for the memory is sufficient for most of situations. On the other hand choosing memory word size is very important and directly affects the overall performance of the system. All the examples are run with 8 or 16 bits word size, but this can be increased at the risk of increasing the complexity of the system.

B. Memory Interface Module (MIM)

This module is responsible for communicating with memory. It receives new members from the Fitness Module and outputs member strings to the Read Module. The rate of writing in the system is much lower than the rate of reading and because of that, the priority of writing is higher than the priority of reading. Both Fitness and Read Modules have to use special handshaking protocols for communicating with the MIM for writing and reading to the memory.

There are two populations in the memory during processing. The Read Module reads the members from the current population and the Fitness Module writes new members into a new population. The Memory Interface Module will organize this arrangement.. At the end of a generation, the MIM swaps the population pointers so that Read Module reads from the current population and the Fitness Module starts writing in the new population.

C. Read Module (RM)

This is the most important module in the GAP. Indeed this module limits the overall performance in the system and is a bottleneck in the GAP.

This module continuously reads the Memory Module. As soon as the reset signal goes high this module starts reading the memory from the first location and increments the read address to access the next location. At the end of the memory, it resets the pointer to zero to start reading from the top of memory. This module delivers each input member to the next module with a handshaking protocol.

D. Selection Module (SM)

The Selection Module receives the members and selects two members to be manipulated. One method of selection we have used involves scanning the memory and forming a running total of the fitness of all the members encountered until this total exceeds a pre-calculated threshold value. The threshold for selection is calculated as follows [Goldberg89, Davis91]:

fitness_threshold = random(1.0) * $\sum$(all fitness values)

At this point the unit selects the corresponding string for crossover and continues to select the second string in the same manner. When two members have been selected they will be passed to the Crossover Unit. This algorithm favours the selection of fit members over weak members as required by the genetic algorithm selection rule.

E. Crossover Module (CM)

Crossover is an operation which swaps corresponding substrings of two members. We have tested a variety of crossover operations including multipoint and uniform crossover [Davis91] but the following description applies only to the simplest case of single point crossover. The Crossover Unit receives two strings from the Selection Module. These strings are concatenated for convenience. A random number is generated and compared to a given global crossover rate to determine if crossover is to be carried out. If this condition is satisfied then another random integer (1...length) will be used to select a crossover point. Then the right hand segment of each string is exchanged to form two new strings. The crossover rate can be adjusted for testing.

F. Mutation Module (MM)

Mutation is an operation which randomly flips one or more bits in a member string of the population. This occurs at a fixed low probability per bit called the mutation rate.

The Mutation Unit receives strings from the Crossover Unit. A random number will be generated for each bit in the string and compared with the preset bit-mutation rate. In each case where the generated number exceeds the mutation rate the corresponding bit is flipped.

G. Fitness Module (FM)

The purpose of the Fitness Module is to evaluate the population members mated by the Crossover/Mutation Module and insert them into the new population. The fitness evaluation should be completed in one clock cycle for optimum performance of this processor. If this cannot be achieved then parallel FM units could be used. This is much more efficient than any software implementation, including those on parallel machines. Because fitness evaluation is often the bottleneck of a GA, this is another advantage of implementing a GA in hardware.

When the FM receives a pair of members from the Crossover/Mutation Module, (CM, MM) it evaluates their fitness's and then writes the new members and their fitness to the appropriate memory location with the help of the Memory Interface Module. After sending the new members to memory, the FM waits for the CM to send other members. The FM also maintains a running sum of the fitness values for reporting to the Selection Module after each generation. The Selection Module will then scale down that sum with a random value and use the scaled sum in the selection process.

Additionally, the Fitness Module maintains a record of how many generations remain and how many members still need to be chosen in the current generation. When the last generation is complete, the FM notifies the Memory Interface Module of GA completion.

Since the FM does more than just evaluate the new population members, it may be desirable to implement another module to do only fitness evaluation (e.g. Control Module). Thus the FM supports the ability to attach to it an optional external Control Module which can perform the actual fitness evaluation of each member. Upon receiving a new member, the Control Module is expected to evaluate this member in one clock cycle and then make the fitness available to the FM. If no Control Module is attached, the FM evaluates the members with an internal default fitness function. Use of an external Control Module allows for implementation of more complex fitness functions.

H. Control Module

The Control Module accepts a genetic string, calculates the fitness value and returns the result to the Fitness Module. The actual process of evaluation of fitness for each string will depend on Control Module. One possible configuration which would suit the application of a GAP to some engineering problem such as instrument tuning is shown in Figure 5. The Transducer (e.g. a DAC or a state machine) converts the genetic string into suitable signals for the Device Under Test (DUT) which is the system to be optimised. A Measurement Unit converts and evaluates the responses of the DUT and delivers the result as a digital fitness value (e.g. an ADC).

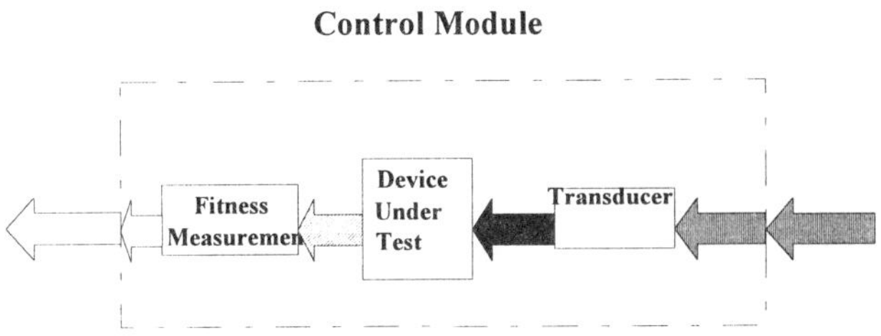

Figure 5 . Typical Control Module

The Control Module should complete the fitness evaluation within one clock cycle from setting up the input value. Actually it is possible to extend this time, but it will decrease the overall performance of the GAP. It means the GAP is suited to fast response environments.

I. The sequence of operation

In summary the operational sequence of the GAP in Figure 4 is as follows.

1. After all the parameters have been loaded into the shared memory, the Memory Interface Module (MIM) receives a starting signal from the front end. The MIM acts as the main control unit of the GAP and is the GAP's sole interface to the Memory Module

2. The MIM notifies the Fitness Module (FM), Crossover Module (CM), Mutation Module (MM), and the Read Module (RM) that the GAP is to begin execution. Each of these modules requests it is required parameters from the MIM, which fetches them from the appropriate places of the shared memory.

3. The Read Module starts the pipeline by requesting population members from the MIM and passing them along to the Selection Module.

4. Whenever the SM receives a new member M from the RM, it determines if M is one of the members it wants to select at this time (the selection process is described in more detail below). If not, it waits for a new member to be sent by the RM. Otherwise, it stores M and checks if another member M' has already been selected. If M' has been selected, then M and M' are sent to the Crossover/Mutation Module for further processing. Otherwise the SM keeps working until M' has also been selected. Once the pair is sent, the SM resets itself and restarts the selection process.

5. When the Crossover/Mutation Module receives a selected pair of members A and B from the SM, it decides whether to perform crossover and mutation based on random values sent from the random number generator. When done, the new members A' and B' are sent to the Fitness Module for evaluation.

6. The Fitness Module takes the two new members A' and B' from the CM and MM and evaluates them. After evaluation, the FM writes the new members to memory through the MIM. The FM also maintains some state variables for the GAP (total fitness and number of generation). These variables are used by the RM to select new members and by the FM to determine when the GAP is finished.

7. The above steps continue until the FM determines that the current GAP run is finished. It then notifies the MIM of completion which in turn shuts down the GAP modules and sends the stop signal to the front end.

IV - A GAP MODEL IN VHDL

A model of the genetic algorithm processor has been constructed using the VHDL (VHSIC Hardware Descriptive Language). VHDL provides programming facilities which permit hardware to be modelled initially at the behavioural level using a block structured approach [Coelho89]. When the behavioural model has been verified, the engineer can implement a structural model describing the internal structure of each block. In a final step this structure can be synthesised in a known hardware technology.

The whole model was written in the VHDL and simulated using Mentor Graphics software. There is a great deal of difference between a model that can be written and simulated using the full power of the VHDL language and a model which has been written to be synthesised in hardware. As an example of this, a Memory Module can easily be used in an unsynthesized model but must be separated for synthesis.

In the first step we simulated the behaviour of the model without regard for synthesis. As parts of the model are converted for synthesis it is possible to mix and test a combination of synthesizable and unsynthesizable units. For the test examples we have synthesised all units except the Memory Module and Control Module which remained in an unsynthesized form during the simulation studies.

V - SIMULATION OF THE BEHAVIOUR

The whole model was tested for performance and comparison with standard genetic algorithm models. Firstly the model was tested with four standard static problems from the DeJong test suits [DeJong75]. Later the VHDL models were tested with the dynamic system described earlier.

A. Example 1 : Static Fitness Function

We give here a brief explanation of the four standard fitness functions used to test the model.

1 Sphere: A unimodal function that is three dimensional and has one minimum at (0,0,0):

$$F_1(\vec{X}) = \sum_{i=1}^{3} x_i^2$$

2 Rosenbrock's saddle: A devilish two-dimensional function that has a curving valley that fools many optimization algorithms into halting prematurely and returning a point that is not the global optimum:

$$F_2(\vec{X}) = (100 * (x_2 - x_1^2)^2 + (x_1 - 1)^2)$$

3 Step: A five-dimensional function that has long, flat surfaces surrounded by discontinuities. Simple hill-climbing algorithms often become stuck on these flat surfaces:

$$F_3(\vec{X}) = \sum_{i=1}^{5} [x_i]$$

4 Quartic: A 30-dimensional function in which evaluation is modified by Gaussian noise. Thus, two evaluations of the same point typically return different values:

$$F_4(\vec{X}) = \sum_{i=1}^{30} i x_i^4 + Gauss\,(0,1)$$

These functions were used to show the optimization

capabilities of the hardware genetic algorithm. Then we compared the results from our hardware implementations with results from software implementations using Genesys Package [Bäck92]. For all the functions, we used a population of 32 individuals and the same probability of mutation and crossover. We used the roulette wheel selection method for selecting the new members.

The results of comparisons are given in Figure 6, which shows the evaluation of the error value of the best point averaged over 10 runs as a function of the number of evaluations. We don't expect that the hardware approach can outperform the software algorithm in any way. However these tests indicate that the hardware is capable of reaching a similar optimum point in same number of generations.

The results shown in the graphs are started in each case with a randomly generated set of members. The starting population is normally different for each curve. However the graphs show only data from hardware and software simulation which proceeds from an identical starting point.

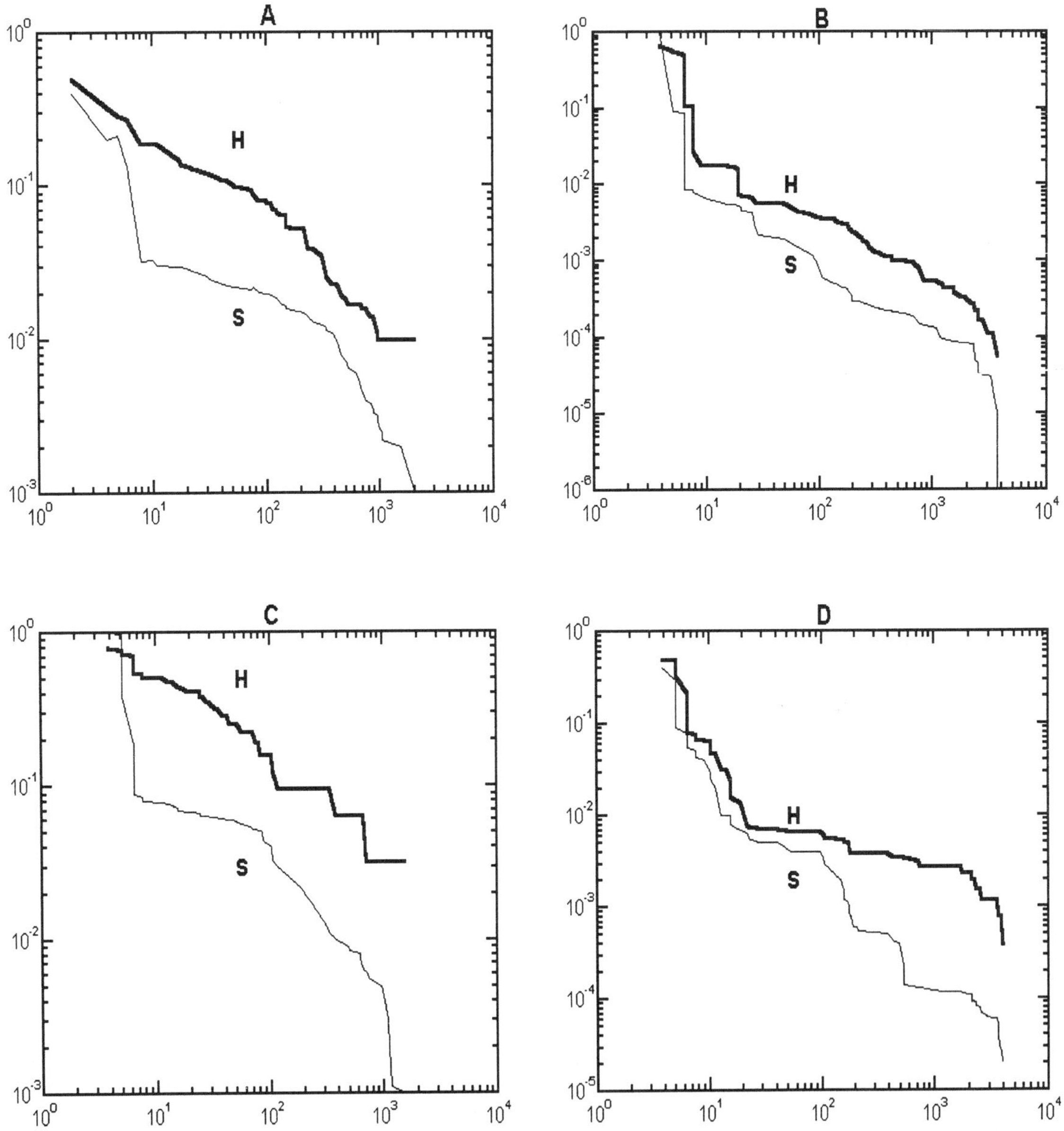

Figure 6 . The results of static functions (S = Software H = hardware) The error value of the best individual versus number of function evaluations A = Sphere, B = Rosenbrock's saddle, C = Step, D = Quartic

B. Example 2 : Dynamic Fitness Function

We have also attempted to apply the hardware model of the genetic algorithm to a dynamic system. We selected the resistor optimization task (Figure 2) to show the performance of our model in a non-static environment.

This example demonstrates the need for a different approach to optimization. For static fitness functions all members of the population eventually merge to the optimum point at the end of a successful search cycle. If the optimum point then moves as a function of time, the GA cannot respond efficiently if there is insufficient genetic diversity in its members. One approach to this situation is to increase the mutation rate thus adding some randomness to the population.

The fitness function for this example can be defined as:

$$F(R_l, R_s) = R_l * R_s * V^2 / (R_l + R_s)^2$$

For our example the value of V is equal to 32 volts. Figure 7 shows the problem space of this function. The solution is always RL=RS and RL changing with frequency f .

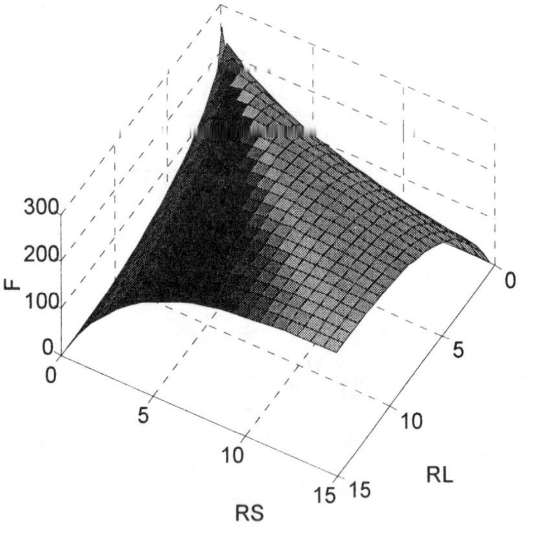

Figure 7 . Problem space of dynamic fitness function

In the genetic algorithm, we used these parameters:

Probability of Mutation = 0.01
Probability of crossover = 0.9
Population Size = 32
Selection Method = Roulette Wheel

Each new generation is created from the current population. The number of generations should be large enough to make sure the genetic algorithm can find the solution in one time period (1/f). We selected 64 for number of generations but it seems that 32 is sufficient.

In the first test we used a one point crossover algorithm and in the second one we changed it to uniform crossover and finally in the third test we changed coding routine to Gray coding [Davis91] (with uniform crossover). The results of the tests averaged over ten runs are shown in Table 1. We ran the simulation for 75 changes in RL.

Test Number	Algorithm	Number of Changes in RL	Incorrect Answers	Error Percentage
1	One Point Crossover	75	15.4	20
2	Uniform Crossover	75	5.2	7
3	Gray Code	75	4.5	6

Table 1 . The results of the dynamic fitness function

The table shows in all cases the genetic algorithm hardware can find the solution with high reliability. It also shows that uniform crossover is better than one point crossover and using Gray code increases the efficiency of the system.

VI - CONCLUSION

In this paper we have described a model of a digital processor which embodies a universal genetic algorithm. We have demonstrated the performance of this model on a static and dynamic optimisation task. Tests reveal that this search can be completed with high reliability at reasonable speed. The next step is to investigate how to add parallelism to the GAP computations.

REFERENCES

[Ackley89] D. H. Ackley , *"A Connectionist Machine for Genetic Hillclimbing"*, Kulwer Academic Publishers, Boston 1989.

[Bäck92] T. F. Bäck , Hoffmeister, *"A User's Guide to Genesys 1.0. Software Package Documentation"*,1992.

[Coelho89] D. R. Coelho , *"The VHDL Handbook"*, Kulwer Academic Publishers, Boston 1989.

[Davis91] L. Davis , *"Handbook of Genetic Algorithms "*, International Thomson Publishing , New York 1991.

[DeJong75] K. DeJong , "*Analysis of the Behaviour of a Class of Genetic Adaptive Systems*" Ph. D. thesis, University of Michigan, Ann Arbor, 1975.

[Goldberg89] D. E. Goldberg , "*Genetic Algorithms in Search, Optimization and Machine Learning*" , Addison-Wesley Co, 1989.

[Holland92] J. H. Holland , "*Genetic Algorithms*", Scientific American, July 1992.

[Kirkpatrick83] S. Kirkpatrick , C. D. Gelatt , M. P. Vecci , "*Optimization by Simulated Annealing*", Science, Vol 220, May 1983.

[Salami94] M. Salami , G. Cain , "*The Quest for a New Computing Architecture Based on Genetic Algorithms*", Proceedings of the Electrical Engineering Congress 1994 (EEC94) , Sydney, Australia.

MODELLING TEAMS AND TEAM TACTICS IN WHOLE AIR MISSION MODELLING*

Gil Tidhar
Australian Artificial Intelligence Institute,
171 LaTrobe Street,
Melbourne, Australia
gil@aaii.oz.au

Mario Selvestrel†
Aeronautical and Maritime Research Laboratory, DSTO
506 Lorimer Street,
Fishermens Bend, Australia
Mario.Selvestrel@dsto.defence.gov.au

Clinton Heinze
Aeronautical and Maritime Research Laboratory, DSTO
506 Lorimer Street,
Fishermens Bend, Australia
Clinton.Heinze@dsto.defence.gov.au

ABSTRACT

The problem of whole air mission modelling is part of a larger problem which is the problem of simulating possible war-like scenarios in the air, sea, and on land. In such modelling systems one is required to model the behaviour of various actors and the resources that are available to them. One aspect of this problem is the modelling of a group of actors as a team and then modelling the coordinated behaviour of such a team to achieve a joint goal.

In the domain of air mission modelling the actors are pilots that control aircraft and their behaviour is referred to as tactics. In this paper we present the approach we adopted in modelling teams and team tactics as part of the development of the Smart Whole AiR Mission Model (SWARMM) for the Aeronautical and Maritime Research Laboratory.

1 INTRODUCTION

Modelling the behaviour of teams is a problem that concerns many analysts who are attempting to model the behaviours of groups of humans and also concerns researchers in Distributed Artificial Intelligence (DAI) who are attempting to model the behaviour of groups of artificial agents. Such attempts include building modelling systems for business processes [Fox93] and power grid monitoring [Jen93]. The problem of modelling the activity of teams of artificial agents [GK93, KLR+92] is a combination of two sub-problems: the first is the modelling of the team itself [Tid93, Wer90] and the second is the modelling of the team activity [Fox81, JM92, KLR+92].

The problem of modelling teams and team behavior includes a wide range of sub-problems. Such sub-problems include the problem of representing teams with a variety of organizational structures and providing a representation of team behavior that allows the implementation of a variety of communication, coordination, and control mechanisms. This problem is further enhanced when considering real-time systems that are embedded in a dynamic environment.

Here we describe the approach taken to the modelling of teams and team behaviour as part of the modelling of whole air missions. This work is part of the development of the Smart Whole AiR Mission Model (SWARMM) [Aer94] system. The purpose of the SWARMM system is to simulate the physics of whole air missions, simulate the pilot reasoning involved in such missions, and to interface with advanced visualization software to enable better understanding of whole air missions.

Whole air mission modelling comprises a set of activities related to the flying of a combat mission by a group of aircraft. The type of aircraft selected for a particular mission will depend first and foremost upon the type of mission that is to be flown. The type of tasks relevant to whole air mission modelling include:

Air Defence - The use of aircraft to defend an area against an attack from the air.

Attack - The use of aircraft to attack a ground target.

Sweep - The use of fighter aircraft to clear a path for an incoming strike force.

Escort - An attachment of fighter aircraft which closely accompany the attack aircraft to defend them against attack by enemy aircraft.

*This work was partially supported by the Cooperative Research Centre for Intelligent Decision Systems, Melbourne, Australia.

†The author is currently at Telstra Applied Technologies, Jindalee Project, 22 Winterton Road, Clayton, Australia.

We refer to the behaviour of the aircraft as *tactics* and to the behaviour of a group of aircraft as *team tactics*. Teams are represented as a set of sub-teams that have adopted structures that describe their inter-team relationships. These relationships include an allocation of organizational and functional responsibilities. The team tactics are modelled as plans that are executed by the individual members of the team. The choice of the team tactics depends on the task to be achieved, the particular situation, the structure adopted, and the availability of communication facilities.

The tactics of individual pilots involve carrying out a sequence of sub-tasks or maneuvers. Each of these tasks can further be divided into smaller sub-tasks. For example, if the pilot decides to go into attack she first needs to find the bearing to the nearest range target; having determined this she needs to change course towards it. Changing the course might involve a sequence of maneuvers. Thus, the formalism must be able to capture sequential, parallel, iterative, and non-deterministic actions.

At any instant the pilot might be able to employ a number of different tactics but may be executing only one of them. Also, while executing a particular tactic the pilot might have already decided (i.e., intended) to execute other tactics some time in the future. Thus, the representational formalism needs to distinguish between the pilot having, executing, and intending a particular set of tactics [GL86, RMSM92].

A combat scenario usually involves more than one aircraft from the same side. Aircraft are organized into tactical units and have specific roles assigned to them. For example, Howlett [How91] shows how a mission to defend an air base might be accomplished by four sector CAPs (Combat Air Patrols) and two back-up CAPs. Any mission involving multiple aircraft is accomplished by adopting team tactics. One must be able to hierarchically decompose the team tactics based on the organization of pilots. Team tactics give rise to notions such as mutual beliefs, joint goals, and joint intentions which are the beliefs, goals, and intentions shared by multiple pilots or teams of pilots.

The dynamic nature of whole air mission scenarios means that aircraft can dynamically reorganize themselves. For example, when two aircraft from different groups are shot down, the remaining aircraft of each group may combine together to form a single group. Dynamic reorganization of aircraft may also force dynamic reassignment of roles. The representational formalism for team tactics must be capable of dynamic reorganization and dynamic reassignment of roles.

The underlying system used for modelling the teams and implementing the team tactics is the Distributed Multi-Agent Reasoning System (dMARS) [Aus94]. Previous work on this problem has included the development of a demonstrator system [RMSM92] that has investigated the ability to model team tactics using the Procedural Reasoning System [GL86]. In that work teams were represented as special agents that did not correspond to any particular aircraft or pilot and that executed the team tactics by instructing the sub-teams to perform particular sub-tasks. This has allowed only for a centrally coordinated type of team behaviour. In this work we are implementing an operational system and have taken a different approach that allows for a distributed control of team behaviours.

Different types of coordinated behavior have also been implemented using the Soar architecture [LJN94, LNR87] although these experiments have focused primarily on the coordination required between two fighter aircraft flying in formation. Furthermore, this work has not considered the coordination required in a large scale air mission and across the command hierarchy.

In Section 2 of this paper we describe the underlying technology used in the development of the SWARMM system and a typical scenario that is modelled. In Section 3 we describe the approach taken in modelling teams and the way teams are implemented and in Section 4 we describe the approach taken in modelling different team tactics and the way these tactics were implemented. We conclude this paper in Section 5 with a short discussion.

2 THE UNDERLYING TECHNOLOGY

The problem of modelling whole air missions has been divided into two components (or sub-problems) each modelled (or solved) using different technology. The first component is modelling the behavior of the physical elements and systems that are involved in whole air missions. These include the aircraft, the radar systems, the weapon systems, and the communication systems. The modelling of the physical systems also includes the modelling of the pilot's body, e.g., eyesight, and the effect of various maneuvers on the pilot's physical state. To model this component we used a Fortran-based computer model called Piloted Air Combat AUStralia (PACAUS) [Ste93].

The second component is modelling the reasoning processes that are part of the pilot. These processes include the process of determining the current situation (referred to as *Situation Awareness*) and the process of choosing and executing the best tactics for the current situation (referred to as *Tactics Selection*). To model this component we used the Distributed Multi-Agent Reasoning System (dMARS) [Aus94].

In the model adopted we view the combination of the pilot's reasoning and the aircraft's physical systems as a single entity which we will refer to as an *aircraft*. We assume that the pilot's reasoning processes receive input from other components/sensory equipment (referred to as the *sensed world*) and provides instructions to other components/equipment (referred to as *pilot response*).

2.1 PACAUS

The Piloted Air Combat AUStralia (PACAUS) system is a time stepped simulation of combat between many aircraft from two opposing sides. In excess of fifty thousand lines of Fortran code comprise the models of the hardware and their tactical use. An advanced graphical post-processing program allows three dimensional visualisation of the combat [Mus93].

Input files allow the user to specify the initial scenario and to define certain flags to control the use of the weapons systems, the radar systems, and the aircraft themselves.

There are several aerodynamic models of combat aircraft. The on-board systems such as the radar, the Radar-Warning-Receiver (RWR), and counter-measure facilities such as chaff, flares, and Electronic Counter-Measures (ECM) are modelled by the system. There are simulations of a variety of weapons including fully active and semi-active missiles, infra-red heat-seeking missiles and guns. The PACAUS program has been utilized for several research programs for the Royal Australian Air Force and is currently being expanded in several important areas.

2.2 DISTRIBUTED MULTI-AGENT REASONING SYSTEM (dMARS)

The Distributed Multi-Agent Reasoning System (dMARS) is an agent-oriented, distributed, real-time system. It provides a representational framework and reasoning mechanisms for implementing agents.

Each agent is composed of a set of beliefs, goals, plans, and intentions. The beliefs of dMARS agents provide information on the state of the environment and are represented in a first-order logic. For example, the belief or fact that the range from aircraft `WARLOCK` to an aircraft `BOGGY1` is 40 miles can be represented by the statement `(range WARLOCK BOGGY1 40)`. Variables are denoted by the character `$`, e.g., `$target` denotes that variable `target`.

The goals of dMARS agents are descriptions of desired tasks or behaviours. In the logic used by dMARS, the goal to achieve a certain condition C is written as `(! C)`; to test for the condition is written as `(? C)`; to wait until the condition is true is written `(^ C)`; to assert that the condition is true is written as `(=> C)`; and to retract a condition is written as `(~> C)`.

Plans are declarative procedural specifications that represent knowledge about how to accomplish given goals or react to certain situations. Each plan consists of a body, an invocation condition, and a context condition [GL86]. The set of plans in a dMARS application system also includes meta-level plans, that is, information about the manipulation of the beliefs, goals, and intentions of the dMARS agent itself.

The body of a plan can be viewed as a procedure or a tactic. It is represented as a graph with one distinguished start node and one or more end nodes. The arcs in the graph are labelled with the sub-goals to be achieved in carrying out the plan. The invocation condition describes the events that must occur for the plan to be executed. Usually, these events consist of the acquisition of some new goals (in which case, the plan is invoked in a goal-directed fashion) or some change in system beliefs (resulting in data-directed invocation) and may involve both. The context condition describes contextual information relevant for the execution of the plan.

The intention list contains all those tasks that the system has chosen for execution, either immediately or at some later time. An intention consists of some initial plan, together with all the sub-plans that are being used in attempting to successfully execute that plan. At any given moment, the intention list of an agent may contain a number of such intentions, some of which may be suspended or deferred, some of which may be waiting for certain conditions to hold prior to activation, and some of which may be meta-level intentions. Only one intention can be executed at any given moment and the choice of that intention depends on the state of the world and the priority of that intention.

In some applications, it is necessary to monitor and process many sources of information at the same time, e.g., simulating a number of pilots. To facilitate this, dMARS was designed to allow several agents to run in parallel. Each agent has its own database, goals, and plans, and operates asynchronously relative to other agents, communicating with them by sending messages.

2.3 A TYPICAL SCENARIO

Let us consider two opposing forces. Red Team is planning a strike mission to destroy a ground target within Blue Team's territory. Red Team assembles a *package* of aircraft operating in several different roles. There is a group of sweepers to clear a path ahead of the strike aircraft, there are escort aircraft to accompany the strike aircraft and of course there are the strike aircraft themselves. These three distinct elements have a single goal: to successfully attack the ground target which has been designated for them. They each have assigned responsibilities within the mission that require communication and interaction within the group.

Blue Team will have aircraft operating in the air-defence role protecting their airspace from the incursion by Red Team. These aircraft will either be launched from an airbase when it becomes apparent that an attack is imminent or, if hostile actions have been present for a number of days, the aircraft may be flying patrols over an area where an attack is expected.

The hierarchy of command within the team exists in a flexible dynamic way to allow the team to operate at several levels and to split and reform as the situation dictates [Sha85]. Within the operation of a standard mission there will be aircraft performing different tasks. An escort aircraft may accompany a strike aircraft as its wingman. A pair of lower performance fighter aircraft might accompany a pair of higher performance fighter aircraft to give the illusion of four high performance fighters.

Each situation may require the use of a different command and control structure. Thus the sub-teams will adopt different command and control roles within the team. The different mission goals adopted by the team may require the sub-teams to adopt functional responsibilities with respect to the conduct of the mission. Thus an aircraft may have both a command and control role (e.g., a leader) and a functional role (e.g., an escort). This two-fold responsibility impacts upon the way in which tactics are chosen and the way in which the mission is conducted.

3 MODELLING TEAMS

The aircraft in whole air missions are identified as *singletons* and have unique names. Aircraft are teamed together into pairs, groups, and packages and again each team has a unique name. As the name *pair* indicates, pairs are teams of two aircraft. Groups are teams made up of pairs and/or singletons. Packages are teams made up of groups and/or pairs and/or singletons. For each such team the various teams of which it is made up are referred to as its *sub-teams*.

Each of the sub-teams is assigned at least one role in the team.[1] The role identifies the sub-team's relationships with other sub-teams and the responsibilities it has towards the various functions of the team. We identify two types of structures that are imposed on the team and that correspond to two types of roles. The first is an organizational structure that defines the Command and Control functions in the team, and which is completely hierarchical. We refer to the roles in this structure as *organizational roles*. The second is a functional structure that defines the functional expertise and responsibilities in achieving the task that the team is set to achieve, and which is completely "flat". We refer to the roles in this structure as *functional roles*. This model of a team is similar to the model described by Tidhar [Tid93].

Typically, the teaming and naming of aircraft is done prior to the mission in a briefing session as is the role assignment. Due to the dynamic nature of the domain, teams can be dismantled and re-formed dynamically. As teams change their structure(s) in response to the situation, roles can be dynamically re-assigned.

3.1 ROLES

In the organizational structure we identify two types of roles, a *leader* and a *wingman*. Each team has only one sub-team as the leader but there can be several sub-teams as a wingman although typically there is also only one wingman. The leader sub-team is responsible to make all the decisions for the team and to instruct or inform the wingman on particular decisions or information. The wingman is responsible for following and obeying the leader and providing it with information that may assist it in making decisions.[2] Each sub-team is aware of the other sub-teams and their roles and reacts to changes to the world according to its roles.

Functional roles correspond to the way the team achieves its tasks and the responsibilities of each of the sub-teams towards achieving the task. The choice of functional roles depends on the tasks that the team is set to achieve. The way the assigned role determines the responsibilities of the sub-team assigned to this role is via the team tactics (see Section 4).

Role assignment is part of the process of team formation. The assignment of roles typically depends on the skills and capabilities of the sub-team. These in turn depend on the hardware setup of the various aircraft in the sub-team, e.g., aircraft type, weapon systems, etc., and the tactical knowledge held by the pilots. When a team is assigned a role all its sub-teams are aware of this assignment and can act accordingly.

Example 1

Consider the team `WARLOCK-12` that is part of the RED team and which has two singleton sub-teams, `WARLOCK-1` and `WARLOCK-2`, that are the `LEADER` and `WINGMAN` respectively. Given that `WARLOCK-12` has to fly in formation towards a particular target, the leader, `WARLOCK-1`, will determine and fly along a course. The wingman, `WARLOCK-2`, will simply follow `WARLOCK-1` while flying in some predetermined disposition. In the case of any observed unique events, `WARLOCK-2` will communicate its observations to `WARLOCK-1` and wait for a command.

The pair `WARLOCK-12` is teamed with the pair `JESTER-12` to form the group `PYTHON`. `WARLOCK-12` is the `LEADER` and `JESTER-12` is the `WINGMAN`. In the context of the team `PYTHON`, the teams `WARLOCK-12` and `JESTER-12` are also assigned the functional roles `ATTACK` and `ESCORT` respectively. The group `PYTHON` is teamed with the singleton `GRIZZLY` to form the package `THUNDER`. In the context of the team `THUNDER` the teams `PYTHON` and `GRIZZLY` are assigned the functional roles `STRIKE` and `TANKER` respectively.

□

Note that `THUNDER` has only functional roles. This means that the coordination and decision making processes are implicit in the team tactics and that each sub-teams is responsible to inform other sub-team of any change to the tactics. This type of behaviour corresponds to the notions of commitment cf. Cohen and Levesque [CL91] and responsibility cf. Jennings and Mamdani [JM92].

[1] Each sub-team may be assigned more then one role, but it is the responsibility of the designer to ensure that a sub-team is not assigned two conflicting roles as is the case in real missions.

[2] Note that the leader of a team can be a sub-team and hence that sub-team's leader will make the decisions for the whole team and so on and so forth until the leader is a singleton.

3.2 IMPLEMENTATION

In dMARS the agent's beliefs are implemented as relations (or predicates) in a relational database. Since each sub-team in a team has at least one role the team itself can be represented as a relation between the team, the sub-teams, and the role that the sub-team has been assigned in the team. We refer to this relation as `role-in-team`. The only teams that do not have sub-teams are aircraft. Such teams are identified with the predicate `singleton` (e.g., `(singleton WARLOCK-1)`).

Each agent is aware of its name (via the predicate `myname`) and can hence deduce its membership in different teams. Basically, if the agent has been assigned a role in a team then it is a member of that team. If that team has been assigned a role in another team then the agent is also a member of the other team. Not only can the agent deduce its membership in a team, it can also deduce the roles it has been assigned in that team. This knowledge allows the agent the adapt its behaviour as specified in the team tactics.

There are two additional aspects of a team that are required for the purpose of specifying the behaviour. The first is the set of members, that is all the aircraft that are either a sub-team of the team or members of any of the sub-teams. The second is the size of a team, that is the number of members. Both these aspects are implemented as functions that deduce this information from the relations `role-in-team` and `singleton`.

Example 2

Let us describe the beliefs of the `WARLOCK-1` pilot. The team `THUNDER` is comprised of the singletons `WARLOCK-1`, `WARLOCK-2`, and `GRIZZLY` and also the aircraft that are part of the team `JESTER-12`. The team `WARLOCK-12` has only organizational roles and is modelled with the predicates `(role-in-team WARLOCK-12 WARLOCK-1 LEADER)` and `(role-in-team WARLOCK-12 WARLOCK-2 WINGMAN)`

In the context of the team `PYTHON` the sub-teams are assigned both organizational and functional roles and the team is modelled with the predicates:

```
(role-in-team PYTHON WARLOCK-12 LEADER)
(role-in-team PYTHON WARLOCK-12 ATTACK)
(role-in-team PYTHON JESTER-12 WINGMAN)
(role-in-team PYTHON JESTER-12 ESCORT)
```

The sub-teams of the team `THUNDER` are assigned only functional roles and the team is modelled with the predicates `(role-in-team THUNDER PYTHON STRIKE)` and `(role-in-team THUNDER GRIZZLY TANKER)`.

□

4 MODELLING TEAM TACTICS

Team tactics are used to ensure that the team works together as a coherent entity. There are two main types of methods to control and coordinate the team behaviour, namely centralized control and coordination, and distributed control and coordination. Distributed control and coordination, in turn, can be implicit or explicit.

The simplest method to control a team is through centralized control and coordination. With this method there is a team member who makes all the decisions and gives orders to the other team members. The team members themselves have no need to know why they are carrying out actions or how the actions contribute to the overall team tactics. Team members under centralized control blindly follow orders given by the team leader. For example, an air defence controller vectors a pair of fighters onto a certain course. The fighters do not necessarily need to understand why they are flying in that direction and obey the order.

On the other hand, team members under distributed control have an understanding of the team's task, the role of each team member, and how the team works together. These team members can cooperate by observing the world (implicit coordination) or through communication messages (explicit coordination).

Implicit coordination involves no communication between team members and it occurs when the members of the team observe the world in order to determine what action should be taken and when. Many team tactics used in whole air missions use implicit coordination. Typically, a team has practiced its standard procedures and tactics together until their execution has become instinctive. The team members do not need to explicitly communicate in order to perform the steps of a given tactic. Through practice the team knows, without communication, the role of individual team members and how each member fits into the overall tactic used to complete the task. During a briefing before a mission begins a team will discuss and confirm the particular tactics applicable to the mission. The mission then proceeds with each individual of the team executing the particular tactic implicitly defined by their role in the team and by the situation in which they find themselves.

Explicit coordination involves communication between team members to coordinate their actions. However, unlike centralized control where team members must follow instructions received, with explicit coordination the messages provide information which is evaluated before any action is taken. In the case of air combat, explicit coordination can occur through data links between aircraft or, more commonly, through radio messages.

Each form of control has its benefits and problems. A team member under centralized control has a relatively easy role with little reasoning to do, it is always told what to do. But, if it is not explicitly told what to do it cannot do anything. When using implicit control each team member has its own view of the world and confusion can occur when

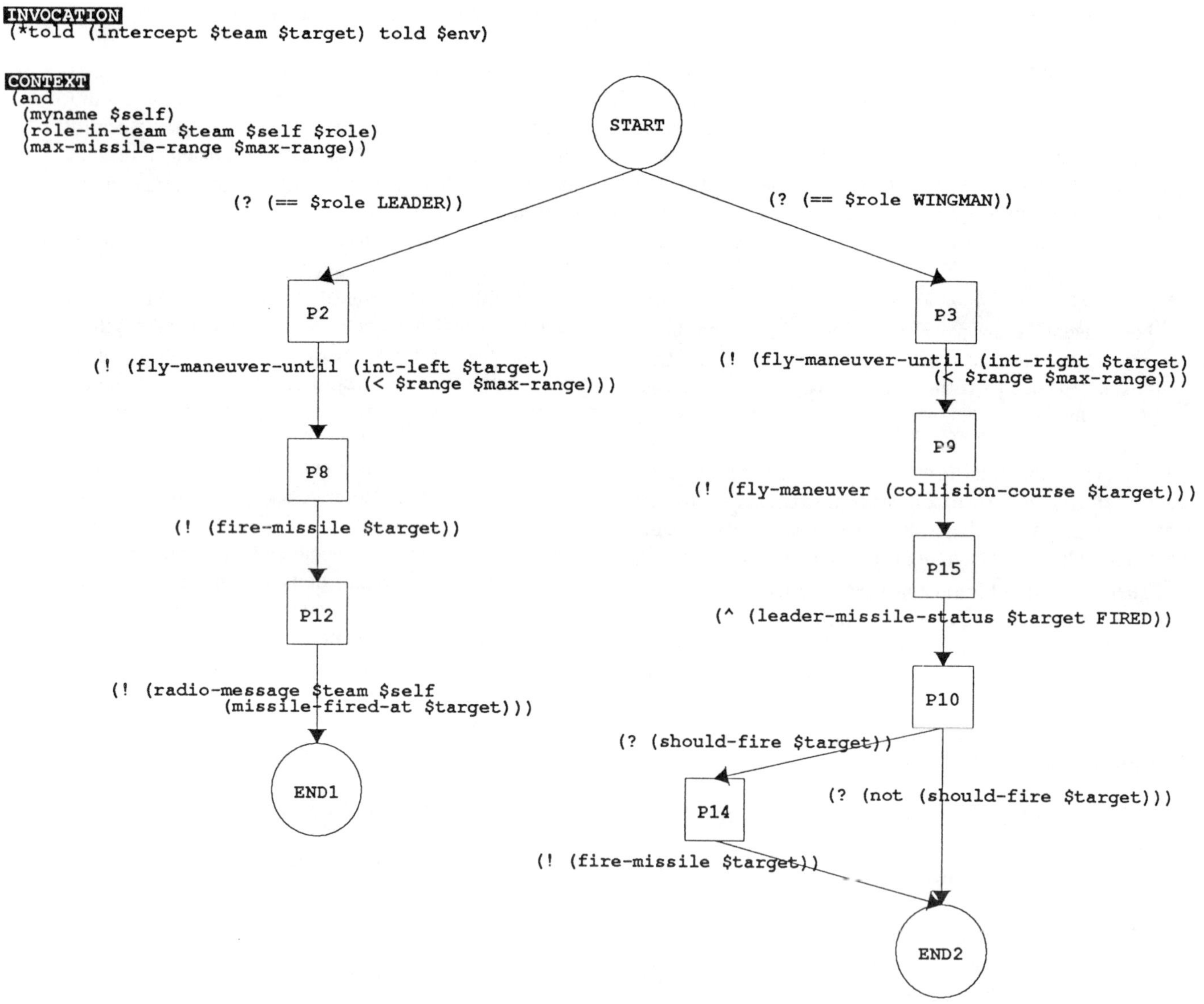

Figure 1: Team tactics for intercepting a target.

team mebers hold conflicting views of the world. Explicit coordination is more reliable than implicit coordination because information is explicitly sent between members of the team. However, it depends on a large number of communication messages and confusion can occur if expected messages are not sent between team members.

In general, all three methods of control need to be used by tactics. Centralized control is necessary when the aircraft under control is relying on the controller to provide information which is otherwise unavailable, e.g., a controller providing intercept instructions for a target that the aircraft cannot detect. Implicit coordination is used to limit exposure to the enemy through radio emissions. Every time a radio message is transmitted there is a possibility that the sender may be detected, this can make explicit coordination through radio and data link dangerous. However, when precise coordination is required, e.g., to inform team members that a missile has been fired, explicit coordination is used.

Example 3

In practice all three methods of control and coordination can be used within the one tactic. To illustrate this we will examine a pincer intercept plan (Figure 1). The pincer intercept involves the team, with two members, forming into two distinct but cooperating elements in order to intercept an enemy aircraft: the leader element which attacks the target from the left and the wingman element which attacks the target from the right. The various stages of the intercept are coordinated using different methods. The following briefly describes the five stages of a pincer intercept:

- The order to commence a pincer intercept is received by the team members.

- The leader and wingman split to obtain lateral separation from the target.
- Once the range to the target is less than the missile's maximum range, the leader begins the missile firing procedures and the wingman changes course towards the target.
- When the leader fires the missile a radio call is made to indicate a missile has been fired.
- Once the leader has shot, the wingman decides whether it should also fire at the target.

□

4.1 MODELLING TEAM TACTICS IN dMARS

Tactics are modelled as plans within the dMARS environment. Each agent has plans which define the tactics available to the agent. In order to model coordinated team tactics, plans are written as sets defining the procedures to be used by each member (or agent) within the team. Each agent of the team executes the portion of the tactic which is relavant to itself. The plans or portion of a plan which must be executed by each team member can be differentiated through the context or by branching within the plan. For example, if a team plan has two members (leader and wingman), the part of the plan that is relavant to the leader can be determined by testing if the current agent is the leader in the context condition, e.g., `(role-in-team $team $self LEADER)`, or branching within the body of the plan, e.g., `(? (== $role LEADER))` (see Figure 1).

All three methods of coordination and control discussed above can be modelled using dMARS plans. The confusion aspects associated with the different tactics discussed above carry through to the dMARS models of the tactics. For example, in real whole air missions implicit coordination can leader to errors in executing a tactic because decisions are based on incorrect data; the same errors can be observed in a dMARS plan.

4.2 CENTRALIZED CONTROL AND COORDINATION

Centralized control and coordination can be modelled by having an agent react to an incoming message without any need for processing or evaluating the message. Commencing the pincer intercept plan is an example of centralized control. The plan is invoked through a told event, e.g., `(*told (intercept $team $target))`. There is no decision making process at the beginning of the plan that gives the pilot agent a choice to follow the intercept command or to ignore it. The first arc of the plan just determines which branch of the plan is relavant for the leader, e.g., `(? (== $role LEADER))`, and which is relevant for the wingman, e.g., `(? (== $role WINGMAN))`.

4.3 DISTRIBUTED IMPLICIT CONTROL AND COORDINATION

Implicit coordination is modelled by examining changes in the world model without communication between the team members. Typically, implicit coordination involves flying a manoeuvre until some geometric observation is made, such as reaching a predefined range to a target.

The initial stage of the pincer intercept uses implicit coordination. The leader and wingman begin by flying different manoeuvres. The leader flies an intercept from the left, e.g., `(int-left $target)`, while the wingman flies an intercept from the right, e.g., `(int-right $target)`. Both agents maintain this manoeuvre until the range to the target is less than the maximum missile range, e.g., `(< $range $max-missile-range)`. The arc between nodes `p2` and `p8` in Figure 1 corresponds to the action of the leader, and the arc between nodes `p3` and `p9` to that of the wingman.

At no time do the leader and wingman need to communicate with each other to achieve the coordination. The coordination occurs because both agents are observing the world and know when they should change manoeuvres, e.g., they change manoeuvre when the required range condition is achieved.

4.4 DISTRIBUTED EXPLICIT COORDINATION

Explicit coordination can be modelled by using communication messages. Such messages convey information about the state of the sub-teams and request information. However, an agent receiving a communication message with explicit coordination evaluates the effect of the information before commencing an action, unlike centralised control where agents blindly obey the command messages. This corresponds to different speech acts as described by Searl [Sea75].

The agents performing a pincer intercept use explicit coordination to determine when to fire a missile. The leader sends a message to the wingman indicating that the leader has fired a missile, e.g., `(! (radio-message $team $self (missile-fired-at $target)))`. The wingman, on receiving the message from the leader, will assert that the leader's missile was fired. The tactic that was waiting for this information, e.g., `(^ (leader-missile-status $target FIRED))`, will now evaluate if it should also fire a missile at the target. If the wingman decides to fire it issues a fire command, e.g., `(! (fire-missile $target))`. In this case the wingman was not ordered to fire, but was informed that a missile was fired and from the operational procedures knew that it should now determine if it should fire a misile.

5 CONCLUDING REMARKS

With an increasing number of expert systems being widely available, a typical industrial environment may include multiple expert systems that are required to cooperate and coordinate their activity. One can view such a group of systems as a team and use control and coordination mechanisms used by humans and adapted to the automated scenario. A similar problem is the problem of modelling the behaviour of groups of humans either in a simulation environment or for the purpose of management and coordination of the group (e.g., business process management).

One of the main problems of modelling the behaviour of groups of humans or coordinating the behavior of human or artificial agents is the problem of representing their behaviour as a team. These problem include the problems of representing the team and acting as a team. In particular the problem of modelling and using different types of control and coordination methods. This is further emphasized when the team is required to work with limited time and communication facilities.

One example of such a problem is the modelling of aircraft and air-combat pilots that are engaged in whole air missions where the pilots are required to reason and act in a highly dynamic environment. Furthermore, they are required to coordinate their activities and to improve their performance as a group. Communication facilities may not always be available either because of technical limitations or because of operational requirements.

To overcome such problems pilots use a wide range of team structures and coordination methods. Different role assignments allow the pilots to pre-determine the behaviour of the team members that is specified in various team tactics used in varying situations.

In this paper we have described how teams are modelled and how this information is used during tactics selection. We have also shown how different control and coordination mechanisms are implemented as plans in the dMARS system. This model together with the PACAUS system is combined to form the SWARMM system.

As part of this work we have identified two key features that dominate the ability of a group of agents (human or artificial) to coordinate their activity. These are the ability to observe the activities of other agents and use predefined methods for coordination, and the ability to use explicit communication. Communication can be used either for a centralized control structure or a distributed one but as it is a limited resource it can not be used frequently.

It seems then that the tactics employed by air-combat pilots to control and coordinate their activities is typically a combination of different types of control and coordination methods. The SWARMM system is currently in its final stages of development and will be commissioned at the Aeronautical and Maritime Research Laboratory in the coming year for further investigation of whole air mission scenarios for the Royal Australian Air-Force.

ACKNOWLEDGMENTS

We would like to thank Dino Appla, David McIlroy, Anthony Schnelbeck, and Serena Steuart from the Aeronautical and Maritime Research Laboratory and Anand Rao and David Morley from the Australian Artificial Intelligence Institute for their contribution to the work described here.

References

[Aer94] Aeronautical and Maritime Research Laboratory, DSTO, Port Melbourne, Australia. *Smart Whole AiR Mission Model (SWARMM): Technical Description*, 1994.

[Aus94] Australian Artificial Intelligence Institute, Melbourne, Australia. *Distributed Multi-Agent Reasoning System (dMARS): Technical Description*, 1994.

[CL91] P. Cohen and H. Levesque. Teamwork. Technical Report 504, SRI International, Menlo Park, CA, 1991.

[Fox81] Mark S. Fox. An organizational view of distributed systems. *IEEE Transactions on Systems, Man, and Cybernetics*, 11(1):70–80, January 1981. (Also published in *Readings in Distributed Artificial Intelligence*, Alan H. Bond and Les Gasser, editors, pages 140–150, Morgan Kaufmann, 1988.).

[Fox93] Mark S. Fox. Enterprise integration. *Personal communication*, February 1993.

[GK93] B. Grosz and S. Kraus. Collaborative plans for group activities. In *Proceedings of the 13th International Joint Conference on Artificial Intelligence*, pages 367–373, Chambry, France, 1993.

[GL86] M. P. Georgeff and A. L. Lansky. Procedural knowledge. In *Proceedings of the IEEE Special Issue on Knowledge Representation*, volume 74, pages 1383–1398, 1986.

[How91] M. Howlett. A description of the criteria used and the process of gameplan selection for air to air intercepts. Technical Report (Working paper), Studies and Simulation Group, Aeronautical Research Laboratory, Melbourne, Australia, 1991.

[Jen93] N. R. Jennings. *Cooperation in Industrial Multi-Agent Systems*. World Scientific Publishing Company, 1993.

[JM92] N. R. Jennings and E. H. Mamdani. Using joint responsibility to coordinate colaborative problem solving in dynamic environments. In *Proceedings of the Tenth National Conference on Artificial Intelligence*, pages 269–275, San Jose, CA, 1992.

[KLR+92] David Kinny, Magnus Ljungberg, Anand Rao, Elizabeth Sonenberg, Gil Tidhar, and Eric Werner. Planned team activity. In *Proceedings of the Fourth European Workshop on Modelling Autonomous Agents in a Multi-Agent World, MAAMAW '92*, Viterbo, Italy, 1992.

[LJN94] J. E. Laird, R. E. Jones, and P. E. Nielsen. Coordinated behavior of computer generated forces in TacAir-Soar. In *Proceedings of the Fourth Conference on Computer Generated Forces Behavioral Representation*, May 1994.

[LNR87] J. E. Laird, A. Newell, and P. S. Rosenbloom. Soar: An architecture for general intelligence. *Artificial Intelligence*, 33(3), 1987.

[Mus93] R. Muscatt. Combat display program - specification and user guide. Working Paper, Air Operations Division, Aeronautical Research Laboratory, Melbourne, Australia, 1993.

[RMSM92] A. Rao, D. Morley, M. Selvestrel, and G. Murray. Representation, selection, and execution of team tactics in air combat modelling. In Anthony Adams and Leon Sterling, editors, *Proceedings of the 5th Australian Joint Conference on Artificial Intelligence*, pages 185–190. World Scientific, November 1992.

[Sea75] J. R. Searl. *A Taxonomy of Illocutionary Acts*, volume 7 of *Minnesota Studies in the Philosophy of Science*. University of Minnesota Press, Minneapolis, 1975.

[Sha85] R. L. Shaw. *Fighter Combat Tactics and Maneuvering*. US Naval Institute Press, 6th edition, 1985.

[Ste93] S. Steuart. Pacaus - overview and source code description. Working Paper, Air Operations Division, Aeronautical Research Laboratory, Melbourne, Australia, 1993.

[Tid93] G. Tidhar. Team oriented programming: Preliminary report. Technical Report 41, Australian Artificial Intelligence Institute, Melbourne, Australia, 1993.

[Wer90] E. Werner. Cooperating agents: A unified theory of communication and social structure. In L. Gasser and M. N. Huhns, editors, *Distributed Artificial Intelligence: Volume II*. Morgan Kaufmann Publishers, 1990.

REPRESENTATION ISSUES FOR DESIGN TOPOLOGICAL OPTIMIZATION BY GENETIC METHODS

P.J. Gage and I.M. Kroo
Department of Aeronautics and Astronautics
Stanford University
Stanford, California
Email: peetah@leland.stanford.edu

ABSTRACT

Standard genetic algorithms operate on a fixed parameterization of the design space, which must be fully specified before optimization begins. Variable-complexity encodings can extend the automated search capability provided by a genetic optimizer. This paper describes the use of shape grammars to provide flexible description of candidate designs. Extended genetic strings, which encode several copies of each variable, are also introduced. Constraint activity is monitored as the genetic string is decoded, to regulate expression of the appropriate copy of each variable. An example application demonstrates that hese extended, variable-length encodings permit discovery of novel topological concepts.

INTRODUCTION

Genetic algorithms have been applied to a wide range of optimization tasks, and are effective in multi-modal and non-smooth domains, where calculus-based search methods are ineffective [GOL89a, GRE87, SCH89, BEL91]. A standard genetic algorithm optimizes the values of a fixed set of parameters, but the best parameter set is often not known *a priori*. Hence, an algorithm that can vary the parameter set during optimization has been developed. This variable-complexity algorithm is particularly appropriate for design studies, where it is common to start with a simple representation and progress to more detailed descriptions which use more design variables. Both standard and variable-complexity genetic algorithms are discussed in Section 2.

When the number of parameters used to describe a design is allowed to vary, exhaustive representation of all possible components is unnecessary. More flexible encodings, describing only the features included in a particular configuration, are helpful in topological design, where the number of components is optimized (rather than just the size and shape of components). A formal method for generating topologies is provided by shape grammars [STI80], which use transition rules to describe changes to a simple initial design. The complexity of the topology is directly related to the number of transition rules encoded in the genetic string. Shape grammars are described in Section 3 of this paper.

Decoding of a genetic string that contains shape transition rules is similar to embryological development. The decoding of each new piece of the string depends on the structure that has been built by prior decoding. This developmental nature can be exploited when constraint activity changes during optimization. The genetic code can be extended to include several copies of each variable, and expression of the appropriate copy can be controlled by constraint activity at the time of decoding. Further details of this extended encoding are provided in Section 4. An example of the use of a shape grammar and an extended encoding is provided in Section 5, where results are presented for the topological optimization of an aircraft wing.

GENETIC OPTIMIZATION

Genetic algorithms are designed to mimic evolutionary selection[GOL89a]. A population of candidate designs is created, and each design is assigned a fitness ranking, which is based on its performance in minimizing an objective and in satisfying constraints. Candidates of relatively high fitness are likely to defeat low-fitness designs in a competition to participate in the production of a new generation. Features of the superior designs are then recombined in later generations. After many generations, and many recombinations, the optimal design may be located.

The high fitness individuals do not actually survive across generations, but a description of their features is propagated. Each individual is represented by a 'genetic' string, which is a coded listing of the values of the design variables. The entire string is analogous to a chromosome, with genes for the different features (or variables). When individuals are selected to be parents for offspring designs, their genetic strings are recombined in a crossover operation, so that the new designs have elements of two earlier designs. A mutation operation also allows modification of elements of the new

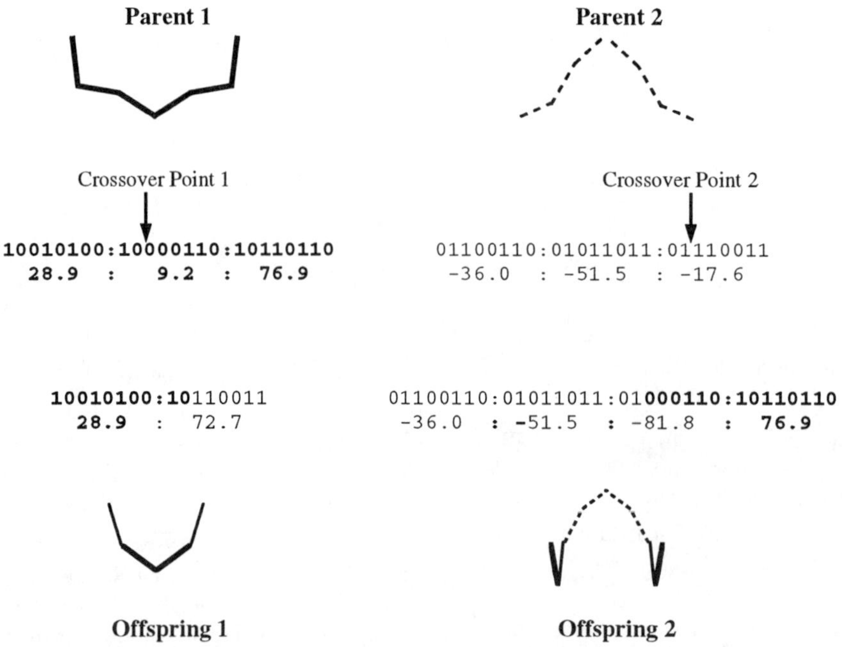

FIGURE 1. Modified Crossover Operator.

individual so that it may include new features that were not present in either parent.

The effectiveness of genetic algorithms depends on useful correlations between parts of the genetic string (genotype) and the performance of the individual it represents (phenotype). Substrings, or building blocks, which appear in the description of above-average phenotypes are likely to survive into the next generation, even if the genotype is broken up by the action of crossover and mutation. Short, low-order building blocks are retained and combined to form higher-order building blocks, with the process repeating over many generations until the best design is found. Promising features of different candidates can be recombined to produce improvements in complete designs.

A standard genetic algorithm operates on a fixed set of design variables. This limitation requires that a description of all possible elements must be explicitly included in the genetic string throughout the optimization process. A modifed genetic algorithm has been developed, to permit the number of design variables to change during optimization [GAG93]. This variation is achieved by modification of the crossover operator, so that the crossover point can be in a different position for each parent, and the offspring have genetic strings of different length. This new operator is similar to that used in genetic programming [KOZ92]. It is illustrated in Figure 1, for a wing design problem, with the dihedral of each wing element concatenated to form the genetic string.

Although variable-length encodings are unusual in genetic search, several investigators have been attracted by their expressive power. S.F. Smith [SMI80] used variable-length strings for development of rule sets for game-playing. He extended Holland's schema theorem [HOL75], and developed a hyperplane analysis to show that promising building blocks are appropriately retained and recombined when string length is not constant. Koza [KOZ92] developed the genetic programming algorithm for automatic generation of computer programs. He cites the empirical evidence of successful applications in a variety of fields as proof that genetic adaptation of variable-length strings is a valid search mechanism. Messy genetic algorithms [GOL89b, GOL90] use variable-length strings to avoid deception that can arise when promising building blocks have a large defining length in a chosen (fixed) encoding.

The modified genetic algorithm is effective because it permits the use of genetic encodings that change length during optimization. Successful features can be identified in simple designs described by few variables, and then included in designs of greater complexity. This requires a general and flexible representation language, in which reasonable designs can be described by different numbers of variables. Shape grammars form one suitable class of representation scheme. They are described in the next section.

SHAPE GRAMMARS

Shape grammars provide a formal method for describing complex topologies in terms of primitive elements

[STI80]. Specific grammars must be developed for different design domains, but each grammar includes four basic features: a set of shapes, a set of labels, shape transition rules, and an initial shape. A sequence of transition rules can be applied to generate complex assemblies of simple components. Labels are used to instantiate special features of each shape, such as size, material density or sites where additional shapes may be added.

Gero et al. have used a shape grammar in conjunction with a standard genetic algorithm to design efficient structural beam sections [GER94]. Their encoding permits evolution of the grammar itself (because the set of shape transition rules is implicitly included in the genetic string). Beam section shapes are evolved simultaneously, as the genetic operators also affect the sequence of shape transition rules encoded in the string. Final beam section shapes are limited by the fixed length of the genetic string, which forces a specified number of shape transition rules to be applied. When a variable-complexity encoding is used, promising shape transition rules can be efficiently identified when rule sequences are short. Subsequently, the high-performance rules can be recombined in longer sequences to produce complex beam section shapes.

Reddy and Cagan employed a shape grammar to represent topologies of structural trusses [RED93]. They seek the optimal topology by starting with the simplest possible shape, and then applying shape transition rules to produce more complex topologies. A simulated annealing algorithm is used for optimization, because the design space is non-convex. At each iteration of the algorithm, a single shape transition rule (randomly selected from all rules in the shape grammar) is appended to the sequence which describes the current design. This 'shape annealing' is an evolutionary process, but the only operator is random mutation. A genetic method, which also includes the powerful crossover operator, can provide the capability for more efficient search. A shape grammar has been used with the variable-complexity genetic algorithm, and complex truss geometries were successfully developed [GAG94a].

EXTENDED ENCODING AND REGULATORY GENES

When individuals of different complexity are allowed in the population, and there is a trend toward increasing complexity, it is possible for constraints that are initially inactive for the most promising individuals to become active in later generations. When variables are selected without regard to an inactive constraint, they may converge to values that are inappropriate when the constraint is active. Without diversity in the values for that variable, the population is unable to adapt to the new environment created by the change in the set of active constraints.

Biological precedent suggests an approach by which adaptation can be encouraged [GOU77]. Only a small fraction of the genetic material in biological systems is used to construct the organism. The locations along the string where decoding starts and stops are controlled by regulatory genes, which respond to environmental influences when determining whether their piece of the genetic string should be decoded or not. A switch in activity of even a single regulatory gene can have profound influence on the final organism.

There are two important aspects of this system to be incorporated in the decoding of variable-length genetic strings: unexpressed sections, and a switching mechanism affected by environmental influences to control the expression and suppression of different sections. In the simplest form of this scheme, the genetic string is extended to carry two values for each design variable. Selection of the copy to be expressed is controlled by constraint activity. Initially, the first value is expressed, and selection of promising building blocks proceeds as usual. There is no selection pressure on the unexpressed values, so they remain randomly distributed. When expression of the first value causes a constraint to become active, the value is ignored, and selection operates on the second value. The first value is retained, because it can be expressed in future generations if it is read when the constraint is not active. The extended genotype is able to produce different phenotypes in different environmental circumstances. The structure built from early sections of the string influences the environment in which later sections of the genetic string are decoded.

APPLICATION: TOPOLOGICAL DESIGN OF A NON-PLANAR WING

The goal in this optimization example is to minimize the drag of a wing that is constrained to lie within a box of specified span and height.

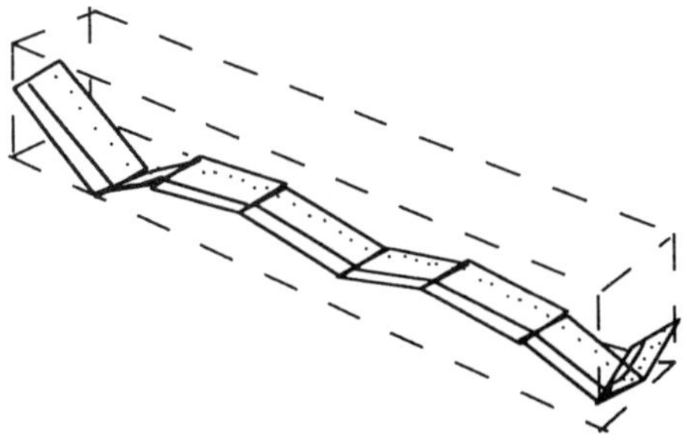

FIGURE 2. A wing must lie within a box of specified span and height.

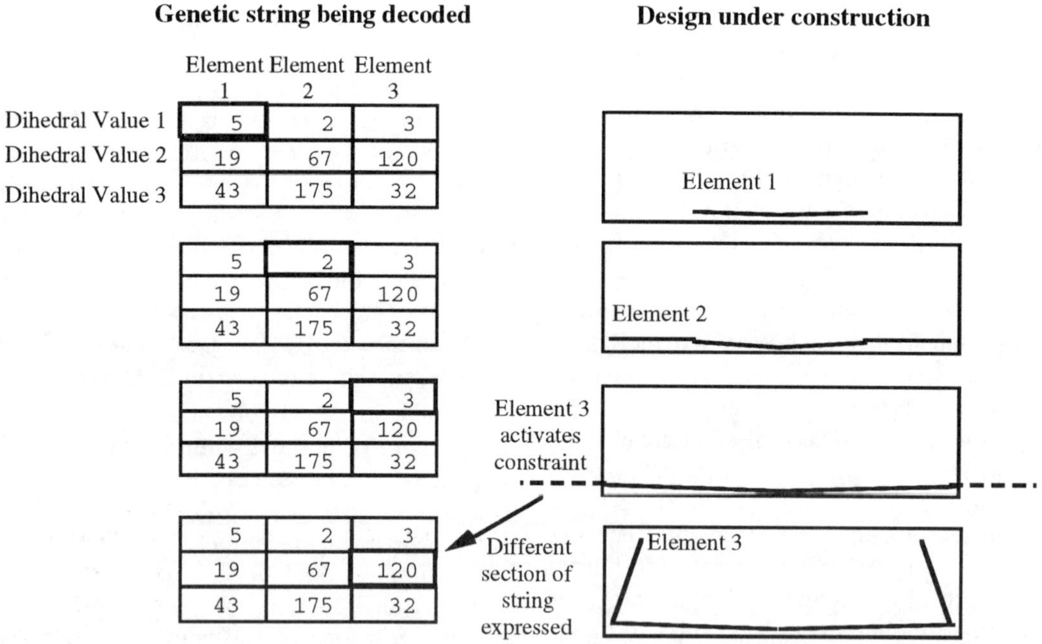

FIGURE 3. Extended genetic string and regulation of expressed sections.

This task is motivated by recent interest in highly nonplanar geometries for very large aircraft, for which span is limited by airport space restrictions. An analytic solution is known for the case of induced drag only (the wing extends around the perimeter of the box) [VON35], but no simple result is available when parasite drag is also considered. Calculus-based optimizers are ineffective in this non-convex design space [GAG94b]. The results presented here indicate that a genetic method can be successful, but the representation issues discussed in this paper strongly influence the quality of the solution.

A shape grammar is used to describe the wings. Primitive shapes are flat rectangular surfaces. Labels specify span and dihedral angle (inclination from the spanwise horizontal direction) for each component of the wing, and indicate where new components can be added. There are two classes of shape transition rules: the last element can be split in two pieces, or a new element can be added at its tip. Rule instances in each class describe the span and dihedral of the element being added. The initial shape is simply a point at the wing root (which may be considered a rectangular surface with zero span). A sequence of transition rules is applied, and the wing is constructed element by element. A standard genetic algorithm produces wings with a fixed number of elements, but the complexity of wings produced by the variable-complexity algorithm depends on the number of transition rules included in the genetic string.

The initial population produced by the variable-complexity algorithm includes designs with few elements, which do not violate the span and height constraints. These feasible designs are likely to be selected for reproduction. In later generations, unequal crossover produces more complex encodings, and a 'growing' wing experiences a change in environment when geometric constraints become active. To ensure that there is sufficient diversity to handle activation of the span and height constraints, the genetic string can be extended to include three copies of each design variable, in the manner described in the previous section.

Figure 3 shows how the string is decoded to construct a wing. The left side of the figure shows the encoding of dihedral values for each element (values of the span variables are omitted, to simplify the figure). The variable that is currently being decoded is indicated by a bold border. On the right is a front view of the wing being constructed, with the span and height limits marked by the exterior box. Wing elements are shown in bold, and labelled as they are added. When no geometric constraint is active, the first copy of each variable is expressed. The span constraint is encountered when the third element is added to the wing, so the second dihedral value is used. If the addition of further elements led to the activation of the height constraint, the third alternative for dihedral value would be decoded.

The key difference between the first and second alternative values for dihedral is that they are influenced by different selection pressures. Dihedral Value 1 is close to zero for all wing elements in Figure 3, because selection has favored elements of maximum span when geometric constraints are inactive. The values shown

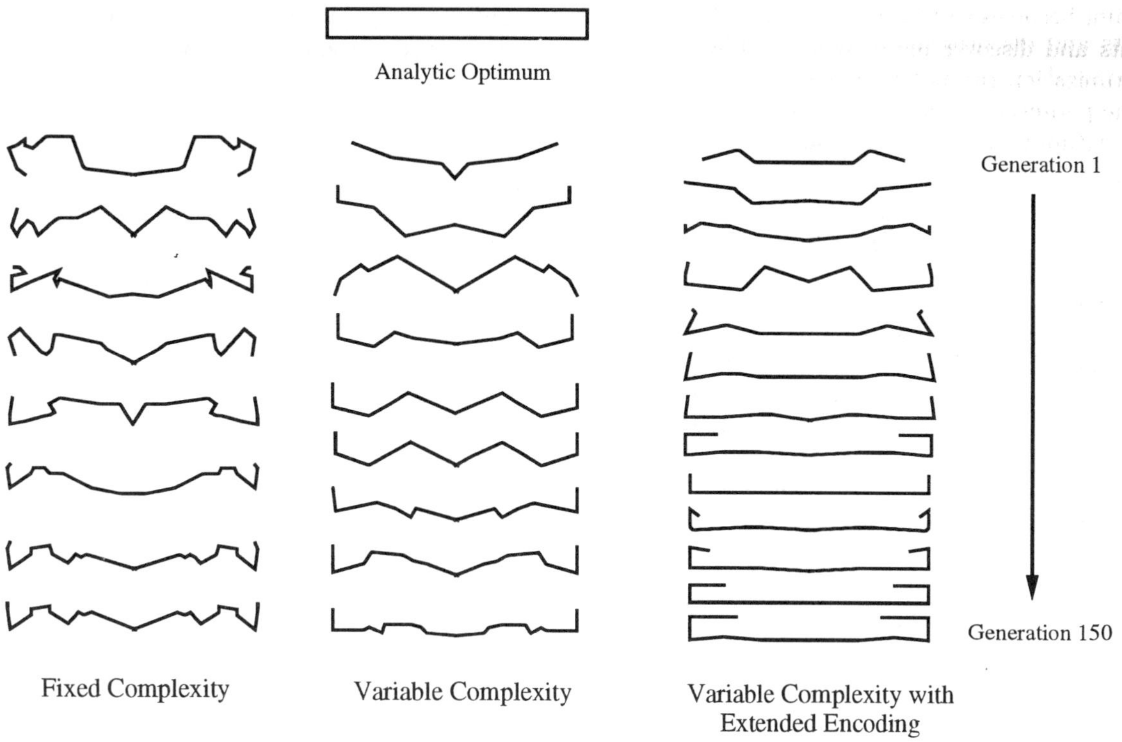

FIGURE 4. Wing optimization results for different genetic encodings.

for Dihedral Value 2 are still random, because selection did not influence this section of the string while it remained unexpressed. Subsequent generations will favor large dihedral values here, because they are more likely to produce feasible designs. This scheme enables selection pressures to operate on different parts of the genetic string in different environmental circumstances.

The importance of this encoding arrangement is illustrated by the comparison of results for standard and extended genetic strings, shown in Figure 4. Front elevations of the wings are displayed, with the best design in the initial population shown at the top of each column, and the best final design appearing at the bottom. Designs from intermediate generations included to demonstrate that the shapes evolve gradually. In this example, the semi-span of the wing is constrained to be less than 20 units, and the height must be less than 4 units. Up to 12 elements may be used to construct the half-wing (the wing is symmetric about the centerline), and each element has an arc-length between 1 and 8 units.

The first column of Figure 4 displays the optimization history for a standard genetic algorithm, which is forced to describe half-wings using 12 elements. The elements in the randomly-generated initial population have an average arc-length of 4 units, so the average half-wing has an arc-length of 48 units, to fit inside a box of semi-span 20, and height 4. The optimizer struggles to find any wings which fit inside the box. Elements of low span and moderate dihedral are favored. The selection pressure to satisfy the geometric constraints opposes the pressure to minimize drag. When the constraints have been satisfied using all available elements, it is difficult to allow further extension of the wing. Any extension can only be achieved by increasing the span of an existing element, which is likely to produce constraint violation.

For the variable-complexity algorithm, the average element in the initial population also has 4 panels. However the average half-wing has only 6 elements, and a total of 24 panels to fit within the box. It is much easier to satisfy the geometric constraints, and consequently there is more freedom to work on drag minimization. The second column of Figure 4 displays a history for the variable-complexity algorithm using a standard encoding. Diversity is quickly lost, and there is no dihedral value that can cope with the height constraint. Winglets are developed, and are retained throughout the optimization history, but winglet extensions are never discovered. There is never strong preference for flat elements, although dihedral values between 0 and 90 degrees are favored.

When there are redundant copies of each variable (Column 3 of Figure 4), constraint activity can cause expression of different copies. Diversity is retained in the third copy of the dihedral variables until the height constraint becomes active, so it is possible to add more elements and discover more complex designs. When the optimization run is terminated, a wing which follows the perimeter of the box has been developed. Its drag is within 0.5% of the analytically-determined optimum.

CONCLUSIONS

The variable-complexity algorithm extends the automated search capability provided by genetic optimization, because it avoids the requirement for a fixed parameterization of the design space. Shape grammars provide a representation scheme which can exploit this flexibility. They describe designs in terms of primitive elements, and complexity is directly related to the number of shape transition rules used in the description.

An extended encoding scheme, with several entries for each variable, aids the optimizer when the influences of the objective and the geometric constraints can be isolated. Expression of different sections of the string is controlled by environmental factors. Selection works on one part of the string when constraints are inactive, and at different points when constraints are activated. Distinct strategies are evolved to cope with changing environmental circumstances. This extended encoding scheme is an important development for evolutionary optimization in constraint-bound domains.

A shape grammar and an extended encoding were used in conjunction with a variable-complexity genetic algorithm to perform a wing optimization task. Near-optimal topologies were discovered in a domain where standard genetic algorithms and gradient-based optimizers are ineffective. This success demonstrates that these representation schemes should be further investigated in more complex applications.

ACKNOWLEDGEMENTS

This work was funded in part by the NASA Langley and NASA Ames Research Centers. Their support is gratefully acknowledged.

REFERENCES

[BEL91] Belew, R., Booker, L., (Editors), Proceedings of the Fourth International Conference on Genetic Algorithms, Morgan Kaufmann, 1991.

[GAG93] Gage, P., Kroo, I. "A Role for Genetic Algorithms in a Preliminary Design Environment" AIAA 93-3933, AIAA Aircraft Design, Systems and Operations Meeting, Monterey, CA, August 1993.

[GAG94a] Gage, P., Kroo, I., Sobieski, I., "A Variable-Complexity Genetic Algorithm for Topological Design", AIAA 94-4413 AIAA/NASA/USAF/ISSMO Symposium on Multidisciplinary Analysis and Optimization, Panama City, FL, Sept 7-9, 1994.

[GAG94b] Gage, P., "New Approaches to Optimization in Aerospace Conceptual Design", PhD thesis, Stanford University, 1994.

[GER94] Gero, J.S., Louis, S.J., Kundu, S., "Evolutionary Learning of Novel Grammars for Design Improvement", Artificial Intelligence for Engineering Design, Analysis and Manufacturing (1994), 8, 83-94.

[GOL89a] Goldberg, D., *Genetic Algorithms in Search, Optimization, and Machine Learning*, Addison Wesley, 1989.

[GOL89b] Goldberg, D.E., Korb, B., Deb, K. "Messy Genetic Algorithms: Motivation, Analysis and First Results", TCGA Report 89003, May 1989.

[GOL90] Goldberg, D.E., Deb, K, Korb, B. "An Investigation of Messy Genetic Algorithms", TCGA Report 90005, May 1990.

[GOU77] Gould, S.J., *Ever Since Darwin: Reflections in Natural History*, Norton, 1977.

[GRE87] Grefenstette, J.J., (Editor), Proceedings of the 2nd International Conference on Genetic Algorithms, Lawrence Erlbaum Associates, 1987.

[HOL75] Holland, J. *Adaptation in Natural and Artificial Systems*, University of Michigan Press, 1975.

[KOZ92] Koza, J.R., *Genetic Programming: On the Programming of Computers by Means of Natural Selection*, MIT Press, 1992.

[RED93] Reddy, G.M., Cagan, J. "Optimally Directed Truss Topology Generation Using Shape Annealing", DE-Vol. 65-1, Advances in Design Automation - Volume 1, ASME 1993 p749-759.

[SCH89] Schaffer, J., (Editor), Proceedings of the Third International Conference on Genetic Algorithms, Morgan Kaufmann, 1989.

[SMI80] Smith, S.F. "A Learning System Based On Genetic Adaptive Algorithms", PhD thesis, University of Pittsburgh,1980.

[STI80] Stiny, G. "Introduction to Shape and Shape Grammars", Environment and Planning B, 7, 1980, pp 343-351.

[VON35] von Karman, T.; Burgers, J. "Airfoils and Airfoil Systems of Finite Span", Vol II of Aerodynamic Theory, div. E, ch. IV, sec. 17, W.F. Durand, ed. Springer (Berlin) 1935.

Fuzzy Logic and Control

APPLYING A MODEL OF THE FUZZY CONTROL SOFTWARE DEVELOPMENT PROCESS IN PRACTICE

Pekka Isomursu, Tapio Rauma, and Seppo Kemppainen
VTT (Technical Research Centre of Finland) Electronics
P.O. Box 1100, FIN-90571 Oulu, Finland
Email: Pekka.Isomursu@vtt.fi

ABSTRACT

We first introduce a model of the fuzzy control software development process. We then show the validity of the model by discussing its use in real-world applications. We concentrate on features that best characterize the model, including managerial aspects, prototyping, and parameterization (a type of information hiding).

1. BACKGROUND

The theory of fuzzy sets [Zad65] can be applied to a wide variety of problems. A common denominator for these problems is their inherent vagueness or uncertainty. So far, the most important application area of the theory has been fuzzy control. A fuzzy logic controller (***FLC***) is a real-time expert system, implementing expertise that cannot easily be expressed in PID parameters or differential equations but rather in situation-action rules. In this paper we do not discuss the theory behind FLCs in detail. If needed, we encourage the reader to refer to other material such as [Lee90], [Zim91] and [Dri93].

While software can be developed without a model of the development process, having such model can make the development easier and, in some cases, be the crucial factor that enables successful completion of the system. No detailed models for the development of FLC software can be found in the literature. Models do exist for general software development, but for better results with FLCs we should recognize the special needs of this software category and tailor the model accordingly. The importance of such models is rapidly growing along with the increasing industrial interest in building FLCs.

To fill in this void we introduced a detailed model of the FLC software development process in [IKR94]. We based the model on and applied it to the development of FLCs in industrial applications for companies in the fields of process control (e.g. Imatran Voima, Kemira, Outokumpu Kokkola Zinc), machine automation (Kone Elevators), and telecommunications (Nokia Mobile Phones). These applications include temperature control of live steam at a power plant [IR94], control of granule size at a fertilizer plant [IT93], and controls of electronic and mechanical devices [Rau94].

In an earlier paper [IKR94] we specified the model in detail but did not discuss its practical use. In this paper our goal is to show validity of the model by discussing its use in real-world applications. We show how concepts such as parameterization and prototyping work in practice. Since certain parts of the model -- such as content and structure of documents -- follow the standard practices of general software development rather well, we do not discuss them in detail. Instead, we concentrate on features that best characterize the model and/or reflect the specific nature and the state of the art of FLCs.

2. INTRODUCTION TO THE MODEL

2.1 WHY A MODEL?

To develop successful industrial systems, problems must be solved that are often neglected in pure research and academic systems. These problems include cost-justifying a project, integrating the delivered system into an existing environment, and managing the user's and the development team's expectations. A good development model helps in coping with these factors. With FLCs, a development model is useful in the following ways:

- The development of the FLC becomes faster, especially for novice developers.
- Clear goals can be set with the work divided into subgoals. This makes better known the point of the project at which things should be done. The system can be developed as contract software.
- Overall, the development process is more structured, better understood, and easier to control. This helps

minimize risks: cost, development time, and efforts can be estimated and controlled better. Software quality improves.

- It is easier for companies to make the decision to use fuzzy logic because they can better manage the development process.
- Having a model helps even when one has to deviate from it; it forces developers to analyze the reasons for doing so.

The development model can be seen as a third, highest view of the design process (see Figure 1). The other two observation levels are the Development Methods & Tools level and the Design Parameters level. In the design of the FLC, typically only the two lower levels are considered and the model level ignored. However, reciprocal actions between all three layers should be considered: we need a model that depicts the life-cycle of an FLC so that we can better understand what to do and when to do it in the development process. When implementing the model, we need to decide what development methods or tools to use in each step of the development. When we choose a certain method or tool it restricts the selection of design alternatives we can use.

2.2 The Model

The model of the FLC software development process exploits general software development paradigms -- exploration, prototyping, evolution, the waterfall model, and the spiral model (more on these in, e.g., [Roy70], [Agr86], [Boe88], [Hum90], and [Why93]) -- selecting suitable features from each according to the needs of FLCs. The top level of the model is depicted in Figure 2.

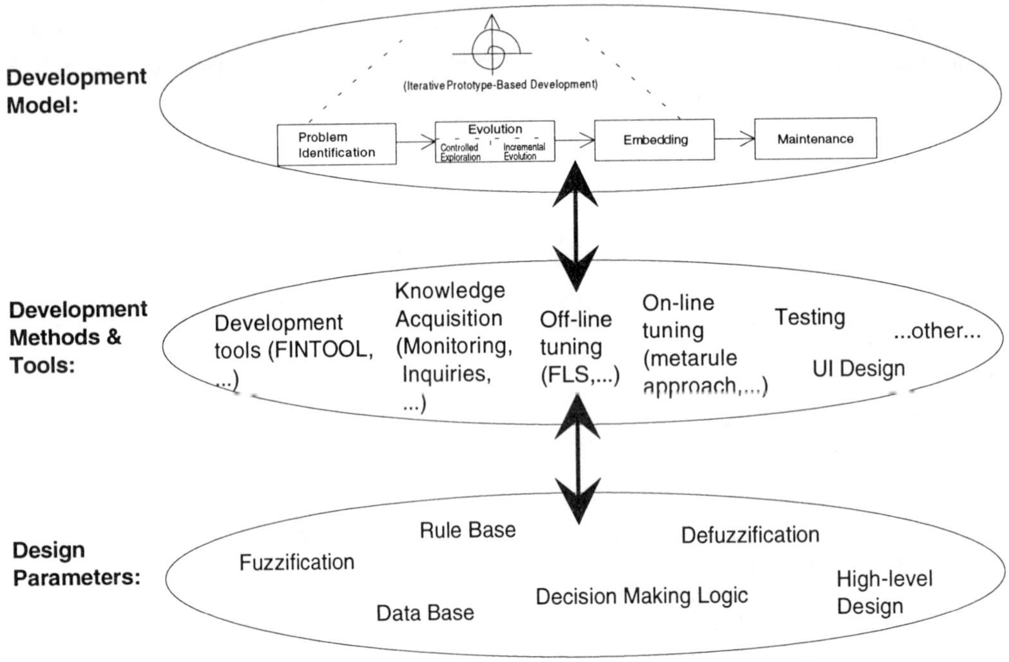

Figure 1. Three views of the software development process of FLCs

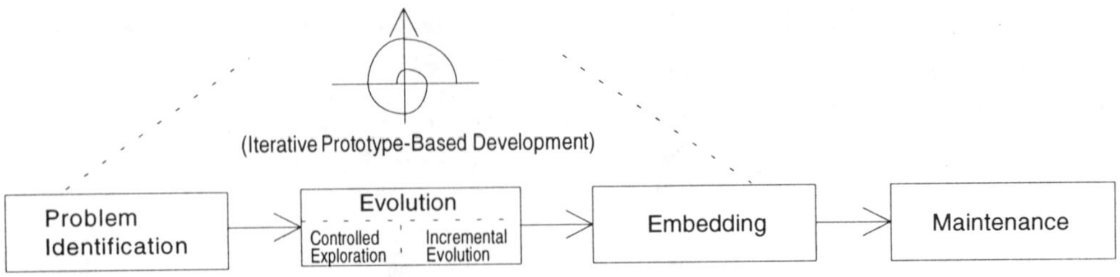

Figure 2. FLC software development model -- top level

To be able to meet *managerial needs* such as risk analysis, time schedules, and contract software development, the top level of the model is sequential, following the waterfall concept ([Roy70]). Inside the linear top level structure of the model, the development process can be seen as a spiral; the model is based on incremental development of prototypes. The prototypes can be grouped according to their role. *Demonstration Prototypes* give application domain experts an understanding of how an FLC works and what kind of knowledge they need to provide. Demonstration Prototypes also help determine whether fuzzy logic is a suitable approach to the problem in hand. *Exploration Prototypes* enable exploration of alternative FLC designs in a controlled fashion. *Evolution Prototypes* help incremental development, enhancement, and validation of essential system features.

Except for Exploration Prototypes, the prototypes are not thrown away; the next prototype is an extension of the previous one. Rather than being separate programs, different Demonstration and Evolution Prototypes are different development stages of the same program. Calling the stages "prototypes" emphasizes the different role that each stage has in the development.

Before embedding the FLC in its final environment, complex fuzzy design parameters are hidden from the user (i.e. *parameterized*). This is done so that the people who will maintain the finished FLC (operators and other application domain experts, maintenance personnel, users of fuzzy logic based consumer goods, etc.) need not understand fuzzy logic. Instead, they can operate in terms familiar to them.

Tables 1 through 4 and Figures 3 through 6 give a summary of the stages below the top level. These will be discussed in more detail via a case study in the next chapter. A more detailed description of the model can be found in [IKR94].

Table 1. Features of the Problem Identification stage

Start criteria:	Interest in fuzzy logic. Reason to believe that fuzzy logic might help solve a problem.
Participants:	Fuzzy logic expert(s). Application domain expert(s). Manager(s) from client.
Tasks:	Specify problem. Create Demonstration Prototype 1. Create Demonstration Prototype 2. Identify design alternatives for the FLC. Select development platform for Evolution stage. Analyze risks, estimate time and cost of development.
Documents:	Requirements specification. Project plan for further development (including time schedule, budget, estimation of risks, FLC design alternatives).
End criteria:	Problem identified and specified. Risks for further development analyzed.

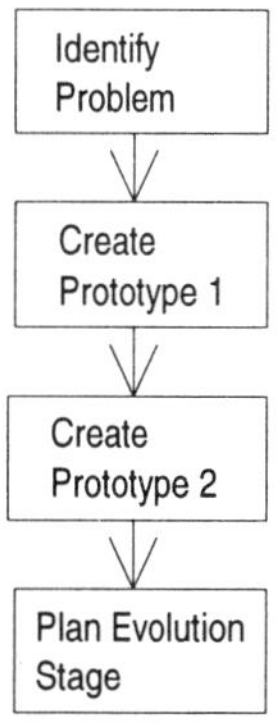

Figure 3. Phases in the Problem Identification stage

Table 2. Features of the Controlled Evolution stage

Start criteria:	Problem identified. Project plan & requirements specification completed.
Participants:	Fuzzy logic expert(s). Application domain expert(s). Manager(s) from the client.
Tasks:	Explore design alternatives. Incrementally create FLC with all essential features. Test and tune to validate essential features.
Documents:	Physical design specification (includes fuzzy design parameters). Enhanced project plan for Embedding and Maintenance stages.
End criteria:	Functionality of essential features has been verified (but not thoroughly tested). Project plan updated (risks, time, budget,...).

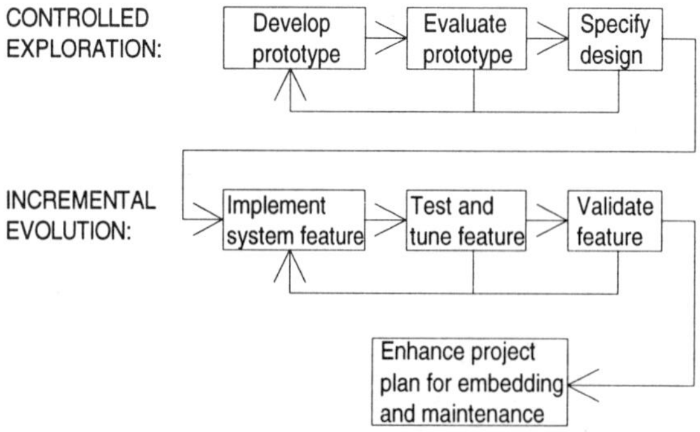

Figure 4. Phases in the Controlled Evolution stage

Table 3. Features of the Embedding stage

Start criteria:	Essential features of FLC validated.
Participants:	Fuzzy logic expert(s). Application domain expert(s). Manager(s) from the client.
Tasks:	Parameterization of FLC (i.e. hiding of complex fuzzy design parameters). Embedding of system in target environment. Off/on-line testing and tuning. Overall system tests. Evaluation of FLC performance.
Documents:	Test report. Operation manual. Service manual. Maintenance plan.
End criteria:	Acceptance tests done. Final product ready. Manuals ready.

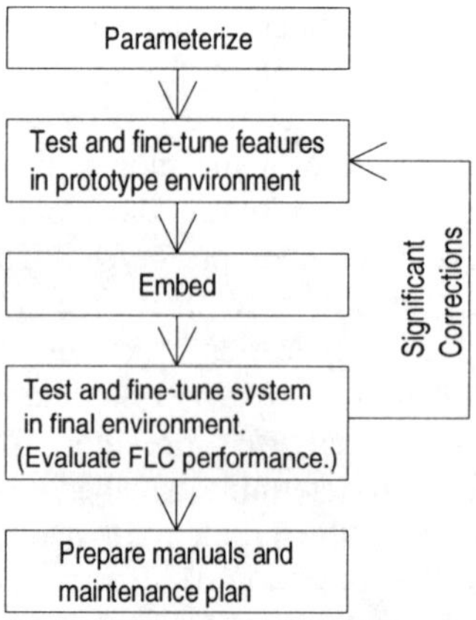

Figure 5. Phases in the Embedding stage

Table 4. Features of the Maintenance stage

Start criteria:	Final product and its documentation ready. Acceptance tests done. The FLC is reliable and in some respect better than earlier solutions.
Participants:	Operators and other application domain experts. Servicemen. (Use of fuzzy logic experts to be avoided as much as possible.)
Tasks:	Operate system and monitor its performance. Maintain system as needed.
Documents:	Operation report. Service report. Documentation of the changes made.
(End criteria:	A better technique is found and implemented.)

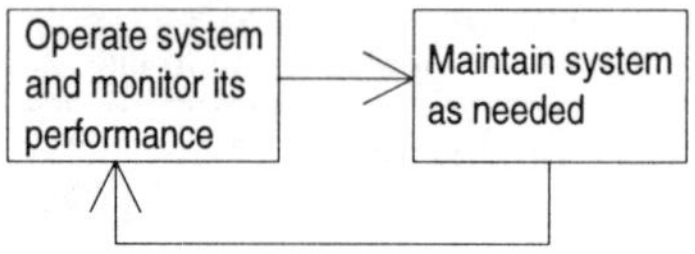

Figure 6. Phases in the Maintenance stage

3. APPLYING THE MODEL IN PRACTICE

We next discuss from a practical point of view the development of industrial FLCs using the model. We mainly concentrate on an FLC we have developed for controlling a chemical process at Outokumpu Kokkola Zinc. We discuss the development of other applications when they deviate from the experience we have had with this process. In the system description we concentrate on the structure of the implemented FLC. We do not discuss in detail the process controlled at Outokumpu because, from the point of view of the development model, it is rather irrelevant what the controlled process is.

3.1 SYSTEM DESCRIPTION

Figure 7 shows the structure of the MIMO (many-in-many-out) FLC that is currently in operation at Outokumpu. The FLC consists of three different modules. Their tasks are as follows:

FLC0 Basic Fuzzy Logic Controller

FLC1 Fuzzy Logic Module based on fuzzy meta rules for automatic tuning of the rule output membership functions of FLC0

FLC2 A fuzzy meta rule module that uses secondary (i.e., in a sense less essential) process parameters to correct the output of FLC0 in certain situations.

We use the following rule format in both the main rule base and meta-rules:

$$R_n: \quad IF\ x_1\ is\ A_{1n}\ AND\ x_2\ is\ A_{2n}\ \ldots\ AND\ x_k\ is\ A_{kn}\ THEN\ y_1\ is\ B_{1n} \qquad (1)$$

where A_{in} and B_{in} are linguistic fuzzy labels for variables x_i and y_i used respectively in rule R_n. The inference method used is max-product and the defuzzification method center-of-area (see [Lee90] for a more detailed discussion of the methods). We have previously used this modular structure in the development of other FLCs ([IR94], [Rau94]). Modularity makes rule derivation, tuning, and maintenance easier because the FLC can be developed one part at a time. [IRK94]

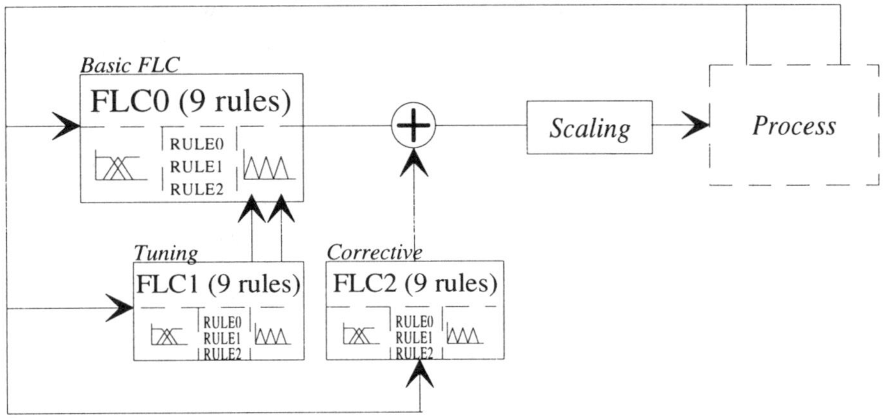

Figure 7. Structure of a modular FLC at Outokumpu Kokkola Zinc

3.2 Applying the Model

Top level:

In order to be able to control the development process from managerial point of view (time, cost, risks, etc.) it is divided into four sequential stages (see Figure 2). At the end of each stage an analysis is made to decide whether or not to move on to the next stage. This makes division of the development process into subprojects easy and the system can be developed as contract software.

In the Outokumpu case, the Problem Identification stage was a separate feasibility study that lasted one man-month. Based on its results (demonstration prototypes, risk estimation, requirements specification, etc.), it was decided to go on and develop the actual FLC. A new project was established that covered the Evolution and Embedding stages. At the end of this second project, a fully functional FLC had been implemented to control the actual real-world process. After using the FLC (i.e. after executing the Maintenance stage) for a while, a need arose for enhancement of the FLC. Thus, a third development project was established to do the necessary corrections to parameterization and the structure of the FLC as discussed below.

As was discussed in Section 2.2, inside the linear top level the development process can be seen as a spiral where the FLC is incrementally developed. We call different stages of the development different prototypes to emphasize the different role of the developed software (or pseudo code) at each stage. Exploration prototypes are an exception in that they form a group of independent programs that are used for exploration and then thrown away, except for one that is used for the development of Evolution Prototypes and finally the final FLC.

The spiral has cycles at two levels. The stages depicted in Figure 2 can be considered as master cycles; they include careful consideration of matters such as project plan, requirements specification, and risk analysis. The development of an individual prototype inside a stage is a minor cycle; it does not include systematic analysis of all development aspects unless a special reason arises to do so.

We next discuss each stage separately. For each stage, the reader may wish to refer to the corresponding descriptions given in Tables 1 through 4 and Figures 3 through 6 which provide the essential features of each stage in a nutshell.

Problem Identification stage:

The first step in the Problem Identification stage (Table 1 and Figure 3) is to identify exactly what the problem is. In the Outokumpu case this was easy to specify since the application experts had a specific problem in mind and the behavior of the process with respect to its inputs and outputs was well known. This is not always the case. In another project where control of a mechatronical device was enhanced with fuzzy logic [Rau94], problem identification took several months. The client could identify what the problem was but was unable to define what parameters cause it and how. Thus, before any FLC could be created, extensive experimentation and modeling of the process behavior had to be done. Also, in some cases the client may only have an idea that fuzzy logic might be of help to some of their problems and wants to test this technology on some unspecified problem. It is then the job of a fuzzy logic expert to obtain sufficient information from application experts and other sources to be able to specify a suitable problem.

The next step is to create a demonstration prototype (Demonstration Prototype 1) of the FLC. After discussing the problem with application experts, a few fuzzy rules were given to them in a written form. Because the rules are related to the actual problem, the experts can more easily relate to the kind of knowledge an FLC rule base needs from them and how an FLC works. This significantly lessens the acceptance threshold of this new technology and helps application experts to start "thinking fuzzy". [IKR94]

During the Problem Identification stage the prototype was incrementally developed as knowledge acquisition from the application domain experts, process monitoring data, and other sources continued. Prototype development helps the fuzzy logic expert in knowledge acquisition and writing both a project plan and a requirements specification. The documents prepared should include time schedule, budget, and risk estimation for further development.

The Problem Identification stage also includes identification of design alternatives for the FLC. Although much of the design can typically be determined at the Problem Identification stage, some alternatives often remain that should be analyzed more carefully. These potentially good designs should be pointed out at this stage and studied further at the beginning of the next (Controlled Evolution) stage. The design alternatives can be grouped as follows:

- ❑Fuzzy design parameters: fuzzification/defuzzification strategies, data base, rule base, decision making logic. A more detailed list is given in [Lee90].
- ❑High-level design: an FLC can be a pure fuzzy logic controller or, more typically, a hybrid. It may include an adaptation mechanism or some other special structure. The prototype may be connected to the real process off-line, on-line, or, more commonly, not at all.

In the Outokumpu case, the main design alternatives proposed were to have either a "standard" FLC using max-product inference [Lee90], or a predictive FLC following the ideas in [Yas85]. Another issue was whether to have one fuzzy rule base or a modular structure.

The final form of the prototype at Problem Identification stage is called Demonstration Prototype 2. It helps convince managers as well as developers that fuzzy logic is useful for the problem (or that it is not), so that a decision can be made whether or not to start the actual development project. In Outokumpu case, Demonstration Prototype 2 consisted of an FLC that could be run on a PC under a commercial fuzzy software development tool. Its operation was demonstrated with simulated data. This was sufficient to demonstrate how the final system would operate in certain situations, and point out what still should be done before the system would be ready.

As mentioned earlier, in the FLC for a mechatronic device [Rau94] the problem was harder to identify and, accordingly, the demand for the performance of Demonstration Prototype 2 was higher. In this case, Demonstration Prototype 2 also ran on PC but was already connected to run the actual target device. This was due to the nature of the problem: running the application in the real environment was the only way to verify that fuzzy logic could help in the problem.

Controlled Evolution stage:

The Controlled Evolution stage (Table 2 and Figure 4) has two main phases:

- ❑***Controlled Exploration:*** Based on the design alternatives identified in the project plan, explore FLC design by building "quick and dirty" prototypes. These Exploration Prototypes are thrown away except for one that is accepted as the basis for further development of the FLC.
- ❑***Incremental Evolution:*** Based on the selected design, incrementally develop all features of the system according to the requirements specification.

After some exploration, it soon became evident that the conventional inference approach would suit the Outokumpu case better than the predictive FLC. In this case the Controlled Exploration phase was quite brief. In other development projects we have had, this phase has often been longer. In an FLC for a telecommunications application (Nokia Mobile Phones), a number of design alternatives were tried out, the result being a hybrid system where in different modes of operation different fuzzy inference mechanisms and rule bases are used.

In the Incremental Evolution phase, a single fuzzy rule base was incrementally developed that could handle the process quite well. A corrective module was then added to it as shown in Figure 8.

If the FLC needs to be adaptive, it is often practical to first develop a non-adaptive prototype, validate its functionality, and only then add the adaptation mechanism. The mechanism is added as a new feature during the Incremental Evolution phase. The tuning module in Figure 7 that adapts the operation of the basic FLC during disturbances was constructed this way, although only at the maintenance stage because the need for the additional module was only then recognized.

During the Controlled Evolution stage the FLC was developed in a PC environment. Before the end of this stage, the FLC prototype was tested sufficiently to verify that the essential features of the FLC functioned (or could be made to function) properly. A list of possible risks, caused mainly by the lack of testing in the final environment, were updated in the project plan.

Embedding stage

The goal of the Embedding Stage (Table 3 and Figure 5) is to make the created prototype into a product. The main tasks are parameterization, testing and fine-tuning of system features, the actual embedding, and overall system testing and fine-tuning.

By *parameterization* we mean that the fuzzy design parameters should be made either invisible to the end user or easy to modify. This is because the end users cannot be expected to understand fuzzy logic. Possible end users include operators, other application domain experts, and service personnel who in one way or another may have to use the FLC. We have used three different approaches in parameterization:

- ***User Interfaces and the technologies related to them***
  We only show users the parameters that they need to know in order to understand and judge the operation of the FLC. The visible parameters should be presented to the users in an as intelligible a form as possible.
  In the Outokumpu case, the system was first tuned by fuzzy logic experts. After reaching a high level of tuning, certain parameters were identified that could be used to simplify fine-tuning. A tuning interface was created that shows only these selected parameters (Figures 8 & 9). Some flexibility is lost but the fine-tuning process becomes easier and more intuitively appealing.
  Figure 8 shows the relationship between the tuning interface and the actual FLC. The parameters that can be changed are the scaling factor of the FLC output, and the location of two output membership functions used in the basic rule base. (This interface was further improved during the Maintenance stage as described later in this paper.)
- ***Linguistic information about the tuning procedure***
  Linguistic information includes manuals and other written instructions or explanations about the operation of the FLC and the meaning of its parameters. These instructions could be implemented as part of the user interface.
  Writing good linguistic information is not as simple as it may seem; the effect of the chosen tuning parameters may be so complicated that it is hard to find clear, easy-to-use instructions for how they should be tuned. This emphasizes the importance of finding the right parameters for the parameterization. In the Outokumpu case, this lead to the development of a new tuning interface during the Maintenance stage.
- ***Automatic adjustment of design parameters***
  Automatic adjustment of design parameters (e.g. via a self-adaptation mechanism) lessens the need for manual fine-tuning. The tuning module in Figure 7 is an example of this approach.

The FLC was thoroughly tested, both in the development environment and in the target environment (an Alcont automation system). As depicted in Figure 6, the embedding was a gradual operation. After testing in prototype environment (PC), the still unfinished FLC was tested in target environment. Using the experience gained, further testing and fine-tuning was done on the PC, and again in target environment. Gradually, more and more of the work was performed in target environment.

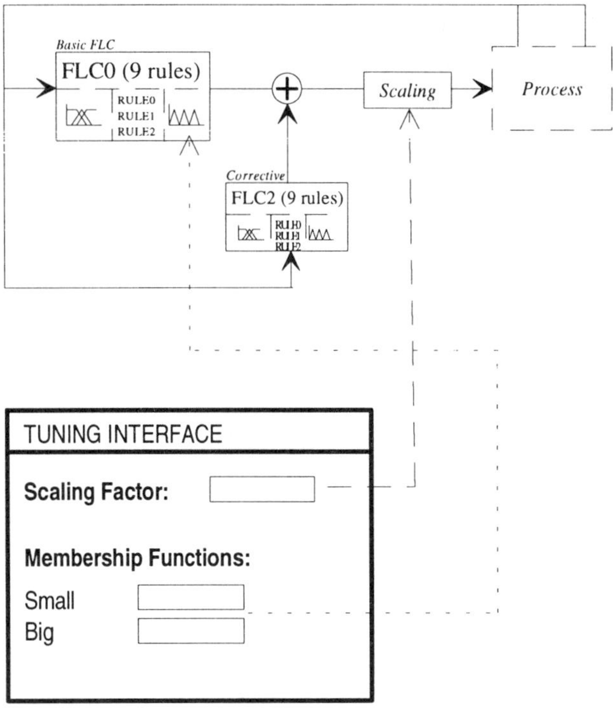

Figure 8. Structure of the FLC after the Controlled Evolution stage and the original tuning interface, created during the Embedding Stage

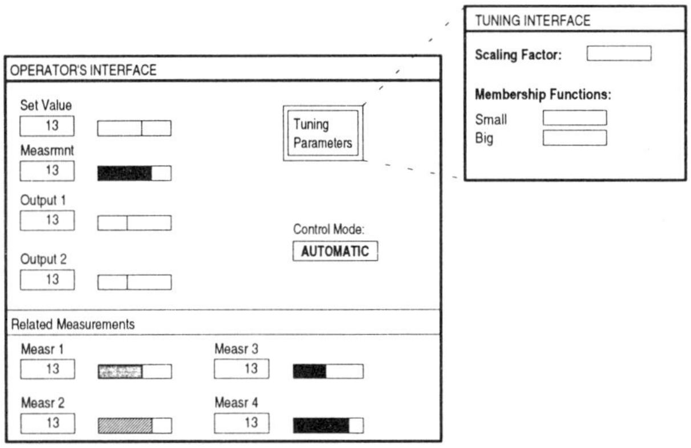

Figure 9. Operator's interface and the original parameterized tuning interface for the FLC, created during the Embedding Stage

How soon one can test in the target environment depends on the nature of that environment. In the Outokumpu case the process is very slow. If the FLC executes incorrectly an operator has plenty of time to switch the process to manual (or other non-fuzzy) control. In faster processes poor functioning of the FLC may be immediately disasterous. In such cases extensive tests have to be done in simulated environments before embedding.

The transformation from the Embedding stage to the Maintenance stage was gradual. The FLC was first run and fine-tuned in the target environment under a fuzzy logic expert's supervision. After sufficiently long test runs without need for further corrections or tuning, the system was taken into permanent use under the process operators' supervision. Operation instructions were written for both normal operation as well as for the tuning interface.

Maintenance stage:

Table 4 and Figure 6 show the main features of the Maintenance stage. In software development, the term "maintenance" denotes not simply an error-correction activity but may involve changes in requirements, design, and implementation, or it may highlight the need for further system testing. We can classify software maintenance as follows [Har92]:

❑***Adaptive Maintenance:*** Maintenance due to changes in the system environment.

❑***User Support:*** Answering questions from the user or correcting user misunderstanding of the operation of the system. The software itself and its documentation will not be modified.

❑***Corrective Maintenance:*** Correction of previously undiscovered system errors.

❑***Evolutive Maintenance:*** Adding new functionality to the software, in response to new or changed requirements.

❑***Perfective Maintenance:*** Changes demanded by the user that improve the system in some way without changing its functionality.

❑***Preventive Maintenance:*** Improving the maintainability of the software.

❑***Anticipative Maintenance:*** Anticipating problems with the use of the software.

The whole development process or parts of it may have to be repeated during the Maintenance stage as modifications are incorporated in the software. In Outokumpu case, one major maintenance operation had to be executed. During the use of the FLC, a need arose for a more intelligible tuning interface. As perfective maintenance, a new interface and new parameterization of the system variables were created (Figure 10). The operation corresponds to a repetition of the Controlled Evolution and Embedding stages.

While the original tuning interface worked rather well, its use of membership functions was somewhat cumbersome, even when textual explanation was provided of their meaning and use. A more intuitively appealing approach for the users is to use terminology which they are used to. The fuzzy membership functions in the user interface were changed to two new parameters: Performance goal and Maximum allowed change rate. A

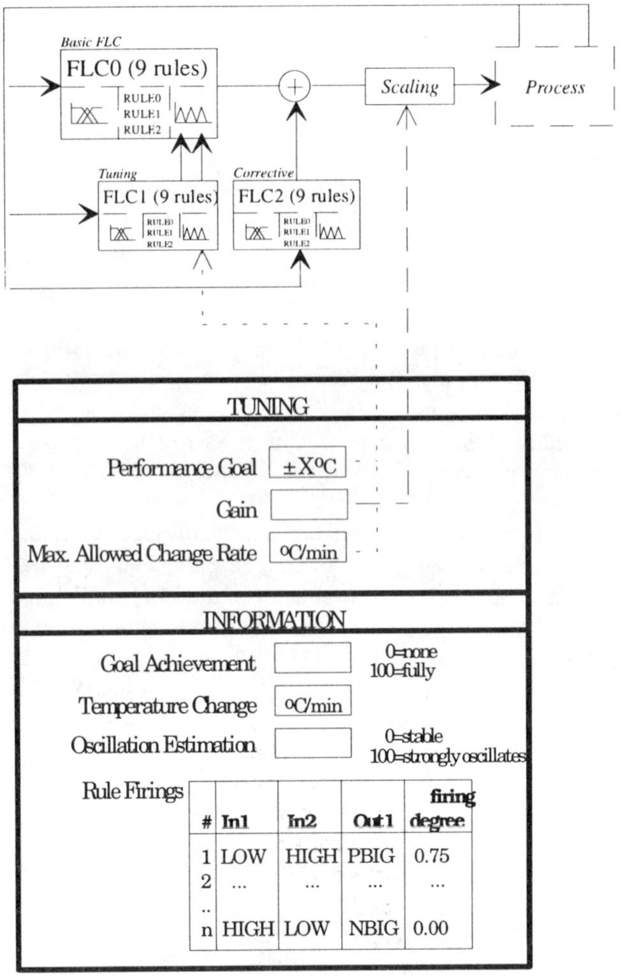

Figure 10. The improved, better parameterized operator interface for the final FLC

new module based on fuzzy meta-rules was added to the FLC that could intelligently interpret the given values into changes in the FLC operation. Information was added to the interface that helps the user estimate the FLC's performance (see Figure 10).

The FLC at Outokumpu has been taken into use and it has worked well. It has been in use for about 95% of the process operation time.

4. CONCLUSIONS

We have introduced a model of the fuzzy control software development process. The model exploits general software development paradigms -- exploration, prototyping, evolution, waterfall model, and the spiral model -- selecting suitable features from each according to the needs of FLCs.

To be able to meet *managerial needs* such as risk analysis, time schedules, and contract software development, the top level of the model is sequential. Inside the linear top level structure of the model, the development process can be seen as a spiral; the model is based on incremental development of *prototypes*.

We can identify three groups of prototypes: Demonstration, Exploration, and Evolution Prototypes. *Demonstration Prototypes* provide application domain experts understanding of how an FLC works and what kind of knowledge the experts need to provide to complete the system. Demonstration Prototypes also help determine whether fuzzy logic is a suitable approach to the problem at hand. *Exploration Prototypes* enable exploration of alternative FLC designs in a controlled fashion. *Evolution Prototypes* help incremental development, enhancement, and validation of essential system features.

Through *parameterization* we intend that the fuzzy design parameters should be made either invisible to the end user or easy to modify. Via a real-world example we have demonstrated how parameterization works in practice.

We have successfully applied the model to the development of FLCs in various industrial applications. In different applications, different parts of the model are emphasized. For example, depending on the control problem, the length of the Problem Identification stage can vary from a few days to a few months. Or, if the basic structure of the system becomes clear in the Problem Identification stage, the use of Exploration Prototypes is brief or may even be skipped.

In our experience, the model depicts the FLC software development process well. If, however, for some reason one has to deviate from the model, it still is helpful; having the model forces analysis of reasons for the deviation.

REFERENCES

[Agr86] Agresti, W.: What Are the New Paradigms? In: Agresti, W. (ed.) New Paradigms for Software Development, IEEE Computer Society Press, Washington D.C., 1986, pp. 6-10.

[Boe88] Boehm, B.: A Spiral Model of Software Development and Enhancement. IEEE Computer, May 1988, pp. 61-72.

[Dri93] Driankov, D., Hellendoorn, H. & Reinfrank, M. 1993. An Introduction to Fuzzy Control. Berlin: Springer-Verlag. 316 p.

[Har92] Harjani, D.-R. and Quelle, J.-P.: A Process Model for the Maintenance of Large Space Systems Software. Proceedings, IEEE Conference on Software Maintenance, Orlando, FL, November 1992, pp. 127-136.

[Hum90] Humphrey, W.S.: Managing the Software Process. Reading, MA, Addison-Wesley, 1990, 494 p.

[IKR94] Isomursu P., Kemppainen S., Rauma T.: A Model of Fuzzy Control Software Development Process. Proceedings, Second European Congress on Fuzzy and Intelligent Technologies (EUFIT'94), September 20-23, 1994, Aachen, Germany, pp. 335-341.

[IR94] Isomursu P., Rauma T.: A Self-Tuning Fuzzy Logic Controller for Temperature Control of Superheated Steam. Proceedings, IEEE World Congress on Computational Intelligence/FUZZ-IEEE'94, June 26 - July 2, 1994, Orlando, FL, pp. 1560-1563.

[IRK94] Isomursu P., Rauma T., Kemppainen S.: A Modular Approach to the Design of Fuzzy Logic Controllers. Proceedings, Seventh International Conference on Systems Research, Informatics & Cybernetics (ICSRIC94), August 15-21, 1994, Baden-Baden, Germany, Vol. II, pp. 85-90.

[IT93] Isomursu P., Tuomivaara S.: Design and Tuning of a Fuzzy Logic Controller. Proceedings, Sixth International Conference on Industrial & Engineering Applications of Artificial Intelligence and Expert Systems (IEA/AIE 93), June 1-4, 1993, Edinburgh, Scotland. pp. 469 - 475.

[Lee90] Lee, C. C.: Fuzzy Logic in Control Systems: Fuzzy Logic Controller, Parts I and II. IEEE Transactions on Systems, Man, and Cybernetics, Vol.20, No.2, March/April 1990, pp. 404-435.

[Rau94] Rauma, T.: Controlling the Starting of an Electromechanical Device with a Fuzzy Logic Controller. (In Finnish) Diploma thesis, University of Oulu, Department of Electrical Engineering, Oulu, Finland, 1994, 54 + 18 p.

[Roy70] Royce, W.W.: Managing the Development of Large Software Systems. Proceedings, Western Conference (WESCON), August 1970, pp. 1-9.

[Why93] Whytock, S.: The Development Life-Cycle. In: Thomé, B. (ed.) Systems Engineering -- Principles and Practice of Computer-Based Systems Engineering. John Wiley & Sons, 1993, 394 p.

[Yas85] Yasunobu S., Miyamoto, S.: Automatic Train Operation System by Predictive Fuzzy Control. In Sugeno, M. (ed.): Industrial Applications of Fuzzy Control, 1985, pp.1-18

[Zad65] Zadeh, L.A.: Fuzzy Sets. Information and Control, Vol. 8., 1965, pp. 338-353

[Zim91] Zimmermann H.-J.: Fuzzy Set Theory And Its Applications. Second, Revised Edition. Kluwer Academic, The Netherlands, 1991, 399 p.

DEVELOPMENT OF A FUZZY LOGIC CONTROLLER AS A POWER SYSTEM STABILISER USING HIERARCHICAL STRUCTURE

J.Shi
Department of Electrical and Electronic Engineering
Victoria University of Technology
P.O.Box 14428, MMC, Melbourne, 3000, Australia
E-mail: joanne@cabsav.vut.edu.au

ABSTRACT:

A fuzzy logic controller (FLC) is designed as a power system stabiliser due to its simplicity and robustness. The input and output membership functions were adjusted according to an evaluation index in order to achieve the optimal performance. The initial FLC design with a complete filled rule table (49 rules) can be reduced using neural network techniques to improve the dynamic performance. However, the simplified FLC does not work well when the power system is subject to a large disturbance such as a three-phase fault. The FLC with hierarchical structure consists of a multilayer controller which incorporates rule based FLCs for direct control at the lower level. The proposed FLC with hierarchical structure has improved the system performance under both small and large perturbations at various operating conditions.

1 INTRODUCTION

One of the most important power system stability problem is the low-frequency oscillations of the interconnected electric power systems. The oscillations may be sustained for minutes and grow to cause system separation, if no adequate damping is available. The application of the state-of-the-art power system stabiliser (PSS) for improving stability has received much attention [LS81, AF77, DC88]. The most widely used conventional PSS is the lead-lag compensator where the gain settings are fixed and determined for a particular operating condition. This means that conventional PSS can provide optimal performance only for that particular operating point. Since the operating point of a power system drifts as a result of continuous load changes or unpredictable major disturbances such as a three-phase fault, fixed gain conventional PSS can not provide the best response in real time. Although general parameters can be decided for a conventional PSS according to a particular range of operating conditions, the design procedure is very complex. For multimachine systems, individual machines may have different requirements of damping torque and synchronising torque, and multi-machine multi-mode power system stabilisation techniques must be sought. These constraints made the design of the conventional PSS even more complex. A self-tuning PSS has been employed to adapt the stabiliser to maintain good dynamic performance over a wide range of operating conditions [HL87, CMH88]. However, self-tuning PSS is difficult for realisation because it requires parameter identification, state observation and feedback gain computation which are very time consuming.

Lately, simpler alternative control schemes for stabilisation of power systems were proposed, they are the fuzzy logic based power system stabiliser (FLBPSS) [HS91, HL89, HMH91, SHK94] and the traditional fuzzy logic controller (FLC) [SHK94*]. These schemes are very suitable for on-line control due to their lower computational burden and simplicity. A comparison study between a FLBPSS and a FLC [SHK94*] shows that the FLBPSS is more suitable due to its robustness. The FLBPSS has been successfully implemented in real-time for a one-machine infinite bus and a two-machine infinite bus power systems [SHK94*].

This paper presents the development of FLC with hierarchical structure to improve the FLC performance in the power system under both small and large disturbance conditions at various operating conditions.

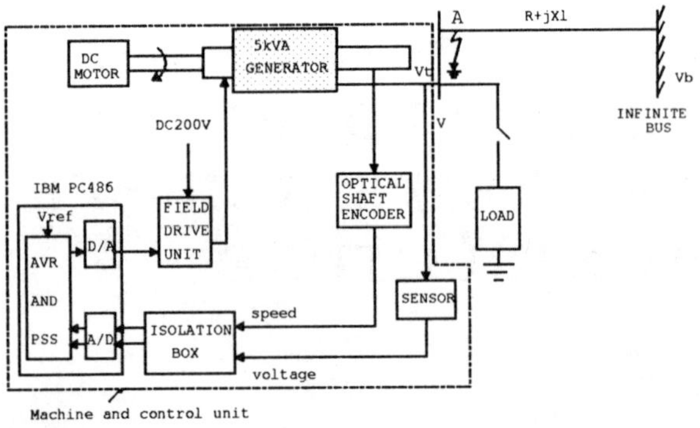

Figure 1: System configuration of single-machine infinite bus power system

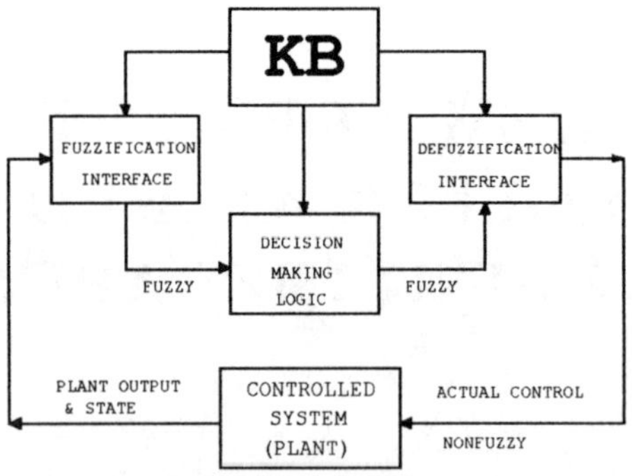

Figure 2: Basic configuration of a fuzzy logic controller

2 ONE-MACHINE INFINITE BUS POWER SYSTEM

The system configuration for one-machine infinite bus system is shown in Figure 1. Since this power system has an oscillating frequency (measured experimentally) of approximately 1.5Hz, the sample time for the digital AVR and the FLBPSS can be chosen as less than 60 milliseconds. A sampling time of 25 milliseconds has been chosen in the present work. The parameters of the synchronous generator and transmission lines are given in APPENDIX.

3 PROPOSED FUZZY LOGIC CONTROLLER

3.1 Introduction to Fuzzy Logic Controller (FLC)

A key part of the FLC is a set of linguistic control rules related by the dual concepts of fuzzy implication and the compositional rule of inference. The FLC provides an algorithm which can convert the linguistic control strategy based on expert knowledge into an automatic control strategy. Figure 2 shows the basic configuration of a FLC, which comprises of four principle components: fuzzification interface, knowledge base, decision-making logic, and defuzzification interface [Lee90].

3.2 Design of the FLC as a PSS

The design of the FLC as a PSS can be separated into the following four steps:

- Step 1) Definition of the input and control variables, i.e. determine which states of the process should be observed and which control actions are to be taken. When applying the FLC as a PSS, generator speed deviation $\Delta\omega$ and acceleration $\Delta\dot{\omega}$ can be observed and have been chosen as the input signal of the FLC. The control variable is the output from the FLC.
- Step 2) Fuzzification strategies and interpretation of a fuzzification operator (fuzzifier), i.e. to transform crisp data into fuzzy sets. Figure 3 shows the membership function for the FLC inputs.
- Step 3) Design of the rule base. The decision table can be defined as follows:

  1. If change of speed and acceleration are 'zero' (in the sense of the linguistic labels), the control applied is 'zero' (also in the sense of the respective linguistic label of control).
  2. If change of speed is 'positive big' and the same as the acceleration, a 'positive big' control has to be used to reach the state in which change of speed and acceleration are reduced to zero. (decelerating control is achieved by applying a positive stabilising signal to the excitation loop).
  3. If change of speed is 'negative big' and the same holds for the acceleration, the control is set to generate negative value (i.e., negative big). (accelerating control of the study unit can be achieved by applying a negative stabilising signal to the excitation loop).

  ⋮

Table 1 shows the complete decision table. The abbreviation used for linguistic variables above mean:
PB: Positive Big, **PM**: Positive Medium, **PS**: Positive Small, **ZR**: Zero
NS: Negative Small, **NM**: Negative Medium, **NB**:

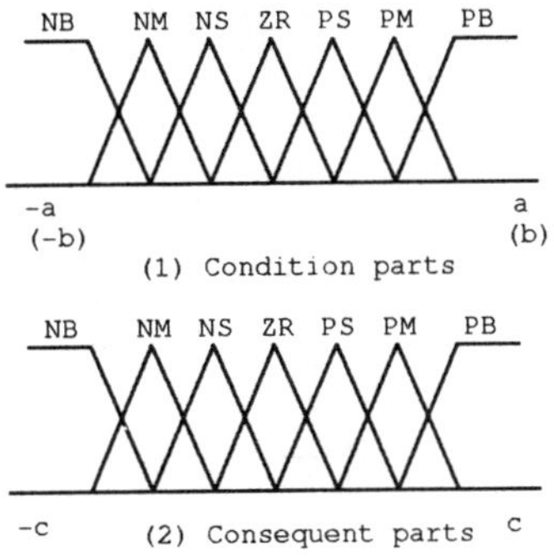

Figure 3: Input and output membership functions for the FLC

Negative Big

- Step 4) Determination of the stabiliser output by using defuzzification. The output membership function is also shown in Figure 3. The centroid and means of maximum defuzzification methods were used in this design.

4 OPTIMISATION OF THE DESIGNED FLC

In order to achieve the best performance for the proposed FLC, the membership functions, rules or defuzzification method have to be tuned according to a performance index. Two aspects of evaluating the PSS performance are maximum transient deviation and settling time. The performance index J is defined as in equation (1):

$$J = \sum_{k=1}^{M} [t_k \Delta\omega(k)]^2 \qquad (1)$$

In the above index, the speed deviation $\Delta\omega(k)$ is weighted by the discrete time $t_k = k\Delta T$. M in the above index stands for the total data number.

$\Delta\omega$				$\Delta\dot{\omega}$			
	NB	NM	NS	ZR	PS	PM	PB
PB	**ZR**	**PS**	**PM**	**PB**	**PB**	**PB**	**PB**
PM	**NS**	**ZR**	**PS**	**PM**	**PM**	**PB**	**PB**
PS	**NM**	**NS**	**ZR**	**PS**	**PS**	**PM**	**PM**
ZR	**NM**	**NM**	**NS**	**ZR**	**PS**	**PM**	**PM**
NS	**NB**	**NM**	**NS**	**NS**	**ZR**	**PS**	**PM**
NM	**NB**	**NB**	**NM**	**NM**	**NS**	**ZR**	**PS**
NB	**NB**	**NB**	**NB**	**NB**	**NM**	**NS**	**ZR**

Table 1: Decision table

4.1 Fine tuning of the designed FLC membership function and the choice of defuzzification method

By adjusting the width of the membership function, the performance of the FLC has been improved. However, the membership function adjustment based on trial and error is very time consuming. In order to improve the efficiency, the choice of a parameter to evaluate the width selection of the membership function is needed. The absolute value of the width is not proper for this goal as it does not compare the width of the separate membership function with the number of classes and the universe of discourse. Marsh [Mot92] has proposed two indices which meet this demand: the overlap ratio and the overlap robustness. These indices evaluate the width of the membership functions through the overlap of two adjacent functions. The idea is very fruitful because it allows to compare the scope of the separate membership function with the universe of discourse and the number of classes.

Marsh [Mot92] considered only linear membership functions and these indices work very well in this case. Their disadvantage is that although they can be easily calculated for membership functions with the limited discourse area, they can not be calculated for functions with the infinite discourse area for example the exponential (Gaussian) membership function. Moreover, the first index does not depend on the membership function shape at all and the second one depends on the shape in the overlap area only.

For these reasons the index of the Whole Overlap Ratio [RS93] is proposed in equation (2):

$$WOR = \frac{\int_{-\infty}^{\infty} MIN(\mu_1(x), \mu_2(x))dx}{\int_{-\infty}^{\infty} MAX(\mu_1(x), \mu_2(x))dx} \qquad (2)$$

where $\mu_1(x)$ and $\mu_2(x)$ are two adjacent membership functions.

The evaluation formulas for this index and some usual membership functions (MF) are given in Table 2.

Where $F(x) = \frac{1}{\sqrt{2\Pi}} \int_{-\infty}^{x} exp(\frac{-u^2}{2})du$ in Table 2 is the cumulative standard normal distribution function.

Centroid and means of maximum defuzzification methods are used for the FLC and the results are also compared by using the above performance index.

Table 3 shows the performance index corresponding to changes of both input and output membership functions WOR in response to a 20% step increase in mechanical torque T_m. When the centroid defuzzification method was used, it can be seen that the FLC gives optimal performance when the WOR is approximately

MF	Formula
Linear	$\mu(x) = 1 - abs(\frac{x-m}{\delta})$
Exponential	$\mu(x) = exp[-(\frac{x-m}{\delta})^2]$
Quadratic	$\mu(x) = max[0, 1 - (\frac{x-m}{\delta})^2]$
MF	Whole Overlap Ratio
Linear	$\frac{1}{\frac{2}{(1-\frac{m_2-m_1}{2\delta})^2}-1}$
Exponential	$\frac{1}{F(\sqrt{2}\cdot\frac{m_2-m_1}{2\delta})} - 1$
Quadratic	$\frac{\frac{2}{3}-\frac{m_2-m_1}{2\delta}+\frac{1}{3}(\frac{m_2-m_1}{2\delta})^3}{\frac{2}{3}+\frac{m_2-m_1}{2\delta}-\frac{1}{3}(\frac{m_2-m_1}{2\delta})^3}$

Table 2: Some usual membership functions and evaluation formulas

WOR%	30.95	24.27	18.79	14.29
J% (Centroid)	*1.05*	*0.78*	*0.75*	*0.69*
J% (Means)	*1.05*	*0.78*	*0.75*	*0.69*
WOR%	13.48	12.72	10.58	7.6
J% (Centroid)	*0.69*	*0.69*	*0.71*	*0.74*
J% (Means)	*0.69*	*0.69*	*0.71*	*0.74*

Table 3: The performance index corresponding to changes in WOR (20% step change in mechanical torque)

from 12.72% to 14.29%. The same results were obtained with means of maximum defuzzification method which are also shown in Table 3.

Table 4 shows the performance index corresponding to changes of both input and output membership functions WOR in response to a three-phase fault at point A as shown in Figure 1 sustained for 125 milliseconds. Means in Table 3 and 4 stands for means of maximum defuzzification method.

From Tables 3 and 4 it can be seen that the two defuzzification methods give the same results under small perturbation and means of maximum defuzzification method performs better in the case of a large disturbance condition. Considering both the small and large perturbation, the best WOR has been chosen as 14.29% and the defuzzification method chosen is means

WOR %	30.95	24.27	18.79	14.29
J% (Centroid)	*40.24*	*40.00*	*39.48*	*39.51*
J% (Means)	*40.24*	*39.89*	*39.39*	*39.40*
WOR%	13.48	12.72	10.58	7.6
J% (Centroid)	*39.58*	*39.78*	*39.77*	*39.81*
J% (Means)	*39.43*	*39.48*	*39.51*	*39.61*

Table 4: The performance index corresponding to changes in WOR with different defuzzification methods (three-phase to ground fault)

WOR %	30.95	24.27	18.79	**14.29**
J% ($H = 1.2s$)	*1.11*	*0.70*	*0.68*	**0.62**
J% ($H = 0.8s$)	*1.01*	*0.84*	*0.82*	**0.77**
J% ($\tau'_{d0} = 0.414s$)	*1.26*	*0.93*	*0.90*	**0.81**
J% ($\tau'_{d0} = 0.276s$)	*0.87*	*0.65*	*0.63*	**0.58**
WOR %	10.58	7.6	5.1	3.2
J% ($H = 1.2s$)	*0.65*	*0.71*	*0.85*	*24.60*
J% ($H = 0.8s$)	*0.78*	*0.80*	*0.86*	*2.18*
J% ($\tau'_{d0} = 0.414s$)	*0.84*	*0.89*	*1.02*	*3.83*
J% ($\tau'_{d0} = 0.276s$)	*0.60*	*0.62*	*0.70*	*6.96*

Table 5: The performance index corresponding to changes in WOR (with parameter variations)

OC	P0	P1	P2	P3
P (pu)	*1*	*0.5*	*0.8*	*1.2*
pf (lag)	*0.85*	*0.95*	*0.95*	*0.95*
OC	P5	P6	P4	
P (pu)	*1.5*	*1* ($X_e = 0.48pu$)	*1* ($X_e = 0.32pu$)	
pf (lag)	*0.95*	*0.85*	*0.85*	

Table 6: Operating conditions

of maximum.

4.2 Sensitivity to nominal parameter variations

The nominal operating point of the system is considered with real power P of 1pu at power factor of 0.85 (lagging). The sensitivity test was performed by changing the nominal system parameters (inertia constant H and the d-axis circuit time constant τ'_{d0}) by ±20%..

Table 5 shows the performance index corresponding to changes of both input and output membership function WOR in response to a 20% step increase in mechanical torque T_m with system parameter changes. It is seen that the WOR value chosen is insensitive to system parameter changes.

4.3 Robustness to operating condition variations

In order to test the robustness, various operating conditions have been applied. Those operating conditions (OC) are shown in Table 6, where P0 is the nominal operating condition as used before and operating conditions P5 and P6 corresponding to ±20% variations in the transmission line inductance.

Table 7 shows the performance index corresponding to changes of both input and output membership function WOR in response to a 20% step increase in mechanical torque T_m under different operating conditions. It shows that the WOR value chosen works

WOR %	30.95	24.27	18.79	**14.29**
J% ($P0$)	1.05	0.78	0.75	**0.69**
J% ($P1$)	1.33	1.07	1.04	**0.97**
J% ($P2$)	0.97	0.73	0.70	**0.64**
J% ($P3$)	0.96	0.67	0.64	**0.58**
J% ($P4$)	5.01	2.02	1.92	**1.70**
J% ($P5$)	1.17	0.87	0.84	**0.77**
J% ($P6$)	1.02	0.76	0.73	**0.67**
WOR %	10.58	7.6	5.1	3.2
J% ($P0$)	0.71	0.74	0.85	4.70
J% ($P1$)	1.00	1.04	1.14	1.81
J% ($P2$)	0.66	0.70	0.80	1.72
J% ($P3$)	0.60	0.64	0.75	37.53
J% ($P4$	1.75	1.86	2.33	96.50
J% ($P5$)	0.79	0.83	0.93	47.62
J% ($P6$)	0.69	0.73	0.83	2.45

Table 7: Performance index corresponding to changes in WOR under different operating conditions

$\Delta\omega$				$\Delta\dot{\omega}$			
	NB	NM	NS	ZR	PS	PM	PB
PB							
PM			**ZR**	**PM**	**PM**	**PB**	
PS		**NS**	**ZR**	**PS**	**PM**	**PB**	
ZR		**NM**	**NM**	**ZR**	**PM**	**PM**	
NS		**NB**	**NM**	**NS**	**ZR**	**PS**	
NM			**NM**	**NM**	**NS**		
NB			**NB**	**NB**			

Table 8: The decision table for the simplified FLC

WOR%	30.95	24.27	18.79	14.29
J% (49-rule)	1.05	0.78	0.75	0.69
J% (24-rule)	0.65	0.46	0.45	0.42
WOR%	10.58	7.6	5.1	3.2
J% (49-rule)	0.71	0.74	0.85	24.60
J% (24-rule)	0.42	0.43	0.46	2.18

Table 9: The performance index corresponding to changes in WOR with the 49-rule and 24-rule FLC

well under various operating conditions. Similar results were obtained for large disturbance condition, such as a three-phase to ground fault.

4.4 Rules reduction for FLC using neural networks

There is an increasing amount of literature dedicated to the use of neural networks for learning rules from input/output data. An example for obtaining the rules to stabilise an inverted pendulum is given by Kosko [Kos92]. A human operator is taking actions in order to move the pendulum to the vertical position. The set of input (angular deviation and its derivative) and output (operator's action) pairs are the data fed into a neural network to perform Adaptive Vector Quantisation (AVQ). The obtained rules in correlation with predefined membership functions are used to define a fuzzy system for which the control surface is approximately the one used by the human operator in stabilising the pendulum.

Patterned after this approach, TILGen [Tog91] is a tool that automatically generates a fuzzy expert system from a set of inputs and required outputs. TILGen is a combination of two techniques: a Neural Network to learn the input-output function described by the set of inputs and outputs, and a technique to decode the neural network to generate the rules for the fuzzy expert system.

Comparing power system stabilisation with the inverted pendulum example, there is no human operator performing the stabilisation of the power system under disturbance conditions. Thus no data can be derived for a predefined control surface. An initial FLC was designed by filling in all the cells of the decision table expressing the knowledge about the system. Here, a neural network is used to simplify the initial FLC structure. The input signals which were generated by applying disturbances such as a 20% step change in mechanical torque T_m and a 20% step change in reference voltage V_{ref} to the power system have been applied to the original 49-rule FLC to obtain the output control signal. Thus the input/output pairs were not obtained from a human-based control but from the excitation of the initial FLC with output signals (speed and acceleration) from the power system . Using this input-output data, TILGen generated the 24 rules decision table presented in Table 8 for the simplified version of the FLC.

Table 9 shows the performance index corresponding to changes of membership function WOR in response to a 20% step increase in mechanical torque T_m under the nominal operating condition with the original 49-rule FLC and the reduced 24-rule FLC. It is seen from Table 9 that the 24-rule FLC performs better than the original 49-rule FLC. The explanation for this is that the starting FLC was not shaping an optimal control surface since some of the cells of the starting FLC are an extrapolation/interpolation of the cells filled in by experience. However, the 24-rule FLC does not perform well under large disturbance condition. The following section describes the development of FLC with hierarchical structure to improve the FLC overall performance.

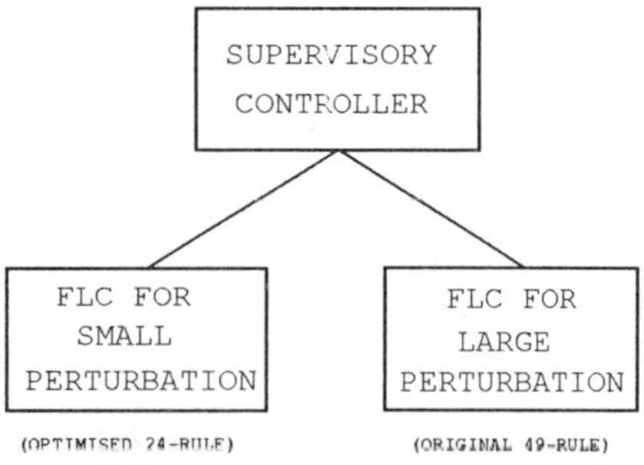

Figure 4: Structure of the hierarchical fuzzy controller

5 DEVELOPMENT OF THE FLC AS A PSS USING HIERARCHICAL STRUCTURE

The hierarchical fuzzy logic control structure consists of a multi-layer controller which incorporates rule based FLCs for direct control at the lower level. Higher level supervisory control is used for supervising the operation of the direct controllers. Typical tasks of the supervisory controller are monitoring the power system and the controller for determining the operating modes, assignment of initial controller and adaptation scheme. There are two FLCs at the lower level, one is responsible for large perturbation at various operating conditions whereas the other for small perturbation. Fuzzy logic for upper level (expert system approach to supervisory control) determines which FLCs to choose. Figure 4 illustrates the structure of the hierarchical fuzzy controller.

The supervisory controller monitors the value of the speed deviation. When the power system is subject to a large perturbation, the value of the speed deviation is very high and vice versa. If the value of the speed deviation is high, the supervisory controller switches the bottom layer controller to the FLC with 49 rules in order to cover all the operating range with full rule table. If the value of the speed deviation is low, the supervisory controller choses the FLC with 24 rules to obtain the optimal performance.

Fuzzy controller	P0	P1	P2	P3
J%(49-rule FLC)	*0.69*	*0.97*	*0.64*	*0.58*
J%(24-rule FLC)	*0.42*	*0.64*	*0.39*	*0.35*
J%(Hierarchical FLC)	*0.42*	*0.64*	*0.39*	*0.35*
Fuzzy controller	P4	P5	P6	
J%(49-rule FLC)	*1.70*	*0.77*	*0.67*	
J%(24-rule FLC)	*0.89*	*0.48*	*0.40*	
J%(Hierarchical FLC)	*0.89*	*0.48*	*0.40*	

Table 10: Comparison of the performance index of the FLC and the Hierarchical FLC (small perturbation)

Fuzzy controller	P0	P1	P2	P3
J%(49-rule FLC)	*2.48*	*1.62*	*1.58*	*2.30*
J%(24-rule FLC)	*1.78*	*1.15*	*1.14*	*1.64*
J%(Hierarchical FLC)	*1.78*	*1.15*	*1.14*	*1.64*
Fuzzy controller	P4	P5	P6	
J%(49-rule FLC)	*7.01*	*2.78*	*2.38*	
J%(24-rule FLC)	*4.03*	*1.96*	*1.74*	
J%(Hierarchical FLC)	*4.03*	*1.96*	*1.74*	

Table 11: Comparison of the performance index of the optimised 24-rule FLC and the hierarchical FLC (small perturbation)

6 COMPARISON STUDY OF THE ORIGINAL 49-RULE FLC, THE OPTIMISED 24-RULE FLC AND THE FLC WITH HIERARCHICAL STRUCTURE

6.1 Comparison of the performance index

Table 10 shows the performance index with the original 49-rule FLC, the optimised 24-rule FLC and the hierarchical FLC for different operating conditions with a 20% step increase in mechanical torque T_m. Table 11 shows the performance index with the original 49-rule FLC, the optimised 24-rule FLC and the hierarchical FLC for different operating conditions with a 20% step increase in mechanical torque T_m followed by a 20% step change in reference voltage V_{ref}. From these tables it can be seen that FLC with hierarchical structure provides the same performance as the optimised 24-rule FLC and much better performance than the original 49-rule FLC.

Table 12 shows the performance index with the original 49-rule FLC, the optimised 24-rule FLC and the hierarchical FLC for different operating conditions un-

Fuzzy controller	P0	P1	P2	P3
J%(49-rule FLC)	*39.48*	*13.58*	*27.23*	*48.46*
J%(24-rule FLC)	*425.33*	*10.03*	*21.57*	*12794*
J%(Hierarchical FLC)	*37.05*	*10.01*	*24.22*	*68.67*
Fuzzy controller	P4	P5	P6	
J%(49-rule FLC)	*73.17*	*27.92*	*52.77*	
J%(24-rule FLC)	*126.99E29*	*27.92*	*71.05*	
J%(Hierarchical FLC)	*65.69*	*27.92*	*71.05*	

Table 12: Comparison of the performance index of the optimised 24-rule FLC and the hierarchical FLC (large perturbation)

der a three-phase fault. From this table it can be seen that the system becomes unstable under operating conditions P0, P3, and P4 when only the optimised 24-rule FLC is in operation. The FLC with hierarchical structure not only stabilises the system under all operating conditions but also improve the system performance in comparison with the system performance obtained by using 49-rule FLC only.

6.2 Comparison of the dynamic response

6.2.1 Comparison of the system response under small disturbance condition

Performance of the proposed FLC with hierarchical structure for operating condition P1 under a 20% step change in mechanical torque T_m is shown in Figure 5. It can be seen that the FLC with hierarchical structure provides the same response as the 24-rule FLC and much better response than the original FLC. Similar results were obtained for different operating conditions which are not shown here.

6.2.2 Comparison of the system response under large disturbance condition

Performance of the proposed FLC for operating condition P5 under a three-phase to ground fault at A (Figure 1) sustained for 125ms is shown in Figure 6. The FLC with hierarchical structure provides better dynamic response under a three-phase fault disturbance condition than the original 49-rule FLC and the optimised 24-rule FLC. Similar results were obtained for different operating conditions. As an example, Figure 7 shows the speed deviation corresponding to a three phase to ground fault under the nominal operating condition P0 with the original 49-rule FLC, the optimised 24-rule FLC and the proposed hierarchical FLC.

From Tables 10, 11 and 12 and the dynamic response it can be seen that the FLC with hierarchical structure

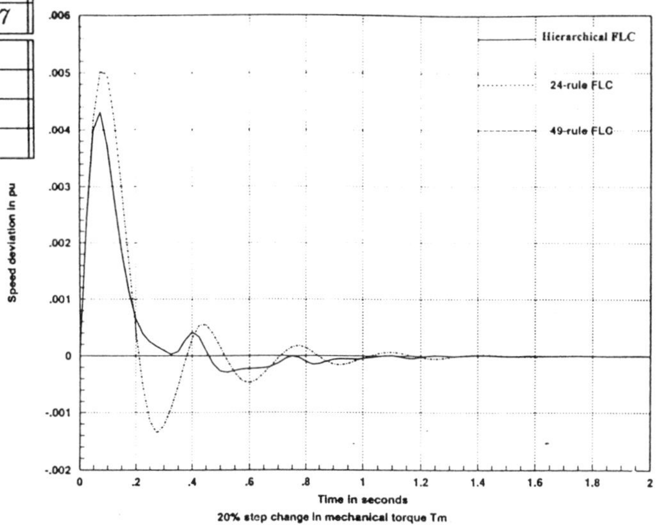

Figure 5: Speed deviation corresponding to a small perturbation of a 20% step increase in mechanical torque for operating condition P1 with the original 49-rule FLC, the optimised 24-rule FLC and the proposed hierarchical FLC

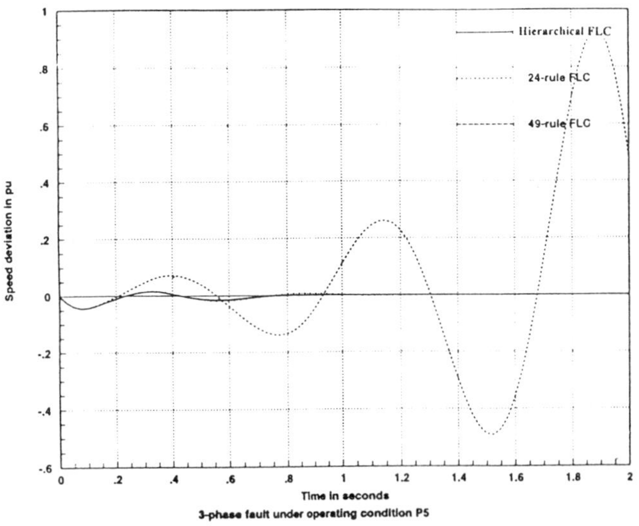

Figure 6: Speed deviation corresponding to a large disturbance of a three phase to ground fault under operating condition P5 with the original 49-rule FLC, the optimised 24-rule FLC and the proposed hierarchical FLC

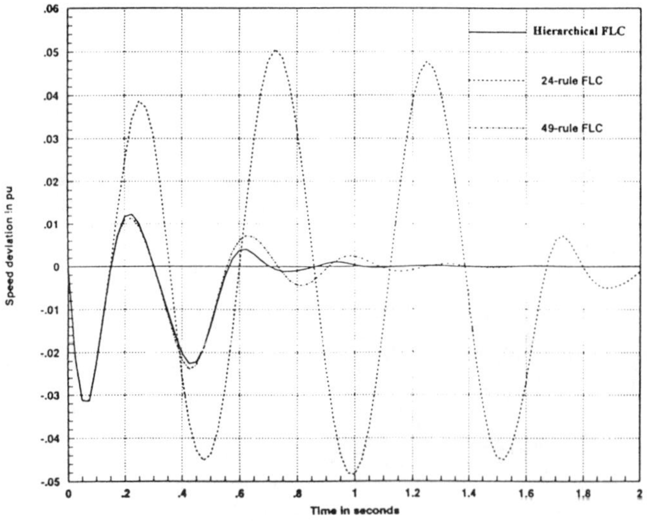

Figure 7: Speed deviation corresponding to a large disturbance of a three phase to ground fault under operating condition P0 with the original 49-rule FLC, the optimised 24-rule FLC and the proposed hierarchical FLC

provides better performance than the original FLC under both small and large disturbance conditions. The FLC with hierarchical structure provides the same performance as the optimised 24-rule FLC under small disturbance condition, however, it provides much better response than 24-rule FLC under large disturbance condition.

7 CONCLUSIONS

The development of the FLC with hierarchical structure to improve the stability of power systems has been presented in this paper. The input and output membership functions of the FLC were adjusted according to an evaluation index in order to achieve the optimal performance. The initial FLC design with a complete filled rule table (49 rules) has been reduced using neural network techniques to improve the dynamic performance. The FLC with hierarchical structure has been developed, which consists of a multi-layer controller incorporating rule based FLCs for direct control at the lower level. Simulation results show that the FLC with hierarchical structure has improved the system performance under both small and large perturbations at various operating conditions.

References

[LS81] E.V., LARSEN and D.A., SWANN: Applying Power System Stabilisers,*IEEE Trans.*, PAS-100, pp. 3017-3046, 1981.

[AF77] P.M., ANDERSON and A.A., FOUAD: Power System Control and Stability. *Iowa State University Press, Ames, Iowa*, 1977.

[DC88] F.P., DEMELLO and C., CONCORDIA: Concepts of synchronous Machine Stability as Affected by Excitation Control,*IEEE Trans.*, PAS-88, pp. 316-329, 1969.

[HL87] Y.Y., HSU and K.L., LIOU: Design of Self-tuning PID power system stabiliser for synchronous Generator. *IEEE Trans.*, EC-2, pp. 343-348, 1987.

[CMH88] A., CHANDRA, O.P., MALIK and G.S., HOPE: A Self-tuning Controller for the Control of Multi-machine Power Systems , *IEEE Trans.*, PWRS-3, (3), pp. 1065-1071, 1988.

[HS91] T., HIYAMA and T., SAMESHIMA: Fuzzy Logic Control Scheme for On-line Stabilisation of Multimachine Power System. *Fuzzy Sets and Systems* 39, pp.181-194, 1991.

[HL89] T., HIYAMA and C.M., LIM: Application of Fuzzy Logic Control Scheme for Stability Enhancement of a Power System, *IFAC Int. Symp. on Power Systems and Power Plant Control*, Seoul, Korea, pp. 613-616, August 1989.

[HMH91] M.A.M., HASSAN, O.P., MALIK and G.S., HOPE: A Fuzzy Logic Based Stabilizer for a Synchronous Machine,*IEEE Trans., EC*, Vol.6, No.3, pp. 407-413, September 1991.

[SHK94] J., SHI, L.H., HERRON and A., KALAM: Design and Implementation of a Fuzzy Logic Based Power System Stabilizer, *AMSE Periodicals: ADVANCES IN MODELLING and ANALYSIS, C. AMSE Press*, Vol. 42, No.4, pp.39-52, 1994.

[SHK94*] J., SHI, L.H., HERRON and A., KALAM: Optimization of Fuzzy Controllers as Real-time Power System Stabilizers, *International Journal of Engineering Applications of Artificial Intelligence*, Vol.7, No.5, p.545-558, 1994.

[Lcc90] C.C., LEE: Fuzzy Logic in Control Systems: Fuzzy Logic Controller-Part I,II, *IEEE Trans. Syst.Man Cybern.*, Vol.SMC-20,No.2 pp. 404-435, 1990.

[Mot92] Motorola: Fuzzy Logic Educational Program. *Center for Emerging Computer Technologies, Motorola Inc.*, 1992.

[RS93] L., REZNIK and J., SHI: Simulation Study of the Choice of a Membership Function Model for Control Applications, *The 2nd International Conference on Modelling and Simulation*, Melbourne, Australia, Vol.1, pp.209-217, 12-14 July 1993.

[Kos92] B., KOSKO: Neural networks and fuzzy systems: a dynamic systems approach to machine intelligence, *Prentice Hall*, 1992.

[Tog91] Tilgen User's Manual, *Togai Infralogic Inc.*, 1991.

APPENDIX

Single-machine infinite bus system generator unit data:
d-axis synchronous reactance, $x_d = 1.027pu$
d-axis transient reactance, $x'_d = 0.479pu$
q-axis synchronous reactance, $x_q = 0.489pu$
Time constant of d-axis circuit, $\tau'_{d0} = 0.345s$
Inertia constant, $H = 1s$
Synchronous speed, $\omega_0 = 314rad/s$

The equivalent transmission line resistance and inductance are:
$R_e = 0.02pu$
$X_e = 0.4pu$

AN ON-LINE DESIGN SYSTEM FOR FUZZY CONTROLLERS

Jobst H. Schmalenbach and Simon D. Hill
Department of Robotics and Digital Technology,
Monash University, Wellington Road, Clayton 3168, Australia.
E-mail: jhs@rdt.monash.edu.au and sdh@rdt.monash.edu.au

ABSTRACT

This paper describes a novel system for designing and evaluating the performance of fuzzy-logic controllers. The novelty derives from the fact that the executable machine code for the fuzzy processor is generated and down-loaded into the fuzzy processor as the designer modifies the rules, membership functions, and rule weightings of the controller design. The example given is of design software running on a Unix engineering workstation development generating and loading code into a custom-built VME-based fuzzy processor.

1 INTRODUCTION

There are numerous examples of fuzzy-logic control incorporated in appliances and equipment used in daily life. Fuzzy logic is used in 35mm camera focusing systems by Minolta and Canon. Canon has found that their fuzzy-logic controlled camera succeeds in focusing on the subject 96% of the time compared to a 76% success rate using a non-fuzzy focusing system [Zad92]. There are other examples, [Tan91, Wal89, D.M92, J.M94, NAS92], but one of the best-known large-scale uses of fuzzy-logic control is the Sendai subway in Japan, [MF93]. The trains can stop within seven centimeters of the desired point on the platform, change velocity so smoothly that passengers do not need to hold onto the straps, and require 10% less energy than conventionally-controlled trains.

The analytic formalism for fuzzy logic is given in [Zad65] and the application of fuzzy-logic theory to control has been pursued in [Zad72, Lee90, YKSH91]. The attractions of fuzzy logic as applied to control are that is a simple, easy to implement technology. It is able to translate imprecise knowledge of human experts into rules that can be processed by digital computers, and valid outputs can be generated for inputs whose values were not predicted in the design phase of the controller.

"Fuzzy Logic is not the absolute solution to every problem in the world of control. But it certainly has its place and could help in combination with other methods to solve some of the problems in control." Lotfi Zadeh in [Zad92].

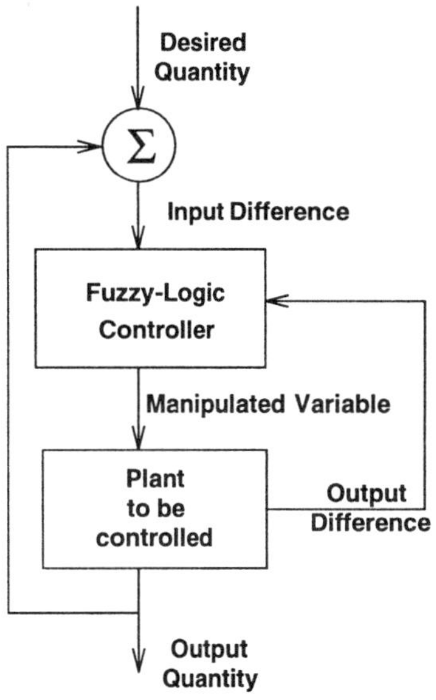

Figure 1: General fuzzy-controller structure

2 OVERVIEW

Fuzzy controllers can be implemented on general-purpose computers, but one of the problems of this approach is the significant computation requirement in the de-fuzzification step [MH95]. The only option is for singleton-consequents, not finite-area membership functions, in a fuzzy-control program running on a general-purpose computer. It is limitations of this nature that has led a number of semiconductor manufacturers to release single-chip devices dedicated to fuzzy processing [OMR91a, Unk94, Tog92].

Figure 1 depicts a block diagram of the general structure used in controlling an arbitrary plant. For simplicity, the illustrated plant has a single input and a single output.

The fuzzy-logic-controller block could be constructed as a software program running on a general-

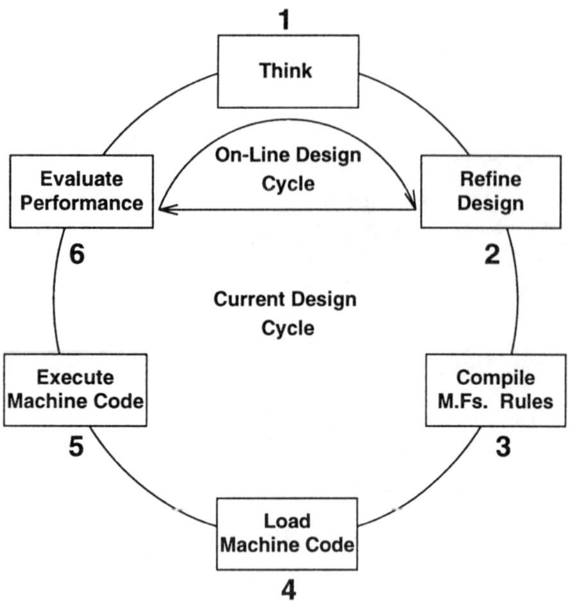

Figure 2: Fuzzy-controller design cycle

purpose computer or could be a special purpose fuzzy-logic controller. The complete controller would need to include hardware interfaces, such as analog-to-digital and digital-to-analog converters, to the measured output quantity and the manipulated variable. The output-difference quantity could be a direct measurement from a sensor, in which case it would be the derivative of the output, or could be synthesised by a differencing operation in the controller software.

Designers faced with the problem of creating a fuzzy-logic controller need to address the problem of partitioning the input difference and output difference quantities, choosing the shape of the membership functions associated with each of the partitions, choosing the shape of the membership functions used in the output, choosing the defuzzification strategy, selecting the number of rules, selecting the antecedent and consequent components for each rule, and choosing the weighting that each rule contributes to the output.

Currently, designers have software tools available to aid the design process [OMR91b, Fuz91, Tog93]. All these tools run on PC platforms either using DOS-based graphics or using the graphical interface features of MS-Windows. The output of the tool is either general-purpose language code, typically C-language instructions or, machine-code instructions for the target special-purpose fuzzy processor. The output is then compiled, for the C-language output, and executed by the general-purpose computer, or, for the machine-code instructions, loaded into the fuzzy-processor and executed. This design-load-execute cycle is illustrated in Figure 2. Current practice is for the designer to follow all six steps explicitly.

The system described in this paper shortens the design cycle from six to only three steps. This is achieved by compiling the membership functions and the rules, along with the inherent weightings, *as the design is refined.* The designer is only aware of the graphical representation of the membership functions, rules, and weightings and is oblivious to the fact that the design software is regenerating the machine code for every modification of the design in step two. The focus of the designer is not distracted by the explicit compilation, loading and execution steps of the current practice. With this on-line system, the designer can concentrate on the fruits of the design, that is, the effect of the design changes on the performance of the control system depicted in step six of Figure 2.

All current fuzzy-controller design systems have PC-based software and PC bus hardware. An additional idea of the on-line design system described here is to to decouple the choice of software-development platform from the choice of hardware execution platform. The key to this decoupling is to utilise computer-bus interface adaptors, [BiT92], which are available, for example, for SBus, VME, MultiBus, TurboChannel, NuBus, MicroChannel, EISA, and ISA buses. The designer can work with the software platform of choice and execute the fuzzy-controller instructions on a hardware platform that suits the particular application.

3 IMPLEMENTATION EXAMPLE

The implementation described in this paper is illustrated in Figure 3 and is an example of a Unix software-development platform communicating with the fuzzy-control hardware that is mounted in a VME-bus card cage. The fuzzy-control hardware has been purpose-built to utilise the OMRON FP-3000 fuzzy processor, [OMR91a], in a VME hardware environment.

3.1 FP-3000 DIGITAL FUZZY PROCESSOR

The FP-3000 appears as a co-processor to the host CPU. The numeric values that specify the membership functions, rules, and rule weightings are written to the FP-3000 as if it were host memory. At the end of each inference cycle, the FP-3000 raises an interrupt line and the host CPU services the interrupt by reading the FP-3000 memory location that holds the defuzzified output value.

The FP-3000 can be used in two different configurations, either single-chip mode or expanded mode. Single-chip mode supports 29 rules each having five antecedents and three consequents. Expanded mode has up to three groups of 128 rules.

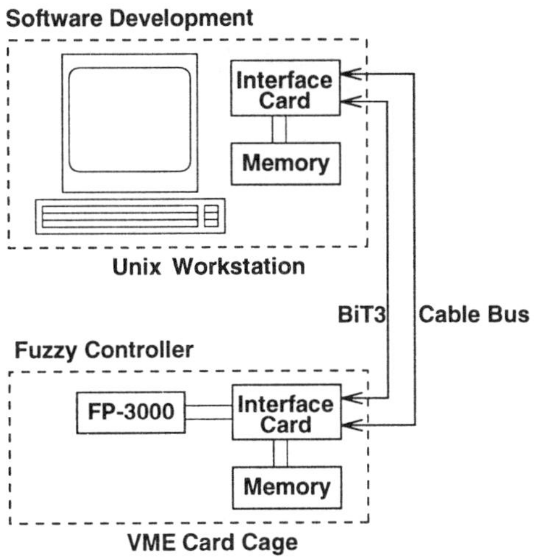

Figure 3: An on-line fuzzy-controller design system

3.1.1 RULE FORMAT

Rules describe how the output data should be calculated by defining the relationship between antecedents and consequents.

Table 1 illustrates the use of rules in the FP-3000. The INx are inputs to the antecedents, A,B,C,D,E,F,G are membership function labels and OUTx are inputs to the consequents.

Consider rule 2 in Table 1. If input 1 is in the range of the B label (antecedent membership function), and input 3 falls in the range of the C label, and input 4 is in the range of the A label, and input 5 is in the range of the A label then this rule will fire. The firing of the rule creates a contribution to output 1 from consequent label A, and creates a contribution to output 3 from consequent label B. Each of the contributions to the output(s) are scaled by the given weight of 0.7. The actual output quantity is calculated by defuzzifying the contributions from all the consequent labels.

In general, for a set of active antecedent labels in a given rule, the value used in the consequent processing is the *minimum* value from the set. If multiple rules have the same consequent label, then it is the *maximum* contribution from that set of consequent label values that is used as the contribution to the output.

3.1.2 MEMBERSHIP FUNCTIONS

Antecedent membership functions subdivide the fuzzy-processor input domains and therefore control the processor's knowledge of the input data. Consequent membership functions partition the rule-space output domain and, through the weighting and defuzzification steps, define the processor's output variables.

Label names referred to in the rules of Table 1 are describing the particular membership function that is used in either the antecedent or the consequent processing. In a given rule there is a maximum of eight antecedents, each selecting from seven membership functions, and a maximum of two consequents each selecting from seven membership functions. The FP-3000 chip allows triangular, trapezoidal, or singleton membership functions for the antecedent processing, but restricts the consequent membership functions to singletons. The main reason for this restriction is to reduce the computation required when using the center-of-gravity method of defuzzification.

3.1.3 DEFUZZIFICATION

The FP-3000 uses two methods to determine the output value of the processor, this is the manipulated variable of the controlled plant as shown in Figure 1.

Center of Gravity Method computes the center of gravity of the singletons that have been defined by the firing of any of the rules. These singletons are weighted by the given rule weight to yield a resulting height, μ_i, and are located at a point in the output domain represented as x_i.

$$\text{output} = \frac{\sum_{i=1}^{n}(x_i * \mu_i)}{\sum_{i=1}^{n} \mu_i}$$

If the denominator is zero, the FP-3000 indicates the exception via an error bit in the processor status byte.

Maximum Height Method The defuzzified output is simply chosen as the maximum-valued singleton in a given output domain. If there is a set of singletons that all have the maximum value, then a tie-breaking mechanism is used to select a single member from that set. The designer nominates to select either the left-most or right-most singleton from the equal-height set.

3.2 SOFTWARE

The software written for this project is an example of a windowed graphical user interface. By using a platform-independent graphics library, [Sma93], the same source code can be compiled and run under the X-Windows system on Unix-based machines as well as under MS-Windows on PC-style machines.

This fuzzy-control design software has three basic functions. Provide a graphical interface for the design of the membership functions and rules, to translate those rules and membership function shapes into fuzzy-processor machine code, and to communicate with the memory-mapped fuzzy processor.

Table 1: A rule set for the FP-3000

Rule No.		Antecedent		Consequent	Weight
1	IF	In1=A & In2=A & In3=B & In4=A & In5=A	THEN	Out1=A, Out2=B	0.5
2	IF	In1=B & In3=C & In4=A & In5=A	THEN	Out1=A, Out3=B	0.7
3	IF	In3=B & In3=A	THEN	Out3=B, Out2=A	0.8

Output 1 uses center of gravity method
Output 2 uses maximum-height method (left priority)
Output 3 uses maximum-height method (right priority)

Design of antecedent membership functions starts by specifying the required number of antecedents and the domain limits by text entry in the main-menu window Figure 5. Then continues by opening the antecedent window, Figure 6, and using the mouse to establish the start and end points of each straight-line segment of the required membership function shape.

A similar sequence applies for the consequent design but, since consequents are constrained by the FP-3000 design to be singletons, it is only necessary to mark the location of a given singleton in the consequent window Figure 7.

Design of the fuzzy-processing rules is a matter of specifying the number of rules in the main menu window, Figure 5, then opening the rule development window, Figure 4, and marking the labels that will be the active membership functions for each input and each output. The weights and the defuzzification strategy are also choices to be made in the rule window.

The menu approach was chosen to define the rules rather than a text-parsing method because it shortens the design time since the rule-translation step is eliminated, and the rule-window state-machine code enforces the underlying constraints that would otherwise be checked by rule-parsing code.

Following these initial design steps, the fuzzy controller can be started by the run buttons in the main menu. There are multiple run buttons to activate one of the three possible rule sets in the FP-3000. The designer is then free to evaluate the performance of the controlled system and modify the membership functions, rules, and weightings, to iterate along the path to the final design. This design cycle is shown as steps 1, 2, and 6 in Figure 2.

For every change in the design parameters, the software translates the updated membership functions, rules and weights into fuzzy-controller machine code and down loads that code into the FP-3000 rule space. All this compilation and loading activity takes place without intervention by the designer who is free to concentrate on the control-system performance.

4 CONCLUSION

A combined software and hardware system that enables control-system designers to evaluate fuzzy-control strategies on line has been described. The particular implementation detailed in this paper, of Unix-based development software that interacts with VME-based fuzzy-processor hardware, Figure 8, provides a concrete example of the concept. One design decision made in this work was to utilise a windows-independent graphics library so that the graphical-user-interface design software could be ported to a variety of hardware platforms supporting different windowing environments such as X-Windows and MS-Windows.

In general, fuzzy processors are only commercially available in either stand-alone configurations with serial interfaces or for the ubiquitous PC-bus interface. This work demonstrates that it is possible, and practical, to build fuzzy-processing hardware for an arbitrary bus structure that suits the needs of the control-system application and that the designer is not constrained by the hardware formats supported by the fuzzy-processor manufacturer.

On-line evaluation of fuzzy-control performance can only be achieved by operating the fuzzy processor as a memory-mapped peripheral to the CPU running the design software. Since bus-to-bus adaptors are available for all popular Unix and PC computers, it is realistic to couple the best available software-development platform to a fuzzy processor mounted on the target-hardware back plane.

The target hardware would include a general purpose, maybe a single-board, CPU runing the control-system software. This CPU would manage any interfaces to external hardware, such as analog-to-digital converters, and interact with the fuzzy processor as a peripheral. It would be running a real-time, interrupt-driven kernel to support, as a minimum, constant-interval sampled-data control. One outcome of this work is to demonstrate that fuzzy controllers can be designed and evaluated in real time. That is, the designer can make changes to the shape, number, and location of antecedent and consequent membership functions, change the structure of the rules linking the antecedents and consequents, and modify the weighted contribution of each rule – and instantaneously see the effects of those changes on the controlled plant.

References

[BiT92] BiT3 Computer Corporation, Minneapolis, MN. *BiT 3 Product Summary*, 1992.

[D.M92] Richard D.Marchbanks. A fuzzy control approach to a two-axis mirror laser beam. *University of New Mexico*, 1992.

[Fuz91] Fuzzy Systems Engineering, Poway, CA. *The Manifold Editor*, 1991.

[J.M94] Jack J.McCauley. A fuzzy logic torque server. *The C users Journal*, 1994.

[Lee90] Chuen Chien Lee. Fuzzy logic in control systems: Fuzzy logic controller — parts I & II. *IEEE Transactions on Systems Man and Cybernetics*, 20(2):404–435, March/April 1990.

[MF93] Daniel McNeill and Paul Freiberger. *Fuzzy Logic*. Bookman Press, Melbourne, 1993.

[MH95] Tim T. McFarland and Simon D. Hill. A comparison of fuzzy and computed-torque control of a robot arm. In *Proceedings of the Eighth International Conference on Industrial and Engineering Applications of Artificial Intelligence and Expert Systems*, Melbourne, June 1995.

[NAS92] NASA, editor. *Proceedings of the Conference of the North Amercian Fuzzy Logic Society*. NASA, Puerto Vallarta, 1992.

[OMR91a] OMRON Corporation, Kyoto. *FP-3000 Digital Fuzzy Processor: User's Manual*, 1991.

[OMR91b] OMRON Corporation, Kyoto. *FS-10AT Fuzzy Inference Software*, 1991.

[Sma93] Julian Smart. *User Manual for wxWindows 1.50, A Portable* C++ GUI Toolkit. University of Edingburgh, December 1993.

[Tan91] Andrew Tanzer. Why fuzzy logic is good in business. *Science and Technology*, 1991.

[Tog92] Togai. *Fuzzy Logic: From Software to Silicon*. Togai Infralogic, Irvine, CA, 1992.

[Tog93] Togai. TILShell 3.0, 1993. Togai Infralogic.

[Unk94] Unknown. Stand-alone fuzzy controller, Summer 1994. Embedded Systems Programming Product News, P. 43.

[Wal89] Larry Waller. Fuzzy logic, it's comprehensive, it's practical and its commercial. *Electronics*, 1989.

[YKSH91] T Yamashita, R Katoh, S Singh, and T Hori. Stability analyis of fuzzy control system applying conventional methods. In *Proceedings of IECON 91*, pages 1579–1584, 1991.

[Zad65] Lotfi Zadeh. Fuzzy sets. *Information and Control*, 8(3):338–353, June 1965.

[Zad72] Lotfi Zadeh. A rationale for fuzzy control. *Journal of Dynamic Systems, Measurement and Control*, 94:3–4, March 1972.

[Zad92] Lotfi Zadeh. Fuzzy logic controllers. IEEE Fuzzy Logic Video Series, 1992.

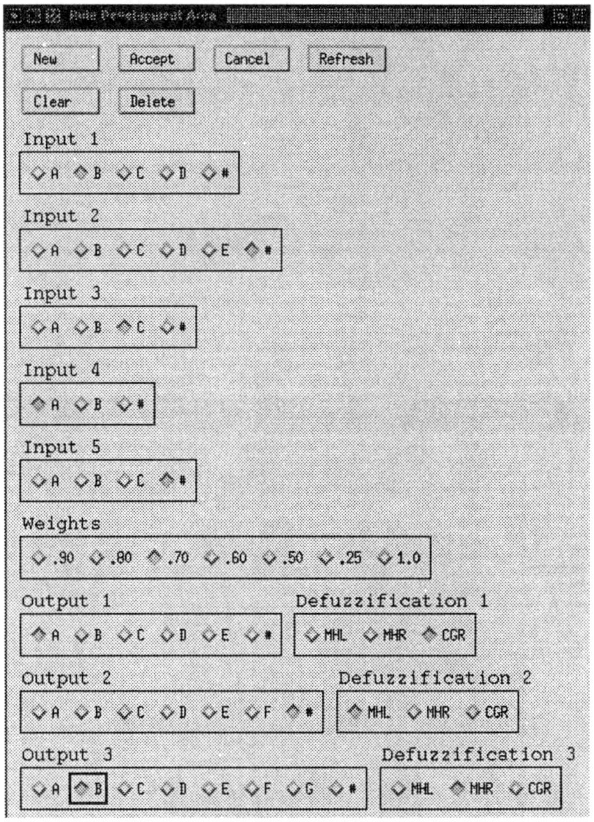

Figure 4: Rule development window

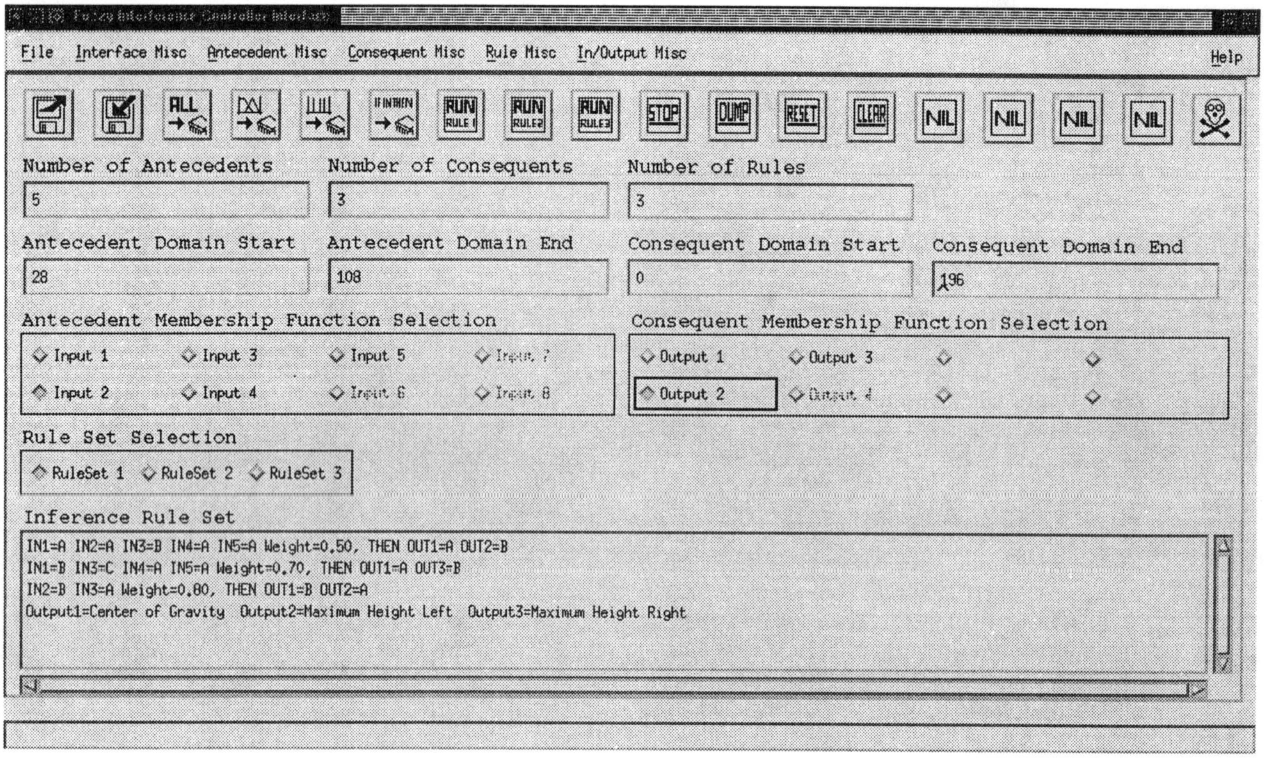

Figure 5: Main menu window

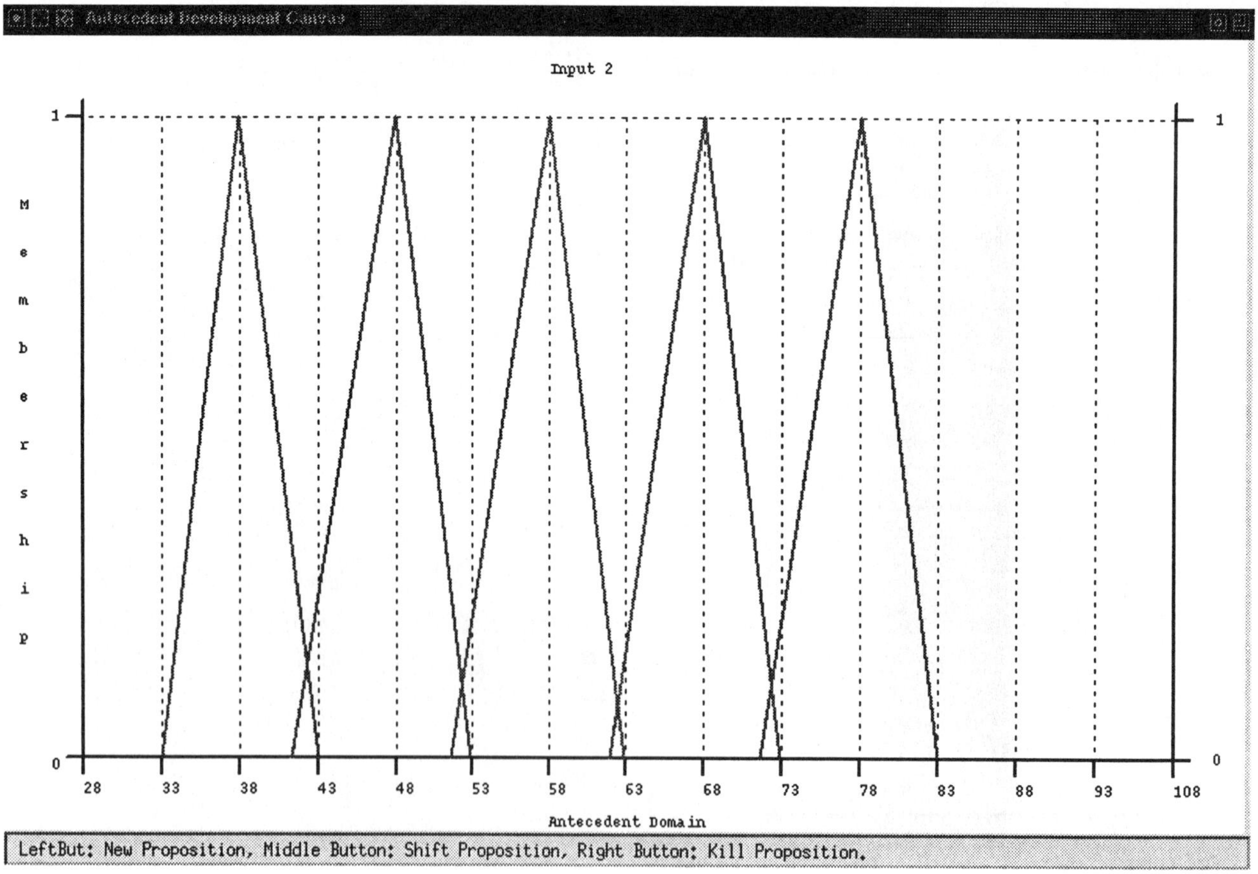

Figure 6: Antecedent development window

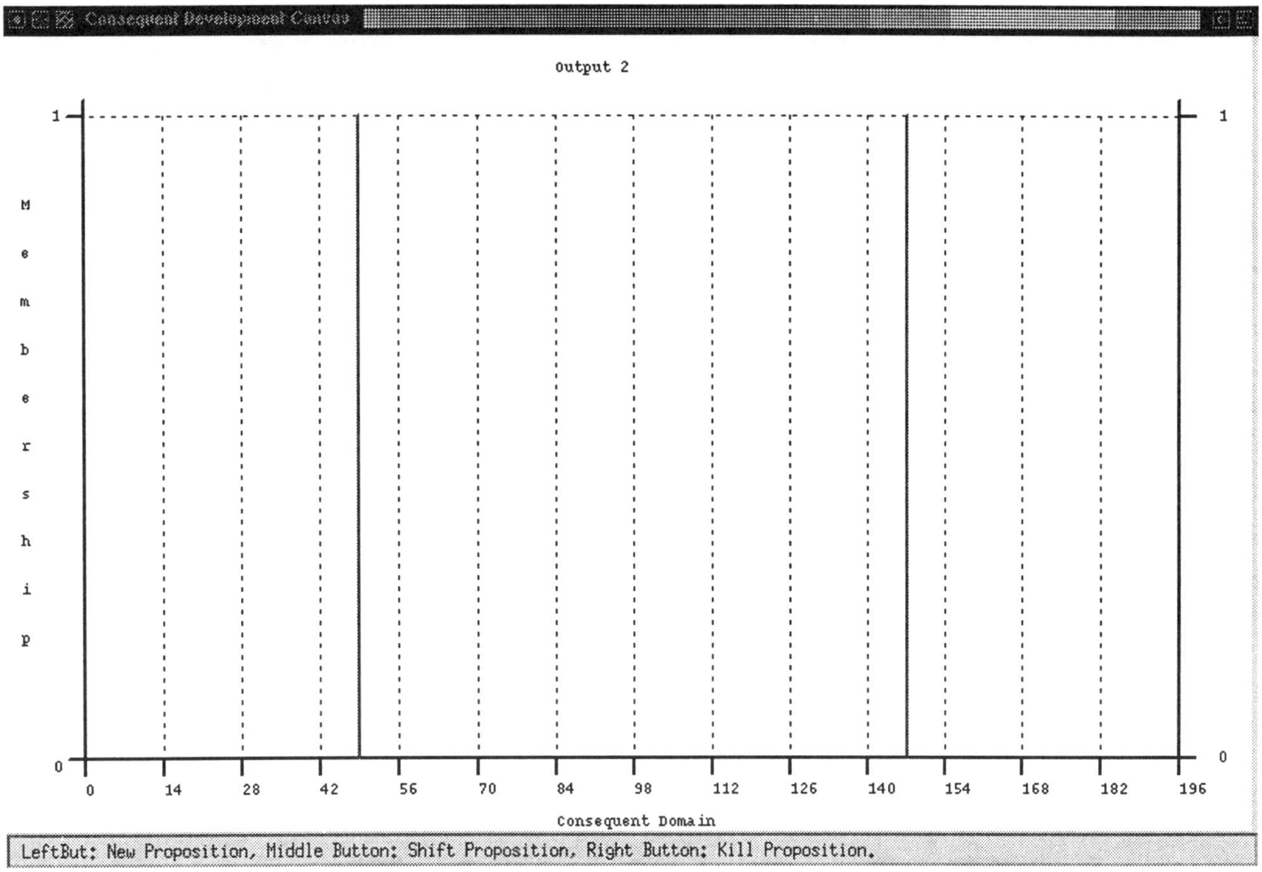

Figure 7: Consequent development window

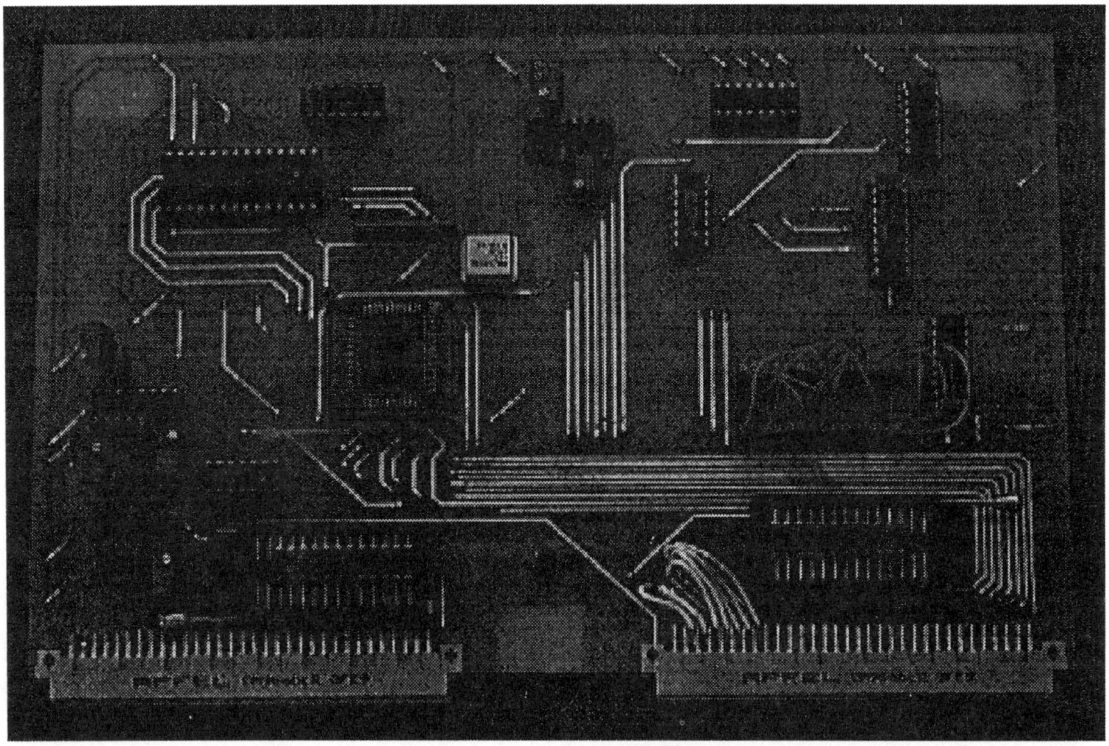

Figure 8: FP-3000 VME board

A FUZZY LOGIC APPROACH TO OPTIMISE INVENTORY

L. F. Sugianto and W. Mielczarski
Centre for Electrical Power Engineering,
Dept.of Electrical and Computer Systems Engineering,
Monash University, Clayton 3168, Victoria, Australia.

ABSTRACT

The need for decision supporting tools in managing large inventory items has been fulfilled by the development of an Intelligent Decision Support System (IDSS). IDSS has been tailored to assist managers or engineers in power plant inventory optimisation.

Despite the superior performance of the system, the accuracy of results still depends on the information supplied. This is a major obstacle to the applicability of IDSS since incomplete historical records are often encountered, not to mention the difficulty faced when the data is to be obtained indirectly from other industries or manufacturers of similar plant item.

Fuzzy Logic approach has been proposed to overcome problems arising from data uncertainty. The paper presents an overview of IDSS and outlines a research direction whereby a fuzzy logic approach would be embraced to enhance the performance of IDSS.

INTRODUCTION TO THE PROBLEM

Researchers in the area of management science have attempted solving the problem of inventory optimisation in many ways. Various commercial software packages implementing different techniques are available in the market today. Approaches range from a highly structured Economic Order Quantity (EOQ) and Re-Ordering Point (ROP) formula [AF79, FW88]; to the application of Artificial Intelligent techniques, such as expert systems. Obviously, they have been designed and developed to address different aspects of inventory management.

A benchmark to evaluate the performance of power stations is the triplet Reliability-Availability-Maintainability (RAM) indices. As far as the electricity supply is concerned, it is desirable to ensure a high availability level. Thus, an inventory decision support system which aims to both optimising inventory and ensuring high availability level, will be an eminent advantage to the power industry.

INTELLIGENT DECISION SUPPORT SYSTEM

Figure 1 below shows the structure of IDSS developed by the authors. The crucial module is the database, which supplies the system with inventory data; such as the price of the plant items, the stock on hand and the delivery time; and the reliability data, such as Mean Time Between Failure (MTBF), Mean Time To Repair (MTTR), operating hours and power station available time. IDSS employs Dynamic Programming to configure the optimal inventory holding. While the availability of the power plant is formulated to be the objective function, the constraint is given by an assumed inventory budget.

An inventory policy module is used as a simulation tool, delivering an optimal policy determined by Dynamic Programming. A user can also propose his own inventory policy, which would then be analysed by IDSS. As an output, IDSS would produce its optimised version under the same constraint, and compares the two versions.

A cost evaluation program measures an increase or decrease of an availability level caused by a particular decision and "translates" it in terms of dollar value. It is a means to indicate profits or losses. A filter is used to provide flexibility of IDSS. It enables a user to specify a

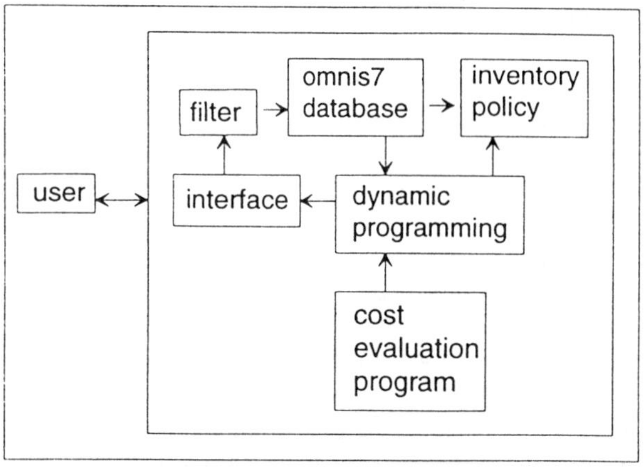

FIGURE 1 IDSS master diagram

type of plant items included in simulation. The interface allows the communication between modules and a user.

Two IDSSs developed for different power systems have been implemented. The first pilot project has been carried out for a Boiler Feed Pump system. The feed pump plant is a subsystem of power generating unit of a base load, coal fired power station. The second, larger scale IDSS has been constructed for a peaking unit, gas turbine power station. The outcomes of both projects are very promising. The cost reduction of approximately 30% of inventory expenses is possible.(Note: see [MS92, Sug92, SM94] for results and detail works.)

DATA UNCERTAINTY IN IDSS

A major dilemma with regards to applicability of IDSS is credibility and accuracy of input data. Input parameters such as MTBF and MTTR, used in both projects have been assumed to be known.

There parameters were collected from historical records. In reality, however, the inventory records do not have all the data needed by IDSS, mostly because a number of failures in power stations is very small. When such a situation arises, a way out is to approximate the unknown data using other databases that may contain "similar" items. There are many levels of "differences" or "similarities" between an item in the system considered and a "similar" item in an external database. These "differences" can be expressed in terms of membership functions and a Fuzzy Logic approach could be used to cope with the fuzziness in information supplied. In conclusion, Fuzzy Logic approach is proposed as the future direction of the development of an advance support system which could tackle the problem of data uncertainty.

THE PROPOSED STRUCTURE

Figure 2 shows the master diagram of the proposed system. The methodology consists of a sequence of procedures which process the fuzzy data before they arrive at the optimal allocation module. To begin with, the data collected from an external database was analysed by comparing the "similarities" and "differences" of the item to the item whose data is to be collected. Appropriate values for membership functions are allocated, indicating the degree of uncertainty or level of similarity. At the next stage, the data with their membership values are aggregated to produce a single fuzzy data. Then, the data can be used for the objective function calculation. Both the data and the membership values are calculated simultaneously to produce a value with an induced membership value. After that, these two elements undergo a transformation to aggregate them into a single value. Having been transformed, the data proceeds to the optimisation module. The outcome of the whole process is an optimised inventory policy, indicating the number of spare parts which should be kept in stock, together with confidence level values. The total cost and total availability also have their own induced confidence level.

Determination of membership function

Fuzziness enters the system through the uncertainty in the data, due to the lack of historical records. There are three ways of determining membership functions. Firstly, if failure records provide statistically significant

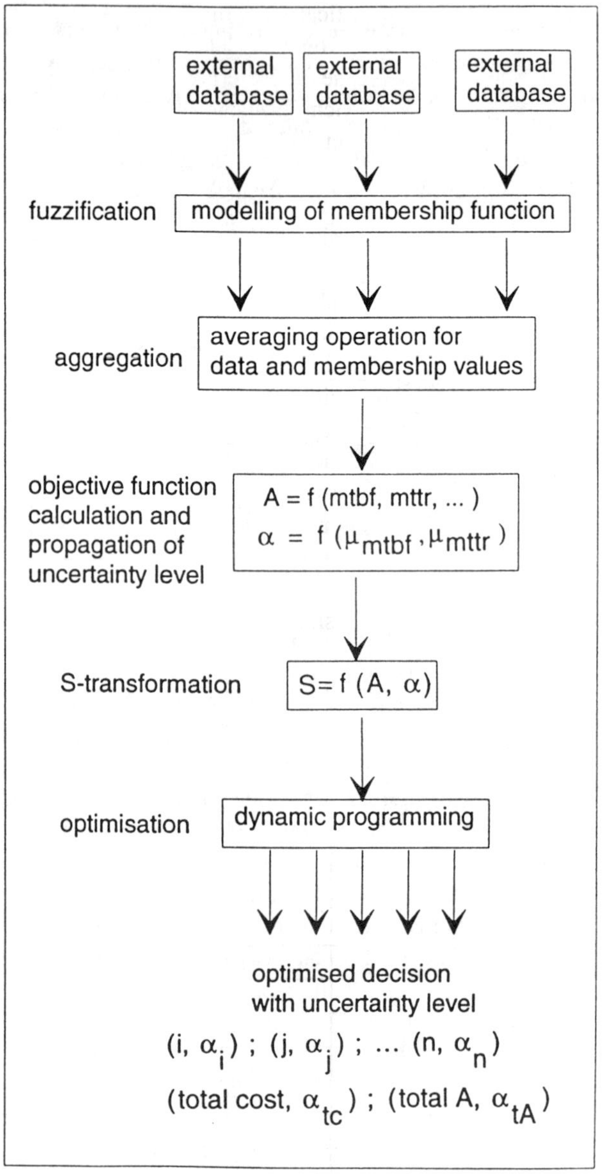

FIGURE 2 Master diagram: Fuzzy Logic Approach in IDSS for Inventory Management.

information, the membership function is determined as a constant with the value "1", expressing the highest confident level. Secondly, if there are no records or records do not provide relevant information, two options can be applied. The first one is to obtain the unknown data from experts who can indicate a possible failure rate with a level of confidence in their judgements. The second method assumes the use of external databases and in this case, membership functions are determined with the use of mathematical operators expressing similarities between the item considered and other "similar" items in external databases. The paper presents details on the later approach.

Realising that the precise information, i.e. the data which satisfies the specification completely, might not be available, the emphasise on this data collection may be laid on the similarity in the characteristic of the plant item. Firstly, parameters which influence the performance characteristic of the plant item must be identified. Such parameters can be:

- manufacturer
- model type or size of the plant item
- year of manufacture
- type of power plant
- operating condition
- environmental impacts.

Secondly, a membership function should be suitably modelled with the use of specified operators. "Similarity" parameters can be combined with different coefficients to imply their relative importance. For instance, the fact that the manufacturer and the model type have more influence on an item performance compared to the environmental impacts, can be reflected by allocating a larger coefficient for those factors and a smaller coefficient for the later factor.

A generalised model can be formulated as follows:

$$\mu_{MTBF}(x) = \sum_{i=1}^{N} \alpha_i \exp\left(\frac{-abs(m_i - x_i)}{k}\right) \quad (1)$$

where

$$\sum_{i=1}^{N} \alpha_i = 1$$

and

$\mu_{MTBF}(x)$ - the membership function of x, influenced by all parameters

N - number of parameters influencing the performance characteristic of the plant item

α_i - coefficient or weighing factor

m_i - specified data

x_i - collected data

k - constant imposing certain condition.

Equation (1) is explained by a case example. The following is an illustration to model a membership function for the parameter determined as Mean Time Between Failure (MTBF) of an electric clutch, with the use of equation (1). MTBF of an electric clutch is to be obtained from an external database. The specification of the electric clutch is as follows:

- Manufactured by company "S" (m_1 = "S")
- Size = "7" (m_2 = "7")
- Year of manufacture :1985 (m_3 = "1985")
- Operated in a peaking unit (m_4 = "peaking").

Table 1 lists 10 data collected from external database. The first electric clutch belonging to Power Station "1" has the following specification :

- Manufactured by company "S" (x_1 = "S")
- Size = "7" (x_2 = "7")
- Year of manufacture :1985 (x_3 = "1987")
- Operated in a peaking unit (x_4="intermediate").

The "similarities" and "differences" between the two electric clutches can be determined by comparing their specifications.

Table 2 on the next page defines the "mapping" between the collected data. As well, let α_i = {0.35, 0.35, 0.2, 0.1} be the set containing the weighing coefficient of the influencing factors.

Table 1 MTBF data collected from external databases

Power Station	Manufacturer	Model type (size)	Year	Operating condition	MTBF (year)	μ_{MTBF}
1	S	S-7	1987	Intermediate	3	0.94
2	GE	GE-5	1988	Base	2.4	0.77
3	S	S-8	1988	Peak	4.1	0.90
4	S	S-4	1988	Peak	3.5	0.90
5	ABB	ABB-3	1983	Base	2.7	0.73
6	S	S-5	1975	Base	2.8	0.78
7	EM	EM-4	1989	Peak	4	0.87
8	EM	EM-5	1985	Peak	3.5	0.90
9	ABB	ABB-3	1986	Intermediate	3	0.76
10	GE	GE-4	1986	Base	2.9	0.89

Table 2 The look-up table for interpreting the collected data in the case example

i (parameter)	m	k	Imposed condition
Manufacturer	S	1	if $x \neq m$, m=1, x = 0
Model type - size	S-7	10	S-7 implies m = 7
Operating condition	Peak = 1	10	Peak =1, Inter.=7, Base = 10
Year	1985	40	k - technological change in manufacturing & age of item

For the "manufacturer" factor, $m_1 = x_1$, therefore the exponential value in equation (1) diminished and the first component of the summation given in equation (1) becomes equal to α_1, since $\alpha_1 = 0.35$. Both electric clutches were size "7" and again, the exponential value in equation (1) diminished and the second component of the summation = α_2 = 0.35. Due to the differences in the year of manufacture, α_3 would be penalised by the exponential part of equation (1): $m_3 = 1985$, $x_3 = 1987$. Knowing $\alpha_3 = 0.2$ and k = 40, the component of the summation was calculated to be 0.19. Having defined "peaking" = 1 and k = 10 in Table 2, $\alpha_4 = 0.1$ from the assumed set, the last component of the summation was calculated to be 0.05. Thus these four components produced by the influencing factors produces a membership value $\mu_{MTBF} = 0.94$. Thus, MTBF = 3 years, collected from Power Station "1", is 94% "suitable" to be taken into consideration.

The possibility distribution for MTBF in the case example, as shown in column 7 of Table 1, was presented in Figure 3.

Aggregation of collected data

At the next stage, the collected data should be aggregated to a single value. Several defuzzification strategies have been introduced by the Fuzzy Logic researches [Ang92, Ber92]; such as the Tsukamoto's defuzzification method, the Center of Area (COA) method and the Mean of Maximum (MOM) Method. However, when the collected data undergo a defuzzification process, the uncertainty level attached with the collected data will disappear.

The problem in propagating the membership values is analogous to multi-criteria decision making. Hence, averaging operators may appropriately be used in this context. Common averaging operators such as the weighted and unweighted arithmetic or geometric mean are adequate models in aggregating decisions [TZ+79, Zim91]. For the purpose of simplicity, the geometric mean will be employed as the averaging operator :

$$L(\mu_{A_1}, \ldots, \mu_{A_n}) = \frac{\sum_{i=1}^{n} \mu_{A_i}}{n} \qquad (2)$$

Objective function calculation

At this stage, the data needed by the optimisation module are ready to be used. An availability model for a base load operation is given by

$$A = \frac{(MTBF, \alpha)}{(MTBF, \alpha) + MTTR} \qquad (3)$$

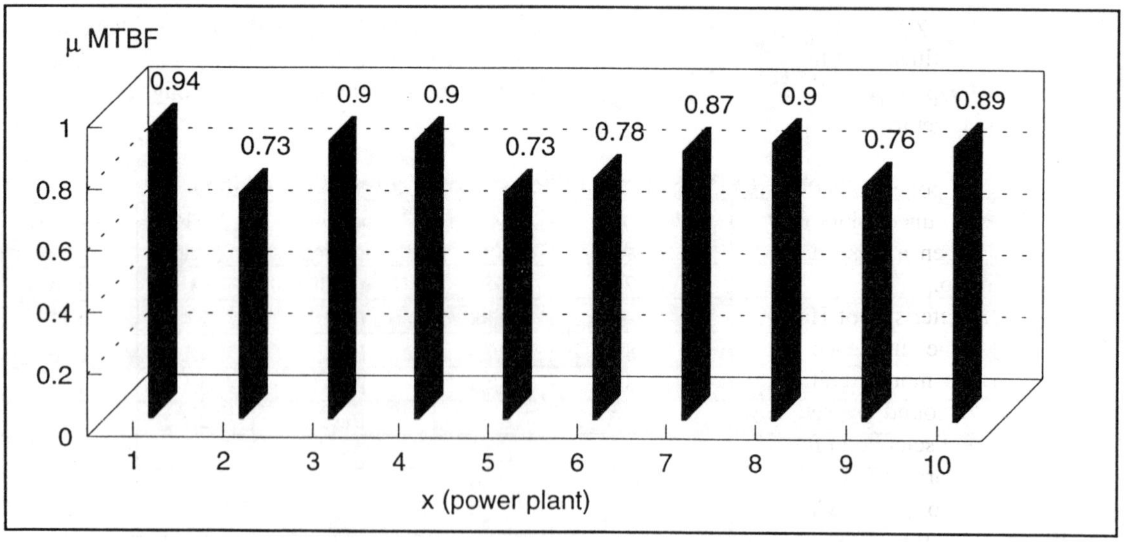

FIGURE 3 Graph of the membership function of the collected data.

where
MTBF = Mean Time Between Failure
α = confidence level of the MTBF value
MTTR = Mean Time To Repair.

Assuming that MTTR data is crisp, uncertainty level of the availability model is derived straight forwardly from the uncertainty level of MTBF. Cases when fuzzy data must be used in multiple parameters, conjunction operations could be used to propagate the confidence level.

Influence of confidence level on data

S-transformation proposed in this section is necessary in order to control the influence of confidence level on data value. The four proposed models are :

- $S_1 = A\alpha^k$
- $S_2 = \dfrac{A\alpha^k}{A + \alpha^k - A\alpha^k}$
- $S_3 = \dfrac{A\alpha^k}{\chi + (1-\chi)(A + \alpha^k - A\alpha^k)}$
- $S_4 = \beta A + (1-\beta)A\alpha$

where
A = data, $0 \le A \le 1$
α = confidence level, $0 \le \alpha \le 1$
k = constant to set the influence of the confidence level, $0 \le k \le \infty$
χ = constant to control the overall influence of the confidence level, $0 \le \chi \le 1$
β = constant to set the influence of the confidence level in the linear relationship, $0 \le \beta \le 1$

As far as the interpretation of these transformation operators is concerned :

- S_1 behaves exponentially. As k increases, the influence of the uncertainty factor becomes more significant. When a crisp data is available, k is assigned to zero.
- S_2 has a characteristic of Hamacher product. As k increases, the influence of the uncertainty factor becomes more significant. An interesting phenomena is found by setting k to 0.5. (The details will be discussed in future paper)
- S_3 is has a characteristic of Hamacher product (extension version). The additional of constant χ is to control the overall influence of the uncertainty factor. A higher value of χ results in a higher degree of influence of the uncertainty level to the data value. In other words, χ is provided to control a "secondary" influence of the uncertainty factor.
- S_4 behaves linearly. Setting β = 1eliminates the influence of the uncertainty factor.

It should be noted that setting k = 1 in S_1 is equivalent to setting $\beta = 0$ in S_4, and the transformation behaves simply like an algebraic product.

Fuzzy optimisation

The process of optimal allocation is carried out according to Dynamic Programming method. In addition, the uncertainty, expressed by the membership functions need to be propagated by transforming the data value and the uncertainty level into a single value. This value would then be compared in obtaining an optimum allocation result. An optimal policy generated by IDSS consists of the total inventory cost, the availability of the overall plant considered and an inventory policy showing how many items of particular spare parts should be kept in the store. Further, each of these values should have a membership value to indicate the level of confidence.

CONCLUSION

This paper presents a Fuzzy Logic approach to cope with uncertainty in decision making. A specific case of application which has been discussed is the propagation of uncertainty level in IDSS for power station inventory optimisation.

REFERENCES

[AF79] A.Alonso, Jr. and R.R. Fernandez. Ordering system ups productivity lower inventory. *IE Journal*, 34-38, 1979.

[Ang94] P. Angelov. A Generalized Approach to Fuzzy Optimization. *International Journal of Intelligent Systems*, 9: 261-268, 1994.

[Ber92] H.R.Berenji. Fuzzy Logic Controllers. In *An Introduction to Fuzzy Logic Applications in Intelligent Systems*, 69-96, 1992, Kluwer Academic Publisher.

[FW88] M.Francis and Webster Jr. A back-order version of the Wagner-Whitin discrete demand EOQ algorithm by Dynamic Programming, *Production and Inventory Management Journal,* First Quarter, 1-5, 1988.

[MS92] W.Mielczarski and L.F.Sugianto. Dynamic Programming in decision support system for reduction of inventory costs. *Proceedings of the 2nd International Conference on Automation, Robotics and Computer Vision*, Singapore, 1992.

[Sug92] L.F.Sugianto. *Intelligent Decision Support System for Inventory Management*, Engineering Thesis, Curtin University, Western Australia, 1992.

[SM94] L.F.Sugianto and W.Mielczarski. *Intelligent Decision Support System for Jeeralang Gas Turbine Station*, Report no.1994-R-2, Centre for Electrical Power Engineering, Monash University, Australia, 1994.

[SM94] L.F.Sugianto and W.Mielczarski. Intelligent Decision Support System for Inventory Management, *Proceedings of the Australasian Universities Power Engineering Conference (AUPEC) 1994*, South Australia, 3:688-693, 1994.

[TZ+79] U.Thole, H.J.Zimmermann and P.Zysno. On the suitability of minimum and product operators for the intersection of fuzzy sets, *Fuzzy Sets and Systems*, 2: 167-180, 1979.

[Zim91] *Fuzzy Sets Theory and its applications*, Kluwer Publishers Ltd., Boston, 1991.

Image

KNOWLEDGE REPRESENTATION FOR MODEL-BASED IMAGE PROCESSING IN MEDICINE

Tatjana Zrimec*† and Claude Sammut*

*School of Computer Science and Engineering
University of New South Wales, Sydney, Australia 2052

†Faculty of Electrical Engineering and Computer Science
University of Ljubljana, Slovenia

Email: tatjana@cse.unsw.edu.au

ABSTRACT

This paper presents a method for modelling the human cerebral vasculature by reconstructing the cerebrovascular system from variety of imaging modalities. Image processing is assisted by a knowledge-based model of the brain's anatomy and blood vessels. The model stores symbolic, structural and geometrical information. A frame representation was used to implement the model since it provides a good platform for comprehensive description of important anatomical features and for establishing communication between the image data and the symbolic knowledge.

INTRODUCTION

Computer assisted 3D visualisation of the human cerebro-vascular system is aimed at helping in the diagnosis and treatment of a variety of brain disorders. The task of this project is to reconstruct and visualise the human cerebro-vascular system from the partial information collected from a variety of medical imaging instruments. The key element in our approach to image processing and image fusion is a knowledge-based model of the cerebral vasculature.

A number of studies have noted the importance of using high-level knowledge (Selfridge 1981, Brinkley 1988, Sawyer 1992) and anatomical knowledge (Kobashi 1992, Inaoka 1992, Robinson 1991, Hill 1991) in medical image processing. Model-based methods have used anatomical information encoded as symbolic or iconic models to facilitate image processing (Baldock 1992, Gerig 1991, Rake 1987). Most of these methods have been used for the reconstruction of the coronary arteries (Rake 1987, Delaere 1991, Stansfield 1986). However there has been little research on the problem of reconstructing the much more complicated cerebral vasculature (Gerig 1993, Barillot 1985, Hill 1991, Hall 1993).

Partial information collected from x-ray angiograms and MR angiograms is used to reconstruct the cerebrovascular tree. X-ray angiograms are 2D projections of the vascular structure. The problem of 3D reconstruction using this information is under-determined and various computational techniques are employed to approximate the missing information. MR angiograms produce a volumetric representation as a set of 2D slices. Further processing is required to extract and represent the imaged structure.

Although both kinds of images contain 3D information about the anatomical structure under investigation, an expert in radiology is needed to extract, combine and interpret the information contained in them. The radiologist overcomes the problem of missing or noisy information by using knowledge of the imaged anatomy.

Many researchers have claimed that if automatic reconstruction methods are to be successful, they must use model-based strategies in much the same way as radiologists do (Baldock 1992, Gerig 1991, Rake 1987, Gerig 1993). Consequently, we use knowledge about the anatomy and the structure of the blood vessels to supply missing information and guide information extraction. A model of the cerebral vasculature was designed to store all relevant information. The model consists of symbolic, structural and geometrical information represented by frames. Spline surfaces are used for 3D graphical visualisation of the model. The frame representation provides a comprehensive description of the important cerebrovascular forms and also provides a link between the image data extracted from the images and the symbolic knowledge.

MODELLING THE HUMAN CEREBRAL VASCULATURE

One of the basic requirements for an anatomical

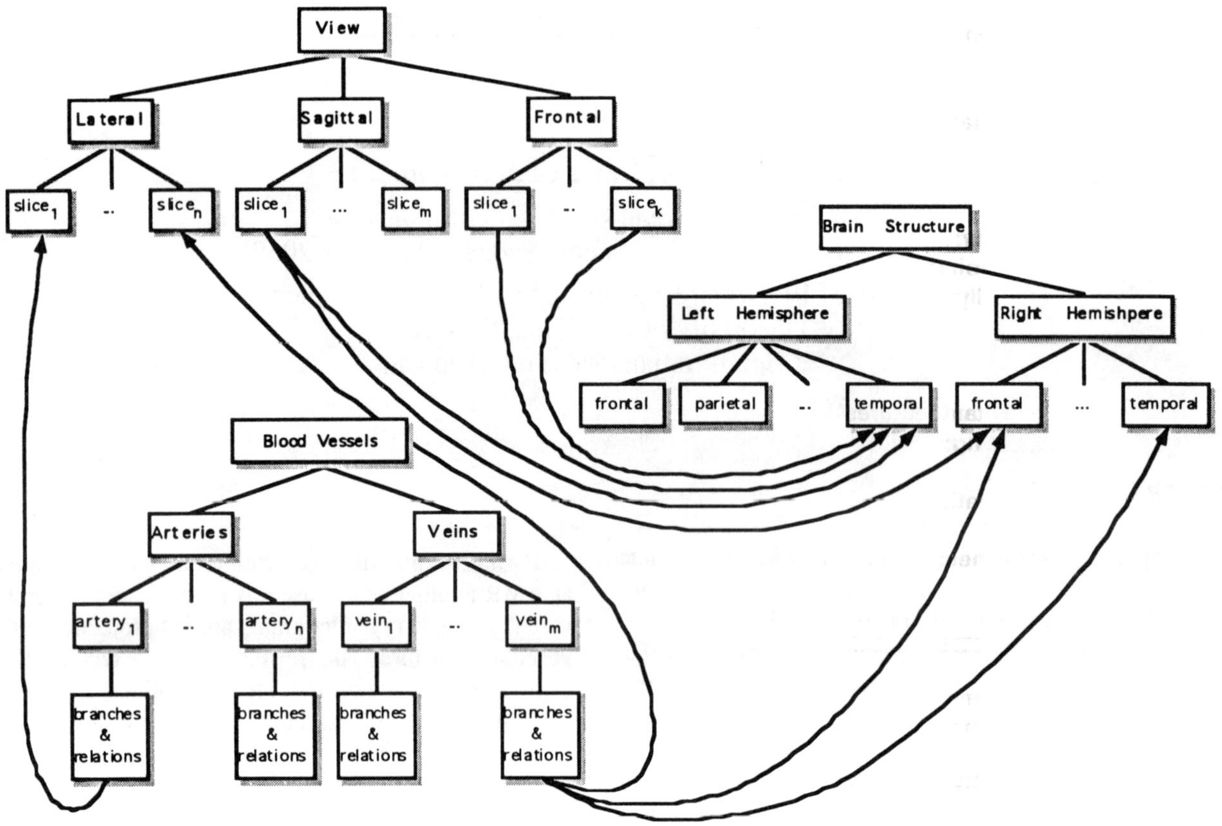

FIGURE 1 Frame Hierarchy

model is the ability to encode the shape of anatomical features and to include normal cases and variations. The model must be easy to understand, must allow easy validation and must be applicable to different imaging modalities (Baldock 1992). Further, models to assist image processing should include knowledge at both the image level and the anatomical level. With these requirements in mind, the first step in building the model is to determine what is to be represented and how.

We deal with four main sources of information:

- A symbolic description of the human anatomy from textbooks (Nolte 1993).
- Structural information of a human brain, depicted in the form of an atlas, e.g. Talairach atlas (Talairach 1988).
- A 3D volumetric representation of the blood vessels of the patient's brain, from a set of slices of Magnetic Resonance Angiograms (MRA).
- Several X-ray angiograms of the patient's brain in a form of 2D projections.

Representation of the information in these sources dictates that the model contains a symbolic description of the anatomy and the structure and geometry of the vasculature of the brain. It is also necessary to store information about spatial relationships among the vessels and other brain features.

We have chosen a frame system to implement the model (Minsky 1975, Goldstein 1977, Horn 1992). The central idea of frame theory is that a large amount of knowledge can be stored in a library of frames, which are packets of knowledge that provide descriptions of typical objects and events. Objects are represented by a set of attributes (also called slots) and their associated values. Frames may describe instances or classes. Class frames describe the generic properties of an entire class. Instance frames describe individual objects. Relations between classes and instances are expressed by special *isa* slots which indicate that an instance is a member of a class or a generic frame is a subclass of another class. Frames can be viewed as one way to structure objects in a hierarchy. Instance frames or generic frames may inherit the properties of generic frames higher up in the "isa" hierarchy.

Demons, are procedures associated with a slot. Different demons are triggered automatically, depending on how the slot is accessed. For example, when a new value is added to a slot, the "if added" demon is automatically invoked by the frame system.

Figure 1 gives an overview of the knowledge-base, organised as a frame hierarchy. The three main types of frames represent brain structure, the vasculature and the

name:	CCA (Comon Carotid Artery)
isa_generic:	artery
size:	large
width:	5 mm
anterolaterally:	crossed by
	superior thyroid vein
	middle thyroid vein
	jugular vein
laterally:	internal jugular vein
posterolaterally:	vagus nerve
medially:	larynx
	pharynx
	trachea
	esophagus
important features:	embedded in the carotid sheath
begining:	from brachiocephalic location, at the right sternoclavicular joint
termination:	at the upper border of the thyroid cartilage
	below the angle of the mandible
branches:	ICC (internal common carotid)
	ECC(external common carotid)

name:	LCCA (Left Comon Carotid Artery)
isa_generic:	CCA
begining:	from arch of the aorta
	in the superior mediastinum

name:	RCCA (Right Comon Carotid Artery)
isa_generic:	CCA
begining:	from brachiocephalic artery
	behind the right sternoclavicular joint

FIGURE 2 The description above represents an idealised Common Carotid Artery which also inherits all the properties of a generic artery. There are two Common Carotid Arteries in the cerebral vasculature, these are represented by their own generic descriptions. Note that they inherit all the properties of the Common Carotid Artery because the isa_generic property states is a kind Common Carotid Artery.

different views provided by the imaging systems. These three hierarchies contain links to each other which show the relationships between the different structures. For example, a frame representing a brain structure will contain slots indicating which vessels supply and drain that structure. The frames for those vessels have corresponding slots to indicate which structure they supply or drain. We will now describe these three frame types in further detail.

Vasculature Frames

The following slots are used to describe a blood vessel:

- the frame's identifier and the long name of the blood vessel;
- topological information, such as average length, average width, possible variations;
- geometrical information in the form of 2D shapes and sets of 3D points;
- where the vessel begins;
- where it terminates;
- important features (i.e. landmarks for radiologists);
- branches or tributaries;
- variations expressed as a percentage of cases;
- image features, such as as expected intensity, best x-ray view for the vessel;
- relations such as, anterior, lateral, anterolateral, posterolateral, posterior, medial, proximal, dorsal, in-front, behind, left-of, right-of.

The geometrical information slot in the blood vessel frame is used for displaying the vascular tree graphically. Automatic visualisation of the symbolic model enables visual inspection of the results making communication with the physicians easier while the model is being developed.

An example of the frame representation is shown in Figure 2 which contains a description of the Common

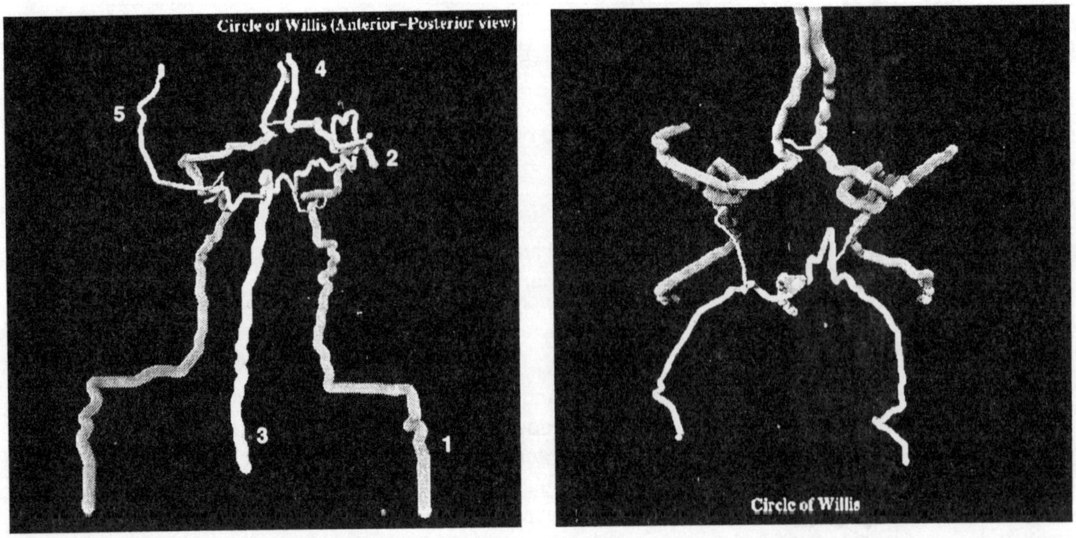

FIGURE 3a shows the anterior posterior view of the Circle of Willis. FIGURE 3b shows the Cephalocaudal view. The vessels are the two Internal Carotid Carotid arteries (1), the two Middle Cerebral arteries (2), the Basilar artery (3) formed by the two Vertebral arteries, both Anterior Cerebral arteries (4) and both Posterior Cerebral arteries (5).

Carotid Artery. We have used English expressions for the purposes of illustration but, in practice, the descriptions are in the form of symbolic expressions that can be interpreted by a program.

Brain Structure Frames

The relationship between brain features and the vasculature is established by the feeding vessels since the vessels supply or drain a particular part of the brain. These are organised as a hierarchy with the brain at the root with branches into large structures such as the left and right hemispheres. The brain is divided into two hemispheres. Each hemisphere branches into four main lobes continuing down to individual brain features. Each lobe is further subdivided. This hierarchy is shown in Figure 1. This part of the model is constructed from a coplanar stereotactic atlas of the human brain which is commonly used by radiologists and surgeons.

Each brain feature contains *supply* and *drain* slots which indicate the blood vessels which supply and drain that feature. As we mentioned earlier, there are corresponding slots in the frames representing those blood vessels so that it is easy to determine which features they supply or drain.

View Frames

A symbolic representation of the slices depicted in the Talairach atlas stores knowledge about spatial organisation. This information is similar to the information captured in MRI slices.

Different sets of slices represent views from three different angles. Slices in each set are grouped according to their approximate location, e.g. 'front or back', 'left or right', etc. This segregation assists in matching slices with the structural representation of the brain. A simplified version of this is shown in Figure 1.

The links between view frames and other kinds of frames indicate on which slices we would expect to find a particular brain feature or vessel and which vessel is expected in a particular slice. The frames also indicate which features surround each vessel.

IMPLEMENTATION

The model needs a coordinate system so that image data can be matched with anatomical knowledge. The coordinate systems was determined using information from MRA data and the Talairach atlas. In the human brain there are some structures, known as a basal brain lines, which can be located and which serve as the basis for a system of reference. The Talairach atlas allows the location of brain features which cannot be directly visualised or identified and must be identified in relation to other anatomical features, such as the basal brain lines. A three dimensional, proportional grid system is established according to the maximal dimensions of the brain in the three planes of space. This system adapts itself to a brains of all dimensions.

We have used the atlas to determine the coordinate system of the Circle of Willis which lies at the base of the brain and is formed by the main arteries of the brain vasculature. The Anterior and Posterior Communicating

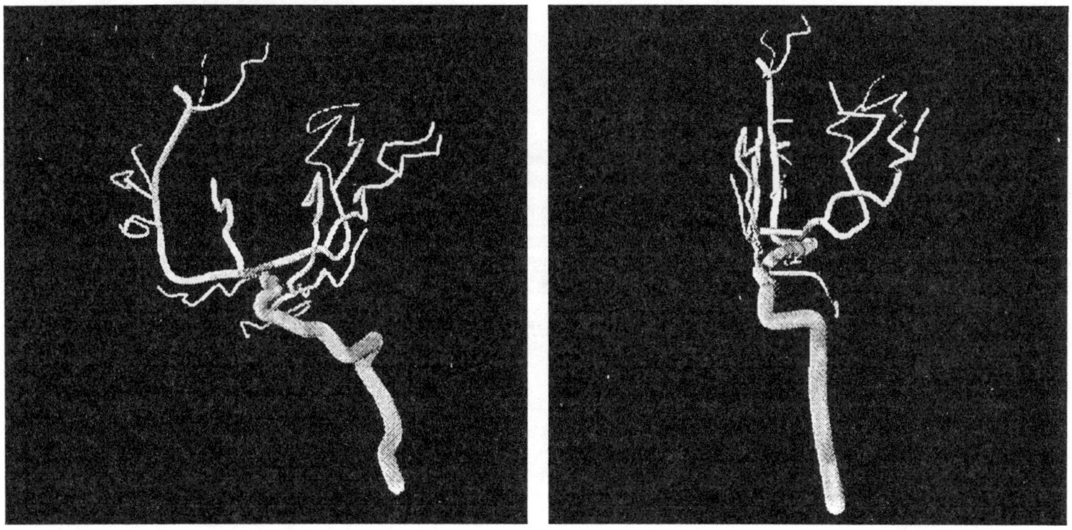

FIGURE 4 Different views of the model of the Internal Carotid Artery

arteries make connections between the basic blood supply of the brain the two Internal Carotid Arteries and the two Vertebral Arteries. The coordinate system of all arteries is determined with respect to the Circle of Willis.

The blood vessels of the Circle of Willis were extracted from a patient's MRA data in a form of 3D skeleton. The MRA data consists of 60 slices, with slice thickness 0.62 x 0.62 x1.5 mm. The vessel thickness was initially determined from anatomical atlases and latter it was corrected by a neuroradiologist. A special graphical environment was developed to visualise the vasculature. A graphical representation of the Circle of Willis is shown in Figure 3.

We have implemented frames for the Common Carotid Artery with all its named branches. We have used FRESH (Sammut 1994) for the frame implementation. The FRESH language is designed as an environment for building knowledge-based systems. It includes features from Prolog and Lisp and Frames. The FRESH system is implemented in C and it allows the use of C. This facility is very useful since the low-level vision processing is implemented in C.

The graphical representation of the model of the Internal Carotid Artery, one of the main branches of the Common Carotid Artery, is shown in Figure 4. This artery is one of the main vessels for supplying blood to the brain, the eyes, the forehead and part of the nose.

USING THE MODEL

In order to make the model operational we must define a set of slots for the blood vessel frames that describe the features extracted from images. During feature extraction an instance frame is created and compared with the frames of the model in order to recognise the object.

The feature extraction system, currently under development, uses low-level image processing techniques to trace blood vessels. This generates line segments, organised as a tree structure, where each line segment corresponds to a blood vessel segment. A frame representation of the blood vessel segments, is built incrementally. The thick vessels are extracted and labelled before proceeding to the finer vessels. The process of using the vascular model during feature extraction and labelling is illustrated in Figure 5.

The tracing algorithm is far from perfect and may miss some blood vessels or add spurious line segments. The reference model of the brain can be used to "fill in" the missing anatomy of the patient's model and filter out noise.

CONCLUSION

We have used a frame representation because it is well suited to "expectation-driven" programming. By this we mean that frames provide an easy way of identifying properties that our knowledge tells us should be present. If these properties are not evident, then executable procedures attached to the properties can be automatically invoked to determine the value of a property. Thus, if we have identified a particular blood vessel and its frame contains information about branches, procedures attached to the branching information can begin a search for those branches in predicted locations.

We have implemented the algorithms for segmentation and vessel tracking. We have also constructed a model of one of the main branches of the cerebral vasculature and we are currently working on the frame

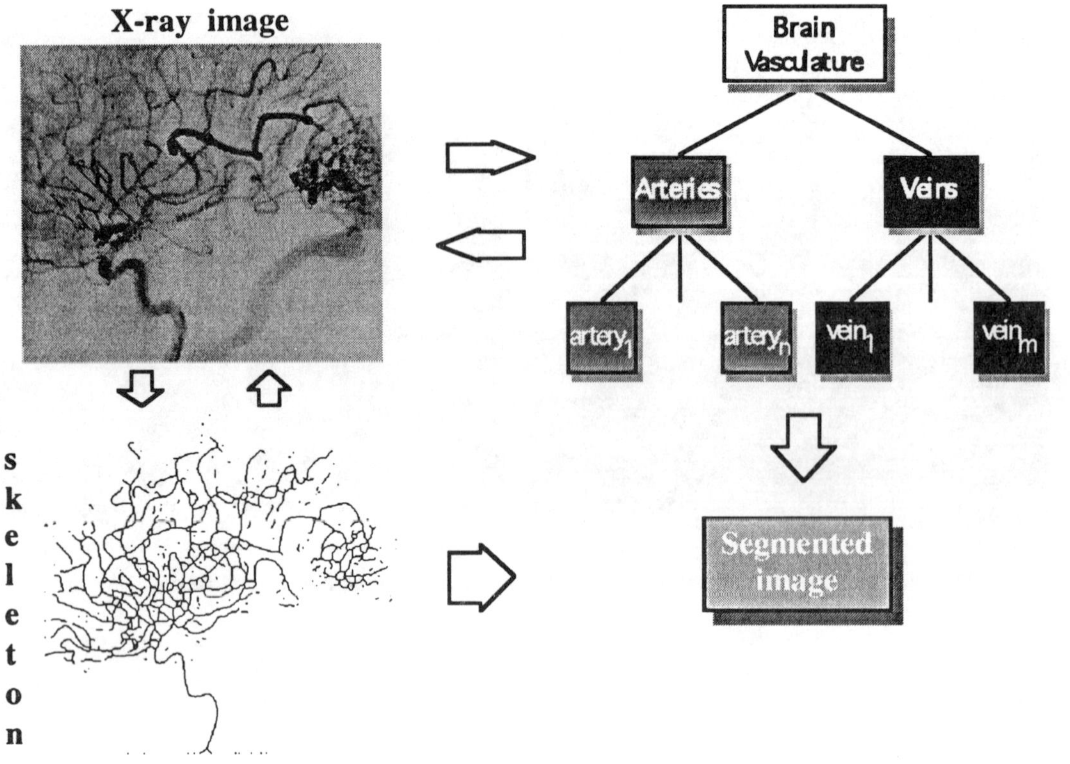

FIGURE 5 Using the model

matching procedures.

Therapy and planning require more information then simply the presence of a disease. 3D reconstruction and visualisation provides information about the vascular topology and it can be used in surgical planning and vascular treatment, to avoid damaging main arteries.

ACKNOWLEDGEMENTS

Thanks to Dr. Geoff Parker for his valuable assistance in constructing the model. This work was supported by the Slovenian Ministry of Science and Technology.

REFERENCES

Baldock R. A. (1992) .Trainable models for the interpretation of biological images. Image and Vision Computing, 10 (6), 444-450.

Barillot, C., Gibaud, B., Scarabin, J.-M., & Coatrieux, J.-L. (1985). 3D Reconstruction of Cerebral Blood Vessels. IEEE Computer Graphic and Algorithms, December, 13-19.

Brinkley, J. F. (1988). Representing Biologic Objects as geometric constraint networks. In AAAI Spring Symposium Series: Artificial Intelligence in Medicine, 7-8.

Delaere, D., Smets,C., Marchal, G., Suetens, P. (1991). Knowledge-based system for the three-dimensional reconstruction of blood vessels from two angiographic projections. MBEC North Sea: Medical physics and imaging, 27-36.

Gerig, G., Kikinis, R., Kuoni, W., Schulthess, W. K,Kuber, O. (1991) Semiautomated ROI Analysis in Dynamic MR Studies. Part I: Image Analysis Tools for Automatic Correction of Organ Displacements, Journal of Computer Assisted Tomography, 15 (5), 725-732.

Gerig, G., Koller, T,. Szekely, G., Brechbuhler, C., Kubler, O. (1993). Symbolic Description of 3-D Structures Applied to a Cerebral Vessel Tree Obtained from M R Angiography Volume Data. Information Processing in Medical Imaging, 13th International Conference, IPMI'93, Arizona, USA, June, 95-111.

Goldstein, I.P. Roberts, B. (1977). Using Frames in Scheduling. In Proc. Fifth Int. Joint Conf. on Artificial Intelligence, August, 257-284.

Hall, P.M., Watt, A., McGewgor, J., Walton, L. ,Bregval, U.,(1993). A Model of a Collection of Vasculature, CS-93-09, The University of Sheffield.

Hill, D. L. G., Hawkes, D. J., & Hardingham, C. R. (1991). The use of anatomical knowledge to register 3D blood vessel data derived from DSA with MR

Images. In SPIE Image Processing, 1445, 348-357.

Horn,W. (1992). Utilizing Detailed Anatomical Knowledgefor Hypothesis Formation and Hypothesis Testing in Rheumatological Decision Support, Deep Modelsfor Medical Knowledge Engineering, E. Keravnou, Elsevier Science Publishers B.V, 27-50.

Kobashi, M., & Shapiro, L. G. (1992). Knowledge-Based Organ Identification from CT Images. In *SPIE Medical Imaging VI: Image Processing*, 1652, 544-554.

Minsky, M. (1975). A framework for representing knowledge. In Winston, P (ed). The Psychology of Computer Vision, McGraw-Hill, New York, 211-277.

Nolte, J. (1993). The Human Brain an Introduction to its Functional Anatomy, Mosby-Year Book, Inc., Missouri.

Noriko, I., Suzuki, H., Fukuda, M. (1992). Hepatic Blood Vessel Recognition Using Anatomical Knowledge. In SPIE Medical Imaging Vl: Image Processing, 519-513.

Rake, S.T., Smith, L.D. (1987). The interpretation of X-ray Angiograms using a Blackboard Control Architecture. Computer Assisted Radiology, International Symposium CAR'87, 681-685.

Robinson, G.P, Colchester, A.C.F., Griffin, L.D., (1993) Model Based Recognition of Anatomical Obiects from Medical Images. In Inormation Processing in Medical Imaging, 13th International Conference, IPMI'93, Arizona, USA, 197-211.

Sammut, C.A. (1994). FRESH Programmer's Manual, School of Computer Science and Engineering, University of New South Wales, Australia.

Sawyer, G., Mason, D. C., Hindley, N., Johnson, D. G., Jones-Parry, I. H., Oddy, C. J., Pike, T. K., Plassars, T., Rye, A. J., de Salabert, A., Serpico, B., & Wielogorski, A. L. (1992). MuSIP Multi-Sensor Image Processing system. Image and Vision Computing, 10 (9), 589-609.

Selfridge, P.G. et al. (1981). Organ Detection in Abdominal Computerised Tomography Scans: Application to the Kidney. Computer Graphics in Image Processing, 15, 265-278.

Stansfield, S.A. (1986). ANGY a Rule-Based Expert System for Automatic Segmentation of Coronary Vessel from Digital Substracted Angiograms. IEEE Trans on Pattern Analysis and Machine Intelligence, March, 8 (2),188-199.

Talairach, J., Tournoux, P (1988). Co-Planar Stereotaxic Atlas of the Human Brain, 3-Dimensional Proportional System: An Approach to Cerebral Imaging, Time Medical Publishers, Inc. Georg Thieme Verlag.

A NEW ALGORITH FOR HANDLING CONTINUOUS-VALUED ATTRIBUTES IN DECISION TREE GENERATION AND ITS APPLICATION TO DRAWING RECOGNITION

Wei Lu, Masao Sakauchi
3rd Dept., Institute of Industrial Science, University of Tokyo,
7-22-1, Roppongi, Minato-ku, Tokyo, 106 Japan
Email:luwei@sak.iis.u-tokyo.ac.jp

ABSTRACT

In this paper, we present a new algorithm for handling continuous-valued attributes in decision tree generation. Our algorithm aims at reducing the efforts for finding the optimal cut value and is very effective when handling attributes that have close-to-uniform distributions. The formal proof of the proposed criterion for cut point selection is given. The proposed algorithm is also applied to rule acquisition for drawing recognition. Experimental results show that efficiency of evaluating continuous-valued attributes for cut value selection is further improved when compared to existing methods.

INTRODUCTION

ID3[Qinlan86] is a powerful and popular algorithm for decision tree generation in empirical learning from examples. When given a set of pre-classified examples expressed in terms of attributes and classes, ID3 forms the decision tree by recursively selecting attributes that locally minimize the information entropy of the classes in any subset. It has been adopted and extended in many machine learning systems[Qinlan88], [Cheng88], [Clark88]. The initial algorithm can only handle nominal attributes, and extended algorithms can handle continuous-valued attributes. One of the problems with this algorithm is that when the number of data is very large and most of the data are continuous, the cut values to be evaluated increase dramatically. Fayyad has proposed a heuristic for reducing the number of cut point candidates[Fayyad92]. But when data has uniform distribution, the reduction becomes less significant. In this paper, we propose two improved heuristics to further reduce the cut point candidates and ensure optimal result at the same time. We also give a formal proof of the heuristic and apply the algorithm to a real-world problem, that is rule acquisition for drawing recognition. Since our main purpose is to improve the efficiency of the construction of decision trees, we choose ID3 for evaluation because of its simplicity and its similarity with other algorithms.

In the following sections we will first discuss the formation of decision trees and handling of continuous-valued attributes. Then we will discuss the proposed new heuristics and the formal proof. Finally we will present an application of the proposed algorithm to rule acquisition for recognition drawings.

DECISION TREE AND ITS HANDLING OF CONTINUOUS-VALUED ATTRIBUTES

In empirical learning from examples, a group of examples $\{E_i\}$ expressed in terms of a set of attributes $\{A_j\}$ and their classifications $\{C_k\}$ are presented to the learning system. The learning system then constructs a decision tree which can uniquely classify the whole group of examples. ID3 is proposed as an algorithm for simple and efficient construction of decision trees. It recursively divides examples into subsets according to their attribute values until every subset contains examples belonging to a single class. There are cases when such divisions are difficult. But since there have been many proposals for this problem and our interest is on the efficiency of constructing decision trees, we will only concentrate on the ideal situation.

When continuous-valued attributes are present, they are typically discretized into subranges. The most common method of discretization is to partition the value range into two subranges. In this case, a threshold value T is determined; all examples that have attribute values smaller than T are partitioned to left branch of the decision tree, and those greater to the right. Such a threshold T is usually called a *cut point.*

Let S be the subset of the examples under consideration, K be the total number of classes, $P(C_k, S)$ be the proportion of examples in S that have class C_k, the *class entropy* of the subset S is defined as[Qinlan86]:

$$Ent(S) = -\sum_{i=1}^{K} P(C_i, S) log(P(C_i, S)) \quad (1)$$

When the subset S is divided into S_1 and S_2 at cut point T its *class information entropy of the partition included by T* is defined as[Fayyad92]:

$$E(A, T; S) = \frac{|S_1|}{N} Ent(S_1) + \frac{|S_2|}{N} Ent(S_2) \quad (2)$$

The basic heuristic for decision tree construction is to find the cut point that minimizes the class information entropy defined above. It does not guarantee optimal trees, but has been proven to be very effective by its various practical applications.

One of the main problems with this cut point selection heuristic is its computation cost, especially when the number of data is very large and attributes are continuous-valued. Since real-world problems for machine learning systems usually contain large number of data, and repeated trial and error process is necessary for fine tuning, more efficient ways of searching for the best cut points are greatly expected. A heuristic(hereafter called *heuristic 1*) proposed in reference [Fayyad92] can reduce the number of cut points to be evaluated by equation 2, and at the same time ensure the *true* best partition of attribute range. The heuristic 1 for determining the potential cut point when attribute A_i is under consideration can be summarized as follows:

1. sort values of attribute A_i of all examples in ascending order
2. find the sub-ranges in which the inclosed examples belong to one single class
3. choose all the borders between the sub-ranges as cut point candidates

It has also been proven in reference [Fayyad92] that theoretically the class information entropy defined by equation 2 only takes its minimum value on the border points described above. Empirical evaluation shows that heuristic 1 can reduce the number of cut points by a ratio of 30% to 85%.

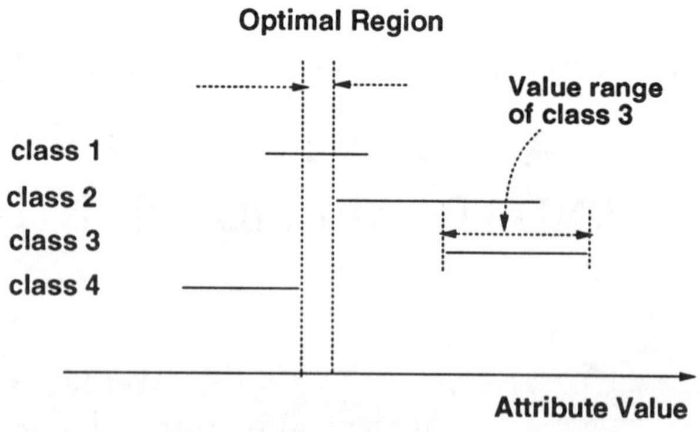

FIGURE 1: Optimal Partition Region

AN IMPROVED ALGORITHM FOR CUT POINT SELECTION

Improved Algorithm

When the number of borders defined in the previous section is k and the total number of examples is N, the number of cut points by heuristic 1 is $k-1$ (where usually $k-1 \ll N$). But when values of an attribute have uniform distribution, which means the values are evenly distributed within a certain range, the number of borders will increase along with the degree of overlapping among classes. In the worst case, after sorting the values in ascending order, all the examples have left and right neighbors of different classes, which makes the number of cut point candidates to be $N-1$, the same as without heuristic 1.

When attributes have uniform distribution, the number of examples is proportional to the length of the value range. we have done some simulations with presumed data sets and the theoretical calculation results of class information entropies for all possible cut points are given in Fig.2. The results show that the entropy always takes minimum values at the border of value ranges of each class. Intuitively, when cut points falls on the borders, the partition result should generate an easier to understand partition, since such cut points can divide one class completely into one side of the partition. We will prove in next section that when the values have uniform distribution, the class information entropy of partitioned result always takes minimum value on the borders. Therefore, as long as the value distribution is close to uniform distribution, the minimum value of class information entropy of partition should always fall around the boarders of value ranges.

From the above results, we propose a new heuristic which examines less cut point candidates and at the same time ensures the best partition of attributes. It can be summarized as follows:

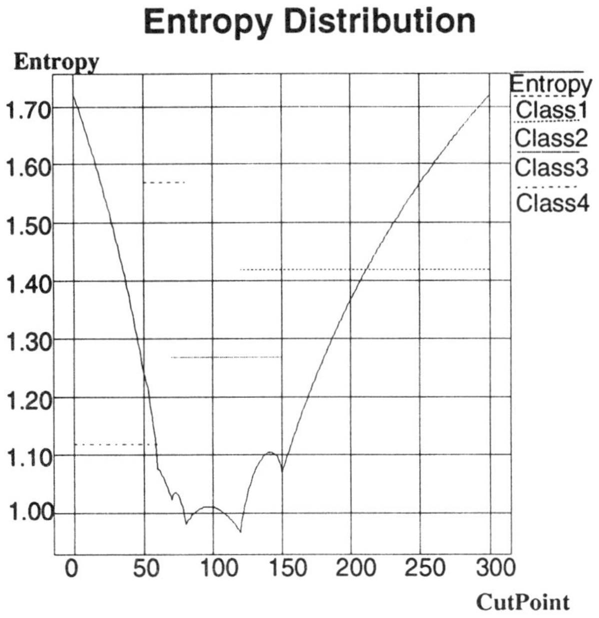

FIGURE 2: Distribution of Entropies

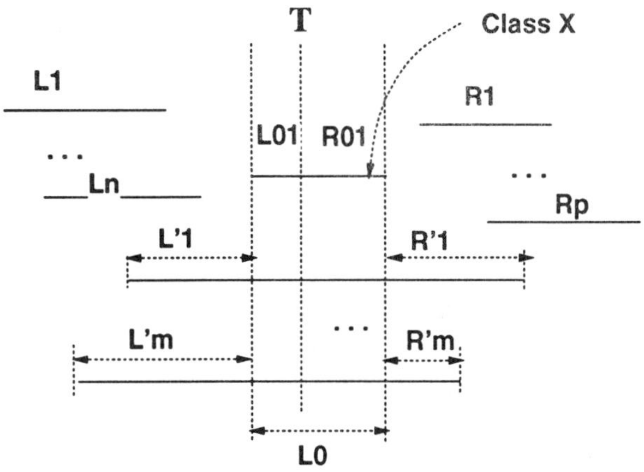

FIGURE 3: Totally Overlapped Area

heuristic 2 when the value ranges of attribute A_i of classes $\{C_k\}$ are $\{[min_k, max_k]\}$, the boundary between $v1$ and min_k, v_2 and max_k are cut point candidates. Here v_1 and v_2 satisfy the following conditions:

1. $v_1 < min_k, v_2 > max_k$
2. v_1 and v_2 belong to different classes from those of min_k and max_k respectively
3. there is no other examples within $[v_1, min_k]$ and $[max_k, v_2]$

By the same simulations, we found that when the ranges are not too unbalanced, the total minimum of entropy alway fall in one of the *optimal partition regions* (Fig.1) or is close to it. Therefore we propose another heuristic as follows:

heuristic 3 when the value ranges are not over unbalanced and there exists optimal partition regions, candidates determined by heuristic 2 can be further limited to those within the optimal partition regions

By heuristic 3, at least two classes can be divided completely into different side of the partition and therefore can be expected to generate more meaningful partition. Our experiments in next section show that with our real-world problem, while for some cases heuristic 3 causes a few extra nodes or leafs(but generates more balanced trees), it can also generate better partition than heuristic 1 can.

In the worst case when there are no candidates determined by heuristic 2 and 3, which is generally unlikely, the heuristic 1 will be used to determine cut point candidates.

Formal Proof of the Proposed Heuristics

First we assume that the cut point falls within the value range of class X and this range is either totally enclosed in some other ranges or has nothing to do with the rest, as shown in Fig.3. For convenience, let $L = \sum_{i=1}^{n} L_i + \sum_{i=1}^{m} L'_i, R = \sum_{i=1}^{p} R_i + \sum_{i=1}^{m} R'_i, N = L + R + L_0, R_{01} = L_0 - L_{01}$.

When the values are of uniform distribution, equation (2) becomes:

$$Ent(A, S; T) = -\frac{L + (1+m)L_{01}}{N}($$
$$\sum_{i=1}^{n} \frac{L_i}{L + (1+m)L_{01}} log(\frac{L_i}{L + (1+m)L_{01}}) +$$
$$\sum_{i=1}^{m} \frac{L'_i + L_{01}}{L + (1+m)L_{01}} log(\frac{L'_i + L_{01}}{L + (1+m)L_{01}}) +$$

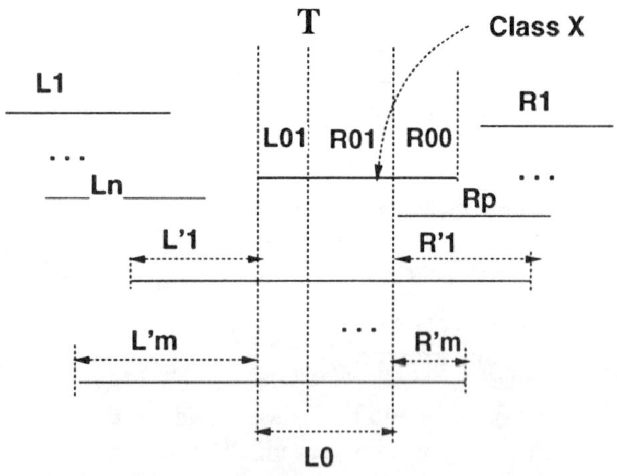

FIGURE 4: Partially Overlapped Area

$$\frac{L_{01}}{L+(1+m)L_{01}}log(\frac{L_{01}}{L+(1+m)L_{01}}))$$
$$-\frac{R+(1+m)R_{01}}{N}($$
$$\sum_{i=1}^{p}\frac{R_i}{R+(1+m)R_{01}}log(\frac{R_i}{R+(1+m)R_{01}})+$$
$$\sum_{i=1}^{m}\frac{R'_i+R_{01}}{R+(1+m)R_{01}}log(\frac{R'_i+R_{01}}{R+(1+m)R_{01}})+$$
$$\frac{R_{01}}{R+(1+m)R_{01}}log(\frac{R_{01}}{R+(1+m)R_{01}})) \qquad (3)$$

By rearranging equation (3), we can get:

$$Ent(A,S;T) = -\frac{1}{N}(\sum_{i=1}^{n}L_i log(L_i) -$$
$$(L+(1+m)L_{01})log(L+(1+m)L_{01})+$$
$$\sum_{i=1}^{m}(L'_i+L_{01})log(L'_i+L_{01})+L_{01}log(L_{01})+$$
$$\sum_{i=1}^{p}R_i log(R_i)-(R+(1+m)R_{01})log(R+$$
$$(1+m)R_{01})+\sum_{i=1}^{m}(R'_i+R_{01})log(R'_i+$$
$$R_{01})+R_{01}log(R_{01}) \qquad (4)$$

We now take the first and second derivatives of equation (4) with respect to L_{01} and get:

$$Ent(A,S;T)'$$
$$= -\frac{1}{N}(-(1+m)log(L+(1+m)L_{01})+$$
$$\sum_{i=1}^{m}log(L'_i+L_{01})+log(L_{01})$$
$$-(1+m)log(R+(1+m)R_{01})+$$
$$\sum_{i=1}^{m}log(R'_i+R_{01})+log(R_{01})) \qquad (5)$$

$$Ent(A,S;T)''$$
$$= -\frac{1}{N}(\frac{-(1+m)^2}{L+(1+m)L_{01}}+\sum_{i=1}^{m}\frac{1}{L'_i+L_{01}}+\frac{1}{L_{01}}$$
$$-\frac{(1+m)^2}{R+(1+m)R_{01}}+\sum_{i=1}^{m}\frac{1}{R'_i+R_{01}}+\frac{1}{R_{01}}) \qquad (6)$$

Next, we prove by induction that $Ent(A,S\ T)'' < 0$.

Let SUM_k be the value of $N \cdot Ent(A,S\ T)''$ when $m=k$.

$$SUM_1 = -\frac{L^2}{L'_{01}(L'_1+L_{01})(L'_1+2L_{01})}-$$
$$\frac{R^2}{R'_{01}(R'_1+R_{01})(R'_1+2R_{01})}$$

Therefore $SUM_1 < 0$. Next assume $SUM_k < 0$,

$$SUM_{K+1}$$
$$= -\frac{(1+k+1)^2}{L+L'_{k+1}+L'_{01}}+\sum_{i=1}^{k+1}\frac{1}{L'_i+L_{01}}+\frac{1}{L_{01}}$$
$$-\frac{(1+k+1)^2}{R+R'_{k+1}+R'_{01}}+\sum_{i=1}^{k+1}\frac{1}{R'_i+R_{01}}+\frac{1}{R_{01}}$$
$$= -\frac{(2+k)^2}{L+L'_{k+1}+(k+2)L_{01}}+\frac{(1+k)^2}{L+(1+k)L_{01}}+$$
$$\frac{1}{L'_{k+1}+L_{01}}-\frac{(2+k)^2}{R+R'_{k+1}+(k+2)R_{01}}+$$
$$\frac{(1+k)^2}{R+(1+k)R_{01}}+\frac{1}{R'_{k+1}+R_{01}}+SUM_k \qquad (7)$$

For convenience, let $a = L'_{k+1}+L_{01}, b = L+(k+1)L_{01}, c = R'_{k+1}+R_{01}, d = R+(k+1)R_{01}$.

$$SUM_{k+1} = -\frac{(b-(k+1)a)^2}{ab(a+b)}-\frac{(d-(k+1)c)^2}{cd(c+d)}+SUM_k$$

From the assumption, $SUM_K < 0$、and because $a > 0$、$b > 0$、$c > 0$、$d > 0$, we have $SUM_{k+1} < 0$. Therefore $Ent(A,S;T)'' < 0$. So far we have proven that Ent(A,S;T) is downward convex and hence always takes its minimum value at $L_{01} = 0 or L_{01} = L_0$, which are the borders of the value range[Luenberger73].

Similarly, for the situation shown in Fig.4, equation (2) becomes:

$$Ent(A,S;T) = -\frac{L+(1+m)L_{01}}{N}($$
$$\sum_{i=1}^{n}\frac{L_i}{L+(1+m)L_{01}}log(\frac{L_i}{L+(1+m)L_{01}})+$$

FIGURE 5: An example of map drawing

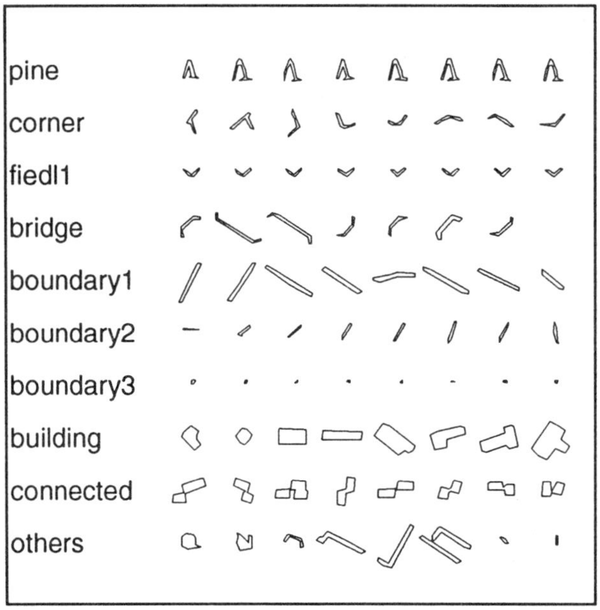

FIGURE 6: Typical Symbols with Distortion

$$\sum_{i=1}^{m}\frac{L'_i+L_{01}}{L+(1+m)L_{01}}log(\frac{L'_i+L_{01}}{L+(1+m)L_{01}})+$$
$$\frac{L_{01}}{L+(1+m)L_{01}}log(\frac{L_{01}}{L+(1+m)L_{01}}))$$
$$-\frac{R+(1+m)R_{01}+R_{00}}{N}($$
$$\sum_{i=1}^{p}\frac{R_i}{R+(1+m)R_{01}+R_{00}}log(\frac{R_i}{R+(1+m)R_{01}+R_{00}})+$$
$$\sum_{i=1}^{m}\frac{R'_i+R_{01}}{R+(1+m)R_{01}+R_{00}}log(\frac{R'_i+R_{01}}{R+(1+m)R_{01}+R_{00}})+$$
$$\frac{R_{01}+R_{00}}{R+(1+m)R_{01}+R_{00}}log(\frac{R_{01}+R_{00}}{R+(1+m)R_{01}+R_{00}})) \quad (8)$$

In this case, it can also be proven by deduction that $Ent(A,S;T)'' < 0$. Since other cases are the combination of the above mentioned two, it has been proven that class information entropy always takes its minimum value at one of the range borders.

APPLICATION TO RECOGNITION OF LINE DRAWINGS

To evaluate the proposed heuristics, we have applied them to a real-world problem, that is rule acquisition for line drawing recognition. We have chosen map drawing as our experimental subjects. The consequential purpose of the recognition system is to convert map information drawn on paper into machine readable abstract data. One major step of the recognition process is to classify the symbols into corresponding classes. Fig.5 shows an example of map drawing vectorized through a scanner and a vectorization system developed by author's group[Lu88]. Fig.6 shows some typical symbols with vectorization distortions. The symbols are designed in such way that they are visually distinguishable without much confusion. Therefore, theoretically it is possible to recognize them with comparatively small number of concise production rules. On the other hand, the values of the attributes overlap with each other, and after vectorization, most of the symbols are distorted in various ways. Besides, shape and dimension of some symbols such as building, bridge and corner vary within certain ranges. Therefore, tremendous trial and error efforts are required for writing production rules for recognition of such objects. Furthermore, whenever there is any change in vectorization conditions, symbol specification and so on, the same efforts become necessary again. As a result, a more flexible way of building drawing reconization system is necessary.

To ease the effort of recognition rule generation and maintenance, we have developed a prototype of recognition system with the feature of empirical learning. This system can acquire recognition rules through interactive operations. An operator only needs to specify examples of every class from the display with mouse and then the system will generate recognition rules from the decision trees constructed by ID3 algorithm. There have been some researches on application of machine learning technologies to image processing and machine vision[Maza91],[Terry93]. But none of them is on ID3-like algorithms. We can show by our experiments that ID3 like algorithm is also effective in the generation of recognition rules for graphical sysmbols. Our approach to rule generation by empirical learning is also very effective when a new drawing is to be recognized and has some new patterns. In this case the recognition rules can be adjusted only by specifying the new pattern and letting the system rebuild the rules. When the objects to be recognized are expressed in continuous-valued attributes, the distribution of values are close to uniform distribution. Therefore this is also a good case for verifying the effectiveness of the proposed heuristics.

The learning system is implemented on Sun Sparc workstation with C++ and Prolog language and X-window system. The total size for decision tree generation is about 1200 lines in Prolog and 1500 in C++.

A map drawing is first digitized by using an image scanner. Then the line segments are detected and converted to line vectors. After vectorization, all the symbols are converted into polygons, and among many shape attributes, represented by 11 shape attributes as follows:

area A the area of the polygon P.

convex area A_c the area of the convex polygon of P.

length L total length of the edges of P.

num Num total number of edges of P.

orientation O orientation of the elongation axis[Yang91].

width W width of the circumscribing rectangle obtained along the elongation axis.

height H height of the circumscribing rectangle obtained along the elongation axis.

convexness R_c A/A_c. the closer to 1, the closer to a convex polygon.

epitomization R_ep L^2/A. the closer to 1, the more integrated around the center

rectangle likeness R_r A/A_r. the closer to 1, the closer to a rectangle.

elongation R_el H/W. the larger, the more elongated along elongation axis.

There are 4940 symbol candidates in the example drawings. 12 cases are chosen for classification. The learning is realized by the repetition of: (a) specification of positive/negative examples, (b) generation of recognition rules, and (c) verification of the generated rules until all the symbols are recognized correctly. The average number of repetition is about 8. We have carried out experiments in three ways: (1)evaluation of heuristic 2, (2)evaluation of heuristic 3, and (3)evaluation of the sensitivity of both heuristics. When

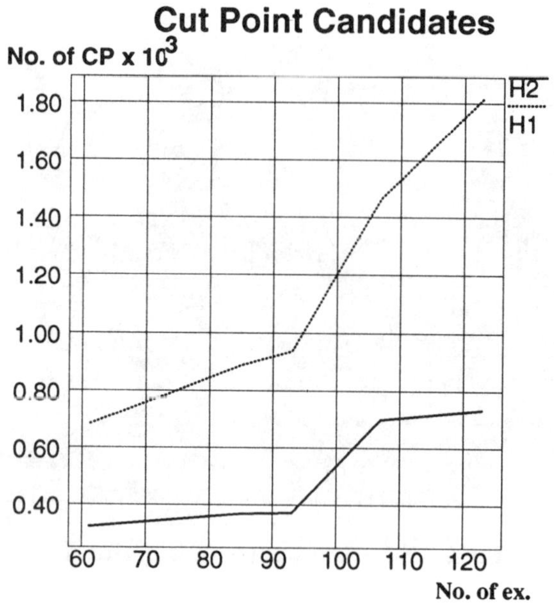

FIGURE 7: Noumber of cut point candidates

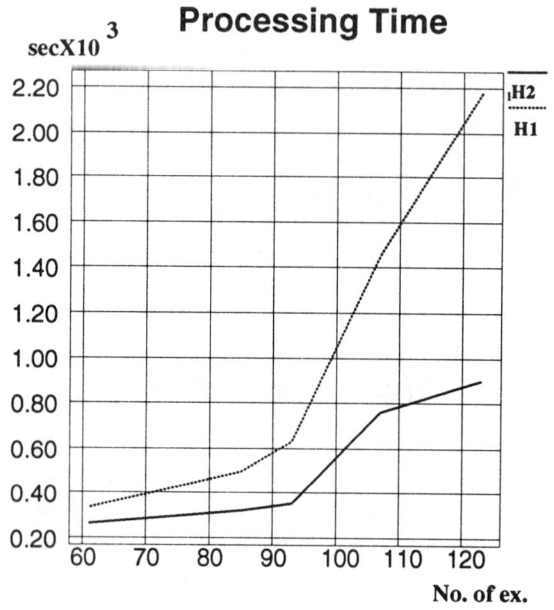

FIGURE 8: Processing time including overhead

data set	ex. no.	class no.	number of cut points			processing time (sec.)		
			H1	H2	H2&H3	H1	H2	H2&H3
data1	61	7	680	326	173	336	204	161
data2	85	7	887	370	217	558	323	283
data3	93	7	936	374	217	630	363	320
data4	107	10	1468	736	348	1454	760	550
data5	123	10	1815	732	316	2177	900	648

TABLE 1: Experimental Result 2

only heuristic 2 is used, the decision trees generated are completely the same as when using heuristic 1. One group of typical experimental results are shown in Fig.7 and Fig.8. Fig.7 shows the number of cut point(CP) candidates to be evaluated for best partition. The axis of abscissas is the number of examples during each repetition stage, the axis of ordinates is the number of CP evaluated by the proposed algorithm and heuristic 1. Fig.8 shows the actual processing time, including overhead. The axes of abscissas and ordinates are the number of example data and the processing time in seconds respectively. The dramatic changes are because of introduction of examples from new classes. The processing time is relatively long, because currently the Prolog has been developed by author's group to realize flexible general-purpose hypermedia development environment and currently has no compiler. It is estimated that the processing time can be at least 50 times faster after compilation. Here, while processing time depends on implementation, we have implemented the system in such a way that only the criterion for CP selection is different. From the result we can see that by heuristic 2 only, the number of CP candidates is further reduced by 52% to 60%.

Table 1 shows the result of using both heuristics 2 and 3 together with other results for comparison. Five typical data sets obtained during the learning process are used. The results show that the number of CP candidates is further reduced by 47% to 56%, making the total saving to be 74% to 82%. The decision trees generated have some differences from those generated by heuristic 1, but most of the trees have the same have the same number of leaves and nodes are the same except one. The decision tree for the last set of data has one node and one leaf less than when using heuristic 1 or 2. For the remaining cases such as data1, data4 and data5, the results are the same as before. Therefore, heuristic 3 can be said to be effective and does not affect the overall performance.

To evaluate the effect of increase in the number of examples on the CP candidates, other sets of examples with more similar examples are tested. Here "similar example" means that the corresponding attribute values are within the same range. For comparison, we also choose 5 sets having almost the same type of examples as those in experiment 2, with the total example numbers being 211, 235, 243, 257 and 273 respectively. The results are shown in Table 2. "Type1" and "type2" mean example data sets used in experiment 1 and experiment 2 respectively. The change in CP candidates determined by heuristic 2 and 3 is always lower than 22% (actually only one 22% at the beginning, the rest are below 10%, whereas that by heuristic 1 is always greater than 25%. This result indicates that the proposed heuristics are less sensitive to changes in number of examples, as long as the value ranges do not change significantly. Actually most of the changes in the number of CP candidates are caused by changes in the structure of decision trees. Since by heuristic 2, as long as the value range borders are duplicated, the candidates should be always $2k-2$, where k is the number of classes in the sub set of examples.

The decision trees in Table 2 have the same number of leaves and nodes except those for data 5. By heuristic 2 only the result has one leaf and node less than that by heuristic 1; by heuristic 2 and 3, the result has one leaf and node more but the maximum length is reduced by one, which means a more balanced tree.

Finally, the complexity of the data set used in our experiments is evaluated by comparing the number of evaluated CP candidates decision trees and savings of CP with those of other systems, since the actual data sets are not available. The total number of cut points evaluated by heuristic 1 at the final stage is about 2000 to 3000. This is about the average when compared with the data sets used in reference(500 to 5000)[Fayyad92]. The saving of CP by heuristic 1 is about 50% to 70%. This is also about the average of those of the data sets mentioned in the same reference (33% to 84%). Therefore when judged by the degree of attribute value's overlapping, this rule acquisition problem can be considered to be of average complexity.

Class no.	heuristic 1		heuristic 2		heuristic 2&3	
	type1	type2	type1	type2	type1	type2
data1	680	1175	326	403	173	181
data2	887	1273	370	404	217	215
data3	936	1290	374	370	217	229
data4	1468	1835	700	722	348	348
data5	1815	2361	732	760	316	355

TABLE 2: Experimental Result 3

CONCLUSION

In this paper, we proposed a new algorithm of handling continuous-valued attributes for decision tree construction. It is effective when the attribute valules have close-to-uniform distributions. Under the condition of uniform distribution, the proof for the main heuristic is also given. Experimental results have shown that for real-world data, the algorithm can reduce the number of cut point candidates by 72% to 82%, without affecting the construction of decision trees. The constructed decision trees sometimes have some differences by proposed heuristics, sometimes with a few more leafs/nodes and sometime less. This proves that the basic heuristic by class information only does not necessarily ensure optimal trees. If the efficiency of decision trees is crucial, at the final stage of learning, both algorithms can be applied to find the better result so that the best inference efficiency during application can be realized.

Since the main purpose of the experiment is to evaluate the efficiency of decision tree construction, we have selected the ID3 algorithm for its simplicity. Our future work will be: (a)evaluation of the algorithm with other real-world data and (b)verification of applicability of the proposed heuristics to other ID3 like algorithms. Since most of the other systems rely on similar cut point selection criterions, good results can be expected.

REFERENCES

[Qinlan86] J. R. Quinlan: "Induction of decision trees ". Machine Learning 1, pp.81-106, 1986

[Qinlan88] J. R. Quinlan: "Decision Trees and Multi-Valued Attributes", Machine Intelligence, 11, pp.305-318, 1988

[Cheng88] J. Cheng, U. M. Fayyad, K. B. Irani et al: "Improved decision trees: A generalized version of ID3". Proceedings of the Fifth International Conference on Machine Learning, pp.100-108. San Mateo, CA: Morgan Kaufmann, 1988

[Clark88] P. Clark and T. Niblett: "The CN2 induction algorithm". Machine Learning, 3, pp261-284, 1988

[Fayyad92] U. M. Fayyad: "On the Handling of Continuous-Valued Attributes in Decision Tree Generation", Machine Learning 8, pp-87-102, 1992

[Luenberger73] Luenberger, D. G. : "Introduction to linear and nonlinear programming". Reading, MA: Addison-Wesley, 1973.

[Lu88] W. Lu, Y. Ohsawa and M. Sakauchi:"A Database Capture System for Mechanical Drawings Using an Efficient Multi-dimensional Graphical Data Structure" Proc. of ICPR, Vol.1, pp.266-269, 1988

[Maza91] Maza, M.: "A Prototype Base Symbolic Concept Learning System", Proc. of the 8th International Workshop on Machine Learning, pp.41-45, 1991

[Terry93] Terry C. and W Bischof: "Learning Structural Descriptions of Pattern and Objects". Proc. of Asian Conference on Computer Vision, pp.23-25, 1993 (Japan)

[Yang91] H. Yang and A. Kundu: "2-D Shape Classification Using Hidden Markov Model", IEEE Transaction on Pattern and Machine Intelligence, Vol.13, No.11, 1991

VISUAL NONLINEARITY CHARACTERIZED AND GOAL DRIVEN IMAGE SEGMENTATION APPROACH

Zhang Tianxu
Institute of Pattern Recognition and Artificial Intelligence
Huazhong University of Science and Technology
Wuhan 430074, P. R. China

ABSTRACT

A new computational model and an algorithm for image segmentation are proposed based on the spatially adaptive principle of human vision and the relevant hypotheses about object recognition. This is a two-stage process of image segmentation. First, initial segmentation is realized with the bottom-up segmenting algorithm, followed by the goal- driven segmenting algorithm to improve the segmentation results concerning certain regions of interest. Experimental results show that, compared with some conventional and gradient-based segmenting methods, the new method has the excellent performance of extracting small objects from the low contrast images of natural scenes with a complicated background.

INTRODUCTION

Despite their considerable progress, the computational theory and algorithm of early vision as the foundation of machine vision are far from perfect [Mar82]. One of the important problems is how to correctly segment a scene into different meaningful regions, that is, how to perform image segmentation. For instance, in object recognition, it is first necessary to distinguish likely objects or object components from the background before it is possible to recognize an object. An important segmenting method is thresholding, which consists of two approaches of depending/not depending on the bimodality of histogram. In addition, thresholding can be subdivided into single thresholding and multiple thresholding (including the spatially variable thresholding) [Sah88]. For their automaticity and adaptability, the method that does not depend on the bimodality of histogram and the spatially variable threshold method have captured widespread attention [Ots79,Kit85,NR79,YB89]. In Ref. [YB89], it is assumed that the object border possesses high intensity gradient, thus using the image gradient (edge) as an indicator for determining the spatially variable threshold so as to solve the problem of image segmentation under poor and nonuniform illuminating conditions. The method has already been successfully used in the segmentation of workpiece images with a simple background. The problem that has thus cropped up is whether or not the determination of the spatially variable threshold must take the gradient or edge features as a test precondition. Secondly, when segmenting the image of an object with a complicated background (e. g. the images of an outdoor scene), usually many parts of the background have a high gradient while the object boundary (in part or as a whole) may have a low gradient. Hence, the basic assumption about the rationality of the above-mentioned method will be affected, running the risk of segmenting many pixels in the background as objects or losing some of the weak ones.

On the other hand, it is interesting to note that for the threshold methods discussed in currently available literature, the almost perfect perceptibility of human vision has not been made full use of. Hence, when they are used in a complicated and changing environment, good performance cannot be guaranteed. For instance, Otsu has proposed a discriminating criterion for segmentation, by which the optimal threshold should be determined [Ots79]. But in the following situations, this method is often incapable of correctly performing segmentation: (1) when the images of natural scenes are of low contrast; (2) the objects of interest are very small; (3) the background brightness and object brightness are spatially variable. These are difficult problems often encountered in actual applications of computer vision.

What is worth mentioning is that investigations on using the knowledge-based AI technique to improve the initial segmentation results obtained with the prevailing segmenting methods are daily increasing [BH90,TSJ91]. It seems that the combination of the data-driven method with knowledge-driven method is a promising approach. However, its ideas should be broadened and before it can be successfully applied to different complicated scenes, many difficult algorithmic problems have to be solved.

With a view to fairly satisfactorily solving these problems, the nonlinearity and adaptability principles of the perception of light intensity variation by the biological visual system are introduced into researches on image segmentation. In addition, the knowledge-based AI technique is also used for the purpose of developing an object image segmenting method with adaptive characteristics and of relatively stable performance to meet the application need for reliable extraction of small objects from a complicated background.

THE MODEL

In view of the extraction of small objects from a complicated background, this model makes it its first and foremost task to prevent objects or object components from getting stuck together to the background after segmentation, paving the way for subsequent object recognition.

The Otsu's discriminating criterion has attracted attention of many researchers for its advantages and disadvantages. Suppose an image contains object and background. The threshold t divides every pixel $\{f(x, y)\}$ into two classes, the dark region C_1 or bright region C_2. Then the within-class variance σ_W^2 and between-class variance σ_B^2 are functions of t.

$$\sigma_W^2(t) = a_1\sigma_1^2 + a_2\sigma_2^2, \qquad (1)$$

$$\sigma_B^2(t) = a_1a_2(\mu_1 - \mu_2)^2, \qquad (2)$$

$$\sigma_W^2 + \sigma_B^2 = \sigma_0^2, \qquad (3)$$

where a_j is the probability of occurrence of class C_j, $\Sigma_{j=1}^{2} a_j = 1$, μ_j and σ_j^2 are the mean and variance of class C_j; μ_0 and σ_0^2 are the mean and variance of the image. The optimal threshold $\hat{t}$ is chosen so that the between-class variance becomes maximum or, equivalently, the within-class variance becomes minimum.

$$\sigma_B^2(\hat{t}) = \mathrm{MAX}\{\sigma_B^2(t)\} \qquad (4)$$

$$\sigma_W^2(\hat{t}) = \mathrm{MIN}\{\sigma_W^2(t)\} \qquad (5)$$

Segmentation can be regarded as a process of perceiving the object and its background. Let us examine Eq. (4) and its physical meaning as well as the problems that exist with this equation as a perceiving criterion. Letting $\Delta\mu = |\mu_1 - \mu_2|$, we can write Eq. (4) as

$$\sigma_B^2(\hat{t}) = \mathrm{MAX}\{a_1(t)a_2(t)\Delta\mu^2(t)\} \qquad (6)$$

Without losing generality, we shall assume that the object is C_1 and the background C_2. Then $\Delta\mu$ can be regarded as the absolute brightness difference between object C_1 with average brightness μ_1 and background C_2 with average brightness μ_2. Equation (6) depends on area a_1 of C_1 and a_2 of C_2 and absolute brightness difference $\Delta\mu$.

An important experimental result of vision physiology and psychology is the statement that in different background brightnesses μ, absolute brightness difference $\Delta\mu$ does not excite the same variation the visual system can perceive. Its discriminating ability decreases with an increase in the background brightness. If such a discriminating ability is represented with a just noticeable brightness difference $\Delta\mu_T$, then $\Delta\mu_T$ will increase with an increase in the background brightness [Buc80]. Apart from the scotopic region, there also exist DeVries-Rose and Weber regions in the relation of $\Delta\mu_T$ with background intensity μ.

$$\Delta\mu_T = \alpha_1\sqrt{\mu}, \qquad (7)$$

$$\Delta\mu_T = \alpha_2\mu. \qquad (8)$$

where α_1 and α_2 are appropriate coefficients.

Obviously, the computational model Eq. (6) is not in agreement with the nonlinear response of the visual system (If C_1 and C_2 are considered to be a background to each other, it can be assumed that $\mu = \mu_2$ or $\mu = \mu_1$). Equation (1) is a measure of within- class nonuniformity as well as a variation of brightness. From the expression for class variance

$$\sigma_j^2 = \sum_{f(x,y)\in C_j} \{f(x,y) - \mu_j\}^2 p_j(f) = \sum_{f(x,y)\in C_j} \Delta\mu_j^2 p_j(f) \qquad (9)$$

where $p_j(\cdot)$ is the probability distribution of the pixel brightness in class C_j and $\Delta\mu_j = |f(x,y) - \mu_j|$ is the absolute brightness difference between the pixels in class C_j and the class average brightness μ_j. It is clear that if Eq. (5) is taken as a criterion, the perception of nonuniform will be dependent only on the average absolute brightness difference. Hence the same problem as with Eq. (6). For this reason, it is necessary to redefine the within-class and between-class variances so that the calculation of the related statistics will be dependent on visual nonlinearity.

In the following redefining process, reference has been made to the definition of contrast in psychophysiology, that is, contrast is the ratio of difference in luminance of an object μ_1 and its immediate surrounding μ, i. e. $C = \frac{|\mu_1 - \mu|}{\mu} = \frac{\Delta\mu}{\mu}$. The absolute brightness difference in the above-mentioned different forms (equivalent to $\Delta\mu$ in the equation) should be modified to relative brightness difference similar to contrast. But this form of contrast only reflects the visual properties in the Weber region. So we should further consider the visual nonlinearity and the global adaptability properties represented by DeVries- Rose region and Weber region as a whole. Furthermore, depending on the need of convenient and easy calculation, we can choose certain monotonically increasing functions $h(\cdot)$, $h_j(\cdot)$ related to the average brightness (e. g. object average brightness, background average brightness, the image average brightness) to represent μ in the equations. The average brightness μ_0 of the image regions to be segmented and the brightness threshold μ_T will be used as control parameters to establish a certain association of analogy between the calculation of the relevant quantities and the DeVries-Rose region or the Weber region.

Definition 1 The relative nonuniformity measure r_j^2 of class C_j is

$$r_j^2 = \left\{\frac{\sigma_j}{h_j(\mu_j,\mu_0)}\right\}^2, \quad (j = 1,2) \tag{10}$$

Definition 2 The within-class relative variance r_W^2 is

$$r_W^2 = a_1 r_1^2 + a_2 r_2^2. \tag{11}$$

Definition 3 The relative brightness difference r is

$$r = \frac{|\mu_1 - \mu_2|}{h(\mu_1,\mu_2,\mu_0)} = \frac{\Delta\mu}{h(\mu_1,\mu_2,\mu_0)}. \tag{12}$$

Definition 4 The between-class relative variance r_B^2 is

$$r_B^2 = a_1 a_2 r^2 = \sigma_B^2 h^{-2}(\mu_1,\mu_2,\mu_0). \tag{13}$$

In the above equations, $h_j(\cdot)$ and $h(\cdot)$ are monotonically increasing functions related to the average brightnesses μ_1, μ_2 and μ_0. It can be seen from Eqs. (10) and (11) that the within-class nonuniformity perception depends on the average brightness. Similarly, it can be seen from Eqs. (12) and (13) that the perception of discriminating the object from its background is dependent on their average brightnesses. The problem is what form of function should be taken for $h_j(\cdot)$ and $h(\cdot)$. According to Eqs. (7) and (8) and the need of actual calculation, a function of the following form is presented:

$$h_j(\mu_j,\mu_0) = \begin{cases} \alpha_1\sqrt{\mu_j} + \beta_1, & \mu_0 < \mu_T \\ \alpha_2\mu_j + \beta_2, & \mu_0 \geqslant \mu_T \end{cases} \tag{14}$$

$h(\cdot)$ can take the following two forms:

$$h(\mu_1,\mu_2,\mu_0) = \begin{cases} \alpha_1\sqrt{\mu_2} + \beta_1, & \mu_0 < \mu_T \\ \alpha_2\mu_2 + \beta_2, & \mu_0 \geqslant \mu_T \end{cases} \tag{15}$$

$$h(\mu_1,\mu_2,\mu_0) = \begin{cases} \alpha_1\sqrt{\mu_1} + \beta_1, & \mu_0 < \mu_T \\ \alpha_2\mu_1 + \beta_2, & \mu_0 \geqslant \mu_T \end{cases} \tag{16}$$

where μ_T is an appropriate mean brightness threshold, $\alpha_1 > 0$, $\alpha_2 > 0$; β_1 and β_2 are appropriate constants; $h_j(\cdot) > 0$, $h(\cdot) > 0$; the optimal threshold $\hat{t}$ should satisfy the new criterion defined as follows:

$$r_B^2(t) = \text{MAX}\{a_1(t)a_2(t)r^2(t)\} = \text{MAX}\left\{a_1a_2\left(\frac{\Delta\mu}{h(\mu_1,\mu_2,\mu_0)}\right)^2\right\}. \tag{17}$$

The following theorem can be proved to be true.

Theorem 1 When a single threshold t is used to segment a discrete image $f(x,y)$, the functions defined by Eqs. (14)-(16) are monotonically increasing functions of threshold t.

What features does the new criterion have? Let (a_1,μ_1) and (a_2,μ_2) be characteristic vectors of object C_1 and background C_2, respectively. It can be seen that the maximum value of Eq. (17) is a result of the joint action of four factors, namely, a_1, a_2, $(\mu_1-\mu_2)^2$ and $h^{-2}(\cdot)$. The optimal threshold $\hat{t}$ has been associated with $h(\cdot)$, or the average brightness of the background and image. The value of Eq. (17) decreases with an increase of $h(\cdot)$, showing that the discriminating ability has been weakened. This is in conformity with the principle of visual nonlinearity and Eq. (6) is but a compromise between the former three items. In the event the object of interest C_1 is in reality rather small (that is, a_1 is small, then the maximum value of Eq. (6) may lead to a wrong threshold, causing too many background pixels to be classified as object pixels.

In many practical applications, although the object is discriminable from its neighboring partial background, yet the brightness of the background and object varies with the spatial position (e. g. multiple objects appear in different positions of an image). Such a variation can be caused by a number of factors. For instance, the spatial variation of the reflectivity of materials itself or the difference in the view- point may bring about nonuniform background and object brightness. The said variation can also be induced by the nonuniform illumination in the image forming process. For this reason, it is necessary to construct different models and algorithms for different situations.

For instance, for the spatial variation of the gray level of image due to poor and nonuniform illumination, Yanowitz segmented the image of a workpiece with the threshold surface obtained from the intensity gradient of the image. The method is highly effective for indoor images with simple background.

In contrast, for images of an outdoor scene with a complicated background and containing small objects, besides the variation due to the uncontrollability of natural lighting, there is the spatial variation of the gray level of image due to the variation of the spatial reflectivity of the background and the objects themselves. A natural background can also possess a high intensity gradient. Problems will crop up if we rely on the gradient. Obviously, the initial recognition of the background and object should also become an important step in the improvement of segmentation. Therefore, effort should be devoted to investigations on new methods different from Yanowitz'.

Let the position of an object be (x,y), whose average brightness is $\mu_1(x,y)$. Its adjacent background can be expressed as $\mu(x,y)$. Thus, the brightness difference $\Delta\mu(x,y) = |\mu_1(x,y) - \mu(x,y)|$ is a function of the coordinates. According to Eqs. (7) and (8), the just perceivable brightness difference may also be expressed as a function of the spatial position:

$$\Delta\mu_T(x,y) = \begin{cases} \alpha_1 \sqrt{\mu(x,y)}, & \text{DeVries-Rose region} \\ \alpha_2 \mu(x,y). & \text{Weber region} \end{cases} \tag{18}$$

It can be seen that a single threshold is in general incapable of properly segmenting all objects from their background at once. It is necessary to use the spatially variable threshold. One method consists in partitioning an image plane into equal subblocks, on which the threshold will be determined according to the bimodality of the histogram [NR79]. But when bimodality is not satisfied, the method will become ineffective. If other criteria such as Otsu's are used on the subblocks, the new problem confronted is that some of the subblocks may almost all become part of the background or the objects, thus violating the prerequisite for using the criterion. Conversely, the human vision possesses superior spatial adaptation. For example, conscious or unconscious ocular movements to scan an entire frame of image and to focus the line of vision on a subregion that may contain object of interest so that the discriminating ability will not only be adapted to the entire frame of image but also be adjusted and varied with different subregions. This spatial adaptability has a certain characteristic of a knowledge- guided active perception. With a view to enabling a segmenting method to possess such adaptability in a certain degree, the following model for segmentation calculation is investigated to facilitate the extraction of small objects from a complicated background. The following knowledge or rational assumptions are used for the said model and algorithm based on the above- mentioned investigation on visual nonlinearity.

(1) The objects in space are nonextensive and compact;

(2) The objects take up very small space in the entire scene;

(3) The internal gray level of the objects or object components is relatively uniform as is that in a partial background;

(4) There is sufficient difference in the gray level or contrast of objects or the visible surfaces (in part or as a whole) relative to their neighbor or the immediate partial background. This implies that a certain region of an appropriate size and containing objects and their partial neighboring background should in general possess sufficient nonuniformity;

(5) The segmented region that takes up a rather large area of an image is usually part of the background;

(6) A segmented region of a complicated shape is not very likely an object.

Knowledge of the 1st and 2nd items will permit an algorithm to divide an entire frame of image into a set of subregions, whose areas depend on the size of the objects. Knowledge of the 3rd and fourth items will yield the concept of the relevant appropriate subregions and appropriate partition of image regions to be discussed later. Here, it can be assumed that those subregions with sufficient nonuniformity may contain both objects and their background. (Of course, certain nonuniform parts of the background may be mistaken for having objects and are thus contained in certain subregions.)

Let Ω represent an image region to be segmented; Ω_i a subregion of Ω. Then we have

Definition 5 Ω_i is a subregion appropriate for segmentation in Ω provided that its relative nonuniformity measure satisfies

$$r_i^2 = \left(\frac{\sigma_i}{h_i(\mu_i,\mu_0)}\right)^2 > r_T^2,$$
$$h_i(\mu_i,\mu_0) = \begin{cases} \alpha_1 \sqrt{\mu_i + \beta_1}, & \mu_0 < \mu_T \\ \alpha_2 \mu_i + \beta_2, & \mu_2 \geqslant \mu_T \end{cases} \tag{19}$$

where μ_i, σ_i^2 are the mean and variance of the subimage in Ω_i and r_T^2 is the threshold of the relative nonuniformity measure.

The prerequisite for using the segmentation criterion is given in this definition, that is, the nonuniformity condition. or the object/background condition a certain to-be-segmented subregion should satisfy so as to prevent the subregion from completely becoming part of the background or completely becoming part of the object.

Definition 6 $\{\Omega_i,\ i=1,2,\cdots,\ I\}$ is a partition of Ω appropriate for segmentation provided that an arbitrary Ω_i in it satisfies Definition 5 and

$$\Omega = \bigcup_{i=1}^{I} \Omega_i \ . \tag{20}$$

Definition 7 Suppose the two classes of regions obtained by segmentation according to threshold t_i in the appropriate subregion Ω_i are C_{i1} (dark region) and C_{i2}, (bright region) and the average brightnesses are μ_{i1} and μ_{i2}, respectively. The regions obtained by segmentation in their neighbor Ω_k according to threshold t_k are C_{k1} and C_{k2} and the means are μ_{k1} and μ_{k2}, respectively. t_i and t_k are said to be compatible provided that in the regions after segmentation

$$\mu_{k2} - \mu_{i1} \geqslant d,\ \mu_{i2} - \mu_{k1} \geqslant d, \tag{21}$$

where $d > 0$ is an appropriate threshold of compatibility.

The criterion function for obtaining spatially variable threshold is set to be

$$R(\{t_i\}) = \sum_{i=1}^{I} r_{Bi}^2(t_i), \tag{22}$$

where $r_{Bi}^2(t_i)$ satisfies Eq. (13) and is defined to be on an appropriate subregion Ω_i. Thus the problem is boiled down to seeking an appropriate subregion set $\{\Omega_i, i=1, 2,\cdots I\}$ and an optimal threshold set $\{t_i^*, i=1,2,\cdots,I\}$ such that under the constraint of Eq. (21), Eq. (22) takes the maximum value.

$$R\{(t_i^*)\} = \mathrm{MAX}\left\{\sum_{i=1}^{I} r_{Bi}^2(t_i)\right\} \tag{23}$$

In order to obtain the optimal segmentation of an object

of interest, it is necessary for certain appropriate subregions to contain the objects of interest as intact as possible and their partial background. The above-mentioned calculating model, however, cannot guarantee this despite its attempt to do so. Certain pixels of an object may get lost, or it is also likely that certain pixels of the object's neighboring background may be allocated to an object after segmentation. These will cause defects or distortion of the shape of an object or object components. Therefore, after obtaining the result of an initial segmentation of an image using the above-mentioned calculating model, there should be an initial object testing followed by a process of getting new appropriate subregions at each candidate object and performing resegmentation. In this process, depending on the requirements of the task, knowledge of the fifth and sixth items will be made use of in the segmenting method while neglecting most of the regions in the image. Knowledge of the 1st, 2nd, and fourth iterms will be used to designate a small number of regions as objects for resegmentation so as to improve the segmenting effect. The object testing criterion is fairly simple as only the relative area and the complexity of shape of the initial segmentation results will be referred to.

THE ALGORITHM

It is usually difficult to obtain the optimal threshold set specified in Eqs. (19) through (23) because it is not easy to write the said computational model in an analytic form of the threshold set. Furthermore, the number, shape, position and size of the appropriate subregions are all undetermined and need be dynamically developed. For this reason, the following simplified algorithm with global and local adaptability to the variation of image brightness has been investigated and implemented as the first stage of a two-stage image segmentation to obtain a satisfactory, if not necessarily optimal, initial image segmentation.

Stage 1:

Step 1 Calculate the statistic parameters μ_0, σ_0^2 and the histogram of the image to be segmented, set the threshold μ_T and determine the function form of $h_j(\cdot)$ and $h(\cdot)$.

Step 2 Determine the size of the initial subregion $\Omega_i(0)$ according to the minimum area A_0 that the object to be extracted may take up in the image, that is, $A_{min} = k_0 A_0$, where we take $k_0 = 2 \sim 3$ in our experiments. Partition the image plane Ω into equiarea initial subregion set $\{\Omega_i(0), i=1,2,\cdots,I(0)\}$.

Step 3 Set the initial nonuniformity measure threshold $r_T^2(0)$ and examine whether $\Omega_i(0)$ is an appropriate subregion or not according to Eq. (19). A subregion that has not passed the examination should be annexed to its neighboring subregion to form a larger one for further examination until Eq. (19) is satisfied or the area of the subregion reaches the specified maximum A_{max}. An appropriate partition $\{\Omega_i(1), i=1,2,\cdots,I(1)\}$ of Ω will be obtained through the simplified growing process.

Step 4 An initial estimate of the optimal threshold set $\{\hat{t}_i(1)\}$ is obtained from the optimal threshold $\hat{t}_i(1)$ of each $\Omega_i(1)$ according to Eq. (17) with the numerical method. Examine whether the region compatibility with Eq. (21) is met. Suppose L pairs have not passed the examination and their mean is $\{(\mu_{i1}, \mu_{k2}), l(i,k)=1,2,\cdots,L\}$. Adjust the initial estimate within a definite range to generate a new threshold set $\{t_i^*(1), i=1,2,\cdots,I(1)\}$ so that the following compatibility measure will reach maximum.

$$S(\{t_i^*(1)\}) = \text{MAX} \sum_{l(i,k)=1}^{L} (\mu_{k2} - \mu_{i1}) \qquad (24)$$

A "Partially Recursively Segmentable" label should be attached to each segmented region and set the recursive segmenting count K=1.

Step 5 Set the nonuniformity measure threshold $r_T^2(K) < r_T^2(K-1)$. According to Eq. (19) examine the nonuniformity of the two classes of labeled regions of the (k-1)-th level. Those that have not passed the examination will have their labels cancelled while those that have will have their optimal thresholds found according to Eq. (17) and the compatibility of the segmentation results will then be checked. Those that are not compatible will also have their labels cancelled and no further segmentation will be made for them.

Step 6 $K \leftarrow K+1$. Repeat Step 5 till the specified count K_0. Now the threshold set is denoted by $\{t_i^*(K_0), i=1,2,\cdots,I(k_0)\}$.

Stage 2:

The data sets to be operated are the original image and the binary picture of the initial segmentation.

Step 1 Perform contour tracking of the binary regions after the initial segmentation and label them. Obtain the number of regions n_0 and set the cyclic variable $n \leftarrow n_0+1$.

Step 2 $n \leftarrow n-1$; if $n=1$, Go To Step 7.

Step 3 Calculate the features of the n-th region: the area A_n, the perimeter P_n, the perimeter to area ratio e_n, and the centroid (x_n, y_n).

Step 4 If $A_n > A_T$ or $e_n > e_T$, then give up this region and Go To Step 2.

Step 5 Generate the appropriate rectangular subregion Ω_n on the original image in accordance with Eq. (19) with the centroid (x_n, y_n) as the center. If this window contains quite a few other binary segmented regions (See the initial results of segmentation), give up this window and Go To Step 2; otherwise

Step 6 Find the optimal threshold of Ω_n in accordance with Eq. (17), perform resegmentation and Go To Step 2.

Step 7 Stop.

EXPERIMENT

The above-mentioned algorithm has successfully been used in the low- level visual calculating module for machine vision. The results of experiments on a group of images of a low contrast natural scene with different types of histogram and containing multiple objects show that the effect of separation of the object images from their background is evidently better than that with certain conventional segmenting algorithms. The experiments were made on a VAX 11/ 785 image processing system. Some examples are given below.

Three segmenting methods are compared in Fig. 1, in which (a) shows the original image, (b) the results of segmentation using Otsu's discriminating criterion, (c) the results of segmentation using Yanowitz's method. The original image has a complicated background, in which the caterpillars and barrels of two vehicles are discernible to the human eye. But neither of the above-mentioned two methods can separate the objects from their background. Figure 1 (d) shows the results of segmentation by the spatially variable threshold calculating model and algorithm proposed in this paper. The caterpillars and barrels on the two vehicles are fairly successfully separated from their background, the effect being quite close to that with the human eye.

In Fig. 2, (a) and (b) show an original image and the final results of using the initial object examination and resegmentation with the non-object region left out. It can be seen that three small objects have already been separated fairly intact from their background.

In the above examples, the selection of the form of the weight function $h_i(\cdot)$ is determined by the average brightness μ_0 of the whole image. μ_0 is rather small in the examples, so we take $h_i(\cdot)=\alpha_1\sqrt{\mu_i+\beta_1}$. If μ_0 is rather great, we can take $h_i(\cdot)=\alpha_2\mu_i+\beta_2$, embodying the global adaptability.

CONCLUSIONS

Theoretical analysis and experimental results show that the said model and algorithm are rational and correct. They differ from certain spatially variable threshold methods given in relevant literature in (1) The spatial variable threshold is made to correspond to an appropriate partition of image region and appropriate subregions, both being dynamically formed in the execution of the algorithm; (2) The edge or gradient information has not been taken as a prerequisite for indicating the variable threshold; (3) the goal-driven process of resegmentation. Of course, an extension and greater improvement of the method and the model are worth further consideration.

ACKNOWLEDGEMENT

This work was supported by the National Natural Science Foundation of China and Wang Kuan-Cheng Education Foundation.

REFERENCES

[Mar82] D. Marr. Vision, W. H. Freeman and Company, San Franciso, 1982.

[Sah88] P. K. Sahoo. A Survey of Thresholding Techniques, Comput. Vision Graphics Image Process., 41:2(1988), 233-260.

[Ots79] N. Otsu. A Threshold Selection Method from Gray-Level Histogram, IEEE Trans. Syst. Man Cybern., 9:1(1979), 62-66.

[Kit85] J. Kittler. On Threshold Selection Using Clustering Criteria, IEEE Trans. Syst. Man Cybern., 15:5(1985), 652-655.

[NR79] Y. Nakagawa and A. Rosenfeld. Some Experiments on Variable Thresholding, Pattern Recognition, 11: 3 (1979), 191-204.

[YB89] S. D. Yanowitz and A. M. Bruckstein. A New Method for Image Segmentation, Comput. Vision Graphics Image Process., 46:1(1989), 82-95.

[BH90] B. Bhanu and R. D. Holben. Model-Based Segmentation of FLIR Images, IEEE Trans. Aerosp. and Electr. Syst., 20: 1 (1990), 2-11.

[TSJ91] J. Ton J. Sticklen, and A. K. Jian. Knowledge-Based Segmentation of Landsat Images, IEEE Trans. Geosc. and Remote Sens., 26:2(1991), 222-231.

[Buc80] G. Buchsbaum. An Analytical Derivation of Visual Nonlinearity, IEEE Trans. Biomed. Engin., 27:5(1980), 237-242.

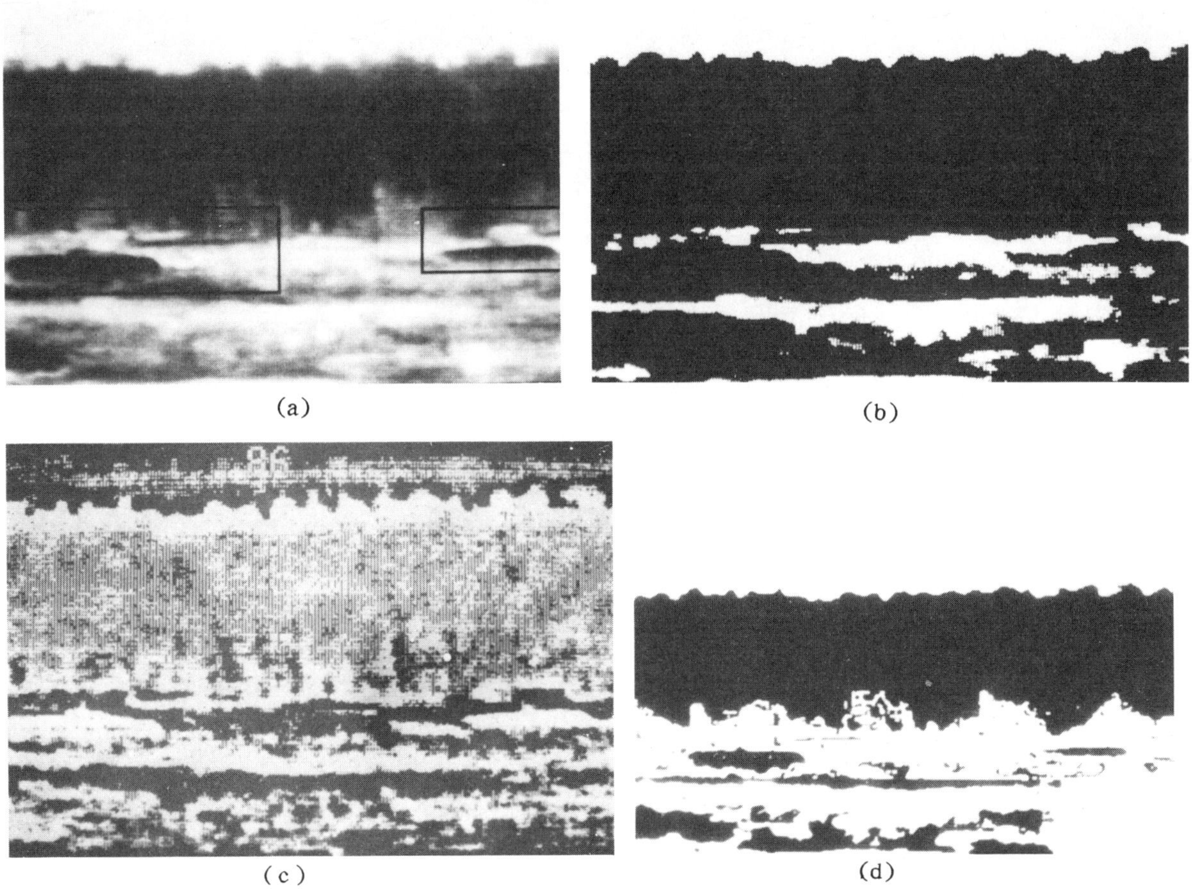

FIGURE 1 A comparison between three segmenting methods.

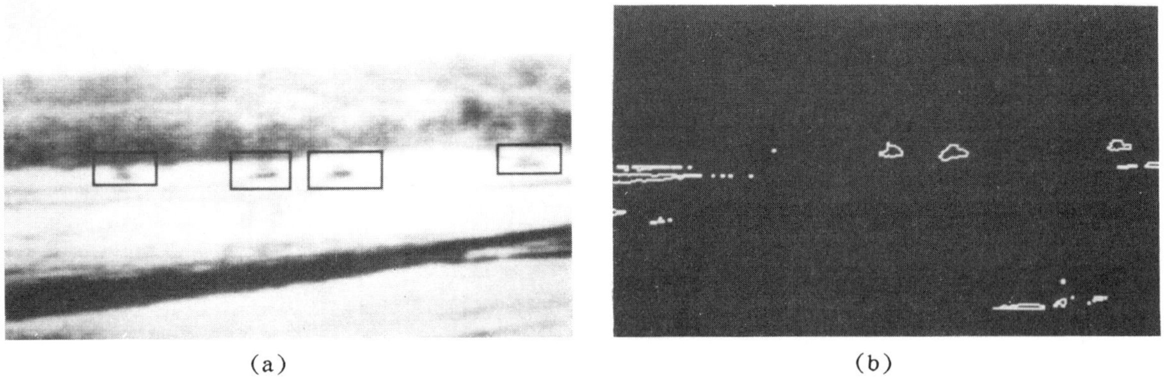

FIGURE 2 Experimental result of goal-driven resegmentation

SCREEN-BASED 3D OBJECT RECOGNITION AND ASSEMBLY

A.E. Samuel[1]
Department of Mechanical and Manufacturing Engineering
The University of Melbourne, Australia

M. Chao[2]
Department of Mechanical Engineering
Monash University, Clayton, Australia
Email: aes@homer.mame.mu.OZ.AU

ABSTRACT

There are several approaches to understanding the way humans solve general problems, the most noted is Newell and Simon (1972). However, recent work has looked at variations of more complex problem-solving models, such as human interaction with computers. Many of these rule-based system models involve complex cognitive processes based on verbal protocols.

Three sets of specific CAD models were designed for the physical world tests. These three sets of models have been designed with a specific goal in mind. They vary in complexity to test the aptitude of the subjects in relation to their abilities to recognise features and patterns in order to detect the correct constructs. In order to explore the protocols how people solve complex problems of object construction and recognition on the computer screen, a powerful mouse driven assembly software (Screen-Assembler) was developed. The Screen-Assembler supports component rotation (-180∞— +180∞) in any of the working areas. This paper describes our initial findings and performance measures with the Screen-Assembler.

1 INTRODUCTION

Verbal protocol analysis has been widely used to capture the processes of problem solving behaviour. Early research studied humans solving well-defined problems such as cryptarithmetic and Tower of Hanoi problems, but it also extended to chess (Chase and Simon, 1973) (De Groot, 1966) (Ericsson and Simon, 1980, 1984) (Newell and Simon, 1972). Other research has focused on continuous and discrete process control tasks. Verbal protocol analysis often becomes necessary when subjects are required to interact with such semantically complex tasks, since there is little overt activity to observe. The principle activities are covert cognitive processes such as thinking, planning, estimating, predicting, scheduling, and judging. Very often, the only way to get access to these processes is to ask the subject to represent the contents of thought verbally (Sanderson, James and Seidler, 1989).

This work explores some of the cognitive difficulties involved in CAD model construction. It addresses mental transformations in the physical world, and the nature of cognitive loads when constructing physical models. It also investigates the transport of skills from physical world model construction to a screen world construction process. The strategic long term objective of this work is the need to reduce the lead time needed to learn and retain the programme constraints associated with specific CAD modellers.

A more immediate objective was to investigate the problem solving protocols associated with CAD model construction. It was hoped that this investigation will lead to the development of measures associated with the cognitive loads on the CAD model generator. The specific aim was to develop measures of cognitive load associated with feature-based mental transformations. Moreover, the work addresses measures of cognition in both the physical world and the screen world.

2 PERFORMANCE MEASURES FOR VERBAL PROTOCOLS

The objective in performing this investigation is to evaluate the thinking rules (protocols) used by subjects in constructing CAD models. Verbalising the thinking process during construction is thought to provide "snapshots" of the subject's thinking and solution strategies. SHAPA, a software verbal protocol analyser (Sanderson et al., 1989) is used as the major analysis tool for this work. SHAPA allows detailed "means-ends" evaluation of problem solving protocols. It is an open architecture analyser which allows the user to decide the basic atomic components of the protocol. Statements, or predicates, must be identified as distinct, meaningful (semantic), elements of the verbalised problem solving process.

[1] A/Professor
[2] Research Fellow

Moreover, it is found that it is necessary to augment the verbal protocols by observing moves on the video tape to identify the next move made by the subject. In each case the subjects were asked to confirm the augmented protocols. This procedure involved viewing the construction process on the video tape and asking the subject at each move "why did you make this move?". Although this procedure is time consuming, the results reflect the whole construction protocol much more closely than the verbal records alone would do.

3 SUBJECTS AND TASKS

During the investigation, two subjects were concentrated on, namely one experienced in CAD or geometric construction, an engineering Ph.D student (subject P) and one not so experienced (subject M), an arts masters student. This investigation took place in two streams. The first stream investigated the assembly of CAD models from constructive solid geometry (CSG) primitives on the computer screen. The second stream investigated the assembly of physical models. Before real tests started a preliminary was used to assimilate the subjects with think-aloud.

Both subjects' completed the following tasks :

Somacube (physical construction) warmup test

SetA (physical and screen construction)

SetB (physical and screen construction)

SetC (physical construction)

Three sets of models were designed and used for the test programme: they are arbitrarily denoted Set A, Set B and Set C. By designing Set A with a central piece, we attempted to lead the subjects to the correct solution, thus allowing for a minimum amount of reasoning. Figure 1 shows one of models in Set A.

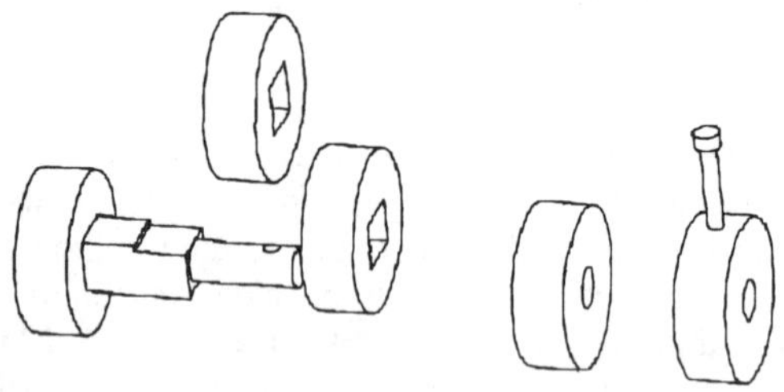

Figure 1 One of the four models in model Set A

Set B has no such central piece and therefore provides no apparent clues to the final solution. Moreover, further complications have been added, to this second set of models, by making some pieces similar to others at first sight. Generally, Set B is more complicated than Set A but can be solved with the careful study of features on each piece.

Set C is based on sixteen components chosen from Set B and then modified in an effort to imply the existence of a reachable solution whereas that is not the case. Our hope in generating Set C was to further expose the solution strategies used by the subjects. It was expected that, since no "solution map" existed, the eventual behaviour of both subjects would degenerate to symptomatic, trial-and-erro search procedures. In this way we hoped to delineate even more clearly between the behaviour of our two subjects.

4 THE SCREEN-ASSEMBLER

In order to explore the protocols how people solve complex problems of object construction and recognition on the computer screen, a powerful mouse driven assembly programme (Screen–Assembler) was developed.

Any motion of a rigid body in three dimensions can be considered as the sum of a rotation about a particular axis, the screw axis, and a translation along that axis (Coxeter, 1961). In the screen world, with its inherent space limitations, it makes sense to concentrate on rotations only. The Screen–Assembler was written in C running on an Iris 4D graphics workstation. The Screen-Assembler is able to accept any 3D object which is constructed from typical mechanical primitives, such as blocks, cylinders, spheres, cones, etc. Figure 2 shows a typical message window in the Screen–Assembler, which presents a screen divided into the following five areas:

- Components menu – where all the components to be assembled are shown;
- Selected component – a place to view the component which is to be added next;
- Assembly space – for components assembly;
- Two sliders - one for rotating components, the other for rotating the assembly.

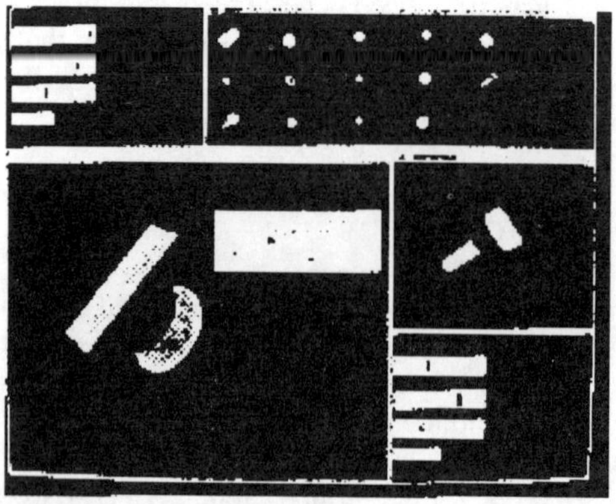

Figure 2: Typical message window of Screen–Assembler

5 SCREEN MODEL TEST PROCEDURE

In the screen assembly experiments, the subjects were given the component pieces from two models in any one set (A or B) – the "components menu". They were asked to assemble them into a complete cylindrical assembly (SetA), or an assembly block (Set B). This was their target in the problem space. In order to identify patterns in the solution strategies of the subjects, as we did in the preliminary and physical CAD model tests, we examined the rates at which the models were solved and also the rates at which the individual pieces were considered by subjects during the construction. Figure 3 shows the flowchart of actions performed by subjects during screen assembly.

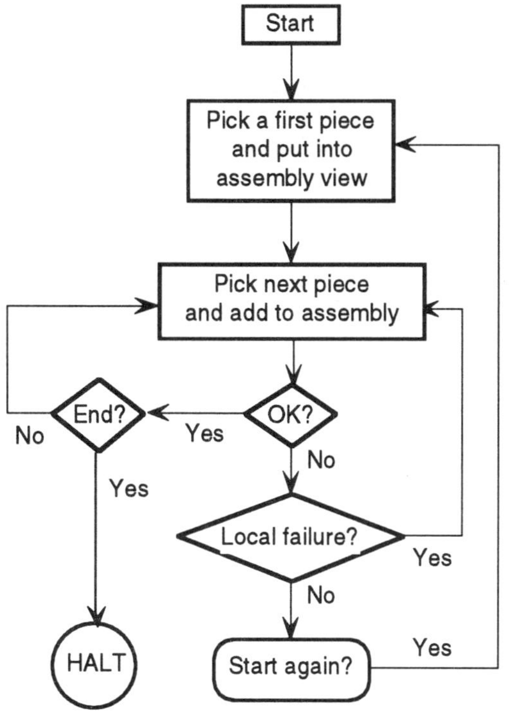

Figure 3: Flow-chart of the screen assembly procedure

Model Set A

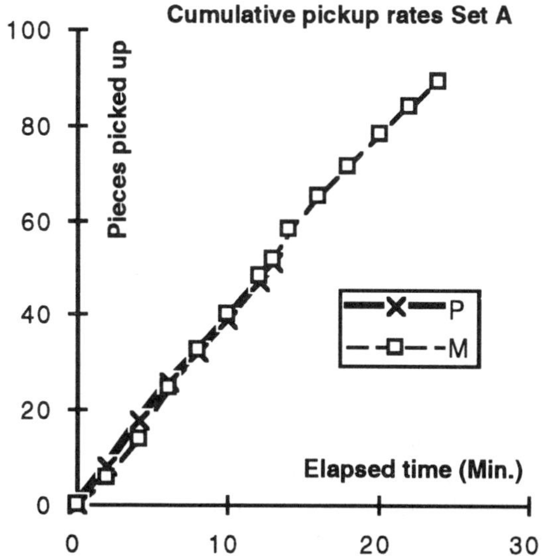

Figure 4: Cumulative pick-up rates during Set A construction

Figure 4 shows the cumulative pick-up rates for the two subjects during the construction of CAD model Set A. Figure 5 shows the relative construction rates of Subject P compared to Subject M.

Clearly subject P was able to reach the solution in considerably less time than subject M, yet, as Figure 4 indicates, both subjects have almost the same average pickup rate (4 pieces per minute). The main difference is in feature recognition speed and the effectiveness with which each subject makes use of individual pieces picked up during construction.

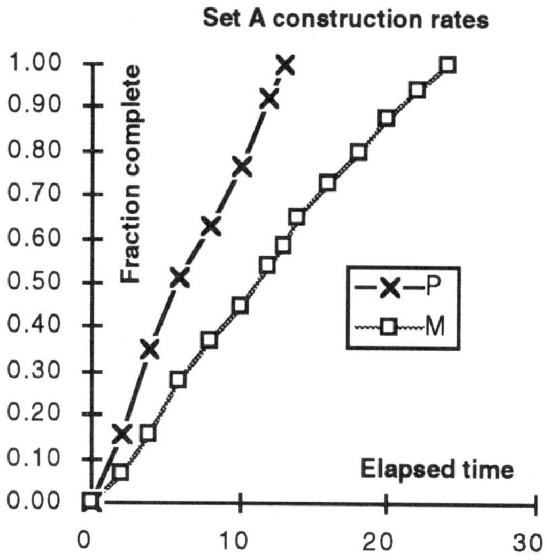

Figure 5: Relative construction rates of the two subjects

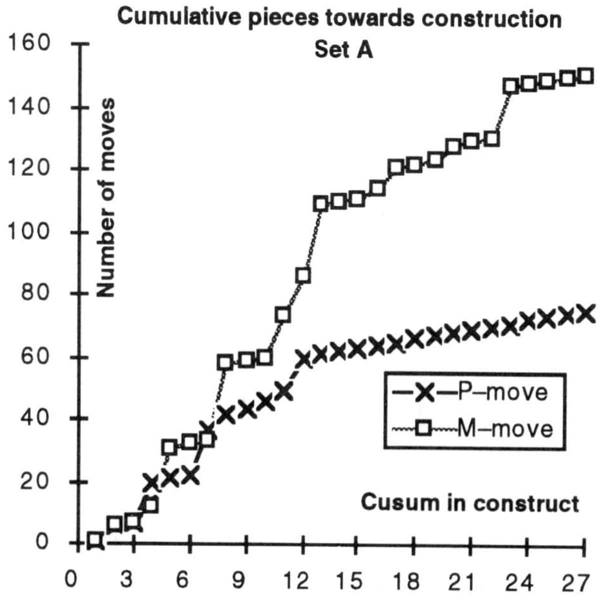

Figure 6: Cumulative pieces in progress towards construction–Model set A

Figure 6 shows the cumulative number of pieces in a given construct as the subjects proceeded towards a solution. The comparison indicates the significant differences between the two subjects, not only in the total moves needed to reach a complete construct, but also in the feature of many more single piece moves for Subject M compared to Subject P.

Model Set B

The cumulative pick-up rates during the construction of CAD model Set B for both subjects are shown in Figure 7. Figure 8 shows the relative construction rates for the two subjects. Figures 8 and 9 show the cumulative number of pieces in progress towards the solution of two subjects. The plot indicates that subject M spent much more time than subject P. On the average subjects P and M again, have almost the same picking up rate (3 pieces per minute).

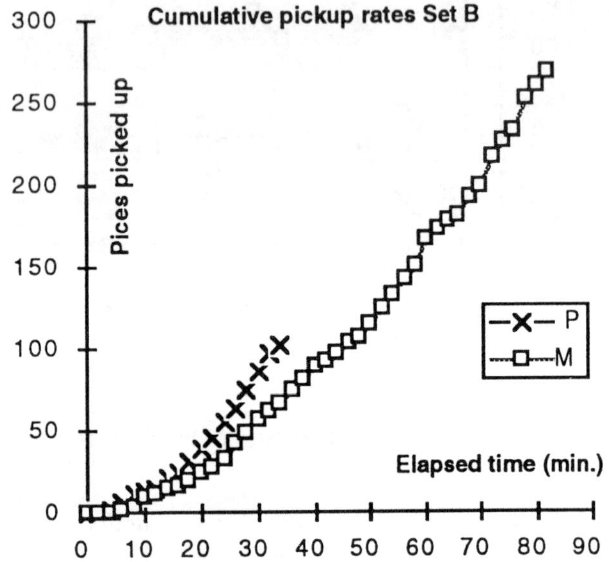

Figure 7: Cumulative pick-up rates during Set B construction

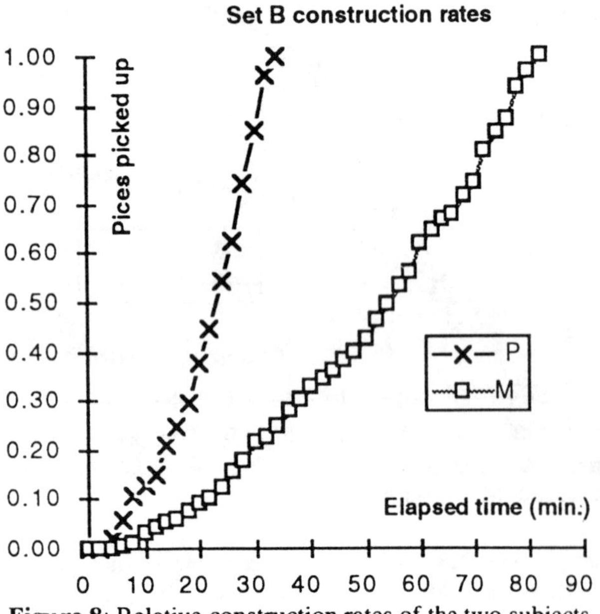

Figure 8: Relative construction rates of the two subjects

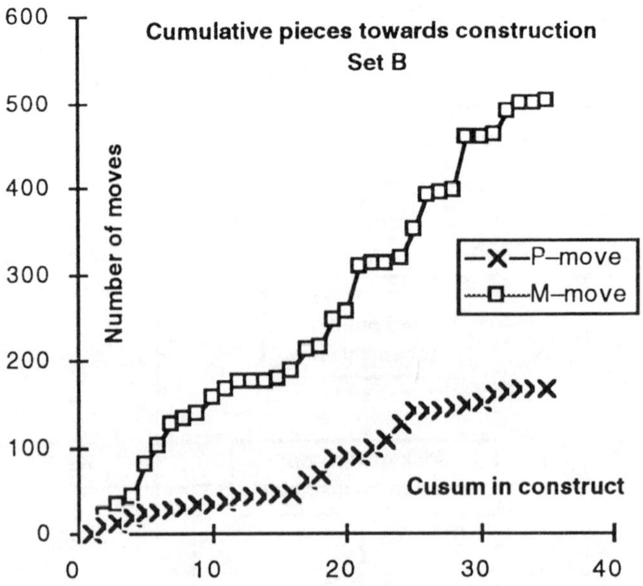

Figure 9: Cumulative pieces in progress towards construction–Model set B

6 PERFORMANCE COMPARISONS

Move counts

Clearly, as Figures 5 to 9 indicate Subject P was considerably more effective in constructing both sets of models. Moreover, as the cumulative constructions (Figures 6 and 9) Subject P made significantly less numbers of single backtrack moves than did Subject M. The overall effect was a significantly reduced number of moves to construction in the screen world. As a first measure of construction efficiency, we calculate the fraction of useful moves made during a construction. Table 1 gives these results.

Verbal protocol analysis

The objective in performing this investigation was to evaluate the thinking rules (protocols) used by subjects in constructing CAD models both in the physical and screen world. Again, verbalising the thinking process during construction was thought to provide "snapshots" of the subject's thinking and solution strategies. A software verbal protocol analyser, SHAPA (Sanderson et al., 1989) was used as an analysis tool for this work. SHAPA relies on identifying suitable predicates which establish the formal aspects of thinking that the subjects apply to their tasks. Identifying the proper predicates is the most difficult task in developing a useful protocol analysis procedure. After watching the video tape-records of screen-based CAD model constructions, we found that we could isolate four clear predicates out of six which were used in the preliminary tests, namely:

Table 1

First measure of construction efficiency

	Subject	Moves to construct	Total moves	Construction efficiency
Model Set A	P	27	75	(27/75) = 36%
	M	27	151	(27/151) = 18%
Model Set B	P	35	167	(35/167) = 21%
	M	35	501	(35/501) = 7%

Table 2

First and second order transition matrix analyses(Partial table)

	Model Set A								Model Set B			
	Subject P				Subject M				Subject P		Subject M	
	Successor				Successor				Successor		Successor	
Predecessor	SM	CS	CM	TOT	SM	CS	CM	TOT	SM	TOT	SM	TOT
NP	1			1			1	1	1	1	1	1
SM	20	8	27	55	6	27	14	47	20	76	9	140
CS	6		5	11	18	7	13	38	24	32	63	142
CM	29	3	12	44	24	4	3	31	32	37	68	83
TOT	56	11	44	111	48	38	31	117	77	146	141	366
NP, SM			1	1	1			1		1		1
SM, SM	7	1	11	19	1	3	2	6	3	19		8
SM, CS	4		4	8	14	5	8	27	21	29	43	106
SM, CM	17	2	8	27	9	2	3	14	24	27	17	25
CS, SM	5		1	6	1	11	6	18	10	24	2	63
CS, CM	4		1	5	11	2		13	6	8	16	26
CM, SM	8	7	14	29	4	13	6	23	7	32	47	53
CM, CS	2		1	3	1	1	2	4	3	3	7	68
CM, CM	8		4	12		3		3	2	2	4	10
TOT	55	11	44	110	42	40	27	109	76	145	136	360

- NEW PARADIGM (NP) *(eg: "Okay, I start with an internal part, seems like ....");* In general NP implied that the subject recognised that the construction process had a major flaw and needed a "restart";
- SUCCESS MOVE (SM) *(eg: "Um, maybe that one, yes.");* SM means that a positive forward move has been made, and the subject is on the right track to a solution;
- CONSTRAINT (CS) *(recognition that one single piece needs to be replaced; eg: "No, I don't think so, no, it's got a square in it, that's not right.")* ;
- COMMENT (CM) *(eg: "That shouldn't go because it's got an indent there and a square there ....")*; CM doesn't add anything positive to the construct, it is simply "thinking out aloud";

Transition matrix analyses (TMA)

Table 2 shows the first and second order transition matrix analyses generated by SHAPA for model Set A. This type of analysis examines the transitions between predicates to determine (if at all possible) any pattern of behaviour during the construction process. First order analysis simply examines the transitions between single predicates, while second order transitions examine the transitions from pairs of predicates.

The most significant finding from the TMA is the proportion of times we find SM–SM transitions and SM–SM–SM transitions. These type of transitions represent two and three level moves towards the successful completion of the construction. For a construction process with a plan, we would expect to find a greater number of such transitions than with an unplanned, trial-and-error based, construction process. Clearly, Table 2 indicates that Subject P had many more such transitions than did subject M.

Table 3 gives a summary of the first and second order transitions for successful moves (SM).

Table 3
Summary of first and second order transitions of SM

Model Set A						
Subject	SM–SM	Total	%	SM–SM–SM	Total	%
P	20	56	36	7	55	13
M	6	48	13	1	42	2
Model Set B						
P	20	77	26	3	76	4
M	9	141	6	0	136	0

There is a noticeable reduction in performance with model Set B compared to Set A. This is due to both the extra pieces (35 compared to 27) and the greater complexity of Set B. However, the overall performance in two and three level successful moves of Subject P significantly exceeds the performance of Subject M.

Lag sequential analyses (LSA)

During protocol analysis, the subject's verbalised thinking, coded as a series of predicates (NP, SM, CS, CM), is recorded in a sequence. The LSA technique uses one predicate as a "key" and calculates the occurrence of other predicates relative to the "key" predicate in the sequence.

Figure 10 shows LSA for the key SM for the two model sets A and B. The ordinate has been scaled with respect to the total, number of moves made by the two subjects. The scale f% is SM moves made as a fraction of total moves. These figures again highlight both, the apparent differences in complexity between the two models, as well as the clear differences in move planning between the two subjects.

Predicate Cycle Analysis (PCA)

Finding regularities in behaviour sequences is a powerful heuristic for pattern recognition, avoiding the intensive computation often needed to identify and compare patterns. Predicate cycle frequency analysis provides a report of actually occurring sequences of predicates, rather than formally defined sequences as in transition matrix analyses (Fisher, 1988; Sanderson et al., 1989). Predicate cycle frequency takes one predicate as a key, and works out the sequence of predicates that fall between the *n*th instance of the key predicate and its *n+1* th instance. The final result of the routine is a listing of all existing cycles, the number of times each occurred, and the sequence in which they occurred. In the screen assembly tests we found no discernible pattern of cycles repeated.

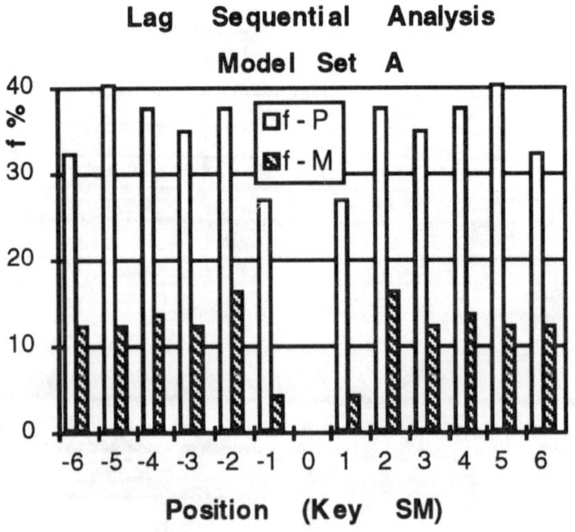

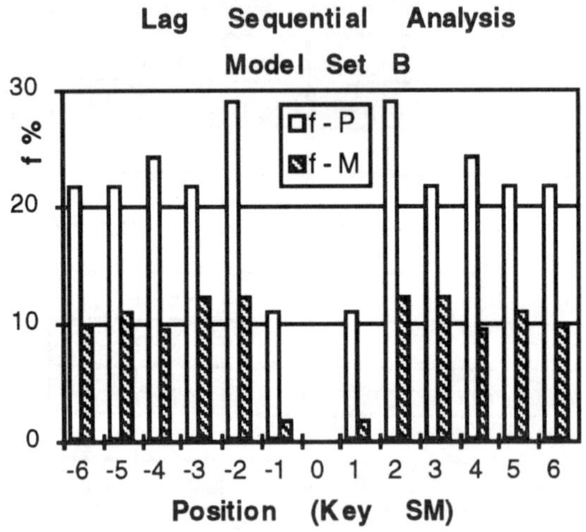

Figure 10: LSA for the two models

7 DISCUSSION

The two subjects used in this experiment had vastly different training backgrounds. One subject, P, had considerable training in 3D visualisation, as well as in the construction of technical equipment. The other subject, M, had no such training. The initial "warm-up" intended to overcome any basic environmental strangeness in using the computer screen. Since both subjects were postgraduate students at Melbourne, albeit in different faculties, one could assume similar levels of basic intelligence, based on the intake scoring system. Nevertheless, even bearing the differences in background in mind, there appeared to be a considerable difference in the problem solving approaches used by the two subjects.

Naturally, verbal protocol analysis is the only plausible approach to evaluating problem solving strategies where there are no visual records of progress through problem space. In the paradigms of protocol analysis presented by Sanderson et al. there is clear evidence of judgment and planning. In our problem space judgment was never used by the problem solvers. This result is due to several factors, some of which have been described by Rasmussen (1986) in dealing with topographic search problems. These problems, just like the model construction problems, are akin to navigating to some known target without a map. Rasmussen notes that:

"If, for instance, a topographic search turns out to be unsuccessful and fails to result in a local search...., the preferred decision is to repeat the search by another parameter. If this search proves unsuccessful too, there is a pronounced tendency to return to a search performed earlier. This seems however to be a repetition with careful judgments of the observations rather than a more careful evaluation of the actual faulty function."

Rasmussen is describing a search for a fault in a complex electrical system. The two types of symptomatic search procedures used for fault diagnosis in such systems are pattern matching and hypothesis testing. Both of these types of search procedures have been observed in the construction of the screen models. Our problem is similar in concept to a symptomatic search problem.

Subject P used some hypothesis about the nature of the model construct. This is exemplified by the number of SM–SM and SM–SM–SM moves made as well as the relatively smaller total number of moves made in solving the model. In contrast to this approach, Subject M used many more single piece moves, with many backtracks, indicating a weaker grasp of feature based mental transformations than that of Subject P. Subject P was able to think ahead to more moves than Subject M (ref. Table 3).

Examining the progress of construction (Figures 4 to 9) and the verbal protocols indicates that Subject P used hypothesis testing (some geometric map or plan) to do the constructions, doggedly continuing with a given plan until a deadlock was clearly reached. In contrast to this Subject M, less experienced in geometric transformations, used a more heuristic approach, we suspect symptomatic search, backtracking many times. We offer the following propositions:

Proposition 1:

Feature based three-dimensional model construction is a problem solving process with strong similarities to topographic search problems. The problem solving process exhibits behavioural problems such as "the law of least resistance" noted by Rasmussen.

Proposition 2

Constructive geometry training is a distinct advantage for formulating hypotheses about feature based three-dimensional model construction.

Proposition 3:

Hypothesis based pattern matching (topographic search) is more economical than symptomatic searches.

8 CONCLUSIONS

This paper describes the results of the screen assembly experiment involving the construction of two sets of CAD models (Sets A and B). Both verbal and videotape records are reported. Verbal protocol analysis is used to analyse the substance of the verbal problem solving approaches. Video tape records were used to augment the verbalised protocols and to allow counting of operational moves performed during model construction. These results of the CAD models experiment give us considerable insight into the mental processed used in three dimensional transformations.

Considerable work is needed to establish the reasons behind the results observed in this paper. Our initial conjectures offer a useful avenue for further explorations

9 ACKNOWLEDGMENTS

The authors would like to thank David Hook and Bernie Kirby, Department of Engineering Computer Resources at the University of Melbourne, for providing generous assistance to us during the development of the "Screen-Assembler" software used in the screen assembly experiments.

REFERENCES

Chase W.G. and Simon, H.A. (1973) *Perception in chess.* Cognitive Psychology, (4): 55-81.

Coxeter, H.S.M. (1961) *Introduction to geometry.* Wiley, New York.

De Groot, A.D. (1966) *Perception and memory versus thought.* In B. Kleinmuntz (Ed.), Problem solving research, methods and theory, John Wiley & Sons, Inc., 22-28.

Ericsson, K.A. and Simon, H.A. (1980) *Verbal reports as data.* Psychological Review, (87): 215-251.

Ericsson, K.A. and Simon, H.A. (1984) *Protocol analysis.* The MIT Press, Cambridge, Massachusetts.

Fisher, C. (1988)*Advancing the study of programming with computer-aided protocol analysis.* In Olson, G.,

Newell, A. and Simon, H.A. (1972) *Human problem solving.* Prentice-Hall, Inc., NJ.

Rasmussen, J. (1986)Information processing and human-machine interaction, North Holland, N.Y.

Samuel, A.E. and Chao, M. (1994) *Cognitive protocols and complexity in feature-based transformation.* Proceedings of IFIP on Feature Modelling and Recognition, 839-852.

Sanderson, P.M., James, J.M. and Seidler, K.S. (1989) *SHAPA: An interactive software environment for protocols analysis.* Ergonomics, (32): 1271-1302.

Soloway, E. and S. Sheppard (Eds.), Empirical Studies of Programmers. pp.198-216, Ablex Publishing Co: Norwood, NJ.

KNOWLEDGE - BASED OBJECT RECOGNITION SYSTEM

Girija Chetty , Dr. Narendra Deshpande
School of Engineering and Technology , Deakin University ,
Geelong , Victoria 3217 , Australia.
Email : girija@deakin.edu.au , narendra@deakin.edu.au

ABSTRACT

Automatic recognition of an object involves several stages such as low level processing, model formation, features extraction and finally model matching. Several algorithms and procedures are now available for each one of the above stages and these tend to be highly complex as well as context dependent, because of which a suitable choice for a given application is rather difficult for a non-expert in this discipline.This paper reports a semantically integrated system developed for recommending the appropriate algorithms for various stages in object recognition. The system is functionally divided into four different modules: Low Level Vision Expert (LLVE), Model Selection Expert (MSE), Geometrical Reasoning expert (GRE), and Question and Answer Module (QAM). The reported work here gives the details of Low Level Vision Expert module based on semantic integration of image processing procedures paradigm. Given an image, the system suggests a suitable image processing procedure, allows the suggested procedure to be evaluated on the image and suggests alternate choice in case of unsatisfied results.

INTRODUCTION :

One of the important elements of achieving unmanned flexible manufacturing is automatic recognition of the components. This area has invoked enormous research interest in the past decade and a number of researchers have focussed the attention on the general problem of locating and identifying the object on the scene. The task of object recognition involves several computational steps as illustrated in Figure1. Each one of the steps acts on the image information available and outputs a more refined data needed to recognize an object in the scene. The first step mainly does the low level processing, for instance, smoothing the image using image filtering operations to achieve a variety of objectives such as reducing the noise, enhancing the image by contrast improvement etc. The next step of processing involves converting the available information into a suitable chosen solid object model. Third step is to extract the distinguishing features of the object which can then be matched with the corresponding features of the reference objects or object models to establish a match.

A variety of algorithms have been developed for the implementation of the above steps, but each one of them is best suited for one class of problems.The appropriate choice is difficult to make for a non-expert.The problem can become further compounded by non-compatibility of the algorithms for different stage. Certain considerations govern the appropriate choice and as such it is possible to lay down the rules governing the choice.The recent ACM surveys [Paul92, Arman93], and some of the review papers [Jolion94, Rosenfield93], give a thorough review of the earlier work done in this domain, outline various procedures and algorithms available for the various stages of the object recognition and suggest the appropriate circumstances for their applications. In view of the above problem, implementation of the knowledge based system for selecting the algorithms for these stages of object recognition has been proposed and reported here. Next section gives a brief decription of the object recogntion task and explains the problems associated with industrial parts recognition.

OBJECT RECOGNITION TASK

Object recognition can be defined as the task of finding and labelling parts of a two dimensional (2-D) image of a scene that correspond to objects in the scene.

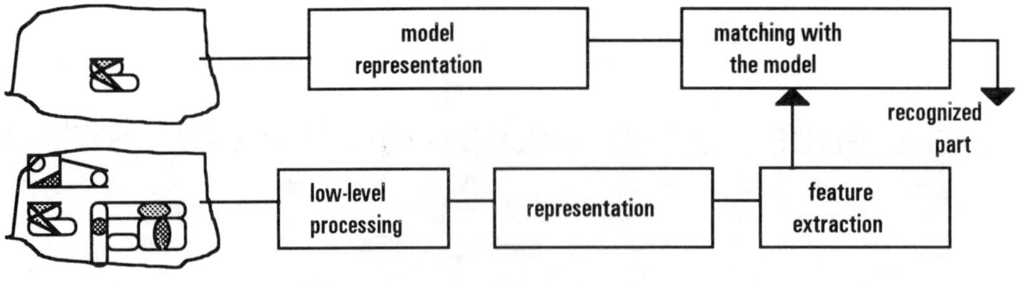

FIGURE 1 Block diagram of an object recognition system

In the flexible manufacturing context, most industrial parts recognition systems can be considered as model based systems in which recognition involves matching the input image with a set of predefined models of parts. The goal of such systems is to precompile a description of each of a known set of industrial parts, then to use these object models to recognize in a image each instance of an object and to specify its position and orientation relative to the viewer. Typically a model includes shape, texture, and context knowledge about the occurence of such objects in a scene.

In an industrial environment the following types of constraints and properties that distinguish this problem domain are usually found [Farshid93].

- The number of parts in a given domain is usually small (1-50).
- Parts may be exactly specified, with known tolerances on particular dimensions and features.
- Parts often have distinctive features (e.g; holes and corners), which are commonly found on many different types of parts.
- In scenes containing multiple parts, there are a number of possible configurations (e.g; touching parts, overlapping parts, and parts at arbitrary orientations with respect to one another and the camera).

Major factors limiting the competence of current recognition systems for complex industrial parts are :

- Low dimensionality in spatial representation and description of parts.
- Simple objects can only be recognised only if presented against a high contrast background with no occlusion, by extracting simple 2-D features and matching them against 2-D object models.
- Lack of higher dimensional spatial descriptions (e.g, 3-D volumetric representations) and their associated matching and feature extraction algorithms restrict the system's capabilities to a limited class of objects observed from a few fixed viewpoints. The ability to recognize a wide variety of rigid parts independent of view-point demands the ability to extract view-invariant 3-D features and match them with features of 3-D object models.
- Another problem is the lack of descriptions of surface characteristics of industrial parts. Without using properties of the surface, many recognition tasks cannot be accomplished by machine vision .

Thus, it can be concluded that the procedures for object recognition will be highly dependent on both the particular application and its intended level of accomplishment.

Three central issues arise as a consequence of the problems mentioned above;

- What features should be extracted from an image in order to describe physical properties and their spatial relations in a scene adequately ?
- In what form should features be combined into object models such that this description is appropriate for recognizing all objects in the given class ?
- How should the correspondence or matching be done between image features and object models in order to recognize the parts in a complex scene ?

These issues have been addressed by many researchers and a growing number of studies have been conducted investigating various approaches to machine recognition of industrial parts, resulting into a huge set of algorithms. These object recognition strategies are described in brief in next section.

Inspite of the rich literature available on the various algorithms, most of the research activities clustered around the development of prototype systems for certain specific industrial applications. These studies are concerned with providing pragmatic solutions to current problems in industrial vision. Some of them show the adequacy of image processing techniques and the availability of technology needs for practical automation systems. Others are concerned with the development of more general parts-recognition algorithms that work well in less controlled environments [Arman93, Chin86, Clement93, Ikeuchi88, Jolion94, Matsuyuma89, Paul92, Rosenfield.

feature extraction) - normally referred to as bottom - up approach and vice versa - normally referred to as top - down approach. A few of this type have been reported, but in a totally different domain. In most of the cases the application domain was aerial image understanding, astronomy etc.[Clement93, Matsuyuma89]. The next section discusses various object recognition categories.

CATEGORIES OF OBJECT RECOGNITION SYTEMS

Object recognition systems can be categorised on the basis of [Chin86] :

- Their dimensionality of spatial description as 2-D, 2-1/2-D, 3-D(Figure2). Associated with each category are issues related to feature-extraction, modeling and matching .
- The computational strategies used for various stages [Paul92]. Based on the computation procedures used for a particular stage of recognition, object recognition task can be classified according to:
  1. Their suitability for complex image data.
  2. Their suitabilty for complex models.
- Complexity of the Image Data: The signal to noise ratio in a digital image is an appropriate measure of complexity. Higher the noise, more complex the data and we need special computational methods to distinguish the correct objects.
- Complexity of the model : This corresponds roughly to the levels of detail in the model. A simple model defined by a simple criterion like shape template.

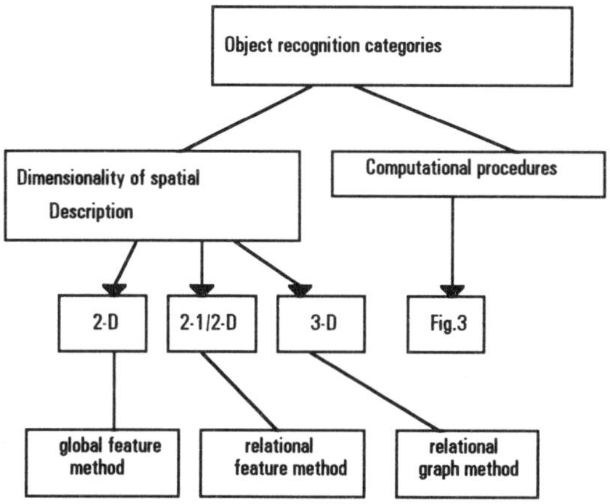

FIGURE 2 Categories of object recognition systems

These four major classes of computational strategies are as shown in Figure3 and can be further classsified as shown in Figure4.

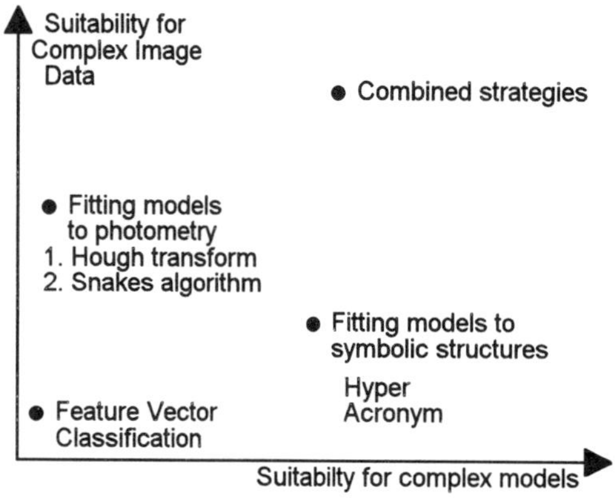

Fig 3 : Classification space for various recognition procedures based on various computational methods

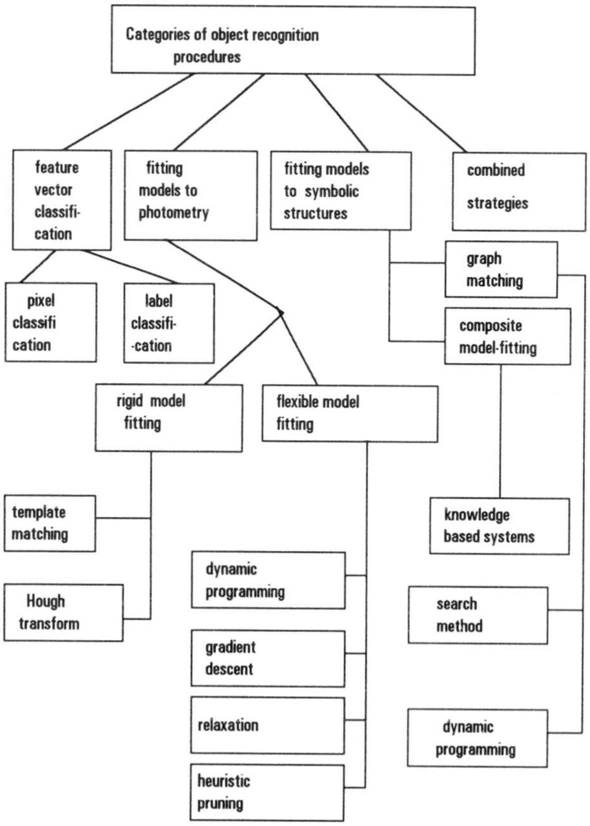

FIGURE 4 Categories of object recognition based on computational complexity

These categories described correspond to a high level vision stage and requires an appropriate choice of the procedures for the low level vision stage of object recognition, based on a particular application domain and a particular computational strategy chosen. Thus the major difficulty is picking up a right computational strategy and the associated low-level parameters. Under such circumstances, a consultative expert system for solving the above mentioned problem is desirable. But the development of such a system requires an integration of different image processing procedures with higher level vision tasks, which is an equally difficult problem. The next section addresses the integration problem and describes integration methodology used for the system proposed by us.

KNOWLEDGE BASED INTEGRATION

The integration of a set of existing programs for any application is a very important and difficult task. The major difficulty of the integration phase is that most of the time modules are developed independently by different specialists, and for different types of images. Knowledge of the use of these modules is not included in the code. Three different types of integration (physical, syntactical and semantic) can be distinguished. More precisely ;

- physical integration of programs is understood as the adaptation of the algorithms to a specialised hardware;
- syntactical integration of programs is integration of a set of programs in a command language or an interface system, it is the integration of the syntactical information needed to run the programs;
- semantic integration of programs is the integration in a system of knowledge of the goals of the programs, knowledge of conditions under which they are applicable, and knowledge of relations between the programs.

Usually, image processing systems are composed of a set of programs or subroutines which can be used directly or through a command language. These systems only provide a physical integration of the procedures or at the best a syntacical integration, that is, the use of abstract commands instead of operating system or programming language syntax.

Since semantic integration of programs includes domain specific knowledge from the expert, it is possible to deal with the problem of robustness of the image processing to noisy input data and of flexibility of the image processing according to variations in the conditions of utilization.

Various methodologies, generally based on artificial intelligence techniques, have been investigated to develop intelligent image processing systems. Such systems make use of intelligent algorithms, intelligent interpretation systems. A classification can easily be performed according to the nature and extent of the knowledge these systems contain explicitly.

- Intelligent algorithms express in detail how programs work and describe explicitly the internal mechanisms of the algorithm. This kind of knowledge is interesting for innovation, generalisation, or educational purposes .One can mention as examples of such intelligent algorithms some interesting work as reported in [Brzakovic91].
- Intelligent interpretation systems contain explicit knowledge on the modelling of objects in the world. An example of such interpretation systems is reported as ACRONYM, which uses geometrical modelling of the objects and CLASSIC, which uses a semantic modelling of classes, and the system reported by [Ikeuchi88], which generates an object recognition program from a 3D-CAD model of the object.
- Semantically integrated systems contain (or have to contain) all the knowledge needed for the selection and the use of programs seen as black boxes. Some algorithms in the library can of course perform object recognition, and they can even be intelligent; but they are considered by the integration system like a black box. In such semantically integrated systems, what programs do (their goal) and when (under which conditions) are expressed explicitly. Work in this domain is recent [Matsuyuma89, Vogt86, Hendler90, Brzakovic91, Ikeuchi88, Clement93].

For the system reported here, we focussed on this semantic integration paradigm with procedures specific to industrial parts recognition task. The next section describes the general architecture of the proposed Semantically Integrated Knowledge-based Object Recognition System.

THE SYSTEM ARCHITECTURE

The architecture of the proposed system consists of following four reasoning modules :

1) Geometrical Reasoning Expert [GRE] for spatial reasoning(matching stage),
2) Model Selection Expert [MSE] for selecting the model of the parts,
3) Low-level Vision Expert [LLVE] for segmentation stage,

4) Question and Answer Module [QAM] for intelligent user interface .

The block schematic of system configuration is as shown in the figure5 here :

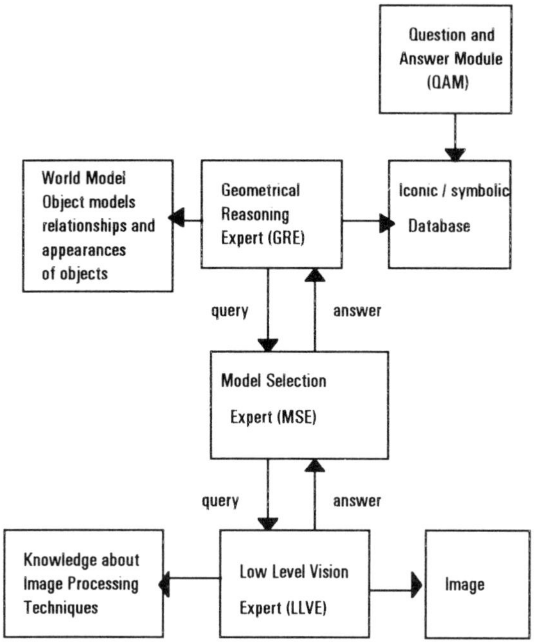

Figure 5 Organization of the system

The former three modules (GRE, MSE and LLVE) are analysis modules to construct the description and recognize the object and the last (QAM) realizes the interactive information retrieval facility to guide through the user and examine the adequacy of the procedures suggested by first three modules (GRE, MSE and LLVE). The three independent experts mentioned above require three different types of knowledge: scene domain knowledge, image domain knowledge, and knowledge about the mapping between the scene and the image. Each expert performs reasoning using its corresponding type of knowledge and communicates with others via simple question and answer protocols. That is, GRE uses scene domain knowledge for spatial reasoning among objects, MSE reasons about appearances of objects in the image based on knowledge about the mapping, and LLVE performs image segmentation using image domain knowledge.

As the type of knowledge for three modules is different we need different model of knowledge for each of the analysis modules outlined above and so is the model of reasoning and control between different modules for efficient integration .

In this paper we present the model of reasoning and control of knowledge for combining the image processing procedure for the low level vision expert [LLVE] module. The methodology used for other two analysis modules and the question and answer module [QAM] will be reported at a later stage .

MODEL OF REASONING FOR LLVE

The reasoning method used for the development of knowledge based integration [Clement93], involves different phases, summarized as follows:

1. Selection of substeps or goals: for example, to reach a certain goal, one has first to perform a segmentation, then to select the biggest region, then to describe its shape .
2. Selection of operators, algorithms, or methods: for example, for image segmentation three methods can exist while, in a particular context, we choose the operator seg1;
3. Initialization of the input arguments: for example, set threshold t = 50 (for grey values between 0 - 255) and input image im21;
4. Effective execution of programs: for example run the command " seg im21-t 50 ", which will perform the segmentation of the image im21 using the threshold value of 50;
5. Test of the results (satisfactory or not) : for example if the size is below 30, the segmentation is too fine .
6. New execution of programs with other values of parameters: for example, if the segmentation is too fine, increase t by 10% ;
7. If the results are still not correct ,selection of other methods : for example, if the operator seg1 fails then use the operator seg2 .

For achieving the above mentioned reasoning methodology we use two techniques:

1. Planning for integration , and
2. Control of Execution for Integration.

Planning for Integration : In this section, the question is; which planning technique is adequate for semantic integration of progrmas? In problem solving, planning is to find a plan for achieving a goal according to a given state of the world. A plan is defined as an ordered list of actions to perform.In general, three main techniques have been developed in planning [Barr82] :

a) Nonhierarchical,
b) Hierarchical, and
c) Skeletal

While the first two methods concentrate on the reduction of conflicts in the plan steps, the last one emphasizes expertise of the specific domain. The third method, also called script based planning, assumes that abstract or skeletal plans exist. A skeletal plan contains the basic

steps to solve a problem. First the best skeleton is selected, then refined .
In image processing, for a low level vision task very few sensitive conflicting subgoals exist ; so the first two techniques are not necessary. As we hold knowledge, at several levels of abstraction, of typical processing methods, a skeletal technique providing a refinement mechanism is appropriate. Another point is the handling of real world data; we have to deal with unreliable or uncomplete information; the state of the world (initial , final, or even intermediate) cannot be described exhaustively. So we have to use a technique based on the description of the actions to be performed and not on the different possible states of the world. Given these prerequisites a skeletal planning technique proves to be adequate for the planning phase of the low level vision module. A plan can eventually be modified during a replanning phase by replacement or insertion of actions.

Control of Execution for Integration:
We are interested here, in defining the type of control of execution which is best adapted for the integration of programs. Four main types of control of execution [Hendler90] have been reported :
1) Prediction of failures and planning of additional actions which wait until the world corresponds to what is expected (example: waiting for a light to turn green before crossing a street);
2) Prediction of failures and integration in the plan of tests and corrective actions;
3) Prediction of failures and integration in the plan of tests and alternatives;
4) Integration of the planning process and control of the execution process.

In LLVE module development context, we do not have to handle secure actions as predicting exactly the behaviour of an image processing operator on an image is not possible. So, a trial and error strategy is need, as well as interleaving of plannig and control of execution processes.

In conclusion, two main tasks have to be performed as shown in Figure 7. First, a planning phase reasons about an appropriate plan to guide the processing of the given images. This step determines the list of instructions to run according to the request of the user; this is done by using a progressive refinement strategy. Second, the control of execution of the plan is done through trial-and-error experiments. The quality of the results obtained from the input images is checked (evaluation of the results), and in case of unsatisfactory results, a mechanism adapts the values of the parameters of the image processing algorithms (adjustment of parameters). The sequence of operators can also be modified using a replanning mechanism if needed.

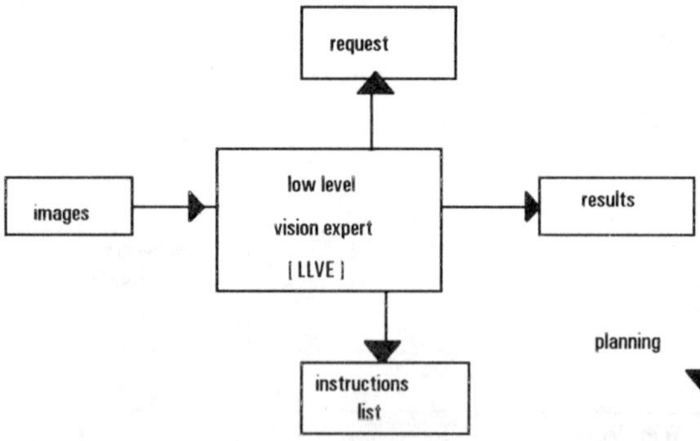

FIGURE 7 Reasoning for integration for LLVE module

For realizing semantic integration of programs with above mentioned model of reasoning for the LLVE module, we need to develop the model for knowledge representation. The next section describes the model of knowledge used for LLVE module.

MODEL OF KNOWLEDGE FOR LLVE

In order to model knowledge of integration and to facilitate the building of knowledge bases, the important concepts like goals, operators, requests, context etc. have to be found.
First, when we process an image, we always have in mind an objective or **goal** to reach. This notion of goal is close to that of the task we need to serve. In the LLVE context the term *goal* represents an image processing functionality. This functionality or objective can be reached by a program or complex processing. For instance: to smooth an image is the goal: in the same way, to count objects in an image is another one. In fact, we distinguish this abstract notion of goal from that of available operators :

Operators are actions which can be performed. They contain specific knowledge to solve a given goal. An operator can be a particular program (it will be referred to as a primitive operator) or a particular sequence of processings (it will be referred to as a complex operator). For example, the primitive operator *median3* is a program implementing a median filter with window 3x3, while the complex operator *count-objects1* is a succession of several processings (smoothing, segmentation, sorting and enumeration).

The **context** is a description of the input data, their conditions of acquisition, the application domain, and even semantic information on the supposed contents of the scene. For instance, a context can be: c-part-recog: application domain is industrial vision, the optical insrtument is a CCD camera, the digitizer is a standard one, scenes are indoor scenes .

A problem in vision is expressed not only by its goal (or functionality), but also by the data to be processed and the eventual constraints on the results to be obtained. This can be termed as a request.

A **request** states which goal has to be reached. The data of the particular case to work on (for example, the name of the input image), the required quality for the results, and the context. A request is thus a query to solve a goal for a specific case. For instance a request, a request can be r-cont : contours detection of the image "view 1"; another one can be r-region: region segmentation of the image "view 2", the size of the image must be greater than 2500 pixels, and the context is "fms-cell".

Thus, we have several phases concerning model of reasoning for integration: how to choose among methods, how to initialze input arguments, how to execute programs how to evaluate results, and eventually how to modify the processing with the determination of new input values. From this description, we can distinguish four notions: choice of operators, evaluation of results, initialization of parameter, and adjustment of parameters

- By the notion of **choice**, we mean the knowledge of how to select, among all the available operators, the operator(s) which is(are) most pertinent, according to the data and the context. For example : if the images have a high contrast, use a contour-based segmentation operator.
- By the notion of **evaluation**, we mean the knowledge of how to assess the results provided by the selected operator after its execution, taking into account constraints expressed in the request. For example: if the number of the segmented regions is greater than 100, the segmentation is too fine .
- By the notion of **initialization**, we mean the knowledge of how to initialize values of input arguments. Various mechanisms are required, as default values or computational methods depending on the values of the other arguments. For example : for the parameter window size is 3 .
- By the notion of **adjustment**, we mean the knowledge of how to modify values of parameters after a negative evaluation .For example : if the segmentation is too fine, increase the value of the threshold th-sigma .

We can thus summarize the model of the knowledge as follows :

The user submits a request which relates to a goal, a set of input data, and a specific context. A goal can be solved by several operators, choice knowledge is used to select an operator among them. An operator can be a program (primitive operator) or it can have a proper decomposition into subrequests (complex operator). Before execution of the operator, initialization knowledge sets the values of its parameters. The results of the execution are estimated by evaluation knowledge. Adjustment knowledge modifies the values of the parameters in case of unsatisfactory results.

The details of the model of reasoning and knowledge for integration for the system reported in this paper is described in the next section.

IMPLEMENTATION

- Implementation of the knowledge model : We have seen that various knowledge must be captured by the system. This knowledge can be classified into two classes

1. The **static** or descriptive knowledge on which the system reasons, such as: the default value of the parameter 'th' is 0.5, or the operator 'med' knows how to perform a filtering;
2. The **dynamic** knowledge used to perform inferences, for instance: if the result image is too noisy then increase the size of the filter.

These two kinds of knowledge are quite different, and so to facilitate the expression of the knowledge by the expert, we propose two different formalisms: **frames** for descriptive knowledge, and the **production rules** for the dynamic knowledge. Moreover, the use of frames permits us to structure the knowledge base by taking into account the relations existing between the various concepts; in particular we have chosen to partition the production rules into small rule bases located in the frames they are concerned with. Such composite architectures are efficient to solve specialized tasks, in which frames and production rules have specific roles .

In case of the system reported here, the static or descriptive knowledge is implemented with frames,i.e;goals, operators, context, requests, and arguments; the heuristic criteria concerning choice, evaluation, initialization, and adjustment are implemented with production rules .

Figure 8 shows the relations between the frames and the rule bases. A request (for instance, request-00) is related to a particular goal (here-goal 1) : this goal can besolved by several operators (Operator-1, Operator-2, and Operator-3 in this case). A decomposition of a complex operator (Operator-2 or Operator-3) is a tree of requests to other subgoals. Each goal and each operator have two private rule bases: one rule base on choice and one on evaluation for a goal; one rule base on initialization and one on adjustment for an operator.

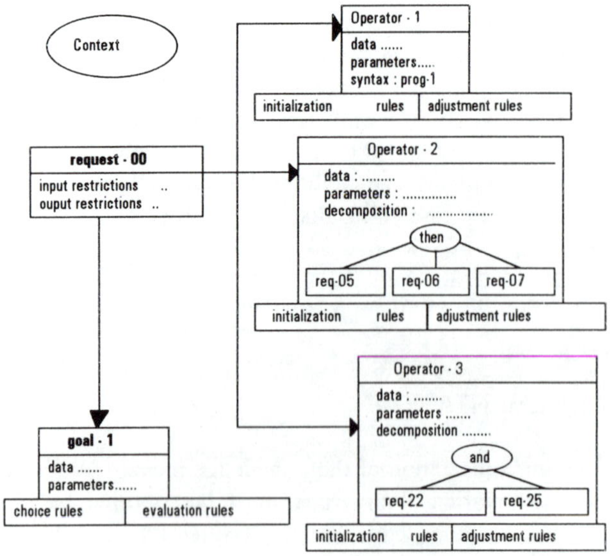

FIGURE 8 Relation between frames and rules for the proposed system

- Implementation of the Reasoning model : The control structure proposed for the reasoning model has five components: a pilot, a planner, a parameter settler, a rule interpreter (a general one called by the previous components), and an interface to activate external programs. The pilot has the role of a supervisor and of a controller: it controls the execution of a tree of requests and decides to call other components for specific tasks. The algorithm presented in Figure 9 controls the execution of a tree of requests. Of course, to process the initial request, this algorithm works on a tree reduced to one request.

Step 1 (performed by the planner) is a global matching verifying that there exists at least one operator to solve the goals of the requests. Step 2 (performed by the pilot) selects one request to process (when the tree is reduced to one request, this step is immediate). Step 3 (performed by the planner) eliminates invalid operators, and performs a classification of the valid ones using the choice rules. Step 4 (performed by the pilot) selects one operator among the valid ones. Before execution of a chosen operator (step 5), the parameter settler determines the values of its parameters : using initialization rules for the first attempt, using adjustment rules for the next ones. If the operator is a primitive one, the execution is done by the interface with the programs ; in case of a complex one, this algorithm is recursively called to control the execution of the tree of requests of its decomposition . Then, according to the request, the evaluation of the results (step 6) is achieved automatically using the evaluation rules, or by graphic presentation of the results to the user who evaluates them interactively. In case of unsatisfactory results, another execution is performed after adjustment of the input parameters of the operator, or after choice of another operator .

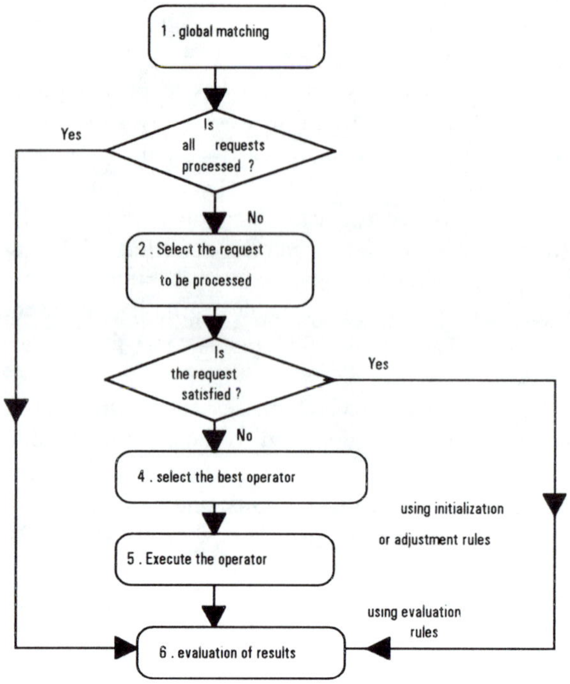

FIGURE 9 Algorithm for a tree of requests

A prototype of the above proposal was implemented on syntetic images and is explained in the next section .

RESULTS AND CONCLUSIONS

The prelimary tests were conducted on synthetic images to test the adequacy of LLVE module proposed. The current knowledge base for the LLVE prototype consists of 15 goals, at which are attached 19 operators .Among these operators 14 are primitive ones (programs) and five are complex ones (with decomposition into steps). Fifty production rules are attached to those frames : there are six choice rules and 10 evaluation rules attached to the goals ; 14 adjustment rules, and 20 initialization rules are attached to the operators .This knowledge base could be extended by adding new alternative operators both for extraction of primitives and matching, with their corresponding rules. Some of the results in which, based on input image, LLVE choses the suitable filter and executes the operator is shown here.

In the input image with gaussian noise the LLVE suggests the weiner filter and executes the operator and ,in case of salt and pepper filter it suggests a median filter and executes it with a default window size of 2x2.

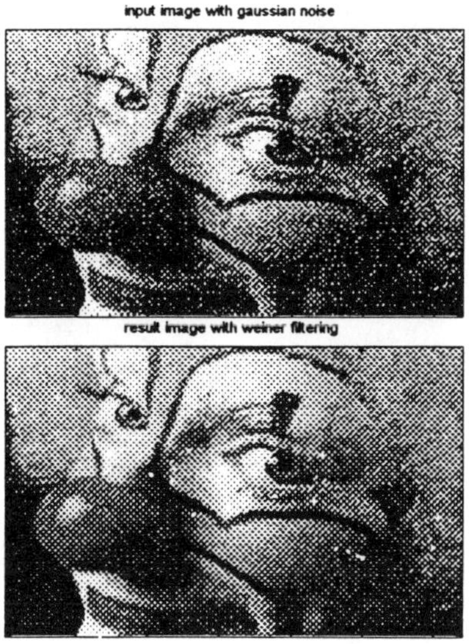

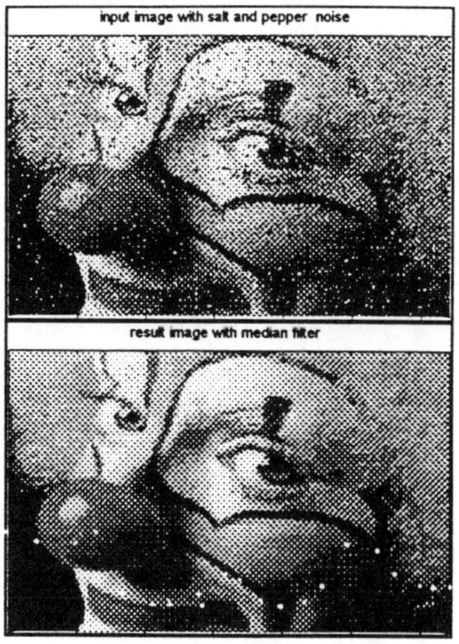

An example of the image segmentation sequence by LLVE is also included here.

Goal Frame : linear-segment
Goal Properties :
Cost-Allowed(Symbolic) : Medium
Window (co-ordinate) :
{(96 70) (182 120))

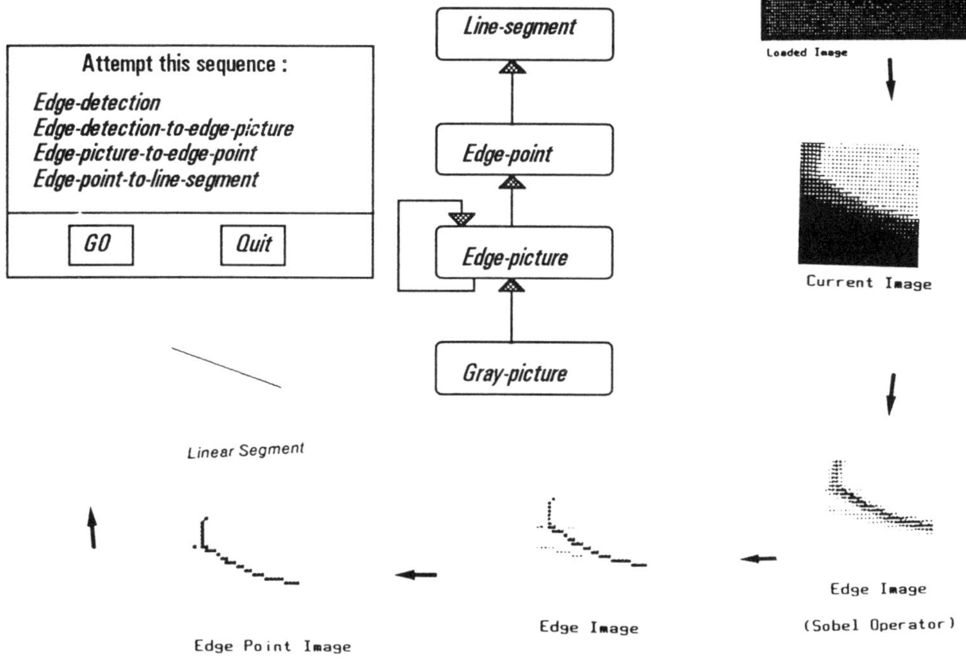

Example of image segmentation

REFERENCES

[Arman93] Farshid Arman and J.K.Agarwal, *Model based Object recognition in Dense Range image - A Review.* ACM Computing surveys, Vol. 25, No.1, March 1993.

[Barr82] A.Barr ,P.R.Cohen and E.A.Feigenbaum . *Handbook of Artificial Intelligence* Pitman, London, New York , 1982 .

[Brzakovic91] D.Brzakovic,R.Patton et.al, *Rule based multi template edge detector,* CGVIP:graphical models and image processing, Vol.53,No.3, pp 258-268, May,1991.

[Chin86] R.T.Chin and C.R.Dyer, *Model -Based Recognition in Robot Vision* ACM Computing surveys, pp.67-108, March 1986.

[Clement93] Veronique Clement and Monique Thonnat, *A Knowledge-Based approach to integration of image processing procedures,* CGVIP :Image Understanding, Vol.57, No.2, pp.166-184, March 1993.

[Hendler 90] J.Hendler .A.Tate and M.Drummond *AI planning:Systems and techniques* AI Magazine, Summer pp 61-77, 1990.

[Ikeuchi88] K.Ikeuchi and T.Kanade*.Automatic generation of object recognition algorithms.*, imProc.IEEE,Vol. 78, No.8, pp1016-1035, 1988.

[Jolion 94] Jean-Michel Jolion, *Computer Vision Vision Methodologies*, CGVIP: Image understanding, Vol.59,No.1,pp.53-71,January,1994.

[Matsuyuma89] T.Matsuyuma.*Expert systems for image processing: Knowledge-based composition of image analysis processes*, Computer vision graphics and image processing. Vol. 48, pp 22-49, 1989.

[Paul92] Paul Seutens, Pascal Fua et.al. *Computational strategies for Object Recognition. ACM Computing surveys,* Vol.24,No.1,March 1992.

[Rosenfield93] Azriel Rosenfield, *SURVEY, Image analysis and Computer Vision : 1993.* CGVIP:Image Understanding,Vol.59 ,no.3 ,pp.367-404,May,1994.

[Vogt86] R.C.Vogt.*Formalized approaches to image algorithm development using mathematical morphology*, in Proceedings, VISION'86, DETROIT 1986.

Knowledge Acquisition

A KNOWLEDGE-BASED METHODOLOGY FOR DEVELOPING KNOWLEDGE ACQUISITION TOOLS

Chih-Cheng Chien and **Cheng-Seen Ho***
Information Technology Lab., Telecommunication Lab., MOTC
No. 9 Lane 74, Hsin-Yi Road Sec. 4, Taipei, TAIWAN, 10656
Email: alex%ltpsuna@tlrouter.motctl.gov.tw

*AI Lab., Department of EE, National Taiwan Institute of Technology
43 Keelung Road, Section 4, Taipei, TAIWAN
Email: csho@et.ntit.edu.tw

ABSTRACT

Knowledge acquisition is one of the major bottlenecks in the development of expert systems. How to develop interactive knowledge acquisition tools easily and efficiently to solve the bottleneck is a recent focus of research in knowledge acquisition. **KAME** is a knowledge-based methodology for rapidly developing specifications for knowledge acquisition tools. It views the development of a knowledge acquisition tool as a primitives construction process which edits a specific knowledge acquisition structure according to the user requirements and generates specifications of a specific knowledge acquisition tool according to the application domain symbols and their attributes.

KEYWORDS: Knowledge Acquisition, Software Methodology, Knowledge Engineering

1. INTRODUCTION

The process of knowledge acquisition is technically challenging and time consuming [Gru89]. It is one of the major bottlenecks in the development of expert systems. Some researches try to attack the bottleneck based on the modeling of problem solving methods and domain models to develop knowledge acquisition tools [Ben85, Esh88, KNM85, Kli88, MM89, Off88]. The other knowledge acquisition tools try to do knowledge acquisition by integrating a variety of methodologies or utilities into a single workbench [AB87, Boo87a, DRM87]. These knowledge acquisition tools are almost domain-specific and difficult to be applied to other domains. It also poses problems to integrate different knowledge acquisition mechanisms in a single knowledge acquisition tool.

We propose a knowledge-based methodology, **KAME**, which tries to attack the above problems and solve the difficulties of knowledge acquisition process by providing a rapid prototyping method for developing knowledge acquisition tools. This methodology views the development of a knowledge acquisition tool as a primitives construction process which edits a specific knowledge acquisition structure according to the user requirements and generates specifications of a specific knowledge acquisition tool according to the application domain symbols and their attributes. In the following sections we will introduce the methodology, the construction process, the primitives used to construct tool specifications, and our conclusions.

2. THE KAME METHODOLOGY

KAME is a methodology based on a primitives set and a tool construction process. The former contain a lot of useful primitives for knowledge representations, problem solving strategies, and knowledge acquisition. They can be used to reconstruct a specification of specific knowledge acquisition tools. The latter is defined to support the operation of the primitives set so that the construction of the specification for a domain-specific knowledge acquisition tool becomes a straight forward job. FIGURE 1 illustrates the methodology, especially on the process of editing and generating a specification of a specific knowledge acquisition tool.

The five types of primitives in the primitives set are problem solving primitives, acquisition primitives, interaction primitives, organization primitives, and relationship primitives. The *problem solving primitives* are basic problem solving techniques that can be used to reconstruct problem solving methods to solve application tasks. The *acquisition primitives* identify the required types of domain knowledge for the problem solving primitives and acquire the knowledge for representation by the representation schemas. They direct the process of knowledge acquisition. The *interaction primitives* are guided by the corresponding acquisition primitives. They specify how to interact with the domain expert by providing proper interaction interfaces according to the types of domain knowledge to be acquired during the knowledge acquisition process. The *organization primitives* specify how to organize the domain knowledge of the application task. The *relationship primitives* specify the relationships among the concepts involved in the domain model represented by the organization primitives. In addition, the methodology also provides four basic *verification functions* for verifying the acquired knowledge bases.

The tool construction process guides the users to edit a specific knowledge acquisition structure and then generates of a specification for the specific knowledge acquisition tool. The edited specific knowledge acquisition structure is domain-independent. Applying domain information (i.e., domain symbols with related attributes) into the specific structure will generate specific knowledge acquisition tools for different domains. A knowledge engineer who intends to derive specifications for specific knowledge acquisition tools through this process must be familiar with the concepts of problem identification, task conceptualization, and the meanings of the primitives in the primitives set.

3. THE PRIMITIVES SET

The *primitives set* contains five types of primitives which are derived by analyzing generic tasks, manual knowledge acquisition methods, and existing knowledge acquisition tools.

First, by applying task analysis technique to *generic tasks* [BC87, Cha86, CJS92] we can obtain several primitive problem solving techniques. Each problem

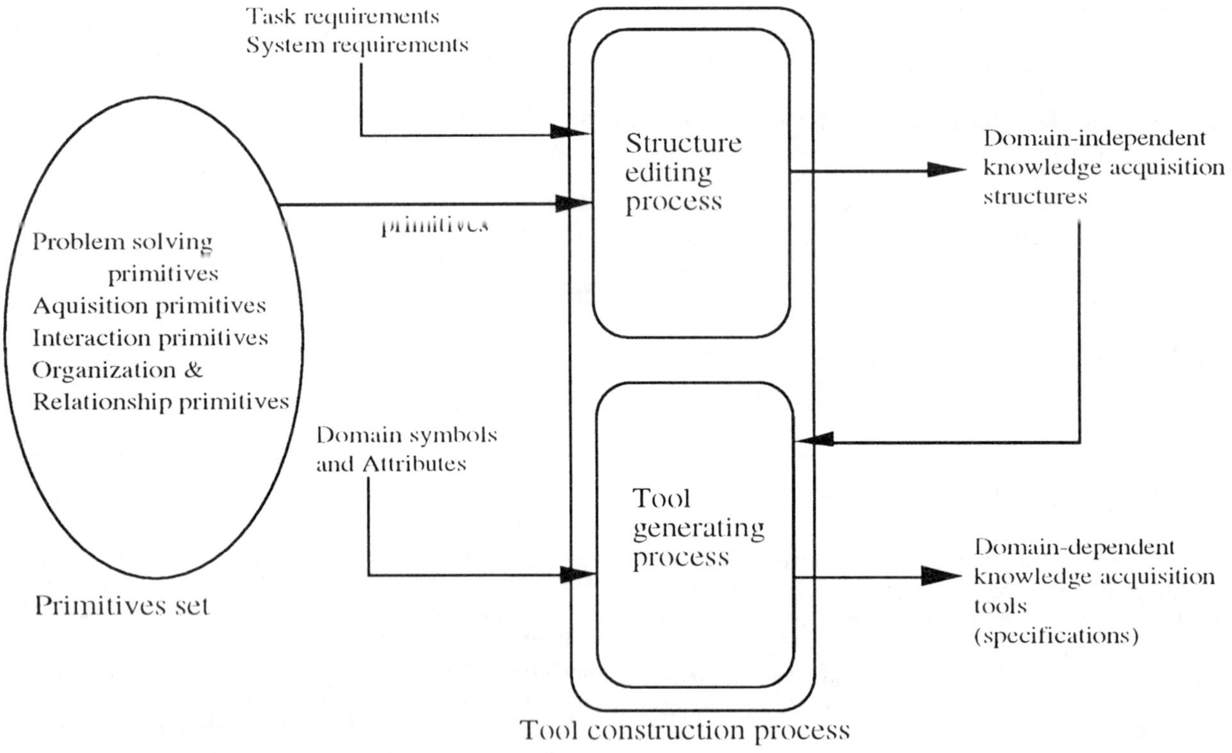

FIGURE 1 The **KAME** methodology

solving primitive is defined by a general function description containing some specialization factors for specializing the primitive, i.e. the primitive is reusable. Thus, a problem solving primitive, if properly specialized according to specific domains, can be used in a specific subtask. Knowledge engineers, who understand how to perform a task through performing corresponding subtasks, can reconstruct a problem solving method from these specialized primitives to solve the task. **KAME** contains twelve problem solving primitives: proposing, checking, revision, restriction, validation, abstraction, refinement, matching, covering, differentiation, remedy, and description.

By applying domain analysis technique to analyze the domains applicable by a generic task, we can derive *organization* and *relationship* primitives that are capable of representing the generic concepts of the domain knowledge models. These primitives can thus be used to construct diverse knowledge models. The organization primitives is used to structure domain knowledge and the relationship primitives is used to define how pieces of domain knowledge are semantically related. We have derived three organization primitives (i.e., *hierarchy*, *network*, and *table*) and twelve relationship primitives (i.e., *constraint*, *contribute-to*, *revision*, *cause*, *condition*, *part-of*, *is-a*, *function*, *abstraction*, *definition*, *element*, and *construct* links).

Acquisition primitives that are responsible for identifying correct types of domain knowledge can be derived by analyzing the problem solving primitives and the representation schemas, because this analysis reveals what types of knowledge are needed and what schemas are appropriate for representing the domain knowledge. For instance, we define *constraint proposing* as an acquisition primitive, since it can be used to acquire the constraint knowledge for the *checking* and the knowledge can be represented by the *constraint* link primitive in a *network* organization primitive. Each acquisition primitive also contains a specialization factor for specialization. For example, we can specialize the *parameter proposing* acquisition primitive into *acquiring the hardware components and their contents* and *acquiring the software components and their contents*. Each specialized acquisition primitive will contain several domain-specific symbols delivered by the knowledge engineer in the tool generating process so that it knows what domain knowledge to acquire. **KAME** contains twelve acquisition primitives: scope setting, syntax describing, data categorizing, solution subdividing, data mapping, data listing, data demarcating, cure describing, parameter proposing, constraint proposing, fix proposing, and format proposing.

We can analyze current manual knowledge acquisition methods and existing knowledge acquisition tools to derive basic *interaction primitives*. These interaction primitives can be specialized with the attributes of the domain symbols to form specific interaction methods for specific knowledge acquisition tools. The interaction methods will be guided by the specialized acquisition primitives to properly communicate with domain experts. Commonly used interaction methods employed in [Boo87b, BG88, Boo89, BG90, GB88, KMY+87, KB87, Mar88, Mar90, MH89] include *concept editor*, *grid constructor*, *menu procedure*, *question/answer procedure*, and *table driven procedures*. Because the concept editor can be replaced by the integration of the grid constructor with the question/answer procedure, **KAME** defines]four interaction primitives: table-driven procedure, menu-driven procedure, grid-driven procedure, Q/A procedure.

From the analysis of existing knowledge acquisition tools and knowledge acquisition methods, we can derive a set of basic verification functions to verify acquired *intermediate* knowledge base. They include primitives for *completeness*, *consistency*, *convergency*, and *subsumption* checks. These verification functions will be automatically included during the construction of specific knowledge acquisition tools.

4. THE TOOL CONSTRUCTION PROCESS

Recall that the construction process contains a domain-independent structure editing process and a domain-dependent tool generating process. In the first process the knowledge engineer determines the task type, the problem solving method, and the domain model. Based on these information he chooses appropriate primitives to construct a specific knowledge acquisition structure. The knowledge engineer then delivers domain-related information into the specific structure to form a specific knowledge acquisition tool in the second process. The domain information contains domain symbols and their attributes. The generated knowledge acquisition tool is a specification which needs to be translated into executable codes for specific implementation environments.

4.1 STRUCTURE EDITING PROCESS

In this process the knowledge engineer determines how to choose the primitives from the primitives set to construct a specific structure to fit a specific domain class. FIGURE 2 illustrates the selection process. Explanations follow.

1.) Task requirements:

a.) What is the task type of the application problem?
We can classify all application problems

into eight task types including classification, debugging, diagnosis, interpretation, construction, design, planning, and scheduling. Normally, the knowledge engineer knows the problem solving strategies if the task type of the application problem is properly identified. Therefore classifying the application problem is the very first decision.

b.) Which problem solving strategies will be used to solve the task?
For each task type there may exists a proper problem solving method to solve it. For example, classification task can be solved by heuristic classification, debugging task can be solved by cover-and-differentiate, construction task can be solved by propose-and-revise, etc. It is necessary for the knowledge engineer to determine which problem solving method will be used to solve the task. The problem solving method can then be constructed from the problem solving primitives in the primitives set or selected from the following pre-defined problem solving methods: heuristic classification method, cover-and-differentiate method, cover-and-classify method, classify-and-interpret method, propose-and-revise method, describe-and-revise method, and restrict-and-revise method

c.) Which domain model will be used to represent the knowledge for the task?
It is essential to determine a domain model to represent the domain knowledge combining with the determined problem solving method to solve the task for developing a specific knowledge acquisition tool. The domain model can be constructed by choosing a proper organization primitive with several relationship primitives in the primitives set. **KAME** also proposes the following pre-defined domain models: causal-model, structural-model, construct-model, functional-model, fault-model, geometric-model, and dependency-mode.

d.) What categories of domain knowledge need to be acquired?
When the problem solving method with corresponding problem solving primitives are identified, basic knowledge types of the domain knowledge are almost clear; i.e., each primitive represents a type of knowledge to be solicited. Thus, the knowledge engineer needs to determine appropriate acquisition primitives according to the selected problem solving primitives. These acquisition primitives will be used to identify and acquire the proper type of

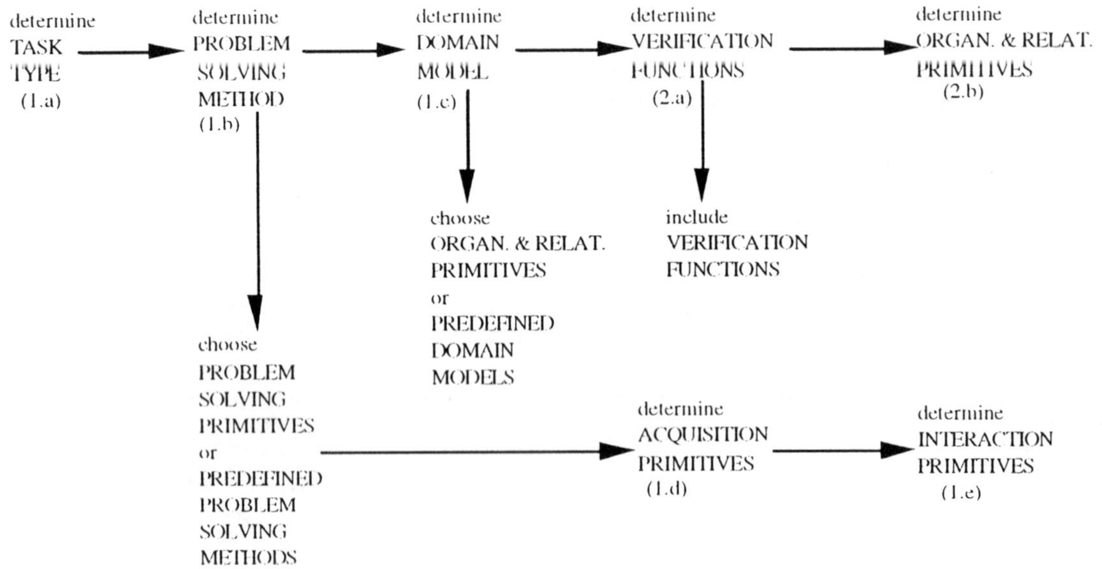

FIGURE 2 Structure editing process of **KAME**

domain knowledge for the problem solving method. Recall that **KAME** proposes twelve acquisition primitives.

e.) How to communicate with domain experts to acquire knowledge?
Recall that **KAME** proposes four types of interaction primitives for the knowledge engineer to construct proper interaction methods to communicate with domain experts. Interaction primitives works with acquisition primitives. As a matter of fact, an acquisition primitive may invoke several interaction primitives to interact with domain experts in different ways.

2.) System requirements:

a.) What knowledge verification functions are needed?
The knowledge engineer determines what types of knowledge verification are necessary. Recall that **KAME** supports four verification functions to verify the acquired knowledge base.

b.) Which knowledge representations are needed for representing the target knowledge base?
The target knowledge representations, e.g., rules, frames, etc., need to be specified. It is dependent on the operation environment of the acquired knowledge bases. For example, we want to develop a generic knowledge acquisition shell on MS-Windows and choose Nexpert Object as the operation environment for the acquired knowledge bases. It must to select *rules* and *frames* as the target knowledge representations, since there are two dimensions (i.e., object dimension and rule dimension) existed in Nexpert Object. Frames is used to represent the object dimension. Rules is used to represent rule dimension.

4.2. TOOL GENERATING PROCESS

In the tool generating process the knowledge engineer

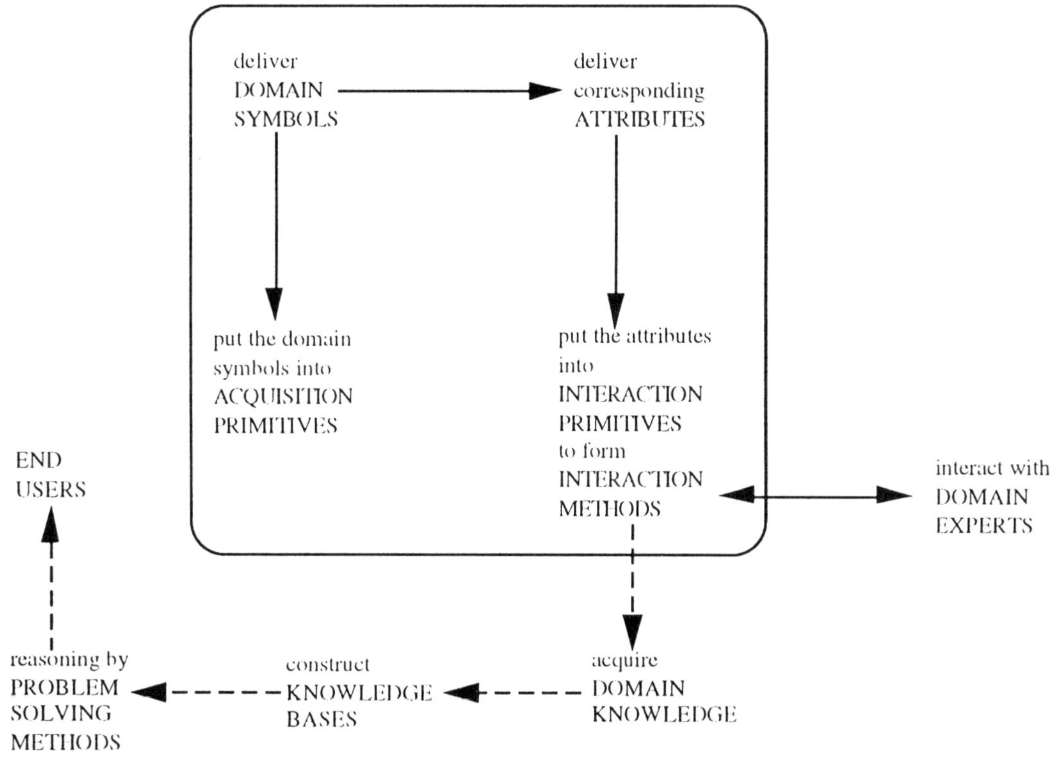

FIGURE 3 Tool generating process of **KAME**

delivers domain symbols and their attributes into the acquisition primitives and interaction primitives, representively, of the constructed specific knowledge acquisition structure to form a specific knowledge acquisition tool. FIGURE 3 illustrates this process.

Domain symbols are domain terminology that are used to describe domain concepts. According to the categories of domain symbols the knowledge engineer delivers them into appropriate acquisition primitives. These acquisition primitives then contain basic and background information about the types of domain knowledge to be acquired. They thus direct the knowledge acquisition process. To support a most suitable interaction method for each symbol, the essential attributes of each delivered domain symbol are delivered by the knowledge engineer into the properly selected interaction primitives to form the interaction methods, which then know what and how to acquire. The acquisition primitives now can guide the interaction methods to interact with domain experts to acquire domain knowledge. The acquired domain knowledge will be represented by the representation schema and incrementally forms an intermediate knowledge base.

5. CONCLUSIONS

We have described the **KAME** methodology for helping knowledge engineers to develop domain-specific knowledge acquisition tools. **KAME** is a knowledge-based methodology for rapidly developing specifications of knowledge acquisition tools. In **KAME**, the development of a knowledge acquisition tool is viewed as a construction process of primitives. That is, it supports a tool construction process for guiding the construction of interactive knowledge acquisition tools from a primitive set that contains five types of primitives to construct a specific knowledge acquisition tool. We summarize the important concepts related to the construction process as follows:

1.) A specific knowledge acquisition tool contains specifications about task requirements and system requirements.
2.) A task contains a domain model and a problem solving method.
3.) A problem solving method may be selected from a set of pre-defined problem solving methods or constructed from proper problem solving primitives.
4.) A problem solving primitive may involve several acquisition primitives to acquire required knowledge.
5.) An acquisition primitive may invoke several interaction primitives to interact with domain experts.
6.) A domain model can be represented by a pre-defined model or an organization primitives with proper relationship primitives.
7.) A specific knowledge acquisition tool may involve several knowledge base verification primitives.
8.) It is necessary to deliver essential domain concepts (symbols) into the acquisition primitives and to deliver corresponding attributes of each domain symbol into the interaction primitives in order to construct a specific knowledge acquisition tool.

The reason why we divide the tool construction process into structure editing process and tool generating process operations will be made clear in below. First, from the viewpoint of software engineering, the structure editing operation is called the requirement analysis phase in the **KAME** methodology. The knowledge engineer determines task requirements and system requirements of an application in this phase to edit a specific knowledge acquisition structure. The structure is *domain-independent* and *reusable*. That is, it can be applied in different applications which are solved by the same problem solving strategy and represented by the same representation schemas. The tool generation operation is called the design phase in the **KAME** methodology. In this phase, the knowledge engineer delivers domain symbols and their attributes into a specific knowledge acquisition structure to form a specific knowledge acquisition tool. The tool is *domain-dependent* and can only be used for a specific application.

A prototype of generic knowledge acquisition shell [CH94] is developed to support **KAME**. The generic shell contains six modules including tool editor, tool generator, tool interpreter, knowledge processor, file manager, and window-based user interface. These modules support all activities in **KAME**. Currently, we are deriving other requirements, e.g., validation functions, learning functions, etc. to extend the methodology and the shell.

REFERENCES

[AB87] G. Abrett and M. Burstein, "The KREME knowledge editing environment," *International Journal of Man-Machine Studies*, vol. 27, no. 2, pp. 103-126, 1987.

[Ben85] J. S. Bennett, "ROGET: A knowledge-based system for acquiring the conceptual structure of a diagnostic expert system," *Journal of Automated Reasoning*, vol. 1, pp. 49-74, 1985.

[Boo87a] J. H. Boose, "Expertise transfer and complex problems: Using AQUINAS as a knowledge acquisition workbench for expert systems," *International Journal of Man-Machine Studies*, vol. 26, no. 1, pp. 3-28, 1987.

[Boo87b] J. H. Boose, "Uses of repertory grid-centred knowledge acquisition tools for knowledge-based system," *Proc. of 2nd AAAI Workshop on Knowledge Acquisition for Knowledge-based system*, Banff, Canada, 1987.

[BG88] J. H. Boose and B. R. Gaines, *Knowledge Acquisition Tools for Expert Systems*. CA: Academic press, 1988.

[Boo89] J. H. Boose, "A survey of knowledge acquisition techniques and tools," *Knowledge Acquisition*, vol. 1, pp. 3-37, March 1989.

[BG90] J. H. Boose and B. R. Gaines, *The Foundation of Knowledge Acquisition*. CA: Academic Press, 1990.

[BC87] T. Bylander and B. Chandrasekaran, "Generic tasks in knowledge-based reasoning: The 'right' level of abstraction for knowledge acquisition," *International Journal of Man-Machine Studies*, vol. 26, pp. 231-244, 1987.

[Cha86] B. Chandrasekaran, "Generic tasks in knowledge-based reasoning: high-level building blocks for expert system design," *IEEE Expert*, vol. 1, pp. 23-30, 1986.

[CJS92] B. Chandrasekaran, T. R. Johnson, and J. W. Smith, "Task-structure analysis for knowledge modeling," *Communications of the ACM*, vol. 35, no. 9, pp. 124-137, September 1992.

[CH94] C. C. Chien and C. S. Ho, "A Primitives-Based Generic Approach to Knowledge Acquisition," *Knowledge Acquisition*, vol. 6, pp. 215-242, 1994.

[DRM87] J. Diedrich, I. Ruhmann and M. May, "KRITON: A knowledge acquisition tool for expert systems," *International Journal of Man-Machine Studies*, vol. 26, no. 1, pp. 29-40, 1987.

[Esh88] L. Eshelman, "MOLE: A knowledge-acquisition tool for cover-and-differentiate systems," in S. Marcus (Eds.), *Automating Knowledge Acquisition for Expert Systems*. MA: Kluwer Academic, 1988, pp. 37-80.

[GB88] B. R. Gaines and J. H. Boose, *Knowledge Acquisition for Knowledge-Based Systems*. CA: Academic press, 1988.

[Gru89] T. R. Gruber, *The Acquisition of Strategic Knowledge*. San Diego, CA: Academic Press, 1989.

[KNM85] G. Kahn, S. Nowlan, and J. McDermott, "Strategies for knowledge acquisition," *IEEE Trans. on Pattern Analysis and Machine Intelligence*, vol. 7, no. 5, pp. 511-522, September 1985.

[KMY+87] A. Kawaguchi, R. Mizoguchi, T. Yamaguchi, and O. Kakushi, "SIS: Shell for interview systems," *Proc. of IJCAI-87*, pp. 359-362, 1987.

[KB87] C. M. Kitto and J. H. Boose, "Selecting knowledge acquisition tools and strategies based on application characteristics," *Proc. of 2nd AAAI Workshop on Knowledge Acquisition for KBS*, Banff, Canada, 1987.

[Kli88] G. Klinker, "KNACK: Sample-driven knowledge acquisition for reporting systems," in S. Marcus (Eds.), *Automating Knowledge Acquisition for Expert Systems*. MA: Kluwer Academic, 1988, pp. 125-174.

[Mar88] S. Marcus, *Automating Knowledge Acquisition for Expert Systems*. MA: Kluwer Academic, 1988.

[MM89] S. Marcus and J. McDermott, "SALT: A knowledge acquisition language for propose-and-revise systems," *Artificial Intelligence*, vol. 39, no. 1, pp. 1-37, May 1989.

[Mar90] S. Marcus, *Knowledge Acquisition: Selected Research and Commentary*. MA: Kluwer Academic, 1990.

[MH89] K. L. McGraw and K. Harbison-Briggs, *Knowledge Acquisition Principles and Guidelines*. NJ: Prentice-Hall Inc., 1989.

[Off88] D. Offutt, "SIZZLE: A knowledge-acquisition tool specialized for the sizing task," In S. Marcus (Eds.), *Automating Knowledge Acquisition for Expert Systems*. MA: Kluwer Academic, 1988, pp. 37-80.

MULTIVARIATE LOGIT MODELS AND NEURAL NETWORKS AS KNOWLEDGE ACQUISITION TOOLS FOR EXPERT SYSTEMS

Martin Lukanowicz, Martin Natter
Department of Industrial Information Processing
Vienna University of Economics
and Business Administration
Pappenheimgasse 35, A-1200 Vienna, Austria/Europe
e-mail: martin.lukanowicz@wu-wien.ac.at, martin.natter@wu-wien.ac.at

ABSTRACT

This paper presents the design of a hybride waste water treatment expert system and its knowledge acquisition methods. The knowledge base can be generated by a multivariate logit model or by neural networks and the transformation of such models into rules.

A multivariate logit model is built by the steps "model generation" and "model reduction" and uses a nonlinear estimator. Scenarios estimated by experts or values measured in a purification plant taken from the operating database can be used as input for the multivariate logit model.

This work proposes a connectionist model (that is, a neural network) of a waste water treatment plant for forecasting and controlling the content of nitrate (NO_3-N) which is a typical pollutant. In an empirical study, we use the operating database of a purification plant for estimating the model. For this purpose, backpropagation — a least square procedure that combines the method of gradient descent with dynamic feedback — is applied. To prevent violations of legal limit values (e.g. NO_3-N < 15mg/l), warning and alarm variables are added to the model.

1 INTRODUCTION

This paper describes KLEX — a hybride waste water treatment expert system. KLEX is intended for the purification plant operator's support and should assist him with the:

- localization and correction of troubles
- optimal regulation of the purification process
- detection of foreseeable troubles
- forecast of his actions ("what-if"-analysis)

The KLEX knowledge base consists of:

- the theoretical basic knowledge of the purification process and waste water treatment technology (from literature, e.g. [OEWWV89])
- knowledge gathered from university experts and practitioners
- the plant configuration data
- knowledge generated by auto-learning methods

Logit models and neural networks are used to avoid "bottlenecks" in knowledge acquisition. Equally, divergent expert opinions of a specific problem can be modeled with the aid of such techniques. The difficulties encountered in knowledge acquisition range from the inability of certain experts to explain their line of reasoning to the simple fact that they are too busy, or even unwilling, to cooperate (see [Trippi&Turban90]).

2 LEARNING WITH MULTIVARIATE LOGIT MODELS

Auto-learning is achieved by estimating an econometric model. From a multivariate logit model, it is possible to generate expert system rules based on the following data:

- a database constructed from scenarios estimated by experts; scenarios generated randomly or by experts from information on the relations between dependent and independent system variables (see [Lukanowicz92]). The experts must only specify all independent variables and dependent system variables (i.e. all data/information they need to estimate a variable).

- values taken from the operating database existing in a purification plant.

Now let us formulate a multivariate logit model:

$$P_{i_k} = \frac{1}{1 + exp^{-\left(\beta_{i_k} + \sum_{j=1}^{n_{i_k}} \sum_{l=1}^{m_j} Y_{jl} * \beta_{i_k jl}\right)}} \quad (1)$$

P_{i_k} ... conditional probability for dependent variable i/interval k
β_{i_k} ... intercept for dependent variable i/interval k
n_{i_k} ... number of declaratory independent variables for dependent variable i/interval k
m_j ... number of intervals of independent variable j
i_k ... subscript for dependent variable i/interval k
j_l ... subscript for independent variable j/interval l
Y_{jl} ... Boolean variable which corresponds to independent variable j/interval l
$\beta_{i_k jl}$ ... parameter describing the relation between dependent variable i/interval k and independent variable j/interval l

To estimate the model parameters, we proceed as follows:

1. Metric variables are subdivided by experts (e.g. legal limits) or statistical methods (e.g. percentiles) into intervals. Each interval is represented by a Boolean variable.

2. In the *model composition phase*, we use the following heuristic:
We checked all pairs of dependent and independent variables able to decrease the sum of squared errors (SSE)[1]. By means of a "normed fit" index, we can decide which independent variables are to be included in the multivariate model. The "normed fit" index measures the percentage of improvement (deterioration) of the SSE by changing to a model with more (less) parameters:

$$\Delta = \frac{SSE_R - SSE_U}{SSE_R} \quad \text{with} \quad 0 \leq \Delta \leq 1 \quad (2)$$

SSE_R is the sum of squared errors of the restricted (intercept) model and SSE_U of the unrestricted model. We select all independent variables with $\Delta \geq 0.02$.

3. We estimate the parameters of the multivariate model generated with the heuristic described in step 2 using a Ridge Regression [Chow85] with a modified version of the Marquardt/Levenberg algorithm [Gallant87, Judge80].

4. In the *model reduction phase*, we eliminate all insignificant independent (input) variables, that is all parameters which are not distinct from zero on a significance level of $\alpha = 0.05$ (t-test). As asymptotically normally distributed t-value we use:

$$\frac{\delta}{\sqrt{VAR_\delta}} \quad \text{with} \quad \delta = \beta_i, \beta_{ij} \quad (3)$$

5. Estimation of the final model.

6. Transformation of the final model into rules (see section 3).

[1] The squared error between the estimated and the empirical dependent variable:
$SSE = \sum_{i=1}^{n} (P_0 - Y_0)^2$

3 KNOWLEDGE BASE AUTOMATICALLY GENERATED FROM A MULTIVARIATE LOGIT MODEL

We will use an example to illustrate the interpretation as rules of the multivariate logit model. The following partial knowledge base shows the rules extracted from the part of the multivariate logit model describing the dependencies of the variable "ammonium nitrate" (NH_4-N) found in the outlet of the waste water treatment plant. For estimating the model, we use values from the operating database of a purification plant.

In our example, the dependent variable "ammonium nitrate" was subdivided into four intervals — that means that we need four rules. Each interval can be interpreted as a linguistic variable (e.g. "low", "normal", "above average", and "high") with a membership value of 1 for all values in the range of the interval. Independent variables are also subdivided into intervals; for these we use a membership value of 1, but membership functions of any distribution are allowed.

If there is no information on the values of the independent variables, the probability of each rule may be deduced by inserting into equation 1 the intercept parameter from the "parameter" column shown in table 1. In this case, the probabilities for rule $\mathcal{R}1$ to $\mathcal{R}4$ are 0.12, 0.24, 0.19, and 0.15, and the rule with the highest probability was used (i.e. rule $\mathcal{R}2$: the NH_4-N content in the outlet of the waste water treatment plant is "normal").

In case more information is available — e.g. "the load of NH_4-N in the waste water is normal" (subdivided into three intervals) —, then the probability of rule $\mathcal{R}2$ rises to 0.43.

Also rule $\mathcal{R}1$ can "win the game", if the NH_4-N level in the waste water is "very high" and there is no oxygen deficit, the nitrite in the outlet is "low", and there is no loading from the mud treatment line. The probabilities for rule $\mathcal{R}1$ to $\mathcal{R}4$ are 0.40, 0.24, 0.19, and 0.30. The distribution of these probabilities can be used as an additional criterion for arriving at a conclusion.

Expert system consultation can be "optimized" by sorting in descending order the absolute values of the

Table 1: Automatically generated rules for NH_4-N

rule	dependent variable		independent variable	parameter	Q#
$\mathcal{R}1$	NH_4-$N_{out} \leq 0.81$	IF		-2.004	
			NH_4-$N_{w-w} \leq 29.0$	0.603	8
			$O_2 \leq 2.0$	-0.647	7
			NO_2-$N_{out} \leq 0.12$	1.017	1
			NO_3-$N_{mud} = 0$	0.564	9
$\mathcal{R}2$	NH_4-$N_{out} > 0.81$ and ≤ 1.5	IF		-1.172	
			NH_4-$N_{w-w} > 29.0$ and ≤ 40.0	0.878	2
$\mathcal{R}3$	NH_4-$N_{out} > 1.5$ and ≤ 3.325	IF		-1.464	
			waste-water pH ≤ 7.75	0.673	6
$\mathcal{R}4$	NH_4-$N_{out} > 3.325$	IF		-1.702	
			NH_4-$N_{w-w} > 40.0$	0.855	3
			$O_2 \leq 2.0$	0.686	5
			NO_3-$N_{mud} > 0.1$ and ≤ 0.3	0.794	4

logit model parameters (R.H. column in table 1). By doing so, it is possible to inquire for the values of independent variables with descending influence over the dependent variable. Queries can be stopped if pending answers are unable to change the conclusion (in contrast to other inference engines). The multivariate logit model parameters are similar to certainty factors for rules and conditions (see[Shapiro83]). Contrary to the MIN/MAX-method (Lukasiewicz's multivalued logic), we use the logit model (conditional probabilities) for the combination of conditions.

4 LEARNING WITH CONNECTIONIST MODELS

In contrast to a multivariate logit model, a connectionist model can process metrically, ordinally, and nominally scaled variables (which means that metric variables need not be subdivided into intervals, and fewer parameters need to be estimated). [Hruschka93] pointed out that connectionist models with sigmoid activation functions are a generalization of multivariate logit models (neural networks without hidden layers).

As input data we use the same operating database as for the logit model. Variables are scaled into the $[0;1] \in \Re$ interval. As the waste water remains in the revival basin for approximately 3 days (see [OEWWV89]), in our model we consider the actual outlet pollutant concentration and that of the previous two days. Interventions by the purification plant operator have a delayed effect. Therefore, in our neural network we use for a dependent system variable two output units (one for time t and one for time $t+1$). These variables forecast the metric value of the NO_3-N content in the purification plant outlet. To prevent violations of legal limits (e.g. NO_3-N $<$ 15mg/l), warning and alarm variables are added to the model.

Our connectionist model consists of three parts:

- a directed graph $G = \{U, E\}$ with nodes U (units) separated into layers. Edges E are allowed only between adjoining layers. The first layer is called input layer (IL) and is followed by one layer with hidden units referred to as hidden layer (HL). The last layer consists of output units and is called output layer (OL).

- a matrix of weights. $w_{ij} \in W$ measures the influence exerted between node i and node j. If no connection exists between two nodes, then $w_{ij} \equiv 0$.

- backpropagation [Werbos74, Rumelhart&McClelland86]: a method used to manipulate the state of units and weights.

4.1 Backpropagation (bp)

A connectionist model can be used to model a functional relation as follows:

$$y_o = f\left(\sum_{\forall h} w_{oh} g\left(\sum_{\forall i} w_{hi} x_i\right)\right) \quad (4)$$

$$f(z), g(z) = \frac{1}{1+\exp^{-z}} \quad (5)$$

For each input variable x_i, we determine an input unit and for each output variable y_o an output unit. In the next step, we calculate the activation of all units (layer by layer) using equation 4. The activation of a unit is a function of the sum of weighted inputs into a unit less its bias (logit model: intercept). As transfer function we use the logistic function (equation 5). The step forward stops at the OL, and we compare the output from network y_o with the empirical values t_o. The difference corresponds to error $e_o = t_o - y_o$. The objective is to change weights w_{ij} of the network so as to minimize the squared error sum for observations p and output units o.

$$E = \frac{1}{2}\sum_{po}(t_{po} - y_{po})^2 \rightarrow \min! \quad (6)$$

Minimizing is done layer by layer by backpropagation of the error. The weights are changed using the first derivative of the objective function E:

$$\Delta w_{ij} = -\eta \frac{\delta E}{\delta w_{ij}} \quad (7)$$

η denotes the learning rate and lies in the interval $[0;1] \in \Re$. In the *initialization phase*, all weights are set to small random values. The forward and backward phases are executed stepwise until the objective function can no longer be improved significantly.

4.2 Evaluated models

We have studied two modeling approaches:

Model with latent variables ($\mathcal{M}1$): Hidden units are not fully connected with input units. The number of interactions is restricted. All hidden units are connected only to a particular group of correlated input variables. The correlations are derived from the theoretical knowledge of the waste water treatment process. This approach allows the interpretation of hidden units. Each hidden unit has a distinct meaning (a concept or principle, e.g. environmental influences). This meaning cannot be measured directly, but it affects the output values. Such variables are commonly called "latent variables".

In the initial model, we used 11 latent variables derived from expert knowledge. In a *model reduction phase*, insignificant links were removed. (Methods for model reduction are presented in [Natter&Lukanowicz93]). We then obtained the final model with 106 parameters, a SSE of 142.4, and a cross-validation rate (CVR) of 159.3 (see [Stone74]).

Nonparametric regression model ($\mathcal{M}2$):
To ensure that the restricted structure of model $\mathcal{M}1$ contains all important dependencies, we tried another modeling approach. Starting with one fully connected hidden unit (all input and output units connected with the hidden unit), we added to the model more hidden units step by step until the cross-validated sum of squared errors could no longer be improved significantly. This method is equivalent to a nonparametric regression (see [White89]).

We found that a model with two hidden units improved the results. A model with three hidden units must be rejected as the cross-validation indicates "overfitting". Adding a second hidden layer does not improve the results. The model with two hidden units underwent a reduction process, and we obtained a model with 92 parameters, a SSE of 130.1, and a CVR of 143.0.

In [Towell&Shavlik91], a neural network is constructed from rules generated by experts. Then, the network is trained with standard neural learning methods allowing knowledge refinements. Refined rules are extracted from the network. Our experience shows that experts may be on the wrong track (see model $\mathcal{M}1$) if interdependencies between variables are complex and input variables are highly correlated. Therefore, it is better to start with a new model and then try to interpret the hidden units. In our example, the two hidden units correspond to the two biological processes "nitrification" and "denitrification".

5 KNOWLEDGE BASE AUTOMATICALLY GENERATED FROM A CONNECTIONIST MODEL

In a neural network, every hidden unit has a distinct meaning. A hidden unit represents a (latent) concept or principle. The hidden layer adds hierarchy to the represented knowledge. This hierarchy is mapped into rules.

Figure 1 shows a network and its representation as a knowledge base. A dependent variable H is determined by concepts $L1$ and $L2$ (first hierarchy). Each latent concept is determined by a distinct set of input variables (second hierarchy). To decide what the logical "*op*" operator (the operator between two conditions) stands for is difficult but if the method described in section 3 is used, this is not necessary. The

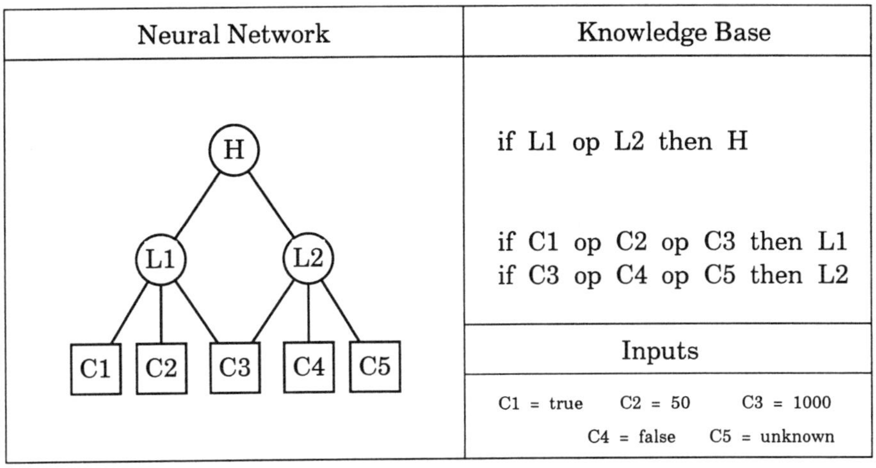

Figure 1: Translation of a neural network structure into a knowledge base

dependencies for the output variables can be calculated using equation 4. If logical operators are really needed, the methods presented in [Towell&Shavlik91] should be used.

Because of the hierarchy and scaling of the metric input values, "optimizing" the consultation of the expert system is not as easy as shown in section 3. To decide which independent variables have the most influence over the dependent variable, one has to analyze the sensitivity of the output value after changing one input value. The sensitivity analysis is done by varying one independent variable for all observations (the input variable is multiplied by e.g. 0.7, 0.8, 0.9, 1.1, 1.2, and 1.3).

6 ACCURACY AND BENEFITS OF THE PRESENTED METHODS

In our empirical study, we use measured values (31 variables and 420 days) taken from the operating database.

Applying the multivariate logit model, we were able to correctly forecast between 40 and 58 percent of the pollutants found in a waste water treatment plant (i.e. BSB_5, CSB_5, NH_4-N, NO_3-N, NO_2-N, and PO_4-P). Furthermore, in 20 to 30 percent the forecast interval of the dependent variable adjoined the empirical interval. In cases of incorrect forecast, the interval probabilities are similar and their distribution should be used as extra safety criteria. The benefit offered by this method is an algorithm that allows fast estimates.

The comparison between the metric empirical value and the forecast value of the connectionist model shows that the network is able to accurately approximate peaks and troughs, and seasonal effects. The comparison of actual and forecast warning and alarm variables indicates a high fault detection level.

The output values range in the interval $[0;1] \in \Re$. This amplitude shows how critically the model assesses the situation. The analysis of the warning and alarm variables is shown in table 2. We use a threshold function for releasing an alert (only values ≥ 0.5). If the threshold value is lower, the model response is more "cautious".

The benefit of this method is a very stable bp-algorithm and the possibility of processing mixed scaled variables.

7 CONCLUSION

This paper presents and empirically validates two nonlinear models for forecasting the pollutant load in the outlet of a waste water treatment plant.

The first model — a more traditional multivariate logitmodel — can be estimated with the aid of a fast nonlinear estimator. The variables are to be subdivided into intervals so that the influence between more variables must be checked. We propose a heuristic for selecting independent variables. The multivariate logit model can easily be translated into a set of rules. The consultation of the expert system can be "optimized" by sorting the input variables queried by their descending influence on the dependent variable.

For the second model — a neural network — we use backpropagation to estimate the parameters. We forecast the metric values and alert levels for nitrate found in the outlet of a purification plant. Two modelling approaches, i.e. (1) correlations between variables derived from the theoretical knowledge of the waste water treatment process and (2) a nonparametric regression model are presented. Mapping of the network structure into a set of rules is possible.

Table 2: Analysis of warnings and alarms forecast by model $\mathcal{M}2$

warnings (W)				alarm (A)			
empirical	forecasted	time	time	empirical	forecasted	time	time
warning phase		(t)	(t+1)	alarm phase		(t)	(t+1)
W	W	101	99	A	A	8	11
	W	35	34		A	1	1
W		29	31	A		13	10
$\sum$ warnings = 130		136	133	$\sum$ alarms = 21		9	12

REFERENCES

[Chow85] Chow, G.C.: *Econometrics*, 2nd Edition, McGraw-Hill, Auckland, 1985.

[Hruschka93] Hruschka, H.: *Determining Market Response Functions by Neural Network Modeling — A comparison to Econometric Techniques*, in: European Journal of Operational Research, Vol. 66, 1/1993.

[Gallant87] Gallant, A.R.: *Nonlinear Statistical Models*, John Wiley & Sons, New York, 1987.

[Judge80] Judge G.J.; Griffiths, W.E.; Carter-Hill, R.; Lee, T.C.: *The Theory and Practice of Econometrics*, John Wiley & Sons, New York, 1980.

[Lukanowicz92] Lukanowicz, M.: *Integration eines multivariaten Logit-modells in das Kläranlagenexpertensystem* KLEX *als induktive Lernkomponente*, Operations Research Proceedings 1991, Springer-Verlag, Berlin Heidelberg, 1992.

[Natter&Lukanowicz93] Natter, M.; Lukanowicz M.: *Ein konnexionistischer Ansatz zur Prognose und Steuerung von Ablaufschadstoffen einer Kläranlage*, in: Operations Research Proceedings 1992, Springer-Verlag, Berlin Heidelberg, 1993.

[OEWWV89] ÖWWV: *Vortragsunterlagen zum Klärwärter Grundkurs des Österreichischen Wasserwirtschaftsverbandes*, Wien, 1989.

[Rumelhart&McClelland86] Rumelhart, D.E.; McClelland, J.: *The PDP Research Group: Parallel Distributed Processing: Explorations in the Microstructure of Cognition*; Volume 1: Foundations, MIT Press, Cambridge MA, 1986.

[Shapiro83] Shapiro, E.Y.: *Logic Programs With Uncertainties: A Tool for Implementing Rule-Based Systems*, in Proceedings of the Eighth International Joint Conference on Artificial Intelligence (ijaci-83), Volume 1, 1983.

[Stone74] Stone, M.: *Cross-Validatory Choice and Assessment of Statistical Predictions*; in: Journal of the Royal Statistical Society, Series B, 1974.

[Towell&Shavlik91] Towell, G.G.; Shavlik, J.W.: *The Extraction of Refined Rules from Knowledge-Based Neural Networks*, Working Paper 91-4, University of Wisconsin — Madison, 1991.

[Trippi&Turban90] Trippi, R.R.; Turban, E.: *Autolearning approaches for building expert systems*, Computer Opns Res., Vol 17. No.6, Pergamon Press, 1990.

[Werbos74] Werbos, P.: *Beyond Regression: New tools for prediction and analysis in the Behavioral Sciences*, Ph.D. thesis, Harvard University, 1974.

[White89] White, H.: *Connectionist Nonparametric Regression: Multilayer Feedforward Networks Can Learn Arbitrary Mappings*, Working Paper, Department of Economics, University of California, San Diego, 1989.

TELSTRA'S AVCORD : A NOVEL APPROACH TO LARGE SCALE KNOWLEDGE ACQUISITION AND RULE PROCESSING

M. J. Cannard
Rapid Response, Information Technology Group, Telstra Corporation Ltd (Telecom Australia)
25/570 Bourke St, Melbourne, Victoria, Australia 3000

KEYWORDS

Knowledge Acquisition, Rule Processing, Expert Systems, Knowledge based systems

ABSTRACT

Conceptualising and developing knowledge based systems has historically been challenging. This is especially so of those systems that face complex decision structures that result in large numbers of decision rules and corresponding actions. This paper is a description of two novel approaches that were used in the construction of a large knowledge based system named AVCORD. The first approach involves a technique used for knowledge acquisition on a large scale using an English orientated medium. The second is the rule processing technique that converts the decision information into a form that can be compared to the acquired data. The AVCORD knowledge based system checks and validates 40,000 daily orders initiated by Telstra's telephone sales representatives. In its first release the system uses over 11,000 decision rules and is projected to be using over 120,000 rules in its final form.

INTRODUCTION

Customer orders initiated incorrectly are costly to any business, and the telecommunications industry is no exception. Dissatisfied customers are lost to competitors, order re-working can be costly, and employee morale can be affected as a result. In Telstra, the nature of the order entry process is quite complex with special relationships and charging rules between different products. Complicating this process is that Telstra has over fifty thousand products which can change to cater for new marketing strategies and new technologies. A knowledge based system addressing the forty thousand or so daily initiated orders was needed to meet specific criteria. It had to be highly flexible, fast , cover a large number of rules, and most importantly, the rules had to be checked easily by the order entry experts.

This paper presents a knowledge based system named AVCORD - Automated Validation of Customer Order Record Details. This system has been implemented successfully in Telstra and targets the ten most error prone product groups of which three were addressed in its first phase. The scale of the project coupled with three month time limitation required a change in direction from the traditional approaches of most knowledge based systems. This paper presents two of those different approaches. Firstly a rapid but flexible knowledge acquisition process in an English orientated medium is presented. This approach removed the need for the traditional knowledge engineer or intermediary person who translates the experts rules into the language of the developers. Secondly an unusual method of rule processing is described which has enabled this system to cope with the acquired knowledge and great number of decision rules produced.

KNOWLEDGE ACQUISITION METHOD

One of the major problems of knowledge acquisition is in the matching of what the experts know, to something that the software developer can use [Gru88][Tur92] . The developer's goal should be to elicit information in a medium that is understood by both the expert and the development engineers [Gru88][WN88]. It was through the achievement of this goal that made the AVCORD project a success.

Early in the project development, it was found that English rules provided by the designated order entry expert were not compatible with what the software engineers could deliver. This is hardly surprising as the expert did not have a computing background. What would normally happen at this time is that a knowledge engineer would be employed to act as an interpreter between the expert and the software developers[Tur92][You89][Zah93]. The AVCORD project, however, had neither the time nor the funding to allow such an inclusion. These constraints necessitated an approach that would cover a great many rules quickly while coping in an environment that contained no intermediary between the expert and the developer. The following is a description of the knowledge acquisition technique used.

Each of the expert's English rules were broken down into what was perceived to be the individual question attributes. Each of these attributes were made headings of a number of columns of a decision matrix. An extra column was added in at the end of the matrix for each of the rules outcomes. Having established a basic rule structure, the experts rules could be mapped to this matrix without losing it's English readability. The development team found that using an English based acquisition structure allowed the experts to relate to the process at a level that they found understandable and comfortable. Once the structure was finalised , a compatible medium was chosen to support the knowledge acquisition structure and the number of rules was chosen. The spreadsheet application MS Excel was chosen, as it was suitable for replicating the matrices and was capable of interaction with the development tools chosen.

The manufacture of rules starts by including every possible combination of the question attributes. Once these are listed in the matrix, the action conditions are then mapped to each decision rule. To add flexibility to this mapping process, the actions are given in numeric form rather than in English. Each of these numeric actions are then specified in an action table which shows the English equivalent. Representing each action by a number allows rules with a number of actions to be represented by a series of numbers rather than a long English statement. A long English statement introduces complexity implications, for both expert and developer, in uniqueness and redundancy.

The number of rules eventually reached eleven thousand which was after pruning and grouping processes. Without each of these compacting processes the number of permutations and combinations of the rules would have been over forty eight million.

The expert became so confident of the knowledge acquisition process that the developers found that they needed only to set the acquisition structure and let the experts take care of the rules and actions. This allowed the development team to move back to the job of building the system, knowing that the system will receive it's decision tables from a known structure and from MS Excel spreadsheets.

The knowledge acquisition technique used forced each of the experts to take on the role of knowledge engineer. Consequentially the experts had now gained a real understanding and feel for the application. Later in testing, the benefits of such a situation soon became apparent to development team. When the decision tool identified missing rules or duplications, the experts immediately pin-pointed and rectified the problem with little or no help from the development team. A further

TABLE 1 Simplified Example of a Single Level Decision Table and Corresponding Action Table

Example Decision Table					Example Action Table	
Is Exchange digital?	Order Type	State	Action		Action No	Description
Yes	O	VIC	[1],[4],[3]		[1]	Replace ‘a’ with ‘b’
Yes	N	VIC	[1]		[2]	Remove product
Yes	O	NSW	[2],[4]		[3]	Modify rate to show 0
Yes	N	NSW	[2]		[4]	Insert correct tariff
No	O	VIC	[5]		[5]	Non digital - Remove
No	N	VIC	[5]		“	“
No	O	NSW	[5]		“	“
No	N	NSW	[5]		“	“
“	“	“	“		“	“

benefit of an English orientated structure was the capability of sending or sharing the tabulated decisions with other experts who would contribute their own opinions and needs of the decision rules. By midway into the AVCORD development the software developers had relatively little input and even little interest in the decision rule making process. As long as the structure was specified and the number of decision tables were known, the experts could include as many rules as they pleased - the honesty of the decision tool would soon determine whether those rules were correct.

RULE PROCESSING

The knowledge acquisition process produced several MS Excel spreadsheets each with English structures and rules ready for use in the decision tool. Development tools generally require knowledge to be expressed at a lower and more detailed level than what humans need for decision making [Tur92]. To address the knowledge level 'gap' a software technique was developed to translate what was available from the order, to a suitable form that could be compared to the English structured decision tables. This method is now described.

Each of these spreadsheets were loaded into a database using the Excel column headings as the corresponding attribute names of the database table. The decision tool, however, is built using a software development language that does not support or understand English. The Telstra order entry process, also, does not give the English answers what the rule attributes require. For example, if

FIGURE 1 Example of Rule Processing - Translation and Query. Compare to TABLE 1

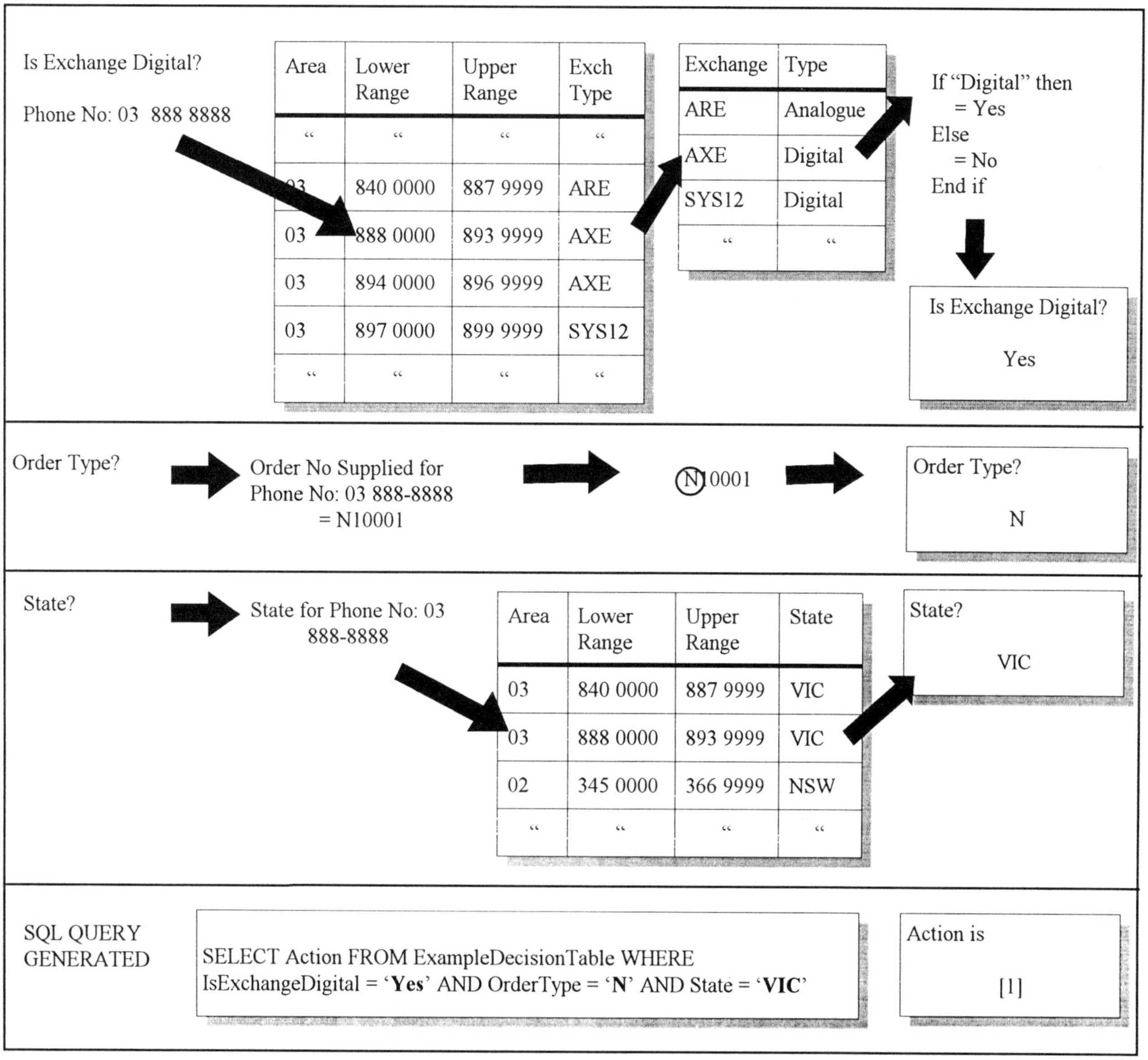

one rule attribute question asked if the order's telephone number was connected to a digital telephone exchange, the order does not have a field that supplies a 'yes' or a 'no'. To interpret this particular answer, from what is available in the order, the telephone number would firstly need to be extracted from the order fields. A search for the extracted number is then conducted in a special exchange lookup table which, as part of it's makeup, includes the name and configuration of the telephone exchange that the number is connected. The search returns the type of exchange which in turn is interpreted as being digital or analogue. The answer can now be converted to a 'Yes' or 'No' and is now able to be compared to the original decision attribute of the decision tables. Each decision attribute is subjected to this translation procedure until all the decision attributes have a recognisable answer to each question. In simpler questions the answers supplied may be a simple 'Yes' or 'No', or it may be one of a list of possible conditions such as an answer to the question of 'what type is present?' and the interpretation supplied might be 'O & N types', 'P Type' , 'R Type' etc.

Once each translation has taken place a SQL query can be pieced together using each of the decision attribute questions as the table attribute identifiers, and each of the interpreted answers as the condition of that attribute. This query is then run against the expert tables that have been loaded into the database from Excel. Each rule has a corresponding action or actions which are returned and can result in either the forwarding of the decision process to another decision table, or it may be the final decision outcome.

This decision process could be likened to pattern matching. Each query built represents a single identifiable pattern. The corresponding expert tables contain only one instance of that pattern which is supplied an action or actions. If more than one of those patterns are identified then there is a mistake in the experts spreadsheet and it will need to be updated. Similarly if a pattern is missing then there is also a mistake in the experts spreadsheet. By the nature of standard SQL querying methods, multiple or missing answers are instantly recognisable as all matched instances are returned by the query.

RESULTS

The number of rules for AVCORD has past the eleven thousand mark for the first three product groupings. This number is to increase to an estimated one hundred and twenty thousand as the order product groupings checked reach the final figure of ten. The large number of rules forced the rules to be spread across individual product group tables. Each of these product groups may include between one or more tables. Simpler product groups only require a single table level, while more complex product groups require three or more levels of pruning tables. Pruning tables allow certain decision outcomes to be quickly channelled and filtered through certain tables instead of having every possible outcome included in one table. A high level table may rule that a certain decision has reached conclusion and by-pass the other decision tables. Other rulings may require that to be able to make a decision all table levels of the product group are to be involved. Smaller tables and better focused decision making allows for rapid fire decision making rather than the slow process of querying exceptional large tables that include every possible combination of action outcomes.

Grouping similar decisions is an obvious way of reducing query time. The earlier example of checking whether the phone number on the order is connected to a

TABLE 2 Example of Rule Grouping. Compare to TABLE 1

Example Decision Table					Example Action Table	
Is Exchange digital?	Order Type	State	Action		Action No	Description
Yes	O	VIC	[1],[4],[3]		[1]	Replace ‘a’ with ‘b’
Yes	N	VIC	[1]		[2]	Remove product
Yes	O	NSW	[2],[4]		[3]	Modify rate to show 0
Yes	N	NSW	[2]		[4]	Insert correct tariff
No	na	na	[5]		[5]	Non digital - Remove
“	“	“	“		“	“

digital exchange can have similar outcomes. If the answer to the digital exchange question is 'No' and the product group demands that it can only be ordered if a digital exchange is present then the outcome will be "Non digital exchange - Remove product x and contact customer'. This answer will be the case in all situations where the digital exchange action is 'No'. All of the 'No' rules can therefore be combined into one rule with all other question attributes replaced with the value 'Na' for not applicable. The SQL query from above must now take on an alternate query field containing 'Na' for each attribute. The query now searches for not only the answer to the decision attribute condition, but also the case where that attribute contains 'Na'. Unique patterns still need to exist but there will be a substantial saving in the number of rules to process.

DISCUSSION

The AVCORD project has been in operation for some time now and has already surprised many of its critics. Many believed that the scope was too large, the rules too complex, and by the time anything was delivered the products and rules would have changed. Delivery time from initiation to implementation was just under three months. The project team included between three and five people at any one time including two full time Telstra order entry experts. Products and rules did change as the project was being developed but the flexible approach of the knowledge acquisition technique allowed the project to continue with little or no impact to the development. Some of the product group areas were found to be quite complex but the decision accuracy rate is still over 98%. This percentage is steadily increasing as the last of the rule quirks and system inconsistencies are ironed out. Accuracy is an important factor of the AVCORD system as the results of the decision making directly impact Telstra's customers.

This system provides a number of outputs and reports. Two thirds of all incorrect orders are output to an automated process that fixes the order - removing what is wrong, manipulates and adds what is correct - all without human intervention. The automated 'fix up' process is part of the AVCORD system and uses software agents to act upon the outcomes and directions supplied by the decision tool. While these agents are part of the AVCORD system they are quite separate to the decision tool. The remaining third of the incorrect orders are orders that need special attention and possible customer contact. These orders are automatically electronically mailed to the appropriate parties who are responsible for the repair of those orders.

All incorrect orders are measured statistically and the results are also mailed automatically to the appropriate supervisory and management areas. These reports contain the number of errors, the types of errors, where they are occurring and what product groups are most at fault. This information is then used to assess and improve the quality of the initial entry of orders at the telephone sales staff level. This may be through better order checking, re-training, or improved entry methods.

The direct savings made in correct product billing procedures and time in order re-work is quite startling especially when comparing to the cost of the system. However, the improvement of customer satisfaction with the correction of incorrect orders is worth many times more than any of the direct savings.

The success of AVCORD is attributed to the ability of the system to draw the experts closely into the development process with the use of the English orientated spreadsheet matrix system. While this is not the most ideal setting to develop an application in, it does acknowledge that as in any knowledge based system, the expert is by far the most important component [Gru88]. It is in the opinion of the author that it does not matter how technically advanced the system is, unless the expert is at the centre of the project development, then the project will suffer deliverable failures. It is up to the developers to acknowledge the experts worth and build an environment that suits the expert rather than themselves [CH88].

CONCLUSION

The size, scope and time frames of the AVCORD project necessitated a change of thinking in the normal approach to knowledge acquisition and to rule processing.

The knowledge acquisition approach presented, describes a novel approach involving a move to a structure and a medium that was based in English.. This approach has brought the experts to the centre of the system's development. The entire environment is set to bend to the needs of the experts rather than the traditional approach of converting the knowledge to a machine recognisable format. The knowledge acquisition medium promoted quick and flexible production of many thousands of rules that were still in a form that was understandable to the experts. The experts closeness to the project was the main key of success of the AVCORD project.

The rule processing approach was a technique devised to handle the unusual input from the data acquisition phase. Once the expert is part of the development environment, it is up to the rule processor to be able to translate the experts decision information into a form that can be used by the decision tool. The rule processing technique makes full use of available database facilities and standard query mechanisms that allow for easy update and flexibility. The process relied on a pattern matching technique that allowed rapid and accurate decisions to be made on thousands of possible rules and outcomes.

The AVCORD system is built using unusual yet simple techniques. These techniques produced a system which was inexpensive, built in a short time frame, used readily available tools and, most importantly, implemented successfully. The system is currently in production, validating the forty thousand daily orders initiated by the Telstra telephone sales representatives.

REFERENCES

[CH88] Cleal, D.M. & Heaton, H.O. ' Knowledge Based Systems: Implications for Human-Computer Interfaces' John Wiley & Sons, Chichester, 1988

[Gru88] Gruber,T.R. 'The Acquisition of Strategic Knowledge' John Wiley & Sons, New York, 1988

[Ols92] Olson,D.L. ' Decision Support Models and Expert Systems' Macmillan Publishing Company, New York, 1992

[SC82] Sprague,R.H. & Carlson, E.D. 'Building Effective Decision Support Systems', Prentice Hall Inc., Englewood Cliffs, 1982

[Tur92] Turban, E. 'Decision Support and Expert Systems : Management Support Systems' Macmillan Publishing Company, New York, 1992

[WN88] Walters,J. & Nielson, N.R. 'Crafting Knowledge-Based Systems: Expert Systems made realistic' John Wiley & Sons, New York, 1988

[You89] Young, L.F. 'Decision Support and Idea Processing Systems', W,C. Brown Publ., Dubuque, 1989

[Zah93] Zahedi,F. 'Intelligent Systems for Business: Expert Systems with Neural Networks' Wadsworth Publ. Company, California, 1993

The Chunking Engine: Closing the Gap Between Formal and Informal Design Specifications

Mihaly Lenart* Ana Pasztor**

*University of Kassel, Dept. of Architecture, CAD–Zentrum, Diagonale 12, 34109 Kassel, Germany
e–mail: michael@architektur.uni-kassel.de

**Florida International University, School of Computer Science, Miami, Fl.33199, USA
e–mail: pasztora@scs.fiu.edu

Abstract

The purpose of a design specification is to describe the goals, parameters, and constraints of a given design task. The better the description, the easier it is to find solutions. In fact, given sufficient information, the specification process might lead to concrete solutions and we talk about "executable specification". Part of the specification is collecting and processing information. Information gathering or knowledge acquisition is difficult because of the gap between different mental images or models of the involved parties and of communication deficiencies. Here, we propose a method to alleviate such problems and automate the elicitation process. The implementation of this method consists of a Chunking Engine. At the heart of the Chunking Engine lies a specification instrument, which is a task and domain independent tool for structuring the elicitation process. The Chunking Engine is responsible for obtaining "testable" specifications, i.e. specifications whose elements are in a common knowledge–base, are quantifiable, or calibrated. Testable specifications lead to formal specifications, which in turn, give rise to implementations or computer programs which generate design objects.

1 Design Specification

The pivotal role of design in product development has been widely recognized by industry and government agencies. Although the cost of design is relatively low compared to the cost of manufacturing and marketing, the impact of design on the production costs, product quality, and marketability is enormous. Therefore spending on design related research has been on the rise in the last decade. Another reason for the increased spending is that although design tasks are inherently difficult to solve, some initial success coming out of AI and Cognitive Science related research raised high hopes for new and efficient ways of dealing with complex design tasks.

One of the difficulties in design is capturing the scope, goals and parameters of a design problem. This activity is called the *specification* which is the foundation of any design process. Its purpose is to describe desirable features, requirements and limitations of the product as well as of the design and manufacturing process. In particular, the specification describes *functional*, *structural* and *behavioral* features of the design product [RG94]. It might also specify modalities for carrying out the design process or constraints for the manufacturing [Kellogg et al.89]. In general, the more *determinate* the description, the easier it is to find solutions. In fact, given sufficient information, the specification process might lead to concrete solutions and we talk about "executable specification" (see e.g. [LPP93]). While less details leave room for creativity, at some point in the design process the specification has to become determinate in order to obtain solutions.

Specification is necessary for the following reasons:

- Without specification there is no goal or direction for the design process. Even though specification is often abstract and intentionally vague, it makes goals and parameters explicit in order to avoid misinterpretations.
- Without specification chances are great to get lost in details. There are no guidelines, no feedback as to which tasks have been accomplished and which comes next. One has no overview of the entire design process.
- Without specification verification of the correctness of emerging solutions is impossible.
- Without specification the impact of each task on the entire process cannot be determined, therefore, it is difficult to develop efficient design strategies. Bottleneck problems or deadlocks cannot be eliminated.
- Without specification the coordination of teamwork is almost impossible. Specification provides means to clearly define, delegate, and create interface between subtasks, as well as avoid redundancies.

There are two kinds of design specifications: formal and informal. *Formal specifications* use formal languages for describing design goals, parameters and constraints. They were first introduced in software design but got eventually adopted in other design areas, e.g. in VLSI–design, as well. *Informal specifications* are the well–known and commonly used specifications in a natural language, such as English, or rather a subset of a natural language used by a given profession. Informal specifications are also codes, guidelines, and regulations as they can be found in the professional literature. Each specification starts out informally. Formal specification, on the other hand, is needed for the following reasons:

- Informal specifications are ambiguous.
- Verification of correctness of solutions is difficult without formal methods.
- With formal specification the solution of implementational tasks becomes easier.
- Without formal specifications organizational tasks can be solved only up to a certain degree of complexity.

However, formal specifications are derived from informal ones and are usually not easily accessible for professionals involved in the design process. According to [Fraser et al.94]: "... only minimal guidelines are provided for eliciting and structuring the requirements before they are expressed in the formal notation [Jon90]". An important part of the specification is collecting and processing information. Information gathering or knowledge acquisition is difficult because of the gap between different mental images or models of the involved parties and of communication deficiencies. This gap is often due to different modes of thinking associated in many cases with each kind of specification, namely formal mode and "narrative" mode, respectively [Kor88]. In narrative thinking goals are less clearly defined than in formal thinking, and models may be lifelike even if their elements are not likely. Even if the user has a clear mental image of the design object, there is still a gap due to the discrepancy between these mental images on one hand, and the vagueness of the verbal specification of the design object, on the other ([JL83] p. 158). Here, we describe a method to alleviate such problems and automate the elicitation process. Our aim in general is to close the gap between formal and informal specifications by developing knowledge or information elicitation methods. In other words, our purpose is to develop strategies and routines for eliciting and structuring design knowledge or information so that we can a) *automate* the design specification *process* and b) derive *formal specifications* from informal ones.

Within this framework, we develop a task and domain independent elicitation instrument providing a structure for the specification and guidelines for the elicitation process

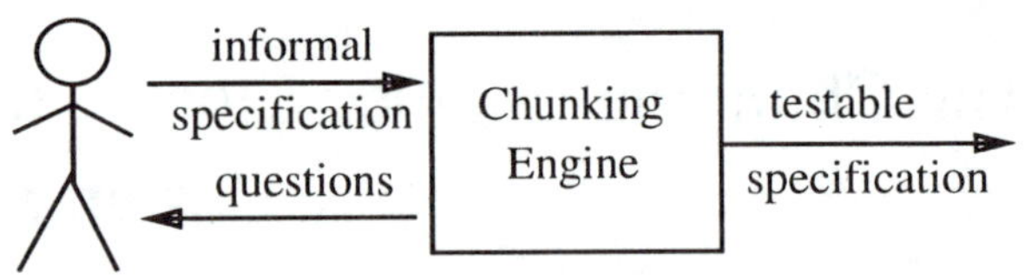

Figure 1: Functional schema of the Chunking Engine.

through constant feedback by asking questions. In particular, we design and implement the *Chunking Engine*, a software tool for systematically refining design specifications. The Chunking Engine takes as input an informal design specification (e.g. verbal protocol, textbook [Eri84], or training manual) and transforms it into a "testable specification" (see Figure 1). Testable specifications correspond to the *atomic objects* of [Alexander et al.88] or *surrogates* in data base terminology [Cod79].

Based on this specification, one can develop a formal specification and turn it into an executable specification. The transformation is done by chunking the input specification into more and more specific information, using existing (cf. [Gar88] section 2) and new knowledge elicitation techniques. Our plan is to use structured interviews ([Ler56], [SN68] and [Zei81]), protocol analysis [Eri84], and hypermedia tools ([Wel89], [KLB]) on one hand, and to develop and utilize new techniques, such as computerized mind mapping based on the concept of generalized mind maps [Buz91], on the other. Information is accepted and output only if it is "testable", i.e. if it is element of a common knowledge base, is quantifiable, or calibrated. For example, the input "tall chair" will be rejected and returned to the user for further specification of "tall" in terms of exact numbers like 3 ft.

2 Verbal vs. Non–verbal Specification

A basic problem seems to be the discrepancy between specification and design, which is due to the different objectives of these two tasks. The primary aim of a specification is to describe "what" the design object is supposed to do, and design is mainly about "how" this can be achieved. In engineering design this discrepancy occures as the one between the verbal specifications and the geometric description of the design object. Specifications are almost always given in verbal form. On the other hand, the dominant representation of the design object is drawings or blueprints, even though the object has many non–geometric features, as well. The designer's task is to recover the design object from the verbal specification. The Chunking Engine's task is to, at least partly, automate this process.

Even if not in the details, cognitive scientists agree that humans use a variety of modalities to represent knowledge.

" ...human knowledge is represented in both visual and verbal modalities ([Vek81], [And84], [Pai86] — quoted in [LWM]). According to Vekker "object names and properties are verbally represented while spatial, logical, and temporal relationships are more likely to be visually represented in a person's cognitive model" [LWM]. Even those who believe that thinking is symbol manipulation agree that "...different types of representation are logically distinguishable at some level of analysis, and, moreover, ... they exist as different options for encoding information ..." ([JL83], p. 146).

People often have a mental image of the design object to be specified, rather than a verbal representation. But specifications are typically verbal in nature. According to Johnson–Laird, "*A picture may be worth a thousand words, but a proposition is worth an infinity of pictures*"[1]. ([JL83] p. 158). The task of the Chunking Engine, then, is to capture the *specificity* of the original image the user had in mind. It helps to elicit more *determinate* representations of the design object with the goal of narrowing down the choice of images of this object in the designer's mind. Johnson–Laird [JL83] p. 160 describes experiments proving that people tend to form a mental model from a spatially determinate description (specification), but fail to do so from an indeterminate one. For indeterminate descriptions, people often come up with models consistent with the description, but which the specifier did not have in mind. Even better, they might form a whole *set* of models [Ioe94], flipping mentally back and forth from one to the other, unable to settle down on any, finally coming to a gestalt or schema.

3 A Specification Example

The example we use for demonstrating the underlying ideas is the design of a nut–cracker. We have a hardware manufacturer with a production line of household dishes and a R&D Department. At some point the upper level management decides to include nut–cracker into the production line because they did not have it yet and market analysis shows new marketing potentials for this particular product. The task is handed down to the head of the R&D Department to specify the new product for a small designer team that has to come up with new, competitive design solutions.

The first specification might be like this: "We need to expand our production line with nut–crackers. You have about four weeks to come up with a new product. The design of the nut–cracker should be simple and unique, and the price competitive." It is needless to say that this specification is vague, providing only a few guidelines for the designer. In order to clarify what exactly is expected from the design process, requirements will often be divided into functional, structural, behavioral, etc. requirements.

[1] emphasis added

Such a classification helps not only to structure, but also to refine the requirements. Here is a possible structuring and refinement of our example:

- *Functional requirements*: The nut–cracker needs to be operated by hand without any power sources. It needs to break nut shells easily (i.e. fast by applying minor force) without destroying the nut inside. It should have a clean, esthetic, 'form–follows–function' appearance.
- *Structural requirements*: It should have a small number of parts, preferably all of the same material. It should also have a simple mechanism and an ergonomic design.
- *Behavioral requirements*: The product needs to be easy to assemble, to clean, and to store. It should be wear and corrosion resistant.
- *Design requirements*: Two or three prototypes have to be delivered within four weeks with estimates for the manufacturing costs. The design is supposed to be unique, i.e. distinguishable from any other product on the market.
- *Manufacturing requirements*: All parts have to be manufactured by the company, except standard parts, such as screws, bolts, nuts, etc. Manufacturing costs need to be minimized by using parts already manufactured by the company or parts that can easily be manufactured on existing machines. The setup time for the production should not take longer than one week and from the start within three weeks a full production line of 200 nut–crackers per day should be possible.

This specification needs further refinements. The stepwise refinements are usually carried out by the designer team by trying to generate designs based on this specification (similarly to revising mental models) and running into new problems. Every time a new problem surfaces, the question to be answered is whether this problem is due to insufficient specification or is it because design solutions need to be further investigated. Especially in the early design stage, it is almost always the specification that needs to be modified or refined because of the lacking knowledge and vagueness of preliminary specifications. This requires some questioning to clarify and complete the specification. Here are some questions that can help to refine the above specification:

What does manual operation mean?
What does it mean to break nuts easily?
What does it mean to keep the inside of the nut intact?
How many parts are considered to be a small number?
Are the cleaning and assembling tasks related?
etc.

The purpose of such questions is to help the designer to elicit all the information needed for the design process. Why are the first specifications almost always insufficient and need further refinements? Why is the elicitation an integral part of the specification process? Here are the main reasons:

- Specifications usually assume some knowledge about the design domain. This knowledge is either common knowledge, such as general properties of a nut, or environmental knowledge, such as manufacturing parameters known to the company's employees or to the designer team. The problem of capturing this knowledge and using it for reasoning is referred to in AI as the frame problem. To make such knowledge explicit is important for design automation but is also helpful for traditional design in order to avoid misunderstandings.
- The first specifications are fuzzy or ambiguous. When we first characterize the objectives of the design task, we either don't have a clear image of the result or we don't want to limit the design by cutting off potentially interesting solutions beyond the narrow scope of our specification. An example is to include a lever–arm mechanism for the nut–cracker that would eliminate many interesting, perhaps ingenious solutions.
- Due to the limited knowledge at the beginning of the design process, specifications often contain contradictory requirements. Such requirements are immanent because of the complex and intricate nature of the relations between the features of a design object. For example, manufacturing costs are related to the shape and the material of the product because they have an impact on e.g. the number of times the product has to go through the milling process. At the early specification stage, however, it is difficult to see all the connections between the various requirements.
- Besides assumed (common or other) knowledge not contained in the specification explicitly, specifications are often incomplete regarding important features. This incompleteness can be on purpose because of the limitation danger mentioned above. For example, we might not want to specify the size or the color of the nut–cracker leaving them to the designer's discretion.
- Specifications also need the flexibility of allowing changes at any stage of the design process. An example is to loosen up the functional requirements by specifying the function of the nut–cracker as a nut breaking device as just one function, leaving additional functional requirements, such as e.g. bottle opener, open. This might lead to the emergence of more valuable design solutions.

Even though for these reasons the elicitation or questioning is perhaps the most important part of the specification process, currently there are no universal methods or guidelines to structure the elicitation process. In particular:

- the questioning is carried out in an ad hoc manner,
- there is no verification method for the completeness of the specification, i.e. whether the specification contains all the information necessary for solving a particular design task, and
- there is no way to tell whether the designer captured the intended specification, i.e. whether the mental models of the designer and of the specifier correspond to each other.

For this reason we want to formalize (and partially automate) the *specification process.* We provide an instrument for *structuring the design process* independently of the design task and domain in such a way that continuous dialogue between the participants in the design process (user, designer, manufacturer, etc.) result in a specification that is *complete* and *testable.*

4 Why Testable Specification?

Usually, specification is carried out hierarchically at different abstraction levels. Any specification can be refined indefinitely. However, there are two limitations regarding the extent of a specification:

- The level of common background or knowledge of the involved parties, and
- the relevance of the specification depth for the solution of a problem.

The more one knows about a problem domain, the less detailed the specification needs to be for solving a task. At the same time, a specification needs only to include details relevant for the solution. The relevance of details for the specification means that details not pertinent for the task should not be included. An example is car design, where the color of the car might be an important feature for the buyer but entirely irrelevant for the designer.

Given a specification, i.e. a description of a task or artifact, how can we make sure that our specification will be understood correctly and the specified task will be carried out satisfactorily? Simply by asking questions, i.e. by information or knowledge elicitation. We need to make sure that the mental image of the design object that the designer creates in his or her mind is "clear enough" for carrying out the specified task. "Clear enough" means that the specification provides enough information to produce a satisfactory outcome or artifact. The specification is successful if the specification process leads to successful production or expected results. Upon completion of the design task the product or result will be evaluated. After evaluation, the specification might get modified and the process repeated until a satisfactory product emerges.

At each stage of the process the specification can be clarified further indefinitely by asking questions about more and more details. For obvious reasons, however, the elicitation process has to get to a halt at some stage. While the relevance of the details for a given task is relative simple to determine, it is hard to tell the kind and extent of the knowledge or world–model the participants of the design process have in common. People, even with the same background, have different experiences. Their common knowledge is based on individual experiences from which they develop abstractions or concepts. The process of developing abstractions is also individual. Therefore, common knowledge does not exclude having different individual models of the world. For example, the concept of cold or hot depends on individual perception, not to speak about concepts like beauty or harmony. Such concepts have a different meaning even for people with the exact same cultural and educational background. The exact meaning of such concepts cannot be determined only relatively by *calibrating* individual perceptions. There are, however, universally accepted concepts. These fall into two categories: common knowledge, and quantifiable entities (see more about these categories in Section 6). Therefore, the Chunking Engine accepts a specification only if it is common knowledge, quantified, or calibrated.

Each specification has two parts: a hidden, common knowledge part and an explicit part describing relevant features and/or requirements. To what extent the explicit part spells out details, depends on the purpose of the specification. If the purpose of the specification is to restrict the outcome as much as possible, then the specification has to spell out the task to the last detail. However, if our purpose is to allow a large number of creative solutions, then the specification needs to be generic. An example for the former is instruction manuals or government regulations. Design specifications belong usually to the latter.

5 A Specification Instrument

Our goal is to develop a specification instrument to guide and structure the elicitation process. In our case, the specification is provided partly by the company's management, partly by the designer team. In general, the specification source is the entire design environment. In fact, specification is often the result of negotiations between involved parties.

For the sake of brevity, we discuss here only functional aspects of the design specification. Each design object has

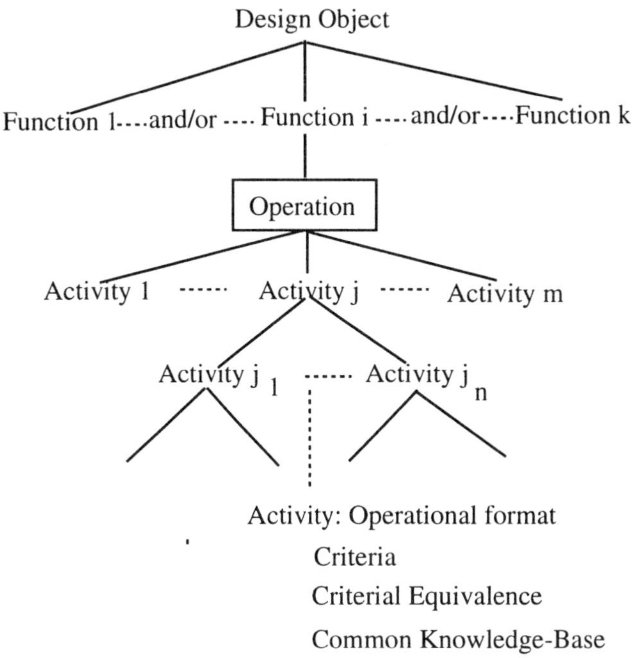

Figure 2: The concept of the specification instrument.

one or more well–defined functions, e.g. the function of the nut–cracker is to break nuts. The first step in the design process is to establish these functions by asking questions about functions the user required from the design object. While our nut–cracker has only one such function, a kitchen machine has several functions, such as "mixing_ingredients", "kneading_ingredients", "cutting_ingredients", etc. These are "or–functions" since each of them is itself an independent function of the kitchen machine. Example for "and–functions" are the ice–maker's "freeze_water" and "disperse_ice" functions.

Each function has three parts: input, operation, and output. Input and output are characterized by user defined criteria. We can elicit these criteria by asking questions about expected features of the input and output. In our example, the input is whole nuts and the output is nut kernels separated from their shell. All criteria will be chunked down into their *criterial equivalences* (see [CGL85]) until all criteria are testable.

The operation consists of actions which have to be performed in order to get from the input to the output. To elicit these actions, we ask the designer the question: "What actions do you need to take in order to design an object performing the operation o?". Often these actions have to be chunked down into subactions until we obtain atomic actions (see Figure 2).

To find out whether an action a_i is atomic, we ask the question: "How do you go about performing a_i?" If action a_i is not atomic, then this question results in a new set

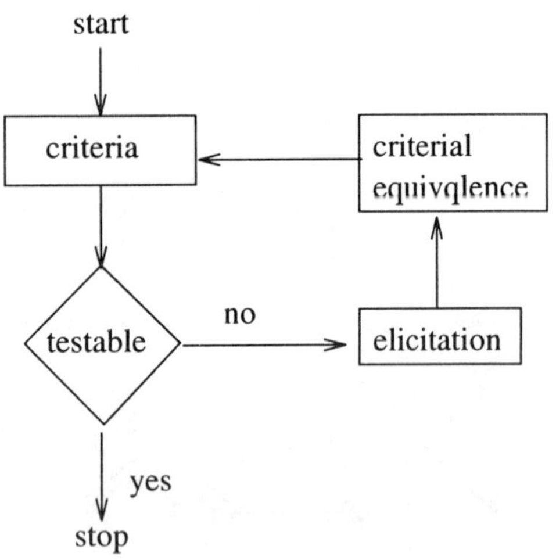

Figure 3: Algorithm for producing testable criteria.

of subactions of action a_i. We can find out whether the list of actions is complete by the question: "What other action do you need to perform in order to design an object performing the operation o?". In our example, the operation is "break_nut". To design an object which performs this operation, the designer has to perform actions like find breaking methods (e.g. by hand, hammer, etc.), choose one or develop a new one, design a device for breaking nuts, etc.

Input, output, as well as each action have an *operational format* (see [CGL85], pp 207). An operational format consists of several variables, whose values are obtained through elicitation. Similarly to [LWM], the elicitation process is based on the structure of human cognition through the use of recognized cognitive categories as defined in [CGL85]. Test *criteria* and *criterial equivalences* are only two of these categories. Criteria are the restrictions the design object has to satisfy. However, to satisfy these restrictions, they must be testable. The algorithm represented in Figure 3 produces testable criteria. Some further variables are: *reference category* (the data base used to evaluate one's criteria), *cause–effect*, and *compelling category and subordination* (which specifies which of the tests result in behavior), some of which overlap with those used in [LWM].

A criterion is testable if it is quantified, calibrated or is part of the common knowledge–base. If a criterion is not testable, it is replaced by it criterial equivalence (see [CGL85] pp 226), i.e. by criteria which explain the original criterion. Questions to elicit criterial equivalence of a criterion x are:

> What does x mean?
> How would you describe x?
> How do you know when an object/action satisfies x?

Elicitation example:

> Q: What does it mean: "The shell has to be easily removable after breaking"?
> A: The shell is easy to remove if it was broken into few pieces.
> Q: What do you mean by "few pieces"?
> A: Not more than five.

The input and output criteria are usually specified by the user (e.g. the input criterion of nut size, or the output criterion of recovering the kernel in one piece), while action criteria are specified by the designer and/or elicited from the user by the designer (e.g. the criterion that the device should be operated manually might be a choice of the designer or the user). The goal is to draw the attention of the user to specification details s/he was not aware to start with, but which are necessary for the designer to recover the user's mental image of the design object.

6 Testing Specifications

Elicitation means completing or modifying the given specification by specifying missing or unsatisfactory values and features. The specification process comes to a halt when all features of the design object are captured and are testable. Testable means that each feature is either common knowledge, quantified, or calibrated.

The Chunking Engine recognizes *common knowledge* by looking it up in the knowledge–base. Since the same word or name might refer to different concepts, the database might have more than one entry by the same word or name. In this case the Chunking Engine will ask the specifier to choose the right one. An example is "nut" that might be a kind of food, characterization of a person, or a machine part, depending on the context.

A feature is *quantified* if it has a standard (universally accepted) numeric measure assigned to it. Such features are e.g. height, weight, and volume; and the measures meter, kg, and cubic centimeter. In some cases the feature can be non–numerical and the measure be absent or just a unit. The measure is absent if we have simply elements or units, such as the number of legs of a chair.

A feature needs to be *calibrated* if it depends on the specifiers's individual experience, taste, or judgment. For example, esthetical values may vary greatly from individual to individual and the only way to specify them is by calibrating the specifier's value system. Calibration means to match or adjust the specifier's and the designer's value systems or scales. When the waitress in a restaurant asks you whether you want the dish you have just ordered mild, hot, or very hot, the only way to find out what this means is to ask how the dish you consider mild (hot, or very hot) is considered by the waitress according to her own scale. In other words, the specifier has to assign his or her own values to a concept or object the other party knows well. The concept of calibration is related to rating grids [BB88] and scaling/clustering [CM88] used for knowledge acquisition.

In our example, we can calibrate the breaking force of the nut–cracker, rather than measuring it in e.g. joule, by asking the specifier to assign numbers to physical efforts which both the specifier and the designer (Chunking Engine), are familiar with. These values, such as e.g. breaking a hard-boiled egg, or driving a nail into the wall, have to be compared with the task at hand, i.e. breaking a nut. There are two calibration methods: either we have enough common references in order to make a close enough match, or we have only a few — but at least two — references and a function for calculating the values for new cases. An example for the first method is to say that "we need twice as much force to break a nut than to break an egg". If we assign the value 5 to the force of breaking the egg then we can assign the value 10 to the force needed to break a nut. With the second method we have a scale function e.g. $y = 2x$, and assuming that for example the specifier assigns $x = 1$ to the force needed to break an egg and $x = 10$ to the force needed to drive a nail into the wall, the calibrated force values would be 2 and 20, respectively. If the specifier finds that the force needed to break a nut is $x = 3$ on this 1 to 10 scale, then the calibrated force value will be 6.

7 Implementation

In the following we describe a prototype system whose implementation in Prolog is currently underway. The prototype analyses specifications and generates questions based on the concept of criterial equivalence, however, it does not incorporate yet the above described specification instrument. Specifications are represented using frames such that each frame corresponds to a concept (class) or an object. The slot values are features or frames that are either given by the initial specification or have to be generated during the refinement/chunking process. For example, the "nut_cracker" frame represents a particular household machine whose features are derived from the classes of "device", "machine", and "household_machine".

nut_cracker
operator|ck: person
has_parts|ck:
mode_of_operation|ck: manual
function|ck: break_nut
target_object|ck: nut
price|qu: low
weight|qu:
manufacturer|ck:

The refinement/chunking process assigns values to the slots and evaluates them whether they are testable. The kind of testable features depends on the slot type indicated by "ck" for common knowledge, "qu" for quantified and "ca" for calibrated features. It is possible to assign more than one test type to a slot, such as "color|qu|ca:". This means that the color–feature can be specified either by quantified values, e.g. spectral numbers, or by calibrating individual perceptions. We can also assign ranges to the slots such that slot values have to be not only of the given type, but also within the given range. If the slot value does not confirm with the type, then further refinement/chunking is needed. This provides either a new value, or creates a new frame. For example the slot value "person" is a frame in the database with the following slots:

person
sex|ck:
age|qu:
profession|ck:
weight|qu:
height|qu:

Since none of the slots have a value, the Chunking Engine requests the specifier to provide them. Upon the following instatiation the new slot values will be evaluated.

person
sex|ck: male female
age|qu: adult
profession|ck: irrelevant
weight|qu: irrelevant
height|qu: irrelevant

The values "male" and "female" are common knowledge found in the database. The slot "age" requires quantified values, such that a new value is requested until the new value is a (whole) number within the predefined range. The value "irrelevant" indicates that the current slot has no bearing on the solution and can be ignored. The value "break_nut" in slot "function" of the frame "nut_cracker" requests the generation of a new frame:

break_nut
force|qu:
actor|ck:
object|ck:
method|ck:
time|qu:

This frame is not in the database. However, it can be created from existing frames, such as "break_object".

The system's architecture contains three parts: the evaluator, the knowledgebase, and the specification database (see Figure 4).

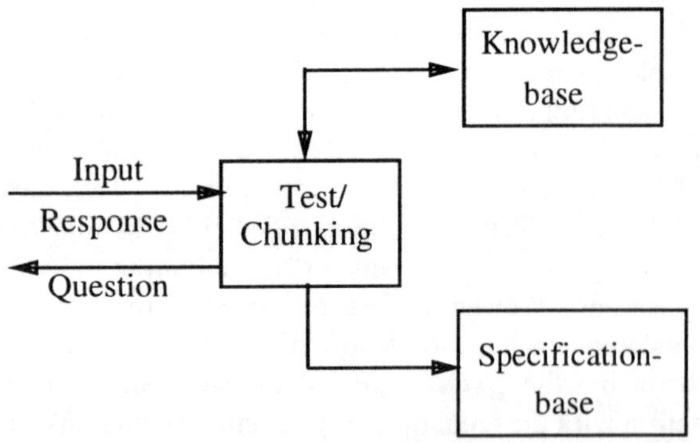

Figure 4: The schema of the system's architecture.

The *evaluator* searches the specification database for uninstantiated slots and non–conforming values. Upon finding non–conforming values, the evaluator searches the knowledgebase for adequate entries. In case of uninstantiated slots or finding no adequate entry, the evaluator requests information from the specifier. Based on this information, the evaluator creates a new frame or new slot values. Frames and values are stored in the *knowledgebase*. The specification is stored in the *specificationbase* and is constantly updated by the evaluator. For example the frame "person" is in the knowledgebase before starting the specification process, for it represents a generic concept related not only to the specification task. Upon finding the slot value "user|ck: person" in the specification, the frame "person" is added to the specification database. The knowledgebase contains also values representing common knowledge concepts, such as material, color, object name, etc. The evaluator requests further specification as long as it cannot find matching entries in the knowledgebase. New general knowledge entered by the specifier will be added not only to the specification but also to the knowledgebase.

8 Conclusion

We described here a method and a tool, the Chunking Engine, for eliciting information in order to automate the design specification process. In particular, we are in the process of developing an instrument for structuring and eliciting information pertinent to solving design tasks. Our questioning method is domain and task independent, however, it provides guidelines to elicit information for *any* design *task* in *any domain*. The elicitation process is supported by computer programs. Currently, we implement the Chunking Engine, a tool that generates *testable* specifications from informal ones.

References

[Alexander et al.88] James H. Alexander, Michael J. Freiling, Sheryl J. Shulman, Steven Rehfuss and Steven L. Messick: Ontological Analysis: An Ongoing Experiment, Knowledge Acquisition Tools for Expert Systems, Knowledge Based–Systems, Vol. 2, eds. J.H. Boose and B.R. Gines, Academic Press, 1988, pp. 25–37.

[And84] Anderson, J.R.: Representational Types: A Tricode Proposal, Technical Report #82-1, Office of Naval research, Washington D.C., 1984.

[BB88] John H. Boose and Jefrey M. Bradshaw: Expertise Transfer and Complex Problems: Using AQUINAS as a Knowledge–Acquisition Workbench for Knowledgebased Systems, Knowledge Acquisition Tools for Expert Systems, Knowledgebased Systems, Vol. 2, eds. J.H. Boose and B.R. Gines, Academic Press, 1988, pp. 39–64.

[Buz91] Buzan, Tony: *Use Both Sides of Your Brain*, A Plume Book, 1991.

[Cod79] Codd, E.F.: Extending the Database Relational Model to Capture More Meaning, ACM TODS 4:4, December 1979, pp. 397–434.

[CM88] Nancy M. Cooke and James E. McDonald: The Application of Psychological Scaling Techniques to Knowledge Elicitation for Knowledgebased Systems, Knowledge Acquisition Tools for Expert Systems, Knowledge Based–Systems, Vol. 2, eds. J.H. Boose and B.R. Gines, Academic Press, 1988, pp. 65–82.

[Gar88] Chaya Garg–Janardan and Gavriel Salvendy: "A Conceptual Framework for Knowledge Elicitation", in *Knowledge Acquisition Tools for Expert Systems*, eds.: J.H. Boose and B.R. Gaines, Academic Press, 1988, pp. 119–129.

[CGL85] Leslie Cameron–Bandler, David Gordon and Michael Lebeau: *The Emprint Method. A Guide to Reproducing Competence*, Moab, Utah: Real People Press, 1985.

[Eri84] Ericcson, K.A., Simon, H.A.: *Protocol Analysis*, MIT Press, Cambridge, MA, 1984.

[Fraser et al.94] Fraser, Martin D., Kumar, Kuldeep, Vaishnavi, Vijay K.: Startegies for Incorporating Formal Specifications, Communications of the ACM, Oct., 1994, pp. 74–84.

[Jon90] Jones, C.B.: *Systematic Software Development Using VDM*, 2nd ed., Prentice–Hall, Englewood Cliff, NJ, 1990.

[Ioe94] Ioerger, Thomas R.: The Manipulation of Images to Handle Indeterminacy in Spacial Reasoning, Cognitive Science, vol. 18, nr. 4, 1994, pp. 551–594.

[JL83] P.N. Johnson–Laird: *Mental Models*, Harvard University Press, 1983.

[Kellogg et al.89] Charles Kellogg, Robert A. Gargan, Jr., William Mark, James G. McGuire, Michael Pontecorvo, Jon L. Schlossberg, Joseph W. Sullivan, Michael R. Genesareth, and Naringer Singh: The Acquisition, Verification, and Explanation of Design Knowledge: SIGART Newsletter, April 1989, No. 108, Knowledge Acquisition Special Issue, pp. 163–165.

[KLB] Danny Kopec, Larry Latour, and Michael Brody: Towards An Expert/Novice Learning System with Application to Infectious Disease, SIGART Newsletter, April 1989, No. 108, Knowledge Acquisition Special Issue, pp. 140–143.

[Kor88] Jim Kornell: Formal Thought and Narrative Thought in Knowledge Acquisition, Knowledge Acquisition for Knowledge Based Systems, Knowledge-based Systems, Vol. 1, eds. J.H. Boose and B.R. Gines, Academic Press, 1988, pp. 35–44.

[LWM] Juliana S. Lancaster, Christopher R. Westphal, Karen L. McGraw: A Cognitively Valid Knowledge Acquisition Tool, SIGART Newsletter, April 1989, No. 108, Knowledge Acquisition Special Issue, pp. 152–154.

[LPP93] M. Lenart, P. Padawitz, A. Pasztor: Formal Specification for Design Automation, IFIP Working Group 5.2, Formal Design Methods for CAD workshop, proceedings, June 1993, Tallinn, Estonia, to appear at North Holland Publ.

[Ler56] Lerner, D.: Interviewing Frenchmen, American Journal of Sociology, 61, pp.187-194, 1956.

%bibitem[Miller et al. 60]Miller et al. 60 Miller, G., Galanter, E.,

[Pai86] Paivio, A.: *Mental Representations: A Dual Coding Approach*, Oxford University Press, New York, 1986.

[RG94] Rosenman, M.A., Gero, J.S.: The what, the how, and the why in design, Applied Artificial Intelligence, 8:199–218, 1994.

[SN68] Sjoberg, G. and Nett, R.: *A Methodology for Social Research*, Harper and Row, New York, 1968.

[Vek81] Vekker, L.M.: Methal Processes: The Agent, the Experience, Action and Consciousness, Vol 3, Leningrad University Press, 1981 (In Russian)

[Wel89] Wells, Tracy L.: Hypertex as a Means for Knowledge Acquisition, SIGART Newsletter, April 1989, Number 108, Knowledge Acquisition Special Issue, pp. 136-138.

[Zei81] Zeisel, J.: Inquiry by Design: Tools for Environment–Behavior Research. Brooks Cole, Monterey, CA, 1981.

KARDS: HYBRID KNOWLEDGE ACQUISITION FOR A SECURITY RISK MODEL

Elizabeth Gurrie[†], Joachim Diederich[§], Alan Tickle[†,§] and Alison Anderson[†,¥]

† School of Information Systems; ¥ Information Security Research Centre;
§ Neurocomputing Research Centre
Faculty of Information Technology, Queensland University of Technology
PO Box 2434 Brisbane Qld Australia 4001
E-mail:gurrie@fit.qut.edu.au

ABSTRACT

The development of knowledge based systems for evaluating computer security, or any specific aspect of this, such as transient vulnerability, has so far foundered on the problem of constructing the objects that are at risk - data, aggregations of data, information, situations etc. This knowledge is organisational and cultural, not technical, and has to be captured anew for each risk study. The paper reports on progress in an Information Security Risk Centre project to develop a hybrid Knowledge Acquisition tool for identifying those "risked objects" from a variety of organisational sources including human experts, printed documents, passive databases and generalised security expertise so that they can be accumulated in a "Risk Data Store" (RDS) needed for security evaluation. ***KARDS*** *(Knowledge Acquisition for RDS), a prototype hybrid knowledge acquisition tool, has been implemented for this purpose. Knowledge acquisition strategies used by this prototype include:- semi-automated interview of domain experts; semi-automated verification of acquired knowledge; heuristic building of object-relationships; laddering - a semi-automated approach to object-relationship verification; database schema analysis - a heuristic approach to identifying information assets held within an organisation's existing databases; the configuration of a neural network to learn new object-relationships.*

INTRODUCTION

In the heterogeneous expertise required for computer security evaluation, it is possible to distinguish a large body of formal and well-documented knowledge encompassing, for example, cryptology, formally secure models, multilevel databases and intrusion detection. The achievement of this knowledge is to permit specification of how secure particular information systems *within the scope of that knowledge base* are required to be. Examples of such systems are military or accounting applications, both of which are subject to well-understood minimum security constraints. In everyday general purpose computing, however, security evaluation draws on a less well-defined subdomain that is context-dependent, extremely application-oriented, and to a large extent home-grown, deriving from the organisation's mission as reinterpreted through the specialised knowledge of experienced local IT staff. Without this background knowledge of the significance of the information itself, it is impossible to evaluate the danger posed to the system's security features by the many environments in which it functions. Yet this is exactly what is required in vulnerability assessment, and in estimation of the risk contingent on that vulnerability.

The fundamental knowledge acquisition problem of risk analysts is therefore not so much locating and testing security features as identifying exactly what is at risk. The risked "objects" can appear in different guises to different system stakeholders. Sometimes information itself has intrinsic value, and could be stolen for profit, e.g. ex-directory telephone numbers. Sometimes the information only has value if known in conjunction with other information (the householder's list of insured items *and* the household address); or known by a particular person, or known at some specific date. Again, the information object can appear to be a business function, as in "daily transactions"; or even an incident, either positive - "we must conclude this contract at all costs on Friday at 4 pm" - or negative: "my worst fear is having to explain why we couldn't meet the deadline". A further level of confusion is introduced by users' perceptions regarding physical files and their corresponding logical entities, e.g., a multi-relation interface with the appearance of a single large file. The analyst's most difficult task is to reduce these redundantly expressed and sometimes contradictory statements to a common representation able to be mapped to information system components and their security enhancements - files, passwords etc.

This paper reports on a project aimed at developing a prototype interface for identifying, characterising and interrelating such illdefined objects in the specialised but informal domain of computer security risk modelling. Firstly we summarise the background to the project, showing how the plan for KARDS arose from successive failures to automate security risk modelling, and analyses of these failures. An account of the architecture of KARDS and of its implementation features follows. Finally, we discuss the current status of the work and immediate plans for its further development, evaluation and integration into its parent risk model.

BACKGROUND

Since its appearance as required system management practice some twenty years ago [NBS74], security risk analysis has been roughly contemporaneous with the spread of knowledge based solutions, and has generated numerous attempts in this genre [B&J88; Wal90; Gil90]. Intensifying requirements for regular and reliable vulnerability and risk analysis derive from many sources - audit, insurance, security certification initiatives and data privacy assurance are only a few. However, existing automated models can supply only unconvincing results, in return for timeconsuming data collection [And91].

The many reasons for the failure of these models derive from the same cause: inadequate means of customising the perceived knowledge base to include organisational values and priorities. Although dependent on information technology, these values develop acyclically with it, outlive any particular implementation of it, and are largely unique to the organisation. Hence they must be expressed as statements about objects meaningful to business rather than in semantics familiar to IT security specialists. The existing models concentrate on making accessible the codified general knowledge available in any textbook. As the examples in the introduction show, no general model will succeed without separating the context data from the general rule knowledge.

This is the objective of KARDS: recognition of contextually-determined objects and their characterisation through hybrid knowledge acquisition from heterogeneous data sources, essentially business experts with user level IT expertise supported by existing documentation. The role of KARDS is to acquire and document "information assets" of interest to a specific business, along with their real-life characteristics and interrelationships. KARDS was designed to act as a knowledge filter for a larger model, the Risk Data Store (RDS) [ALT93]. Its role in the RDS, which is outside the scope of this paper, is to supply asynchronous update of the information assets register within the RDS itself.

In knowledge acquisition, the method required is determined by the type of knowledge which has to be represented by the application. The knowledge may be elicited from a human expert, from text which has been supplied in a computer readable form or may be contained within a passive database. Given this combination of factors, it is evident that the type and source of the knowledge will influence the selection of a knowledge acquisition method. A hybrid Knowledge Acquisition tool such as KRITON [DRM87] appears to suit the requirements of this project and has influenced the direction of this applied research. The acquisition tool must be able to generate a variety of knowledge representations since *more than one type of representational format can be combined to represent the knowledge required to support the inferencing demanded by a variety of tasks* [C&W89]. Further, before knowledge can be acquired by a semi-automated tool, the tool must be tuned to the domain in its use of vocabulary and documentation.

This research seeks to achieve knowledge acquisition through a customised hybrid interface featuring semi-automated interviewing, passive database schema analysis, heuristic-supported object relationship analysis, and knowledge verification by example using neural networks. Each of these is a task that, in a manually-executed project, would represent a taxing information-gathering phase of risk analysis, performed by a highly-paid specialist.

THE FUNCTION OF KARDS

Individual knowledge acquisition methods result in restricted capabilities to elicit knowledge. A hybrid approach to knowledge acquisition reduces these restrictions. Hybrid knowledge acquisition is a theory guided approach which attempts to overcome the deficiencies of individual knowledge acquisition strategies. *...no single acquisition method will be powerful enough to overcome the so called knowledge-acquisition bottleneck in knowledge engineering* [DRM87].

Risk analysis is recognised to suffer from the same bottleneck [And91; W&C92; Bas93]. Industry and emerging standards have identified a need for intelligent risk analysis tools [CEC91], capable of modelling the system environment efficiently. Knowledge acquisition strategies must be able to assist, but only if they can adapt to the variety of knowledge sources used in the risk analysis process.

In many application areas, domain knowledge is not only available in the form of rules or facts, as provided by an expert or the analysis of relevant documents, but also in the form of data or cases which represent prior experiences in similar situations. It is natural to combine knowledge-driven and example-based acquisition methods. Neural networks are candidates for "learning by example" acquisition techniques. Neural networks support data-analysis and feature extraction methods, i.e. they discover previously unknown relations and attributes in data sets. These relations can enrich the object knowledge acquired by other KARDS knowledge acquisition techniques.

It is important to recognise that any learning or acquisition method includes an "algorithmic bias." That is, the learning or acquisition technique will create a representation which can be used for classification and further learning, but the classification behaviour will be biased in a certain direction, depending on the nature of the representation. Since neural networks form different representations to other KARDS acquisition methods, it is natural to integrate neural networks in KARDS in order to control and possibly cancel the algorithmic bias in the symbolic knowledge acquisition techniques.

THE FEATURES OF KARDS

The design of a knowledge system should anticipate the acquisition process [G&C88]. KARDS is an implementation of this philosophy. By providing a front-end acquisition tool to the RDS the knowledge engineer is supported in acquiring knowledge from a hybrid set of sources.

Gruber et al present three principles for acquisition design. [G&C88 pp132-134]

1. *Design task-level representational primitives to capture important domain concepts defined by the expert.* This ensures the expert is not going to be restricted in defining their knowledge by knowledge-base implementation issues.
2. *Design explicit, declarative representational primitives.* Familiar "form-filling" user-interface technology can be used to assist the expert in expressing knowledge explicitly and declaratively. This allows the expert to examine the assumptions underlying control.
3. *Design representations at the same level of generalization as the experts' knowledge.* This principle emphasises that experts should not be forced to generalise nor should they be required to provide information not available to them.

These design principles have been taken into account during the design and development of the KARDS architecture.

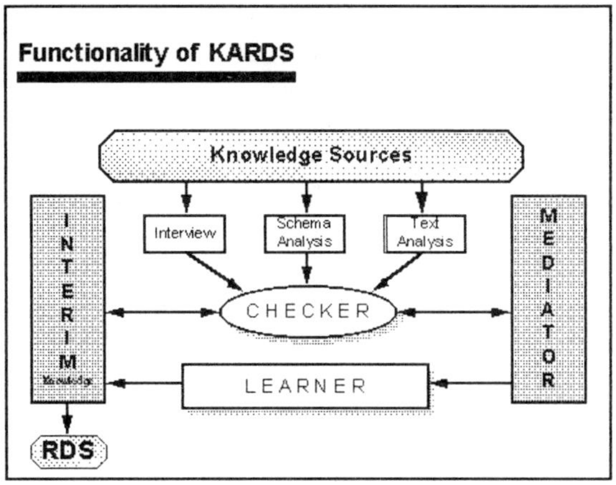

FIGURE 1 - The Architecture of KARDS

The KARDS architecture is designed to support the acquisition of knowledge from a variety of sources using a variety of elicitation methods. As knowledge components are acquired, the *Mediator* assists the expert in developing a conceptual model of their approach to and interest in risk information. As knowledge is acquired and mediated, it is verified by the *Checker*. Mediation and checking allow the development of an *Interim Knowledge Base*, an integration of the different conceptual models offered by the various stakeholders contributing to the knowledge acquisition task. The *Learner* supports the principle of knowledge-guided elicitation. Pre-elicited knowledge serves as examples which can be used by the *Learner* to derive new knowledge. Finally, the interim knowledge is encoded in a form suitable for the RDS.

The following knowledge acquisition strategies are available within KARDS.

Semi-Automated Interview: A combination of interview techniques are available within KARDS. The primary method uses the repertory grid technique. This technique is based on Kelly's personal construct theory [Kel55]. Repertory grids guide the interviewee by focusing on a particular aspect of their knowledge thus assisting to identify those concepts which build that knowledge. Each concept is considered and features are extracted which help to differentiate one concept from another. This concept/feature structure provides a basis for further revision of the concepts [S&G88; DRM87; Boo93].

Part of the interview process involves a preliminary level of knowledge verification. Knowledge is checked for completeness and apparent importance. This is achieved via an optional scoring process. The scores can be used to determine the level of importance each concept/feature plays within the knowledge construct and as evidence that one concept can be differentiated from another [Har89]. If two concepts cannot be differentiated by their scores, this indicates either redundancy or incompleteness within the knowledge.

The focus provided by KARDS within the interview process is a "security incident". Since the purpose of KARDS is to assist in identifying "risked-objects", a focus is required to allow the expert to define required changes in security. A "security incident" helps to demonstrate any potential loss or gain in security. The expert is asked to identify an incident. Then, via the repertory grid process, a description of the knowledge underlying the expert's understanding of that incident is systematically elicited, thus contributing additional knowledge to the overall information security model for the organisation.

Passive Database Schema Analysis: Most organisations have valuable assets in the form of databases. Knowledge about these assets is relevant to an organisation's information security model and is important to the risk analysis process. KARDS provides a fully automated schema analysis feature. By analysing a database schema, information assets can be identified and automatically made known to the *interim* knowledge base within KARDS.

Heuristic Establishment of a Relationship Hierarchy: A set of heuristics is designed into KARDS which predict apparent relationships between objects identified via interview process or database schema analysis. These relationships form the basis of an initial relationship hierarchy. This is vitally important to the RDS since it provides the paths by which statements can be derived about the impact on the organisation as the result of some information security event.

Graphical Representation and Modification of Relationship Hierarchy: The relationship hierarchy can be presented graphically to the expert to assist them in verifying the knowledge structures being acquired. The representation used is an directed acyclic graph [Smi87]. Each vertex of the graph represents a KARDS object which has been elicited via interview or schema analysis. The vertices can be selected, or "opened" for inspection. Each arc between vertices represents a relationship which has been previously derived. A graphical interface allows the user to change these relationships if desired.

Laddering - Semi-automated verification of Relationship Hierarchy: A secondary interview technique is included to provide an alternative method to verify the relationship hierarchy. The relationship heuristics may not identify relationships between objects which have been acquired via different knowledge elicitation techniques, or the relationships established heuristically may not best represent the expert's opinion of how the objects are linked. Therefore a verification method is required. Laddering [DRM87; B&B88] is a strategy which systematically guides the expert to consider an object in relationship to other objects which have been derived. It may be that objects acquired by different acquisition methods are identical. The laddering process allows this to be identified and modifies the knowledge relationship structure appropriately.

Configuration and Training of a Neural Network: KARDS investigates the use of neural networks to learn previously unknown relationships from extant knowledge, i.e. use knowledge which has already been captured to guide the acquisition of new knowledge. The neural network used implements a Rumelhart architecture [Die91] which is a feed-forward, fully connected network. The neural network is configured at runtime according to the number of examples that have been derived from other elicitation methods. The Baum-Haussler Rule is used to configure the network [B&H89].

Interim Knowledge Representation: The knowledge acquired by any of the acquisition methods described above is stored in an interim knowledge base which is accessible by the RDS. This interim knowledge is continually updated and restructured as new knowledge is acquired. At this stage of our work, the knowledge generator [DRM87] required to generate appropriate representations for the RDS has not been completed but is designed to generate different representations suitable for the distributed knowledge-base environment in which the RDS operates.

FUTURE RESEARCH DIRECTIONS AND ENHANCEMENTS TO KARDS

- Extended use of captured knowledge for the guidance of new knowledge acquisition;
- Implementation of a security related thesaurus for improved heuristic building of object relationships;
- Analysis of computer based documents for extraction of security related knowledge;
- Passive Database knowledge discovery of possible vulnerabilities within the data;
- Integration and evaluation of KARDS in the RDS architecture.
- Evaluation of KARDS against those components of existing risk analysis tools which identify information assets. The evaluation is required to determine if KARDS more accurately identifies sufficient and necessary risked information objects.

CONCLUSION

The current implementation of KARDS demonstrates the feasibility of adapting a hybrid knowledge acquisition framework to a specialised informal domain. Prototype development indicates that capture of the illdefined and context dependent objects of the risk analysis can successfully be reduced to a computable representation. This can occur without loss of meaning and without pre-empting the level of abstraction or user expectation of the conceptualisation process.

REFERENCES

[And91] **Anderson, A.** (1991) "Comparing Risk Analysis Methodologies" in **Lindsay, D T & Price, W L (Ed)** *Proceedings of IFIP TC11 Seventh International Conference on Information Security: Creating Confidence in Information Processing, IFIP Sec '91 Brighton, UK, 15-17 May 1991* Elsevier Science Publishers, North-Holland.

[ALT93] **Anderson, A.M., Longley, D. & Tickle, A.B.** (1993) "The Risk Data Repository: a novel approach to security risk modelling" *Proceedings IFIP Sec '93*, Deerhurst, Canada.

[Bas93] **Baskerville, R.** (1993) "Information Systems Security Design Methods: Implications for Information Systems Development" *ACM Computing Surveys*, v25 n4 Dec pp 375-414.

[B&H89] **Baum, E.B. & Haussler, D.** (1989) "What Size Net Gives Valid Generalization?" *Neural Computation* 1, pp151-160.

[B&J88] **Bonyun, D & Jones, G** (1988) "An Expert Systems Approach to the Modelling of Risks in Dynamic Environments" *Proceedings 1988 Computer Security Risk Model Builders Workshop* Denver, Colorado, May 1988.

[B&B88] **Boose, J.H. and Bradshaw, J.M.** (1988) "Expertise transfer and complex problems: using AQUINAS as a knowledge-acquisition workbench for knowledge-based systems" in **Boose, J. & Gaines, B.,** *Knowledge Acquisition Tools for Expert Systems: Knowledge-based Systems Vol 2* Academic Press Ltd.

[B&G88] **Boose, J. & Gaines, B.,** (1988) *Knowledge Acquisition Tools for Expert Systems: Knowledge-based Systems Vol 2* Academic Press Ltd.

[Boo93] **Boose, J. H.** (1993) "A survey of knowledge acquisition techniques and tools" in **Buchanan, Bruce & Wilkins, David (Eds)** *Readings in knowledge acquisition and learning: Automating the Construction and Improvement of Expert Systems*, Morgan Kaufmann Publishers, San Mateo, California.

[CEC91] **Commission of European Communities (CEC)** (1991) *Information Technology Security Evaluation Criteria (ITSEC) - Provisional Harmonized Criteria Version 1.2*, Brussels, June.

[C&W89] **Conway, T. & Wilson, M.** (1989) "Psychological Studies of Knowledge Representation" in **Ringland, G.A. & Duce, D.A. (Eds)** *Approaches to Knowledge Representation - An Introduction* Research Studies Press Ltd.

[Die91] **Diederich, J.** (1991) "Re-Learning in Connectionist Semantic Networks" in **Morik, K, Bergandano, F & Buntine, W (Eds)** *IJCAI-91 Workshop: Evaluating and Cnaging Representaion in Machine Learning* Sydney, Australia.

[DRM87] **Diederich, J., Ruhman, I. & May, M.** (1987) "KRITON: a knowledge-acquisition tool for expert systems" *International Journal of Man-Machine Studies* 26 pp29-40.

[Gil90] **Gilbert, I.** (1990) "Automated Risk Management Software Tools" *Proceedings 3rd International Computer Security Risk Model Builders Workshop* Santa Fe, New Mexico, August 1990.

[G&C88] **Gruber, T.R. & Cohen, P.R.** (1988) "Design for acquisition: principles of knowledge-system design to facilitate knowledge acquisition" in **Boose, J. & Gaines, B.,** *Knowledge Acquisition Tools for Expert Systems: Knowledge-based Systems Vol 2* Academic Press Ltd.

[Har89] **Hart, A.** (1989) *Knowledge Acquisition for Expert Systems* Kogan Page, London.

[Kel55] **Kelly, G.** (1955) *The Psychology of Personal Constructs* Norton, New York.

[NBS74] **National Bureau of Standards (USA)**, *Guidelines for ADP Physical Security and Risk Management*, Washington.

[S&G88] **Shaw, M.L. & Gaines, B.R**. (1988) "KITTEN: Knowledge initiation and transfer fools for experts and novices" in **Boose, J. & Gaines, B.,** *Knowledge Acquisition Tools for Expert Systems: Knowledge-based Systems Vol 2* Academic Press Ltd.

[Smi87] **Smith, Harry** (1987) *Data Structures: Form and Function* Harcourt Brace Jovanovich Inc., Orlando, Florida.

[Wah90] **Wahlgren, G.** (1990) "Survey of Computer Aided Risk Analysis Packages for Computer Security" *Working Paper: 90-16 System Integrity and Information Security research centre,* Department of Computer and Systems Science, Royal Institute of Technology, Stockholm, Sweden.

[W&C92] **Wahlgren, G. & Carroll, J.** (1992) "General Systems Theoretic Model of InfoSecMan" *Proceedings of Workshop on Information Security Management IFIP, Technical Committee 11, Working Group 11.1, Carlton Hotel Singapore, May 1992.*

EXPERT-TEACHING: AN APPROACH TO COMPUTER-AIDED MECHANICAL ENGINEERING DESIGN

Daniel P.J. Goodburn, David D. Harris and Peter H. Sydenham
School of Electronic Engineering, University of South Australia, Levels Campus,
The Levels, South Australia, 5095.
Email: 9311408w@lux.levels.unisa.edu.au

ABSTRACT

This paper discusses a search for the requirements for creating a knowledge-based support system for mechanical engineering design. Attempts at design support in the past have used expert systems to capture design knowledge. This technique has not overcome the well-known difficulties of knowledge acquisition and the inability of experts and end-users to develop and/or maintain expert systems. As a result expert systems in general have become somewhat unpopular in the world of commercial Computer-Aided Design (CAD) software. It is proposed that, in the case of design, some of these problems can be overcome if tools can be provided to designers to help them develop applications themselves to support their work practices. Further these applications should be targetted at the detail design stage. The paper discusses two studies of CAD users to support these claims. Finally, an illustrative example is given describing how a system incorporating the above ideas might work.

INTRODUCTION

In principle, knowledge-based systems should be highly beneficial to the CAD user. However, although there has been much research into developing Intelligent CAD (ICAD) systems, very little of this research has found its way into commercial CAD systems. Many research applications which have been reported in the past decade or so are based on expert system technology [BC84, CKK93, For93, HA88]. The traditional expert system application is a computer program which captures the knowledge of human experts and makes it available for use by a larger number of non-experts (for example see Moseley et al. [MC92]).

This paper discusses some problems with the expert system approach to developing ICAD systems and indicates an approach which, we believe, addresses these issues.

PROBLEMS WITH EXPERT SYSTEMS

Expert Systems have received a mixed reception in practice. In many business applications they have been very successful and well-received. In technical areas, although there are certainly many highly successful applications, their success has been more limited. In the research community, and especially in relation to design support, they have met with indifferent success, and enthusiasm towards expert systems across research disciplines has declined recently [BMZ92, Met92, HW93]. We believe that, in attempting to apply expert system technology to CAD, two problems predominate - Knowledge Acquisition, and the problems, for non-programmers, of writing knowledge-based applications.

Knowledge Acquisition.

Knowledge acquisition is recognised as being a difficult and time-intensive process [BB87, FW93, Gai89]. There are several methods for acquiring knowledge about the problem domain. Puppe et al. [PG92] discuss three methods for knowledge acquisition:-

Indirect knowledge acquisition is when one or more knowledge engineers elicit information from one or more domain experts. This is also known as *knowledge elicitation* and is probably the most common method of knowledge acquisition.

Direct knowledge acquisition, where the experts formalise their knowledge by themselves.

Automatic knowledge acquisition, where knowledge is transformed automatically from already existing knowledge.

Sometimes a combination of these methods may be used. The knowledge acquired is represented using a knowledge representation scheme (eg. rules, frames, semantic nets) before being programmed into a computer.

The problems of knowledge acquisition have been widely reported. Bachmann et al. [BMZ92] report on the selection of knowledge sources. The cooperation of the expert is essential, and the success of the system could be jeopardised if full cooperation is not given [MC92, FW93]. Differences of opinion between experts or in vocabulary must be overcome [Gai89]. Further difficulties can arise through lack of communication between the knowledge engineer and the expert [BMZ92].

In order to aid the knowledge engineering process, tools have been developed which automate aspects of knowledge acquisition and representation [Cle92]. Other tools have been developed to help the knowledge engineer to structure the knowledge acquisition interview [FW93]. The purpose of these tools is to improve the efficiency of the knowledge engineering process, decreasing time and costs.

The difficulties experienced in knowledge acquisition are obviously related to the complexity of the subject area. This perhaps explains at least part of the reason for the success of expert systems in business applications, where rule-sets are often quite small and readily determined. The complexity and subjectivity of technical areas make for very difficult construction of expert systems.

Generation of Knowledge Based Systems

Expert system shells to aid the development of expert systems are available commercially, and are often marketed as being capable of making the development of expert systems an exercise well within the ability of a user with no prior programming skills. This type of comment is usually based on the fact that rules can be entered as normal English language sentences. This kind of knowledge representation has been criticised by Mettrey [Met92] for its sacrifice of expressiveness and pattern-matching flexibility. Bachmann et al. [BMZ92] are also critical of expert system shell use, as the inference engines in these shells incorporate mechanisms which are only appropriate for the particular domains for which they were intended. The survey suggests that those expert systems that are successful were developed using specially-made shells.

Once the expert system has been built there is still much development which needs to take place before the system will be accepted in the workplace. End-users will need to learn how to use the system. The introduction of the new technology will also affect the normal work practices, which will have to change in order to accommodate the new system. Users may not be familiar with the logic of the system and thus will not accept it [BMZ92]. As with any system, bugs will be found, which will slow down the process of integration. These teething problems must be anticipated and accepted by the developers and users alike.

THE EXPERT USER/TEACHER

Expert systems have traditionally been viewed as tools to carry the expertise of an expert for use by a non-expert. We believe that a new paradigm - expert systems developed by the expert for use by the expert - has much to recommend it.

Consider the case where the user of the expert system is quite capable of doing the work without the expert system. The user is an expert in the domain of the expert system. In that case, why use an expert system? The answer is simply to save time, as a well-constructed expert system should be able to carry out the task much faster than the human. However, with the user as the expert, many benefits arise:-

- The user will be able to observe the work of the expert system and check that it appears to have achieved a sensible result;
- The user will be able to teach the expert system new knowledge;
- The user will have confidence in the system.

This approach at once eliminates the need for the knowledge engineer by placing the knowledge acquisition process in the hands of the user. It can therefore be done at the convenience of the user, adding only knowledge the user considers important, allowing incremental upgrading of the knowledge, and eliminating human errors of understanding.

SIMPLIFIED DEVELOPMENT TOOLS

The removal of the knowledge engineer from the development process can only be justified if the experts are then able to develop the system themselves. Bachmann et al. [BMZ92] describe situations where experts did have success in developing tools themselves. Further evidence of this can be found in the study by Nardi and Gantt [NG92] in which they describe CAD users nicknamed "gardeners and gurus". This is discussed in detail later in this paper. The authors believe tools such as these, which are aimed at helping local developers to build expert systems, will be the key to the success of future CAD systems.

It is expected that the expert will still be required to perform his normal work during the development of the system. Therefore the development of the system must take place intermittently during the designer's work. Knowledge should be added incrementally during the performance of a task, and the system must be operable almost immediately so that the validation can take place in the work environment. The key is that development software must be easy to use, given that its use will be intermittant.

Need for mechanisms for knowledge transformation

Although the designer has the domain expertise, the designer is not in general a computer programmer, and although computers are common in the design workplace these days, which is to say that the designer is likely to be computer literate, the computer does not normally support the main tasks usually performed by designers. This will mean that ways must be found to assist the designer to externalise his know-how, and to represent this in the computer.

Machine-learning is one path which has been followed to aid the developers of expert systems during the knowledge acquisition stage [Cle92]. The authors believe that this avenue is still too immature to elicit all the knowledge required to develop useable expert systems. The acquisition of knowledge and representation of knowledge within the computer will still be in the hands of humans. With the above arguments that the designer-expert should be the knowledge engineer, we have coined the term *expert-teaching*, to emphasise the fact that the onus in solely on the designer to develop the expert system application. To aid the designer, researchers into ICAD must concern themselves with providing tools to facilitate the development process.

THE PRACTICAL CAD USER

One feature which we have found remarkable in much of the reported work on Intelligent CAD research is the very few projects in which the work of real CAD users has been studied. Another is the lack of research into the product documentation and communication task (which is what CAD supports) to identify areas in which intelligence in the CAD system will be of real benefit - i.e. where it will be cost effective. We have undertaken study of designers at work, and have drawn upon other research into this activity to identify what would provide the designer with real benefits, and what is hindering this in current approaches.

Who uses CAD?

Several researchers have reported that CAD systems do not support the early stages of design i.e. the conceptual and embodiment stages [SSJ89, JT91, Joh91, UWC90]. This raises the question as to what stages current CAD systems do support in the design process. In order to obtain a clearer understanding of this, four CAD users in a large engineering company were interviewed to gain an idea of who uses CAD within the company, and how they use it. Of the four, three use the CAD system exclusively throughout their work. One of these is the CAD manager of the company. The fourth CAD user is a conceptual designer who only makes use of the system occasionally.

Of the four CAD users interviewed only the conceptual designer had heard of expert systems, and had thought about how they could help him in his design work. The other CAD users were unfamiliar with the concept of expert systems and had not seen them in use. The CAD manager had seen value in expert systems as a mechanism by which "corporate knowledge" generated by an expert could be captured and retained by the company after the expert had moved on to other parts of the company, or indeed left the company. The remaining CAD users felt that they could have little to contribute to the development of expert systems to aid their work. They considered that they just entered the design as directed by the conceptual designer or manufacturing engineer, and that they themselves used no knowledge relevant enough for incorporation into an expert system - a point of view with which the authors disagree!

A video recording of a fifth CAD user was made while he worked on a typical CAD task. The recording shows that this person uses considerable expertise which was attributed to his experience and familiarity with the CAD system and the performance of the task. This observation brings into question the comments described above by the CAD users concerning the lack of relevance of their knowledge.

Although there is considerable research currently going into high levels of conceptual design, and the potential for tools to support people working at this level, at present (and, we believe, for some time into the future) the major users of CAD systems will be those charged with taking a design concept and designing its functional details. This falls into Brown and Chandrasekeran's "Level 3 Design" [BC84], where routine selection is being made from a range of well-understood options. In our study of the CAD users it was interesting to note that people working at this level often did not feel that they were "designing". Nevertheless, designers readily acknowledge that the time taken in detailing the concepts, and even more in detailing updates and changes, is very much more than in concept development, and it is the area where CAD is most productive. On a cost-benefit, or time-savings approach, this is the obvious target for knowledge-based support tools.

Harris [Har94] points out that there is a clear mapping between the needs of the detail designer and "Level 3 Design". The expertise required in support tools is routine and low-level but time-consuming to apply manually. Targeting support tools at this level of design also supports the concept of *end-user programming* [Nar93] which is a major supporting concept for our work.

Gardeners and Gurus

Potentially, knowledge-based support tools should be very useful, particularly in detail development/CAD where the time-savings of using such tools would be the greatest. But despite this apparently worthwhile application, knowledge-based tools to support detail design are rarely successful. Some reasons for this are given above. Nevertheless there is evidence that CAD systems incorporating knowledge-based tools can succeed.

Nardi and Gantt [NG92] describe a study of communities of CAD users where some users have become local developers within their local CAD user communities and support their colleagues by customising environments and programming applications using macros, or more sophisticated programming languages such as C or AutoLISP. They are also familiar with the the platform under which the CAD system runs and are able to mediate between the system administrators and the CAD users. The local developer looks out for code being developed by other CAD users in order to make useful code available to the whole group and to ensure standardisation and elimination of bugs. The activity of the local developers directly contributes to the productivity of the whole CAD community, and where this has been recognised by management, the local developer has been given a formal position by management. This formal local developer position is referred to as a *gardener* (they help to "grow productivity" by nurturing fellow employees).

The reasons for the success of gardeners are :-

- The gardener is familiar with the domain of the CAD users as well as their everyday patterns of work and group interaction. The gardener also has good programming knowledge and communication skills, and can act as an intermediary for the CAD and Information Technology people, and the CAD users and management.
- CAD users are more comfortable with CAD systems knowing that support is available to help with any problems they might have.
- Gardeners are given leave to concentrate on "tinkering" with the computer to research methods for improving productivity.
- Managers can be assured that program applications will be standardised and that CAD users will concentrate on their domain-related tasks, leaving the "tinkering" to the gardeners.
- The gardener comes from the rank and file: the gardening skills develop after familiarisation with the domain and CAD system. Gardeners know the frustrations of CAD usage and so show concern and empathy towards their fellow CAD users. They produce code with a very clear picture in mind of the group's needs and preferences.

To ensure the success of gardeners and local developers, CAD systems must provide the flexibility required by gardeners and local developers to enable them to enhance and customise CAD systems to the needs of the group.

The current approach

This study shows that CAD applications are useful when developed by people who know not only the domain but are familiar with the specific needs of the CAD community they are supporting. Communities of CAD users adapt themselves and their work practices in order to improve the productivity of their CAD usage. Their ability to do this is determined by their level of programming skill and the flexibility of the CAD system, which must allow for the needs of end-users by providing facilities to enhance the system via customisation tools and programming language interfaces. A feature of the adaptability of CAD users is the possible evolution of local developers who are able, with their advanced programming and communication skills, to develop applications to aid all CAD users and improve productivity.

Given the support of knowledge-based development tools there is no reason why the local developer could not program knowledge-based systems. The local developer already has a deep knowledge of the domain and work environment and an understanding of the needs and frustrations of CAD users. These are all the necessary qualities of the knowledge engineer. Furthermore the local developer has first hand knowledge of the problems experienced by CAD users due to his experience as a CAD user. The local developer is therefore the ideal person to program knowledge-based applications.

THE PROPOSED SYSTEM

A computer system is being developed to explore these ideas and their implications. The system consists of two modules:

- A drafting module
- A knowledge base module for viewing, entering and modifying knowledge about the design.

The drafting module provides various drawing tools to aid the designer in producing schematic and detailed drawings of designs. The knowledge base window allows the designer access to hierarchical representations of the objects drawn in the drafting window, as well as rules relating to the specification of the objects. Objects and rules may be defined and modified in this module. The module may also suggest specifications of various objects of the design based on the rules entered by the designer.

It is important to note that the system represents an integration of current CAD drafting facilities with the facilities provided by expert system technology. Expert systems are not seen as a replacement to current CAD systems, but as a way of enhancing their usefulness to the CAD user. To this end work is currently being done to integrate knowledge-based techniques with the QikDraw Advanced Solids Modeller.

Illustrative example - Flange specification

To illustrate the proposed system, consider the example of a designer working on the design of a pipe system. One of the tasks that the designer would be required to perform often during this work is specifying flanges to join pipes in the system. The knowledge sources used for this purpose would be the problem specification, standards for flanges, catalogs from suppliers, and most importantly his past experience. To specify a flange the designer would need to specify the values of the parameters of the flange: the nominal size, flange type, flange face, material, and the flange table (which determines the dimensions of the flange). These values could also be considered to be outputs of an expert system and could be represented as leaves of a tree hierarchy. To the branches of the tree the designer would attach the rules used to specify each value of the tree. Rules could be added at leisure, and could be associated with parameters of other objects relevant to the design.

By selecting a pipe represented on the drafting window and then selecting **Flange** from the **Tool** menu, a dialog box will appear from which the designer may choose to perform several actions. If the computer has enough knowledge entered already, it may suggest a specification based on its available information. This may only be a previously specified flange, eg. if the designer had only specified one previous flange, or it could be suggested based on the rules and the information already specified by the designer. For instance, if the designer has already specified the size of the pipe to which the flange is to be joined, the system may conclude that the diameter of the flange must appropriately suit this pipe size, if the designer has entered

Flange_Size = Selected_Pipe.Pipe_Size

as a rule to determine the Flange Size. The designer may accept the suggested flange, or re-specify the flange. This re-specification would be based on the designer's rules of specification of flanges. Each re-specification of a parameter value is an opportunity for the designer to enter a new rule, a rule which the designer uses to make that re-specification. Therefore the entry of the rule is relevant to the task at hand. The designer verifies rules each time the computer makes a correct specification of the flange, according to the designer's existing knowledge.

CONCLUSIONS

This paper commenced with the hypothesis that knowledge based systems should be useful to the CAD user, but that successful applications were rare. Nevertheless, successes do exist where the well-known problems of knowledge engineering are addressed. It is concluded that there are identifiable requirements for a successful linking of KBS and CAD, and that these include:-

- The KBS must be developed for, and in the environment of, the individual CAD user.
- The KBS should target those CAD applications which are most time consuming, easily defined and repetitive. This occurs in the area of detailing, not in conceptual design.
- KBS support tools should be developed for use by the expert (not the non-expert) user.
- The KBS must be readily teachable by the user, and this teaching function must be sufficiently simple and intuitive to cater for the fact that its use may be intermittent and infrequent.

Work is currently in hand to develop such a KBS support tool, integrated with the QikDraw Advanced Solids Modeller.

ACKNOWLEDGEMENTS

Mr. Goodburn is supported by the Australian Research Council under an APRA(I) scholarship, with QikDraw Systems as the industrial collaborator.

Access to QikDraw's Advanced Solids Modeller program, and senior programmer Dayong Zhang is greatly appreciated.

The assistance and cooperation of staff at Britax Rainsfords Pty. Ltd. is gratefully acknowledged.

Part of this work was carried out at the LUTCHI Research Centre, Loughborough University of Technology, UK. The assistance provided by Professor Ernest Edmonds and Ms. Linda Candy as well as many others within LUTCHI is gratefully acknowledged.

REFERENCES

[BB87] J.H. Boose and J.M. Bradshaw. Expertise Transfer and Complex Problems: Using AQUINAS as a Knowledge Acquisition Workbench for Knowledge-Based Systems. *International Journal of Man-Machine Studies*, 26:3-28, 1987.

[BC84] D.C. Brown and B. Chandrasekaran. Expert Systems for a Class of Mechanical Design Activity. *IFIP Working Party Conference Knowledge Engineering in Computer-Aided Design*, Budapest, Hungary, 1984.

[BMZ92] R. Bachmann, T. Malsch and S. Ziegler. Success and Failure of Expert Systems in Different Fields of Industrial Application. In H.J. Ohlbach, editor, *Proceedings of the Sixteenth German Conference on Artificial Intelligence (GWAI-92)*, pages 77-86, Bonn, Germany, August/September 1992. (Published in *Lecture Notes in Computer Science*, volume 671, Springer Verlag).

[CKK93] Y. Chon, K.I. Kim and K. Kim. A Knowledge-Based System for Centrifugal Fan-Blade Design. *Engineering Applications of Artificial Intelligence*, 6:425-435, 1993.

[Cle92] R.P. Clement. Learning Expert Systems by Being Corrected. *International Journal of Man-Machine Studies*, 36:617-637, 1992.

[For93] G.A. Forgionne. IMDS: A Knowledge-Based System to Support Concurrent Engineering at Westinghouse. *Expert Systems With Applications*, 6:193-202, 1993.

[FW93] X. Feng and T.A. Weber. Knowledge Acquisition Advisor (KA^2): An Interactive Knowledge Acquisition Tool for Expert System Development. *Engineering Applications in Artificial Intelligence*, 6:507-518.

[Gai89] B.R. Gaines. Integration Issues in Knowledge Support Systems. *International Journal of Man-Machine Studies*, 31:495-515.

[HA88] M.N. Huhns and R.D. Acosta. Argo: A System for Design by Analogy. *IEEE Expert*, 3: 53-68, 1988.

[Har94] D.D. Harris. Knowledge Based Tools to Support the Measurement System Designer. M.Eng Thesis, University of South Australia, Faculty of Engineering, 1994.

[HW93] A. Horsfall and D. Wilde. AI Application Survey. In *Artificial Intelligence in Australia Research Report 1993*, pages 56-66. Australian Computer Society, Australia, 1993.

[Joh91] A.L. Johnson. Designing by Functions. *Design Studies*, 12: 51-57, 1991.

[JT91] A.L. Johnson and A.C. Thornton. Towards Real CAD. *Design Studies*, 12:232-236, 1991.

[MC92] L.G. Moseley and M. Cartwright. The Development of an Expert System for Operational Use in the Selection of Industrial Adhesives. *Engineering Applications in Artificial Intelligence*, 5:319-328, 1992.

[Met92] W. Mettrey. Expert Systems and Tools: Myths and Realities. *IEEE Expert*, 7:4-12, 1992.

[Nar93] B.A. Nardi. A Small Matter of Programming: Perspectives on End User Computing. MIT Press, Cambridge, Massachusetts, U.S.A., 1993.

[NG92] B.A. Nardi and M. Gantt. Gardeners and gurus: Patterns of Cooperation Among CAD Users. In *Proceedings of CHI '92*, pages 107-117, Monterey, California, U.S.A. May 3-7, 1992.

[PG92] F. Puppe and U. Gappa. Towards Knowledge Acquisition by Experts. In F. Belli and F.J. Rademacher, editors, *Proceedings of the Fifth International Conference in Industrial and Engineering Applications of Artificial Intelligence and Expert Systems (IEA/AIE-92)*, pages 546-555, Paderborn, Germany, 1992. (Published in *Lecture Notes in Computer Science*, volume 604, Springer Verlag).

[SSJ89] M.A. Sinclair, C.E. Siemieniuch and P.A. John. A User-Centred Approach to Define High-Level Requirements for Next-Generation CAD Systems for Mechanical Engineering. *IEEE Transactions on Engineering Management*, 36:262-270, 1989.

[UWC90] D.G. Ullman, S. Wood and D. Craig. The Importance of Drawing in the Mechanical Design Process. *Computers and Graphics*, 14: 263-274, 1990.

Knowledge-Based Systems

A DECISION MATRIX APPROACH FOR CONSTRUCTING MULTIPLE KNOWLEDGE BASES SYSTEM

Ning Shan and Xiaohua Hu
Department of Computer Science, University of Regina
Regina, Saskatchewan, Canada S4S 0A2
e-mail : {ning,xiaohua}@cs.uregina.ca

ABSTRACT

Recently, multiple knowledge bases system has been paid attention by people. In the practical point of view, the result produced by multiple knowledge bases system is more valuable than that produced by single knowledge base system, since the former involves more cost and intelligence than the latter.

In this paper we present a new approach to construct multiple knowledge bases system using rough set theory. Multiple knowledge bases system can be formulated precisely and in an unified way within the framework of rough sets theory. Our approach is based on the decision matrix method which has multiple learning and incremental learning capability. Our method first computes all the minimal knowledge rules from the knowledge representation system, then constructs a minimal knowledge bases corresponding to each reduct of the knowledge representation system. The computation of all minimal knowledge rules and reducts is reduced to the problem of simplifying a group of associated Boolean expressions. Our study shows that decision matrix approach provides an efficient and effective mechanism to construct a multiple knowledge bases system in a dynamic environment.
keyword: machine learning, knowledge-based systems

1 INTRODUCTION

Today, expert systems have been successfully developed in widely real-world applications where intelligent decisions have to be made. Knowledge base is an important component of expert systems and will directly affect the efficiency of a decision making process. Usually, knowledge base can be represented as a set of decision rules or a decision tree. This kind of knowledge base can be derived from human experts or a collected data. Most of the time the collected data is so huge that it is beyond the human experts' ability to analyze without automated analysis techniques. The useful information usually has complex patterns and relationships hidden in the collected data. In the past decade, the analysis and extraction of useful information from a collected data has been a subject of active research in machine learning area.

Rough sets as a nonstatistical methodologies for data analysis was introduced by Pawlak (Paw91, Paw82). It provides powerful tools for data analysis and knowledge discovery from imprecise and ambiguous data and derivation decision rules from them. So far, the theory of rough sets has been successfully applied in many areas, such as knowledge acquisition, forecasting and predictive modeling, knowledge base system, and data learning (Slo92, Zia94). The rough sets theory assumes the representation of data set in a *decision table* form which is a special case of a knowledge representation system. Rows of this table correspond to objects, and columns correspond to attribute. A number of algorithms and systems have been developed based on rough sets theory which may induce a set of decision rules from a given decision table, and may use induced decision rules to classify future examples. Most of them are attempting to find and select the best minimal set of decision rules that use only a minimal subset of attributes from the given data table. A minimal set of decision rules as a knowledge base represents the minimum knowledge necessary to support the classification of the given data.

A single knowledge base system which utilizes a single minimal set of decision rules to classify future examples may lead to mistakes, because each minimal set of decision rules represent different domain of the knowledge representation system and has different criteria. Recently, in order to enhance the accuracy of expert system, the concept of redundant or multiple knowledge bases emerged. The idea is to generate several knowledge bases instead of one knowledge base for the classification of new object, hoping that the combination of answers of multiple knowledge bases will result in better performance. Typically one object is classified with several rules in the multiple knowledge bases system, and the decisions are then combined to obtain the final conclusion. Many research results illustrated that such multiple rules, if appropriately combined during classification, can improve the classification accuracy (Kon92).

In this paper, we present a decision matrix approach to construct a multiple knowledge bases system in a dynamic environment. This approach combines the results of our previous works (Hu94, Zia93, Sh94). The maintenance of knowledge base in a dynamic environment is an important problem in many applications. The current knowledge base would have to be changed when a new piece of information is delivered with a new object. Incremental learning system has the significant capability to change the knowledge base in a dynamic environment. The decision matrix method has the multiple learning and incremental learning capability.

2 ROUGH SETS

In this section we present the concept *reduct* of rough sets theory. In the process of data analysis in a knowledge representation system, the main computational effort of rough sets is associated with the determination of the reducts. The concept of reduct plays a very important role in rough set theory, and is adopted by many application systems.

2.1 Knowledge Representation Systems

The basic component of a knowledge representation system (KRS), is a set of objects. The knowledge representation system provides information about the real world phenomena. Knowledge is understood as the ability to classify objects. A decision table is a special case of knowledge representation systems. The decision table is a finite set of decision rules which specify what decisions (actions) should be undertaken when some conditions are satisfied. A knowledge representation system is defined as follows,

Let $S =< OBJ, C, D, VAL, f >$ be a *knowledge representation system*, where OBJ is a non-empty set of objects (i.e., $OBJ = \{obj_1, obj_2, ..., obj_n\}$), C is a non-empty set of condition attributes, and D is a non-empty set of decision attributes. We have $A = C \cup D$ which is the set of all attributes and $C \cap D = \emptyset$. Let $VAL = \bigcup_{a \in A} VAL_a$, where for each $a \in A$, VAL_a is a finite attribute domain with at least two elements and the elements of VAL_a are called values of attribute a ($a \in A$). f is an information function such that $f(obj, a) \in VAL_a$ for every $a \in A$ and $obj \in OBJ$. Every object which belongs to OBJ is associated with a set of values corresponding to the condition attributes C and decision attributes D.

OBJ	S	H	E	C	CLASS
obj_1	0	0	1	0	0
obj_2	1	0	2	1	1
obj_3	1	1	1	0	0
obj_4	0	2	1	1	1
obj_5	1	2	1	0	1
obj_6	1	0	1	0	0
obj_7	1	2	2	1	1
obj_8	0	0	2	1	1

Table 1: An example of a knowledge representation system.

Let X be a subset of OBJ, we say that X is a concept if, and only if, $f(obj_i, d)$ has the same value of decision attribute $d \in D$ for any $obj_i \in X$. Any family of concepts which forms a partition of OBJ will be referred as *knowledge about* OBJ.

Let B be a nonempty subset of A, let obj_x, obj_y be members of OBJ, and let R be an *equivalence relation* over OBJ ($R = OBJ \times OBJ$). We define a binary relation, called an *indiscernibility relation* and defined as $IND(B) = \{(obj_x, obj_y) \in R : \forall a \in B \;\; f(obj_x, a) = f(obj_y, a)\}$. We say that obj_x and obj_y are indiscernible by a set of condition attributes B in a knowledge representation system iff $\forall a \in B$, $f(obj_x, a) = f(obj_y, a)$. The indiscernibility relation partitions OBJ into *equivalence classes*. Equivalence classes of the relation R are called *elementary sets* in an *approximation space* $Apr = (OBJ, R)$. Any finite union of elementary sets is called a *definable set* in Apr. The empty set is also an elementary set. For any object $obj_i \subseteq OBJ$, the equivalence classes of the relation R containing obj_i will be denoted $[obj_i]_R$.

Let X be a subset of OBJ, the *lower approximation* of X in Apr, denoted by $\underline{Apr}(X)$, is the set $\underline{Apr}(X) = \{obj_i \in OBJ | [obj_i]_R \subseteq X\}$ The *upper approximation* of X in Apr, denoted by $\overline{Apr}(X)$, is the set $\overline{Apr}(X) = \{obj_i \in OBJ | [obj_i]_R \cap X \neq \emptyset\}$ The $\underline{Apr}(X)$ is the union of all elementary sets contained in X. For any $obj_i \in \underline{Apr}(X)$, the obj_i can be classified certainly belonging to X. The $\overline{Apr}(X)$ is the union of all elementary sets overlapping with X. For any $obj_i \in \overline{Apr}(X)$, the obj_i can be classified as possibly belonging to X.

Table 1 shows an example of a knowledge representation system. That is, the universe of discourse OBJ consists of eight objects, $OBJ = \{ obj_1, obj_2, ..., obj_8 \}$. Each object is described by a set of condition attributes $C = \{$ S, H, E, C $\}$, with attribute values $VAL_S = \{ 0, 1 \}$, $VAL_H = \{ 0, 1, 2 \}$, $VAL_E = \{1,2\}$, and $VAL_C = \{ 0, 1 \}$. The set of values $VAL_{CLASS} = \{ 0, 1 \}$ of the decision attribute D represents the set of concept descriptions which are to be learned based on the attribute values of C. In our terminology, the concept is a subset of objects with a particular value of a decision attribute. All objects belonging to the concept are said to be positive, whereas all objects outside the concept are negative.

2.2 Attribute Reduction Techniques

Attribute reduction techniques aim at finding minimal subsets of attributes, each of which has the same discriminating power as the whole attributes. A minimal subset of attributes is a reduct, denoted as RED, that cannot be reduced further without affecting the essential information.

Let $S =< OBJ, C \cup D, \{VAL_a\}_{a \in A} >$ be a knowledge representation system $(A = C \cup D)$, let C^* denote the collection of equivalent classes of the relation $IND(C)$ and D^* be a family of equivalent classes of the relation $IND(D)$. The $POS(C, D)$ is a union of lower approximation of all elementary sets of the partition D^* in the approximation space $Apr = (OBJ, IND(D))$,such as

$$POS(C, D) = \bigcup_{X \in IND(D)} \underline{Apr}(X).$$

Definition 2.1 The set of condition attributes C is *independent* with respect to the set of decision attributes D if, for every $P \subset C$, $POS(C, D) \neq POS(P, D)$; otherwise C is said to be *dependent* on D.

Definition 2.2 The *degree of dependency* between the condition attributes and the decision attributes D is denoted as $\gamma(C, D)$ and defined as

$$\gamma(C, D) = \frac{card(POS(C, D))}{card(OBJ)},$$

where *card* denotes set cardinality.

Definition 2.3 The set of attributes P $(P \subseteq C)$ is a *reduct* of attributes C which satisfies the following conditions:

$$\begin{aligned} POS(P, D) &\neq POS(P', D),\ \forall P' \subset P \quad \text{and} \\ \gamma(P, D) &= \gamma(C, D). \end{aligned}$$

Attribute reducts, denoted as $RED(D)$, are the minimal subsets of condition attributes C with respect to decision attributes D, none of the attributes of any minimal subsets can be eliminated without affecting the essential information. These minimal subsets can discern decision classes with the same discriminating power as the full set of condition attributes. Any of the reduct sets, $RED_i \in RED(D)$, can be used instead of the original system S. For each reduct, we can derive a reduct table from the original knowledge representation system by removing those attributes which are not in the reduct. For example, in Table 1, $RED(D) = \{HE, HC\}$ (we will discuss how to compute reducts in Section 3), then we can generate two reduct table from the original system as shown in Table 2 (a) and (b).

3 A DECISION MATRIX APPROACH FOR COMPUTING ALL MINIMAL DECISION RULES

3.1 Decision Matrix

A decision matrix can be used to compute all minimal sets of decision rules and the reduct sets of a knowledge representation system S. It provides a way to generate the simplest set of decision rules, while preserving all essential information. The approach to generation of rules as presented here is based upon the construction of a number of Boolean functions (Sko92, Zia93, Sh94) from decision matrices.

Before we define the concept of a decision matrix we will assume some notational conventions. That is, we will assume that all positive and negative objects are separately numbered with subscript i (*i.e.*, $i = 1, 2, ...\gamma$) and j (*i.e.*, $j = 1, 2, ...\rho$) respectively. To distinguish positive from negative objects we will use superscripts V and $\sim V$, for instance, obj_i^V versus $obj_j^{\sim V}$ for the class "V".

Definition 3.1 A decision matrix $M(S) = (M_{ij})$ of a knowledge representation system $S =< OBJ, C \cup D, \{VAL_a\}_{a \in A} >$ is defined as

$$M_{ij} = \{(a, a(obj_i^V)) : a(obj_i^V) \neq a(obj_j^{\sim V})\}$$

The set M_{ij} contains all attribute-value pairs (*attribute, value*) which are not identical between obj_i^V and $obj_j^{\sim V}$. In other words, M_{ij} represents the complete information distinguishing obj_i^V from $obj_j^{\sim V}$.

The set of minimal decision rules $|B_i|$ for a given object obj_i^V $(i = 1, 2, ...\gamma)$ is obtained by forming the Boolean expression

$$B_i^V = \bigwedge_j \bigvee M_{ij}$$

where $\bigwedge$ and $\bigvee$ are respectively generalized conjunction and disjunction operators.

The Boolean expression called a decision function B_i^V is constructed out of row i of the decision matrix, that is $(M_{i1}, M_{i2}, ...M_{i\rho})$, by formally treating each attribute-value pair occurring in the component M_{ij} as a Boolean variable and then forming a Boolean conjunction of disjunctions of components belonging to each set M_{ij} $(j = 1, 2, ..., \rho)$.

The decision rules $|B_i^V|$ are obtained by turning such an expression into disjunctive normal form and using the absorption law of Boolean algebra to simplify it. The conjuncts, or prime implicants of the simplified decision function correspond to the minimal decision rules (Sko92, Zia93). By treating each of the classes as a target concept, a set of minimal decision rules can be computed for each of classes.

OBJ	H	E	Class
obj_1, obj_6	0	1	0
obj_2, obj_8	0	2	1
obj_3	1	1	0
obj_4, obj_5	2	1	1
obj_7	2	2	1

Table 2 (a)

OBJ	H	C	CLASS
obj_1, obj_6	0	0	0
obj_2, obj_8	0	1	1
obj_3	1	0	0
obj_4, obj_7	2	1	1
obj_5	2	0	1

Table 2 (b)

Similarly, by treating the complement of the class "V" as a target concept, a set of decision rules can be computed for each object of the class "$\sim V$" using the same approach. Some of the decision functions obtained from decision matrices given in Figure 1 and derived from the information contained in Table 1 are presented in *Example 1*.

Once all the decision rule sets $|B_i^V|$ have been computed, a set of all minimal decision rules $RUL(|V_d|)$ for the concept $|V_d|$ corresponding to the decision value V_d ($|V_d| = \{obj \in OBJ : d(obj) = V_d, d \in D, V_d \in VAL_d\}$) is given by

$$RUL(|V_d|) = \bigcup |B_i^V| \qquad (i = 1, 2, ...\gamma)$$

For computing the set of reducts of a knowledge representation system, we will introduce the concepts of the *phantom* decision function $\tilde{B}_i^V$ and the *reduct* function $F_{RED(V)}$. A phantom decision function $\tilde{B}_i^V$ is a Boolean expression defined by the conjunction of all Boolean expression $\vee\tilde{M}_{ij}$ of row i in the given decision matrix, where $\vee\tilde{M}_{ij}$ represents the disjunction of the only attribute names (does not contain the value of attributes) of the component M_{ij}. So that we have the following formula:

$$\tilde{B}_i^V = \bigwedge_j \bigvee \tilde{M}_{ij} \qquad (j = 1, 2, ..., \rho)$$

Informally speaking, A phantom decision function $\tilde{B}_i^V$ is a similarity of a decision function except for the elements of Boolean expression without the value of attributes. One can directly derive the result of a phantom decision function $\tilde{B}_i^V$ from the result of a decision function B_i^V, it is just eliminate the values of attributes in the prime implicants of the result.

The reduct function $F_{RED(V)}$ is a Boolean function constructed by the conjunction of all phantom decision function $\tilde{B}_i^V$ in the decision matrix. So that we have the following equivalence,

$$F_{RED(V)} = \bigwedge_i \tilde{B}_i^V$$

or

$$F_{RED(V)} = \bigwedge_i (\bigwedge_j \bigvee \tilde{M}_{ij})$$

The set of reducts, denoted as $RED(|V_d|)$, is obtained by performing the multiplications and applying the absorption law of Boolean algebra over the Boolean expression $F_{RED(|V_d|)}$. The conjuncts, or prime implicants of the result of the reduct function, are the whole set of reducts for the target concept V_d in a given knowledge representation system.

Example 1: *Figure 1 depicts two decision matrices obtained from the knowledge representation system given in Table 1. Each cell (i, j) in a decision matrix is a collection of attribute-value pairs distinguishing row i of the target class from column j of its complement.*

Based on these decision matrices we can obtain the following decision functions B_i^0 (i = 1, 2, 3) from the class '0' decision matrix (and similarly, we can obtain B_i^1 (i = 1, 2, ...5) from the class '1' decision matrix).

Class '0' decision functions are shown in Figure 2.

The $\bigcup |B_i^0|$ corresponds to all the minimal decision rules RUL for the class '0' of the knowledge representation system shown in Table 1:

$$(S = 0) \wedge (C = 0) \rightarrow (CLASS =' 0')$$
$$(H = 0) \wedge (E = 1) \rightarrow (CLASS =' 0')$$
$$(H = 0) \wedge (C = 0) \rightarrow (CLASS =' 0')$$
$$(H = 1) \rightarrow (CLASS =' 0')$$

Similarly, we can obtain the set of all minimal rules for the class '1':

$$(E = 2) \rightarrow (CLASS =' 1')$$
$$(C = 1) \rightarrow (CLASS =' 1')$$
$$(H = 2) \rightarrow (CLASS =' 1')$$

Now, let us compute the reduct function for the class '1' and class '0', such that

$$F_{RED(0)} = \bigwedge \tilde{B}_i^0 \qquad (i = 1, 2, 3)$$
$$= (H \wedge E) \vee (H \wedge C)$$
$$F_{RED(1)} = \bigwedge \tilde{B}_j^1 \qquad (j = 1, 2, 3, 4, 5)$$
$$= (H \wedge E) \vee (H \wedge C)$$

So that we can obtain the sets of reducts for the class '0' and the class '1',

$$RED(0) = \{HE, HC\}; \qquad RED(1) = \{HE, HC\}$$

j		1	2	3	4	5
i	OBJ	obj_2	obj_4	obj_5	obj_7	obj_8
1	obj_1	(S,0)(E,1)(C,0)	(H,0)(C,0)	(S,0)(H,0)	(S,0)(H,0)(E,1)(C,0)	(E,1)(C,0)
2	obj_3	(H,1)(E,1)(C,0)	(S,1)(H,1)(C,0)	(H,1)	(H,1)(E,1)(C,0)	(S,1)(H,1)(E,1)(C,0)
3	obj_6	(E,1)(C,0)	(S,1)(H,0)(C,0)	(H,0)	(H,0)(E,1)(C,0)	(S,1)(E,1)(C,0)

(a) A decision matrix for class '0'

j		1	2	3
i	OBJ	obj_1	obj_3	obj_6
1	obj_2	(S,1)(E,2)(C,1)	(H,0)(E,2)(C,1)	(E,2)(C,1)
2	obj_4	(H,2)(C,1)	(S,0)(H,2)(C,1)	(S,0)(H,2)(C,1)
3	obj_5	(S,1)(H,2)	(H,2)	(H,2)
4	obj_7	(S,1)(H,2)(E,2)(C,1)	(H,2)(E,2)(C,1)	(H,2)(E,2)(C,1)
5	obj_8	(E,2)(C,1)	(S,0)(H,0)(E,2)(C,1)	(S,0)(E,2)(C,1)

(b) A decision matrix for class '1'

Figure 1: Decision matrices for Table 1

$$B_1^0 = ((S,0) \vee (E,1) \vee (C,0)) \wedge ((H,0) \vee (C,0)) \wedge ((S,0) \vee (H,0)) \wedge ((S,0) \vee (H,0) \vee (E,1) \vee (C,0)) \wedge ((E,1) \vee (C,0)) = ((S,0) \wedge (C,0)) \vee ((H,0) \wedge (E,1)) \vee ((H,0) \wedge (C,0))$$

$$B_2^0 = ((H,1) \vee (E,1) \vee (C,0)) \wedge ((S,1) \vee (H,1) \vee (C,0)) \wedge ((H,1)) \wedge ((H,1) \vee (E,1) \vee (C,0)) \wedge ((S,1) \vee (H,1) \vee (E,1) \vee (C,0)) = (H,1)$$

$$B_3^0 = ((E,1) \vee (C,0)) \wedge ((S,1) \vee (H,0) \vee (C,0) \wedge ((H,0)) \wedge ((H,0) \vee (E,1) \vee (C,0)) \wedge ((S,1) \vee (E,1) \vee (C,0)) = ((H,0) \wedge (E,1)) \vee ((H,0) \wedge (C,0))$$

Figure 2: Class '0' decision functions.

3.2 An Incremental Learning Algorithm

An important objective of an incremental learning algorithm is the maintenance of the knowledge base in a dynamic environment. Here we present a algorithm which can processes new objects by modifying current decision matrices and decision rules RUL instead of regenerating a new set of decision rules. Each decision rule set for the knowledge representation system is maintained during the process, so that it can be updated during the generation of the set of decision rules. The addition of a new object results in the following knowledge base update algorithm:

Incremental Learning Algorithm

Step 1. Determine which target concept $|V_d|$ the new object belongs to. If the new object does not belong to any of the current target concepts, then define a new target concept and add it to the set of current target concepts. Next, consider two class decision problem, class "V" equivalent to $|V_d|$ and class "$\sim V$" equivalent to $OBJ - |V_d|$. Let (M_{ij}^V) be the current decision matrix for this problem. If there is only one class then stop processing.

Let us denote the new object as obj_k^V according to convention described in Section 3.1.

Step 2. Create a new row of decision matrix (M_{kj}^V): $(M_{kj}^V) = \{(a, a(obj_k^V)) : a(obj_k^V) \neq a(obj_j^{\sim V}), obj_j^{\sim V} \notin |V_d|\}$.

Step 3. Compute the decision function of obj_k^V

$$B_k^V = \bigwedge_j \bigvee M_{kj}^V$$

and the associated set of decision rules $|B_k^V|$.

Update the current set of decision rules $RUL(|V_d|)$ to become

$$RUL'(|V_d|) = RUL(|V_d|) \cup |B_k^V|$$

Step 4. Modify the reduct function $F_{RED(V)}$ of the current concept,

$$F'_{RED(V)} = F_{RED(V)} \bigwedge (\bigwedge_j \vee \tilde{M}_{kj}^V)$$

and the associated set of reducts $RED(|V_d|)$.

Step 5. For each target concept $W_d \neq V_d$, alter the current decision matrix (M_{ij}^W) by creating a new column of decision matrix: $(M_{ik}^W) = \{(a, a(obj_i^W)) : a(obj_i^W) \neq a(obj_k^{\sim W}), obj_i^W \in |W_d|\}$, where $obj_k^{\sim W}$ is the new object which belongs to $OBJ - |W_d|$ in this case.

Next, change the current decision functions B_i^W to become $B_i^{W'}$ by computing and simplifying the Boolean expression

$$B_i^{W'} = B_i^W \bigwedge \bigvee M_{ik}^W$$

i	OBJ	S	H	E	C	CLASS
6	obj_9	1	0	2	0	1

Table 3: A new object

and then update the rule set by taking the union of the associated rule sets, that is

$$RUL'(|W_d|) = \bigcup |B_i^{W'}|$$

Finally, update the current reduct function $F_{RED(W)}$ to become the new reduct function $F'_{RED(W)}$ by performing and simplifying the following Boolean expression

$$F'_{RED(W)} = F_{RED(W)} \bigwedge (\vee \tilde{M}_{ik})$$

and then update the set of reducts $RED'(|W_d|)$ by turning the prime implicants of the above result of Boolean expression.

Example 2: *In Example 1, we obtained all minimal decision rules RUL for the knowledge representation system shown in Table 1. Now, let's consider adding a new object as shown in Table 3. Suppose that the new object obj_9 belongs to class '1'. To update the knowledge base, we create the following new row in the decision matrix for class '1'*

$$M^1_{6,1} = (S,1)(E,2) \quad M^1_{6,2} = (H,0)(E,2) \quad M^1_{6,3} = (E,2)$$

and compute the decision function B^1_6 for the new object:

$$B^1_6 = \quad ((S,1) \vee (E,2)) \wedge ((H,0) \vee (E,2)) \wedge ((E,2)) = (E,2)$$

By adding the rule corresponding to the term (E, 2) *to the set of rules obtained in the Example 1 we obtain the following updated set of rules for class '1':*

$$(E=2) \rightarrow (CLASS='1')$$
$$(C=1) \rightarrow (CLASS='1')$$
$$(H=2) \rightarrow (CLASS='1')$$

Now, we compute the new set of reducts for the class '1' by updating the reduct function obtained in the Example 1 with conjunction of the new elements of $\bigvee \tilde{M}^1_{6,j}$ (i.e., $j = 1,2,3$) or with the phantom decision function $\tilde{B}^1_6 = (E)$, so that

$$\begin{aligned} F'_{RED(1)} &= F_{RED(1)} \bigwedge (\bigwedge_j \bigvee \tilde{M}_{6,j}) \qquad (j=1,2,3) \\ &= F_{RED(1)} \bigwedge \tilde{B}^1_6 = ((H \wedge E) \vee (H \wedge C)) \wedge (E) \\ &= H \wedge E \end{aligned}$$

Thus, the updated set of reducts of class '1', $RED'(1)$ is

$$RED'(1) = \{HE\}$$

To update the rules for class '0', we create a new column in the decision matrix for class '0'

$$M^0_{1,4} = (S,0)(E,1) \quad M^0_{2,4} = (H,1)(E,1) \quad M^0_{3,4} = (E,1)$$

and the we will consider updating the decision functions B^0_i $(i = 1,2,3)$ as derived from Figure 1 (a).

$$\begin{aligned} B^{0'}_1 &= B^0_1 \wedge ((S,0) \vee (E,1)) \\ &= ((S,0) \wedge (C,0)) \vee ((H,0) \wedge (E,1)) \\ B^{0'}_2 &= B^0_2 \wedge ((H,1) \vee (E,1)) = (H,1) \\ B^{0'}_3 &= B^0_3 \wedge ((E,1)) = ((H,0) \wedge (E,1)) \end{aligned}$$

From the above computation, we obtain the new set of minimal decision rules for class '0'.

$$(S=0) \wedge (C=0) \rightarrow (CLASS='0')$$
$$(H=0) \wedge (E=1) \rightarrow (CLASS='0')$$
$$(H=1) \rightarrow (CLASS='0')$$

Similarly, we consider to update the set of reducts for the class '0' by performing the reduct function $F_{RED(0)}$ obtained in Example 1 conjuncts with the elements of $\bigvee \tilde{M}^0_{i,4}$ (i.e., $i = 1,2,3$). Thus,

$$\begin{aligned} F'_{RED(0)} &= F_{RED(0)} \bigwedge_i \bigvee \tilde{M}_{i,4} \qquad (i=1,2,3) \\ &= ((H \wedge E) \vee (H \wedge C)) \wedge (S \vee E) \wedge (H \vee E) \wedge (E) \\ &= (H \wedge E) \end{aligned}$$

So that we obtain the set of reducts $RED'(0)$ as

$$RED'(0) = \{HE\}$$

4 BUILDING A MULTIPLE KNOWLEDGE BASES SYSTEM

4.1 The Concept of Multiple Knowledge Bases System

Rule generation is an important phase for data analysis and knowledge discovery from the knowledge representation system. Different algorithms and approaches will induce different minimal knowledge bases which, may or may not, use the same condition attributes from the given knowledge representation system. A minimal knowledge base employs only the information necessary to represent the given data set without loss of essential information. In other words, a minimal knowledge base is a set of decision rules without any redundant attributes and attribute values. It is very hard to say which one is better than the others in the real applications. Depending on the criterion, One minimal knowledge base can be more useful than another, that employs different information. From the predictive point of view, we believe that the multiple knowledge bases systems will be better than single knowledge base systems.

We have introduced the concept of multiple knowledge bases system by considering to employ more than one knowledge bases (Hu94). A multiple knowledge bases system may have more than one expert and there may be more than one knowledge bases associated with each expert. It could be partitioned into sub-systems based on the decision attributes. Each of partitioned sub-systems corresponds to one expert. A structure of the multiple knowledge bases system is shown in the Figure 3.

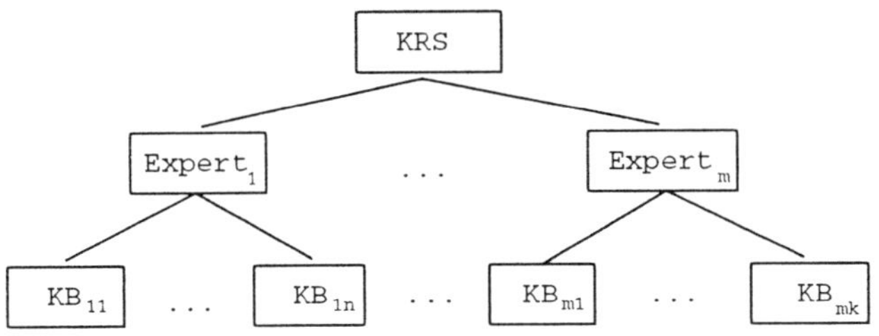

Figure 3: structure of Multiple Knowledge Bases Systems

4.2 Multiple Knowledge Bases System with Decision Matrix

One can use different algorithms and systems to generate several different knowledge bases from a given knowledge representation system, and embed these knowledge bases into a expert system to form a multiple knowledge bases system. Different knowledge bases are taken into account in the problem solving phase. This method does not have the incremental learning capability. When new information is expected to become available on a knowledge representation system, it has to regenerate the knowledge bases from the newly organized knowledge representation system. The process of regeneration can be costly when the knowledge representation system increases in growth. For machine learning problems in a dynamic environment, it would be preferable to accept new information incrementally, without needing to regenerate a new knowledge base.

Our approach uses decision matrix method to construct a multiple knowledge bases system defined above. The constructing task is to find multiple knowledge bases that can be used to predict the class of an unseen object as a function of its attribute values. Each knowledge bases is done by taking into account the same type of theory (rough sets theory) and represents different domain of the universe. One reduct corresponds to a minimal knowledge base in this paper. We first use decision matrix to compute all the minimal decision rules of the knowledge representation system and then construct different minimal knowledge bases corresponding to different reducts. A minimalized knowledge base corresponding to a reduct is a set of decision rules which is *fully covered* by the attributes of a reduct. The fully cover means that all the condition attributes used by the decision rules is also the attributes of the reduct table.

Let $RUL_{min} = \{r_1, r_2, ..., r_k\}$ be the set of all minimal decision rules generated by decision matrix method and let $RED = \{RED_1, RED_2, ..., RED_i\}$ be the set of knowledge bases (attribute reducts). A minimal knowledge base referred to RED_i ($RED_i \in RED$) is denoted by $RUL_{min}(RED_i)$ and defined as $RUL_{min}(RED_i) = \bigcup\{Cond(r_k) \subseteq Cond(RED_i) : r_k \in RUL_{min}\}$, where $Cond()$ is the set of attribute names.

For example, in *Example 1*, we have the set of reducts $RED = \{HE, HC\}$ with respect to decision attribute. According to the above definition, the minimized knowledge bases corresponding to reducts "H, E" and "H, C" on the class '0', and to reducts "H, E" and "H, C" on the class '1' are the following sets of decision rules extracted from all minimal decision rules:

The minimal set of decision rules for reduct "H, E" on the class '0' is

$$(H = 0) \wedge (E = 1) \rightarrow (CLASS =' 0')$$
$$(H = 1) \rightarrow (CLASS =' 0')$$

The minimal set of decision rules for reduct "H, C" on the class '0' is

$$(H = 0) \wedge (C = 0) \rightarrow (CLASS =' 0')$$
$$(H = 1) \rightarrow (CLASS =' 0')$$

The minimal set of decision rules for reduct "H, C" on the class '1' is

$$(E = 2) \rightarrow (CLASS =' 1')$$
$$(H = 2) \rightarrow (CLASS =' 1')$$

The minimal set of decision rules for reduct "H, C" on the class '1' is

$$(C = 1) \rightarrow (CLASS =' 1')$$
$$(H = 2) \rightarrow (CLASS =' 1')$$

5 DISCUSSION AND CONCLUSIONS

An approach for constructing multiple knowledge bases system, which is based on rough set theory, was presented. The approach is based on the decision matrix method which has multiple learning and incremental learning capability. The core of the multiple knowledge bases system for generating decision rules and computing the reducts has some advantages over traditional approaches, it dynamically construct the rules and reducts on the learning process.

There are several advantages of using decision matrix method. First, the decision matrix method has the ability of multiple learning to concurrently generate multiple knowledge bases. It can compute all minimal sets of decision rules and the reducts of a knowledge representation system.

Second, the decision matrix method has the ability to incremental learning new knowledge bases in a dynamic environment. The new information (objects) can be processed by modifying the current knowledge bases instead of regenerating new knowledge bases. Most of known incremental learning algorithms based on generating a single knowledge base and can not concurrently generate multiple knowledge bases.

Third, the main computation of rules and reducts is reduced to the problem of simplifying a group of associated Boolean expressions. Since the simplification problem can be easily decomposed into a number of disjoint subproblems by independently simplifying parts of the whole formula, our approach seems to be most suitable for implementation in multiprocessor systems.

References

PAW91 Pawlak, Z. Rough Sets: Theoretical Aspects of Reasoning About Data. Kluwer Academic Publishers, 1991.

PAW82 Pawlak, Z. *Rough Sets.* International Journal of Information and Computer Science 11(5), pp.341-356, 1982

KON92 Kononko, I. and Kovacie, M. *Learning as Optimization: Stochastic Generation of Multiple knowledge.* Proceeding of the 9th International Workshop on Machine Learning (ML92), Aberden,Scotland, July 1-3, pp. 259-262, 1992

HU94 Hu, X. and Shan, N. *Rough Set and Multiple Knowledge Bases.* Proceeding of the 7th Florida AI Research Symposium, Pensacola Beach, FL, USA, 1994, pp. 255-258

SLO92 Slowinski, R (ed.) Intelligent Decision Support: Handbook of Applications and Advances of Rough Sets Theory. Kluwer Academic Publishers, 1992

ZIA94 Ziarko, W. (ed.) Rough Sets, Fuzzy Sets and Knowledge Discovery. Springer-Verlag, 1994.

SKO92 Skowron, A. and Rauszer, C. *The Discernibility Matrices and Functions in Information Systems.* In Slowinski, R. (ed.) Intelligent Decision Support: Handbook of Applications and Advances of Rough Sets Theory, Kluwer Academic Publishers, 1992.

ZIA93 Ziarko, W and Shan, N 1993. *A Method For Computing All Maximally Gene ral Rules in Attribute-Value Systems*, Computational Intelligence: An International Journal, to appear.

SH94 Shan, N. and Ziarko, W. 1994. *Data-Based Acquisition and Incremental Modification of Classification Rules*, Computational Intelligence: An International Journal, Special Issue on Rough Sets and Knowledge Discovery (to appear).

MODELLING HYBRID RULE/FRAME-BASED EXPERT SYSTEMS USING COLOURED PETRI NETS

Simon C. K. SHIU*, James N. K. LIU, Daniel S. Yeung
Department of Computing, Hong Kong Polytechnic University
Hung Hom, Kowloon, Hong Kong
*Email : csckshiu@hkpucc.polyu.edu.hk

ABSTRACT

High level Petri Nets have recently been used for many AI applications, particularly for modelling traditional rule-based expert systems. The major effect is to facilitate the analysis of the knowledge inference during the reasoning process, and to support the system verification which increasingly becomes an integral part of expert system development. Nevertheless, there is not much attention being put on systems other than the traditional ones. In this paper, we described an approach to model hybrid (rule- and frame-based) expert systems using Coloured Petri Nets and the concept of controlled state tokens. The analysis of the proposed model is by constructing and examining the reachability tree spanned by the knowledge inference. Such methodology has an implication for supporting the verification process in hybrid systems.

Keywords : Artificial Intelligence, Expert System Modelling, Verification, Coloured Petri Nets

INTRODUCTION

Expert Systems (ES) have reached the stage where they are implemented and used in a wide variety of organizations and industries, a selection of operational expert systems in US, Europe, Canada and the Far East can be found in [Lie91, Zar91, SC91 and LYM+91]. There is increasing need for expert systems validation and verification (V&V) because erroneous advice may lead to invaluable economic loss and even fatal loss of life in some domain applications. Traditionally, attention has been concentrated in using verification techniques to tackle rule-based systems [SSS82, NPL+85, Bea90, Eve91, LD91]. However, these techniques exhibit a limited range of applicability. They could not cope with the kind of hybrid expert systems (HES), rule-based plus frame-based, which many of today's expert systems are developed [OO93, Dur94]. The use of this hybrid approach integrates the power of organizing data objects in a class hierarchy and reasoning about the objects through user pre-defined logical associations. This advantage accounts for many popular expert system developing software, such as ADS, ART, EXSYS EL, KAPPA-PC, KBMS, Nexpert Object, Level5 Object, ProKappa, ReMind, which combine some sort of frame-based representation with a rule-based inference engine. Although this approach benefits from the advantages of both representation techniques, it complicates the V&V of the expert systems. In this paper, we describe an approach based on Coloured Petri Nets(CPN) proposed by [Jen92] plus the concepts in State Controlled Petri Nets (SCPN) proposed by [Liu91] to model hybrid expert systems. It is through the analysis of the CPN that behavioural properties of the hybrid expert system can be verified and modified, thus enhancing the quality and reliability of the developed expert system.

Traditionally, there are a few approaches in modelling expert systems, such as [Cha91]'s Normal Form approach, [CS87, Van91]'s Decision Table Method, [Lan90]'s Incidence Matrix Method, [deK86]'s Truth Maintenance Systems, [CCS90, SC88, PS91]'s Generic Rule Systems. One of the major criticism of the above techniques is that none or very little consideration is given to allow for the dynamic checking of the knowledge inference. On the other hand, CPN, can support a formal description for modelling systems which consists of concurrent and synchronous activities. Besides, they also have a graphical representation and a well-defined semantics allowing for dynmaic analysis of the modelled system. In this paper, a contribution is made to the modelling of hybrid rule/frame-based expert systems. We will introduce an approach based on Coloured Petri Nets plus the concept of state tokens for the representation of knowledge inference in a HES. We will examine the transition sequences and check against the properties of the network in CPN for HES modelling. The paper is organized in five main sections. The first section gives the introduction, the second section describes a practical example of a hybrid expert system, and the third section gives the definitions of CPN and illustrates how the HES example can be modelled by CPN. Section four discusses the methods for analyzing the CPN and the implications of our approach to support formal verification of the systems. The last session gives the conclusion and discussion.

A HYBRID EXPERT SYSTEM EXAMPLE

We adopt a simplified version of a personnel selection expert system which has currently been used in Hong Kong [Hue93] for illustrating the HES modelling by our proposed CPN methodology. The aim of this system is to find out, among all the clerks in the organization, who should be promoted to senior clerk. The organization's employee data structure is represented in a frame-based hierarchy as shown in Figure 1 and details of relevant frames in the hierarchical structure are given below.

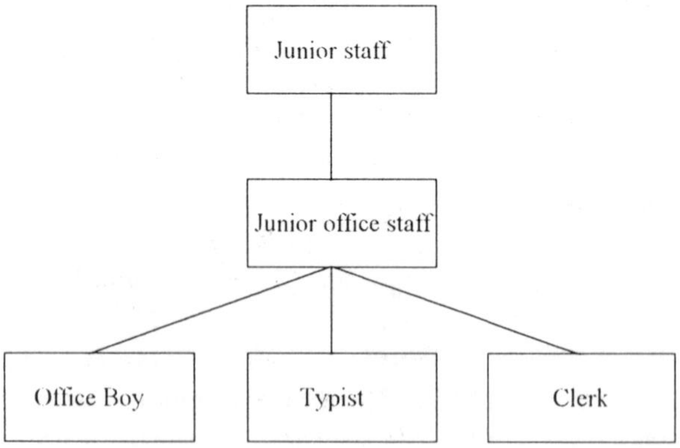

Figure 1 : The Frame Hierarchy

A Junior Staff Frame :

Slot Name	Value	Type	Demons
Job Title	Junior Staff	String	
Office Hours	09:00 17:00	Time	
Qualification Requirement	Five HKCEE	String	
Salary Pay Scale	1 to 10	String	
Father Frame	Staff		
Son Frame	Junior Office Staff		

A Junior Office Staff Frame :

Slot Name	Value	Type	Demons
*Job Title	Junior Office Staff	String	
*Office Hours	09:00 17:00	Time	
*Qualification Requirement	Five HKCEE	String	
*Salary Pay Scale	1 to 10	String	
Department	General Sec.	String	
Annual leave	21 days	Integer	
Father Frame	Junior Staff		
Son Frame	Clerk, Typist and Office boy.		

* denotes slots inherited from parent frame

A Clerk Frame :

Slot Name	Value	Type	Demons
*Job Title	Clerk	String	
Clerk Name		String	
Address and Telephone		String	

HKID#		String	IF process HKID# THEN Privilege is Local ELSE Privilege is Overseas.
Privilege		Local/ Overseas	
Sex		M/F	
*Office Hours	09:00 17:00	Time	
*Qualification Requirement	Five HKCEE	String	
*Department	General Sec.	String	
*Salary Pay Scale	1 to 10	String	
Present Salary Point		Integer	Present Salary Point must between 1 to 10 inclusive.
Years of Service		Integer	
*Annual leave	21 days	Integer	
Leave taken		Integer	
Leave balance		Integer	Leave balance = Annual leave - Leave taken
Knowledge of Work		G/M/L (Good, Medium, Low)	
Acceptance of Responsibility		G/M/L	
Organization of Work		G/M/L	
Initiative		G/M/L	
Relations with Colleagues		G/M/L	
Relations with Public		G/M/L	
Expression on Paper		G/M/L	
Oral Expression		G/M/L	
Supervisory Skills		G/M/L	
Leading Skills		G/M/L	
Performance		G/M/L	
Experience		G/M/L	
Ability		G/M/L	
Quality of Services		G/M/L	
Seniority		G/M/L	
Promotion		Yes /Wait /Reject	
Father Frame	Junior Office Staff		
Son Frame	NIL		

* denotes slots inherited from parent frame

It is noted that a frame is equivalent to a data structure with various types declarations, (or an object with different attributes). Demons are declared as methods or procedures within the frame. In the above example, the three frames are Class frames. Each individual clerk's information is captured by the creation of a clerk frame instance. The data value of Clerk Name, Sex, Address...etc are input via the user interface. The data values and demons in the slots with a * are inherited from the parent frame, the data value of Privilege and Leave balance are updated by firing the demons in HKID# and Leave balance. The data values for slots between Knowledge of Work and Leading Skills inclusively are input by the individual clerk's supervisor at the beginning of the inference process. The data value of Performance, Experience, Ability, Quality of Services and Seniority are being inferred by the execution of the rules pre-defined earlier by the personnel manager of the organization. The goal is to find out the data value of the slot Promotion, which can be inferred by forward chaining or backward chaining within the rule sets. (Over 100 rules were constructed for the original expert system based on the Multiple Criteria Decision Model [Tay84]). Typical rules are as follows:

Rule 1 : IF Quality of Services is Good
AND Seniority is Good
THEN Promotion is Yes.

Rule 2 : IF Quality of Services is Good
AND Seniority is Medium
THEN Promotion is Wait.

Rule 3 : IF Privilege is Local
AND Years of Service is more than 5 years
AND Present Salary Point is 10
THEN Seniority is Good.

Rule 4 : IF Performance is Good
AND Experience is Good
AND Ability is Good
THEN Quality of Service is Good.

In order to model the above HES, we have identified the need to provide an enriched modelling language which must be able to represent the hierarchical framed data structure, the individual frame instance, the various data types of each slot in a frame instance, the demons attached to individual slots, the separate rule sets which the truth of predicates are determined by firing the rules using the value in the slots. Details are as follows.

MODELING THE HYBRID EXPERT SYSTEMS USING COLOURED PETRI NETS

Definition of CPN

A Coloured Petri Net can be defined as a tuple CPN = (Σ, P, T, A, N, C, G, E, I) satisfying the requirements below:

Σ = a finite set of non-empty types, called colour sets.
P = a finite set of places.
T = a finite set of transitions.
A = a finite set of arcs such that : $P \cap T = P \cap A = T \cap A = \emptyset$.
N = a node function. It is defined from A into $PxT \cup TxP$.
C = a colour function. It is defined from P into Σ.
G = a guard function. It is defined from T into expressions such that: $\forall t \in T$: $[Type(G(t)) = \mathbb{B} \wedge Type(Var(G(t))) \subseteq \Sigma]$
E = an arc expression function. It is defined from A into expressions such that : $\forall a \in A : [Type(E(a)) = C(p(a))ms \wedge Type(Var(E(a))) \subseteq \Sigma]$.
I = an initialization function. It is defined from P into closed expressions such that $\forall p \in P$: $[Type(I(p)) = C(p)ms]$.

The set of colour sets determines the types, operations and functions that can be used in the net inscriptions. It is assumed that each colour set has at least one element.

The places, transitions and arcs are described by three sets P, T and A which are required to be finite and pairwise disjointed.

The node function N maps each arc into a pair where the first element is the source node and the second the destination node. The two nodes have to be of different kind (ie one of the nodes must be a place while the other is a transition). Several arcs may be allowed to link between the same ordered pair of nodes.

The colour function C maps each place, p, to a colour set C(p). This means that each token on p must have a token colour that belongs to the type C(p).

The guard function G maps each transition, t, to an expression of type Boolean, i.e., a predicate. All variables in G(t) must have types that belong to Σ . A guard is allowed to be a list of Boolean expressions [Bexpr1, Bexpr2..etc]$=\mathbb{B}$. This means that the binding must fulfill each of the Boolean expression in the list.

The arc expression function E maps each arc, a, into an expression which must be of type C(p(a))ms. This means that each evaluation of the arc expression must yield a multi-set over the colour set that is attached to the corresponding place.

The initialization function I maps each place, p, into a closed expression which must be of type C(p)ms, ie a multi-set over C(p).

Illustrative Example

The Personnel Selection Expert System can be modelled by CPN as follows :

First, each framed data structure is represented by a compound colour set, and each frame instance is represented by a token in that set. For instance, there are fifteen sets of non-empty types or colour sets being used in the illustrated example, ie Σ=AA,BB,...OO:

Color AA = string; (all text strings)
Color BB = with Local | Overseas; (colours explicitly specified)
Color CC = with Male | Female;
Color DD = time; (date)
Color EE = integer with 0..10;(between 0&10)
Color FF = integer;
Color GG = with Good | Medium | Low;
Color HH = with Yes | Wait | Reject;
Color II = list AA with 4; (a list of four strings)
Color JJ = list AA with 3;
Color KK = list FF with 5;
Color LL = list GG with 15;
Color MM = with Clerk | Typist | Office Boy;
Color NN = product II * BB * CC * DD * JJ * KK * LL * HH; (all tuples (i,b,c,d,j,k,l,h) where i∈II, b∈ BB,....h∈HH)
Color OO = with Yes | No; (for state token, if the value is Yes, it denotes that the predicate is true, else if the value is No, the negation of the predicate is true.)

Var i:II; var b:BB; var c:CC; var d:DD; var j:JJ; var k:KK; var l:LL; var h:HH; var clerk : NN; (var denotes variable declaration which introduces one or more variables. Here we have one variable, clerk, which is with colour NN. We may use var clerk1, clerk2, clerk3 : NN for declaring three different clerks for example.)

Secondly, places in the CPN are taken to correspond to predicates of the production rules which are pre-defined earlier by the user, and the transitions in the CPN correspond to the implications of the rules. The transition operations are represented by the arc expression functions. (eg. Rule 1 can be represented in CPN as shown in Figures 2a & 2b:)

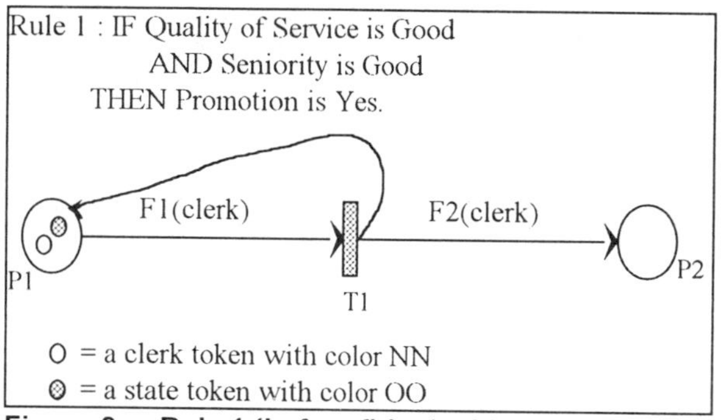

Figure 2a : Rule 1 (before firing) with an input token Clerk and a state token in P1 with 'Yes'

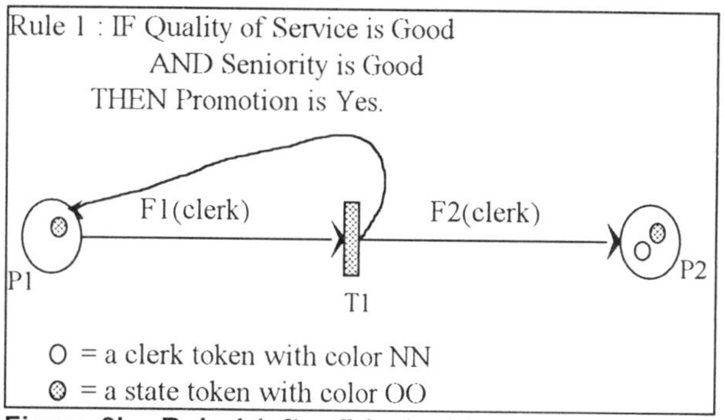

Figure 2b : Rule 1 (after firing) with an output token Clerk and a state token in P2 with 'Yes'

P1 and P2 are Places with colour type NN, T1 is a Transition which is enabled iff the input arc expression F1(clerk) is evaluated to be true. If F1(clerk) is true then T1 is fired, it implies that Rule 1 is executed. The token clerk will be removed from P1 and a new token clerk will be created in P2 with new data values determined by the output arc expression F2(clerk). A state token with 'Yes' value will also be created in P2 for further inference (if any). In order to preserve the state of the predicate in P1, a state token marked with 'Yes' is created in P1 via the self-loop. In the example F1(clerk) is an arc expression function which evaluates if the data value of variable l[14][1] and l[15] are equal to "Good", and if true, it will assign the value "Yes" to variable h. After a sequence of firings has been exhaustly executed, we can examine the attributes in the token clerk for consideration for promotion to senior clerk.

Similarly, a demon can also be represented by a Places/Transition tuple. (eg. the demon attached to slot HKID# is given in Figure 3 prior to its execution)

---

[1] l[14] means the 14th element in the list variable l:LL with color Good | Medium | Low.

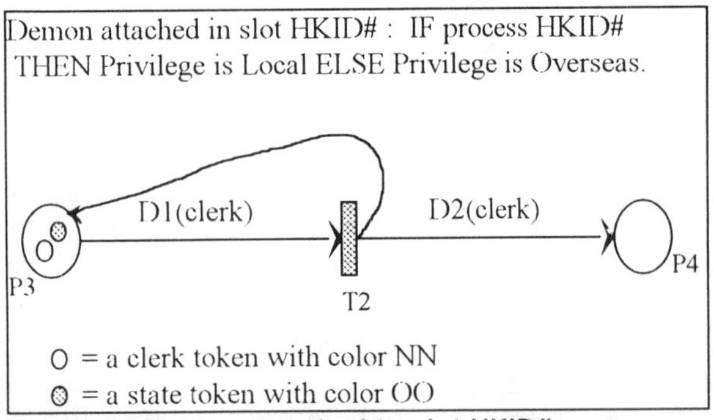

Figure 3 : Demon attached to slot HKID#

ANALYSIS OF COLOURED PETRI NETS

The major analysis technique, within the context of expert system verification, is the use of reachability tree which represents the reachability set of the CPN (or occurrence graph in Jensen's terminology). The basic idea behind is to construct a tree/graph containing a node for each reachable marking and an arc for each occurring binding element. In expert system verification, it refers to exhaustively exploring all the possible interactions of predicates within the model. From a given initial state, all possible transitions are generated, leading to a number of new states. This process is repeated for each of the newly generated states until no new states are generated. In our example, by constructing and examining the reachability tree using one particular clerk token (equivalent to an initial marking of the CPN because different clerks have different values in their attribute slots), we can detect any irregularities of the predicate places. Obviously such a tree/graph may become very large even for a small CPN. Therefore, for simplicity reason, without taking any transition conditions or transition operations into consideration, we can concentrate our analysis by enabling a specific transition and then check the reachability set for any irregularities of the associated predicate places. In a hybrid rule/frame based expert system, irregularities may come from the rule sets or the frames or both. Typically, there may be redundant rules or frames, dead end rules, missing rules or frames, misplaced slots and frames, duplication of frames, subsumptions of rules, auxiliary rules or frames, circular rule sets, contradiction rules, inconsistent rules. The checking of the above irregularities can be done exhaustively or heuristically by adequately initiation of the sequence of transitions and closely examining the reachability markings. The problems can be located through the trace of the sequence of transitions which may provide alternative or multiple marking effects.

CONCLUSION AND DISCUSSION

In this paper, we have described an approach to model hybrid rule/frame based expert system using Coloured Petri Nets and the concept in State Controlled Petri Nets. The frame data structures are represented using colour tokens. The data value of each colour token may be of an arbitrarily complex type, thus enabling various classes of frames be represented. The rules and demons are represented by a Place/Transition tuple with a self-loop in order to preserve the state of the predicate. The analysis of the CPN is by constructing and examining the reachability tree to detect irregularities of the predicate places.

In a pure frame-based expert system, reasoning is by comparing descriptions of incoming facts with the frames in the knowledge base, and retrieving the class frame that best matches the situation. The main inference mechanism or strategy for applying general information to specific instances is inheritance. This reasoning mechanism is rather limited in practical situations. In a pure rule-based expert system, reasoning is by firing a sequence of rules using incoming facts. Although this method is simple and useful, complex domain knowledge could not be represented. Our approach is useful to model systems that combined rule/frame based representation techniques.

Future work will include formalizing our approach and developing of algorithms to detect irregularities in the HES. We would also like to investigate further the capability of the methodology to handle fuzzy systems and the complexity involved against the traditional approaches.

REFERENCES

[Bea90] Beauvieux, A. A General Consistency Checking and Restoring Engine for Knowledge Bases. In *Proceedings of the 9th European Conference on Artificial Intelligence,* Stockholm, Sweden, 1990.

[CCS90] Chang, C.L., Combs, J.B. and Stachowitz, R.A. A Report on the Expert Systems Validation Associate (EVA). *Experts Systems with Applications,* Vol. 1, No. 3, pp219-230, 1990.

[Cha91] Charles, E. Checking Knowledge Bases for Inconsistencies and other Anomalies. In *Workshop Notes from the Ninth National Conference in Artificial Intelligence,* AAAI-91, Knowledge-Based Systems Verification, Validation and Testing, 17 July, Anaheim CA. 1991.

[CS87] Cragen, B.J. and Steudel, H.J. A Decision Table Based Processor for Checking Completeness and Consistency in Rule-Based Expert Systems. *International Journal of Man-Machine Studies,* Vol 26, pp633-648, 1987.

[deK86] de Kleer, J. An Assumption-Based TMS. *Artificial Intelligence* 28, pp127-162, 1986.

[Dur94] Durkin, J. *Expert Systems: Design and Development,* Macmillan Publishing Company, pp12-23, pp711-771, 1994.

[Eve91] Evertsz, R. The Automated Analysis of Rule-based Systems Based on their Procedural Semantics. In *Proceedings of the 12th International Joint Conference on Artificial Intelligence*. Sydney, Australia, 1991.

[Hue93] Huen, H.S.M. *A Prototype Decision Support System for Assessing The Claims for Promotion of Clerical Officers in the Hong Kong Civil Service.* M.Sc. Dissertation, Department of Computing, Hong Kong Polytechnic University, 1993.

[Jen92] Jensen, K. *Coloured Petri Nets : Basic Concepts, Analysis Methods and Practical Use, Volume 1*, Springer-Verlag, 1992.

[Lan90] Landauer, C. Correctness Principles for Rule-Based Expert Systems. *Expert Systems with Applications,* Vol 1, No. 3, pp291-317, 1990.

[LYM+91] Lee, J.K., Yeung, D.S., Mizoguchi R. and Narasimhalu D. *Operational Expert System Applications in the Far East*, Published by Pergamon Press, 1991.

[Lie91] Liebowitz, J. *Operational Expert System Applications in the United States*, Published by Pergamon Press, 1991.

[LD91] Liu, N.K. and Dillon, T.S. An Approach Towards the Verification of Expert Systems Using Numerical Petri Nets. *International Journal of Intelligent Systems,* 6, pp. 255-276, 1991.

[Liu91] Liu, N.K. *Formal Description and Verification of Expert Systems.* Ph.D. dissertation, Department of Computer Science and Computer Engineering, School of Mathematical and Information Sciences, La Trobe University, Bundoora, Victoria, Australia, 1991.

[NPL+85] Nguyen, T.A., Perkins, W.A., Laffey, T.J. and Pecora, D. Checking an expert system knowledge base for consistency and completeness. In *Proceedings of the 9th International Joint Conference on Artificial Intelligence.* Los Angeles, CA. 1985.

[OO93] O'Keefe, R.E. and O'Leary, D.E. Expert System verification and validation: a survey and tutorial, *Artificial Intelligence Review* 7, pp3-42, 1993.

[PS91] Preece, A.D. and Shinghal, R. DARC: A Procedure for Verifying Rule-Based Systems. In *Expert Systems World Congress Proceedings,* Liebowitz, J. (ed) Vol. 2, Pergamon Press, pp971-979, 1991.

[SC88] Stachowitz, R.A. and Chang, C.L. Verification and Validation of Expert Systems. *Tutorial note at AAAI-88.*

[SC91] Suen, C.Y. and Chinghal, R. *Operational Expert System Applications in Canada*, Published by Pergamon Press, 1991.

[SSS82] Suwa, M., Scott, A.C. and Shortliffe, E.H. An approach to verifying completeness and consistency in a rule-based expert system. *AI Magazine*, pp16-21, 1982.

[Tay84] Taylor, R.N. *Behavioural Decision Making*, Scott Foresman & Company, 1984.

[Van91] Vanthienen, J. Knowledge Acquisition and Validation Using a Decision Table Engineering Workbench. In *Expert Systems World Congress Proceedings,* Liebowitz, J. (ed), Pergamon Press, Vol 3, pp1861-1868, 1991.

[Zar91] Zarri, G.P. *Operational Expert System Applications in Europe*, Published by Pergamon Press, 1991.

HANDLING COMMUNICATIONS IN CONCURRENT KBS

A. Bokma, M. Huiban and A. Slade

School of Computing
and Information Systems
University of Sunderland
Sunderland, United Kingdom

S. A. Jarvis

Artificial Intelligence Systems
Research Group
University of Durham
Durham, United Kingdom

ABSTRACT

The paper addresses the issue of specifying and verifying communications in Distributed Artificial Intelligence (DAI) Systems, taking the Consensus method for the development of large-scale Agent-Based Systems as a test case. In developing distributed systems care needs to be taken to deal appropriately with the issue of communication between distributed components which are vital to the performance of the overall system. This paper proposes the use of guidelines to help in the selection of appropriate communication links and to ensure the correct implementation of reliable communication links between cooperating agents.

1 INTRODUCTION

There has been a noticeable shift from large mainframes towards networked hardware architectures in recent years. This requires a different approach to the construction of large systems to run on these new platforms. Another important development has been the increased need for faster and more intelligent systems. Systems of this kind still need to be engineered to the same rigorous standards expected of systems developed for more traditional implementation paradigms. The Consensus method is a step in that direction. In the process of specifying DAI Systems (KBS) it is important that the analysis and design of distributed solutions as well as their reliable and robust implementations should be dealt with satisfactorily. The performance of the system does not just depend on the performance of individual components, but also on the speed and reliability of the communication links between components. In monolithic systems which are designed in a modular fashion, communication of data between different modules is relatively easy to handle, as it does not usually require inter-process communication and network services provided by the operating system. Distributed systems are much more vulnerable in this respect, and in order to successfully specify and implement them, careful thought needs to be given to the selection of appropriate types of communication links and their implementation.

In this paper we give guidelines to help in the individual selection of appropriate types of communication links for a distributed architecture. Guidance is also given for the specification and verification of processor interconnection. Our emphasis is based on the Consensus method which has been used for the specification of large, concurrent Knowledge Based Systems in a safety critical environment. The Consensus method, which is explained in the next section, includes both a specification technique and a life-cycle model, drawing on some existing approaches such as [Hat88] [Cra89] [Boe88] as well as learning from other relevant developments in the field, such as [Ack92] [Tay89] [Sta87] [Fje92]. The method specialises in the development of large real-time applications, that build concurrent knowledge based components into an agent-based architecture, covering the complete life-cycle from concept through to operation. In the sections which follow, we discuss the importance of choosing correctly between different types of communication links which may exist between these different agents. The formal specification of such links is an essential component in the life-cycle, if we are to consider the reliability of the system architecture supporting our distributed system.

The Consensus method, is the result of research into software development methods and approaches, and has been refined through practical experience gained from applying the method to the development of a large-scale Air Traffic Control application with British Aerospace and Cambridge Consultants, UK. The method has also been used in two further projects, IRAS and MIRAM [Ira93] [Mir93] who are concerned with condition monitoring and preventive maintenance in manufacturing environments.

2 THE CONSENSUS METHOD

The Consensus method is concerned with the development of large parallel Knowledge Based Systems. This involves the development process, from conception through to operation, including the requirements definition, the design of the solution and the implementation, culminating in a working system. A successful method has to be comprehensive, yet compact, while being intuitive and easy to use and effective enough to deal with large complex applications.

The requirements analysis is carried out with the help of the Requirement Model. This uses a top-down structured analysis approach to analyse the requirements iteratively, by starting from an abstract level and proceeds to analyse the requirements down to the most specific level, where they are readily implementable. The requirements are expressed in a hierarchical set of data and control flow diagrams, together with associated documentation to record the description and detailed specification of the depicted components. The system design process is performed with the help of the Architecture Model. The Architecture Model also uses a structured approach with the goal of mapping the requirements expressed in the Requirement Model into a network of distributed knowledge based components that exhibit a high degree of concurrent behaviour. Again the approach is top-down and iterative.

In the Consensus method the requirements and system architecture are developed in conjunction and the two models together constitute the System Specification. This ensures that the final system takes account of both the desired functionality and possible restrictions. It is important in the specification process to balance the desired functionality of the system with the constraints imposed by the implementation platform and thereby force the specification process to pursue only feasible solutions. This includes both platform constraints and the need of structuring the requirements in terms of a distributed knowledge based system. By establishing a direct mapping, where the Architecture Model analyses and structures the Requirement Model to produce a specific system architecture, the tasks of validation and verification are substantially facilitated. The direct mapping between the Requirement and Architecture Models facilitates the task of validation, by allowing the analyst to inspect that all requirements have been incorporated in the architecture and how this has been accomplished. The analyst can also observe directly the trail from user requirements into specific components implementing them and to assess whether they have been implemented correctly. For the architecture Consensus uses a distributed, collaborative problem solving paradigm, where systems are built as networks of medium grain-sized knowledge based agents (KBAs), which engage in a collaborative problem solving activity. The process of structured analysis produces a network of primitive processes. As the grain-size of the structured analysis is finer than that of the final network of KBAs there is a relatively large set of primitive processes which have to be accommodated in a few KBAs. The goal of the Architecture Model is to group these processes into a number of KBAs, in a way which ensures that the individual KBAs can work efficiently and concurrently, while requiring a minimum of communication with other KBAs, to achieve the overall goal of the system.

In Consensus each KBA is an autonomous KBS, which contains a local blackboard, local knowledge sources, a local controller and communication channels. The local blackboard contains data and results of processing. Access to it is restricted to the local knowledge sources and controlled by the local controller. Each KBA has a set of local knowledge sources which perform the tasks assigned to that KBA. They operate solely on the information contained in the local blackboard and may request the local controller to communicate with other KBAs to transfer data. The local controller is in complete control of processing at a local level within the agent. It schedules the execution of local knowledge sources which operate on the local blackboard and initiates communication through the communication channels module with other KBAs on behalf of its knowledge sources, as well as receiving messages and posting them to the local blackboard.

3 THE SPECIFICATION OF COMMUNICATION CHANNELS

It is the purpose of the system Architecture Model: to identify a suitable set of agents which together constitute the system; to define the information flow between the agents in the system; and to specify the channels on which the information flows. In order to achieve this, the process of design takes the decomposition of DFDs from the Requirements Model and performs a number of analyses to group the transforms in a way that results in highly cohesive KBA. This should display a low degree of coupling as well as displaying a high degree of concurrency. To illustrate this, Figure 1 shows a sample DFD with a superimposed grouping into KBAs:

Those flows which connect processes or data-stores which reside in different KBAs become communication channels, and are therefore distinguished from flows within a KBA, which are handled internally. Given that the analyst has decided on a candidate system architecture, he needs to analyse the properties of each communication channel, and:

- consider, for each communication channel, the nature and amount of information that will be transferred
- determine the most suitable type of communication link for that channel

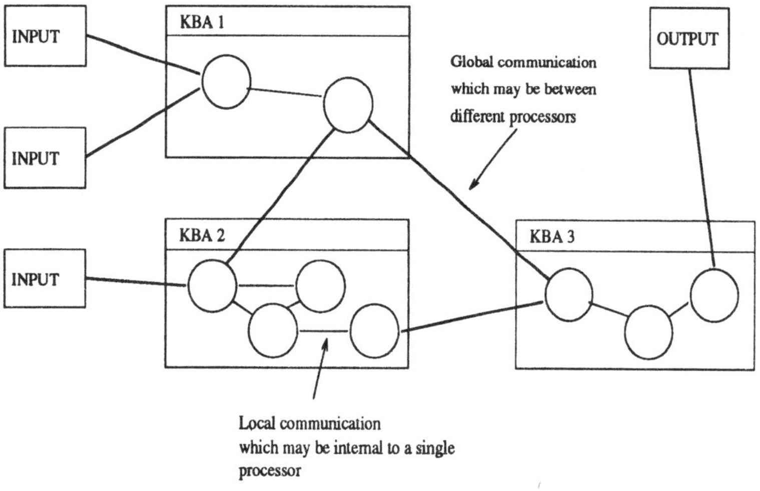

Figure 1: Sample DFD with Superimposed KBA Diagram

- calculate the total amount and rate including overheads incurred for the type of communication link
- consider candidate allocations of KBAs over the given network of processors, including any constraints
- consider the expected load on the processors and network connections and whether the network and its processors can adequately support the system.

It is essential that the choice of communication link between two KBAs suits the requirements identified. The choice of an incorrect implementation could lead to a system which will not quickly or reliably transfer messages between two key parts of the system, or would place emphasis and resources on communication which rarely took place. In the next section a formal approach to the analysis of agent interconnection is presented, there is a clear emphasis on reliability in the process, formal specifications and verifications are used. In choosing a suitable communication link, the Consensus process identifies a numbers of requirements, these requirements and other system constraints are modelled mathematically and the behaviour patterns between agents are formally clarified.

4 THE SPECIFICATION OF COMMUNICATION LINKS FOR KBAs

Processor interconnection quickly produces highly complex systems, involving a wide variety of possible interactions, and communications between many different constituent components. Much attention has been paid to developing methods which will help to clarify complicated interconnections; informal approaches have proven to be notoriously unreliable and the use of Formal Design Techniques (FDTs) is now widely accepted. The use of FDTs also enables the exploitation of specification reuse; specifications of components can be combined and reused in other parts of the system.

A method based on Communicating Sequential Processes (CSP) [HiJ95] [HiJ93] is proposed, a notation for the analysis of communication patterns of cooperating programs or objects. Its language of programs, or *processes*, may be used to provide algorithmic descriptions of system behaviours and it also includes semantic models which make it possible to reason about system properties.

The following section includes a brief description of the rôle of algebraic semantics in the analysis of communication behaviour and demonstrates how this can be applied to an example using three selected models of communication links between processors. The choice of how and when to use these communication links is explored with their requirements, the latter of which is identified within the unambiguous framework of mathematical specifications.

Analysing safety properties

The semantic models of CSP define a refinement relation between algorithm specifications and the immediate levels of design through which we pass to an implementation. Using CSPs semantic models the safety properties of a communication link can be analysed. The intended behaviour of a system can be specified and a proof of correctness established between a behavioural specification and an algorithm specification. The consistency between what we say the communication link will do, in terms of the messages passing between the sending and receiving KBAs, and the specification itself is proved.

CSP offers a number of rules and conventions which will be used in the verification of the communications link [Jar92]. Like any mathematical proof, not all of the rules of CSP need to be used, or even understood, instead it is sufficient to select one or two such rules which suit the particular processor interconnection. The selection of a small subset of the rules is regarded as a domain specific language which is suited to the particular task at hand; this retains an approach which is mathematically sound yet versatile and amenable to simple use and understanding.

Consider, for example, the CSP semantic rule which we shall use most often,

$$\frac{P \textbf{ sat } S(tr) \;\wedge\; Q \textbf{ sat } T(tr)}{P\,{}_A\|_B\, Q \textbf{ sat } S(tr\lceil A\checkmark) \wedge T(tr\lceil B\checkmark)}$$

This rule states that any properties (S and T) on the order of messages guaranteed by some component of the system (the processors P and Q for instance) must still be true of the whole system;
and its corollary

$$\frac{P \textbf{ sat } S(tr)}{P\,{}_A\|_B\, Q \textbf{ sat } S(tr\lceil A\checkmark)}$$

any constraint imposed by a component is a constraint upon the whole system.

Before a proof is begun it is useful to introduce functions such as '<u>before</u>' to express the ordering of messages. A message 'a' is defined as being before message 'b' when

$$a \;\underline{before}\; b \;\cong\; tr = tr_0^\frown\langle b\rangle^\frown tr_1 \;\Rightarrow\; \langle a\rangle \;\underline{in}\; tr_0$$

We also obtain an inference rule expressing the transitive behaviour of the '<u>before</u>' function.

$$a \;\underline{before}\; b\;(tr) \;\wedge\; b \;\underline{before}\; c\;(tr) \;\Rightarrow\; a \;\underline{before}\; c\;(tr)$$

tr describes the *trace* of a communications channel, the sequence of messages which have been transmitted at any one time. This trace is simply represented as a sequence.

It is common too for communication events to be specified in an algebraic format. For example, $i.COM.m$ represents processor i inputting message m to the communications channel COM. The messages may have their own format consisting of headers, content, checksums etc. However, at this stage it is not important to consider all of these details, rather, the sequence of messages is more important than their actual content. The symbol $*$ is used to represent all possibilities, for example:

$$*.a.m = \{i.a.m \mid i.a.m \in \Sigma\}$$

Σ represents the set of all possible communication events.
Such rules are also used in the specification of the communications medium.

The specification of the communications medium

There may exist many different requirements for a communications medium, each of these requirements can be specified in turn. Consider the simple requirement which states that any message which comes out of the communications medium, at a processor, must have previously been input to the medium from some other processor:

$$tr = tr_1 \frown \langle i.OUTPUT.m\rangle \frown tr_2 \Rightarrow$$
$$\exists\, k \bullet \langle k.INPUT.m\rangle \in tr_1$$

or

$$k.INPUT.m \;\underline{before}\; i.OUTPUT.m$$

If message m is output to a processor i, then there must be some processor k at which m was previously input.

This may seem to state the obvious and indeed it does, though with this type of definition it is possible to produce unambiguous answers to any questions which may arise in the requirements analysis. The above specification does allow one to assume that the communications medium will not output spurious messages for instance.

Modelling the system requirements

Using the CSP approach exactly what messages will pass across the communications channel can be modelled. For instance the messages output at KBA i labelled as originating at KBA j must appear in the same order in which they are sent to, or generated by, KBA j

$$i.OUT.j \;\leq\; j.IN$$

The second system requirement ensures throughput of messages via the communications system. It specifies that there will be no more than M messages which have been input but not yet output.

$$\#IN - \#i.OUT \;\leq\; M$$

Mapping the requirements onto the processor architecture

<u>Communication link 1 : Unrestricted Simplex</u>

Consider an initial example of a communication link in which data can flow between KBAs in one direction only (unrestricted simplex). There are a number of simple assumptions that must be made to support the specification and which in turn govern the suitability of this communications link to a particular communication problem; The assumption that both the transmitting and receiving processors are always in a ready state is made, and that they are able to store any number of messages at a single time. The simplification that the processing time can initially be ignored is also taken, and that the communications channel is reliable, that is, information is never corrupted or lost.

Two distinct processors, a sender and a receiver, are modelled. The sender KBA loops continuously, accepting messages from the local controller, formatting the messages into an outbound frame, and passing the resulting message to the communications channels. A ? is used as an infix operator to denote input to the RHS expression, similarly the infix ! operator denotes output to the LHS expression.
A single execution of the sender can be defined by:

$$S1 = (FromLocalController?buffer \rightarrow ToCommunicationChannel!Format(buffer) \rightarrow SKIP)$$

where *SKIP* is a process that always terminates successfully, and the function *Format* formats an outbound communication. The sender process may then be defined as continuous executions of the process *S1*, i.e.

$$SENDER1 \mathrel{\hat{=}} S1 \,;\, SENDER1$$

The receiver KBA can be modelled as simply and may be defined similarly; it receives messages (assumed to be undamaged) from the communications channel, strips the data part from the message (using the function *Info*), and passes it on to the local controller. It then loops, waiting to receive another message.

$$R1 = (FromCommunicationChannel?r \rightarrow ToLocalController!Info(r) \rightarrow SKIP)$$

$$RECEIVER1 \mathrel{\hat{=}} R1 \,;\, RECEIVER1$$

The communications link is then mathematically defined as the parallel composition of a SENDER1 and a RECEIVER1 process. The two agents are labelled i and j, and the protocol is defined as:

$$COMMLINK1 \mathrel{\hat{=}} i : SENDER1 \parallel j : RECEIVER1$$

Communications link 2 : Stop-and-Wait

Consider now how a slightly more complicated communications link based on the Stop-and-Wait message passing principal.

The processes *S1* and *R1* need not have been defined as above, but could have modelled the sender and receiver KBAs directly using explicit recursion. However, by separating out the definitions they may be reused in a more realistic communications link.

The assumption above, that the receiving KBAs can process incoming messages infinitely quickly, can now be dropped. In doing this, the specification must now ensure that the sending KBA does not flood the receiving KBA with incoming messages, by sending messages faster than the receiver can process them.

A general solution in such a case is to have the receiving KBA provide feedback to the sending KBA. Once the receiver passes a message to the local controller, it sends a dummy message back to the sending KBA, indicating that it is ready to receive more data. The sending processor will behave as before, but now must wait for an acknowledgement from the receiving KBA before it accepts another message for transmission:

$$SENDER2 \mathrel{\hat{=}} S1[ToComChannelS/ToComChannel]; (FromLocalControllerS?goAhead \rightarrow SENDER2)$$

$$RECEIVER2 \mathrel{\hat{=}} R1[FromComChannelR/FromComChannel]; (ToLocalControllerR!goAhead \rightarrow RECEIVER2)$$

ComChannel is used instead of *CommunicationChannel* in the above expressions for brevity. In order to produce verifiable specifications in CSP, channels are modelled to operate in one direction only. A second channel is added for transmission of the acknowledgement, though in an implementation a single physical link may be used; this does not effect the correctness of the design. To avoid confusion, an "S" or an "R" is added to the CSP channel names to indicate the sending or receiving KBA, the definitions of *SENDER2* and *RECEIVER2* are suitably renamed.
As before, the communications link is defined as the parallel composition of the sender and receiver KBAs:

$$COMMLINK2 \mathrel{\hat{=}} i : SENDER2 \parallel j : RECEIVER2$$

The synchronisation between the sender and receiver KBAs is provided implicitly by the hand-shaking on the communication channels used; the specification does not need to use additional wait functions.

Communication link 3 : Positive Acknowledgement and Retransmission

In the previous section the assumption that the communications medium was completely reliable was adopted. In reality a communications channel may be used between KBAs where messages and acknowledgements are often lost, or corrupted requiring retransmission. In order to specify a safe system in such a case, there must be some way for the receiving KBA to distinguish whether it is seeing a message for the first time, or whether it is a retransmitted message.

The simple way to achieve this is to have the sender put a sequence number in the header of each message it sends. The receiver can then inspect the frame to determine whether it is a new frame or a duplicate which may be discarded.

As before, the model of the next communications link involves the sending KBA waiting for a positive acknowledgement from the receiver before transmitting the next data item. For this reason, a model using a one-bit sequence number is sufficient.

The sender records the sequence number of the next message to be sent in *NFTS*. Similarly, the receiver records the sequence number of the next message it expects to receive in *FE*.

As before, the sender receives a message from the local controller, formats this information, and passes it on to the communications channels. It then waits on the arrival of a corresponding acknowledgement message.

If after a certain length of time, the sending KBA has not received an acknowledgement, then it assumes that the original message never arrived, or that the acknowledgement was lost, and therefore transmits a duplicate copy of the message. However, the time interval that the sender waits must be such as to allow for the propagation of the original message to the receiver, for the receiver to process that message and for an acknowledgement to be transmitted back. This time interval may be crucial in the system, by specifying such a trait it allows the system to be proved correct before the details are mapped onto a physical implementation.

If an acknowledgement is received, then the sender increments the sequence number *NFTS* using the modulo 2 addition operator *2Inc*, ready to transmit the next message. When the receiving processor receives a message, it checks its sequence number. If this is the same as *FE*, then an acknowledgement is sent. Alternatively, if the message is a duplicate, then it is ignored and not passed to the local controller.

The new specification is merely expansions of those which were addressed previously. Now, however, the record sequence numbers are recorded in process subscripts.

The communication link is then defined as the parallel composition of labelled *SENDER3* and *RECEIVER3* processes, with *FE* and *NFTS* both set to 0:

$$COMMLINK3 \mathrel{\widehat{=}} i : SENDER3_0 \parallel j : RECEIVER3_0$$

Verification of communication link specifications to system specification

It can be proven mathematically that the specification of the communication links meet the system specification seen earlier, that the system implemented is 'safe' with respect to the requirements imposed. The proof can take a similar incremental approach as the models of the communication links themselves.

The verification, for the Unrestricted Simplex link for example, is simple: *SENDER1* receives messages on *FromLocalController* and transmits these to its output channel *ToCommunicationChannel*. As a result, at any instance, the trace of messages on *ToCommunicationChannel* will be a prefix of those on *FromLocalController*. In addition, the *SENDER1* process cannot input any further messages until it has output the current message; as a result, there can never by more than one message more on *FromLocalController*:

(*1*) $ToCommunicationChannel \leq^1 FromLocalController$

Similarly, the *RECEIVER1* process inputs from *FromCommunicationChannel*, and must output to *ToLocalController* before it inputs again:

(*2*) $ToLocalController \leq^1 FromCommunicationChannel$

In the first instance the assumption was made that the transmission medium was such that there was never any corruption, messages were never lost, and consequently that all messages would arrive in the order in which they were sent. This implies that:

(*3*)
$FromCommunicationChannel \leq ToCommunicationChannel$

From (*1*) to (*3*), above:

$$ToLocalController \leq^1 FromCommunicationChannel$$
$$\leq ToCommunicationChannel \leq^1 FromLocalController$$
$$\Rightarrow ToLocalController \leq FromLocalController$$

This result means that the sequence of messages passed on to the local controller at the receiving KBA will always be a prefix of those received from the local controller at the sending KBA. Since the assumption of infinite buffering space and that processing time is negligible has been made, the number of elements by which the traces differ is not important.

Thus, it has been proven that the first communications link model meets the system specification of the communications link between these two KBAs. The mapping between this specification model and a physical implementation is then trivial; the reader will be aware that the communication link models were in fact manufactured physical networking communication models. The specification is refined directly to a hardware implementation and in doing so the transfer of the requirements identified at the level of the Requirements Model and the Architecture Model to a particularly suited communications implementation is made simple.

$$SENDER3_{NFTS} = (FromLocalController?buffer \rightarrow ToCommunicationChannelS!Format(NFTS, buffer) \rightarrow WOA_{NFTS,buffer})$$

$$WOA_{NFTS,buffer} = (ACK-MESSAGE_{NFTS} \triangleright^{t_0} RETRANSMIT_{NFTS,buffer})$$

$$ACK-MESSAGE_{NFTS} = (FromCommunicationChannelS?goAhead \rightarrow SENDER3_{inc(NFTS)})$$

$$RETRANSMIT_{NFTS,buffer} = (ToCommunicationChannelS!Frame(NFTS, buffer) \rightarrow WOA_{NFTS,buffer})$$

$$RECEIVER3_{FE} = (FromCommunicationChannelR?r \rightarrow RECEIVE_{r,FE})$$

$$RECEIVE_{r,FE} = \begin{array}{l} (ToLocalController!Info(r) \rightarrow \\ ToCommunicationChannelR!goAhead \rightarrow RECEIVER3_{2Inc(FE)}), \\ \text{if } seq(r) = FE \\ (ToCommunicationChannelR!goAhead \rightarrow RECEIVER3_{FE}), \\ \text{if } seq(r) \neq FE \end{array}$$

Figure 2: Specification of a Positive Acknowledgement and Retransmission protocol

5 CONCLUSION

The Consensus method fills a gap in the current field of software engineering, providing a comprehensive and integrated method for the development of large agent-based systems. The use of this method in the development of an Air Traffic Control application has shown that agent-based systems can be successfully applied to systems which are relevant to current needs. The method covers the whole process from the statement of the users need, to a detailed specification and through to implementation. The gradual progression from basic user requirements to a more refined specification with the help of a diagrammatic approach allows the customer to verify their requirements. The direct mapping from detailed requirements to system architecture and process specification also facilitates validation. There is a strong emphasis on sound software engineering principles in the system development process.

The application of Air Traffic Control is exemplary for a range of possible applications in the domain of aerospace and the experience gained by the project shows that Agent-Based Systems have considerable potential to provide solutions in this area. There are a number of reasons for using distributed systems of this kind, one of which is the fact that it is difficult to conceive using mainframe computers, say, on board an air- or space-craft. Given high processing requirements and the need to keep weight to a minimum, the use of a number of networked processors would therefore appear to be much more appealing. Considering that systems on board of an aircraft, for example, have a number of distinct functions to fulfil, such as (amongst others) manoeuvring the craft, life-support systems and communication with ground-based systems, it seems sensible to divide a system to combine these functions into entirely separate sub-systems which could run on different processors and communicate with each other as and when necessary. This could be combined with a separate control system to supervise the proper functioning of these separate sub-systems, and which would take appropriate action when problems arise. Given the need to avoid malfunctioning at all cost, such a distributed system could also be made to be more resilient than a monolithic system as the malfunctioning of a sub-system would leave the other sub-systems still operative and enable the control system to take corrective action to re-establish full functionality. In addition, given stringent real-time requirements the lack of performance of one sub-system need not necessarily affect the performance of other sub-systems.

The development of distributed systems is dependent on the interconnection between cooperating components. High-performance and reliability in the system can be engineered with the mathematical analysis of these communication links and the adoption of FDTs such as those proposed in this paper. It is essential in the specification of safety critical systems to be able to formally demonstrate the correct implementation of the user requirments. In doing so situations in which deadlock and livelock occurring in our communications are avoided. Multi-processing requires us to consider two tasks; the first is the decomposition of a problem onto a number of smaller agents, the second is the communication between these agents, effective systems will be built with both these in mind.

Acknowledgements

We would like to acknowledge the work of M. G. Hinchey of the University of Cambridge, United Kingdom, for his dedicated and enthusiastic approach to the applications of CSP and for the use of joint research which he and Stephen Jarvis had previously done together.

References

[Ack92] A Anjewierden, B Wielinga & N Shadbolt: *Supporting Knowledge Acquisition: The AC-Knowledge Project*, ESPRIT-92 Knowledge Engineering

[Boe88] B W Boehm, *A Spiral Model of Software Development and Enhancement*, IEEE, May 1988

[Bok94] A Bokma Et al., *The Consensus Method, Final Methodology Report D17*, Technical Report School of Computing and Information Systems, University of Sunderland, UK, February 1995

[Cra89] I D Craig, *The Cassandra Architecture, Distributed Control in a Blackboard System*. Ellis Horwood, 1987, ISBN 0-7458-0579-5

[Fje92] R Fjellheim, T Pettersen & B Chriastoffersen, *REAKT Application Methodology Overview*, ESPRIT-IIP5146 REAKT, Computas Expert Systems A.S, October 1992, CX Doc. CXN92125

[Hat88] D Hatley & I Pirbhai, *Strategies for Real-time System Specification*, Dorset House, 1988

[HiJ93] M G Hinchey & S A Jarvis, *An Incremental Approach to Network Specification and Verification*, In Proceedings SVNC'93, 4th Silicon Valley Networking Conference, Santa Clara, CA, 13-16 April 1993, pp 123-132, Maple Press, 1993

[HiJ95] M G Hinchey & S A Jarvis, *Concurrent Systems: Formal Development in CSP*, McGraw-Hill International Series in Software Engineering, 1995

[Ira93] *Integrated Computer Aided Reliability Analysis and Real-time Fault Location Using Component and System Modelling (IRAS)*, BE4250, Brite-EuRam II Synopses of Current Projects 1993, Commission of European Communities ISBN 92-826-4454-5

[Jar92] S A Jarvis, *Verification of a reliable network protocol*, Oxford University Computing Laboratory, Programming Research Group, TR-2-93, September 1992.

[Mir93] *Machine management for Increasing Reliability, Availability and Maintainability (MIRAM)*, BE4302, Brite-EuRam II Synopses of Current Projects 1993, Commission of European Communities ISBN 92-826-4454-5

[Roy70] W Royce, *Managing the Development of Large Software Systems: Concepts and Techniques*, Proceedings of WestCon, August 1970, (Chapter 3)

[Sta87] *The STARTS Guide to methods and software tools for the construction of large real-time systems*, NCC Publications, Hobbs Southampton, 1987

[Tay89] R M Taylor (Ed.), *System Evolution - Principles and Methods*, KBS Centre of Touche Ross Management Consultants, London, December 1989

FETAL HEALTH DECISION SUPPORT SYSTEM

John Falconer *, Peter Summons % and Hernan Villanueva #
* Dept of Reproductive Medicine, University of Newcastle,
E-Mail: johnfal@minsky.newcastle.edu.au
% Dept of Management, Information Systems Group, University of Newcastle,
E-Mail: mgpfs@cc.newcastle.edu.au
Dept of Electrical and Computer Engineering, University of Newcastle,
E-Mail: hev@ee.newcastle.edu.au
Callaghan, NSW, Australia, 2308.

ABSTRACT

A diagnostic instrument to report on the well-being of a fetus in utero is being developed at the University of Newcastle. The major goal of the instrument is to give an index to clinicians of any potential disease and extend present instrumentation by giving a dynamic index which can show any progressive decline in well-being of the fetus. This may contribute to the clinicians decision as to whether early delivery of the fetus or leaving it in utero enables the best chance of survival and reduces the risk of morbidity. The instrument incorporates three main components: an Expert System that forms the heart of a Decision Support System for the clinician, software modules that take input from fetal Heart Rate recordings and Colour Doppler Ultrasound raw data and furnish processed results to the Decision Support System, and finally a database that stores current results from the software modules regarding the fetus of interest and historical data about the fetus and its parents. This paper gives a review of the overall design of the instrument and the current progress.

Although low birthweight babies are a relatively small percentage of deliveries (<1%) they account for the majority of the neonatal mortality and morbidity. Despite improvements in antenatal care in the developed countries of the world prenatal mortality is still 12.8/1000 (and morbidity is higher since the survival of very low birthweight babies has improved). Our current instrumentation is able to detect major fetal disorders (such as growth retardation or neurological defects) but is very poor for monitoring the short-term progress of babies (for instance during labor). Improvements in prenatal health care are also of great economic benefit due to the high cost of neonatal intensive care.

The Fetal Health Decision Support System (FHDSS) is based upon cooperative work from several departments of the University of Newcastle including Electrical and Computer Engineering, Management and Reproductive Medicine. Fetal data is being obtained from the John Hunter hospital in the form of ultrasound recordings and recordings of the Fetal Heart Rate.

A major aim of the research is to produce a single tool that is capable of giving rule-based decisions regarding the well-being of a fetus and predictions of its future health using non-intrusive measurements of fetal data as well as historical data regarding the fetus and parents. The tool will be a Fetal Health Decision Support System (FHDSS) that can be used by clinicians when dealing with a mother and fetus, or by educators training clinicians. The tool comprises an Expert System, an external database and software modules pertaining to processing of input data. The software modules will be directed towards pre-processing of data to the Expert System, eg image enhancement and processing of Ultrasound data, calculation of a correlation dimension metric from Fetal Heart Rate (FHR) recordings, morphological parameter extraction such as fetal head diameter based upon estimation of measurements from Ultrasound data and calculation of derived parameters such as fetal weight based upon these estimates. It will utilise a knowledge base from an Expert System and will allow interaction between the clinician and the system. This may take the form of a "what-if" situation where a simulator provides a scenario of fetal health for various introduced pathological conditions.

Fetal parameters, in isolation, have not yet been proven as positive indicators of fetal outcome. Another major aim of the research is to identify multi-variable indicators or combinations of parameters that may indicate the state of well-being of a fetus. These fetal parameters should be easily obtained and, if possible, be available via non-intrusive means on the fetus.

A schematic of the system is shown in Figure 1. The dashed lines indicate components of the system that are currently being changed.

INPUTS TO THE SYSTEM

Fetal Heart Rate (FHR) is one of the biometric parameters being obtained at present. It is non-intrusive on the fetus and can be routinely measured in examinations of the mother and fetus. Even though it has been widely accepted as a gross indicator of fetal health for at least 100 years, it has not proved a reliable measure when used in isolation from other factors and as yet no study has proven that an improvement in the outcome of a fetus can be based on predictions from FHR recordings. We suggest that the FHR needs to be used in conjunction with other metrics for an overall prediction of fetal health. A study being conducted into the analysis of the FHR with chaos techniques suggests that indexes such as the correlation dimension can give single numeric values which correlate with fetal well-being and may be used a summary metric to input to the FHDSS as a fetal health predictor metric [BVF94] [BDG+94].

A Doppler Colour Ultrasound unit is used to provide fetal data. This data is processed to generate metrics for the fetus that may ultimately provide a fetal model. The data is obtained "raw", ie it is obtained before it has been filtered or processed by the ultrasound system. This approach ensures that the data to our system is not affected by any FFT processing that the ultrasound machine may use, which might result in limitations on metrics derived by the software modules of the FHDSS. It also allows us to store and reconstruct data for later processing that may be required as the system evolves. It does, however, impose an initial consideration as to the selection of the database used for the development of the FHDSS - that it be capable of storing large amounts of image data in record format. This may not necessarily a constraint for the final FHDSS database.

EXPERT SYSTEM AND KNOWLEDGE BASE

Gensym G2 was chosen as the Expert System for development of the knowledge base for the FHDSS. G2 is a powerful hybrid expert system, combining rule-based and object-oriented methods for knowledge representation [Zah93]. Although its primary roles to date have been in real-time manufacturing and production systems, it possesses a simple graphical interface for the user and offers the capacity to interface to external C or FORTRAN software modules and to interact with external databases. An internal simulator also allows modelling of discrete and continuous event systems. G2 was also currently available, and supportable, to the University of Newcastle. The FHDSS is being implemented on a Digital (DEC) Alpha System running under an OSF platform.

Hilal and Soltan [HS93] suggest that a prototyping style of development methodology would be the most appropriate for our system. This is implemented within the overall framework of the Knowledge Analysis and Decision Support (KADS) methodology, which has been demonstrated as a practical strategy for the design of Knowledge Based Systems [Hic+89].

During the development period, the knowledge base will be sparse and will contain details of "standards" regarding "normal" fetus development. This will allow design and testing using a prototyping strategy for initial cases.

Individual parameters regarding fetal state can be tested in a "static" sense, eg a head diameter of the fetus under study would be compared to a charted range of values indicating a "normal" fetal head diameter. This allows simple interface design for the clinician and validation of out-of-range situations. Variations will then be added, eg cases where the fetus is a part ofa twin pair - which alters the accepted "normal" range of values for an individual fetus. Once selected independent parameters are working within the scope of the FHDSS, combinations between the individual parameters will be studied.

An inherent problem associated with the application area is the uncertainty of the figure for the chronological development of the fetus. This leads to difficulties in matching the fetal subject parameters against established class parameters for fetal development.

One of the potential problems in any medical decision system is that it may lead to legal liability. This would be especially true in a closed-loop control system. Our system is not closed-loop, except as far as the clinician provides any feedback path. The system presents information to the clinician who is free to accept or reject any of the indications from the FHDSS (completely or with modification), ie the system under development offers only a service to the clinician user.

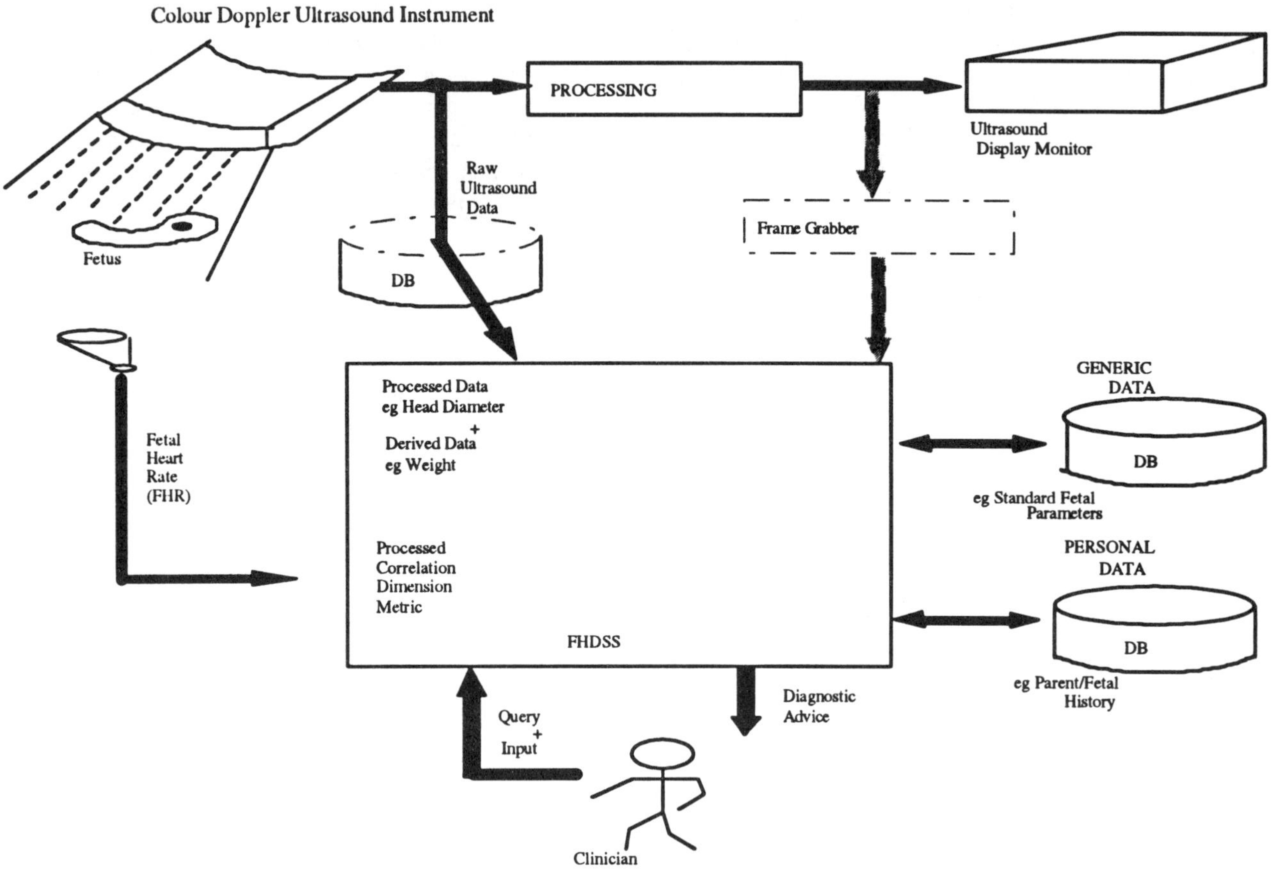

FIGURE 1. Fetal Health Decision Support System

Krause et al [KFO+93] indicate potential legal liabilities for the implementation of a Medical Decision Support System. They state that most of the liabilities mentioned are able to be avoided by full and clear detail in the specifications and limits of the scope of the system. They outline that a liability that may arise from its future use is: "an expert system furnishes incorrect information...". This may arise from a number of factors, and should be considered in a DSS design. One of these factors is stated as: "correct data has been presented to the user, but in a misleading format;...". Given that the design of the final interface will have to be implemented in a normal systems fashion, involving the designer and potential user group interacting with an evolving prototype, an advantage of the Expert System chosen is that it has a simple, easily understood user interface that is capable of presenting results graphically.. A shortfall is that G2 has no easy mechanism for reporting on the basis for its decisions, ie reporting the rules and search path that finally produce a given decision. This may not be relevant to the potential user group, but must be clearly stated in its specifications.

Another advantage of the expert system is that fuzzy logic rules can be incorporated to allow for situations where degrees of confidence are placed on a decision.

SOFTWARE MODULES

The software modules form data collection and processing functions for the Expert System. At the time of writing, few modules have been implemented. One module processes FHR recordings and generates a correlation dimension metric relating to fetal well-being. Another module processes image data obtained from the ultrasound machine. The module displays the ultrasound image on the alpha monitor and provides edge detection on selected regions. The focus of this module is currently on a selected heart chamber. From this module, we can currently estimate an image of the chamber and will later derive methods to extract direct and calculated metrics such as diameter and volume from the image data. At present, the data for this module is obtained by a frame grabber in the alpha machine, ie after it has undergone processing (including FFT) within the ultrasound machine. The next stage of development will take data that has not undergone processing by the ultrasound instrument and port it directly to the alpha system module. A hardware/software interface has been implemented to accomplish the transfer and display of ultrasound wand data directly to an IBM PC, and an interface has been designed for the alpha, but not implemented at the time of writing. Preliminary software has been tested to generate a 3 Dimensional image of the fetus from the ultrasound images. This module may, in the future, provide the basis for part of the main output data display for the system.

DATABASE

As yet no database has been chosen for development. Investigations are being conducted to find a database that can run under OSF (such as Postgress) to confirm whether they can interface easily to G2. Questions as to the suitability of an Object-Oriented database or a Relational database as being the most appropriate for the final system are still unanswered at the time of writing.

The database will be required to store details of previous metrics for the fetus as well as historical data regarding the parents of the fetus. Details of fetal history would include items from the software modules as well as details derived from the Expert System on previous examinations. The history of fetal measurements allow chronological trends of the fetus under consideration, such as estimated weight or head diameter, to be charted and immediately available to the clinician. This historical fetal data may also be used as an indicator or predictor as to what is "expected" under "normal" conditions, for the current measurement, using the simulation capacity of G2. This would then be compared to the current readings to notify the clinician of abnormal deviations. Details of parental history may also serve as indicators or "flags" to the clinician that could modify the interpretation of fetal deviations, eg if the mother was a diabetic or drug user, fetal measurements would be interpreted within this context and not modelled on the assumption of a "normal" pregnancy.

The database should initially be capable of storing image data captured from ultrasound. The clinician should have the ability of data replay and identification of scenes of interest, allowing the possibility of enhancement or further processing of these regions of interest.

SYSTEM OUTPUTS

The outputs from the FHDSS should be clear and quickly understood. Ideally graphs or charts will indicate some of the outputs. The expert system allows for layering of icons or schematics so that the clinician can explode a schematic symbol to a lower layer containing greater levels of detail. This gives the clinician the capacity to quickly and easily examine specific measurements, derived metrics or decisions to the level of detail appropriate.

CONCLUSION

The FHDSS has been outlined with discussion of its present state of development and some factors influencing the expectation and choice of components.

REFERENCES

[BDG+94] Borserio, D., Duggan, L., Giles, W., Villanueva, H & Falconer, J. Correlation Dimension - a Useful Technique for Analysis of Fetal Heart Rate. *Proc. Soc. Study. Fetal Physiology* 21, A4, 1994.

[BVF94] Borserio, D., Villanueva, H, Falconer, J. Chaotic Analysis of Fetal Heart Rate Variability IEEE Engineering, Medical & Biology International Conference, Maryland, USA. Nov, 1994.

[Hic+89] Hickman, F.R. et al. *Analysis For Knowledge-Based Systems: A Practical Guide to the KADS Methodology* John Wiley & Sons, New York, 1989.

[HS93] Hilal, D,K., Soltan, H. Towards a Comprehensive Methodology for KBS Development *Expert Systems*, pp75-91, Vol 10, No 2, May, 1993.

[KFO+93] Krause, P., Fox, J., O'Neil, M., Glowinski, A. Can we Formally Specify a Medical Decision Support System? *IEEE Expert*, pp56-61, June, 1993.

[Zah93] Zahedi, F. *Intelligent Systems for Business: Expert Systems with Neural Networks* Wadsworth Publishing Company, Belmont, California, 1993.

USE OF COMMONKADS IN THE DEVELOPMENT OF KNOWLEDGE BASED SYSTEMS FOR THE UK NAVAL NUCLEAR INDUSTRY

Colin Cadas and Ken Burns
Rolls-Royce and Associates Ltd
PO Box 31
Derby UK DE24 8BJ

ABSTRACT

This paper describes work carried out by RRA (Rolls-Royce and Associates) in the production of IKBSs (Intelligent Knowledge Based Systems) which support the operation of the UK Naval fleet of nuclear powered submarines. In particular, the use of the CommonKADS methodology is reported.

IKBS applications have been developed by RRA since 1987. Applications typically have been in the real-time domain in diagnosis, alarm and warning rationalisation, and control. In most cases demonstrators are built which are trialled on one of a number of Naval nuclear power plant simulators. As IKBS technology matures and becomes more acceptable, and IKBS trials on simulated submarine environments show continued success, the reality of submarine manoeuvring rooms complete with operator support systems based on IKBS technology moves closer.

Issues currently being addressed are primarily concerned with software quality, verification and validation, maintainability and reusability. To improve the IKBS software development process, current work to build an IKBS control system follows the CommonKADS methodology. The CommonKADS methodology has been developed under the Esprit programme, a programme funded by the CEC (Commission of European Communities). It is anticipated that adopting a more rigorous approach to IKBS development such as CommonKADS will lead to significant improvements in the areas mentioned above, and subsequently increase customer confidence and user acceptance.

Three real-time IKBSs are described in this paper. The paper also describes in detail one of these applications which has been developed under the CommonKADS methodology. The results and recommendations of using CommonKADS are also described.

INTRODUCTION

Rolls-Royce and Associates Ltd (RRA) has been involved in design, procurement and support of nuclear power plant for the UK Royal Navy for over 30 years.

The work of the company encompasses a broad range of engineering technology, and one aspect of this work is to provide instrumentation for controlling, operating and maintaining the plant. Like most process plant operators, attempts to reduce manpower and de-skill tasks have led to alternative means being sought to provide the necessary level of behaviour and skill. Over the past few years interest in the development and application of expert systems has led to a range of IKBS demonstrators being produced in areas such as fault detection, alarm/warning rationalisation and control. It has been a principal aim of IKBS development to solve the cognitive overload problem, and provide operators with timely, concise and relevant information.

This paper describes three applications developed. In addition, the results of adopting the CommonKADS methodology for IKBS development are described in some detail.

INTELLIGENT KNOWLEDGE BASED SYSTEMS - IKBS

Expert systems, or IKBS, (these two terms are often used interchangeably even though there are subtle distinctions between them), is a rapidly maturing technology now entering its second generation. IKBSs are concerned with the representation and execution of expertise and knowledge, in addition to the procedural code found in more conventional software. They typically combine a number of disciplines including computer science, psychology, engineering and ergonomics. Expert systems gained in popularity during the 1980s and successful examples of first generation systems include MYCIN [BS84] for diagnosis and treatment of blood infections and RI [Mc82] - a system for configuring VAX computers.

First generation expert systems were standalone, off-line systems used interactively in areas like diagnosis and decision support. With the increase of hardware capability and development of IKBS technology, second generation systems are emerging which are more powerful, have substantially improved GUIs (Graphical User Interface) and can reason on-line, in real-time. Several systems operating in real-time process control have now been successfully deployed, e.g. LINKMAN [Dti90]. SURVEX, in addition to being a real-time application of IKBS is also integrated with conventional software systems.

It is not the intention of this paper to give a detailed account on the subject of IKBS, or expert systems. The interested reader is directed to Jackson [Jac86], Waterman [Wat86] and Charniak[ChMc] for further information, although there are several other good introductory texts available.

IKBS DEVELOPMENT METHODS

Since IKBS is a relatively new technology, and one dealing with knowledge, expertise and inference, standard software lifecycle models have always been seen as inappropriate. In the 1980s, concerns were mostly with technical issues such as inference engines, knowledge representation, knowledge elicitation and search space - and there were few real applications. Towards the end of the 1980s, as technical problems were better understood, computing power had increased, and, more importantly, industry had begun to exploit the technology, real applications began to emerge and the focus of research shifted towards verification and validation, and later formal methodology.

Even so, most applications of IKBS today have been developed in an ad hoc way, using rapid prototyping as the principal development method. The resulting systems, including those developed by RRA, suffer from a common set of problems:

(i) Verification and Validation (V&V) is difficult. The expert knowledge is often represented only in the prototype software. This is invariably difficult to: read, trace back to source knowledge and test. In some cases, acceptance of the system can only be achieved through trialling and testing of the prototype. This will test the side effects of the knowledge representation, not the knowledge itself.

(ii) Modification is difficult. An in-depth understanding of the prototype is required before changes to the knowledge representation can be made. Additional documentation is required if the prototype is to be transferred to a third party.

(iii) Each system is a new system. Reusability of development methods and knowledge is difficult between projects - each new IKBS starts again from scratch.

At RRA, we are now adopting structured methods for IKBS development, viz. CommonKADS. Such a structured development methodology offers many advantages over a prototyping approach:

(i) Implementation independent models representing the knowledge and knowledge processes. V&V is possible. By constructing a model of the underlying knowledge structures and problem solving methods, the expert is able to review the knowledge representation.

(ii) Reduced development costs. Elicited knowledge is represented is explicitly modelled and categorised which greatly improves reusability, traceability and maintenance. Further, many mistakes in the knowledge representation are discovered earlier in the development lifecycle.

(iii) Standard development methodology. By adopting an ISO 9001 compliant methodology, customer confidence is improved through improved quality assurance techniques

SURVEX (SURVEILLANCE EXPERT)

In recent years, modern process plants have increased automation and installed surveillance equipment whilst reducing the level of manpower operating and monitoring the plant.

As a result, some of the local watchkeeping tasks have been transferred to control room operators, and the data reduction and warning filtering expertise inherent in local plant operators has been lost, whilst an additional workload has been placed upon operators. The surveillance systems installed to date have therefore been less usable than anticipated. The root cause for this can be traced to the reduced manning philosophy employed in modern process plants. It has been assumed that an increase in automation and installation of surveillance equipment can compensate adequately for this reduction.

Previously, tasks that were carried out routinely, such as local instrumentation logs, were carried out at hourly intervals by watchkeepers in the various machinery spaces. For trained and experienced watchkeepers, this provided trend analysis. In addition, there would be a large degree of subjective monitoring, in which the human senses of vision, sound, touch and smell would be automatically used to draw attention to any piece of equipment not behaving in the expected manner. Deviations would be reported to the control room on an "as necessary and appropriate" basis. This ensured minimum intrusion for operators. Watchkeepers were, therefore, the earliest form of surveillance and warning system.

It is important to realise that the introduction of automatic surveillance systems has transferred some of these local watchkeeping tasks to control room operators. Cycle monitoring times have been reduced and the channel count has increased. Most significant is that the data reduction and warning filtering expertise inherent in local plant operators has been lost. Table 1 compares the surveillance characteristics for the three most recent designs of surveillance systems.

Whilst the first design generation consisted of human expertise in the form of watchkeepers, later generations consisted of simple comparison between measured parameters and set warning levels. Table 1 shows a tenfold increase in the channel count, an increase of 3600 in warning cycle times, a decrease in the functional sophistication and a transfer of operational responsibility. A perusal of the trade journals indicates that these problems are typical of most process plants.

Problems arising from this situation include:

(i) Cognitive overload suffered by operators. This is particularly evident during emergency or transient conditions. The large channel count and re-direction of the surveillance operations task means that operators are presented with large quantities of raw, uninterpreted data which would have previously been filtered by knowledgeable watchkeepers.

(ii) Simplistic surveillance model. Channel warning levels are set for normal operating conditions. This means that during other plant conditions, spurious warnings can be raised. For example, during plant start-up, operations staff are flooded with low temperature and low pressure warnings. So many warnings are raised, that it is impossible to tell whether any really useful information regarding "live" warnings - those relevant to the operational characteristics at the time - is being transmitted. Useful information is therefore simply lost or ignored. Even during normal operations, many channels have acceptable values which vary depending on the operational status of the plant equipment - this lack of dynamic behaviour on channel warnings can also lead to spurious warnings.

(iii) Poor design of user interface. VDU listings are limited to a few warnings with no prioritisation or grouping. The number of parameters which could actually be in warning at any one time is an order of magnitude greater than can actually be viewed (even if an operator could actually absorb so much information). In addition, single screen displays mean that only one display can be viewed at a time and graphical information has poor scaling capabilities for viewing history graphs of channels.

(iv) The design of software based surveillance systems has taken little or no account of the

Table 1 Surveillance Characteristics For Different Design Generations.

PERFORMANCE PARAMETER	DESIGN GENERATION		
	FIRST	SECOND	THIRD
NUMBER OF CHANNELS	250 - 400	500+	2500+
WARNING TIME	1 hour	1 second	1 second
FUNCTIONAL CHARACTERISTICS	Human Expertise	Simple Comparison	Simple Comparison

operators mental model of the plant. The result is that surveillance systems have no idea of how important any particular warning is with respect to any other, so that during a particular plant state, warnings on equipment that may have little relevance to an operator assume the same importance as warnings on equipment of major interest to the operator at the time.

The solution being achieved for submarine power plant is to introduce IKBS into surveillance to replace lost expertise, i.e. to return to a situation in which operators receive small amounts of high quality information rather than large amounts of low quality information. This has employed a user-centred approach to integrate IKBS employing rapid prototyping, with existing surveillance technologies. This approach is particularly important when viewed in the long term, as the channel count is expected to increase further, and the approach is compatible with reduced manning policies whilst maintaining levels of expertise.

A 2 year programme of work has recently completed to develop the IKBS, called SURVEX (SURVeillance EXpert) [CM+93]. SURVEX is implemented using a real-time expert systems software package called G2 [Moo91], marketed by Gensym in the USA. SURVEX is currently undergoing functional and operational trials within the environment of the SMCTD (see later). The IKBS reasons with all incoming plant data, identifies machinery state, and using this as a basis, prioritises and filters the information. This information is presented to the operator via significantly improved HCI (human computer interaction).

Future generations of SURVEX will then show further incremental improvements providing an architecture which will evolve in time incorporating increased amounts of expert knowledge.

ALDACS (AUTOMATIC LEAK DETECTION AND CLASSIFICATION SYSTEM)

Diagnosis of nuclear power plant malfunctions is a very complex task, with much reliance placed on operators' knowledge, training and experience. The task is complicated by plant component interactions, stress, and most importantly the large number of alarms which could overwhelm the operators during an emergency. Plant safety is of paramount importance, and although no serious incidents have occurred, incidents such as that at Three Mile Island reinforce the aim to provide all possible means to ensure safe operation.

Between 1987 and 1991, an expert system called ALDACS [BM90] was developed at RRA to detect and classify leaks. During 1991, a real-time version of ALDACS was produced using G2. Trials were then carried out with ALDACS connected to a submarine simulator facility. This successfully demonstrated that ALDACS could detect leaks faster than human operators.

In 1993, a series of prolonged ALDACS trials were conducted to compare the performance of ALDACS against trainee manoeuvring room operators in realistic operational circumstances. ALDACS performed well by detecting over 90% of the leaks encountered during the trial period. The trials also identified a number of improvements which have been implemented during 1994.

The success of ALDACS, and the development of expert systems technology within RRA, has led to the application of IKBS to other areas including SURVEX. Current development includes integrating ALDACS with SURVEX.

FEED (FEED WATER CONTROL EXPERT SYSTEM)

This system differs from the others in one main respect - the methodology employed in its production. FEED has been developed using CommonKADS - using the Catalyst project described later. The results of the Catalyst experiment are extremely important, because if it fulfils its expectations it will help to facilitate an important objective for all the applications described in this paper - that of implementation onboard a nuclear powered submarine

FEED is a demonstrator IKBS for Steam Generator (SG) feedwater control (described below). The aims of the FEED project are to:

(i) Show improved control beyond the operational power envelope of the existing conventional control system.

(ii) Reduce the number and degree of feedwater regulating valve movements.

(iii) Demonstrate a real-time closed loop control IKBS architecture using the SMCTD.

The feedwater control system is required to maintain a constant mass of water within the SG. An SG is a boiler which converts heat generated by the submarine nuclear reactor into steam used by the steam turbines for propulsion and power generation (see Figure 1). The controller moves the feedwater regulating valve (FRV) to adjust the feedwater flow returning to the SG. Feedwater flow is maintained to balance the output steam flow and the water level against a calculated program level.

As well as showing improved control, FEED is required to operate beyond the operational power envelope of the current controller. Also, FEED is to automate some of the manual feedwater drills and procedures. This is to be achieved using only the limited plant parameters available to the current controller.

CommonKADS has been used to develop a 'current' expertise model of the current controller. A 'future' expertise model has also been developed which describes the proposed FEED control philosophy (see Figure 2). This contains both strategic and application knowledge. The strategic knowledge describes a generic control system which monitors the environment and applies the most appropriate control strategy. The application knowledge describes each control strategy and the selection criteria.

To date, the current expertise model has been implemented using G2 on a SUN IPX and trialled using the SMCTD. This was to prove that the expertise model and G2 was able to achieve control at least as well as the current controller. The future expertise model is now being implemented in G2.

EVALUATION OF IKBS - SMCTD

All the IKBSs described have undergone trialling or evaluation by connecting to a variety of facilities. These include specific submarine simulators as well as a large test plant located in Scotland. More recently, a purpose built facility called the SMCTD (Submarine Machinery Control Technology Demonstrator) has been commissioned. The SMCTD consists of a Gould computer connected via Ethernet and Sun workstation link to a Charles Rivers graphic processor. This provides generic simulation of a generic PWR (Pressurised Water Reactor) reactor plant together with associated propulsion and electrical power generating equipment.

Through the Ethernet link various IKBSs can be demonstrated and trialled. A passive MR (Manoeuvring Room) mock-up is currently available. This is in the process of being semi-automated by the addition of the SURVEX system.

State of the art Virtual Reality software enables the rapid prototyping of machinery areas for real-time interactive geography training, maintenance training and design. Similarly, the SMCTD possesses a facility

Figure1 - Steam Generator Schematic

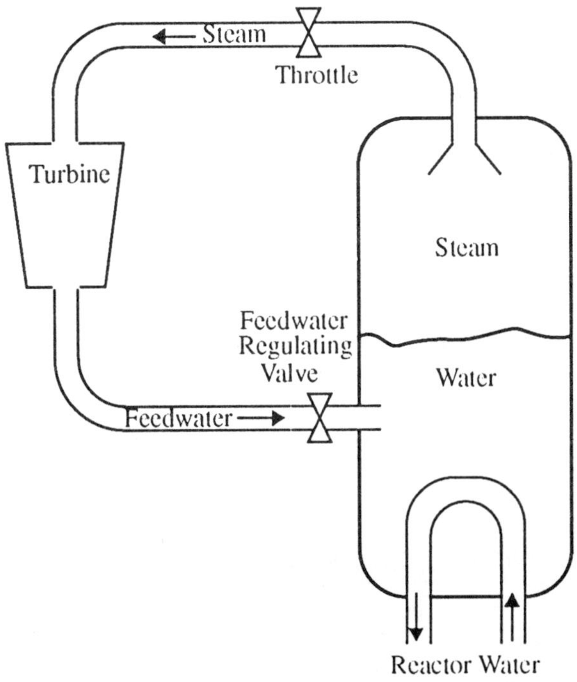

Figure2 - FEED Inference Structure

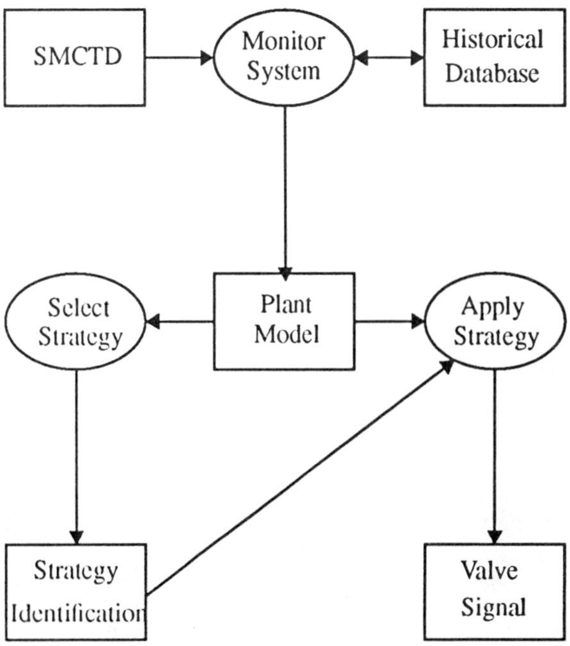

for developing multimedia applications such as interactive panel Intermediate Trainers and Data Retrieval Systems.

As well as the above, the SMCTD is intended as a trials platform for the following:

Assessment of the operability and physical arrangement of workspaces and surfaces using reconfigurable panels allowing design changes at decreased cost.

Cognitive and physical workload assessments for operator tasks under realistic and controlled environments.

Evaluation and confirmation of manning policies and strategies for efficient operation of equipment.

FUTURE FOR IKBS

The development of several demonstrator IKBSs has been described. It is in no-one's interest to leave the demonstrators in their present state. All the IKBSs have shown that, individually, they solve technical problems and perform well. However, prior to installation onboard a submarine, several issues need to be addressed:

(i) Several stand alone IKBSs each occupying physical space and requiring screen output is not realistic.

(ii) V&V, ARM, quality and reusability need to be significantly improved

Current work concentrates heavily on these two issues. To address the first issue, work is ongoing to provide mechanisms by which all the IKBSs can be integrated together, and with existing surveillance and SCADA packages. To address the second issue, the CommonKADS methodology is being adopted. This, it is hoped, will lead to better quality, more maintainable and reusable software. The remainder of this paper describes the results and recommendations of using this methodology.

CATALYST

Catalyst is an ESSI project funded by the CEC (Commission of European Communities). Catalyst aims to introduce and monitor the uptake of the CommonKADS methodology, and associated tool support, into IKBS development.

The Catalyst project will demonstrate the benefits of adopting a structured IKBS development methodology. The uptake of CommonKADS into the existing IKBS development strategy is to be achieved through:

(i) An initial assessment of existing IKBS development practices.

(ii) Training and adoption of CommonKADS.

(iii) Develop the FEED application (see below) using CommonKADS. Monitor the adoption and evaluate the experiences of using CommonKADS.

(iv) Final assessment to objectively asses the changes made to IKBS development practices.

On completion of the Catalyst programme we expect to see improvements in IKBS development and project management. These will be evident from:

(i) Improved maintainability, traceability and readability of design.

(ii) More rigorous and effective V & V.

(iii) Reduced future development costs through knowledge reusability.

(iv) Increased customer confidence and quality assurance.

COMMONKADS

CommonKADS is a framework for IKBS project management, IKBS development and knowledge representation [SW+94].

CommonKADS is a suite of six interrelated abstract modelling tools. Central to the CommonKADS methodology is the Expertise Model which describes the problem solving knowledge to perform a task. It attempts to produce an implementation independent model which is equivalent to the current system or human expert. This is in contrast to conventional AI development techniques which attempt to reproduce the same behaviour.

An expertise model is composed of three layers or categories of knowledge:

(i) Domain layer which describes the concepts and structure of the problem domain.

(ii) Inference layer which describes how to use the domain knowledge.

(iii) Task layer which describes the tasks and task

structure to perform the application goal.

Supporting the CommonKADS model set is a cyclic risk based project management system for IKBS development. The project is broken into a number of Project Management Activity Cycles (PMAC) timed to coincide with project milestones and deliverables. During each PMAC cycle, a process of review, risk, plan and monitor is adopted. This identifies the cycle products, analyses the project and cycle risks and plans mitigating actions to reduce the threat of each risk. Progress against the plan is monitored throughout the cycle.

Tools are available to aid expertise modelling. At RRA, we have adopted the KADSTOOL marketed by ILOG. Use of a tool for expertise modelling is not essential but does offer automated verification, improved traceability and a drawing tool which understands expertise model structure.

RRA EXPERIENCES OF COMMONKADS

One of the most striking impressions when first encountering CommonKADS is the size and complexity of the documentation. This is something of which CommonKADS professionals seem perversely proud. Out of 6 models, only the expertise model (by far the most important model in CommonKADS) has been covered so far. The document supporting the expertise model runs to 760 pages, and in total Rolls-Royce and Associates has collated 2350 pages of documentation. We are assured that the documentation for the remaining models is more concise.

In addition to the size of the documentation is the language. Like most AI technologies, CommonKADS is couched in its own terminology which to the novice can appear difficult or confusing, particularly as certain words have particular meanings

For any organisation on a limited budget, an incremental and pragmatic approach to adopting CommonKADS is recommended. Simple guidelines to CommonKADS in layman's terms would be useful. It is likely that more case studies like this one will lead to increased understanding of the theory and practice of using CommonKADS.

The PMAC has been useful in providing traceability, when a threat to the project materialized. However, this has led to Rolls-Royce and Associates amending the way in which risk analysis is conducted during the PMAC. Generally, the way in which threats are assessed is to allocate each a numeric likelihood and a numeric impact, and multiply the two together. Based on the results, those risks with the highest scores are tackled to reduce the risks. During one cycle, a threat which had registered as having high impact, but low probability, did not have a score high enough to be assessed in detail. When the threat occurred, it was a surprise since it had not been identified as a significant potential problem. Now, in addition to the highest overall scores, those threats having high impacts are reviewed for that cycle, and included as "possible nasties" even if the likelihood is low. This has proved to be a useful approach, and one more likely to lead to success.

Initial expertise model development suffered from a number of problems which were a consequence of an incomplete understanding of the modelling process - i.e. the role of the model and the role of each of the layers. This led to models being discarded due to lack of relevance to the application field. Also, models were discarded because the terminology was interpreted in different ways by different individuals - this led to a failure to agree on an understanding of the model.

CommonKADS comes with a library of reusable inference structures, but since this is incomplete it is largely a matter of luck if one finds something useful. For RRA, the library inference structure for control was non-existent, and indeed the libraries seem poorly equipped for most real-time applications. A recommendation for the CommonKADS community would be to have a framework for expanding the libraries as industry fills in the gaps. The wider issue of modifying CommonKADS to reflect the experiences of industry, this experiment included, is also left to the CommonKADS community to consider.

It is clear that, as applications are developed, particularly in similar areas e.g. diagnosis, or similar domains, then more and more of the models will become reusable. This is much more true than the current situation in which knowledge tends to become embedded in the application. However, to some extent reusability and, more importantly, maintainability, relies on the continued availability of the chosen commercial tool. Should the tool become obsolete, then this will cause reversion to paper models, or reworking the model into another tool. Both reusability and maintainability are as yet untested, but it seems likely that this will be more successful than the current methods.

The use of semantic networks and domain networks, as prescribed by the graphical version of the CML (Conceptual Modelling Language) is considered a useful representation of large amounts of knowledge compared with a simple textual description. However, problems have arisen in assuring that the reader ascribes the meaning that the author intended. Significant effort has been required to reduce the amount of ambiguity that may be contained in the graphical CML.

Some aspects of expertise modelling are not easily understood by experts. This, it is believed, is because there is great emphasis on trying to produce reusable, or generic descriptions from pre-defined role types. It may be better (at least initially) to adopt a more pragmatic model which is more understandable by the expert. Also, the inference layer is particularly difficult to read - there is a strong temptation to read it like a flow chart, which it strongly resembles. There is an implied compromise here between making models readable by CommonKADS experts, and training a novice to be able to read an existing model.

To date, CommonKADS expertise modelling has proved to be a time consuming process. This is in no doubt due in part to the learning process involved. No conclusions can be made at present as to whether CommonKADS expertise modelling will actually increase development costs once CommonKADS is an established working practice.

CONCLUSIONS

A number of real-time IKBSs for the UK Naval Nuclear Industry have been described in the areas of diagnosis, control, accident management and alarm/ warning rationalisation. In addition, the simulator facility for IKBS evaluation has also been described.

The future plans for the IKBSs has been outlined and the remaining problems discussed. The solution to some of these probles lies in the adoption of an accepted methodology. CommonKADS, a European funded methodology has been adopted, and a number of conclusions and recommendations are made surrounding its use and adoption.

ACKNOWLEDGEMENTS

All the work described here relating to the adoption of CommonKADS development has been funded by the CEC, and has made the production of this paper possible.

All work relating to the development of IKBS applications has been funded by the UK MoD. Particular thanks go to Lt Cdr Jon Heley and Lt Cdr Nick Walker for the continued support for the work and publication of this paper.

REFERENCES

[BM90] Brown, D. and Mayfield, T. P. "ALDACS - A Knowledge-Based System to Assist Power Plant Operation" Research and Development in Expert Systems VII Cambridge University Press, ISBN 0 521 40403 7. 1990.

[BS84] Buchanan, B.G. and Shortliffe, E.H. (Eds) "Rule - Based Expert Systems - The MYCIN Experience of The Stanford Heuristic Programming Project" Addison - Wesley Publishing Company 1984.

[CM+93] Cadas, C., Bowskill, J., Mayfield, T. and Clarke, C. "An IKBS Approach To Surveillance For Naval Nuclear Submarine Propulsion". Proceedings of the 2nd ASME/JSME Nuclear Engineering Joint Conference, Volume 2. ISBN 0-7918-0637-5. 1993.

[ChMc] Charniak, E. and McDermott, D. "Introduction To Artificial Intelligence" Addison - Wesley Publishing Company.

[Dti90] DTI "Real - Time Process Control: Improved Efficiency" DTI Booklet ISBN 0 11 515221 0 1990.

[Jac86] Jackson, P. "Introduction To Expert Systems" Addison - Wesley Publishing Company 1986.

[Mc82] McDermott, J. "RI - A Rule - Based Configurer Of Computer Systems" Artificial Intelligence VOL 19 pp39 - 88. 1982.

[Moo91] Moore, R.L. "The G2 Real - Time Expert System" Expert System Applications For The Electric Power Industry. EPRI conference proceedings. SEPT 9-11 1991.

[SW+94] Schreiber, G; Wielinga, B; Hans Akkermans; Walter Van de Veld; Robert de Hoog. "CommonKADS A comprehensive methodology for KBS development" 1994.

[Wat86] Waterman, D.A. "A Guide To Expert Systems" Addison - Wesley Publishing Company 1986.

KNOWLEDGE CONSTRAINTS

John Debenham
Key Centre for Advanced Computing Sciences,
University of Technology, Sydney, PO Box 123, NSW 2007, Australia.
Email: debenham@socs.uts.edu.au

ABSTRACT

Knowledge constraints are proposed as a tool for maintaining the integrity of knowledge-based systems. Knowledge-based systems applications are frequently constructed as a knowledge base, or rule base, combined with a relational database. The most expensive of these components to build and maintain in such a combination is typically the knowledge base. However, constraints, which are an integral feature of database applications, are seldom mentioned in connection with knowledge. One class of knowledge constraints provides the basis of a practical and effective mechanism for protecting a knowledge-based system against update anomalies. Another class of knowledge constraints, coupled with the referential integrity constraint, together provide the foundation for a procedure to substantially automate the maintenance process. This procedure is highly efficient if the system's rules have been normalised. An experimental knowledge-based systems design and maintenance tool based on this approach has been built and trialed in a commercial environment.

INTRODUCTION

A key issue in the management of knowledge-based systems is the effective maintenance of the "knowledge base", or "rule base", component of that system. The relative importance of this issue increases with the degree of "volatility" of the system. *Volatile* systems are systems which are subject to a high rate of change. The derivation of effective techniques for the maintenance of knowledge-based systems in which the knowledge base itself is highly volatile is a major research problem.

A major collaborative research project between the University of Technology, Sydney and the CSIRO Division of Information Technology has addressed the effective maintenance of knowledge-based systems. The early results of this project are summarised in [Deb89]. Recent work in this project has focused on two issues; first, the development of a unified framework for conceptual modelling in which the "data", "information" and "knowledge" in the application can all be represented entirely as "items" in a single formalism [Deb92], and second, the development of classes of constraints for knowledge which can protect the knowledge base effectively against the introduction of update anomalies. Early results on the development of knowledge constraints were reported in [Deb94b]; more recent results are reported here. A key product of this collaborative research project has been the development of a complete methodology for the management of knowledge-based systems. This methodology is supported by a Computer Assisted Knowledge Engineering tool, or CAKE tool, called "The Knowledge Analyst's Assistant". An experimental version of this tool has been constructed [DL90] and has been trialed in a commercial environment.

Constraints play two distinct roles in our approach:

- constraints are attached to the representation of the application with the intention of preventing the introduction of update anomalies, and
- constraints are employed by our representation of the maintenance procedure with the intention of making that procedure more efficient.

Constraints which are attached to the representation of the application with the intention of preventing the introduction of update anomalies are called *pragmatic constraints*. Constraints which are employed by our representation of the maintenance procedure with the intention of making that procedure more efficient are called *referential constraints*.

Pragmatic constraints are of two distinct types:

- constraints which are attached to each item, these are:
  - the item value constraints, and
  - the item set constraints.
- constraints which are attached to the conceptual model itself.

Constraints which are attached to each item are called the *item constraints*. Constraints which are attached to the

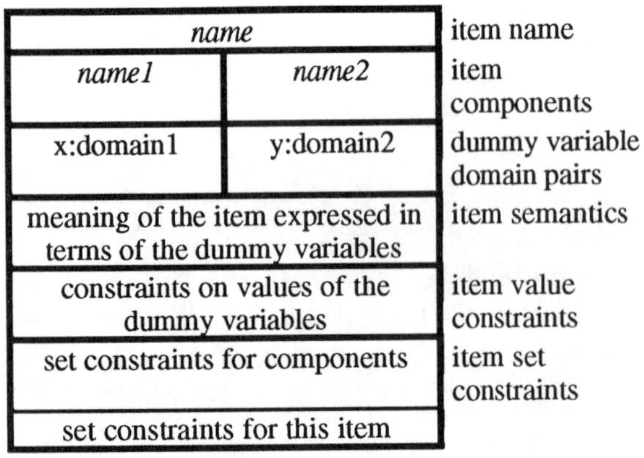

Figure 1 The i-schema for representing items

conceptual model are called the *model constraints*. For the purpose of illustrating our discussion only we will assume that the knowledge in the application is to be implemented using Horn clause logic [Kow91], and that the data and information are to be implemented using a relational database which is tightly coupled with the logic [Gra89], [Ito91], [NH88], [Twi89]. Note that we are not promoting Horn clause logic as a practical language for implementing knowledge. As far as ease of maintenance is concerned, it should not matter what language is used to actually implement knowledge as long as the knowledge has been modelled rigorously and provided that this model of the knowledge is subsequently maintained, and is then used to drive the system maintenance process [CSEM91].

CONCEPTUAL MODELLING

The kernel of our approach to the design of rule-based knowledge systems is the construction of a "conceptual model". The *conceptual model* is a complete, non-functional, implementation-independent description of the application. Our conceptual model may *either* be constructed explicitly by representing each thing in the application *or* implicitly by representing the way in which a representation of each thing in the application can be constructed. The implicit conceptual model prevents the introduction of unnecessary redundancy and is preferable, but the explicit model is easier to describe. We will introduce "items" as a uniform formalism for representing things explicitly in the conceptual model, and will introduce "objects" as "model building" operators which can be used to represent things implicitly. Each item in the explicit conceptual model may be constructed from the implicit conceptual model by applying a sequence of the object operators to simpler items.

Items

Items are a formalism for describing the things in an application and have three important properties: items have a uniform format no matter whether they represent data, information or knowledge things; items incorporate two powerful classes of constraints, and a single rule of normalisation can be specified for items. The "meaning" of an item is called its *semantics*. We either present items graphically as "i-schema" or formally as λ-calculus expressions. The *i-schema* notation for representing items graphically is shown in Figure 1.

For example, an application may contain up to one hundred "parts" which are identified by "part numbers" which are constrained to lie between 1,000 and 9,999. Also, for example, the application may contain "cost-prices" which are identified by positive "dollar-amounts". In which case the data item named *part* could be as shown in Figure 2. Further, the application could contain an association whereby each *part* is associated with a *cost-price*. This association could be subject to the value constraint that parts whose part-number is less that 1,999 will be associated with a cost price of no more than $300. This association could be subject to the set constraint that every part must be in this association and that each part is associated with a unique cost-price. This association could be represented by the information item named *part/cost-price*; this item is also shown in Figure 2. Rules, or knowledge, can also be represented naturally using i-schema. As items inherit all the facts about their components knowledge items are very clumsy and space prevents us from including an example of a knowledge item here. The *part/cost-price* item may be presented in the λ-calculus notation as:

$$
\begin{aligned}
&\textit{part/cost-price}[\\
&\quad \lambda xy\bullet[\text{costs}(x\text{:part-number},\\
&\qquad y\text{:dollar-amount})]\bullet,\\
&\quad \lambda xy\bullet[\ (1000 < x < 9999) \wedge (0 < y) \wedge\\
&\qquad ((x < 1999) \rightarrow (y \le 300))\ ]\bullet,\\
&\quad (\text{Uni}(\textit{part}) \wedge\\
&\qquad \text{Can}(\text{cost-price}, \{\text{part}\}))_{\textit{part/cost-price}} \wedge\\
&\qquad (\text{Card} \le 100)_{\textit{part}}\ ]
\end{aligned}
$$

We will present a single rule of normalisation for items [Deb93b]. From this single rule the five normal forms for relational database may be derived as well as a comprehensive set of new normal forms for knowledge.

Combining Items

We have just introduced items which are capable of representing data, information and knowledge things, [Ito91]. Formally an item is a named triple; in addition a set of representations of physical things, ie a set of labels, is associated with some items. Thus in its formal λ-calculus presentation an item has the form:

$$A(\text{S},\text{V},\text{C})$$

where S is an expression called the *semantics* of A, V is an expression called the *value constraints* of A and C is called the *set constraints* of A.

part
∅
x:part-number
'x' is a part-number
1000<x<9999
≤100

part/cost-price	
part	*cost-price*
x:part-number	y:dollar-amount
'x' is a part-number and 'y' is a dollar-amount and costs(x,y)	
1000<x<9999	0<y
x<1999 → y≤300	
≤100	
∀	
----------------	o

Figure 2 i-schema for the items '*part*' and '*part/cost-price*'

Two items may be combined in the following way which is called the *join* of the two items. Given two items:

$$A(\, S_A, V_A, C_A \,) \text{ and}$$
$$B(\, S_B, V_B, C_B \,)$$

Suppose that S_A has n variables, ie A is an n-adic item; and that S_B has m variables, ie that B is an m-adic item. Some of the components of A and B may be identical. Suppose that we identify k pairs of components of A and B which are identical, where $k \geq 0$. Let E be an ordered set of components where each is one of these identical pairs of components of A and B; note that E may be empty. To ensure that the definition is well defined we insist that the components in the set E occur in the same order as they do as components of A. Suppose the components from item A (or B) which are in the set E are expressed in terms of a total of p variables. Let A^* be an n-adic item which is identical to item A except for the order of its variables; the last p variables in A^* are those variables in A which belong to the components of A in the set E. Let B^* be an m-adic item which is identical to item B except for the order of its variables; the first p variables in B^* are those variables in B which belong to the components of B in the set E. Let π be a permutation which turns the ordered set of variables of A^* into the ordered set of variables of A. Let π' be a permutation which turns the ordered set of variables of B^* into the ordered set of variables of B. Suppose that 'x' is an (n - p)-tuple of free variables, 'y' is a p-tuple of free variables and 'z' is an (m - p)-tuple of free variables. Then the item with name $A \otimes_E B$ is called the *join* of A and B on E and is defined to be the item:

$$(A \otimes_E B)[\lambda xyz\bullet[S_A(\pi(x,y)) \wedge S_B(\pi'(y,z))]\bullet,\ \lambda xyz\bullet[V_A(\pi(x,y)) \wedge V_B(\pi'(y,z))]\bullet,\ C_{A \otimes_E B}\,]$$

where $C_{A \otimes_E B}$ is defined as follows. Suppose that C_A is an expression of the form $c_A \wedge G$ where c is that part of C_A which carries the subscript 'A' and G is that part of C_A which carries subscripts other than 'A'. Likewise suppose that C_B is an expression of the form $d_B \wedge H$. Then:

$$C_{A \otimes_E B} = (c \wedge d)_{A \otimes_E B} \wedge (G \wedge H)$$

Note that the set E is *a* set of identical pairs of components of A and B. In practice E is frequently the set of *all* identical pairs of components of A and B in which case $A \otimes_E B$ is simply written as $A \otimes B$.

Using the rule of composition ⊗, knowledge items, information items and data items may be joined with one another regardless of type.

We say that an item I is *decomposable* into $D = \{I_1, I_2, .., I_n\}$ if

- I_i has non-trivial semantics for all i,
- $I = I_1 \otimes I_2 \otimes \ldots \otimes I_n$, where
- each term in this composition contributes at least one component to I.

If item I is decomposable then it will not necessarily have a unique decomposition. We will see that the concept of decomposition provides the foundation for our theory of normalisation.

Objects

A knowledge item contains embedded within it the full specification of its constituent components, and thus items are unnecessarily clumsy. We now introduce the notion of an "object", [Deb94a], [Gra89]. Objects are model building operators which have four important properties: objects have a uniform format no matter whether they represent data, information or knowledge things; objects incorporate powerful classes of constraints; objects enable items to be built in such a way as to inherit the characteristics of their components, and a single rule of normalisation can be specified for

name		object name
(tuple1)/type1	(tuple2)/type2	tuple type pairs
meaning of the object expressed in terms of the dummy variables		object semantics
constraints on values of the dummy variables		object value constraints
set constraints		object set constraints

Figure 3 The o-schema for representing objects

objects such that an implicit conceptual model built with normal object operators will contain only normal items. As for items, we either present objects graphically as "o-schema" or formally as λ-calculus expressions. The *o-schema* notation for representing objects graphically is shown in Figure 3.

Each object is an operator which turns n items into another item for some value of n. Further, the definition of each object will presume that the set of items to which that object may be applied are of a specific "type". In this context the *type* of an m-adic item is determined both by whether it is a data item, an information item or a knowledge item and by the value of m. The *argument type* of an n-adic object is an n-tuple which specifies the types of the n items to which that object may be applied. Each of the n elements in an argument type will be "free" or "fixed". A *free argument type* is denoted by $\mathbf{X}^n$ and indicates that the object may be applied to any type of n-adic item and thus simply specifies the arity of that item. For example, if an object has argument type ($\mathbf{X}^2$, $\mathbf{X}^2$, $\mathbf{X}^1$) then it may be applied to any 2-adic item, followed by any other 2-adic item which is followed by any 1-adic item. A *fixed argument type* is denoted by $\mathbf{D}^n$ (standing for "data"), $\mathbf{I}^n$ (standing for "information") or $\mathbf{K}^n$ (standing for "knowledge") and indicates that the object may be applied to an n-adic item of the nominated type and thus simply specifies both the arity of each argument and whether that argument should be a data item, an information item or a knowledge item.

It is interesting to note that the formal definition of an object is similar to that of an item. An object is a named, typed triple ***A***[E,F,G], where:

- ***A*** is a unique *object name* that, by convention, is written in bold italic script;
- E is an n-argument typed expression called the *object semantics*;
- F is an n-argument typed expression called the *object value constraints* and
- G is an n-argument typed expression called the *object set constraints*.

The object *value constraints* is an n-argument typed expression which is satisfied by the labels which belong to the value set of any item constructed by applying that object to a set of items. The object *set constraints* is an n-argument typed expression constructed from the

sell
(x)/$\mathbf{D}^1$
'x' is a selling price
x≠0
∅

costs	
(x)/$\mathbf{X}^1$	(y)/$\mathbf{X}^1$
'x' costs 'y'	
x<1999 → y≤300	
∀	
----------------	O

Figure 4 o-schema for objects '***sell***' and '***costs***'

primitives "Card", "Uni" and "Can" which represents structural constraints on that object and will be illustrated in the examples which follow.

Data objects provide a representation of sub-typing. For example, the application could contain a special type of "price" called "selling prices". In which case, this notion could be represented by the data object ***sell*** of argument type ($\mathbf{D}^1$) as shown in Figure 4. Further, the application could also contain an association "costs" whereby some things arc said to "cost" something. This association could be subject to the value constraint that things whose identifying number lies between 1,000 and 1,999 will cost no more that $300. This association could also be subject to the set constraints that all members of the value set of the first argument must be in this association and the value of the first argument functionally determines the value of the second argument. In which case, this association could be represented by the information object ***costs*** of argument type ($\mathbf{X}^1$, $\mathbf{X}^1$) as shown in Figure 4. Consider the item instance of the object ***costs***:

costs(*part, cost-price*)

This information item has already been given the name *part/cost-price*, and is identical to the *part/cost-price* item described above.

Further the application could contain the association that some things are derivable from other things by the application of a single mark-up factor. In which case this association could be represented by the knowledge object ***mark-up-rule*** of argument type ($\mathbf{I}^2$, $\mathbf{I}^2$, $\mathbf{D}^1$) as shown in Figure 5. The use of objects as operators enables us to decompose items into an object operator sequence. For example, consider:

[part/sale-price, part/cost-price, mark-up] =
mark-up-rule(*part/sale-price,*
part/cost-price, mark-up) =
mark-up-rule(***sells-for***(*part,* ***sell***(*price*)),
costs(*part,* ***cost***(*price*)), *mark-up*) [E]

mark-up-rule		
(u,v)/I^2	(x,y)/I^2	(z)/D^1
u = x → v = z × y		
u = x → v > y		
∀	∀	
------------------------------------		o
o	------------------------------------	
---------------	o	---------------

Figure 5 o-schema for object '***mark-up-rule***'

Thus we see how the knowledge item [E] may be represented as the application of a knowledge object to a data item and two information objects each of which in turn are applied to two data items. Note that this has all been achieved in a single framework in which no distinction is made between data, information and knowledge. Also note that objects are capable of representing value and set constraints in a uniform way.

The definition of "object joint" is very similar to the definition of item join given above. Just as for items, the notion of object join provides the basis for the decomposability of objects which in turn provides the foundation for our theory of the normalisation of objects. Thus we may normalise the implicit conceptual model.

Normalisation of Knowledge

A key goal of the normalisation of conventional models of data and information in traditional database technology is to prevent update anomalies [Da86]. Likewise the normalisation of our conceptual model is intended to obtain a model of a knowledge systems application that is actually designed to support the maintenance process by reducing the occurrence of update anomalies [Deb94b], [WKL89].

We now give one principle for normalisation from which we may derive all the classical normal forms for information as well as some new normal forms for information and a comprehensive suite of normal forms for knowledge. This single principal of normalisation applies to all items and objects no matter whether they represent data, information or knowledge things. Thus the problem of defining the concept of "normal" for a knowledge system is dealt with in one simple principle.

> Principle NF. Given a set of items or objects discard those which are decomposable.

A conceptual model of a knowledge systems application is said to be *normal* if the items and objects in it are not decomposable. For simplicity, the following discussion is phrased in terms of items in the explicit conceptual model.

Suppose that item I has the three components A, B and C. It is interesting to consider the different ways in which the item $I = I(C, B, A)$ can be decomposed into two sub-items I_1 and I_2. These different ways are categorised by the different ways in which functional associations are present in I. It is not difficult to show that the different ways in which functional associations are present lead precisely to the classical normal forms:

3NF $I((C, B) \Leftarrow A) = I_2(C \Leftarrow B) \otimes_{\{B\}} I_1(B \Leftarrow A)$

2NF $I(C \Leftarrow (B, A)) = I_2(C \Leftarrow B) \otimes_{\{B\}} I_1(B, A)$

BCNF $I(B \Leftarrow (C, A)) = I_2(C \Leftarrow B) \otimes_{\{B\}} I_1(B, A)$

4NF $I(C, B, A) = I_2(C, B) \otimes_{\{B\}} I_1(B, A)$

5NF $I(C, B, A) = I_3(A, C) \otimes_{\{C\}} I_2(C, B) \otimes_{\{B\}} I_1(B, A)$

It should be stressed that the classical normal forms noted above apply equally well to knowledge as to data or information [Deb92]. Thus we see that the classical normal forms provide a complete characterisation of the ways in which an item of three components may be decomposed.

It is interesting to note that by considering decompositions of items into more than three components we can derive new normal forms. We now consider the decomposition of items of four components. For example:

$$I((D, C) \Leftarrow (B, A)) = I_2(D \Leftarrow (C, B)) \otimes_{\{B, C\}} I_1(C \Leftarrow (B, A))$$

This rule of normalisation is not equivalent to any of the classical forms introduced above. For example, consider the item whose semantics is defined by rule [a] [Kow91]:

part/profit(x, y) ← part/cost-price(x, w),
  part/mark-up-factor(x, u),
  y = w × (u - 1) [a]

Buried within this rule is the sub-rule that "the sale price of a part is the cost-price for that part multiplied by the part's mark up factor". This sub-rule could represented as the semantics of another item:

part/sale-price(x, y) ← part/cost-price(x, z),
  part/mark-up-factor(x, w), y = (z × w) [b]

Now it should be clear that rule [b] is buried within rule [a], and that rule [a] should thus be re-expressed as:

part/profit(x, y) ← part/sale-price(x, z),
part/cost-price(x, w), y = z - w [c]

This is an example of the use of the new rule of normalisation given above.

MAINTENANCE

We establish a *coupling relationship* between two items in the conceptual model if modification of one of these items could require that the other item be checked for correctness if the consistency of the conceptual model is to be preserved. Managing the maintenance of knowledge-based systems is substantially concerned with the management of the complex coupling relationships between the items in the conceptual model. It may be shown that these coupling relationships are of four distinct kinds. First, *duplicate relationships* which are relationships between two items which actually represent the same real fact to some extent. In other words, a real fact has been represented, at least in part, in more than one place. Second, *component relationships* which are relationships which represent the syntactic structure of the application. Third, *equivalence relationships* which are relationships between two items whose semantics are logically equivalent. Fourth, *sub-item relationships* which are relationships between two items one of whose semantics logically implies the other's semantics; it may be shown that sub-item relationships may be reduced to sub-type relationships.

Duplicate relationships

Duplicate relationships are relationships between two items which actually represent the same real fact; these relationships can be eliminated by the application of a systematic analysis method and by normalising the conceptual model. Established database technology includes techniques to ensure that data models are normal in this sense. In general terms, the normalisation of the conceptual model attempts to ensure each real fact is represented in the model in one place and in one place only. [Deb89] describes a comprehensive set of normal forms for knowledge; these new normal forms are quite compatible with the established normal forms for information [Da86]. [Deb92] presents a powerful notion of "normal" in the form of one simple principle from which *all* normal forms, including the normal forms for knowledge, may be derived. It is important to appreciate that if the first form of coupling relationships have been removed during the normalisation process then the rules will all be logically independent.

We have seen that items will be classified as data, information or knowledge. Thus if a fact is represented twice then the items which represent it will either be classified in the same way (as data, information or knowledge) or they will be classified differently. Duplications caused by two items which are classified differently may be prevented by a systematic analysis; duplications caused by two items which are classified in the same way can be removed by normalisation. Normalisation is an expensive process, particularly the normalisation of knowledge items, thus we prescribe that normalisation should only be applied to the highly volatile sections of the knowledge base. Indicative relative costs and benefits in normalising items are shown in Figure 6.

Consider, for example, three different representations of the fact "the mark up rate for pencils is 35%". This fact could be represented by attaching the tuple ('pencil', '35%') to the value set of the information item "*part/mark-up-rate*"; in which case the fact is an information thing. The same fact could also be represented by including the clause:

part/sale-price('pencil', y) ←
part/cost-price('pencil', z), $y = z \times 1.35$

in the semantics of the knowledge item "*[part/sale-price, part/cost-price]*" in which case the fact is a knowledge thing. The same fact could also be represented by attaching the label '35%' to the value set of the data item '*pencil-mark-up-rate*'; in which case the fact is a data thing. Thus we see that the same real fact can be correctly implemented in the data dictionary, the database or in the knowledge base. The danger in this situation is realised when the mark-up rate for pencils should happen to change; in which case, all three representations would have to be correctly altered to preserve consistency. We have argued extensively in [Deb89] that duplications caused by representations of facts which are classified

Data items	Information items	Knowledge items	Normalisation cost	Maintenance cost
Non-normal	Non-normal	Non-normal	Nil	Very high
Normal	Non-normal	Non-normal	Low	High
Normal	Normal	Non-normal	Medium	Medium
Normal	Normal	Normal	High	Low

Figure 6 The costs and benefits of normalisation

differently can be avoided by constructing a conceptual model. If the fact in the above example had been represented as an information item then the formal design superstructure would prevent the explicit representation of "mark-up-rate" as data or knowledge.

Component relationships

The second form of coupling relationship are the component relationships which are representations of the syntactic structure of the application. These component relationships are implemented as "component links" in our CAKE tool "The Knowledge Analyst's Assistant".

For example, in item [E] above there is a component relationship between this knowledge item and the items "*part/sale-price*", "*part/cost-price*" and "*mark-up*". Furthermore there is a component relationship between the information item "*part/sale-price*" and the data items "*part*" and "*sale-price*". These component relationships are shown in Figure 4. To understand the vital role that these component relationships play in the maintenance process suppose that item [E] is implemented by the following clause:

$$\begin{aligned}&\text{part/sale-price}(x, y) \leftarrow\\ &\qquad \text{part/cost-price}(x, z),\\ &\qquad \text{is-the[mark-up-rate]}(w),\\ &\qquad y = z \times w \qquad\qquad [F]\end{aligned}$$

That is suppose that in our implementation, "*part/cost-price*" is represented as a relation named "part/cost-price", and "*mark-up*" is represented as a population named "mark-up", and suppose that "part/cost-price" and "mark-up" are actually stored. Suppose that "*part/sale-price*" is implemented by the relation named "part/sale-price" which is deduced using clause-group [F] as required. Now suppose that we are advised that "wheel nuts for tractors now sell for $9.36". Suppose that 'wheel nuts' have an part number '1470', then the tuple ('1470', '9.36') should now be in the value set of "*part/sale-price*". Then a systematic execution of this maintenance operation might follow the links shown in Figure 4. First we note that there is a discrepancy with the relation "part/sale-price" and check that '1470' is a label associated with the population 'part' and that '9.36' is a label associated with the population 'sale-price'; suppose that they are. Then we check the rule by which this relation is derived, ie group [F]. Suppose that clause [F] is still correct then we check the other component relationships from item [E]; suppose we check the component relationship to item '*part/cost-price*' first. Here we find that the relation part/cost-price contains the tuple ('1470', '6.78'). Suppose that we note that the '6.78' is in fact incorrect and should be '7.80', so we make this change and note that '7.80' is now a member of the 'cost-price' population. Note that throughout the execution of this maintenance operation we have followed the component relationships. However, if we are to be sure that this maintenance operation has been completely effected then there is more to do than to follow the links shown on Figure 7. Note that a new 'cost-price' has been introduced, namely '7.80'; thus all relations in which 'cost-price' is a domain will have to be checked for correctness and then, if any of those relations have to be modified, all clauses in which those relations occur will also have to be checked for correctness. Furthermore, any changes to those clauses will lead in turn to the checking of the correctness of the constituent relations and populations within those clauses, and so on and so on.

A naive approach to knowledge base maintenance is to pursue each and every component relationship from any knowledge, information or data item that has to be modified during the execution of a particular maintenance operation. This naive approach can be very costly and can result in a substantial proportion of the knowledge base being searched and checked for correctness during each maintenance operation. As we will see, a particular class of constraints can be employed to prune this search considerably. Note that the management of the component links is just one of a comprehensive range of functions performed by "The Knowledge Analyst's Assistant".

CONSTRAINTS

Constraints for data and information are well understood and are implemented in relational database managements systems [Da86] as a matter of routine. Further, we advocate the use of constraints on knowledge as a key to the construction of maintainable knowledge-based

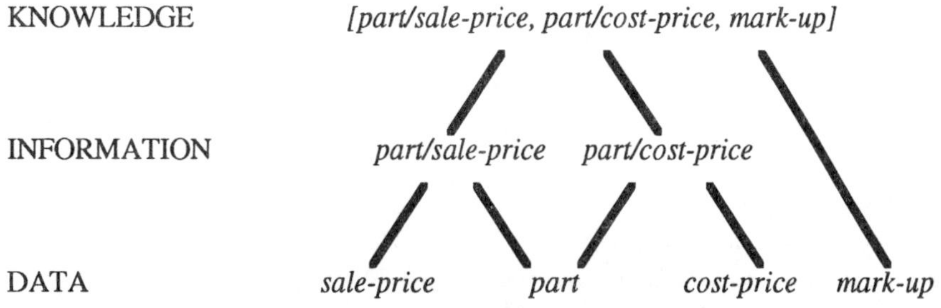

Figure 7 The component relationships for knowledge item [E]

systems particularly when the application is highly volatile. As we will see, we apply constraints to knowledge items and consequently to their implementation as clauses. An important interest of ours is the partial mechanisation of the knowledge base maintenance process, see [DL90] for a description of the CAKE tool "The Knowledge Analyst's Assistant". We will see that constraints are essentially of two kinds. First, "pragmatic" constraints which are constraints specified by the knowledge analyst to preserve the integrity of the knowledge when it is subsequently subjected to maintenance operations. Second, "referential" constraints which are constraints employed by our Knowledge Analyst's Assistant software to prune the search of the knowledge base which would otherwise follow an exhaustive search to ensure the preservation of referential integrity.

Pragmatic Constraints

Pragmatic constraints are attached to the representation of the application with the intention of preventing the introduction of update anomalies. Pragmatic constraints are constraints on the conceptual model, and are thus inherited by the internal model. We identify two important types of pragmatic constraint, the "item constraints" and the "model constraints". The *item constraints* are constraints which are attached to each item, these are the *static item constraints* which are constraints on how an item may be expressed, and *dynamic item constraints* which are constraints on how an item may be modified. The *model constraints* are constraints which are attached to the conceptual model itself.

Item Constraints. Static item constraints are constraints on how a knowledge item may be expressed. The static item constraints are actually contained within each item as the "value constraints" and the "set constraints". For example, consider the item [E], this item could incorporate the value constraint that if [E] is used to calculate the sale-price from cost-price, or vice versa, for any 'part' then the sale-price must be greater than the cost-price. In addition, knowledge item [E] could incorporate the set constraints:

Uni(*part/sale-price*)

which requires that in the value set of [E] all values of the value set of item "*part/sale-price*" must occur. This could be implemented as the requirement that any clause-group implementing [E] which derives the tuples for the predicate "part/sale-price", such as the clause [F], must derive *all* such tuples. Knowledge item [E] could incorporate the set constraint:

Can(*part/sale-price*, {*part/cost-price*, *mark-up*})

which requires that in the value set of [E] the values associated with the components *part/cost-price* and *mark-up* functionally determine the values associated with the component *part/sale-price*. This could be implemented as the requirement that any implementation of [E] should derive *all* the tuples for the predicate "part/sale-price" from the predicates "part/cost-price" and "is-the[mark-up]"; clause [F] is such an implementation.

Dynamic item constraints are constraints on how an item may be modified. An example of a dynamic item constraint on item [E] is the requirement that if item [E] is modified for any reason then the values in its value set associated with the component "*sale-price*" may not vary by more that 10% on the original values.

Model Constraints. Model constraints are constraints on the conceptual model. As an example of a model constraint we describe an "information model constraint". Consider any sets of tuples which could form the value sets of items "*part/sale-price*", "*part/cost-price*" and "*mark-up*"; for example the tuples shown in Figure 5. Note that these tuples should be consistent with the knowledge represented in the knowledge item [E]. Furthermore, note that given an inference engine it is a simple matter to demonstrate automatically that these tuples are consistent with that knowledge item. An example of an information model constraint on the knowledge item [E] is the requirement that that knowledge item remains consistent with any particular information model such as the model shown in Figure 8. This constraint requires that item [E] remains consistent with that particular information model. In practice, we advocate the use of hand-coded, simple but non-trivial information models as the basis for information model constraints. Information model constraints are simple, powerful and effective constraints on knowledge. They are especially useful in applications where the knowledge is subject to a high rate of change and the essential meaning of the information model is fairly stable.

Referential Constraints

In the maintenance example discussed above we showed that the component relationships can be employed to form the basis of an algorithm for assisting with the complete execution of maintenance operations. It should also be clear that by blindly following every component relationship in this way a very substantial proportion of the physical model will have to be checked during the execution of each maintenance operation. We will now

part/sale-price

part-number	dollar-amount
1234	1.44
2479	3.07
5312	0.90
2643	4.08

part/cost-price

part-number	dollar-amount
1234	1.20
2479	2.56
5312	0.75
2643	3.40

mark-up

mark-up-factor
1.2

Figure 8 A simple information model

show that "referential" constraints may be employed to substantially prune this search space in practice.

The raw method for effecting a maintenance operation is to follow every component relationship from an item that has been modified to identify other items as candidates for modification. *Referential constraints* are constraints which simply state that a particular component relationship need not be followed during the complete execution of a maintenance operation. For example, if we have the referential constraint that "the value set of the information item *part/cost-price* is *fixed* in knowledge item [E]" then this means that if the tuples of the relation part/cost-price are modified then it is not necessary to follow the link to the clauses which implement item [E]. In practice this amounts to saying that "the validity of item [E] is invariant of the contents of the value set of the component *part/cost-price* item", and thus this referential constraint is in fact a static constraint on item [E]. Thus this constraint prunes the component relationship *from* the information item *part/cost-price to* the knowledge item [E] in the instance when the value set of item *part/cost-price* is modified.

The referential constraint just considered has the effect of pruning a component relationship *from* an information item *to* a knowledge item. It is interesting to note that many component relationships *from* knowledge items *to* their constituent data or information items can also be pruned. For example, suppose that the information item *part/cost-price* is implemented by a relation 'part/cost-price' which is not a real relation, in other words it is not actually stored but its values deduced when required using the clause:

part/cost-price(x,y) ← part/type(x,z),
type/cost-price(z,y)

which is an implementation of the semantics of the knowledge item:

[part/cost-price, part/type, type/cost-price] [G]

then the link from knowledge item [E] to its component information item *part/cost-price* is replaced by a link from [E] direct to item [G]. Thus we see that if a knowledge item contains an information item as a component, and that information item is implemented by a relation which is not actually stored then the component relationship from the knowledge item to that component information item is replaced by a link to the knowledge item which is used to derive that information item. In other words, we only need follow links to the information and data levels in the case of items which are implemented by relations and populations which are actually stored. As we noted in the sample modification to knowledge item [E] above, this also helps to restrict the extent of the search.

CONCLUSION

Constraints for knowledge have been promoted as a tool for protecting knowledge-based systems from the introduction of update anomalies during the execution of maintenance operations. An extensive taxonomy of knowledge constraints has been proposed. We have seen that one class of knowledge constraints provides the basis of a practical and effective mechanism for protecting an expert system against the introduction of update anomalies. Another class of knowledge constraints, coupled with the referential integrity constraint, together provide the foundation for a procedure to substantially automate the expert systems maintenance process. This procedure is highly efficient if the knowledge has been normalised.

REFERENCES

[CSEM91] Compton, P., Srinivasan, A., Edwards, G., Malor, R. & Lazarus, L. "Knowledge Base Maintenance without a Knowledge Engineer", in proceedings Expert Systems World Congress, J. Liebowitz (Ed), Pergamon Press 1991.

[Da86] Date, C.J. "An Introduction to Database Systems" (4th edition) Addison-Wesley, 1986.

[Deb89] Debenham, J.K. "Knowledge Systems Design", Prentice Hall, 1989.

[Deb92] Debenham, J.K. "Modelling Knowledge Systems", in proceedings Third International Conference on Database and Expert Systems Applications DEXA'92, Valencia, Spain, September, 1992, pp488-493.

[Deb93a] Debenham, J.K. "Decomposition of Four Component Items", in proceedings Fourth International Conference DEXA93, Prague, September 1993, pp457-460.

[Deb93b] Debenham, J.K. "Normalising Knowledge Objects", in proceedings Second International Conference on Information and Knowledge Management, Washington, November 1993, pp335-343.

[Deb94a] Debenham J.K. "Objects for Knowledge Modelling", in proceedings World Congress on Expert Systems, Estoril, Portugal, January 1994.

[Deb94b] Debenham J.K. "Normalisation and Maintainable Knowledge", in proceedings 7th International Conference on Systems Research, Informatics and Cybernetics, Baden-Baden, Germany, August 1994.

[DL90] Debenham, J.K. and Lindley, C. "The Knowledge Analyst's Assistant: A Tool for Knowledge Systems Design", in proceedings of the 4th Australian Joint Conference on Artificial Intelligence, Perth, 21-23 November 1990, pp 343-354.

[Gai87] Gaines, B.R. "Foundations of knowledge engineering", in (M.A. Bramer, Ed.) "Research and Development in Expert Systems III", Cambridge University Press, 1987.

[Gra89] Gray, P.M.D. "Expert Systems and Object-Oriented Databases: Evolving a New Software Architecture", in "Research and Development in Expert Systems V", Cambridge University Press, 1989, pp 284-295.

[Ito91] Ito, H. "Interface for Integrating a Knowledge-Based System and Database Management System Using Frame-Based Knowledge Representation", in proceedings Expert Systems World Congress, J. Liebowitz (Ed), Pergamon Press 1991.

[Kow91] Kowalski, R.A. "Logic Programming in Artificial Intelligence", in proceedings International Joint Conference on Artificial Intelligence, Sydney, August 1991.

[NH88] Napheys, B. & Herkimer, D. "A Look at Loosely-Coupled Prolog/Database Systems", in Proceedings Second International Conference on Expert Database Systems, George Mason University, 1988.

[Ste87] Steels, L. "Second Generation Expert Systems", in (M.A. Bramer, Ed.) "Research and Development in Expert Systems III", Cambridge University Press, 1987.

[Sto87] Stonier, T. "What is Information", in (M.A. Bramer, Ed.) "Research and Development in Expert Systems III", Cambridge University Press, 1987.

[Twi89] Twine, S. "Towards a Knowledge Engineering Procedure", in "Research and Development in Expert Systems V", Cambridge University Press, 1989, p 90-102.

[WKL89] Walker, A., Kowalski, R., Lenat, D., Soloway, E. and Stonebraker, M. "Knowledge Management", in (L. Kerschberg, Ed.), "Proceedings from the Second International Conference on Expert Database Systems", Benjamin Cummings, 1989.

A KNOWLEDGE-BASED STRATEGY FOR THE RE-ASSOCIATION OF FRAGMENTED SENSOR REPORTS

Keith P. Mason
Electronics and Surveillance Research Laboratory
Defence Science and Technology Organisation
P.O. Box 1500
Salisbury SA 5108
AUSTRALIA

ABSTRACT

This paper discusses a knowledge based strategy to assist in the task of identifying and tracking radar emitters from radar signals intercepted by a passive sensor system. The novelty of this strategy is that it utilises knowledge of the performance (and in particular the limitations) of the sensor processing algorithms as well as knowledge of environmental data (such as from geographic information systems) and knowledge of radar emitter characteristics to re-associate reports of radar emitters which have been fragmented by the sensor system. An illustration of this strategy from a laboratory software prototype is provided in this paper.

1 INTRODUCTION

The passive detection of radar signals and the identification and tracking of the platforms that carry them is an important source of surveillance information. Radars operate in one of two modes, either pulsed or continuous wave, and it is the processing of pulsed signal waveforms that are discussed in this paper.

Sensors deployed for detecting radar emitters commonly report pulsed signal intercepts containing parameters representing the time of arrival (toa), direction of arrival (doa) and signal parameters including radio frequency (rf) and pulse width (pw). Analysis of the pulses from a single emitter will allow other parameters to be derived, such as pulse repetition interval (pri), measures of agility of pri and rf (that is, whether pri or rf are constant or change over time), and the emitter's scan pattern. The values of an emitter's parameters can be used as a classifier of the type of emitting radar.

The signals intercepted by the sensor system's antenna will often comprise a superposition of emissions from several radars active in the environment. These signals will be interleaved in time of arrival order. For measurements of the derived parameters of the individual radars these signals need to be de-interleaved into streams containing pulses only belonging to a single radar emitter. Many algorithms for de-interleaving have been proposed (a selection of de-interleaving techniques are described in [4]), however none so far has been without weaknesses. The weaknesses can lead to the fragmentation of the pulses belonging to a single emitter into multiple reported pulse trains, or lead to the false merging of pulses from distinct emitters into a single reported pulse train. In this paper one particular de-interleaving algorithm known as the sequence search [8] is used for illustration.

This paper presents a strategy that uses knowledge of weakness of sensor processing algorithms to assist in emitter identification. A laboratory software prototype is being developed to evaluate the benefits of this strategy for future operational systems.

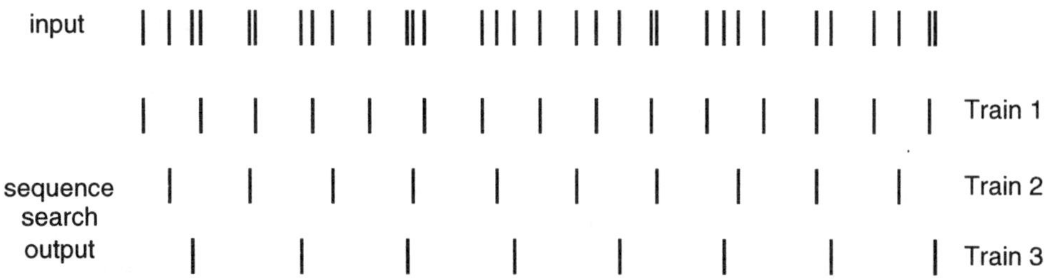

FIGURE 1 Sequence Search Applied to an Interleaved Sequence.

More information about radar systems can be found in [7], and on using radar intercepts for surveillance in [4]. Real time signal interpretation is discussed, for example, in [3]. Other knowledge-based approaches to the radar emitter domain can be found in [1], [2], [5], [6] and [9].

The next section details the problem of fragmentation of sensor reports of radar emitters. Section 3 describes the knowledge-based strategy used to repair fragmentation errors. Section 4 provides some examples illustrating this strategy. Section 5 provides a conclusion to this paper.

2 FRAGMENTATION

The sequence search tries to establish pulse trains for which pulses fall within narrow windows around a constant pri pulse train template. This works well when applied to constant pri pulse trains. However, when applied to a pulse train from an emitter with agile pri, the sequence search will find components of the pulse train that match the template, and report each separately, fragmenting the sequence of pulses from the emitter into several pulse train reports. This is illustrated below.

The sequence search starts by choosing two pulses separated by at least a set small time interval. It tries to use these two pulses to establish a constant pri pulse train. If a pulse train (with length greater than a threshold of say 8 pulses) is found, it is removed from the input data and the process is repeated. If a pulse train is not found then a different pair of pulses in the interleaved sequence is chosen (according to a strategy that gradually increases the time between the selected pulses) and the process is repeated. Figure 1 shows an input pulse train consisting of three interleaved constant pri pulse trains and each of the component pulse trains found by the sequence search. More details of the sequence search algorithm can be obtained from [8].

When faced with an emitter whose pri slowly slides linearly from one value (say t) to another, the sequence search will find pulses at 2t, 3t, 4t etc from the initial pulse as the small amount of slide is within the tolerance range of the search. However at some point the accumulated slide will be out of the tolerance range of the search, so terminating the search and reporting the pulses already found as a pulse train. The process will then start again, and will find another pulse train with pri slightly different from the previous train. This pattern will continue until the end of the slide, where the pri jumps back to the start again and the whole pattern begins over again. This is shown in Figure 2. Thus the sequence search fragments the pulse train into a sequence of short pulse trains, rather than giving the desired single report containing all the pulses of the emitter.

There are a variety of emitter behaviours that include agile pri. These all result in fragmented reports from de-interleaving techniques such as the sequence search. As well as the sliding pri case described above, dwell and switch emitters, staggered pri emitters and interrupted pulse sequence emitters are described below and in the discussion of the re-association strategy in section 3.

Dwell and switch emitters are characterised by dwells of a number (say 80) pulses that have a constant pri and then switch to another pri for another dwell of (say 70) pulses, and so on, through a sequence of dwells. A sequence search de-interleaver will fragment the pulse train into separate reports for each dwell.

An interrupted pulse sequence emitter consists of a series of constant pri pulse trains, each with the same pri, separated by a gap larger than the pri. A sequence search de-interleaver will fragment the emitter into a series of reports, one for each component pulse train.

A staggered pri emitter continually cycles through a set of pri values (called stagger intervals). The number of stagger intervals is typically from 2 to several hundred. The cycle time of the stagger is called the frame interval. A staggered pulse train can be also be viewed as a number of time interleaved pulse trains with the same pri (equal to the frame interval) but off-set in start time. The sequence search is likely to find one pulse train for each stagger interval with pri equal to the frame interval but with off-set start times (that is, the sequence search de-interleaves the constant pri pulse train components of the stagger).

Since the sequence search uses a pri increasing from a small value, when processing a pulse train from a staggered pri emitter, a lot of potential pulse trains will

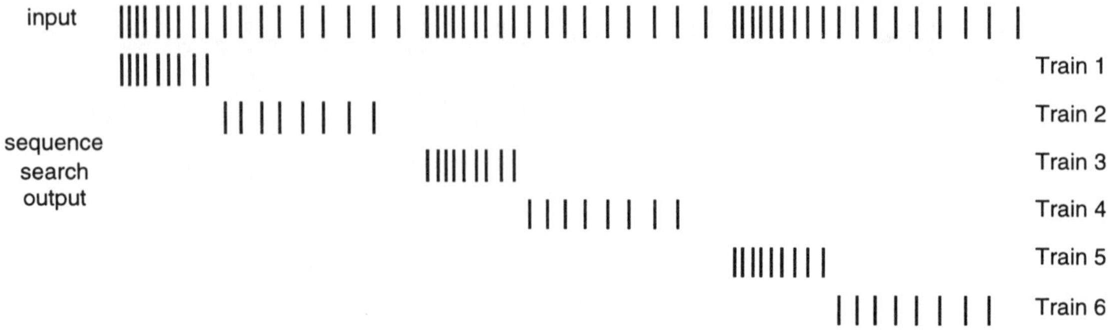

FIGURE 2 Sequence Search Applied to a Sliding PRI Emitter.

have been tested before searching for trains with the pri of the frame interval. It may be that some of these searches are able to match some pulses of the stagger as a constant pri train because the matching process includes a tolerance. This adds to the complexity of the fragmentation, as shown in Figure 3 where the first train found by the sequence search has a pri of one third of the frame interval.

It is clear from these examples that fragmentation of sensor reports complicates the process of postulating emitter entities from pulse trains reported by sequence search style de-interleavers.

3 STRATEGY

The approach of this paper to the re-association of pulse train fragments uses a generate and test paradigm. A library of emitter characteristics is used to generate candidates for each pulse train. Each candidate is then evaluated by comparing the expected fragmentation pattern for it with the reports from the de-interleaver. Knowledge of fragmentation patterns used in the evaluation process can be obtained analytically or empirically. The evaluation process also considers operational constraints (such as geographic location) on the candidates. The evaluations are then ranked to produce an interpretation of which emitters are active in the environment. The ranking process is weighted to minimize the number of emitters in the preferred interpretation.

3.1 Candidate Generation

The library of emitter characteristics is of central importance to the generation process of emitter candidates. The library function must be able to provide a list of emitter candidates that could produce a pulse train with the measured primary parameters of rf and pw, and the derived parameter of pri. The library must also be able to provide information about the fragmentation pattern that would be expected to be produced by the de-interleaver for each emitter candidate.

For example, a library search that returns a dwell and switch emitter also needs to return information on the emitter's other dwells, including sequencing. A search that returns a staggered emitter will need to return all of the stagger intervals. Interrupted pulse sequence emitter candidates should be reported with information on the duration of the interruption, and the number of pulses per fragment. A sliding pri emitter candidate requires characteristics of the pri variation to be returned. An emitter that only emits constant pri pulse trains needs to be identified as such by the library function.

3.2 Candidate Evaluation

An evaluation is performed for each candidate retrieved by the library search. The evaluation process will depend on the nature of each candidate and will measure the extent that the fragmentation pattern expected for a candidate is observed in the output from the de-interleaver. A library of emitter characteristics is also consulted to determine whether a candidate has any operational constraints, such as being at a certain geographic location, and these constraints are also considered as part of the evaluation.

Candidate evaluation yields a metric consisting of class and number, and also a list of pulse trains that supports the candidate. The number is a measure of the supporting evidence for the particular candidate class. For example, *dwell(9)* indicates that 9 subsequent fragments have been found to support the dwell and switch candidate, and *stagger(82%)* indicates that fragments covering 82% of the stagger intervals have been reported. The evaluation number is set to zero if an operational criterion associated with a particular candidate is not met.

Since constant pri emitter candidates exactly match the template of the sequence search they are not expected to be fragmented by a sequence search (excluding long term affects such as scan pattern), and so there is nothing to seek amongst a de-interleaver's output to support such a candidate. The default evaluation of a constant pri emitter therefore is *simple(1)*.

However, due to the possibility that a fragment pulse train from a complex emitter could be recognised as a

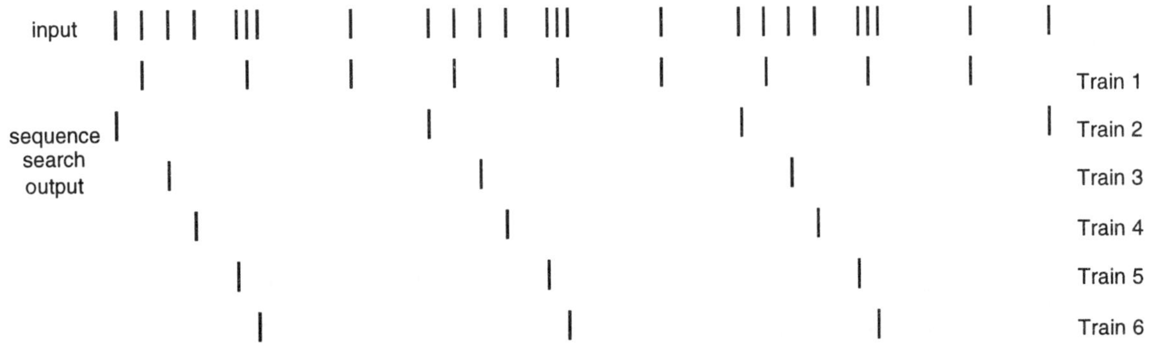

FIGURE 3 Sequence Search Applied to a Staggered PRI Emitter.

(spurious) constant pri emitter, it is prudent to look further to minimize the number of falsely postulated emitter entities. For example, if an emitter has previously been established with the same parameters (such as from an earlier scan), then this is additional evidence that supports a constant pri pulse train truly characterising an emitter (rather than being a spurious fragment), and so the evaluation in this case is *simple(2)*. Otherwise if there are multiple reports from the de-interleaver on the same bearing then it may be the case that fragmentation has occurred, and the candidate is ranked as *simple(0.5)*, indicating that care should be taken to check fragmentation possibilities before postulating a constant pri emitter on the basis of this pulse train report.

3.3 Candidate Ranking and Emitter Track Initiation

The aim of candidate ranking is to choose between multiple interpretations of the signal environment. That is, to choose between the various ways the pulse train reports from the de-interleaver could be associated into groups that corresponded to emitter entities. It is the approach of the strategy of this paper to attempt to rank candidates so that the smallest number of emitters is postulated. For example, a set of pulse trains might be interpreted as being six dwells from a dwell and switch emitter, or six emitters of constant pri and short duration. This ranking strategy prefers to postulate a dwell and switch emitter entity with six pulse trains assigned to it, rather than to postulate six emitters each with one pulse train assigned.

The ranking strategy eliminates from further consideration any candidate with a zero evaluation number and, except for simple candidates, any candidate which does not reach a (heuristic) threshold for sufficient evidence to establish the candidate class. For example, a dwell and switch emitter candidate has a threshold of 3 dwell fragments, a stagger emitter candidate has a threshold of 70% of total intervals observed, interrupted pulse train candidates and sliding pri candidates have a threshold of 4 pulse train fragments.

As candidates which have achieved their threshold are supported by several pulse train fragments, a potential for ambiguity exists if one or more pulse trains provide support for more than one of these candidates. However constraints (such as dwell and switch fragments being contiguous) restrict the cases where cross class ambiguities can believably exist. One case of ambiguity, for example, is where fragments from a stagger happen to have the same characteristics as an interrupted pri emitter. Such ambiguities can be resolved by a fixed order ranking strategy of processing dwell and switch candidates, then stagger, slide, interrupted, and lastly simple candidates.

Ambiguities of pulse train association may remain within a class, but these can be resolved by the amount of evidence supporting each candidate. The candidate with the largest evaluation number in the class is chosen first and an emitter entity is created for it and its supporting trains. The process is repeated with the remaining fragments and their candidates. It is unlikely the evaluation numbers would be equal for candidates referring to different but overlapping associations of pulse trains if the library is complete. Thus this process assigns (without ambiguity) the pulse trains to an emitter entity (the identity of which may be ambiguous).

When the above process comes to consider pulse trains which have only simple candidates, emitter entities should be created for those which have evaluation numbers of 1 or 2. However for those with an evaluation of 0.5 there is the possibility that they are spurious fragments as there are other emitters on the same bearing and it is prudent to attempt to associate them with established complex emitters. The analysis of fragmentation can be quite complex and the details of this association is beyond the scope of this paper.

The main requirement for the above strategy to be successful is that the de-interleaver reports for fragmented pulse trains will have enough components for the evaluation function to return values above the thresholds for the emitter types active in the environment.

4 ILLUSTRATION

A laboratory sequence search de-interleaver yields reports of the format:

train(index,rf,pw,min_pri,av_pri,max_pri,
doa,start_toa,stop_toa).

The following is an example set of eight pulse train reports from this de-interleaver:

```
train(1,9000,2.0,1330,1330,1330,53,445462,499992),
train(2,9000,2.0,1130,1130,1130,53,364832,442802),
train(3,9000,2.0,1040,1040,1040,53,291942,363702),
train(4,9000,2.0,970,970,970,53,223972,290902),
train(5,9000,2.0,1230,1230,1230,53,138132,223002),
train(6,9000,2.0,1490,1490,1490,53,34092,136902),
train(7,9000,2.0,1810,1810,1810,53,22,32602), and
train(8,9000,2.0,9898,9898,9898,53,1800,40000).
```

Using a library of emitter characteristics, a laboratory prototype of the strategy described in this paper generates candidates of the form:

cand(index,train_index,emitter,type,constraints_list). The following candidates could be generated for the above pulse train reports:

```
cand(1,1,ajax_c3.2,dwell,
        [1630,1810,1490,1230,970,1040,1130,1330]),
cand(2,1,acme_123,simple,[]);
cand(3,2,see_far,interrupted,[4000]).
cand(4,2,ajax_c3.2,dwell,
        [1330,1630,1810,1490,1230,970,1040,1130]),
cand(5,2,vigilant_v23,simple,[]),
cand(6,3,vigilant_v1225,simple,[]),
cand(7,3,ajax_c3.2,dwell,
        [1130,1330,1630,1810,1490,1230,970,1040]),
cand(8,4,ajax_c3.2,dwell,
        [1040,1130,1330,1630,1810,1490,1230,970]),
cand(9,4,ajax_c1.3,dwell,[1810,1490,1230,970]),
cand(10,5,ajax_c3.2,dwell,
        [970,1040,1130,1330,1630,1810,1490,1230]),
cand(11,5,ajax_c1.3,dwell,[970,1810,1490,1230]),
```

cand(12,5,vigilant_v23,simple,[]),
cand(13,6,ajax_c3.2,dwell,
[1230,970,1040,1130,1330,1630,1810,1490]),
cand(14,6,ajax_c1.3,dwell,[1230,970,1810,1490]),
cand(15,6,acme_123,simple,[]),
cand(16,7,ajax_c3.2,dwell,
[1490,1230,970,1040,1130,1330,1630,1810]),
cand(17,7,ajax_c1.3,dwell,[1490,1230,970,1810]),
cand(18,7,acme_123,simple,[]), and
cand(19,8,long_sight,simple,[]).

These candidates show, for example, that pulse train 6 (which has a reported pri of 1490) has 3 candidates: (candidate 13) ajax_c3.2 which is a dwell and switch emitter with 8 dwells of pri (in sequence) of 1230, 970, 1040, 1130, 1330, 1630, 1810 and 1490, (candidate 14) ajax_c1.3 which is a dwell and switch emitter with 4 dwells of pri (in sequence) of 1230, 970, 1810 and 1490, and (candidate 15) acme_123 which is a constant pri emitter.

Each candidate is evaluated by looking for the expected fragments for the candidate type being reported by the de-interleaver. Evaluations are structured as eval(candidate_index,score,[train_list]) where train_list is are the pulse trains that support the candidate. Evaluation of the above yields:
eval(1,dwell(0),[1]),
eval(2,simple(0.5),[3]),
eval(3,interrupted(0),[2]),
eval(4,dwell(1),[1,2]),
eval(5,simple(0.5),[2]),
eval(6,simple(0.5),[3]),
eval(7,dwell(2),[1,2,3]),
eval(8,dwell(3),[1,2,3,4]),
eval(9,dwell(0),[4]),
eval(10,dwell(4),[1,2,3,4,5]),
eval(11,dwell(1),[4,5]),
eval(12,simple(0.5),[5]),
eval(13,dwell(5),[1,2,3,4,5,6]),
eval(14,dwell(2),[4,5,6]),
eval(15,simple(0.5),[6]),
eval(16,dwell(6),[1,2,3,4,5,6,7]),
eval(17,dwell(3),[8,5,6,7]),
eval(18,simple(0.5),[7]), and
eval(19,simple(0.5),[8]).

These show, for example, that for pulse train 6, candidate ajax_c3.2 evaluated as dwell (5) and is supported by trains 1, 2, 3, 4, 5 and 6; candidate ajax_c1.3 evaluated as dwell(2) and is supported by trains 4, 5 and 6; candidate acme_123 evaluated as simple(0.5) and is supported by train 6.

Ranking is used to choose between emitter candidates, and so to postulate emitter entities. The ranking process starts with dwell and switch candidates, and chooses candidate 16 as it has the highest evaluation number, dwell(6). An emitter entity is created for this candidate, and assigns its 7 supporting pulse trains. This leaves only one pulse train to consider. It has only one candidate with an evaluation of simple(0.5). As there is no possible fragmentation of the previously postulated emitter that could account for this pulse train, a second emitter entity is created. Emitter entities in the prototype have the form emitter(index,identity,type,rank,train_list), and the preferred interpretation of emitters active in the environment is:

emitter(1,ajax_c3.2,dwell,6,[1,2,3,4,5,6,7]), and
emitter(2,long_sight,simple,0.5,[8]).

5 CONCLUSION

This paper has presented a knowledge-based strategy to re-associate fragmented emitter reports from a de-interleaver based on the sequence search. This strategy has been demonstrated for several emitter types. The method could be extended to other emitter types and to other de-interleaving techniques.

ACKNOWLEDGEMENTS

The author thanks Dr. David Bovar and Dr. Stephen Elton for their helpful comments on this paper.

REFERENCES

[1] Mason, K.P. (1992). *Information Fusion in a Knowledge-Based Classification and Tracking System.* Industrial and Engineering Applications of Artificial Intelligence and Expert Systems. Lecture Notes in Artificial Intelligence, volume 604, pages 666-675. Springer Verlag.

[2] Mason, K.P. and Hood, S.T. (1989). *KBESM - A Prolog Blackboard System.* Australian Computer Journal, volume 21, number 3, pages 100-107.

[3] Hood, S.T. and Mason, K.P. (1987). *Knowledge-based Systems for Real-time Applications.* In Applications of Expert Systems, R. Quinlan (ed), pages 102-119, Addision Wesley.

[4] Wiley, R.G. (1993). *Electronic Intelligence: The Analysis of Radar Signals.* 2nd Edition. Artech House.

[5] Byrne, C.D., Miles, J.A.H. and Lakin, W.L. (1989). *Towards knowledge-based Naval Command Systems.* 3rd IEE International Conference on Command, Control, Communication and Management Information Systems, pages 33-42.

[6] Roe, J., Cussons, S. and Faltham, A. (1990). *Knowledge-based Signal Processing for Radar ESM Systems.* IEE Proceedings, Vol. 137, Part F, No. 5, pages 293-301.

[7] Skolnik, M.I., (1981). *Introduction to Radar Systems.* 2nd edition McGraw Hill.

[8] Mardia, H.K. (1989). *New Techniques for the De-interleaving of Repetitive Sequences.* IEE Procedings, Part F, Communications Radar Signal Processing, vol136(4), pages 149-154.

[9] Marshall, C.B. and Simons, P.R. (1990). The Application of Real-time AI Techniques to Pulse Train De-interleaving. Military Microwaves '90. Conference proceedings, London, July 11-13,1990.

OPINION TRANSITION MODEL UNDER DYNAMIC ENVIRONMENT: EXPERIMENT IN INTRODUCING PERSONALITY TO KNOWLEDGE-BASED SYSTEMS

Masakatsu Ohta, Toshiyuki Iida, and Tsukasa Kawaoka
NTT Communication Science Laboratories
1-2356 Take, Yokosuka-Shi, Kanagawa, 238–03 Japan
E-mail: ohta@nttkb.ntt.jp

ABSTRACT

A dynamic model that permits an entity to respond to changes in its environment by modifying its opinions is proposed. The goal is to reproduce the decision making ability of humans; knowledge-based systems can be structured to express the personality characteristics of *adaptability* and *homeostasis*. The balance between these properties is simply controlled by three parameters. Simulations confirm: (1) a system weighted toward Public Self-Consciousness responds to changes in its environment by readily changing its opinions, on the other hand, weighting for Private Self-Consciousness stiffens its resistance to changing its opinion; and (2) the strength with which an opinion is held becomes most unstable immediately before it is altered.

1 INTRODUCTION

Up to the present, many expert systems have been developed and used to solve problems. However, even though limited to small scale problems, these systems fail to attain any reasonable level of performance due to the lack of common knowledge. Recently, many researchers believe that very large-scale knowledge bases will be the answer. For example, the Cyc project of MCC is building a common sense knowledge base to make an intelligent system [Lenat89]. EDR (Japan Electronic Dictionary Research Center) accumulates common knowledge on vocabulary and usage [Yokoi93]. One property of common knowledge is its dependence on the environment, such as the society. For example, in Japan common knowledge includes "Mt. Fuji is tall". However, people living in a country with many mountains much higher than Mt. Fuji would probably not hold the same opinion. This implies that system adaptability is the key to achieving better system performance. If the system's environment changes, the system must modify its opinions accordingly. The obvious approach is to establish a set of environments and the corresponding sets of opinions. Changing the environment automatically changes the opinions. Unfortunately, it seems impossible to adequately classify environments in advance. Moreover, this approach assumes the existence of some authority to tell the system that the environment has changed.

As a new approach, this paper suggests that the support rate of opinions be used as the input triggering opinion changes. This approach is inherently more flexible because environments are not explicitly differentiated and the system is allowed to express some attributes of human personality. The system attributes are *adaptability* and *homeostasis*. The former indicates the ability to readily change opinion while the latter suggests system resistance to change. The proposed model is based on an analysis of the way in which humans form common knowledge. Three parameters are established that allow us to set the balance between these two characteristics. The result is that system response, both system output and system adaptability, can be freely changed to meet a wide range of situations.

The rest of this paper is organized as follows. In Section 2, assuming that the opinions supported by the majority is indicative of the environment, we examine how opinions can change to conform to common knowledge. Section 3 describes a dynamic model for opinion transition. Moreover, we analyze the stability of opinion conviction. Section 4 introduces numerical examples on the qualitative behavior of the model: the effect of the three parameters on the transition flexibility; and changes in stability before and after the transition. Finally, concluding remarks are given in Section 5.

2 PERSONALITY AS EXPRESSED BY CONFORMITY

As a public opinion reflects the society, the opinion supported by most people reflects the environment to which they belong. Therefore, by altering its opinions to those supported by the majority, the system can adapt itself to any environment; classification of the environment is not needed. This is also true if the environment changes in a dynamic fashion. The question is how best to represent the way in which the system changes its opinions. The actions of humans in forming and conforming to common knowledge was analyzed for this purpose.

The influence of personality on conformity has already been studied [Froming81]. This personality attribute is associated with two major characteristics: *Public Self-Consciousness* concerned with one's overt self-aspect such as appearance, impression to others, and so on; and *Private Self-Consciousness* concerned with one's covert self-aspect such as motives, thoughts, feeling, and so on [Fenigstein75]. These characteristics are relatively independent of each other and exert quite different effects on conformity. People high in public self-consciousness conform to the majority more easily than those low in public self-consciousness, on the other hand, people high in private self-consciousness override the pressure from others [Froming81]. Therefore, a system that reproduces public and private self-consciousness would appear to have a personality and flexibly respond to a dynamic environment.

3 OPINION TRANSITION MODEL

3.1 TRANSITION DYNAMICS

We developed a nonlinear dynamic model that changes its opinions by analyzing the decision making behavior of humans.

(1) *Adaptability* :
A person's conviction in the accuracy of an opinion may change over time as input is received over time from the environment (FIGURE 1).

(2) *Nonlinearity* :
The process of changing an opinion involves hysteresis, i.e. it is a nonlinear phenomenon.

The system replicates these two characteristics. Public self-consciousness weakens the system's conviction in its own opinions and makes it more susceptible to the opinions of its environment. Private self-consciousness strengthens the system's convictions so that it changes opinion less readily. We establish a nonlinear model

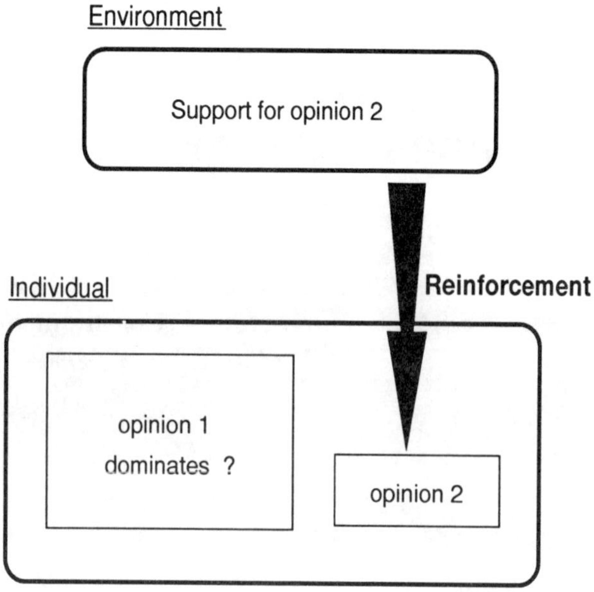

FIGURE 1 Adaptability to Environment

that allows these characteristics to be easily controlled using three simple parameters.

The model uses the following nonlinear differential equation. Note that for simplicity we assume there is only one topic, for which there are several possible opinions.

$$\frac{dX_i}{dt} = \alpha\,(P_i - X_i) + \beta \sum_{j \neq i} (X_j W_{ji} - X_i W_{ij}), \quad (1)$$

where W_{ij}, which represents the relative attractiveness of opinion j compared to i, is a nonlinear function written as

$$W_{ij} = 1 \,/\, [\, 1 + exp\,\{\, \gamma\,(X_i - X_j)\}\,]. \quad (2)$$

P_i is the support rate for opinion i and $\boldsymbol{P} = (P_1, P_2, \cdots)$ represents the opinions of the environment. $X_i(t)$ indicates the certainty with which the system supports opinion i (afterward, written as X_i). P_i and X_i must satisfy the following conditions

$$\sum_i P_i = 1, \qquad \sum_i X_i = 1. \quad (3)$$

W_{ij} denotes the transition rate of certainty factor from opinion i to j. The first term in the right side of eq.1 addresses public self-consciousness and allows interference from the environment to make X_i equal to P_i. Parameter α (> 0) denotes the weight of this interference. The second term addresses private self-consciousness.

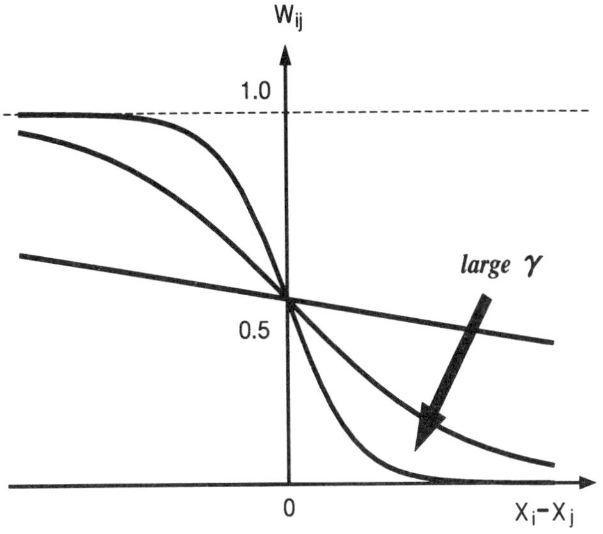

FIGURE 2 Transition Rate of Certainty Factor

Irrespective of the environment, the system's conviction for opinion i becomes, over time, stronger than its conviction for opinion j if X_i is larger than X_j. W_{ij} is the measure of preference for j over i, so it decreases X_i by $\beta X_i W_{ij}$; however, due to negative feedback acting on X_j, X_i increases by $\beta X_j W_{ji}$. Parameter β (> 0) determines the strength of this feedback. W_{ij} contains parameter γ (> 0) which is a measure of decision conviction. As shown in FIGURE 2, increasing γ increases the system's confidence in holding one opinion over time. This model defines the system's opinion as that with the largest certainty factor. It is obvious that if α is large, the system changes opinions easily. If β and γ are large, the system's conviction in its opinions becomes stronger over time.

3.2 PERTURBATION IN CONVICTION

After establishing a system it is important to monitor its internal state. For each topic it is possible to estimate the stability of the system's conviction by assessing changes in $(X_1, X_2, \cdots)$. If the system is close to changing its opinion, even small changes in the environment can lead to large changes or perturbations in $(X_1, X_2, \cdots)$. If the system is stable, the perturbation diminishes with time, so that the opinion does not change. This can happen either when the environment reinforces the system's opinion or when the disagreement of the environment is not strong enough to overcome the system's self-conviction. If the system is unstable, the perturbation strengthens and the system may change its opinion. An interesting case occurs when two opinions have equal conviction. The perturbation caused by external noise can trigger unpredictable selections [Nicolis89].

System stability can be estimated from the stability analysis of the solution of eq.1. In order to simplify the analysis, we consider only the stability of the stationary solution. The current internal state is expressed as

$$\begin{aligned} \boldsymbol{X}(t) &= (X_1^S + u_1(t),\ X_2^S + u_2(t), \cdots) \\ &= \boldsymbol{X}^S + \boldsymbol{u}(t), \end{aligned} \tag{4}$$

where $\boldsymbol{X}^S = (X_1^S, X_2^S, \cdots)$ is the stationary solution of eq.1 and $\boldsymbol{u}(t) = (u_1(t), u_2(t), \cdots)$ is the perturbation. We rewrite eq.1 as

$$\frac{dX_i}{dt} = F_i(\boldsymbol{X}), \tag{5}$$

and substitute eq.4 into 5 assuming that the perturbation is very small (thus we examine local stability), yielding the linearized equation

$$\frac{du_i}{dt} = \sum_j L_{ij} u_j, \tag{6}$$

where L_{ij} is written as

$$L_{ij} = \left. \frac{\partial F_i}{\partial X_j} \right|_{\boldsymbol{X}=\boldsymbol{X}^S}. \tag{7}$$

Eq.6 is a linear differential equation with constant coefficients and its solutions include $u_i(t) = u_i(0)\, exp(\lambda t)$, where $u_i(0)$ is the initial perturbation and λ is an eigenvalue of the following matrix

$$L = \begin{pmatrix} L_{11} & L_{12} & \cdots \\ L_{21} & L_{22} & \cdots \\ \vdots & \vdots & \ddots \end{pmatrix}. \tag{8}$$

Therefore, if all real parts of the eigenvalues are negative, $\boldsymbol{u}(t)$ diminishes with time, so the perturbation does not affect $\boldsymbol{X}^S$: $\boldsymbol{X}^S$ is asymptotically stable, else $\boldsymbol{u}(t)$ amplifies with time, so $\boldsymbol{X}(t)$ moves farther away from $\boldsymbol{X}^S$: $\boldsymbol{X}^S$ is unstable. From this analysis, it is clear that $\boldsymbol{X}^S$ becomes more unstable as an opposing opinion becomes more dominant than the system's opinion, and eventually returns to a stable state after the transition occurs.

4 NUMERICAL EXAMPLES

We performed numerical experiments to confirm the qualitative behavior of the model under the following conditions: (1) a transition occurs between two opinions (thus $P_2 = 1 - P_1$, $X_2 = 1 - X_1$), and (2) the environment changes more slowly than the certainty factors. From condition 2 we can take $\boldsymbol{P}$ as

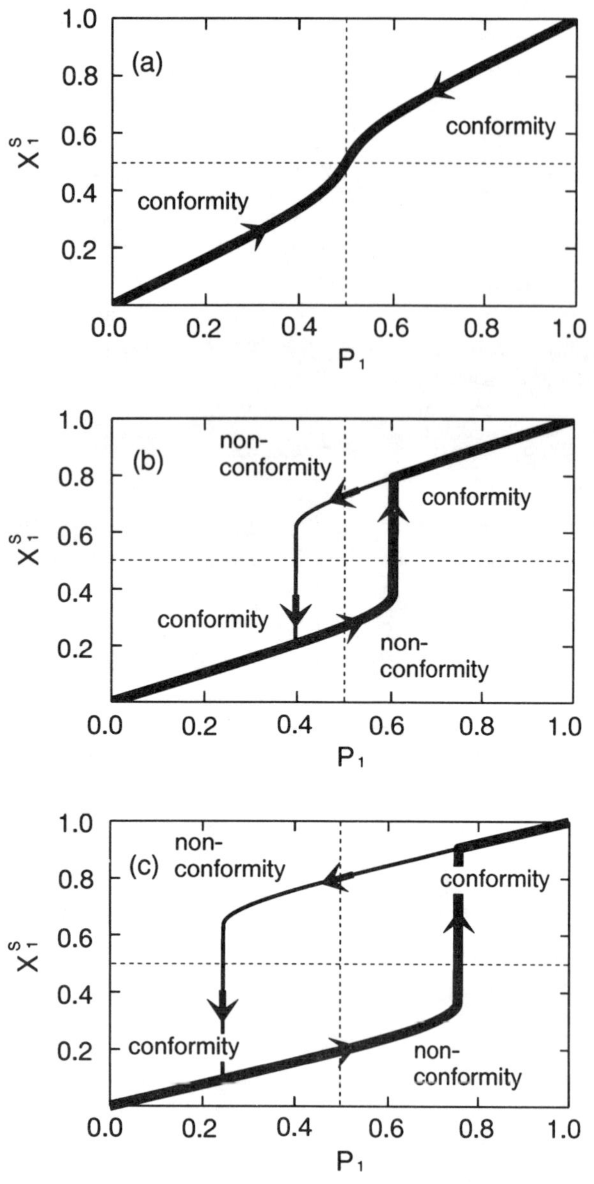

FIGURE 3 Public Self-Consciousness Dependency on α. α values are:(a)0.20, (b)0.05, and (c)0.03; β=0.05 and γ=7 in all plots.

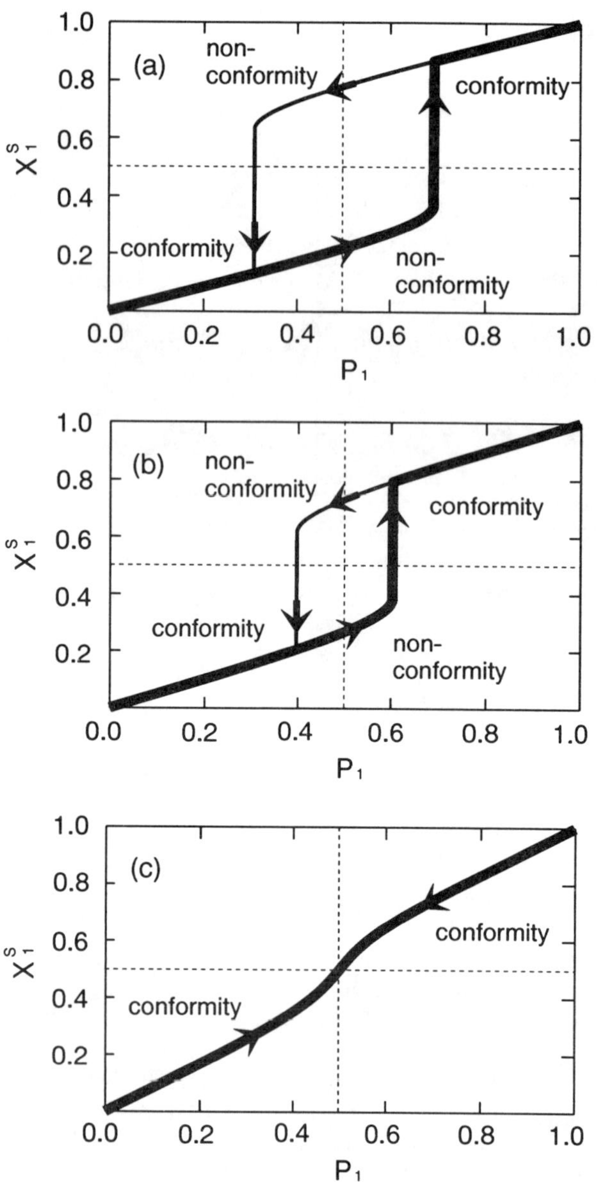

FIGURE 4 Private Self-Consciousness Dependency on β. β values are:(a)0.07, (b)0.05, and (c)0.01; α=0.05 and γ=7 in all plots.

constant while the certainty factors change while satisfying eq.1. We calculated the stationary solution of eq.1 at each $\boldsymbol{P}$ numerically, while changing $\boldsymbol{P}$ from $(0,1)$ to $(1,0)$ and back to $(0,1)$ gradually. At each $\boldsymbol{P}$ except $(0,1)$, the stationary solution given at the former $\boldsymbol{P}$ was used for the initial value of $\boldsymbol{X}(t)$. At $\boldsymbol{P} = (0,1)$, $\boldsymbol{X}(0) = (0,1)$ was used. We performed experiments using many different parameter values and found the same qualitative behavior. Here we examine some typical examples.

With regard to public self-consciousness, FIGURE 3 shows the behavior of the model as a function of α. As shown in the figure, opinion transitions occurred more easily at large values of α, the actual transition occurring at $P_1 = 0.5$ where the two opinions had equal dominance. However, as shown in FIGURE 3(b), for small value of α, the present opinion was kept even after the dominance between the opinions reversed and hysteresis was found in the behavior due to the nonlinear dynamics of the model. The transition from opinion 2 to 1 occurred at $P_1 \approx 0.6$, and at $P_1 \approx 0.4$ for the reverse transition. As α

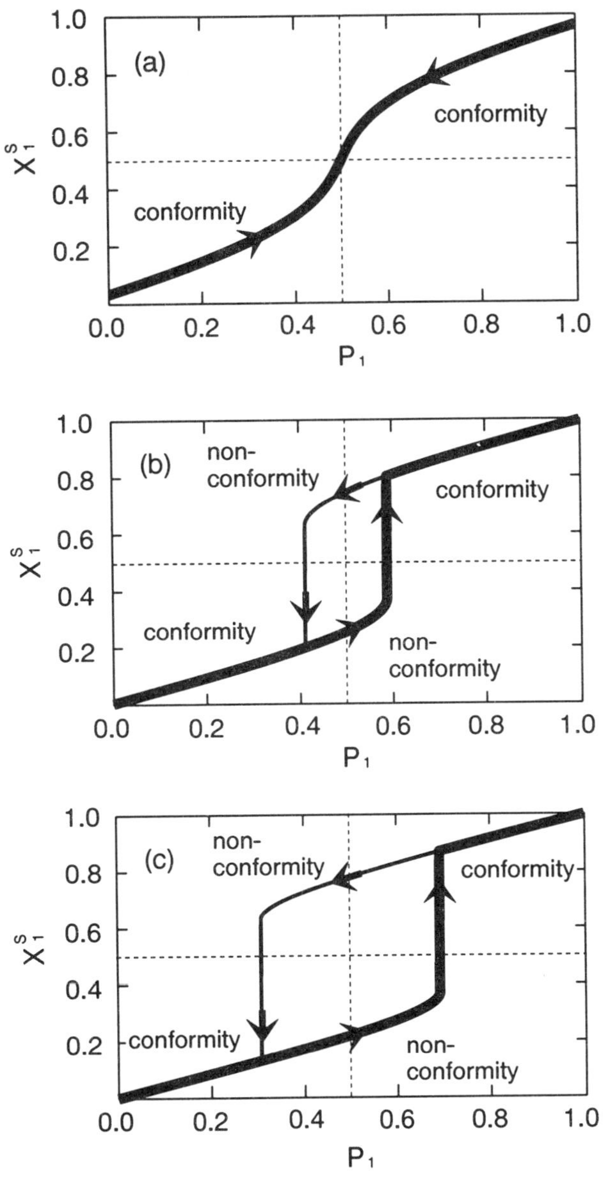

FIGURE 5 Private Self-Consciousness Dependency on γ. γ values are:(a)3, (b)5, and (c)7; α=0.05 and β=0.07 in all plots.

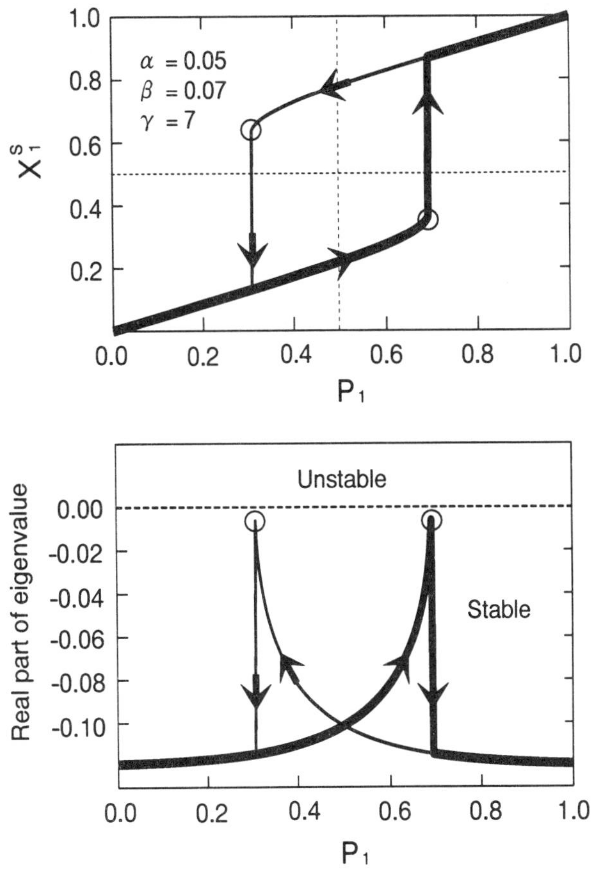

FIGURE 6 Stability of the Stationary Solution

becomes smaller, the system resists any transition and the hysteresis effect becomes stronger (FIGURE 3(c)). This behavior matches that of people high in public self-consciousness; they readily adopt the opinions of others.

With regard to private self-consciousness, FIGURES 4 and 5 show its dependency on β and γ, respectively. As β decreases, transitions occur more easily even if γ is large and hysteresis decreases (FIGURE 4). As γ increases, the transition is resisted and hysteresis becomes more prominent (FIGURE 5). This behavior matches that of people who are self-confident; they endure the pressure from others and retain their own opinions. These examples suggest that a system with large α, and small β and γ is adaptable; and one with small α, and large β and γ is homeostatic.

FIGURE 6 shows the stability of the stationary solution at each $\boldsymbol{P}$ for which the hysteresis phenomenon was found. Since the eigenvalue of L had negative real part at each $\boldsymbol{P}$, the system's opinion is not affected by very small perturbations: it is asymptotically stable. However, as the opposing opinion strengthens, the real part approaches zero and the internal state becomes unstable. When a transition occurs, stability is recovered. This corresponds to human behavior: under strong external pressure to change opinion, the person becomes more uneasy about his present position, and is relieved by adopting the majority's opinion. It should be noted that the stability decreases rapidly, even though the certainty factors do not change much. This rapid change of stability alerts us to the fact that a transition is not far away.

5 CONCLUSIONS

An attempt to introduce personality, though much simplified, to knowledge-based systems was made. We described a dynamic model for opinion holding that permits different opinions to be held in different environments. The model is based on the concept of conforming to the majority's opinion which is often seen in human societies. Numerical examples confirmed that the model is suitable for reproducing the decision making ability of humans. The model's qualitative behavior can be easily controlled by three parameters: the parameters may be set constant to yield an entity that adopts different opinions flexibly, or an entity that is confident in its own opinions. One interesting idea is to change the parameters under some circumstances to make the system unstable in order to reach global solutions. In future work, we will investigate these parameter controls and apply the proposed model to an intelligent database system [Iida94].

ACKNOWLEDGMENT

This study was performed at NTT Communication Science Laboratories as part of research into artificial intelligence with the goal of common knowledge to solve problems. We would like to thank the members of The Very Large Knowledge Base Research Group. Special thanks to Shigeo Shimada for an instructive discussion.

REFERENCES

[Fenigstein75] Fenigstein, A., Scheier, M.F., and Buss, A.H., Public and Private Self-Consciousness: Assesment and Theory, *Journal of Consulting and Clinical Psychology*, 43, pp.522–527, 1975.

[Froming81] Froming, W.J. and Carver, C.S., Divergent Influences of Private and Public Self-Consciousness in a Compliance Paradigm, *Journal of Research in Personality*, 15, pp.159–171, 1981.

[Iida94] Iida, T., Shimada, S., Ohta, M., and Kawaoka, T., Artificial Intelligence for Semantic Understanding, *Proceedings of the IFIP Congress 94*, 2, pp.137–142, 1994.

[Lenat89] Lenat, D.B. and Guha, R.V., *Building Large Knowledge-Based Systems: Representation and Inference in the Cyc Project*, p.372, Addison-Wesley, Massachusetts, 1989.

[Nicolis89] Nicolis, G. and I. Prigogine, *Exploring Complexity: An Introduction*, Freeman & Co., New York, 1989.

[Yokoi93] Yokoi, T., From Electronic Dictionaries to Very Large-Scale Knowledge Bases, *Proceedings of the MT Summit IV*, pp.47–62, 1993.

A SYSTEM SUPPORTING THE HUMAN DIVERGENT THINKING PROCESS BY PROVISION OF RELEVANT AND HETEROGENEOUS PIECES OF INFORMATION BASED ON AN OUTSIDER MODEL

Kazushi Nishimoto, Shinji Abe, Tsutomu Miyasato, and Fumio Kishino
ATR Communication Systems Research Laboratories,
2-2, Hikaridai, Seika-cho, Soraku-gun, Kyoto, 619-02, Japan
Email : knishi@atr-sw.atr.co.jp

ABSTRACT

The authors have been trying to construct a system that stimulates the divergent thinking process by making an "artificial outsider" attend a brainstorming session in a teleconferencing system with realistic sensations. In this paper, we discuss what kinds of information can break human fixed ideas and how an outsider can extract such information. Based on the result of this discussion, we propose an "Outsider model" for getting such information and construct a prototype system. Moreover, by classifying the whole information space into four regions from objective and subjective viewpoints, we discuss which region stimulates the human imagination and propose an evaluation method for determining whether a piece of information can effectively stimulate the human imagination. We experiment using a prototype system and carry out evaluations based on the evaluation method. The results show that the prototype system can extract information capable of stimulating the human imagination.

1 INTRODUCTION

We have been researching a teleconferencing system with realistic sensations made possible by virtual reality technologies[Kis91]. The concept of this system is shown in Figure 1. The main feature of this environment is that through the use of 3-D display techniques, the images of the participants are rendered in stereoscopic mode so that they appear in 3-D and look real to viewers from the real space. The objects appearing in the virtual space are all synthesized by using computer graphics techniques so that the virtual space appears, as realistically as possible, as an extension of the real space. Moreover, the participants in the virtual space are displayed in such a way that their gazing directions all converge on the same focus of attention, further enhancing the sensation of a real conference room atmosphere. Another feature of this virtual teleconferencing room is the capability of displaying, in the virtual space, the object that is being discussed in 3-D mode. For example, an image of an automobile can be displayed in 3-D in the virtual space as shown in Fig. 1. Furthermore, direct manipulation of the virtual object is also possible during the conference, giving an extra degree of flexibility for articulation during the conference. It therefore has great potential in making people generate new ideas and concepts.

As one example, the authors have been trying to combine the teleconferencing system with realistic sensations with a

FIGURE 1 A concept of the teleconferencing system with realistic sensations

system supporting human creativity[Nis93]. The goal of this research has been to construct a system that stimulates the divergent ways of thinking of all participants by making a virtual participant synthesized by a computer and a database attend a brainstorming session as an "artificial outsider". This makes the system listen to all opinions and enables it to provide timely pieces of information that take account of the opinions.

In section 2, we discuss a concept supporting human divergent thinking and propose an "outsider model". In section 3, we explain a prototype system based on the outsider model. In section 4, we discuss a methodology capable of evaluating supporting systems for the human divergent thinking process and propose an evaluation method. In section 5, we show a subjective experiment using the prototype system and discuss its efficiency based on the evaluation method.

2 A METHOD FOR SUPPORTING THE HUMAN DIVERGENT THINKING PROCESS BY AN ARTIFICIAL OUTSIDER

2.1 Discussion about the function of an outsider in the divergent thinking process.

The purpose of the human divergent thinking process is to make a problem clear and to collect relevant pieces of information [Kun93]. Therefore, it would be effective if we could gather pieces of information having various relevance to the problem in question.

Brainstorming[Osb63] is a familiar way of supporting this process and experience tells us that it is effective for not only experts but also for an outsider to attend a brainstorming session. A brainstorming session serves to have the participants collect various ideas. However, a deadlock often occurs when all participants are experts of the same domain, i.e., fixed ideas exist. Therefore, it is expected that heterogeneous information provided by an outsider would provide a stimulus to the imagination of experts, leading them to associate more information of various types. That is, the function of the outsider is to break the others' fixed ideas by providing some pieces of heterogeneous information.

As a simple example, we will explain a method of extracting heterogeneous information and a mechanism capable of breaking fixed ideas to a group. Here, we consider the way in which people see a pattern on a full moon. Objectively speaking, the same pattern can be seen by everyone. However, a Japanese may say "A rabbit is on the moon and she is making MOCHI (rice cakes)," and a Moroccan may say "The former king is riding on a horse." Thus, even if we see the same pattern, we individually understand it in different ways. We assume that such information processing is constructed by the following steps:

(1) Getting a part possibly having some feature from a given pattern.

(2) Associatively expanding the part using one's own knowledge.

(3) Extracting a known pattern resembling the expanded pattern.

In this process, by getting a different part in step (1) and/or by associatively expanding it in a different way in step (2), heterogeneous information from a different viewpoint can be extracted even if the same information is given.

It should be noted that all of the data processed in the experiments in this paper are text data. We regard text as a pattern constructed by words. The basic method for retrieving text data in our experimental system is based on the method mentioned above. However, text has meaning. Therefore, we should process it by considering the meaning, different from the way we would process meaningless patterns. If not, information (text) extracted by the system would diverge too much and there would be no relevance in many cases. Therefore, when constructing an artificial outsider system based on the above concept using text data, it is necessary to include some process capable of understanding/handling the meaning of the text data.

Next, we will explain how the heterogeneous information breaks one's set of fixed ideas. In this example, most Moroccans do not know MOCHI. Therefore, they cannot imagine "A rabbit making MOCHI on a full moon". On the other hand, most Japanese do not know the figure of the former king of Morocco. Therefore, they cannot think of the pattern as "The former king of Morocco riding a horse." However, if a Japanese tells a Moroccan how to make MOCHI, or a Moroccan shows a picture of the former king to a Japanese, the Moroccan may understand how a Japanese sees the moon and the Japanese may understand how a Moroccan sees it.

In such a process, having one's viewpoint changed is essential. For a Moroccan, he/she must have his/her ordinal viewpoint changed, obtain other parts of the pattern and associatively expand the newly-obtained part using the image of a rabbit making MOCHI. It is hard for people to change their own viewpoints. However, once someone succeeds in changing such viewpoints for them, their scope can be expanded and they can get out of their fixed ideas. In this manner, heterogeneous information makes one's viewpoint change and breaks his/her set of fixed ideas.

2.2 An outsider model

We propose the outsider model shown in Figure 2 in

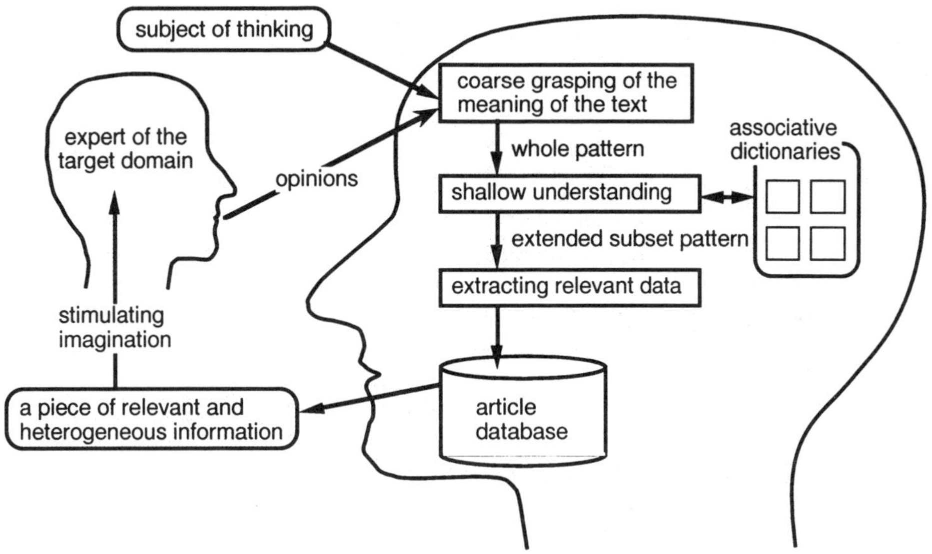

FIGURE 2 An outsider model: Listening to experts' opinions, an outsider coarsely grasps their meaning, shallowly understands them and extracts a piece of information which the outsider has. Such a piece of information can be not only relevant but also heterogeneous, stimulates the experts' imagination and breaks their fixed-ideas.

order to extract information having not only some relevance to the subject of thinking, but also some heterogeneousness. An outsider provides pieces of information through the following process.
(1) Coarsely grasping the meaning of the text: Superficially grasping the meaning of the content of other's opinions.
(2) Obtaining a shallow understanding: Based on the grasped meanings, getting a shallow understanding using only personal knowledge.
(3) Extracting relevant data: Based on the shallow understanding, extracting data appearing to have some relevance, using personal knowledge.
Every step corresponds to a step of the process of association from meaningless patterns as we mentioned in section 2.1, but there is a difference in that this model allows the grasping and understanding of meaning.

2.2.1 Coarsely grasping the meaning of the text

This step gets a set of words that possibly holds the meaning of the text. We call this set of words a "whole pattern", corresponding to the whole pattern on the full moon.

We assume that a set of words not forming a sentence can coarsely represent the meaning of the original text. That is, each word included in the set restricts the ambiguities of other words mutually and, as a result, the set represents a concrete meaning. On the other hand, the outsider-model does not include a step for deep understanding. Therefore, the whole pattern may include not only so-called keywords which have a direct relation but also some words having only an indirect relation. Such indirect relation may introduce heterogeneousness.

2.2.2 Shallow understanding

This step associatively modifies the whole pattern by using some associative dictionaries. In this step, depending on the associative dictionary, different words are selected and expanded. We call the modified "whole pattern" an "expanded subset pattern" below.

In this step, we regard associating some relational words from a given word as a kind of understanding. For example, when someone explains about “an apple”, by giving some relational words recalled associatively (e.g., red, fruit, jam, Newton, etc.), he/she tries to make the meaning correct. Other words appearing with “apple” in the same text restrict ambiguities mutually and make the whole meaning clear. For example, if “planet” appears with “apple”, then “Newton” may be strongly recalled from the set of words associated with “apple”, and it will be assumed that the text is about physics or astronomy.

However, it is easy to see that this understanding method by associative expansion gives different results depending on knowledge. With similar knowledge, people could associatively expand in almost the same way. However, with heterogeneous knowledge, people would associatively expand in different ways. Consequently, if we prepare an associative dictionary that is constructed by heterogeneous information/knowledge, the expanded subset pattern will not only have some relevance but will also be heterogeneous at the same time. If we use an associative dictionary constructed based on specific domain knowledge, we can give the character of an expert of another domain to the artificial outsider. Using such a mechanism, this model can grasp a given pattern from a different viewpoint and shallowly understand it. This process is the most important part of this model.

2.2.3 Extracting relevant data

According to the result of the previous step, the outsider refers to his/her own knowledge and provides a piece of information which is inferred to have all possible relevance. For selecting a certain piece of information, the degree of resemblance between each whole pattern of data stored in the database and the expanded subset pattern must be calculated.

3 STRUCTURE OF THE PROTOTYPE SYSTEM

Based on the outsider model, we have constructed a prototype system. Figure 3 shows the software structure and the process flow of the system. The software consists of a parser, an associative memory module, and a database manager with an article database. Each module corresponds closely to each step of the outsider model.

3.1 Overview of the process flow

Before this system can be used as an artificial outsider, associative dictionaries and article databases have to be constructed. First, an article is input into the parser. The parser analyzes the article and modifies it into a keywords vector. The keywords vector is input into the associative memory module. This module associatively memorizes it and generates an associative dictionary. At the same time, the database manager registers the article with its keywords vector, if needed.

When this system is used as an artificial outsider, a user inputs an opinion into the system. The opinion is analyzed by the parser and modified into a keywords vector. This vector corresponds to the "whole pattern". The keywords vector is input into the associative memory module, and then this module recalls a certain keywords vector using some associative dictionaries. The recalled vector corresponds to the "expanded subset pattern". The database manager calculates the mutual correlation between the recalled vector and the keywords vector of each article stored in the article database. Then, after referring to the user's history, which contains information on which articles had been given to the user, an article with a high mutual correlation value and not yet given to the user is provided to the user.

3.2 Details of each module

3.2.1 Parser

This module divides the input text into words, decides the part of speech of each word and modifies each word to a regular form, e.g., a verb is modified into its present tense and a noun is modified into its singular form. As a result of this analysis, a word and part-of-speech list is generated. Next, a subset word list that includes only words of specified parts-of-speech is made. Using this subset list, a keywords vector is generated. The keywords vector K_j of the article A_j is generated as follows:

$$K_j = (\delta_1, \delta_2, \delta_3, \ldots\ldots, \delta_i, \ldots\ldots, \delta_m) \quad (1)$$

$$\delta_i = \begin{cases} 1 & (w_i \in A_j) \\ 0 & (w_i \notin A_j) \end{cases} \quad (2)$$

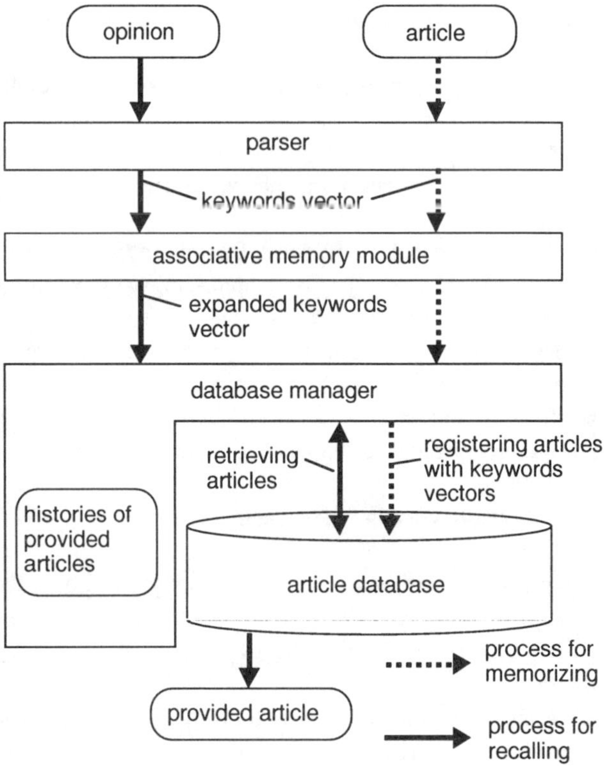

FIGURE 3 Software structure of the prototype system based on the outsider model: There are two process flows, one is for memorizing input articles and the other is for recalling by using input opinions.

where m is the total number of different kinds of words the system has processed, and w_i is the i-th word of the different kinds of words. The words included in the subset list are selected by their order of appearance in the text, and the number is restricted.

3.2.2 Associative memory module

We applied the Associatron[Nak72] to the associative memory method. Memorization is performed by generating an associative memory matrix. n keywords vectors are memorized as follows:

$$M = \sum_{j=1}^{n} K_j K_j' \tag{3}$$

where K_j is a vertical vector, the operator ' means the transposition of the vector and M is an associative memory matrix. All vectors we use below are vertical vectors.

Recalling is done as follows:

$$R = \phi_\theta \left\{ \phi_0(M)Q \right\} \tag{4}$$

where M is the associative memory matrix, Q is a keywords vector generated from an input opinion, R is a recalled keywords vector and ϕ_θ, ϕ_0 are the following quantizing functions:

$$\phi_0(x) = \begin{cases} 1 ; x > 0 \\ 0 ; x = 0 \\ -1 ; x < 0 \end{cases} \quad \phi_\theta(x) = \begin{cases} 1 ; x > \theta \\ 0 ; -\theta \leqq x \leqq \theta \\ -1 ; x < -\theta \end{cases} \tag{5}$$

These quantizing functions can also be applied to a vector and a matrix. In such a case, each element of the vector or the matrix is quantized.

The threshold value θ of ϕ_θ is determined so as to make the number of set-1-elements of a recalled vector equal or near to the number of set-1-elements of each memorized vector.

3.2.3 Database manager

This module calculates the mutual correlation r_j between the recalled vector R and the keywords vector K_j of the j-th articles as follows:

$$r_j = \frac{K_j' \cdot R'}{m_R} \times \frac{K_j' \cdot R'}{m_{K_j}} \tag{6}$$

where m_R is the number of set-1-elements of the recalled vector, m_{K_j} is the number of set-1-elements of the vector K_j and the operator · means the inner product of the vectors.

Then, the article whose keywords vector has the highest mutual correlation value is retrieved from the article database. However, this module has a history showing which articles have already been provided to which user. Referring to them, only those articles not having been provided yet are provided.

4 DISCUSSION ABOUT AN EVALUATION METHOD AND AN EVALUATION EXPERIMENT

4.1 Analysis on how various kinds of information stimulate the human imagination

Figure 4 shows how we classify the whole information space when the subject of the thinking is given. The whole space is classified into two regions: the outer region or "Objectively no relevance" region and the inner region or "Objectively some relevance" region. The line o dividing these two regions is the border line of the objective relevance limit. Region 4, on the outside of line o, includes pieces of information having no relevance to the subject of thinking.

The pieces of information inside of line o are classified into two regions by the user's subjectivity: the outer region or "Subjectively no relevance" region, and the inner region or "Subjectively some relevance" region. The line s dividing these two regions is the border line of the subjective recognition limit of the relevance. Region 3, which is between lines o and s, includes pieces of information having some relevance to the subject of the thinking. However, the user cannot recognize the relevance for certain reasons.

The pieces of information inside of line s are classified into two regions: The information in region 1, the inner region, has already been recalled and its relevance is known. Meanwhile, the information in region 2, the outer region, has not been recalled and will not until the information is given; but as soon as it is given, the user notices the relevance rather easily. The line a dividing these two regions is the border line of the recalled limit.

According to the classification above, information stimulating the user's imagination and/or creativity must be included in regions 2 and 3. The information in region 2 is "Overlooked information". The user can easily notice the relevance, however, he/she overlooks it until it is given. That is, such information makes the user notice the overlooked relevance directly. Therefore, it efficiently breaks the user's set of fixed ideas as soon as it is given. The information in region 3 essentially includes some relevance. However, the user cannot find/recognize them clearly. Therefore, it is impossible with such information to expect a quick and direct effect different from the information in region 2. However, once the user notices the hidden relevance, his/her imagination can be expanded and the generation of some new ideas can be expected.

On the other hand, the information in region 4 has

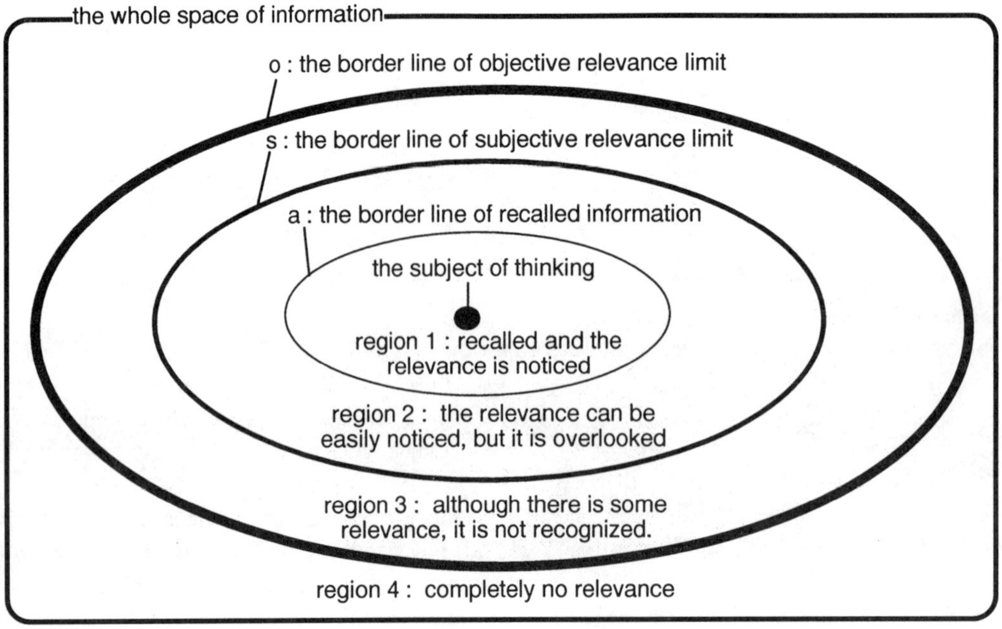

FIGURE 4 Classifying the whole information space from a objective and subjective viewpoint based on a given subject of thinking : There are three border lines and four regions. The border line a and s are border lines of fixed-ideas. So as to break them, it is efficient to provide pieces of information included in region 2 and/or region 3.

completely no relevance. Therefore, it is hard to expect that such information will stimulate the user's imagination. Furthermore, although the information in region 1 is very significant as basic information for generating new ideas, it is hard to expect that such information will stimulate the user's imagination.

As a result, we can regard the set of fixed ideas as the lines a and s. Then, by extracting and showing information included in region 2 and/or region 3, the user can find his/her own set of fixed ideas and break the set.

4.2 Evaluation method for pieces of information provided by a supporting system for the human divergent thinking process

To evaluate in what region a provided piece of information is included, we propose a subjective evaluation method. From the following two points of view, subjects are to evaluate each piece of information:

1) Relevance: How much does the provided piece of information relate to the subject of the thinking.

2) Heterogeneousness: To what extent could the subject not expect (predict) such a piece of information to be provided.

Intuitively speaking, relevance and heterogeneousness are opposing concepts. Therefore, regarding to a subject of the thinking, if we provide various pieces of information for subjects to evaluate from these two viewpoints and then plot the evaluation results on a graph like that shown in Figure 5, we can expect the plots to be distributed almost exclusively along line l and in the upper-right region of line l. It is hard to suppose that a piece of information can be distributed in the lower-left region of line l.

The relevance of the pieces of information included in region 1 of Fig. 4 has been learned in advance by the subjects and they expect/predict to be provided such pieces of information. Therefore, when such pieces are provided, they are evaluated as having high relevance. However, the subjects are not surprised at seeing them. Consequently, they are also evaluated as having low heterogeneousness. Therefore, region A of Fig. 5 corresponds to region 1 of Fig. 4.

The pieces of information included in region 2 of Fig. 4 have some evident relevance and the subjects can easily notice it. Therefore, the pieces are evaluated as having high relevance. However, the subjects have not expected/predicted to have them provided and they are surprised at the fact that they have overlooked such an evident relevance. Consequently, they also evaluate such pieces as having high heterogeneousness. Therefore, region B of Fig. 5 corresponds to region 2 of Fig. 4.

The pieces of information included in region 3 of Fig. 4 ordinarily have some relevance, but the subjects cannot evidently recognize it and sometimes determine that there is no relevance. Consequently, if such pieces are provided,

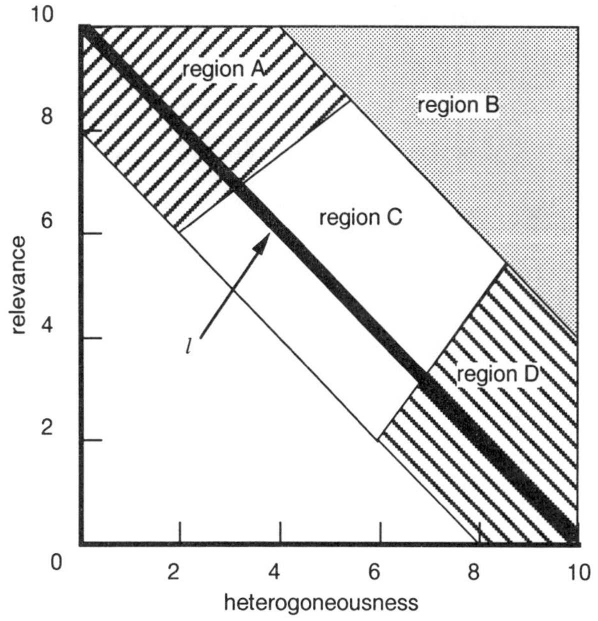

FIGURE 5 An expected distribution of pieces of information evaluated by relevance and heterogeneousness: Region B corresponds to region 2 of Fig.4, and region C and region D correspond to region 3 of Fig.4.

they are evaluated as having low (or sometimes no) relevance and high heterogeneousness. Therefore, region 3 of Fig. 4 corresponds to regions C and D of Fig. 5.

As pieces of information included in region 4 of Fig. 4 objectively have no relevance, of course the subjects evaluate that they have no relevance and very high (complete) heterogeneousness. Therefore, region 4 of Fig. 4 corresponds to region D of Fig. 5. Consequently, if we implement a system that provides only pieces of information having some relevance from an objective viewpoint, we can determine the pieces of information evaluated to be included in region D as information included in region 3 of Fig. 4.

According to the evaluation of the provided pieces of information, we can say that the system has the ability to stimulate human imagination if the provided pieces are properly distributed in region B, C or D. If many pieces are distributed in region B, the system can quickly and directly break the user's set of fixed ideas, or if many pieces are distributed in region C and/or region D, the system is not able to stimulate so quickly and directly, but by thinking of the relevance of the provided information, the subjects may find some amazing relevance and get new ideas.

5 SUBJECTIVE EVALUATION EXPERIMENT OF A PROTOTYPE SYSTEM BASED ON THE OUTSIDER MODEL

5.1 Subjective evaluation experiment and results

To make an article database and an associative dictionary, we used articles from CLARINET news, which is a commercial Internet news network. The number of processed articles was 1465. The "whole pattern" of each article was made to include only nouns and unknown words. The number of set-1-elements of each keywords vector was restricted to 50 maximum. The total number of different kinds of processed words was 12637.

In the experiment, the following subject of the thinking/interest/opinion was input to the system.

Subject of the thinking: Computer market
Interest: The future of Apple Macintosh.
Opinion: I think Windows will be in place of the Macintosh, even if the Macintosh's performance is raised dramatically by the adoption of the PowerPC chip.

In this case, the distinction of subject/interest/opinion actually has no meaning. This distinction is used in other functions of the system.

Using the two algorithms mentioned below, six articles for each algorithm (twelve articles in total) are extracted. The number of subjects is eight. By providing three articles per algorithm, six articles in total to each subject, the subjects subjectively evaluate them based on their relevance and heterogeneousness. The subjects are not informed of what algorithm is used to extract each article. The articles are provided as hard copies to the subjects.

The two algorithms are as follows.

(1) Algorithm based on the outsider model
(2) Algorithm that directly calculates mutual correlations.

In other words, algorithm (2) is "based on the outsider model but without a step for shallow understanding". In algorithm (2), the "whole pattern" of the input text is not expanded by associative dictionaries and is directly calculated with the "whole pattern" of each article. Comparing the evaluation results of algorithm (2) and algorithm (1), we expect to find the effect of the step for shallow understanding, which can be the most significant step for getting heterogeneousness.

There is a possibility that identical articles may be extracted by both algorithms (1) and (2). However, all twelve extracted articles were different in this experiment.

The subjects read the provided articles and evaluated their relevance and heterogeneousness on a scale of from 0 to 10. Figure 6 shows the evaluation results. For example,

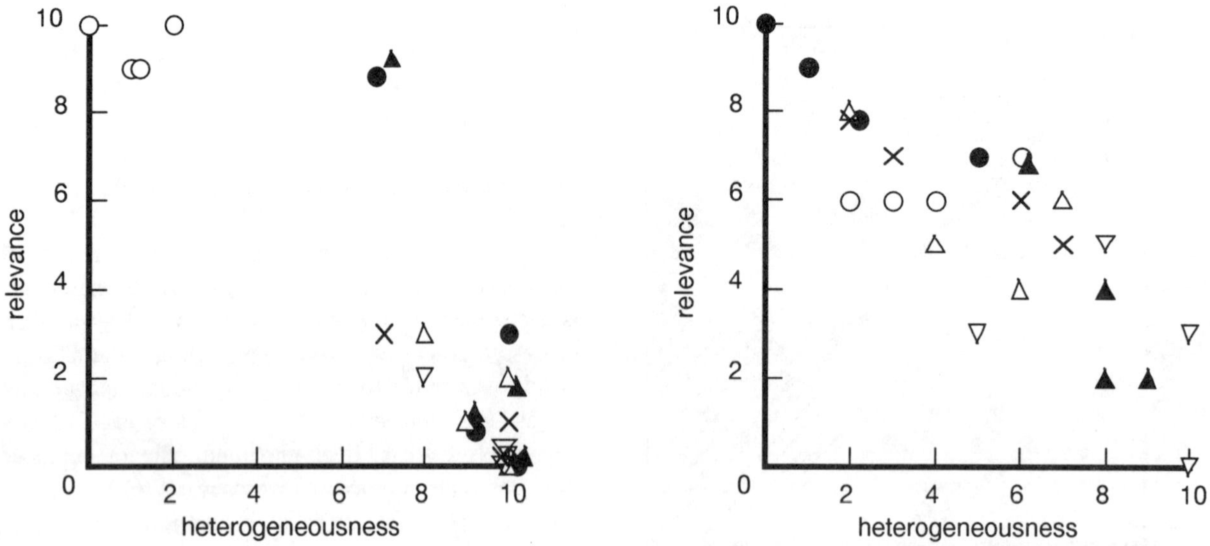

(1) using the alogorithm based on the outsider model

(2) using the algorithm directly calculates mutual correlations.

FIGURE 6 Results of a subjective experiment : Without shallow understanding step, algorithm (2), the experimental system extracts pieces of information mainly included in region C of Fig. 5. Based on the outsider model, algorithm (1), extracted pieces of information are more heterogeneous and less relevant than algorithm (2). And it becomes possible to get pieces included in region B of Fig. 5 using algorithm (1). It is hard to get such information without shallow understanding.

in Fig. 6 (1), ○ shows the evaluation results for a certain article by four subjects. The article represented by ○ in Fig. 6 (1) is different from the article represented by ○ in Fig. 6 (2).

5.2 Discussion

In Fig. 6, the results of algorithm (2) are almost entirely distributed along line l of Fig. 5, especially in region C. According to the results of algorithm (2), information included in region 3 of Fig. 4 can be extracted. On the other hand, the results of algorithm (1) are mainly distributed in region D. That is, algorithm (1) also mainly extracts information included in region 3 of Fig. 4, however, these pieces of information are more heterogeneous than those with algorithm (2).

In algorithm (2), the articles are extracted by using only words included in the input text and higher r_j articles include more of them. Therefore, the relevance of high r_j articles always appears directly. And as r_j becomes low, the article comes to include fewer keywords. Consequently, the relevance becomes lower while the heterogeneousness becomes higher. Therefore, using algorithm (2), the extracted articles usually distribute along line l of Fig .4 and it is difficult to get articles included in region B of Fig. 5.

On the other hand, in algorithm (1), as the articles are extracted by using not only the words included in the input text but also the words expanded from them, the relevance can become indirect and a little ambiguous. Therefore, the subjects have more difficulty recognizing their relevance than with algorithm (2). This is the reason for the low relevance/high heterogeneous results of algorithm (1). However, the prototype system is designed to extract only pieces of information which have some relevance. Therefore, such articles must include a certain (indirect) relevance. Consequently, even if they are evaluated as being included in region D, they are included in region 3, and not 4 of Fig. 4.

If someone recognizes this hidden indirect relevance, as soon as it is provided, they evaluate the articles as having both high relevance and high heterogeneousness. Therefore, using algorithm (1), information included in region B can be gotten. Such information is hard to obtain by algorithm (2). As a matter of fact, two articles represented by ● and ▲ are evaluated as being included in region B by one of the subjects. It is assumed that this subject succeeded in finding the hidden relevance of the two articles. This result shows the fact that these two articles include a certain relevance although the other subjects evaluate them as being included in region D of Fig. 5. Therefore, if they can also find the hidden relevance, it is expected that they will evaluate them as being included in region B of Fig. 5 and may get new ideas by the pieces of information.

In conclusion, based on the outsider model, we can get not only low relevance /high heterogeneous information but also high relevance/high heterogeneous information.

6 RELATED WORK

The Colab[Ste87], a well known collaborative system, provides a face-to-face meeting room to users. It supports a brainstorming session by providing some useful tools. However, it does not provide any information to users. The AA1[Hor94] also provides a space for brainstorming to users. It restructures an information space input by users. Therefore, it actively provides unnoticed structural information to users. However, it is different from our system in terms that the AA1 does not provide other new relevant pieces of information.

The Metaphor machine[You87] aims to support the human creative thinking process by providing metaphors. The system automatically generates metaphors by using a relational database method. The Keyword Associator [Wat90] aims to support the human creative thinking process by providing relational keywords associated from input words using some associative dictionaries. The concepts of these systems to support human creativity by providing some heterogeneous pieces of information resemble our system. However, the method for getting/giving heterogeneousness is different from ours. The metaphor machine gives users only metaphors. The keyword associator basically gives users only associated words, not sentences or text. And they do not process the meaning of input information.

7 CONCLUSION

In order to support the human divergent thinking process by synthesizing an artificial outsider and making it attend a brainstorming session in a teleconferencing system with realistic sensations, we have discussed what kinds of information can efficiently break human fixed ideas and how an outsider can extract such information. We have shown that if some pieces of information extracted from a different viewpoint are given to people and if they notice the difference of viewpoints, they can change their own viewpoint and break through the frame of their fixed ideas.

To get such pieces of information, we proposed an "Outsider model", consisting of a "Coarsely grasping of the meaning of text" step, a "Shallow understanding" step and an "Extracting relevant data" step. In the shallow understanding step, which is the most significant step of this model, by associatively expanding words included in an input text of opinions using some associative dictionaries, pieces of information from a different viewpoint can be gotten. We also constructed a prototype system based on this model. On the other hand, by classifying the whole information space into four regions from objective and subjective viewpoints, we discussed which region stimulates human imagination. We concluded that pieces of information which have some unnoticed relevance are efficient for stimulating people's imagination. We proposed an evaluation method for determining in which region a certain piece of information is included and whether the piece can effectively stimulate human imagination by evaluating the relevance and heterogeneousness of the piece. We experimented using the prototype system and carried out evaluations based on the evaluation method for eight subjects. The results showed that the prototype system can extract pieces of information capable of stimulating human imagination. Moreover, the system can extract pieces of information which are not only low-relevant/high-heterogeneous but also high-relevant/high-heterogeneous by the function of the shallow understanding step. It is difficult to get such information without this step.

In this paper, we used only one associative dictionary made from newspaper articles and showed a system capable of extracting relevant and heterogeneous pieces of information using only it. However, this system cannot control and change its own viewpoint. Therefore, we have been trying to control the viewpoint of the system at will by using plural associative dictionaries, each of which is made from articles of a different domain and combining them. By this method, we expect to get "Artificial experts of different domains". We are also planning to experimentally apply such an artificial outsider or expert-of-a-different-domain to a teleconferencing system and to evaluate the system's efficiency.

REFERENCES

[Hor94] K. Hori, *"A System for Aiding Creative Concept Formation"*, IEEE Trans. on Systems, Man, and Cybernetics, vol. 24, no. 6, pp. 882-894, June 1994.

[Kis91] F. Kishino, *"Communication with Realistic Sensations through Integrated Multi-media Environment"*, Proc. WS on Integrated Comprehension and Generation System, pp. 49-54, December 1991.

[Kun93] S. Kunifuji, *"A Survey on Creative Thinking Support Systems and the Issues for Developing Them"*, Journal of Japanese Society for Artificial Intelligence, vol. 8, no. 5, pp. 552-559, September 1993.

[Nak72] K. Nakano, *"Associatron - A Model of Associative Memory"*, IEEE Trans. on S.M.C., SMC-2, 3, pp. 381-388, July 1972.

[Nis93] K. Nishimoto, K. Mochizuki, T. Miyasato, and F. Kishino, *"A supporting system for human creativity: Computer aided divergent thinking process by provision of associative pieces of information"*, Proc. Vienna Conference on Human Computer Interaction (VCHCI '93), pp. 437-438, September 1993.

[Osb63] A. Osborn, *"Applied Imagination: Principles and Procedures of Creative Thinking"*, Scribner's, New York, 1963.

[Ste87] M. Stefik, G. Foster, D. G. Bobraw, K. Kahn, S. Lanning, and L. Suchman, *"Beyond the chalkboard: Computer support for collaboration and problem solving in meetings"*, Commn. ACM, vol. 30, no. 1, pp. 32-47, 1987.

[Wat90] I. Watanabe, *"IdeaEditor: Overview"*, Research Report HAS-RR-90-2E, HAS-SIS, FUJITSU LIMITED, 1990.

[You87] L. F. Young, *"The Metaphor Machine: A Database Method for Creativity Support"*, Decision Support Systems, vol. 3, no. 4, pp. 309-317, 1987.

Machine Learning

POLYGONAL INDUCTIVE GENERALISATION SYSTEM

D.A.Newlands and G.I.Webb
School of Computing and Mathematics,Deakin University,
Geelong, Victoria 3217, Australia.
Email:doug@deakin.edu.au,webb@deakin.edu.au

ABSTRACT

Classification learning has been dominated by the induction of axis-orthogonal decision surfaces. While induction of alternate forms of decision surface has received some attention in the context of decision trees, this issue has received little attention in the context of decision rules. An inductive learning algorithm has been developed which creates arbitrarily shaped concepts. Results from a prototype implementation demonstrate that the approach performs well on target concepts that are not readily represented by long, flat decision surfaces.

INTRODUCTION

The complexity of the representation of an hypothesis produced by a given computational learning system is a function of both the hypothesis language and the target concept. If the concept being learned cannot be concisely represented, it will have to be approximated, possibly by a large number of inappropriately shaped decision surfaces. The typical axis-orthogonal decision tree representation of the concept shown in Figure 1 is an example of the problem. Systems which represent concepts in disjunctive normal form propositional calculus or similar form and which use the natural attributes of the learning situation (decision trees [Ces87], covering algorithms [Mich86], [Web92]), are implicitly using hyper-rectangles to represent the concept in the instance space. If the concepts to be learned have axis-parallel boundaries, the representation will be concise, using a small number of large decision surfaces. However, concepts with straight but non-axis-parallel or curved boundaries will only be approximately represented by a large number of small decision surfaces. Systems for the induction of concepts with oblique boundaries have been described including oblique decision trees [Mur94] and synthesised attributes [Yip94, Quin93, Ren90, Blo91, Bre94, Mat89, Meh84, Mich86, Pag90, Utg91, Wne94] but most have limitations when concept boundaries are not linear. Statistical and nearest neighbour methods of concept formation can be regarded as representing concepts by hyperspheres or similar shapes e.g.CLUSTER [Mich83], UNIMEM [Leb87] and COBWEB [Fis87]. Instance-based learning, derived from nearest neighbour classifiers [Aha91], does not maintain a generalisation of instances but stores instances and examines them to make classifications. Salzberg [Sal91] describes Nested Generalized Exemplar theory which is currently implemented using hyper-rectangles but, in principle, can be implemented using other shaped decision regions.

All of the above methods avoid the n-dimensional geometry of the learning area: the hyper-rectangle based methods by being able to examine each axis separately since surfaces are orthogonal to axes and the statistical methods by imposing symmetry on the situation using Pythagorean measures on the instances. This article examines the use of arbitrarily-shaped n-dimensional solids to represent decision regions. The quality of the representation of hypotheses in this framework should not depend on how well the geometry of the actual con-

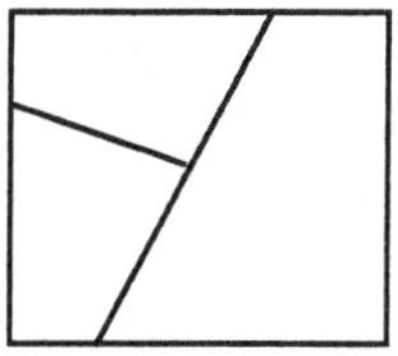

Target concepts represented by two decision surfaces.

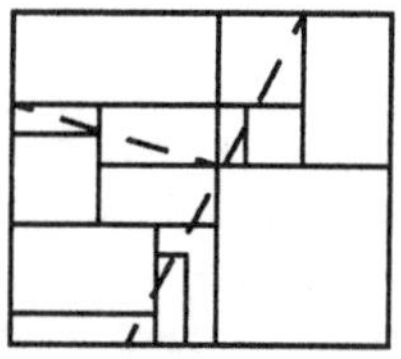

Approximation that might be formed with axis-orthogonal decision surfaces

FIGURE 1 A complex representation of a simple concept

cept fits the hypothesis language of the inductive system and thus one would expect high quality hypotheses and representations over a wider range of tasks than methods with an implicit reliance on a regular polygon structure.

POLYGONAL INDUCTIVE GENERALISATION SYSTEM(PIGS)

The proposed system will construct hyper-polygons, each representing (part of) a concept, using a successive generalisation algorithm such as employed by GOLEM [Mug90], DLG [Web92a] and Multiple Convergence [Mur87]. Each hyper-polygon will use instances as vertices, but no internal vertices or edges will be represented. New instances which are internal to an appropriate (i.e. belonging to the correct concept) hyper-polygon will be regarded as being covered. New instances which are not so covered will be attached to the nearest polygon of the correct concept by performing a minimal generalisation which does not lead to any negative instances being covered. If no generalisation is possible, a new polygon will be started.

Generalisation of a polygon and point will be effected by inserting edges between the new instance and all vertices which are directly visible from the new point and which are one edge from the point on the polygon which is nearest to the new instance. The algorithm for generalisation of a hyper-polygon is shown below.

```
generalise-polygon(POLYGON p,
                   POINT new-instance)
    FIND nearest vertex, V0, of polygon p
    FIND all vertices,V1...Vn, connected to V0
    IF all V1...Vn are visible from new-instance
              REPLACE coordinates of V0 with
                        those of new-instance
    ELSE
              DELETE invisible vertices V1...Vn
              CREATE new vertex, Vn+1,
                        containing new-instance
              CONNECT new vertex,Vn+1, to
                        V0,V1....Vn
              DELETE connections from V0 to
                        V1...Vn
    ENDIF
END.
```

If there is a subsequent problem with covering of negative instances, the generalisation can be rolled back. This form of generalisation, when there are no hidden vertices, does not lead to any increase in storage size since no new vertex is created; only the location of a preexisting one is altered.

In a continuous instance space, the measure of minimality of generalisation should be in terms of the volume of instance space enclosed since the representation is purely geometric and not restrained to be at particular angles to the axes. In methods where the representation is at a symbolic level, least generalisation, as discussed by Plotkin [Plo70], in terms of the hypothesis language is reasonable but subsumes larger, rectangular volumes of instance space than does polygonal generalisation. The actual generalisation method is, therefore, considerably more conservative than hyper-rectangle based methods and would lead one to expect fewer false positives than with other generalisation techniques. Post-processing on the decision regions (hyper-polygons) should permit extraction of higher level hypotheses by fitting large regular shapes to the regions, using mathematical techniques to select among possible large shapes and using regularities in one area to complete other areas.

THE PROTOTYPE

This initial implementation will be in 2-dimensions since all situations are readily visualisable in 2-dimensions. In particular then, a concept, in the prototype, will be a set of surfaces; each surface will be a set of lines and each line will be a pair of vertices. An instance of a concept is a point within one of the surfaces representing that concept. Classifying an instance requires identification of which surface(s) it lies within.

The learning task is to construct concepts when presented with attribute vectors consisting of pairs of continuous numerical values. Non-numeric attributes, missing values, noise and counter-examples are not considered here as the objective is to gain insight into the operation of a new generalisation technique which has characteristics not found in hyper-rectangle based methods. For the same reason, dealing with overfitting and the computational complexity of of the prototype algorithm are not central to this work.

COVER AND GENERALISATION

An instance will be *covered* by a concept if it lies within one of the surfaces of the concept. A simple approach to cover and a desire to minimise storage requirements require a little care in generalisation so that the polygon has no internal edges. Consider the left part of figure 2 where points a, b, c have already been generalised to form a surface, P is a new positive instance of the same concept and N is a negative instance. If the generalisation is done as at the right in figure 2, the internal lines xa and xc will cause the cover algorithm to malfunction. (Generalisation bpc is not permitted because it would cover the negative instance!

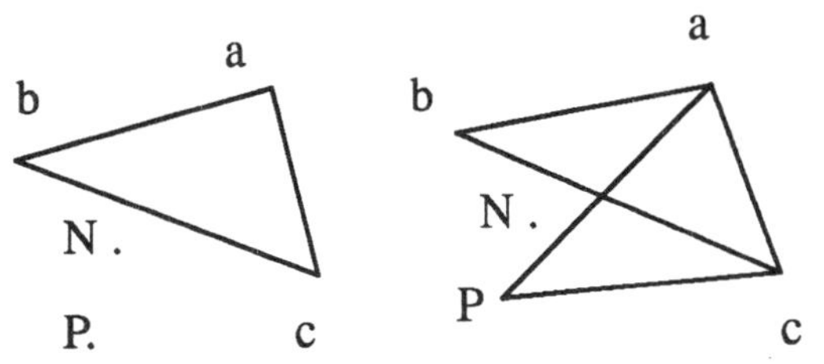

FIGURE 2 Faulty generalisation

The algorithm for generalisation of a concept is

```
generalise(POINT newpoint,
        CONCEPT concept)
    FOR all polygons in concept
        FIND nearest vertex to newpoint
        PUT polygon, vertex and distance
            in possible-list
    ENDFOR
    ORDER possible-list on distance
    PUT empty polygon last in possible-
        list to guarantee generalisability
    FOR each polygon-vertex-distance-tuple
            in the possible-list
        generalise-polygon (polygon,newpoint)
    IF polygon now covers any negative point
        roll back generalisation
    ELSE
        RETURN from successful generalisation
    ENDIF
    ENDFOR
END.
```

SPIKING

The reason for sorting the polygons before generalisation is to avoid the situation where an instance is generalised onto an inappropriate, distant region. Such effects almost always create very narrow "spikes" as, to be viable, they must not cover any negative instances. However spiking has been observed to occur in two other situations. Early in the induction phase when there are few, or no, valid (i.e. having 3 or more points in the 2-d case being studied) regions, two points which are from different regions of the same concept may get joined and if a third point near one of these is processed then a spike from one region to another will result. This can be accepted if no negative instance in the training set contraindicates this generalisation. However, it is undesirable as, when applying a classifier to classify previously unsighted objects, if an instance falls within the spike, it will be classified positive to both the concept represented by the spike and to the concept of the area through which the spike passes. This problem can be side-stepped by a constraint on the length of the new sides of the newly generalised area either relative to the pre-existing edges of the newly generalised area or in absolute terms. Clearly the optimal length limit is less than an actual concept width but the geometry of concepts is not available before the induction process so some heuristic scheme has to be used. Preprocessing the data, which will be seen later to have some attractions, might allow estimation of the average inter-concept distance and this could be used to form an absolute length limit. The other reason for needing a length limit is in the case of the actual concept being quantised and there not being any negative instances e.g. two bands with a "forbidden" region between them. Now, if the data starts with 2 instances in one band and one in the other, generalisation would occur across the "forbidden" area between them. While it is true that the classifier could never get any instance wrong because none could occur in the "forbidden" area, there would be no possibility of extracting correct higher level formulations of the actual concept.

EVALUATION

The Conservation Law of Generalisation Performance [Sch94] states that no learning algorithm can, in general, obtain higher generalisation performance than any other. In this context, it is incumbent upon the researcher presenting a learning algorithm to identify the types of learning problems for which it might be expected to obtain high generalisation performance. By escaping the constraints of axis-orthogonal decision surfaces, PIGS should enjoy an advantage over systems restricted to axis-orthogonal decision surfaces when learning concepts that cannot be readily represented by such surfaces. With respect to oblique decision trees, the relatively short lines developed by PIGS should give it an advantage when the target concept cannot be well approximated by long, straight decision surfaces. PIGS, however, is unsuited to learning tasks where closeness in the instance space is not predictive of class. With this in mind, it is expected that PIGS will perform well on a wide variety of concept shapes since there is no bias towards a particular geometry although it will not perform as well as methods which are tailored to the type of concept being learned.

To evaluate these assumptions, comparisons between PIGS, OC1 (oblique decision trees [Mur94]) and C4.5

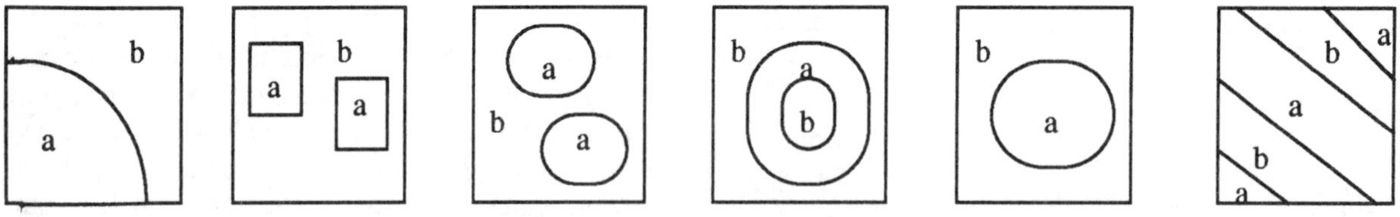

quad squares discs polo-mint circle POL

FIGURE 3 Test concepts

(axis-orthogonal decision trees [Quin93]) were performed on a range of artificial data sets ranging from "squares" where one would expect a decision tree to perform best since it will automatically produce straight edges of the correct orientation, to "POL"[Mur94] (parallel oblique lines) where an axis orthogonal decision tree would do less well than, say, an oblique decision tree [Mur94], to various curved concepts where all straight-edged decision trees should do less well and PIGS should be superior. The test concepts shown in figure 3 were used for experimentation. Each dimension is in the range [0..16]. Training sets consisted of 1600 randomly generated points and test sets of 400 not drawn from the training set. Fifty training and test set pairs were generated for each of the test concepts (except POL, only 30) and presented to PIGS, OC1 and C4.5. When developing rules, the maximum edge length allowed was the default for PIGS, 6. When applying the rules developed by PIGS, if no rule applied to an instance, the instance was inferred to belong to the nearest concept using a simple nearest neighbour technique. In no case did two or more contradictory rules cover an instance. The accuracy of PIGS was compared to OC1 and C4.5 using a 2-tailed, matched-pairs t-test. The results are shown in table 1 and table 2. The mean and standard deviation of the accuracy of each system is shown together with the test statistic and the probability that the outcome is by chance.

It can be seen that PIGS provided significantly better performance than C4.5 and OC1 on all concepts with curved geometry. On the "squares" data set, C4.5 does very well as one would expect but OC1 (not set to prefer axis-orthogonal surfaces) does not do significantly better than PIGS. On the "POL" data set, OC1 does significantly better than PIGS, as expected, but C4.5 does significantly worse as it is badly biased for this type of concept where there are no axis-orthogonal components. While testing PIGS it was observed that:-

1] the number of polygons per concept was typically 2 to 5 and this variation seems to be a function of the order in which instances are seen.

2] in more than 93% of runs, no instance was classified as belonging to two classes. This suggests that the simple, absolute value restriction on the size of new sides is reasonably successful in stopping spiking.

3] a number of items, on average 6%, lay outside the decision regions produced by PIGS. Clearly, the edges of concepts are over-specialised and there will always be

Table 1:

Concept	PIGS	PIGS	OC1	OC1	Statistics	
	Mean	St.Dev.	Mean	St.Dev.	t value	p
squares	98.90	1.107	99.02	0.848	0.8915	0.3770
quad	99.55	0.467	99.38	0.429	-3.0645	0.0035
circle	99.09	0.834	98.17	1.080	-8.0501	0.0000
discs	98.84	1.002	98.22	1.134	-4.9240	0.0000
polo	98.15	1.838	97.43	1.672	-3.5298	0.0009
POL	98.22	0.684	99.18	0.636	6.8737	0.0000

interstices between concepts. Consequent upon this, we note that it is errors in the nearest neighbour technique for unclassified points which produces the majority of the false positive and negative outcomes, *not* PIGS itself.

CONCLUSIONS

PIGS obtains significantly better predictive accuracy than C4.5 on every domain except "squares" where it was expected to be inferior. PIGS also obtains better predictive accuracy than OC1 on all curved concepts and is not significantly worse for "squares". OC1 does perform better on "POL" but this is not unexpected given the learning bias of OC1. The results give a clear practical demonstration of the consequences of conservation of generalisation performance. This performance from the prototype justifies continuing with future plans. As well as extending PIGS to n-dimensions, other areas to be investigated include

1] preprocessing the data with the objective of having clumped data at the front to enable seeding of good concepts in the induction process which should minimise spiking without any ad hoc constraint.

2] preprocessing the data with the objective of having well separated points at the front to form large concepts early and minimise the amount of induction to be done by having more points covered early in the process.

3] postprocessing the concepts to aggregate overlapping polygons to reduce the amount of computation in subsequent classification of instances.

3] postprocessing the concepts to smooth their surfaces to reduce their over-specialisation and to minimise the interstices between concepts.

4] postprocessing the concepts to construct higher level hypotheses from regularities in the polygons and their placement in instance space, e.g. having three equally spaced concepts, two of which are spherical and one of which is poorly represented, one might induce that the odd one should also be spherical; having four identically shaped, regularly spaced polygons, one might induce some kind of repetitive law.

PIGS has demonstrated the feasibility of induction of oblique decision surfaces within a covering algorithm. While this prototype implementation is restricted to two-attribute domains, the approach is, in principle, extensible to any number of dimensions. The excellent results obtained by this prototype demonstrate great potential.

REFERENCES

[Aha91] Aha,D.W. and Kibler,D. and Albert,M.K Instance-Based Learning Algorithms. *Machine Learning*, 6,1991, 37-66.

[Blo91] Bloedorn,E. and Michalski, R.S.: Data-Driven Constructive Induction in AQ17-PRE. *Proceedings of the Third International Conference on Tools for AI*, 1991, San Jose, CA, 30-37.

[Bre94] Breiman,L. and Friedman,J.H. and Olsen,R.A.and Stone,C.J.: Classification and Regression Trees. *Wadsworth statistic/ probability series*, 1994.

[Ces87] Cestnik,B. and Kononenko,I. and Bratko,I.: ASSISTANT 86: A Knowledge-Elicitation

Table 1:

Concept	PIGS	PIGS	C4.5	C4.5	Stats	
	Mean	St.Dev.	Mean	St.Dev.	t value	probability
squares	98.81	1.167	99.69	0.354	5.8082	0.0000
quad	99.52	0.414	98.82	0.758	-8.3494	0.0000
circle	99.07	0.813	98.31	1.193	-6.8948	0.0000
discs	98.84	1.002	98.16	1.093	-6.4078	0.0000
polo	98.59	1.245	97.27	1.883	-8.3943	0.0000
POL	98.22	0.684	94.40	1.063	-16.572	0.0000

Tool for Sophisticated Users. *Progress in Machine Learning;Proc.of EWSL 87*, 1987.

[Fis87] Fisher,D.: Knowledge Acquisition Via Incremental Concept Clustering. *Machine Learning*, 2, 139-172.

[Leb87] Lebowitz,M.: Experiments with Incremental Concept Formation:UNIMEM. *Machine Learning*, 2, 2, 103-138.

[Mat89] Matheus,C. and Rendell,L.A.: Constructive Induction on Decision Trees. *Proceedings of the International Joint Conference on Artificial Intelligence.*, 1989, 645-650.

[Meh84] Mehra,P. and Rendell,L.A. and Wah,B.W.: Principled Constructive Induction. *Proceedins of the Eleventh International Joint Conference on Artificial Intelligence*, 651-656.

[Mich83] Michalski,R.S.: Learning from Observtion: Conceptual Clustering. In *Machine Learning: an Artificial Intelligence Approach*, 1983, Springer Verlag.

[Mich86] Michalski,R.S. and Mozetic,I. and Hong,J. and Lavrac,N.: The Multi-purpose Incremental Learning System AQ15 and its Testing and Application to three Medical Domains. *Proceedings of the Fifth National Conference on Artificial Intelligence*, 1986, 1041-1045.

[Mug90] Muggleton,S. and Feng,S.: Efficient Induction of Logic Programs. *Proc. First Conference on Algorithmic Learning Theory*, 1990, Tokyo, 1-14 .

[Mug93] Muggleton,S.: Inductive Logic Program ming: Derivations, successes and shortcomings. *Proceedings of European Conference on Machine Learning*, Springer Verlag , 1993, 1-37.

[Mur87] Murray,K.S.: Multiple Convergence: An Approach to Disjunctive Concept Acquisi tion. *Proc. Tenth International Joint Confer ence on Artificial Intelligence*, 1987, Morgan Kaufman, Los Altos, 297-300.

[Mur94] Murthy,S.K. and Kasif,S. and Salzberg,S.: A System for Induction of Oblique Decision Trees. *Artificial Intelligence*, 2, 1994, 1-32.

[Pag90] Pagallo,G. and Haussler,D.: Boolean Feature Discovery in Empirical Learning. *Machine Learning*, 5.

[Quin93] Quinlan,J.R.: C4.5 Programs for Machine Learning, Morgan Kaufman, 1993, San Mateo, California.

[Plo70] Plotkin,G.D.: A Note on Inductive Generalisation. In B.Melzer & D.Michie (Eds.), *Machine Intelligence*, 5, 1970, Edinburgh University Press, Edinburgh, 153-163

[Ren90] Rendell,L.A. and Seshu,R.: Learning Hard Concepts through Constructive Induction: Framework and Rationale, 1990, *Computational Intelligence.*

[Sal91] Salzberg,S.: A Nearest Hyperrectangle Learning Method. *Machine Learning*, 6, 251-276.

[Sch94] Schaffer,C.: A Conservation Law for Generalisation Performance. *Machine Learning Conference*, 1994.

[Utg91] Utgoff,P.E. and Brodley,C.E.: Linear Machine Decision Trees:COINS. Tech. Report 91-10, Univ. of Massachussetts, Amherst,MA.

[Web92] Webb,G.I.: Man-Machine Collaboration for Knowledge Acquisition. *Proc. Fifth Australian Joint Conference on Artificial Intelligence*, 1992, World Scientific, 329-334.

[Web92a] Webb,G.I.: Learning Disjunctive Class Descriptions by Least Generalisation, Tech. Report C92/9, Deakin University, Geelong, Victoria 3217, Australia.

[Wne94] Wnek,J. and Michalski,R.S.: Hypothesis-Driven Constructive Induction in AQ17-HCI. *Machine Learning* , 14, 2, 139-168 .

[Yip94] Yip,S.P. and Webb,G.I.: Empirical Function Attribute Construction in Classification Learning. In C.Zhang, J.Debenham & D.Lukose (Eds.), *Artificial Intelligence: Sowing the Seeds for the Future*, World Scientific, Singapore, 29-36.

A RULE DISCOVERY METHOD BASED ON APPROXIMATE DEPENDENCY INFERENCE

Atsuhiro Takasu[1] and Moonis Ali[2]

[1]Research and Development Department
National Center for Science Information Systems
Otsuka, Bunkyo-ku, Tokyo 112, Japan
Email: takasu@rd.nacsis.ac.jp

[2]Department of Computer Science, Southwest Texas State University
San Marcos, Texas 78666-4616, U.S.A.
MA04@academia.swt.edu

ABSTRACT

Functional relationship among attributes in relational databases has wide a variety of applications, as a typical rule, appearing in databases. It can be also regarded as a constraint for database design, query optimization, intelligent query processing, and so on. This paper presents a functional rule discovery system, named Profile, which discovers a set of functional relationships through three steps: generalizing values in the objective relation, inferring approximate functional relationships, and transforming the inferred relationships into presentation form. The functional relationship inference problem has combinational difficulty with respect to the number of attributes. This paper discusses Profile, especially focusing on its inference algorithm, and shows the feasibility of the system by some experiments.

INTRODUCTION

Recent computer systems accumulate various kinds of data, such as, experimental data, performance observation data, logs for computer and network managements, and so on. Although these raw data are expected to include valuable information, their utility is limited as they are not fully analyzed. In order to utilize those data, systems that perform either automatic or computer-aided analysis are required. This motivates many researchers to study *knowledge discovery in databases* (KDD [Sha92, CT93] or *data mining*).

Current KDD research is focused on relational databases, although some researchers have begun to discuss this problem in more complexly structured databases such as object-oriented and fulltext databases. Typical regularities handled in KDD are concept description and attribute relationship. As an example of concept description, let us consider a relation about a person's sport preference, represented with a schema (*name, nationality, age, preference*) in Table 1, and assume that the relation is clustered by the nationality. Then, the concept description will be a list of sport preferences for each country such as {*baseball,tennis,football* } for the U.S.A., {*ice hockey,tennis* } for Canada, {*soccer, cricket*} for the U.K., and {*soccer, tennis*} for Germany. In discovering concept description, databases are regarded as a set of instances of concepts. Tuples are summarized into classes corresponding to concepts by projecting them onto some attributes and/or generalizing values, and then, the summarized data is transformed into a concept description. For the above example, the concept description is obtained by projecting the data onto *nationality* and *preference*.

As for the attribute relationship, correlative and functional relationships are handled mainly. The latter one is different from the former one in distinguishing the causation and the covariation. Żyktow [ZB91] proposed a KDD method for correlative relationship using χ^2 test which discovers, for example, the relationship between *nationality* and *preference*. Note that there is no direction between attributes in this relationship (cf. functional relationship is described below). On the other hand, functional relationships are further classified into statistical, exactly functional and approximately functional ones. For statistical relationship, Cohen [CBCA93] proposed methods based on regressive analysis. Exactly functional relationship (or *functional dependency* (FD)) means that a set X of attributes functionally determines a set Y of attributes. FD is originally introduced to discuss the normal form of relations for database design and applied to various problems of relational databases, such as, integrity constraint maintenance, schema integration in heterogeneous databases, as well as database design. In general, *nationality* does not functionally determine *preference* since there are exceptions for this relation, although this kind of regularities is useful. Approximately functional relationship represents this kind of regularity, that is, one which holds for almost all data, even when it is not complete. Several measures of approximation (e.g., [SM93]) are proposed.

We have developed an approximately functional rule discovery system named Profile which extracts all functional relationships whose measure of approximation is larger than the given parameter. When the approximation level is 1.0, Profile enumerates all of exactly functional relationships (i.e., FDs). Profile challenges three problems. The first one is about the generalization of data. A database usually contains sparse data which prohibits systems from discovering useful rules. By generalizing data appropriately, we can obtain useful rules. The second one is about time efficiency of an inference mechanism. The problem space of the functional rule inference consists of any subset of all attributes, which is exponential with respect to the number of attributes. Therefore, the discovery system should provide a practically usable algorithm. The third problem is about the output form of rules. The discovery system possibly infers too many rules whose size can also be exponential with respect to the number of attributes. Therefore, it should provide some mechanism to transform rules to easily understandable forms.

In this paper, we describe Profile focusing on the second problem, that is, the time efficiency problem. The remainder of this paper is organized as follows. Section 2 contains basic definitions about the approximately functional relationship. In Section 3, rule discovery mechanism of Profile is described and in Section 4, an efficient inference algorithm is presented. Experimental results about the feasibility of the algorithm are shown in Section 5. Conclusions are given in Section 6.

NOTATION AND DEFINITION

In this section, we briefly review the concept of FD (see [Ull88] for details), then we define the approximate dependency inference problem.

A *relation schema* U is a set (sequence) $(A_1, \cdots, A_n)$ of attributes, and for each attribute A_i, there is an associated domain D_i. A *relation* R over U is a subset of $D_1 \times D_2 \times \cdots \times D_n$. Each element t of R is called a *tuple*, with $t[A_i]$ denoting the i'th attribute value of tuple t. For attributes $A_{i_1}, A_{i_2}, \cdots, A_{i_k}$ such that $i_1 < i_2 < \cdots < i_k$, $A_{i_1}A_{i_2}\cdots A_{i_k}$ denotes the subset (subsequence) $(A_{i_1}, A_{i_2}, \cdots, A_{i_k})$ of U. For a subset $A_{i_1}A_{i_2}\cdots A_{i_k}$ of U, $t[A_{i_1}A_{i_2}\cdots A_{i_k}]$ denotes the subtuple $(t[A_{i_1}], t[A_{i_2}], \cdots, t[A_{i_k}])$ of t.

An FD over U is an expression $X \rightarrow A$, where $X \subseteq U$ and $A \in U$ holds, and means that for any relation R over schema U,

$$(\forall t, s \in R)(t[X] = s[X] \rightarrow t[A] = s[A]) \quad (1)$$

holds. For a relation R, an FD $X \rightarrow A$ is said to be *consistent* with R if formula (1) holds for the relation R. For an FD f, $lhs(f)$ and $rhs(f)$ denote left hand side and right hand side of f, respectively.

[Definition 1] For a set R of tuples and an FD f, *approximation level* of f for R is defined as $1 - \frac{|v|}{|R|}$, where v satisfies the following conditions:

(1) f is consistent with $R - v$,

(2) $\frac{|v|}{|R|}$ is the minimum among sets satisfying the condition (1).

In this definition, v represents minimum exceptions of tuples about f, and the approximation level is defined as the ratio of the size of the maximum consistent tuple set over $|R|$.

[Example 1] Let us consider an FD $Nationality \rightarrow Preference$ over a relation in Table 1, then a tuple set $\{1, 2, 4, 6, 9\}$ satisfies the conditions in Definition 1 and the approximation level is 0.5. Note that there are other tuple sets satisfying the conditions (e.g., $\{1, 3, 4, 6, 9\}$), however, the approximation level is same for any tuple set that satisfies the conditions.

Approximately functional dependency (AFD) of R denoted by $X \xrightarrow{a} A$ is an FD with the approximation level a for R. $X \xrightarrow{1.0} A$ stands for an FD $X \rightarrow A$. For a relation R and a real number a, an AFD $X \xrightarrow{b} A$ is said to be *approximately consistent* if $b \geq a$ holds.

[Definition 2] For a relation R over a schema U, approximation level a, and an attribute $A \in U$, if an FD

Table 1: An Example Relation of Sport Preference

Id	Name	Nationality	Age	Preference
1	Robert	U.S.A.	25	football
2	John	U.S.A.	39	tennis
3	Jane	U.S.A.	55	baseball
4	Mary	Canada	28	ice hockey
5	Michael	Canada	38	tennis
6	Jimmy	U.K.	22	soccer
7	Kate	U.K.	49	cricket
8	Richard	U.K.	58	cricket
9	Boris	Germany	32	soccer
10	Georg	Germany	35	tennis

$X \rightarrow A$ is approximately consistent, and for any proper subset Y of X, $Y \rightarrow A$ is approximately inconsistent, then $X \rightarrow A$ is *α-minimal*, while if $X \rightarrow A$ is approximately inconsistent, and for any proper superset Y of X, $Y \rightarrow A$ is approximately consistent, then $X \rightarrow A$ is *α-maximal.*

[Example 2] For a relation in Table 1 and approximation level 0.95, an FD $Age\ Preference \rightarrow Nationality$ is α-minimal, because its approximation level is 1.0, whereas approximation levels of FDs $Age \rightarrow Nationality$ and $Preference \rightarrow Nationality$ are 0.9 and 0.7, respectively.

MIN(R, A, a) and MAX(R, A, a) respectively stand for α-minimal and α-maximal FDs whose right hand side is A.

Since for a relation R over a schema U, a property

$$X \xrightarrow{a} A, Y \xrightarrow{b} A, Y \supset X \Rightarrow a \geq b \tag{2}$$

holds by the definition of approximation level, either MIN(R, A, a) or MAX(R, A, a) is not empty for any relation R.

Approximate dependency inference problem is defined as:

Instance : A relation R over a schema U, an attribute $A \in U$, and real number a

Problem : Find MIN(R, A, a).

[Example 3] For the relation of Table 1, let us consider AFDs whose right hand side is *Preference*. Then, as described in Example 1, the approximation level for attribute *Nationality* is 0.5, whereas the approximation levels for attributes *Age* and *Name* are 1.0. These FDs are α-minimal. Therefore, MIN$(R, Preference, 0.5)$ is

$Nationality \xrightarrow{0.5} Preference$

$Name \xrightarrow{1.0} Preference$

$Age \xrightarrow{1.0} Preference$,

Similarly, MIN$(R, Nationality, 0.8)$ is

$Name \xrightarrow{1.0} Nationality$

$Age \xrightarrow{0.9} Nationality$

$Preference \xrightarrow{0.7} Nationality$.

SYSTEM DESCRIPTION

Profile extracts functional rules from databases according to the three parameters: a number of ranges for discretizing continuous data, classification hierarchies for categorical data, and a set of covariance attributes. As shown in Figure 1, Profile processes databases by the following three phases:

1. generalizes values in the given databases using the first and the second parameters,
2. applies an approximate dependency inference algorithm for each covariance attribute,
3. transforms inferred rules to dependency graph and decomposes the given relation according to the rules.

Databases usually contain attributes having sparse values, which means that their attribute values differ in many tuples of the database. For example, attributes *Name* and *Age* in Table 1 have sparse values. This sparsity of attribute values prohibits discovery systems from extracting useful rules. Generally, when causation attributes contain sparse values, a system tends to extract unuseful rules. For example, the approximation level of an FD $Age \rightarrow Nationality$ is 1.0 in Table 1, not because its functional relationship is strong, but because every attribute value of *Age* is different. On the other hand, the sparsity in the covariance attribute tends to cause loss of important rules. For example, although an FD $Preference \rightarrow Age$ seems to have a functional relationship, the approximation level of this FD is low.

Profile uses two kinds of generalization methods for categorical and continuous data. Domains of continuous attributes are generalized by discretizing them into the given number of ranges. Profile uses information loss as a criterion of discretization, that is, for the given number n of ranges, it discretizes continuous domain into n ranges as the information loss is minimum by a discretization algorithm similar to Wong's algorithm [WC87]. For example, let us consider to discretize the domain of *Age* in Table 1 into three ranges, namely young, middle, and old. Then, the discretized ranges whose information loss is minimum are $(0, 35.5]$, $(35.5, 44]$, and $(44, \infty)$.

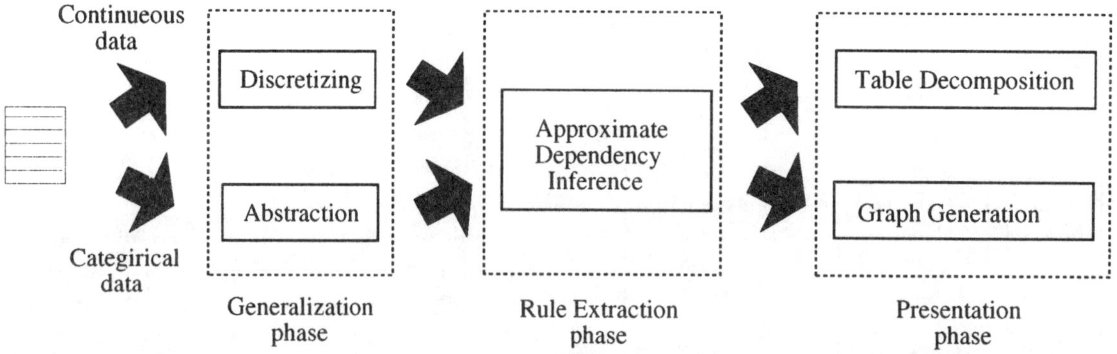

Figure 1: Process Outline

Table 2: A Generalized Relation

Id	Nationality	Age	Preference
1	North America	young	football
2	North America	middle	tennis
3	North America	old	baseball
4	North America	young	ice hockey
5	North America	middle	tennis
6	Europe	young	soccer
7	Europe	old	cricket
8	Europe	old	cricket
9	Europe	young	soccer
10	Europe	middle	tennis

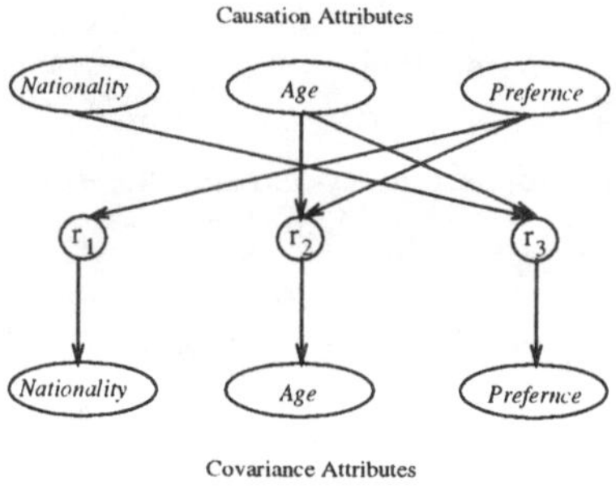

Figure 2: An example of Dependency Graph

Profile generalizes domains of categorical attributes with classification hierarchy [DT93] which represents abstraction of attribute values of each element of the domain. For example, in the domain of *Nationality*, U.S.A and Canada are abstracted to North America, while U.K. and Germany are abstracted to Europe.

Profile discards attributes whose domains are still sparse after the generalization phase. In Table 1, for example, *Name* is discarded. The generalized relation of Table 1 is shown in Table 2. In this generalized data, the approximation level of unuseful FDs $Age \rightarrow Name$ decreases to 0.7, whereas the approximation of $Preference \rightarrow Age$ increases to 1.0.

Approximate dependency inference phase is the most important part of Profile. In this phase, all FDs whose approximation levels are larger than the given one and right hand sides are included in the given covariance attributes. Let us assume that the given approximation level is 0.8 and covariance attributes are *Preference* and *Age*, then the following AFDs are inferred from the relation in Table 2:

$$\begin{aligned} r_1&: \quad Preference \overset{0.9}{\rightarrow} Nationality \\ r_2&: \quad Preference \overset{1.0}{\rightarrow} Age \\ r_3&: \quad Nationality\ \ Age \overset{0.9}{\rightarrow} Preference\ . \end{aligned}$$

In the functional rule extraction from databases, combinatorial difficulty with respect to attributes is most difficult hurdle in both theoretical and practical viewpoints. We discuss this problem in the next section and present an efficient approximate dependency inference algorithm.

The inference phase produces a set of rules in the form of $X \overset{a}{\rightarrow} A$. This form is appropriate for other systems such as database schema design tools. However, for presentation to human users, the rules should be transformed to more easily understandable forms. Profile transforms AFDs into two kinds of rule presentations. One is a dependency graph, where each rule $r : X \rightarrow A$ is represented with a directed graph (N, A) where nodes N are $X \cup \{A\} \cup \{r\}$ and arcs A is $\{(B, r) \mid B \in X\} \cup \{(r, A)\}$. Figure 2 depicts a dependency graph for the rules r_1, r_2 and r_3.

Another form is a decomposed relation of the given one. For each extracted rule $X \rightarrow A$, Profile generates a relation which is a projection of the given relation onto the attributes XA. This relation is an extensional representation of the rule. If all attributes in a rule are continuous, it may be replaced with an intentional representation such as an function $A = f(X)$.

INFERENCE ALGORITHM

In this section, we present an efficient approximate dependency inference algorithm. In the following discussion, we assume schema is $A_1 \cdots A_n A_{n+1}$ and consider to find MIN(R, A_{n+1}, a). MIN(R, A_{n+1}, a) and MAX(R, A_{n+1}, a) are respectively abbreviated to MIN and MAX. Since the right hand sides of FDs are fixed to A_{n+1}, FDs are represented only with their left hand sides, i.e., a set of attributes. For an attribute set X, $\overline{X}$ denotes an attribute set $A_1 A_2 \cdots A_n \backslash X$. When discussing MIN, Hasse diagram constructed with $A_1 A_2 \cdots A_n$ like Figure 3 is helpful to overview the problem space where nodes correspond to FDs. Nodes in the Hasse diagram and FDs are used interchangeably. Note that by the formula (2), if some node X is an approximately consistent FD, every ancestor of X is approximately consistent, and if X is approximately inconsistent, every descendent is approximately inconsistent.

The difficulty of the approximate dependency inference problem is in its huge problem space which consists of FDs $\{X \mid X \subseteq U\}$, whose size is 2^n, for a schema $U = A_1 A_2 \cdots A_{n+1}$. Theoretically, the approximate dependency inference problem is the same in polynomiality as the dependency inference problem [MR92] which consists of enumeration of all exactly functional dependencies, enumeration of minimal transversals of hypergraphs, which is described later, and identification of positive Boolean functions [BI93]. Mannila [MR92] proved that even the size of output (i.e., MIN) can be exponential both in the numbers of attributes and tuples which represent the input size of the dependency inference problem, therefore, we cannot expect any polynomial approximate dependency inference algorithm with respect to input size. It is open problem whether these problems have polynomial algorithm with respect to output size.

When the output size is exponential with respect to the input size, we can estimate algorithms more appropriately, by measuring the algorithm with the output size. When obtaining MIN, we also need to obtain MAX simultaneously in order to show that the obtained approximately consistent FDs are sufficient. Therefore, in order to measure the complexity of algorithms, we use the following four parameters:

- the number of attribute denoted by n,
- the number of tuples in R denoted by m,
- the number of α-maximal FDs denoted by $\overline{m}$,
- the number of α-minimal FDs denoted by $\underline{m}$.

Approximate dependency inference algorithm consists of tuple operations which calculate the approximation level of the given FD in R and attribute operations which search α-minimal FDs in the problem space (e.g., Figure 3). The approximation level of $X \rightarrow A$ can be calculated by sorting R with the key XA. Theoretically, the attribute operation is hard, however, practically the efficiency of the tuple operation is also important because the number of tuples is far more larger than the number of attributes. We measure the efficiency of approximate dependency inference algorithm by the tuple and the attribute operations that are respectively represented with

- the number of sorts carried out in the procedures,
- the number of nodes visited by the procedure.

[Definition 3] For a set *min* of approximately consistent FDs and a set *max* of approximately inconsistent FDs, an FD X is *undeducible* if X is not a subset of any element in *max* and X is not a superset of any element in *min*. An undeducible FD which is not a subset of any undeducible FD is referred to as *maximally undeducible*.

Note that the maximally undeducible FDs are maximal ones whose approximate consistency cannot be determined either from *min* or *max* by the formula (2).

[Example 4] For an attribute set $ABCDE$ with right hand side fixed to E, suppose approximately consistent and inconsistent FDs are $\{AB, AC, AD\}$ and $\{BD, A\}$, respectively. Then, the undeducible FDs are $\{C, BC, CD, BCD\}$, and a set of maximally undeducible FDs are $\{BCD\}$. Figure 3 shows approximately consistent (slightly hatched), approximately inconsistent (thickly hatched), and undeducible FDs (no hatched).

We can obtain MIN by repeating the following procedure until there is no undeducible FD:

- find a maximally undeducible FD X,

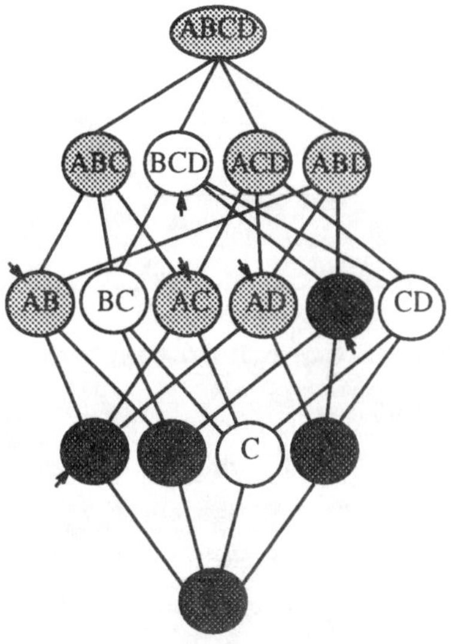

Figure 3: An example of Undeducible FDs

- search an α-minimal FD from X.

The maximally undeducible FDs are related to minimal transversals of a hypergragh [Ber89]. For a set V, a hypergraph is defined to be a set of subsets of V. If we regard the attribute set $A_1 A_2 \cdots A_n$ as V, MIN, MAX, min, and max can be regarded as hypergraphs. For a hypergraph H, transversal of H is a subset X of V such that

$$(\forall Y \in H)(X \cap Y \neq \emptyset)$$

holds, and minimal transversal Z is one such that any subset of Z is not transversal. Note that for a transversal X of a hypergraph H, $\overline{X}$ is not a superset of any element in H, that is, if H is a set of approximately consistent FDs, the approximate consistency of $\overline{X}$ can not be derived from H. Let MT be a set of minimal transversals of the hypergragh MIN, then MAX is represented as $\{\overline{X} \mid X \in MT\}$. Similarly, for a minimal transversal set MT of min, maximally undeducible FDs are represented as

$$\{\overline{X} \mid X \in MT, (\forall Y \in max)(\overline{X} \not\subseteq Y)\}. \qquad (3)$$

For example, minimal transversals of $min = \{AB, AC, AD\}$ in Figure 3 are A and BCD. Then, $\overline{A} = BCD$ is maximally undeducible FD, while $\overline{BCD} = A$ is already known as an approximately inconsistent FD. Therefore, we can obtain maximally undeducible FDs by calculating minimal transversals of min.

For the approximate dependency inference, it is sufficient to calculate a maximally undeducible FD for min and max, if it exists, otherwise to indicate that no undeducible FD exists. Let min be $\{X_1, X_2, \cdots, X_k\}$ and consider a set of transversals then a set

$$\{A_1 A_2 \cdots A_k \mid A_i \in X_i\}. \qquad (4)$$

Then, for any minimal transversal T for min, there exists an attribute set Y in the set (4) and $T \subseteq Y$ holds. Therefore, we can obtain a minimal transversal by generating the set (4). A minimal transversal is obtained from an attribute set Y by reducing attributes from Y and checking whether it is not superset of any element in min. We denote this procedure with $minimal(Y)$.

An algorithm calculating maximally undeducible FDs for min and max is depicted in Figure 4. The third argument of the procedure MaxUndeduce is initialized to an empty set and keeps the attribute set $(A_1, A_2, \cdots, A_i)$ at the ith recursion.

```
procedure MaxUndeduce(min, max, X)
  begin
(1)   if min = ∅ then
(2)     if X is not subset of any element in max then
(3)       return minimal(X̄);
(4)     else
(5)       return null;
(6)     endif
(7)   endif
(8)   select an element Y from min;
(9)   if X ∩ Y ≠ ∅ then
(10)    return MaxUndeduce(min\{Y}, max, X);
(11)  for all A ∈ Y do
(12)    MaxUndeduce(min\{Y}, max, AX);
  end
```

Figure 4: Algorithm for Maximally Undeducible FDs

Lemma 1 The procedure MaxUndeduce finds a maximally undeducible FD for min and max if any exists, otherwise it returns **null**. The space complexity is $O(|max| + n|min|)$.

Proof

The procedure MaxUndeduce recursively generates an attribute set in the set given in (4). If there exists maximally undeducible FD for min and max, MaxUndeduce finds it at line (3) from a transversal X of min. Otherwise, it returns **null**, because any minimal transversal is a subset of some element in (4) and MaxUndeduce generates a transversal in (4) until the transversal satisfies the condition given in (3).

The depth of recursion of MAXUndeduce is at most $|min|$ and at each recursion it uses space $O(n)$. MaxUndeduce does not update max. The procedure *minimal* calculates a minimal transversal with the space $O(|min| + n)$ which is smaller than the other part of MaxUndeduce. Therefore, the space complexity of MaxUndeduce is $O(|max| + n|min|)$. □

Unfortunately, this algorithm is not polynomial with respect to the size of MIN. The usability and executional limitation of this algorithm are discussed in a later section.

For a maximally undeducible FD, an α-minimal FD can be obtained by a simple depth-first search depicted in the Figure 5 which uses two global variables max and min. At line (1), for the consistency check, it first checks whether X is a subset of some FD in max. If X is not a subset of any FD in max, it calculates the approximation level of X.

procedure FindMin(X)
  begin
(1) **if** X is approximately inconsistent **then**
(2) $max := max \backslash \{Y \in max \mid Y \subset X\} \cup \{X\}$;
(3)   **return false**;
(4) **endif**
(5) **for all** $A \in X$ **do**
(6)   **if** FindMin($X \backslash A$) is **true then**
(7)   **return true**;
(8) $min := min \cup \{X\}$;
(9) **return true**;
  end

Figure 5: Algorithm to Find an α-Minimal FD

Lemma 2 FindMin(X) finds an α-maximal FD with one sort or finds an α-minimal FD with at most n^2 times of sort. The size of max increases at most n^2.

Proof

Since one attribute is removed from the attribute sets at line (6) at each recursive call, the depth of recursion is, at most, n. The number of loops at (5) to (7) is, at most, $|X|$. Therefore, this procedure always stops.

By the definition of the maximally undeducible FD, all superset of X is approximately consistent. Hence, if maximally undeducible FD X is approximately inconsistent, X is α-maximal. In this case, the procedure sorts the relation only once, and finds an α-maximal FD at line (1).

Since X is an undeducible FD, X is not superset of any FD in min. If X is approximately consistent, it has at least one new α-minimal FD in its descendent including itself. In this case, this procedure clearly finds an α-minimal FD at line (8). The depth of the recursion of this procedure is at most $|X| \leq n$, and the loop at line (5) to (7) is repeated at most $|X| \leq n$. Since the procedure is sufficient to find one α-minimal FD, backtrack does not occur from approximately consistent FDs. Therefore, the number of recursive calls is at most n^2. For each call in this procedure, the sort is carried out, at most, once at line (1). Hence, the number of sorts carried out in this procedure is, at most, n^2.

For each recursive call of this procedure, at most, one approximately inconsistent FD is added to max at line (2). By the above discussion, the size of max increases, at most, n^2. □

Finally, we show a dependency inference algorithm in Figure 6 where min and max are global variables.

procedure InferFD
  begin
(1) $min := max := \emptyset$;
(2) **while** ($X :=$ MaxUndeduce($max, min, \emptyset$))
      $\neq$ **null do**
(3)   FindMin(X);
(4) **end**
  end

Figure 6: Approximate Dependency Inference Algorithm

Theorem 1

(1)InferFD finds MIN and MAX.

(2)The number of sort carried out in this algorithm is, at most, $n^2\underline{m} + \overline{m}$.

(3)Space complexity of this algorithm is $O(n^3\underline{m} + n\overline{m} + m)$.

Proof

This procedure clearly stops for any relation. When this procedure stops, by the lemma 1, there are no undeducible FDs. Furthermore, $(\forall X, Y \in min)(X \not\subset Y \wedge Y \not\subset X)$ and $(\forall X, Y \in max)(X \not\subset Y \wedge Y \not\subset X)$ hold. Therefore, min and max are MIN and MAX, respectively.

For each call, FindMin finds one α-minimal FDs or one α-maximal FDs. Therefore, the FindMin is called $\underline{m} + \overline{m}$ times. By the lemma 2, the number of sorts carried out is, at most, $n^2\underline{m} + \overline{m}$.

Since the global variable min keeps α-minimal FDs and size of each FDs is, at most, n, the size of min

is $n\underline{m}$. Approximately inconsistent FDs are added to max only in the procedure FindMin, and its size is, at most, n^2 by the lemma 2. Therefore, the size of max is at most $n^3\underline{m} + n\overline{m}$. The space complexity of sort is $O(m)$. Since the depth of the recursive call for the procedures is n, other spaces are negligible. Hence, the space complexity is $O(n^3\underline{m} + n\overline{m} + m)$. □

FEASIBILITY OF DEPENDENCY INFERENCE

In this section, we discuss the feasibility and effectiveness of the presented algorithm through several experiments. As shown in Theorem 1, the space complexity of the algorithm is small enough. Hence, we concentrate on the execution time in the following discussion.

In the experiments, we use Armstrong relation [Fa82, BDFS84] to make experimental relations. For a set F of FDs, an Armstrong relation is such that it satisfies any FD in F^+ and does not satisfy any FD which is not in F^+, where F^+ stands for the closure of F [Ul188].

For a number n of attributes, a number $\underline{m}$ of α-minimal FDs, and a number r of tuples, we make experimental data as follows:

1. generate a set F of α-minimal FDs by repeating the random generation of an FD (i.e., a subset of $A_1A_2\cdots A_n$) until they are not a subset of any other FDs and their size is $\underline{m}$,
2. construct an Armstrong relation R of F whose size is r by the algorithm in [BDFS84].

Note that the number of α-maximal FDs depends on the generated α-minimal FDs. Algorithms are implemented in C and experimented on HP9000/750.

The first experiment is on the number of rf-maximal FDs. In this experiment, for any combination of $\underline{m}$ and n in $\{10, 15, 20, \cdots, 50\}$, we generate 50 data sets. For each data set, we measure the number of MAX, the number of sorts carried out, and execution time of the attribute operation. Figure 7 (a) and (b), respectively, depict the number of sorts and the attribute operation time for the number of α-maximal FDs, when $n = 30$ and $\underline{m} = 10$. As shown in Figure 7 (a), the number of sort is proportional to the number of rf-maximal FDs, which is claimed in Theorem 1. Figure 7 (b) shows that the attribute operation time is almost proportional to the number of rf-maximal FDs. These

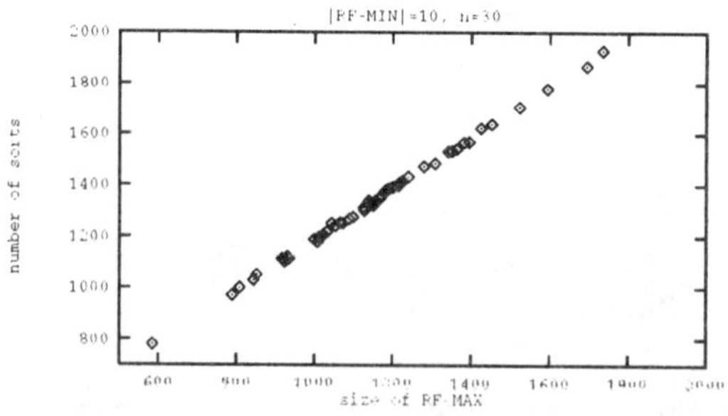

(a) the number of sorts

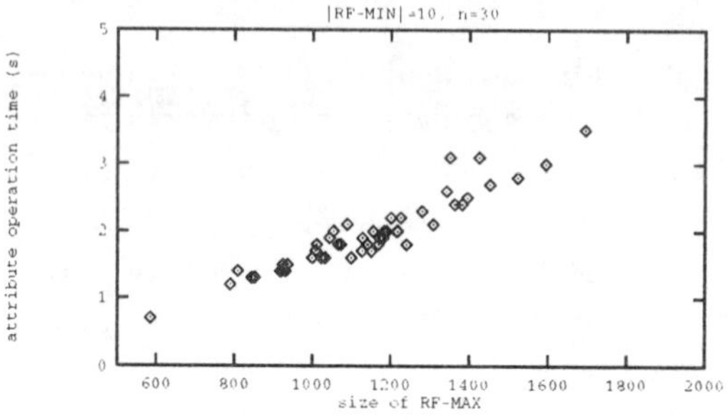

(b) attribute operation time

Figure 7: Characteristics for α-Maximal FDs

tendencies are observed for all combinations of n and $\underline{m}$ in this experiment.

The second experiment is on the number of α-minimal FDs. In this experiment, n is fixed to 20, and for each $\underline{m}$ in $\{10, 15, 20, \cdots, 50\}$, we generate 50 data sets and then measure average number of sorts and attribute operation time. The result is depicted in Figure 8. Figure 8 (a) shows that the number of sorts is proportional to $\underline{m}$ as claimed in the Theorem 1, while Figure 8 (b) shows that the attribute operation time is almost proportional to $\underline{m}$. From Figure 7 (b) and Figure 8 (b), execution time for the procedure MaxUndeduce is supposed to be almost propotional to $\overline{m}$ and $\underline{m}$, although the polynomiality of the procedure is not theoretically guaranteed.

The last experiment is on the number of attributes. This experiment is the same as the previous experiment except for varying n instead of $\underline{m}$, which is fixed to 10. We first discuss the number of sorts. Figure 9 (a) depicts the number of sorts for n, while Figure 9 (b) depicts the number of sorts to find α-minimal FDs, i.e., total number of sorts minus |MAX|. Although the Theorem 1 claims the number of sorts is $O(n^2)$ at the worst case, Figure 9 (b) shows that it is almost proportional to n. Figure 9 (c) depicts the attribute operation time, and shows that the performance is sharply deteriorated when the number of attributes is around

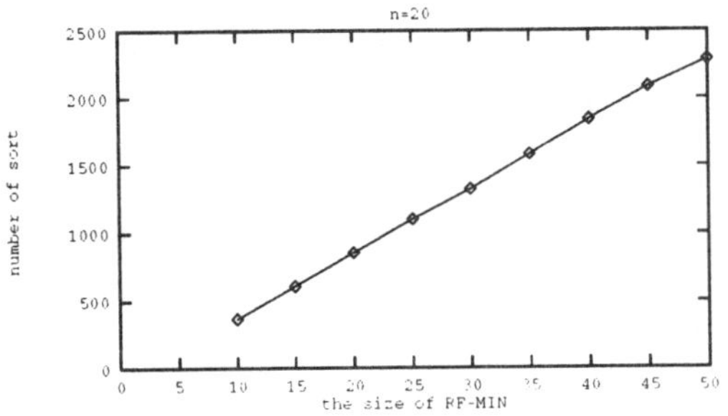

(a) the number of sorts

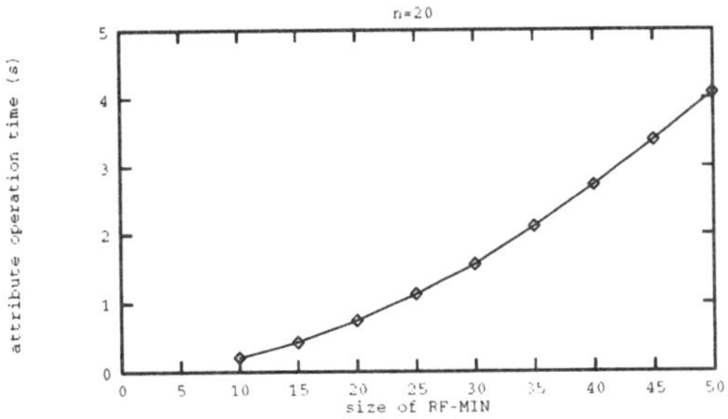

(b) attribute operation time

Figure 8: Characteristics for α-minimal FDs

50. This is due to the increase of the number of α-maximal FDs as shown in Figure 9 (a) and (b), which consequently increases the tuple operations in the procedure FindMin. Figure 7 (a), Figure 8 (a), and Figure 9 (b) show that the order of the number of sorts is nearly $O(n\underline{m} + \overline{m})$. Since the dependency inference problem needs to calculate both MIN and MAX, the presented algorithm is optimal as for the number of sorts.

By these experiments, the presented algorithm can be applied to relations whose number of attributes is up to 50. Practically, the number of attributes is not so large, and this limitation will not be serious.

CONCLUSIONS

In this paper, we presented a rule discovery system Profile that extracts approximately functional rules from databases. In order to generate useful rules, it first generalizes database data by using information loss as criterion to discretize continuous data and classification hierarchies for categorical data. Time efficiency is a key point for the approximate dependency inference. In this paper, we presented an efficient algorithm which is constructed by transforming a portion of the dependency inference problem to the enumeration of minimal transversals of hypergraphs. The feasibility of the presented algorithm is checked by some experiments and shown that the algorithm is applicable to the relation whose attribute is up to 50.

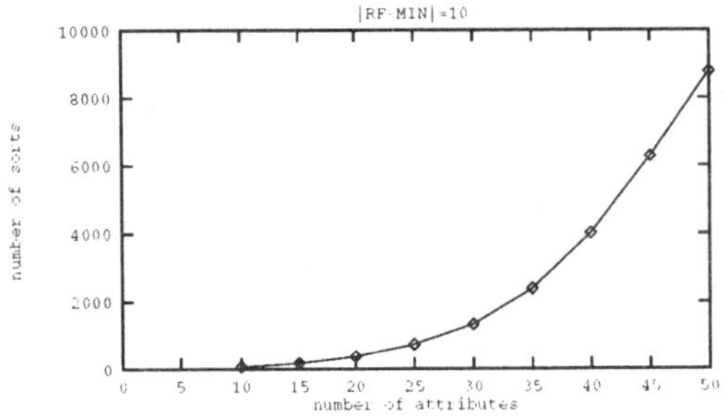

(a) the number of sorts

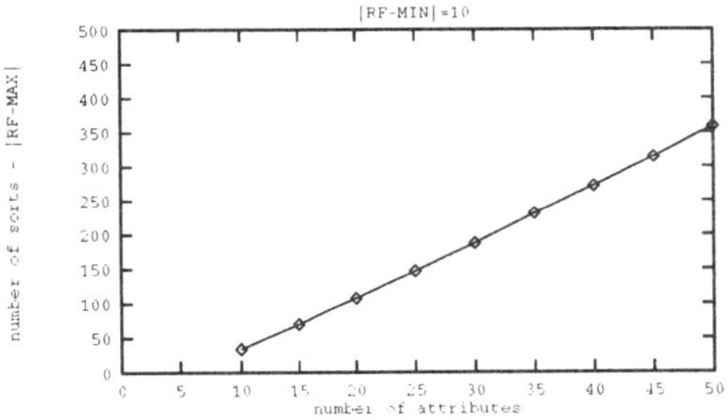

(b) the number of sorts - the number of α-maximal FDs

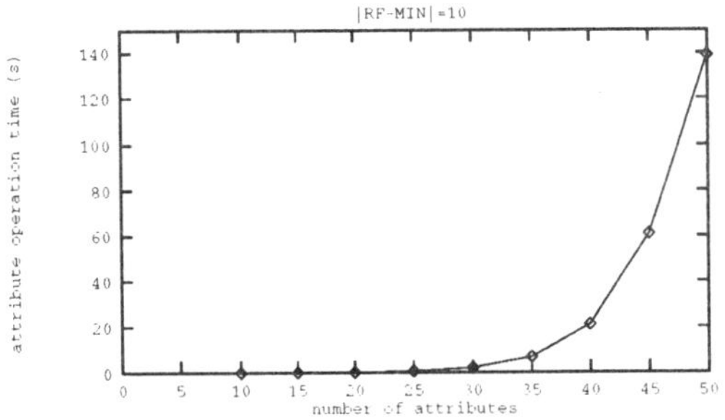

(c) the Attribute Operation Time

Figure 9: Characteristics for the number of Attributes

From a theoretical viewpoint, the most challenging future problem is to clarify whether the approximate dependency inference problem has a polynomial algorithm. Since there are several related problem, whose polynomiality is equivalent to this problem, clarifying the polynomiality of the approximate dependency inference simultaneously solves related open problems.

By sampling data, we can apply the presented methodology to larger databases which consist of millions of tuples. We have developed a probably approximately correct learning model for the dependency inference problem and discussed the sampling effect in

[AT93]. By extending this model, we will be able to apply this methodology to large databases.

The relationship between data generalization methods and the obtained rules are an interesting and important problem. Although Profile uses two kinds of generalization methods at present, we can control the discovered rules by given alternative generalization methods and generalization levels. This is the another future research direction.

References

[AT93] T. Akutsu and A. Takasu. Inferring Approximate Functional Dependencies from Example Data. In *Proceedings of AAAI workshop on Knowledge Discovery in Databases*, Washington, D.C., U.S.A., July 1993.

[BDFS84] C. Beeri, M. Dowd, R. Fagin, and R. Statman. On the Structure of Armstrong Relations for Functional Dependencies. *Journal of the Association of Computing Machinery*, 31:30–46, 1984.

[BI93] J. C. Bioch and T. Ibaraki. Complexity of Identification and Dualizatio of Positive Boolean Functions. *Technical Report Rutcor Research Report RRR25-93*, Rutgers University, 1993.

[CT93] N. Cercone and M. Tsuchiya *eds.* Learning and Discovery in Knowledge-Based Databases. *IEEE Transactions on Knowledge and Data Engineering*, 5(6), 1993.

[CBCA93] P. Cohen, L. Ballesteros, A. Carlos, and R. St. Amant. Automating Path Analysis for Building Causal Models from Data: First Results and Open Problems. In *Proceedings of AAAI workshop on Knowledge Discovery in Databases*, Washington, D.C., U.S.A., July 1993.

[DT93] V. Dhar and A. Tuzhilin. Abstract-Driven Pattern Discovery in Databases. *IEEE Transactions on Knowledge and Data Engineering*, 5:926–938, 1993.

[Fa82] R. Fagin. Horn Clauses and Database Dependencies. *Journal of the Association of Computing Machinery*, 29:952–985, 1982.

[MR92] H. Mannila and K.-J. Räihä. On the Complexity of Inferring Functional Dependencies. *Discrete Applied Mathematics*, 40:237–343, 1992.

[Ber89] C. Berge. *Hypergraphs - Combinatorics of Finite Sets.* Elsevier Science Publishing Company, North-Holland, 1989.

[SM93] G. Piatetsky-Shapiro and C. J. Matheus. Measuring Data Dependencies in Large Databases. In *Proceedings of AAAI workshop on Knowledge Discovery in Databases*, Washington, D.C., U.S.A., July 1993.

[Sha92] G. P. Shapiro *eds.* Special Issue on: Knowledge Discovery in Data- and Knowledge Bases. *International Journal of Intelligent Systems*, 7(7), 1992.

[Ull88] J. D. Ullman. *Principles of Database and Knowledge-base Systems - volume 1.* Computer Science Press, 1988.

[WC87] A. K. C. Wong and D. K. Y. Chiu. Synthesizing Statistical Knowledge from Incomplete Mixed-Mode Data. *IEEE Transaction on pattern Analysis and Machine Intelligence*, 9:796–805, 1987.

[ZB91] J. M. Żyktow and J. Baker. Interactice Mining of Regularities in Databases. In G. P. Shapiro and W. J. Frawley, editors, *Knowledge Discovery in Databases.* AAAI/MIT Press, Cambridge, MA, 1991.

INTELLIGENT AIR TRAVEL AND TOURIST INFORMATION SYSTEMS

Bing Liu

Department of Information Systems and Computer Science
National University of Singapore
Lower Kent Ridge Road
Singapore, 0511
Republic of Singapore

E-mail: liub@iscs.nus.sg

ABSTRACT

Traveling and touring has become an important part of modern life. Planning of such activities, is, however, often time consuming and frustrating. In this paper, a prototype computer system is described that is aimed at helping in these situations. Specifically, the system helps in two types of planning. The first type is the travel planning, i.e., before we leave, we need to make travel arrangements. The second is the sightseeing planning, i.e., when we arrive at our destination, we need to get some advice and to plan our sightseeing so that we can make best use of our time there. To deal with the first problem, two modified *k*-shortest path algorithms and a double search technique are proposed. To deal with the second problem, a machine learning technique is employed to make the system learn to give sightseeing advice. A branch and bound technique is also used to sequence sightseeing activities.

1. INTRODUCTION

Frequently, we encounter two problems when we travel. The first problem is that before traveling, we need to know which flights to take and where to changeover. If there are direct flights that go to our destination, a decision is relatively easy to make. But when there is no direct flight that goes to our destination, it will be quite difficult to make a decision. The second problem is that when we arrive at our destination, we need to get some tourist advice and to plan our sightseeing so that we can make best use of our time.

To solve the first problem, we normally consult a travel agency and ask them to make the travel arrangement. However, they may not be able do it satisfactorily because they do not fully understand our constraints and preferences. Very often, we may need to call them many times and vice verse in order to make the final travel plan. This final plan may still not be the best for us because both sides do not have sufficient information about the other.

The second problem is even more frustrating because we may know nothing about the new place and our time of stay is limited. Typically, we will ask a hotel staff or a tourist information officer to give us some suggestions or advice. Although they may be quite helpful, they cannot be at our service all the time. As the previous problem, the hotel staff or the tourist information officer may not fully appreciate our interests and constraints. Most of the time, they will give some general advice, which may not be suitable for us.

Recent years, some computer tourist information systems have been built in some cities. These systems mostly provide the information about various tourist attractions. They do not give specific advice or suggestions to tourists and do not help tourists to plan their sightseeing.

In this project, we are building a system to help in these situations. Currently, a prototype system has been implemented, which has two sub-systems, namely, a travel planner (called Travel-Planner) and an intelligent tourist information system (called Tour-Info).

The travel planning problem is similar to the shortest path problem [Deo84, Dij59, Shi79, Ski87]. However, there are differences. The network used by the travel planning problem is different from that of a normal shortest path problem. Travel planning also has time constraints, which are not present in the shortest path problem. We have designed some algorithms (based on the *k*-shortest path algorithms in [Ski87]) and techniques for its solution.

The intelligent tourist information system consists of

three modules, information browser, tourist advisor, and sightseeing planner. Information browser allows the tourist to freely browse though all the tourist information. Sightseeing planner helps the tourist to sequence his/her sightseeing if he/she selects his/her own places to visit. This module uses a Branch and bound technique. Tourist advisor gives advice to the tourist on what to visit with his/her limited time. This module employs a machine learning technique to learn from past experiences and uses the knowledge gained to give advice.

The paper is organized as follows: In the next 3 sections, we first present Tour-Info as we believe it is more significant. Section 2 gives an overview. Section 3 discusses the tourist advisor and Section 4 discusses the sightseeing planner. Travel-Planner is presented in Section 5. Section 6 provides a discussion. Section 7 describes the related work and Section 8 concludes the paper.

2. THE TOUR-INFO SUB-SYSTEM

In this section, we give an overview of the Tour-Info system. Figure 1 shows the general architecture. A brief description of each sub-system is given below:

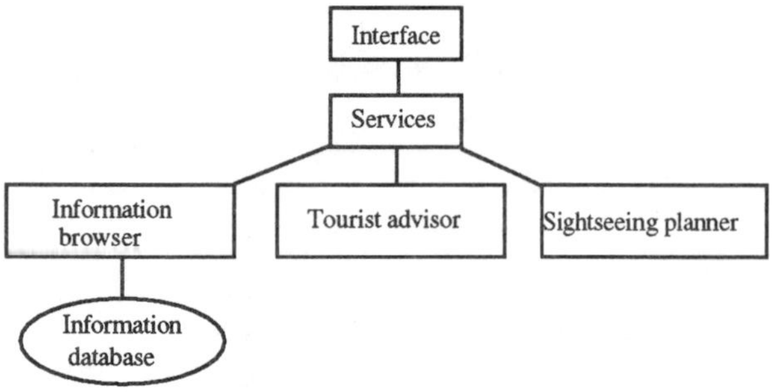

Figure 1. Tour-Info system

Interface: It is for the user to issue his/her requests and for displaying system results.

Services: It introduces various services provided by the system.

Information browser: It allows the user to freely browse the tourist information available, and to select the places he/she wishes to visit.

Tourist advisor: It gives advice to the user on what to visit with his/her time. This module first asks for some general information about the user, and then suggests a number of sightseeing plans for him/her to choose.

Sightseeing planner: When the user does not like the advice or suggestions given by Tourist Advisor, he/she can choose the places he/she wishes to visit by using Information Browser. Then, this module can be used to sequence the visits. Unlike Tourist Advisor, Sightseeing Planner does not ask for any user information.

Information database: It contains all the tourist information.

In the next two sections, we will discuss Tourist Advisor and Sightseeing Planner in more details. Information Browser will not be described as such systems are already available in some places. It is also not the focus of our project.

3. TOURIST ADVISOR

When tourists come to a new place, they normally have little knowledge about the place. Typically, they would like to get some advice or suggestions on what to visit with their limited time. Then, they would most likely to ask a tourist information officer who is knowledgeable about the place and also experienced with tourists. They will then follow that advice or explore some more information to make their decision. However, such services may not always be readily available and the advice given may not suit individual needs. Tourist Advisor is designed to help in this situation, i.e., to play the role of a tourist information officer.

3.1. Machine learning is needed for giving advice

Giving advice is a very complicated task. The system not only needs to know various tourist attractions, but also what might be interesting to different groups of tourists. For example, people of different age groups may like different things. A family with children may be interested in the places that are attractive to children. People from different countries may also like different things. For a human tourist information officer, these kinds of knowledge are gained from years of experiences.

In order to build a computer system to provide such a service, the obvious approach is to implement an expert system, which embeds all the knowledge of some experienced tourist information officers. However, the problem with a typical expert system is that it cannot handle changes. Any change will need manual modification of the knowledge base, which is very difficult most of the time. For example, a new tourist spot may be built, and as time goes by people may become interested in different things because of new trends and fashions. This means that the system needs to learn to keep up with the constant change.

Another important issue is that if a system is to be installed in each hotel (or hotel room), then the advice given will very much depend on the location of the hotel. For example, a tourist would usually like to go to a night spot that is not too far away from his/her hotel. This kind of information cannot be embedded in a general system because different hotels are at the different locations. Their surrounding facilities are also quite different. Furthermore, different hotels may also cater for different classes of tourists. For example, expensive hotels are catered for people who are relatively rich. These people may have different interests and spending powers from those staying in cheap hotels. So, advice given is location dependent and/or hotel dependent. It will be very costly to build a system and to keep modifying it for each individual hotel. Thus, learning capability becomes essential for an advisory system.

In our prototype system, Tourist Advisor uses a machine learning technique for classification. It is called the decision tree induction [Qui86, Qui93]. We did not implement a new decision tree induction system in this project. Instead, we use the C4.5 system [Qui93] in our system to perform the learning task.

3.2. Decision tree induction and C4.5

Decision tree induction is perhaps the mostly used learning technique in practice. C4.5 is its most recent implementation. Essentially, C4.5 takes a dataset and induces a decision tree for classifying the examples in the dataset. The decision tree is then used to produce a set of rules for classification of new data. In our context, it uses past cases as dataset to derive some general rules. These rules are then used as knowledge to give advice to tourists. The following is a brief overview of the technique in C4.5.

The dataset used by C4.5 is a two dimensional table. Each column represents an attribute and each row represents an example case. Each example is assigned a class beforehand. Thus, each example in the dataset is of the following format (attribute-*i*-value is the value for the *i*th attribute of the example, and class-*j* is the class that the example belongs to):

attribute-1-value attribute-2-value ...
attribute-i-value ... attribute-n-value class-j

A decision tree basically divides the data according to different attribute values and the classes of the data. It is generated by setting the root node to be the set of examples; partitioning the examples in the root node, thus creating child nodes; and recursively partitioning each of the child nodes to get the next levels in the tree. For each node, C4.5 selects an attribute and creates a branch for each value of the attributes appearing in the subset of examples corresponding to the node. This process of specialization is then recursively applied to each non-leaf node.

The algorithm for decision tree generation requires the specification of four steps [Qui86]. Each step represents a decision that constitutes parts of the strategy that a decision tree generation algorithm adopts in exploring the space of possible trees.

Step 1. Select an attribute to use in partitioning the examples at a node

Step 2. Choose a particular partition of the examples based on the attribute of Step 1.

Step 3. Decide whether to stop partitioning a node

Step 4. Assign a class to a node deemed to be a leaf in Step 3.

3.3. Tourist Advisor sub-system

Tourist Advisor provides two kinds of advice. One is on sightseeing, i.e., what places to visit and in what sequence. The other is on individual items, such as restaurants, night spots, etc. We will discuss them below in turn.

The module that provides sightseeing advice works as follows:

Step 1. The system requests for some general information about the tourist (see below).

Step 2. This information is used to match some rules (the rules are produced by the learning system) in the knowledge base to give some alternative sightseeing plans. The tourist can then choose a plan from them. The plan that has been chosen and the rule that produces the plan will accumulate credits.

Step 3. If the tourist does not like any of the suggested plans, he/she can turn to the Information Broswer and Sightseeing Planner modules to select and plan his/her own places to visit. After he/she has decided a sightseeing plan, it will be recorded as an example case with the information collected about the tourist.

Step 4. When enough cases that do not use suggested sightseeing plans have been collected, the machine learning system will be used to generate a new set of rules. This represents that old knowledge is no longer valid or suitable. However, those old plans and rules that have sufficient credits will be converted into cases to be used together with the new cases by the learning system to generate the new knowledge. Those old rules and/or plans that are seldom used will be discarded.

The working of this module can be characterized as a continuous process of learning, using the learnt knowledge to give tourist advice, learning again, and so on.

In our current prototype system, the information required from the user is as follows:

- Age group: the age group of the tourist

- Sex: male or female
- Country: home country of the tourist
- Companion: traveling alone or in group, e.g., couple, and with children
- Length of stay: half day, one day or two days
- Historical sites: whether the tourist is interested in the historical sites
- Fauna: whether the tourist is interested in the zoo, bird park, etc.
- Theme parks: whether the tourist is interested in theme parks
- Sea sports: whether the tourist is interested in sea sports
- Shopping: whether the tourist is interested in shopping
- Culture shows: whether the tourist is interested in cultural shows

The classes for the dataset are the possible sightseeing plans.

For different places this list could be different because different places may have different tourist attractions.

A question may be asked: where does the initial set of rules (with sightseeing plans) come from? The system does not really need the initial set of rules. It can actually accumulate some cases first and then generate the initial set of rules. In this case, the system will not give advice at the beginning, although it still requests for information from the tourist. Alternatively, the initial set of rules could be provided by a human expert.

Now let us have an example. Suppose, a user gives the following input (some attributes may be left blank):

Age group:	30
Sex:	male
Country:	UK
Companion:	wife and children
Length of stay:	0.5 (days)
Historical sites:	yes
Fauna:	yes
Theme parks:	yes
Sea sports:	no
Shopping:	
Culture shows:	yes

The system will try to use this set of information to match some rules in the rule base to produce a set of plans. A sightseeing plan could be like this: Leave the hotel at 8:30, (1) go to Underwater World (30 minutes travel and 1 hour visit), (2) go to Bird Park (30 minutes travel and 2 hours visit), (3) back to the hotel at 1:00pm. Notice that the tourist will be asked when he/she can leave and when he/she wishes to come back. The times giving for traveling and visiting in the plan are not so important but only as a reference. The tourist can adjust them to suit his/her situation. The present system has not included the public transportation information. It assumes that the tourist uses a car.

We now turn to the module that provides the other type of advice, i.e., giving advice on specific items, such as restaurants and night spots. Typically, there are hundreds of restaurants and night spots in a big city. It is very difficult for a tourist to choose. This sub-system will help him/her to narrow down to a small subset so that he/she can make his or her choice easily. The attributes (or the information about the tourist) used for providing restaurant advice is like this:

- Age group: The age group the tourist is in
- Sex: Male or female
- Country: The tourist's home country.
- Companion: Traveling alone or in group (e.g., friends, couple, or with children)

For a different item, the set of attributes may be different. The work of this module is similar to the one above. Due to the space limitation, we will not discuss it here.

4. SIGHTSEEING PLANNER

In the previous section, we presented Tourist Advisor that gives suggestions and advice to tourists. However, some tourists may not like the advice given. Then, they may choose to select their own places to visit. They can browse though the tourist information to make their selections. After that, they need to sequence their visits, i.e., which one place to go first and which next, and so on, so that they can cover as many places as possible without wasting time. Figure 2 shows an example, where the tourist wishes to visit three places, P1, P2, P3 (the background is the map of Singapore).

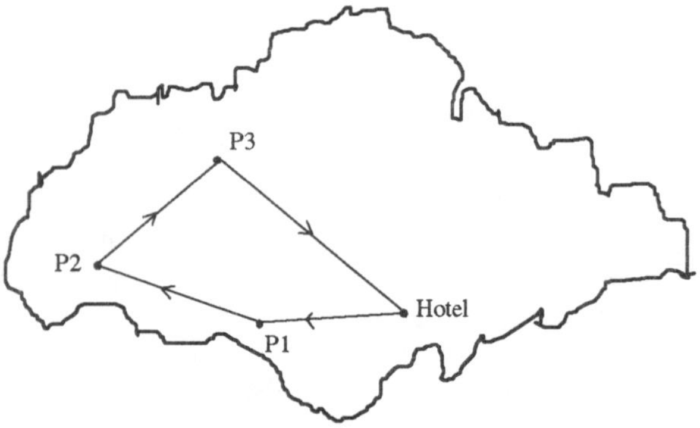

Figure 2. An example sequence problem

This sequencing problem is a constrained traveling sales man problem. The constraints are opening hours of each place and the time a tourist needs to stay there.

In our system, a branch and bound technique is used.

This technique is sufficient because the number of places one can visit within a day (each day has a different plan) is quite limited. It is normally around 1-6. For a large traveling sales man problem, more heuristics may be needed for efficient solution generation.

5. TRAVEL PLANNER

We now turn to the travel planner system. The travel planning problem may be modeled as finding k-shortest paths between two nodes in a weighted and directed network $G = (N, A)$ with a node set N and an arc set A. Each arc has 5 components, (i, j, d, e, n), where i and j are two adjacent nodes, d is the start time at node i and e is the ending time at node j, and n is the arc identification number. Each arc $(i, j, d, e, n) \in A$ also has a length $d_{ij\text{-}n} \geq 0$. In order to use an arc, the minimal changeover time constraint must be satisfied, i.e., the end time of the previous arc on the path must be smaller than the starting time of the present arc by a certain fixed time (called minimal changeover time constraint). K-shortest paths are desired between two nodes s and t that satisfy the earliest departure time constraint at node s or the latest arrival time constraint at node t.

In the air travel context, the nodes are airports. Arcs are the flight services available. Each of these services has a flight number (which is corresponding to the arc identification number), a departure time and an arrival time. The length of an arc could be the cost in terms of money or traveling time or a changeover. K-best solutions are required because it gives the user choices, from which he/she can select the one that is most suitable to him/her. In the traveling domain, there are normally three criteria to minimize, namely, cost (in terms of money), traveling time, and the number of changeovers. People normally would like to have a travel arrangement that is cheap, takes a shorter time and also has fewer changeovers.

This travel planning problem is very similar to the standard shortest path problem [Shi79]. Shortest path problem is modeled as finding the shortest path between two nodes s and t in a weighted and directed network $G = (N, A)$ with a node set N and an arc set A. Each arc $(i, j) \in A$ also has a length (or weight) $d_{ij} \geq 0$. Figure 3 shows an example network for the shortest path problem.

The flight network is different. The key difference is that in a normal directed and weighted network, there is only one arc going in the same direction between any two adjacent nodes, while in our case there could be many arcs going in the same direction. Each arc also has its own length. This situation is caused by the fact that different airlines offer similar services. Figure 4 shows a typical situation between two adjacent nodes in a flight network.

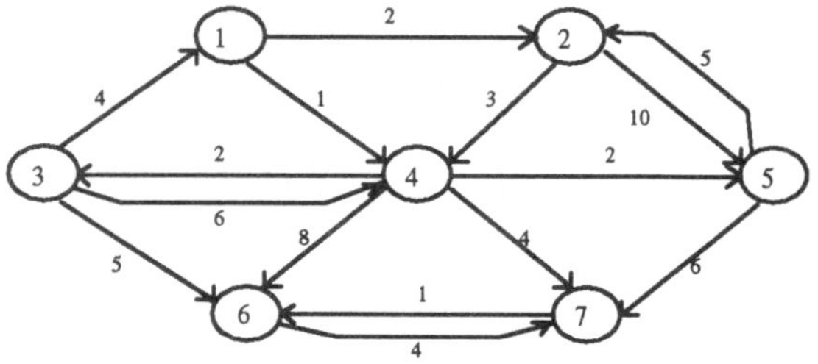

Figure 3. An example network

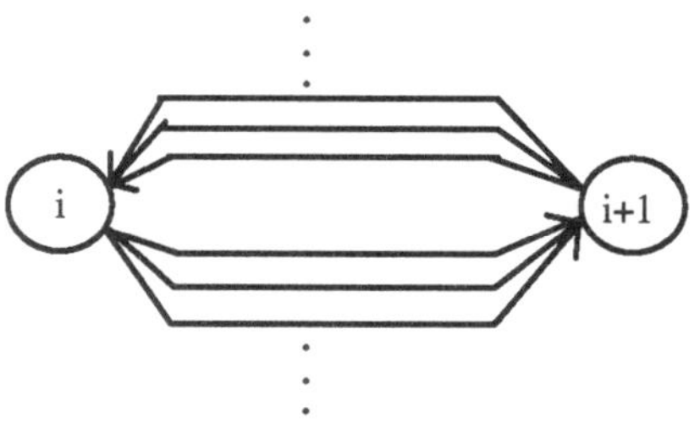

Figure 4. The typical situation in an air transportation network.

Another difference is that in the traveling domain time is important, which is not present in a normal shortest path problem. Despite the differences, we can borrow some of the ideas from the shortest path problem and generalize some shortest path algorithms for our purpose.

The shortest path problem has received considerable attention in the past three decades. Many algorithms have been devised for its solution [Deo84, Dij59, Gal86]. Dijkstra algorithm [Dij59] is perhaps the earliest and also one of most efficient algorithms for the problem. Most of the other algorithms are variations of Dijkstra algorithm. K-shortest path algorithms have also been well studied [Shi79, Ski87]. The k-shortest path algorithms are similar to those algorithms for finding only one path, except that they keep k solutions at each node. There are two types of k-shortest path algorithms, one based on Dijkstra algorithm and the other based on A* [Ski87]. A* may be used when the network satisfies the condition that the estimated distance from the current node to the destination node is always less than or equal to the real distance between the two nodes (see [Pea88] for more details). This algorithm can reduce search space substantially [Pea88] because of its directional heuristics.

In this work, we have generalized the k-shortest path algorithms in [Ski87] to solve our travel planning problem. The modified algorithms are presented below.

Every node i is assigned a k-vector of labels $L_i = \{l_{i1}, l_{i2}, ..., l_{ik}\}$ with $l_{if} = (c, a)$, where c is the current length from s to i, and a the arrival time. The entries of L_i are listed in an increasing order according to c. The components of this vector are the values, either temporary or permanent, of the shortest, second shortest, up to the kth shortest path lengths from the starting node s to the node i. To efficiently manage the labels we maintain a priority queue (by using a binary heap [Sed88]) $Q = \{q_1, q_2, ..., q_r\}$ whose entries are the smallest, temporary components of each node's k-vector. An information list P is maintained at each state of the algorithm to indicate which nodes still have temporary components left in their k-vector. The earliest departure time at s, which is given by the user, is *startt*. (In the case that the user gives the latest arrival time at t, the same algorithms can still be used, but will search backward from t, instead of search forward from s).

K-shortest path algorithm can be stated as follows (similar to the one in [Ski87]):

Step 1. Initialize $L_s = \{(0, startt), (\infty, 0), (\infty, 0), ..., (\infty, 0)\}$. For all $i \neq s$, set $L_i = \{(\infty, 0), (\infty, 0), ..., (\infty, 0)\}$. Insert l_{s1} into Q. Make all labels temporary and place all nodes on P.

Step 2. Set $N = q_1$ and call its associated node i^*. For each arc S originating from i^* (with its terminating node j and identification number n) that satisfies the minimal changeover time constraint, if $R = N(c) + d_{i^*j\text{-}n} + w(N, S)$ (where $N(c)$ is the current length of q_1, and $w(N, S)$ is the waiting length) is smaller than at least one component of node j's k-vector, then retain the k smallest values among the values $\{l_{i1}, l_{i2}, ..., l_{ik}\}$and $(R, S(a))$ (where $S(a)$ is the arrival time of arc S) as node j's k-vector. If R is smaller than node i's entry in the priority queue, replace node j's entry by $(R, S(a))$ and reorder the queue Q.

Step 3. Make permanent the component L_{i^*} associated with the value N. If there are no more temporary components for node i^*, remove node i^* from list P and its associated value from the priority queue. If the above holds and $i^* = t$, STOP; L_t contains the k-shortest path lengths from node s to node t. Otherwise, replace node i^*'s entry in the priority queue by its smallest remaining temporary component and reorder the queue.

Step 4. Go to Step 2.

A* version of the algorithm can also be devised. It makes use of the available knowledge about the shortest air distance (or the estimated shortest traveling time or the estimated cheapest cost) between the source and the destination. In the above algorithm, Q is ordered according to the distance so far from the source. For the A* version, we will use the distance so far from the source plus the air distance from the current node to the destination. In order to modify the above algorithm to the A* version, Step 2 needs to be changed to Step 2A*. We use A_{it} to denote the air distance from i to t. We then have:

Step 2A*. Set $N = q_1$, and call its associated node i^*. If $N(c) > 0$, set $T = N(c) - A_{i^*t}$. Otherwise $T = N(c)$. For each arc S originating from i^* (with its terminating node j and identification number n) that satisfies the minimal changeover time constraint, if $R = T + d_{i^*j\text{-}n} + w(N, S)$ is smaller than at least one component of node j's k-vector, then retain the k smallest values among the values $\{l_{i1}, l_{i2}, ..., l_{ik}\}$and $(R, S(a))$ as node j's k-vector. If $E = min\{l_{im}(c), m = 1, 2, ..., k | l_{im}$ is temporary$\}$ $+ A_{it}$ is smaller than node i's entry in Q, replace node i's entry by $(E, S(a))$ and reorder the queue Q.

We will not analyze these two algorithms as they are very similar to the ones in [Shi79, Ski87]. The only differences are that we need to check the minimal changeover time constraint and to add changeover waiting length into the total length of each path.

Both algorithms above can be used to find k-shortest paths in terms of money cost. A* cannot be used to find solutions in terms of number of changeovers because it is impossible to know the minimal changeovers required from one location to another except using 0. This reduces the A* version of the algorithm to the Dijkstra version.

Using the above algorithms alone, however, is not sufficient for finding the k-shortest paths in terms of traveling time, which refers to the period between the time the user departs from the source and the time he/she arrives at the destination. There are typically two situations when the user is concerned about the traveling time. The first one is that the user can only leave at a specific time at the source s (we call this the earliest departure time) and the other is that the user must arrive at the destination t at a specific time (we call this the latest arrival time). They are actually the same problem but require search in different directions, one using forward search and the other backward search. Now let us see why the above algorithms are not sufficient. We use the first case as an example. The problem is that the algorithms always keep k-earliest arrival times at each node in relation to the user specified earliest departure time rather than the departure times of the flights at the source. Then, it is possible to depart later by taking a later flight and wait less time at each changeover point, but also reach the destination at the same time. To solve this problem, we propose a double search technique to produce k-best solutions in terms of traveling time.

The technique works as follows (for the case that the user specifies the earliest departure time): Firstly, a forward search is performed producing *k*-best solutions in relation to this earliest departure time. Then, it searches backward from *t* using the real latest arrival time among the *k* solutions as the base to produce the final results. Likewise, when the user specifies the latest arrival time, the system will first search backward and then search forward.

In the above algorithms, we only considered the minimal changeover time constraint, i.e., the minimal waiting time at a node. However, in practice, people may not like to wait for too long either, i.e., there may also be the maximal waiting time constraint. When this constraint is imposed, the above algorithms may not produce the *k*-best possible solutions. This does not affect the solutions using traveling time as the measure because the double search technique produces the *k*-best solutions. If maximal waiting time constraints cannot be satisfied at some nodes, it means it is impossible to satisfy them. However, it will affect the solutions that using cost or the number of changeovers as the measure since time is not a measure here but a constraint. The problem is that some or all of the current *k*-best solutions at a node may not satisfy a maximal waiting time constraint later on, but some of those outside the *k* solutions may be able to (which have been discarded). In this case, backtracking is required, which is very time consuming.

Our current solution to the problem is to search for more solutions, i.e., $k+n$ (n is a fix integer value) solutions when the user asks for only *k* solutions so as to reduce the effect of the problem situation. Only *k* solutions will be reported to the user even if more than *k* solutions are found. Theoretically, this approach still does not guarantee to found the *k*-best solutions. But for practical use, we believe it is sufficient.

6. DISCUSSION

Currently, both Tour-Info and Travel Planner are prototypes implemented on 486 PC using the C language. So far, not much work has been done about the interface. It is still text based. We understand for a practical system a sophisticated interface must be built.

We have two purposes in mind in building the Tour-Info system. The first one is to provide an integrated tourist information system, which not only provide information, but also help the tourist to do his/her sightseeing planning and give sightseeing advice. The second one is to study how to give advice by using machine learning techniques.

Tourist Advisor and Sightseeing Planner have been tested with tourist information data of Singapore. The results are very promising. We believe Sightseeing Planner is already satisfactory. Tourist Advisor still needs some more work both in terms of research and development. Using machine learning techniques to give tourist advice is a new approach. The current experimentation is still in its early stages. There are also quite a number of factors that have not been considered in our system, e.g.,

1. If many people do follow the advice, some places may have very few visitors, while some others may be quite crowded.
2. The system is only catered for new tourists, i.e., they have never been to the place before. To give advice to those who are not the first timers, more work needs to be done.
3. How to give advice to tourists on ad hoc events, such as important shows or culture events, which cannot be included in learnt rules.

These and other issues, such as interaction between the user and the system and understanding the user's needs, require careful modeling. They will be the focus of our future work

The purpose of building Travel Planner is to model the problem and design some algorithms for its solution. We believe our algorithms and the double search technique can be readily used in a practical system. So far, many tests have been carried out with the algorithms using two randomly generated networks. One has 40 nodes and 4290 arcs, and the other has 90 nodes and 11,789 arcs. The algorithms are very efficient. The results always return in less than 1.5 seconds for *k* up to 5.

A number of options are possible in the travel planning system. The user can fix the earliest departure time at the source or the latest arrival time at the destination, and ask the system to produce the appropriate solutions. The user can also choose his/her preferred airlines to travel, and ask the system to suppress those airlines he/she does not like to fly with.

7. RELATED WORK

Decision tree induction is one of the earliest machine learning technique that discovers rules classifying a dataset into many classes. This technique and its representative system C4.5 (and its earlier version ID3 [Qui86]) are perhaps the most widely used machine learning technique and system respectively. They are mostly used in knowledge discovery in databases [Pia91]. We believe this is the first time it is applied in the domain of giving tourist advice.

Regarding the travel planning, [Dah91] reported a system for rail-air travel at New-Delhi in India. The algorithm they use is a modified $A*$ algorithm, which is able to take into consideration of the additional knowledge about the rail network. The system works in two steps: Firstly, it uses the modified $A*$ algorithm to find the optimal path. Then, it uses some heuristic strategies to identify the possible changeover points. The final route is evaluated on the basis of distance, number of changeover points, time required to travel and the number of trains available between the changeover points. They did not report about air travel planning, except stating the system also handles domestic air travel with Indian airline. Our system works differently. Firstly, our system works in only one step. We believe that their method will not be appropriate for international air travel because it is very difficult to find the path first and then identify the changeover points (which is suitable for rail travel). In the international air travel, a route depends on changeover points, and departure and arrival times of airlines, rather than the other way around. Secondly, our system produces multiple (alternative) solutions that allow the user to make his/her selection.

[Bos89] reported a system called Global-Trotter, an intelligent flight itinerary planner. The system is modeled on travel agent/client interactions. Its main strength is in reasoning about the user requirements and constraints. Finding the best route is not the major concern as it only displays the flights for the user to select. This is different from the purpose of our system. Travel Planner's main task is to find the *k*-best itineraries for the user. However, we see the two systems as complimentary. In our future work, we will add their facilities to our system.

8. CONCLUSION

In this paper, we discuss a prototype system for helping travelers and/or tourists in their travel and sightseeing planning. We have emphasized on two modules, namely, Tourist Advisor and Travel Planner. In Tourist Advisor, a machine learning technique is used to make the system learn to give advice. This approach is very promising. It solves many problems with a normal expert system as it is able to learn and to change with time and trends. However, we do not claim that the current technique used is the best or even sufficient for the purpose of giving advice. Instead, we see the current prototype system as the beginning. There are still a number of problems need to be solved. Some of them have been discussed in Section 6. Our future work will focus on solving those problems.

Travel Planner helps the user to make his/her travel arrangement. We proposed some new algorithms and techniques. Our aim is to build a system that can be used by general public to plan their travels. It may also be used as a tool for travel agencies. Our future work for this sub-system will be focused on providing more flexibility to the user, and to add the reasoning capability as the one in Global-Trotter.

REFERENCE

[Bos89] P. K. Bose, M. R. Padala, "Globe-Trotter: An Intelligent Flight Itinerary Planner," IEEE Expert, pp. 56-64, 1989.

[Bro85] L. Brownston, R. Farrell, E. Kant and N. Martin, *Programming Expert Systems in OPS5*, Addison-Wesley Publishing Company, 1985.

[Deo84] N. Deo and C. Pang, "Shortest Path Algorithms: Taxonomy and Annotation," *Networks,* vol 14, pp. 257-323, 1984.

[Dah91] V. Dhawan, V.S Rajput and R.G.S. Asthana, "Intelligent Rail-air Travel Planner," Proceedings of CAIA-91, pp. 412-418, 1991.

[Dij59] E. W. Dijkstra, "A Note on Two Problems in Connexion with Graphs," *Numerische Mathematik*, vol. 1, pp. 269-271, 1959.

[Gal86] G. Gallo, and S. Pallottino, "Shortest Path Methods: A Unified Approach," *Mathematical Programming Study*, vol. 26, pp. 38-64, 1986.

[Hay83] F. Hayes-Roth, D.A. Waterman, and D.B. Lenat (eds), *Building Expert Systems*, Addison-Wesley, 1983.

[Pea88] J. Pearl, *Heuristics*. Addison-Wesley, Reading, MA, 1988.

[Pia91] G. Piatetsky-Shapiro and W.J Frawley, eds, *Knowledge Discovery in Databases,* Cambridge, MA; AAAI/MIT, 1991.

[Qui86] R. Guinlan, "Induction of Decision Trees," Machine Learning, 1(1), pp.81-106, 1986.

[Qui93] R. Quinlan, *C4.5: Programs for Machine Learning*, Morgan Kaufmann Publisher, 1993.

[Sed88] R. Sedgewick., *Algorithms*, Addison-Wesley Publishing Company, Chapter 11 pp. 145-163, 1988.

[Shi79] D. Shier, "On Algorithms for Finding the k-Shortest Paths in a Network," Networks, vol 9, pp. 195-214, 1979.

[Ski87] C.C. Skiscim, and B.L. Golden., "Computing k-Shortest Path Lengths in Euclidean Networks," Networks, vol 17, pp. 341-352, 1987.

THE FLEXIBILITY OF AN UNSUPERVISED NEURAL NETWORK FOR MACHINE PART CLASSIFICATION

C.K. Lee, C.H. Chung
Department of Electronic Engineering, Hong Kong Polytechnic,
Hung Hom, Hong Kong.
Email:Encklee@hkpcc.hkp.hk

ABSTRACT

In this paper, we provide an application on the learning adaptation of an unsupervised neural network for machine part classification. The network we shall use is based on an unsupervised learning algorithm called learning by experience (LBE) [LY93b]. Here, we modify the network, so that whenever it encounters a memory full case, it solves the problem by releasing the constraint. Hence, it provides the flexibility for machine part classification when the preset acceptance level is stringent. Simulations results to illustrate the effectiveness of this approach will be included.

INTRODUCTION

Machine part classification is a necessary step where the process will classify the good parts into different categories and reject any parts which are not within a certain tolerance. Usually, for manufacturing purpose, we like to have some certain flexibilities when performing this task, for example, we may prefer not to have the number of categories and the bounds of the tolerances fixed first, as these factors will vary from day and day production needs. Hence, here we propose to use an unsupervised neural network, the LBE network, for such purpose.

However, like other unsupervised learning neural networks, when the number of input patterns to the network is larger than the pre-defined number of output neurons, and a new input pattern is not similar with all the previously stored patterns, the LBE network [LY93a] will become memory full. Therefore, the network cannot recognize any new input pattern and it envisages the memory catastrophe problem.

Here, we propose an approach to cope with the memory full effect, which we call it a constraint release. The threshold - a measure between the mean square error and the weights connected to the output neurons - will automatically be increased whenever the network becomes memory full. Also, we use *fuzzy rule sets* [Kos92a] to control the release of the threshold.

As mentioned previously, when this network applied to machine part classification, we shall face the following problem. Considering in a production line that produces machine parts such as screws and nuts, it is needed to classify good machine parts and locate defective parts, whereas this recognition job can be done by the LBE network. Moreover, we also want to classify good machine parts into different categories of some fixed acceptance levels. Hence, when there are too many parts falling below the acceptance level of each category, one cannot have a successful classification, because there is no more output neurons for category assignment. The phenomenon occurs when we set the acceptance level too tight or there are some problems in manufacturing the parts. Anyway, it will subsequently cause the network to become memory full. In such situation, we have two alternatives: The first one is that we can reject all these components; while the second one is that we can release the constraint (acceptance level) governing the classification so as to accept these components. Of course, when we have set the acceptance level too tight, it is better to release the constraint. Here, we propose a neural system which allows such constraint release. This practice allows us to accept more good parts, rather than to reject them, mainly because we have set an unrealistic level in the first place. Therefore, in this paper, we propose an unsupervised neural network which can automatically adjust this release.

An unsupervised network is used because we assume that a set of training data is unavailable, that is, the number of categories and the acceptance level of each category. Also, we want the network to alert us that we have not set a pragmatic acceptance level in the first place. An supervised network cannot perform this job satisfactorily as it always needs a training set [Kos92b]. But, when we want to release the constraint, we face the problem - what is the suitable level of such release. Here, we propose to use a fuzzy logic approach to handle this problem because we feel that it is difficult to model the practical situation and hence to calculate the appropriate level.

Currently, a well-established unsupervised learning neural network, ART [CG87] can be chosen for this job, where the vigilance can be varied for the similarity measure. However, it needs modifications for a successful implementation. Fuzzy logic approach [YF94] is a good candidate whenever we encounter a situation of uncertainty and this uncertainty is difficult to model. Ghosh & Pal [GPP93] has provided a good study where they applied fuzzy logic to a neural network for object extraction, however, they did not describe the situation where similarity measure of the neural network is adaptively changed. Xu, Krzyzak and Oja [XKO93] has given a critical analysis of applying competitive learning for clustering analysis, however, they have not provided the situation when clustering radii are dynamically changed due learning. In this paper, we attempt to provide an integrated approach when a neural network encounters a memory full situation. Beni and Liu [BL94] has proposed a fuzzy clustering method, however, they have considered the change of clustering radii.

The arrangement of the paper is as follows. In Section 2, we describe the structure of the LBE network and its learning algorithm. Section 3 will provide a method of image compression of the machine parts classified by the network. We need it because we want to reduce the number of neurons and hence the complexity of the network. In Section 4, the considerations and problems when the constraint is released are stated. Section 5 will provide a full picture of this paper, where the application of *fuzzy rule sets* will be elaborated and formulated. In Section 6, simulations will be presented to illustrate the feasibility of this constraint release adaptation. Also, an application to machine part classification is given. While Section 7 gives some judgements of this approach and its future development.

LEARNING BY EXPERIENCE (LBE) NETWORK

1 Structure of the LBE network

Unlike other unsupervised learning neural networks, the main idea of LBE network is not how to update the weights of the winner, but instead, is how to use *memory* to let output neurons know whether the applied input pattern belonging to the stored patterns or not. In fact, it employs the competitive learning as the updating rule for the winner.

First, a memory is added to each output neuron to store the strength, SS_i. This strength is used to compare the current input pattern with the stored patterns. As analog signals are applied to the network, the strength function is represented by the *mean square error* between the input pattern and the weights. Therefore, the lower the strength, the closer is the 2 compared patterns. The structure of the LBE network is shown in Fig. 1.

Basically, it is a two-layer, feed forward type network. The lower layer contains the input neurons and the upper layer consists of the output neurons. The connections among the lower layer and the upper layer store the weights, W_{ij}, which in turn store the learnt patterns. The connections within the upper layer indicate that each output neuron receives SS_i, as the inhibitory input, from all the other output neurons. Also, each output neuron has a memory to store its experience. Once an output neuron has become the winner, the current strength SS_i will be stored into its memory. This memory is used to determine whether an applied pattern is associated to this output neuron.

When an input pattern is applied to the network, X_j will be equal to P_j, the input pattern. Then S_i will be built up through the network connections. When S_i is obtained, it will be compared with the strength stored in its internal memory M_i to work out SS_i. Finally the output neuron which has the minimum strength will become the winner, and its internal memory and the weights will be updated only.

2 The Learning Algorithm:

Fig. 1 The structure of the neural network

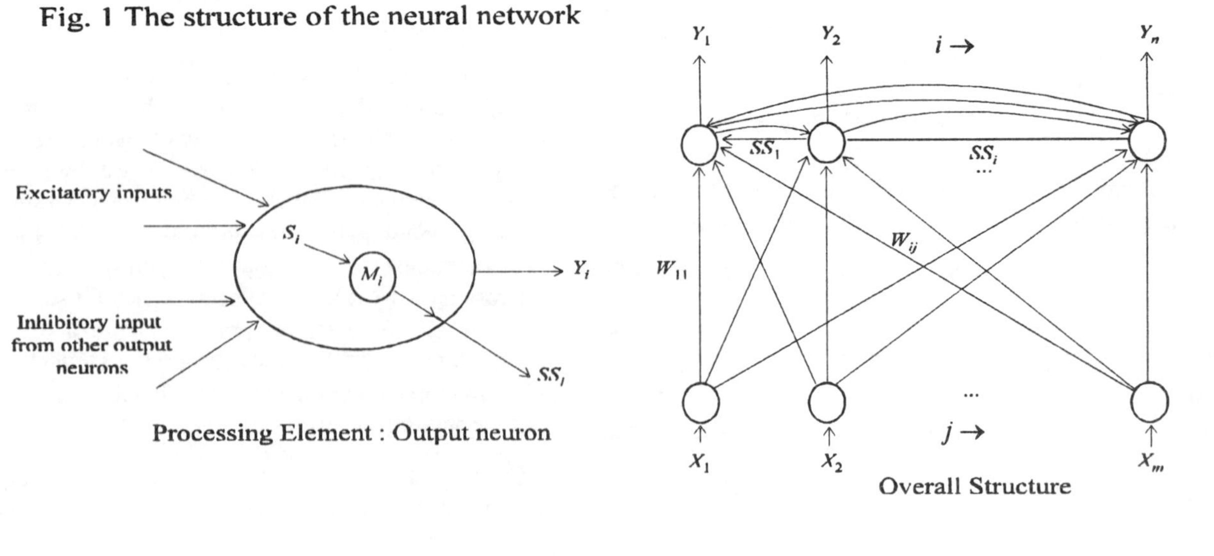

Step 1- the initialization phase

All weights, W_{ij}, are set to 0. All memory locations, M_i, are set to $(N+1)$, where N is the number of input neurons. The *learning rate* can be set to any value from 0 to 1. The *mean square error* can be set to any value from 0 to 1. This is the maximum allowable error between two input patterns to be classified to the same class.

Step 2 - the weight updating phase

Apply the pattern to the input neuron X_j of the network, then S_i is calculated as

$$S_i = \sum_{j=0}^{M} \frac{(W_{ij} - X_j)^2}{N} \quad (1)$$

The final strength SS_i, for competition within the upper layer is obtained as

$$SS_i = \begin{cases} S_i & \text{if } S_i \le M_i + \textit{mean square error} \\ \infty & \textit{otherwise} \end{cases} \quad (2)$$

The winner is searched among all the output neurons, and it is the one which has the minimum strength. Two cases will occur. They are: (1) *the normal case* -- the winner with SS_i not equal to ∞. This means the input pattern is recognized; (2) *the memory full case* -- the winner with SS_i equal to ∞. This indicates that there is no pattern stored in this network matched with this pattern and the network has no more unused output neuron to store this new pattern, which is a memory catastrophe.

For the normal case, the internal memory and the weights of the winner are updated as:

If $M_i = N + 1$ then

$$W_{ij} = X_j, \quad (3)$$

else

$$W_{ij} = W_{ij} + L \times (X_j - W_{ij}) \quad (4)$$

where i = index of the winner

L = learning rate

Next, we compute the S_i again and use these new weights, W_{ij}, to update the memory M_i as

$$M_i = S_i \quad (5)$$

and the output is given as

$$Y_i = \begin{cases} 1 & \text{if } i = \textit{index of winner} \\ 0 & \textit{otherwise} \end{cases} \quad (6)$$

IMAGE COMPRESSION

For machine part classification, we need to have images of the machine parts first. Usually for the simplicity of implementation, the images of the machine parts are thresholded into binary values first. For a reasonable resolution of the parts, the size of the binary images will be very large, which is usually in 256 × 256. In our case, as for illustration purpose, we assume that each image is represented by a 24 × 24 binary bitmap. Moreover, for the 2 different views of each part, we need 2 × 576 bits. Hence, the size of the LBE network will have 2 × 576 input neurons, and as a result, its size is large and it takes time to be trained. Therefore, we need to compress the image before inputting to the LBE network.

In our approach, a simple method is employed as the purpose of this paper is not to investigate the image compression techniques. Therefore rather controversially, we convert the original binary image into a gray-scale image. Hence, each block of the original image (size $n \times n$, where n is no. of pixels) is changed to a gray scale value -- dividing the number of black dots by the block size. The block size can be changed, which sometimes depends on the size of the original image. For example, for a 4 × 4 block, the original 24 × 24 binary image can be represented by 36 gray scale values. These 36 gray scale values are then input to the LBE network for classification. We use this approach to take the advantage that the LBE network can accept analog data so as to reduce its complexity. However, we also provide an investigation of this simple image compression on the recognition effects of machine parts.

THE PROBLEMS OF WHEN RELEASING THE CONSTRAINT

As mentioned in Section 1, we shall encounter a memory full case when the classes of the input patterns are larger than the number of output neurons. Therefore, the network cannot recognize any new input pattern which is different from the stored patterns and it causes the memory catastrophic effect.

One way to cope with the memory catastrophic effect is to increase the threshold - the *mean square error*, whenever the network becomes memory full. This threshold is used to control the allowable error between the current input pattern and the patterns stored in the output neurons. A new input pattern is not recognized because it is not similar to any stored patterns. The measure of similarity is based on the threshold and the mean square error between the patterns - stored and input. Hence, the input pattern will be classified as the same class learnt by that output neuron. Therefore, if the *mean square error* is larger than the threshold set by the network, this input pattern will not be classified, and the network will become memory full. In other word, we can increase the threshold of the network to increase the maximum allowable error of the network. As a result, we can classify some input patterns that are not classified before, which is equivalent to say, we have increased the learning capability of the network.

Although we can increase the threshold to allow the network to classify some input patterns that are not classified previously, but such practice will introduce another problem. For two very similar input patterns that have already been stored in the network, when we increase the threshold, we will also increase the allowable error of one output neuron. As the allowable error increases, those two similar patterns will be grouped into one particular output neuron. This can be illustrated as follows. In Fig. 2a, there are two input patterns represented in a 2-dimensional space. If the network is configured with 2 output neurons only, these two patterns will be stored in 2 output neurons as shown in Fig. 2b, and each circle corresponds to that cluster which belongs to that output neuron. As the threshold is being increased, the cluster corresponding to the output neuron will be increased accordingly. When the threshold becomes sufficiently large, the other input pattern will fall into its cluster as shown in Fig. 2c. Then the two patterns will be classified into one output

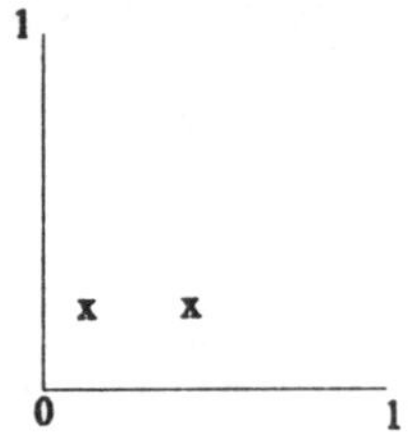

Fig.2a The 2 input patterns.

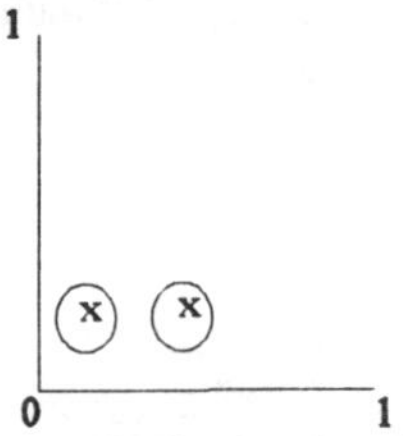

Fig. 2b The clustering of output neuron 1 and 2

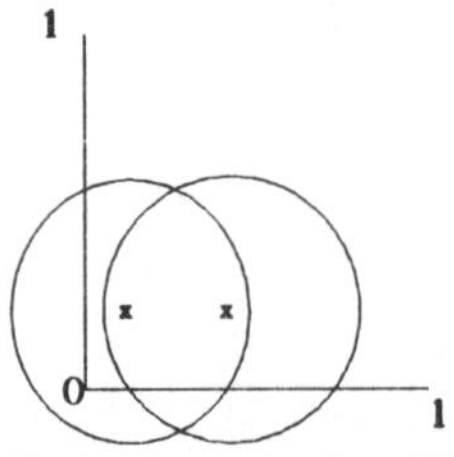

Fig. 2c The overlapping of the clusters after releasing the threshold

Fig. 2 The causes of instability.

neuron. If these two patterns should not be classified as a group, the network is considered to be unstable. In such a way, the previously stored pattern is wrongly classified to other output neuron after releasing the threshold. When the network becomes unstable, we have to restore the network to its previous state, and we have to reject the input pattern which makes the network become unstable. We can avoid the network to become unstable by simply not increasing the threshold at the very beginning when the network becomes memory full. But in such a case, we cannot increase the learning capability of the network. If the two input patterns can be grouped together, we can ignore the unstable behavior of the network. Therefore, we need the human decision to determine if the threshold should be released or not. In this paper, we simply increase the threshold when the network becomes memory full. Moreover, we assume that when any previously stored pattern is grouped to other output neuron, the network is merely considered to be unstable, and we need to restore the network previously.

Apart from instability problem, there are other problems concerning of releasing the threshold. First, one may ask: To what extent should the threshold be released? In what step-size should the threshold be released? In this paper, we try to resolve these problems by adopting a fuzzy rule approach. Here, a fuzzy threshold control algorithm is incorporated in the network to perform this job, and the block diagram of the modified LBE network is shown in Fig. 2d.

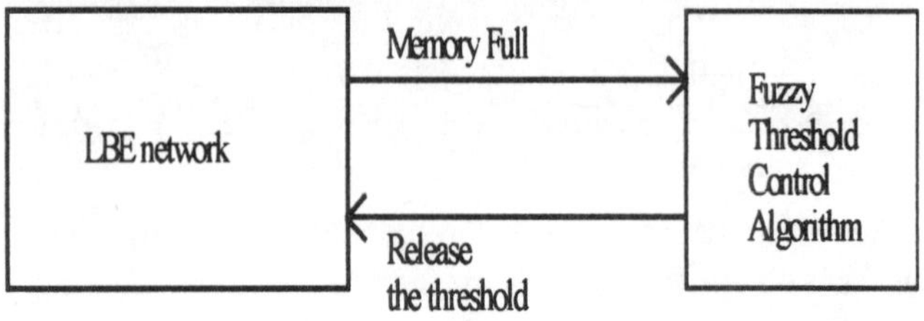

Fig. 2d The network with the fuzzy threshold control.

From the above, we can see that the network becomes unstable when we increase the threshold. This reveals from the fact that the cluster area belonging to one particular output neuron increases as we increase the threshold. When the network has successfully classified the first input pattern, the first output neuron is occupied. As new input patterns are applied to the network, they will be compared with the pattern stored in the first output neuron. If a new input pattern falls within the cluster of the first output neuron, it will be classified into the same output neuron. If it is outside the cluster of the first output neuron, and there is an unused output neuron, it will then occupy that particular output neuron. If there is no other unused output neuron, then the network will becomes memory full. The cluster radius is determined by the threshold of the network. As the threshold is increased, the cluster area is enlarged. If the two patterns stored previously are closely enough, then the network will classify them together. Unfortunately, the network will treat this situation as unstable even though these two patterns can be classified as the same group. This is an disadvantage of the clustering method using our network and it happens in any unsupervised clustering methods.

FUZZIFICATION OF THE VARIABLES

First, we list the abbreviations used as below:

ZE: ZERO

SM : SMALL

ME: MEDIUM

LA: LARGE

VL: VERY LARGE

PS: POSITIVE SMALL

PM: POSITIVE MEDIUM

PL: POSITIVE LARGE

NS: NEGATIVE SMALL

NM: NEGATIVE MEDIUM

NL: NEGATIVE LARGE

There are 3 factors that we need to monitor when we want to release the amount of the threshold. They are: (1) the current mean square error, (2) the frequency of memory full occurrence, and (3) the instability of the network. We use these factors as input fuzzy variables to an inference engine to determine the amount of threshold release. Here, we construct the fuzzification of memberships of these 3 factors first.

1. Current Mean Square Error

The current mean square error ranges from 0 to 1. A large value reveals that the input pattern deviates much from the stored patterns. Therefore, we can change the new threshold to a greater extent. The current mean square error is divided into 5 fuzzy set values, namely ZERO, SMALL, MEDIUM, LARGE, VERY LARGE, and its membership function is depicted in Fig. 3. Here we use trapezoidal fuzzy set values. The length of the upper and lower base is 0.1 and 0.4 respectively, and for each adjacent fuzzy set value, there is an approximate 17% overlapping. This is the same for the other two input values: the frequency of memory full occurrence and instability.

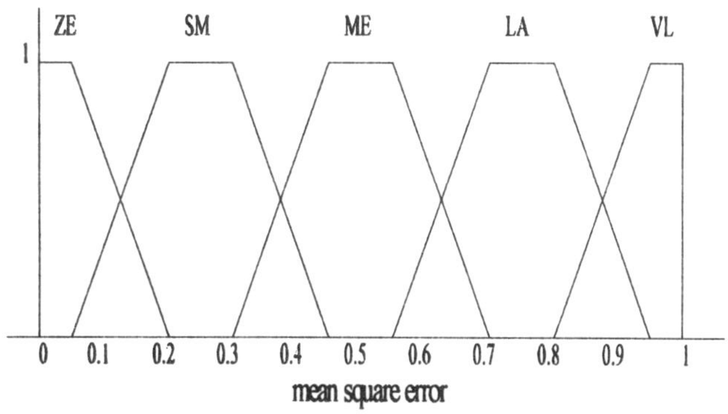

Fig. 3 Membership graph for the mean square error

2. Frequency of Memory full occurrence

The fuzzy threshold control algorithm will determine the new threshold after 10 iterations of learning. The choice of a 10-iteration period is to ease off the calculation of the frequency of memory full occurrence. We normalize the frequency of memory full occurrence so that it ranges from 0 to 1. A counter is assigned to count the number of input patterns which has caused the network to become memory full. After that, the frequency of memory full occurrence is equal to this counter value divided by 10. Also, this frequency is divided into five fuzzy subsets namely ZERO, SMALL, MEDIUM, LARGE, and VERY LARGE, with the membership functions as shown in Fig. 4.

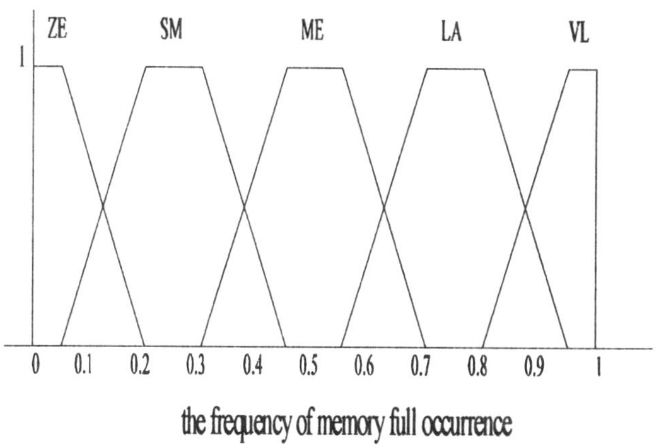

Fig. 4 Membership graph for the frequency of memory full occurrence

3. Instability

Before we can investigate the instability effect, we assign the network an extra memory which size is equal to the number of input patterns. Hence, we can keep track with the class relationship between input patterns and output neurons, which indirectly determines the instability of the network after releasing the threshold. If any previously stored patterns are wrongly classified into other output neuron when we release the threshold, then the network is considered to be unstable, and the threshold has to be reduced to restore the network to its previous state. Here, the instability ranges from 0 to the number of input patterns. For simplicity, it is normalized by dividing the number of stored patterns that are mapped to different classes by the number of previously stored input patterns. A 0 means no stored pattern is re-mapped to another class; whereas an 1 indicates that all stored patterns are re-mapped to other classes. Again, it is divided into 5 fuzzy subsets with the membership functions as shown in Fig. 5.

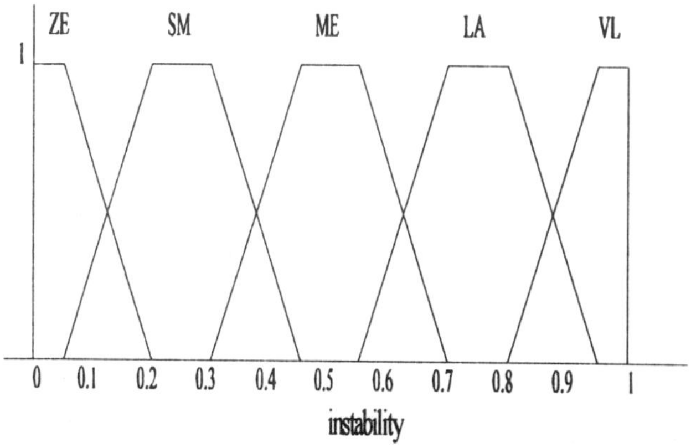

Fig. 5 Membership graph for instability

4. Percentage change in threshold of the network

From the above 3 input fuzzy set variables, an inference engine can determine the percentage change in threshold. The percentage change in threshold is ranging from -100% to +100%. From Section 4, we see that the network becomes unstable when the cluster center of one output neuron falls within the cluster region of other output neurons. In other words, the amount of threshold that can be increased is determined by the distance between the center of the cluster of each output neuron. Therefore, the percentage change of the threshold is normalized to the distances among the input patterns. In Section 6, from the simulation results, we observe that we can increase the threshold to a greater extent if the distances among the input patterns are larger. Here, in order to have the sufficient precision, we divide it into 7 fuzzy set values namely: NEGATIVE LARGE, NEGATIVE MEDIUM, NEGATIVE SMALL, ZERO, POSITIVE SMALL, POSITIVE MEDIUM, and POSITIVE LARGE. Fig. 6 shows its membership functions. From the membership graph, the lengths of the upper and lower base are 20% and 40% respectively. There is an approximate 18.75% overlapping for each adjacent fuzzy set value, and the centers are -82.5%, -50%, -20%, 0%, +20%, +50%, and +82.5% respectively. These values are used in calculating the fuzzy centroid.

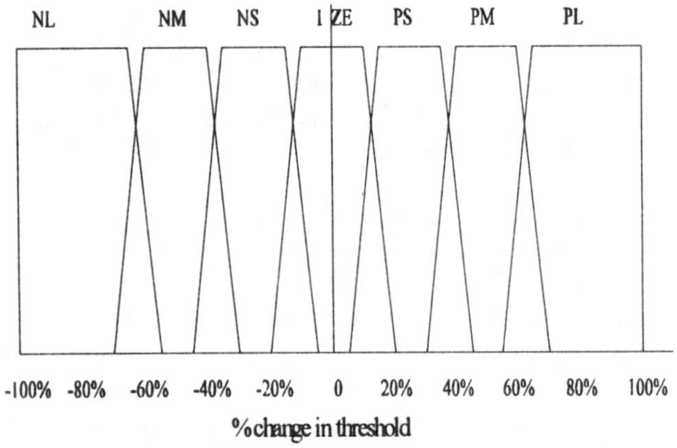

Fig. 6 Membership graph for the %change in threshold

After defining the ranges of the fuzzy variables, we have to build the rule matrix of the inference engine. The constructions of the different rules are elaborated as below.

1. The stability of the network is considered to be the most important one for any learning algorithm. Therefore, if any one of the previously stored patterns is re-mapped to other classes after releasing the threshold, the threshold must be tightened to restore the network to its stable state. Hence, if the instability is NOT ZERO, the percentage change of threshold can be worked out as Table 1.

Table 1 Instability vs. % change in threshold.

Instability	SM	ME	LA	VL
%change in threshold	NS	NM	NM	NL

As observed from Table 1, if the degree of instability is small, we can restore the network to its stable state by decreasing the threshold by a small amount. If the degree of instability is very large, we must decrease the threshold by a much larger extent.

2. If the instability is ZERO, then the percentage change in threshold can be obtained from the two variables, the mean square error and the frequency of memory full occurrence. We tabulate the rule matrix as in Table 2.

Table 2 % change in the threshold.

%change in threshold		The frequency of memory full occurrence				
		ZE	SM	ME	LA	VL
mean square error	ZE	ZE	PS	PM	PL	PL
	SM	ZE	PS	PM	PM	PL
	ME	ZE	PS	PM	PM	PL
	LA	ZE	PM	PM	PL	PL
	VL	ZE	PM	PL	PL	PL

If the frequency of memory full occurrence is zero (ZE), there is no need to change the threshold since all input patterns are successfully classified. For a small frequency of memory full occurrence, and a small mean square error, we can change the threshold by a small extent (PS). For a large (LA) frequency of memory full occurrence, it reveals that many input patterns cannot be classified. Therefore, we need to change the threshold by a greater extent (PL).

With the above 25 rules, the inference engine first finds out the scalar activation value w_i for each rule. From these 25 activation values w_i, we can then work out the output fit vector, and the percentage change in threshold is calculated using a defuzzification algorithm - the centroid of the output fit vector.

SIMULATIONS

1 Test images

First, we construct the synthetic images of some machine parts. Initially, they are represented by 24 × 24 binary bitmap, but finally they are compressed into different ratios. We perform these steps as we want to illustrate the effects of compression on the network. In addition, for a 24 × 24 binary bit-map, the LBE network needs 576 input neurons, which takes time to train.

First, we compress the original 24 × 24 binary images into 6 × 6 gray-scale images. Hence, we divide the image into blocks of 4 × 4 image. For every block of 16 pixels, we calculate the gray-scale value by counting the number of

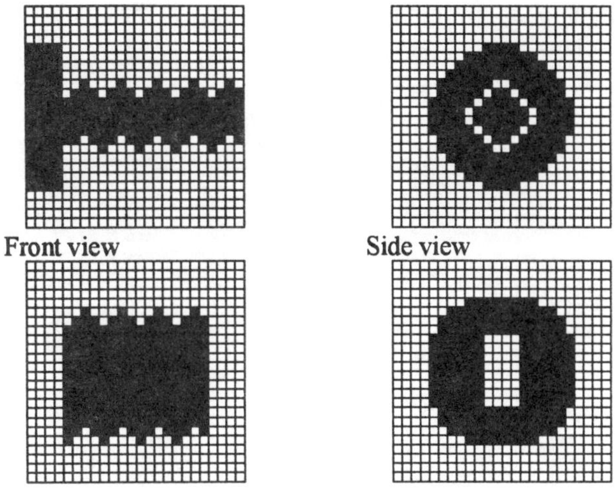

Fig. 7 Non-defective machine parts

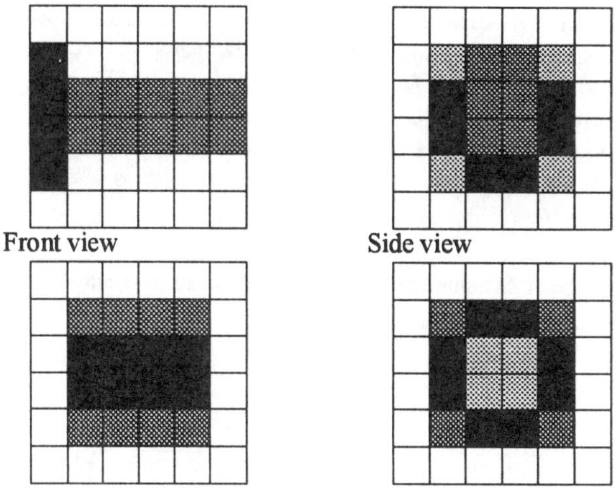

Fig. 8 Defective machine parts after compression.

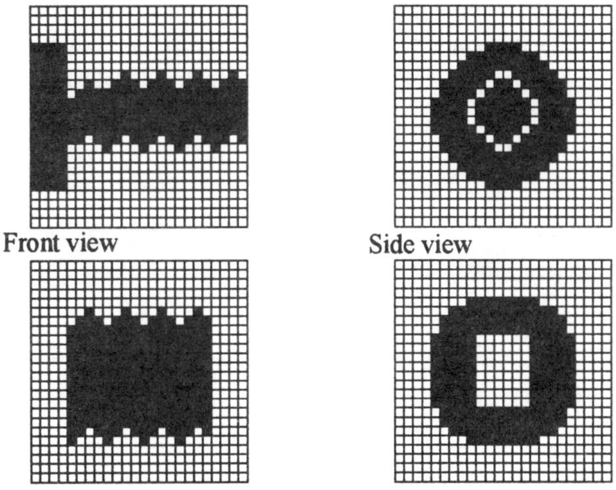

Fig. 8 Defective machine parts.

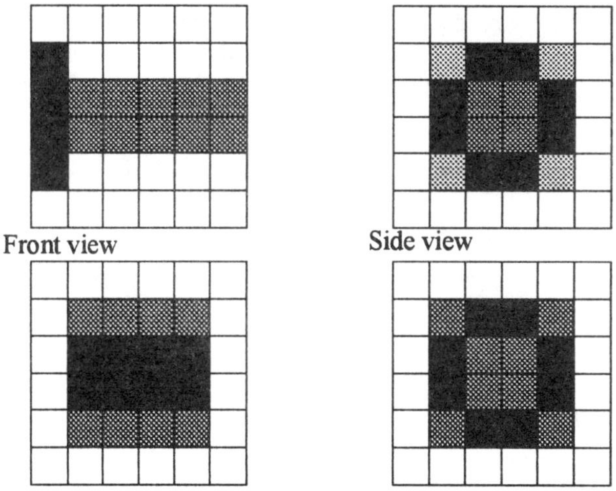

Fig. 10 Non-defective machine parts after compression.

black pixels in that block and dividing that number by 16. The result is the required gray-scale representation of that block. Hence,

$$\text{gray-scale value} = \text{no. of black pixels} / 16 \qquad (7)$$

As a result, the original 24×24 binary image can be represented by 36 gray-scale values. Therefore, the size of the LBE network is reduced from 576 to 36 input neurons. Fig. 7 shows two examples of non-defective machine parts and Fig. 9 shows two defective machine parts. While Figs. 8 and 10 show their corresponding gray-scale images after compression.

2 steps

Now, we apply these compressed images of good and defective machine parts to the network for classification. Also, another 15 uncompressed images as shown in Fig. 11 (at the end of this paper) are used. For simulation purposes, the front and the side views of them are treated separately and are both applied to the network for classification.

As observed from Fig. 11, image 7 is identical to image 3. In addition, image 1 to 4 is the non-defective machine parts, whereas image 5 to 8 are the defective parts as shown in Fig. 7 to 10. Image 9 and 10 are defective parts which are slightly different from image 8, and 7 respectively. Image 11 and 15 are defective parts which both deviate from image 5. Image 12 is identical to image 1. Moreover, image 13 is identical to image 4 and image 14 is identical to image 5. Therefore, they are in only 7 classes as tabulated in Table 3.

Table 3 Classifications of the images.

Class	1	2	3	4	5	6	7
Image	1, 12	2	3, 7, 10	4, 13	5, 11, 14, 15	6	8, 9

Thus, the LBE network is configured to have 7 output neurons and 36 input neurons. The initial threshold is set to 0.000001 and the number of iterations is 200.

3 Results

Because of the small threshold, the input image 1 to 8 will use up all the output neurons, so the network becomes memory full when a new different image arrives. The fuzzy algorithm releases the threshold until iteration 170 with threshold 0.003553. Beyond iteration 170, the threshold remains unchanged, and the result of classification is tabulated as Table 4.

Table 4 Assignments of the images to the output neurons.

Image	1	2	3	4	5	6	7	8	9	10	11	12	13	14	15
output neuron	1	2	3	4	5	6	3	7	7	3	5	1	4	5	5

The results of simulation is expected. Also, we plot the threshold change and the changes in weights connected the input neurons and the output neurons, which in effect, illustrate the learning process and hence the occurrence of the stability here. Fig. 12 shows the change in threshold against the number of iterations, while Fig. 13 shows the summed weight product of each output neuron against the number of iterations.

Fig. 12 Threshold vs. no. of iterations

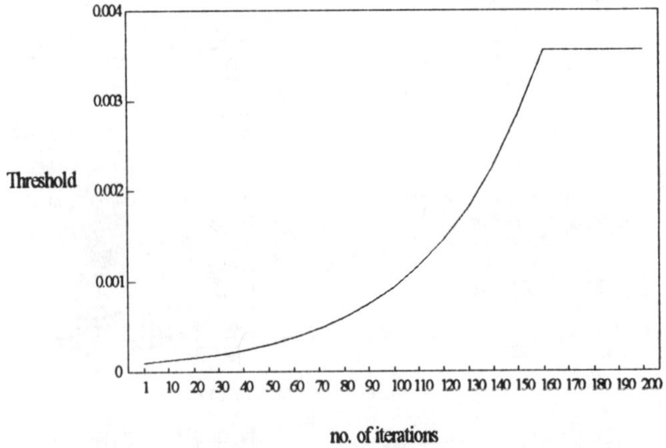

As revealed from Fig. 13, the summed weight products of output neurons 1, 2, 4, 6 and 7 remain unchanged and these

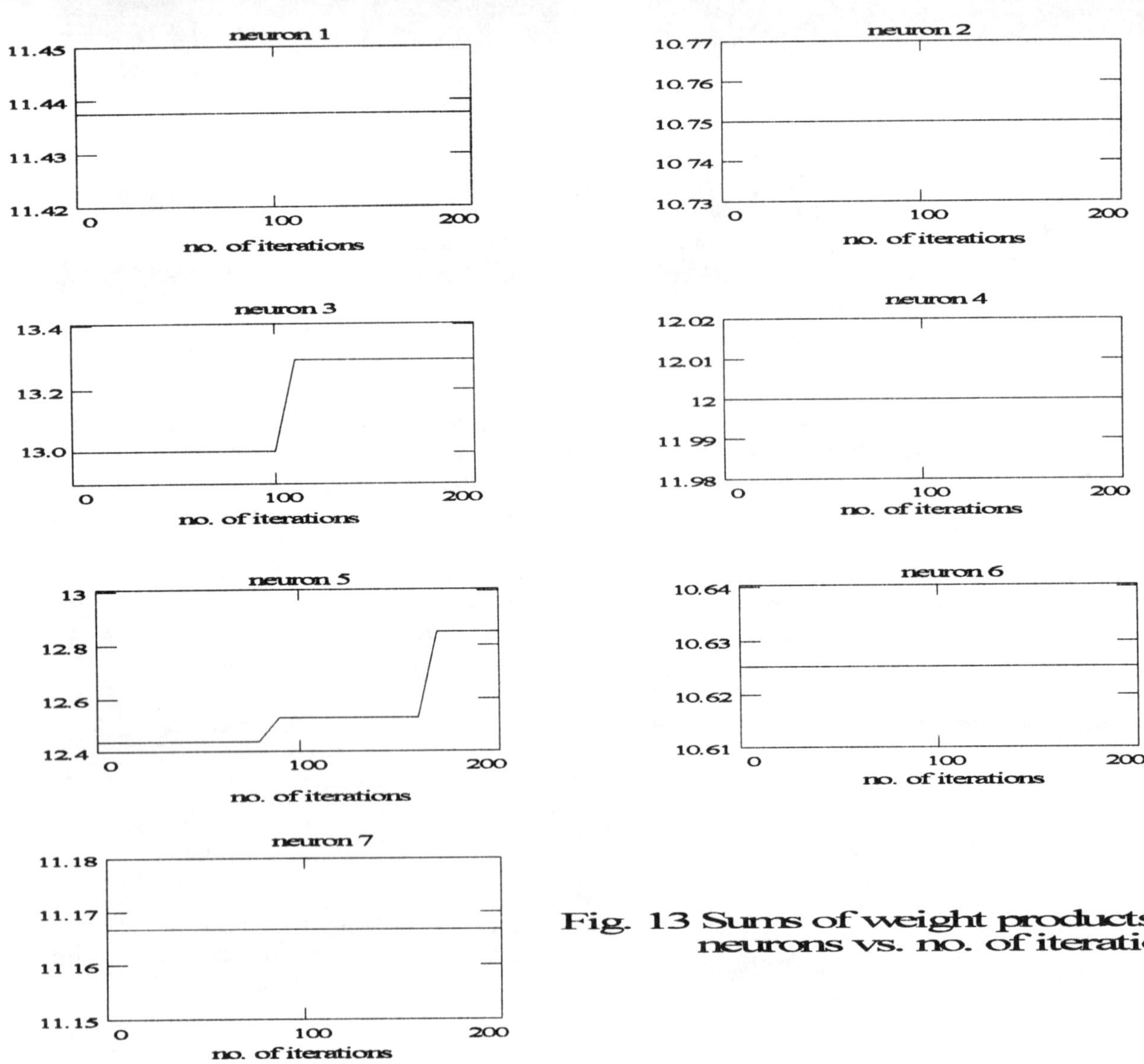

Fig. 13 Sums of weight products of neurons vs. no. of iterations

output neurons have settled down within the first 10 iterations. For output neuron 3, it has to settle down at about iteration 110 with the summed weight product equal to 13.2875. For output neuron 5, its response becomes steady at about iteration 170 with a summed weight product equal to 12.845834.

When the threshold is increased too much, some previously stored pattern will be wrongly classified to other output neurons, and the LBE network is considered to be unstable. To show the effect of the fuzzy threshold control algorithm in the case when it becomes unstable, it is configured to have only 2 output neurons, and 3 images as shown in Fig. 14a was input to the network for classification. These 3 images are represented as 3 points on the *xy*-axes. And the no. of iterations was configured to be 300.

Hence, from the simulations we observed the following phenomenon. Initially, image 1 and 2 occupy output neurons 1 and 2 respectively, and image 3 is unclassified. Thus, the fuzzy threshold control algorithm increases the threshold. The circle shown in Fig. 14b represents the corresponding cluster belonging to that output neuron, and the size of the circle is proportional to the threshold of the network. As the threshold increases, the corresponding cluster region of each neuron will increase. When the area of the cluster enlarge to a certain extent, the other image will fall into it. Therefore, both images will be wrongly classified into the same class. This happens when the threshold becomes 0.261955. The

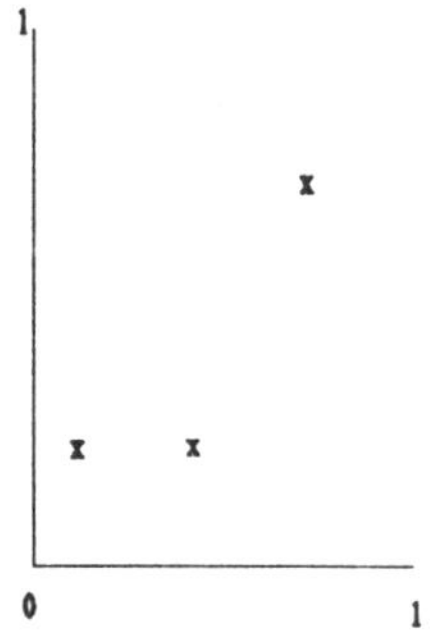

Fig. 14(a) The 3 input images.

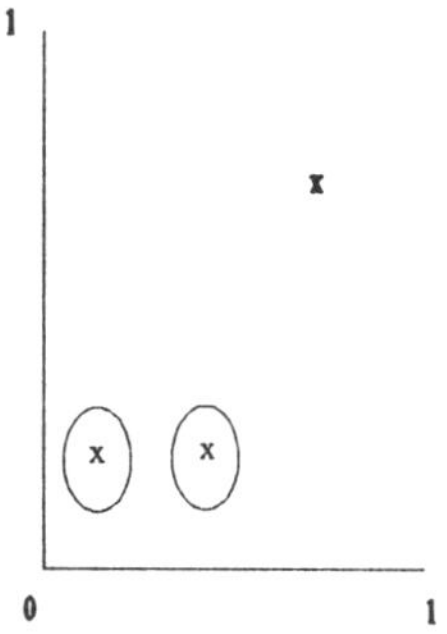

Fig. 14(b) The clustering of the 3 images.

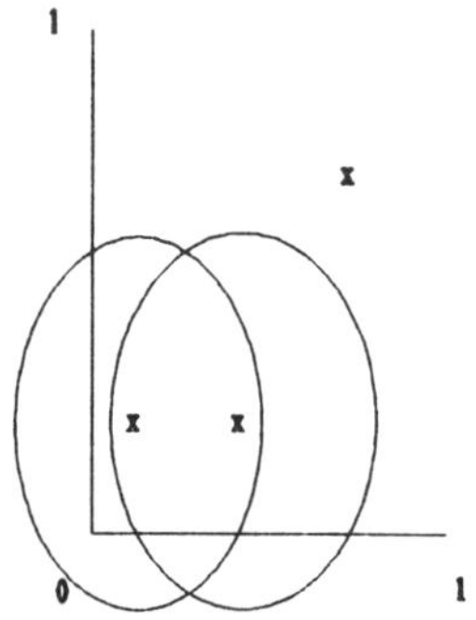

Fig. 14(c) The overlapping of the two clusters with threshold = 0.261955

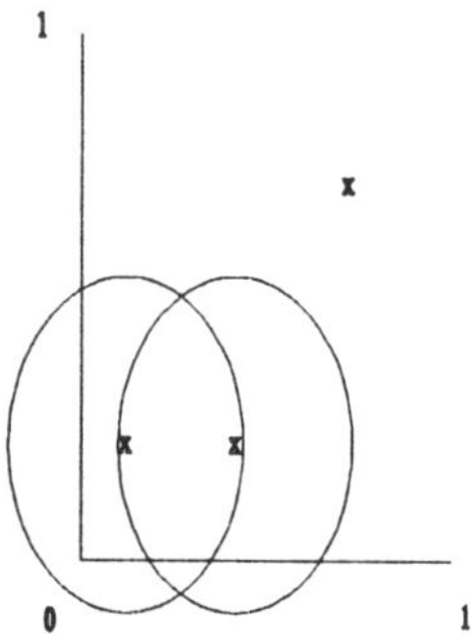

Fig. 14(d) The overlapping of the two clusters with threshold = 0.181913

Fig. 14 The clustering of incoming images.

situation becomes that image 1 and 2 occupy output neuron 1 and image 3 occupies output neuron 2. This is shown in Fig. 14c. Therefore, the fuzzy threshold control algorithm then reduces the threshold, and the network has to learn all the images again. Beside reducing the threshold, the algorithm also resumes the original internal weights and strengths stored in the network just before the network becomes unstable. The algorithm continues to reduce the threshold until the network restores to a stable state in which image 1 and 2 occupy output neuron 1 and 2 respectively and image 3 being unclassified. Afterwards, the threshold will be increased again, but this time a smaller step size is used. However, the network becomes unstable again when the threshold becomes 0.181913 as shown in Fig. 14d. The algorithm then restores the network to a stable state, and the network will increase the threshold again with even a smaller step size. For these 3 images, the network will flip between stable and unstable states until the learning process was terminated by the no. of iterations configured at the very beginning. The result of the simulations shows that image 1 and 2 occupy output neuron 1 and 2, whereas image 3 is still unclassified. In the practical situation, it means that the part corresponding to image 3 has to be rejected as it does not fit any of these 2 clusters.

The network will become unstable when the cluster correspond to each output neuron cross over with other output neuron. The cluster of each output neuron is proportional to the threshold of the network. Therefore, the stability of the network is determined by the threshold. To investigate how much the threshold can be increased, we configure the network to have 1 output neuron with 2 input images. The threshold is being increased until the network becomes unstable. The threshold is then compared with different sets of images with different distance between the 2 image. The results are tabulated in Table 5.

Table 5 Distance between images vs thresholds.

distance	0.1414	0.3	0.4243	0.583	0.781
threshold	0.010451	0.051345	0.148386	0.252257	0.428836

We can see from the result that, as the distance between the two image is large, the network can allow a larger threshold before it becomes unstable. Therefore, we can determine how much the threshold can be increased from the difference between the input images.

To investigate the effect of image compression on the performance of the LBE network. We have applied image 1 and 5 into the LBE network for classification using different compression ratios. That is the original image is divided into blocks of 3 × 3, 4 × 4, 6 × 6, 8 × 8 and 12 × 12. The network is configured to have only one output neuron. Therefore, we can compare the mean square error between two images using different compression ratios. The obtained *mean square errors* are shown in Table 6.

Table 6 Compression ratios vs. mean square errors.

Compression ratio	3x3	4x4	6x6	8x8	12x12
Mean square error	$1\text{x}10^{-5}$	$1.3\text{x}10^{-5}$	$2.9\text{x}10^{-5}$	$2.7\text{x}10^{-5}$	$5.4\text{x}10^{-5}$

From the results, as the compression ratio increases, the mean square error also increases. Hence, we have to compromise between the LBE network size and the error introduced by image compression. Therefore, we use a 4 × 4 block unit or a 16:1 ratio in the data compression so as to keep a good balance between the network size and a smaller error.

CONCLUSION

In this paper, we have illustrated the application of fuzzy logic in an automatic selection of the threshold releasing value. With this fuzzy threshold releasing algorithm, we can eliminate the memory full effect and hence we can increase the learning capability of the network. In addition, we have illustrated its application for machine part recognition when there are many parts.

Beside using an automatic threshold releasing algorithm, the constraint releasing algorithm can follow an user-set profile, in which the releasing criteria come from the manufacturing history of a product. Hence, in the next step, we shall investigate this effect on this aspect.

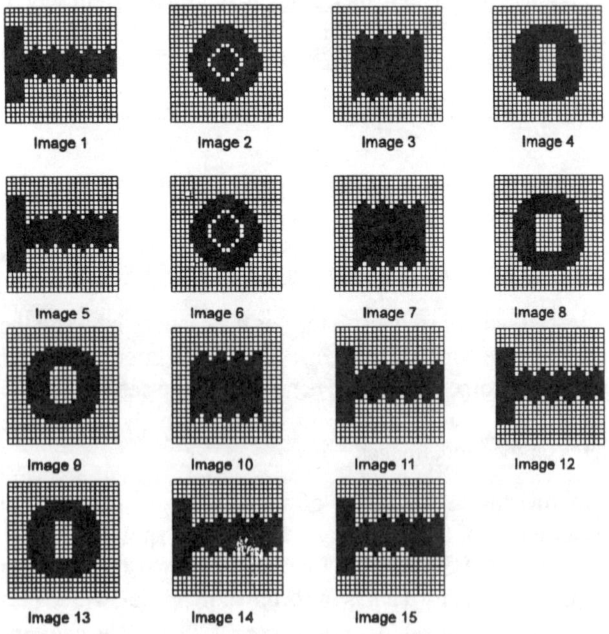

Fig. 11 Uncompressed images for recognition.

References

[LY93a] C.K. Lee, K.F. Yeung, "An Unsupervised Learning Algorithm for Character Recognition and Classification," *WCNN'93*, 11-15 July, 1993, vol2, pp. 427-430, Portland, USA.

[Kos92a] B. Kosko, *Neural Networks and Fuzzy Systems, A Dynamical Systems Approach to Machine Intelligence*, Prentice Hall, 1992.

[Kos92b] B. Kosko, *Neural Networks for Signal Processing*, Prentice Hall, 1992.

[LY93b] C.K. Lee, K.F. Yeung, "Waveform Recognition and Classification Using an Unsupervised Network," *IJCNN'93-Nagoya*, October 25-29, 1993, Japan.

[CG87] G. A. Carpenter, S. Grossberg, "ART2: Self Organization of Stable Category Recognition Codes for Analog Input Patterns," *Applied Optics*, vol. 26, pp. 4919-4930, 1987.

[YF94] R. R. Yager, D. P. Filev, *Essentials of Fuzzy Modeling and Control*, Wiley-Interscience, 1994.

[GPP93] A. Ghosh, N. R. Pal, S. K. Pal, "Self-Organization for Object Extraction Using a Multilayer Neural Network and Fuzziness Measure," *IEEE Trans. Fuzzy Systems*, vol. 1, no. 1, pp. 54-68, 1993.

[XKO93] L. Xu, A. Krzyzak, E. Oja, "Rival Penalized Competitive Learning for Clustering Analysis, RBF Net, and Curve Detection," *IEEE Trans. Neural Networks*, vol. 4, no. 4, pp. 636-649, 1993.

[BL94] G. Beni, X. Liu, "A Least Biased Fuzzy Clustering Methods," *IEEE Trans. Pattern Analysis and Machine Intelligence*, vol. 16, no. 9, pp. 954-960, Sept. 1994.

THE INTENSITY OF IMPLICATION, A MEASUREMENT LEARNING MACHINE

Laurent Fleury **Yann Masson**
IRESTE UNIVERSITY OF NANTES
La Chantrerie, CP 3003, 44087 Nantes cedex 03, France
☎ 33 40 68 30 00 - Fax 33 40 68 30 66
e-mail : fleury@silr.ireste.fr

PerformanSe
3, rue Racine, 44000 Nantes, France
☎ 33 40 73 93 96 - Fax 33 40 69 61 59

ABSTRACT

This study presents a machine learning method based on examples extracted from a database. It has been carried out for an industrial field of application, the constraints of which have obliged us to give a measurement of the intensity of implication between the premise and the conclusion of a rule. The intensity of implication is an index which makes it possible to answer the question "What is the probability of having an implication between two propositions ?". Adapted to the field of symbolic learning, the intensity of implication enables to measurement the probability of a rule such as : if *premise* then *conclusion*. This index seems to be particularly adapted to the field of learning. It resists noise, converges with the size of the sample, eliminates coarse rules, can be used within the framcwork of an incremental algorithm. In that article, we analyse it in detail, compare each of its attributes with that of conditional probability.

INTRODUCTION
INDUSTRIAL ACCOUNT

PERFORMANSE is a firm which designs expert-systems for the psychological analysis of behaviour. In order to be relevant, it has to follow a policy of global analysis of the individual instead of reacting to a specific component irrespective of the others. That is the reason why PERFORMANSE has initially turned to the technique of expert-systems.

The original knowledge base was first conceived by HUMAN EDGE SOFTWARE corporation, a firm located in PALO ALTO, founded in 1984 by Doctor James H. Johnson. Being both a computerist and a specialist in psychology, he was the source of the products of the firm [GJM+77, GJW78, GJW79, GJK80]. The results have been a series of software programs using the technique of expert-systems and dealing with the multiple aspects of human behaviour in a professional environment: communication, leadership, management of human resources in a firm.

PERFORMANSE has brought its products to France and since 1986, it has been interested in developing and extending the knowledge base in order to make a reference tool for the management of human resources within firms and French consulting services. Nowadays, as we shall see later, its industrial needs have led them to become more involved in machine learning.

The main package of PERFORMANSE is composed of three expert-systems (amounting to thousands of rules), one of them for self evaluation, another one for the evaluation of someone else, and the third one, which complements the firsts two, is a tool for vocational guidance. The way these expert-systems work is standard. The expert knowledge is conveyed in rules and the knowledge concerning the individual under evaluation is obtained from an initial questionnaire.

PERFORMANSE is now widely established in France and so, through its privileged customers, has at its disposal a set of evaluations, distributed in several databases which are centralised in order to constitute a learning set.

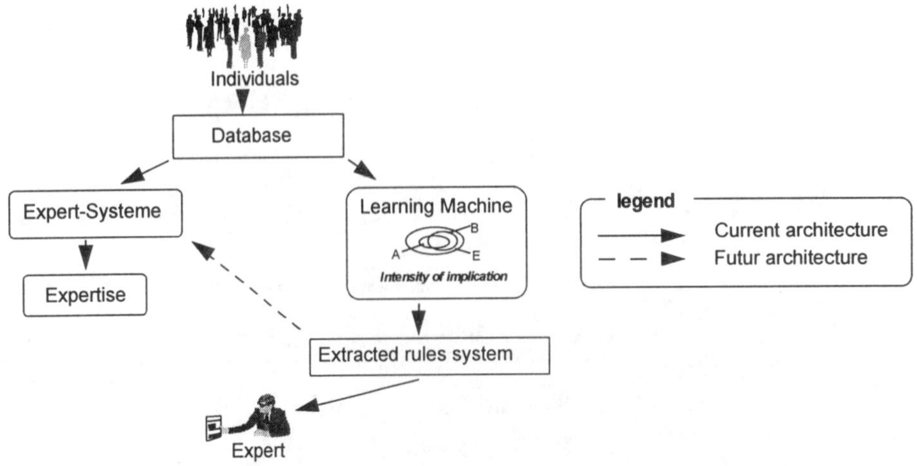

- Figure n°1 : -

PURPOSE AND CONSTRAINTS OF AUTOMATIC LEARNING

The method which is used nowadays to assess the importance of a component is that of scoring. That approach, according to the experts is too restrictive. Can we indeed be satisfied with the IQ to appraise an intelligence which may also be synthetic, intuitive, creative ? One way to compensate for the limitations of scoring is to find, among individuals of a same class, specific configurations. By getting involved in machine learning we want to achieve several objectives. The first one is to provide a qualitative validation for a tool which already exists, in a field where quantitative validation is the only one which is acknowledged. The second one is to provide the expert with a feed back of his expertise, in order to corroborate his views or lead him onto another direction.. Eventually, and this is our most ambitious goal, we want to find out the behaviour of a group of individuals. There will be lots of applications then. We will be able to define some distinctive configuration of a professional activity and determine a component of the corporation culture.

We have started our research with the study of some existing methods. There are too many algorithms for concept acquisition to make an inventory of them without forgetting any [CMM83] [GD86] [Gan87, Gan88] [KT93] [MST94] [CN88] [BMP+90]. Very quickly, we found it necessary to place that mechanism between the expert and the expert-system (knowledge acquisition). That is the reason why we have to obtain some knowledge that can be both understood by the expert and used by the expert system. So we have been interested in the search for the logical and statistical regularities within a learning space.

Of course, we have taken a particular interest in the measurement index of the quality of an implication existing in various articles. For instance : the lack of counter-examples, conditional probability, ratio of examples and counter-examples, entropy.

As we have already said, we would like to extract a set of rules usable by an expert-system, in other words, we are looking for non-isolated rules. Entropy, often used for decision trees, does not answer our needs.

The description of the evaluated individuals, involves a certain lack of precision which is difficult to assess. As a matter of fact, depending on the components used in the initial questionnaire a few contradictions may appear. That is the reason why these would be a high waste of information if we limited ourselves to implications without counter-examples.

Eventually such is our latin sensitiveness that the way some choices are defined, may influence the answer of the individual to the detriment of the component it is trying to measurement. The following example brings out the choice of item n°1 in a proportion of 82% :

1- What I say pledges my word as much as what I write.

2- For me, only what is written means obligation.

As we will see in detail later, when a whole population is described with a component which is always true, process learning, based on conditional probability or the ratio of examples over counter-examples, results in a system of rules the majority of which concludes that that same component is true. This is nonsense because, if a proposition is not discriminating, it gives little information, even though the acquired system of rules seems to say the contrary.

For all those reasons, we have tried to build up an index which will be able to answer our needs and eliminate incoherences induced by certain other indices.

PROBLEMATIC FORMALISM

Some learning systems work as follows. We have at our disposal a set $E = \{e_1, e_2, \ldots, e_n\}$ composed of examples in the field of what we want to study. An example e_i of E is distinguished by a conjunct of propositions. We call $d(e_i)$ the description of e_i : $d(e_i) = d_1 \wedge d_2 \wedge \ldots \wedge d_p$ Let

$D = \{d_1, d_2, \ldots, d_k\}$ be the set of propositions which may characterise an example of *E*. Then, we look for all the conjuncts between the propositions in the form of implication such as *Premise* $\Rightarrow$ *Conclusion* in which *Premise* is a conjunct of propositions from *D* and *Conclusion* a proposition from *D*.

A rule is considered as logical (Figure n°2) if it does not allow any counter-examples. In terms of sets, a rule is logical if the sub-set *PREMISE* of the examples from *E* for which *Premise* is true is strictly included in the subset *CONCLUSION* of the examples from *E* for which *Conclusion* is true

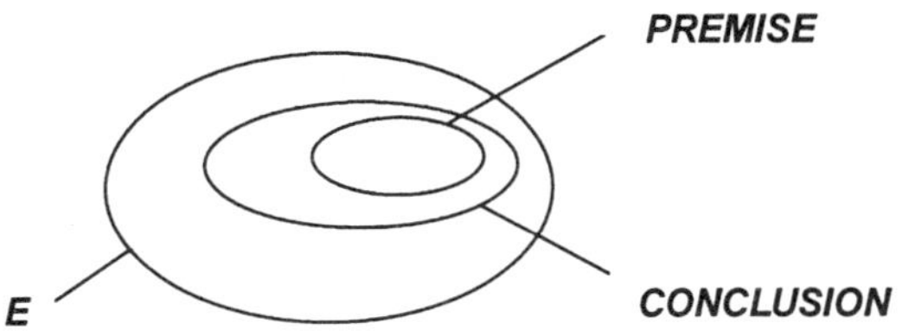

- Figure n°2 : A logical rule -

When set *E* is large and/or noisy, it can be interesting to generate rules which allow a certain number of counter-examples. Then we refer to statistical rules (Figure n°3).

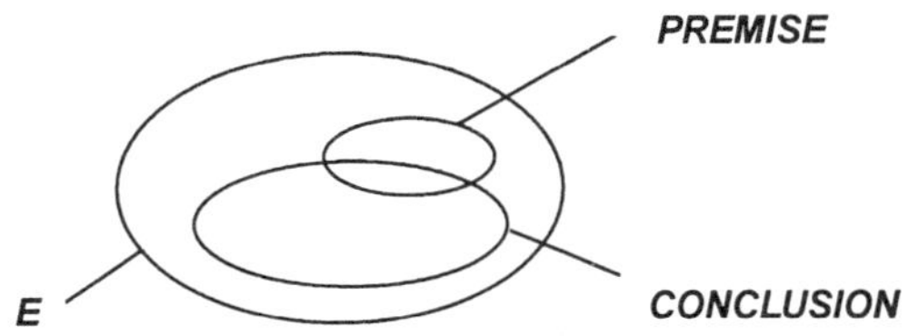

- Figure n°3 : A statistical rule -

In that case, we have to ask ourselves a few questions : Which conditions are necessary to accept a "quasi-implication" from *Premise* to *Conclusion* ? Can we associate a measurement with the "quasi-implication" *Premise*$\Rightarrow$*Conclusion*, its value being an index of the quality of this "quasi-implication" ?

The intensity of implication, a notion resulting from the work of R. Gras and A. Larher [GL93] based on probability theory, gives an answer to these two questions. Before using this index for the purpose of learning concepts, we will make a detailed presentation of it, after a non-exhaustive study of proposed solutions in various articles. Finally we will present it in detail.

PRESENTATION OF CURRENTLY USED INDICES

In order to make a synthetic comparative study of different indices which are found in various articles, we will present them here gathered in homogeneous families.

In order not to complicate index formulas and according to the formalism given at the end of this article, we will present them in order to give an estimation of the implication quality $a \Rightarrow b$. Let *A* be the sub-set of examples for which the proposition *a* is true. Let *B* be the sub-set of examples for which proposition *b* is true. The set $A \cap \overline{B}$ then corresponds to the sub-set of examples for which the implication $a \Rightarrow b$ is denied.

⇨ When a rule (if *a* then *b*) verifies a maximum of examples and a minimum of counter-examples in *E*, we can consider it pertinent. So, the GENREG system generates a rule only if it recognises a number of counter-examples smaller than a given threshold α and a number of examples greater than a threshold β [Ral91].

$$Card(A \cap \overline{B}) \leq \alpha$$
$$Card(A \cap B) \geq \beta$$

<u>Note</u> *: The thresholds given above (α and β) refer to particular knowledge of the studied field. So they will be given by the user.*

⇨ The system described in [SS91] uses a similar approach. The learning process is achieved through successive generalisation of examples. The only implications for which the ratio of the number of examples over the number of counter-examples is greater than a given threshold are preserved : $\frac{Card(A \cap B)}{Card(A \cap \overline{B})} \geq \alpha$

<u>Note</u> *: This index is commonly used in the learning fields because it also allows to propose a measurement for the implication of a conjunct of propositions toward a proposition as the discriminant ability of a proposition for a set of examples in relation to a set of counter-examples [HT91].*

⇨ Conditional probabilities (conditional probability of a given b) are often widely used too, as a measurement of the quality of a rule : $P(b|a) = \frac{Card(A \cap B)}{Card(A)}$

This index, as has been written before, comes directly from the definition of the conditional probability. Some methods use it as it is, and accept the existence of an implication provided it is higher than a given threshold [STT92]. Other methods use it in a different way, or mixed with other indices [DM91]. We will present one of these methods below :

- Ganascia [Gan87, Gan88] defines an index : the plausibility of a rule. He uses mainly the definition of the conditional probability to calculate it. Let us call $Pl(a \Rightarrow b)$, the plausibility of the uncertain rule $a \Rightarrow b$. We have : $Pl(a \Rightarrow b) = 2.\Pr(b|a) - 1$

This formula is meant to take its values between -1 and 1, -1 referring to the false, 0 the lack of information and 1 the true.

⇨ The entropy is a notion highly used in learning machine mainly for decisions trees [Qui86, Qui87]. Let us cite J_measure, index employed in the ITRULE algorithm [GS89].

$$J(B;A) = p(a) \times \left[p(b|a) \times \log\left(\frac{p(b|a)}{P(b)}\right) + p(\bar{b}|a) \times \log\left(\frac{p(\bar{b}|a)}{p(\bar{b})}\right) \right]$$

After this overview of existing solutions, we are going to expose the concept of intensity of implication in the next paragraph.

INTRODUCTION TO THE CONCEPT OF INTENSITY OF IMPLICATION

PRINCIPLE OF THE INTENSITY OF IMPLICATION

As the implication $a \Rightarrow b$ is denied in the sub-set $A \cap \bar{B}$, instead of considering A and B as fixed, we prefer to let them vary independently and randomly in set E. Let us call these two random subsets X and Y. These sub-sets have the same cardinal as A and B. Then we compare the cardinal of $A \cap \bar{B}$ (fixed) with that of $X \cap \bar{Y}$ (random variable), supposing there is no link inclusive between X and Y, or inductive between x and y.

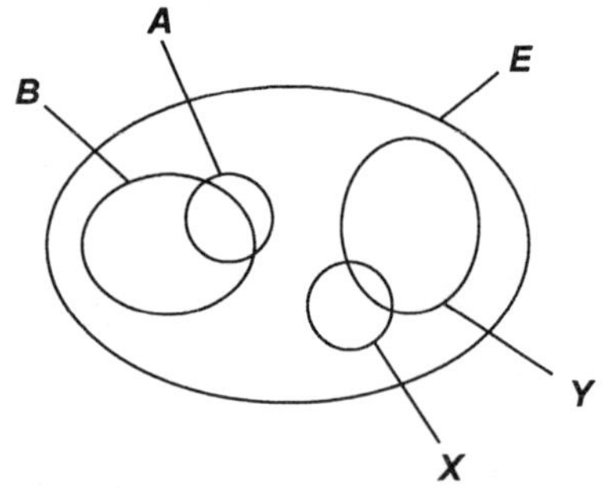

- Figure n°4 : X and Y vary at random in E -

If the cardinal of the set $A \cap \bar{B}$ is improbably small compared with the one we could expect from the distribution of the cardinals of $X \cap \bar{Y}$, we will accept a quasi-implication, $a \Rightarrow b$. The probability of $a \Rightarrow b$ is measurement by the complement on one of the measurements of the probability inferred from the comparison between the aleatory cardinal of $X \cap \bar{Y}$ and the one of $A \cap \bar{B}$.

> ☞ Intuitively, the value of the intensity of implication measurements the degree of statistic astonishment of the size of $A \cap \bar{B}$, considering the sizes of A, B and E, and assuming there is no link a priori between A and B [GL93].

DEFINITION OF THE INTENSITY OF IMPLICATION

Considering 2 parts X and $\bar{Y}$, of respective cardinals $Card(A)$ and $Card(\bar{B})$, we define the random variable $Card(X \cap \bar{Y})$ which we describe below.

× The random variable $Card(X \cap \bar{Y})$ obeys a Binomial distribution because :

(i) the random variable can occupy only two complementary states :

- to be described with the conjunct of propositions $x \wedge \bar{y}$
- not to be described with the conjunct $x \wedge \bar{y}$

(ii) the random variable has a constant probability of taking one of the two states

× In addition, it is proved that the random variable $Card(X \cap \bar{Y})$ obeys the Poisson distribution with the parameter λ. Its formula is given below :

$\Pr\left[Card(X \cap \bar{Y}) = k\right] = \frac{\lambda^k}{k!} . e^{-\lambda}$, and $\lambda = n.p(a).p(\bar{b})$, with :

$p(a) = \frac{Card(A)}{Card(E)}$ and $p(\bar{b}) = \frac{Card(\bar{B})}{Card(E)}$.

- $n = Card(E)$ and,
- $p(a).p(\bar{b})$, *the frequency of the conjunct of propositions* a *and* $\bar{b}$ *supposed independent.*

× In short, the random variable $Card(X \cap \bar{Y})$ obeys the Poisson distribution with mean λ, with : $\lambda = \frac{Card(A).Card(\bar{B})}{Card(E)}$.

> ☞ The quality of implication is even better if the number of counter-examples is smaller than the expected number, in other terms if the quantity $\Pr(Card(X \cap \bar{Y}) \leq Card(A \cap \bar{B}))$ is small. This is the observation "smallness" of $Card(A \cap \bar{B})$ from $Card(X \cap \bar{Y})$ which is taken as a basic pointer of the quasi-implication $a \Rightarrow b$. The implication intensity is then defined with the function φ_1 :
>
> $\varphi_1(a,\bar{b}) = 1 - \Pr\left[Card(X \cap \bar{Y}) \leq Card(A \cap \bar{B})\right]$ with
>
> $$\Pr\left[Card(X \cap \bar{Y}) \leq Card(A \cap \bar{B})\right] = \sum_{i=0}^{Card(A \cap \bar{B})} \frac{\lambda^i}{i!} . e^{-\lambda} \quad (3)$$

× Let $Q(a,\bar{b})$ be : $Q(a,\bar{b}) = \frac{Card(X \cap \bar{Y}) - \lambda}{\sqrt{\lambda}}$ (4)

Let $q(a,\bar{b})$, be the observation of $(a,\bar{b})$:

$$q(a,\bar{b}) = \frac{Card(A \cap \bar{B}) - \lambda}{\sqrt{\lambda}} \quad (5)$$

In the cases where the cardinalities of sets A, B and E are so that $\lambda \geq 3$, the probability calculation of the random variable $Card(X \cap \overline{Y})$ according to the Poisson distribution with parameter λ can be made from the variable $Q(a,\bar{b})$ obeying the normal distribution with 0 and variance 1 ($N(0,1)$).

☞ Then, it is the "smallness" of the observation $q(a,\bar{b})$ from $Q(a,\bar{b})$ which has been kept as a basic indicator for quasi-implication $a \Rightarrow b$. The intensity of implication is defined from the index :

$$\varphi_2(a,\bar{b}) = 1 - \Pr\left[Q(a,\bar{b}) \le q(a,\bar{b})\right] \qquad (6)$$

where : $$\varphi_2(a,\bar{b}) = \frac{1}{\sqrt{2\pi}} \int_{q(a,\bar{b})}^{+\infty} e^{-\frac{t^2}{2}} dt \qquad (7)$$

As a conclusion :

We can say that the implication $a \Rightarrow b$ can be admitted at a level of confidence $1-\alpha$ if and only if :

- If $\lambda \le 3$, then

$$\varphi(a,\bar{b}) = \varphi_1(a,\bar{b}) = \left[1 - \sum_{i=0}^{Card(A \cap \overline{B})} \frac{\lambda^i}{i!} . e^{-\lambda}\right] \ge 1 - \alpha \qquad (8)$$

- If $\lambda \ge 3$, then

$$\varphi(a,\bar{b}) = \varphi_2(a,\bar{b}) = \left[1 - \Pr\left[Q(a,\bar{b}) \le q(a,\bar{b})\right]\right] \ge 1 - \alpha \quad (9)$$

BEHAVIOUR OF THE INTENSITY OF IMPLICATION

SENSITIVENESS OF THE INTENSITY OF IMPLICATION TO EXPANSION OF SETS

The intensity of implication increases with the representativeness of the population samples which are studied.

Example : Let us look for the regularities in physical features of 10 people. We will be more particularly interested in the implicative link between the two propositions *Red-Haired* and *Blue-Eyed.*

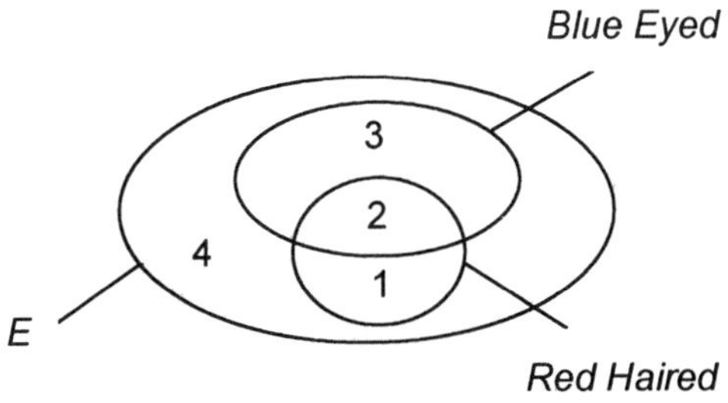

- Figure n°5 : One sample not really representative -

× Let us calculate the conditional probability :

$$P(BlueEyed|RedHaired) = \frac{Card(RedHaired \cap BlueEyed)}{Card(RedHaired)} = \frac{2}{3} = 0{,}67$$

× With $Card(RedHaired) = 3$ and $Card(\overline{BlueEyed}) = 5$, we have : $\lambda = 3 \times 5 / 10 = 1{,}5 \le 3$, the calculation of the intensity of implication becomes :

$$\varphi(RedHeared, \overline{BlueEyed}) = 1 - \sum_{i=0}^{1} \frac{\lambda^i}{i!} . e^{-\lambda} = 1 - \left[e^{-1,5}.(1 + 1{,}5)\right] = 0{,}44$$

The reduced size of the sample does not allow us to conclude that there is an implication from *Red-Haired* towards *Blue-Eyed.* Let us multiply the sample size by 10 (Figure n°6) and keep the *E* set proportions

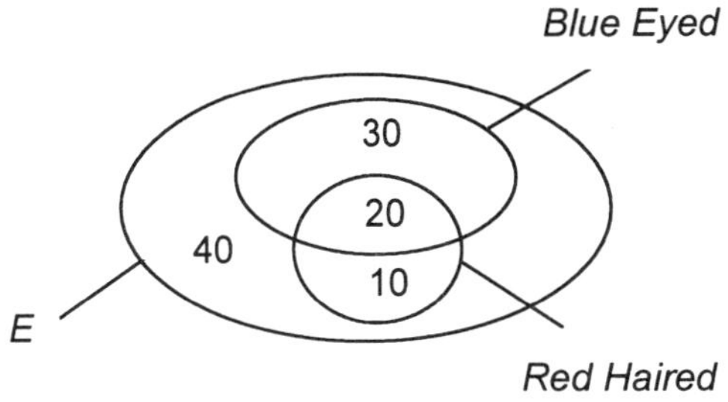

Figure n°6:Sets *E, Red-Haired, Blue-Eyed, Raid-Haired∩Blue-Eyed-*

× One more time, we calculate the intensity of implication.

$$P(BlueEyed|RedHaired) = \frac{Card(RedHaired \cap BlueEyed)}{Card(RedHaired)} = \frac{20}{30} = 0{,}67$$

With $Card(RedHaired) = 30$ and $Card(\overline{BlueEyed}) = 50$, we have : $\lambda = 3 \times 5 / 10 = 1.5 \le 3$, the intensity of implication calculation becomes:

$$q(RedHaired, \overline{BlueEyed}) = \frac{10 - \frac{30 \times 50}{100}}{\sqrt{\frac{30 \times 50}{100}}} = -1{,}29 \text{ then}$$

$$\varphi(RedHaired, \overline{BlueEyed}) = \frac{1}{\sqrt{2\pi}} \int_{-1.29}^{+\infty} e^{-\frac{t^2}{2}} dt = 0{,}90$$

This time, the sample is large enough to conclude that there is an implication *Red-Haired* to *Blue-Eyed.* We can notice that the conditional probability of *Blue-Eyed* given that *Red-Haired* remains unchanged.

More generally, if the manipulated cardinalities are such that $\lambda \ge 3$, and if we multiply the *E, A, B, A∩B* set cardinalities by a coefficient ω, $q(a,\bar{b})$ is multiplied by $\sqrt{\omega}$:

$$q(a,\bar{b}) = \frac{\omega.Card(A \cap \overline{B}) - \frac{\omega.Card(A).\omega.Card(\overline{B})}{\omega.Card(E)}}{\sqrt{\frac{\omega.Card(A) \times \omega.Card(\overline{B})}{\omega.Card(E)}}} = \sqrt{\omega}.\frac{Card(A \cap \overline{B}) - \frac{Card(A).Card(\overline{B})}{Card(E)}}{\sqrt{\frac{Card(A).Card(\overline{B})}{Card(E)}}}$$

☞ The intensity of implication seems to be particularly adapted to algorithms which deal with disparate volumes of data.

BEHAVIOUR OF INTENSITY OF IMPLICATION FOR A POPULATION WIDELY DESCRIBED WITH A PROPOSITION

Example : A lucky walker picks up 100 mushrooms (Figure n°7). We consider the set of white mushrooms and the set of toxic mushrooms : does there exist an implicative link between the propositions *White* and *Toxic* ?

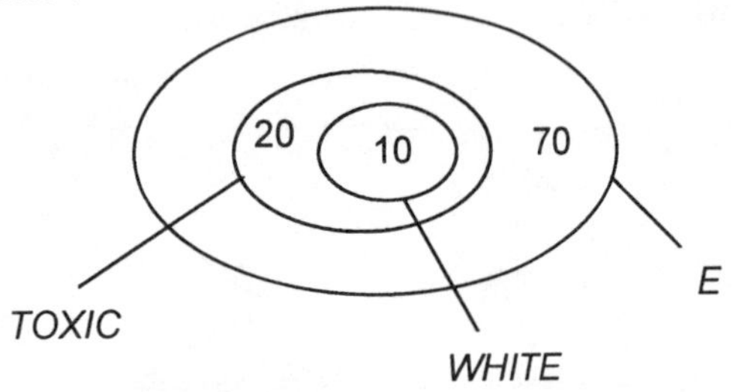

- Figure n°7 : *TOXIC* is small compared with *E* -

× Let us calculate the conditional probability.

$$P(Toxic|White) = \frac{Card(TOXIC \cap WHITE)}{Card(WHITE)} = \frac{10}{10} = 1$$

× With $Card(White) = 10$ and $Card(\overline{Toxic}) = 70$, we have : $\lambda = 10 \times 70/100 = 7 \geq 3$, the intensity of implication becomes :

$$q(White, \overline{Toxic}) = \frac{0 - \frac{10 \times 70}{100}}{\sqrt{\frac{10 \times 70}{100}}} = -2{,}64 \text{ then}$$

$$\varphi(White, \overline{Toxic}) = \frac{1}{\sqrt{2\pi}} \int_{-2.64}^{+\infty} e^{-\frac{t^2}{2}} dt = 0{,}99$$

Example : The walker makes the experiment again in another forest (Figure n°8).

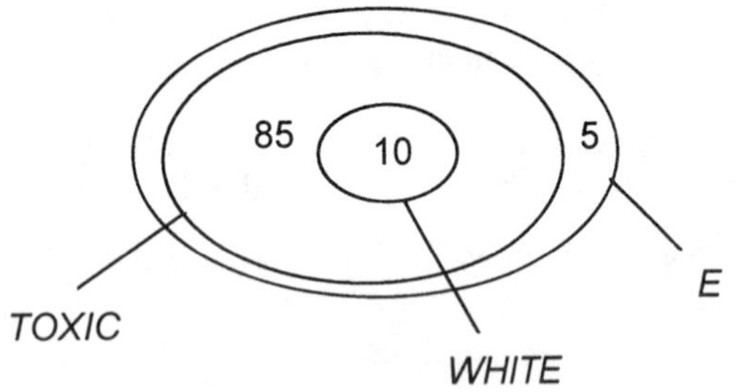

- Figure n°8 : *TOXIC* is nearly equal to *E* -

× Let us calculate the conditional probability.

$$P(Toxic|White) = \frac{Card(TOXIC \cap WHITE)}{Card(WHITE)} = \frac{10}{10} = 1$$

× With $Card(White) = 10$ and $Card(\overline{Toxic}) = 5$, we have : $\lambda = 10 \times 5/100 = 0{,}5 \leq 3$, the intensity of implication becomes :

$$\varphi(White, \overline{Toxic}) = 1 - \sum_{i=0}^{0} \frac{\lambda^i}{i!} . e^{-\lambda} = 1 - \left[e^{-0{,}5}\right] = 0{,}40$$

The cardinal of the set *TOXIC* (*Card(TOXIC)*=95) is great compared to the one of *E* (*Card(E)*=100), so a mushroom picked at random among the set *E* is likely to belong to the set *TOXIC* (Pr(*Toxic*)=0,95). The probability that all the elements of the set *WHITE* (*Card(White)*=10) belong to the set $\overline{TOXIC}$ is not negligible ($\Pr(Card(WHITE \cap \overline{TOXIC}) = 0) = 0{,}60$). So we cannot reasonably conclude in favour of the existence of an implicative link between the propositions *White* and *Toxic*. That is why the intensity of implication associated to the rule *White* ⇒ *Toxic* is low.

If the sets *A* and *B* are fixed, the intensity of implication decreases when set *B* is directed towards set *E*. This does not apply to the conditional probability which remains unchanged. In the case when *B*=*E*, the following property has been proved :

For $Card(E)$ and $Card(A)$ fixed, we have : $\lim_{Card(B) \to Card(E)} q(a, \overline{b}) = 0^-$ and $\lim_{Card(B) \to Card(E)} \varphi(a, \overline{b}) = 0{,}5^+$

☞ This behaviour is very interesting in a context of symbolic learning. If a proposition *b* is true for most of the examples of the learning set, some systems will produce too many rules of the type *Premise*⇒*b*. The use of the intensity of implication makes it possible to solve that problem efficiently.

BEHAVIOUR OF THE INTENSITY OF IMPLICATION IN A NOISY FIELD

We are going to study the variation of $q(a, \overline{b})$ with that of $Card(A \cap \overline{B})$. Let us consider a real situation for which we know *Card(E)*, *Card(A)*, $Card(\overline{B})$ and $Card(A \cap \overline{B})$. We have :

$$q(a, \overline{b}) = \frac{Card(A \cap \overline{B}) - \frac{Card(A).Card(\overline{B})}{Card(E)}}{\sqrt{\frac{Card(A).Card(\overline{B})}{Card(E)}}} \leq \beta \text{ and } \varphi(a, \overline{b}) \geq 1 - \alpha$$

What error of measurement on $Card(A \cap \overline{B})$ does the implication still allow (with 1-α, level of confidence), even though *Card(E)*, *Card(A)*, and $Card(\overline{B})$ remain constant. In other words, what is the sensitiveness of the implication to a variation of $Card(A \cap \overline{B})$?

It has been proved that the implication is still valid ($\varphi(a, \overline{b}) \geq 1 - \alpha$) for any variation *k* of the size, such as :

$$Card(A \cap \overline{B}) \leq k \leq \beta \sqrt{\frac{Card(A).Card(\overline{B})}{Card(E)}} + \frac{Card(A).Card(\overline{B})}{Card(E)} - Card(A \cap \overline{B})$$

Example : Let us study the influence of the consumption of coffee on the success to exams. We have a sample of 100 students. The rule *Coffee⇒Success* is only kept when its intensity of implication is superior to the threshold $1\text{-}\alpha=0{,}95$.

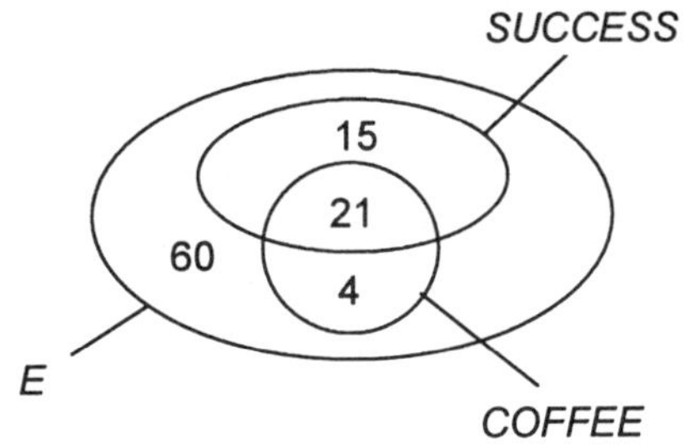

- Figure n°9 : Sensitiveness of $\varphi(Coffe, \overline{Success})$ to the variation of $Card(Coffee \cap \overline{Success})$ -

With $Card(Coffee) = 25$ and $Card(\overline{Success}) = 64$, we have : $\lambda = 25 \times 64 / 100 = 16 \geq 3$, the intensity of implication calculation becomes :

$$q(Coffe, \overline{Success}) = \frac{4 - \frac{25 \times 64}{100}}{\sqrt{\frac{25 \times 64}{100}}} = -3 \text{, then :}$$

$$\varphi(Coffe, \overline{Success}) = \frac{1}{\sqrt{2\pi}} \int_{-3}^{+\infty} e^{-\frac{t^2}{2}} dt = 0{,}99$$

$$-4 \leq k \leq -1{,}65\sqrt{\frac{25 \times 64}{100}} + \frac{25 \times 64}{100} - 4 \Leftrightarrow -4 \leq k \leq 5$$

$Card(Coffe \cap \overline{Success})$ can evolve between 0 and 9 without $\varphi(Coffe, \overline{Success})$ getting under the threshold $1\text{-}\alpha=0{,}95$.

☞ According to the learning parameters linked with the field, it is possible to measurement the resistance to noise of extracted implications. Concerning our example, the generated rules are very noise-proof.

DEVELOPMENT PROSPECTS WITHIN AN INCREMENTAL SYSTEM

We find the possibility of measuring the sensitiveness of implication to size variations for $Card(A \cap \overline{B})$ particularly interesting within an incremental system.

One of the questions asked by Ganascia [Gan87] is : given the high cost of an incremental construction, is it better to model the learning behaviour of those who are strong minded and learn slowly but surely, or those who are more brilliant but more inconstant and who generalise quickly at the risk of being constrained to question again their knowledge which has been acquired too quickly.

Concerning this problem, we think we can give an answer. Considering that for every implication, we can give the number of counter-examples beyond which it will be questioned again, we will consider this number as the maximum not to exceed without questioning again the acquired rule system.

☞ If for every rule, we combine the maximum limit of counter-examples limit of not to be exceeded (cf §3.3.), the addition of an example which transgresses one rule of the system or more, does not necessarily imply the reconstruction of a new rule system. Thus, the cost of generating a new system is spared.

CONCLUSION

This study dealt with the question of the implication measurement from the premise to the conclusion of a rule. The intensity of implication is an index which allows to generate strong rules and whose robustness we know. It is efficient in a noisy environment and can be used within an incremental algorithm.

We have developed a symbolic learning system for the generation of rules guided by the intensity of implication : the FIABLE system (Force d'Implication pour l'Apprentissage Basé sur Les Exemples). It is written in C++ language and runs on PC. FIABLE system has enabled us to make an experimental comparison (by simulation) between the intensity of implication and the various indices currently used in the field of symbolic learning. We have also compared the results from FIABLE and those obtained from ID3 system and CHARADE. This comparison has corroborated the idea that the intensity of implication is an excellent measurement of the pertinence of a rule.

This method is now used by experts of PERFORMANSE in order to work on a specific group of individuals. Especially, these experts have been able to show up that it exists several profiles for one job, whereas the present propensities assert the contrary, namely that only one profile satisfies one job.

REFERENCE

[BPM90] Bisson G., Porter R., Mooney B., A Knowledge Based Generalizer, 7ème International Conference on Machine LearningEdition Morgan Kaufamnn, Austin Texas 1990

[CN88] Clark P., Niblett T. The NC2 induction algorithm, Machine Learning, 3. p 261-283 1988

[DM91] Diday E. Mennessier M.O."Analyse symbolique pour la prévision de séries chronologiques pseudo-périodiques", Induction symbolique et numérique à partir de données, p 179-192, Cépaduès-éditions, 1991.

[Gan87] Ganascia J.G. "AGAPE et CHARADE : deux techniques d'apprentissage symbolique appliquées à la construction de bases de connaissances", Thèse d'état, Université Paris-Sud, Mai 1987.

[Gan88] GANASCIA J.G., "CHARADE : A rule system learning intelligence" (IJCAI), Milan, Italy, Août 1987

[GS89] Goodman R.M., Smyth P."The induction of probabilistic rules set - the itrule algorithme" Proceedings of the sixth international workshop on machine learning Splatz, B, ed p 129-132, San Mateo, CA Morgan Kaufmann 1989

[GL93] Gras R., Larher A. "L'implication statistique, une nouvelle méthode d'analyse de données", Mathématiques, Informatique, et Sciences Humaines n°120

[GD86] Guignes J.L., Duquenne V. Mathématiques et sciences humaines 24 ème année n°95, Familles minimales d'implications informatives résultant d'un tableau de données binaires" p 5-18 1986

[HT91] Ho Tu Bao, Tong Thi Thanh Huyen "A method for generating rules from examples and its application "Symbolic Numeric Data Analysis And Learning, p 493-504, Nova Sciences Publishers 1991

[GJM+77]Glannetti R.A., Johnson J.H., Miller D.A., Williams T.A."Design for an online computerized system for MMPI interpretation" Behaviour Research Method and instrumentation, 9.(2) p117-122 1977

[GJW78] Glannetti R.A., Johnson, Williams T.A."A self-contained micro-computer system for psychological testing" Behaviour Research Method and instrumentation, 10. p 576-581 1978

[GJW79] Glannetti R.A., Johnson, Williams T.A. "Psychological Systems Questionnaire : An objective personality test designed for on-line computer presentation, scoring and interpretation" Behaviour Research Method and instrumentation 11(2), p 257-260 1979

[GJK80] Glannetti R.A., Johnson, Klingler D.E., Williams T.A."The reliability of diagnoses by technician computer and algorithm"Journal of Clinical Psychology, 36. p 447-451 1980

[KT93] Kodratoff Y., Tecuci G. "Techniques of design and DISCIPLE learning apprentice" knowledge acquisition and learning, Kaufmann p 655-668 1993

[MST94] Michie D., Spiegelhalter D.J., Taylor C.C."Machine Learning, neural and statistical classification", Ellis Horwood 1994

[CMM83] Carbonnell J.G., Michalski R.S., Mitchell T.M.A theory and methodology for inductive learning An Artificial Intelligence Approach, Edition Springer Verlag, , p 83 - 134 1983

[Quin86] Quilan J.R. "Induction of decision trees", Learning Machine 1, p81-106 1986

[Quin87] Quinlan J.R."Generating Production Rules from Decisions Trees" The 10 th International Conference on Artificial Intelligence 1987, p 304-307

[Ral91] Ralambondrainy H. "Apprentissage dans le contexte d'un schéma de base de données", Induction symbolique et numérique à partir de données, p 241-255, Cépaduès-éditions, 1991.

[STT92] Schektman Y. Trejos J. Troupe M."Un générateur de règles floues à partir de bases de données volumineuses", 3èmes journées "symboliques-numériques", p 121-143, Mai 1992, Paris.

[SS91] Sebag M. Schoenauer M."Un réseau de règles d'apprentissage", Induction symbolique et numérique à partir de données, p 241-255, Cépaduès-éditions, 1991.

APPENDIX
FORMALISM

e_i Example from E.

d_j Proposition of D.

A Sub set of examples from E for which a is true.

B Sub set of examples from E for which b is true.

X Random sub-set of E for which the cardinal is $Card(A)$.

Y Random sub-set of E for which the cardinal is $Card(B)$.

$a \Rightarrow b$ Implication where a and b are two propositions from D.

$Card(E)$ Cardinal of set E.

$Card(A)$ Cardinal of sub set A.

$Card(B)$ Cardinal of sub set B.

$A \cap \bar{B}$ Sub-set of counter examples of the implication $a \Rightarrow b$.

$Card(X \cap \bar{Y})$ Random cardinal of sub set $X \cap \bar{Y}$: obeys Poisson distribution with mean λ.

$Q(a,\bar{b})$ Random variable $Card(X \cap \bar{Y})$: obeys Poisson distribution with mean 0 and variance 1

$q(a,\bar{b})$ Observation of $Q(a,\bar{b})$.

$\varphi(a,\bar{b})$ Intensity of implication of the implication $a \Rightarrow b$.

$N(0,1)$ Normal distribution.

$1\text{-}\alpha$ Level of confidence of the implication $a \Rightarrow b$.

$D = \{d_1, d_2, \ldots, d_k\}$ Set of propositions which can characterise an example from E.

$E = \{e_1, e_2, \ldots, e_n\}$ Set learning.

$d(e_i) = d_1 \wedge d_2 \wedge \ldots \wedge d_n$ Description of e_i : conjonction of propositions from D.

Premise$\Rightarrow$*Conclusion* Implication where *Premise* is a conjonction of propositions from D and *Conclusion* a proposition from D.

PREMISE Sub-set of examples from E for which *Premise* is true.

CONCLUSION Sub-set of examples from E for which *Conclusion* is true.

Man-Machine Interface

APPLICATION OF VISUAL PSYCHOPHYSICS TO THE DESIGN OF AN ENGINEERING DIAGRAMS DISPLAY SYSTEM

Laurent Ponthieu
Electricité de France, Direction des Etudes et Recherches*,
1 av. du Général de Gaulle, F-92141 Clamart, France.
Email: laurent.ponthieu@der.edf.fr

ABSTRACT

Electrical Diagrams are read by technicians to troubleshoot complex electric systems. A diagram is typically made of one hundred pages, each featuring dozens of symbols and text captions. Results from research in *visual attention and visual search* suggest that the technician might benefit from the help of a computer as early as *reading* the diagrams is concerned. The principle of a *hybrid visual system* is to display the information in a manner that best exploits the abilities of the human visual system. As a preliminary step towards this goal, we experimented with two visual tasks that are part of reading the diagrams. Experimental data show firstly, that appealing presentations do not necessarily mean efficient search, secondly, that a presentation grounded on knowledge about human visual attention can dramatically improve the search performance. The conclusion is that the diagram reading problem is likely to benefit from the tools and the ideas of visual search.

INTRODUCTION

The goal

It is not easy, even for the skilled technician, to troubleshoot an electrical system documented through a hundred of wiring diagrams. Professionals don't complain about the complexity of troubleshooting in itself; they only think that the system is too large, and consequently that the diagrams are too easy to become lost in.

The HYVIS project (Hybrid Visualization System for Technical Diagrams) aims at designing an electronic display system that would improve the efficiency and the reliability of troubleshooting by making the diagrams easier to read and to apprehend. HYVIS is both a design technique and an evaluation technique for technical displays. HYVIS applies ideas and experimental procedures originated in psychophysical research on human visual search and attentional performance.

HYVIS operates under the following constraints, or guidelines:

- The layout of the diagrams is defined by engineering standards that must be respected. However, some variations or additions are allowed, as long as the original shapes and layout are preserved.
- The technician is driving the inquiry; the computer is only providing assistance for diagram reading and understanding. This project does not address automated diagnosis and rather concentrates on image perception and understanding, as the place from where everything starts.
- The amount of user input, for a certain task category, should be kept to a minimum depending on both the human capability for that task ("if the human can do it, then the computer must not help") and on the rate at which such a task changes ("if the task changes often, then it should require no input at all").

The material

The Electrical Diagram in figure 1 shows the implementation of an automaton. Such a system is made with simple dedicated components that exchange information through wires. Usually, the system is too large to fit on a single page, so it is broken into up to a hundred pages, linked by off-page reference signs.

Troubleshooting takes place after a failure condition is notified. The technician starts from components listed in the failure notification and explores the electric paths. The goal is to find a component whose normal behavior contradicts the current observations. It is more or less like walking into a maze: one must find a way through the diagrams and remember any previous inference. This gives rise to several kinds of tasks, only one of which

*This work has been carried out while the author was a Visiting Fellow at the Visual Psychophysics laboratory, Center for Ophthalmic Research, Harvard Medical School, Brigham & Women's Hospital, 221 Longwood Ave., Boston, MA 02115, USA.

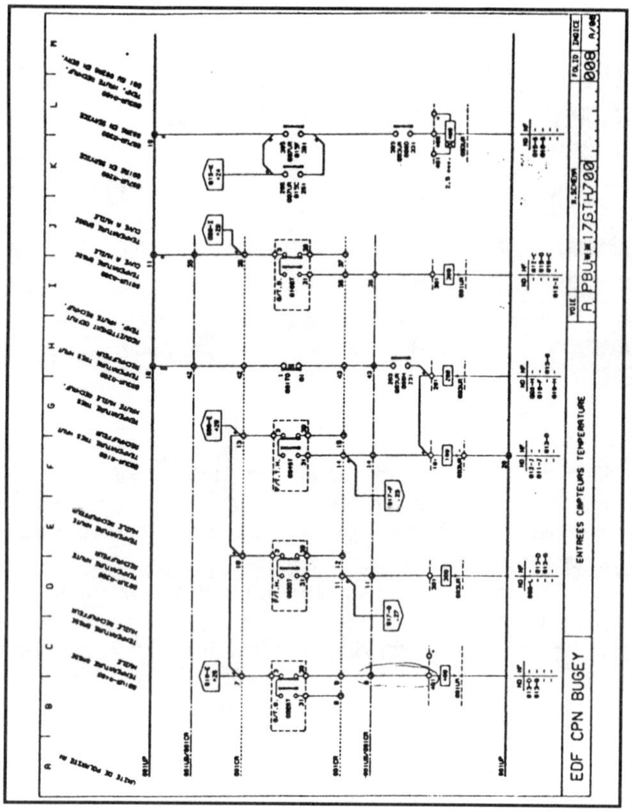

Figure 1: One page from a real Electrical Diagram. On the top, one can see the coordinate system (letters A through M) then several captions that describe the components. On the bottom, there are the page title and reference number. The central part shows the wires (horizontal and vertical solid lines) and the components. The horizontal dotted lines are related to the physical location of the components and can be ignored here. The layout tries to align the components on a virtual grid. In this example, from top to bottom and left to right, the components are: only off-page arrows (AR) — first line; three sensors (SE), one push-button (PB), one sensor again, then two relay contacts (NORC) – second line; AR and NORC – third line; six relay magnets (RM) – last line.

(Individual Search Tasks, or IST for short, [Pon93]) is considered here. IST use explicit graphical information apart from any understanding of the electric system. The search goal is fully known at the beginning of the task, and it does not change as the task is performed. Examples are: find a component by name, from index information, look for a specific reference in a caption, or in a series of captions, consider all the components on the same path.

The tools

Visual Search has explored the deployment of attention when searching for a target item among distractor items within an image. Experimental data shows that *visual features*, such as color or orientation, can *attract attention* towards the target item faster than a sequential examination of the items would. The attraction strength depends both on how many features distinguish the target from the distractor, and on the number of shared features between the target and the distractors. Under Guided Search model [WCF89, Wol94], each feature turns to be an independent source of information for guidance. Featural information is assumed to be available preattentively, and it has been proved that several of these sources can be processed at the same time.

When the target is defined by a unique feature (such as search for a red 'X' among green 'X's), search is said to be "parallel" because the target can be found in a time that slowly increases with the number of distractors that surround the target [TG80]. Guidance is very strong and Reaction Time (RT) is said to be independent of Set Size (SS).

Searches for conjunctions (such as search for a red vertical bar among red horizontal and green vertical bars) are more difficult. There is no single feature that defines the target and information from color and orientation must be combined. However, as both the red locations and the vertical locations are more likely to contain the target than the other locations, the search is not random. Attention is "guided" by both color and orientation information [WFC89]. As a result, RT rises moderately with set size (between 5 and 20 ms/item for target-present displays, less than double for target-absent displays).

When no featural information is available (e.g., a search for a letter T among letters L), attention is deployed from item to item in an apparent random manner, and search is said to be "serial" [BJ83, TG80]. In that case, Reaction Time rises steeply with Set Size (more than 20 ms/item for target-present displays, about double for target-absent displays).

Towards a Hybrid Display Solution

HYVIS looks for a display procedure that helps the performance of any ISTs. If the only ISTs were to search for symbols from categories A or B, it would be helpful, and maybe enough, to draw all symbols from A in red and all symbols from B in green. As color is one of the visual features that allow for parallel search, the items from these categories would be found and distinguished in no time. But considering there are many more than two symbol categories (and inside one category, it is sometimes important to distinguish the individuals), it is much preferable to use several "fast-processing" or "effortless" features (orientation, size, frequency, ...) at the same time.

This is one of the main ideas that define a hybrid search system: arrange, even dynamically as the task changes, so that the target item is salient in comparison with the undesired, distractor items. Among the techniques that can contribute to that purpose, only one, *feature recoding*, is experimented with here. It consists in modifying the graphical attributes of an object so that the search for it can make more use of preattentive information processing.

Two experiments

Each of these experiments addresses a specific Individual Search Task that occurs in real situations. They are "Reaction Time versus Set Size" experiments: the subjects were asked to tell whether one symbol (the target) was present or absent in images that could contain other symbols (the distractors); the reaction time (RT) was measured and evaluated with respect to the image size (or set size); the mean RT for each set size is roughly a linear function of the set size; the slope of that function characterizes the task complexity. Experiment 1 evaluates search for a symbol defined by shape only, while Experiment 2 deals with shape *and* label. The latter task is the most frequent IST in the real world. But since label reading is not undertaken before a component from the right category is found, Expt 1 studies shape search alone, and Expt 2 studies label search with little shape search involved.

The task complexity is measured on standard diagrams, then one or more variations are tested. Expt 1 shows, through a negative and rather unexpected result, that the subjects can easily sort the information by type so as to examine only the relevant subparts of the symbols. Expt 2 reproduces that result even more surprisingly, then shows that a label can be efficiently recoded by means of shape and color, providing an insight on to what a hybrid system can be.

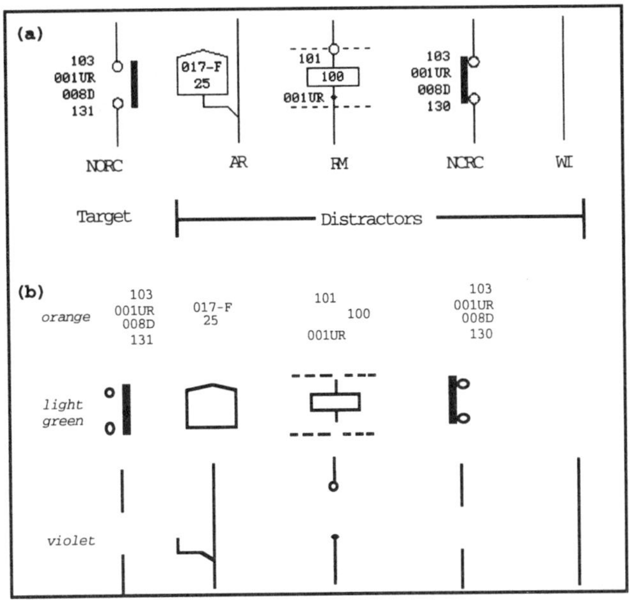

Figure 2: Models of real components used for Expt 1. The target is a "normally open relay contact" (NORC). The distractors are: "off-page connector" (AR), "relay magnet" (RM) and "normally closed relay contact" (NCRC). (a) The symbols are drawn white on black for the *standard shape* task. (b) In the *colored shape* condition, three contrasting (triadic complementary) colors are used: orange for the labels, light green for the component body, violet for the wires and the plugs.

EXPERIMENT 1 - SEARCH BY SHAPE

Methods

Models of real components were created for the purpose of this experiment (see figure 2). As much as possible, the original proportions of the symbols were respected, as one can notice by comparing with figure 1 where a real diagram is shown. The vertical wires, abutting to the symbol from top and bottom, simulated an inter-symbol connection when symbols were stacked in a column. The target symbol was always a "normally open relay contact" (NORC). The distractors were chosen from three categories: the "off-page connectors" (AR), the relay magnets (RM) and the "normally closed relay contacts" (NCRC). In the real diagrams, components from these four types occur with approximately the same relative frequency, but they occur along with other, less frequent, categories that were not used in this experiment. For this reason, and in order to add space to the diagrams, vertical wires (WI), as high as any symbol, were added as distractors. Images of size 2x2, 3x4 or 4x5 items (row x column) were formed by a uniform random choice from the distractor set. The items were vertically aligned to form 2, 4 or 5 columns of connected components (see figure 3). For half of the images, one distractor was replaced by the target. The target locations were equally likely over the whole image. In comparison with real diagrams, these simulated ones reproduced vertical alignment, connectivity and approximate relative frequency of symbols. They did not reproduce the horizontal connections, the dotted lines, the frame nor the comments that are visible on figure 1.

The target item (NORC) is very similar to one distractor item (NCRC, see figure 2): they are the only ones with the same large vertical bar. The other distractors show more difference in shape, and symbol WI, being a simple vertical line aligned on the general vertical layout of the image, probably does not even count as a distractor. But still it is difficult to define a "set size": does one symbol count for one or for several items? As no convincing answer can be given to that question without further work, it is assumed that the set size is equal to the number of locations on the screen multiplied by 3/4 (to account for WI's no-role), that is 3, 9 and 15. The "actual" slopes can then only be smaller than the ones reported here.

The purpose of the *standard shape* task was to measure the base complexity of a search by shape when target versus distractor similarity is high. The symbols were drawn in white on a dark background (see figure 2(a)). The only difference between the target and the (principal) distractor was that the target had the vertical bar to the right of the wire, while the distractor had it to the left. It is likely, but not proven, that the symbol has to be

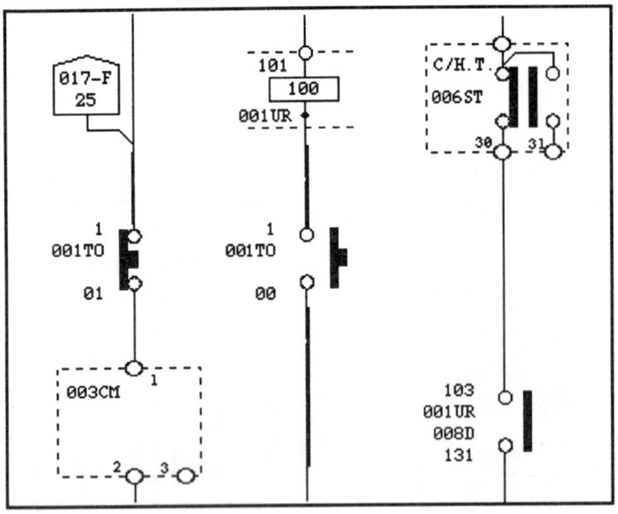

Figure 3: A simulated diagram used in the experiments. The vertical alignment of the symbols is usual in real diagrams. Real diagrams can bear horizontal connections that were not simulated here. The diagrams used in the experiments featured between 4 and 20 symbols.

considered as an assembly of subparts and the subparts' relationships be examined.

In the *colored shape* condition, three contrasting colors were used (triadic complementary): orange for the labels, light green for the component body, violet for the wires and the plugs (see figure 2(b)). This coloring scheme aims at making easier the apprehension of the components' subparts, and at reducing the number of locations to be examined (labels are not informative for this task).

The subjects were shown a simulated diagram (in the style of figure 3) and were asked to tell whether the target was present or not in that display. The target item and the distractor items were shown before the experiment started; they remained the same throughout the session. After each trial, the subjects were given feedback on their answer through the display of a small sign in the center of the screen. This forced the subjects to fixate at the screen center between each trial. The display field was 23 deg X 17 deg and the target could appear at any (equally probable) location within this field. The images were kept on display until an answer was given. The experiments were run on a MacIIci running VSearch software [EOR90]. The subjects were unaware about the goals of the experiment. All subjects had given informed consent. Ten subjects ran 330 trials for each task. The first 30 trials of each task were discarded as practice and only the correct answers were kept for further analysis. Each subject performed all the tasks, but the order under which the tasks were executed was varied.

Results and discussion

The results are presented in figure 4 (recall that the slopes could be overestimated due to the choice in counting the number of items). Contrary to what was expected, the two tasks do not show any obvious difference. Both the target and the blank slopes (averages of individual by-subject slopes) are not significantly different between the two conditions (d.f.=9, $|T|=0.82$ and $p=0.44$ for the target slopes, $|T|=0.51$ and $p=0.62$ for the blank slopes), and so are the intercepts. The slopes average to 26 ms/item and to 98 ms/item for target and blank trials respectively. The blank slope to target slope ratio averages to 3.6 for the *standard shape* task, it averages to 4.6 for the *colored shape* task. They are both significantly different from 2 (d.f.=9, $|T|>4.5$ and $p<0.002$), which is the theoretical value for a serial self terminating search [CW94].

The values of the target trial slopes suggest that the searches are "serial" with a rate of approximately 50 ms/item (this is twice the target slope). High slope ratio values could mean that the subject is extremely cautious when the target is absent (note however that the error rate is about 1 %). The fact that the *colored shape* task is not more efficient than the *standard shape* task suggests that the spatial cues can be treated rather independently from the color cues. This can be instanced either by saying that the spatial cues are sufficient to decompose items into subparts (stimuli like these ones are hard to conceive as wholes and any search strategy is likely to involve some sort of item decomposition), or by saying that the occurrence of multiple heterogeneous colors does not lower the search performance. Whatever the reason for that effect, the practical conclusion is that coloring an image, even if the coloring scheme is seductive, even if this leads to more "comfortable" searches, does not always provide better performance.

However, the subjects judged the *colored shape* task as more "comfortable" than the *standard shape* task, and this

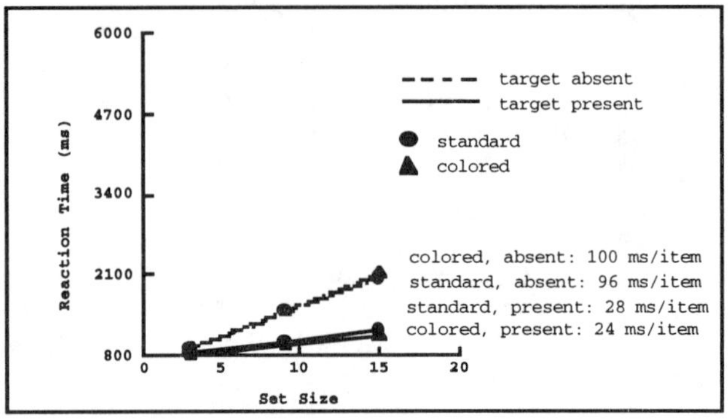

Figure 4: Results for the *standard shape* and for the *colored shape* tasks (experiment 1). Although the subjects reported more comfort for the second task, the slopes are not significantly smaller.

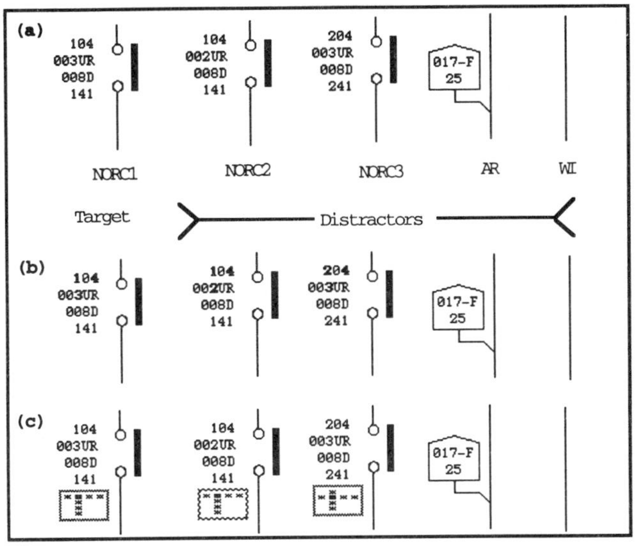

Figure 5: Models of real components used for Expt 2. The target is a "normally open relay contact" (NORC1, label "104/003UR"). The distractors are: two other relay contacts (NORC2, label "104/002UR", or NORC3, label "204/003UR") and an "off-page connector" (AR). (a) The symbols are drawn white on black for the *standard label* task. (b) In the *colored label* task, the subparts of the labels that served identification were drawn in bright green (boldface in this figure). (c) In the *recoded label* condition, the symbols were again entirely drawn in black and white, but a small figure was added next to each. The figure recodes the label information in a purely graphical way.

first one yields a smaller target slope (even if non significantly smaller). Is this the conjunction of a purely subjective feeling (due to the use of color) with a random experimental effect, or is coloring really making the image easier to read, while at the same time adding complexity to some part of the search process? More control experiments are needed to address this question.

EXPERIMENT 2 – SEARCH BY LABEL

Methods

This experiment used the same methods as the previous one, unless otherwise noted. The subjects were asked to find a specific relay contact in an image that contained other nearly identical relay contacts as distractors (see figure 5(a)). The target symbol was always a "normally open relay contact" (NORC1, label "104/003UR"); the distractors could be: another relay contact (NORC2, label "104/002UR", or NORC3, label "204/003UR") or an "off-page connector" (AR). Vertical wires (WI) were added for reasons of spacing. As the target and the distractors differed only by their labels, the tasks described here can be considered as the most complex ones with respect to what can happen in real situations.

The purpose of the standard label task was to measure the base complexity of a search by shape and label when target versus distractor similarity is the highest. The symbols were drawn in white on a dark background (see figure 5(a)). This task amounts (not talking of the arrow that has a label but an irrelevant shape) to finding the contact labeled "104/003UR" among other contacts labeled "104/002UR" and "204/003UR".

Although a label is made of four lines of text, the first two were enough in this example to distinguish the target from the distractors. More, only three characters inside these two lines sufficed (this corresponds to a real case where the items would share enclosing units, or locations within their specific units). These three characters were written with a contrasting, unique color for the *colored label* task, in order to evaluate the effort devoted to extracting the relevant information within a symbol. The language that must be read here is quite simple: five symbols (0,1,2,3,4), four words (104, 204, 002, 003), and the words can be ordered in only four ways. Such a simple task might lead to large effects of training. Despite that apparent simplicity, how would it compare to the equivalent task where the information is purely graphical, and not textual at all? In the *recoded label* task, quite all the information contained in the label is recoded with colors and shapes, through a small figure that is added next to each contact. The figure is made with a thick rectangle that encloses a matrix-like area (see figure 6). The rectangle border color represents the unit digit (2 or 3) of the relay *main unit* number (002 or 003), according to the standard scheme in electrical engineering (black=0, brown=1, red=2, orange=3, yellow=4, green=5, blue=6, violet=7, gray=8, white=9). Only the unit digit is represented, for it is the most likely to vary within one page, and because coding more than ten numbers is difficult if not useless. Inside the rectangle, one can see that one row and one column are highlighted. The row number represents the *sub-unit* number within the main unit, which is also the hundred number in the top line of the label. The column rank represents the *contact* position within the sub-unit, which is related to the rightmost digit in the top line of the label (first position is 3, second is 4, etc.). The intersection of the row and the column is drawn with a brighter color, so that both the lines and their intersection are visible.

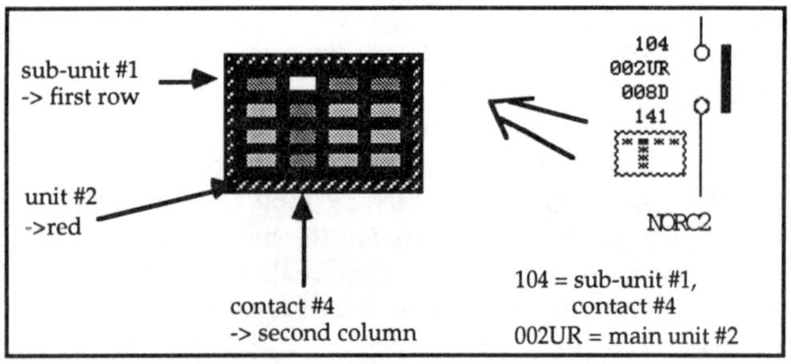

Figure 6: In the *recoded label* task, the component label is represented with a geometrical figure. The enlargement shows the closest black and white equivalent to the actual figures that were used. A color represents the *main unit* number, a shape represents the *sub-unit* and the *contact* number.

Why such a representation? First, when stimuli tend to be complex conjunctions of simpler graphical features, some sort of hierarchical organization makes the search easier. This is especially true when the conjunction occurs within the same feature (such as the search for a figure that has a red *and* a green part). In that case, organizing the stimulus as a part surrounded by a whole (or anything else that can create a graspable object with subparts) can make the search much easier, as easy as a two-feature conjunction search [WFH*94]. As color and differences in color were used here to code for several informations, the stimuli subparts were organized in shapes (row and column), then in shape conjunctions (intersection of row and column), then in a part and a whole. This representation is also intended to keep the drawing as close as possible to the mental model of the component (this is intended for domain specialists). The real (operating) relay contacts (numbered 3, 4, 5, 6) are arranged in rows where each element of the row gets activated at the same time by the same command unit. Four independent rows (numbered 100, 200, 300, 400) are grouped on the same relay unit. A relay unit is merely a box that gives its name (001UR, 002UR, ...) to its subcomponents. This is where the column by row by box color design comes from.

Results and discussion

The results are presented in figure 7 (recall that the slopes could be overestimated due to the choice in counting the number of items). The slopes for the *standard label* task are 190 ms/item and 360 ms/item for target present and target absent trials respectively. The slopes of the *colored label* task, although smaller, at 170 ms/item and 320 ms/item, are not that different: the improvement (average drop of the slope values from the *standard label* task) is only 11 % for both trial types, and only the blank trial slope is significantly smaller. On the other hand, the *recoded label* task is very different: its slopes are 66 ms/item and 140 ms/item, which represents an improvement of 64 %. The blank slope to target slope ratio is in neither task significantly different from 2 (d.f.=9, |T|>1.1, p≥.2 for each task). The error rates are 2.3 %, 1.6 % and 0.9 % for the *standard, colored* and *recoded* tasks respectively. The only significant difference between the error rates is that the *recoded task* error rate is smaller than any other. The intercepts (of the slope line with the RT axis) do not show statistical evidence for a dependency with the task (ANOVA: F(2,18)=0.33, p=0.72).

The steep slopes and the 2:1 slope ratio suggest that all these tasks are serial self-terminating searches. Despite the high slope values, the error rate is very small, typically between 0 and 5 %. This is quite surprising for these difficult tasks, but can be explained by saying that the subjects performed an exhaustive search among well-isolated items. Thus the items were inspected and rejected in turn, with a very small "revisiting rate" as they could easily be tagged. Supporting this argument, the slope ratio is the highest for the *recoded label* task: because the subjects relied on more information for reaching the target, they were more cautious when the target was not found.

Just as in Expt 1, increasing the salience of the significant information does not help a lot (compare the *standard* and the *colored label* tasks), although subjects reported more comfort. It is even more striking here, as the spatial cues are dimmer: the interesting subpart is embedded in strings where some irrelevant characters are found, whereas in the previous experiment, the subpart stood by itself. The small amount of improvement that still exists could reflect the time saved in accessing the digits.

The performance increase due to the introduction of the recoding figures is clear. Both the slopes and the error rates are three times smaller (we won't discuss the error rates as they are very small). The subjects spontaneously described the stimuli as a colored box with a cross-like shape inside. As one can imagine from figure 5, the *recoded label* task

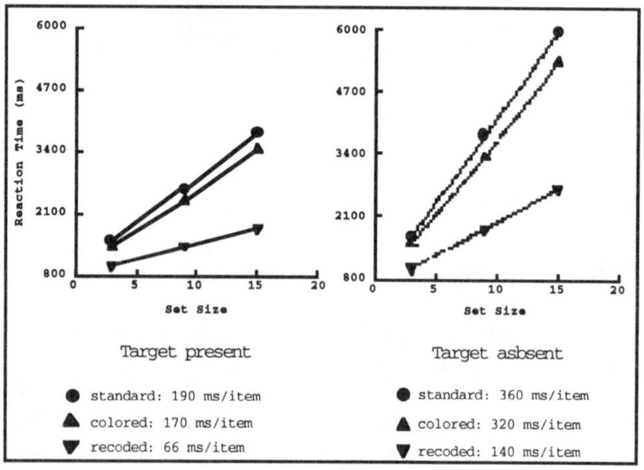

Figure 7: Results for the *label* tasks (experiment 2). The *standard* and the *colored label* tasks are not very different. The *recoded label* task shows a strong increase in performance.

amounted to search for the orange box with the horizontal line on the top of the matrix, or, alternatively, for the orange box with the bright smaller rectangle on the top line. The color of the surrounding box eliminates half of the distractors; the shape inside eliminates the other half but the target. The search becomes a color by shape search where the color information does not occur at the same place as the shape information.

Note that the *recoded label* design might also help a (much harder) Composite Search Task as "search for all the contacts that are parts of the same bigger unit", which is quite frequent in troubleshooting.

GENERAL DISCUSSION

When troubleshooting a large electric system, the technician must find its way through an equally large set of wiring diagrams, where the symbols and the pages look very much alike. The HYVIS project (Hybrid Visualization System for Technical Diagrams) aims at designing an electronic display system that would, within the frame of this task, make the diagrams easier to read and to apprehend. Among the "tools" that HYVIS can rely on, results from experimental work on *visual search* seem to be very useful. Under that light, HYVIS needs to present a display that best exploits the abilities of the human visual and attentional system.

The experiments reported here dealt with models of real diagrams. Two symbol search tasks were examined and one or several variants were measured for each (in terms of the subject's performance). The tasks had been identified during a preliminary study of the troubleshooting activity. They consist in looking for a specific symbol (representing an electrical device) among a variety of others. The variants were designed after ideas originating in experimental observations with more or less simple stimuli. The purpose of the variants was to measure the improvement, in the search performance, caused by a better usage of the visual system abilities.

The first experiment was a search for a symbol among other symbols that differed only by the way their *geometric subparts* were arranged (a shape similarity case). The base condition was monochrome; in the variant condition, the distinctive subparts were drawn with a contrasting color. Results show that the subjects' performance was the same accross both conditions. This means that the subjects can easily restrict their attention to the relevant subparts of the symbols. In the second experiment, the distracting symbols and the target symbol shared everything but the *label string*. The base condition was monochrome; in the variant condition, the distinctive substrings were drawn with a contrasting color. Results once again show that the subjects' performance was the same under both conditions. It means that the subjects can easily restrict their attention even to substrings of letters within the symbols. These, in fact, are negative results with respect to what was expected. More surprisingly, the subjects always felt more confortable in the colored conditions, although their actual performance was not improved. The practical conclusion is that simple ideas, such as highlighting the relevant information, as promising they seem to be, might not improve the actual performance at all.

On the other hand, with a second variant, the second experiment shows that the search performance can be dramatically improved when using more elaborate attentional abilities of the visual system. When the discriminating text is *recoded* with a combination of color and shape, the task complexity gets divided by three (and thus becomes a reasonably difficult task). This improvement might be linked with observations performed with hierarchical stimuli [WFH*94].

The benefit of this study is twofold: first, we experimented with the design of a hybrid visual system for an industrial application, and we also provided some strong support to the initial belief that visual search was a good background to start with (human visual system knowledge, experimental methods).

Acknowledgements

I thank Dr. Jeremy M. Wolfe for inviting me to perform this study in his lab. I thank Alexander Bilsky for conducting the experiments. I thank Dr Patricia O'Neill for teaching me how to analyze the experimental data (all interpretation errors are mine). I thank all the above for their many very helpul comments and suggestions.

REFERENCES

[BJ83] Bergen, J. R. & Julesz, B. Parallel versus serial processing in rapid pattern discrimination. *Nature*, 303:696-698, 1983.

[CW94] Chun, M. M. & Wolfe, J. M. Just say no: How are visual searches terminated when there is no target present? *Submitted.* 1994.

[EOR90] Enns, J. T., Ochs, E. P. & Rensink, R. A. VSearch: a MacIntosh software for experiments in visual search. *Behaviour Research Methods, Instruments and Computers*, 22:118-122, 1990.

[Pon93] Ponthieu, L. Computer-Aided Search for Technical Diagrams. 1 - Problem Overview. *Electricité de France, Direction des Etudes et Recherches, Technical Report*, HI23/94-006, 1994.

[TG80] Treisman, A. & Gelade, G. A feature-integration theory of attention. *Cognitive Psychology*, 12:97-136, 1980.

[TS85] Treisman, A. & Souther, J. Search asymmetry: A diagnostic for preattentive processing of separable features. *J. of Exper. Psych.: General*, 114:285-310, 1985.

[Wol94] Wolfe, J. M. Guided Search 2.0: A revised model of visual search. *Psychonomics Bulletin & Review*, 1(2):202-238, 1994.

[WCF89] Wolfe, J. M., Cave, R. C. & Franzel, S. L. Guided Search: An alternative to the Feature Integration Model for Visual Search. *J. of Exper. Psych.: Human Perception and Performance*, 15:419-433, 1989.

[WFH*94] Wolfe, J. M., Friedman-Hill, S. R. & Bilsky, A. B. Parallel processing of part-whole information in visual search tasks. *Perception & Psychophysics*, 55 (5):537-550, 1994.

Model-Based Reasoning

MTDM — AN ACTIVE MODEL FOR DRAWING UNDERSTANDING

Wei WU, Wei LU, Masao SAKAUCHI
3rd Dept., Institute of Industrial Science, University of Tokyo,
7-22-1, Roppongi, Minato-ku, Tokyo, 106 Japan
Email:wuwei@sak.iis.u-tokyo.ac.jp

ABSTRACT

In this paper, we present a new framework of active object-oriented model, MTDM(Matching Tree Driving Model), which drives the matching procedure for drawing understanding. To meet the requirements of multitarget, multipurpose environment, MTDM represents the features of recognition targets with two kinds of objects. The static features of targets are declared as objects independent of the matching process. For every recognition target, a data structure named **matching-tree** is defined to drive the matching procedure of the target. For obtaining a highly structural representation style, MTDM represents the abstract features of targets by members of objects. A prototype system of drawing understanding system has been developed to evaluate the suitability of MTDM for a multitarget and multipurpose environment.

1 INTRODUCTION

As an important component of multi-media systems, a lot of research has been conducted for the development of drawing and image understanding systems. Many successful results have been reported. Amongst many other issues, it is important for drawing understanding system to represent knowledge about recognition targets. A lot of research centered on the description of shape for recognition targets according to their outlines [Stein92, Bobick92, Nishida92, Ao92]. These kinds of models are insufficient to represent the compositional structure of recognition targets. Other research concentrated on model representations which depict the compositional features of targets in a more formal, structural way. Most of the systems developed are for specific fields and incorporate pattern recognition techniques according to the characteristics of their targets. For example, Dori's model [Dori89] for dimension of engineering machine drawing consists of syntactic/geometric descriptions. T.Fan and G.Medioni [Fan89] present 3D objects with a graph whose nodes correspond to the surfaces of target and links representing the relationships between the surfaces. Kise and Momota[Kise93] represent the structure of the documents in a tree structure. The object-oriented technology has been also been applied to model representations. For example, Joseph's model[Joseph92] declares the properties of basic elements such as dimensions, labels and hatching and the construction relationship around drawing in object-oriented style to achieve a highly structured description. In general, the above models have to take different description methods when the recognition targets are different. To realize more general model representation, Pasternak[Pasternak93] introduced an object-oriented model which is based on line segments and points in input drawing data. A compound target is defined by the definition of the geometrical characteristics among the basic elements such as angle, near and linkage. A reasoning engine then interprets the models and retracts corresponding information from the input drawing data. However, this model representation has the following deficiencies:

1. Definition is based on the basic data elements and no abstract characteristics of targets. Therefore, it is difficult to handle the distortion of input data.

2. All geometrical features have to be declared using basic elements and the structure of targets becomes difficult to understand.

3. In many cases, the structure of targets is described based on the repeating of substructure. Hatching, coil and so on are such examples. This model is weak in describing these kinds of targets.

In reviewing the relevant works, we understood that the model for targets must not only define what the targets are but also specify how to obtain them from the input drawing data which contains many noises. Moreover, a model must have the capacity to represent various kinds of static features of targets on abstract levels.

In this paper, we present an active model named MTDM(Matching Tree Driving Model) which can be used in a multitarget and multipurpose drawing understanding environment. MTDM is composed of two types of object-oriented style declarations. One is for static features of targets such as geometrical and structural features. The other for dynamic characteristics such as matching processes and matching control. The reasoning processes are driven by MTDM models. The MTDM describes the features of recognition targets based on the abstract characteristics, making it more flexible for absorbing noises, describing various kinds of properties for recognition targets. For easier understanding, reasoning driving declaration is based on a tree structure.

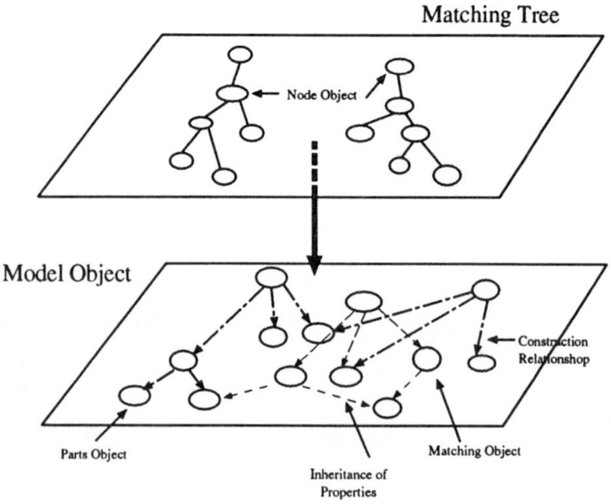

FIGURE 1: The composition of MTDM

2 THE COMPOSITION OF MTDM

Knowledge for drawing understanding are shape, properties of targets, restriction rules and so on. They are usually represented in different ways. Model of target is abstract and structural description for the knowledges. MTDM is an active model that describes what the target is by the declaration of compositional and geometrical features, and how to obtain the target by matching control declarations that are based on the compositional declaration. Matching control declaration describe the dynamic action of models and are used to drive the reasoning engine of drawing understanding system. The composition of MTDM is shown in Fig. 1. "Object" in Fig. 1 means the declaration of targets in object-oriented style, with details as follows.

1. The lower layer is composed of objects called **model-objects** that describe the static characteristics of targets and knowledge about geometrical and spatial relationships. Depending on the contents, the object describing, model-object can be divided into :
   (a) Parts-object: Declaring the features of target, attributes inheritance relationship among the objects, and so on.
   (b) Matching-object : Declaring restriction rules, concepts of spatial relationship, and so on.
2. The upper layer shows a tree structure based on the declaration of model-object. The tree structure shows the control sequence in driving the matching procedure of target and is named **matching-tree**. It is defined by the declaration of **node object** of the tree.

2.1 Composition of Model Object

With the model-object declaration, features of target, spatial relationship and the rules set about drawing can be represented in object-oriented style. The parts-object declares features of recognition targets. It is independent of the reasoning process and matching methods. Matching-object shows which drawing rules or spatial restrictions can be used in the target matching procedure. For example, the commonly used geometrical parts-objects are; line, arc, rectangle, and so on. The commonly used matching-objects are; connecting, affecting, neighboring, near and so on. The definition of spatial restrictions are based on the abstract features of parts-object, such as; the features with type of position, the elongation axis of object and so on. All objects declared will be stored in a model base. A model-object contains the following information:

1. Member declaration: recognition result contains not only labelling of the input drawing data, but also the abstract expression of the corresponding target. The abstract features of targets are described by member declaration of model-object in MTDM. A member provides a reference name to a feature, whose value is implemented dynamically. For example, an arc model can have abstract features of center and end positions, which are declared by members of **strPosition, endPosition** and **centerPosition** respectively. In the process of arc-matching, the coordinates of the center and end positions are computed by the matching process and implemented in related member. The computation results show the po-

sitions of corresponding points when a set of vectors are considered as an arc. The position may not be any point in the input data. Another example is arrow. Recognition target of arrow has three members of **headPos, tailPos**(specifing the typical positions of head) and tail of the arrow, and **body**. Using the abstract members of **headPos** and **tailPos**, the distortions can be absorbed.

2. Method declaration: Describes the usage of model-objects. For example, a matching-object corresponding to the spatial relation of "near" will have a function of matching data near a specified object. The parts-object of arc have a function to compute the center and end points of the presumed arc according to the related vector set. The usages of model object are declared by the method declaration statements in MTDM.

3. Inheritance declaration: Describes the inheritance relation between model-objects. The members and methods of the declared object will be inherited by the current object. For example, the member and method declarations for polygon are as follows:

$$Member : verticeslist;$$

$$Method : display, computing - area;$$

a rectangle model that inherits properties from polygon will have the member attributes of "vertices list" and methods declared in the polygon model.

2.2 Composition of Matching Tree

A matching-tree is a tree structure for declaring the control sequence of matching processes and is activated by a **firing-parts** which is one of the leaf nodes. The definitions of the various nodes of a matching-tree are as follows:

1. Root node of the matching-tree: Corresponds to a parts-object relates to the recognition target. Satisfaction of the node means the success of the target matching procedure.

2. Internal node of the matching-tree: Corresponds to a matching step in the reasoning procedure and relates to a matching-object. The information about firing control of methods that are declared in the matching-object and tracing control to the next matching step is declared. Controlling declaration is represented by the declaration of acting conditions and acting methods.

3. Leaf node of the matching-tree: Represents the components of root target, corresponds to a parts-object. It shows the matching targets of every matching step and the composition of the recognition target described in the root node.

A matching-tree relates to one of the matching sequence for the recognition target. When there exist multiple matching sequences for a recognition target, multiple matching-trees will be built in accordance with each matching sequence. A node in the matching-tree is related to a model-object, which includes the information of the model-object and matching control declaration. The detailed declaration format will be presented in Section 3. The target-matching process starts from firing-parts, matches other components of the target in a sequence specified by the matching-tree. Fig. 2(a) shows an example of matching tree for recognition of arrow symbol and rectangle. the node marked with "*" indicates firing-parts. By tracing the tree structure in depth first order, the nodes are fired one by one. When a node is fired, the acting conditions are evaluated, and if successful, the acting methods are executed. As the example shown in Fig. 2(a), for a short line with an opening end point, the elements with the same properties near it are matched and the angle between them are computed. An arrow mark is matched if the angle is less than the given threshold. Since an arrow mark acts as the firing-parts of the arrow, the matched arrow marks will then start an arrow matching procedure. Driven by the matching- object of "between", a line object between the arrow mark is matched. If the line object is matched, an instance related to the arrow will be created. Fig. 2(b) shows two models for driving the matching procedure for rectangle. In a perfect situation, a rectangle is composed of four right-angle objects. Therefor the corresponding model specifies that if right angles form closed loop, then the loop is rectangle. The second model shows how to get rectangle in a noisy situation. For loops that have right angles, by comparing the rate of the loop's area to the area of its extent box, an rectangle can be determined. In fact, the first model causes an ineffective check of loops. It is only an example to show that a target can be matched according to its characteristics by various models. The last model is more general than the first one.

3 DECLARATION OF MODEL

A model is declared by an object-oriented language named OMDL (Object-oriented Model Declaration Language). The OMDL is composed of two declaration sections of object-model declaration and matching-tree declaration described in Section 2.

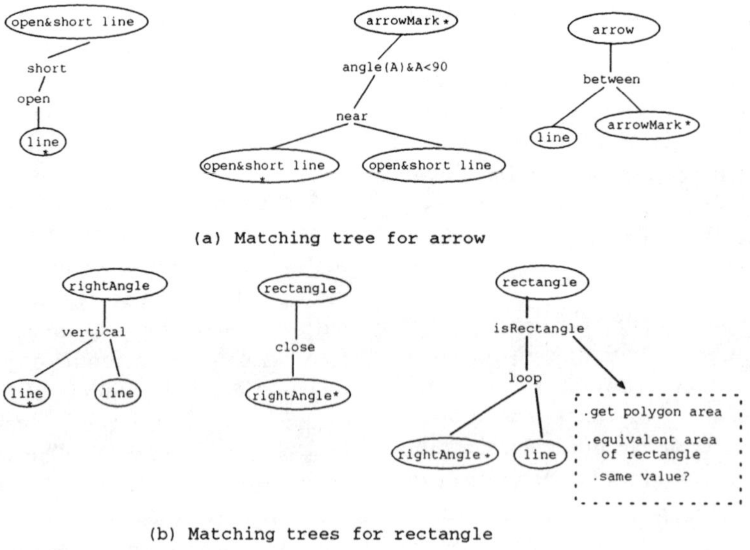

FIGURE 2: Declaration example of matching tree

3.1 Model Object Declaration

In this declaration section, the features, usage and the information about matching action are declared in object-oriented style. Model-objects are stored in the model base of the drawing understanding system. All of the model-objects that are used in matching-tree must be declared in this section. Using OMDL, model-objects are declared in the following format.

```
element{
  name : <object name>;
  kind_of : <name of the object
                  to be inherited>;
  member : <member1>,<member2>, ... ;
  method : <method1>,<method2>, ... ;
};
```

The "name" item specifies the ID of object and the "kind-of" item shows the name of object to inherit from. "member" item describes the reference name of features of the recognition target. The "method" item shows the reference name of method that represents the usage of the object. A built-in object named "TOP" is declared at the top of the inheritance structure. This object contains commonly used fundamental methods, such as

```
display: displays result with
             specified color
getMember: refers to the member
             specified
create:  creates new recognition
             result instance
elongation: computes the elongation
             axis of recognition result
```

and so on. All objects have a member named "degree" which refers to the degree of certainty of recognition results. An implicit method named "degree" for computing the certainty degree of matching results is called automatically whenever an object is created. The "TOP" model has a "degree" method that always returns a value of 1, when there is no declaration of "degree" method for a model object, it inherits the "degree" method declared in the "TOP" object . As an example, the model for arrow stated in Section 2 becomes:

```
element{
  name : arrow;
  kind_of : top;
  member : tailPosition, topPosition;
  method : degree, getPosition;
};
```

By this declaration,

1. The name of model-object is declared as "arrow"
2. The arrow has two members for its head and tail positions.
3. Two methods are defined. The method of "degree" computes the certainty degree of arrow by the rate of the length of the mark and the length of the line object related to the arrow. "getPosition" method evaluates the equivalent position of the arrow head from the components.

3.2 Declaration of Matching Tree

Based on pre-declared model-object, matching-trees that drives the target matching procedure can be declared. A matching-tree is implemented by declaring the structure of tree nodes. The information that must be declared is relation among nodes representing tree structure and the matching control knowledge. The following is the node declaration format.

```
m_tree{
  model : <model name>;
  name : <name of related model-object>;
  parent: <name of parent node according
                to the tree structure>;
  key : <firing part specification>;
  weight : <weight coefficient>;
  trace_condition : <tracing control
                          declaration>;
  act_condition : <action control
                          declaration>;
};
```

Where, the meaning of each item is as follows:

1. Model : Specifies the ID of model.
2. Name : Specifies the name of related model-object. The node is identified by the pair of the "model" item and the "name" item.
3. Parent : Specifies the parent node of tree structure. Nodes that have the same parent will be fired according to the order of declaration
4. Key : Declares whether the node acts as a firing-parts or not.
5. Weight : Shows the weight coefficient of certainty degree of related model-object, and is decided by experiment.
6. Trace-condition : Tracing control of matching tree is declared in the following format.

   ```
   ([ci1,...cin], command),
                  ( ... ), ...
   ```

   The item $[ci1, ..., cin]$ represents a set of conjunctive conditions such as *exist*(member name) and *notExist*(member name) which checks whether the specified member is implemented. The tracing predicate specified by *command* will be executed when the conditions are satisfied. An empty condition specification means unconditional execution of tracing action. Currently the prototype system now has the following tracing predicates.

   (a) *exit* : reasoning terminates. System will than continue to create instances of matched results.

   (b) *quit* : Terminates the reasoning procedure but do not create result instance.

   (c) *fail* :Indicates the failure and starts a back-tracking procedure.

   (d) *continue* : Traces the next node according to the declaration of matching tree.

7. act-condition : Declares matching actions and their execution conditions when the corresponding node in a matching tree is reached.

   ```
   ( [ci1,...,cin ],
     method(object, parameter1, ...)),
   ( .... )
   ```

The items of $[ci1, ..., cin]$ is that the same as in the trace-condition item. "method" is a message sending sequence, with the arguments declared in the model-object declaration section. The first parameter "object" is a model-object name, specifying the destination of the message. Parameters that follow "object" are member names declared in the matching tree, including members declared in nodes and the name of nodes. For the parameters list, the unified members are treated as input arguments, the others output. A matching method tries to unify the output arguments according to the input ones. Moreover, the method returns a status that shows whether the matching action has been successful. The predicate ***status*** is used to check the status.

Models described in OMDL are then translated to object-oriented Prolog running on GOLS [Satoh92].

4 DRIVING REASONING PROCEDURE BY MODEL

The reasoning engine driven by the matching-tree is implemented in a drawing understanding system named OO–Mudams [Wu93]. The drawing data is first converted to basic elements of line, curve and linkage. The basic elements and intermediate results matched by models are stored in the result space, a result data base of OO–Mudams. Reasoning engine works on this result space.

A reasoning procedure consists of several inference cycles. Whenever a new result acting as a firing-parts for certain models is created in the result space, an inference cycle will be started, whereby all the models related to the firing-parts are fired one by one according to the control sequence declared in the matching-tree. An inference cycle terminates after all related models have been fired. Fig. 3 shows the matching procedures for dimension.

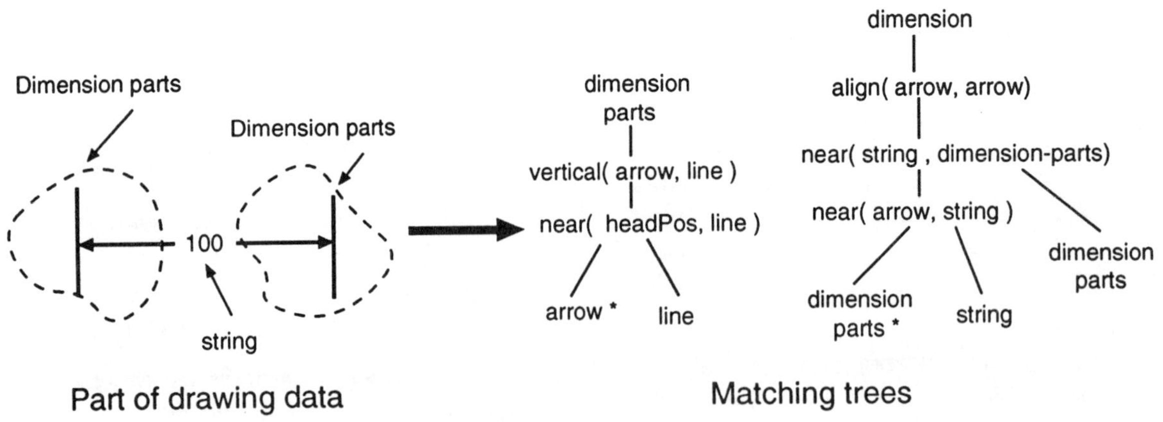

FIGURE 3: Example of matching dimension

In this case, the arrow is the firing-parts for the recognition target of dimension-parts. After an arrow is matched, the matching-tree corresponding to the dimension-parts will be fired. According to the structure of matching-tree and the matching action described in the internal nodes, the lines around the head position of arrow are searched first. Then, among the lines found, the ones vertical to the arrow are passed to the leaf node of "line". If the matching is successful, a new result of the dimension-parts will be created. In the succeeding inference cycles, the newly created dimension parts results will in turn fire the dimension model, which matches dimensions according to the follwoing matching sequence:

1. searches string objects around the tail position of arrow.
2. searches other dimension-parts which are near the searched string
3. checks the alignment relationship between the firing-parts and the searched dimension-parts

Internal nodes of the matching trees show what members will be matched. Several possible matching results are returned from the matching method and stored in a content stack related to each member. When the root node of a matching-tree is satisfied or a matching process causes a failed status, backtracking process is started to find other results according to the content stack of the members. Using this backtracking, all potential matching results that relate to a matching-tree will be searched. In this way, matching-tree can easily declare targets such as hatching, coil, that have structures described by the repeating of a special substructure.

5 EXPERIMENTAL RESULT

To evaluate the suitability of MTDM description for a multitarget and multipurpose environment, we have developed a prototype system and applied it to several types of typical engineering drawings. The system has been termed OO–Mudams [Wu93] and has been implemented on a multi-media platform GOLS [Satoh92]. The programming languages on this platform are object-oriented PROLOG and C++. Several experiments has been conducted to verify the following aspects:

- Same model applied to different types of drawings
- Flexible construction of model according to the characteristics of the recognition targets
- In a noisy environment, covering as many possibilities as possible with less models

The recognized objects in the experimental results shown in all the figures are printed in thick lines.

First, a model is built for recognition of the distorted arrow symbols in engineering machine drawings. Fig. 4 shows some typical distortions of arrow symbol.

Since the linkage situation based on line segments changes with distortion, we have constructed a model for matching arrows by using the following geometrical characteristics of the arrow.

1. The typical features of arrow head are two short lines with an opening end point.
2. The angle between the two arrow head mentioned above is less than 90 degrees.
3. Distortions concentrate on a small area near the head of arrow.

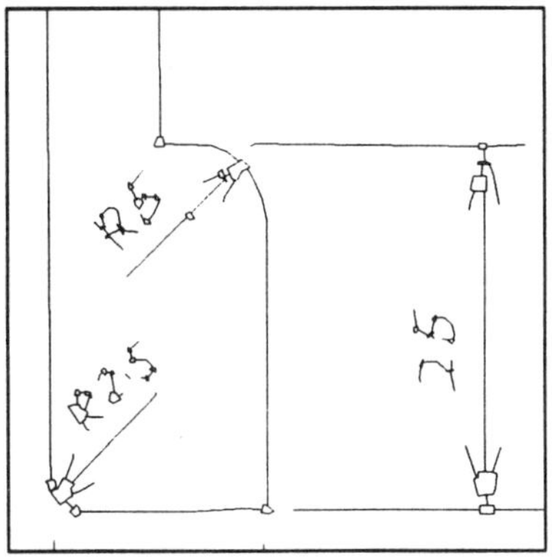

FIGURE 4: Typical distortion of arrow

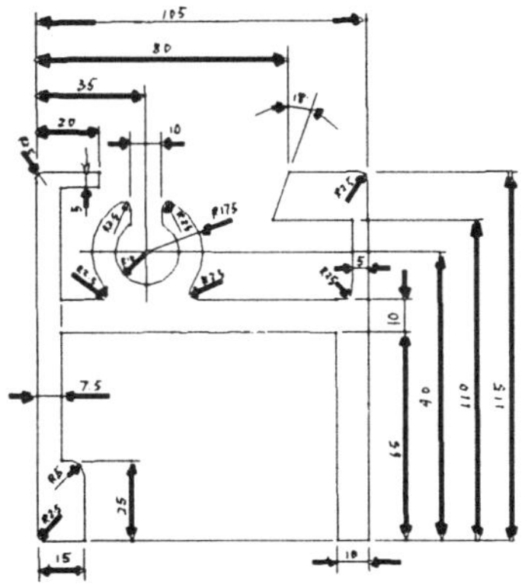

FIGURE 5: Result of arrow matching

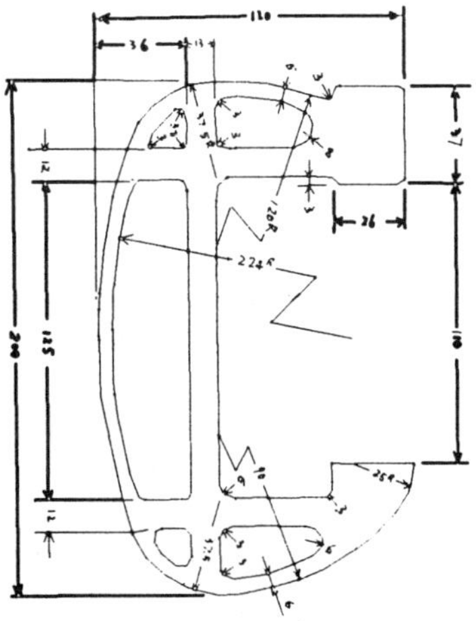

FIGURE 6: Result of dimension matching

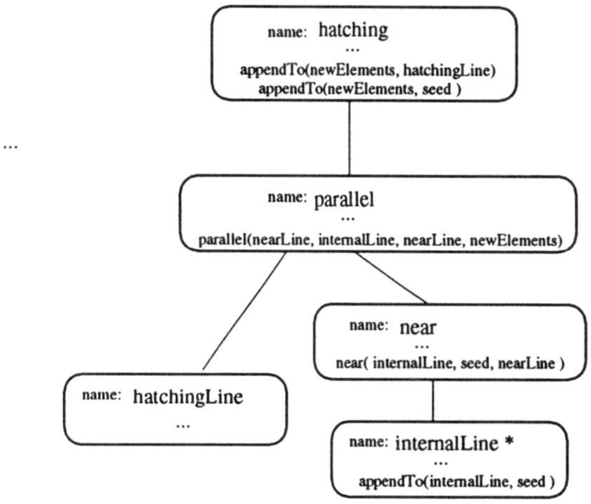

FIGURE 7: Structure of matching tree for hatching

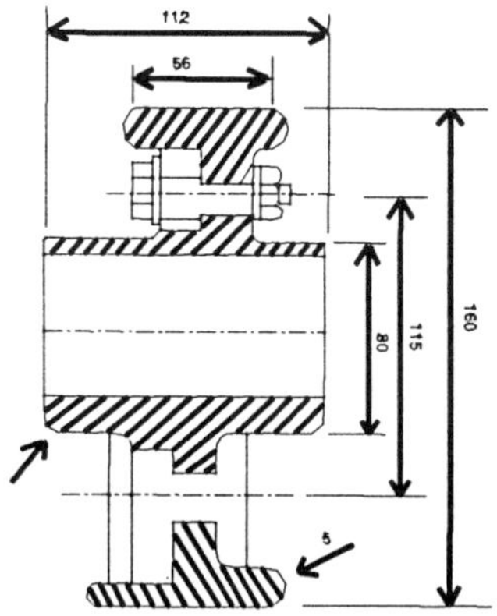

FIGURE 8: Matching results of hatching

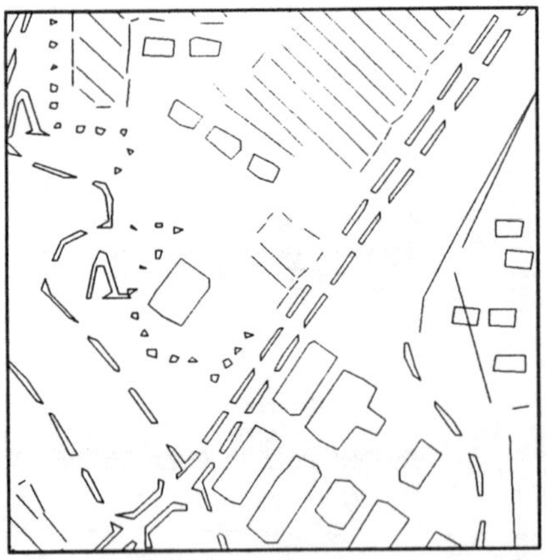

FIGURE 9: Typical distortions of boundary symbols

FIGURE 10: Recognition of symbols in map

4. The line related to arrow is between arrow heads.

Fig. 2(a) shows the matching-tree that drives the reasoning engine to match the above geometrical characteristics of arrow. The result is shown in Fig. 5. The input drawings have 136 arrows symbols, with 70 containing distortions and noises, 13 arrow symbols lacking the mark features. The 122 arrow symbols were correctly recognized without error. All the rest were rejected. Fig. 6 is the application results of the model described in Fig.3 to machine drawing.

The second experiment is for recognition of hatching in engineering machine drawing and map drawing. Hatching is defined as a set of parallel lines that are enclosed in a polygon. This restriction rule indicates how to construct hatching from lines. Fig. 7 shows a matching-tree structure that matches hatching by backtracking. The matching-tree intermediate member *seed* controls the backtracking process. The member *hatchingLine* refers to the list of *internalLine* satisfying the restriction of hatching. The matching-tree tries to maximize the contents of *hatchingLine*.

Fig. 8 is the application result of the above model to machine drawing.

The last experiment is for recognition of symbols in map drawing. Fig. 9 shows an application result of the recognition of boundary and building symbols. When components of recognition targets are very small, they are sensitive to noise. As a result, it is difficult to define models for every possible situation. Therefore, for recognition of symbols, a learning system developed by our group has been used to obtain a classification space for symbols [Lu93] according to their shape parameters [Pavlidis80, Kita92]. The geometric parameter space has 21 dimensions such as direction, size, convexity and so on. The decision trees obtained from the learning system is mapped to matching trees. For the boundary symbols, 933 correct results and 12 error results were obtained from 974 symbols and the recognition results are shown in Fig.10 .

6 CONCLUSION

In this paper, we have described an active model for drawing understanding. A model contains two kinds of information, one for describing the static characteristics of target such as geometrical features and composition attributes, the other for driving the matching procedure. And as the reasoning procedure is driven by models, matching of features can be implemented in appropriate sequences and algorithms by defining corresponding matching trees that are suitable to the recognition target. Since the matching tree only describes how to match features of target, the same model object that describes the static characteristics of target can be applied to different matching trees. Experimental results have shown that the proposed model is suitable for building multitarget and multipurpose systems. In future works, a larger number of drawings will be applied to the OO–Mudams, in order to find a set of more appropriate predicates for describing geometrical features and relations. The experimental results show that it is difficult using only one model(recognition methods) to handle all the distortions and noises. Currently, experiments are being conducted to evaluate the ability to absorb more noises using less models.

REFERENCES

[Stein92] F.Stein and G.Medioni, "Structural Indexing: Efficient 2-D Object Recognition", IEEE Trans. PAMI, VOL.14, No.12, 1992

[Bobick92] Aaron F.Bobick, "The Representation Shape Paradigm of Concurrent Evolving Object Descriptions", IEEE Trans. PAMI, VOL.14, No.2, 1992

[Nishida92] H.Nishida and S.Mori, "Algebrsaic Description of Curve Structure", IEEE Trans. PAMI, VOL.14, No.5, 1992

[Ao92] G.Ao and Y.Chi, "Structural Tree Representation of Outlines and Its Application to Object Recognition", IEICE Trans., VOL. J75-D-II, No.3, 1992

[Dori89] Dov Dori, "A Syntactic/Geometric Approach to Recognition of Dimensions in Engineering Machine Drawings", COMPUTER VISION,GRAPHICS,AND IMAGE PROCESSING 47, 271-291, 1989

[Fan89] T.Fan and G.Medioni, "Recognizing 3-D Objects Using Surface Descriptions", IEEE Trans., PAMI, VOL.11, No.11, 1989

[Kise93] Koichi Kise and Ken-Ichi Momota, "Hypothesis Driven Document Image Understanding with Knowledge about Layout and Content", Vol.34, Trans. IPSJ, 1993

[Joseph92] S. H. Joseph, T. P. Pridmore, "Knowledge-Directed Interpretation of Mechanical Engineering Drawings", IEEE Trans. PAMI, VOL.14, NO.9, 1992

[Pasternak93] Boris Pasternak,Bernd Neumann", Adaptable Drawing Interpretation Using Object-Oriented and Constraint-Based Graphic Specification ",ICDAR'93,1993

[Wu93] W.Wu and M.Sakauchi, "A Multipurpose Drawing Understanding System with Flexible Object-Oriented Framework", ICDAR '93, 1993

[Lu93] W.Lu, W.Wu and M.Sakauchi, "Learning of Structural Information for Mechanical Drawings", IEICE, National Conference, Japan, 1993.3

[Pavlidis80] T.Pavlidis, "Algorithms for Shape Analysis of Contours and Waveforms", IEEE Trans. VOL. PAMI-2, NO4, 1980

[Kita92] N.Kita, "Similarity Measurement of 2D Shape Based on Concentric Circular Features", ICICE Trans. VOL. J75-D-II, No.9, 1992

[Satoh92] T.Satoh and M.Sakauchi, "GOLS:A Hypermedia Platform Which Can Process Television Pictures", Technical Report of ITE, Japan, 1992.12

Natural Language

Making Requirements Specifications Accessible *via* Logic, Language and Graphics: A Progress Report

Jeremy Pitt* **and Jim Cunningham**
Department of Computing, Imperial College of Science, Technology, and Medicine,
180, Queen's Gate, London, SW7 2BZ, UK.
Email: {jvp,rjc}@doc.ic.ac.uk URL: http://medlar.doc.ic.ac.uk/~jvp/

Abstract

Natural language software tools may have an important role in making requirements specifications more accessible. Possible tools include text processors to support requirements elicitation, and text generators to support requirements validation. The current paper reports on our progress in developing a natural language generation system, integrating this tool with a graphical interface and an automated reasoning system, and applying it in the domain of requirements validation. The resulting synthesis of logic, language and graphics is an important first step in developing an intelligent assistant to support a designer in both requirements elicitation and validation.

1 Introduction

The requirements specification for any system, written in some combination of diagrams and natural language, may be difficult for a designer or manager to reason about, just because of its complexity. An equivalent specification, written in a formal language, can also cause problems. It may be difficult to create, and for those untrained in the formal language, it may still be difficult, if not impossible, to read and understand. Furthermore, there may be interactions between separate parts of the specification which produce results that are not only inconsistent with some original intention, but are also not apparent until the specification is 'evaluated' in some way. Even then, the symbolic results of this evaluation may not be easily understood.

There is scope, therefore, for making a formal specification more 'accessible', in two senses: by making it both easier to create and easier to animate. This is a potentially important application for natural language software tools, for example text processors to support requirements elicitation, and text generators to support requirements validation.

In [PCK94], we introduced a system which provided cooperative answers to natural language email queries. This system used a text processor to convert queries into logical form, which were then evaluated against a knowledge base. The answers, still logical statements, were then translated back into natural language by a discourse generator. In this paper, we report on our progress in using the discourse generator for another application – requirements validation. This tool is integrated with a graphical interface and an (another) automated reasoning program, to produce an interactive system which can assist a designer with requirements validation.

The general technique we are using for requirements validation is a form of symbolic execution called *logical animation*. In section 2, we discuss the idea of logical animation, as originally presented in [CCB90], and the advantages of the logical calculus we are now using to implement this idea. In the following section, section 3, we briefly describe our text generation tool and its application, in conjunction with an interactive graphical interface, in supporting logical animation. Throughout, we will illustrate the ideas by reference to the blocks world [GN87]: however, this is a simple, exemplary scenario, and in section 4.2 we consider how we intend to support reasoning about a specification in a more complex scenario, a robotic factory. This scenario is the focus of a case study in the last year of the CEC Medlar II Project (Esprit 6471).

Our overall intention is to use the discourse processor for requirements elicitation, so that our two natural language software tools would work in tandem as illustrated in figure 1. Although we have a working discourse understanding system (as described in [PCK94]), its application for requirements elicitation is less advanced than our application of the discourse generation system for requirements validation. In section 5 we outline our approach to the former activity, and draw some further conclusions.

2 Logical Animation

The motivation for logical animation is to deduce and display a logical model of a formal requirements specification [CCB90]; in this paper, we will use as our working example a specification of the blocks world presented as a logical theory of Modal Action Logic [Mai87].

*Supported by CEC Esprit Project GOAL (Esprit 6283) and CEC Esprit BRA Medlar II (Esprit 6471).

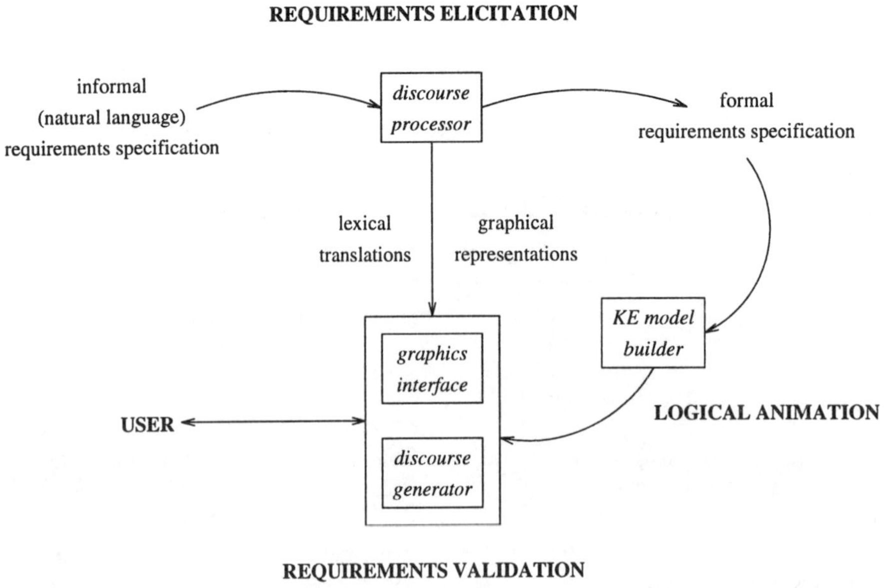

Figure 1: *Logic, Language & Graphics in Requirements Engineering*

One deduction method for finding a logical model of such a specification is a modal form of semantic tableaux [Fit90], which is especially suitable for automation and providing interaction with the user. The original system used the 'standard' tableau rules, however, for the following reasons our implementation uses the free variable **KE** calculus [DM94, PC95]:

Efficiency: **KE** is different from the tableau method, although the **KE** proof procedure is related to tableau with lemmata plus a certain formula selection strategy. However, **KE** does offer a substantial (space) improvement because the only branching rule decreases the size of the search space;

Model building: an essential feature of **KE** is that all complete open branches are mutually inconsistent. This facilitates our search for minimal models in comparison with the tableau method, which can lead to a redundant enumeration of models (i.e. some of the models may be subsumed by others). This redundancy may not be visible in small examples, but can have a dramatic effect on the enumeration process when lengthier sets are involved;

Behavior explanation: The standard tableau method is essentially a 'consistency checker', i.e., given a set of formulas Γ (of an axiomatic theory) and a single formula α, it implements a function f such that:

$$f(\Gamma, \alpha) = \begin{cases} \text{yes}, & \text{if } \alpha \text{ is consistent with } \Gamma, \\ \text{no}, & \text{otherwise} \end{cases}$$

The problem with the standard tableau method is that not all the rules used to implement f can be easily related to 'traditional' forms of reasoning (Swartout [Swa83] reports a similar problem using a theorem prover based on resolution). For example, the branching rule for implication derives its justification from the truth-tables for implication in classical logic (only), but is hard to explain to those untrained in formal logic, and hard to motivate the use of the rule in building a model. The explanation of a proof involving such steps requires additional effort (e.g. by translating a tableau proof into natural deduction steps). **KE**, on the other hand, uses as its analytic rules for implication, *precisely* those rules which correspond to 'practical' forms of reasoning, such as modus ponens and modus tollens. Even the single branching rule of **KE** is related to the fundamental logical notion of classical bivalence, i.e. every formula is either true or false. All of these rules are intuitive, easily justified, and, it is expected, will be easier to explain as a reasoning step in animating a specification and explaining its behaviour.

The KE Model Builder we have implemented (formal details are included as an Appendix) breaks down each formula in the database (i.e. the specification) in order to build a **KE**-tree in which each branch is an alternative way of satisfying the database, i.e., a logical model in the Herbrand Universe.

Having done this, the intention of logical animation is to provide the user with detailed information about the actual state of the system. In practice, this means listing all positive extensions of the predicates that can be derived from the (intensional and extensional) database; semantically, we are trying to build a *minimal model* [Hin88] using the **KE** proof procedure. The condition of the **KE**-tree when the deductive process stops can be as follows:

- each branch is closed by an inconsistent pair A and $\neg A$. The original database is then inconsistent;

Sorts:	*BLOCK*
Actions:	*Unstack:* $BLOCK \times BLOCK$
	Stack: $BLOCK \times BLOCK$
Agents:	*user*
Predicates:	$OnTable \subseteq BLOCK$
	$On \subseteq BLOCK \times BLOCK$
	$Clear \subseteq BLOCK$
Axioms:	$\forall x.OnTable(x) \Rightarrow \forall y.\neg On(x,y)$
	$\forall x.OnTable(x) \vee \exists y.On(x,y)$
	$\forall x.Clear(x) \Rightarrow \forall y.\neg On(y,x)$
	$\forall x.Clear(x) \vee \exists y.On(y,x)$
	$\forall x \forall y.On(x,y) \wedge Clear(x) \Rightarrow \mathbf{per}(user, Unstack(x,y))$
	$\forall x \forall y.\neg On(x,y) \Rightarrow \neg\mathbf{per}(user, Unstack(x,y))$
	$\forall x \forall y.OnTable(x) \wedge Clear(x) \wedge Clear(y) \wedge x \neq y \Rightarrow \mathbf{per}(user, Stack(x,y))$
	$\forall x \forall y.\neg Clear(x) \Rightarrow \neg\mathbf{per}(user, Stack(y,x))$
	$\forall x \forall y.\mathbf{per}(user, Unstack(x,y)) \Rightarrow [user, Unstack(x,y)]OnTable(x)$
	$\forall x \forall y.\mathbf{per}(user, Stack(x,y)) \Rightarrow [user, Stack(x,y)]On(x,y)$

Figure 2: *Modal Action Logic specification for the Blocks World*

- there are alternative complete open branches, which may be disregarded either because they do not introduce new information, or because they introduce a non-minimal model, or because they add incompleteness;
- there is only one complete open branch which represents the actual state of a model for the intensional database.

Clearly, it is only the latter case which provides the user with information about the actual state of the system being animated, so we restrict our attention to minimal models and disregard branches which introduce incompleteness.

To give an example, consider the Modal Action Logic specification of the blocks world presented in figure 2.

Given only the first four axioms above and the fact $\forall x.OnTable(x)$ (i.e. the initial state whereby all the blocks are on the table), the output from the KE Model Builder is the **KE**-tree shown in figure 3, in which the only open branch contains the positive literals *OnTable(X4)* and *Clear(X7)*, where *X4* and *X7* are free variables, indicating that all the blocks are clear.

However, if we perform a stack action, e.g. to stack the green block on the blue block, so that the postcondition *On(green,blue)* is true, a **KE**-tree can be developed as illustrated in figure 4.

At this point, as no rules apply and the branch will not close, so we apply a *frame rule*, by which we carry over those aspects of the state of the system which are unchanged by an action to the new state. Here, we add all the (positive) facts which hold in the current state and are consistent with the facts (i.e. positive and negative literals) on the branch. The KE Model Builder then continues making derivations, and if it now stops with an open branch, the positive literals on that branch can be collected and this (minimal model) now represents the new state. In this case we will add the facts that the red block is clear and on the table, the green block is clear, and the blue block is on the table. Derivation continues and an open branch is obtained, with these same facts.

3 Discourse Generation

There is evidence from cognitive science [LS87] that the optimal way of conveying maximal information is vision. Hence we look for a graphical presentation of our logical model. However, our experience with the tool of [CCB90] was that, with complicated displays, looking for the effects of changes became a bit like a "spot the difference" game, with the added complication that the previous display was no longer available for comparison. Therefore, we resolved to use natural language to draw attention to details.

Our text generation system consists of four modules: a parser, an inference system, the generator, and a post-processor. These modules are organized serially, and their operation is as follows. The first module parses logical formulas and converts them into a Prolog structure to be used by the inference system in the second module. This module transforms the semantic structure output from the first module into a syntactic structure based on a traditional grammar of English [QGLS85]. To do this the inference system exploits a variety of heuristics, e.g. actions are treated as dynamic verbs, predicates as prepositions or adjectives, determiners are chosen according to the quantifiers and *where* in the logical formula the quantified variable is introduced, etc. The third module is the surface generator, which is a Prolog logic grammar run 'in reverse' to produce the actual English sentences. The final module is a post-processor, which is used to analyse the sentences and improve the

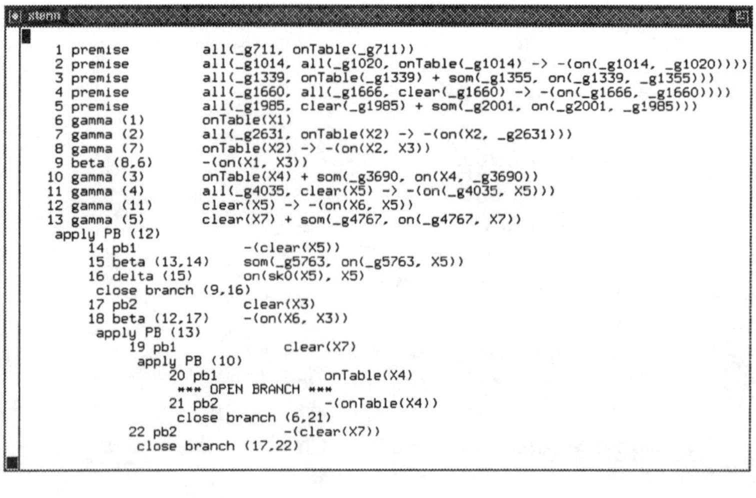

```
 1 premise          all(_g711, onTable(_g711))
 2 premise          all(_g1014, all(_g1020, onTable(_g1014) -> -(on(_g1014, _g1020))))
 3 premise          all(_g1339, onTable(_g1339) + som(_g1355, on(_g1339, _g1355)))
 4 premise          all(_g1660, all(_g1666, clear(_g1660) -> -(on(_g1666, _g1660))))
 5 premise          all(_g1985, clear(_g1985) + som(_g2001, on(_g2001, _g1985)))
 6 gamma (1)        onTable(X1)
 7 gamma (2)        all(_g2631, onTable(X2) -> -(on(X2, _g2631)))
 8 gamma (7)        onTable(X2) -> -(on(X2, X3))
 9 beta (8,6)       -(on(X1, X3))
10 gamma (3)        onTable(X4) + som(_g3690, on(X4, _g3690))
11 gamma (4)        all(_g4035, clear(X5) -> -(on(_g4035, X5)))
12 gamma (11)       clear(X5) -> -(on(X6, X5))
13 gamma (5)        clear(X7) + som(_g4767, on(_g4767, X7))
 apply PB (12)
     14 pb1              -(clear(X5))
     15 beta (13,14)     som(_g5763, on(_g5763, X5))
     16 delta (15)       on(sk0(X5), X5)
      close branch (9,16)
     17 pb2              clear(X3)
     18 beta (12,17)     -(on(X6, X3))
      apply PB (13)
          19 pb1              clear(X7)
           apply PB (10)
               20 pb1              onTable(X4)
                *** OPEN BRANCH ***
               21 pb2              -(onTable(X4))
                close branch (6,21)
          22 pb2              -(clear(X7))
           close branch (17,22)
```

Figure 3: **KE**-*tree for Blocks World in the initial state*

	axioms 1–4	
fact	*On(green,blue)*	
1	*OnTable(X1)* $\rightarrow \neg$*On(X1,Y1)*	*Rule G, axiom 1 (twice)*
2	*OnTable(X2)* $\vee$ *On(X2,sk0(X2))*	*Rule G, Rule D, axiom 2*
3	*Clear(X3)* $\rightarrow \neg$*On(Y3,X3)*	*Rule G, axiom 3 (twice)*
4	*Clear(X4)* $\vee$ *On(sk1(Y4),X4)*	*Rule G, Rule D, axiom 4*
5	$\neg$*OnTable(green)*	*Rule B, fact and 1*
6	$\neg$*Clear(blue)*	*Rule B, fact and 3*

Figure 4: *Development of* ***KE****-tree after* stack(green,blue)

quality of the output.

In processing a formal specification, the first two modules do additional work by constructing all the necessary lexicons for 'talking about' that specification in English: some interaction with the user may be required at this point, e.g. to translate predicates like *OnTable*. When the system is initialized the axioms of the specification are processed by these two modules. This enables an extensional database, which is a description of the system's initial state, to be converted into English after being processed by the entire system. Subsequent calls to the system are made with a set of (positive and negated) logical formulas which describe the change of state as a result of doing a particular action. These formulas pass through all four modules, and a natural language description produced at the end.

At this point we should stress that this natural language generation system was designed as a demonstration of feasibility only. Clearly, it is very limited in comparison with state-of-the-art systems (for a review, see [HZ93]), in that we have had to make too many simplifying assumptions, restricted the domain of application, and resorted at times to efficacious heuristics rather than general principles. However, the system is a useful stepping stone towards a more powerful system which exploits *graphical dialogues* and *multimedia generation*.

Graphical dialogues were used in the CEC Esprit ACORD project (see [LKM89]). Here, natural language was combined with graphics in the interface to a knowledge based system. The flow of information between the graphics systems and the dialogue manager fulfilled two functions. Firstly, it set up and maintained a semantic mapping whereby pictures were interpretable; and secondly, it caused pictures, whose interpretation was understood, to appear and change, and could register (and react to) a user interaction with respect to a picture.

Using this system, a user could construct a graphical representation of objects in his/her system. For example, in the ACORD demonstrator this included trucks, cars, cities, and loads. The user could define icons, and accompany this by a deictic instruction to interpret such-and-such an icon as a certain object. The user could also relate a natural language description to a graphical event, for example by issuing a natural language command such as "interpret this action as *'the car goes to the city'*", and accompanying this command by a graphical action, e.g. clicking on a car icon and dragging it to a city. The system could then use these descriptions to describe any graphical changes it makes during logical animation.

This process of system initialization via graphical dialogues is also a good source of input for multimedia generation, i.e. the generation of natural language and graphics. A good example of such a system is COMET [MEF+90], which was designed to generate explanations for system maintenance and repair. This system highlighted the requirement for extra underlying knowledge resources in order to relate the graphics to text, and also the need for different components each contributing to the overall output, and a sophisticated method for coordinating the interactions of these components to produce coherent output.

Considerable research and development effort lies between our current prototype and realizing a system that incorporates the elements described above, but our first experiments and prototype natural language generation system have demonstrated three general points:

- the use of natural language generation to supplement graphical representations, to call attention to specific details, is important and necessary. We would intuitively expect this: for example, art gallery guides do not explain details of paintings merely by blowing up the details; these are instead described in natural language;
- the use of traditional grammar can be applied to generating as well as to processing natural language. Previously [PCK94], we used traditional grammar in our discourse processing system; the generator shows that the same principles can be useful for structuring text for generation, at least at the sentential level;
- the use of logic as the basis for generation is highly promising. It may well be that an adequate knowledge representation system for generating English may be closer to the predicate calculus than other formalisms such as semantic networks, which were invented to capture the meaning of natural language.

In fact, it is our contention that the representation of natural language semantics is best conducted through the medium of multi-modal logic, which is being actively researched in the Medlar II Project. Multi-modal logics, we argue, are sufficiently rich and expressive to represent the full range of meanings of events, actions, and behaviour required for our applications, and, in addition to the usual formal advantages of a well-defined semantics, completeness and consistency, these logics are powerful enough for direct symbolic execution, as in logical animation.

4 Demonstrators

4.1 The Blocks World

A prototype system has been implemented in order to develop and evaluate the ideas in a simple test domain, i.e. the blocks world. The prototype implementation integrates three components: the KE Model Builder, the Natural Language Generator, and a Graphics Interface, which allows menu-driven user interaction. The operation of the system follows two phases: initialization and animation. The sequence of processes performed by each component of the system, and in response to user interactions, is as follows:

Initialization:

- the KE Model Builder applies the tableau rules to the specification plus a given extensional database to create the initial state of the system;
- the Natural Language Generator processes the specification to construct the lexicons, and describes the world model in English;
- the Graphics Interface draws the world model, computes from the specification the permissible actions in this state, and makes these available via a menu (the *Action Menu*);

Animation: repeat until the user selects *Quit*:

- the user selects an action from the *Action Menu*;
- the KE Model Builder starts new tableau with the axioms of the specification and the postconditions of the action. When no further **KE** rules apply on an open branch, the *frame rule* is invoked, and the **KE** rules re-applied until a minimal model is constructed;
- the Natural Language Generator computes the changes from one state to the new state represented by the minimal model. English descriptions of these changes are generated and displayed;
- the Graphics Interface updates the graphical representation and determines the permissible actions in the new state. The *Action Menu* is updated dynamically and the user can now select a new action.

As discussed above, the blocks world in its initial state is as displayed and described in the left hand screen dump in figure 5. After selecting the action *Stack green on blue*, the new situation is as shown, and the changes are as described, in the right hand screen dump. (For those watching in black and white, the red block is on the left, the green one in the middle, and the blue one on the right.)

4.2 The 'Car Painting Robots'

To give a more realistic picture of a potential application, we briefly review a case study that is the focus for work in the last year of the CEC Medlar II project, and is the test environment for the tools described in this paper. Consider then the way that model building could be used in designing a 'factory of the future'. Figure 6 shows a robotic car painting workplace with various spatial constraints. Initially there are a large number of cars in one workspace (E1) each to be painted a selected colour by one of the robots. A robot is only permitted to paint a car in a place which is free of neighbouring objects, so cars must be moved accordingly. E4 is another space for storing cars, but eventually all cars must be painted and returned to E1.

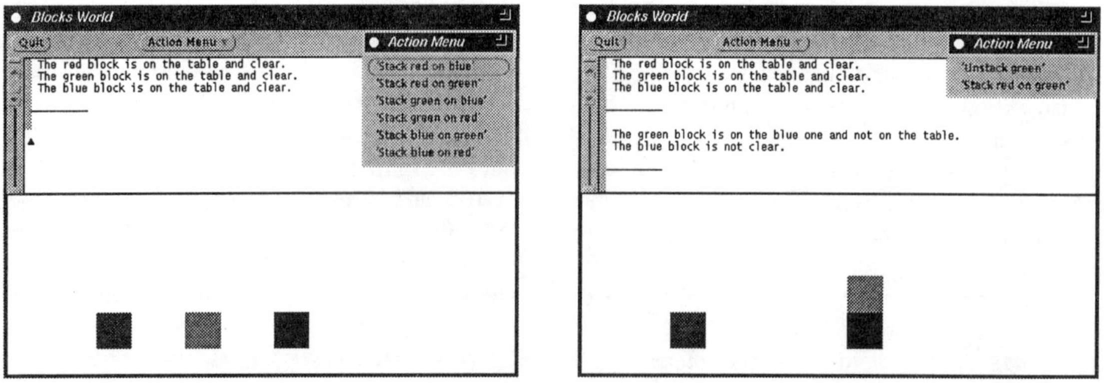

Figure 5: *Interactive Graphical Displays for the Blocks World*

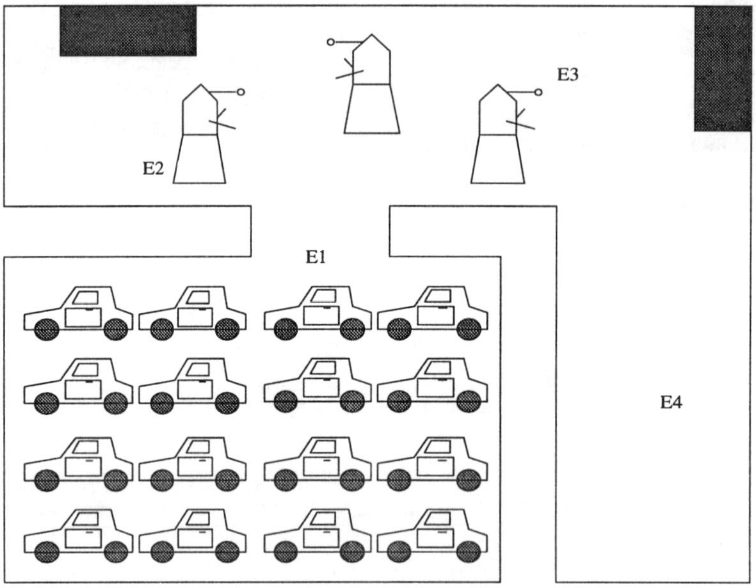

Figure 6: *A Robot Car Painting Factory*

As sketched, there are a myriad of unresolved questions which we will not go into save to indicate the design space of solutions. It is clear that special geometric reasoning is needed, but the major open questions concern the capability of the robots. This in turn determines the extent to which knowledge and planning is distributed, and the structure of the communication system. Even without autonomous robots there is a considerable problem in finding some partial ordering of tasks which will eventually achieve the goal situation. Having autonomous agents makes the conception of a plan more difficult, but does not necessarily complicate the computation of a solution, because with many agents eager to work the search space for tasks of each individual agent can be much smaller. Indeed there are indications that more agents can make the synthesis task more, not less tractable.

Now consider what would happen if we had specified the logical behaviour of each robot, with or without a global planing agent. Whatever our specification, we need to gain confidence in it. One way is to prove some theorems about it, this is traditional deduction. We would like to prove that the goal is eventually achieved, that robots don't crash, respray cars or anything else we would not like (see [AC91] for an indication that this sort of proof is not fantasy.) Unfortunately we cannot think of everything, and nor will robots. We could gain more confidence and validate the whole system by simulating it and interacting with the simulation. Such a simulation would be a restricted form of closure computation, where we circumscribe the class of consequences which we admit, for which we can use the model building techniques described above. Moreover, if we want to interact with the robots, and get them to explain themselves and justify why they are doing certain actions in a certain order, then some form of two-way natural language communication is highly desirable.

5 Further Work and Conclusions

5.1 Requirements Elicitation

The general idea of using discourse understanding for requirements elicitation was motivated by the TELL system [HSE87], in which an English specification was translated one sentence at a time into a formula of temporal logic, each sentence being treated as a separate semantic unit. However, representing the 'meaning' of individual English sentences as isolated expressions of a logic put heavy restrictions on the English syntax, and limited the specifier's 'freedom of expression'. Simple direct translation into a collection of logical axioms could not reflect the thematic structure of, or interaction between discontiguous parts of, the English specification.

In an attempt to overcome these problems, we have been considering the application of our discourse understanding system [PCK94] to an informal requirements specification written in natural language. The intention is not to produce a formal requirements specification automatically, but to provide automated 'intelligent assistance' for the system designer. This Intelligent Assistant would include a discourse understanding system, and would enable the designer to investigate the components and the behaviour of selected parts of the system piece by piece (cf. [Cox94]). The analysis would therefore give feedback for establishing and subsequently validating the formal specification, by providing useful information relating to:

- initial identification of agents, actions and system axioms;
- interactive construction of the graphical representation and lexical paraphrasal of system predicates, for use in the graphics interface and text generation programs to be used in validation (cf. [LKM89]);
- early indication of inconsistencies via 'incremental compilation': not only are axioms identified but their effect can be tested by constructing a partial model of the system.

There are, though, still great lacunae in both theory and practice between what we can handle and what we must be able to handle. Consider the following extract from the description of the blocks world in [GN87]:

> ...there are just three blocks. Each block can be somewhere on the table or on top of exactly one other block. Each block can have at most one other block immediately on top of it. ...for all blocks x [and] y ...If x is on y and x is clear, then, after an *unstack* operation, x is on the table and y is clear ...

Ideally, the first sentence would be recognized as a definition of *sorts*, the second and third as axioms, and the fourth as an action, which an 'intelligent' assistant could 'animate' by giving appropriate arbitrary values to x and y and building a model. However, many issues are immediately raised in trying to automatically process the discourse above, even before an attempt to extract the sorts and axioms and animate them (for example, that in this context 'clear' means 'having no blocks on'; the vacuous meanings of 'just' and 'somewhere'; etc.). The situation is exacerbated when we consider less tightly constrained, more 'real-life' examples, and a working solution now way well lie in the direction of 'controlled language' (i.e. disallowing such vague expressions) rather than ever more sophisticated text processors.

5.2 Final Comments

In this paper, we have described methods for model building with free variable **KE**, logical animation based on model building, and discourse generation not just for stating the results of reasoning, but, in conjunction with graphical descriptions, also to help *explain* the results of reasoning (a major challenge that we are currently investigating is to how explain the proofs that led to these results). This research has illustrated the ideas that:

- natural language software tools, plus graphical interfaces, used for both input and output (cf. [LKM89, RK94]) can help to alleviate complexity, and can be used to make requirements specifications more accessible;
- the synthesis of logic, language and graphics is important for enabling interactive 'intelligent systems' to reason about change, to explain changes, and to show the effects of change.

Though we still have some way to go in scaling up and integrating our prototypes, we believe this combination of Artificial Intelligence technologies can accompany the static illustration of specifications – as in, for example, a recent application of virtual reality – to enable a more dynamic interaction between a system designer and its specification to take place *via* an 'intelligent assistant'.

References

[AC91] W. Atkinson and J. Cunningham. Proving Properties of a Safety-Critical System. *IEE Software Engineering Journal*, 6(2):41–50, 1991.

[CCB90] M. C. Costa, R. J. Cunningham, and J. Booth. Logical Animation. In *Proc. of 12th International Conference on Software Engineering*, pages 144–149, Nice, 1990. IEEE Computer Society Press.

[Cox94] B. Cox. Communicating conceptual integriy in distributed systems through intelligent assistance. *Omega*, 22(2):113–122, 1994.

[DM94] M. D'Agostino and M. Mondadori. The Taming of the Cut. *Journal of Logic and Computation*, 4:285–319, 1994.

[Fit90] M. Fitting. *First-Order Logic and Automated Theorem Proving*. Springer-Verlag, 1990.

[GN87] M. Genesereth and N. Nilsson. *Logical Foundations of Artificial Intelligence*. Morgan Kaufmann, 1987.

[Hin88] J. Hintikka. Model Mimimization – An Alternative to Circumscription. *Journal of Automated Reasoning*, 4, 1988.

[HSE87] H. Horai, M. Saeki, and H. Enomoto. Specification-based software development system Pure TELL. Research Report 73, Fujitsu Ltd, International Institute for Advanced Study of Information Science, 1987.

[HZ93] H. Horacek and M. Zock, eds. *New Concepts in Natural Language Generation*. Pinter, 1993.

[LKM89] J. Lee, B. Kemp, and T. Manz. Knowledge-based graphical dialogue: A strategy and architecture. In *Esprit-89 Conference Proceedings*. Kluwer, 1989.

[LS87] J. Larkin and H. Simon. Why a diagram is (sometimes) worth 10,000 words. *Cognitive Science*, 11:65–99, 1987.

[Mai87] T. Maibaum. A Logic for the Formal Requirements Specification of Real-Time Embedded Systems. Alvey Forest Project Deliverable R3, Dept. of Computing, Imperial College, 1987.

[MEF+90] K. McKeown, N. Elhadad, Y. Fukumoto, J. Lim, C. Lombardi, J. Robin, and F. Smadja. Natural language generation in COMET. In R. Dale, C. Mellish, and M. Zock, editors, *Current Research in Natural Language Generation*, pages 103–138. Academic Press, 1990.

[PC95] J. Pitt and J. Cunningham. Theorem proving and model building with the calculus **KE**. To appear in the Bulletin of the IGPL, 1995.

[PCK94] J. Pitt, J. Cunningham, and J. H. Kim. Cooperative answering to natural language email queries. In F. Anger, R. Rodriguez, and M. Ali, editors, *Proc. 7th Intnl. Conf. on Industrial and Engineering Applications of Artificial Intelligence and Expert Systems*, pages 273–281. Gordon and Breach Science Publishers, 1994.

[QGLS85] R. Quirk, S. Greenbaum, G. Leech, and J. Svartvik. *A Comprehensive Grammar of the English Language*. Longmans, 1985.

[RK94] R. Rajagopalan and B. Kuipers. The figure understander: A system for integrating text and diagram input to a knowledge base. In F. Anger, R. Rodriguez, and M. Ali, editors, *Proc. 7th Intnl. Conf. on Industrial and Engineering Applications of Artificial Intelligence and Expert Systems*, pages 273–281. Gordon and Breach Science Publishers, 1994.

[Swa83] W. Swartout. The GIST Behavior Explainer. In *Proceedings of AAAI-83*, pages 402–407, Washington, 1983. Kaufmann.

Appendix: KE Specification

A formal specification of the **KE** rules used in our implementation is given in figure 7.

A branch of a **KE**-tree is represented by a 4-tuple $(\Delta,\Lambda,\Gamma,n)$ where Δ is the set of unanalysed formulas and γ formulas that have been analysed x times, Λ is the set of analysed formulas, Γ is the set of formulas that have been analysed $x + 1$ times, and n is the number of times more that the γ rule can be applied on the branch.

Something akin to Smullyan's unifying notation is used to represent formulas more compactly:

α	α_1	α_2
$a_1 \wedge a_2$	a_1	a_2
$\neg(a_1 \rightarrow a_2)$	a_1	a_2^c
$\neg(a_1 \vee a_2)$	a_1^c	a_2^c
$\neg\neg a$	a	

β	β_1	β_2
$\neg(b_1 \wedge b_2)$	b_1^c	b_2^c
$b_1 \rightarrow b_2$	b_1^c	b_2
$b_1 \vee b_2$	b_1	b_2

η	η_{11}	η_{12}	η_{21}	η_{22}
$e_1 \leftrightarrow e_2$	e_1	e_2	e_1^c	e_2^c
$\neg(e_1 \leftrightarrow e_2)$	e_1	e_2^c	e_1^c	e_2

so that in Rules **A**, **B**, **Bs** and **E** the following schemata are used, where $i \in \{1,2\}$ and if $j = 1$ then $k = 2$ and vice versa:

$$\text{Schema A: } \frac{\alpha}{\begin{matrix}\alpha_1\\ \alpha_2\end{matrix}} \qquad \text{Schema B: } \frac{\begin{matrix}\beta\\ \beta_j^c\end{matrix}}{\beta_k}$$

$$\text{Schema Bs: } \frac{\begin{matrix}\beta\\ \beta_i\end{matrix}}{} \qquad \text{Schema E: } \frac{\begin{matrix}\eta\\ \eta_{ij}\end{matrix}}{\eta_{ik}}$$

The proof procedure used in the KE Model Builder is shown in figure 8.

Move Literal:

$$\frac{\Delta \cup \{l\}, \Lambda, \Gamma, n}{\Delta, \Lambda \cup \{l\}, \Gamma, n}$$

Closure:

$$\frac{\Delta, \Lambda \cup \{l, l^c\}, \Gamma, n}{\Box}$$

Rule A:

$$\frac{\Delta \cup \{\alpha\}, \Lambda, \Gamma, n}{\Delta \cup \{\alpha_1, \alpha_2\}, \Lambda \cup \{\alpha\}, \Gamma, n}$$

Rule Bs:

$$\frac{\Delta \cup \{\beta\}, \Lambda, \Gamma, n}{\Delta, \Lambda \cup \{\beta\}, \Gamma, n} \qquad \text{if } \beta_i \in \Delta \cup \Lambda \cup \Gamma$$

Rule B:

$$\frac{\Delta \cup \{\beta\}, \Lambda, \Gamma, n}{\Delta \cup \{\beta_k\}, \Lambda \cup \{\beta\}, \Gamma, n} \qquad \text{if } \beta_j^c \in \Delta \cup \Lambda \cup \Gamma$$

Rule E:

$$\frac{\Delta \cup \{\eta\}, \Lambda, \Gamma, n}{\Delta \cup \{\eta_{ik}\}, \Lambda, \Gamma, n} \qquad \text{if } \eta_{ij} \in \Delta \cup \Lambda \cup \Gamma$$

Rule D:

$$\frac{\Delta \cup \{\exists x.\phi\}, \Lambda, \Gamma, n}{\Delta \cup \{\phi[x/sk_i(X_j)]\}, \Lambda \cup \{\exists x.\phi\}, \Gamma, n}$$

for sk_i a new skolem constant, X_j the free variables in ϕ

Rule G:

$$\frac{\Delta \cup \{\forall x.\phi\}, \Lambda, \Gamma, n}{\Delta \cup \{\phi[x/X_i]\}, \Lambda, \Gamma \cup \{\forall x.\phi\}, n-1} \qquad \text{if } n > 0, \text{ and for } X_i \text{ a new free variable}$$

PB:

$$\frac{\Delta, \Lambda, \Gamma, n}{\Delta \cup \{\beta_i\}, \Lambda, \Gamma, n \mid \Delta \cup \{\beta_i^c\}, \Lambda, \Gamma, n} \qquad \text{if } \beta \in \Delta \text{ or } \eta \in \Delta \text{ (and let } \beta_i = \eta_{ij})$$

Restart Gamma:

$$\frac{\Delta, \Lambda, \Gamma, n}{\Delta \cup \Gamma, \Lambda, \varnothing, n} \qquad \text{if } n > 0, \text{ and there is no } \gamma \in \Delta$$

Figure 7: *Formal Specification of free variable* **KE**

1 *if one of rules* ***move literal, A, Bs, B, E, D*** *or* ***G*** *applies to the first formula in* Δ*, apply it;*
  1.1 *if* ***closure*** *applies,* end with close*;*
  1.2 *otherwise repeat from* 1*;*
2 *if there is a* β *or* η *formula in* Δ*, apply* ***PB****;*
  2.1 *process both branches (i.e. start from* 1 *with each branch):*
3 *if* $n > 0$ *apply* ***restart gamma*** *and repeat from* 1*;*
4 *if* $n = 0$, end with open.

Figure 8: *Proof Procedure for the* KE Model Builder

ESTIMATING FUZZY PHONEME SIMILARITY RELATIONS FOR CONTINUOUS SPEECH RECOGNITION

Hajin Yu, Sung-Joo Kim, and Yung-Hwan Oh
Dept. of Computer Science, KAIST
373-1, Kuseong Yuseong Taejon 305-701 Korea
Email : hjyu@adam.kaist.ac.kr

ABSTRACT

In this research, the fuzzy similarity relation of phonemes is evaluated and it is applied to continuous speech recognition. We propose a method to evaluate the similarity of each pair of Korean phonemes based on the similarity of articulation of phonemes. The similarities of place and manner of articulations of phoneme pairs are estimated and then the results are combined to calculate the similarities of phonemes by using some fuzzy operations. We derive an algorithm to apply the result to word spotting in continuous speech, and test the algorithm in a train ticket reservation system.

1 Introduction

Human speech is very ambiguous, but humans can communicate with it without any difficulty. The languages themselves can be ambiguous in grammar, a word can have several different meanings, and the phonemes are not always pronounced clearly. To deal with the vague data in computers, the computers should be able to deal with the fuzziness in the similar way as humans do. In this point, fuzzy concept can be a good way of solving many ambiguity problems in speech understanding.

We have built an expert system for continuous speech recognition using fuzzy reasoning[YOYM93]. Figure 1 shows the structure of the system. In the system, speech is first segmented into subword units, and then the phonemes are recognized by using fuzzy rules. As the result, we can get a string of phonemes. Then, the phoneme string is matched with the words in the word dictionary. The phonemes are, however, not always recognized correctly, and some of the phonemes can be recognized as the similar phonemes. Some phonemes in human speech are confused very often, and even humans can not discriminate all the phonemes exactly in conversation. Thus, the machine may not be able to find the word that match exactly with the phoneme string, the result of phoneme recognition.

As a solution of the problems described above, we can store all the variations of the pronunciations in the dictionary. In that case, the size of the dictionary becomes too large, which makes the development of the system too expensive. Moreover, we can hardly predict all the variations that may occur in reality. However, if we can find a word that has the similar phonemes with the recognized phonemes, we do not have to describe all the variations in the dictionary. When we can not search the word that match exactly with the recognized phoneme string, we can search the word that is similar to it. To find similar phoneme string, we have to know which phonemes are similar to each other. Hence, we have to calculate the similarity between the phonemes.

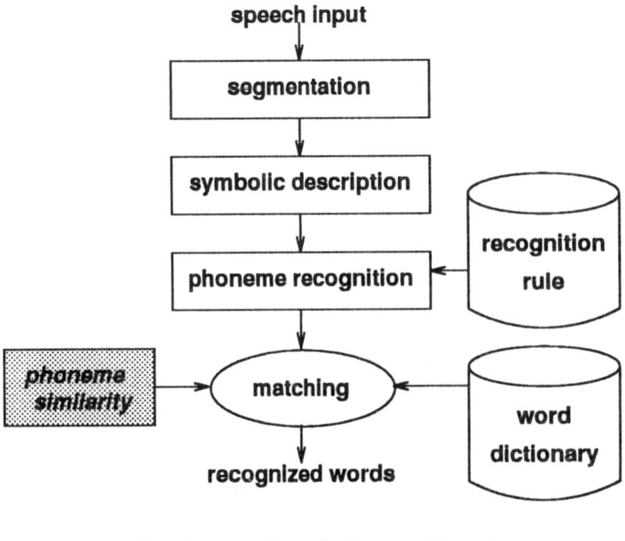

Figure 1 Continuous Speech Recognition System

We use 33 Korean phonemes in our system. Hence, we have to make a table of the similarities with more than 500 pairs of the phonemes. The problem here is that the number of entries of the table is too large to estimate. If an expert of phonetics just determines the similarities of the phonemes and fill the table, the result can be inconsistent, that is, we can not judge all the similarities of the phonemes fairly. We know which phonemes are similar, but we can not score it in a proper way. Another approach to estimate the phoneme similarities may be calculating the distance of the parameters of the phonemes. However, it is not always possible to segment phonemes exactly from speech data, and to collect phoneme speech data from various context.

In this research, we propose a method to evaluate the similarity relations of the phonemes based on the articulatory features which constitute the phonemes. We decompose the phonemes into articulatory features, and then we evaluate the similarities based on the similarity of the articulatory features by using fuzzy operations. We applied the result to word spotting in continuous speech recognition. This paper is organized as follows. Section 2 briefly Reviews the articulatory features of speech. Section 3 presents the idea of evaluating the phoneme similarities based on the basic features. Section 4 describes the algorithm to apply the result of Section 3 to word recognition in continuous speech and the results of the experiment. The summary and conclusions of the paper are given in Section 5.

2 Articulatory Features of Phonemes

We can discriminate the phonemes using the articulatory features. Vowels are usually described in terms of three features [Lad82] : (1) the height of the body of the tongue, (2) the front-back position of the highest part of the tongue, (3) and the degree of lip rounding. The articulatory features of consonants are [Pic80] : (1) the manner of articulation, (2) the voicing feature, (3) the place of articulation along the vocal tract.

In our study, we describe the vowels and the consonants in the same way to deal with them consistently. We divide the basic articulatory features of the phonemes into place of articulation and manner of articulation. The height and position of the tongue in the articulations of vowels are treated as the place of articulation as in the articulations of consonants. The voicing feature is considered as a factor of manners, and vowel type is regarded as a kind of manner of articulation. The rounding of a vowel is not included in the two parts, but it gives an additional effect to the similarities of the vowels as explained in section 3.3.

2.1 Place of Articulation

The parts of the vocal tract that can be used to form sounds are called articulators[Lad82]. Some articulators come close together causing the disturbance to the airstream in the production of phonemes. In order to form consonants, the airstream through the vocal tract must be obstructed in some way. The place of the obstruction is called place of articulation. For example, in a consonant sound /g/, the back of the tongue is raised so that it touches the velum. Vowel sounds may be classified into front vowels and back vowels by means of the position of the highest point of the tongue. They are in turn may be characterized by the height of the tongue. For example, the vowel /a/ is classified as low back vowel.

In our study, we describe phonemes in terms of horizontal places and vertical places. Horizontal place is the position in the vocal tract where the articulation occurs, and vertical place is the degree of the closing of the articulator or the height of the tongue. Table 1 shows Korean phonemes classified according to the horizontal and vertical places of articulations. In the table, (R) and (U) describe if the vowels are rounded or unrounded.

2.2 Manner of Articulation

The shape of the vocal tract changes dynamically in time to produce a consonant. Therefore, consonants can be classified according to the movements of articulators. For example, in the production of an affricate /ʧ/, the tongue comes up first to make contact with the back part of the alveolar ridge to form a stop closure. This contact is then slackened so that there is a friction at the same place of articulation [Lad82]. Each of the change of the articulators is called event, and the manner of articulation can be described by series of the events. For example, the manner of articulation of the phoneme /ʧ/ is the series of the events,

$$\text{closure} \rightarrow \text{friction} \rightarrow \text{aspiration}$$

Table 1 Korean phonemes

	labial	alveolar	palatal	velar	pharyngeal
obstruction	ㅂㅃㅍ: plosive ㅁ:nasal	ㄷㄸㅌ: plosive ㄴ:nasal ㄹ:lateral ㅅㅆ:fricative	ㅈㅉㅊ: affricate	ㄱㄲㅋ : plosive ㅇ :nasal	
high		ㅣ		ㅜ(R) ㅡ(U)	ㅏ
middle		ㅔ			ㅓ(U) ㅗ(R)
low		ㅐ			

Table 2 Similarities of Horizontal Places of Articulation

	LAB	ALV	PAL	VEL	PHA
LAB	1.0	0.3	0.1	0.0	0.0
ALV	0.3	1.0	0.8	0.5	0.1
PAL	0.1	0.8	1.0	0.5	0.3
VEL	0.0	0.5	0.5	1.0	0.5
PHA	0.0	0.1	0.3	0.5	1.0

LAB : labial VEL : velar
ALV : alveolar PHA : pharyngeal
PAL : palatal

Table 3 Similarities of Vertical Places of Articulation

	obstruction	high	middle	low
obstruction	1.0	0.1	0.0	0.0
high	0.1	1.0	0.6	0.3
middle	0.0	0.6	1.0	0.7
low	0.0	0.3	0.7	1.0

3 Evaluating the similarities

The phonetic values of phonemes are decided by place and manner of articulations. Hence, we can evaluate the similarities of phonemes from the similarities of the articulatory features of the phonemes described in Section 2. To evaluate the similarity of the places of articulations, we can consider the distance between the articulation places. We can evaluate the similarity of the manners of articulations by using the similarity of the acoustic events that constitute the articulations.

3.1 The similarity of place of articulation

Table 2 and 3 show the similarities of vertical and horizontal places of articulations. We can simply fill the table by using the distance of the places. For example, labial and pharyngeal are at a longest distance from each other. The place similarity can be defined to be a function that maps two phonemes A and B into the interval from 0 to 1. Let $S_{horizontal}$(A,B) and $S_{vertical}$(A,B) be the similarities of the places of articulations of phonemes A and B, in horizontal and vertical respectively. Then using the table, we can define

$$S_{place} : A \times B \rightarrow [0,1] \tag{1}$$

$$S_{place}(A,B) = S_{horizontal}(A,B) \wedge S_{vertical}(A,B) \tag{2}$$

Here, minimum is used for the conjunction. The reason why we use the conjunction of the two value is that, humans feel two phonemes to be different if one of the two places are different. The vertical places of the two vowels /i/ and /a/ are the same, but the horizontal places of them are different, and humans can discriminate them easily.

For example, the horizontal place of the phoneme /ʧ/ is palatal, and the horizontal place of the phoneme /s/ is alveolar. The similarity of palatal and

Table 4 Similarities Between Acoustic Events

	CN	CL	NC	LC	FL	FR	AS	VO
CN	1.0	0.7	0.2	0.2	0.1	0.0	0.0	0.0
CL	0.7	1.0	0.2	0.2	0.2	0.1	0.1	0.0
NC	0.2	0.2	1.0	0.7	0.6	0.0	0.3	0.6
LC	0.2	0.2	0.7	1.0	0.9	0.0	0.3	0.6
FL	0.1	0.2	0.6	0.9	1.0	0.0	0.2	0.5
FR	0.0	0.1	0.0	0.0	0.0	1.0	0.5	0.1
AS	0.0	0.1	0.3	0.3	0.2	0.5	1.0	0.5
VO	0.0	0.0	0.6	0.6	0.5	0.1	0.5	1.0

CN : closure without explosion FL : flap
CL : closure FR : fricative
NC : nasal closure AS : aspiration
LC : lateral closure VO : vowel

alveolar is 0.8, and the vertical place is the same, because the two phonemes are consonants, which makes the similarity to be 1.0. Then the similarity of places of the two phonemes becomes the minimum of the two, 0.8.

3.2 The similarity of manner of articulation

The similarity of the manners of articulations can be evaluated from the similarity of the acoustic events. The similarity of the acoustic events can be evaluated based on the acoustic resemblances such as the state of the vocal tract, the degree of obstruction and the sonority of the events. Table 4 shows the similarity of the events.

A phoneme consists of different number of events, so we use dynamic matching[SN85] to compare the series of the two events of the phonemes. Let $a_1, a_2, ..., a_I$ and $b_1, b_2, ..., b_J$ be the event series of phonemes A and B, respectively. Then we find the nonlinear expansion of one of the event series which maximize the sum of the similarity of the matched pairs. The manner similarity between A and B is defined by S_{manner}(A,B),

$$S_{manner} : A \times B \rightarrow [0,1] \tag{3}$$

$$S_{manner}(A,B) = \max_F \sum_{k=1}^{K} S_{event}(c_k)/K, \tag{4}$$

where c_k = (i,j) is the combination of a_i and b_j for $1 \leq i \leq I$ and $1 \leq j \leq J$, $S_{event}(c_k)$ the similarity

		/s/	
		friction	aspiration
/ʧ/	closure	**0.1**	0.1
	friction	**1.0**	0.6
	aspiration	0.6	**1.0**

Figure 2 Dynamic matching of the series of acoustic events

between the two events a_i and b_j, F represents a series of c_k, $c_1 = (1,1)$ and $c_K = (I,J)$. The relation between c_k and c_{k-1} is limited as

$$c_k = c_{k-1} + (1,0), \tag{5}$$
$$= c_{k-1} + (0,1), \tag{6}$$
$$= c_{k-1} + (1,1) \tag{7}$$

This relation satisfies the monotonically increasing condition of i and j.

In the previous example, the phoneme /ʧ/ is composed of the three events : closure, friction and aspiration, and the phoneme /s/ is composed of the two events : friction and aspiration. Figure 2 shows the example of matching the two events series. In the figure, the real values are the similarities of the pairs of events S_{events}, and the matched events pairs are marked by boxes. As the result, the similarity of the manner of articulation is the sum of the three similarities of the pairs divided by 3, which becomes,

$$S_{manner}(/ʧ/, /s/) = (0.1 + 1.0 + 1.0)/3 = 0.7 \tag{8}$$

Here, we used average of the similarities of the events to calculate the similarity of the manner of articulation, assuming all the events characterize the property of a phoneme equally.

3.3 The similarity of phonemes

The similarity of two phonemes is the weighted sum of the two similarities, place similarity and manner similarity. The similarity of two phonemes A and B is a function $S_{phoneme}$(A,B) that maps the two phonemes into the interval from 0 to 1.

$$S_{phoneme} : A \times B \rightarrow [0,1] \tag{9}$$

$$S_{phoneme}(A,B) = (1-t) \times S_{place}(A,B) + t \times S_{manner}(A,B)) \tag{10}$$

where, $0 \leq t \leq 1$

$t = 0.8$ if A,B are consonants

$t = 0.5$ otherwise

When the two phonemes A and B are consonants, the manner similarity is given more weight than the place similarity, since humans often confuse the phonemes with the same manner of articulation. In the previous examples, the similarity of the two phonemes becomes,

$$\begin{aligned} S_{phoneme}(/\text{ʧ}/,/s/) &= (1-t)\times 0.8 + t\times 0.7 \\ &= 0.2\times 0.8 + 0.8\times 0.7 \\ &= 0.72 \end{aligned} \quad (11)$$

In case of two vowels, the roundings of the vowels give an additional score in the similarity. Only two vowels are rounded in Korean vowels. The rounded vowels are very similar to each other, and they are distinguished clearly from other vowels. Hence, the similarity of two vowels are multiplied by a factor k which is decided according to the roundings of the two vowels :

	Rounded	Unrounded
Rounded	1.5	0.1
Unrounded	0.1	1.0

The similarity of the two vowels A and B becomes :

$$\begin{aligned} S_{vowel}(A,B) &= 1 \wedge k((1-t)\times S_{place}(A,B) \\ &\quad + t\times S_{manner}(A,B))) \end{aligned} \quad (12)$$

4 Application to word spotting

We applied the result of the similarity table to word spotting in continuous speech. We have built an expert system for speaker-independent continuous speech recognition [YOYM93]. The output of the phoneme recognizer is a series of phonemes without any information about the boundaries of words. The task here is to find words from the dictionary which has the same or similar phoneme transcription in the recognized phoneme series. Since it is not always possible to find a word that match exactly with the recognized output, we use the phoneme similarity relation derived in this study. The similarity between a registered word in the dictionary and a part of the phoneme recognition result is calculated by using dynamic matching [SM90] as the algorithm in figure 3. In the algorithm, the matching is based on the phoneme similarities, whereas most of conventional algorithms are based on the distances of the parameters in frames [DPH93]. The algorithm can find the start and end points of words in one scan. In the matching paths, the local constraint considered by Sakoe and Chiba [SM90] is used, but there is not a weight difference of the paths. Figure 4 shows an example of searching words in a sentence. In the figure, the values in the tables show the similarities between the phonemes in the registered word and the phonemes in the input sequence. The matched path is represented by boxes.

Using this algorithm, we can find a word that included in a phoneme sequence quickly. We first consider the computational requirement of the algorithm in comparison with conventional frame-based dynamic time warping (DTW) algorithm. Suppose that the total number of frames of the input sentence is F, and the average number of frames in a word is S, and there are W words in the dictionary, and there are R templates for each word. Then, to match the input sentence and the words by using conventional frame-based DTW, we need $F \times S \times W \times R$ frame comparisons. By contrast, in our system, the input is segmented first and the phonemes are recognized in advance. Recognizing the phonemes needs smaller amount of computation than recognizing the words, since the length of a phoneme is much smaller than that of a word. Suppose that the average number of the frames in the recognition units is M, the total number of units in the dictionary is U, the number of phonemes in a sentence is T, the number of phonemes in a word is N. Then, we need just $M^2 \times U$ of frame comparisons for phoneme recognition by usual DTW and T × N phoneme comparisons for the proposed word search algorithm. Since M is relatively small (about 1/5 of the number of frames in a word), and the number of units is less than 1,000, the number of frame comparison is small and independent of the vocabulary size. The computations for the proposed algorithm is much cheaper than the frame comparisons, since the computations in phoneme comparison are scalar comparisons which are calculated in advance, whereas the calculations in frame comparison are vector comparisons.

Both the proposed algorithm and the conventional DTW algorithm is implemented on a workstation, and it is tested with a continuous Korean speech corpus. Table 5 shows the analysis conditions for the experiments. In the experiments, the number of templates for a reference word for the conventional DTW is five when the word detection rate and false alarm

Variables :

$x_0, x_1, \ldots, x_{T-1}$: recognized phoneme sequence

$w_0, w_1, \ldots, w_{N-1}$:

phoneme sequence of a word in the dictionary

S(t,i) : similarity between x_t and w_i

start_point :

estimated start point of the word in the input

end_point :

estimated end point of the word in the input

n : length of the path

S_{word} : word similarity

Step 1 - Initialization

$\delta_t(0) = 0, 0 \leq t < T$

$\delta_1(0) = 1.0 - S(0,0)$

$\delta_i(0) = \delta_{i-1}(0) + 1.0 - S(0, i-1), 2 \leq i \leq N$

Step 2 - Recursion

For $1 \leq i \leq W, 1 \leq t < T$ **begin**

$$\delta_1(i) = \min \left\{ \begin{array}{l} \delta_{t-1}(i-1) \\ \delta_t(i-1) \\ \delta_{t-1}(i) \end{array} \right\} + 1.0 - S(i-1, j) \qquad (13)$$

$\Psi_t(i)$ = (one of the three paths selected in 13)

end

Step 3 - Find end point

end_point = $\underset{1 \leq t < T}{\text{argmin}} \quad \delta_t(W)$

Step 4 - Backtracking for start point

(t,i) = (end_point, W)

n = 0

A = 0

while $1 < i$ **and** $0 < t$ **begin**

A = A + S(t,i)

n = n + 1

(t,i) = $\Psi_t(i)$

start_point = t

end

Step 5 - Evaluating word similarity

S_{word} = A / n

Figure 3 Spotting a word in a phoneme sequence

Recognized Phonemes : "nailbusaemculbal"

Registered Word : "neil" (tomorrow)

	n	a	i	l	b	u	s	a	e	m	c	u	l	b	a	l
n	1.0	0.4	0.4	0.8	0.3	0.3	0.3	0.4	0.3	0.9	0.3	0.3	0.8	0.3	0.4	0.8
e	0.3	0.6	0.8	0.3	0.1	0.5	0.2	0.6	1.0	0.3	0.1	0.5	0.3	0.1	0.6	0.3
i	0.4	0.6	1.0	0.4	0.2	0.5	0.2	0.6	0.8	0.4	0.2	0.5	0.4	0.2	0.6	0.4
l	0.8	0.4	0.4	1.0	0.3	0.3	0.3	0.4	0.3	0.6	0.3	0.3	1.0	0.3	0.4	1.0

Word Similarity = (1.0+0.6+1.0+1.0)/4 = 0.90

Registered Word : "busan" (name of the city)

	n	a	i	l	b	u	s	a	e	m	c	u	l	b	a
b	0.3	0.1	0.2	0.3	1.0	0.1	0.5	0.1	0.1	0.4	0.7	0.1	0.3	1.0	0.1
u	0.3	0.5	0.5	0.3	0.1	1.0	0.2	0.5	0.5	0.3	0.1	1.0	0.3	0.1	0.5
s	0.3	0.2	0.2	0.3	0.5	0.2	1.0	0.2	0.2	0.2	0.7	0.2	0.3	0.5	0.2
a	0.4	1.0	0.6	0.4	0.1	0.5	0.2	1.0	0.6	0.3	0.2	0.5	0.4	0.1	1.0
n	1.0	0.4	0.4	0.8	0.3	0.3	0.3	0.4	0.3	0.9	0.3	0.3	0.8	0.3	0.4

Word Similarity = (1.0+1.0+1.0+1.0+0.6+0.9)/6 = 0.92

Figure 4 Examples of Fuzzy Word Spotting

Table 5 Analysis conditions for the experiments

Task	train ticket reservation
Vocabulary size	50 words
Sampling rate	16 kHz
Parameter	melcepstrum
Number of speakers	10
Training data	200 sentences (1580 words)
Test data	200 sentences (1210 words)
Phoneme Recognition	knowledge-based

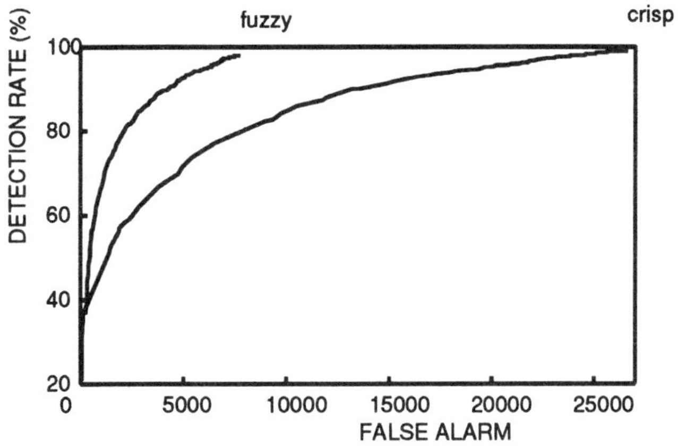

Figure 5 Recognition result

is almost the same with the proposed algorithm. In that condition, the processing time of frame-based DTW was about sixty times of that of the proposed algorithm. The processing time of the proposed algorithm is almost twice as long as real time on Sparc-10 workstation.

Figure 5 shows the recognition rate in comparison with the crisp matching. Here, the recognition result of crisp matching means that the phoneme similarity is set to zero when it is less than 1.0, that is, a word is matched only when all the phonemes in a word are matched exactly in the sentence. As the result shows, the word detection rate is higher when we use the fuzzy matching than that when we use the crisp matching, in the condition of the same number of false alarms.

5 Conclusion

We evaluated the similarity of each pair of 33 Korean phonemes, and used the result to improve the continuous speech recognition system. The phonemes are described by acoustic features, place of articulation and manner of articulation. The similarities of the acoustic features of the phonemes are evaluated, and then the similarities of the phonemes are calculated based on the similarities of the basic features of the phonemes by using fuzzy operations. The result of the similarity table is applied to word spotting in continuous speech with the task of train ticket reservation. As the result, the similarities of the phonemes evaluated by the proposed method have been very useful in searching words in continuous speech. We plan to do some experiments comparing the similarities with the distances of various speech parameters.

References

[DPH93] John R. Deller, John G. Proakis, John H. L. Hansen, *Discrete-time processing of speech signals,"* Macmillan Publishing Company, 1993

[Lad82] Peter Ladefoged, *A Course in Phonetics,* Harcourt Brace Jovanovich, Inc., 1982

[Pic80] J. M. Pickett, *The Sounds of speech communication,* University Park Press, Baltimore, 1980

[SM90] Harvey F. Silverman and David P. Morgan, "The application of dynamic programming to connected speech recognition," IEEE ASSP Magazine, July 1990, pp. 6-25

[SN85] Shuzo Saito, Kazuo Nakata, *Fundamentals of Speech Signal Processing,* Academic Press, 1985

[YOYM93] Hajin Yu, Yung Hwan Oh, Yoichi Yamashita, Riichiro Mizoguchi, "Fuzzy Expert System for Continuous Speech Recognition", Proceedings of '93 Korean/Japan Joint Conference on Expert Systems, 1993, pp 951-965

SPEECH WORD RECOGNITION WITH BACKPROPAGATION AND FUZZY-ARTMAP NEURAL NETWORKS

Lipo Wang
School of Computing and Mathematics
Deakin University
662 Blackburn Road
Clayton, Victoria 3168 Australia
Email: lwang@deakin.edu.au

ABSTRACT

We present the initial results on speaker-independent speech word recognition using a modified version of the QuickProp network, a back-error-propagation variant, and the Fuzzy-ARTMAP network, an ART variant. The TIMIT speech database is used for training and testing. While both networks achieve results better than existing results on speech phonetic recognition, the Fuzzy-ARTMAP network runs much faster, but yields generally inferior results, in comparison to the modified QuickProp network.

INTRODUCTION

Neural networks have shown some promise in speech processing lately [e.g., BDFK92, DG91, MZ91, WC91, DRO91, BF90] and have yielded comparable or better results in comparison with the hidden Markov model (HMM) approaches [e.g., LH89, PC91]. These work are done using phonemes in the TIMIT speech database [ZSG90]. Speaker-independent phonetic recognition accuracy ranging from 40%-80% has been reported.

In this brief paper, we present our initial results on speaker-independent word recognition, also using the TIMIT speech database. We shall use two types of neural networks, i.e., a modified version of the QuickProp network [Fah88], a backpropagation [RM86] variant, and the Fuzzy-ARTMAP network [CG92], an ART [CG86] variant. We deal with speech words, instead of phonemes. Specifically, we use speech data for the 11 words in the following sentence: "She had your dark suit in greasy wash water all year."

SPECIFICATIONS AND RESULTS

The training sets consist of speech data recorded from a total of 400 speakers from 8 main dialect regions in the United States [ZSG90]. The testing sets consist of speech data from 230 speakers not included in the training set.

Since the number of data in a digitised speech recording is usually very large, e.g., it includes about 60,000 data points per sentence, preprocessing of the raw data is necessary before they are suitable for neural networks. We did the following preprocessing with all training and testing data:

(1) The speech signal is high-frequency pre-emphasized with transfer function $(1 - 0.95/z)$;
(2) The speech signal is then windowed using a 32 msec Hamming window, with a 10 ms frame spacing;
(3) The magnitude spectrum is computed using a 512 point Fast Fourier Transform for each frame;
(4) The spectrum is then log amplitude scaled, frequency warped with a bilinear transform with a coefficient of 0.6;
(5) Fifteen sepstral coefficients are then computed over a frequency range of 150 Hz to 6000 Hz.

We have modified the quick-prop network [Fah88] to enable it to output "unclassified" or "don't know" answers. Specifically, we have added the following rules during testing after training. When a testing pattern is presented to the network after training, we calculate a sum of squared error for each output node. If the sum is less than a preset error threshold, the test pattern is considered to be "classified", and it can be either "correctly classified" or "incorrectly classified". Otherwise the test pattern is "unclassified". We note that this classification error threshold is not necessarily equal to the error threshold for training to stop.

After training the two networks with the preprocessing training data, we test the networks with the preprocessed testing data. Both the recognition accuracy and the computer time are recorded. Our work is carried out in a Convex C240 Super Computer. The results are presented in Table 1.

Note that the "time" used above refers to the "user time", rather than the "real time" or the "system time".

The Fuzzy-ARTMAP results in the table are obtained using a 2-voter scheme [CG92]; however, we note that similar results are obtained with a 1-voter scheme.

In the present work, we use a three-layer feed-forward backpropagation network. We chose the number of output nodes to be the same as the number of classes, i.e., 11. We chose the number of input nodes to be the same as the largest number of input data point for a speech word after preprocessing, which depends on dialect region and the sex of the speakers. We use 400 hidden units for all cases, which yields good results compared with other choices for the number of hidden units. We chose the classification threshold ERROR=0.2 and the error threshold for training to stop TOTALERRORT=10.0. We let the maximum number of epox be 400.

CONCLUSIONS

We present the initial results on speaker-independent speech word recognition using a modified version of the QuickProp network, a backpropagation variant, and the Fuzzy-ARTMAP network, an ART variant. The TIMIT speech database is used for training and testing. While both networks achieve results better than existing results on speech phonetic recognition, the Fuzzy-ARTMAP network runs much faster, but yields generally inferior results, in comparison to the modified QuickProp network.

REFERENCES

[BDFK92] Y. Bengio, R. De Mori, G. Flammia, and R. Kompe. Global optimiztion of a neural network-hidden Markov model hybrid. IEEE Trans. on Neural Networks, 3: 252-259, 1992.

[BF90] D. Bijl and F. Fallside. A speech application of a probabilistically conditioned neural network. in C.P. Tsang (ed.) Proc. 4th Australian Joint Conf. Artificial Intelligence: 265-273, 1990.

[CG86] G. A. Carpenter and S. Grossberg. Neural dynamics of category learning and recognition: Attention, memory consolidation, and amnesia. in Brain Structure, Learning, and Memory, J. Davis, R. Newburgh, and E. Wegman (eds.), AAAS Symposium Series, 1986.

[CG92] G.A. Carpenter, S. Grossberg, N. Markuzon, J.H.

Table 1 Performance Comparison on Speech Recognition

Dialect Region & Sex of Speaker	Fuzzy-ARTMAP Classification Accuracy Overall Correct	Fuzzy-ARTMAP Computer Time in Seconds	Modified QuickProp Classification Accuracy Correct in Classified (Overall, Unclassified)	Modified QuickProp Computer Time in Seconds
1 f	81.8%	56.0	90.9% (68.2%, 25.0%)	16343.6
1 m	83.1%	104.3	96.9% (81.8%, 15.6%)	45386.5
2 f	94.3%	77.9	98.6% (80.7%, 18.2%)	30077.9
2 m	87.9%	201.2	96.6% (86.9%, 10.1%)	142878.5
3 f	100.0%	56.5	100.0% (78.8%, 21.2%)	19190
3 m	96.8%	205.9	98.3% (90.9%, 7.5%)	128075.5
4 f	92.0%	73.5	97.1% (75.6%, 22.2%)	12322.9
4 m	94.3%	239.5	98.8% (89.8%, 9.1%)	171333.0
5 f	90.9%	106.1	100.0% (82.6%,17.4%)	38334.3
5 m	90.9%	221.7	98.8%(86.1%, 12.8%)	118969.3
6 f	97.0%	43.1	92.3% (72.7%, 21.2%)	11654.1
6 m	85.2%	93.3	94.7% (81.8%, 13.6%)	33369.6
7 f	94.3%	59.5	98.6% (83.0%, 15.9%)	17742.8
7 m	86.1%	177.3	95.9% (85.5%, 10.9%)	134079.0
8 f	84.8%	26.1	88.9% (72.7%, 18.2%)	4687.4
8 m	84.1%	56.5	97.1% (76.1%, 21.6%)	17037.4

Reynolds, and D.B. Rosen. Fuzzy ARTMAP: A neural network architecture for incremental supervised learning of analog multidimensional maps. IEEE Trans. Neural Networks, 3: 698-713, 1992.

[DG91] M.R. Davenport and H. Garudadri, A neural net acoustic phonetic feature extractor based on wavelets. IEEE Pacific Rim Conference on Communications, Computers and Signal Processing (May 9-10, 1991), 449-452 (1991)

[DRO91] V. Digalakis, J.r. Rohlicek, and M. Ostendorf. A dynamical system approach to continuous speech recognition, IEEE Conf. Proc. CH2977-7/91/0000-0289: 289-291, 1991.

[Fah88] S. Fahlman. Faster-Learning Variations on Back-Propagation: An Empirical Study. in Proceedings of 1988 Connectionist Models Summer School, Morgan Kaufmann, 1988.

[LH89] K.-F. Lee and H.-W. Hon. Speaker-independent phone recognition using hidden Markov models. IEEE Trans. Acoustic, Speech, and signal Processing, 37: 1641-1648, 1989.

[MZ91] H.M. Meng and V.W. Zue. Signal representation comparison for phonetic classification, IEEE Conf. Proc. CH2977-7/91/0000-0285: 285-288, 1991.

[PC91] D.J. Pepper and M.A. Clements. On the phonetic structure of a large hidden Markov model. IEEE Conf. Proc. CH2977-7/91/0000-0465: 465-468, 1991.

[RM86] D.E. Rumelhart and J.L. McClelland. Parallel Distributed Processing: Explorations in the Microstructure of Cognition, I and II, MIT Press, Cambridge MA, 1986

[WC91] J.-X. Wu and C. Chan. Recognition of phonetic labels of the TIMIT speech corpus by means of an atificail neurla netowrk. Pattern Recognition, 24: 1085-1091, 1991.

[ZSG90] V. Zue, S. Seneff, and J. Glass. Speech database development at MIT: TIMIT and beyond. Speech Comm., 9: 351-356, 1990.

Planning and Scheduling

MAINTAINING QUALITY AND QUANTITY PRODUCTION CONSTRAINTS IN A MULTI-AGENT BATCH SCHEDULING ENVIRONMENT

Jim Butler
Knowledge-Based Systems Group, CRA Advanced Technical Development
Bentley, W.A., Australia
Email: jlb@perth.atd.cra.com.au

ABSTRACT

This paper describes a multi-agent architecture and agent language for solving distributed batch production scheduling problems where the satisfaction of quality constraints on the final product is of paramount importance. The problem of rescheduling in the face of unexpected events is addressed through the use of multi-stage negotiation with search as a coordination mechanism for a team of scheduling agents.

1.0 INTRODUCTION

There are numerous production scheduling domains where the production operations are characterised by the combination or blending of batches of product into a final product. The problem is to determine a blending schedule of batches to satisfy customer demand. Demand is defined in terms of a quantity and quality of final product at a requested time. Quality is measured by the percentage of constituent elements composing the product. There are a number of factors contributing to the problem complexity in these domains. The first one is related to the distributed nature of the problem. Often the products to be blended may be drawn from a number of different and possibly physically distributed production processes. When the production scheduling of the batches at each production process is considered together with the blending, the problem's complexity increases. One of the reasons for this is that the processes involved in producing the product may be quite different, and thus have different operational constraints and costs associated with production.

The second contributing factor to problem complexity is the inclusion of quality constraints in production requirements. The production of batches and subsequent combination of batches into target quantities of final product at specified times, where quality constraints on levels of constituent elements are met, is the essence of the scheduling problem in these domains. If quality were not an issue, the problem would be much simpler, since quantities from multiple production processes could be blended in any manner which satisfied the total production quantity requirement.

The dynamic aspects of this class of problem become evident when one of the sources of produced resource fails to deliver a planned quantity at the planned quality. There may be many reasons for the failure, ranging from an unexpected operational breakdown to error in estimating the percentage of constituent elements in a batch. An example of the latter cause may be found in the mine scheduling domain. In the mining of iron ore there is uncertainty associated with the estimation of impurities. When a batch of ore is mined and is to be blended with other batches, more accurate sampling may determine that the quality is not as high as expected. If blending occurs as planned, the quality of the final product will be below specification (assuming all other quality estimates are correct). Therefore, rescheduling must take place, possibly involving a number of production processes, to ensure that the quality constraints are met.

In this paper, an approach for solving the above problem is proposed. It addresses the issues of quality constraints and reactive scheduling requirements and its foundation is upon research carried out in the areas of multi-agent planning and constraint programming. In Section 2, using the constraint programming paradigm, a representation for quality constraints in production scheduling is presented. In Section 3, a distributed solution to the rescheduling problem is outlined and Section 4 concludes this work.

2.0 MODELLING QUALITY CONSTRAINTS

From a production scheduling perspective, when an activity is carried out, resources may be required, provided, consumed or produced. A resource is required when an activity involves the resource (or part thereof) during some part (or all) of its duration and the resource is available for use in other activities when it is no longer required by the original activity. Conversely, a resource is said to be provided when it is made available (possibly for use by other activities) during some part of the execution of the activity. A resource is consumed when the quantity available after the activity has completed is less than that before the activity started. Finally, a resource is produced when the execution of the activity results in more of the resource being available.

A model for a single process which produces a resource is shown in figure 1. This resource is subsequently used as a consumable resource in a blending operation which aggregates this production with those of other processes to meet production objectives. This process has operational constraints which are modelled as required resources with capacity constraints. The required resources are represented by their availabilities which can vary over time. Typically, these required resource constraints correspond to the labour and equipment needed for operations.

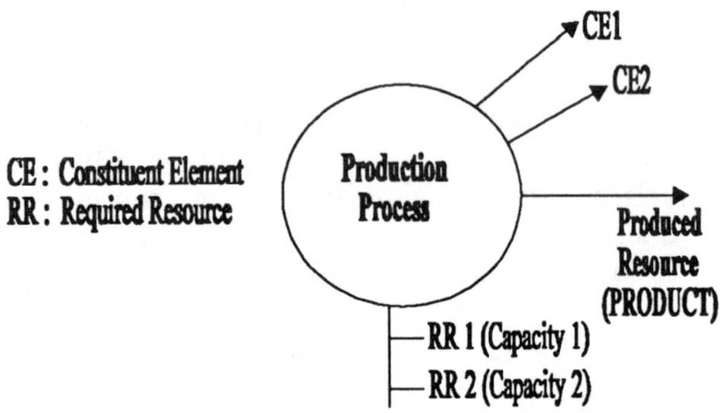

FIGURE 1 Single production process.

From the point of view of this process, a resource is *produced* in the execution of batch activities. A batch activity corresponds to the operation of the production process over some period of time. Each batch has a known quantity and an estimated quality. To model the production at this process, we say it produces a certain quantity of a resource called PRODUCT. PRODUCT is composed of a number of constituent resources, which are the elements of the product. This is modelled by assigning a produced resource to each constituent element of interest in PRODUCT. The levels of constituent elements in PRODUCT may be constant or may vary, depending upon the application domain.

Production requirements, in terms of quantity and quality, are specified for a number of production periods. The quality constraints are specified by the minimum and/or maximum amount of the element which can be produced. Since the quantity of PRODUCT may not necessarily be known beforehand, it is actually necessary to constrain the quantity of constituent elements to be percentages of PRODUCT. With constraint programming tools like ILOG's Schedule, capacities of resources are represented as tables or arrays of constrained variables [Le Pape, 1993]. Thus, one can directly post constraints between the variable corresponding to the quantity of PRODUCT at targeted times and the variable corresponding to the quantity of constituent elements. This ensures the maintenance of the quality constraints of production requirements.

2.1 Example

The modelling may be clarified with an example. In open pit iron ore mine planning and scheduling, one of the problems is to determine the extraction sequence of mining blocks. These blocks may be either iron ore or waste. Currently, techniques such as dynamic programming [Dowd and Onur, 1992], mixed integer programming, and linear programming (LP) [Mann and Wilke, 1992] are being investigated for applicability in open pit mine planning and scheduling. While LP appears best to solve the blending problem, according to [Mann and Wilke 1992], it has not satisfactorily solved the following problems:

- precedence constraints;
- considering equipment movements across the pit; and
- optimising multiple period planning.

The constraint programming approach outlined in this example does not suffer from these drawbacks.

Within a constraint-based formulation of open pit iron ore mine scheduling, there are temporal precedence constraints associated with the physical relationships of the blocks (a block cannot be mined before all blocks above it are mined). Each block is characterised by its mass, grade (percentage of iron ore) and a number of quality parameters (percentages of certain impurities). Certain amounts of required resources (equipment and people) are necessary to mine a block whether it is iron ore or waste and there are capacity constraints on these resources corresponding to their availabilities. Produced resources include the overall product, iron, and a resource for each impurity. If a block contains iron ore, it produces a resource corresponding to its mass, an iron resource of quantity equal to its mass x grade, and impurity resources equal to mass multiplied by impurity percentages. There are production requirements for given time intervals which translate into minimal and maximal capacity constraints on these produced resources. Suppose that there is a production requirement to produce 500,000 tonnes of ore by the end of a production period at a grade of at least 60% and with a maximum silica level of 3% and maximum alumina level of 2%. Then the following constraints would be posted for the time corresponding to the end of the production period:

- the amount of product is constrained to be at least 500,000 tonnes at the end of the period;
- the amount of iron is constrained to be at least 60% of the amount of product;
- the amount of silica is constrained to be at most 3% of the amount of product;
- the amount of alumina is constrained to be at most 2% of the amount of product.

By posting these constraints, the solution space for block extraction sequences will decrease due to constraint propagation involving the temporal precedence constraints between the blocks and the required and produced resource constraints of the blocks. Subsequently, search techniques may be used to compute a complete schedule which satisfies all constraints and optimises a defined objective.

3.0 RESCHEDULING

As mentioned earlier, there may be a high degree of uncertainty associated with the estimates of the level of constituent elements in the batches of each production process. Remembering that each batch fits into a blending plan which has precise quality constraints, it becomes clear that there is a strong requirement for a mechanism to take corrective action when it is discovered that a production process is producing batches with qualities significantly different from the estimates. There are at least three approaches to satisfying this requirement. The first is to attempt to reschedule at the local production process level. For example, in open-pit mining, if the sampled quality starts to degrade, local rescheduling could result in the operations being moved to a deposit whose quality is expected to be higher than that currently being mined. There are advantages and disadvantages with this approach. The search space is smaller when a production process is considered in isolation, however there is also the danger that a solution may not be found. Even if a solution is found, there may be a lower cost solution when other production processes are considered.

Another approach is to consider the schedules of *all* production processes as a global schedule and, accounting for the operational constraints of each production process, compute a new global schedule. This has the advantage of reaching an optimal solution, but suffers computationally due to the increased size of the search space.

The third method is to make the problem-solving control distributed by equipping the decision support system of each production process with communication and coordination abilities. The result is a collection of scheduling agents, with an agent responsible for each production process. A solution is reached through the internal problem-solving techniques of each scheduling agent, together with their collaboration abilities. A problem which originated at one production process can then be resolved by rescheduling at another production process (or possibly a number of processes). A compromise between computation cost and goodness of solution is achieved with this method.

The work upon which this paper is based uses both the first and third approaches. The second approach is not practically feasible when there is a requirement for quick rescheduling. If a solution cannot be found locally (approach one), help is sought externally from the scheduling agents responsible for other production processes (approach three). However, the scope of this paper is confined to a discussion of the third approach. The techniques used for local scheduling (approach one) involve heuristic search or branch and bound in conjunction which constraint propagation.

3.1 Distributed Rescheduling Architecture

Figure 2 illustrates an organisation of scheduling agents into a team for collaborative schedule maintenance. Clearly, an agent needs to have knowledge of its local constraints, which include the resource capacities within its domain and its production requirements which contribute to the joint production requirements upon the team of agents. The notion of a joint production requirement is similar to that of joint responsibility in [Jennings and Mamdani 1992], but more so to that of a joint persistent goal discussed in [Levesque, Cohen and Nunes 1990]. The definition of a joint persistent goal stipulates that when a single agent believes that the goal is impossible, it informs the other team members of this belief, whereby consensus is reached as to whether or not the team believes that the goal is unattainable. This stipulation has its place in joint production requirements as follows. If a single scheduling agent becomes aware of an unexpected event which renders its local production requirement impossible to achieve, it first confers with the other team members to determine, based on the collective knowledge of all agents, if the joint production requirement is now impossible to achieve. In practice, if a single agent is unable to reschedule to accommodate the unexpected event, it is highly likely that corrective action may be taken at other agents to maintain the joint commitment to the joint production requirement.

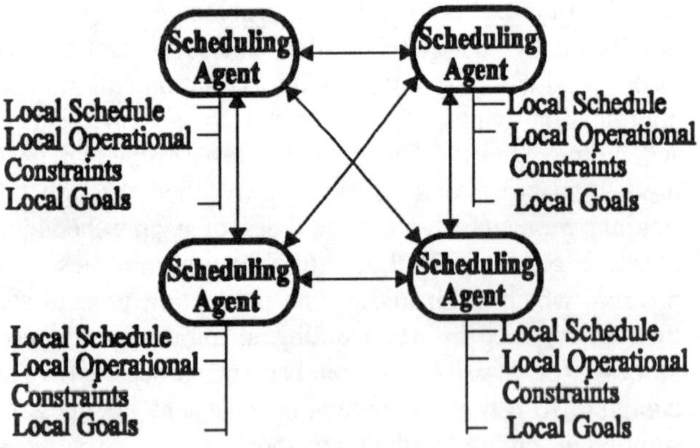

FIGURE 2 Team organisation of scheduling agents.

The nature of the inter-agent communications which maintain joint commitment and thus maintain quality constraints reflects the underlying mechanism for agent coordination. [Durfee, Lesser and Corkhill, 1989] classify approaches to achieving coordination into negotiation, functionally-accurate cooperation, organisational structuring, multi-agent planning, sophisticated local control and theoretical frameworks. An example of negotiation in multi-agent production scheduling is the contract net [Smith, 1980] used in [Butler and Ohtsubo 1992] and [Parunak 1987]. The contract net is appropriate in that domain because there is a predefined task decomposition. In the domain under consideration here, however, the problem is different in that a task is the amount and quality of resource which will be produced at an agent. This is not determined beforehand, but is a result of the distributed problem-solving process.

The coordination mechanism used for this problem domain draws on both distributed multi-agent planning [Georgeff, 1984] and multi-stage negotiation [Conry, Meyer and Lesser, 1988]. In multi-agent planning, it is necessary for each agent to have models of the other agents. This model may take the form of the agents' knowledge, plans, or goals, depending on the level of reasoning about other agents required. In our domain, it is sufficient to model other agents solely in terms of their goals, where a goal is to satisfy a production requirement for a specific amount and quality of produced resource by a certain time. An agent does not need to know how the resource will be produced at other agents, just the amount, quality and time of production.

In multi-stage negotiation, goal decomposition, while not the intended purpose of the negotiation process, is a natural result of it. At any given time, all agents in our domain will have a set of goals to which they are committed. Each goal is the satisfaction of a production requirement by a given time and together with the goals of other agents for the same time, contributes to a joint goal which is related to customer orders for a certain production period. There may be an agent who experiences an unforseen event and determines it is no longer able to satisfy one of its goals. In the communication protocol described below, this agent can request assistance from other agents to attempt to maintain commitment to the joint goal through a re-decomposition of the joint goal. When problem solving has concluded through a staged set of interactions among the agents, this goal will have been decomposed into a number of local goals for the agents which have contributed to the solution. These local goals are merged with the agents' previous goals for the production period, giving each agent a new goal for that production period.

3.2 Inter-Agent Communication Language

The language with which agents can communicate was designed to be as simple as possible while at the same time capturing the expressiveness necessary for this problem domain. The messages which can be sent between agents may be divided into two categories. One category includes all messages which can be involved in a rescheduling session while the other includes those messages related to agents maintaining consistent models of other agents. This second category may be used as a precursor to a rescheduling session, when an agent may want to obtain and reason about information contained within other agents to determine which agents it believes would be appropriate to participate in the session. The issue of team selection, however, is beyond the scope of this paper. The messages belonging to the first category are the following:

- **Exact Production Request** : Directed to any number of agents, indicating that the sending agent requires assistance in producing a specified quantity and quality of resource by a certain time.
- **Approximate Production Request** : Same as above except that the sending agent welcomes responses in which the quantity and quality are not necessarily the same as that requested. It makes provision for the local preferences of responding agents.
- **Production Refusal** : From one agent to another, in response to a Production Request, indicating a refusal to provide any assistance in production. Normally reserved for situations where the agent requires much assistance and is unable to give any itself.

- **Production Offer** : From one agent to another, in response to an Exact Production Request or Approximate Production Request, indicating an offer to produce a certain quantity and quality of resource by a certain time. In the case of response to an Approximate Production Request, the quantity and/or quality may be different to that requested. Additionally, the cost of the extra production to the offering agent, is also included.
- **Offer Accept** : From one agent to another, in response to a Production Offer, indicating acceptance of the offer.
- **Offer Reject** : From one agent to another, in response to a Production Offer, indicating rejection of the offer.

There are a number of noteworthy points regarding this collection of messages. Firstly, the reason that the offered quantity and quality of resource may be different in a Production Offer to that requested in the Approximate Production Request is related to the operating constraints and current schedule of the responding agent. While the agent may not be able to satisfy the request exactly, its cooperative disposition encourages it to attempt to give some assistance. Additionally, if other agents make Production Offers to the initiating agent, the combined effort, after a number of communications, may satisfy the initial request in terms of quantity and quality of produced resource.

Secondly, with both an Exact Production Request and an Approximate Production Request message in the collection, there are two broad classes of strategies which may be developed. With a rescheduling session that is initiated with an Exact Production Request message, the strategy will be such that the initiating agent will dispatch successive varying Exact Production Request messages to a number of other agents. Based on reasoning about the cost components of the responses, the agent will determine the quantity and quality components of subsequent requests, with the aim of reaching a lowest cost solution across all the agents. In essence, this class of strategy is centrally directed despite the fact that it occurs among a group of distributed agents. Alternatively, with a rescheduling session that is initiated with an Approximate Production Request message, a receiving agent is given the opportunity to vary the quantity or quality requested in a manner which suits its current state. For example, an agent which has a preference for maintaining a smooth utilisation of required resources, such as equipment operators, may increase the quantity requested in its response if it believes this will result in a more even distribution of work across its resources. This class of strategy is more distributed in its nature than the previous class. In the next section two instances of each class of strategy will be compared.

The second category of messages currently includes the following:

- **Request Goal** : From one agent to another, requesting the current production target of the receiving agent.
- **Inform Goal** : Directed to any number of agents (usually either broadcast to all when a change occurs, or sent to a single agent in response to a Request Goal message), informing of the current goal.

While not documented here as formal messages, there are also communication primitives to ensure that only one problem at a time is dealt with. Clearly, it is advantageous to combine multiple problems which may all be solved in a single distributed problem solving session, however this will be left to future research.

3.3 Agent Coordination Strategies

The language for communication between agents outlined above may be used to implement a wide range of coordination strategies. Two classes of strategies were outlined previously and in this section an instance from each class will be presented in the form of examples.

The first example uses an instance of the distributed-oriented class of strategies. Suppose a collection of agents share a joint goal of satisfying a given production requirement. Each agent has its own plan (schedule) for achieving its individual goal. This goal is the satisfaction of a local production requirement which contributes to the satisfaction of the joint goal. The local scheduling which occurred to satisfy the goal was based upon certain assumptions about resource availabilities. In the actual execution of the schedule, one of those assumptions may become invalidated due to an equipment breakdown or due to staff becoming unavailable. In this situation, the agent, say Agent1, may try to reschedule to determine a new schedule which continues to meet the production requirements, including quality constraints, for the local goal. If this is not possible, Agent1 may send an Approximate Production Request to all other agents (Agent2, Agent3, Agent4). The quantity of the request is 80 and the quality (assuming a single constituent element) is 90%, which stipulates 90% x 80 = 72 units of the constituent element in the batch. In response to the Approximate Production Request, Agent3 may be under-resourced and would thus respond with a Production Refusal message. However, Agent2 and Agent4 may have excess capacity but not enough to satisfy the request individually. Suppose Agent2 performs local "what if" rescheduling and responds with a Production Offer of 65 units at 95% quality and Agent4 similarly responds with 30 units at 85% quality. Upon receiving these offers, Agent1 calculates that if it were to accept both offers, it would result in 95 units at 92% ((65 x 95%) + (30 x 85%) / (65 + 30)). If Agent1 believes that any variation in quantities of produced resource will not affect the qualities of the respective agent offers, it may choose to respond with an Exact Production Request to Agent2 of 40 units at 95% and one to Agent4 of 40 units at 85% to achieve a total of 80 units at 90%. If both Agent2 and Agent4 respond with Production Offer messages with quantities and qualities equal to those of the respective request messages, then Agent1 can respond with Offer Accept messages to both and the problem is solved.

There are many simplifications in this example. Firstly, cost is not considered. This may be valid, however, when *a* solution is desired as contrasted with the optimal solution. Secondly, the assumption that agents will only vary quantities and not qualities in Production Offers, together with the fact that only one constituent element was considered, reduces the solution space to a single solution (two algebraic equations with two variables). The solution would not be so trivial with significantly more agents involved compared to the number of constituent elements, however this example was chosen merely to demonstrate a strategy.

An instance of the centrally-directed strategies will now be presented through an example. This strategy uses a form of search, where the initiating agent repeatedly queries a set of agents using Exact Production Request messages and uses the costs to determine the quantities to include in the subsequent query. Again assume Agent1 runs into trouble satisfying its goal for the current production period due to an unpredicted event. If it cannot continue commitment to the joint goal to which its goal contributes through local rescheduling, Agent1 may attempt to enlist help from the other agents (Agent2, Agent3, Agent4). Again, suppose the required production assistance is a quantity of 80 and quality (assuming a single constituent element) of 90%. Before sending Exact Production Request messages, it may initially send Request Goal messages to all agents to get a rough idea of the quality of resource that is planned for production. In response to this, suppose the agents respond with Inform Goal messages from which Agent1 deduces that Agent2 is currently producing at a quality of 87%, Agent3 at 89% and Agent4 at 92%. Agent1 then sends each agent an Exact Production Request message with a quantity equal to the total quantity of 80 and qualities which correspond to the respective Inform Goal messages. Agent1's strategy is to try to determine the nature of the cost functions for the different agents, in order to minimise the total cost of the extra production. To do this, it iteratively queries the agents of production costs with the quantity component converging towards a solution. With the Production Offer message contents from the three agents Agent1 can set up a set of linear equations which satisfy the quantity and quality constraints. By dividing the cost by the quantity a linear cost coefficient can be obtained for each agent. An objective function for the total cost can then be set up in terms of the cost coefficients and variable quantities for each agent. Solving the set of equations to minimise the total cost will instantiate the quantity variables. A new set of Exact Production Request messages is then sent out with the new quantities. If the costs which return in the Production Offer messages are different to those obtained from multiplying the cost coefficients by the new quantities, the new costs are then used to calculate new cost coefficients and a new set of equations are solved. This process continues until the cost coefficients are differing less than some specified amount. At this time Offer Accept Messages are sent to all agents and a solution has been reached.

4.0 CONCLUSIONS

A formulation for the batch production scheduling problem which includes quality constraints has been developed within the context of the constraint programming paradigm. Additionally, a multi-agent architecture for maintaining production requirements, including both quantity and quality constraints has been outlined. An inter-agent communication language was designed and two strategies for achieving coordination of agents in multi-agent rescheduling were outlined and contrasted. The distributed strategy is suited to situations where a satisfactory solution is required and local preferences which may not be reflected in the cost should be considered. The centrally directed strategy is useful when a minimum cost solution is strongly desired.

Future research will be directed towards the investigation of other strategies for coordination and experimental analysis where parameters such as bandwidth, speed, and quality of solution are measured. These are the typical parameters used to measure the efficiency of distributed problem-solving techniques.

ACKNOWLEDGEMENTS

The author would like to thank Richard Kowalczyk and Gil Tidhar for their valued contributions to this research.

REFERENCES

BUTLER, J and OHTSUBO, H (1992) ADDYMS: Architecture for Distributed Dynamic Manufacturing Scheduling. In Artificial Intelligence Applications in Manufacturing. MIT Press: 199-214.

CONRY, S E, MEYER, R A and LESSER, V R (1988) Multistage Negotiation in Distributed Planning. In BOND, A H and GASSER, L (Eds) Readings in Distributed Artificial Intelligence. Morgan Kaufman, San Mateo, CA: 367-384.

DOWD, P A and ONUR, A H (1992) Optimising Open Pit Design and Sequencing. In KIM, Y C (Ed) Application of Computers and Operations Research in the Mineral Industry. Port City Press, Baltimore: 411-422.

DURFEE, E H, LESSER, V R and CORKHILL, D D (1989) Trends in Cooperative Distributed Problem Solving. In *IEEE Transactions on Knowledge and Data Engineering*. 1(1): 63-83.

GEORGEFF, M (1984) A Theory of Action for Multi-agent Planning. In *Proceedings of the National Conference on Artificial Intelligence*: 121-125.

JENNINGS, N R and MAMDANI, E H (1992) Using Joint Responsibility to Coordinate Collaborative Problem Solving in Dynamic Environments. In *Proceedings of the Tenth National Conference on Artificial Intelligence*. San Jose, CA: 269-275.

LE PAPE, C (1993) A Universal Constraint-based Representation of Time Tables: Benefits and Costs ... and Benefits. In *Proceedings of the AAII-SIGMAN Workshop on Knowledge-based Production Planning, Scheduling and Control*, IJCAI-93. Chambery, France.

LEVESQUE, H J, COHEN, P R and NUNES, J H (1990) On Acting Together. In *Proceedings of the Eighth National Conference on Artificial Intelligence*: 94-99.

MANN, C and WILKE, F L (1992) Open Pit Short Term Mine Planning for Grade Control - a Combination of CAD Techniques and Linear Programming. In KIM, Y C (Ed) Application of Computers and Operations Research in the Mineral Industry. Port City Press, Baltimore: 411-422.

PARUNAK, H V D (1987) Manufacturing Experience with the Contract Net. In HUHNS, M (Ed) Distributed Artificial Intelligence, Research Notes in Artificial Intelligence. Pitman: 285-310.

SMITH, R G (1980) The Contract Net Protocol: High Level Communication and Control in a Distributed Problem Solver. In *IEEE Transactions on Computers*. C-29(12):1104-1113.

OPTIMAL FLIGHT PATHS FOR AIRCRAFT : A LITTLE KNOWLEDGE GOES A LONG WAY

Mario C. Selvestrel and Simon Goss
DSTO Air Operations Division
PO BOX 4331, Melbourne, Victoria, Australia 3001
Email : simon.goss@dsto.defence.gov.au

Abstract

This paper presents a method for generating optimal flight paths across defended terrain. Constraints to be considered include those due to the aircraft's aerodynamic limits and the avoidance of obstacles. The problem was formulated in a state-space representation and trajectories produced using A* heuristic search. As A* is exponential in space requirements it is often deemed impractical to determine optimal solutions to real problems that have a large search space. Unlike previous work where solution optimality was given up to gain computational feasibility, in this study, optimality of solution was preserved by using novel heuristics, searching over regions of the map rather than through individual grid cells of the map, and intelligent pruning of inferior states.

1. INTRODUCTION

This paper presents a method for generating optimal flight paths across defended terrain. The scenario of interest is that of a tactical aircraft on a low-altitude terrain-following/terrain-avoidance mission. For such a mission, the pilot typically tries to minimise the chance of detection by flying very close to the ground. Proximity to the ground enables the pilot to take advantage of radar shadows provided by hills and ridge lines so that the aircraft is not visible to ground based radars.

A digital terrain map can be viewed as an integer array specifying the altitude of each cell in metres. Each element in the array corresponds to a terrain patch approximately 100 by 100 metres square.

Route planning for low altitude involves detailed trade offs amongst conflicting requirements. The requirements are modelled by constraints on the possible motion. A multi-objective cost function balances the conflicting objectives.

Constraints to be considered include those due to the aircraft's aerodynamic limits and the avoidance obstacles. The trajectories generated must be flyable by an aircraft. For example, an aircraft cannot instantaneously change direction. It must follow a minimum radius of turn that is governed by the aircraft's aerodynamic characteristics. The solution considers hard obstacles, such as the ground, which cannot be passed through, and soft obstacles, such as threats, which can be passed through at increased cost.

Conflicting requirements include trade offs between fuel usage and exposure to threats. For example, to reduce fuel usage one would like to minimise manoeuvring and travel in a straight line at constant altitude and speed. However, to minimise exposure to threats a winding route between hills, behind ridges, and through valleys may be best.

The trajectories generated minimised a cost function equal to the weighted sum of three components. The first component examined the aircraft altitude along the path; the second considered the distance travelled by the aircraft; and, the third considered threats to which the aircraft was exposed. Minimisation of the cost function encouraged the resultant path to be a trade off between short paths over high cost regions and longer paths over lower cost regions.

The problem was formulated using a state-space representation in which changing state corresponds to moving the aircraft through a predefined manoeuvre.

The search algorithm chosen to generate the aircraft flight path was based on the A* heuristic search algorithm. The heuristics determined the way in which the state space is explored. High quality heuristics lead to a reduced search space and faster solutions. The A* algorithm guarantees to find an optimal path, according to a given function, when a path exists and the heuristics used are admissible.

Unlike previous work using A* where solution optimality was given up to gain computational feasibility, optimality of solution was preserved by using novel heuristics, searching over regions of the map rather than through individual grid cells of the map, and intelligent pruning of inferior states.

1.1 PRIOR APPROACHES

Several approaches have been applied to determining optimal flight paths (Krozel 1988, Hwang and Ahuja 1992). However, few have attempted to simultaneously address all the issues outlined above.

Voronoi diagrams have been used to determine paths for aircraft in two dimensions (Krozel and Andrisani 1990) and three dimensions (Aurenhammer 1991). The methods allow paths to be determined quickly, but fail to adequately model the physics of flight or soft obstacles. The rates of turn required by an aircraft to follow the Voronoi diagram vertices are often unobtainable in practice.

Geometric solutions using two dimensional splines (Cheng 1990), or three dimensional Free Space Representation (Rao and Arkin 1989), have addressed the physics of motion but have limited ability to deal with soft obstacles.

Several algorithmic approaches have been applied (Mitchell 1988, Schwartz and Sharir 1988) but these often require special properties between the digital terrain map and the physics of motion. The most common requirement is that the aircraft must be able to move from its current grid cell to any of its adjacent eight cells in one step. This movement may be possible for slow moving aircraft, such as helicopters, but is unsuitable for high speed aircraft which require a large turning radius.

Control theory methods (Menon *et al.* 1991) have been developed but these tend to be computationally expensive.

Several researchers have attacked the problem using state space search or graph search techniques. It is generally accepted that the A* algorithm is one of the most efficient methods for searching graphs. Unfortunately it is hindered by exponential requirements in storage space (Nilsson 1982). In order to avoid these problems, non optimal solutions have been accepted. These were found by using inadmissible heuristics (Bate and Stanley 1988), coarse to fine refinement methods (Sudkamp and Courte 1989) or by pruning of paths believed to be inferior.

The approach presented here is based on A* search, but unlike previous work, optimality of solution was maintained by using novel heuristics, searching through regions on the map rather than individual grid cells and intelligent pruning of paths known to be inferior.

2. PROBLEM FORMULATION

The problem was formulated using a state-space representation in which changing state corresponds to moving the aircraft through a predefined manoeuvre. The aircraft is permitted to fly straight, turn left or turn right (FIGURE 1).

Each manoeuvre corresponds to an arc joining states in the search space (FIGURE 2).

The search graph was generated dynamically and there was no pre-defined representation of the search space. This representation allows the search space to be generated independently of the digital terrain map and physics of motion.

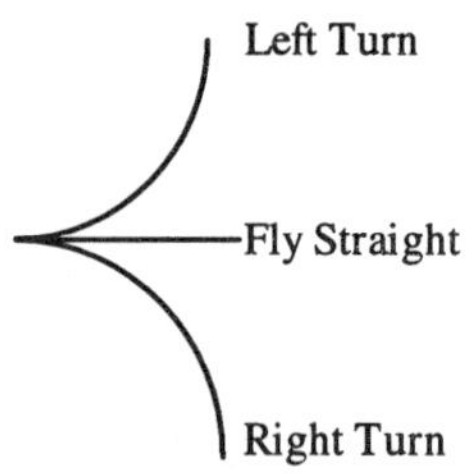

FIGURE 1 Valid Aircraft Manoeuvres

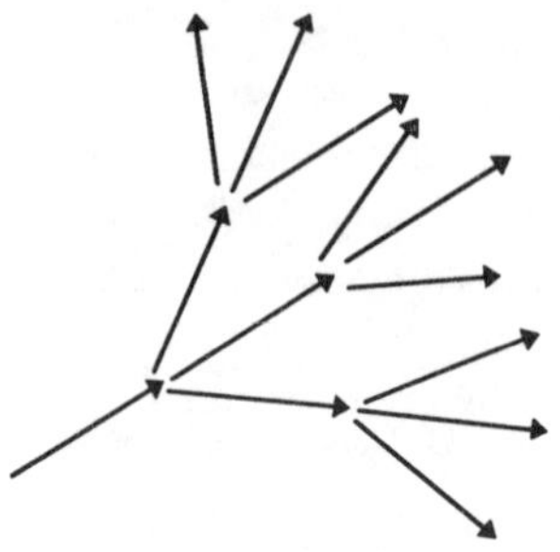

FIGURE 2 A Generated Search Tree

The cost associated with traversing each arc of the graph is a function of the terrain crossed, threats encountered, and the manoeuvre required to cause the change of state. Independence between these components allows different terrain representations, aerodynamic models, search algorithms, and cost functions to be interchanged easily.

3. MODELLING OF THE PHYSICS

The aircraft is modelled by a simple three degree of freedom point mass aircraft model. The aircraft is permitted to move for a fixed period of time at a fixed speed. In the horizontal plane the aircraft is allowed to travel straight ahead, or to maintain a constant acceleration turn to the left or right (FIGURE 1). The aircraft's elevation above the terrain is determined by a simple terrain following-algorithm (Asseo 1988). For the purpose of this work the aircraft speed is 300 knots (approximately 150 m/s), turns are of 3 g and the duration of the manoeuvres are two seconds.

The aircraft's position and heading are maintained in real space (ie physical space). To allow the use of state space search techniques they are then mapped onto a quantised space. The quantised position is described by the digital terrain map grid cells. The aircraft heading is quantised to the nearest 22.5 degrees, that is, there are sixteen quantised headings. States within the same grid cell having the same quantised heading are considered identical. The use of this mapping prevents the accumulation of quantisation errors which often occur with state space search techniques (Mitchell 1988). The basic three manoeuvres result in a search graph with a branching ratio of three (FIGURE 2).

4. THE COST FUNCTION

The cost function comprised the weighted sum of three parts: an elevation component, $C_{elevation}$; an exposure to threat component, C_{threat} ; and a distance travelled component, $C_{distance}$. The ratio between the three weighting factors determines the nature of the path generated. A short straight path is encouraged by giving the distance travelled component a weighting significantly higher than the other weightings. To ensure a path which stays low and avoids high ground the elevation and threat components are given large weightings.

The most significant component of the cost function is that due to aircraft elevation. It is generally accepted that in the absence of specific threat information low flight over low ground is a sensible policy (Wendel et al. 1982). This cost is modelled by the integral of the effective elevation as shown in FIGURE 3.

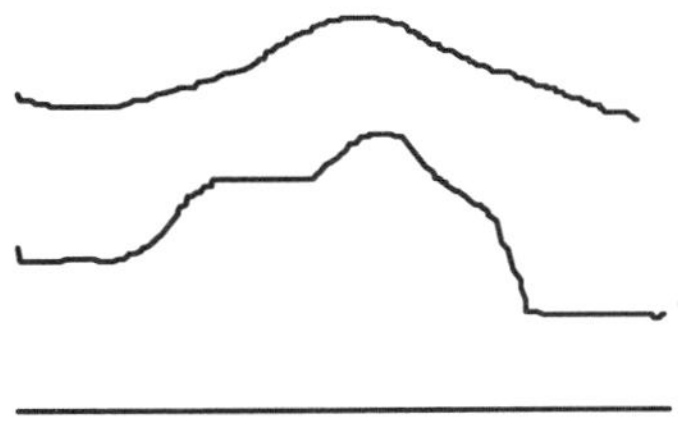

FIGURE 3 Elevation Cost Function

Threats, such as exposure to radar sites, should be avoided to maximise the protection available to an aircraft The threat cost was modelled by increasing the effective elevation of the aircraft when it was exposed to a threat. The effective elevation is used only in the cost function and does not alter the movement of the aircraft. Line of sight calculations and weapon ranges are used to determine if the aircraft is exposed to threats.

To prevent excessively long paths it is necessary to consider the overall distance travelled by the aircraft. This tends to be the least significant component of the cost function.

The suitability of specific cost functions is an operations research issue currently being studied.

5. REDUCING THE SEARCH SPACE

As A* is exponential in space requirements (Nilsson 1982) it is often deemed impractical to determine optimal solutions to real problems that have a large search space.

For example, using the simple three manoeuvres described above over a relatively short path of 150 steps would, in the worst case, require in the order of 10^{71} states to be expanded. Empirical studies across a number of terrain maps using A*, with Euclidean distance as the heuristic, showed the number of states examined to be less than 10^6, an effective branching ratio of less than 1.1 for a given path length.

The search space examined by A* can be reduced by a number of techniques. However, the solutions found may be non optimal. Such solutions can be obtained using inadmissible heuristics (Bate and Stanley 1988) or using a coarse to fine method (Sudkamp and Courte 1989). On the other hand, optimal solutions can be found in a reduced search space, by using good heuristics to guide the search or reducing the path length (ie the number of steps needed to reach the target state) by modifying the operators used to dynamically generate the graph.

The search space can also be reduced by pruning states which are believed to be inferior. This results not only in the saving of storage space for the current state but also for those of its future descendants. If the states pruned are known not to be on the optimal path the solution found will be optimal.

This study is restricted to methods leading to optimal solutions.

5.1 FINDING GOOD HEURISTICS

It has been shown (Pearl 1984) that the closer the estimated heuristic cost is to the actual cost the less nodes will be expanded by the A* algorithm. Pearl (1984) has developed equations to illustrate the effect of various heuristics on the A* algorithm. However, the equations require the analyst to make judgements about the relative and absolute errors of the heuristics used. Unfortunately such parameters are generally too difficult to estimate with any accuracy.

To obtain an empirical understanding for the impact of heuristics on the size of the search space a simple experiment was carried out. Using the Euclidean distance heuristic, h, a number of paths were determined using A*. The algorithm was then re-run with the heuristic reduced by some amount and the number of nodes expanded recorded. For example, a solution was found using h as the heuristic cost estimate, then a solution was found using $0.9h$, $0.8h$, $0.7h$, and so on. Under some conditions a 20% reduction in the heuristic value would result in a 300% increase in the number of nodes expanded. Such observations provided considerable incentives to develop good heuristics.

Good heuristics cause a reduction in the effective branching ratio of the search space. However, in practice it is difficult to develop good heuristics that will reduce the search space without incurring significant computational costs. Typically, the more accurate a heuristic is, the more effort is required to calculate it.

5.1.1 EUCLIDEAN DISTANCE HEURISTIC

The simplest heuristic that can be used in motion planning is the Euclidean distance between the current and goal state. This heuristic assumes the vehicle can instantaneously change direction and move in a straight line at minimum terrain cost from the current position to the goal position with no exposure to threats. This Euclidean distance heuristic ignores several key factors including:

- The physics of motion. An aircraft cannot instantaneously change its direction. It must follow some minimum radius of turn governed by its performance parameters.
- The terrain over which the aircraft moves. If the goal position is in a high cost region, then assuming the aircraft can travel at the lowest terrain cost on the terrain map for the complete journey is an unnecessarily low underestimate.
- Exposure to threats.

5.1.2 PATH OF MINIMUM LENGTH HEURISTIC

The Euclidean distance heuristic can be improved by considering the aircraft radius of turn when estimating the total distance to be covered by the aircraft (FIGURE 4). This improved the distance component of the heuristic but not the elevation or threat components. In practice this heuristic successfully reduced the number of nodes expanded, but took longer to return a solution because of the geometric calculations required.

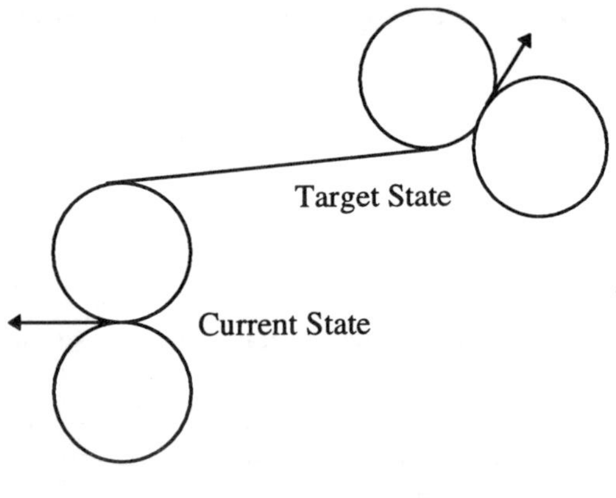

FIGURE 4 Considering Radius of Turn

5.1.3 CONSIDER TERRAIN ELEVATION

Observation of the dynamics of the A* search showed that the simple Euclidean distance heuristic performed exceptionally poorly when the goal state was located in a high cost region. In such cases the heuristic grossly underestimated the cost to completion and thus the algorithm spent significant time exploring low cost areas away from the goal state. This suggested that more emphasis needed to be placed on the costs caused by terrain elevation. This corresponds to looking ahead in the search space in order to help guide the search, the theoretical benefits of which have been illustrated by Kaindl and Scheucher (1992).

The solution adopted was to emphasise the terrain elevation in the heuristic. The cost estimate due to terrain elevation can be increased by considering the terrain the aircraft must fly over in travelling between the current state and goal state.

Radial heuristics from goal state point of view

The poor performance was most pronounced when the goal state was in a high cost region. Particular emphasis was placed on the terrain surrounding the goal. This was done by defining concentric regions extending radially from the goal and determining the minimum cost to cross the regions. The minimum cost to reach the goal was then bounded from below by the sum of the minimum costs of crossing each region (FIGURE 5).

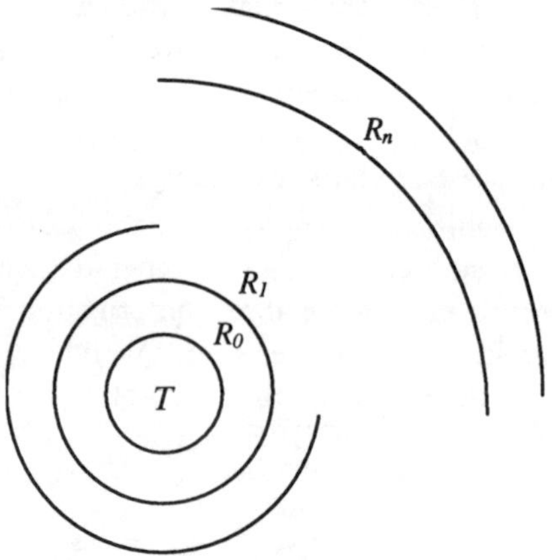

FIGURE 5 Considering Effects of Terrain Near the Goal

A brief description of the technique follows:

1. Define regions, R_i bounded by concentric circles around the goal state.
2. Determine the minimum cost, $c_{min}(R_i)$, obtained when crossing the region.
3. The lower bound cost of moving from the current state, a distance d from the goal, to the goal state is the sum of the minimum cost of each region that must be crossed to reach the goal state, therefore $h = \sum_{i=0}^{d} c_{min}(R_i)$.

This heuristic performs exceptionally well when the goal state is in a high cost region, since it results in an estimate that is substantially higher than the lower bound calculated by the simple Euclidean distance heuristic.

Radial heuristics from goal and current state point of view

In an attempt to improve the radial heuristic defined above, an improved heuristic was developed that considered not only the regions near the goal state but also those near the current state. Thus, if the current state was in a high cost region, future movements required to leave the region would be reflected in the heuristic.

As discussed above, to move from the current state to the goal state the aircraft must cross all the regions radiating from the goal state to the current state. It must also cross all the regions radiating from the current state to the goal state. The overall minimum cost of crossing the regions must be the larger of the two respective minima (FIGURE 6).

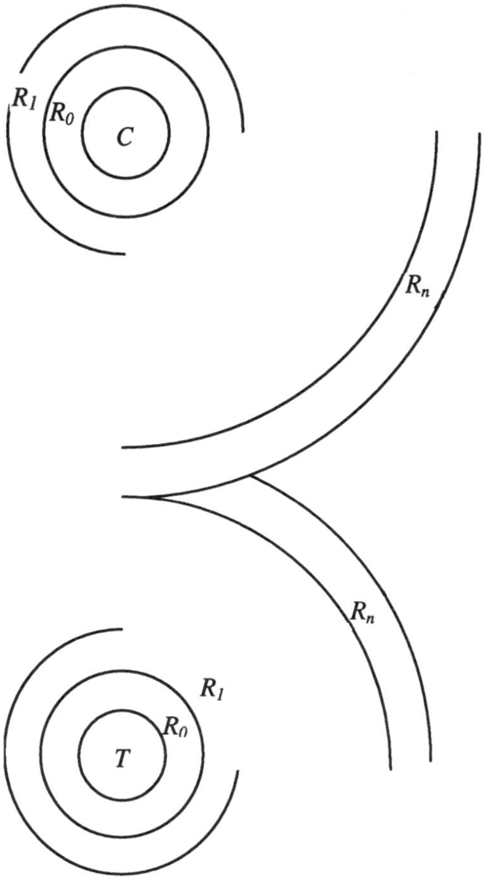

FIGURE 6 Considering Effects of Terrain Near the Current and Goal State

A brief description of the approach follows:

1. Define regions, Rt_i, bounded by concentric circles around the goal state.
2. Determine the minimum cost, $c_{min}(Rt_i)$, obtained when crossing the region.
3. Define regions, Rc_j, bounded by concentric circles around the goal state.
4. Determine the minimum cost, $c_{min}(Rc_j)$, obtained when crossing the region.
5. Determine the region Ro_i, where Rt_i and Rc_j overlap. If the distance between the initial and goal states is l, then $i + j = l$.
6. Determine the minimum cost, $c_{min}(Ro_i)$, of the overlapping regions. This corresponds to the greater of the minimum costs $c_{min}(Rt_i)$ and $c_{min}(Rc_{l-i})$: $c_{min}(Ro_i) = \max(c_{min}(Rt_i), c_{min}(Rc_{l-i}))$.
7. The lower bound cost of moving from the current state, a distance d from the goal, to the goal state is the sum of the minimum cost of each region that must be crossed to reach the goal state, therefore

$$h = \sum_{i=0}^{d} c_{min}(Ro_i) .$$

Although this led to a higher cost estimate than the previous radial heuristic it provided an inconsistent heuristic due to discontinuities in the estimate. The inconsistent h, caused non-optimal solutions to be generated by the A* algorithm.

5.2 SEARCH THROUGH REGIONS

Examining the nodes expanded by A* showed that significant effort was required to generate and store nodes which lay in regions of uniform cost. The cost in moving through such a region is constant per unit distance travelled, regardless of the direction travelled. In such cases the heuristics discussed above were unable to differentiate between *good* and *inferior* child states. This inability effectively led to an exhaustive search in regions of uniform cost.

The notion of regions has been used extensively in two-dimensional binary maps (Holmes and Jungert 1992). A map described by regions is often described as a *tile world* representation. Tile world representations are especially valuable in binary maps where large areas of the search space can be classified as obstacles and do not need to be searched. This is not of direct benefit to the weighted region representation chosen for the current study, but can, nevertheless, be used to guide the search.

The problem caused by exhaustive search of the region is not so much because of expanding the nodes in the region, but rather because of the storage requirement for the nodes within the region. When considering a region only the entry and exit states to the region are significant, as illustrated in FIGURE 7. The intermediate states caused by manoeuvres between entry and exit states can be described by one arc in the search graph and do not need to be stored. This arc must have a parameter associated with it which describes the manoeuvres executed within the region. Nevertheless, the space required by this parameter is significantly less than the space required to store many states.

Another significant benefit of searching through regions is that an accurate representation of areas which should be searched is implicitly built. The costs of nodes in regions which have been searched play a significant role in determining the order in which nodes from the open queue will be expanded. Nodes on the boundary of expensive regions tend to enter the open queue at positions far from the front of the queue and may never be expanded. On the other hand, nodes on the boundary of low cost regions are likely to enter the open

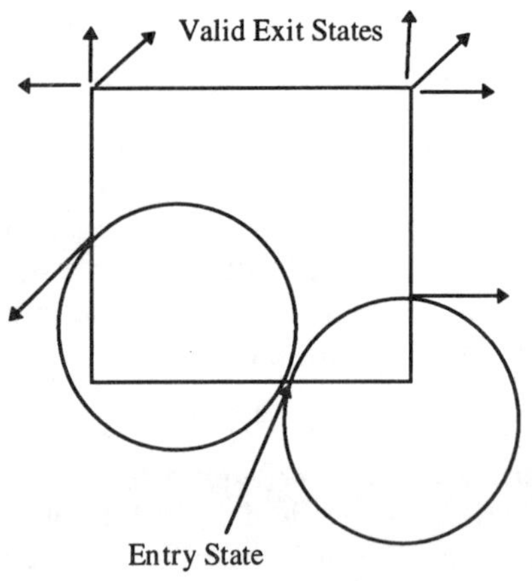

FIGURE 7 Searching Through Regions

queue near the front of the queue and are likely to be expanded quickly.

Searching through regions has the effect of reducing the overall path length, as only one step is required to cross a region that may otherwise require many basic manoeuvres to cross. However, for every entry point to the region there can be several exit points, increasing the branching ratio.

The use of a tile world representation in a weighted region map is extremely sensitive to terrain profile. Tiles represent areas of uniform terrain cost. At the finest granularity, each grid cell on a digital terrain map can be considered a tile. However, to be useful a tile must cover an area that would need several steps to traverse. Tiles which can be crossed in only one or two steps do not provide the savings discussed above. There is little benefit in using a tile world representation in areas of undulating terrain, where the sizes of tiles are small. On the other hand, areas such as plateaus, lakes or wide valleys, where a tile covers a large area of the map, benefit significantly from such representations.

To overcome this dilemma each digital terrain map has a list of regions associated with it. When new states are generated a quick test is made to determine if the state lies within a region. If it does the region is exhaustively searched, otherwise the search continues as usual. For the purpose of this study regions are defined by squares obtained by using a regular quadtree (Foley *et al.* 1990) to describe the terrain map. Further improvements could be obtained by using techniques where regions are defined by more general perimeters such as polygons. A general polygon can cover the same area as several quadtree elements.

5.3 PRUNING

Pruning is the act of removing states from the open and closed list. The benefits of pruning states from the search space have been described by several authors (Pearl 1984, Korf 1988 and Zhang and Korf 1993). Pruning can take several forms. The easiest method is to store only the best *n* states in the open list at any time. States which exceed the limit are removed from the open list without examination. This form of pruning, although quick to execute and easy to implement, may result in non-optimal solutions. States that are on the optimal path can initially enter the open list far from the lists head, and be pruned before they can rise to the head of the list.

However, a simple modification to the A* algorithm allows states, known not to be on the optimal path, to be pruned without affecting the optimal solution. The technique relies on having a known solution path while the A* algorithm is running. The known solution provides an upper bound on the cost of an optimal path. The heuristic cost estimate determined by the A* algorithm provides a lower bound on the cost of an optimal path. Therefore any state with a heuristic cost estimate greater than a known solution cannot be on the optimal path. The state generated to achieve this condition can be removed without affecting the optimality of the solution.

FIGURE 8 presents the pruning A* algorithm. It consists of a typical A* implementation with slight modifications. Step [4] involves calculating a cost from the initial state, *s*, to the target state, *t*, and assigning it to *min_cost*. The minimum cost of a known path (not necessarily optimal) between the starting state and the goal state is now *min_cost*. Step [10] calculates the cost of a path, $c(s,t)$, between *s* and *t* which passes through *n*. If this path costs less than *min_cost*, *min_cost* is set to $c(s,t)$ (steps [11][12]). Step [13] frees the space occupied by states in the open and closed queues which are known not to be on the optimal path. Step [16] prevents states with a heuristic cost greater than *min_cost* entering the open or closed lists.

As with search through regions, this form of pruning is sensitive to terrain profile. It performs especially well when low ground is found between the current state and the goal state. In such cases *min_cost* tends to be relatively low and inferior paths which attempt to explore high cost areas are quickly pruned. On the other hand, if the goal state is surrounded by high cost areas, *min_cost* will generally be larger than $f(n)$, and most states will be entered into the queue, reducing the benefits of pruning. This is a limitation on the available heuristics, not the pruning method.

However, as calculating the cost to goal takes negligible computational resources compared to queue management, it is a useful technique to use all the time. A useful side effect of this form of pruning is to give A* the ability to behave as an *anytime* search algorithm, where the current minimum cost path is known at all times.

```
[1] procedure pruning_A*
[2]  O = {s}
[3]  C = {}
[4]  min_cost = c(s,t)
[5]  while O ≠ {}
[6]    choose state n in O such that
       f(n) is a minimum
[7]    remove n from O
[8]    if n = t
[9]      return path that led to n as
         optimal path
[10]     calculate c(n,t), and
       c(s,t) = g(n)+c(n,t)
[11]     if c(s,t) < min_cost
[12]       min_cost = c(s,t)
[13]       remove all states n in
         O and C with
         f(n) > min_cost
[14]     for each state n' generated
       from n
[15]       calculate h(n'), g(n'), and
         f(n') = g(n') + h(n')
[16]       if f(n) < min_cost
[17]         if n' ⊄ O and n' ⊄ C
[18]           add n' to O
[19]         if n' ⊂ O' as n"
[20]           if f(n') < f(n")
[21]             remove n" from O
[22]             add n' to O
[23]         if n' ⊂ C as n"
[24]           if f(n') < f(n")
[25]             remove n" from C
[26]             add n' to O
[27]     add n to C
[28]   return failure
```

FIGURE 8 The A* Algorithm Modified To Prune Paths Known Not To Be On the Optimal Path

6. RESULTS

This section applies to the above techniques to a number of pathological terrain models and presents a qualitative description of how the techniques perform. The terrain models illustrate the best and worst case conditions for the techniques described above. The terrain models examined include: a plane; the goal located a the top of a hill (ie a cone (FIGURE 9)); the goal located on a plane surrounded by a ridge (FIGURE 10); and the goal located on a plane surrounded by a ridge, with the ridge broken by a saddle (FIGURE 11).

Flat terrain

Both the Euclidean distance and radial heuristics performed well over flat terrain because both heuristics are near ideal.

Flat terrain is effectively one large region containing both the start and goal node which negates the benefits illustrated above.

Pruning is wasteful because it uses CPU time which could otherwise be used by the heuristics. This is a unusual case because of the accurate heuristics.

T on a Hill

The Euclidean distance heuristic is poor and a significant amount of effort is spent searching the low cost ground rather than climbing the hill.

The radial heuristic performs well as it provides a near perfect heuristic.

Using regions with the Euclidean distance heuristic improves performance as the low cost region is explored quickly. The benefit of using regions is not present with the accurate radial heuristic.

Pruning performs poorly in this example because no low cost path exists between the goal and other states, that is the hill must be climbed to reach the goal.

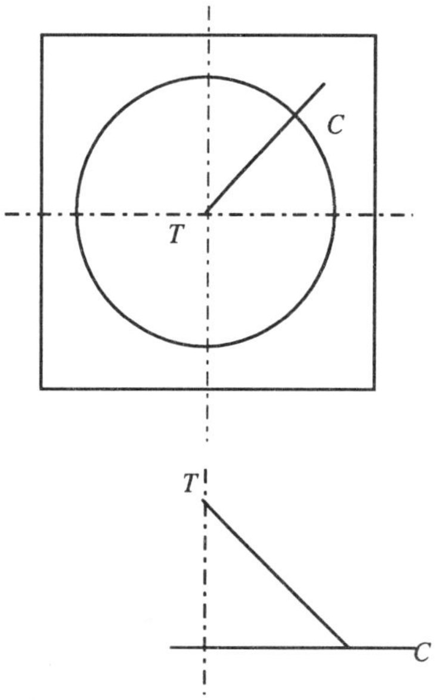

FIGURE 9 ***T*** **on a hill**

T surrounded by a ridge

The Euclidean distance heuristic is poor and a significant amount of effort is spent searching the low cost ground rather than climbing over the ridge.

The radial heuristic performs well as it provides a near perfect heuristic.

Using regions with the Euclidean distance heuristic improves performance as the low cost region is explored quickly. The benefit of using regions is not present with the accurate radial heuristic.

Pruning performs poorly in this example because, once again, no low cost path exists between the goal and other states, that is the ridge must be climbed to reach the goal.

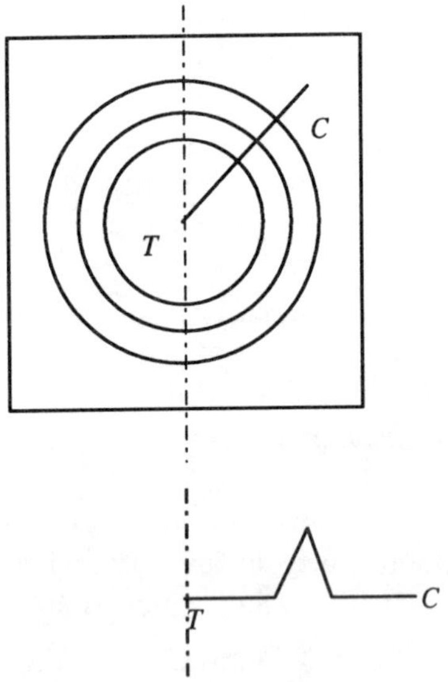

FIGURE 10 ***T* surrounded by a ridge**

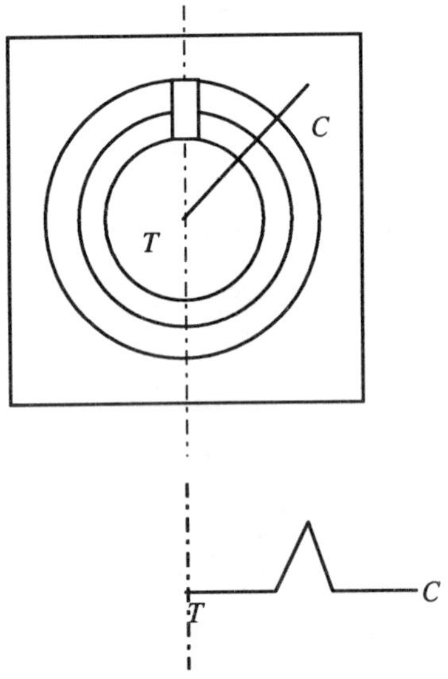

FIGURE 11 ***T* surrounded by a ridge broken by a saddle**

T surrounded by a ridge broken by a saddle

In this case the Euclidean distance and radial heuristic are equal. The radial heuristic must consider all the terrain surrounding the goal because the optimal path may circle the ridge completely until the saddle is found.

Using regions helps expand the low cost regions quickly, thus leading to a faster solution.

Pruning performs well in this example. Once the saddle is found a relatively low upper bound to the cost of the path is also found. This upper bound prevents many states entering the queues.

To determine the effectiveness of the techniques, optimal flight paths were determined over a real digital terrain map for a region approximately 100 km square (1000 by 1000 grid cells). The terrain examined contained several significant features (valleys, ridge lines, lakes and rolling hills). Several starting and target positions were used to expose the algorithm to the terrain features from various orientations. This is required as each technique is sensitive to terrain profile. For example, a valley leading to a target may prove useful, whereas one found across a path to the target provides negligible benefits.

Table 1 summarises empirical results obtained using a SUN sparcSTATION 2 with 32 Mb of RAM and 100Mb of swap space. The results are averaged out over 32 runs.

Table 1 Summary of Result

	Run Time (min)	Nodes expanded	Path length (km)
Euclidean Heuristic	94	1 102 456	193
Consider Heading	106	985 157	193
Search Regions	54	845 987	137
Radial Heuristic	45	534 167	193
Radial Heuristic with Regions	39	498 576	127
Radial Heuristic with Regions and Pruning	21	387 466	127

7. CONCLUSIONS

In practice the radial heuristic proved most useful to determine optimal paths over digital terrain maps. More sophisticated heuristics such as considering the radius of turn required to obtain the target state heading given the current heading proved computationally expensive without significantly reducing the search space.

Examination of various terrain maps show there is little benefit in using a tile world representation in mountainous areas with many small valleys and ridges, whereas areas with plateaus, lakes or wide valleys can benefit greatly from such a representations.

Pruning is also sensitive to terrain profile. Pruning performs poorly if the goal is surrounded by high cost regions. Pruning performs best when a low cost path passes through high cost terrain. However, as calculating the cost to goal requires negligible computational resources compared to queue management it is a useful technique to use all the time. A useful side effect of this form of pruning is to give A* the ability to behave as an anytime search algorithm.

Using a dynamic representation to switch between searching through regions and normal search combined with a radial heuristic and pruning has proved useful to determine optimal solutions for problems which could otherwise not be solved.

8. REFERENCES

Asseo, S. (1988), Terrain Following / Terrain Avoidance Path Optimisation Using The Method of Steepest Descent. Proceedings of the IEEE National Aerospace and Electronic Conference, pp. 1128-1136, Dayton, Ohio.

Aurenhammer, F. (1991), Voronoi Diagrams - A Survey of a Fundamental Geometric Data Structure, ACM Computing Surveys, Vol 23, No 3, September 1991, pp. 333-405

Bate, S. and Stanley, K. (1988), Heuristic Route Planning: An Application To Fighter Aircraft, Proceedings of the IEEE National Aerospace and Electronic Conference 1988, pp. 1114-1120, Dayton, Ohio.

Cheng, V. (1990), Concept Development of Automatic Guidance for Rotorcraft Obstacle Avoidance, IEEE Transactions on Robotics and Automation, Vol 6, No 2, April 1990, pp. 252-257

Foley, J.D., van Dam, A., Feiner, S.K. and Hughes, J.F. (1990), *Computer Graphics - Principles and Practice*, 2nd Edition, Addison-Wesley, USA.

Holmes, P. and Jungert, E. (1992), Symbolic and Geometric Connectivity Graph Methods for Route Planning in Digitised Maps, IEEE Transactions on Pattern Analysis and Machine Intelligence, Vol 14, No 5, May 1992, pp. 549-565

Hwang, Y.K. and Ahuja, N. (1992), Gross Motion Planning - A Survey, ACM Computing Surveys, Vol 24, No 3, September 1992, pp. 219-291

Kaindl, H. and Scheucher, A. (1992), Reasons for the Effects of Bounded Look-Ahead Search, IEEE Transactions on Systems, Man and Cybernetics, Vol 22, No 5, September/October 1992, pp. 992-1007

Korf, R.E. (1988), Real-Time Heuristic Search: New Results, Proceedings of the 7th National Conference on Artificial Intelligence, AAAI88, 1988, pp. 139-144

Krozel, J. (1988), Search Problems in Mission Planning and Navigation of Autonomous Aircraft, Master of Science thesis in Aeronautics and Astronautics, Purdue University

Krozel, J. and Andrisani, D. (1990), Navigation Path Planning for Autonomous Vehicle: Voronoi Diagram Approach, Journal of Guidance and Control, Vol 13, No 4, November - December 1990, pp. 1152-1154

Mitchell, J. (1988), An Algorithmic Approach to Some Problems in Terrain Navigation, Artificial Intelligence, Vol 37 (1988), pp. 171-201

Menon, P., Kim, E. and Cheng, V. (1991), Optimal Trajectory Synthesis for Terrain Following Flight, Journal of Guidance and Control, Vol 14, No 4, July-August 1991, pp. 807-813

Nilsson, N., Principles of Artificial Intelligence, Springer-Verlag, 1982

Pearl, J. (1984), HEURISTICS: Intelligent Search Strategies for Computer Problem Solving, Addison-Wesley

Rao, T. and Arkin, R. (1989), 3D Path Planning for Flying/Crawling Robots, SPIE Vol 1195, Mobile Robots IV, (1989), pp. 88-96

Schwartz, J. and Sharir, M. (1988), A Survey of Motion Planning and Related Geometric Algorithms, Artificial Intelligence, Vol 37 (1988), pp. 157-169

Sudkamp, T. and Courte, J. (1989), Selection of Subdomains in Hierocratic Path Generation, SPIE Vol 1095, Applications of Artificial Intelligence VII, (1989), pp. 426-434

Weimann, A., Scheibe, D. and Schröder, P. (1988), Terrain Masking and Threat Avoidance Using Land Mass Data, Proceedings of National Aerospace and Electronic Conference 1988, pp. 540-545, Dayton, Ohio.

Wendel, M.J., Katt, D.R. and Young, G.D., *Advanced Automatic Terrain Following/Terrain Avoidance Control Concepts Study*, Proceedings of the IEEE National Aerospace and Electronic Conference 1982, pp. 1366-1372

Zhang, W. and Korf, R.E. (1993), Depth-First Vs. Best-First Search: New Results, Proceedings of the 11th National Conference on Artificial Intelligence, AAAI93, pp. 769-775, Washington DC.

EXPERT SYSTEM FOR SHIP TRANSPORTATION PLANNING

Yugo Taka*, Satoshi Fukumura* and Zhi Min Xie**
* Information Systems Department, Kawasaki Steel Corporation,
Chiba, Japan, 261-01.
** Infrastructure Systems Division, Kawasaki Steel Systems R&D Corporation,
Chiba, Japan, 261-01.
Email: taka@system.kawasaki-steel.co.jp

ABSTRACT

This paper describes an expert system for ship transportation planning which provides efficient transportation schedules for steel products transported by ship from steelworks to distribution stations nationwide. This scheduling problem can be described as a combinatorial optimization problem, and has the characteristics of large scale, many constraints, multi-purposes, and combined problems typical of real scheduling problems. A practical heuristic scheduling algorithm was developed to solve this problem by acquiring human expert knowledge and analyzing the essence of the problem. A comprehensive coastwise transportation administration system was also developed to enable the best use of the planned schedules. As a result, the system has reached practical operation, reducing wasteful waiting time for ships.

INTRODUCTION

The manufacturing industry has responded to the diversifying and advancing needs of customers by such means as multi-kind and small-quantity production, and short time to delivery. The requirements that both production departments and distribution departments need to fulfill are increasing, such as just-in-time delivery, high-frequency delivery, and reduction of delivery lead time. Under these circumstances, it is important to develop more advanced distribution control systems, and especially to improve the planning functions in these systems.

Kawasaki Steel Corporation has been developing for several years systems for planning distribution tasks which require a skilled human expert's knowledge and judgement by applying expert system technology [FY91, IF91, IYF+91]. One of these is a complex and large-scale expert system for ship transportation planning which provides efficient transportation schedules by ship for steel products and has now been put into practical operation.

A number of research projects have applied the AI approach to the development of practical systems for shipment scheduling. For example, the system developed by Itoh et al. [IIH+91] and the system developed by Hayase et al. [HIM+93] plan loading and unloading schedules at berths in a works producing oil products. The system developed by Fukumura et al. [FY91] also falls into this category, although it is concerned with steel products. The system developed by Takenaka et al. [TMS+89] plans cargoes and service schedules for ships for chemical raw materials. These systems deal with many constraints and complex problems, but the system described in this paper deals with a more difficult problem, in that it involves the selection and composition of shipping products, assignment of ships, and berth planning in loading and unloading ports.

The next two sections of this paper describe the ship transportation planning problem and its characteristics. A practical scheduling algorithm to solve this problem is then presented. The next section describes the configuration of the whole system, which is evaluated finally.

THE SHIP TRANSPORTATION PLANNING PROBLEM

The problem addressed by this system is planning efficient transportation schedules for steel products by ship from steelworks to distribution stations nationwide. FIGURE 1 shows the steelworks and the main distribution stations of Kawasaki Steel Corporation. There are more than one hundred distribution stations involved, the number of ships reaches approximately one hundred, and the quantity of products for shipping constitutes about two thousand records.

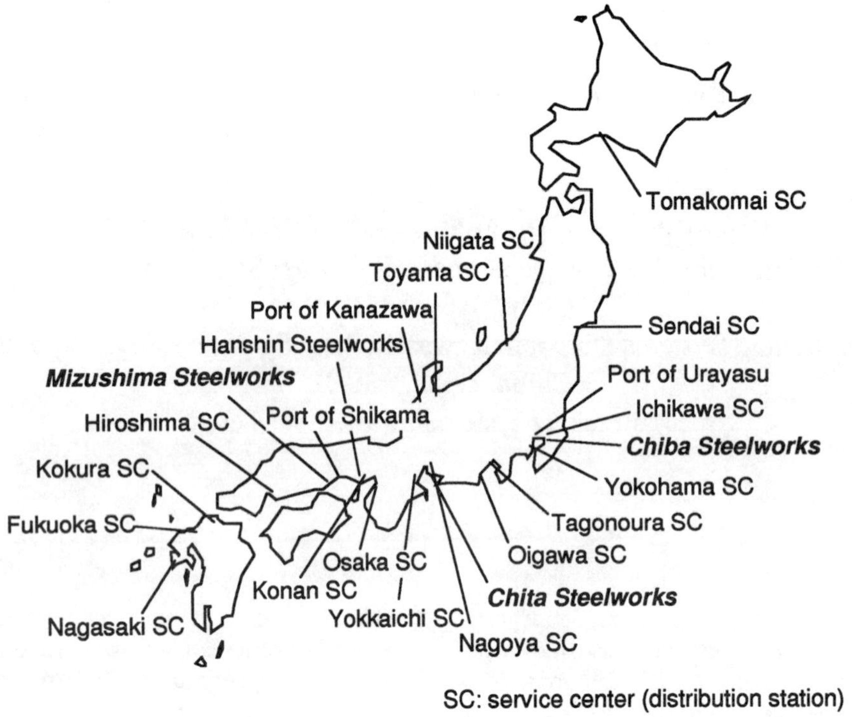

FIGURE 1 Steelworks and main distribution stations of Kawasaki Steel Corporation

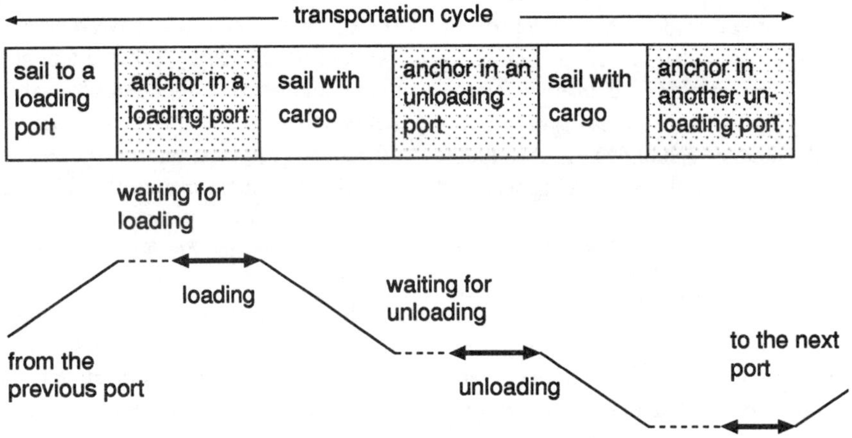

FIGURE 2 Ship transportation cycle

FIGURE 2 shows the ship transportation cycle consisting of the following stages: (1) sailing to a loading port (steelworks) with no cargo, (2) anchoring or tying up to load cargo, (3) sailing to the unloading port (distribution station) with the cargo, (4) anchoring or tying up to unload the cargo. A cycle may contain some of the stages more than once if loading or unloading in more than one port is needed. After unloading all the cargo, the ship sails to one of the loading ports to load the next cargo, repeating the cycle.

The items of the ship transportation schedule to be planned consist of the following:

(1) Making up the transporting lot
Cargo (referred to as a transporting lot) is made up by selecting products for shipping within the deadweight tonnage of the ship involved.

(2) Assignment of a ship
A ship is assigned to each transporting lot, depending on the cargos and destinations.

(3) Loading berth planning
The cranes and time period for loading the transporting lot at the berth of each loading port is planned.

(4) Unloading berth planning
The cranes and time period for unloading the transporting lot at the berth of each unloading port is planned.

The period to be planned is from two to five days later at the loading port. Schedules corresponding to transportation lots loaded in the loading ports are planned for daily basis, covering approximately twenty ships per day.

The main purpose of this planning is to achieve efficient utilization of ships, and to reduce transportation costs. But, the following indexes need to be considered:

(1) to keep the appointed date of delivery,
(2) to achieve the required shipping quantity.

The latter index involves minimizing the idle time of cranes in the loading ports. The following indexes are concerned with efficiency:

(3) to maximize the loading rate of a ship,
(4) to minimize the waiting time of a ship at an unloading port.

The latter index is particularly important, because it speeds up the turnaround of ships and enables products to be shipped by fewer ships.

A complete outline of the ship transportation planning problem is shown in FIGURE 3.

CHARACTERISTICS OF THE PROBLEM

Assuming that transporting lots and the ships to be loaded are given, making loading and unloading plans can be described as a job shop scheduling problem, if the series of activities from the start of loading a transporting lot to the end of unloading it is regarded as a job, if each loading or unloading step is regarded as an operation, and if each crane is regarded as a machine.

This problem has the following characteristics:

(1) There may be more than one crane by which each loading or unloading operation can be done.
(2) There may be no precedence constraints between unloading ports, that is, the order of operations may be arbitrary.
(3) The number of working cranes at the same time is given, but which ones are working is not specified.
(4) Standby time corresponding to the sailing time (or berthing or crane shifting time) is necessary between operations.
(5) While loading ports operate for 24 hours a day, most unloading ports operate only during the daytime. Moreover, unloading ports tend to be influenced by such constraints as no work on holidays, the capacity of warehouses, and double berthing.

In a practical operation, the system must compose transporting lots and assign ships. The following characteristics must be considered to compose transporting lots.

(1) The selection of products for shipping can be done quite freely, except for those whose appointed date of delivery is close.
(2) There are various possibilities for transporting lots with regard to the size of ship and combination of destinations and kinds of products for shipping.
(3) Although such selection of products for shipping and the composition of transporting lots influences the evaluation of schedules, the optimal combination varies according to the constraints imposed by the ships, loading ports, and unloading ports.

To assign a ship to a transporting lot, the starting time of loading must be considered as well.

These things indicate that the composition of transporting lots, assignment of ships, and making loading and unloading berth plans are closely related with each other, and that good schedules cannot be obtained if they are treated individually. However, it is almost impossible to resolve all of them as a single problem because of their enormous number of combinations. Hence, it is necessary to devise a practical scheduling algorithm.

SCHEDULING ALGORITHM

Ship transportation planning can be described as a combinatorial optimization problem, and as stated above, has the characteristics of large scale, many constraints, multi-purposes, and combined problems typical of real scheduling problems. In purely practical terms, it is only necessary to find a satisfactory solution within a given time rather than the truly optimal solution. Therefore, it is not realistic to solve this problem only by mathematical methods such as the branch-and-bound method, so that a heuristic method should be effective. A practical scheduling algorithm was thus developed by acquiring human expert knowledge and analyzing the essence of the problem.

The point of this problem is that transporting lots which correspond to jobs in a typical scheduling problem are not given, and that, although their composition is important for the results, an evaluation of their true quality is difficult unless all schedules are planned. With this in mind, the following algorithm mainly based on how jobs should be composed was developed.

(1) Generate many candidates for jobs.
(2) In order to evaluate these candidates, solve provisionally a partial problem most closely related to the evaluation.
The estimated values obtained from this evaluation enable priority to be assigned among the candidates. The premises on which the partial problem is solved are treated as constraints to solve other partial problems.
(3) Select a candidate for the job which has the highest priority.
(4) Plan partial schedules for the selected job.

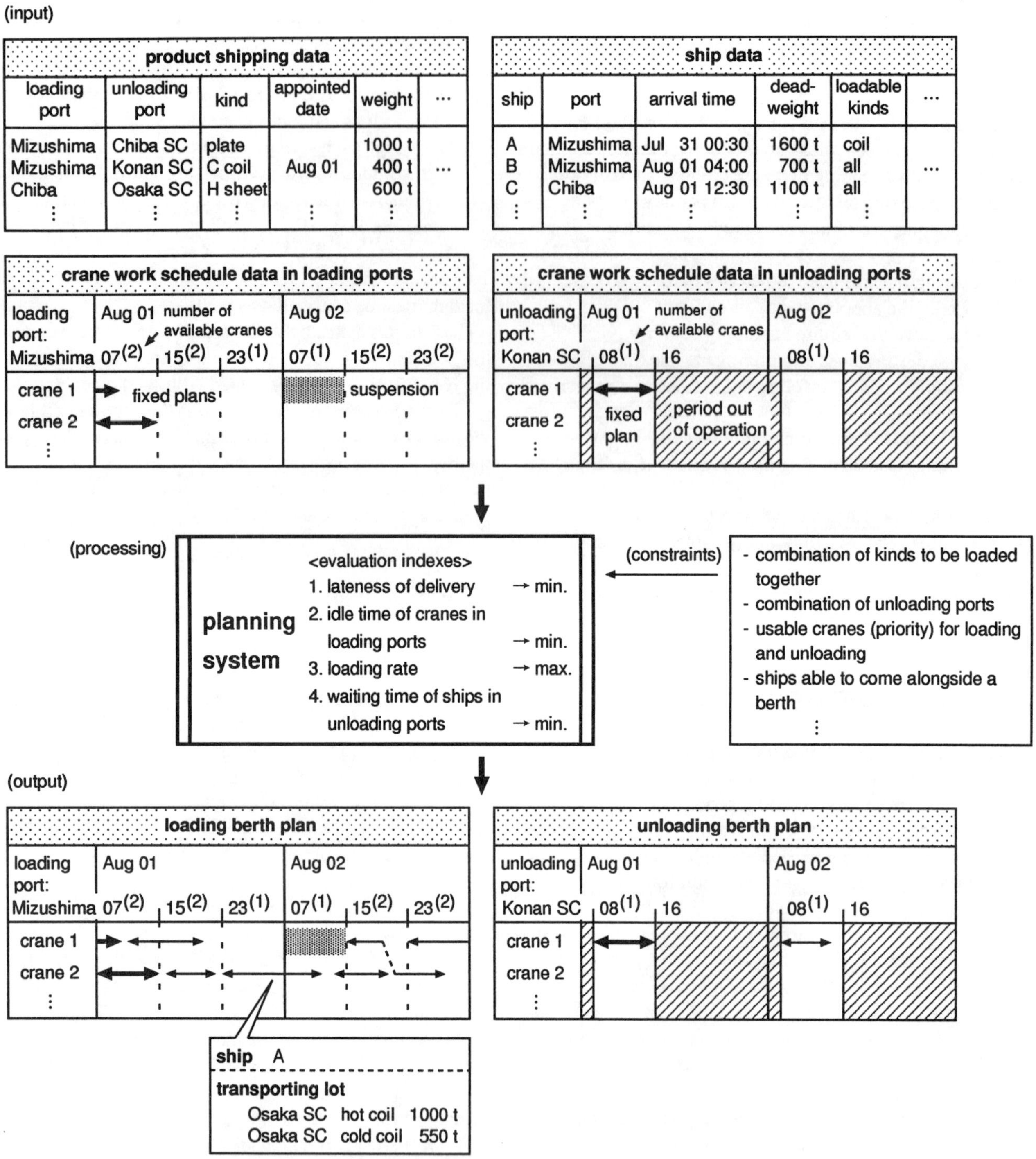

FIGURE 3 Outline of the ship transportation planning problem

Each partial step in scheduling process has the function of planning the optimal partial schedule for jobs already planned and newly selected. Re-scheduling for jobs already planned is needed for this purpose.

(5) Return to (3).

In this way, suitable schedules for the whole plan can be produced. This procedure is illustrated in FIGURE 4 and is described next in detail.

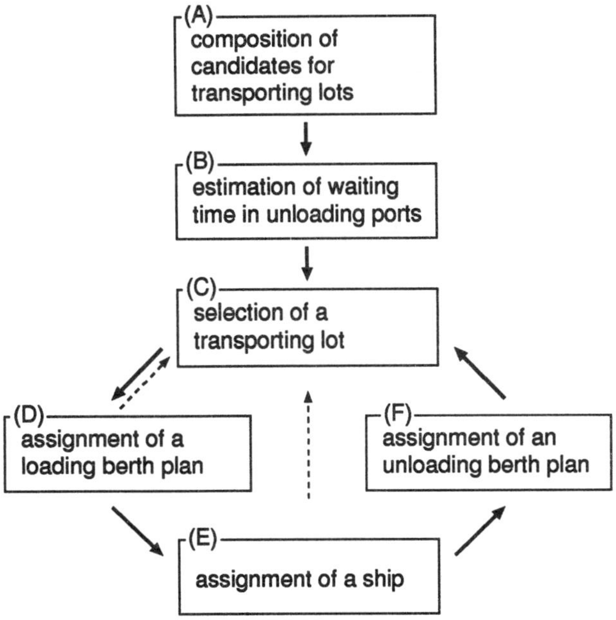

FIGURE 4 Scheduling procedure

A. Composition of Candidates for Transporting Lots Compose allowable candidates for transporting lots, which consist of several sizes of ship and various combinations of destinations and kinds of products for shipping.

B. Estimation of Waiting Time in Unloading Ports Estimate the waiting time in unloading ports of each candidate for transporting lots in a similar way to making unloading berth plans. Since the loading time, the unloading time, and the sailing time for each lot can be roughly calculated from its components, the waiting times (days) and the optimal arrival times at the unloading ports can be estimated in the way shown in FIGURE 5.

The waiting time in the unloading ports is used to set the priority of transporting lots, and the optimal arrival times at the unloading ports is used for the constraints of making loading berth plans by calculating back to the optimal departure times from the loading ports.

C. Selection of a Transporting Lot Select candidates for transporting lots one by one in the defined order of priority. The first priority is given to those which have strictly appointed dates of delivery. The second priority is given to those which have high values calculated from the waiting times in unloading ports and from loading rates.

Thus, the evaluation indexes of the appointed date of delivery, the loading rate, and the waiting time in unloading ports are considered. The idle time of cranes in the loading ports is reduced by repeating the selection of transporting lots until the idle time become sufficiently small or until no unselected transporting is left.

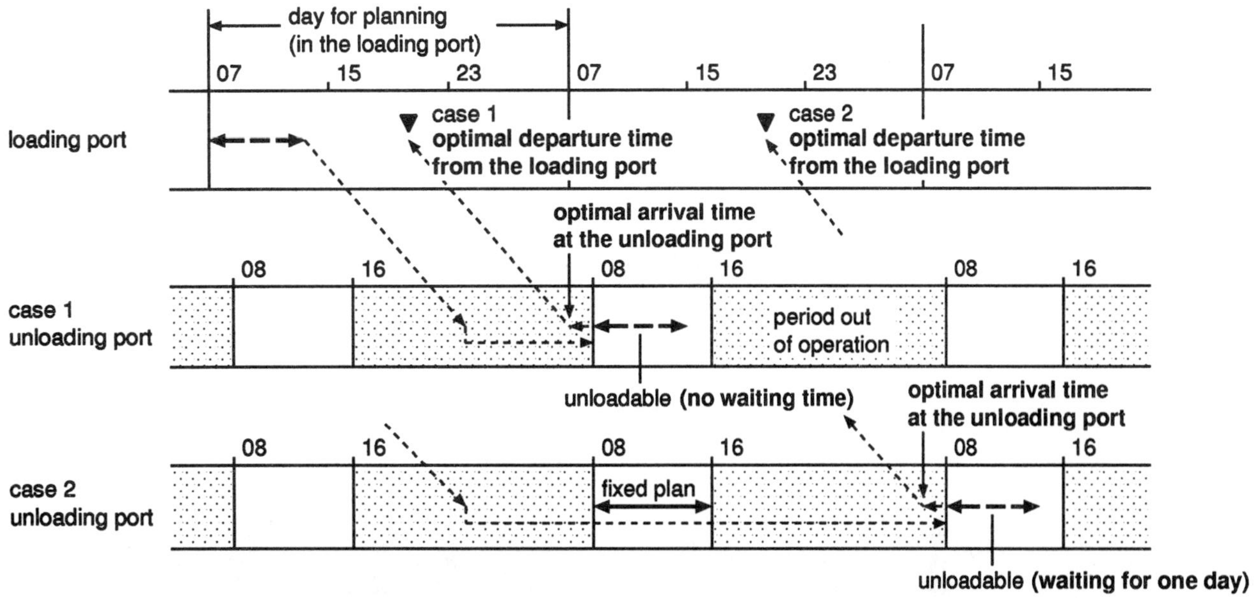

FIGURE 5 Method for estimating the waiting times in unloading ports

D. Assignment of a Loading Berth Plan In order to minimize the waiting time in an unloading port, it is necessary to bring the end-of-loading as close as possible to the optimal departure time from the loading port when making a loading berth plan. Generally, arrival at an unloading port later than the optimal time is not desirable, so the optimal departure time from the loading port can be regarded as being like the constraint of the appointed time. Therefore, making a loading berth plan can be treated as a scheduling problem to keep the appointed time and to bring jobs as close as possible to the appointed time.

Decide a crane (or cranes) and a time period for loading the selected transporting lot, considering the constraints of available cranes and the number of workable cranes, and adjusting for duplication with other lots. When it is not possible to put the transporting lot in the permissible position, return to the selection of a transporting lot, and select the next best one.

E. Assignment of a Ship Assign a ship to the selected transporting lot, whose arrival is in time for loading and which satisfies such constraints as the loadable kinds of products and destinations for sailing to. When all assignable ships have already been assigned to other transporting lots, try to exchange assigned ships.

F. Assignment of an Unloading Berth Plan Calculate the sailing time to the unloading port(s), select a crane (or cranes) and decide the time period for unloading the selected transporting lot.

Compare the resulting waiting time for unloading with the estimated time. If the former is longer than the latter because of duplication with other lots, cancel the schedules for the selected transporting lots, re-evaluate the estimated waiting time for unloading, and return to the selection of a transporting lot. Alternatively, if the schedules for the selected transporting lots are acceptable, return to the selection of a transporting lot and plan the next one.

FIGURE 6 shows an example of the schedules planned in this way.

The scale of this expert system is as follows: the rule part is 9K lines with 400 rules (using an expert system tool developed in the company); the other part is 95K lines (most of which is written in C language).

CONFIGURATION OF THE WHOLE SYSTEM

In order to apply a scheduling system to practical work and to use it effectively, peripheral systems supporting it

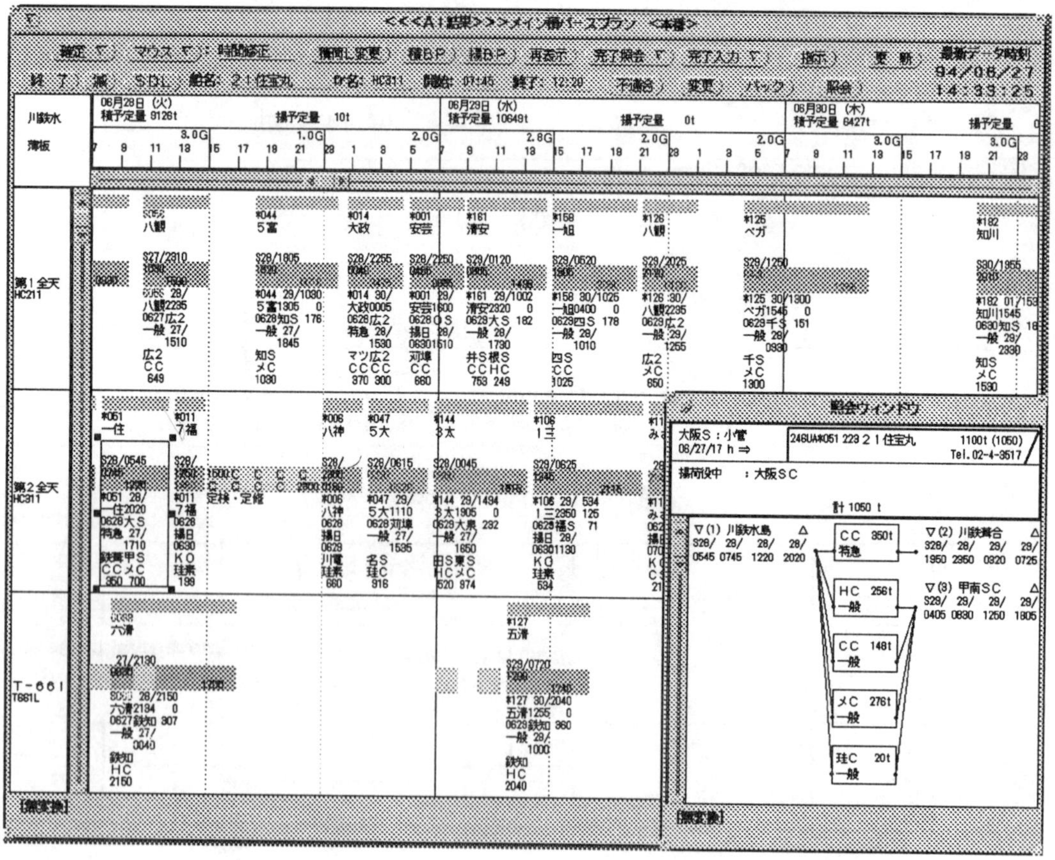

FIGURE 6 Example of the output for a lodaing berth plan

are indispensable, such as those collecting precise input data, those providing an easy-to-use human interface, and those supporting the execution of schedules.

This system has been developed as a subsystem for a comprehensive coastwise transportation administration system. FIGURE 7 shows the configuration of the whole systems and FIGURE 8 shows the configuration of the network. It has the following characteristics:

(1) The present position of each ship can be grasped by on-board GPS (Global Positioning System) on a notebook-type personal computer connected with host computers through the marine wireless telephony system. A function for forecasting the precise arrival time of each ship was also developed.
(2) A new company-wide distribution database was developed to hold the movements of ships as just stated, the progress of loading and unloading, and various other information including the planned schedules.
(3) A graphical user interface for reference and modification of the schedules was also developed. By using work stations at the steelworks and offices and connecting them through a network, adjustment among steelworks and offices can be done at the same time.

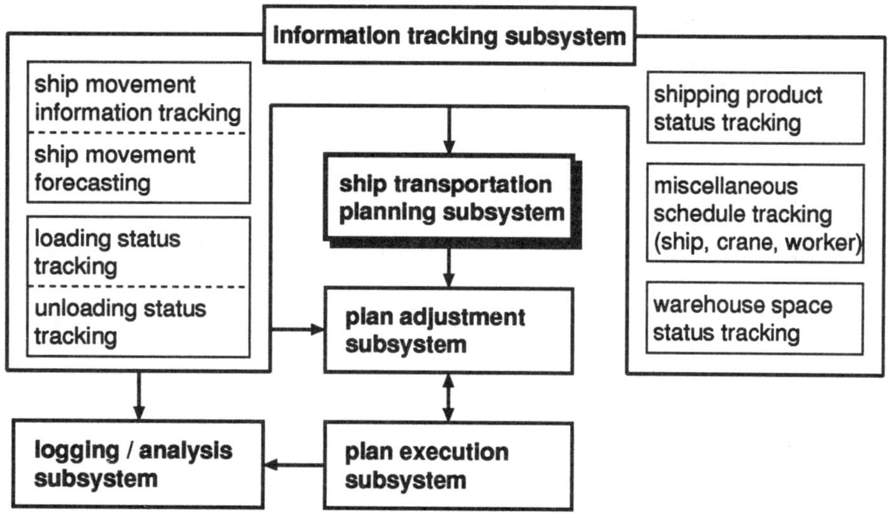

FIGURE 7 Configuration of the whole system

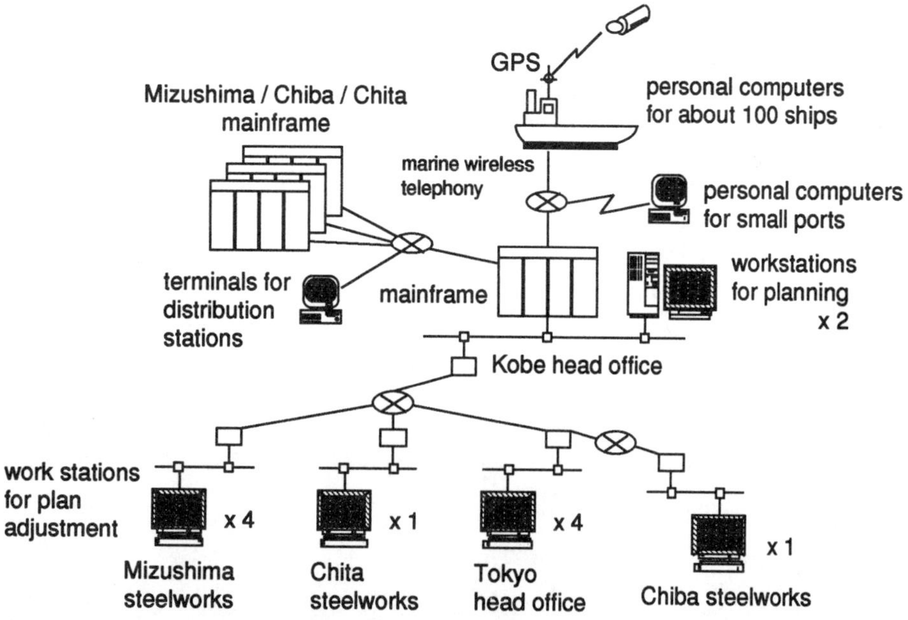

FIGURE 8 Configuration of the network

EVALUATION OF THE SYSTEM

This system was put into practical operation in December 1993, and daily work has been performed since based on schedules planned by this system.

The effects of this system are as follows:

(1) It is now possible to plan schedules by considering the total picture of ship transportation, these formerly having been planned separately by personnel in the loading ports, unloading ports, and ship allocation section to give only partial optimization. As a result, the waiting time in unloading ports has been reduced, which contributes to reducing the number of ships necessary for transportation.

(2) By reducing the planning time and improving the precision of the forecast data, it is now possible to plan schedules spanning up to five days. Consequently, arrangements for loading and unloading have been smoothed.

(3) By sharing the database and the schedules on a company-wide basis, the facility for discussing the solutions to various problems from plans to operations has been built up.

CONCLUSION

A ship transportation planning system having the characteristics of large scale, many constraints, multi-purposes, and combined problems, was developed by applying a practical scheduling algorithm. To support this, a comprehensive administration system was also developed. The system is now in practical operation, and more efficient transportation by ship has been achieved to prove the effectiveness of the system.

It can be concluded that developing a scheduling system to solve a practical problem is not easy, but it proved effective to analyze the essence of the problem, to apply human expert knowledge, and to design a scheduling algorithm accordingly.

REFERENCES

[FY91] S. Fukumura and E. Yamakawa. An Application of Expert Systems to Shipment Planning. *Asia-Pacific Journal of Operational Research* 8, 1991.

[HIM+93] T. Hayase, K. Itoh, I. Matsumoto and H. Innami. OSSES: Oil Shipment Scheduling Expert System for Oil Shipment Yard (in Japanese). *Journal of Japanese Society for Artificial Intelligence* 8 (1), 1993.

[IF91] K. Iritsuki and S. Fukumura. Planning Expert system using Multiple World Mechanism. In *Proceedings of the 4th International Conference on Industrial & Engineering Applications of Artificial Intelligence & Expert Systems*, Hawaii, June 1991.

[IIH+91] K. Itoh, M. Ishizuka, K. Harada, S. L. Chung and T. Enkawa. A Scheduling System for Berth Assignments Based on a Hierarchical Constraint-Directed Search (in Japanese). *Journal of Japanese Society for Artificial Intelligence* 6 (1), 1991.

[IYF+91] K. Iritsuki, E. Yamakawa, S. Fukumura, K. Nogami and J. Ikuta. Application of Expert System to Material Handling System (in Japanese). *Kawasaki Steel Giho* 23 (3), 1991.

[TMS+89] T. Takenaka, K. Maejima, F. Sekine and J. Mino. A Development of Shipping Scheduling System (in Japanese). In *Proceedings of the 3rd Annual Conference of Japanese Society for Artificial Intelligence*, Tokyo, Japan, July 1989.

AI PLANNING FOR HAZARD ACTION RESPONSE

Todd Mansell **Elizabeth Sonenberg**
Department of Computer Science
The University of Melbourne
Parkville, Victoria 3052, Australia.
emails: tmm,eas@cs.mu.oz.au

Grahame Smith
CRC for Intelligent Decision Systems
723 Swanston St,
Carlton, Victoria 3053, Australia.
email: gsmith@cs.mu.oz.au

ABSTRACT

This paper describes the application of the U-Plan planning system to what is described here as the hazard action response domain. U-Plan is a planning system that constructs quantitatively ranked plans given an incomplete description of the world. The final plan produced by U-Plan includes a method for choosing between planning alternatives after the acquisition of additional information or evidential support. U-Plan has been successfully applied to an air combat domain, and preliminary results suggest the hazard action response domain a suitable application for the U-Plan approach.

1 INTRODUCTION

Traditional AI Planning has been described as composing a course of action that can transform the world from a given initial state to a desired goal state. Traditional planning systems assume the availability of complete knowledge about the environment, and the effect an action will have when taken in that environment. Similarly, the traditional planner assumes that the initial state of the world does not change from the time that planning begins, till the time the plan executor commences the implementation of the set of actions.

Planning under uncertainty can loosely be described as the composition of a course of action that will achieve the goals presented given uncertain and/or incomplete information, and/or where the effect of the operators on the world are not known with absolute confidence. This paper describes the application of U-Plan (an AI planning system) to the problem of combating a chemical spillage and/or fire.

1.1 U-Plan

A major fact of planning with incomplete and uncertain information about the environment is that it is not possible to construct one initial state that precisely and unambiguously represents the world. U-Plan uses a possible worlds representation, where the available initial information is used to construct every possible initial state of the world. Associated with each possible world is a numerical measure of belief specifying the degree to which the evidence supports each possible world as the one that represents the true state of the world.

A hierarchical approach to planning is used as it significantly reduces the search space by first planning at abstract levels, and then expanding these abstract plans into more detailed plans. At the highest abstraction level strategic decisions are made, while at the lowest levels of abstraction, tactical decisions about how best to implement the strategy, are made. In support of hierarchical planning, each possible world is described at a number of predefined abstraction levels, allowing decisions to be made using a state representation at an equivalently detailed level of abstraction.

U-Plan utilises a set of (predefined) goal reduction operators that encode how a planning goal is reduced by the application of that operator. What results is a planning hierarchy tree where the goals are broken up into subgoals by the goal reduction operators. This allows us to first make the strategic decisions, which then guides all other decisions down to the tactical implementation of the subgoals. The reduction operators are expressed at various levels of abstraction. The planning algorithm uses the same abstraction level for both operator and possible world. A measure of expected fulfilment [Man94a,b] is used when selecting which operator to apply next.

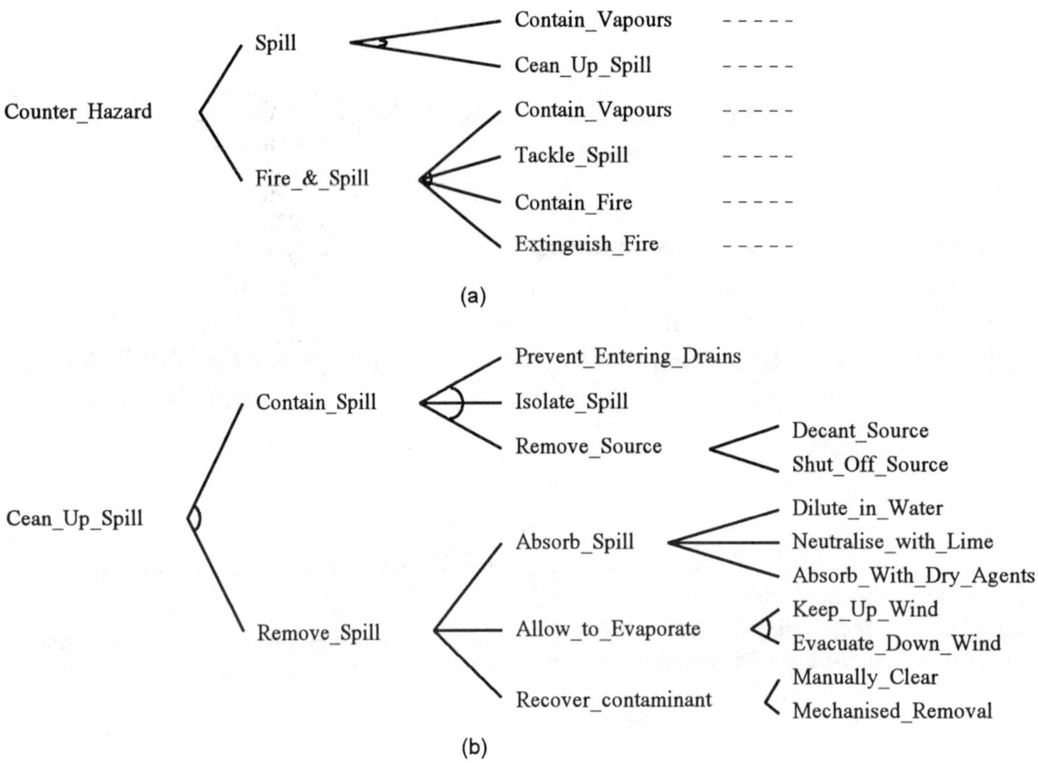

FIGURE 1: (a) The strategic portion of the abstraction hierarchy for the HAR domain. (b) A detailed view of the strategic Clean_Up_Spill action.

U-Plan constructs a plan for one possible world at a time, the first plan being constructed for the possible world with the greatest likelihood of representing the true world. Before subsequent plans are constructed, the suitability of reapplying an existing plan to this new possible world is assessed. Associated with each plan is the possible worlds it works for. If a plan partially works for some other possible world (e.g. the strategy works but some of the detail is different), then that part of the plan is used for this possible world, and planning continues from where the plan failed. When a plan exists for every possible world, the operator order of all the plans are combined to obtain a single planning tree that branches when the operator execution order differs. At this point the ability to acquire additional knowledge is used. At each branch, a knowledge acquisition operator can be inserted to determine which action in the planning tree to carry out next.

This planning algorithm has been applied to an air combat domain [Man94b]. In this domain, U-Plan demonstrated the ability to consistently produce a super-plan for the set of possible worlds in less time than it would take to produce a plan for each individual world.

Central to planning using U-Plan is: the set of states are represented at several abstraction levels; the selection of reduction operators is not purely state based but dependent on a calculation of expected fulfilment; the system will attempt to acquire additional knowledge when it is advantageous to do so, and an attempt is made to apply an existing plan to more than one possible state.

2 THE HAZARD ACTION RESPONSE DOMAIN

U-Plan is intended for use in emergency response type situations where a solution is likely to be required before complete and precise information is available (i.e., using the available uncertain or incomplete information). A typical emergency response domain suited to the U-Plan approach is responding to hazardous chemical spills and/or fires, the so called *hazard action response* (HAR) domain. When confronted with the spillage of a hazardous chemical, a number of uncertainties are likely in the initial information gathered.

Often, the exact type of chemical spilled is not always immediately known. This means the properties of the chemical may need to be estimated or may only be partially known. For the HAR domain, the type of information likely to be imprecise are (1) type of hazard represented by the material, (2) form the material may take, (3) flammable nature of the material, (4) active nature of the material, (5) location of fuel for the fire, (6) source of fuel for the fire, (7) location of the object causing the spillage, and (8) object that is leaking. However, evidence may be available, allowing an imprecise picture of the situation to be constructed. For example, if a building is on fire and explosions are occurring, one knows the chemical is explosive, but one does not know the precise threat that chemical poses (i.e., toxic, carcinogenic, etc).

Fire-fighters are regularly forced to act using information suffering these critical shortcomings. They must gather as much information as possible and formulate a response, with the knowledge that additional information must be acquired before some of the more critical decisions may be taken. An example of part of an abstraction hierarchy for the HAR problem is given in figure 1. These actions were selected with the aid of a fire officer familiar with the procedures when dealing with these types of situations[1].

The fire-fighter will prepare a response to the spillage based on what is known about the spill and what equipment is available to fight the spill. To do this, a representation of the assets available to fight the spill must represented. In U-Plan this information may be assumed to be known to the planner. Such information represented here is the type of units available to combat the spill (eg Hoses, Foam, Sand, etc), and the protective equipment available for the fighters (eg, Chemical Splash Suit, Breathing Apparatus, etc).

The plan of action produced by the senior officer will be influenced by all these pieces of information. A number of strategies are available for selection based on their level of risk and reaction to certain chemicals. One might choose to contain the spread of the fire before attempting any further action. This may be achieved by preventing the spread of the spillage by containment. Alternatively, by cutting off the source of a leak, one may remove the fuel of a fire, thereby gaining control. Many of the tactical actions are common to a number of higher level strategies, but are used to achieve a different overall result. For example, the action *Water_Spray* may be used to contain vapours produced by a chemical, or can be used as a barriers to prevent embers spreading to other buildings, assets or regions.

A large number of plan strategies as well as tactical implementations exist in this domain. Most importantly, a number of different plan solutions exist for each possible world.

3 FRAMING THE PROBLEM

In order to apply U-Plan to the HAR problem, certain aspects of the domain had to be represented in the form used by the planning system. Some of the steps taken include working with an expert to identify how problems are solved in the domain, identifying the type of information likely to be uncertain, and writing the U-Plan operators.

3.1 Identifying the Available Information

The first problem to be faced when constructing a new domain of application for U-Plan is to identify the type of information (available at plan time) that is likely to he uncertain, and the type of information likely to be precisely known. For the hazard action response domain, the type of information likely to be imprecise are (1) type of hazard represented by the material involved, (2) form the material may take, (3) flammable nature of the material, (4) active nature of the material, (5) location of fuel for the fire, (6) source of fuel for the fire, (7) location of the object causing the spillage, and/or (8) object that is leaking.

Some information in the hazard action response domain is assumed to be known with complete accuracy. In our representation of the domain, precise information includes the types of available fire fighting equipment (in our case, say *Fire_Hose, Foam, Alcohol_Foam, Sand, Carbon_Dioxide, Lime,* etc), and the available personal protection equipment (*Chemical_Splash_Suit, Gas_Tight_Suit, Breathing_Apparatus,* and *Gloves_&_Boots*).

3.2 Possible Worlds

U-Plan assumes an incomplete model of the world is all that is available, and uses a set of initial possible states (P-state) to describe what might be true of the world. A P-state is a complete description of one possible world using propositional statements. Each P-state is described hierarchically with n levels of abstraction, where n is domain dependent and selected during knowledge engineering. Each of these levels is a complete description of a world at a specified abstraction level. The highest level of abstraction gives a coarse description of the world. The lowest level gives a detailed view of the world. Intermediate

[1] The steps outlined by our expert have been modified (where required) to simplify exposition.

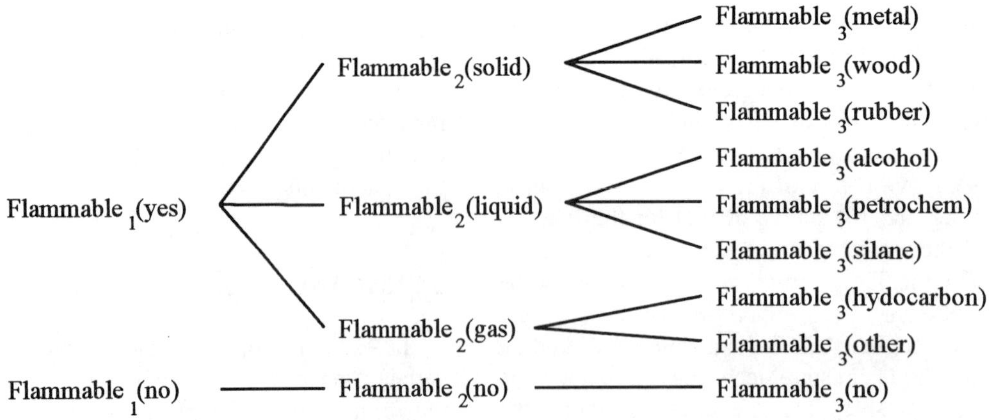

FIGURE 2: The compatibility relation describing the flammable nature of the material.

levels provide the description required to make a smooth transition between both extremes. In the HAR domain three levels of abstraction are used.

The utilisation of representations of possible worlds at differing levels of abstraction is intended to assist the logical propagation of strategic decisions being based on strategic information; and conversely tactical decisions being based on tactical information.

Mapping functions (called compatibility relations) are used to represent the interrelationships between the elements within the frames at the next highest or lowest level of abstraction. A compatibility relation must exist for every frame from one level of abstraction to the next, and compatibility relations can only link the abstraction levels directly above or below them.

The compatibility relations describing the degree to which the chemical is thought to be *Flammable,* is given in figures 2. The highest level of abstraction details whether the chemical is thought to be flammable or not. This level of information is used at the most strategic level of planning. The intermediate level (2) discerns whether the chemical is a solid liquid or gas. Finally, the material is categorised into more detailed functional groupings.

3.3 Reduction Operator

Planning operators represent actions that the system may perform in the given domain. The role of an action is to change the state of the world, the aim of an operator is to represent how applying that action will change the system's view of the state of the world. U-Plan uses reduction operators to give alternative methods for achieving the goal at a lower level of abstraction, the tactical level reduction operators describe the direct effects of an action on the P-state. These are SIPE-like operators [Wil88] where the closed world assumption is implemented, and hierarchical planning used. What follows is a description of the information contained in a U-Plan operator (with two descriptive representations of operators presented in figure 3):

- The necessary preconditions slot contains information that must be true of the world before the operator can be applied. This slot includes propositions that must be true of the world but are unable to actively achieved, or are prohibitively expensive to do so. Alternatively, propositions that represent aspects of the environment that are to be observed and not changed are included. For example, when dealing with a chemical fire a precondition may be to act in a certain way if the chemical is oil. The necessary preconditions provide a method of avoiding the possibility of adding oil to the fire to enable the action.
- Satisfiable preconditions designate propositions that must be made true of the world before the action can be applied. This is usually some tactical detail that should be true of the world before application of the operator. An example of a precondition that can be satisfied is the *Contain_Spill* operator (see fig. 3) that requires the planner to request *Sandbags* before this action can be achieved.
- Plot identifies the lower level actions that should be taken to achieve the goal of this action. As we see in figure 1, the *Contain_Spill* operator is achieved by the further application of three lower level reduction operators (*Prevent_Entering_Drains, Remove_Source, and Isolate_Spill*). These operators are placed in the plot section of the *Contain_Spill* operator (see fig. 3).

At the lowest level of abstraction, the reduction operators plot describes the effects that action has

Name:	Contain_Spill
Level:	Moderate
N-Precond:	
S-Precond:	Sandbags
Plot:	Prevent_Entering_Drain ∧ Remove_Source ∧ Isolate_Spill.
Probability:	Spill_Size ∧ Location ∧ Source_Location
Postconditions:	Nil
Planfail:	Backtrack

Name:	Shut_off_Source
Level:	Low
N-Precond:	Source_Location(known)
S-Precond:	Spill_Source_Access
Plot:	Spill_Flow(off)
Probability:	Spill_Source_Access ∧ Source_Location
Postconditions:	Spill_Flow(off)
Planfail:	Backtrack

FIGURE 3: Qualitative representation of two hazard action response operators.

on the state of the world. The *Shut_Off_Source* operator will include in its plot a proposition that informs U-Plan that the spillage flow has been stopped (in figure 3 this concept is represented by the proposition *Spill_Flow(off)*).

- The probability slot of a reduction operator summarises how different features of the domain will affect the likelihood of the operator successfully achieving the goals. This information is only qualitatively represented to date, hence only the propositions likely to alter the operator's probability have been inserted so far. The precise conditional probabilities must be gathered by either historic data, or a survey of expert opinion. For example, the probability of operator *Douse_With_Water* is conditionally dependent on the size and location of the spillage. The probability of a reduction operator is only conditioned on the state of the world, as operators are assumed to be independent of each other [Man94c].

When adding this information to the operators, the degree of abstraction for the information required to represent these details for the reduction operator should be considered, and a qualitative description included in the abstraction slot. For example, the *Shut_Off_Source* operator is considered to require detail state description, where the *Contain_Spill* warrants a more abstract representation of state information.

4 PLAN REAPPLICATION

U-Plan applies plan reapplication in an attempt to determine if a plan generated for one initial P-state can be adopted for another initial P-state. The desired result being fewer plans than the number of initial P-states. This is implemented by attempting to reapply plans generated for one initial P-state to other initial P-state. This is accomplished by first generating a plan for the initial P-state with the greatest support, then apply it to the next highest supported state to see if it achieves the goal function. Similarly, the rest of the initial P-states, with a support and plausibility over some threshold, are selected to see if the plan works in their situation. The initial P-states the plan works for are recorded and this information used in plan selection.

A plan is reapplicable if all the reduction operators in the plan (that are not redundant) have their preconditions met under the new initial P-state, and when applied result in the goal state being achieved. If a plan, during reapplication, fails due to the unsuccessful application of an operator, that plan is not entirely discarded. U-Plan will attempt to use the part of the plan that was successful and planning continues from the point where the plan failed. The desire is to construct plans with the same or similar strategies by reusing, at least part of, the plan at the high level of abstraction. When more than one plan partially works for a new initial P-state the best plan [Man94b] is used.

5 SUPER-PLANS

Once plans exists for all the P-states, with support and plausibility above some threshold, a single super-plan is constructed. This is achieved by merging the set of plans constructed for the set of initial P-states, that is applying identical operator sequences and branching at the point where plans differ. At each branch in the super-plan a knowledge acquisition operator is added, attaining the information required to select which

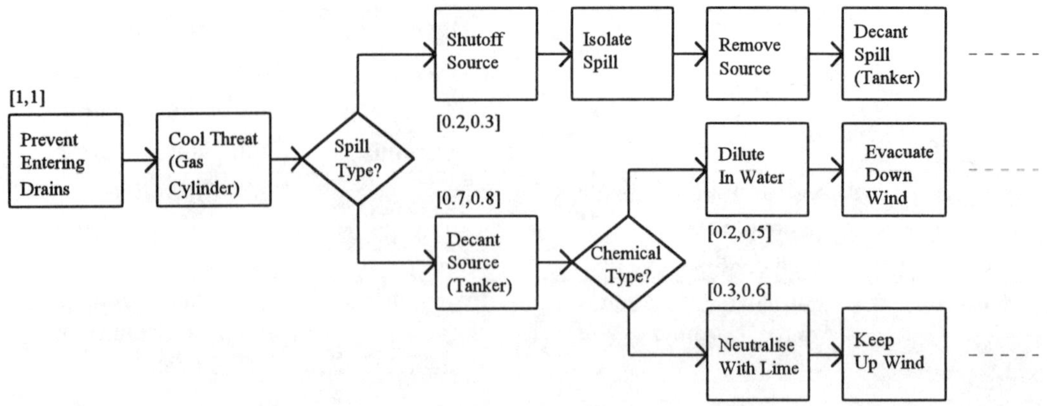

FIGURE 4: A super-plan outlining three simple plans hazard action responses.

action in the super-plan to apply next. The case may arise when the required information to differentiate between alternative branches is not available. In this case, the selection is based on the degree of evidence supporting for each branch of the super-plan, (see [Man94b] for more detail). This final step of producing a super-plan is an important part of presenting useful coarse of action that could be applied by the system.

For example, figure 4 depicts part of a super-plan generated for 5 initial P-states that reused 3 different plans. In this super-plan operators identical to all plans (i.e., *Prevent_Entering_Drains*, and *Cool_Threat(Gas_Cylinders)*) are identified as being able to be applied immediately. Following this, additional information about the *Spill_Type* is sought before we can determine how to stop the spillage continuing. Similarly, the type of chemical spilled may be required before committing to a particular clean-up strategy. At both these branch points, a quantitative measure of the weight of evidence supporting each branch is provided, should the additional information be unavailable.

In most cases, U-Plan produces a super-plan in less time than it would take a traditional planner to produce one plan for every possible world [Man94b].

6 CONCLUSION

This paper outlines the proposed application of U-Plan (a system for planning given an uncertain environment and incomplete information) to the HAR domain. The system represents the incomplete and uncertain description of the environment using a set of possible worlds. Each of these possible worlds contains a representation of the world at a number of abstraction levels. This enables the planning system to make high level (strategic) decisions, based on a high level representation of the world. The planning system's ability to fit part or all of an existing plan to a number of possible worlds, has the potential to produces fewer plans than one for each possible world.

U-Plan is intended for use within emergency response type domains where actions should proceed while information is being collected; or complex environments where complete information is unobtainable. Preliminary results suggest that the HAR domain is well suited to the U-Plan system as plan reapplication appears viable, resulting in fewer plans than one for every possible being produced.

REFERENCES

[Man94a] T. M. Mansell, Quantitative Operator Selection for Planning Under Uncertainty. In The Second International Conference on Artificial Intelligence Planning Systems, Chicago, 311-316, 1994.

[Man94b] T. M. Mansell, *An Approach To Planning Given Uncertain and Incomplete Information*. PhD Thesis, The University of Melbourne, Australia, 1994.

[Man94c] T. M. Mansell, An Approach To Planning Given Uncertain and Incomplete Information. *Computational Intelligence* (In Submission), 1994.

[Wil88] D. Wilkins, *Practical Planning: Extending the Classical AI Planning Paradigm*. Morgan Kaufmann, Los Altos, California, 1988.

HEURISTIC SCHEDULING IN A DYNAMIC PRODUCTION ENVIRONMENT: A PRACTICAL CASE

C. L. Chang
Department of Business Systems, Monash University
Clayton Campus, VIC, Australia, 3168
Email: chaolin.chang@fcit.monash.edu.au

ABSTRACT

The management of a tyre manufacturing company, call it company A, is aware of the need for an efficient production scheduling system to alleviate their production problems. The scheduling problem is an intractable one so an optimized schedule is not the objective. However, the scheduling system to be implemented is expected to produce a feasible schedule within an acceptable time, say one minute to allow the production planner to evaluate alternative schedules, plan forward, and maximize the utilization of resources on the shop floor. In addition to these, many practical issues in implementing the scheduling system, that may not be directly related to the scheduling algorithm but play a vital role in the success of the implementation, are discussed.

INTRODUCTION

Company A experiences two typical problems in the shop floor:

1. Work in progress inventory is piling up on the shop floor and
2. Safety stock is constantly kept high.

The management identifies one key factor that is causing these two problems — lack of an effective scheduling tool for the shop floor.

The production process is divided into three divisions: tube moulding, tread pressing and tyre manufacturing. Currently, production scheduling in each division is handled manually. However, the shop floor is a very dynamic environment and the scope and the pertinent variables in scheduling far exceed any human scheduler's capabilities. The human scheduler is only able to produce a rough schedule; the actual shop floor operation still relies on constant monitoring by the human scheduler.

Consequently, the human scheduler is so busy with the daily operation and co-ordinating the shop floor that he/she cannot perform scheduling effectively, and material planning is not considered. Because the material consumption cannot be planned, the work in progress inventory needs to be kept high to avoid out-of-stock events.

This case study is to develop a computer scheduling system for the tube moulding division based on the author's Very Fast Production Scheduler (VFS) technique [Chang, et. al. 1994]. Because the management requested that jobs to be finished as early as possible, forward scheduling was used.

VFS uses job-oriented heuristic [Hastings and Yeh, 1990]. Job-oriented heuristics schedules take a job as the basic unit of schedule. All operations in a job are scheduled before the next job is processed. In addition, VFS takes into account job precedence and technology constraints between operation.

The distinct feature of VFS is its computational efficiency. Typically, it can schedule a hundred jobs in about a second of CPU, during which job precedence, technology constraint and other practical constraints are considered.

There are basically two steps involved in VFS:

1. Find a job loading sequence based on determined constraints
2. Load jobs according to the job loading sequence

The second step in VFS is often an iterative process in which heuristics are employed to produce a feasible schedule.

Figure 1: The production process for a tube.

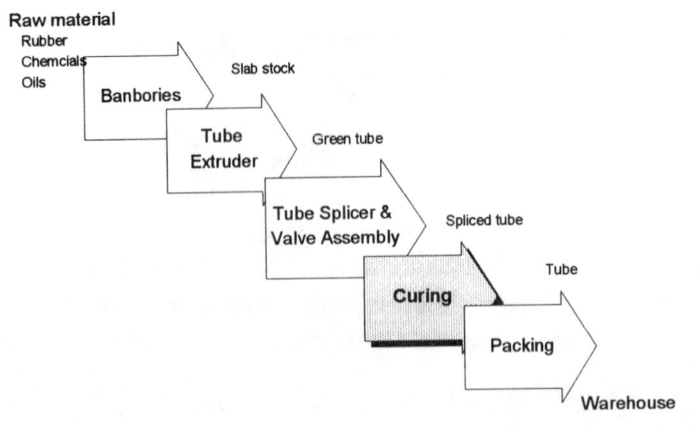

The specific objectives of the scheduling system are:

1. Meeting the due dates of jobs and
2. Minimising the numbers of times moulds are changed.

THE PRODUCTION PROCESS

Figure 1 presented the full production process of a tyre's liner tube. The production starts from the Banbories where all raw materials, such as rubber, chemical, and oils, are mixed. The intermediate product after Banbories, called slab stock, then goes through the tube extruder to produce the green tubes.

The green tubes are then transferred to the Pressing division where tubes are spliced and assembled with a valve. The green tubes go through the curing process to stiffen them. Once curing is completed, the tubes become the finished product, and are then sent to the warehouse for packaging and storage. This type of production is basically make-to-stock with characteristics of a job shop.

This case study takes the curing process as the starting point. The curing process starts with a mould in a press called a suite. The action of removing a current fitted mould from a press and fitting in another mould is called mould change. At present, mould changes is the major reason of lost productivity. Time required for a mould change depends on what suite is changed from and to.

After a suite is fitted, a spliced green tube is inflated and placed in the press, which is then closed for a specified time. After a tube is cured the press is opened. The curing time for a tube, called cycle time, depends on the suite. In most cases, the possible suite for making a tube ranges from one to tens of choices.

Figure 2 gives an example of possible suites in tube production. In this case, this tube is made by either press TR1 or AB1. Three possible moulds can be used on TR1 and two on AB1. The cycle (curing) time for all suites in TR1 is 9 minutes and a mould change requires 480 minutes. The cycle time for two suites of AB1 is 25 minutes and a mould change will require 300 minutes. During a mould change, no tube are produced.

DIFFICULTIES OF SCHEDULING PRACTICE

Monthly forecast of tubes are issued ten days before the end of each month. Because of seasonal fluctuation on the demand for different tubes, the forecast of tubes differs every month. The demand of one type of tube constitutes a job.

Each job has the following attributes:

•Job number: a unique number assigned to each job to identify the job.

•Job batch size: the monthly demand for one type of

Figure 2: An example of suites for a tube

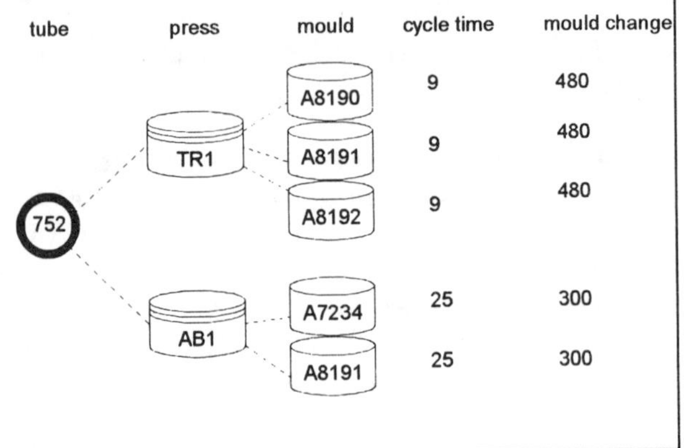

Table 1: Example of a three-day schedule produced by the production planner on 2/11.

PRESS NO	SIZE/ VALVE	CARRY OVER	TOTAL ORDER	SCH 1/11	ACT	SCH 2/11	ACT	SCH 3/11	ACT
1 &2	7.50-16 V274	1410		85	85	85		85	
22	12.00-20 V179	452	250	50	36	50		50	
10 &21	7.60-15 V259	1820	5000	170	160	170		340	

tube.

- Priority: a management assigned value.
- Due date: usually the last day of the following month.
- Size of tube: indicating what type of tube.
- Valve: indicating the type of valve used.

Each job is for a particular size of tube. Because this is a make-to-stock production, the due date for each job is the last day of the following month. There are currently over thirty presses and over eighty moulds on the shop floor.

There are three variables involved in producing a schedule for the shop floor determining the job, press, and mould. The main scheduling task is to determine when and what job to start, on which press, and with which mould. There are four major difficulties to overcome in doing the above scheduling.

Firstly, the resultant schedule has to be sufficiently stable so that it can be used at the operational level as well as at the planning level. The production data is updated daily. Therefore, the scheduling system needs to be able to generate a stable output for the operator to follow. A sensible schedule is the one that makes sense to operators on the shop floor; not to the computer algorithm.

Secondly, a press is often able to cure more than one type of tube. There is often more than one type of mould that can be fitted into one press. As a result, a job often contains a number of possible suites resulting in variations to the cycle and set-up time. The total number of possible suites for all jobs is between 1,500 and 3,000, depending on the number of jobs and types of tubes produced.

Thirdly, the scheduling process needs to take into account what the current set-ups on the shop floor are. If job 100 is scheduled onto press TR1 on day 1, the computer scheduler should continue to schedule job 100 from TR1, although some better alternatives may exist by assuming any desired set-up is available.

Fourthly, some jobs that are issued with a relative large job batch size will require to be run on more than one press at a time, called splitting. Job splitting at present depends largely on the past experience of the production planner. At present, the scheduling process is conducted by a human scheduler based on experience and some guess work. Unfortunately, it is a difficult task for the computer scheduler to fully emulate the experience and flexibility of its human counterpart.

The production planner's general loading principles of jobs is based on the following hierarchy:

1. Jobs are loaded according to the priority assigned by the production manager.

2. Current loading on the shop floor is checked to see if any suite can produce tubes for the job. The job is loaded onto that particular press, if the mould in that press can be used by the job. A mould change may be required, if no current suite can take the job.

3. A job is split when the job quantity is obviously too large for one press.

The production planner lays out the schedule in a tabular format as shown in Table 1.

In the schedule produced by the production planner, jobs are identified by the size of the tube and valve used in the tube as shown in the second column, SIZE/VALVE. Presses are identified by a unique number assigned by the production manager. The type of mould is not included in the schedule. The CARRYOVER column records the remaining demand for that job on the date of the schedule. The TOTAL ORDER column records the issued demand of the job. If a job has a blank TOTAL ORDER, it is a pending job from the previous month. The rest of the columns are the scheduled and actual production layout for each day of the month. The scheduled production quantity is placed in the SCH column and the ACT column records the actual production figure.

In Table 1, the job 7.50-16 V274, which has a carryover of 1410, is to be loaded onto press 1 and press 2 (shown in column 1). As the TOTAL ORDER column for this job is blank, we know that this is an unfinished job from last month's ticket. The job is scheduled onto

1/11, 2/11, and 3/11 for 85 each. Since the schedule date is 2/11, we have a production figure from 1/11 recorded in the ACT column.

LIMITATIONS OF THE MANUAL SCHEDULE

As observed by McKay et al.[1988], a scheduler on the shop floor is often required to dispatch jobs based on incomplete information on a dynamic shop floor. Consequently, the scheduler often uses some simple dispatching rules, e.g. SPT, FIFO, etc, plus intuition to co-ordinate the operations on the shop floor.

The complexity of the scheduling decision grows exponentially as the job size increases and/or time horizon of the schedule expands. Although the production planner contributes a level of flexibility that cannot be matched by any computer algorithm, the human scheduler is unable to evaluate all possible alternative schedules within a reasonable time frame.

The decision to determine what job to load on what press using what mould can be as little as one suite to almost two hundred possible suites. The total suites for all jobs reach almost three thousand possibilities. However, the current schedule, done in a tabular format, does not consider what mould to use, thus simplifying the scheduling decision.

There are currently more than 30 presses on the shop floor. These presses can be categorised into three types: AB, TR, and TB. As the loading of jobs can be easily divided into these three types of presses, the scheduling decision is very much simplified. Consequently, the schedule produced by the production planner is a rough schedule that requires constant monitoring of the job progress. Expedition and overproduction become the guiding rules in the factory. Excess quantity is constantly kept to avoid the possibility of a stock-out situation.

Once a job is loaded onto a press, it is run continuously until the required quantity is produced. In some situations a job cannot be produced by the current press. The production planner will split the jobs onto the second press or even third, and fourth press.

Because of the fact that the production planner can only consider the current and near future loading situation based on past experience, it is not possible to predict whether a job will be finished before the due date, and therefore a mould change is largely based on what should happen next.

PROBLEM DEFINITION

The scheduling decision is to determine, at any given point in time, on what job to start, at what press, using what mould. The resultant schedule should also be sufficiently stable so that it is able to look backwards on the schedule produced one day before and intelligent enough to produce a feasible schedule. Moreover, a job splitting method needs to be incorporated into the computer scheduler.

The main constraint is the availability of presses and moulds that is defined by the work hours using one minute as the unit of measure. The tube division is worked by shifts. Some presses are used 3 shifts per day and others are used 2 shifts per day. The utilisation of presses depends on the demand of tubes. Preventive maintenance, repair, and machine breakdown also need to be considered.

The scheduling decision is a time-critical one, meaning that a new schedule is expected to be produced every morning after the production data from yesterday is entered into the system. The scheduling process should ideally take no more than one minute.

In short, the objectives of the computer scheduler are to

1. meet due dates,
2. keep suite changes to a minimum, and
3. produce a schedule within one minute.

PRACTICAL CONSIDERATION

The computer scheduler produces feasible schedules based on rules embedded in the algorithm and production data given. Nevertheless, this may cause problems when implementing the computer scheduler in the production environment, as the underlying assumptions of the computer algorithms may be violated when the situation on the shop floor changes. Some practical considerations with equal importance as the computer algorithm encountered while implementing the scheduling system are outlined as follows.

The practical problem is dynamic in nature

On the shop floor, production data is updated daily and new jobs are released every 30 day. The computer scheduler needs to be able to schedule forward, based on the last schedule. Schedules are produced continuously and are affected by the schedules produced previously. Also, the scheduling process needs to be sufficiently fast so that the new production data is reflected in the new schedule within minutes. In this case, stability provided by the JOH [White, 1986] becomes an important asset.

A practical scheduling system needs to co-exist with the current production system

The company has been developing a production system in-house. The computer scheduler therefore needs to be able to accept input from the production system and to produce output (schedule) that is readable by the system. The interface between the production system and the computer scheduler is accomplished by using fixed width text files.

The shop floor uses calendar days not day numbers

The computer scheduler uses day numbers to simplify the task of computation but the shop floor uses actual calendar dates and runs 3 shifts per day. The conversion between the day numbers, which are serialised integers, and calendar dates becomes essential.

Capacity fluctuation is commonplace on the shop floor

In addition to usual working hours which constitute standard capacity, there are other factors that may affect the actual available capacity:

a) Rostered-day-off
b) Preventive maintenance
c) Repair
d) Work over-time

Other subtle factors that affects the actual output need consideration also. For instance, a highly motivated operator normally has higher productivity than a less motivated one; an ailing operator may not be as efficient as a healthy one. Thus, an up-time factor is assigned to each press to inflate or deflate specified capacities. This up-time factor is an arbitrary value determined by the production manager.

Computer-human interfaces is the crux of a computer scheduler

If the production planner rejects what the computer scheduler generates, the factory can be even worse off than without the computer scheduler. To integrate the scheduling system, two tasks related to computer-human interface need to be resolved. Firstly, the integrated system needs to be accepted by its potential users, i.e. the human scheduler and the foreman on the shop floor. Next, the scheduling system will require minimum input from its users.

In addition to these practical considerations, it is important that the users have the opportunity to override the schedule produced by the system [Turner 1991].

SCHEDULING APPROACH

The computer scheduling system may be faster than its human counterpart. Nevertheless, it is difficult to dismiss the fact that the factory has been run by the production planner for years. We should not overlook how the production planner performs his/her job before starting to develop a new system. In a way, we are trying to produce a computer scheduler that emulates how the production planner reacts to the practical problem. It is understood from the outset that no matter how sophisticated the computer algorithm is, the flexibility of a production planner is not replaceable.

In addition, the scheduling problem encountered here is an intractable problem, and we do not expect the computer scheduler to provide the same flexibility as that of the production planner. In other words, the computer scheduler is developed to release the production planner from the chore of scheduling. Yet, running of the shop floor is still in the hands of the production planner.

Table 2: A partial list of suites for a job with various attributes.

Job	Remaining	Due Date	Priority	Size	Valve	Mould	Press	Cycle time	Setup time
42	22	30/09/94	3	L235/85/750-16LT	15	A8196	TB 01	9	300
42	22	30/09/94	3	L235/85/750-16LT	15	A8197	TB 01	9	300
42	22	30/09/94	3	L235/85/750-16LT	15	A8201	TB 01	9	300
42	22	30/09/94	3	L235/85/750-16LT	15	A8182	TB 01	9	300
42	22	30/09/94	3	L235/85/750-16LT	15	A8177	TB 01	9	300
42	22	30/09/94	3	L235/85/750-16LT	15	A8191	TB 01	9	300
42	22	30/09/94	3	L235/85/750-16LT	15	A8196	TB 02	9	300
42	22	30/09/94	3	L235/85/750-16LT	15	A8197	TB 02	9	300
42	22	30/09/94	3	L235/85/750-16LT	15	A8201	TB 02	9	300
42	22	30/09/94	3	L235/85/750-16LT	15	A8182	TB 02	9	300
42	22	30/09/94	3	L235/85/750-16LT	15	A8177	TB 02	9	300
42	22	30/09/94	3	L235/85/750-16LT	15	A8191	TB 02	9	300

THE DATA

Table 2 is an abridged list of data that shows the possible suites for a few jobs. A job is identified uniquely by a unique job number, which is recorded in the **job** column. However on the shop floor, a job is normally identified by its **size** and **valve** used. The **remaining** column records the remaining job quantity to be completed. Each job is assigned a **due date** and **priority**. Priority is assigned by the production planning manager with a scale from 1 to 3. A job with a priority value of 1 is an urgent job.

The possible combinations of mould and press for each job are shown in the rest of the four columns: **mould**, **press**, **cycle time**, and **setup time**. On an average schedule, there are about 3000 records on the list.

The current suites and job loading situation are kept in a text file, called the lock file. A sample of the lock file is shown in Table 3. The lock file ensures that the resultant schedule does not deviate from the previous schedule too dramatically and the current setups on the shop floor are taken into consideration.

Available capacity on any particular day for a press is defined as

$$c_{p,t} = u_p \times \begin{cases} 0, \text{ if at time } t \text{ press } p \text{ is scheduled for} \\ \quad \text{maintenance or a rostered - day off.} \\ o_{p,t}, \text{ if } o_{p,t} > 0. \\ s_{p,t}, \text{otherwise.} \end{cases}$$

where

$c_{p,t}$ = actual capacity at time t of press p.

u_p = uptime factor of press p.

$o_{p,t}$ = overtime capacity defined at time t of press p.

$s_{p,t}$ = standard capacity defined at time t of press p.

Table 3: A example of the lock file.

Press	Mould	Job
AB 1	A8187	61
IN 01	A8204	256
IN 02	A8218	289
TR 01	A8206	64
TR 02	A8207	65
TB 01	A8196	44
TB 02	A8201	45

Available capacity in mould is defined as either available or occupied, i.e. a Boolean value is used: 1 - available, 0 - occupied.

$$c_{m,t} = \begin{cases} 1, \text{ if } c_{p,t} > 0. \\ 0, \text{ if } c_{p,t} = 0. \end{cases}$$

where

$c_{m,t}$ = actual capacity at time t of mould m.

The unit of measure for capacity is measured in minutes. In the actual schedule, the time t is expressed in actual calendar date, for example, 20/12/94; and serial day numbers are used internally for the computational efficiency.

The up-time factor is an arbitrary value expressed as a percentage; for example 80%. It reflects the actual output from each press based on past experience from the shop floor foreman.

THE SCHEDULING PROCESS

The target of the scheduling algorithm is to produce a 60-day schedule showing the loading of jobs onto presses and what mould is to be used. The main task at any particular point in time is to first determine what mould will be used with a particular press, i.e. a suite is formed, then jobs that can utilise this suite are loaded until no more capacity is available in the suite.

The scheduling algorithm is based on the Very Fast Production Scheduler [Chang, et al. 1994]. There are basically two steps involved:

1. Determining a job loading sequence by sorting jobs according to such factors as
   a) Due date
   b) Priority of the job
   c) Size of tube
   d) Job quantity
   e) Job number

2. Load jobs onto suites according to the loading sequence determined in step 1. Forward loading will be used because the primary objective is to finish jobs as early as possible.

JOB LOADING SEQUENCE

The job loading sequence is determined by sorting jobs according to the parameters assigned to jobs, such as job status, priority, due date, job batch size, and job number.

Job status: used to distinguish whether a job has already commenced. A new schedule every morning will first schedule those jobs that have commenced back to where it started.

Due date: jobs that have earlier due dates are scheduled first.

Priority: priority is normally used to flag an urgent job.

Size of tube: two jobs may have the same size but different valve. In terms of the production, size is the determining factor of how a tube is made rather than the job number. By incorporating the size of a tube in the determination of the job loading sequence, the same size of tube can be scheduled together.

Job quantity: the larger the quantity of a job, the earlier the job is to be scheduled.

Job number: job number is a unique number assigned to each job by the system. The incorporation of the job number in determining the job loading sequence allows ties to be resolved.

SCHEDULING ALGORITHM

After the job loading sequence is determined, the scheduling algorithm commences with two main stages.

Stage 1 - Load scheduled jobs based on current suites on the shop floor.

Stage 2 - Load job onto press according to the job loading sequence.

Stage 1

A new schedule needs to reflect the current setup and loading on the shop floor. For example, if Mould A8196 is already in Press TB 01, a sensible schedule should try to use the existing setup instead of proposing a mould change. Each mould change can take from 4 hours to a day. It is therefore crucial for the computer scheduler to minimise mould changes.

There are two possibilities in the current loading: one is the current setting used by a particular job; the other is a job finished, but the mould is not removed from the press yet.

Stage 2

Stage 2 is a cyclic process. It stops when there is no more capacity in all suites to make tubes. Based on computational experience, the scheduling process normally takes less than 20 rounds to achieve an acceptable result. One reason for this cyclic process is to split jobs. Since a job can be split at any point, in order to minimise the need for mould changes, a job is loaded onto a setup until its due date. If the job can be finished by the due date, the job is split with the remaining quantity becoming a new job. The job splitting method is in accord with how the production planner splits a job. The following is the algorithm in Stage 2.

ALGORITHM

1. Choose a suite (a press fitted with a mould)
2. If there is any job that can use this setup, proceed to Step 4. If not, proceed to the next step.
3. Change the mould and proceed to Step 2.
4. If the mould is currently used by another press, proceed to Step 3. If not, proceed to the next step.
5. Choose an unfinished job according to the job loading sequence and use this setup.
6. If this job is able to use this setup and more than one tube can be produced, proceed to Step 8. Proceed to the next, otherwise.
7. If this is the last job in the job loading sequence, proceed to Step 9. Otherwise, choose the next unfinished job and proceed to Step 6.
8. Commit capacity required by this job using this setup
9. If this is the last press, terminate Stage 2. Otherwise, select the next press and proceed to Step 2.

There are two subroutines that are equally important in the scheduling algorithm. The first routine — *mould selection* — determines how a new mould is chosen when no jobs can use any suites on the shop floor. The second routine — *job committing* — actually loads the job onto the press.

Mould Selection

A mould list based on the due dates of jobs is constructed for each press. Moulds that are used by jobs with earlier due date are placed on the list earlier than otherwise.

Job Committing

Figure 3: Algorithm for job committing.

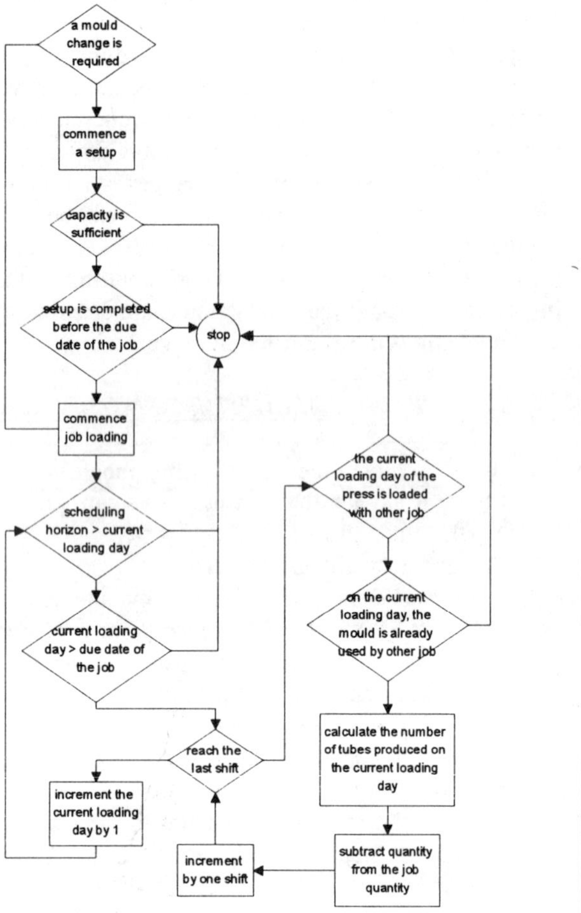

ALGORITHM

1. If there is a mould change, commence a setup. If not, proceed to Step 3.
2. If capacity in this press is sufficient to complete the mould change and the scheduled mould is available , schedule a setup on this press. Then proceed to Step 3. If not, terminate the routine.
3. If the scheduled finish date after the setup is after the due date of the job, flag this job as a late job and terminate the routine.

LIMITATION OF THE COMPUTER SCHEDULER

It is understood from the outset of the project that the scheduling problem on the shop floor is intractable. An optimal schedule is not expected but a feasible one is. There are also limitations to the algorithm. The algorithm was developed specifically for situation of this particular shop floor in which a job is always associated with more than one suite. The algorithm may fail if this assumption is invalid.

THE RESULTS

Table 4 is the schedule produced by the computer scheduler.

The significant advantages of the schedule produced by the computer scheduler over human scheduler are:

1. The computer scheduler schedules moulds.
2. The computer scheduler takes into account the capacity fluctuation.
3. The schedule is more realistic than the one produced by the production planner, since the actual capacity is considered rather than an estimate.
4. The production planner is able to forecast, more accurately, the due date of a job using the computer scheduler.
5. Job splitting is done automatically by the computer scheduler.
6. Current suites on the shop floor are considered in creating the new schedule.
7. The computer scheduler generates the schedule very quickly, in under 10 seconds.
8. The computer scheduler is based on the heuristic of utilising a mould until no other job can use it. Therefore, it avoids unnecessary mould change.

Table 4: The schedule produced by the computer scheduler.

Job	Press	Mould	Remaining	23/10/94	24/10/94	25/10/94	26/10/94	27/10/94
256	IN 01	A8204	291	56	56		R	
241	IN 01	A8214	202			14	R	50
57	TR 03	A8183	118	57	57	4	R	
266	TR 03	A8243	252				R	25
44	TB 01	A8196	3237	88	88	22	R	
54	TB 01	A8196	957				R	88
45	TB 02	A8201	843	88	88		R	
46	TB 02	A8201	79			22	R	
58	TB 02	A8183	151				R	10

CONCLUSIONS

To apply a scheduling system in practice, an effective algorithm is crucial as well as some practical considerations, such as interfacing with the existing system, acceptance of users, etc.

Computer human interface determines the life expectancy of a scheduling system in practice. Without the support of the human scheduler or operator, a scheduling system is nothing more than a computer program.

Optimised solutions are often impractical in a time-critical decision making situation, more so if the job shop is in a dynamic environment. The computational efficiency and quality of the schedule are often more prominent than an optimal schedule. After all, a computer scheduler reacts to a problem domain with a boundary, but the situations on the shop floor may not be within the boundary defined in the computer scheduler. In such cases, human schedulers play a vital role.

Another important aspect of a practical scheduling system is that the current shop floor resource commitment needs to be taken into account while updating a schedule.

In this case study, we have demonstrated the potential of applying Very Fast Scheduling (VFS) technique in practice. The objective of VFS is to provide reliable, realistic, and computational efficient schedules. Though resultant schedules from VFS may not be optimal ones, the users of this case study have shown little interest in pursuing optimal schedules but instead have shown great interest in practical schedules. Moreover, users would like to be able to evaluate alternate schedules, which means that the scheduling system needs to generate a schedule within an acceptable response time.

ACKNOWLEDGEMENT
The author would like to thank Mr. Bruce. A. Van Twest, Business Centre Manager, and Mr. David Burt, Computer Co-ordinator, of Company A for their assistance on the development of the scheduling system.

REFERENCE

1. Chang, C. L., Hastings, N.A.J., and White, C. (1994), "A Very Fast Production Scheduler", *International Journal of Operations and Production Management*, 14(8), pp. 88-101.
2. Hastings, N.A.J., Yeh, C.-H. (1990) "Job Oriented Production Scheduling", *European Journal of Operational Research*, 47, pp. 35-48.
3. McKay, K. N. and Safayeni, Frank R and Buzacott, John A (1988) "Job shop scheduling theory: what's relevant", *Interfaces*, 18(4), pp. 84-90.
4. Turner, J. (1991) "Finite scheduling: MRP II's final frontier", *Manufacturing Systems*,9(2), pp. 62-64.
5. White, C. (1986) *Capacity Constrained Load Sequencing Techniques for Production Management*, unpublished PhD dissertation, Dept. Econometrics and O.R., Monash University.

AUTOMATING PROBLEM SOLVING USING TRANSACTIONAL PARADIGM

V.K.Murthy and E.V.Krishnamurthy
Computer Sciences Laboratory,
Research School of Information Sciences and Engineering,
Australian National University,Canberra,
ACT 0200, Australia
Email: "abk@nimbus.anu.edu.au"

ABSTRACT

This paper deals with automated problem solving using the transactional paradigm on database augmented with associated knowledge sources.The advantages of this approach in the context of human acquisition of concepts as well as object-oriented database system are described. The notion of abstraction in problem solving is then considered and how an object-oriented transaction model can contribute towards an improvement in the Knoblock's method of automated abstraction for problem solving. Examples are provided to illustrate the concept of problem solving through transaction modeling. Finally, a discussion is included on the structuring and control of transactions, based on Brodie-Ridjanovic conceptual model of specification and Brown-Chandrasekaran model for design problem solving, for automating transactions. An extensive up-to-date bibliography is provided.

1.INTRODUCTION

Automatic methods for solving problems, proving theorems, reasoning, planning designing algorithms, implementing office procedures and programming have been widely studied in Artificial Intelligence (AI) for over two decades, see [NS72], [Ni71], [Ni80], [CM85], [Fe90], [Pe94]. Yet this field is still in its infancy and needs a lot of additional research before we expect any maturity in this field. This is a very fascinating area of research since it would also lead to an understanding of acquisition of mathematical concepts and development of analytical thinking in humans. The rapid advances in computer technology, particularly,in the distributed computing area, are providing greater thrust and support in handling this complex automated problem solving more effectively.

In this paper we consider the problem of automating the solution of problem based on automating a set of transactions in the context of a database (and knowledge-base) in order to achieve the desired goal.

2. TRANSACTIONAL PARADIGM

Automating problem solving through transactional paradigm or modeling has several attractive features both from computational [KM91], as well as human problem solving points of view [Gi83], [LL83].

1.Transactions operate in a database and a knowledge-base environment and the operations involved in the syntactic model are simple :
Read (R), Write (W), Commit (C), Abort (A), Exclusivelock(X), Sharelock(S), Exclusiveunlock (X*), Shareunlock(S*),Undo (U), Redo(D).
The R action can be interpreted as a matching or query operation , and the W action can be interpreted as an update or a replacement or a function computation action. Hence, R and W actions can simulate any production or classifier system; this means every computationally solvable problem can be reformulated as a transaction handling problem. Further, locking and timestamping permit to realize concurrent computations that are serializable. Thus poblems belonging to any input domain and output domain can be subjected to any desired operations.

2.We can deal with both passive objects and active processes and achieve a very complex set of computations using the syntactic model of the transaction.

3.Atomicity is ensured during an action and either all or none of a transaction operations are performed.
4.Consistency conditions, error-diagnosing and recovery methods are well-defined with respect to objects and actions.
5.If a transaction completes successfully, it commits and the results remain intact; if the transaction fails it aborts by undoing all the earlier actions to ensure database consistency.
6.The serializability notion to ensure total temporal order is well-defined and so concurrent (or partially ordered) operations can take place using the well-known concurrency control techniques-such as locks,timestamps to indicate priority and obsolescence. and implement temporal logic connectives.
7.Cascade-abort free schedules are well defined and can form the basis for avoiding failure cascade.
8.The recently introduced notion of transactional blackboard paradigm [TT91], [Ja89], [Co89] [EG89] fits very well to support the automated transaction design in a multiprocessor environment. In particular this paradigm would permit concurrent execution, as well as, negotiation and cooperation among subtasks in problem solving.
9.The transaction processing system can be integrated with the operating system and hence higher speed in problem solving can be achieved.
10.When a human solves a problem, the solution process has a similarity to transaction handling problem; for each transaction is an exploratory non pre-programmed real-time procedure that uses a memory recall (Read), acquires a new information and performs a memory revision (Write). Each transaction is also in addition provided with the facility for repair (recovery-Undo) much like the repair process encountered in human problem solving (see Chapter 5, [Gi83]).
11.Also in human problem solving, several independent or dependent information is acquired from various knowledge sources [LL83] and their consistency is verified before completing a step of the solution to achieve each subgoal ; this process corresponds to committing a subtransaction in a distributed transaction processing system, before proceeding to reach the next level of subgoal arranged in a hierarchy. Thus the transactional paradigm to solve a problem is verification -oriented [Pe94].

We now proceed to discuss how automatic problem solving can be carried out through transaction modeling.

3.AUTOMATING PROBLEM SOLVING

All the automatic methods described in the literature (see [Lo78], [BM79], [MW77], [BS89], [Bu83], [Kn91], [St89], [DMM91], [Ro91]) have a common basis: a problem is defined over a space called the '*Problem space* ' that models the real world where the problem lies. This space consists of the following triplet: states , operators and a language that provide their description . In the terminology of category theory, we can call this space as the signature of the problem. The states are defined by a set of conditions (or properties) that accurately describe the relevant features of the world. The operators have the capability to change the conditions or states when they are applied. However, these operators can only be applied when a set of conditions - called preconditions become true; the effects of applying operators are then described by another set of conditions - called post conditions [Pe94].

A problem is specified by a consistent initial state, which describes the initial configuration of the world, and a consistent goal state which describes the desired configurtion corresponding to the solution of the problem. Thus solving a problem amounts to finding a totally or partially ordered sequence of application of operators that can transform a given initial state into the desired final state - the goal, through a sequence of subgoals. In many problems the partial order may not exist among the subgoals and "ingenuity" is needed to break the infinite loops, by "undoing" previous subgoals.

A variety of approaches are currently used to achieve the desired sequencing:
(i) *forward chaining* beginning with the initial state to the goal through a set of successive transformations ,each transformation achieving a subgoal (Ru89] calls this 'data directed');
(ii) *backward chaining* starting from goal to reach the initial state through a succession of subgoals, (Ru89] calls this 'goal directed'];
(iii) *means-end analysis* uses a measure to compute the difference between the goal and current state and choose the operator that can minimize the difference;
(iv) *least commitment approach* in which each partial plan is refined by a proper choice and partial ordering of operators to refine the plan further; this strategy involves making some commitment for each subproblem at each pass, using these commitments to generate additional specifications, **undoing** earlier comitments if required, and repeating the process.

All these approaches can be used for automating problem solving through transactional paradigm.

Also note that the problem solving model used is quite general. The language used for describing states and operators can be of different types and of different complexities [Ja86]:
(i) elementary command languages in imperative form, or
(ii) production rules or
(iii) structured representation using forms -such as frame representation, object oriented representation or
(iv) pictorial or network (graph) form , or
(v) logic (first order predicate calculus) in declarative form.

The choice of a suitable language for description is dictated by its ease of comprehension, ease of programming, ease of datastructure manipulation, versatility, reasoning ability, flexibility for incorporation of new information and modification. Among the different approaches available at present the structured representation - in particular the object oriented representation of the problem space seems to offer several advantages as described in the following section.

4. OBJECT ORIENTED PROCESSING

Object-oriented processing has the following advantages : (see [Ki89], [Ru89], [CM90], [WMG89], [Hu91]):
(i) It provides a collection of objects that can correspond to data or process components.
(ii) It provides a mehanism, called ' *generalization or inheritance mechanism"* that permits objects to be organized in taxonomies in which specialized objects inherit the properties and operations of more generalized objects. Similar classes of objects which share properties and functions can be modelled by specifying a super class which defines the common part, and then deriving the specialized subclasses . Inheritance realizes the relationship 'IS-A'.
(iii) It provides for putting together objects to construct more complex objects through '*Aggregation mechanism* '. Aggregation provides a convenient mechanism for modelling the relationship IS- PART- OF between objects.
(iv) It provides for the '*Association mechanism* 'where a set of objects satisfying specific propertries can be associated together as a complex object. Association provides for the relationship 'covering' or 'IS_A MEMBER_OF'. . Hence the different immediate descendants of a hierarchy need not be mutually exclusive object classes.

The use of the above mechanisms (called abstraction mechanisms) result in the following advantages :
(a) The aggregation operation corresponds to the Cartesian product of sets; the generalization operation corresponds to the disjoint union; and the association operation corresponds to covering or powerset operation. From the programming language viewpoint these three operations respectively correspond to concatenation (sequencing), Selection (Choice) and Recursion (looping) which are the basic domain constructors. Accordingly , there exists an isomorphism between object oriented approach and programming language approach to describe a domain of knowledge and manipulate the objects in the domain. In fact this isomorphism is helpful to provide the denotational semantics of object-oriented programming languages in a way similar to the imperative or functional programming languages.

(b)They provide for encapsulation of very complex behaviour at higher level leaving the details transparent to the representation at that level.
(c)Since the related information is structured together, changes in problem situation can be easily transmitted to the representation.
(d)Since there is no distinction between objects and their behaviour, flexibility of control is achieved.

In view of the above advantages, it is clear that transaction processing on object-oriented database will be useful in automating problem solving. This means we must first be able to provide procedures that can generate abstractions automatically. This aspect will therefore be considered in the next section.
Remark: The use of Entity -relationship model and its integration with Gamma paradigm for the structuring of parallel programs is considered elsewhere [MK95b].

5. AUTOMATING ABSTRACTION

A basic requirement for automating problem solving is automatic abstraction. Thus finding good abstractions is very important for efficient problem solving.With this in mind, Knoblock [Kn91] has attempted automatic abstraction for problem solving. He has identified several useful abstraction hierarchies and provided an automatic approach for generating abstractions based on these properties.Knoblock [Kn91] presents two basic properties of abstraction hierarchies that relate a problem space to the possible abstractions of that problem space:

(i) *Monotonicity* : Structure of abstract solution should be preserved as it it is refined
(ii) *Ordered monotonicity* : This introduces a restriction on the monotonicity; not only the structure is preserved, but also that the space is partitioned such that achieving the properties at one level does not interfere with the properties in a more abstract level.

Essentially, the above two properties avoid the multiple inheritance situation of 'tangled hierarchy' so that each descendent has no more than one strand of ancestors. Such a non-tangled hierarchy permits an ordered flow of events; see Fahlman [Fa87]. However, in many problem solving situations cycles (that resemble deadlocks) can arise and ordered monotonoicity may not be achievable except through the use of ingenuity.

[Kn91] has not utilized the three very powerful basic abstraction mechanisms that are available in Object oriented modelling. We now explain how a combination of Knoblock technique together with the object oriented abstraction mechanisms such as aggregation, association and generalization is helpful in automating transactions and also to solve problems based on transaction paradigm.

6.TRANSACTIONAL PROBLEM SOLVING

We now illustrate the application of transactional paradigm for problem -solving using examples.

6.1.Example
Consider the problem of decomposing a composite number N as a product of powers of primes. The number N can be abstracted as an aggregation and association of suitable powers of primes from 2 to N. Thus a rudimentary approach to solve this problem is to remove powers of 2 by successive division, and construct a consistent data object $2^{a}.x = N$; then remove powers of 3 and form a consistent database $3^{b}.y = x$; then remove powers of 4 (now non-existent) and so on. This is a elementary modification of the Eratosthenes sieve approach; this approach uses forward chaining .Here each time N or the reduced N is read and each transaction constructs a consistent subgoal .
By using the notation iXR for reading object X by transaction i ,iXW for writing object X by transaction i.and iC for commit of i and iA for abort of i,we can express the above process as a schedule (a collection of transactions) thus:
1NR, 1N{:= 2^{a}.x}W, 1C; 2 NR, 2N{:= $2^{a}.3^{b}$.y}W, 2C; and so on.

6.2.Example (from Decker et al [DMM91)
A farmer f wants to move his bag of corn c, a dog d and a goose g from the site X on one side of the river to site Y on another side of the river using a boat B that can carry at most 2 objects. Only f can row the boat ; also g and d, as well as, c and g are not allowed to remain together in X or Y or B in the absence of f. Design a schedule (a set of transactions) to solve this problem using the dataobjects X, B, Y.

Abstraction: The basic idea is to represent and manipulate sets of candidate solutions. The principal operations are extraction of candidate solutions, splitting them and filtering them.The association abstraction serves as a filter and allow only the following possible membership of the subsets of {c,d,f,g} in X,Y :
{ c, d, f, g}, {c, d, f}, {c, f, g}, {d, f, g}, {c, d}, {c,f}, {d, f}, {f, g}, {c}, {d}, {g}, {null }.
The disallowed associations are: {c, d,g}, {c, g}, {d, g}, {f}.
In B only the following membership association is permissible:
B= {c, f}, {d, f}, {f, g}, {f}, {null}
Any other associations are invalid or inconsistent in the database at X, Y or B.

We use the same notation as before: iXR for reading object X by transaction i ; iXW for writing object X by transaction i. and iC for commit of i and iA for abort of i. Initially, the dataobject X is consistent with X= {c, d, f, g} and Y= {null}; eventually our goal is to have X= {null} and Y={c, d, f, g} through a sequence of subgoals (transactions) each leaving the database consistent. We use the least commitment approach that permits undoing earlier commitments , if required.

Transaction 1:
1XR:={X={c,d,f,g}};1BR:={B=null}; 1BW:={ B ={f,g}};note that any other choice of association would make B or X inconsistent at this point of time. Then 1 XW:={ X={c,d}}, and 1 commits. Thus we have:
1XR,1BR,1BW,1XW,1C as the initial schedule.We will use this notation hereafter.

Transaction 2:
2YR:={null}; 2 BR:= {B={f,g}};2YW:={ Y={g}}; 2BW:={B=f}; 2C.
Thus Y is consistent and our goal is to reduce the cardinality of set X denoted by $|X|$ to zero.Since $|X| \neq 0$ we need to perform more transactions.
Transaction 3:
3BR:= {B={f};3XR:={X ={c, d}};
3BW:={B={d, f}} , 3XW:={X={c}}, 3C.
Thus all these actions are consistent.

Transaction 4:
4YR:={Y={g}},4BR:={B={d,f}}; since Y={d,g} is not permissible and Y={d,g,f} will not achieve the goal |X|=0, we have a tricky situation here. Due to inconsistency that can arise we can abort transaction 4 and get back to a situation at commitment of transaction 3 ;this will make no progress since goal is not achieved. In other words, we are now in a situation of an infinite loop.Hence we need to violate an earlier subgoal by taking g back to X and leaving d in Y ; that is we "UNDO" an earlier committed transaction .This is described by: 4YW:={Y={d}}; 4BW:={B={f,g}}; 4C; thus consistency is got.
Remark: Such"UNDO" which violate earlier subgoals can result in a nonserializable schedule . In many problem solving situations only a nonserializable schedule can lead to semantic consistency and there may not exist a serializable schedule that results in semantic consistency [KM91]; see also [Kn91], [LRN86], [K91] for nonserializability of subgoals arising in the Rubik's cube problem.
Transaction 5:
5XR:={X={c}};
5BR:={B={f,g}};5BW:={B={c,f}};
5XW:={X={g}}; 5C.
Note |X|=1;|Y|=1. So there is improvement in the sense |X| is monotonically decreasing,while |Y| is monotonically increasing.This satisfies the means-end analysis.
Transaction 6 :
6YR:={Y={d}}; 6BR:= {B={c, f }};
6YW: = {Y= {c, d}}; 6BW := {B= {f}}, 6C.
Thus the dataobjects X, Y, B are left consistent.
Transaction 7:
7XR: = {X={g}}; 7BR:={f}; 7BW:={B={f, g}};
7XW:={X=null }; 7C.
Transaction 8:
8YR:={Y={c,d}};8BR:={B={g,f}};
8BW:= {B=null}; 8YW:={Y={c, d, f, g}}; 8C.
The goal is achieved.

6.3.Example
The constraints in the above example may be modified by permitting 3 objects in B. Then we need only 4 transactions:
Transaction 1: moves c, d, f from X to B leaving g in X;
Transaction 2: leaves c, d in Y and sends f back to X in B;
Transaction 3: puts g, f in B, leaving none in X;
Transaction 4: unload f, g in Y and achieves the goal.
We can have also have other solutions:
Transaction 1: moves f, d, g from X to B;
Transaction 2: leaves d in Y and moves f, g back to X in B;
Transaction 3: loads c, f, g in B to Y leaving none in X;
Transaction 4 achieves the goal.

6.4. Use of locks and timestamps
The association abstraction is very powerful to describe temporal asociations (as well as non-associations) encountered in assembly problem.It is well known how concurrency and temporal order in association abstraction, can be brought about using locks and timestamps [KM91]. If there is a violation of temporal consistency then the corresponding transaction will be aborted and the assembly left at a consistent state.Several diffferent timestamping and locking techniques are applicable depending upon the problem at hand.This is very vast area and so we will not go into the details.We will illustrate by a simple example the use of locks to control temporal order.

6.5.Example
Consider the vertical bottom-up assembly of blocks a,b,c,d,e given in Figure 1. Let t denote a table -top representing the database on which the required assembly takes place. We represent the association of objects using parentheses. also we use the relation xTy for x on the top of y, and xLy for x on the left with respect to y .

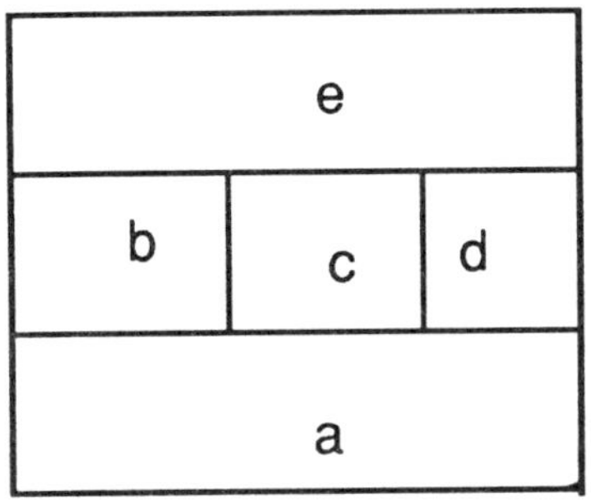

Figure 1

The whole assembly of Figure 1 can be described by : aTt and bTa and (cTa and cLb) and (dTa and dLc) and (eT(b and cLb and dLc)) ; these actions can be realised thus:
Transaction 1:
1(t,a)X (1 locks t and a); 1tR (1 reads t); 1aW (1 places a on top of t) ; 1(t,a)X* (Unlocks t and a); 1 C. [Result: Object a is placed on t].
Transaction 2:
2(a,b)X; 2aR; 2 bW, 2(a,b)X*, 2C.
[Result: b is on top of a]
Transaction 3:
3(a,b,c)X; 3(a,b)R; 3cW; 3(a,b,c)X*, 3C.
[Result:c is on left of b and on top of a]

Transaction 4:
4(a,b,c,d)X, 4(a,b,c)R, 4dW, 4(a,b,c,d)X*, 4C.
[Result: d on top of a and left of c]
Transaction 5:
5(a, b, c, d,e)X, 5(a, b, c, d) R, 5eW, 5(a,b,c,d,e)X*, 5C. {Result: vertical assembly complete}.

We can use the timestamp to control the order in the same way. Also we can use a concurrent approach in which we place a on top of t and concurrently sub -assemble (b and cLb and dLc) and put this subasdsembly on top of a followed by the action e on top of the subassembly (b and cLb and dLc).

We now consider some structural aspects of automated transaction design.

7. STRUCTURING AND CONTROL

From the examples provided in the Section 6, it is clear that automating transaction design for database systems (augmented with associated knowledge sources) would require a complete specification of a set of primitive components and their relations that satisfies a set of constraints . Such an approach to a complete specification is already available in the classic paper by Brodie and Ridjanovic[BR86]. However, in order to computationally realize this approach , we need to structure the Brodie-Ridjanovic approach and devise suitable control strategies. The impetus in this direction has been provided by Brown-Chandrasekaran [BC89] in their excellent monograph. The basic philosophy in Brown - Chandrasekaran design (BCD) rests on improving agreement (reducing inconsistencies and uncertainties) on common viewpoints or plans among a set of specialists (knowledge sources) through negotiation, which consists of a structured exchange of information. Such a negotiation should ultimately lead to shared plans or consistent viewpoints in order to arrive at the required specifications for transaction design. If a consistent viewpoint is not achieved a failure message is generated leading to a situation of 'back to the drawing board'.

Hence we call this paradigm ' a drawing-board paradigm' since it can possibly lead to a beginning afresh situation , when earlier plans have failed to achieve the desired goal. The drawing board paradigm thus consists in starting with an initial version, which is subsequently revised adding consistent viewpoints and eliminating inconsistent view points,as more and more instances are provided. This concept is analogous to the concept of version spaces suggested by Mitchell ,see Ja[86], [GM88] and [St91]. Also there are some similarities between the drawing board paradigm and the transactional blackboard pradigm [TT91] , [EG89], [Ja89], [Co89]. In order to facilitate planning, negotiation, failure detection, failure message generation and recovery, each specialist or knowledge source is structured on Brown-Chandrasehkaran design (BCD) thus :

1. Each specialist has a collection of plans and a plan consists of a sequence of calls to other specialists or tasks. In computer science terms, these calls resemble coroutines (if there is a symmetric calling relation), or subroutines (a master-slave relation or a client server relation) .
2. Each task is expressed as a sequence of steps with interspersed constraints. The tasks have no control on problem solving; as earlier mentioned they are only slaves executing steps satisfying specific constraints (expressed as predicates) and generate error messages when failure strikes.
3. Each step is also further structured. The steps provide tentative values for attributes of the design and check for consistency. If the required constraint fails, error message is generated and passed on to the task level and to the concerned specialist for the change of plans or constraints and choose new tasks. Thus communication between specialist to specialist and to the drawing board are effected through message passing, and communication between specialist and tasks and between tasks and steps are effected through shared memory or message passing .

The above structuring permits failure handling and recovery based on local knowledge. But this may not always be adequate since specialists at different nodes may fail to communicate and global knowledge will be needed for solving many problems.In the present attempt here,we therefore concern only with reliable nodes and links; the failures we are concerned are one of violation of constraints that cause inconsistencies. As a next step we need to consider negotiations among unreliable participants (similar to Byzantine generals problem; we will not discuss this here).

We also assume that before the final specifications are arrived at we work with failure- free shadow or copy or versions of all the available databases; and we execute hypothetical or pseudo-transactions to test the intention of actions for trial-error design. Such a scheme, called 'virtual copy ' was first suggested by Fahlman [Fa87]; it is used in ORION [Ki89].

A hypothetical or pseudo- transaction always aborts; that is no matter what changes are incorporated,they are undone. Thus they provide a mechanism for experimenting with the imaginary effects of 'what-if' changes to the database. Since these changes are not permanently recorded, the user has the freedom to examine the impact of complex changes due to automated transaction design. As advocated by Kim et al [Ki89] we can define transient and working versions. These have the following properties.

Transient (or pseudo version):

1.can be updated by the user who created it

2.It can be deleted by same user

3.New versions can be derived and promoted as a working version.

Working version :

1.Stable and cannot be updated.

2.Can be deleted by owner.

3.A transient version can be created from working version.

4.A transient version can be promoted to a working version.

The scheme proposed above will only permit loose coupling among the database and knowledge sources i.e. the planning system is not integrated and is built on the top of the existing DBMS. With sufficient experience and for specific domains our approach can perhaps lead to the design of tightly coupled systems that provides an integration with the kernel of the operating system of the database management sytem.

8. CONCLUSIONS

The approach described here will be of practical value in real - time consistency driven planning [MK90], robotized assembly tasks, Knowledge - based process planning, automated parallelisation of programs [MK95a], scheduling and for intelligent cooperation in robotics , expert systems ,constraint directed negotiation of resource allocations, seeking cooperative answers in databases [GM88] , knowledge base management, and learning in distributed systems and multi agent environments (see [RO91], [LW91], [DGC91], [SF89], [BGS91], [OSR91], [Ma91]). Further this approach can be of use in modeling human acqusition of mathematical concepts and processes , as well as, analytical thinking [Gi83], [LL83], [LRN86], [K91], [G88]. Accordingly, the integration of transactions and database modifications into logic database programming languages as described by Bottcher [B091] will be of great value to implement this approach practically. We briefly outline Bottcher's approach.

In order to integrate transactions with a logic programming language the following requirements are to be met:

1.The code of several transactions have to be combined into a single logic program;

2.Transaction steps as well as consistency checks are to be programmed in that particular logic language;

3.Backtracking should preserve atomicity of transactions; that is if transactions abort , then undo actions are to be done and committed transactions in the stable database should be left unharmed.

Bottcher [Bo91] has developed a logic language PROTIOS-L based on these concepts for handling transactions .

REFERENCES

[BC89] D.Brown and B.Chandrasekaran, Design Problem Solving, Pitman, London,1989.

[BG91] P.Brazdil, M.Gams, S.Sian, L.Torgo, and W.Va de Velde, Learning in distributed systems and multi-agent environments, in Machine Learning-EWSL-91, Editor:Y.Kodraff, Lecture Notes in Computer Science, Vol.482, Springer Verlag, New York, 1991, pp.412-423.

[BM79] R.S.Boyer and J.S.Moore, A computational Logic, Academic Press, New York, N.Y., 1979.

[Bo91] S.Bottcher, A semantics for the integration of database modifications and transaction brackets in a logic programming language, pp.1-10, in Information Systems and Artificial intelligence: Integration aspects, Editor:D.Karagiannis, Lecture Notes in Computer Science,Vol.474, Springer Verlag, New York, 1991.

[BR86] M.Brodie and D.Ridjanovic , On the design specification of database transactions, in Conceptual Modelling, Chapter 10, M.L.Brodie, J.Mylopoulos and J.W.Schmidt, Springer Verlag, New York, 1986.

[BS89] D.Benyon and S.Skidmore (editors), Automating Systems development, Plenum Press, New York, N.Y., 1989.

[Bu83] A.Bundy, Computer model of Mathematical reasoning, Academic Press, New York, N.Y., 1983.

[CM85] E.Charniak and D.McDermott, Introduction to artificial Intelligence, Addison Wesley, Reading., Mass., 1985.

[CM90] A.F.Cardenas, D.McLeod, Editors, Research Foundations in Object oriented and semantic database systems, Prentice Hall, Englewood Cliffs, N.J., 1990.

[Co89] D.D.Corkill, Design alternatives for parallel and distributed blackboard systems, pp.99- 136, in V. Jagannathan, R.Dodhiawala and L.S.Baum: Blackboard architectures and applications, Academic Press, New York, 1989.

[DG91] A.Delchambre, P.Gaspart, D.Coupez and P.R.Gilbert, Knowledge - based process planning, scheduling and error-recovery in robotized assembly, in Expert Systems and Robotics, Editors: T.Jordanides and B.Torby, pp.619-641, Springer Verlag, 1991.

[DMM91] M.Decker,G.Moerkoytte,H.Muller, J.Posegga, Consistency driven planning, pp.195-209, Lecture Notes in Artificial Intelligence, Vol.541, EPIA 91, Springer Verlag, New York, N.Y., 1991.

[EG89] J.R. Ensor and J.D.Gabbe, Transactional blackboards, in Blackboard Systems, Editors: R.Engelmore and T.Morgan, pp.557-561, Addison Wesley, Reading , Mass, 1989.

[Fa87] S.E.Fahlman,NETL: A system for representing and using real-world knowledge, M. I. T. Press, Cambridge, Mass., 1987.

[Fe90] H.Fetzer, Artificial Intelligence , Its scope and limits, Kluwer Academic Publishers, Boston, 1990

[Gi83] H.P.Ginsburg, The development of mathematical thinking, Academic Press, New York, N.Y., 1983.

[Gi88] A.Ginsberg, Automatic refinement of Expert System Knowledge Bases, Pitman, London, 1988.

[GM88] A.Gal and J.Minker, Informative and cooperative answers in databases using integrity constraints, pp.277-301, in Natural language understanding and Logic Programming II, Editors: V.Dahl and P.Saint-Dizier, Elsevier, New York, 1988.

[Hu91] J.G.Hughes, Object-oriented Databases, Prentice Hall International, New York, 1991.

[Ja86] P.Jackson, Introduction to Expert Systems, Addison Wesley, Reading, Mass., 1986.

[Ja89] V.Jagannathan, Realizing the concurrent blackboard model pp.85-97, in V.Jagannathan, R.Dodhiawala, and L.S.Baum: Blackboard architectures and applications, Academic Press, New York, 1989.

[Ki89] W.Kim et al , Features of the ORION object oriented database system,pp.251-282 ,in Object oriented concepts, databases and applications, Editors:W.Kim, , F.H.Lochovsky, ACM Press,Addison Wesley, Reading, Mass.,1989.

[KM91] E.V.Krishnamurthy and V.K.Murthy, Transaction Processing Systems, Prentice Hall , Englewood Cliffs, N.J., 1991.

[Kn91] C. A. Knoblock, Automatically Generating abstractions for problem solving, CMU-CS-91-120, Carnegie Mellon University, Pittsburgh, 15213, U.S.A.

[Ko91] R.Korf, Problem solving, Pitman, London, 1991.

[La86] J.Laird, P.Rosenbloom and A.Newell, Universal subgoaling and chunking, Kluwer Academic Publishers, Boston, 1986.

[LL83] R.Lesh, and M.Landau, Acquisition of mathematics, concepts and Processes, Academic Press, New York, N.Y., 1983.

[Lo78] D.W.Loveland, Automated theorem proving, North Holland, Amsterdam, 1978.

[LW91] T.Ludwig and B.Walter, Database Systems for the support of knowledge based systems, in Information Systems and Artificial Intelligence: Integration aspects, Editor:D.Karagiannis, Lecture Notes in Computer Science, Vol.474,Springer Verlag, New York, 1991.

[Ma91] N.M.Mattos, An approach to Knowledge base management, Lecture Notes in Computer Science, Vol.513, Springer Verlag, New York, 1991.

[MK90] T.Murata and N.Komoda, Real-time control software for transaction processing based on Coloured safe Petri net model, The Journal of Real -Time Systems, 1, 299-312, 1990.

[MK95a] V.K.Murthy and E.V.Krishnamurthy, Transactional Paradigm for the management of distributed parallelisation of programs, to be published.

[MK95b] V.K.Murthy and E.V.Krishnamurthy, Integrating entity relationship model with Gamma paradigm for structuring parallel program design, to be published.

[MW77] Z.Manna and R.Waldinger, Studies in Automatic Programming Logic, North Holland, Amsterdam, 1977.

[Ni71] N.J.Nilsson, Problem Solving Methods in Artificial Intelligence, McGraw Hill, New York, N.Y, 1971.

[Ni80] N.J.Nilsson, Principles of Artificial Intelligence, Tioga Publishing Co., Palo Alto, California,1980.

[NS72] A.Newell and H.A.Simon, Human Problem Solving, Prentice Hall, Englewood Cliffs, N.J., 1972.

[OS91] E.Oliviera, R.F.Silva and C.Ramos, Intelligent cooperation for robotics, in Expert systems and Robotics, Editors:T.Jordanides and B.Torby, pp.259-275, Springer Verlag, New York, 1991.

[Pe94] P.Pepper, Program development in algebraic setting, in Formal Program development, Lecture Notes in Computer Science, Vol.755, pp.225-261, Springer Verlag, New York, 1994.

[RO91] C.Ramos and E.Oliveira, An efficient approach to planning in assembly tasks, Lecture Notes in Artificial Intelligence, Vol.541, pp.210-221, EPIA 91, Springer Verlag, New York, N.Y., 1991.

[Ru89] D.M.Russinoff, Proteus:A frame based nonmonotonic inference system, pp.127-150, in Object oriented concepts, databases, and applications, W.Kim,F.H.Lochovsky,Editors: ACM Press, Addison Wesley, Reading, Mass.,1989.

[SF89] A.Sathii, M.S.Fox, Constraint directed negotiation of resource allocations, Ch.8,pp.163-195, in Distributed Artificial Inteligence, Vol.II , Editors: Les Gasser and Michael N.Huhns, Pitman, London, 1989.

[Sm94] D.R.Smith, Automating the design of algorithms, in Formal Program development, Lecture Notes in Computer Science, Vol.755, pp.324-354, Springer Verlag, New York, 1994.

[St89] D Steier, Automatic algorithm design with a general architecture for intelligence, CMU-CS-89-128, Carnegie Mellon University, Pittsburgh, 15213, U.S.A.

[St91] A.Strasser, Towards a lattice of knowledge representation systems, pp.200-213, in Lecture Notes in Computer Science, Vol.535, Editors:Jorrand, Ph., Keleman, J, Springer Verlag, New York, 1991.

[TT91] S.Tzafestas and E.Tzafestas, The blackboard architecture in knowledge-based robotics system, in Expert Systems and Robotics, Editors: T.Jordanides and B.Torby, pp.285-319, Springer Verlag, 1991.

[WM89] I.H.Witten,B.A.MacDonald, S.Greenberg, Specifying Procedures to Office Systems : pp.477-500, in D.Benyon and S.Skidmore Editors: Automating systems development, Plenum Press, New York, N.Y., 1989.

VARIABLE PRECISION ROUGH SETS APPROACH TO PREDICTING FACTORS AFFECTING DELAY LIKELIHOODS

Jack David Katzberg, MIEEE, MIEE
Faculty of Engineering
University of Regina
Regina, Saskatchewan, CANADA S4S OA2
TEL: (306)585-4733
e-mail:KATZBERG@ROBINHOOD.ENGG.UREGINA.CA

Pauline Katzberg
89 Haultain Crescent
Regina, Saskatchewan, CANADA S4S 4B4
TEL: (306)586-7544

Wojciech Ziarko
Computer Science Department
University of Regina
Regina, Saskatchewan, CANADA S4S OA2
TEL: (306)585-5213
e-mail:ZIARKO@CS.UREGINA.CA

ABSTRACT

A method of predicting the factors which most effect the likelihoods of specified events from data bases or tables of data is presented. This method is called the Variable Precision Rough Sets Model with Asymmetric Bounds. The data is broken into a limited number of meaningful ranges. The tables of data are then reduced to a minimum set of variables and rules for predicting the likelihood of the specified event. The problem of designing a pacing control system for a steel mill when delays are common is described. A data base of production data is then used to determine the best set of variables for predicting the likelihood of a delay.

1. INTRODUCTION

This work was motivated by a "pacing" problem in the steel industry. The rough sets methodology analyzes data in a data base or table by breaking continuous variables into rough meaningful categories and then using a set theoretic approach to infer the minimum set of variables and rules to predict the outcome of interest. In this paper we determine which variables and what categorization will give good prediction of delay frequency with reasonable complexity. There are many factors that could be used to predict the likelihood of a delay occurring while steel is being rolled by the finishing mill. In addition to product gauge, width, and material specification, the factors include melting practice and possible groupings of the melting practice, measured chemistry, carbon equivalent, rolling practice, finishing mill entry and predicted exit temperatures, predicted peak rolling force, and exit speed. If the effect of all these variables is to be determined then an organized approach is required. The variable precision rough set [Zia93] or probabilistic approximate classification methods [PWZ88] offer such an organized approach. However, the variable precision rough set model assumes we are interested in events with probability greater than 0.5. Normally delay probabilities are less than that. Consequently, the basic model was redefined to make it possible to characterize events whose probabilities fall into any range. We have termed this new methodology, for automated pattern analysis, Variable Precision Rough Sets with Asymmetric Bounds.

In this paper we present the basic definitions and those features of the method required for computation. It has been shown [KZ93] that this generalized model inherits all the basic mathematical properties of the original rough sets model introduced by Pawlak [Paw82, Paw91]. We then describe the steel mill pacing problem that motivated this work. Finally we describe how

Variable Precision Rough Sets with Asymmetric Bounds was applied to determine which variables were most important in predicting delay and the results that were obtained.

2. CLASSIFICATION

For the steel mill data we have analyzed, the data was available in a Steel Quality Information System data base of IPSCO INC. This information has been analyzed to determine the relationship between the variables holding values in specified ranges and the likelihood of delay. The data base reflects the presence of dependencies existing between two groups of system variables: the condition variables and the decision variables. Let

$$C^* = \{C_1, C_2, \ldots C_I\}$$

denote the classification of the set of situations in the data base generated by values of condition variables C. That is, each C_j is the set of elements which have values in the ranges specified for that element of the classification or partition. For our problem the decision variables of interest in the data we extracted are: melting practice, rolling practice, grade, transfer bar temperature, number of rolling passes, width, length, and final product gauge. Similarly, let

$$D^* = \{D_1, D_2, \ldots D_I\}$$

denote the classification of the set of situations in the data base generated by values of decision variables D. In this case there is one decision variable with two values: delayed or not delayed.

We are interested in the frequency of a set defined by a decision value given a set defined by specified values of the condition variables. Let us define the frequency of Y given X to be $p(Y|X)$. Let us assume we have some frequency, F, of Y given X that is of interest. Let us further assume that σ_1 and σ_2 represent a significant increase or decrease in frequency such that if

$$p(Y|X) \geq F+\sigma_1$$

then one decision will result or if

$$p(Y|X) \leq F-\sigma_2$$

then a different decision will be taken. If $p(Y|X)$ is between these limits either the decision may be uncertain or a third action might be appropriate. In the case of delay frequencies in a steel mill different decisions on pacing strategy will result.

As a given situation must either be in the set, Y, or its complement, -Y, and as a set and its complement are disjoint, for the set of situations, X, we can define

$$c(X,Y) = 1 - \text{card}(X \cap Y)/\text{card}(X)$$
$$= \text{card}(X \cap -Y)/\text{card}(X) = p(-Y|X)$$

$c(X,Y)$ is called the *classification factor*. It is the frequency of the complement of Y given X. In a manner analogous to the variable precision rough sets concept, if the frequency of Y given X is above the value of interest then the classification factor will be below some significant lower value

$$c(X,Y) \leq 1 - (F + \sigma_1) \equiv \ell$$

We will associate this lower range of classification factors with the inclusion of the set X within Y

$$Y \overset{\ell}{\supseteq} X \text{ if and only if } c(X,Y) \leq \ell.$$

Similarly if the frequency of Y given X is below the value of interest then the classification factor will be above some significant upper value

$$c(X,Y) \geq 1 - (F - \sigma_2) \equiv u$$

We will associate this upper range of classification factors with the exclusion of the set X from Y

$$X \overset{u}{\not\subset} Y \qquad \text{if and only if } c(X,Y) \geq u.$$

Note that it is only necessary that $0 \leq \ell < u \leq 1$. It is possible for both upper and lower classification factor ranges to be small.

3. APPROXIMATE DEPENDENCY OF ATTRIBUTES

Let Z be a subset of the condition variables, C, whose values induce the classification $Z^* = [Z_1, Z_2, \ldots Z_I]$ of the records in the data base. In addition, let $D^* = [D_1, D_2, \ldots D_J]$ be a classification of decisions or outcomes.

Given a class of decisions $D_j \in D^*$, we can define the lower approximation $\underline{Z}_\ell(D_j)$ of D_j with respect to the partition Z^* with respect to the lower classification factor, ℓ, as

$$\underline{Z}_\ell(D_j) = \{Z_i \in Z^* : \text{for all } Z_i \overset{\ell}{\subseteq} D_j\}$$

Given a class of decisions $D_j \varepsilon D^*$, we can define the negative region approximation $NEGZ_u(D_j)$ of D_j with respect to the partition Z^* with respect to the upper classification factor, u, as

$$NEGZ_u(D_j) = \{Z_i \in Z^* : \text{for all } Z_i \overset{u}{\not\subset} D_j\}$$

We will define the degree of dependency between the subset of condition variables, Z, and the decision, D_j, to be

$$\deg_{\ell,u}(Z, D_j) = \frac{\text{card}\left(\underline{Z}_\ell(D_j)\right) + \text{card}\left(NEGZ_u(D_j)\right)}{\text{card}(U)}$$

where card(U) is the total number of records in the data base. U signifies the universe of interest which in this case is that data base we are analyzing. This degree of dependency is of fundamental importance. It characterizes our ability to predict an outcome likelihood from the accumulated experience recorded in the data base. The degree of dependency is the proportion of those situations in the data base which can be assigned a higher than $1-\ell$ likelihood, of the occurrence of D_j, or lower than $1-u$ likelihood based solely on the values of variables from the set Z.

If we define $N(D_j,X)$ to be the number of elements associated with the decision set, D_j, in the set, X, then

$$N(D_j,X) = card(D_j \cap X)$$

The frequency of D_j, in the lower approximation is

$$p(D_j|\underline{Z}_\ell(D_j)) = \frac{N(D_j,\underline{Z}_\ell(D_j))}{card(\underline{Z}_\ell(D_j))} > 1-\ell$$

For each set, Z_i, within the lower approximation, the classification factor $c(D_j,Z_i)$ is lower than ℓ thus the frequency of $p(D_j|Z_i)$ is higher than $1-\ell$ in each set, Z_i. This implies this must be true for the lower approximation, $\underline{ZD}_j(\ell)$, as it by definition contains only these sets. The better the classification, Z^*, the greater the difference between the frequency of the decision or outcome, D_j, in the lower approximation and $1-\ell$.

The frequency of D_j, in the negative region is

$$p(D_j|NEGZ_u(D_j)) = \frac{N(D_j,NEGZ_u(D_j))}{card(NEGZ_u(D_j))} < 1-u$$

Again, as this is true for each set within the negative region it must be true for the negative region as a whole. Similarly, the better the classification, Z^*, the greater the difference between the frequency of the decision or outcome, D_j, in the negative region and $1-u$.

These frequencies give us additional insight into how well a classification is able to separate the frequency of a given outcome. They are particularly usefull when two or more classifications give a similar dependency.

4. STEEL SHEET AND PLATE PRODUCTION

The computation of delay for a steel mill "pacing control system" motivated this work and has been the first practical application of the Variable Precision Rough Sets with Asymmetric Bounds methodology. To understand this application one must first understand the IPSCO INC.'S production system. The layout of IPSCO INC.'s steel making facility in Regina is depicted in Figure 1. The caster produces slabs 15 to 30 feet long, 38.5 to 74 inches wide and 8 inches thick. The slabs are taken by rail car from the caster to a storage yard, termed the slab yard. If the slabs can not be loaded directly into the furnace from the transfer car they are stored in the slab yard.

The slabs are then loaded onto the walking beam charge table of the reheat furnace by an overhead crane. The operator of the roughing mill activates a slab extractor when he is ready to process the next slab. The slab extractor removes the slab from the discharge position in the reheat furnace and deposits the slab on the roll line. The reheated slab is then rolled to the 2HI roughing mill. The mill is a reversing mill. The slab is reduced from 8" thickness to about 1.2" thickness in 8 to 12 passes. The reduced slab is termed a transfer bar.

The transfer bar is then fed down the roll line to the 4HI finishing mill where it is further reduced to plate or sheet product. The finishing mill is a reversing steckel mill with backup rolls above and below the working rolls which are in contact with the product being reduced. The backing rolls provide support for the smaller diameter working rolls to prevent them from bending. The use of small diameter work rolls is advantageous because the smaller diameter produces greater rolling forces allowing a larger percentage reduction on each pass. The finishing mill has coiling furnaces on either side into which the steel may be coiled to prevent the long thin strip from cooling too much to be easily worked. The finishing mill has descaling sprays which may be used to remove any iron oxide which has formed. The finishing mill rolling time varies from about two and a half minutes for plate to over 10 minutes for some thin gauge strip. Typically the mill will process one slab about every 10 minutes.

After rolling in the finishing mill, sheet product is rolled into coils by the upcoiler. Between the finishing mill and the upcoiler it is water quenched by the laminar flow unit to give the desired coiling temperature.

The pacing control problem in a Steel Production System is the problem of determining when is it most economic to draw the steel from the reheat furnace, given the expected time the present product will be completed in the finishing mill and the associated probabilities of delay, the expected time it will take to prepare the next product on the roughing mill and transport it to the finishing mill and the expected cost of delay associated with that product. Close pacing in this type of operation will improve throughput, consistency and quality. Constant operation will keep temperatures and thus the product uniform. If a delay occurs in the finishing mill after a slab has been pulled, that slab will cool and be more difficult or even impossible to roll. The most economic pacing decision depends on the likelihood of delay for the product currently being rolled by the finishing mill. The first question to answer is how to group the material to calculate the most meaningful delay statistics.

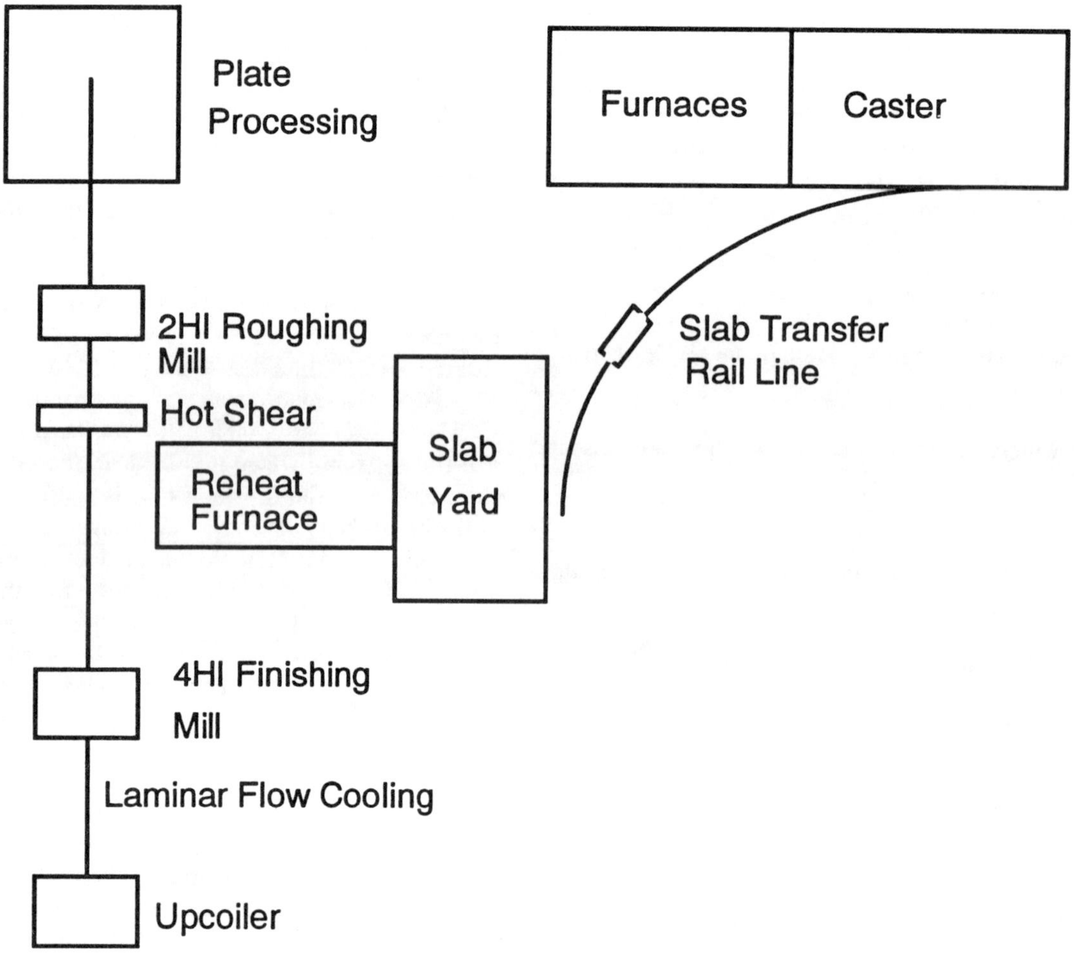

FIGURE 1: The Layout of IPSCO INC.'S Steel Making Facility

5. DATA COLLECTION AND ANALYSIS

Two weeks of production data containing 1301 records have been analyzed. The two week period is two of IPSCO INC.'s normal production cycles and thus insures a typical product mix. IPSCO INC.'s practice is to change the back-up rolls in the finishing mill once a week. Products are scheduled to be rolled based on the condition of these back-up rolls. This data was extracted as a text report from IPSCO INC.'s Steel Quality Information Data Base System. The variables extracted were: the slab identification number, the melting practice code, the rolling practice, the steel grade, the transfer bar temperature, the number of rolling passes on the finishing mill, p, the slab thickness, g_S, the slab width, the slab length, L_S, the final gauge of the finished product, g_f, the production time of the product in the finishing mill, T, the date on which the product was rolled, and the time at which rolling was started on the finishing mill. Records with abnormally slow production were identified as delays. The extracted data was analyzed using dBase IV v2.0.

In order to estimate slow production time it was necessary to first predict a normal production time. The formula used was

$$T = K_R L_f + K_P p + K_D$$

Our estimate of the start-up time, K_D, was 20 seconds. It is the time when the transfer bar is threading onto the first drum, before rolling properly starts. Our estimate of the delay time between passes, K_P, was 8 seconds. It is the time in which the finishing mill reverses. The first term in the expression is the estimate of the time the product is being reduced between the rolls. It is proportional to the length of the finished product, L_f, which we calculate as

$$L_f = \frac{L_S g_S}{g_f}$$

The rolling constant, K_R, is a function of the rolling schedule used. We calculated a separate rolling constant for each steel grade, divided into each of 10 gauge groupings, each of 3 width groupings, and either 5, 7 or 9 passes. We eliminated all groups with less than 4 records. That left us with 1247 records divided into 41 production groups.

To be able to predict production times we first calculated a typical rolling constant for each group. We assumed that not more than 50% of material would ever be delayed, so we could safely use the faster production in computing the rolling constant. We also felt it prudent to eliminate the fastest records as perhaps being atypical. For each record we used the actual production time to calculate its rolling time constant. The software then took the average of the group of rolling time constants in the 25% to 50% band, ordered form lowest to highest, as our typical rolling time constant.

This typical rolling time constant was then used to predict an estimate, T_e, of the normal production time for each piece. If

$$T - T_e > \left\{\left(K_d + K_p T_e\right) or \left(nE_{av}\right)\right\}$$

then that element was given the decision value 'delayed'. Otherwise it was considered 'undelayed'.

In the above equation:

K_d is a constant number of seconds

K_p is a constant percentage of the actual production time.

E_{av} is the average of the estimation errors in the group used to calculate the rolling time constant. It was calculated to determine the quality of our estimates.

n is an integer.

That is, if the amount the actual time exceed the estimated time was less than a specified percentage of actual production time plus a constant, that production was also not considered to be delayed. However, if the production times for a given grouping were very scattered this could be too stringent. So if the amount by which the actual time exceeded the estimated time was less than the multiple n times the expected error, E_{av}, that product was not considered to be delayed. We took n as 3, K_d as 20 seconds and after considerable experimentation chose K_p to be 8%. Using these constants, it was estimated 12% of production was delayed in the two week period under study. This seemed reasonable.

6. ANALYSIS OF DELAY DATA

The dBase IV v2.0 language was used to compute the degree of dependency, the frequency of delays in the lower approximation, and the frequency of delays in the negative region using the formulae of Section 3. Consideration and examination of available data, its consistency, and its availability for prediction suggested that melting practice, rolling practice, original grade, transfer bar temperature, number of finishing mill passes, width, length, and gauge were the best available variables for predicting delay likelihoods. In addition to the previous groupings, slab lengths were broken into two ranges and transfer bar temperatures into four ranges. Various groups of these variables were used to partition the data base of two weeks production data. The outcome set was all production we estimated to be delayed. An upper classification factor, u = 0.92 and a lower classification factor, ℓ = 0.79 were used. This

Table 1: The Degree of Dependency for several possible variables with the upper classification factor u = 0.92 and the lower classification factor = 0.79

Z	$deg_{\ell,u}(Z, D_j)$	$P(D_j \mid \underline{Z}_{\ell}(D_j))$	$P(D_j \mid NEGZ_u(D_j))$
Gauge	0.6541	0.2407	0.0319
Grade	0.7141	0.2359	0.0241
Rolling Practice	0.7417	0.2891	0.0288
Melting Practice	0.7733	0.3125	0.0299
Gauge, Melting Practice	0.7909	0.3498	0.0282
Gauge, Rolling Practice	0.7709	0.3345	0.0309
Gauge, Grade	0.7748	0.2874	0.0255
Melting Practice, Transfer Bar Temperature	0.7502	0.3820	0.0256
Gauge, Melting Practice, Width	0.7748	0.3534	0.0262
Gauge, Melting Practice, Transfer Bar Temperature	0.8186	0.3903	0.0226
Gauge, Melting Practice, Transfer Bar Temperature, Width	0.8255	0.3905	0.0225

corresponds to considering production delayed less than 8% of the time as little delayed and production delayed more than 21% of the time as frequently delayed. Table 1 contains a representative set of results. Computation times for 3 or fewer variables were under 1 minute on a 486DX33 with 8MEG of RAM. For seven variables the computation times went up to 36 minutes.

The best single variable for predicting delay was melting practice with a dependency of 0.77. This is contrary to conventional wisdom which holds that final gauge is the most important variable. It has normally been thought that, as cooling in the final passes greatly increases as gauge decreases and as cold ends of the strip can be difficult to handle, gauge would be the most important variable. It should be noted however that different melting practices tend to be used for different gauge ranges, so melting practice does have some implied gauge information. In addition, in the period we considered there were 35 melting practices and only 10 gauge groupings. The next best was rolling practice with a dependency of 0.74. Comparing these two, note that melting practice has the higher average frequency of delays in the group predicted to have increased delays, which indicates less low delay material is being averaged in.

The best pair of two variables we found was gauge and melting practice with a dependency of 0.79. Note as well that adding variables does not always add monotonically to the dependency. When melting practice and transfer bar temperature are taken together the dependency is actually lower than with melting practice alone. This is explained by noting that frequency of delays in lower bound has increased. This indicates some larger sets have been subdivided. A subset with evaluated delay statistics grouped as delayed increasing the frequency of delays. The rest of the set has been grouped in the boundary between the lower bound and negative regions, thus reducing the dependency.

The combination of melting practice, gauge and transfer bar temperature gave a dependency of 0.82. In other words in 82% of case we can predict whether elevated or less than average delay is likely. It also gives a large separation in the average frequencies. If we predict that a given product is in the elevated delay group, then it will be delayed on average about 40% of the time. If we predict another product is in the reduced delay group it will be delayed less than 3% of the time. This was the best combination we found. Adding further variables, such as width gave a negligible improvement.

7. CONCLUSIONS

In this paper we have shown how the Variable Precision Rough Set Model with Asymmetric Bounds can be used to analyze the dependencies in practical industrial problems. This procedure for automated pattern analysis can be used to determine the best variables and data patterns to categorize the data in terms of the likelihood of a given event. It is assumed the data is available in tabular or data base form. This information is then analyzed to determine the relationship between the variables holding values in specified ranges and the likelihood of the specified event occurring. This approach is suited to problems with random data in which some of the less likely outcomes are important. Such outcomes might be a delay of a given length, the possible discovery of oil, or the concentration of gold in an ore. In such problems we are interested in the likelihood or frequency of a decision variable when predicted by a given vector of condition variables. The approach which we have presented is computationally simple and requires modest computation resources.

We have applied this method to a practical steel industry problem and clearly indicated how the method has been applied. Our methodology has determined that a categorization using only three variables can be used to predict whether a given steel product is likely to be subject to increased or reduced delay frequencies, in most cases.

REFERENCES

[KZ93] J. D. Katzberg and W. Ziarko, "Variable Precision Rough Sets with Asymmetric Bounds", Proceedings of the International Workshop on Rough Sets and Knowledge Discovery RSKD-93, pp. 163-190, Banff, October 12-15, 1993 (published by Springer Verlag, Workshops in Computing Series).

[Paw82] Z. Pawlak, "Rough Sets", *International Journal of Computer and Systems Sciences*, vol. 11, pp. 145-172, 1982.

[Paw91] Z. Pawlak, *Rough Sets - Theoretical Aspects of Reasoning about Data*, Kluwer Academic Publishers, 1991.

[PWZ88] Z. Pawlak, S. M. K. Wong, W. Ziarko, "Rough Sets: Probabilistic Versus Deterministic Approach", *International Journal of Man-Machine Studies*," vol. 29, pp.81-95, 1988.

[Zia93] Wojciech Ziarko, "Variable Precision Rough Set Model", *Journal of Computer and Systems Sciences*, Vol. 46, No. 1, pp. 39-59, February 1993

Call Routing by Simulated Annealing*

Xin Yao and Nitin Kanani
Department of Computer Science
University College, The University of New South Wales
Australian Defence Force Academy, Canberra, ACT, Australia 2600
Email: xin@csadfa.cs.adfa.oz.au

Abstract

Simulated Annealing (SA) is a powerful stochastic search algorithm applicable to a wide range of problems. This paper presents some experiments of applying SA to an NP-hard problem, i.e., call routing in circuit-switched telecommunications networks. The call routing problem considered here can be described as assigning calls to paths in a network so that the number of calls blocked can be minimised. Previously, the problem has been approached by linear programming and some heuristic algorithms. However, none of these algorithms has solved the problem satisfactorily due to the nonlinear nature of the problem. This paper applies SA to the call routing problem in circuit-switched telecommunication networks. We have carried out two sets of experiments. The first set investigates random selection of paths in SA with uniform distribution. The second set studies non-uniform probabilistic selection of paths in SA according to whether a path is a direct one or not and the path capacity available. These two results are also compared with the result generated by a simple heuristic algorithm. Our experiments show that non-uniform probabilistic selection of paths in SA can lead to the least number of calls blocked among the three methods we investigated. We also examine the impact of different probability distributions on the result of routing in this paper.

*This work is supported by an ATERB (Australian Telecommunications and Electronics Research Board) grant and an ADFA Special Research Grant to the first author.

1 Introduction

Call routing is an important issue in circuit-switched telecommunication networks. The problem can be described as follows: Given a network with n nodes (i.e., switching stations), the total number of node pairs is $n(n-1)/2$. For each node pair (i,j), there is a direct *link* with capacity $A_{(i,j)}$. There is also call request (i.e., load request) for this node pair. A *path* for the node pair (i,j) is either a direct one, i.e., the direct link, or an indirect one, i.e., the one which connects the origination i with the destination j via a number of non-direct links. A *route* $R(k)$ for the node pair k is a sequence of paths, $p_1, p_2, \ldots, p_{L(k)}$, such that the traffic (calls) blocked from p_l is overflowed to p_{l+1} for $l = 1, 2, \ldots, L(k)-1$. $L(k)$ is a number depending on k. It can be the same or different for different k's. A *routing scheme* is a set $\{R(k) | k = 1, 2, \ldots, n(n-1)/2\}$ of routes for all node pairs. The aim of a routing algorithm is to find a routing scheme which minimises the number of calls blocked. In other words, we want to route as many calls as possible from their originations to their destinations without exceeding link capacities along the paths. Our formulation of the routing problem is similar to Chang's [1], but rather different from the formulation adopted in some other papers where the probability of call blocking is minimised. The main aim of this paper is to examine the effectiveness of applying SA to call routing.

Call routing in telecommunication networks is extremely complex in the real world because of its dynamic and nonlinear nature. There are numerous factors which can affect the result of routing. To make the problem approachable, following assumptions are usually made.

1. The call connection and disconnection time is very short in comparison with the call holding time, and thus is negligible.

2. Each link consists of n channels. If all n channels are carrying calls, the link is said to be *busy*. If a path consists of more than one link and at least one of them is busy, then the path is busy.

3. Each call can be routed through either the direct path, i.e., the direct link, or indirect paths, i.e., those paths which connect the origination with the destination via a number of non-direct links. A call is *blocked* if all the direct and indirect paths between the origination and the destination are busy. We only consider direct paths and two-link indirect paths in this paper.

It is worth noting that the above assumptions do not imply the routing problem investigated in this paper is simple. On the contrary, the problem is actually NP-hard! That is, there is no polynomial time algorithm which can solve the problem exactly unless P=NP, a very unlikely event.

In addition to minimising call blocking, a second criterion used in measuring the quality of routing is link capacity balancing. If all the calls can be routed successfully, the best routing scheme should be the one which leaves roughly the same link capacity for every link in the network so that congestion would not be easily formed when there are additional calls coming into the network.

Call routing has been studied by many researchers for a long time. Most of them used either linear programming or a heuristic algorithm to minimise call blocking [1, 2, 3]. However, none of these algorithms has solved the problem satisfactorily due to the nonlinear nature of the problem. Linear programming can only find a rough approximation to the actual global optimal solution. This paper applies SA to the call routing problem in circuit-switched telecommunication networks. Section 2 of this paper introduces the basic idea of SA. Section 3 describes the application of SA to the routing problem and our simulation results. The first simulation investigates random selection of paths in SA with uniform distribution. The second simulation studies non-uniform probabilistic selection of paths in SA according to whether a path is a direct one or not and the path capacity available. Section 4 presents the simulation results on an existing 7-node network. Finally, Section 5 concludes with some remarks and discussions.

2 Simulated Annealing

SA is a powerful stochastic search method applicable to a wide range of problems for which little prior knowledge is available. It has been widely used in computer-aided VLSI design, combinatorial optimisation, neural network training, image processing, code design, function optimisation, etc. The most important reason for SA's popularity is its ability to produce high quality solutions for hard problems.

The basic idea of SA comes from condensed matter physics. It is well known in condensed matter physics that a good way to find minimum energy states, called ground states, of complex systems such as solids is to use the annealing technique, in which the system (solid) is first heated to some high temperature, and then slowly cooled down. The system (solid) will reach a ground state if the cooling rate around the freezing point of the system is sufficiently slow. This process can be simulated on computers via abstract models, such as systems of interacting particles with many degrees of freedom.

At each step of the simulation, a new state of the system is generated from the current state by giving a random displacement to a randomly selected particle. The new state will be accepted as the current one if the energy of the new state is no greater than that of the current state, otherwise, it will only be accepted with probability

$$\exp\left(-\frac{E_{new\_state} - E_{current\_state}}{T}\right)$$

where E stands for the energy of the system and T is the temperature. This step can be repeated as many as necessary with a slow decrease of temperature in order to find a minimum energy state. Of course, only finite steps are taken in practical simulations. This simulation procedure was proposed by N. Metropolis et al. [4] and is called the Metropolis procedure.

The analogy between finding minimum energy states in a physical system and finding minimum cost configurations in a combinatorial optimisation problem was independently observed by S. Kirkpatrick et al. [5] and V. Černy [6]. They proposed a new algorithm for solving combinatorial optimisation problems, SA, based on the Metropolis procedure. Since then, research on SA has been booming in both theory and applications.

A combinatorial optimisation problem can be described as: Given a finite configuration space $S = \{X \mid X = (x_1, x_2, \cdots, x_m)\}$, where m is called the dimension of the space, and a cost function $C : S \longrightarrow R^1$,we want to find an optimum configuration $X \in S$, such that $\forall Y \in S, C_X \leq C_Y$. (We only deal with minimisation problems here. Maximisation problems can be treated similarly.) The neighbourhood system of a combinatorial optimisation problem can be defined as

$$\mathcal{N} = \{N_X \mid X \in S, N_X \subseteq S\} \tag{1}$$

where $X \notin N_X$ and $Y \in N_X \iff X \in N_Y$. Obviously, elements in N_X represent the configurations next to X in the configuration space. They are called neighbours of X.

The classical SA (CSA) [5] starts with an initial configuration generated at random. At each step, it selects the next configuration Y from the neighbourhood N_X of the current configuration X. The next configuration will be accepted as the current one if its cost is no greater than that of the current configuration, otherwise it will only be accepted with probability

$$\exp\left(-\frac{C_Y - C_X}{T}\right).$$

This procedure is repeated with a slow decrease of the control parameter T, called temperature, until a sufficiently good solution has been found. CSA can be summarised as Figure 1.

```
select initial configuration X at random;
select initial temperature T;
REPEAT
  REPEAT
    Select Y from N_X at random;
    IF C_Y ≤ C_X
      THEN accept Y as the new configuration
      ELSE accept Y as the new configuration
           with probability
           exp(-(C_Y - C_X) / T)
  UNTIL 'inner-loop stop criterion' satisfied;
  decrease temperature T
UNTIL 'outer-loop stop criterion' satisfied
```

Figure 1: Main steps of classical simulated annealing.

3 Call Routing by Simulated Annealing

It has been shown that SA can converge to global optima independent of the initial solution if certain conditions can be met [7, 8]. For some applications of SA, the initial solution is often generated at random. However, the finite time behaviour of SA still depends on the initial solution. A better initial solution usually leads to a better near optimum using shorter computation time.

We have used a simple heuristic method to generate an initial solution for SA so that better routing schemes can be found quickly. The initial routing scheme is generated according to the ratio between call request and path/link capacity. For each node pair (i, j), let the call request be $D_{(i,j)}$. For each path $p_l(i, j)$, let its capacity be $A_l(i, j)$, for $l = 1, 2, \ldots, L(k)$. $A_l(i, j)$ is the same as the link capacity for direct paths, and is the smaller value between the two link capacities for indirect paths. (Note that we only consider two link indirect paths.) The initial route for node pair (i, j) is a nonincreasing sequence of paths sorted according to

$$A_l(i, j) - D_{(i,j)}.$$

That is, paths with higher remaining capacity, which could be negative, are preferred. In our implementation, the number of paths generated for each node pair is a user adjustable parameter, called *num_paths*, and is restricted to be the same for all node pairs.

After generating the initial routing scheme by the heuristic method, SA is evoked. The cost function for a routing scheme R used by SA is defined as

$$C(R) = \sum_{i,j=1}^{n} \sum_{l=1}^{num\_paths} \left(D_{(i,j)} - A_l(i,j)\right)$$

where n is the number of nodes in the network. That is, the cost is measured by the number of calls blocked or the remaining path capacities. The initial temperature is chosen as the same as the number of call requests, i.e., $T_0 = \sum_{i,j=1}^{n} D_{(i,j)}$. The acceptance function is defined as

$$a_{R_{current}R_{next}}(T) = \min\left\{1, \exp\left(-\frac{C(R_{next}) - C(R_{current})}{T}\right)\right\} \quad (2)$$

where $R_{current}$ and R_{next} are the current and next routing scheme respectively, T is the temperature, and C is the cost function. The cooling schedule used in our implementation is

$$T_{next} = 0.9 * T_{current}.$$

It is well-known that the generation function in SA has a major impact on SA's performance [8, 9]. We have tested two methods of generating the next routing scheme from the current one; the uniform random selection and probabilistic selection. Two different probabilistic distributions have been used in our second experiment.

3.1 Uniform Random Selection Method

This method is simple and has been shown to be effective for some other applications. Roughly speaking, the next routing scheme is generated by selecting randomly two paths in a route which is also chosen at random, and then swap their positions in the sequence. For example, given a routing scheme $\{R(k)|k = 1, 2, \ldots, n(n-1)/2\}$ where $R(k) = \{p_1, \ldots, p_i, \ldots, p_j, \ldots, p_{num\_paths}\}$, the next routing scheme is generated according to the following algorithm:

1. Generate a uniformly distributed random number r between 1 and $n(n-1)/2$ and let $R(r) = \{p_1, p_2, \ldots, \ldots, p_{num\_paths}\}$.
2. Generate two uniformly distributed random numbers i and j ($j \neq i$) between 1 and *num_paths* and assume, without loss of generality, that $R(r) = \{p_1, \ldots, p_i, \ldots, p_j, \ldots, p_{num\_paths}\}$.
3. Generate the new route $R'(r) = \{p_1, \ldots, p_j, \ldots, p_i, \ldots, p_{num\_paths}\}$ from $R(r)$ by swapping the position of p_i and p_j and the new routing scheme by replacing $R(r)$ with $R'(r)$ (Other routes in the routing scheme are unchanged).

It should be noted that the same set of paths with different orders can give rise to different numbers of call blocking because routing between different node pairs is interrelated and a link can be shared by more than one path. After generation of the next routing scheme, Eq.(2) is used to determine whether the routing scheme should be accepted. According to Eq.(2), the routing scheme will always be accepted, i.e., become the new current scheme if it is better than the current one, or else it will only be accepted with a small probability determined by the cost difference and the current temperature. We have carried out some experiments with randomly generated networks using the above method. The result of our experiments with a randomly generated 20-node network is summarised in Table 1.

There are two observations which are quite clear from the above experimental results. First, our simple implementation of SA has improved the results of the heuristic method significantly. Second, the call blocking decreases monotonically as the number of available indirect paths increases. This implies that better utilisation of link capacities can be achieved with more indirect paths.

3.2 Probabilistic Selection Method

The uniform random selection method is simple, but it ignores the current status of a network, i.e., the current path capacity and call request in the network. It is not a very "intelligent" algorithm. For example, for a node pair, a path

No. of Paths	Initial Blocking	Final Blocking
4	2099	2062
8	1785	1754
12	1604	1589
16	1530	1495

Table 1: Results obtained using the uniform random selection method for a 20 node network. The number of paths indicates the number of alternative paths in each route for a node pair. The initial blocking is the result generated by our heuristic method, which is used as the initial solution for SA. The final blocking is the result produced by SA.

with capacity 100 will have the same probability to be assigned the next call as a path with capacity 1. Such a uniform distribution can lead to severe congestion in some parts of the network while there are still some available *link* capacity in other parts of the network which cannot be used. Another disadvantage of the uniform distribution is the long computation time needed to find a path with available capacity to route the next call. The algorithm might have to generate several new routing schemes before one can be accepted.

The probabilistic selection method attempts to make use of the information about the current path capacity and call request in the network in order to generate more promising next routing schemes. It biases towards those paths with more remaining capacity using a probability distribution when assigning calls. It also considers the link capacity balancing problem by distributing calls evenly among all available paths if no blocking occurs, so that congestion is less likely to happen when there are additional calls coming into the network. We have used the following probability distribution in our experiments:

$$Prob(p_l(i,j)) = \frac{\Delta A}{\sum_{l=1}^{n} \Delta A} \qquad (3)$$

where

$$\Delta A = \begin{cases} \beta\left(A_l(i,j) - D_{(i,j)}\right) & \text{if } A_l(i,j) \geq D_{(i,j)} \\ \frac{\alpha}{D_{(i,j)} - A_l(i,j)} & \text{if } A_l(i,j) < D_{(i,j)} \end{cases}$$

where β and α are two coefficients adjustable by the user. For each node pair, its next route $R(k)$ is generated by selecting (without replacement) *num_paths* number of paths from n possible paths according to Eq.(3). The probabilistic selection enables our algorithm to escape from possible local optima while allocating higher probabilities to those paths with more available path capacities.

We have carried out some experiments on the same 20-node network as that used in Section 3.1 and set $\beta = D_{(i,j)}$ and $\alpha = A_l(i,j)$ in our experiments. The results are shown in Table 2.

No. of Paths	Initial Blocking	Final Blocking
4	2099	1700
8	1785	1662
12	1604	1569
16	1530	1506

Table 2: Results obtained using the probabilistic selection method for the 20-node network. The number of paths indicates the number of alternative paths in each route for a node pair. The initial blocking is the result generated by our heuristic method, which is used as the initial solution for SA. The final blocking is the result produced by SA.

Table 2 has shown some interesting results in comparison with Table 1. In general, results produced by the probabilistic selection method are better than those obtained by the uniform random selection method. However, such difference decreases quickly as the number of alternative paths increases. In the case of 16 alternative paths, the uniform random selection method actually generates a better result than the probabilistic selection method. It seems that the probabilistic selection method can make better use of the network information and generate better routing schemes, especially when there are only a few alternative paths available for each node pair. There is a big difference in the number of calls blocked between selecting a path with large capacity and selecting one with small capacity. As the number of alternative paths increases, selection becomes less crucial since most paths will be selected anyway. In our example, it is more important to select probabilistically 4 best paths out of possible 19 paths than to select 16 out of 19 since the difference between two sets of 16 paths is small in comparison with that between two 4-path sets.

3.3 Probabilistic Selection with Bias Towards Direct Paths

It is worth noting that whenever an indirect path is used two links are involved. Hence, it is more costly to use indirect paths than to use direct ones. The third method of generating a routing scheme is to employ probabilistic selection with biases towards direct paths. For each node pair, the route $R(k)$ is generated by selecting the direct path with a much higher probability 0.8 and selecting an indirect path with probability 0.2. The indirect path is selected from a total of $n-2$ indirect ones according to a probability distribution similar to Eq.(3), i.e., l must be an indirect path in Eq.(3). The result of our experiments is summarised in Table 3.

No. of Paths	Initial Blocking	Final Blocking
4	2099	1551
8	1785	1460
12	1604	1459
16	1530	1468

Table 3: Results obtained using the probabilistic selection method which encourages utilisation of direct paths. The number of paths indicates the number of alternative paths in each route for a node pair. The initial blocking is the result generated by our heuristic method, which is used as the initial solution for SA. The final blocking is the result produced by SA.

It is clear from Table 3 that significantly better results have been achieved by our third method. The number of calls blocked is consistently lower than that obtained by our previous two methods. Such improvement is mainly due to the efficient use of links by direct paths.

4 Simulation Results on a 7-Node Network

This section presents our simulation results on an existing 7-node network shown in Figure 2 studied by Herzberg and Bye [2].

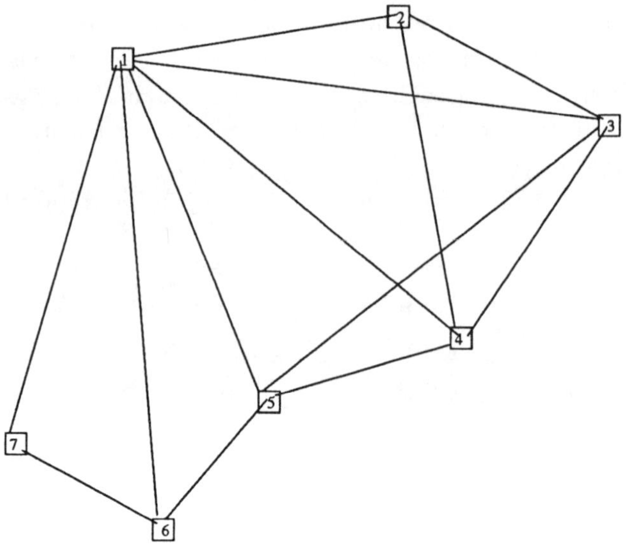

Figure 2: A 7-node network studied by Herzberg and Bye [2].

In order to evaluate the performance of our algorithms, we have carried out experiments with different call requests (i.e., traffic load) in the network. Figure 3 shows the results with different traffic loads, where the link capacity of the network remains the same for all experiments, i.e., it is always 521.

It is worth mentioning that the results shown in Figure 3 were averaged over ten runs of SA. According to the assumptions made in Herzberg and Bye's paper [2], the traffic load of 311 in this paper is roughly equivalent to the traffic load in Erlangs in their paper. In this case, SA with the probabilistic selection method (with or without bias towards direct paths) produces routing schemes with 0 call blocking. The exact results can be seen clearer from Table 4.

The above results were obtained with 4 paths in each route. If we increase the number of paths in each route, a decrease in call blocking is expected. Figure 4 shows the results of SA with probabilistic selection method using different number of alternative paths in a route. The traffic load (i.e., call

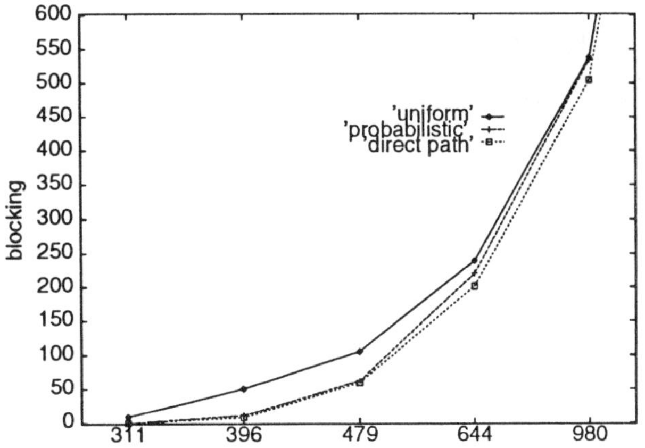

Figure 3: Simulation results on an existing 7-node network with different call requests. The horizontal axis represents traffic load and the vertical axis represents call blocking. The results were averaged over ten runs.

Calls	Blocking 1	Blocking 2	Blocking 3
311	8	0	0
396	51	11	8
479	111	62	59
644	237	220	201
980	541	534	505
1978	1484	1483	1436

Table 4: Simulation results on an existing 7-node network with different call requests. The total link capacity of the network is 521. Blocking 1, 2 and 3 list results generated, respectively, by SA with uniform random selection method, SA with probabilistic selection method and SA with bias towards direct paths. The results were averaged over ten runs.

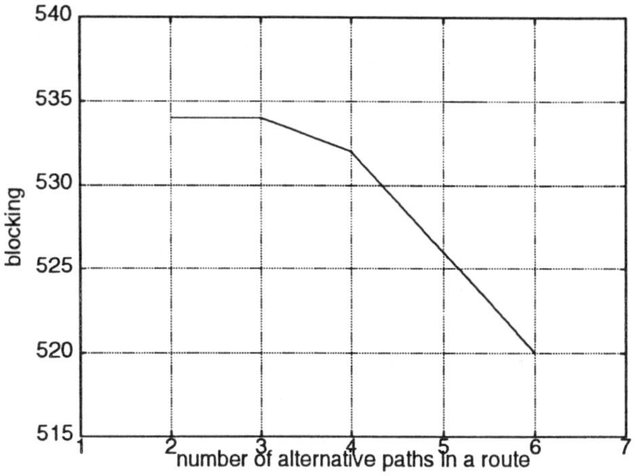

Figure 4: Simulation results of SA with probabilistic selection method using different number of paths in a route. The traffic load (i.e., call requests) is 980 and the link capacity is 521 for all the runs.

requests) is 980 and the link capacity is 521 for all the runs.

5 Concluding Remarks

This paper applies SA to an NP-hard problem, i.e., call routing in telecommunication networks. The advantages of SA include its simplicity, generality and the ability to produce high quality near optimal solutions (i.e., routing schemes). Our approach can be regarded as a two-stage one where an initial solution is first generated by a heuristic method and then further improved by SA. Three methods have been proposed and tested to generate the next routing scheme for SA. Our experiments on a 20-node network and a 7-node network have shown that SA with bias towards direct paths achieves the best performance in terms of minimum call blocking. Our experiments also support the view that introduction of biases into a stochastic algorithm usually increases the algorithm's efficiency greatly while the algorithm can still escape from local optima because of its stochastic nature.

Further improvement can be made to our existing selection methods. Our probabilistic method did not maintain a very strong link between the current and the next routing schemes because the algorithm tends to take long jumps in the search space relatively often, even at low temperatures. We believe the algorithm's performance can be further improved if the probability of taking long jumps can be reduced at low temperatures. A method similar to that proposed previously [8] might be the solution.

Our formulation of the call routing problem in this paper is similar to Chang's [1], but different from some others. It is, however, not difficult to apply our algorithm to other problem formulations. The next step in our study is to apply SA to a more realistic model of call routing in circuit-switched telecommunication networks and to use fast simulated annealing [10, 11].

ACKNOWLEDGEMENT — The authors would like to thank Dr. M. Herzberg for answering our questions regarding his paper [2].

References

[1] F. Chang, "Routing-sequence optimization for circuit-switched networks," *AT&T Technical Journal*, pp.57–63, May-June,1989.

[2] M. Herzberg and S. J. Bye, "Bandwidth management in reconfigurable networks," *Australia Telecommunication Research*, **27**(2):57–69, 1993.

[3] T.-K. G. Yum and M. Schwartz, "Comparison of routing procedures for circuit-switched traffic in nonhierarchical network," *IEEE Trans. on Communications*, **COM-35**(5):535–544, 1987.

[4] N. Metropolis, A. Rosenbluth, M. Rosenbluth, A. Teller, and E. Teller, "Equations of state calculations by fast computing machines," *Journal of Chemical Physics*, **21**:1087–1091, 1953.

[5] S. Kirkpatrick, C. D. Gelatt, and M. P. Vecchi, "Optimization by simulated annealing," *Science*, **220**:671–680, 1983.

[6] V. Černy, "Thermodynamical approach to the travelling salesman problem: an efficient simulation algorithm," *Journal of Optimization Theory and Applications*, **45**:41–51, 1985.

[7] S. Geman and D. Geman, "Stochastic relaxation, Gibbs distribution, and the Bayesian restoration of images," *IEEE Trans. on Pattern Analysis and Machine Intelligence*, **PAMI-6**:721–741, 1984.

[8] X. Yao, "Simulated annealing with extended neighbourhood," *International Journal of Computer Math.*, **40**:169–189, 1991.

[9] X. Yao, "Comparison of different neighbourhood sizes in simulated annealing," In P. Leong and M. Jabri, editors, *Proc. of Fourth Australian Conf. on Neural Networks*, pages 216–219, Sydney, Australia, 1993.

[10] L. Ingber, "Very fast simulated re-annealing," *Mathl. Comput. Modelling*, **12**(8):967–973, 1989.

[11] X. Yao, "A new simulated annealing algorithm," *International Journal of Computer Math.*, **57**(3+4):1–8, 1995.

SCHEDULING MANUFACTURING TASKS CONSIDERING DUE DATES: A NEW METHOD BASED ON BEHAVIOURS AND AGENDAS

Carlos Ramos[1][2], Ana Almeida[1], Zita Vale[2]

[1]Instituto Superior de Engenharia do Porto
Dept. Engenharia Informática
Rua São Tomé, 4200 Porto
Portugal
[2]Faculdade de Engenharia da Universidade do Porto
Dept. Engenharia Electrotécnica e Computadores
Rua dos Bragas 4099 Porto CODEX
Portugal
email: csr@fe.up.pt

ABSTRACT

This paper presents a methodology for finding time intervals for scheduling manufacturing tasks considering deadlines for industrial tasks. This method is an important part of a Decision Support System for dynamic scheduling and involves two phases: the forward influence phase and the backward influence phase. Algorithms for these phases are presented and explained. The method is appropriate to deal with real life problems of industrial Manufacturing Systems.

INTRODUCTION

An industrial task is composed by a set of operations. Operations are assigned to the resources of Manufacturing Systems (e.g. machines, robots, conveyors). To each basic operation performed in one or more product parts we can relate a processing time (this time can also include set up time, time for changing tools, time for transporting parts). By performing several operations on parts we will obtain a product (e.g. a smoke detector). However, orders for manufacturing are not for only one product, being usual to specify a set of N products. Nowadays, it is also usual to associate delivery dates (or deadlines) to the products that the client orders.

The scheduling of Manufacturing Systems is a well known problem for which many contributions have been done. Most of the proposed methods have been tested for job shop scheduling [Bak-74]. However, job shop approaches are too limited since they do not cover all industrial processes, namely those in which the parallel execution of the operations of the same task is possible. Moreover, it has been shown that the problem is NP-hard [GD-79]. Some well known traditional scheduling methods are: the Moore's algorithm; the Lawer's algorithm [Law-73] and the Johnson's algorithm [Joh-54]. Due to the complexity of the problem heuristic approaches have been adopted later, some examples of these being: the Holloway's algorithm [HN-74], the modified Giffler's algorithm [Fre-82], the Adams' algorithm [ABZ-88]; the Gray's algorithm [GD-91] and the Hoitomt algorithm [HLP-93]. However, all these methods are limited to cover cases more complex than traditional job shop, and usually they assume that all manufacturing task orders are available at the beginning. In fact, this is not a real assumption since orders may appear at any time and what is really useful is a tool for supporting decisions to help operators in scheduling new manufacturing orders.

The aim of this work is to define a new policy for a Decision Support System for scheduling operations on the resources of the Manufacturing System considering deadlines for the production of the goods specified in an industrial task. The method to be described uses some concepts of a method for computational task execution on the framework of Assembly Robotics considering temporal deadlines [RO-93].

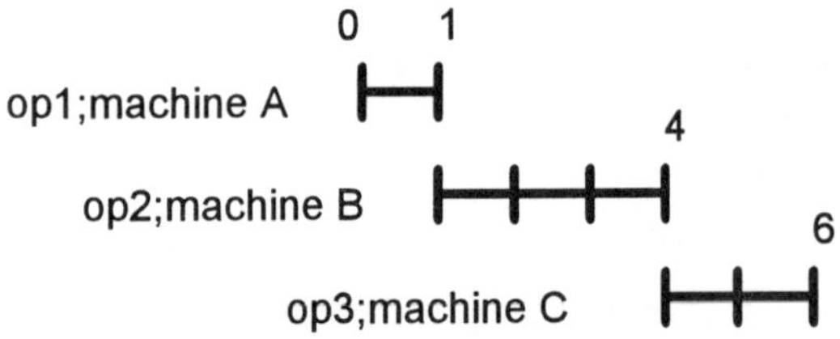

FIGURE 1 Behavior for one product

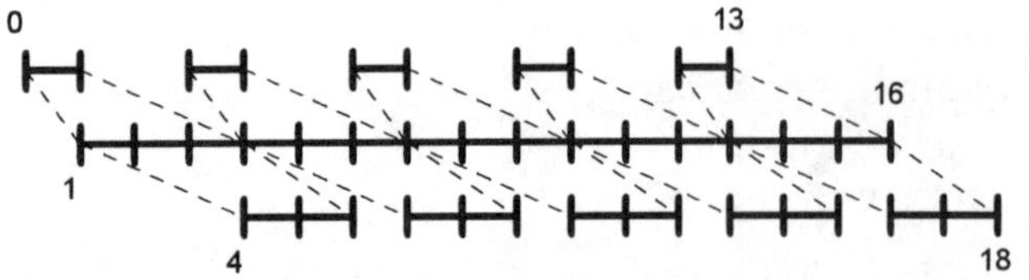

FIGURE 2 Behavior for 5 similar products

BASIC CONCEPTS OF THE METHOD

Let us explain our concept of behavior. A behavior is a well temporised sequence of operations to produce a good. Figure 1 illustrates the behavior for manufacturing a product. For this product 3 operations are necessary: op1, op2 and op3. The operation op1 needs 1 *tu* (time unit) in machine A, operation op2 needs 3 *tu* in machine B and operation op3 needs 2 *tu* in machine C.

However for producing N products we will not spend N*6 *tu*. Figure 2 illustrates that we have overlapped between the manufacturing of products. In this figure we see the manufacturing of 5 products by appending the basic behavior. In fact, after the first product we will be able to deliver a new product every 3 *tu*.

Let us consider Δtp_i as the processing time for operation i and Δtw_i as the waiting time for the next similar processing. For the illustrated example of figure 2 we have:

$\Delta tp_1 = 1\ tu$ $\Delta tw_1 = 2\ tu$
$\Delta tp_2 = 3\ tu$ $\Delta tw_2 = 0\ tu$
$\Delta tp_3 = 2\ tu$ $\Delta tw_3 = 1\ tu$

Considering these values we can calculate the occupation time for each resource, assuming N products to be manufactured, and using the following formula:

$\Delta toc_i = N*\Delta tp_i + (N-1)*\Delta tw_i$

For the example of figure 2 we will have:

$\Delta toc_1 = 5*1 + 4*2 = 13\ tu\ (13-0)$
$\Delta toc_2 = 5*3 + 4*0 = 15\ tu\ (16-1)$
$\Delta toc_3 = 5*2 + 4*1 = 14\ tu\ (18-4)$

Another important value to consider is the time interval from the beginning (or end) of an operation till the beginning (or end) of another operation (the first operation preceeds the second).

For this example we have:

$\Delta tin_{12} = 1\ tu$ $\Delta tfin_{12} = 3\ tu$
$\Delta tin_{23} = 3\ tu$ $\Delta tfin_{23} = 2\ tu$
$\Delta tin_{13} = 4\ tu$ $\Delta tfin_{13} = 5\ tu$

As we will see in section 5 one can use the concept of behaviors to deal with more realistic situations of Manufacturing Systems (e.g. setup times, transport times, queues, parallel execution of the operations of the same task dealing with cases more complex than simple job shop, configuration of the system).

The focus intervals are the time intervals in which the scheduling makes sense. The upper limit of the task focus interval is imposed by the delivery date (or deadline). The concept of deadline can be extended to the operations, for example the deadline of operation op2 (example of figure 2) is the deadline of operation op3 minus 2 *tu* ($\Delta tfin_{23} = 2\ tu$). The lower limit of the focus interval depends on the goal of the scheduling. If we are scheduling for the next week then the lower limit is the beginning of this week, if we are scheduling for immediate manufacturing then the lower limit is the current time plus the expected time for deciding about the scheduling. The concept of lower limit of the focus interval can be extended to the operations. For example, the lower limit of operation op2 is the lower limit for operation op1 plus 1 *tu* ($\Delta tin_{12} = 1\ tu$).

For instance, in the case of figure 2 the focus interval of operations will be put in the following form: (tinf,tsup). If ti is the lower limit for a task and td is the deadline for the same task then the focus intervals for the operations referred on figure 1 and 2 are:

$T1 = (ti\ ,\ td - \Delta tfin_{13})$
$T2 = (ti + \Delta tin_{12}\ ,\ td - \Delta tfin_{23})$
$T3 = (ti + \Delta tin_{13}\ ,\ td)$

Let us assume the following values: ti = 0 and td = 70, then:

T1 = (0,65)
T2 = (1,68)
T3 = (4,70)

An agenda represents the time intervals in which each resource is free, then we can easily know the time intervals in which the resource is in use. Each time interval of the Agenda is represented in the following form (tin_i, tfin_i).

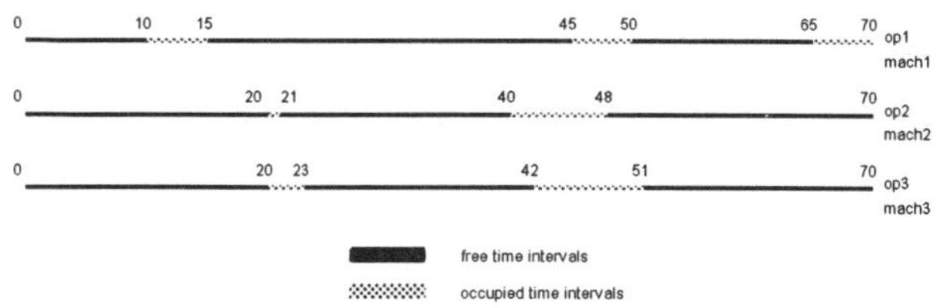

FIGURE 3 Free and Busy Time Intervals

Now, let us consider figure 3, where the free and busy time intervals for three resources (machines 1, 2 and 3) are represented for the time interval (0,70):

The agendas for the free time of these machines in the time interval (0,70) are the following:

$Ag1_{(0,70)}$ = [(0,10), (15,45), (50,65)]
$Ag2_{(0,70)}$ = [(0,20), (21,40), (48,70)]
$Ag3_{(0,70)}$ = [(0,20), (23,42), (51,70)]

AGENDA SET OPERATIONS

Let us define the following set of operations on Agendas:

$Agx_{(tin,tfin)}$: Agenda of a resource x in the interval (tin,tfin)

$\neg Agx_{(tin,tfin)}$: complement of an Agenda x in the interval (tin,tfin)
e.g. $\neg Ag1_{(0,70)}$ = ¬[(0,10), (15,45), (50,65)] = =[(10,15), (45,50), (65,70)]

$Ag|_{\{interval,\{constraints\},result\}}$: submission of all intervals of an Agenda to a set of constraints
e.g.
$Ag|_{\{(ti,tf),\{tf-ti < TL\},()\}}$ deletes intervals with duration inferior to T_L
$Ag|_{\{(ti,tf),\{\},(ti+\Delta T1,\ tf-\Delta T2)\}}$ adds ΔT_1 to ti and subtracts ΔT_2 from tf

$Ag_a \cup Ag_b$: union of two Agendas
e.g.
[(0,10),(15,30),(40,50),(55,70)]∪[(0,12),(16,32),(48,60)]= = [(0,12),(15,32),(40,70)]
notice that the union operates on time intervals and not on members of sets, for instance, the union of (15,30) with (16,32) results in (15,32)

$Ag_a \cap Ag_b$: intersection of two Agendas
e.g.
[(0,10),(15,30),(40,50),(55,70)]∩[(0,12),(16,32),(48,60)] = = [(0,10),(16,30),(48,50),(55,60)]

notice that the intersection also operates on time intervals
for instance, the intersection of (15,30) with (16,32) results in (16,30).

SCHEDULING WITH DEADLINES

Considering the operations on Agendas described before we will explain the basic concepts of our Scheduling methodology. Here, we will consider a relation machine-operation, what means that the index i will be used for operation i and for the machine where the operation is supposed to be performed. In order to find the possible time intervals for scheduling a task we apply a method involving two phases: the forward influence phase and the backward influence phase.

The algorithm for the forward influence phase is the following:

1) obtain the focus interval for operation i : (tc_i ,td_i)
where tc_i = tinf, for i = first (first operation of the task operation chain)
$tc_i = tinf + \Delta tin_{first,i}$, for i ≠ first
and
td_i = td, for i=last (last operation of the chain ; td = deadline for the task)
$td_i = td - \Delta tfin_{i,last}$, for i≠last
and compute the Agenda Ag_i in the interval (tc_i ,td_i)

2) if there is any influence of a previous operation k in the operation i under consideration then the new Agenda Ag_i' is obtained from the intersection of Ag_i and Ag_{ki}, where Ag_{ki} is the forward influence of k on i (see step four).
$Ag_i' = Ag_i \cap Ag_{ki}$
notice that we may have more than one operation k preceding the operation i, in this case the new agenda will be: $Ag_i' = Ag_i \cap Ag_{k_1 i} \cap Ag_{k_2 i} \cap \ldots \cap Ag_{k_n i}$

3) delete from Ag_i' the time intervals with duration inferior to toc_i.
$Ag_i'' = Ag_i'|_{\{(ti,tf),\{tf-ti<toc_i\},()\}}$

4) if operation i precedes operation j, the forward influence of i on j is obtained by adding Δtin_{ij} to the beginning of the time intervals and $\Delta tfin_{ij}$ to the end of the time intervals of Ag_i''.
$Ag_{ij} = Ag_i''|_{\{(ti,tf),\{\},(ti + \Delta tin_{ij},\ tf+\Delta tfin_{ij})\}}$

5) if the last operation has not been achieved then return to step 1 to the next operation of the chain (in section 5.4 we will see how to generate the sequence of operations from an operation graph, even when the parallel execution of operations exists in the same task).

This algorithm allows to deal with the forward influence between machines. Further, we will see that there is also a backward influence.

Let us illustrate the method with the example of figure 3 for the Agendas of the machines and considering the behavior of figure 1 used for manufacturing 5 products (as in figure 2).

First iteration of the method (i=1)

1)Focus Interval = (tinf , td - $\Delta tfin_{13}$) = (0 , 70 - 5) = =(0,65)

Ag_1 = [(0,10),(15,45),(50,65)]

2)since we have not operations preceding op1 then $Ag_1' = Ag_1$

3)toc_1 = 13 *tu*, and deleting the time intervals inferior to toc_1 we obtain Ag_1''

$Ag_1'' = Ag_1' |_{\{(ti,tf),\{tf-ti<13\},(\,)\}}$ = [(15,45),(50,65)]

4)considering the time differences between the beginning and end of operations 1 and 2

Δtin_{12} = 1 *tu*, $\Delta tfin_{12}$ = 3 *tu*, we can see the forward influence of machine 1 on 2

$Ag_{12} = Ag_1'' |_{\{(ti,tf),\{\},(ti + \Delta tin12,\ tf+\Delta tfin12)\}}$ = =[(16,48),(51,68)]

Second iteration of the method (i=2)

1)Focus Interval = (0+ Δtin_{12} , td - $\Delta tfin_{23}$) = =(0 + 1 , 70 - 2) = (1,68)

The Agenda Ag_2 in the interval (1,68) is:

Ag_2 = [(1,20),(21,40),(48,68)]

2)The Agenda Ag_2' is obtained from the intersection of Ag_2 with Ag_{12}

$Ag_2'=Ag_2 \cap Ag_{12}$=[(1,20),(21,40),(48,68)]∩[(16,48),(51,68)] = [(16,20),(21,40),(51,68)]

3) toc_2 = 15 *tu*, and deleting the time intervals inferior to toc_2 we obtain Ag_2''

$Ag_2''= Ag_2' |_{\{(ti,tf),\{tf-ti<15\},(\,)\}}$ = [(21,40),(51,68)]

4) considering the time differences between the beginning and end of operations 2 and 3

Δtin_{23} = 3 *tu*, $\Delta tfin_{23}$ = 2 *tu*, we can see the forward influence of machine 2 on 3

$Ag_{23}=Ag_2'' |_{\{(ti,tf),\{\},(ti+\Delta tin23,tf+\Delta tfin23)\}}$= =[(24,42),(54,70)]

Third iteration of the method (i=3)

1) Focus Interval = (0+ Δtin_{13} , td) = (0 + 4 , 70) = =(4,70)

The Agenda Ag_3 in the interval (4,70) is:

Ag_3 = [(4,20),(23,42),(51,70)]

2) The Agenda Ag_3' is obtained from the intersection of Ag_3 with Ag_{23}

$Ag_3'=Ag_3 \cap Ag_{23}$=[(4,20),(23,42),(51,70)]∩[(24,42),(54,70)] = [(24,42),(54,70)]

3) toc_3 = 14 *tu*, and deleting the time intervals inferior to toc_3 we obtain Ag_3''

$Ag_3''= Ag_3' |_{\{(ti,tf),\{tf-ti<14\},(\,)\}}$ = [(24,42),(54,70)]

4) since op3 is the last operation in the chain there is no forward influence from op3 to another operation

After this procedure we obtain as result the following Agendas:

$AG_{forward1} = Ag_1''$ = [(15,45),(50,65)]
$AG_{forward2} = Ag_2''$ = [(21,40),(51,68)]
$AG_{forward3} = Ag_3''$ = [(24,42),(54,70)]

The Agenda $AG_{forward3}$ is already the final set of scheduling possibilities for scheduling op3 on machine 3. However, for machines whose operations precede the operation to be allocated to machine 3 (op3) it is necessary to see the backward influence. In this phase we must start with the last operation of the chain in direction of the first operations. The algorithm for the backward influence is the following:

1) If operation i precedes operation j then subtract Δtin_{ij} to the beginning of the time intervals of AG_j and subtract $\Delta tfin_{ij}$ to the end of these intervals.

$AG_{ji} = AG_j |_{\{(ti,tf),\{\},(ti-\Delta tin_{ij},tf-\Delta tfin_{ij})\}}$

2) The final AG_i is the intersection of $AG_{forward\_i}$ with AG_{ji}

$AG_i = AG_{forward\_i} \cap AG_{ji}$

notice that we may have more than one operation j preceding the operation i, in this case the new agenda will be: $Ag_i' = Ag_i \cap Ag_{j_1i} \cap Ag_{j_2i} \cap .... \cap Ag_{j_ni}$

3) Repeat steps 1 and 2 for other operations until all precedences between operations have been taken into account.

For the example we are dealing with, we have:

First iteration: influence of machine/operation 3 on machine/operation 2

1) $AG_{32}=AG_3 |_{\{(ti,tf),\{\},(ti-\Delta tin_{23},tf-\Delta tfin_{23})\}}$= =[(24,42),(54,70)]$|_{\{(ti,tf),\{\},(ti-3,tf-2)\}}$ = = [(21,40),(51,68)]

2) $AG_2 = AG_{forward2} \cap AG_{32}$ = =[(21,40),(51,68)]∩[(21,40),(51,68)]= [(21,40),(51,68)]

Second iteration: influence of machine/operation 2 on machine/operation 1

1) $AG_{21}=AG_2 |_{\{(ti,tf),\{\},(ti-\Delta tin_{12},tf-\Delta tfin_{12})\}}$= =[(21,40),(51,68)]$|_{\{(ti,tf),\{\},(ti-1,tf-3)\}}$ = = [(20,37),(50,65)]

2) $AG_1 = AG_{forward1} \cap AG_{21}$ = =[(15,45),(50,65)] ∩ [(20,37),(50,65)]=[(20,37),(50,65)]

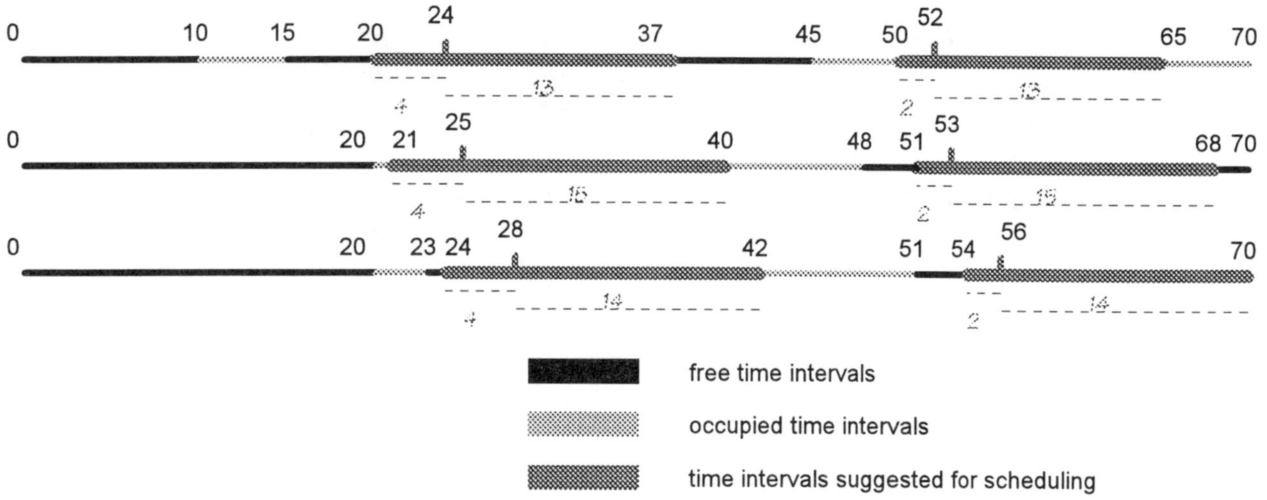

FIGURE 4 Result of scheduling

Figure 4 represents the possibilities of scheduling of operations 1, 2 and 3 on machines 1, 2 and 3 by the method presented in this paper.

By the analysis of figure 4 we can see that two possible time allocations are shown.

Besides the necessary time intervals for scheduling the task, it is shown at the beginning of the intervals an additional time that can be used for shifting the operations (4 *tu* for the first solution and 2 *tu* for the second). The first possibility of time allocation is obtained from the first pair of the lists AG_1, AG_2 and AG_3. For this solution the operation with more constraints is operation 2 limited on machine 2 at the left by the occupied time interval (20,21) and at the right by the occupied time interval (40,48). Notice that the free time interval (23,24) of machine 3 cannot be used since it is only possible to start operation 3 on machine 3 after 3 *tu* (Δtin_{23} = 3 tu). This constraint has been considered during the forward influence of operation 2 on operation 3. Notice also that the first pair of $AG_{forward\_1}$ was (15,45). However, when this free time interval was allocated it was not possible to know about the constraints of operation 2. This is the reason why we need to consider the backward influence. For this specific example, the free time intervals (15,20) and (37,45) of machine 1 cannot be used, the first due to the need to start operation 1 at most 1 *tu* (Δtin_{12} = 1 *tu*) before operation 2, the second due to the need to end operation 1 at least 3 *tu* ($\Delta tfin_{12}$ = 3 *tu*) before operation 2.

A similar analysis can be done to the second possibility of scheduling (now operation 1 will be the one with more constraints).

EXTENDING THE METHOD FOR MORE COMPLEX CASES

At the first sight the method proposed in this paper may appear to have some kind of limitations of the method to deal with real problems of operational scheduling. In this section we will show that the method is well adapted for most of the problems that may appear in real situations.

In the example of the previuos section setup times have not been considered. The setup time is the time needed for preparing a machine for an operation (e.g. load or change a tool on the machine; change the ink in a painting tool). In order to use setup times using the method proposed in this paper, the behaviors of our tasks will incorporate this setup time and some small changes are performed in some calculations:

the occupation time is:

$$\Delta toc_i = tsetup_i + N*\Delta tp_i + (N-1)*\Delta tw_i$$

the time between the start of operations i and j is:

$$\Delta tin_{ij}' = tsetup_i - tsetup_j + \Delta tin_{ij}$$

and a negative time $\Delta tin_{ij}'$ is now allowed.

Usually, it is also necessary to spend some time transporting products from one machine to the next machine in the manufacturing process chain. Some examples of this need are: the loading/unloading of machines performed by robot arms, the transport of products by means of conveyors and the transport of products by means of an AGV. This is the easier case to deal with since transport can be considered as any other operation in the manufacturing process. Machines considered for transport operations will be conveyors, AGVs or robots.

This paper has shown an example in which the products are not stocked after the end of an operation and before another operation. However, this is not mandatory in Manufacturing Systems. Usually, queues may be created. In the example of figure 2 this corresponds to put the operation op1 in the interval (0,5). Notice that this is a behavior different from the illustrated on figure 2. Due to this reason, we have implemented a module able to generate behaviors based on the specifications of the manufacturing operations precedences and times. The methodology for scheduling is the same as described before, the only difference is that some values will change (e.g. $\Delta tfin_{ij}$).

The example we have illustrated here has shown one case in which we have a completely sequential execution of operations for manufacturing the product. However, the presented method is more general, allowing to treat cases in which operations for manufacturing one product are performed in parallel. For this purpose we start with a graph representing a task and the precedences between the operations of the task. In this graph the nodes represent operations and the arches represent the precedences between the operations. This kind of graphs is used in another work that we are developing for generating and choosing plans for sequencing operations for Process Planning [RR-94]. The sequence of Agendas to deal with in the forward influence phase, as well as the values Δtin_{ij} and $\Delta tfin_{ij}$ to be considered can be obtained from the visit of this Precedence Graph. The sequence of Agendas to deal with in the backward influence phase is obtained by the inversion of the sequence for the forward influence. Figure 5 illustrates a Precedence Graph for a Task.

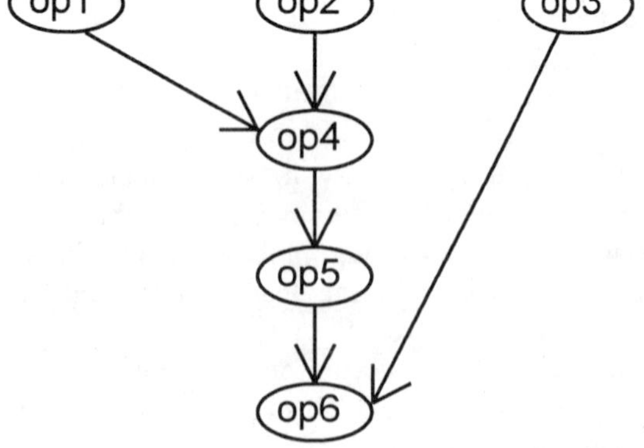

FIGURE 5 A Precedence Graph involving parallelism

In the example of figure 5 the sequence of Agendas to handle in the forward influence phase could be one of the following:

$Ag_1, Ag_2, Ag_3, Ag_4, Ag_5, Ag_6$
$Ag_1, Ag_2, Ag_4, Ag_3, Ag_5, Ag_6$
$Ag_1, Ag_2, Ag_4, Ag_5, Ag_3, Ag_6$
$Ag_1, Ag_3, Ag_2, Ag_4, Ag_5, Ag_6$
$Ag_2, Ag_1, Ag_3, Ag_4, Ag_5, Ag_6$
$Ag_2, Ag_1, Ag_4, Ag_3, Ag_5, Ag_6$
$Ag_2, Ag_1, Ag_4, Ag_5, Ag_3, Ag_6$
$Ag_2, Ag_3, Ag_1, Ag_4, Ag_5, Ag_6$
$Ag_3, Ag_1, Ag_2, Ag_4, Ag_5, Ag_6$
$Ag_3, Ag_2, Ag_1, Ag_4, Ag_5, Ag_6$

Any of these sequence of Agendas can be generated by an algorithm that visits the Precedence Graph, any sequence is good enough for the purposes of the method we have described in this paper.

The configuration of the Manufacturing System is important for the Scheduling of tasks. Sometimes, an operation in the manufacturing chain is much more expensive, in time, than other operations, acting as a "bottleneck" of the task. In this case there is a trend for the waste of the resources incharged of the other operations since these resources wait for the critic operation resource.

In order to solve this kind of problems it is usual to have more than one resource for the critic operation. Let us consider an example of 3 sequential operations in which one spends 4 *tu*, while each of the other two operations spends 1 *tu*. However, with 4 resources (M_a, M_b, M_c and M_d) for the most expensive operation it is possible to deliver one product each 1 *tu* since the 4 resources are working in parallel. This kind of problems is easily dealt with in our system because the module of behavior generation has the capability to generate behaviors for these configurations. Everything happens as if instead of one operation op we had four operations: op_a, op_b , op_c and op_d.

CONCLUSIONS AND FUTURE WORK

In this paper we have presented a methodology for finding time intervals for scheduling manufacturing tasks considering deadlines for the tasks. This method is an important part of a Decision Support System for dynamic scheduling. The original method presented here involves two phases: the forward influence phase and the backward influence phase. Algorithms have been presented for these phases. It has been referred that the method is appropriate to deal with real life problems of industrial Manufacturing Systems.

However, some aspects need to be taken into account in the future development of this Decision Support System. The balance between autonomy and user interaction is one of the most critic aspects in the real use of this kind of systems. A completely autonomous system without any possibility to interact with the operator is useless. The most important part of the expertise about scheduling Manufacturing Systems is not still acquired and remains in the operator. The Decision Support System must interact with the operator giving the possibility to accept his suggestions, even if the computerised method does not agree with him. Notice that he is the final guilty for the good or bad decisions concerning the scheduling.

Another important question is how to decide in which resource a certain operation will be performed, assuming that the operation can be executed in more than one resource. The solution for this kind of problems may involve negotiation between task agents and resource agents. We are developing some work in this topic [Ram-94] and we are envisaging the integration of the work described in this paper with the negotiation protocol we have established.

ACKNOWLEDGEMENTS

This work is partially supported by FLAD (Fundação Luso-Americana para o Desenvolvimento) under Project 443/93 - SFF.

REFERENCES

[ABZ-88] J. Adams, E. Balas. and D. Zawack. The shifting bottleneck procedure for job shop scheduling. Management Science, vol. 34, n. 3, pp. 391-401, 1988.

[Bak-74] K. Baker. Introduction to sequencing and scheduling. John Wiley & Sons Inc, 1974.

[Fre-82] S. French. Sequencing and scheduling: an introduction to the mathematics of job shop. Ellis Horwood Limited Publishers, 1982.

[GD-79] M. Garey and D. Johnson. Computers and Intractability: a guide to the theory of NP completeness. Freeman, 1979.

[GH-91] C. Gray and M. Hoesada. Matching heuristic scheduling rules for job shops to the business sales level. Production and Inventory Management Journal, Second Quarter, 1991.

[GRL-94] S. Gottschlich, C. Ramos and D. Lyons. Assembly and task planning - a taxonomy and annotated bibliography. IEEE Robotics and Automation Magazine, vol. 1, n. 3, 1994.

[HLP-93] D. Hoitomt, H. Luh and R. Pattipati. A practical approach to job-shop scheduling problems. IEEE Transactions on Robotics and Automation, vol. 9, n. 1, pp. 1-13, 1993.

[HN-74] C. Holloway and R. Nelson. Job shop scheduling with due dates and variable processing times. Management Science, vol. 20, n. 9, pp. 1264-1275, 1974.

[Joh-54] S. Jonhson. Optimal two-stage and three-stage production schedules with setup times. Naval Research Logistics Quarterly, vol. 1, n. 1, pp. 61-68, 1954.

[Law-73] E. Lawer. Optimal sequencing of a single machine subject to precedence constraints. Management Science, vol. 19, pp. 544-546, 1973.

[Ram-94] C. Ramos. An architecture and a negotiation protocol for the dynamic scheduling of manufacturing systems. IEEE International Conference on Robotics and Automation, pp. 3161-3166, 1994.

[R0-93] C. Ramos and E. Oliveira. CIARC: a multi-agent community for intelligent robotics assembly systems with real-time capabilities. IEEE International Conference on Systems, Man and Cybernetics, pp. 617-622, 1993.

[RR-94] J. Rocha and C. Ramos. Task planning for flexible and agile manufacturing systems. IEEE International Conference on Intelligent Robots and Systems, pp. 105-112, 1994.

A KNOWLEDGE-BASED SYSTEM FOR AUTOMATED INDUSTRIAL VISUAL INSPECTION PLANNING

Mu G. Jeong and Suk I. Yoo
Department of Computer Science, Seoul National University,
Shilim-Dong, Kwanak-Ku, Seoul, Korea, 151-742
jmk@zeus.snu.ac.kr

ABSTRACT

The essential question to automate various visual inspection tasks in the industrial domain is how to define an object model and develop an inspection program for any given inspection performance, which is called *the industrial visual inspection planning problem* in this paper. The decision-making works required to solve the planning problem are so knowledge-intensive and interrelated and should cope with various performances as well. As an effective approach to the planning problem on two-dimensional domains, a hypothesis-based reasoning model and a decomposed knowledge model are suggested. The suggested approach has been implemented as a knowledge-based system which supports an automated divide-and-conquer problem solving strategy based on an automated reasoning mechanism. The implemented system is illustrated with an application example and compared with the other approaches pursuing different objectives.

1. INTRODUCTION

The essential question to automate the various visual inspection tasks in the industrial domain is how to define an object model and develop an inspection program for any given inspection performance. In general, the inspection performance is given by both a manufactured part containing features to be inspected and inspection constraints specified by time and accuracy. Given an inspection performance, defining an object model involves such works that select the most promising object representation method, set up an image acquisition environment, construct an instance of the selected object representation method from a set of real images, verify the usability of the instance, and then modify or optimize the instance if necessary. On the other hand, given an inspection performance and an object model, developing an inspection program involves such works that determine a sequence of promising operators, set specific parameter values of the chosen operators, verify the usability of the instnace, and then modify or optimize the instance if necessary. These works are called *the industrial visual inspection planning problem(*shortly, *the IVIPP)* in this paper.

Difficulties of the IVIPP come from the fact that the decision-making works to solve the IVIPP are so knowledge-intensive and interrelated between them and should cope with various requirements of inspection performance as well. Because of such difficulties, it has been broadly accepted that human engineers can not perform the decision-making works effectively due to the inconsistency, the cost ineffectiveness, and no way to prove the optimality [Dra93, Chi88].

Thus, a lot of approaches for automation have been studied for decades. They can be classified into the following four groups based on their primary objectives. The first group focuses on only the object model definition problem on the assumption of pre-defined inspections or recognition programs, based on CAD data [MM93] or image data [CM89]. The second group focuses on only the program development problem on the assumption of pre-defined perfect object models, based on the knowledge of built-in operators [Dra93] or pre-defined programs(or strategies) [CT93, Mat89]. The third group considers both problems together but for only one(or very limited) requirement of inspection performance [Chi88]. The approaches of the three groups, however, may not always guarantee the required performance completely or may be useless if the requirement is changed. On the other hand, the fourth group considers the two problems together for various inspection performances. Most approaches in this group surely overcome such limits previously addressed. They, however, can not always guarantee the consistent and optimal solution because the specific steps must be completed by human engineers' decisions as Hedengren's approach [Hed89].

From the survey of related previous works, we may suggest the following five properties of a successful approach to the IVIPP: First, all the decision-making processes to solve the IVIPP should be integrated based on well-trained knowledge of decision dependencies between them. Second, for a required inspection performance, all possible object representation methods and all object model instances by an object representation method should be considered and an optimal object model instance can be determined. Third, given an object model and a required inspection performance, all possible inspection strategies and all inspection program instances of an inspection strategy should be considered and an optimal inspection program can be determined. Fourth, all specific decision-makings should be automatically completed by some automated reasoning methods and usability criteria. And, fifth, the problem solving power should be gradually enhanced by experience.

The primary objective of this paper is to design and implement an automated visual inspection planning system supporting the five properties. For such system, we suggest a knowledge-based approach based on the expert system techniques[BC89] and the blackboard paradigm[BC89] due to the complexity and heterogeneous features of the IVIPP. In addition, the problem scope of the approach is confined to the IVIPP for the two-dimensional features of industrial manufactured parts(shortly, *the 2D IVIPP*). From the confinement of the problem scope, we can assume that first, the work to set up an image acquisition environment would be simplified, second, as long as the image acquisition environment and the object to be inspected are not changed, the quality and contents of all acquired images would be kept same independently of the acquisition time, and third, the performance of the solution of the 2D IVIPP would be independent of the image acquisition environment.

The organization of this paper is as follows: Section 2 and section 3 describe the reasoning model and the knowledge model of our approach respectively. Section 4 describes a knowledge-based system implemented based on the suggested approach. Section 5 shows a problem solving example of the implemented system. Section 6 discusses conclusions of this paper.

2. HYPOTHESIS-BASED REASONING MODEL

The reasoning model of our knowledge-based approach is based on a concept, *hypothesis*. By the term, *hypothesis*, we mean object models and inspection programs not yet verified their usability. The object model hypothesis is classified into *the abstract object model* and *the object model instance*. The abstract object model refers to abstract object representations based on pre-defined symbolic terms, not ready for inspection yet, for example, "*(regional local indirect (total-area, perimeter, centroid-x, centroid-y, hole-number))*". The object model instance refers to a concrete object representation, ready for inspection, as an example for the above abstract object model, "*(1 (50 20 200 220) 2 (10.9, 13.7, 133, 100, 3))*". On the other hand, the inspection program hypothesis is similarly classified into *the abstract inspection program* and *the inspection program instance*. The abstract inspection program refers to the ordered lists of the primitive operators without parameter values, as an example using the object model instance addressed above, "*(capture, windowing, thresholding, regional-indirect-feature-extracting, regional-indirect-feature-matching, decision-making)*". The inspection program instance is the ordered list of the primitive operators having fixed parameter values, as an example using the abstract inspection program, "*((capture 1 0 0 1 1) (windowing (50 20 200 220)) (thresholding 130 256) (regional-indirect-feature-extracting 1 (total area perimeter centroid-x, centroid-y, hole-number)) (regional-indirect-feature-matching 1 1 3) (decision-making))*".

The hypothesis-based reasoning process is described precisely by the following algorithm.

Algorithm Hypothesis-Based-Reasoning-Process:
1. Get a specification of required inspection performance from a human engineer.
2. Based on the object representation knowledge, determine all abstract object model candidates and select a candidate out of them; if failed then return FAILURE if there is no solution by now or return SUCCESS if there is a solution at least by now.
3. Based on the inspection program knowledge, determine all abstract inspection program candidates and select a candidate out of them; if succeeded then go to step 6.
4. Based on the inspection operator knowledge, determine an abstract inspection program; if failed then go to step 2 to get another abstract object model candidate.
5. Based on the inspection operator knowledge, determine an inspection program instance by setting parameter values for the selected operators if necessary; if failed then go to step 3 to get another abstract inspection program candidate.
6. Based on some operators selected at the previous step and human engineer's cooperation, determine an object model instance by constructing specific contents of the abstract object model from a set of images of standard objects.
7. Based on a set of samples for test and their defect information supplied by human engineer, evaluate the performance of the inspection program instance and the object model instance.
8. If the result is good then update all related knowledge with positive effect, otherwise, update with negative effect; go to step 5 to determine another inspection program instance.

End_Algorithm

For object model definition, an image-based method is proposed from the two reasons: First, it can support various object representations such as boundary-based, region-based, and skeleton-based, while CAD-based methods generally support only the boundary-based representation. Second, it does not have such representational disagreement between object models and object instances that CAD-based methods involve.

For inspection program development, a hybrid method which integrates a rule-based hierarchical planning method [BC82] and a case-based planning method [BC82] is proposed. The rule-based hierarchical planning method always finds a plan for any given inspection goal and supports such advantages of conventional hierarchical planning methods. It, however, might be wasteful where similar inspection goals are frequently given because it would repeat same complex searching process for every similar goals. On the other hand, the case-based planning method can save such wasteful computations by using old cases through indexing and support the case-based learning capability. It, however, might not find a solution if the cases accumulated by the time may not have similar cases for a given goal. By integrating the two methods, the hybrid planning method always finds a solution for any inspection goal effectively and supports gradual improvement of problem solving power.

For performance evaluation, a statistical image-based method is proposed, since the performance of industrial visual inspections in real-world refers to the averaged performance after sufficiently large number of applications to different samples of various quality level. The method requires human engineer's assistance for the preparation of samples for test and the acquisition of some real-world information about defects on them.

For optimization, a hierarchical bottom-up method is proposed, since this strategy can search all possible cases with little effect on the current result. As the top-level optimization processes, as shown in the algorithm, of the whole optimization hierarchy, the inspection program instance optimization level triggered at step 7 is the bottom-most level, the abstract inspection program optimization level triggered at step 5 is in the intermediate level, and the abstract object model optimization triggered at step 4 is the up-most level. The optimization is achieved from the lowest to the highest. Similarly, for all levels of the optimization hierarchy, the bottom-up method would be applied.

For learning, a short-term learning and a long-term learning are proposed. The short-term learning performs the information accumulation activities to exclude wasteful cyclic searching for a current problem, while the long-term learning performs the information accumulation to gradually improve the problem solving power for next problems.

For integration of all the decision-makings, a least-commitment policy based on the temporal dependencies to complete each decision-making work of the reasoning process is adopted, since it makes possible to find complete solutions with little wasteful effort.

3. DECOMPOSED KNOWLEDGE MODEL

All the strategies of the reasoning model are based on various kinds of information, called *knowledge*. The notion of knowledge in this paper includes all information given or derived during problem solving process such as factual data, heuristic rules, various strategies, and primitive procedural operations. In addition, a group of knowledge is defined as a meaningful problem solving unit to perform a decision-making step of the reasoning.

Application Domain Knowledge - The application domain knowledge specifies all possible application domains. An application domain is defined by object descriptions and constraints imposed on the objects. A specified application domain is called *the inspection goal*. An inspection goal defines a 2D IVIPP implicitly. For example, given an inspection goal for a mark inspection domain, "*((individual local shape) (1 0.2 30) (98 1 70))*", where the *(individual local shape)* specifies the object to be inspected, the *(1 0.2 30)* specifies a time constraint, and the *(98 1 70)* specifies an accuracy constraint, it define implicitly a 2D IVIPP, "*Plan an inspection program and an object model to inspect the given object, (individual local shape), guaranteeing that the processing time should be in 0.8 sec. - 1.2 sec; the accuracy should be in 97% - 99%; and the accuracy should be considered more than the time by 1.25 times*".

The object of the application domain knowledge refers to a set of symbolic terms which represent real-world 2D features of industrial manufactured parts, such as, "individual" or "relational"; "global" or "local"; "shape", "area", "frequency", etc.. Two object types are considered in this paper, *the individual object* and *the relational object*. The individual object refers to such 2D features without regard to any relationships between them as a local shape, a global shape, area, length, perimeter, radius, etc., while the relational object refers to such 2D features with regard to various relationships between the individual features as distribution, layout, frequency, serial-id, and so on.

The constraint of the application domain knowledge refers to a set of real number for time and accuracy. The time constraint specifies the required time to complete an inspection cycle, the maximum error tolerance, and the relative weight to the accuracy weight, while the accuracy constraint specifies the required statistical probability of the correct decision after testing a number of samples, the maximum error tolerance, and the relative weight to the time weight.

Object Representation Knowledge - The object representation knowledge specifies a set of representation methods and a set of decision rules. Object representation refers to internal descriptions of the object based on image data. In this paper, an object is represented by the three-level abstractness(*the most abstract level*, *the intermediate-level*, and *the least abstract level*) in order to support the least-commitment policy and the bottom-up optimization policy of the reasoning model. The most abstract level represents an object by all possible combinations of the five types: *the boundary-based*, *the skeleton-based*, *the region-based*, and *the relation-based.*

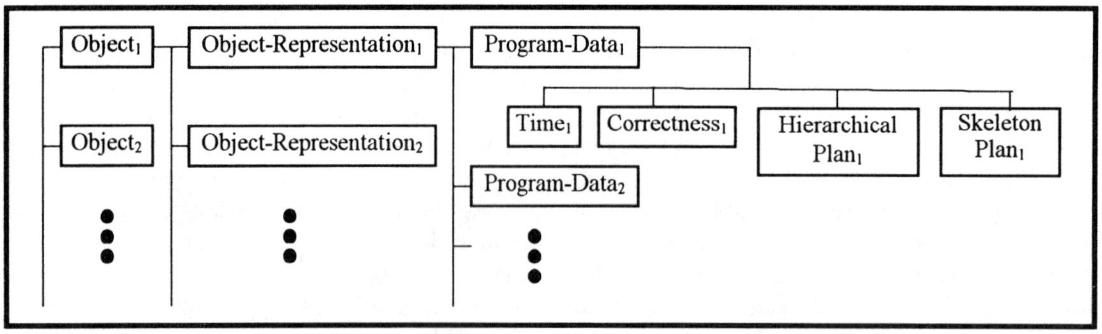

FIGURE 1. The inspection program knowledge

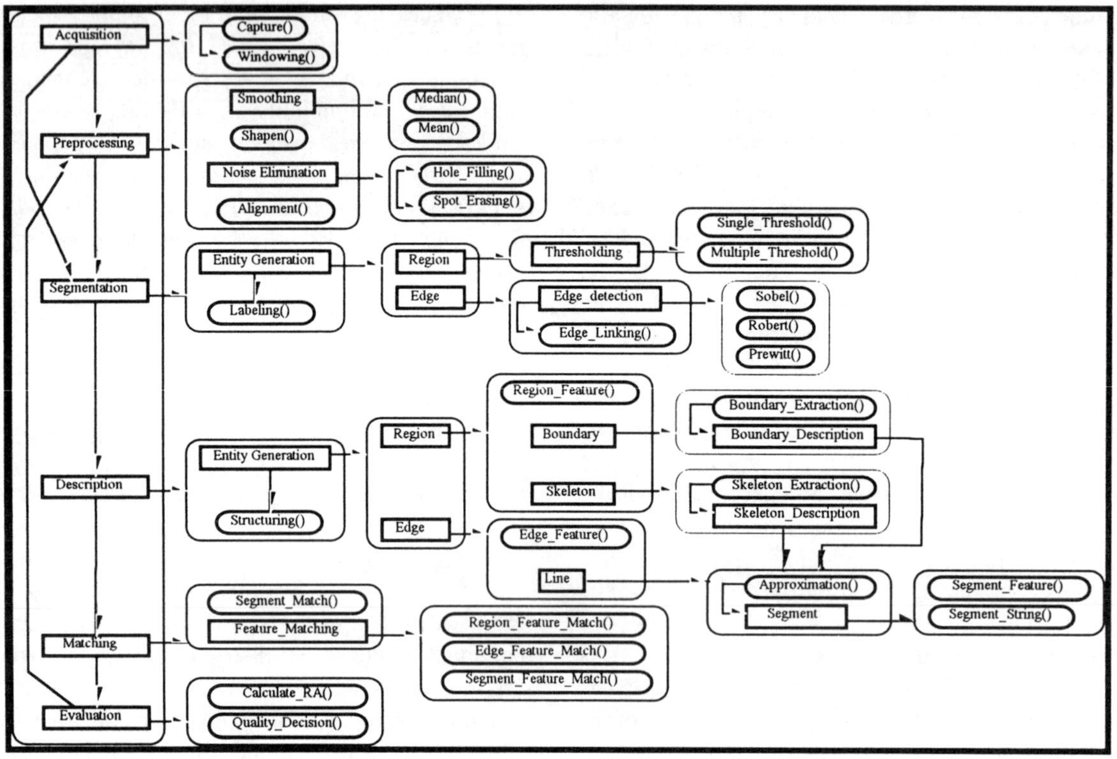

FIGURE 2. The inspection operator knowledge

The intermediate level represents it by the two types: *the direct* and *the indirect*. The direct type means direct representations of the results of the applied primitive operators, while the indirect type means their indirect representations. The least abstract level represents by all the possible results of them abstractly. For example, the least abstract level for the regional-based indirect representation would define items computed from real images by applying some image analysis operators, such as, the number of holes in a blob, the number of pixels of a blob, and so on.

Between the object and the object representation, a set of decision rules can be defined as follows: *If (an object is determined) then (one or more high-level object representations are selected); If (an object AND a high-level object representation is determined) then (a intermediate-level object representation is selected); If (an object AND a high-level representation method AND a intermediate-level object representation are determined) then (one or more low-level object representational are selected).*

Inspection Program Knowledge - The inspection program knowledge represents a set of inspection programs and a set of rules for searching promising inspection program cases. An inspection program is an unit to represent an inspection strategy and can be defined by the five factors: *an object, an object representation, a skeletal plan, a hierarchical plan*, and *its performance data.* All the inspection program cases that have been developed until now are organized into such a hierarchical data structure as shown in Figure 1 for effective indexing.

A set of indexing rules can be defined as follows: *If (object$_i$ is equal to the current object AND object-representation$_i$ is equal to the current object representation AND the expected performance value calculated by step 3 and 4 of the performance evaluation algorithm using the constraints given as inspection goal and the constraints$_i$ of a selected program case is not bad) then (the skeletal plan $_i$ and the hierarchical plan $_i$ is set as a candidate for an abstract inspection program).*

Inspection Operator Knowledge - The inspection operator knowledge specifies a set of inspection operators, a set of relationships between them, and a set of decision rules for parameter setting. The inspection operator is an unit to represent an action composing inspection programs, which is defined by the four components: *an input*, *an output*, *a function*, and *a set of parameters*. The inspection operators, primitive or abstract, are organized into such a hierarchical structure as shown in Figure 2. The hierarchical representation effectively supports the rule-based hierarchical planning method of the reasoning model. All leaf nodes are primitive operators actually implemented by C language. The others are abstract operators.

A set of rules by the knowledge structure can be defined to represent such two types of relationships as all possible relationships between the operators in each group denoted by a larger rounded rectangle as follows: *If (current processing is for programming AND an operation is given AND performance measure is given) then (next operation to perform is selected);*
and all possible expansion relationships between each of the operators in a group and a sub-group as follows: *if (the current processing is for expansion of an abstract operator AND an operation is given) then (related next detail level operators are loaded AND perform another programming mode based on the loaded operators).*

In addition, for every primitive operator which has one or more parameters, a set of rules to decide appropriate values are defined as follows: *If (evaluated performance values are given AND an operator of the inspection program instance previously constructed is given with its parameter value) then (the parameter value is modified by some degree).*

Performance Evaluation Knowledge - The performance evaluation knowledge specifies various performance criteria and the performance evaluation strategy. The criteria includes false-negative error rate, false-positive error rate, accuracy, accuracy performance, time, time performance, and integrated performance.

The performance evaluation strategy is the algorithm, *Image-Based-Statistical-Method.*

Algorithm Image-based-Statistical-Method:

1. For all sample objects numbered N_{so}, the following process is applied: Given i'th sample object including some artificially madden defects about which various data such as the defects number(n_{AEi}), position, and size are known, compute the frequency of false-positive errors(N_{PEi}), the frequency of false-negative errors(N_{NEi}), the inspection time(t_{i*}), and the accuracy(a_{i*}) after applying the hypotheses by N_i times as followings:

$$t_{i*} = \sum_{j=1}^{N_i} t_{ij} \,/\, N_i,$$

$$\text{if } N_{AEi} > 0,\ a_{i*} = 1 - \sum_{j=1}^{N_i} \{(N_{PEij} + N_{NEij}) \,/\, N_{AEi}\} \,/\, N_i;$$

$$\text{else if } N_{AEi} = 0,\ a_{i*} = -\sum_{j=1}^{N_i} N_{PEij} \,/\, N_i;$$

2. Compute the average inspection time(E_t) and the average accuracy(E_a);

$$E_t = (\sum_{i=1}^{N_{so}} t_{i*}) / N_{so},\ E_a = (\sum_{i=1}^{N_{so}} a_{i*}) / N_{so},$$

where N_{so} denotes the total number of sample objects;

3. Compute the performance measure for time(PM_t) and accuracy(PM_a):

$$PM_t = (T - E_t) / Tol_T,\ PM_a = (E_a - A) / Tol_A,$$

where T and Tol_T are time constraints and A and Tol_A are accuracy constraints, given as an inspection goal;

4. Compute an integrated performance measure, PM;

$$PM = W_T \times PM_t + W_A \times PM_a,$$

where W_T and W_C are relative weights, given as an inspection goal, $W_T + W_A = 1$, $W_T \geq 0$, $W_A \geq 0$;

5. If PM< -1 then the performance of current hypotheses is "bad", otherwise since the their performance is evaluated as "good", currently reasoned hypotheses are taken as a solution.

End-Algorithm

Processing Control Knowledge - The processing control knowledge specifies a set of processing units and a set of temporal relationships between them to determine next processing unit. A processing unit refers to a well decomposed knowledge group in this paper. We organize all the processing units to be hierarchical. The processing units for the eight steps of the hypothesis-based reasoning algorithm presented in section 2 are defined as the top level processing units. The lower-level processing units are defined for the sub-steps of each of steps in the top level, so and so. As an example of the second level processing units, for the abstract object modeling step, three processing units are defined for reasoning each of the three level abstractness of the object representation as the object representation knowledge.

The temporal relationships between the processing units in a level are as follows: *if (a processing unit AND the processing result status AND the processing history are given) then (one or more next processing units can be activated).*

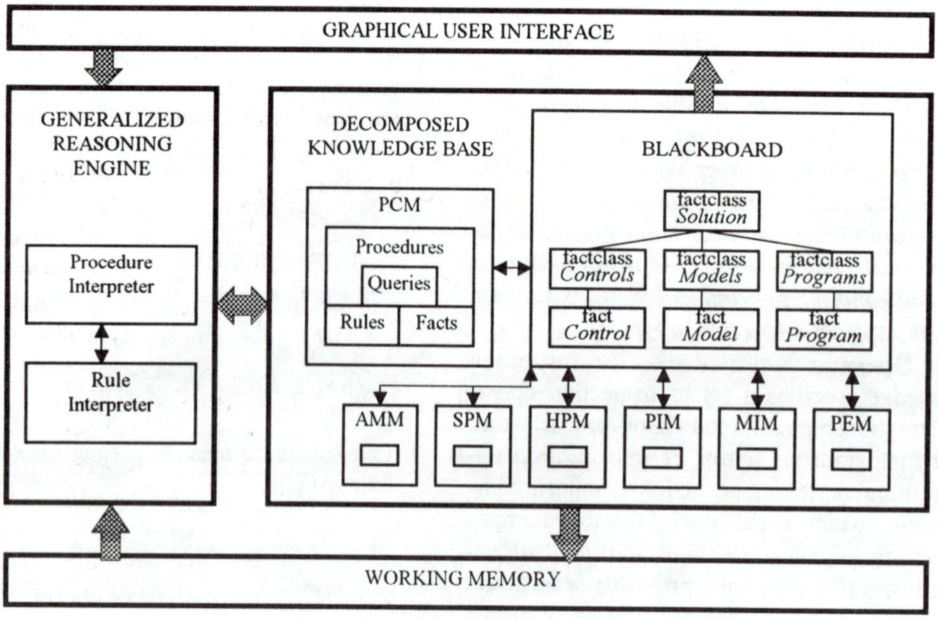

FIGURE 3. The Automated Visual Inspection Planning System

4. IMPLEMENTATION

The suggested hypothesis-based reasoning model consists of various knowledge-intensive steps each of which has a different goal and requires different types of knowledge and solution method. We have thus implemented a knowledge-based system, AVIPS (Automated Visual Inspection Planning System), based on the expert system technique[BC89] and the blackboard paradigm[BC89] that effectively support an automated divide-and-conquer based on an automated reasoning mechanism for a given 2D IVIPP.

The main components of the AVIPS are the decomposed knowledge base(DKB), the working memory, the generalized reasoning engine(GRE), and the graphical user interface(GUI). The knowledge groups and the related reasoning strategies of the suggested approach are implemented as knowledge modules each of which would be little coupled between them and highly cohesive for its goal so that one could be defined independently of the other knowledge modules. They compose the knowledge base of the AVIPS, the DKB. To interpret and apply the knowledge group to a problem state, a goal-directed rule interpreter and procedure interpreter are implemented and integrated as a reasoning engine of the AVIPS, the GRE. Therefore, without effecting on the reasoning engine, the knowledge modules can be defined, modified, and extended effectively. Through the GRE, the problem solving results are shown to users and some necessary inputs are taken from users.

The overall problem solving process in terms of the components are as follows: First, the user gives an inspection goal through the GUI and triggers the problem solving process. By the user's trigger, the PCM(the processing control module) and the blackboard are loaded by the rule interpreter of the GRE. Then, by the GRE based on the PCM, a knowledge module appropriate for current problem state is loaded and applied continuously until all possible solutions would be found.

4.1 Decomposed Knowledge Base

The DKB is composed of the eight knowledge modules as shown in Figure 3. Each knowledge module is implemented by at least one or more types of knowledge bases among the four types of knowledge: a fact base, a rule base, a procedure base, and a query base. All factual data of a knowledge module are represented by *frames*[BC89] and defined in the fact base, the relationships and all heuristics rules are represented by *if-then rules*[BC89] and defined in the rule base, a set of static temporal relations between the actions built in the AVIPS or defined by human engineers are represented by *Symbolic Expression*[WH89] and defined in the procedure base, and all goal formula associated with a fact class and a rule set are represented by the *Symbolic Expression* and defined in the query base. Two types of query, *the out-query* and *the in-query*, are defined. An out-query is just a call specified by a name of a procedure, while an in-query is a definition of a procedure. The blackboard given by a fact base is a global data area referenced or modified by the knowledge modules during their problem solving process. The blackboard represents a solution state at any time. The PCM controls the overall problem solving process to get all possible solutions by continuous activation of an appropriate knowledge module based on current blackboard content.

4.2 Working Memory

The working memory represents a problem state and a solution state of the 2D IVIPP. An out-query makes the procedure interpreter of the GRE form a problem state, while an in-query makes the GRE form a solution state.

The problem space provides the basis to define our all possible problem states, which is defined by $\boldsymbol{P} = \langle F, R, P, Q\rangle$ where $\boldsymbol{P}$ refers to the problem space of the 2D IVIPP, F refers to all fact bases, R refers to all rule bases, P refers to all procedure bases, and Q refers to all query bases. The problem space with specific values at a time, i, is an association of an activated knowledge module and the blackboard state at that time. It is defined by $\boldsymbol{P}_i = \langle F_i, R_i, P_i, Q_i\rangle$ where $\boldsymbol{P}_i$ refers to *the i'th problem state*, F_i refers to all facts of an active knowledge module and the blackboard on the working memory, and R_i, P_i, and Q_i refer to all rules, all procedures, and all queries of an activated knowledge module respectively on the working memory.

On the other hand, the solution space provides the basis to define our all possible partial and full solutions to the problem states. It is defined the blackboard state as $\boldsymbol{S} = \langle C, M, P\rangle$ where $\boldsymbol{S}$ refers to the solution space, C refers to all possible solution states of the process control problem, M refers to a set of all possible solution states of the object model definition problem, and P refers to a set of all possible solution states of the inspection program definition problem. A specified solution space at a time, i, is defined by $\boldsymbol{S}_i = \langle C_i, M_i, P_i\rangle$, called *the i'th solution state*.

4.3 Generalized Reasoning Engine

The GRE forms not only a solution state from a problem state by an in-query but also a problem state from a solution state by an out-query. It consists of a rule interpreter and a procedure interpreter. The rule interpreter deals with rules and facts, while the procedure interpreter treats procedures and queries.

The reasoning characteristics is determined dynamically by the two parameters: *the rule chaining direction*(*RCD*) and *the result type*(*RT*). The RCD value is either *FORWARD* or *BACKWARD*. The FORWARD represents the forward reasoning method, while the BACKWARD denotes the backward reasoning method. The RT value is either *FACT* or *PLAN*. The FACT means that the solution should be a set of facts satisfying the given query, while the PLAN means that the solution should be a set of sequences of facts which can transform an initial state to a solution state. For a specific problem state, the GRE is specialized with specific values before its application.

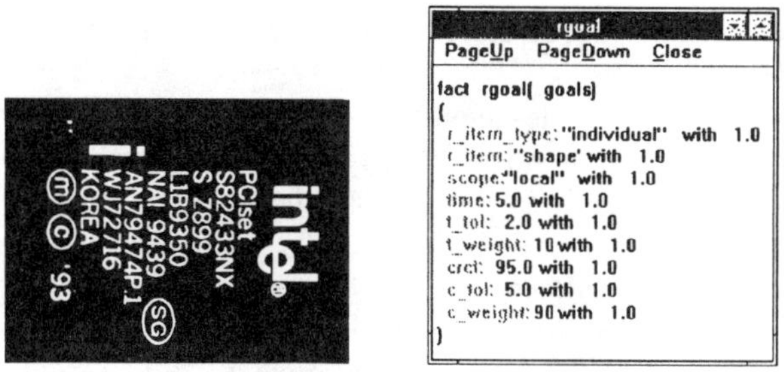

FIGURE 4. An Inspection Goal

5. APPLICATION EXAMPLE

In this section, the implemented system, AVIPS, is illustrated with an application example to solve a 2D IVIPP defined by the inspection goal of an industrial mark inspection domain as shown in Figure 4.

Step 1 : Get an inspection goal from a user - First of all, the AVIPS accepts an inspection goal from human engineer, $rgoal_0$, as shown above. By human engineer's *build* command, an out-query, *pcm* is then called and activated automatically. The *pcm* makes the GRE load the PCM to form the first problem state, $\boldsymbol{P}_0$: $\boldsymbol{P}_0=\langle F_0, R_0, P_0, Q_0\rangle$, where $F_0=\{control_0, rgoal_0\}$, $R_0=\{control\text{-}rules\}$, $P_0=\{\}$, and $Q_0=\{pcm\text{-}query, (amm), (spm), (hpm), (pim), (mim), (pem), (learning)\}$.

Step 2 : Determine an abstract object model - The *pcm* then activates an in-query, *pcm-query* in Q_0: *(query '((control has "pcm")) '() 'control-rules)*, the GRE then finds a rule, *amm-call-rule* from the rule set, *control-rules* in R_0:

rule amm-call-rule(control-rules){
if: (control has "pcm") (pcm level "initial")
then : PROC (amm)1.0 }.

By this rule, an out-query, *amm* in Q_0 is called. The *amm* then makes the GRE load the AMM to form a new problem state, $\boldsymbol{P}_{01}$: $\boldsymbol{P}_{01}=\langle F_{01}, P_{01}, P_{01}, Q_{01}\rangle$, where $F_{01}=\{control_{01}, rgoal_0, model_{01}\}$, $R_{01}=\{reprtype\text{-}rules, itemtype\text{-}rules, item\_rules\}$, $P_{01}=\{(mod\text{-}reprtype), (mod\text{-}itemtype), (mod\text{-}item)\}$, and $Q_{01}=\{reprtype\text{-}iquery, itemtype\text{-}iquery, item\text{-}iquery\}$.

The *amm* activates an in-query, *reprtype-iquery* in Q_{01}: *(query '((am level "reprtype")) '() 'reprtype-rules),* the GRE then finds a rule from the *reprtype-rules*, in R_{01}:

rule shape-rt-rule(reprtype-rules){
if: (am level "repretype") (rgoal r_item "shape")
then : PROC(mod-repr-type "bline" "sline" "region" "hybrid") 1.0 }.

By this rule, as the top-level object representation, four candidates such as *boundary-based("bline"), skeleton-based("sline"), region-based("region"),* and *hybrid-based("hybrid")* are determined, and then the *FOA-*

model-repr-type of the blackboard fact, *model* is set to "*bline*".

The *amm* activates another in-query, *itemtype-iquery* in Q_{01}: *(query '((am level "itemtype")) '() 'itemtype-rules)*, the GRE then finds a rule from the rule set, *itemtype-rules* in R_{01}:

rule direct-rule(itemtype-rules){
if: (am level "itemtype")
PROC(or (= "sline" (get-attr-val 'model 'FOA-model-repr-type)) (= "bline" (get-attr-val 'model 'FOA-model-repr-type)))
then : PROC(mod-item-type "direct")1.0 }.

By this rule, as the intermediate-level object representation, a candidate, "*direct*" is determined, and then the *FOA-model-item-type* of the blackboard fact, *model* is set to "*direct*".

The *amm* activates the last in-query, *item-iquery* in Q_{01}: *(query '((am level "item")) '() 'item-rules)*, the GRE finds a rule from the rule set, *item-rules* in R_{01}:

rule shape-line-item-rule(item-rules){
if: (am level "item")
(model FOA-model-item-type "direct")
PROC(or (= "sline" (get-attr-val 'model 'FOA-model- repr-type)) (= "bline" (get-attr-val 'model 'FOA-model- repr-type)))
then : PROC(mod-item "l-seg-str")1.0 }.

By this rule, as the low-level object representation, the line segment string("*l-seg-str*") is determined, and then the *FOA-model-item* of the blackboard fact, *model* is set to "*l-seg-str*". After all, a new solution state, $\boldsymbol{S}_{013}$=<*control*$_{013}$, *model*$_{013}$, *NULL*> is constructed, as shown in Figure 5.

Step 3: Determine an abstract inspection program by skeletal planning - The $\boldsymbol{S}_{013}$ is a partial solution state of $\boldsymbol{P}_0$, $\boldsymbol{S}_1$. Now, the *pcm* activates the *pcm-query* again, based on $\boldsymbol{S}_1$, the GRE then finds a next rule, *spm-call-rule* from the rule set, *control-rules* in R_0:

rule spm-call-rule(control-rules){
if: (control step "amm")(control status "success")
then : PROC(spm) 1.0 }.

The GRE fails to get an abstract inspection program because the inspection program knowledge is empty.

Step 4: Determine an abstract inspection program by hierarchical planning - The *pcm* activates the *pcm-query* again, the GRE then finds another rule, *hpm-call-rule* from the rule set, *control-rules* in R_0:

rule hpm-call-rule(control-rules){
if: (control step "spm")(control status "failure")
then :PROC(hpm) 1.0 }.

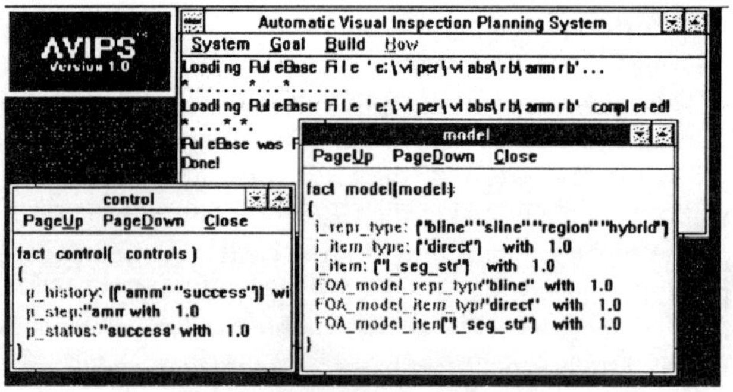

FIGURE 5. $\boldsymbol{S}_{013}$

By this rule, the out-query, *hpm* in Q_0 is called. The *hpm* then makes the GRE load the HPM to form a new problem state, $\boldsymbol{P}_{11}$: $\boldsymbol{P}_{11}$=<F_{11}, R_{11}, P_{11}, Q_{11}>, where F_{11}={*control*$_{013}$, *rgoal*$_0$, *model*$_{013}$, *program*$_0$}, R_{11}={*root-prg-rules, acq-prg-rules, cap-prg-rules, ..., lsstr-mat-prg-rules, expanding-rules*}, P_{11} ={*(insert-partial-plan) (partial-querying abs-op) (get-next-abs-op)*}, and Q_{11}={ *root-partial-planning-query, acq-partial-planning-query, ..., root-expanding-query, acq-expanding-query ...*}.

At first, the *hpm* activates an in-query, *root-expanding-query* in Q_{11}: *(query '((expansion level "root")) '() 'expansion-rules)*, the GRE then finds a rule, *root-expand-rule* from the *expansion-rules* in R_{11} :

rule root-expand-rule(expansion-rules){
if: (expansion level "root")
then : PROC(sequence (loadrb "root-prg-rules") (partial-querying "root")) 1.0 }.

By this rule, a procedure, *partial-querying* in P_{11} activates an in-query, *root-partial-planning-query* in Q_{11}: *(query '((op level "root")) '() 'root-prg-rules)*, the GRE then finds a rule from the rule set, *root-prg-rules* in R_{11}. After the partial planning using the rule set, *root-prg-rules* is completed, the *t-plan* of the blackboard fact, *program* is set to the following value, ("*ig*" (("*acq*" "*seg*" "*des*" "*mat*" "*eva*"))).

The *hpm* then calls a procedure, *get-next-abs-op* in P_{11} to search an abstract operator to be expanded by the breadth-first searching manner, "*acq*". For the selected abstract operator, the previously explained sequence of tasks are performed. In this way, all abstract operators are replaced by a sequence of primitive operators. Finally, $\boldsymbol{S}_{11}$ = <*control*$_{11}$, *model*$_{013}$, *program*$_{11}$> as shown in Figure 6 is generated. From the Figure 6, the *t-plan* and the *s-plan* of the blackboard fact, *program,* represent a hierarchical plan and a skeleton plan each as results.

Step 5: Determine an inspection program instance - The $\boldsymbol{S}_{11}$ is a partial solution state of $\boldsymbol{P}_0$, $\boldsymbol{S}_2$. The *pcm* activates the *pcm-query* again, based on $\boldsymbol{S}_2$, the GRE then finds a next rule, *pim-call-rule* from the rule set, *control-rules* in R_0 :

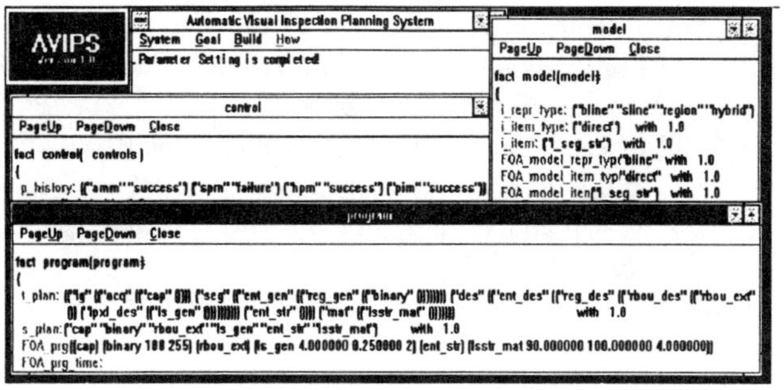

FIGURE 6. S_{11} and S_{21}

rule pim-call-rule(control-rules){
if: (control step "hpm")(control status "success")
then : PROC(pim)1.0 }.

By this rule, an out-query, *pim* in Q_0 is called. The *pim* then makes the GRE load the PIM to form a new problem state, P_{21}: $P_{21}=<F_{21}, R_{21}, P_{21}, Q_{21}>$, where $F_{21}=<control_{11}, rgoal_0, model_{013}, program_{11}>$, R_{21}=*{cap-param-rules, ..., lsstr-mat-param-rules}*, P_{21}=*{cap-param-dialog, bin-param-dialog}*, and Q_{21}=*{cap-param-query, ..., lsstr-mat-param-query}*.

The *pim* then activates an in-query, *cap-param-query* in Q_{21} : *(query '((op is "cap")) '() 'cap-param-rules)*, the GRE then finds a rule, *cap-param-rule01* from the rule set, *cap-param-rules* in R_{21} :

rule cap-param-rule01(cap-param-rules){
if: (op is "cap")
then : PROC(cap-param-dialog) 1.0 }.

By this rule, the procedure, *cap-param-dialog* in P_{21}, is called. The *cap-param-dialog* set the parameter values of *"capture"* operator. In this way, all parameters values are determined.

Finally, $S_{21}=<control_{21}, model_{013}, program_{21}\}$ as shown in Figure 6 is generated. From the Figure 6, the value of *FOA-prg* of the blackboard fact, *program* represents the inspection program instance: *((cap) (binary 188 255) (rbou_ext) (ls_gen 4.0 0.25, 2) (lsstr_mat 90.0 100.0 4.0))*.

Step 6: Determine an object model instance - The S_{21} is a partial solution state of P_0, S_3. The *pcm* activates the *pcm-query* again, based on the S_3, the GRE then finds a next rule, *mim-call-rule* from the rule set, *control-rules* in R_0 :

rule mim-call-rule(control-rules){
if:(control step "pim")(control status "success")
then: PROC(mim)1.0 }.

By this rule, the out-query, *mim* in Q_0 is called. The *mim* then makes the GRE load the MIM to form another problem state, P_{31} : $P_{31}=<F_{31}, R_{31}, P_{31}, Q_{31}>$, where $F_{31}=\{control_{21}, rgoal_0, model_{013}, program_{21}\}$, $R_{31}=\{\}$, P_{31}=*{(model-construction)}*, and $Q_{31}=\{\}$.

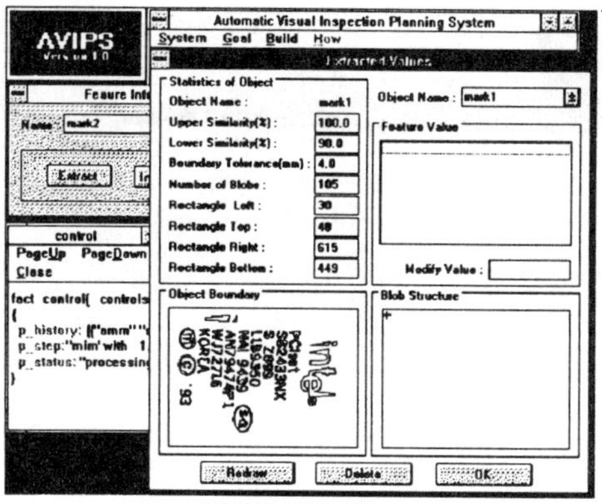

FIGURE 7. S_{31}

The *mim* then calls a procedure, *model-construction* in P_{31}. The *model-construction* generates some interactive tools to a human engineer, by which the human engineer can select some region where required features exist and then processing the determined inspection program instance to extract specific values. Finally, a new solution state, $S_{31}=<control_{31}, model_{31}, program_{21}>$ as shown in Figure 7 is generated. The $model_{31}$ is the object model instance.

Step 7: Performance Evaluation - The S_{31} is a partial solution state of P_0, S_4. The *pcm* then activates the *pcm-query* again, based on the S_4, the GRE then finds a next rule, *pem-call-rule* from the rule set, *control-rules* in R_0 :

rule pem-call-rule(control-rules){
if: (control step "mim") (control status "success")
then : PROC(pem)1.0 }.

By this rule, the out-query, *pem* in Q_0 is called. The *pem* makes the GRE load the PEM to form another problem state, P_{41} : $P_{41}=<F_{41}, R_{41}, P_{41}, Q_{41}>$, where $F_{41}=\{control_{31}, rgoal_0, model_{31}, program_{21}\}$, $R_{41}=\{\}$, P_{41}=*{ (performance-evaluation)}*, $Q_{41}=\{\}$.

The *pem* then calls the procedure, *performance-evaluation* in P_{41}. The *performance-evaluation* applies the hypotheses to 100 image samples which includes various 50 defects. Since the integrated performance, *FOA_prg_total_prfm* of the blackboard fact, *program* is 0.941496, the *status* of the blackboard fact, *control* is set to *"success"*. Finally, $S_{41}=<control_{41}, model_{31}, program_{21}>$ as shown in Figure 8 is generated.

Step 8: Learning - The S_{41} is a partial solution state of P_0, S_5. The *pcm* then activates the *pcm-query* again, based on the S_5, the GRE then finds a next rule, *learning-rule* from the rule set, *control-rules* in R_0:

rule learning-rule(control-rules){
if: (control step "pem") (control status "success")
then : PROC(learning)1.0 }.

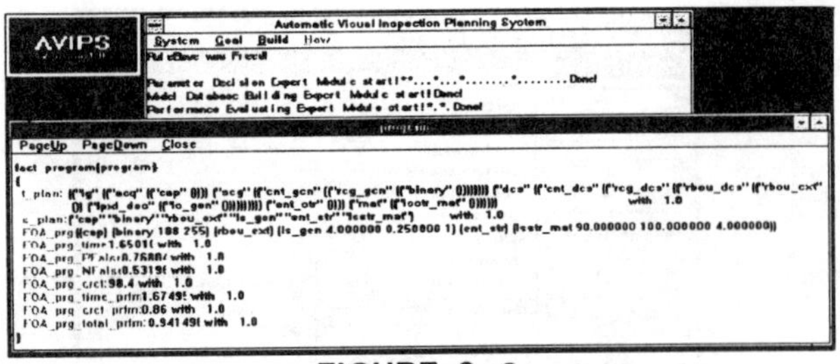

FIGURE 8. S_{41}

By this rule, a procedure, *learning* in Q_0 is called. The *learning* updates the related knowledge with positive effects with the solution, $\boldsymbol{S}_5$.

6. CONCLUSIONS

By now, the industrial visual inspection planning problem, previous works that attempt to solve the problem, and the five essential properties of a successful approach to the 2D IVIPP were addressed, and then, an knowledge-based approach and its implementation were described with an problem solving example.

In this section, the AVIPS are compared with the other approaches addressed in section 1 about various items in Table 1. As shown, the AVIPS may be a successful problem solving environment for the industrial visual inspection planning tasks on the two-dimensional domains, which guarantees consistent, optimal, and complete solutions by the automated, cost-effective, flexible, and intelligent problem solving mechanism.

REFERENCES

[BC82] A. Barr, P. R. Cohen, and E. A. Feignbaum, *Handbook of Artificial Intelligence*, Ptsman, London/New York, 1982.

[BC89] A. Barr, P. R. Cohen, and E. A. Feignbaum, *Handbook of Artificial Intelligence*, Ptsman, London/New York, 1989.

[CM89] Chein-Huei Chen and Prasanna G. Mulgaonkar, "*Automatic Vision Programming*", CVGIP: Image Understanding, Vol. 55, No. 2, March, pp. 170-183, 1992.

[Chi88] Roland T. Chin, "*SURVEY: Automated Visual Inspection : 1980 - 1987*", CVGIP, Vol. 41, pp. 346 - 381, 1988.

[CT93] Veronique Clement and Monique Thonnat, "*A Knowledge-Based Approach to Integration of Image Processing Procedures*", CVGIP : Image Understanding, Vol. 57, No. 2, March, pp. 166-184, 1993.

[Dra93] Bruce A. Draper, "*Statistical Properties of Learning Recognition Strategies*", Proceedings : Image Understanding, April, pp. 557-565, 1993.

[Hed89] Kristinal Hedengren, "*Methodology for Automatic Image-Based Inspection of Industrial Objects*", Advances in Machine Vision, Jorge L. C. Sanz, ed. Springer-Verlag New York Inc., pp. 160-191, 1989.

[MM93] M. Marafet, Sandeep Malhotra and R. L. Kashyab, "*Object-Oriented Intelligent Computer-Integrated Design, Process Planning, and Inspection*", IEEE COMPUTER, pp. 54-65, 1993.

[Mat89] Takashi Matsuyama, "Expert System for Image Processing: Knowledge-Based Composition of Image Analysis Processes", *CVGIP*, vol. 48, pp. 22-49, 1989.

[WH89] Patrick Henry Winston and Bertbold Klaus Paul Horn, *LISP 3rd edition*, Addison-Wesley Publishing Company, 1989.

Table 1. Comparison between the systems for visual inspection planning problem

ITEMS / APPROACH	Decision Scope	Various Goals	Various Object Models	Various Inspection Programs	Automatic Evaluation & Adjustment	Knowledge Based Decisions	Machine Learning	Input Device Planning	3D - Object Support
[CM89]	*OM*	*yes*	*limited*	*no*	*no*	*no*	*no*	*no*	*yes*
[MM93]	*OM*	*yes*	*limited*	*no*	*no*	*yes*	*no*	*yes*	*yes*
[CT93]	*IP*	*yes*	*no*	*yes*	*yes*	*yes*	*no*	*no*	*no*
[Dra93]	*IP*	*yes*	*no*	*yes*	*no*	*no*	*no*	*no*	*yes*
[Mat89]	*IP*	*yes*	*no*	*yes*	*no*	*yes*	*no*	*no*	*no*
[Chi88]	*OM & IP*	*no*	*no*	*no*	*no*	*no*	*no*	*yes / no*	*yes / no*
[Hed89]	*OM & IP*	*yes*	*yes*	*yes*	*no*	*no*	*no*	*no*	*no*
AVIPS	*OM & IP*	*yes*	*yes*	*yes*	*yes*	*yes*	*yes*	*no*	*no*

Note . The symbol *OM* denotes the object modeling and the symbol *IP* denotes the inspection programming.

Qualitative Modelling

AUTOMATED HAZARD IDENTIFICATION BY EMULATION OF HAZARD AND OPERABILITY STUDIES

M. Jefferson, P. W. H. Chung, A. G. Rushton
Department of Chemical Engineering, Loughborough University of Technology, Loughborough, England, LE11 3TU

ABSTRACT

A computer program was developed to perform qualitative analysis of the effect of deviations in continuous process plant. This analysis is performed by determining how deviations propagate using a signed directed graph of the plant. The program is known by the acronym QUEEN, which stands for **QU**alitative **E**ffects **EN**gine. QUEEN can be used as an engine to emulate, in particular, various forms of hazard identification. This paper describes how QUEEN has been used to form the basis of an automated hazard identification tool, emulating conventional hazard and operability studies (HAZOP). The main elements of the hazard identification system are QUEEN and a HAZOP driver for QUEEN to emulate conventional HAZOP. This HAZOP driver is known by the acronym CHEQUER, which stands for **C**omputer **H**AZOP **E**mulation using **QU**alitative **E**ffects **R**easoning. CHEQUER also contains some simple filtering rules.

INTRODUCTION

The identification of possible hazards in chemical plants has become increasingly important. Not only have chemical plants become larger and more complex, but some countries now have regulations requiring that some form of formal hazard identification be carried out. Moreover, environmental regulations have been tightened as the public has become aware of the dangers posed by large chemical plants. In the United States it is also a legal requirement to carry out some form of hazard identification retrospectively for plants already built and operating but which have not previously been subjected to any formal hazard identification (OSHA, 1992).

There are a number of hazard identification techniques available. Over the years many companies have refined particular techniques to suit their own particular needs and to try to make the process of hazard identification more effective and more efficient. However, in general, this has not reduced the time taken to carry out hazard identification. With the increased speed of many other parts of the design process, hazard identification is becoming the log-jam in attempts to speed up the design of new plants still further.

Hazard and operability studies (HAZOP) are one of the most widely used and most respected procedures for hazard identification during design. As with many other techniques for hazard identification, HAZOP is very time consuming.

One possible route for speeding up the identification of hazards in chemical plant is to use computers to identify hazards automatically. Zerkani & Rushton (1993) describe a computer aid for hazard identification. This emulates the HAZOP procedure commonly used in industry. That work has been extended using the emulation of the HAZOP procedure as a base.

HAZOP METHODOLOGY

There are numerous texts that give detailed instructions on how to carry out HAZOP (CIA, 1977; Kletz, 1992; Knowlton, 1992; Lees, 1980). The basic principle is for a group of people to apply certain defined guide words, to a piping and instrumentation diagram (P&ID) or engineering line diagram (ELD), on a line by line basis in an effort to identify causes and consequences of process deviations. The guide words relate to process deviations. The approach taken is outlined in figure 1.

Deviations considered in conventional HAZOP are derived from a combination of guide words and intentions (CIA, 1977). The list of guide words includes NONE, LESS, MORE, PART OF, etc. The combination of guide word and intention gives rise to a process deviation; e.g. if the guide word is MORE and the intention is to heat some liquid up, then the combination of the two gives MORE HEAT. The guide words can be applied to both batch and continuous plant when combined with the appropriate intentions.

QUEEN AND CHEQUER

In developing automated hazard identification, the qualitative modelling of propagation equations has been used as a basis. These propagation equations are analysed by a Qualitative Effects Engine (QUEEN) (Chung, 1993) using a signed directed graph (SDG) representation. The SDG is built up by QUEEN from a plant model and a library of standard continuous plant units, e.g. pumps, vessels, heat exchangers, etc. This model-based approach is also used in other work on automated hazard identification (Venkatsubramanian, 1994).

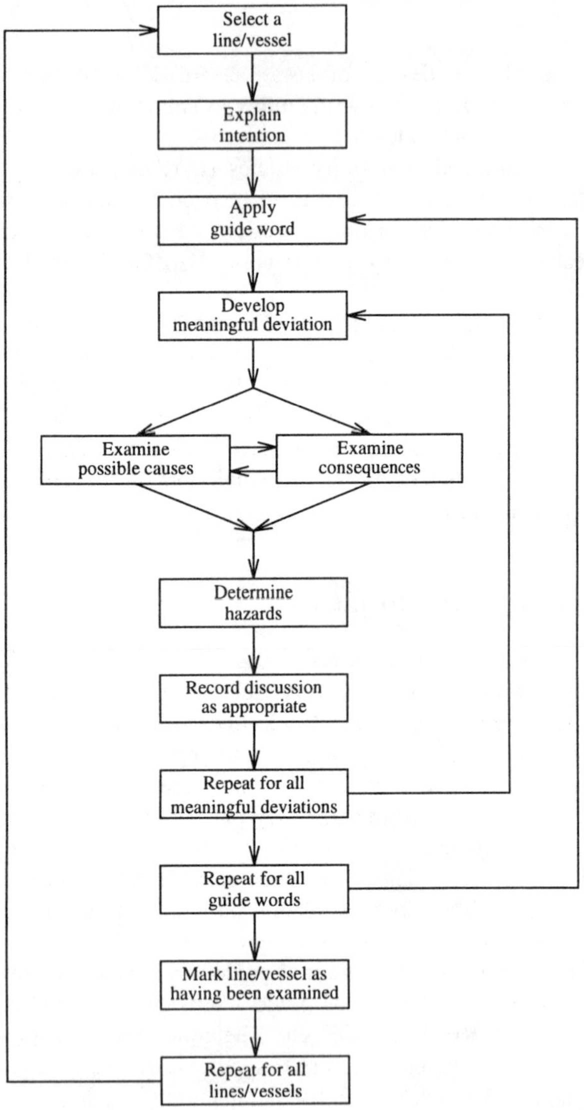

Figure 1. Basic method for conventional hazard and operability studies (HAZOP).

QUEEN is a powerful engine that can be used, with the appropriate drivers, to emulate a variety of hazard identification tasks. These include fault tree analysis and failure modes and effects analysis. Chung (1993) illustrates how QUEEN could be used for carrying out a crude form of fault tree analysis. This paper is concerned with how HAZOP can be emulated as an application of QUEEN. Given the widespread adoption of HAZOP in the chemical industry, work has been concentrated on the emulation of this method of hazard identification.

The main elements of the hazard identification system are QUEEN and a HAZOP driver for QUEEN to emulate conventional HAZOP. The other parts of the system are a component library, a fluid library, a plant model and an internal SDG. The HAZOP driver is known by the acronym CHEQUER, which stands for **C**omputer **HAZOP** **E**mulation using **QU**alitative **E**ffects **R**easoning. CHEQUER also contains some simple filtering rules to enhance the output. The relationship between QUEEN, CHEQUER and other data files used by these programs is illustrated in figure 2. QUEEN is fully described elsewhere (Chung,1993). A description of the CHEQUER driver forms the basis of much of the remainder of this paper. A brief description of the remaining parts of the system is given below.

The fluid library contains information on the physical properties of fluids, including toxicity, flammability and freezing and boiling points at various temperatures and pressures.

The component library contains text representations of the propagation equations that describe the behaviour of process variables in units found in continuous plant. This library also contains representations of how deviations in process variables within units are linked to initiating and terminating events. There are also direct links between initiating and terminating events. Initiating events take the form of faults and terminating events take the form of consequences. The information for propagation equations is stored in a textual representation, using PROLOG syntax, as an arc of an SDG.

The plant model is a description of a plant based on how process units are connected. The units in the plant are references to units in the component library.

Finally the internal SDG is built up by QUEEN from the textual representations of arcs in the component library and the units and their connections specified in the plant model. The SDG representation is used by QUEEN to link process variables. The effect one process variable and the location at which it applies has on another is represented by a value assigned to an arc linking them. This value can take the value of plus or minus one. A plus one represents a positive effect of one process variable on the other. A minus one represents a negative effect of one process variable on the other. The order of process variables is important, the effect is not commutative; i.e. The effect process variable A has on process variable B is not necessarily the same as the effect that B has on A. The textual representation for the effect of flow into a pipe on flow out of a pipe is as follows:

```
arc([in,flow],1,[out,flow]).
```

This represents the fact that there is an arc (of an SDG) signifying that flow in has a direct and positive effect on flow out. The representation for the link between a fault, such as a partial blockage in a pipe, and the deviation, less flow out, arising from this fault is:

arc([fault,'pipe partial blockage'],-1,[out,flow]).

An example of a link between a deviation and a consequence is:

arc([deviation,[morePressure,outlet]],1,
[consequence,'Overpressure rupture']).

This represents the fact that there needs to be an arc linking the deviation 'more pressure' at the outlet of a unit to the consequence 'overpressure rupture' in that unit. Note that not only is the deviation specified but also a point in the unit at which that deviation occurs, in this case the outlet. For units with more than one inlet or outlet this becomes particularly important as a deviation at one point may have a different consequence to a deviation at another point.

Finally an example of a link between a fault and a consequence is:

arc([fault,'leak to environment'],1,
[consequence,'Loss of material']).

HAZOP EMULATION

Figure 3 illustrates the basic method behind the computer emulation of HAZOP by CHEQUER. Some of the detail behind this method is outlined below.

SELECTING A PLANT UNIT

Conventional HAZOP meetings are conducted on a line by line basis. There is no specific guidance given on what should constitute a line. HAZOP team leaders have discretion to decide what constitutes a line based on the complexity of the piping and instrumentation diagram (P&ID) and other factors affecting the level of detail required from the study. This discretion is not exercised in the computer emulation of the HAZOP procedure and it is therefore necessary to define explicitly what constitutes a line. It should also be noted that, when modelling plants using QUEEN, it is not necessary to include the pipework that joins units together. Therefore, the plant as modelled does not necessarily contain any lines in the conventional sense.

The strategy we have adopted is to present the user with two options. Either a 'HAZOP' can be performed on the main items or a 'HAZOP' can be performed on all the items. This is intended to reflect the choice a HAZOP team leader may make on what level of decomposition is required for the HAZOP study. The level of decomposition is determined by the HAZOP team leader when deciding what constitutes a line. A low level of decomposition may provide more detailed results than a high level of decomposition. For CHEQUER a 'HAZOP' on main items is likely to produce less detail than a 'HAZOP' performed on all items. In all cases, the more detail that is required the longer the analysis will take. Main items have been defined to include such things as vessels, pumps, heat exchangers, etc. The items included in this definition can be customised by the user to include units that may have particular significance for the plant under consideration.

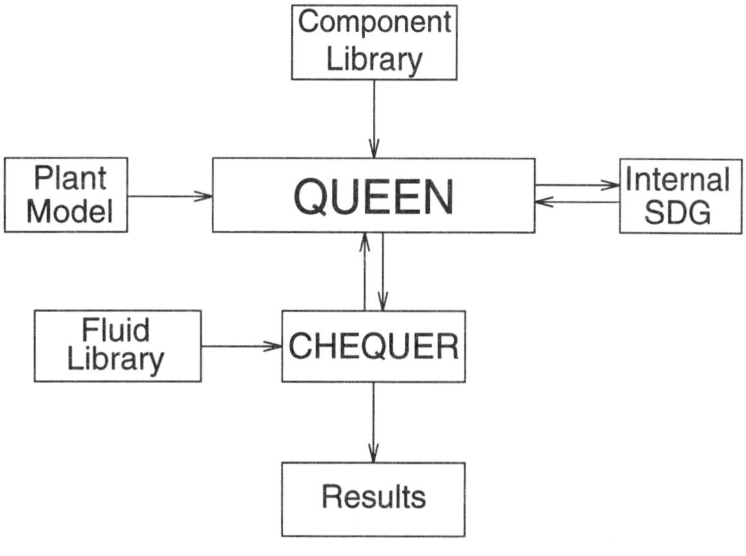

Figure2. Interaction between QUEEN and CHEQUER.

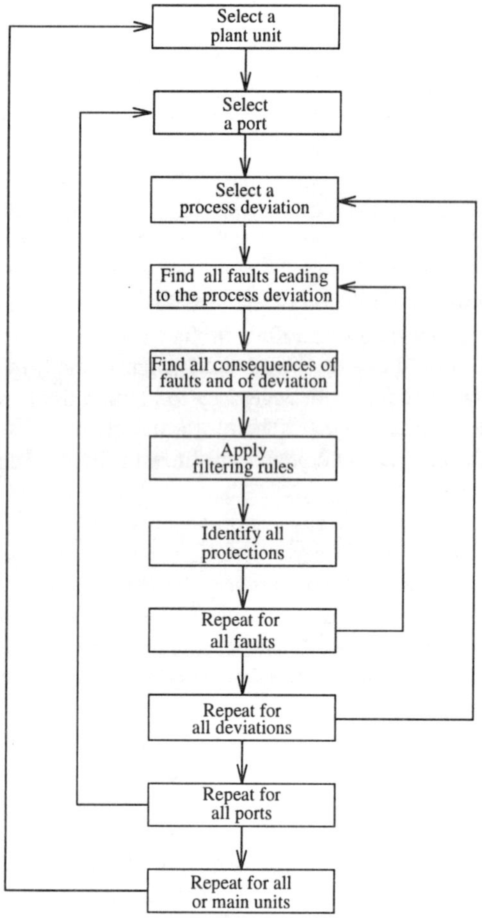

Figure 3. Basic method of CHEQUER

In a conventional HAZOP, the HAZOP team leader will decide the order in which the lines will be studied. A logical order will be determined so that the HAZOP team carries out the study efficiently and effectively with the minimum of duplication. For CHEQUER the amount of duplication does not really present any difficulties, so units are studied simply in the order in which they appear in the model of the plant under consideration.

SELECTING A PORT

Each unit has a number of inlets and a number of outlets and these are referred to as ports. It is these ports which form the connections between units. The inlet port of one unit will be connected directly to the outlet port of another.

When studying just the main items, the outlet of one main item is not necessarily connected directly to the inlet of another main item. Therefore it is necessary to study all the ports associated with all of the main items.

Given the direct connection between inlet and outlet ports, it ought to be possible, when performing a HAZOP of all the units, to study the effects of deviations either at all the inlet ports or all the outlet ports. However, because of the requirement to specify a port when making the link between a deviation and a consequence and the mechanism of searching for consequences that are used, this approach is not possible. All ports of all units must be studied if a complete analysis is to be made.

In both cases then the effect of deviations is considered at all the inlet and outlet ports of the relevant unit. The inlets of each unit are considered before the outlets. This emulates the approach taken in conventional HAZOP when vessels are being studied. All inlet lines of a unit will be studied before any outlet lines are considered.

SELECTING A PROCESS DEVIATION

Conventional HAZOP is a powerful technique and has been developed for use on both continuous and batch processes. QUEEN, however, has so far only been developed to handle continuous processes. This reduces the number of guide words and intentions that need to be handled by CHEQUER. For continuous plant the main process deviations that can be generated from the combination of guide word and intention are:

HIGH/LOW FLOW
NO FLOW
REVERSE FLOW
HIGH/LOW PRESSURE
HIGH/LOW TEMPERATURE
HIGH/LOW LEVEL

The representation QUEEN uses for process deviations combines a plus or minus one with a process variable. In this case, plus one represents a deviation that gives rise to a higher than normal value of the process deviation. Minus one represents a deviation that gives rise to a lower than normal value of the process deviation; i.e. A combination of plus one and the process variable 'flow' represents HIGH FLOW. A combination of minus one and 'flow' represents LOW FLOW. NO FLOW and REVERSE FLOW fall outside the scope of the standard plus or minus one representation, so they are represented by combining plus one with their own particular process variable. NO FLOW is represented by combining plus one with a 'process variable' no flow and REVERSE FLOW is represented by combining plus one with reverse flow. These process deviations are considered in turn for every port of every unit, although there is an exception with HIGH and LOW LEVEL which are applied to vessels only.

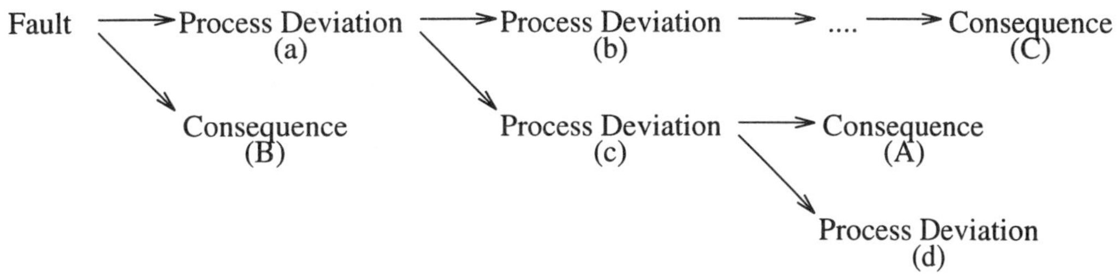

Figure 4. An example of the connection between faults and consequences.

IDENTIFYING FAULTS AND CONSEQUENCES

In conventional HAZOP, having established a possible process deviation from a guide word and intention, the team will simultaneously attempt to identify possible faults that give rise to this deviation and consequences of this deviation. The intention is to come up with a realistic cause-consequence scenario. If no consequences are found then causes are not a problem and if no causes can be found the consequences should never occur. If no realistic cause consequence scenario can be found then the HAZOP team will move on to another process deviation.

The connection between faults and consequences in QUEEN is illustrated in figure 4. In particular it is important to notice that consequences can be linked to both faults and process deviations. Obviously QUEEN can only search in one direction at a time. It is not clear whether there are any advantages in searching in one particular direction first. QUEEN is directed to search backwards first, from a process deviation to a cause. Having established a fault as a cause then a fairly simple method is used for identifying consequences.

One method would be to search forward from every fault. A number of paths can be found by following the process deviations forward from the fault. Following all of these process deviation paths to their ends would allow us to identify all the consequences of any particular fault. This would be very time consuming and would generate a lot of redundant information.

CHEQUER identifies consequences based on two criteria. Firstly, consequences are identified if they are directly linked with the original query made to QUEEN. That is the consequences are directly linked to the process deviation, the unit and the port under consideration. Secondly, consequences are identified if they are directly linked with any faults. So long as all the items in the plant which have consequences linked to them are studied, then all consequences will be found.

Looking at figure 4, if the original query to QUEEN was made concerning process deviation (c), then QUEEN traces back and finds the fault linked to process deviation (a). Having found this fault it looks for consequences. Consequence (A) is directly linked to the original process deviation (c) and consequence (B) is directly linked to the fault, so both are returned by QUEEN. If a further query is made at process deviation (b) then only consequence (B) which is directly linked to the fault is returned.

The above illustrates how CHEQUER handles the identification of consequences when there is only a single fault giving rise to a particular process deviation. When there are multiple faults, then the method that CHEQUER uses to identify the causes and consequences of a particular process deviation is as follows. First it identifies all faults that give rise to the process deviation under consideration. CHEQUER goes through these faults, one at a time, identifying consequences linked either to the fault or to the original deviation. This results in a list of faults and associated with each fault is a list of consequences. At this point some simple filtering rules are applied to cut out unwanted faults and consequences. Finally, CHEQUER identifies possible protections in the form of controllers and other items of equipment that act to reduce or reveal the effects of the process deviation.

FILTERING

Following the identification of causes and consequences, CHEQUER applies some filtering rules to reduce significantly the amount of output it would otherwise generate.

CHEQUER only looks for consequences once a fault has been identified but it is possible for it to return no consequences as a result of the fault. In this case the filtering mechanism rejects the fault and returns to attempting to identify another fault.

Secondly, CHEQUER will identify the same fault over and over again when queried about different process deviations. There may be different consequences owing to the fact that different process deviations are being considered, but on many occasions the same consequences are identified. So once CHEQUER has identified a particular cause consequence scenario, the filtering mechanism will reject any repetitions it comes across.

Finally there is some filtering that is carried out based on the properties of the fluid in any units in which causes or consequences are identified. Certain faults and consequences are only applicable given particular fluid properties. For example the possibility of fire or explosion requires that the fluid be flammable. If QUEEN identifies a cause or consequence that is only appropriate to certain fluid properties then CHEQUER will check whether or not they are relevant. This is done by checking the properties of fluids in the unit where the cause or consequence has been identified. Fluid properties are contained in the fluid library with fluids present in the plant being specified in the plant model.

IDENTIFYING PROTECTIONS

When CHEQUER has identified faults and consequences for a particular process deviation, then it queries QUEEN to find out what equipment exists to act as protections.

The mechanism for this identification by QUEEN has been fully described by Chung (1993).

However, there are two additional rules that CHEQUER applies to make sure that incorrect protections are not included. Firstly, if CHEQUER identifies as a fault the failure of part of a control system and QUEEN includes the controller for this control system in the list of protections, then CHEQUER will delete it. Secondly, if the fault gives rise to a process deviation no flow, then any controllers are presumed to be useless. In this case CHEQUER does not query QUEEN. As an example of why this first rule is needed, consider a control valve failure. If this is identified as a fault, then the controller for that control valve can have no effect, so it is useless as protection. The second of these rules is based on the presumption that most controllers act by having some effect on the flow of a fluid and if there is no flow they can have no effect.

CONCLUSIONS

This paper describes how conventional hazard and operability studies (HAZOP) can be emulated using an expert system. HAZOP is a respected technique, widely used in the chemical industry for hazard identification during the design of process plants. The system developed uses a computer program (QUEEN) as a core with the emulation being driven by a program known as CHEQUER. QUEEN was written as a general tool for performing qualitative analysis of deviations in process plant.

Future development work will be concerned mainly with the filtering rules used. At present they are fairly crude, particularly in the way in which fluid properties are used. Other work is the extension of CHEQUER to be able to handle non-continuous operations. These include the identification of problems that may arise from maintenance, start-up, shut-down, batch or semi-batch operations.

ACKNOWLEDGEMENTS

This research has been carried out with the financial support of the EPSRC (Project ref.: GR/H74629) and the financial and technical support of ICI Engineering Technology, PO Box 13, Runcorn, Cheshire, UK.

P.W.H. Chung is supported by a British Gas and Royal Academy of Engineering Senior Research Fellowship.

REFERENCES

Chemical Industries Association Ltd.; "A Guide to Hazard and Operability Studies".
Chemical Industry Safety and Health Council of the Chemical Industries Association, 1977.

Chung, P. W. H.; "Qualitative Analysis of Process Plant Behaviour". 6th International Conference on Industrial and Engineering Applications of Artificial Intelligence, 1993; pp. 277-283.

Kletz, T. A.; "HAZOP and HAZAN". 3rd Edition, Hemisphere, 1992.

Knowlton, R. E.; "A Manual of Hazard and Operability Studies". Chemetics International, 1992.

Lees, F. P.; "Loss Prevention in the Process Industries". Butterworths, London, 1980.

OSHA; Department of Labour, Occupational Safety and Health Administration; "Process Safety Management of Highly Hazardous Chemicals; Explosives and Blasting Agents; Final Rule". Federal Register, February 1992; pp. 6356-6417.

Venkatsubramanian, V. & Vaidyanathan, R.; "A Knowledge Based Framework for Automating HAZOP Analysis". AIChE Journal, March 1994.

Zerkani, H. & Rushton, A. G.; "Computer Aid for Hazard Identification". 6th International Conference on Industrial and Engineering Applications of Artificial Intelligence, 1993; pp. 102-109.

A SIMPLE AND EFFICIENT METHOD FOR DIAGNOSING EQUIPMENT FAULTS BASED ON QUANTITATIVE PHYSICAL KNOWLEDGE

Hisashi Shimodaira
Research and Development Department, Nihon MECCS Co., Ltd.
4-13-27 Midori-Cho, Koganei-Shi, Tokyo, Japan, 184

ABSTRACT

This paper proposes a method for identifying and locating the origin of a fault of an equipment by using approximate linear quantitative equations in variable increments which are derived from the equations representing the steady state of the equipment and measured values of the variables. The proposed method differs fundamentally from strict numerical simulation methods and the qualitative methods. Although the solution which is obtained by the proposed method is approximate, it is devised so that errors which is contained in them can be reduced. In the proposed method, because only linear equations are dealt with, it has the advantages of being easy to implement and process and of taking much less computation time.

INTRODUCTION

In a fault diagnosis † of the equipment, the origin of the abnormality needs to be detected from the observed abnormal symptoms in the state variables of the equipment. Various methods for assisting in and automating such a task by using knowledge about fault diagnoses have been proposed. Among them, methods reasoning from first-principles based on deep knowledge [IA79, Is89, SV87] which represents the structure and function of the equipment are promising, because they are versatile and can perform diagnoses of novel or unanticipated faults. In the methods based on deep knowledge, the equipment is represented as qualitative or quantitative models. The qualitative methods [DB84] have the problem of combinatoric explosion of possible outcomes and of generating behaviors which do not actually occur. On the other hand, a major disadvantage of the quantitative methods, especially strict numerical simulation methods, is the practical difficulty in solving the equations in real time which are mostly nonlinear and complicated.

In this paper, we propose to employ approximate linear quantitative equations in variable increments, which are derived from the equations representing the steady state of an equipment, to identify and locate the origin of a fault. The proposed method is applicable to any equipment of which state variables are represented by continuous variables.

† In this paper, a fault is defined as a malfunction of the equipment, which is observed as a departure from the acceptable range of a variable at a measuring point. This defines a fault as an abnormality in the equipment.

BACKGROUND OF THE PROPOSED METHOD

For example, the flow rate through a valve is given by

$$Q = CA\sqrt{\frac{2P}{\rho}} \tag{1}$$

where Q is the flow rate, C is the discharge coefficient, P is the pressure across the valve, ρ is the mass density of the fluid, and A is the area available for flow. Differentiating with respect to time t,

$$\frac{dQ}{dt} = C\sqrt{\frac{2P}{\rho}}\frac{dA}{dt} + \frac{CA}{\rho}\sqrt{\frac{\rho}{2P}}\frac{dP}{dt} \tag{2}$$

In the qualitative physics [DB84], confluences are derived from Equation (2) as

$$\partial Q - \partial A - \partial P = 0 \tag{3}$$

where ∂Q represents the qualitative value of dQ/dt. In the qualitative methods, because most confluences can eliminate only the combinations of qualitative values which obviously contradict them, the so-called combinatoric explosion occurs which makes the problem more difficult to solve. The cause of this situation consists in that much information is lost in the process transforming Equation (2) into Equation (3). This also causes the problem of generating behaviors which do not actually occur in the equipment.

The basic idea which is employed in the proposed method is as follows. When a change occurs in the condition of the valve whose state variables satisfy the Equation (1) and then the variables Q, A and P change by ΔQ, ΔA and ΔP, respectively having finite magnitudes after some period, the following approximate equation in these increments needs to be satisfied.

$$\Delta Q = C\sqrt{\frac{2P}{\rho}}\Delta A + \frac{CA}{\rho}\sqrt{\frac{\rho}{2P}}\Delta P \qquad (4)$$

The above equation is derived by representing Equation (1) by the Taylor series about the variable values of the steady state and taking the terms with the first degree with respect to the increment. If we neglect the transient changes, we can estimate the approximate variations of variables by using the above equation. We call this method the variation estimation method.

VARIATION ESTIMATION METHOD

Most equipments are operated in a steady state where the state variables are kept within the acceptable ranges. In this paper, we define a fault as a condition which occurs through continuous changes from a normal steady state due to a malfunction of a part of the equipment and reaches another steady state corresponding to the malfunction.

We use the following equation as an example of equations in variables representing a steady state of the equipment.

$$z = f(x, y) \qquad (5)$$

where x, y and z are the variables representing the state, input, disturbance and output for the equipment. Such an equation representing a steady state can be derived by making the term with the derivative of the state variable with respect to time in the state equation of the equipment reduce to zero. Equation (1) corresponds to Equation (5).

Let variable values of x, y and z in a normal steady state be a, b and c, respectively. When a change occurs due to a malfunction in a part of the equipment and then the variables x, y and z change by Δx, Δy and Δz, respectively having finite magnitudes from the point (a, b, c) to reach another steady state, these increments of the variables need to satisfy the following approximate equation.

$$\Delta z - f_x \Delta x - f_y \Delta y = 0 \qquad (6)$$

where f_x represents $\partial f/\partial x$. The above equation is derived by representing Equation (5) by the Taylor series about a point (a, b, c) and taking the terms with the first degree with respect to the increment. The details are described in APPENDIX 1. Equation (4) corresponds to Equation (6).

We can calculate the variations of variables of the equipment in such an abnormal state from the normal steady state by using a set of equations in variable increments such as Equation (6). The coefficients of the equations in variable increments are calculated by using the measured values of the variables at the measuring points in the normal steady state and the values of some variables which can be calculated by using the measured values of variables with the aid of the equations in variables such as Equation (5). These equations in variable increments are used as constraints. The values of the variable variations in the abnormal conditions are calculated by inserting the values of the variable variations at the measuring points into the constraints and by propagating them over the constraints [Wm88]. For example, when Δx and Δz in Equation (6) are known, then Δy can be calculated.

Because the equation in variable increments such as Equation (6) is an approximate relationship among the variable increments, as the magnitudes of the variable variations become larger, the errors which are contained in the solution become larger. Thus we employ the following method (called mean-coefficient method) to reduce the errors. If the magnitude of a variable variation which is calculated by an equation in variable increments is larger than the prescribed threshold, all the coefficients in the equation are replaced by a mean value of two coefficients which are calculated by using the variable values before and after the change. For example, in Equation (6) when Δx which is larger than the threshold with respect to x is obtained, the coefficient containing the variables x and y is replaced by a mean value of the coefficient for $x=a$ and $y=b$ and that for $x=a+\Delta x$ and $y=b+\Delta y$. Then the variable variation Δx is recalculated by using the equation with the updated coefficients and the improved value of Δx is obtained. Concerning the theoretical meaning of the mean-coefficient method, see APPENDIX 2.

In the variation estimation method, we assume that the magnitude of the variable variation which is calculated from the equation in variable increments is within the range which is determined by the condition where the original equation in variables such as Equation (5) is valid. Therefore if the magnitude of the variable variation exceeds such a limit, the calculated variable variation is trivial. For example, Equation (1) holds under the condition where the valve is working ($0<A<A_{max}$) and is not valid under the condition where it is completely open ($A=A_{max}$) or completely closed ($A=0$). In this sense, the behavior which is obtained by the variation estimation method corresponds to the intrastate behavior in the qualitative physics [DB84].

In the above, we are concerned with the lumped-parameter system in which all the state variables are functions of only time and not coordinates. Because the distributed-parameter system can be approximately transformed into the lumped-parameter system by the difference calculus, we are concerned with only the lumped-parameter system.

FAULT DIAGNOSIS METHOD USING VARIATION ESTIMATION METHOD

When a part of the equipment functions normally, the variable variations which are calculated by the equations in variable increments representing the normal function of the part approximately coincide with the actual variable variations at the part. If a part of the equipment malfunctions, the variable variations which are calculated by the equations representing the normal function of the part do not correspond to the actual variable variations and the differences are observed at the measuring points.

In order to reason about a malfunctioning part, the constraints are divided into groups so that the variable variation at a prescribed measuring point or a check point can be calculated by substituting the variable variations at the other measuring points into the set of constraints in each group. The check point is a prescribed point where the two values of a variable variation are calculated through different routes and compared. In this paper, it is assumed that multiple faults do not occur at the same time in each constraint group.

The malfunctioning part is located through the following two stages.

1) Identifying the constraint group containing a malfunctioning part

In each group, it is assumed that the parts of the equipment function normally. The variable variation at the prescribed measuring point is calculated by substituting the variable variations at the other measuring points into the equations in variable increments representing the normal functions of the parts and compared with the measured one. Or the two values of the variable variation at the check point are calculated in the same way and compared with each other. If both the values do not coincide with each other, we can identify that the constraint group contains a malfunctioning part.

2) Locating the malfunctioning part

At this stage, the malfunctioning part in the constraint group which is identified at the stage 1) is located by the trial-error method. It is assumed that in the constraint group only one part of the equipment malfunctions and the other parts function normally. A malfunctioning part is assumed and the variable variation associated with it is calculated by inserting the variable variations at all the measuring points and at the check point into the constraints. If the calculated value corresponds to the abnormal condition and can explain it according to the knowledge about the fault, we can infer that the part is the origin of the fault. Such a process is repeated until the possible solution is obtained.

When a fault is the abnormal condition which does not concern the original function of the equipment (for example, leakage in a tank), we need to add a term for representing it to the equation in variable increments at the stage 2). For example, if a change $k\Delta u$ (u: a variable representing the fault, k: a proportional constant) occurs in Equation (6) by the fault, then Δu is calculated from Δx, Δy and Δz by the following equation.

$$\Delta z - f_x \Delta x - f_y \Delta y + k \Delta u = 0 \tag{7}$$

Usually most equipments consist of the parts and the parts have their own functions. Therefore we can diagnose a part of the equipment by grouping the the equations (constraints) representing the functions of the part.

A SAMPLE PROBLEM

The proposed method has been applied to the refrigerating equipment which is a part of the air conditioning equipment and the effectiveness has been investigated. The equations in variables for the equipment are shown in Table 1. The equations in variable increments which are derived from the equations in Table 1 are shown in Table 2. Concerning the symbols used, see Table 3. The equations in variable increments are divided into three groups and the group containing a malfunctioning part is checked according to the procedure as shown in Figure 1. Concerning the condenser, on the assumption that the condenser functions normally and thus ΔK_c, Δf_w, Δl_{oc} are equal to zero, Δp_2 is calculated through Equations (2.4), (2.5), (2.6) and (2.12) by using the measured values of Δt_{w1}, Δt_{w2} and ΔW and compared with the measured value of Δp_2. Concerning the evaporator and cold storage, the same procedure as the condenser is used. After the condenser and evaporator are confirmed to function normally, the compressor is checked. On the assumption that the compressor functions normally, ΔN is calculated through Equations (2.4), (2.8) and (2.3) and ΔN is also calculated through Equation (2.2) by using the measured values of Δp_1 and Δp_2, then both the calculated values are compared.

The equipment constants for the equipment are shown in Table 4. These values can be obtained from the design specifications. The variable values at the measuring points in the normal steady state are shown in Table 5. The variable values except those at the measuring points can be calculated in advance by using some equations in Table 1. These variable values are shown in Table 6.

We examined an abnormal condition where the variable values at the measuring points present the variations which are shown in Table 5. In this condition, the condensation pressure is about 10%

Table 1 Equations in Variables for Refrigerating Equipment

$$R = \frac{V}{v_1}\eta_v(h_1 - h_u) \quad (1.1)$$

$$N = AV\frac{\kappa}{\kappa-1}\frac{\eta_v}{\eta_c}p_1\left[\left(\frac{p_2}{p_1}\right)^{\frac{\kappa-1}{\kappa}} - 1\right] \quad (1.2)$$

$$Q_c = R + N \quad (1.3)$$

$$Q_c = cW(t_{w2} - t_{w1}) \quad (1.4)$$

$$t_{wm} = \frac{1}{2}(t_{w1} + t_{w2}) \quad (1.5)$$

$$Q_c = K_c F_c(t_2 - t_{wm}) \quad (1.6)$$

$$\frac{1}{K_c} = \frac{1}{\alpha_r} + \frac{l_{oc}}{\lambda_o} + f_w + \frac{1}{\alpha_w} \quad (1.7)$$

$$Q_e = R \quad (1.8)$$

$$Q_e = K_e F_e(t_r - t_1) \quad (1.9)$$

$$Q_e' = K_w F_w(\theta - t_r) \quad (1.10)$$

$$p_1 = f_p(t_1) \quad (1.11)$$

$$p_2 = f_p(t_2) \quad (1.12)$$

$$h_1 = g_h(p_1) \quad (1.13)$$

$$v_1 = g_v(p_1) \quad (1.14)$$

Table 2 Equations in Variable Increments for Refrigerating Equipment

$$\Delta R + \frac{V}{v_1^2}\eta_v(h_1 - h_u)\Delta v_1 - \frac{V}{v_1}\eta_v(\Delta h_1 - \Delta h_u) = 0 \quad (2.1)$$

$$\Delta N - AV\frac{\kappa}{\kappa-1}\frac{\eta_v}{\eta_c}\left[\frac{1}{\kappa}\left(\frac{p_2}{p_1}\right)^{\frac{\kappa-1}{\kappa}} - 1\right] - AV\frac{\eta_v}{\eta_c}\left(\frac{p_2}{p_1}\right)^{-\frac{1}{\kappa}}\Delta p_2 = 0 \quad (2.2)$$

$$\Delta Q_c - \Delta R - \Delta N = 0 \quad (2.3)$$

$$\Delta Q_c - c(t_{w2} - t_{w1})\Delta W - cW(\Delta t_{w2} - \Delta t_{w1}) = 0 \quad (2.4)$$

$$\Delta t_{wm} - \frac{1}{2}(\Delta t_{w1} + \Delta t_{w2}) = 0 \quad (2.5)$$

$$\Delta Q_c - F_c(t_2 - t_{wm})\Delta K_c - K_c F_c(\Delta t_2 - \Delta t_{wm}) = 0 \quad (2.6)$$

$$\Delta K_c + K_c^2\Delta f_w + \frac{K_c^2}{\lambda_{oc}}\Delta l_{oc} = 0 \quad (2.7)$$

$$\Delta Q_e - \Delta R = 0 \quad (2.8)$$

$$\Delta Q_e - F_e(t_r - t_1)\Delta K_e - K_e F_e(\Delta t_r - \Delta t_1) = 0 \quad (2.9)$$

$$\Delta Q_e' - F_w(\theta - t_r)\Delta K_w - K_w F_w(\Delta\theta - \Delta t_r) = 0 \quad (2.10)$$

$$\Delta p_1 - f_p'(t_1)\Delta t_1 = 0 \quad (2.11)$$

$$\Delta p_2 - f_p'(t_2)\Delta t_2 = 0 \quad (2.12)$$

$$\Delta h_1 - g_h'(p_1)\Delta p_1 = 0 \quad (2.13)$$

$$\Delta v_1 - g_v'(p_1)\Delta p_1 = 0 \quad (2.14)$$

■ Condenser

Measured value: Δt_{w1}, Δt_{w2}, ΔW, Δp_2 Assumption: $\Delta K_c = 0$, $\Delta f_w = 0$, $\Delta l_{oc} = 0$

[Eq. (2.4) → ΔQ_c] ↘

[Eq. (2.6) → Δt_2] → [Eq. (2.12) → Δp_2] ⇔ (Compare) Measured value Δp_2

[Eq. (2.5) → Δt_{wm}] ↗

■ Evaporator and cold storage

Measured value: $\Delta\theta$, Δt_r, Δp_1 Assumption: $\Delta K_w = 0$, $\Delta K_e = 0$

[Eq. (2.10) → ΔQ_e] → [Eq. (2.9) → Δt_1] → [Eq. (2.11) → Δp_1] ⇔ (Compare) Measured value Δp_1

■ Compressor

Measured value: Δp_1, Δp_2 Assumption: Compressor functions normally

[Eq. (2.4) → ΔQ_c] ↘

[Eq. (2.3) → ΔN] ⇔ (Compare) [ΔN ← Eq. (2.2)]

[Eq. (2.8) → ΔR] ↗

Figure 1 Method for Identifying Constraints Containing Malfunctioning Part

higher than the normal one, which is close to the one at which the safety valve starts to operate. According to the Japanese law, it is 15% higher pressure than the normal setting one. In this case, the external conditions such as atmosphere temperature do not change. However the proposed method is applicable to the case where the external conditions in the abnormal state change from those in the normal state.

The values of the coefficients of the equations in Table 2 can be calculated by using the equipment constants, the measured variable values in Table 5 in the normal steady state and the variable values (t_1, t_2, h_1, v_1 and so on) in Table 6 which are calculated by using the measured variable values.

By checking the condenser according to the procedure in Figure 1, the variation of the condensation pressure $\Delta p_2 = 0$ is obtained. Because this value does not correspond to the measured value $\Delta p_2 = 1.06$ kgf/cm^{2}abs, we infer that the condenser may have a malfunctioning part. Then the malfunctioning part is assumed to be the heat transmission of the cooling pipe of the condenser. By calculating the variation of the heat transmission coefficient of the cooling pipe by Equations (2.12) and (2.6) by using the measured values of Δp_2 (1.06kgf/cm^2 abs) and so on, $\Delta K_c = -342$ kcal/m^{2}h℃ (39.3% reduction from the normal value) is obtained. Because the calculated variation is considerably large, the coefficients of the second and third terms of Equation (2.6) are replaced by the mean values of those which are calculated according to the mean-coefficient method and ΔK_c is recalculated. As a

Table 3 References of Major Symbols

R	Refrigerating capacity
N	Compression power of compressor
Q_c	Condensation load in condenser
p_1	Evaporation pressure of refrigerant
t_1	Evaporation temperature of refrigerant
p_2	Condensation pressure of refrigerant
t_2	Condensation temperature of refrigerant
c	Specific heat of cooling water in condenser
W	Flow rate of cooling water
t_{w1}	Temperature of cooling water at condenser inlet
t_{w2}	Temperature of cooling water at condenser outlet
t_{wm}	Mean temperature of cooling water in condenser
K_c	Heat transmission coefficient of cooling pipe
F_c	Area of cooling pipe
λ_o	Heat transfer coefficient of oil film
l_{oc}	Thickness of oil film
f_p	Function for saturated pressure and temperature
t_r	Room temperature in cold storage

Table 4. Equipment Constants

Symbol	Value	Unit
V	200	m³/h
η_v	0.72	
η_c	0.79	
A	1/427	kcal/kgm
κ	1.295	
α_r	5000	kcal/m²h℃
α_w	4000	kcal/m²h℃
f_w	0.0002	m²h℃/kcal
λ_{oc}	0.1	kcal/mh℃
F_c	12	m²
$K_e F_e$	6090	kcal/h℃
$K_w F_w$	2030	kcal/h℃

Table 5. Variable Values at Measuring Points in Normal State and the Variation Values in Abnormal State

Symbol	Variable Values	Variation Values	Unit
p_1	1.94	0.0761	kgf/cm²abs
p_2	11.9	1.06	kgf/cm²abs
t_{w1}	20	0	℃
t_{w2}	24.8	0	℃
W	16600	0	l/h
t_r	-10	0.665	℃
θ	20	0	℃

Table 6. Variable Values in Normal State and the Calculated Variation Values in Abnormal State

Symbol	Variable Values	Variation Values	Unit
Q_c	79500	0	kcal/h
t_2	30	3	℃
K_c	870	-230	kcal/m²h℃
l_{oc}	0.05	0.0395	mm
t_1	-20	0.887	℃
R	60900	-1350	kcal/h
N	18600	1350	kcal/h
h_1	396	0.604	kcal/kg
v_1	0.624	-0.0463	m³/kg
h_u	132	24.0	kcal/kg

result, K_c=-230 kcal/m²h℃ (26.5% reduction) is obtained. If it is assumed that the decrease in K_c is only due to the increase in oil film, Δl_{oc}= 0.0395mm is obtained through Equation (2.7). Next, by checking the evaporator and compressor according to the procedure in Figure 1, the calculated value and the measured one of Δp_1 and the two values of ΔN which are calculated through the different routes coincide with each other respectively. As a result, it can be confirmed that these function normally. Other variable variation values which are calculated by using ΔK_c=-230 kcal/m²h℃ are shown in Table 6.

As the result of the above reasoning, it is inferred that the origin of the fault is the decrease in the heat transmission coefficient of the cooling pipe of the condenser due to the increase in oil film, which causes the higher condensation pressure (temperature) and the reduction in the refrigerating capacity and leads to the rise in the room temperature of the cold storage. These interpretation coincides with the knowledge about the cause and effect of the fault in the refrigerating equipment, so that the above reasoning can be confirmed.

The exact value of the variation of K_c which is calculated by Equation (1.6), is ΔK_c=-245 kcal/m²h℃. The relative error, which is compared with the original value of K_c, in the variation value ΔK_c=-342 kcal/m²h℃ which is calculated without using the mean-coefficient method is 11.1%. On the other hand, that in ΔK_c=-230 kcal/m²h℃ which is calculated by using the mean-coefficient method is 1.7%. This proves that the mean-coefficient method can reduce the errors in the variable variations which is calculated by using approximate equations in variable increments.

This sample problem shows that the proposed method is effective in identifying and locating the cause of the fault in the equipment.

FEATURES AND LIMITATIONS OF THE PROPOSED METHOD

The variation estimation method differs fundamentally from the qualitative methods. In the proposed method, the combinatoric problem which is the most serious disadvantage of the qualitative method is avoided and the quantitative values of variable variations can be easily obtained. These values can be used to perform the fault diagnosis in a simple and efficient way. The variation estimation method also differs methodologically from strict numerical simulation methods, the semi-quantitative simulation method proposed by La86 and the approximate simulation method proposed by QB88, although equations in variable increments may be used in these methods. In the proposed method, the quantitative values of variable variations can be calculated in a simpler way than strict numerical simulation methods.

The most salient feature of the variation estimotion method is that the condition of the equipment after a change can be conjectured by using the linear equations in variable increments instead of using the nonlinear complex equations in variables, although some of the equations in variables need to be used to calculate the coefficients of the equations in variable increments.

The advantages of the variation estimation method that come with dealing with only linear equations are as follows.

1) If we use equations in variables to obtain the solution, because most of them are complex and nonlinear, we need to employ the iteration method such as the relaxation method and that results in more complex implementation and more computation time. In the proposed method, because only linear equations are dealt with, the implementation is easy and the solution can be obtained in less computation time.

2) In some kind of problems, the local constraint propagation fails to assign all values one might desire due to cyclic dependencies among the variables [SS80]. In such a case, because the variation estimation method uses only linear equations in variable increments, the solution can be easily obtained by solving the simultaneous linear equations or by propagating symbolic expressions of unknown variables as well as numerical values [SS80, Wm88].

3) If we use the equations in variables such as those in Table 1, we need to implement multiple equations in the program to calculate efficiently a value from the other variables, each of which is of the form which is solved for a variable (for example, $x=g(y, z)$ and $y=h(z, x)$ for Equation (5)). In the variation estimation method, on the contrary, only one equation of the form such as Equation (6) is sufficient to calculate each variable variation, if we use appropriate programming techniques. This makes the processing in propagation of constraints much simpler and results in reduction of the system size.

In the variation estimation method, because approximate equations in variable increments are used, we need to note that the calculated values contain errors. However, because the errors can be reduced by using the mean-coefficient method as shown in the previous section, the method can be applicable to the case where the variables present considerably large variations.

In addition, the variation estimation method is not valid as mentioned in the previous section for the case where the variable variation exceeds the limit at which the structure of the model changes. However, because most equipments are monitored so that a large change of the state will not occur, the change due to a malfunction may be considerably small. Thus the proposed method can cover many faults in such cases.

In the proposed method, it is assumed that the fault is the steady state corresponding to a malfunction of the equipment. However, because the assumption is approximately satisfied, if the speed of changes of the variables with respect to time is not so large, the proposed method can cover many faults in practice.

Because it is often difficult to obtain quantitative values of variables for detailed parts of the equipment, we need to devise an effective method to form the whole model of the system. The qualitative method has the advantage that the model can be constructed by using qualitative knowledge without using quantitative one. Therefore a method seems to be promising which uses cooperatively the proposed quantitative method and the qualitative method to carry out the diagnosis of the whole equipment.

In the proposed method, the way for dividing the constraints into groups depends on the object of diagnosis and the task of locating a malfunctioning part needs the knowledge about the function and the faults of the equipment. We hope to develop versatile algorithms for automating these tasks.

CONCLUSIONS

In this paper, we have proposed a method for identifying and locating the origin of a fault of an equipment by using approximate linear quantitative equations in variable increments which are derived from the equations representing the steady state of the equipment and measured values of the variables. The proposed method differs fundamentally from strict numerical simulation methods and the qualitative methods. In the proposed method, the combinatoric problem which is the most serious disadvantage of the qualitative method is avoided and the quantitative values of variable variations can be calculated in a simpler way than strict numerical simulation methods. Although the solution is approximate, the error is considerably reduced by using the mean-coefficient method. The most salient feature of the proposed method is that the condition of the equipment after a change due to the fault can be conjectured by using the linear equations. This brings the method the advantages of being easy to implement and process and of taking much less computation time.

We have applied the method to a diagnosis of the refrigerating equipment and showed that the origin of the fault can be identified and located. The proposed method is useful as a simple and efficient one for performing diagnoses of the equipment by using quantitative physical knowledge.

REFERENCES

[DB84] J. De Kleer and J. S. Brown. A Qualitative Physics Based on Confluences. *Artificial Intelligence* 24:7-83, 1984.

[IA79] M. Iri, K. Aoki, E. O'shima, and H. Matsuyama. An Algorithm for Diagnosis of System Failures in the Chemical Process. *Computers & Chemical Engineering* 3:489-493, 1979.

[Is89] Y. Ishida. An Application of Qualitative Reasoning to Knowledge Acquisition in Process Diagnosis : Diagnosis Rule Generation by Qualitative Simulation. *Transactions of the Institute of Electronics* D-II J72-D-II(3):416-426, 1989.

[La86] E. W. Lawrence. Representation Method for Dynamic Causal Knowledge Using Semi-Quantitative Simulation. In *Proceedings of the Fifth World Conference on Medical Informatics*, 180-184, 1986.

[QB88] B. Quipers and D. Berleant. Using Incomplete Quantitative Knowledge in Qualitative Reasoning. In *Proceedings AAAI*, 324-329, 1988.

[SV87] H. R. Steven and V. Venkatasubramanian. Model-based Reasoning in Diagnostic Expert Systems for Chemical Process Plants. *Computers & Chemical Engineering* 11(2): 111-122, 1987.

[SS80] G. J. Sussman and G. L. Steel Jr.. CONSTRAINTS — A Language for Expressing Almost-Hierarchical Descriptions. *Artificial Intelligence* 14:1-39, 1980

[Wm88] Wm Leler. *Constraint Programming Languages.* ADDISON-WESLEY PUBLISHING COMPANY, Massachusetts, 1988, p. 28.

APPENDIX 1 THEORETICAL FOUNDATION OF THE PROPOSED METHOD

The behavior of a system is described in the form:

$$\frac{dx}{dt} = f(x, u, w) \quad \text{(A1. 1)}$$

$$y = g(x, u, w) \quad \text{(A1. 2)}$$

where u denotes the manipulated variables, w the disturbances, x the state variables, and y the output variables. The disturbances w contains variables for representing a fault which does not concern the original function of the equipment (for example, the flow rate of liquid leak in a tank). If the system is in a steady state, because $dx/dt=0$, the variables representing the steady state satisfy the following equations:

$$f(x_s, u_s, w_s) = 0 \quad \text{(A1. 3)}$$

$$y_s = g(x_s, u_s, w_s) \quad \text{(A1. 4)}$$

where suffix s denotes the steady state. At this stage, because the variable value in the w representing the fault which does not concern the original function of the equipment is zero, it does not produce any influence on the steady state.

If a fault occurs in the system and the variables change by Δu, Δw, Δx and Δy, respectively to reach another steady state corresponding the fault, then the following equations need to be approximately satisfied:

$$f_x(x_s, u_s, w_s)\Delta x + f_u(x_s, u_s, w_s)\Delta u + f_w(x_s, u_s, w_s)\Delta w = 0 \quad \text{(A1. 5)}$$

$$\Delta y = g_x(x_s, u_s, w_s)\Delta x + g_u(x_s, u_s, w_s)\Delta u + g_w(x_s, u_s, w_s)\Delta w \quad \text{(A1. 6)}$$

where f_x denotes $\partial f/\partial x$. The above equations can be derived by expanding Equations (A1. 3) and (A1. 4) about the point (x_s, u_s, w_s) by the Taylor series and neglecting the terms with increment which is greater than the first degree.

In this paper, we are concerned with a fault such as a leak of liquid in a tank by which the structure of the system is not changed. Let's suppose that Equations (A1. 5) and (A1. 6) are constructed by assuming that the equipment functions normally without taking the fault into consideration. If a part of the equipment functions normally, the measured values of the variable variations at the part satisfy such equations. However if a part of the equipment malfunctions, the values of the variable variations which are calculated by such equations do not coincide with the measured values at the part. Therefore we can identify the malfunctioning part based on Equations (A1. 5) and (A1. 6) as described in the section of fault diagnosis method using variation estimation method.

APPENDIX 2 THEORETICAL MEANING OF THE MEAN-COEFFICIENT METHOD

We consider the meaning of the mean-coefficient method through the following function z of x:

$$z = f(x) \quad \text{(A2. 1)}$$

The approximate equation in variable increments can be written:

$$\Delta z = f'(a)\Delta x \quad \text{(A2. 2)}$$

where $f' = df/dx$ and Δx and Δz are the finite variations of x and z from $x=a$ and $z=f(a)$ respectively (see Figure A1). Δz is the approximate variation value of z. If we define the approximate error e in Δz as the third term of the Taylor series of $f(x)$ about the point $x=a$, then we obtain

$$e \fallingdotseq \frac{(\Delta x)^2}{2} f''(a) \tag{A2.3}$$

The error e' in the variation which is calculated by using the mean-coefficient method is represented as

$$e' = f(a+\Delta x) - f(a) - \frac{1}{2}\{f'(a) + f'(a+\Delta x)\}\Delta x \tag{A2.4}$$

Substituting the Taylor series of $f(a+\Delta x)$ and $f'(a+\Delta x)$ about the point $x=a$ and neglecting the terms with derivatives greater than the third degree, then we obtain

$$e' \fallingdotseq -\frac{(\Delta x)^3}{12} f'''(a) \tag{A2.5}$$

In order to estimate the approximate relative error, we consider z to be a polynomial of degree 3 in x and take only the term with degree 3 of x as the approximate value representing the magnitude of z at $x=a$. If we take only the terms with the greatest degree of x in Equations (A2.3) and (A2.5), then the relative errors for e and e' which are compared with the original value of z are given by

$$E \fallingdotseq 3\left(\frac{\Delta x}{a}\right)^2 \tag{A2.6}$$

$$E' \fallingdotseq -\frac{1}{2}\left(\frac{\Delta x}{a}\right)^3 \tag{A2.7}$$

According to the calculation of K_c in the sample problem, Equation (A2.7) has given an appropriate value of the relative error ($E' = -0.0093$).

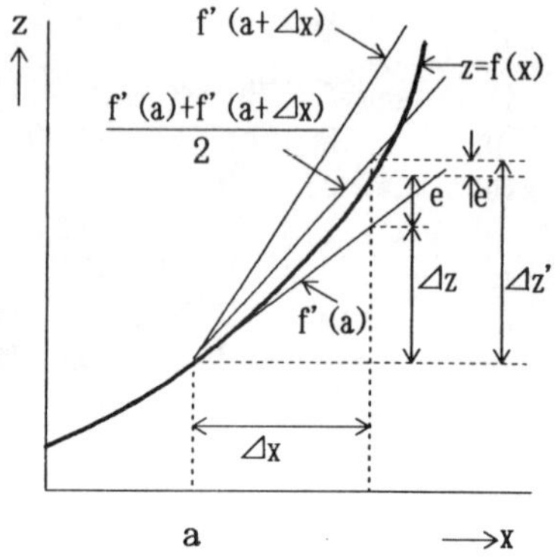

Figure A1 Meaning of Mean-Coefficient Method

Software Engineering

The SEED Project: A Software Environment to Support the Early Phases in Building Design

Robert Woodbury
Department of Architecture
The University of Adelaide
S.A. 5005, Australia

Ulrich Flemming, Robert Coyne
Department of Architecture and Engineering Design Research Center (EDRC)
Carnegie Mellon University
Pittsburgh, PA 15213, U.S.A.

Steven Fenves, James Garrett
Department of Civil Engineering and Engineering Design Research Center (EDRC)
Carnegie Mellon University
Pittsburgh, PA 15213, U.S.A.

The Engineering Design Research Center at Carnegie Mellon University is a National Science Foundation (USA) supported Engineering Research Center.

Abstract. SEED is a collection of software modules intended to support the early phases of building design. The system incorporates the results from two multi-generational research efforts on grammar-based design generation and offers designers a broad range of form generation tools, from interactive construction that is controlled by the designer to the automated generation of design alternatives and case-based design. The system will eventually combine these capabilities with a broad range of analysis, evaluation, and visualization tools.

The SEED modules are based on a shared logic and are accessible through a shared interface. This logic results from a uniform problem-solving view underlying the tasks supported by modules. The key is a uniform problem specification based on functional units which allow a module to participate actively in form generation and support a uniform case indexing and retrieval mechanism across modules and across problem decompositions within a module.

1. Introduction

The SEED project support the early phases in building design (Flemming et al., 1993). Its goal is to provide support, in principle, for the preliminary design of buildings *in all aspects that can gain from computer support*. This includes using the computer not only for analysis and evaluation, but also more actively for the *generation* of designs, or more accurately, for the generation of design representations.

A major motivation for the development of SEED is to bring the results of two multi-generational research efforts focusing on 'generative' design systems closer to practice:

(1) LOOS, a generative system for the synthesis of layouts of rectangles (Flemming et al., 1988; Flemming, 1989), and ABLOOS, which extends the representation and operators underlying LOOS hierarchically and recursively so that (i) a rectangle in a layout can represent a single object or a layout of rectangles and (ii) the generation operators can be used to generate layouts at any level of a hierarchical problem decomposition (Coyne and Flemming, 1990; Coyne, 1991).

(2) GENESIS (Heisserman, 1991; Heisserman and Woodbury 1993), a rule-based system that supports the generation of assemblies of 3-dimensional solids.

It is extremely tedious and time-consuming to generate with current CAD systems an internal representation of a design that is sufficiently detailed to support sophisticated visualization or analysis tasks. The above mentioned systems demonstrate how these deficiencies can be overcome by allowing the computer to play a more active role in form generation. The systems also indicate how the generative operators employed can support the fast exploration of design *alternatives*. SEED intends to incorporate these capabilities and to make them available to designers through an effective interface.

In addition, the project sponsors are interested in support for work with *recurring building types*, that is, building types dealt with frequently in a firm or institution. Such organizations - from housing manufacturers to government agencies - accumulate considerable experience with recurring building types. But current CAD systems capture this experience and support its reuse only marginally. SEED intends to provide systematic support for the storing and retrieval of past solutions and their adaptation to similar problem situations. This motivation aligns aspects of SEED closely with current work in Artificial Intelligence that focuses on *case-based design* (see, for example, Kolodner, 1991; Domeshek and Kolodner, 1992; Hua et al., 1992).

Emphasis is given throughout to the development of a collection of generic capabilities that can be easily adapted to different building types. We start by giving an overview of SEED and its modules in the next section. We then deal with those aspects of the system that enable it to play are more active role in form generation and case-based design: the explicit problem statement (section 3) and generation modes (section 4). Section 5 gives some details about the software engineering process employed and the current implementation.

2. Overview of SEED

We conceive of SEED as a *collection of modules*. This allows us to make local use of various pieces of existing and possibly heterogeneous software and to distribute the development efforts among several teams and over time. It also facilitates the "plugging in" and "plugging out" of individual modules and module versions as the system grows and changes in response to user feedback. On the other hand, each module should appear to the user as part of a unified whole. We attempt to achieve these multiple objectives by basing each module on a common logic, which includes a common style for the interfaces.

In order to arrive at this common logic, we divide the overall design process into distinct *tasks* or *phases*, where a phase is defined by the type of *problem* it addresses and the type of *solution* it generates. For example, the schematic layout phase takes a list of functional units (given in an architectural program) and a context and supports the generation of an organizational scheme and overall building form in which these units are allocated and shaped. A *module* supports work in an entire phase. A common architecture and interface can then be developed for the modules based on the uniform problem-solving view that underlies the definition of the phases they support (Flemming et al., 1993).

Module 1: Architectural Programming
Problem: A project description as it is typically given at the start of a building project: context (including the site); the overall function and size of the building (e.g. an elementary school for 300 pupils); budget etc.
Solution: An architectural program or design brief that expands and elaborates the initial project description, especially the functional and size information in terms of individual functional units and components (e.g. 15 class rooms for max. 30. pupils each etc.) including constraints and criteria on the shape and placement of these units; applicable building codes etc.

Each SEED module offers designers a broad range of form generation capabilities that range from stepwise construction under the designer's control to the automatic generation of design alternatives and case-based design. In order to enable the module to participate actively and effectively in each of these modes, an explicit statement of the problem to be solved in an invocation of a module is needed. Each module therefore has a *problem specification component* that allows designers to specify and modify dynamically the design problem to be solved. In addition, a module has a *generation component* that actively aids designers in finding solutions to the problem specified, and an *evaluation component* that evaluates solutions according to the constraints and criteria identified in the problem specification. Results produced in a module session can be stored and retrieved from a database for use in the same module or in other modules and within the same or across projects. Figure 1 depicts the data flow in a generic SEED module. The following three modules with this generic architecture are planned for the first SEED prototype and currently under development:

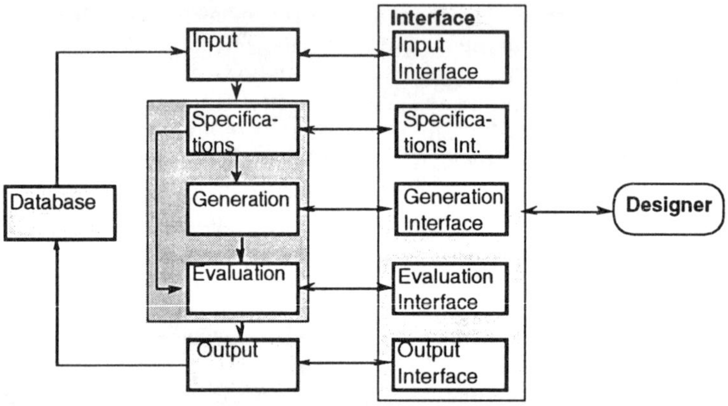

Figure 1: Data flow in a generic SEED module

Module 2: Schematic Layout Design
Problem: A specification of functional components, a context, budget, and other parts of an architectural program
Solution: A layout of the functional units that determines the overall building organization, zones (including the distribution of units on different floors), and form subject to the specifications of the program

Module 3: Schematic Configuration Design
Problem: A schematic layout and programmatic requirements
Solution: A three-dimensional configuration of spatial and physical building components that realizes the given organization subject to the programmatic requirements

The form generation operators of LOOS/ABLOOS are used in module 2 and those of GENESIS are used in module 3. Generation operators for module 1 are yet to be developed.

3. Problem Specification

We attempt to keep the problem specifications in the diverse modules as uniform as possible in order to emphasize the common logic underlying the modules. This uniformity also gives us the chance to develop a uniform indexing and retrieval mechanism to support case-based design across modules: if a solution has been developed with the aid of a module, the associated problem specification can be used to compute an index to be stored together with the solutions so that at a later time, when the same or a different designer defines a similar problem, the resulting problem specification can be compared with the stored indices to retrieve solutions to similar problems (see Flemming, 1994, for details).

In developing such a uniform problem specification, we start with the observation that the description of a design specifies the shape, location and non-geometric attributes of the spatial and physical entities that make up a building, which we call *design units*. Furthermore, each design unit is generally multi-functional, that is, has to accomplish more than one purpose. For example, a room may have to provide sufficient space for the placement and operation of equipment, provide light and acoustic isolation etc. A wall may have to bear loads, provide visual and acoustic privacy, contribute to the illumination of a room through reflecting surfaces etc. We collect all of the requirements that a design unit has to satisfy in a single construct, which we call a *functional unit*. A functional unit is a collection of constraints, criteria and associated procedures that collectively represent requirements a design unit must satisfy.

A functional unit defines a design problem, namely that of shaping, placing and otherwise determining the properties of a design unit that satisfies the respective requirements in a given design context. There is no assumption that this specification is complete at any time in the design process. It is restricted formally to requirements that can be evaluated by the module and taken into account during automated form generation and case retrieval. It is restricted in its content to requirements that appear currently important to the designer. In general, design in SEED proceeds in an 'open world' (Hinrichs, 1992).

A functional unit can contain *constituent functional units*. For example, the functional unit 'floor' may contain constituent 'functional zones', each functional zone constituent 'rooms', and each room constituent pieces of equipment and furniture. Similarly, a functional unit 'gravity-load resisting system' may contain constituent 'structural zones', each structural zone 'structural planes', and each plane 'structural members' like columns and beams or frames. Thus, functional units can be hierarchically decomposed into constituent units, which leads to a parallel decomposition of the resulting design problem into subproblems through several abstraction levels. We mention in passing the strong resemblance between the problem specification in a SEED module in terms of a hierarchical decompositons of functional units and the general AEC reference model GARM (Gielingh, 1988).

A SEED module accepts hierarchically decomposed problem statements and supports the generation and evaluation of solutions at each level. For example, the schematic layout module supports the allocation of zones on a floor, the allocation of rooms within a zone, and the placement of equipment in a room. In this case, the constituent relations imply spatial containment relations, but this need not be the case for other modules. Case-based design is available not only across modules, but also across problem decompositions within a module.

The problem specification component of a module allows designers to specify or remove functional units, to modify the requirements of a functional unit, and to aggregate and disaggregate or otherwise recombine functional units in terms of the constituents they contain. This process is supported by a database of predefined functional units that can be browsed and from which default requirements can be retrieved. We expect that the possibly very intricate process of finding a good problem decomposition will be helped by case retrieval, which may supply not only good alternative starting solutions, but also, and at the same time, good alternative problem decompositions. In general, a SEED module supports not only the exploration of solution alternatives to a given problem specification, but also the exploration of implications of different problem formulations.

4. Generation Modes

Finding solutions to a design problem can occur principally in three basic operations in a SEED module:

1 **Allocate functional units**. The designer selects a set of specified functional units and requests their allocation. Depending on the state of the design, the system creates associated design units and places and shapes them according to the associated requirements, or it discovers that existing design units can take on additional functions and establishes the appropriate associations. This operation underlies, for example, the execution of a task where the designer asks the system to show all feasible ways of placing a certain functional area in a current layout. The use of existing design units arises typically when a design has been developed to a certain degree (for example, as an overall massing scheme) and other decisions have to be taken on it (for example, those dealing with structural support).

2 **Allocate design units**. The designer selects a type of design unit for which no functional unit specification exists. The system generates not only instances of the design units, but also the associated functional units. For example, the designer asks the system to create all columns required in a specific structural zone at all intersections of structural planes in that zone without having specified the functional requirements for these columns before-hand. The system generates these requirements when it creates the design units and associated functional units, which are added to the current problem specification. The designer can inspect them there in order to find out what requirements are specified and can edit them individually (see the preceding section). This operation circumvents the tedium that would be attached especially to repetitive

allocation tasks when they have to rely solely on operations of type 1.

3 **Retrieve past solutions**. The designer asks the system to find good candidate solutions in the case base to solve the problem currently specified. This operation combines aspects of the previous two: The system tries to find solutions that allocate all the functional units that are specified, but such solutions may contain design units associated with functional units that are not yet specified; for example, in trying to find a good layout of given rooms on a floor, module 2 may retrieve a layout that contains not only the specified units but also units like corridors, stairs, utility rooms etc. that are needed for this particular scheme, but have been specified neither as functional nor design units by the designer yet. After retrieval, the new design units become part of the current solution and the associated functional units (which are preserved in the index) become part of the current problem specification, where they are again available for inspection and editing.

These operations are complemented by a range other operations that allow designers to remove design units, edit their attributes etc.

The basic operations can be executed in various modes that range from total designer control through various degrees of automation. An operation of type 1, for example, is completely under designer control when the designer specifies exactly where each selected functional unit is to be allocated in the current design state. The operation is partially automated when the system generates the first, next or all feasible allocation alternatives, but leaves the selection of the current best one to the designer. It is fully automated when the system generates the best allocation according to the current problem specification. Similar modes exist for the execution of the other operations. Note that transitions between these modes within a module are seamless because they are based on the same representation and operators.

We are, at the present time, discussing and experimenting with a range of *navigation aids* that help designers explore various portions of the, possibly rapidly, expanding design space within a SEED module. Figure 2 shows as an example the current version of the *design space window* that belongs to the standard windows brought up when design generation in module 2 commences. The window shows all design states that have been generated so far and the derivational relations between states. Its vertical aixs traces derivation relations among states; its horizontal axis alternatives at a particular level of generation. Designers can scroll through the display to refresh their memory and reactivate any state for further elaboration. Analogous design space windows are provided by other modules.

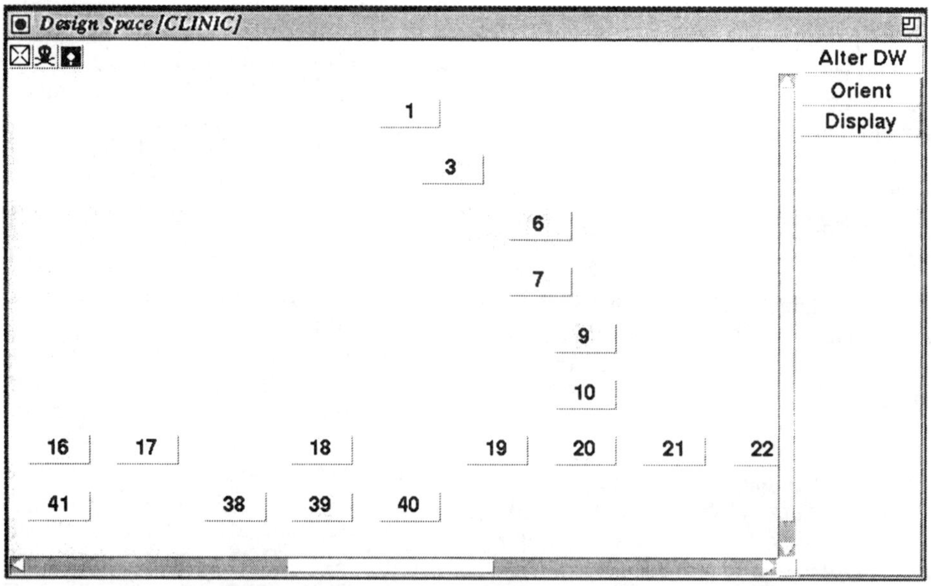

Figure 2: Design space window of module 2

5. Implementation and Current Status

First prototypes of modules 2 and 3 are currently under development using, within the limits of the resources available to us, the object-oriented software engineering approach OOSE (Jacobson et al., 1992). This approach develops a system prototype through five 'models', starting with a high-level functional specification or behavior model through object, design, implementation and test models.

Of these models, the behavior model has been of greatest utility to us to date. It clarifies and determines the desired behavior of the planned system by describing it from the user's perspective through a set of *use cases*. Each use case describes in detail an individual task that a designer can execute with the help of the system; collectively, the use cases provide a complete description of the desired system behavior. Note that the software development process is not expected to be linear: iterations and feedback from more detailed development stages to earlier ones are an integral part of it.

We have completed the behavior specification and design models of modules 2 and 3, to the degree necessary to initiate implementation (see Coyne et al., 1993, for a preliminary description). The implementation of module 2 has been tested by one of our sponsors and signficant refinements to its interface have been made. An initial implementation of module 3 preceded the complete behaviour and design models for that module. This first implemenation served as a demonstration testbed and a vehicle for developing user interface concepts. We are presently implementing a new version of module 3 using a commercial solids modeler and logic programming toolkit. At the time of writing a behaviour model for module 1 is underway.

Our initial design of the SEED implementation model specifies a tripartite structure comprising *view*, *task* and *model* components. The view component presents the system to users and provides low level tools for their interaction with it. The model component is a repository for design problems, solutions and design spaces and a set of operations for changing the state of problems and solutions and, implicitly, design spaces. The task component comprises a set of tasks, each of which implements a use case by linking together view and model elements.

In our first implementations, we faced a situation in which the various parts of the system were best written in different programming languages. The model components were best known to us at the outset of the project because the methodologies underlying each module were developed through the research projects mentioned in the introduction. This meant that we could trade flexibility for rigor and speed; we thus wanted a strongly typed and compiled language for developing these components. Furthermore, we required a language which was readily integrated with other languages and widely supported, leaving only C or C^{++} in the choice set. As module 2 was written from scratch, we chose C^{++} for it. Since we used much of the existing GENESIS code in the initial implementation of module 3, we retained for it its original primary language, C. Part of GENESIS was written in CLP(R), a Prolog superset, available as a compiler/ interpreter written in C (Jaffar et al., 1992). We retained this structure; the model component of module 3 is therefore written partially in both C and CLP(R).

The task and view components were less well-understood, yet are the parts most emphasized in our software engineering methodology. We expected these parts to change as we discovered ways to present end-users with a system that helps and motivates them to solve real design tasks. Thus we wanted a language that permits fast experimentation with interface design. We chose TK/Tcl, which provides the benefits of a user interaction toolkit (the TK part; see Ousterhout, 1991); an interpreted language in which it is easy to treat data as program and vice versa; and a language from which it is easy to call routines in C or C^{++} (Ousterhout, 1990). Since the TK toolkit presumes the X Window system, our basic environment comprises UNIX, X, C, C^{++}, CLP(R), TK and Tcl.

Unfortunately our experiences with TK/Tcl did not fulfill our expectations. While very easy to prototype interfaces with it, the resulting implementations did not scale well, and we found ourselves dealing with increasingly complex code-management and data translations tasks. We prefer not to speculate whether this is inherent in the TK/Tcl design or an artefact of sloppy programming practice on our part. Nonetheless, we are examing our interface environment for future versions of SEED and expect to use an environment more directly related to C++.

We use graphics packages for the display of designs, which have until now been supported in specialized graphics hardware. We have used up to recently the DEC implementation of the PHIGS graphics package, but we have now ported to OPENGL. Lastly, our hardware environment at the outset comprised DEC MIPS stations, to which we have since added DEC Alphas.

Acknowledgments. The development of SEED is sponsored by the US Army Corps of Engineers Construction Engineering Research Laboratories (USACERL), Battelle Pacific Northwest Laboratories, the National Institute of Standards and Technology (NIST), the Engineering Design Research Center (EDRC) at Carnegie Mellon University, the Australian Research Council, the Australian Department of Industry, Science and Technology, the University of Adelaide and Johnson Schmidt Architects.

References

R. F. Coyne (1991) *ABLOOS. An Evolving Hierarchical Design Framework*. Ph.D. dissertation, Dept. of Architecture, Carnegie Mellon University, Pittsburgh, PA

R. F. Coyne and U. Flemming (1990) "Planning in Design Synthesis: Abstraction-Based LOOS," in: *Artificial Intelligence in Engineering V, Vol. 1: Design (Proc. Fifth International Conference, Boston, Mass.)*. J. Gero (ed.), New York: Springer, 91-111

R. F. Coyne, U. Flemming, P. Piela, R. Woodbury (1993) "Behavior Modeling in Design System Development" in *CAAD Futures '93 (Proc. International Conference on Computer-Aided Architectural Design Futures, Pittsburgh, PA)*, U. Flemming and S. Van Wyk (eds.). New York: Elsevier, 335-354

E. A. Domeshek and J. L. Kolodner (1992) "A Case-Based Design Aid for Architecture", in *Artificial Intelligence in Design '92* (J. Gero, ed.), Boston: Kluwer Academic Publishers, 497-516

U. Flemming (1989) "More on the Representation and Generation of Loosely Packed Arrangements of Rectangles", *Environment and Planning B. Planning and Design.* **16**:327-359

U. Flemming (1994) "Case-Based Design in the SEED System", *Automation in Construction* , forthcoming

U. Flemming, R. Coyne, T. Glavin, and M. Rychener (1988) "A Generative Expert System for the Design of Building Layouts - Version 2, " in *Artificial Intelligence in Engineering: Design (Proc. Third International Conference, Palo Alto, Ca)*, J. Gero (ed.). New York: Elsevier, 445-464

U. Flemming, R. Coyne, R. Woodbury (1993) "SEED: A Software Environment to Support the Early Phases in Building Design" in *ARECDAO93 (Proc. IV International Conference on Computer Aided Design in Architecture and Civil Engineering, Barcelona, Spain)* 111-122

W. Gielingh (1988) *General AEC Reference Model.* ISO TC 184/SC4/WG1 doc 3.2.2.1, TNO Report BI-88-150

J. Heisserman (1991) *Generative Geometric Design and Boundary Solid Grammars.* Ph.D. dissertation, Dept. of Architecture, Carnegie Mellon University, Pittsburgh, PA

J. Heisserman and R. Woodbury (1993) "Generating Languages of Solid Models" *Proc. Second ACM/IEEE Conference on Solid Modeling and Applications, Montreal, Canada, May 19-21*

T. R. Hinrichs (1992) *Problem Solving in Open Worlds. A Case Study in Design.* Hillsdale, NJ: Erlbaum Associates

K. Hua, I. Smith, B. Faltings, S. Shih, G. Schmitt (1992) "Adaptation of Spatial Design Cases", in *Artificial Intelligence in Design '92*(J. Gero, ed.), Boston: Kluwer Academic Publishers, 559-575

I. Jacobson, M. Christerson, P. Jonsson, and G. Overgaard (1992) *Object-oriented Software Engineering: A Use Case Driven Approach*, New York: Addison-Wesley

J. Jaffar, S. Michaylov, P. J. Stuckey and R. H. C. Yap (1992) "The CLP(R) Language and System", *ACM Transactions on Programming Languages and Systems (TOPLAS)*, **14**(3):339-395

J. L. Kolodner (1991) "Improving Human Decision Making through Case-Based Decision Aiding", *AI Magazine*, **12**(2): 52-68

J. K. Ousterhout (1990) "Tcl: An Embeddable Command Language", *Proc. 1990 Winter USENIX Conference*

J. K. Ousterhout (1991) "An X11 Toolkit Based on the Tcl Language", *Proc. 1991 Winter USENIX Conference*

SIKT: A Structured Interactive Knowledge Transfer Program

Xindong Wu

Department of Software Development
Monash University
900 Dandenong Road
Melbourne, VIC 3145, Australia

Email: xindong@insect.sd.monash.edu.au

Abstract

Facilitating and formalising the process of specification acquisition in software development is an important problem in the automatic generation of software. This paper presents a structured interactive knowledge transfer program, SIKT, with which subject experts/professionals can interact directly to build executable programs without the aid of computer programmers. SIKT supports the software life cycle from requirements definition through implementation in rule-based systems development. A 'rule schema + rule body' representation, which integrates numeric computation, inexact calculus and logic inference, is adopted to enable the subject experts to develop programs capable of carrying out both numerical computation and logic inference.

1 Introduction

Automatic Programming (AP) is a long-term target for both conventional Computer Science and Artificial Intelligence research. The allure of this goal is based on the recognition that programming has been the bottleneck in the use of computers. Over the last 35 years phenomenal improvements have been made in hardware capability with corresponding decrease in cost. However, software productivity has progressed much more slowly. Despite the development of high level languages (including functional and declarative/logic-based ones), structured methods for developing software, and object-oriented tools, the basic nature of programming remains unchanged. It is still largely an person-centered manual conversion from user requirements/specifications to executable programs, and is error prone and labor intensive.

Program transformation, in which a formal specification is compiled into an implementation, has traditionally dominated AP research [Rich & Waters 88]. With an AP environment, the programmer does not need to worry about *how* to implement his or her program but only *what* to do in the program. The AP environment is expected to automatically or semi-automatically produce the final executable program from the programmer's high level specification. However, there are two serious problems with program transformation research. Firstly, specification acquisition is as difficult as the knowledge acquisition problem in knowledge-based systems development. The specification languages used in existing AP environments are nearly as hard to read and use as high-level/logic-based programming languages. In their effort to be formal, all these languages have scrubbed out many of the mechanisms which make informal languages understandable. Secondly, it is not clear when a program can be usefully transformed [Burstall & Darlington 77, Abadi & Lamport 93]. This problem is still intractable. By imposing a set of transformation rules and applying well-defined sequences of transformation, transformation efficiency can sometimes be achieved. However, the price is that we can preserve only partial correctness of the final programs or restrict the programs that can be produced.

In recognition of the difficult goal of full automation of programming, a lot of research has moved from "automatic" to semi-automatic programming, although program transformation/synthesis still remains an active area in Software Engineering. For example, the KBE-macs research [Waters 85, Rich & Waters 90] at the Massachusetts Institute of Technology in the U.S.A. and the Haskell system [Spinellis 93] developed at Imperial College of Science in Britain, whose power came principally from the ability to construct a program out of the algorithmic fragments/specialised tools stored in a library, and research into techniques editors [Robertson 91] at the University of Edinburgh in Britain, have demonstrated that programming support is a practical way to facilitate programmers. The aim of the SIKT program in this paper is to improve on this previous work by facilitating and formalising the process of specification acquisition. We address the specification acquisition problem in program transformation research and avoid its intractable correctness problem[1] mentioned above. The software life cycle is supported from requirements definition through implementation. We also incorporate symbolic knowledge acquisition techniques developed in Artificial Intelligence research into the development of software to make it capable of carrying out both numerical computation and logic inference. This feature supports a possible way to integrate numerical computation in Software Engineering with symbolic knowledge processing in Artificial Intelligence.

SIKT is a structured interactive knowledge transfer program, implemented in Prolog, with which subject experts/professionals can interact directly to build executable knowledge bases without the aid of computer programmers. The program guides the subject experts in inputing the specifications and domain knowledge of their programs in a top-down, interactive manner, and generates executable knowledge bases from the input from the experts. A knowledge base so produced plus a predefined inference engine from the *KEshell* [Wu 91] forms the final executable program.

[1]This is because we do not use any heuristic transformation rules in our research. Domain knowledge for problem solving is also acquired during interactive specification acquisition. Requirements definition and knowledge are both put in a knowledge base. The knowledge base plus a predefined inference engine from an expert system-like environment forms the final executable program.

2 Related Research

The first successful example of software tools via which experts can interact directly with an expert system is TEIRESIAS [Davis 82]. When an expert identifies an error in the performance of a MYCIN-like expert system, he or she can invoke TEIRESIAS to help track down and correct the error. TEIRESIAS provides facilities for modifying or adding to the knowledge base of the expert system to implement knowledge acquisition. There has been broad agreement that an expert knows more about a field than he or she is able to articulate explicitly. Thus, asking a general question like "Tell me what you know" is often less productive than raising a focused one: "What makes your expert diagnosis different in this case?" TEIRESIAS is a tool which helps the expert augment or modify the knowledge base of rule-based systems. It does not attempt to generate new information on its own. The transfer of expertise by interaction gives a useful mechanism for the study of interactive interface design. However, like a lot of other knowledge base tools (such as KADS [Wielinga *et al.* 92], MORE and MOLE [Marcus 88]) in existing expert systems development and the interactive debugging facilities in high-level languages (*e.g.*, Prolog compilers), TEIRESIAS does not support interactive acquisition of a whole knowledge base.

Inspired by TEIRESIAS and other knowledge-based tools and their problems (see, for example, [Pressburger & Smith 88, Marcus 88, Wielinga *et al.* 92]), the SIKT program is designed and implemented in the expert system shell, *KEshell* [Wu 91], based on a 2-level knowledge representation which integrates numerical and rule-based computations. SIKT can automatically build executable knowledge bases out of direct dialogue with domain experts. As the dialogue process is structurally engineered, a domain expert does not need to know much about knowledge engineering or programming languages. All the expert needs to do is answer the questions asked by SIKT. SIKT builds up a factor dictionary and a reasoning network to describe the logic relationships among the factors. The expert can specify both numeric computation and logical inference during the dialogue.

This paper presents the SIKT program in detail. In the following section, we outline the 2-level 'rule schema + rule body' representation [Wu 94]. In Section 4, we describe the dialogue design between the SIKT program and the expert, and we follow this with an introduction to the *KEshell* system in Section 5. Finally, we summarise in Section 6 our ongoing work based on the SKIT program.

3 Rule Schema + Rule Body: A 2-Level Representation

3.1 The syntax

The grammar used in rule schema + rule body is that of *(factor, value)* pairs as opposed to *(object, attribute, value)* triples in OPS5-like languages [Brownston *et al.* 85].

Definition 1. A *factor*, which has a similar meaning to an attribute in M.1, EXPERT and KES [Harmon & King 85], is a name involved in a domain expertise. It can be a logical assertion, a discrete set variable or a continuous numeric variable.

Definition 2. A rule schema has a rule-like structure with the general form of

IF $E_1, ..., E_n$ THEN A

or IF $E_1 \wedge ... \wedge E_n$ THEN A

where all of $E_1, ..., E_n$ and A are factors. Each E_i $(i = 1, ..., n)$ is called a premise factor and A is called the conclusion factor in the rule schema. The rule schema says that if there are data for all of $E_1, ..., E_n$ then A can be evaluated.

Definition 3. Each rule schema IF $E_1, ..., E_n$ THEN A has a corresponding rule body which contains all the possible expertise to evaluate the value of A (when A is a non-logical variable) and/or the certainty factor of the value/A. All the factors included in the rule body must appear in the rule schema so that when there are data for all of $E_1, ..., E_n$, A can be surely evaluated by the rule body if the expertise inside of the rule body is complete.

Definition 4. *A computing rule* is different from an inference rule in that its RHS is an assignment to a non-logical variable or a certainty factor. For example, *IF A>10, CF(B)>0.4 THEN X=sin(C)+max{0.5, cos(D)}*.

In each rule body, there may be one or more inference rules similar to those in production systems [Barr & Feigenbaum 81, Jackson 90] and/or computing rules for computation. All the rules in a rule body are used to determine the value of the conclusion factor in its corresponding rule schema and/or the certainty factor (CF) of the value/factor. When the conclusion factor is a logical assertion, the rule body can be used to compute the CF of the assertion. When the conclusion factor is a set variable or a numeric variable, the rule body is used both to evaluate the value of the factor and to compute its CF. Thus, the computation of the CF of a factor can be processed in the same way as the evaluation of non-logical factors, both being explicitly expressed in rule bodies. When all the factors in a domain expertise are logical assertions and all the rule bodies have the same rules for computing CFs, the inexact inference then behaves similarly to the standard implementation approach mentioned in the motivations. When all the factors are numeric variables and no uncertainty calculus is needed, all the rule bodies will be used to express computation models and a rule schema plus its rule body is analogous to a procedure or function in conventional programming. Therefore, a knowledge base in this context, which is composed of a number of procedures and functions for numeric computation, plus the inference engine which solves user problems by using the knowledge base and is analogous to a main procedure, can have the same function as a conventional algorithm-based software. This feature of the rule schema + rule body language supports a possible way to integrate Software Engineering with Artificial Intelligence.

Definition 5. A rule schema with its corresponding rule body is called a *rule set*.

A rule set is an independent knowledge unit in

the rule schema + rule body language. It can be described in extended *Backus Naur form* as follows.

<rule set> := <rule set number><rule schema><rule body>
<rule set number> := <integer>
<rule schema> := 'IF' <premise factors> 'THEN' <conclusion factor>
<premise factors> := <premise factor>{','<premise factors>}
<premise factor> := <factor>
<conclusion factor> := <factor>
<factor> := <variable name>|<logical assertion>
<logical assertion> := <predicate(objects)>|<predicate>
<rule body> := (<C-rule>| <I-rule>){<rule body>}
<C-rule> := (<factor>| CF(<factor>))'=' <assignment expression>
<assignment expression> := <value>|<algebraic expression>
<I-rule> := 'IF'<antecedents> 'THEN'<conclusion>
<antecedents> := <antecedent>{','<antecedents>}
<antecedent> := (<factor>| CF(<factor>))<relation-sym> <assignment expression>
<relation-sym> := '>'|'<'|'='|'<>'|'>='|'<='
<conclusion> := <C-rule>
<value> := <integer>|<real> |<symbolic value>|<probability> | <fuzzy degree>

The terms <variable name>, <predicate(objects)>, <predicate>, <algebraic expression> and different kinds of values above are simple enough to be understood.

3.2 Advantages of the language

As compared with classic production systems, the rule schema + rule body language has five main advantages:

1. it can avoid matching all the rules in a knowledge base with the working memory at run time when some piece of data is not available;

Example 1. Suppose we have a rule schema "If A_1, A_2, A_3, A_4, A_5 then B" and all $A_1, ..., A_5$ have a value domain {1,...,1000}, there are at least 1000^5 possible decision-tree like rules with form of or similar to

$$\text{If } A_1 = N_1, A_2 = N_2, A_3 = N_3, A_4 = N_4, A_5 = N_5 \text{ then } ...$$

in total where $N_i \in \{1, ..., 1000\}$ $(i = 1, ..., 5)$. Now if we have a large quantity of data on A_1, A_2, A_3 and A_4 but no data about A_5, only 5 comparisons are needed: Once no data about A_5 is found during matching of the rule schema, all of the rules in the rule body will be neglected without matching.

A linear forward chaining algorithm [Wu 93c] has been designed based on this advantage.

2. the naturally structured organization of knowledge bases in terms of rule sets;

The self-contained format (see Definition 3) of rule sets and their communication channel (the working memory in grammar of $(factor, value)$ pairs with associated certainty factors giving the strength of belief in the values, see Section 5.2) clearly enhance the modularity of the language. One can edit, modify and execute a rule set independently without much attention to other rule sets. Modularity is an important virtue to AI programs as well as Software Engineering because it makes programs easier to understand, explain and modify.

3. the expressive capacity of computation knowledge;

Example 2. The knowledge for solving the general equation $AX^2 + BX + C = 0$ can be expressed in rule schema + rule body as follows.

Rule schema: *IF A, B, C THEN X*

Rule body:

IF A=0, B≠0 THEN X=-C/B
IF A≠0 THEN

X=(-B+$\sqrt{B^2-4AC}$)/(2A)
IF A$\neq$0 THEN
X=(-B-$\sqrt{B^2-4AC}$)/(2A)

4. the flexible expression of uncertainty calculus in rule bodies.

Example 3. In fuzzy multi-objective judgement problems [Wu & Zhang 87], we often need different judging operators, such as $M(\wedge,\vee), M(\otimes,\oplus)$, $M(\wedge,\oplus)$ and $M(\otimes,\vee)$, for different objectives to evaluate their fuzzy degrees. It is easy to define a specific judging operator for each objective in the rule sets where the objective is the conclusion factor.

Also, a domain expert can define CFs as either probabilities or fuzzy degrees, as required. It is possible to couple probability calculus with fuzzy calculus in a knowledge base for inexact inference in rule schema + rule body. However, the coupling depends on the calculus formulae and possible transformation defined in rule bodies. For each factor in a domain expertise, its CF will be computed according to the rule bodies which have it as their conclusion factor.

4 Interactive Knowledge Acquisition

Below we detail the guidance procedure of SIKT based on the rule schema + rule body representation described in Section 3. In the procedure, DFD stands for the defined factor dictionary, where all the factors that have been given evaluation methods are put, and UDF stands for the undefined-factor stack, which contains all the factors whose evaluation methods have not yet been given by the expert.

1. Initialize DFD and UDF to empty sets.

2. Ask the expert to enumerate the problems or output variables for which the program must be able to solve (*e.g.*, names of diseases). Put all these *top factors* into UDF.

3. If UDF is empty, all factors have been defined so goto step 9 for consistency checking and rule body acquisition.

4. Select a factor F from UDF. If F is in DFD, goto step 3.

5. Ask the expert to define F's type: numerical, logical (truth/false) or symbolic (e.g., "good", "ok", and "bad").

6. Ask the expert if users are supposed to provide evidence for F during problem solving. If "yes", F is a *terminal* factor so put F into DFD and goto step 3.

7. Acquire a new rule schema whose conclusion factor is F: Ask the expert to give all the factors F_i that can determine F in a new evaluation method. Put into UDF all the F_i that are not already in DFD or UDF. Add a new rule schema of the form "`IF all of` F_i `THEN F`".

8. Ask the expert if there is any other evaluation method to determine F. If "yes", goto step 7, otherwise put F into DFD and goto Step 3.

9. Check for semantic inconsistencies in the schemata given. There are three types of semantic inconsistencies:

   (a) Dead cycles: Rule schemata "`IF A THEN B`", "`IF B THEN C`" and "`IF C THEN A`" form a dead cycle if none of `A`, `B` and `C` is a terminal factor (*e.g.* a medical symptom) or can be evaluated from other schemata. If a dead cycle is found, the expert will be asked to resolve it.

   (b) Repetition: When there are two rule schemata which are the same, the expert will be asked to integrate the two rule schemata into one.

   (c) Knowledge Insufficiency: If there exists a factor that is neither a top factor (*e.g.* a disease) nor a terminal factor and there is no rule schema which has the factor as its conclusion factor, the rule schemata are incomplete. The expert will be asked to input more information on the factor when it is discovered.

10. For each rule schema acquire its rule body textually from the expert. A rule can be a *computing rule* such as:

```
IF A>10 AND CF(B)>0.4
THEN X=sin(C)+max{0.5, cos(D)}
```

or a *production rule* like those in OPS5-like expert systems.

11. Check for semantic inconsistencies in each rule body. There are three types of semantic inconsistencies in rule bodies:

    (a) Repetition: When there are two rules in a rule body which are the same, one of them will be automatically deleted.

    (b) Redundancy: When there are two rules in a rule body which have the same right-hand side (RHS) but the left-hand side (LHS) of one contains the LHS of the other, the former rule will be automatically deleted.

    (c) Contradiction: When two rules in a rule body have the same LHS but different RHS's, the rules will be displayed to the expert so that appropriate alterations may be made.

12. Let the expert approve the rules that have just been built. If the expert declares that there are still some modifications needed, allow him or her to modify the existing working buffer and call steps 9 and 11 to check consistency. Modifications can be conducted interactively, or via an editor and associated compiler. The editor is a standard text editor like `vi` or `emacs`, and the compiler is based on the 2-level 'rule schema + rule body' representation used above, and incorporates the semantic checking described in steps 9 and 11.

5 *KEshell*: A Rule Schema + Rule Body Based Knowledge Engineering Shell

5.1 System Structure

KEshell is a rule schema + rule body based knowledge engineering shell, with its structure shown in Figure 1.

In the diagram, KB denotes knowledge bases and KBMS is a knowledge base maintenance subsystem. All other modules will be described in detail below.

5.2 Building expert systems in *KEshell*

KEshell has been implemented in Turbo-PROLOG on IBM PC computers. It is a tool suitable for any problems that can be represented in the rule schema + rule body language.

Facts in the working memory are represented as $(factor, value)$ pairs with accompanying certainty factors, which can be either probabilities or fuzzy degrees. Rule schemata and rule bodies in KB represent domain expertise. As the factors in a rule schema are allowed to be variables, it is easy to enter many concrete rules, inference rules and/or computing formulae, in one generic rule schema and then take the rule body as a "look-up" table.

The inference engine in *KEshell* supports both forward and backward chaining with the forward chaining working in linear time [Wu 93c]. When there is a dead cycle in a knowledge base, which means there is some error in the knowledge base that can not be removed, only backward chaining can be adopted. *KEshell* has a special engine in KBMS to detect cycles in knowledge bases.

There are two ways to build knowledge bases. The first is the interactive mode described in Section 4. As shown above, the communication between experts and *SIKT* is quite friendly. An expert does not need to know much about knowledge engineering or programming to build a knowledge base in this mode, just answering all the questions *SIKT* asked. The other way is to use a text editor. The syntax for this editor is the rule schema + rule body language in the *Backus Naur form* plus a declaration of factors' types and domains for those factors which are set variables.

There is a tracing engine in *KEshell* for knowledge base debugging at run time. It first labels all the knowledge has been used during inference, then deletes the knowledge which is useless for concluding the goal(s) even if it was used during the inference and has been labeled, and finally

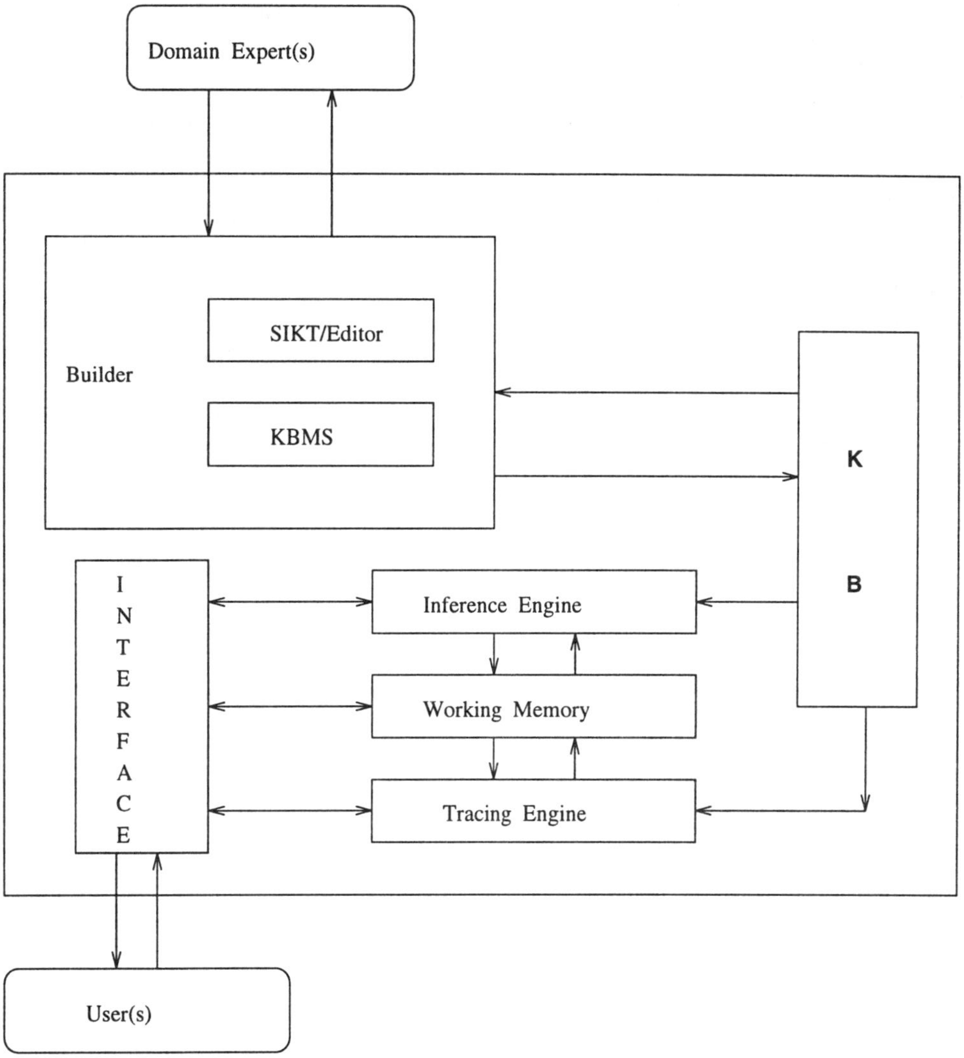

Figure 1: The System Structure of *KEshell*

displays the inference process by using only the still labeled knowledge.

The interface for users is still simpler. Questions concerning evidence are asked, answers are obtained, the chaining process takes place, and the solution(s) are given.

The major feature of *KEshell* is that the shell has supported all the techniques mentioned in Sections 3 to 5 and put them into practical applications. As compared with large system programming languages [Harmon & King 85], such as OPS5, ART and KEE, inexperienced programmers and subject matter experts can interactively build knowledge bases in *KEshell* by calling the *SIKT* module. The price paid for this is that the representation in *KEshell* is not so powerful as in programming languages. For example, users can define control in an OPS5-like rule but they can not do so in *KEshell*.

6 Conclusions

Rule schema + rule body is a representation language that integrates numeric computation, inexact calculus and logical inference. We have presented the representation language and the structured interactive knowledge transfer module (*SIKT*) in *KEshell* in this paper.

SIKT and *KEshell* provide an efficient and comprehensive environment for domain professionals (rather than computer programmers) to build their own application programs. They have wide application to various computing problems, such as diagnosis, pattern recognition, prediction, decision-making and so on. The SIKT program as it stands can only build knowledge bases for *KEshell*, the author and his colleagues have started to extend its capabilities in the following ways:

- Produce an executable program (rather than a knowledge base only) directly from the dialogue with the expert. This will require the inference engine in KEshell to be modified and incorporated into the executable program.
- Provide an extendible library of algorithm fragments and corresponding documentation for the expert to refer to and use. This will enable the expert to build more powerful programs to solve more complex problems.
- Supply a set of machine learning facilities such as inductive algorithms (CN2, C4.5 and HCV) [Wu 93a, Wu 93b] and inductive logic programming (FOIL [Quinlan 90] and GOLEM [Muggleton 92]). The expert can document case studies of his or her skill at work when it is difficult to articulate his or her expertise explicitly.

Acknowledgements

The representation language described in Section 3 is derived originally from Xiong *et al.*'s "rule skeleton + rule body" representation [Wu 90].

References

[Abadi & Lamport 93] M. Abadi & L. Lamport, Composing Specifications, *ACM Trans. on Programming Languages and Systems*, **15**(1993), 1: 73–132.

[Barr & Feigenbaum 81] A. Barr & E.A. Feigenbaum, *The Handbook of Artificial Intelligence*, Vol. I, William Kaufmann, 1981.

[Brownston *et al.* 85] L. Brownston, R. Farrell, E. Kant & N. Martin, *Programming Expert Systems in OPS5*, Addison-Wesley, 1985.

[Burstall & Darlington 77] R.M. Burstall & J. Darlington, A Transformation System for Developing Recursive Programs, *Journal of ACM*, **24**(1977), 1: 44–67.

[Davis 82] R. Davis, Applications of Meta-Level Knowledge to the Construction, Maintenance and Use of Large Knowledge Bases, *Ph.D. Thesis*, Tech. Rept. Stan-CS-76-564, Stanford University, USA, 1976; Reprinted in *Knowledge-Based Systems in Artificial Intelligence*, R. Davis & D.B. Lenat (Eds.), McGraw-Hill, 1982.

[Jackson 90] P. Jackson, *Introduction to Expert Systems*, Second Edition, Addison-Wesley, 1990.

[Harmon & King 85] P. Harmon & D. King, *Expert Systems*, John Wiley & Sons, USA, 1985.

[Marcus 88] S. Marcus, Automating Knowledge Acquisition for Expert Systems, Kluwer Academic Publishers, 1988

[Muggleton 92] S. Muggleton, Inductive Logic Programming, In *Inductive Logic Programming*, S. Muggleton (Ed.), Academic Press, 1992, Chapter 1, 3–27.

[Pressburger & Smith 88] T.T. Pressburger & D.R. Smith, Knowledge-Based Software Development Tools, *Software Engineering Environments*, Chichester Horwood, P. Brereton (Ed.), 1988, 79–103.

[Quinlan 90] J.R. Quinlan, Learning Logical Definitions from Relations, *Machine Learning*, **5**(1990), 239–266.

[Rich & Waters 88] C. Rich & R.C. Waters, Automatic Programming: Myths and Prospects, *IEEE Computer*, **21**(1988), 8: 40–51.

[Rich & Waters 90] C. Rich & R.C. Waters, *The Programmer's Apprentice*, Addison Wesley and ACM Press, USA, 1990.

[Robertson 91] D. Robertson, A Simple Prolog Techniques Editor for Novice Users, *Proceedings of 3rd Annual Conference on Logic*

Programming (Edinburgh, UK, April 1991), G.A. Wiggins, C. Mellish & T. Duncan (Eds.), Springer-Verlag Workshops in Computing Series, 1991, 190–205.

[Spinellis 93] D. Spinellis, Implementing Haskell: Language Implementation as a Tool Building Exercise, *Structured Programming*, **14**(1993), 1: 37–48.

[Waters 85] R.C. Waters, KBEmacs: A Step toward the Programmer's Apprentice, *Technical Report 753*, AI Laboratory, MIT, 1985.

[Wielinga *et al.* 92] B.J. Wielinga, A.T. Schreiber & J.A. Breuker, KADS: A Modelling Approach to Knowledge Engineering, *Knowledge Acquisition Journal*, **4**(1992), 1: 5–53.

[Wu 90] X. Wu, *Constructing Expert Systems*, Press of the University of Science and Technology of China, China, 1990.

[Wu 91] X. Wu, *KEshell*: A "Rule Skeleton + Rule Body" Based Knowledge Engineering Shell, *Applications of Artificial Intelligence IX*, M.M. Trivedi (Ed.), SPIE Press, U.S.A., 1991, 632–639.

[Wu 93a] X. Wu, Inductive Learning: Algorithms and Frontiers, *Artificial Intelligence Review*, **7**(1993), 2: 93-108.

[Wu 93b] X. Wu, The HCV Induction Algorithm, *Proceedings of the 21st ACM Computer Science Conference*, S.C. Kwasny & J.F. Buck (Eds.), ACM Press, U.S.A., 1993, 168–175.

[Wu 93c] X. Wu, *LFA*: A Linear Forward-chaining Algorithm for AI Production Systems, *Expert Systems: The International Journal of Knowledge Engineering*, **10**(1993), 4: 237–242.

[Wu 94] X. Wu, *Rule Schema + Rule Body*: A 2-Level Representation Language, *International Journal of Computers and Their Applications*, **1**(1994), 1: 49–59.

[Wu & Zhang 87] X. Wu & D. Zhang, An Expert System for Multi-Objective Decision-Making—DECISION, *Proceedings of the 1987 International Symposium on Fuzzy Systems and Knowledge Engineering*, Guangzhou, China, 1987, 216–223.

A FRAMEWORK FOR CODESIGN BASED ON FUZZY LOGIC AND GENETIC ALGORITHMS

V. Catania[(1)], N. Fiorito[(1)], M. Malgeri[(2)] and M. Russo[(1)]
[(1)]Istituto di Informatica e Telecomunicazioni e [(2)]Centro di Calcolo
Facoltà di Ingegneria - Università di Catania - Viale A. Doria, 6, 95125 Catania

ABSTRACT

This paper proposes a new approach to solving the problems connected with the simultaneous design of hardware and software components. The idea underlying the paper is to solve the problems of evaluating the performance of the modules by setting up the system as a designer would do. To achieve this we used a tool based on fuzzy logic and genetic algorithms, which allows the performance of single modules to be evaluated without having to actually implement them and also allows evaluation on the basis of imprecise terms. We also show the impact of this approach in the partitiong phase which became feasible.

1 INTRODUCTION

Hardware-Software Codesign is a design discipline whose aim is the simultaneous design of the hardware and software components of a system in order to optimize both the performance and the cost, and to have an easier redesign when needed. Research in this field has been given great incentive by events such as the development of a lot of hardware devices, which has made it possible to implement complex algorithms in hardware, with excellent performance levels and a moderate cost. By contrast, the development of RISC processor technology has allowed the software implementation of several functions which previously had to be implemented in hardware to meet time requirements.

Lastly several methodologies have been proposed for simultaneous design of hardware and software exploiting the advantages of both software and hardware. The term codesign covers different aspects including allocation of hardware and software, evaluation of performance, validation and synthesis [Mic94]. One of the key items is evaluation of the cost of the system and its partitioning. This paper aims to provide a possible solution to the problem mainly in the embedded system area, with a view to drastically reducing design costs and at the same time facilitating implementation choices by faster exploration of the possible implementation alternatives.

To achieve this, we propose a tool which, after a short training phase, is capable of extracting fuzzy rules that allow the performance of single modules to be evaluated without having to actually implement them.

The choice of fuzzy logic [Zad88] is due to the fact that it is capable of specifying the magnitudes observed in fuzzy terms, i.e. in an imprecise way: it does not, in fact, use the traditional concept of membership or non-membership in a given set, but refers to the concept of degree of membership. Thanks to this it is possible to define an object as *very good*, or *slow*, or *cheap*, and so on.

In order to compensate for the low learning capacity of fuzzy logic, a mixed technique based on genetic algorithms and backpropagation was used.

The idea underlying the project, in fact, was to meet two specific requirements. The first is that of providing a work environment which will integrate the tools which are already available for hardware and software system design, thus facilitating the work of the designer. The second, which is the core of our work, is that of estimating the cost of both the modules the system comprises and the system as a whole, before the system is actually implemented.

In order to gain a better understanding of the implications of the use of soft computing techniques, we have limited the field of application to embedded control systems; these computing and control systems are dedicated to an application and may vary widely in size and scope. Embedded systems, which we are studying, fall into the class of reactive systems, that is, systems which react to the environment by executing functions in response to specific external stimuli. If the functions have to be executed within a predefined time they are called real-time systems.

The paper is organized as follows: Section 2 introduces fuzzy logic and genetic algorithms. Section 3 describes the kernel of the tool proposed in the paper: subsection 3.1 describes the tool; Section 3.3 is an example of application of genetic algorithms to a real case and section 3.4 describes the algorithm used by the module which performs the partitioning.

2 FUZZY LOGIC AND GENETIC ALGORITHMS

2.1 Fuzzy Logic

Fuzzy logic is based on the concepts of linguistic variables and fuzzy sets. A *fuzzy set* in a *universe of discourse* $\mathcal{U}$ is characterized by a *membership function* m_f which assumes values in the interval [0,1]. A fuzzy set F is represented as a set of ordered pairs, each made up of a generic element $\in \mathcal{U}$ and its degree of membership $m_f(u)$.

A *linguistic variable* x in a universe of discourse $\mathcal{U}$ is characterized by a set $\mathcal{W}(x)=(\mathcal{W}_{1x}, \ldots \mathcal{W}_{nx})$ and a set $\mathcal{M}(x)=(\mathcal{M}_{1x},\ldots,\mathcal{M}_{nx})$, where $\mathcal{W}(x)$ is the term-set, i.e. the set of names the linguistic variable x can assume, and $\mathcal{W}_{ix}$ is a fuzzy set whose membership function is $\mathcal{M}_{ix}$. If, for instance, x indicates a temperature, $\mathcal{W}(x)$ could be the set $\mathcal{W}(x)=(Low, Medium, High)$, each element of which is associated with a membership function.

The rules governing a fuzzy system are often written using linguistic expressions which formalize the empirical rules by means of which a human operator is able to describe the process in question using his own experience.

In general in a fuzzy conditional rule

if **premise** then **conclusion,**

the premise is made up of a statement in which fuzzy predicates P_j (henceforward also called antecedents) of the general form (X_j is A_j) are combined by different operators such as the fuzzy operators AND and OR; in this case X_j is a linguistic variable defined in the Universe of the Discourse U and A_j is one of the names of the term set of X_j.

The following is an example of a fuzzy conditional rule using such operators:

if $\mathbf{P_1}$ and $\mathbf{P_2}$ or $\mathbf{P_3}$ then $\mathbf{P_4}$,

where:

$\mathbf{P_1}$=(X_1 is A_1), $\mathbf{P_2}$=(X_2 is A_2),
$\mathbf{P_3}$=(X_3 is A_3), $\mathbf{P_4}$=(Y_4 is B_4).

To apply an inference method to the conclusion, it is first necessary to assess the degree of membership θ of the premise, through assessment of the degrees of membership α_i of each predicate P_i in the premise.

The *membership degree* α_i is calculated by assessing the degree of membership of a generic value of X_i in the fuzzy set A_i. If X_i is made up of a fuzzy set, its degree of membership α_i is determined by making an intersection between the fuzzy value of X_i and the fuzzy set A_i and choosing the maximum value of membership; if X_i is a crisp value, its degree of membership in the fuzzy set A_i is made up of the value the membership function of A assumes corresponding to x.

The degree of membership, θ, of the premise can thus be calculated by assessing the fuzzy operations on the predicates (Figure 2-1)

Operator	Meaning
α_i AND α_j	MIN(α_i,α_j)
α_i OR α_j	MAX(α_i,α_j)
NOT α_ι	1-α_ι

Figure 2-1 Fuzzy logic operators

Once the value of θ is known, an inference method can be applied to assess the conclusion. The latter is expressed in the form:

$$(Y_1 \textit{ is } B_1) \textbf{ and } (Y_2 \textit{ is } B_2) \textbf{ and } (Y_3 \textit{ is } B_3) \ldots$$

where: Y_i are linguistic variables and B_i are names belonging to the term set $W(Y_i)$.

In this case the fuzzy operators **and** acting on the output fuzzy predicates (also called consequents) have a different meaning than in the premise. They only express in a single rule a number of rules with the same premise but different conclusions.

Application of the inference process, in our case *center of area* [Lee90], means that the output variable is obtained using the following formula:

$$y = \frac{\sum_{r=1}^{R} \Theta_r C_r A_r}{\sum_{r=1}^{R} \Theta_r A_r}$$

where:

Θ_r is the generic degree of membership of the premise belonging to i-th rule:

C_r and A_r are, respectively, the center and the area of the output fuzzy set belonging to i-th rule.

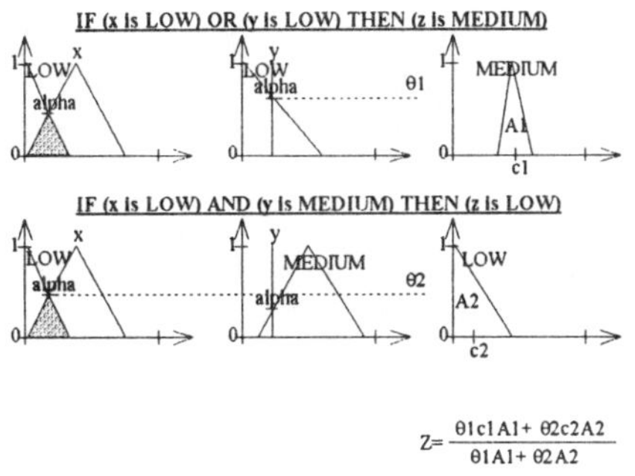

Figure 2-2 Inferential procedure using min-max method

2.2 An Overview of Genetic Algorithms

These algorithms are a sort of artificial evolution of virtual individuals, selected by means of a Fitness Function. In practice, genetic algorithms are a robust way to search for the global optimum of a generic function with several variables; they are very flexible and not sensitive to the classical problem of local optimum. Genetic algorithms are essentially based on three operators: reproduction, crossover and mutation. The flow of a classic genetic algorithm can be simply explained as follows:

P(0) := (P₁,...,Pₙ);
While not (End Condition) do
  Generate P_A from P(t), applying ***reproduction operator;***
  Generate P_B from P_A, applying ***crossover operator;***
  Generate P_C from P_B, applying ***mutation operator;***
  P(t+1) := P_C;
  t := t+1.
end while

The generic operators are summarized in Table 1

Operator	*Operation performed*
Reproduction	The individuals in the population P_A come from the probabilistic selection of the individuals in the population P(t) with the greatest fitness.
Crossover	Applied to two chromosomes (parents) it creates offspring using the genes of both. In most cases a point called crossover is chosen and two offspring are created. The first will have a chromosome with the father's chromosome in the first part and the mother's in the second. The sibling will have the opposite.
Mutation	Each single bit of a generic sibling will be inverted with a probability of p_m.

Table 1 Main operator applied to population

The basic theorem of genetic algorithms guarantees that the individual with the greatest fitness and shortest chromosomes will increase in each generation. This allows a rapid evolution towards individuals with the best degree of fitness.

2.3 Implementation of the Genetic Algorithm

Several genetic algorithms have recently been developed with different implementations of the genetic operators, thus allowing the creation of a new class of algorithms which, although they recall classical genetic algorithms, give better results in a shorter time in terms of convergence towards an optimal solution. One of these is the algorithm used to implement the Fuzzy Performance Estimator.

It is well known that execution of a genetic algorithm leads to genetic homogenization, i.e. the individuals in a population end up by being so alike that no significant improvements between one generation and the next are possible. The attempt is therefore made to find countermeasures which will avoid this problem but at the same time prevent the genetic algorithm from degenerating into a purely random search.

We used a variant of canonical genetic algorithms which always maintains the best solution in the population because it is proved [Rud94] that canonical genetic algorithms will never converge to the global optimum.

2.4 Fuzzy Rules

Let us first discuss the representation of the *fuzzy rules* which will be used by our tool.

Each individual in the genetic population is a set of ***R*** fuzzy rules; in turn, each rule comprises ***I*** inputs (*antecedents*) and ***O*** outputs (*consequents*) represented by an equal number of fuzzy sets connected by the fuzzy operator *and*. The membership functions are Gaussians in the form:

$$\mu(x) = e^{\left[-\frac{(x-c)^2}{2\sigma^2}\right]};$$

which can thus be characterized by the centre ***c*** and the variance ***θ***.

Every element in the grid is therefore an individual whose genetic heritage is made up of the sequence of *I* inputs and O outputs for all the *R* rules, giving a total of $(I+O)R$ fuzzy sets.

Given the nature of genetic algorithms, the number of rules and antecedents in the fuzzy program is minimized and the fitness function is naturally maximized. In order to enhance performance in terms of execution time and minimization of genetic algorithm error [UBKM93] an equivalent neural network has been associated to processing of the fuzzy rules. The resulting network is a three-layered one.

Therefore the mixture of genetic algorithms and neural networks generates a fuzzy inference, i.e. fuzzy rules and tuned membership functions. These results represent the best description of the cost function in terms of fuzzy logic which is represented by the examples used to learn.

3 THE FRAMEWORK

The framework this paper proposes is a complete tool which goes from specification of the system to be implemented to actual physical implementation. Where possible, the tools used were ones available on the market.

The first step towards solving the problem is to find a tool capable of describing the system as a whole before it is split into hardware and software (partitioning). The tool thus has to be capable of representing the system without establishing that a subset has to be hardware- or software-implemented. The description of the system also has to be modular and able to be simulated to provide the designer with parameters he can use to assess his choices.

As there are several commercially available tools whose performance is suitable for our purposes, where possible they were used in the various development stages described below. The novelty of our approach lies in the method used to evaluate the cost of the modules the system comprises and the technique used for partitioning, i.e. distributing the modules between hardware and software.

In general, all the frameworks proposed so far have the following development stages which allow the system to be specified and ultimately implemented:

1. definition of specifications (time constraints, etc.);
2. identification of the target architecture. This step is essential to establish the margins available for mapping the modules onto the hardware or software part of the system;
3. modularization of the specifications;
4. evaluation of the performance of each module implemented both in hardware and in software. This is a basic step towards subsequent partitioning. It should be pointed out that the possibility of assessing the cost of implementation of a module on the basis of its specifications is extremely useful in VLSI projects as well, as it gives the designer useful information about how to achieve possible improvements;
5. partitioning, i.e. hardware or software mapping of the processes;
6. definition of the handshake techniques between hardware and software;
7. implementation, which is usually done by a synthesizer for the hardware part and by compilation of the software part.
8) checking to see whether requirements have really been met. This stage usually involves simulation of the behaviour of the system when stimulated by a set of test patterns, unfortunately this technique does not guarantee that requirements will always be met. Approaches using formal techniques are currently being developed.

Our work aims to solve the problems connected with evaluating the cost of each module in order to achieve correct partitioning and avoid going through the synthesis phase over and over again. It is therefore necessary to define what exactly we mean by cost: we considered a cost function which takes into account the delay introduced by the module and its cost in dollars. In the case of hardware implementation, for example, the cost in dollars is closely linked to the hardware required, while the delay is linked to the software.

The task of the FPE is to assess, or rather estimate, the cost function. So far we have only assessed hardware implementations as it is more difficult to establish the correlation between the parameters which distinguish them and the function itself. As far as software is concerned, the situation is more simple as software costs are usually due to the delay introduced. This can be evaluated using traditional techniques, as the average complexity of the software required is quite low.

Partitioning, as we said above, maps the modules onto the hardware or software part of the system according to the overall cost of the system as a whole. Here we restrict the field of application of the framework to embedded reactive control systems with a well-defined target architecture (see

Figure 3). Our system produces a possible partitioning, i.e. a partitioning which respects the cost constraints imposed.

Phases 4 and 5 are the most difficult to define because they are based on imprecise knowledge of the data and on the designer's ability to interpret it. He can, in fact, expect a particular function to give a good performance when implemented in software, or an interface with the outside world to be sufficiently fast if hardware-implemented, and so on. His final choice falls on the solution which represents a trade-off between cost and efficiency, and he checks that this is true by synthesizing the hardware components and simulating their behaviour, or even by going so far as to build a prototype of the controller. This solution is obviously very costly in terms of time and is not easy to generalize as it mainly depends on the individual designer developing the system. In order to provide real assistance in the development of a design by using the co-design approach, it is necessary to reduce the search for an optimal solution and provide a knowledge base comparable to that of an expert designer. If this knowledge base can then be updated by learning each time the system is used, even further improvement is achieved. To design a controller using this method it is therefore necessary to imitate the behaviour of a human being, in particular his capacity to reason by approximation, i.e. to define and use imprecise assessments such as quite good, very fast, inexpensive, etc. In addition, a designer often has to deal with requirements expressed in linguistic, and therefore imprecise, terms: a device which is *"not too slow and not too large in size"* instead of a device with *"t seconds' delay and a maximum size of x millimetres"*.

3.1 The Structure of the Framework

To insert the high-level specifications of the system the SpeeDChart EFSM formalism was used; although originally conceived for hardware, it proved suitable for specification of the software parts needed in a system of the kind being studied. SpeedChart allows a specification to be made in the form of deterministic extended finite state machines (EFSM) as a code block can be associated with each state. SpeedChart can also provide specification implementations in VHDL. As the SpeeDChart tool does not provide an output in a programming language, a translator was developed from the internal representation into C. The choice of VHDL outputs for the hardware and C for the software allow any kind of synthesizer and CPU to be used for the controller.

The framework comprises the following elements:

- insertion of the specifications of the system to be developed;
- modularization of the specifications;
- pre-partitioning of the system, i.e. identification of the components which have to be in hardware or in software and those whose implementation has not yet been established (codesign components);
- characterization of codesign modules. It is performed by the FPE basing its results on the knowledge database created in a *learning* phase which must be performed on the technology;
- partitioning of the system by a module called a Decision Maker (DM). It puts out a possible implementation;
- synthesis of the components of the partitioning solution found, simulation and assessment. This phase allows updating of the technology database and, by the use of FPE off-line, also updating of the knowledge database.

In the light of the results obtained, it is possible to see whether the partitioning meets the design requirements. If not, another partitioning hypothesis is made by the DM, taking the results of the cosynthesis into account. If the requirements are met, the final implementation of the system can take place. As can be seen, the great advantage of the methodology proposed is that the cosynthesis phase is only repeated if the real cosynthesis results greatly differ from those estimated by the expert system. In practice design starts with insertion of the specification according to the requirements, which is currently done using SpeedChart. When the description of the whole system has been inserted its accuracy is checked, where possible. The designer then divides the specification into functional modules. It should be pointed out that at this stage no choice has been made as regards the type of implementation (hardware or software) for each single module. The next step consists of translating the specification for each module into both VHDL and C. At this point the partitioning, or rather what we call pre-partitioning, of the system begins: the modules are divided into three sets: software, hardware and codesign. The first and second sets comprise modules which necessarily have to be implemented in software and hardware respectively for the requirements to be met. while the third set includes all those modules for which no precise decision has been made. It is on the modules belonging to this latter set that the Decision Maker, using the knowledge previously created by the Fuzzy Performance Estimator, works.

3.2 The Fuzzy Performance Estimator

The Fuzzy Performance Estimator, exploiting genetic algorithms, estimates some fundamental parameters for choice of the modules, on the basis of the semantics of the description in VHDL (hardware). The results of the Fuzzy Performance estimator are the values needed by the Decision Maker to choose a partitioning. For the software counterpart the evaluation is simpler due to the kind of code necessary in the system being studied; therefore we decided to test the FPE only on hardware.

It is a fact [Car91] that the VHDL constructs used to describe a hardware module affect the characteristics of its silicon implementation, especially as regards latency times and the area occupied. Unfortunately it is not possible to quantify the final silicon characteristics, nor is it possible to state with any accuracy which parameters are of importance at each stage: it is only possible to state that the structure of the VHDL code contains information about the *cost* (not only in terms of money) of the final circuit.

In the tool proposed in this paper it was therefore decided to consider almost all the parameters which could be extracted from the VHDL code because the genetic algorithms by themselves do not take into account parameters which are not relevant. The advantage of this approach is that it is independent of the kind of technology since, as mentioned previously, it takes all the significant parameters into consideration and so their weight can vary according to the technology used.

The parameters taken into consideration during our work to assess the hardware modules are the following: the total number of code lines, the width of the bus, the number of processes and blocks accessing the bus, the number of internal signals, the number of ports, the size of the registers, the number of parallel processes, the number of local variables, the number of loops, the maximum length of the loops, the number of conditional assignments, the number of elementary logical instructions, the number of multiplication or division instructions in a loop, the number of addition instructions in a loop, the number of *if* constructs, the number of *case* constructs, the number of synchronization instructions, the number of assignments to variables, and the number of assignments to signals.

At this point the problems to solve are:

- expressing the relationship between the parameters listed and those needed to assess the cost of hardware or software implementation (cost and maximum delay, for example);
- establishing the weight each parameter has in this relationship.

Fuzzy logic, which by its nature tolerates imprecision, is highly suitable to solve this kind of problem. Through *fuzzy conditional rules*, in fact, it allows the formalization of procedures that can only be imprecisely expressed. At the same time, using the concept of *fuzzy sets*, it allows us to handle imprecision regarding the values of the parameters, weighting them through the definition of appropriate *membership functions*.

In our approach the fuzzy sets needed to obtain the fuzzy relationships are generated using genetic algorithms, which allow construction of the fuzzy relationships needed to interpret the parameters chosen for a given technology. They also allow us to eliminate those parameters which, on the basis of the examples chosen, are not significant.

The system obviously depends on the accuracy of the set of examples used to learn the technology, i.e., a preliminary phase is required to gather a set of significant examples the expert system can use to create its database; therefore this database could be updated automatically as the system develops so as to take design variations into account.

3.3 An Example

The Fuzzy Performance Estimator (FPE) is the block which estimates the characteristics of the modules belonging to the codesign set, using exclusively parameters which can be extracted from their high-level specification or implementation in VHDL or C.

In order to check that the FPE is functioning properly, some real cases were chosen, more specifically a certain number of combinatory modules including multi-input adders, subtractors and multipliers, totalling 200 examples which are part of a fuzzy processor currently being designed. These modules were implemented using a 0.7 μm CMOS technology. For the sake of simplicity each module featured the following parameters: width of external bus (ebd), number of VHDL code lines (cln), I/O ports (iop), number of internal operations (nio), number of internal variables (niv), number of conditional constructs (ncc), and number of combinatory operations (nco). These parameters were chosen as they seemed to be the most significant for the examples chosen. An analysis performed on a large number of parameters including, for example, the number of divisors and the number of synchronization instructions showed, in fact, that only those chosen had significant variations in their values. Of course, in assessing further examples it will be necessary to take all the other significant parameters into account.

To allow the FPE to learn this technology, the 200 patterns were subdivided into 160 learning patterns and 40 testing patterns. After about 80000 iterations the confidence parameters reached highly satisfactory values, the standard deviation from the testing patterns being about 10%, while that from the learning patterns was about 5%. The FPE also showed that the number of conditional constructs is not as significant as it was thought to be before the learning program was executed on the basis of analysis of the variations in the values of the parameter.

In practice the complete study of a technology requires identification of a number of *significant* examples and the parameters which appear to affect the values to be estimated. The data collected is supplied to the FPE, which will only be able to *learn* what it requires, eliminating any insignificant input parameters.

3.4 The Decision Maker

The Decision Maker is the block which, on the basis of the costs estimated by the FPE, allocates the modules to hardware or software. Various proposals can be found in literature. It is evident that the choice of the possible partitioning is fundamental. For the process to be used it cannot be confined to proposing just any one of the various possible partitionings, but has to supply the one that appears to be the best. To make this choice, the DM has to be able to analyze the possible solutions as quickly as possible, assessing each single module but without having to make simulations which are usually long and wasteful.

The novelty of our approach is that these choices are made using fuzzy logic which, on the basis of the knowledge obtained by the Fuzzy Performance Estimator data calculated by the assessment module, can extract a possible partitioning. The problem the Decision Maker has to solve is that of finding a partitioning without, however, carrying out an exhaustive search.

As mentioned previously, before the modules are mapped onto the hardware or software part of the system, a target architecture has to be identified; here we refer to the typical architecture of embedded reactive systems, in which there is a precise limit on the number of modules that can be hardware-implemented, given by the size and number of the FPGAs used. The constraint is much less limiting for software, as it depends on the amount of memory available. In this phase the algorithm for scheduling the modules (whether hardware or software) is also developed.

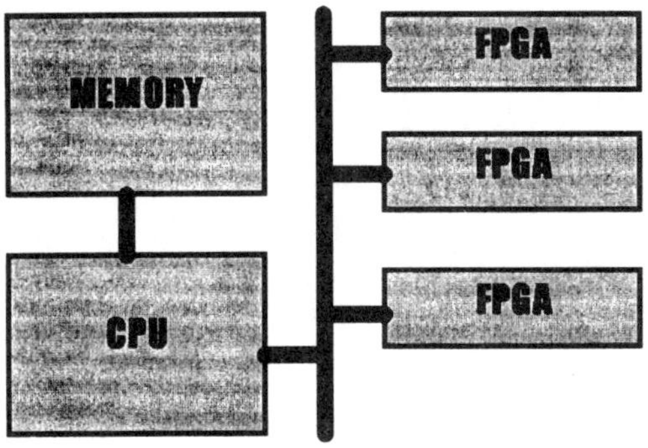

Figure 3: Target architecture.

In our framework the cost of each module is known in fuzzy terms, that is, a fuzzy set is associated with each module, representing the delay and cost in dollars of the module. These estimated values are used by the DM to obtain a possible partitioning.

The DM is essentially based on two fuzzy programs; the first, which we will call isSosftware() tells us if a particular module can be software-implemented, while the second allows us to assess whether the system's requirements, as defined in fuzzy terms, are met (requirementOk()). Communication costs are evaluated, again in fuzzy terms, and are of fundamental importance when it comes to deciding whether a given module is to be hardware- or software-implemented.

Let us now define the parameters used by the fuzzy programs:

M_i = i-th module;

CR_i^t = cost of implementing the i-th module, with t belonging to the set *{hw,sw}*;

CD_i^t = delay introduced by the i-th module, with t belonging to the set *{hw,sw}*;

$CC_{i,j}^{t_i,t_j}$ = cost of communication between module i and module j, where t_i and t_j belonging to the set *{hw,sw}*;

N = total number of modules;

n = number of modules already allocated.

The fuzzy program isSoftware() comprises the following rules:

if $\frac{CR_{tot} + CR_i^{hw} - CR_{max}}{CR_{max}}$ *is negative then* (M_i *is hw)*

if $\frac{CR_{tot} + CR_i^{hw} - CR_{max}}{CR_{max}}$ *is positive then* (M_i *is sw);*

if $\frac{CD_{tot} + CD_i^{sw} - CD_{max}}{CD_{max}}$ *is negative then* (M_i *is sw);*

if $\frac{CD_{tot} + CD_i^{sw} - CD_{max}}{CD_{max}}$ *is positive then* (M_i *is hw);*

if $\frac{\sum_{j=1}^{n} CC_{i,j}^{sw,t_j} - \sum_{j=1}^{n} CC_{i,j}^{hw,t_j})}{n}$ *is positive then* (M_i *is hw);*

if $\frac{\sum_{j=1}^{n} CC_{i,j}^{sw,t_j} - \sum_{j=1}^{n} CC_{i,j}^{hw,t_j})}{n}$ *is negative then* (M_i *is sw).*

The first 4 rules consider the costs due to delay and the cost in dollars of the modules, while the last two evaluate the communication costs.

The values CR_{max} and CD_{max} are threshold values above which the cost and delay are unacceptable.

These programs are used in the DM to search for a possible partitioning. The DM starts from a process chosen at random and, through the isSoftware() function, establishes whether the module is to be software or hardware-implemented. After making this choice it checks to see whether the global requirements, in fuzzy terms, are met. If they are not, the requirementOk() function is used to obtain a different partitioning, until a solution to the problem is reached.

4 CONCLUSIONS

This paper proposes a tool which will allow the designer to (partially) automatize the development of embedded systems by using the codesign approach. The tool takes advantage of both tools which are already available on the market for VLSI CAD and techniques based on soft computing. The main novelty proposed is the use of artificial intelligence techniques to imitate the behaviour of a human in defining the partitioning of the system; choice of the techniques to be used fell on fuzzy logic, which allows for imprecision. We have also shown that the use of genetic algorithms allows cost functions to be obtained from the various modules without excessive computational effort.

Further studies should be devoted to analisys of the Decision Maker in order to optimize the algorithms and to test its effectiveness on a larger set of experiments. Moreover, other rules should be added to take into account the degree of parallelism, the scheduling cost and all the parameters which affects partitioning.

[Mic94] G. De Michelei, *Computer Aided Hardware Software Codesign,* IEEE Micro , august 1994

[Zad88] L.A.Zadeh, *Fuzzy Logic,* IEEE Comp., Mag, Apr 1988.

[Lee90] C.C.Lee, *Fuzzy logic in control system, fuzzy logic controller,* IEEE Transaction on system and amb. contr. vol SMC-20 no. 2 1990

[Rud94] G. Rudolph, *Convergence analysis of canonical genetic algorithms,* IEEE Transaction on neural network vol. 5 no. 1 jan 94.

[UBKM93] S.Uckun, S.Bagchi, K. Kawamura, Y. Miyabe, *Managing genetic search in job shop scheduling*, IEEE expert, oct. 93

[Car91] Carlson, Introduction to HDL Based Design Using VHDL, Synopsis 1991

A METHODOLOGY FOR KNOWLEDGE BASED ENGINEERING OF PARALLEL PROGRAM SYSTEMS

Vladimir D.Ilyin
Department of Methodology for Programming,
Institute for Informatics Problems of Russian Academy of Sciences
30/6, Vavilova St., Moscow, 117900, Russia
E-mail:root@ipiand.ipian.msk.su

ABSTRACT

This paper addresses the problem of creation of an effective methodology for computer aided engineering of parallel program systems (PPS). It describes the approach to computer aided engineering of PPS that is the attempt to solve this problem combining the generalization of our methodology for computer aided construction of non-parallel program systems (NPS) and our automatic parallelizing method designed for construction of PPS. The theoretical foundations of PPS-construction, task knowledge representation language and automatic parallelizing method are presented and described in short.

KEYWORDS

Parallel Program Systems, Computer Aided Construction of Parallel Program Systems, Automatic Parallelizing Method.

1. INTRODUCTION

The problem of program construction is bound up with the automatic parallelizing problem. To create an effective methodology for PPS-construction it is necessary to fix methodological differences between development of NPS and PPS. Besides it is important to ascertain ideas that are effective for development of NPS and PPS as well.

The known approaches to facilitation of safe and efficient programming of a parallel architectures involve the development of following features: languages (such as concurrent PASCAL, ADA, OCCAM, etc.) with their programming environments, parallelizing translators and runtime support environments. The data-flow model of computation is widely used methodological element of these approaches.

The functional model of computation is used in functional languages [Backus78]. This model does not depend upon architectural notions. Functional principles of execution are used in data-flow systems [Dennis74]. Both the functional model of computation and functional principles of execution have influenced on creation of the representation of algorithm to be parallelized (Section 4.).

It is very important for development of safe PPS to create a methodology for computer aided construction of PPS. The purpose of this paper is to present and describe in short the theoretical foundations of methodology for computer aided construction of PPS, OBRAZ language for task knowledge representation and the automatic parallelizing method called MPM.

2. PPS-CONSTRUCTION AS A FORMAL PROCESS

PPS-construction is a formal process based upon both the set of task constructive objects and the family of construction rules. The formal representation of task constructive object is a special graph called task relation graph (G-graph). The family of construction rules is based on both the set of G-graph operations and the set of inference procedures which search subgraphs of G-graphs called solving structures.

PPS-construction is realized in spaces of task constructive objects and consists of such processes as object creation, concretization, specialization, generalization, notion shell replacement and new objects construction. From the formal view point PPS-construction may be considered as the process of inductive set construction.

The parallel program systems development in PPS-construction environment is considered as task constructions building. For this purpose the notion of task constructive object is introduced along with the number of other notions derived from it. They form the complex of notions used in modelling of the process of transition from the problem to be programmed to the program. This complex allows the user to formulate his ideas in terms of tasks at no diffidence and to advance step by step towards the producing of specified program system keeping in mind the meaning of every transition without difficulties.

The task constructive object concept is used to support the "stage to stage" transition from a problem formulation (WHAT-representation of problem being programmed) to a program (final HOW-representation). It is presented by the finite tree where task formulation corresponds to the tree root and programs correspond to the leafs. The intermediate (on the way from formulation to program) HOW-representations of problem (methods and program templates) are necessary for taking into account the implementation specification.

The process of PPS-construction consists of other processes which are performed at different stages of working with the objects in the world of programmable problems. This world is understood as the aggregate of task constructive object spaces.

The construction processes of NPS and PPS have common components. The first of them is the creation of basic objects (b-objects) and it is called p-creation. The specialization of basic objects and the derived ones (p-specialization) is the process of setting of the correspondence between some object and non-empty set of derived objects. Derived object is designed for solving the problem which is a partial case of the problem represented by its base object. Object notion shell replacement (p-replacement) assigns certain problem domain names in accordance with a given lexicon. Making constructions is the process of linking task constructive objects by means of special communication functions. The process of object concretization called p-concretization defines the correspondence between the problem formulation and the methods of its solution, program templates and programs in accordance with implementation specification. There are some additional components in the process of PPS-construction that will be mentioned later (Section 4.).

3. OBRAZ LANGUAGE FOR TASK KNOWLEDGE REPRESENTATION

The conception and formalisms of task knowledge representation are developed for building of task knowledge base of the system of computer aided program construction [Ilyin89]. They form the conceptual basis of OBRAZ language.

OBRAZ is designed to describe notion systems, which contain knowledge that is necessary for supporting the processes of task constructive objects definition, specialization, generalization, construction and concretization (Section 2). The language implements an idea to replace the development of task specifications by the development of the specification of application domain notions system. Interpreting of this notions system specification produces specifications of tasks that are necessary for solving fixed problem. This task specifications and specification of problem solving environment (computational resources, data sizes, etc.) provide all information necessary to build program system.

The central concept of OBRAZ is a notion defined by its name and specification. Notion specification is a set of other subordinate notions (attributes) and relations between them. Language statement usually contains notion definition including such relations as membership and inheritance. Language statements can also describe restrictions for notion value set and equivalence of notions as a special case of restriction.

The notion can have several definitions which express various points of view. Language interpreter considers every definition as a separate notion. User can establish his own view to the problem area selecting the only definition among multiple possible ones.

OBRAZ language is developed to be a tool that supports representation of knowledge that is necessary to build a target-program system producing (computing) values of some notions by values of other notions. This goal have determined that notion model should be based on relations represented by programmable tasks and called task relations. Except task relations the notion specification can contain some relations on notion value sets that are called constraints and are used to define suitable conditions when application of this specification is possible.

The task that formulates a relation on set of notions is also a notion with special meaning. Task specification contains two non-intersecting subsets of attributes called input and output, the aggregation of input and output is called task memory.

The program construction environment has no means for evaluation of task relations stored in its database. This determines the absence of variables in OBRAZ because there is no temporary data to store. The notion specification is close to concept of object class in object-oriented languages. To be briefly we'll sometimes skip in subsequent paragraphs words "specification" or "description", so one should consider terms "notion" and "task" as "notion specification" and "task specification". Text enclosed in /* and */ (multiple lines) is a comment. Text between // and end of line is also a comment.

3.1. Notion representation

The basic structural unit in OBRAZ language is a notion definition that is represented by its name and specification which in turn is a collection of definitions of other notions and relations between them. Specification can be formed from other specifications using inheritance.

Some notions in OBRAZ have built-in specifications. First of all, there are the most general notions such as "text", "number", "set" for which OBRAZ language supports constants and special syntax extensions that increase readability and compactness of notation. There are also notions that are used to build the task based models, such as "task relation", "task memory element", "input", "output", "task graph", "query". All these notions are called predefined.

Every notion in OBRAZ is defined inside the specification of some another notion. If notion A is defined inside notion B then we'll say that B is owner of A and A is attribute of B. This creates membership relations hierarchy that covers the whole notion system. Every notion that can be specified using OBRAZ is the owner of something or the attribute of something or both. Membership relations are transitive. For any notion the owner of its owner is also its owner, and the attribute of its attribute is also its attribute.

3.2. Operations on notion specifications

Notion specifications can be built from other notion specifications by applying to them operations of concatenation, intersection and exclusion. General form of notion definition is

<name> : <expression>;

where <expression> contains operators: "+" – for concatenation, "-" – for exclusion and "*" – for intersection. Operands in expression are specifications presented by explicit set of definitions enclosed in curly braces {} or by names of notions possessing them. The notion is called base notion if its specification is an operand in expression defining other notion which is called derived one.

Concatenation operator "+" adds to its leftside operand all attributes, relations, constraints that are contained in the rightside specification. Notion specification defined by concatenation of some base notions contains specification of any its base. Those attributes that appear in result specification from specifications of other notions after concatenation are called inherited unlike own attributes that were defined inside curly braces.

The exclusion operator "-" selects only those attributes and relations from the leftside specification that are not defined in the right side specification. The latter can contain some notions that are unknown in the left side specification – this notions are ignored. Specification to be excluded can be presented by a notion name or as explicit list of attributes and relations enclosed in curly braces.

The group of two or more specifications separated by intersection operator "*" defines new specification containing only those attributes and relations that are defined in every member of this group.

Example:

```
matrix { rows, columns : integer; };
row_vector : matrix + {
      rows = 1;
      size = columns;
};
column_vector : matrix + {
      columns = 1;
      size = rows;
};
vector : row_vector * column_vector;
```

Result of intersection in this example is equal to definition:

```
vector {
      rows, columns, size : integer;
}
```

Relations defined by operators "=" and "<>" in OBRAZ are more than restrictions on the value sets. Operator "=" means also synonymy, mutual concatenation of specifications and establishing of memory links between elements of task construction. Operator "<>" prohibits the concatenation and making equivalent for the notions it connects and for their derivatives too.

3.3. Task relation

Task relation is a notion whose specification is formed similar to specification of any other notion. The "task relation" notion is predefined in OBRAZ with following description:

```
trel {
      mem;            // general concept of
                      // "memory element";
      input,          // "input" and
      output: mem;    // "output"  elements
};                    // are derived from it
```

This specification is interpreted as follows: there is the "task relation" notion TREL that has attribute "task memory element" TREL.MEM and attributes "task input element" TREL.INPUT and "task output element" TREL.OUTPUT derived from TREL.MEM. To specify some notion X as a task relation one should derive it from TREL:

```
X : trel + { ... };
```

Concatenation X with TREL specification gives to X inherited attributes X.MEM, X.INPUT and X.OUTPUT. These attributes should be used as base notions when specifying real task memory elements:

```
matrix {
      rows, columns : integer;
      element {
            row, column : integer;
            row <= rows;
            column <= columns;
      };
}
reverse_matrix : matrix;
reverse : trel + {
      // "reverse" task relation reverses
      // input matrix A and produces
      // output matrix B
      A : matrix + input;
      B : reverse_matrix + output;
}
```

"Task relation" notion has special built-in derivatives: function, equation, program. The "function" notion differs from inherited TREL notion with the only output element that is implicitly equivalented with the function itself. Task relation is recognized as a function specification when it has the only output memory element with the same identifier as the function one. Task relation presented by function simplifies specification of compound task relations which may be given in functional form:

```
reverse (À : matrix ->
            reverse : reverse_matrix);
```

```
compound_task (À : matrix -> Â : matrix)
{
     multiply (À, reverse (À) -> Â);
}
```

Equation is a task relation where all memory elements are derived from MEM notion, not from INPUT or OUTPUT. The language interpreter dynamically redefines equation's memory according to its current needs during the search of problem solution so that one of the memory elements becomes output and all others form the input. So equation can be considered as compact notation defining multiple task relations in one specification.

The "program" notion is a task relation containing some attributes that allow to describe the process of source or object program code generating. The attribute set structure depends on software and hardware environment used to build the final program system and to execute it. The built-in in OBRAZ "program" notion itself is only template and cannot specify any program code generation process, but there are some notions in construction system knowledge base that are derived from "program" and define proper attributes to describe construction in various special environments.

3.4. Task relation graph

From the formal view point every notion can be considered as task relation graph if it contains some task relations or other task relation graphs which all play the role of graph nodes. The joint set of all memory elements of all tasks in task relation graph forms graph memory. Equivalence of nodes memory elements defines memory intersections creating graph edges. Every group of mutually equivalented notions with at least one task memory element in group forms one graph's memory element. All members of such group are synonyms. Every task memory element that isn't connected with any other notion also forms one graph's memory element.

4. MPM: AUTOMATIC PARALLELIZING METHOD

The new automatic parallelizing method called MPM is developed to support construction of safe and efficient PPS. MPM-parallelizer is designed to transform non-parallel algorithm to parallel one.

MPM-parallelizer is an abstract machine for automatic parallelizing of algorithms represented by task constructive objects (Section 2). Input of MPM-parallelizer consists of non-parallel algorithm representation and runtime environment specification. Its output contains representations of parallel algorithm and runtime environment configuration.

4.1. Task Constructive Automaton

Task Constructive Automaton (TCA) is designed for representation of algorithm to be parallelized. From formal point of view algorithm A is considered as operations set S on which the order relation R is defined (A = <S,R>). If algorithm PA = <SPA,RPA> is derived from algorithm NA = <SNA,RNA> by parallelizing then SPA = SNA and either RPA ≠ RNA or RPA = RNA. If RPA = RNA algorithm NA may not be parallelized.

TCA represents the task constructive object to be executed in parallel environment. It has hierarchical unlimited memory (TCA-memory). Input and output subsets (TCA-inp and TCA-out) are defined on TCA-memory. Intersection between TCA-inp and TCA-out sets is null. If TCA-inp is available (TCA-inp[a]) TCA is ready to be executed (TCA[re]). When TCA[re] obtains computational resource it becomes executable (TCA[e]). Then output elements of TCA[e] become available (TCA-out[a]).

There are two types of TCA: TCA-unit and TCA-block. TCA-unit has one-level unlimited memory. For example single operation of algorithm may be represented by TCA-unit. TCA-block consists of TCA-units that are related by TCA-block memory relation. TCA-block memory elements set is formed by mapping of union of sets of memory elements of TCA-units included. Intersection between sets of TCA-unit and TCA-block memory elements is null. Communication between TCA-block memory and memory of TCA-unit obeys the following assertions.

If the memory state condition of corresponding domain of TCA-block memory is "true" then memory access of TCA-unit to TCA-block is allowed to read from TCA-block and write to TCA-unit memory.

If the memory state condition of TCA-unit output is "true" then memory access of TCA-unit to TCA-block is allowed to read from TCA-unit and write to TCA-block memory. As a result TCA-block memory state is altered.

TCA-block may consist of other TCA-blocks which behaviour is the same as behaviour of TCA-units.

4.2. Parallelizing stages

There may be two different situations of MPM-parallelizing. The first one is outside the PPS-construction system and the second is inside it.

In the first situation non-parallel algorithm is represented by non-parallel source code and parallel algorithm is represented by parallel source code. In the second situation non-parallel algorithm is produced from problem specification and algorithm representation is non-parallel TCA-block. Representation of parallel algorithm is parallel TCA-block.

The first situation differs from the second one by that the automatic parallelizing process includes two additional components: transformation from non-parallel source code to non-parallel TCA-block and transformation from parallel TCA-block to parallel source code.

Thus MPM-parallelizing stages are:

a. Transformation from non-parallel source code to non-parallel TCA-block;

b. Transformation from non-parallel TCA-block to maximum parallel TCA-block;

c. Transformation of the maximum parallel TCA-block to parallel TCA-block implemented in accordance with computational resources specification.

d. Transformation from parallel TCA-block to parallel source code.

The stage b is the main one. This transformation is step by step process that is implemented in accordance the assumption that computational resources of maximum parallel TCA-block are unlimited.

The stage b. obeys the following rules:

i. If on k-step of parallelizing of non-parallel TCA-block there are one or more TCA-units that are TCA[e] then the transformator includes them in k-box of parallel TCA-block (k = 1,2,...).

ii. If on N-step there is none of TCA-units that are TCA[e] then MPM-parallelizing is finished.

iii. Parallel TCA-block is k-box sequence (k = 1,...,N).

5. CONCLUSIONS

At this time PPS-construction system is partially implemented on the basis of PPS-construction methodology. When the system is complete, it may be used for several parallel architectures.

PPS-construction system consists of five interactive components: the PPS-construction knowledge base including PPS-specification editor, the knowledge based parallel programs constructor with automatic parallelizer embedded, the knowledge based runtime support constructor and graphic tools. PPS-construction system produces PPS to be run in specified runtime environment.

ACKNOWLEDGMENTS

The author would like to thank his colleagues at Institute for Informatics Problems of Russian Academy of Sciences for their contributions to the research discussed in this paper. Victor Borisov have contributed to OBRAZ language development and have implemented OBRAZ language interpreter. Alexander Ilyin have contributed to MPM-parallelizing development. Boris Kurov have contributed to the preparing of some benchmark programs to test MPM method. Many others have participated in discussion of PPS-construction methodology.

REFERENCES

[Backus78] J.Backus. *Can programming be liberated from the von Neumann style? A functional style and its algebra of programs.* Comm.ACM, 21 (8), pp. 613-641, 1978.

[Dennis74] J.B.Dennis. *First version of a data flow procedure language.* In B.Robinet (Ed.). Programming Symposium: Proc. Collogue sur la Programmation (Paris, France, Apr. 1974). Lecture Notes in Computer Science, Vol. 19. Springer-Verlag, New York, pp.362-376, 1974.

[Ilyin89] V.D.Ilyin. *Program Generating System.* Nauka, Moscow, 1989.

Spacial and Temporal Reasoning

SPATIAL REPRESENTATIONS AND ROUTE DESCRIPTIONS

Agnès GRYL - Marie-Rose GONÇALVES
LIMSI-CNRS
Paris-Sud University
BP 133
91403 Orsay Cedex
tél: (33 1) 69 85 80 07 - 69 85 80 24
fax (33 1) 69 85 80 88
e-mail: gryl@limsi.fr - goncalve@limsi.fr

ABSTRACT

In this paper, we consider an everyday life use of space – route descriptions – to point out the needs in terms of representation primitives of objects and relations. In a first part, we shortly describe what is a route description (RD) and the experiments we conducted to study how the environment is used to describe routes. In a second part, we present what is needed and what is available in terms of spatial representation and reasoning. Finally, we conclude on the general needs to design a RD system useful for other applications.

INTRODUCTION

Representing and reasoning about space is understanding how spatial knowledge may be represented and how it may be manipulated. From the artificial intelligence point of view, representation of spatial knowledge covers many tasks such as interpretation and generation of natural language, user interfaces design, geographic information systems design, and so on. The processing of space finds in route descriptions (RD) an application that is on the one hand common and usual in everyday life and which, on the other hand, implies to consider a rich variety of spatial knowledge. Thus, in order to design a RD generator as close as possible to human RD, we focused our research on what the elements of space used in RD are, how they are used, what we need and what we get to represent them.

I ROUTE DESCRIPTIONS

For a speaker (S), giving a route description (RD) corresponds to describing a path from a starting point (SP) to an ending point (EP) to a listener (L). This task is very constrained. First, S should elaborate the path between the two points, then he/she should construct the representation of this path, which only contains the useful elements of the environment for the purpose and then he/she should generate the natural language description. At each step, the result should provide L with a useful representation. Thus, the elements, and the lexical items used to express it, should be understandable by L. For example, the landmarks mentioned should be understandable by both L and S without ambiguities on names, localization and characteristics. The ultimate aim of this kind of verbalization is to provide L with a representation of the environment he/she has to move in without too many problems (ideally without any problem). In the actual RDs, many apparent difficulties arising from the use of natural language (ambiguities, implicit information and ellipses) are easily solved by using the context.

In order to point out the existence of a canonical RD structure or at least typical features of RDs, we conducted experiments to extract the different elements used in route descriptions, especially in terms of landmarks (objects of the environments) and actions indicators (use of these objects).

Thus, we conducted two experiments consisting in asking subjects to describe pedestrian RD in urban (Paris) and semi-urban (university campus) environments. The RDs were indirect in the sense that the experimental conditions required S to produce a description for L, who is not physically present and who does not know the environment. Moreover, the subjects were interviewed in a place which is not on the route. All these points suppress deictic references, gesture and so give a kind of complete RD in the way there is no confirmation and interaction.

The first experiment we conducted, consisted in asking 60 subjects for 3 route descriptions in a university campus near Paris [Gry92].

The analysis of the corpus route descriptions led us to conclude that:

- among the different possible landmarks only a few of them seem to be relevant to route description,
- a specific path seems to be preferred to link two places,
- the actions indicators express a few basic concepts,
- it seems to be possible to build a schematic description containing the landmarks used by a majority of people and the basic semantic concepts expressed by the action indicators.

The analysis of the second corpus, consisting in 30 RDs in Paris, collected, under the same experimental conditions, is in progress; but the first results confirm the results obtained by the analyses of the first corpus.

These results allowed us to define what is needed in terms of representation and reasoning. This is developed in the following sections.

II. SPATIAL REPRESENTATION AND REASONING

Defining a spatial representation consists in selecting the entities to represent, choosing the properties of these entities, and finally taking into account the spatial relations occurring among these entities. Because of the multidimensionality of space, we must consider a wider range of notions than those dealing with the unidimensional temporal domain. The selection criteria of a representation are generally imposed by the domain of the associated application. We have to define a system of representation adequate to RDs, that would be developed with respect to the specific needs in the generation of such descriptions.

Two different kinds of spatial representations must be distinguished: those dealing with the qualitative aspect of spatial knowledge and those dealing with the quantitative one. The generally adopted point of view in qualitative approaches consists in only defining as many distinctions as necessary to identify an object, a property or a relation in a given context. A quantitative representation, on the contrary, requires the representation of the value of all the variables according to an a priori fixed unit, whatever the given context is. By their under-determination, qualitative formalisms are the best adapted to the representation of spatial information included in the RDs, since they contain typically qualitative, imprecise, partial and subjective information. We are interested in the different qualitative formalisms defined from the point of view of their applicability to the specific needs of RD generation.

To classify the various qualitative approaches existing, Ligozat [Lig93] and Freksa and Röhrig [Fre93] have suggested many different discriminating parameters. Some of these criteria, listed below, are crucial for our purpose and thus, have to be taken into account when elaborating a useful representation:

- The representational primitives of the objects: the representation can deal either with extended objects (regions in 2D space), or with abstract objects (points, vectors or rectangles).
- The physical properties of the objects: the definition of the representation can, for example, be aware of the possibility of a change in the size or in the shape of the objects.
- The presence or absence of intrinsic orientations of the objects: in some cases, the objects manipulated are regions without orientation, in other cases they are rectangles provided with, for example, a direction of motion.
- The dimension and the nature of space: a 2D or 3D euclidean space, or an abstract topological space.
- The existence or absence of a global orientation of space.
- The relational information represented: topology, distance, orientation.
- The ability to handle partial and uncertain knowledge.

We are interested here in the formalisms that cope with two-dimensional space. In the pedestrian RDs we have collected, we noticed that spatial information may be considered as an orthogonal projection of objects of the environment to an horizontal plane. Even if terms linked to the third dimension (the vertical one) are used (like climbing up or going downstairs), the information retained and used refers rather to discriminating characteristics of the path, in the same way as the mention, for example, of the path texture (a gravel or an asphalt road), than to a modification of the vertical position of the listener (the position before and after the climb). Here, the third dimension of space will be considered as a feature of the object to which is related, in the same way as its shape or its color.

Actually, the choice of the representational primitives of the objects depends on the nature of the relevant spatial information to model.

For example, in the topological approaches of Randell & al. [Ran92] or Egenhofer & al. [Ege90], the basic entities used are regions handled in 2D abstract spaces. The considered aspects are the relations existing among the spatial entities in terms of containment, interior and closure. The relative orientation of the entities does not play any significant role in this case. In the approaches describing the respective orientations of the objects, as opposed to the previous ones, the degrees of connection (contact, inclusion, etc...) between the objects may not be taken into account. The handling of abstractions such as points, vectors or rectangles is more appropriate.

While analyzing the corpus, we noticed that among the wide set of landmarks available in the environment, only few of them were used in RD. Moreover, these landmarks could be distinguished on the one hand by their nature and on the other hand by their ways of utilization. Thus, the underlying representation has to take into account different types of objects (in terms of shape or size for example), and different degrees of precision (according to the context, the same object could be modeled by different abstract objects without loss of information, the nature of the object being relevant or not).

The aim of this section is to determine wether according to the nature of the considered objects, the orientational representations are usable and sufficient to model the knowledge contained in the RD.

Representing objects by points

The point is the simplest spatial object used. This primitive of representation has been developed for example by Freksa [Fre92]. He proposes a qualitative calculus in which the orientation of a point is given related to two others distinct points (which form a vector of reference). The primitive relations of the representation are those corresponding to the disjoint regions obtained by the division of the plane induced by the vector of reference.

Let us take an example of RD's structure of environment (cf. Figure 1). Somebody stands at the starting point and wants to reach building B. We need to represent four streets and three buildings and the relations occurring among them. The choice of the representative points of the different objects in order to keep the structure of the environment explicit or at least inferable is not obvious.

If we decide to represent buildings A, B and C respectively by the points a1, b1 and c1 (as in Figure 1), we lose the information that building B is in front of building C. In another way, if we represent buildings A, B and C respectively by the points a1, b1 and c2, we lose the information that, when walking along building A, the first building that appears at the right, is building C.
Moreover, it is impossible to include in the punctual buildings representation a representation by points for the streets. For example, given the configuration of the streets 1, 2, 3 and 4, we are not able to choose a representative point for each one of the streets in such a way that the layout of the buildings in the street could be represented.

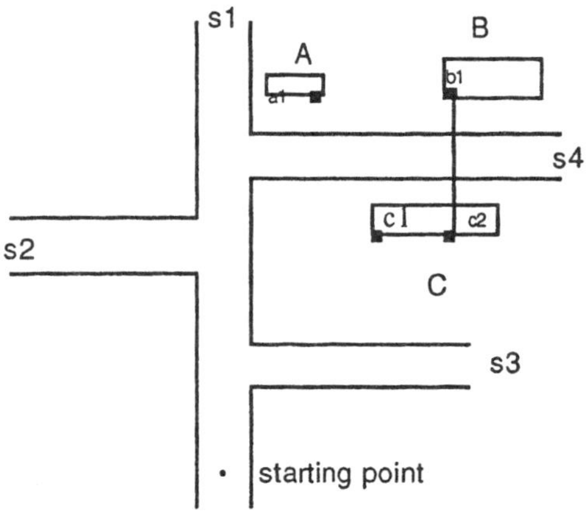

Figure 1. Representation by points.

Hence, representing the structure of environment by punctual entities seems to be insufficient in the case of RDs. Spatial information contained in RDs is often related to notions such as the (qualitative or quantitative) distance among the objects, the relative size of the objects or the area occupied by the objects.

Representing objects by rectangles

Formalisms dealing with rectangular objects provide us with another category of qualitative spatial representations. Allen's work [All83] on temporal intervals was extended to the spatial domain. Relations among objects using Cartesian t-uples of 1-dimensional Allen's relations are defined. We can find such an example in Güsgen's formalism [Güs89]. A Cartesian, extrinsic reference frame is fixed and each object is projected, in an orthogonal way, on the axes of the referential. A spatial relation between two objects is, in that case, the Allen's relation that exists between the projections of the objets on each axis. The underlying algebra is then a product of two copies of Allen's algebra. In this approach, only objects aligned with the axes of the referential or included in a rectangle oriented with the axes can be represented; the rectangular closure of not aligned objects can overlap even if these objects are disjoint.

Consider the example of figure 2 where three streets intersect. The difficulty is to determine a canonical street representation that allows us to represent the crossroads, without loss or gain of surface. If we model the streets by the rectangles 1, 2 and 3, the intersection surface among the three streets is then lost. If a landmark, like a statue, is in the middle of the square, we cannot represent it. It is also possible to represent the streets by the rectangles 1', 2', 3'. But it is still not satisfying because instead of losing surface, we add some to each of the considered streets.

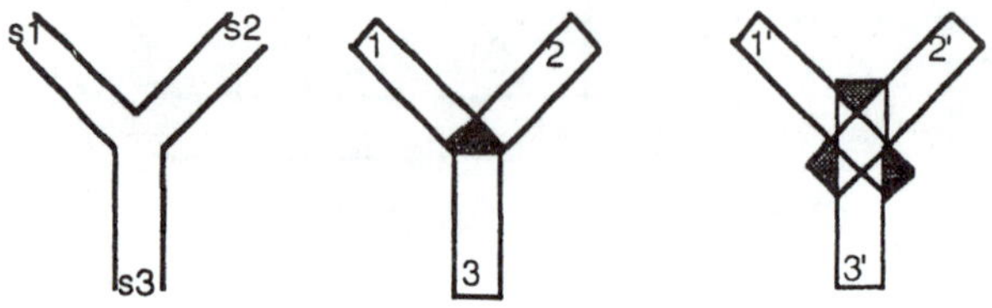

Figure 2. Representation by rectangles.

Nevertheless, such formalisms based on rectangular primitive representation seem to be more adapted to the RD case. Relations between objects are handled without losing information such as relative size or distance. But situations similar to those illustrated by the previous example show the limits of these representations.

Reference frames

As seen in the previous examples of qualitative orientational representation, localizing an object in an environment consists in giving its relative orientation in relation to a second object which is considered as a reference object. Thus, it is necessary to define a reference frame assigning an orientation to the reference object.

Three different kinds of referential are distinguished in relational approaches: the intrinsic, the extrinsic and the deictic reference frames.

A reference frame is called intrinsic when centered on the reference object whose intrinsic properties define the orientation of axes. For example, we commonly consider the front of a building as determined by the location of its principal entrance. Mukerjee and Joe [Muk90] used such reference frames.

A reference frame is called extrinsic when external factors induce the orientation of the reference object. For example, cardinal directions (North, South, East, West) may be used for guidance in a town, or if an object has no intrinsic orientation, the orientation of an intrinsically oriented object in its close proximity may be used. Güsgen [Güs89] and Frank [Fra92] used such reference frames.

The last kind of reference frame is the deictic one. The orientation of a first reference object is defined by the introduction of a second reference object (a point of view). For example, the point of view of the speaker may be used and invoked explicitly by a sentence like "seen from where we are". Freksa [Fre92] and Schlieder [Sch92] use such reference frames.

No reference frame is privileged in RDs. Let us consider the following examples coming from the first corpus.

1. *Turn on your left. You see the restaurant in front of you and on the left of the restaurant you find the library.*
2. *You find a high building, which is tower B and the first building after this tower is building A.*
3. a. *You go straight forward and the library is on your right.*
   b. *Take the corridor inside the buildings.*
4. *The library is near the restaurant.*

Different kinds of reference frame apply to these three descriptions .

A deictic reference frame is considered in sentence 1. An orientation is given to the restaurant by referring to the point of view of somebody looking at it, with respect to which the library is then localized.

An example of an extrinsic reference frame is shown in sentence 2. An external orientation, for instance the geographical orientation of the town, is used. In this case, the front-back axis of building B is then copied from the north-south direction of the town.

Finally, the example given in the sentence 3a is an intrinsic one. The intrinsic orientation of the person (his/her left, right, front and back) determines the reference frame. The location of the library is given related to the position and orientation of the reference object, i.e. the person considered.

Thus, we can conclude that the manipulation of the three reference frames mentioned above is imperative in a human-adapted representation system.

Representing relations between objects

After studying the constraints referring to the objects representation primitives and the reference frames used, we need to define the different kinds of relational information existing in RDs. The three aspects mainly developed in qualitative spatial approaches in order to describe the relative position between two objects are distance, orientation and topology. These three kinds of qualitative relations are used in RDs.

Consider again sentence 1. The given information consists in the orientational position of the library with respect to the restaurant. In sentence 3 b, the relation is a topological one and in sentence 4, the notion of qualitative distance allows to determine the relative position between the two buildings.

Different levels of granularity have been taken into account in the existing qualitative representations. Hernandez [Her91], for instance, considered this aspect. Representation of the orientation is achieved by a sector model: each point corresponding to an object separates the plane in either two, four or eight disjoint sectors, depending on the level of granularity adopted.

Spatial reasoning consists in defining mechanisms allowing reasoning on a given representation. An example is determining procedures to infer implicit knowledge from the knowledge explicitly represented or maintaining the consistence of a database at the time of acquisition and deletion of knowledge.

As seen in the previous section, an adequate representation for RDs should allow us to use different kinds of reference frames to orientate objects in space. Operations to transform between different reference frames should be defined to consider variations of objects orientation criteria, according to the nature of objects or their used context.

Symmetry is also an operation needed. Given a relation between A and B, we have to deduce the relation between B and A. For example, if we know that the post office is on the left of the station, then we want to deduce that the station is on the right of the post office (according to the same reference frame).

The last inference operation we mention in this paper is composition. Given the explicit relation between A and B, and the relation between B and C, we want to be able to deduce the relation (or the set of relations) that can exist between A and C. Composition is an operation defined in all the qualitative spatial formalisms.

As mentioned earlier, among the variety of the existing qualitative approaches, from the point of view of the representational primitives of objects as well as from the point of view of the primitives of manipulation, none is complete enough to cover all the needs of the RDs. Nevertheless, an adequate system for RDs may account and use with benefits some aspects developed in these formalisms.

We need in the same representation to:

- represent different kinds of objects and associated primitives,
- conceptualize an object according to the given context,
- deal with different levels of granularity,
- represent particular features of objects.

All these needs are obviously interrelated. An important aspect not been mentioned yet is the notion of motion in RDs. We need to represent a fictitious movement: the movement of a pedestrian along the RD. We tried to point out the necessity of a hybrid representation of space in order to be able not to restrict ourselves to very specific applications.

CONCLUSION

We collected natural language route descriptions in order to define primitives based on human perception, representation and use of space. We pointed out that, to get as close as possible to the human use of space, it is essential to define systems integrating both different kinds of objects and primitives and different levels of precision. We think that our approach, directly based on a specific but very common use of space, may be useful for other applications such as building scenes under constraints, planification in robotics and transportation and scenes descriptions.

REFERENCES

[All83] J.F.Allen. Maintaining knowledge about temporal intervals. In *Communications of the ACM*, 26 (11), 832-843, 1983.

[Ege90] M.J. Egenhofer & J. Herring. A mathematical framework for the definition of topological relationships. In Brassel K. and Kishimoto H., editors, *Proceedings of the Fourth International Symposium on Spatial Data Handling*, 803-813, Zurich, Switzerland, 1990.

[Fra92] A.U. Frank. Qualitative spatial reasoning about distances and directions in geographic space. In *Journal of Visual Languages and Computing*, 1992.

[Fre92] C. Freksa. Using orientation information for qualitative spatial reasoning. In Frank A., Campari I. and Formentini U., editors, *Theories and Methods of Spatial-Temporal Reasoning in Geographic Space, Proceedings of the International Conference GIS - From Space to Territory*, 162-178, Springer Verlag, Pisa, Italy, 1992.

[Fre93] C. Freksa & R. Röhrig. Dimensions of qualitative spatial reasoning. In Piera Carreté N.and Singh M.G., editors, *Proceedings of the III IMACS International Workshop on Qualitative Reasoning and Decision Technologies - QUARDET'93*, 483-492, CIMNE Barcelona, 1993.

[Gry92] A. Gryl. Opérations cognitives mises en oeuvre dans la description d'itinéraires, Mémoire de DEA, Paris XI, Orsay, 1992.

[Güs90] H.W. Güsgen. Spatial reasoning based on Allen's temporal logic. Technical report TR-89-049, International Computer Science Institute, Berkeley, CA., 1990

[Her91] D. Hernandez. Relative representation of spatial knowledge: the 2-D case. In Mark D.M. and Frank A.U., editors, *Cognitive and Linguistic Aspects of Geographic Space*, 373-385, NATO Advanced Studies Institute, Kluwer, Dordrecht, 1991.

[Lig93] G. Ligozat. Models for qualitative spatial reasoning. In Anger F., Güsgen H.W. and Van Benthem, editors, *Proceedings of the Workshop on Spatial and Temporal Reasoning at the 13th International Joint Conference on Artificial Intelligence IJCAI'93*, 35-45, Chambéry, France, 1993.

[Muk90] A. Mukerjee & G. Joe. A qualitative model f or space. In *Proceedings of AAAI-90*, 721-727, AAAIPress / The MIT Press, Boston, MA, 1990.

[Ran92] D.A. Randell, Z. Cui and A.G. Cohn. A spatial logic based on regions and connection. In *Proceedings 3rd International Conference on Knowledge Representation and Reasoning*, 165-176, Morgan Kaufmann, San Mateo, CA, 1992.

[Sch92] C. Schlieder. Anordnung und Sichtbarkeit. Eine Charackterisierung Unvollständigen Räumlichen Wissens, Doctoral Dissertation, Universität Hamburg, 1992.

TEMPORAL REASONING IN INTELLIGENT APPLICATIONS FOR POWER SYSTEM CONTROL CENTERS

Zita A. Vale, Carlos Ramos

University of Porto
Faculty of Engineering/DEEC
Rua dos Bragas - 4099 Porto Codex - Portugal
Fax: 351-2-2000808
Email: zav@fe.up.pt

ABSTRACT

The results already obtained with intelligent real-time applications in power systems show that Artificial Intelligent techniques are adequate and useful for solving some problems for which traditional programming techniques are not able to provide good solutions.

On the other hand, the duration of this kind of projects is still rather long and the total number of deployed systems all over the world is still very low. This is due to the complex issues that real-time applications for power systems must deal with, among which temporal reasoning plays a key role.

This paper presents the most important temporal reasoning requirements for real-time applications in power systems. It also presents the methods used in some systems that have been developed to deal with temporal reasoning, with particular emphasis on SPARSE, an expert system for fault analysis and service restoration, developed for portuguese Control Centers.

INTRODUCTION

Modern Power System Control Centers must operate and control medium and large electrical networks that usually cover a large geographic area. The accomplishement of these tasks has been made possible due to important developments either in computer and in transmission of information techniques.

Nowadays, complex Power Systems require a very accurate and efficient control what makes Control Centers especially important. These centers are equipped with SCADA (Supervisory Control and Data Acquisition) systems allowing the acquisition of information about the Power System and its transmission to Control Centers in real-time.

Modern Control Centers use several computer applications that include load flow, state estimation, contingency analysis and short-circuit analysis among many others. However, human operators still play the most important role in supervising the Power System and decision making. These tasks can be very complex, especially during emergency situations when several hundreds of messages may be received during one minute and decisions must be taken under great stress and, sometimes, in the absence of most experienced operators. This led to the necessity of developing new applications to assist Control Center operators in decision making.

Due to the symbolic nature of the reasoning involved in operators' tasks, Artificial Intelligence techniques appeared as a promising solution. Some years ago, some of the most important electrical utilities around the world began to develop knowledge-based applications to incorporate in their Control Centers [Tam87]. These applications deal with the information available in the Control Center and help operators in decision making.

However, these applications are very demanding because they must deal with enormous volumes of information in real-time. On the other hand, temporal questions are a key issue in most of Control Center operators' task. These reasons make temporal reasoning a key issue in real-time applications for Power System Control Centers. The available commercial shells do not present characteristics adequate to this kind of applications requiring the development of inference engines from scratch or the use of some alternative methods.

In this paper, we deal with the most important issues concerning temporal reasoning in real-time applications for Power System Control Centers. We refer to the methods used in several systems that have been developed. Temporal reasoning techniques used in SPARSE, an expert system for fault analysis and service restoration, developed for portuguese Control Centers, are also presented.

TEMPORAL REASONING

Temporal Reasoning in Artificial Intelligence

Till 1980, almost every research works based on Artificial Intelligence techniques did not involve time representation. They considered knowledge as independent from time, i.e., a fact that is known to be true at a specific instant, due either to a conclusion of the intelligent system or to an external source of information, is considered as true from then on.

However, almost all real applications involve some kind of temporal reasoning and even those usually treated using Artificial Intelligence techniques: natural language, databases and robotics.

McDermott [McD82], Allen [All83, All84] and Kowalski and Sergot [Kow86] presented some of the first approaches to temporal reasoning. However, their approaches are mainly devoted to narrative interpretation and databases update. Temporal reasoning requirements of real-time applications dealing with a large volume of information with severe time constraints still constitute an open question.

Temporal Reasoning Requirements in Power Systems

As Power Systems are dynamically changing environments, the corresponding information also changes dynamically. This requires dynamically changing knowledge bases and inference engines supporting non-monotonic reasoning. In fact, alarm messages reflect changes in the Power System and the arrival of a message may turn false some of the conclusions previously achieved. The inference engine must be able to test the veracity of previous conclusions according to the most recent information. On the other hand, in what concerns real-time applications, this reasoning must attend to severe constraints and process huge volumes of information in very short periods of time.

The most important points in what concerns temporal issues in artificial intelligence applications for power systems are the following:

- the reasoning involves a lot of temporal questions because conclusions are dependent from the following aspects:
  - the time in which events take place
  - the chronological order of the events
  - the period of time that separates one event from another
  - not only the already known events but also on the events that may or may not happen afterwards
- as information is acquired through an information system that has its own limitations one must consider that:
  - the time-tag of each piece of information always includes some kind of errors (the time corresponding to information time-tagged at the origin of the events includes errors inherent to the acquisition systems and may be furnished with important delays to Control Centers whereas information time-tagged at Control Centers presents a time always with important errors)
  - faults may occur in the transmission system (e.g. a fault in a Remote Terminal Unit) making the consideration of incomplete information necessary; when the system recovers, it may transmit some delayed information that must then be considered and may alter the previously achieved conclusions
  - when the volume of information to be transmitted is too large (e.g. in case of severe incidents) a part of this information may be lost.

All these questions, together with several limitations due to the necessity to guarantee real-time processing, make temporal and non-monotonic reasoning very important in power system intelligent applications.

Temporal Reasoning Methods Used in Power System Applications

The use of Artificial Intelligence techniques in power system applications has been experienced since about 10 years ago [Tam87] and acquired a special importance since 1990 [Liu93, Son93]. Presently, Artificial Intelligence techniques have proved to be adequate to solve some important questions in power systems, especially in power system control. However, Artificial Intelligence techniques, as developed by Artificial Intelligence researchers, could not be directly applied in power system applications, namely in what concerns temporal and non-monotonic reasoning. This meant that the mean duration of the development of power system intelligent applications is rather long requiring important manpower effort.

Presently, there are only a few real-time intelligent applications in power system deployed all over the world. This does represent the success of the last ten years research work in this area but it also shows the severity of the difficulties encountered. As these difficulties are common to a lot of applications, its discussion may be of great help for on-going and future projects.

The several systems developed and described in the literature use widely different methods to deal with temporal reasoning [Kir93].

The system developed for the Consolidated Edison Company of New York, USA, [Sho89, Sho92] uses a Doyle-style truth maintenance system in order to preserve the consistency of its model of the power system [Kir93]. It builds a dependency network that relates the achieved conclusions to the facts that support them. If one fact becomes false, all the conclusions that depend on it are automatically retracted. Automatic retractions are allowed up to five minutes after the first alarm for a particular event.

The system developed for the Northern States Power Company of Minneapolis, USA, [Bij89, Bij91] uses, for this purpose, a situation stack and a larger set of rules [Kir93]. The situation stack is used as a notebook where the system keeps information about unresolved issues and of alarms that have not yet been cancelled. The analysis of this information involves the consideration of aditional rules.

The intelligent alarm processor developed for Regional Control Centers of the network of Electricité de France [Mon89] uses an entirely different approach. It diagnoses the fault and generates an ideal list of alarms for that kind of fault. This list is compared with the list that has been received in order to generate the messages to be presented to operators. When an incident occurs, the alarms are stored during 30 seconds and only then are sent to the inference engine that provides the diagnostic and the ideal list of alarms.

The system developed for the 22kV distribution network of Singapura [Wag92, Bru93] only performs its reasoning when an incident occurs. When the electrical network is in a normal state, it only updates its information when there are changes in the topology. This system considers an area of the network, based on the first message that corresponds to an incident. The diagnostic only begins to be performed when it is considered that all the necessary messages are already stored.

For this purpose, a temporal window of dynamic nature is used. The width of the window depends on the type of the first alarm message received. This width corresponds to the larger period of time during which one must wait for information concerning the incident. This window width may be shortened according to the time interval between two consecutive alarms. On the other hand, if no more information arrives during a pre-defined period of time, the incident is considered to have finished.

All these systems use different methods for considering temporal issues. As this theme is usually not very detailed in the literature, advantages and disadvantages of each method are not clear. On the next section, we will present the system developed for portuguese transmission substations Control Centers - SPARSE - and the involved temporal reasoning.

SPARSE

General Overview

The main goal of SPARSE is to help Control Center (CC) operators, in case of incident, to interpret rapidly the enormous quantity of alarm messages that are generated according to SCADA information and to minimize the amount of non supplied power.

SPARSE is implemented in Prolog and C language and runs on a DECstation 5000/240, under ULTRIX environment. This machine is connected through a Local Area Network Ethernet of duplicate configuration with the two μVAX II machines in the CC. One of these machines is always on-line, supporting SCADA and control functions, while the other is in hot standby.

In order to send to SPARSE the required real-time information, a software application, TTLOGW, has been developed and installed in the μVAX II machines. This application is developed in FORTRAN and includes the following features:

- when the power system is in a normal state, it only sends to SPARSE the messages corresponding to changes in the state of breakers and disconnectors and of the type of operation of plants
- when the beginning of an incident is detected, it sends to SPARSE all the messages (alarms and changes of state), since two minutes before the beginning of the incident and till it receives information about the end of the incident (from SPARSE or from CC operators).

In order to give to SPARSE knowledge about the power system state, TTLOGW sends a "state protocol" when SPARSE starts to run and when μVAX boots. This "state protocol" includes information about the state of all the breakers and disconnectors and of the operating mode used in all substations.

Temporal Issues

The syntax of the rules included in SPARSE Knowledge Base allows to deal with temporal reasoning.

The Left Hand Side (LHS) of the rules may be a disjunction of conjunctions of several conditions. The Right Hand Side (RHS) contains the actions/conclusions to be taken if the LHS is verified. Let us consider the following rule:

```
rule xx: 'EXAMPLE' :
[[C1 and C2 and C3]
or
[C4 and C5]]
==>
[A1,A2].
```

Conditions considered in the LHS (C1 to C5 in this example) may be one of the following types:

L1) a fact whose truth must be proved

L2) a temporal condition.

The actions/conclusions to be taken (A1 to A2 in this example) may be of one of the following types:

R1) generation of facts (conclusions to be asserted to the knowledge base)

R2) killing of facts (conclusions to be deleted from the knowledge base)

R3) interaction with the user interface.

Let us now consider a simplified version of a rule concerning the closure of a line, under fault, by the operator:

```
rule d8: 'THREE PHASE TRIPPING WITH
RECLOSURE'
[[
message(Date1,Time1,[Plant1,Panel1,[Plant2,NL]],
'>>>> TRIP ORDER', 'BEGIN') at T1
and
message(Date2,Time2,[Plant1,Panel1,[Plant2,NL,
'BREAKER']],'BREAKER','OPEN') at T2
and
panel_in_service(Plant1,Panel1)
and
condition(mod_dif_time_less_or_equal (T2,T1,15))
and
message(Date3,Time3,[Plant1,Panel1,[Plant2,NL,
'BREAKER']],'BREAKER','CLOSED') at T3
and
condition(dif_time_less_or_equal(T3, T2, 50))
]]
==>
[
kill_fact(breaker_closed(_,_,_,Plant1,Panel1,
Plant 2,NL),_,T2),
generate_fact(breaker_open(Date2,Time2,trip,Plant1,
Panel1,Plant2,NL),T2),
kill_fact(breaker_open(_,_,_,Plant1,Panel1,Plant2, NL),
_, T3),
generate_fact(breaker_closed(Date3,Time3,tp_reclosure,
Plant1,Panel1,Plant2,NL),T3)
].
```

As we can see from this example, the conditions considered in the LHS of SPARSE's rules may be time-tagged (as the first, second and fifth ones) or not (as the third one). The time-tags considered in the conditions may be used to test temporal relations involving the required time-tags (as the 4th and the 6th conditions). These temporal conditions allow to correctly identify the situation, taking into account temporal issues.

In this example, the 4th condition imposes that the period of time between the event considered in the 1st condition (trip order sent to the breaker) and the event considered in the 2nd condition (opening of the breaker) is not greater than 150 miliseconds. This guarantees that the opening of the breaker is a consequence of the tripping order that it received.

On the other hand, as this 4th condition considers the absolute value of the difference of instants T2 and T1, it allows the tripping order message to arrive after the message that reports the opening of the breaker. Although this may seem a strange situation, because the trip order must always preceed the consequent opening of the breaker, it allows to deal with errors and delays that may occur in the transmission of some messages which may result in inversion in the order of SCADA messages.

The last condition considered in the LHS of this rule imposes that the breaker must be reclosed within a 500 miliseconds period after its opening. If this is the case, we may conclude that it has been closed due to fast reclosure. In fact, protection devices are usually equipped with fast reclosure features that allow to minimize the non supplied power in case of fugitive faults. In the case of a three-phase fault, as considered in this rule, the fast reclosure is usually performed 300 miliseconds after the opening of the breaker. In order to deal with eventual delays in the transmission of the information, we consider 500 miliseconds in this rule.

All these times considered in temporal conditions are based on the technical knowledge about the power system and have been adjusted according to the analysis of a large set of real incidents.

The Inference Engine of our system [Val93c] has been completely developed in Prolog especially for this application. It allows an efficient processing of alarm messages in real time, treating more than 1200 alarm messages per minute.

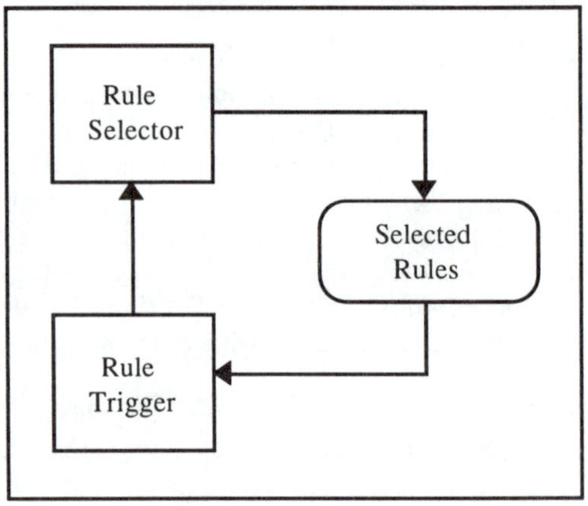

Figure 1 - Inference mechanism

It uses a forward chaining strategy of reasoning that triggers the appropriate rules when a new fact, external (alarm) or internal (conclusion) arrives, in order to derive new conclusions.

The Inference Engine uses metaknowledge in order to select the appropriate rules and guide the reasoning process, improving the system efficiency (figure 1).

For each arriving fact (alarm or conclusion) the Rule Selector selects the rules to be triggered. The Rule Selector generates facts as the following:

trigger(NF,NR,T)

that means that rule number NR should be triggered at instant T, due to the arrived fact number NF.

The metaknowledge used by this module specifies, in each case, the time that the system should wait before triggering the selected rules. This allows the introduction of a delay in the triggering of the rules. This delay allows the system to:

- wait, for some moments, for the arriving of facts that are premises of the rule to be triggered; in this case, there are two possible situations:
  - the delay corresponds to a tolerance allowed in the arriving of the messages (to consider delays in the transmission of the information)
  - the delay corresponds to the time required for the arrival of other messages involved in the same rule (e.g. the maximum time interval between a "trip order" message and a "breaker open" message)
- wait, for a certain period of time, for the arrival of facts that are, in the negative form, premises of the rules to be triggered (if one of the extremities of a line opens, due to tripping, and the other extremity does not open within a defined time interval, the operator is informed about the tripping in the other end).

The inference engine of an alarm processor must deal with complex temporal reasoning, mainly due to the following reasons:

- the sequence of events is especially important for the interpretation of alarm lists
- the meaning of a message depends on what happened before and, in certain cases, on what may happen next.

The Rule Trigger module selects, for the present moment, among the Selected Rules, what is the next one to be triggered. For each triggered rule, the veracity of the premises is verified. If this veracity is proved, the rule is successful and the actions/conclusions of its RHS are taken.

When all the rules selected to be triggered until the present moment have been considered, the Rule Selector considers the facts that have arrived to the system and that have not yet been considered.

In our system, the facts that are found out to be false are not simply erased from the Knowledge Base. They are recorded as "old facts" allowing its later use for explanation facilities. The following fact:

old_fact(504,urg_al('SFA','102'),'23609096','23609100').

means that a fact reporting an urgent alarm in pannel 102 of Ferreira do Alentejo substation (SFA) has become true at instant 23609096 and has become false at instant 23609100.

EXAMPLE

Let us consider the following messages that arrived to Vermoim Substations Control Center on the 8th of May 1993 in consequence of an incident:

17:40:48.655	SVG	204	VERMOIM
URGENT ALARM		BEGIN	
17:40:48.655	SVG	204	VERMOIM
>>>> TRIP ORDER		BEGIN	
17:40:48.656	SEJ	SGER	GEN SERVICES
URGENT ALARM		BEGIN	
17:40:48.656	SEJ	SGER	GEN SERVICES
COMP. FAULT		BEGIN	
17:40:48.664	SPN	205	VALDIGEM II
OTHER ALARMS		END	
17:40:48.670	SRA	119	TRANSF 2
NON URGENT ALARM-B		BEGIN	
17:40:48.697	SED	BAR6	60KV BAR
NON URGENT ALARM		END	
17:40:48.715	SVG	204	VERMOIM
BREAKER		OPEN	
17:40:48.736	SEJ	SGER	GEN SERVICES
URGENT ALARM		END	
17:40:48.736	SEJ	SGER	GEN SERVICES
COMP. FAULT		END	
17:40:48.795	SVG	204	VERMOIM
URGENT ALARM		END	
17:40:48.795	SVG	204	VERMOIM
OTHER ALARMS		END	
17:40:48.795	SVG	204	VERMOIM
>>>> TRIP ORDER		END	
17:40:48.797	SED	SGER	GEN SERVICES
POWER FAULT		BEGIN	
17:40:48.797	SED	SGER	GEN SERVICES
AGENT PRESENT		BEGIN	
17:40:48.797	SED	SGER	GEN SERVICES
URGENT ALARM		BEGIN	
17:40:48.850	SRA	119	TRANSF 2
NON URGENT ALARM-B		END	
17:40:48.910	SRA	123	TRANSF 1
NON URGENT ALARM-B		END	
17:40:48.530	SVM	211	VALDIGEM
>>>> TRIP ORDER		BEGIN	

17:40:48.544 SVM	213	RECAREI II
OTHER ALARMS	BEGIN	
17:40:48.544 SVM	214	TRANSF 1
OTHER ALARMS	BEGIN	
17:40:48.544 SVM	211	VALDIGEM
OTHER ALARMS	BEGIN	
17:40:48.544 SVM	210	TRANSF 3
OTHER ALARMS	BEGIN	
17:40:48.564 SVM	214	TRANSF 1
OTHER ALARMS	END	
17:40:48.564 SVM	211	VALDIGEM
URGENT ALARM	BEGIN	
17:40:48.584 SVM	214	TRANSF 1
OTHER ALARMS	BEGIN	
17:40:48.610 SVG	SGER	GEN SERVICES
OTHER ALARMS	BEGIN	
17:40:48.624 SVG	212	POCINHO II
OTHER ALARMS	BEGIN	
17:40:48.624 SVM	211	VALDIGEM
BREAKER	MOVING	
17:40:48.638 SRR	201	VALDIGEM I
OTHER ALARMS	END	
17:40:48.638 SRR	203	VALDIGEM II
OTHER ALARMS	END	
17:40:48.644 SVG	211	POCINHO I
OTHER ALARMS	BEGIN	
17:40:48.644 SVM	213	RECAREI II
OTHER ALARMS	END	
17:40:48.644 SVM	214	TRANSF 1
OTHER ALARMS	END	
17:40:48.651 SVM	208	TRANSF 4
OTHER ALARMS	END	
17:40:48.651 SVM	205	RECAREI III
OTHER ALARMS	END	
17:40:48.651 SVM	204	TRANSF 6
OTHER ALARMS	END	
17:40:48.651 SVM	202	TRANSF 7
OTHER ALARMS	END	
17:40:48.664 SVM	210	TRANSF 3
OTHER ALARMS	END	
17:40:48.664 SVM	211	VALDIGEM
BREAKER	OPEN	
17:40:48.684 SVG	212	POCINHO II
OTHER ALARMS	END	
17:40:48.684 SVM	211	VALDIGEM
URGENT ALARM	END	
17:40:48.690 SRA	624	RUIVAES I
VOLTAGE ZERO	END	
17:40:48.731 SVM	211	VALDIGEM
>>>> TRIP ORDER	END	
17:40:48.764 SVM	211	VALDIGEM
OTHER ALARMS	END	
17:40:48.784 SVG	211	POCINHO I
OTHER ALARMS	END	
17:40:50.210 SVG	OPA	AUT. OPER.
AUTOMATIC OPERATOR	BEGIN	
17:40:52.188 SRR	208	VERMOIM I
OTHER ALARMS	END	
17:40:52.238 SRR	209	VERMOIM II
OTHER ALARMS	END	
17:40:52.248 SRR	207	VERMOIM III
OTHER ALARMS	END	
17:40:53.810 SVG	SGER	GEN SERVICES
NON URGENT ALARM	BEGIN	
17:40:53.830 SVG	SGER	GEN SERVICES
NON URGENT ALARM	END	
17:40:54. 10 SVG	SGER	GEN SERVICES
OTHER ALARMS	END	
17:41:56. 30 SVG	SGER	GEN SERVICES
NON URGENT ALARM	BEGIN	
17:41:56. 55 SVG	204	VERMOIM
BREAKER	CLOSED	
17:41:56. 70 SVG	SGER	GEN SERVICES
NON URGENT ALARM	END	
17:42:01.110 SVG	OPA	AUT. OPER.
AUTOMATIC OPERATOR	END	
17:43:38. 67 SVM	SGER	GEN SERVICES
TYPE OF OPERATION	MANUAL	
17:44:12.844 SVM	211	VALDIGEM
BREAKER	MOVING	
17:44:12.924 SVM	211	VALDIGEM
BREAKER	CLOSED	
17:44:34.167 SVM	SGER	GEN SERVICES
TYPE OF OPERATION	AUTOM.	
17:44:51.870 SVG	SGER	GEN SERVICES
NON URGENT ALARM	BEGIN	
17:44:52.510 SVG	SGER	GEN SERVICES
NON URGENT ALARM	END	

For this set of messages, SPARSE presents the following conclusions:

17:40:48:530			BEGINNING OF INCIDENT
17:40:48:715			Line SVG-SVM - opened in two ends by single tripping
17:40:48:715	SVG	204	>> Wait service restoration by OPA (1 minute)
17:41:56:055	SVG	204	Breaker closed by OPA
17:41:56:055	SVM	211	>> Close breaker
17:44:12:924	SVM	211	Breaker closed by remote order
17:44:12:924			Line SVG-SVM - service restored
17:44:12:924			END OF INCIDENT

SPARSE enormously reduces the number of messages presented to operators. The original set of messages includes almost 50 messages for a period of 185 ms. SPARSE reduces the number of messages presented to operators during the same period to 3. On the other hand, these messages include not only an interpretation of the incident (as the first and second messages) but also advice for operators (third and fifth messages).

CONCLUSIONS

The results already obtained with intelligent real-time applications in power systems show that Artificial Intelligence techniques are adequate and useful for solving some problems for which traditional programming techniques are not able to provide good solutions.

On the other hand, the duration of this kind of projects is still rather long and the total number of deployed systems all over the world is still very low. This is justified due to the complex issues that real-time applications for power systems must deal with, among which temporal reasoning plays a key role.

This paper presents the most important temporal reasoning requirements for real-time applications in power systems. It also presentes the methods used in some systems that have been developed to deal with temporal reasoning.

It is not yet possible to decide what methods are better because the discussion about this theme is just beginning. The authors hope that this paper provides a good basis for promoting this discussion among power system and Artificial Intelligence experts and to arrive to conclusions that can make the projects involving intelligent real-time applications for power systems less time consuming and more efficient.

REFERENCES

[All83] James F. Allen, "Maintaining Knowledge About Temporal Intervals", Communications of the ACM, vol. 26, n. 11, pp. 832-843, November 1983

[All84] James F. Allen, "Towards a General Theory of Action and Time", Artificial Intelligence, vol. 23, n. 2, pp. 123-154, 1984

[Bij89] R. W. Bijoch, S. H. Harris, T. L. Volkmann, "Intelligent Alarm Processor at Northern States Power", Second Symposium on Expert Systems Application to Power Systems, Seattle, USA, pp. 79-83, 17-20 July 1989

[Bij91] R. W. Bijoch, S. H. Harris, T. L. Volkmann, J. J. Bann, Bruce F. Wollenberg, "Development and Implementation of the NSP Intelligent Alarm Processor", IEEE Transactions on Power Systems, vol. 6, n. 2, pp. 806-812, May 1991

[Bru93] Thomas Brunner, Wolfgang Nejdl, Harald Schwarzjirg, Monika Sturm, "On-Line Expert System for Power System Diagnosis and Restoration", Intelligent Systems Engineering, IEE, vol. 2, n. 1, pp. 15-24, Spring, 1993

[Kir93] Daniel S. Kirschen, Bruce F. Wollenberg, "Intelligent Alarm Processing in Power Systems", Proceedings of the IEEE, vol. 80, n. 5, pp. 663-672, May 1992

[Kow86] Robert Kowalski, Marek Sergot, "A Logic-Based Calculus of Events", New Generation Computing, vol. 4, n. 1, pp. 67-95, 1986

[Liu93] C.C. Liu et al., CIGRE Working Group 38.06.03, "Expert Systems: Development Experience and User Requirements", Electra, n. 146, pp. 30-67, February 1993

[McD82] Drew McDermott, "A Temporal Logic for Reasoning About Processes and Plans", Cognitive Science, 6, pp. 101-155, 1982

[Mon89] Georges de Montravel, Daniel Arrieta, "Pour les Futurs Centres de Conduite de Réseaux Électriques: une Synthèse des Alarmes", Révue Général d'Électricité, n. 1, pp. 33-37, January 1989

[Sho89] Alvin H. Shoop, Steven Silverman, Bhashyam Ramesh, "Consolidated Edison System Operation Computer Control System (SOCCS) Alarm Advisor (SAA)", Second Symposium on Expert Systems Application to Power Systems, Seattle, USA, pp. 84-88, 17-20 July 1989

[Sho92] Alvin H. Shoop, Steven Silverman, "A Real-Time Alarm Processor", Electrical Power & Energy Systems, vol. 14, n. 2/3, pp. 108-113, April/June 1992

[Son93] P. J. M. Van Son et al. (CIGRE Working Group 39.03), "Exploring User Requirements of Expert Systems in Power System Operation and Control", Électra, n. 146, pp. 68-84, February 1993

[Tam87] Y. Tamura et al. (CIGRE Working Group 38.02.07), "An International Survey of The Present Status and the Perspective of Expert Systems on Power System Analysis and Techniques", CIGRE, 1987

[Val93a] Zita A. Vale, A. Machado e Moura, "An Expert System with Temporal Reasoning for Alarm Processing in Power System Control Centers",IEEE Transactions on Power Systems, vol. 8, no. 3, pp. 1307-1314, August 1993

[Val93b] Zita A. Vale, A. Machado e Moura, Albino Marques, "Artificial Intelligence Applications in a Regional Substations Operational Center", 11th Power Systems Computation Conference (PSCC), Avignon, France, 30 August - 3 September 1993

[Val93c] Zita A. Vale, A. Machado e Moura, "Intelligent Alarm Processing in Control Centers", Athens Power Tech - Joint International Power

Conference, Athens, Greece, 5 - 8 September 1993

[Val93d] A. Almeida do Vale, M. J. Dias Gonçalves, Zita A. Vale, "Improving Man-Machine Interaction in Control Centers: The Importance of a Full-Graphics Interface", 11th Power Systems Computation Conference (PSCC), Avignon, France, 30 August - 3 September 1993

[Wag92] Monika Pfau-Wagenbauer, Wolfgang Nejdl, "Integrating Model-Based and Heuristic Features in a Real-Time Expert System for Power Distribution Networks", The Eight Conference on Artificial Intelligence for Applications (CAIA '92), IEEE, Monterey, California, USA, pp. 303-309, 2-6 de March 1992

Tutors and Teaching Systems

INTELLIGENT TUTORING SYSTEM: A KNOWLEDGE BASED APPROACH

Leila Alem
CSIRO Division of Information Technology,
Locked Bag 17, North Ryde, NSW 2113, Australia
leila. alem@syd.dit.csiro.au

ABSTRACT

Good educational software is active, not passive, ie the student ought to be doing something, not just watching. Multimedia offers a serious way of improving simulation based instruction, but to date it has been generally passive in its implementation [Sch93]. A good intelligent tutoring system should be able to support a rich learning interaction by encouraging the learner to take initiative and by providing assistance and explanation, rather than directive tutoring. In this paper we present our approach which employs a "learning by doing with explanation" style of tutoring interaction, implemented using a student model and an instructor model. The tutoring is context sensitive, ie the response of the intelligent tutor is based on a model of the student's interaction with the tutorial. Our approach enables a student to manipulate his/her progress directly. In this paper, examples are given from a prototype system which has been built for sonar operators.

Keywords: Student model, Instructor model, Intelligent Tutoring System, Knowledge Based System

1. INTRODUCTION

This paper presents research done for the design of an intelligent tutoring system for sonar operators. The requirements were for a knowledge based system that provides assistance to the student, that supports a rich learning interaction and that provides on-line control of student progress. For the design of such a system we have adopted a constructive view of learning which aims to encourage the learner to take initiative by providing assistance and explanations rather than directive tutoring. Clancey [Cla87] among others suggests that a good Intelligent Educational System (IES) should be able to give some explanation and guidance about the task activities. While the importance of explanation is generally recognised, tutoring systems often fall short in this respect.

In this paper we present our approach which employs a "learning by doing with explanation" style of tutoring interaction. The instructor model and the student model are presented. The tutoring is context sensitive, ie the responses of the intelligent tutor are based on a model of the student's interaction with the tutorial.

Section two presents the instructor model which is based on the model of the activity where the training is taking place and the relevant tutoring interventions, while section three presents the student model. The use of such models is presented in section four in the context of the design of an intelligent tutor for sonar operators.

2. THE INSTRUCTIONAL MODEL

The questions that must be addressed in designing an intelligent tutor are:
(1) how should the task be structured for the student?
(2) how should the student actions be tracked?
(3) when should the tutor intervene? and finally
(4) how should the tutor intervene?

The purpose of our instructor model is to answer questions 1, 3 and 4 and is composed of three main components:

- The instructor model which is the instructor's representation of the task to be achieved by the student during the training session.
- The tutoring interventions: the explanations, supporting information and advice provided by the tutor.

The interaction model which defines when and how the tutoring interventions are provided to the student during the training session.

2.1 The Instructor Model

The purpose of the instructor model is to describe the knowledge that the system must have in order to help the student to carry out his or her tasks during the training session. It is a representation of the "how to do it?",

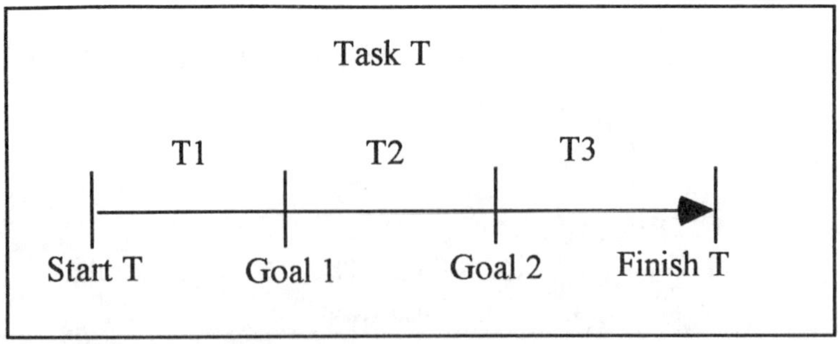

Figure 1. The concept of task.

knowledge in order to get the intended tasks accomplished. The instructor model is based on a generic model of the activity and a tutorial model. The instantiation of the instructor model is then derived by performing the activity initiation based on a specific tutorial.

2.1.1 The generic model of the activity

We use our task model proposed in [AL93/1] as a basis for the representation of the generic model of the activity in which the training is taking place. Many task models have been proposed in the literature as a conceptual entity for modelling: the reasoning process [Cla87], the user activity being expert or non expert [TF88] or the problem solving knowledge [BW89]. In our case, our task model is used to capture the instructor's representation of the task to be performed by the student during the training session. This representation will then be used for monitoring the student's activity and performance, assisting the learning process and finally modelling task-related knowledge.

As a preliminary to the design of our task model a number of requirements were identified:

- the task should be achievable,
- task goals should be identifiable and expressible,
- there should be a number of possible solutions so as to permit the production of alternatives,
- explicit rules and procedures should be involved in the production of the solution and finally,
- the task should support the generation of partial solution.

Our task model contains knowledge about its function (ie the goal or objective of the task), its structure (ie the action plan that enables the execution of the task), its behaviour (ie the status of the task), and the assistance capabilities (ie its related explanations, supporting information and general advice). The latter corresponds to the tutoring interventions.

An activity is modelled by a set of interrelated tasks. A task can be decomposed into sub tasks. Each task corresponds to one goal. How each goal is achieved is determined by the action-plan of the task. The concept of task is shown in Figure 1.

This model has been used in TOITS [AL93/2] an intelligent tutoring system for sonar operators. This application is described in more detail in [AL94, AL93/2]. The purpose of the training activity was to identify vessel classes and is mainly composed of four major tasks: (A) initial tactical evaluation, (B) component parameter, (C) equipment parameter, and (D) classify vessel. Figure 2 shows a subset of the task-structure.

The goal of vessel-identification task is to determine the vehicle class. The vehicle identification process involves the selection of the most likely vessel class or classes from those which are known or are likely to be within the operational region. The initial tactical evaluation is to determine the number of targets to be identified. Once the number of targets is known, then the analysis will continue by attempting to identify each target separately. Task T1 is represented as:

Task: T1
Goal: identify the vehicle classes
Action-plan: [T2, T3, T4]

Task: T2
Goal: determine number of targets
Action-Plan: user_action [TargetNo]
Explanation: "to determine the number of vehicles represented. This may be done by examining the Doppler shifts appearing on the lines."
Reference: Chapter 1 Component Identification
Tool: slop tool

The component analysis for a particular vehicle begins by attempting to determine the number of propeller shafts the vehicle. Once the shaft component has been identified the number of propeller blades can be determined.

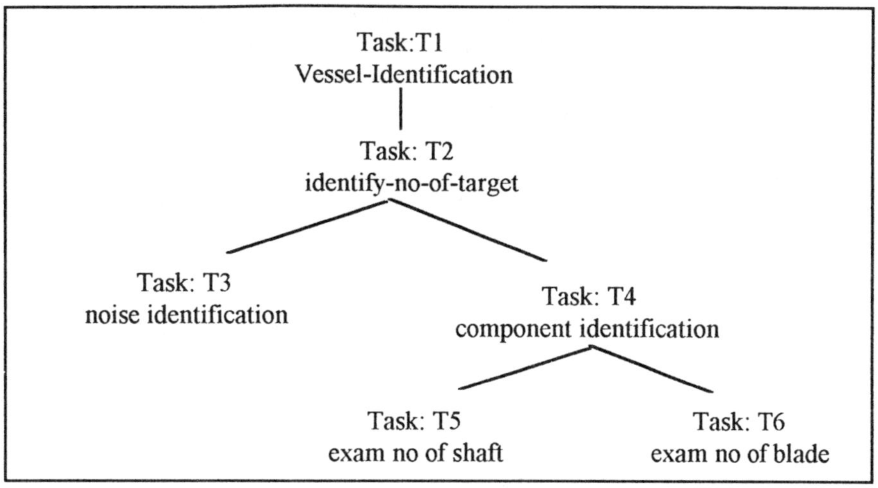

Figure 2. A sub set of the task structure.

Task: T4
Goal: determine shaft and blade components
Action-plan: [T5, T6]

2.1.2 The instantiation of the instructor model

The instantiation of the instructor model is derived by performing the activity initiation in relation to a given tutorial. The activity initiation is processed by the task engine which
(1) instantiates the generic activity model based on a given tutorial, and
(2) processes tasks status of this instance.

Tokens and status are associated with each task to describe the behaviour of the task. The status of the task is represented as one of:

- ready - ready to perform this task,
- wait - waiting for other tasks to be finished,
- undefined - the task might or might not be performed,
- achieved - the task has been successfully preformed.

The task engine is implemented in the M.4^{TM} expert system shell. The task engine schedules the execution sequence of tasks and processes the status of tasks based on a given tutorial. It also describes the sequence of the tasks, in which order and under which conditions the task are activated, and how they interact.

The use of our task modelling approach provides a suitable way of structuring knowledge during the learning process. It models both the student activity in a learning scenario and the reasoning process of the knowledge based system. It also describes the user actions and the interface feedback.

2.2 The tutoring interventions

The tutoring interventions are modelled as assistance knowledge. In the context of the intelligent tutor for sonar operators, the assistance knowledge is acquired from domain experts (ie the instructor) and is modelled as part of the task conceptual model. It is composed of:

- explanations that are of two types: static and dynamic explanation,
- supporting information providing cross references to the relevant sections of an electronic book which contains domain knowledge, and
- recommended tools to be used while performing the task at hand.

Static explanations address the problem of why and how to perform tasks, they are static in the sense that they do not depend on the student course of action.

Dynamic explanations provide strategic help and address the problem of what to do next based on what has been done. They are dynamic in the sense that they are generated depending upon student action. While engaged in a task, student's actions are mapped onto the activity model in order to derive the goal pursued by the student. The assessment of the student's action is used to compute task status and derive context-sensitive tutoring intervention which explains the relationships between tasks, and delivers advice on what should/could be done based on what has to be done by the student.

At any time the student can access the most relevant section of an electronic book which contains the domain knowledge (called the system model). It combines hypertext and multi-media techniques to present interactive domain knowledge to the user/student. The electronic book has been designed in a way that corresponds to the pedagogical structuring of the domain in order to organise the knowledge in a way that it is easy to communicate to the student. The table of contents of this book represents the curriculum, ie the subject and topic to be taught. During the learning process the student may wish to read the section which presents and explains a key concept/subject in direct relation with the task at hand. Some sections of the electronic book are structured around a specific example which allows the student to move from "learning by doing" to "learning by example". In this sense the supporting information also allows the student to change the tutoring style.

The interaction model which defines when and how the tutoring interventions are provided to the student during the training session is defined below.

2.3 The interaction model

There are three forms of interaction, each of them leading to a different tutoring style:

- System initiative which corresponds to a directive style of tutoring, ie the system interrupting the student frequently.
- Mixed initiative which corresponds to an opportunistic coaching style of tutoring, ie the system providing feedback at the right time.
- Student initiative which corresponds to the style of tutoring where the system provides advice and guidance to the student only on request.

We have adopted an approach called "learning by doing with explanations" where the learning process is described in terms of the learner working at the task level, by undertaking tasks while at the higher level the system provides explanations and guidance. As Cumming stated [CS90] learning is not given, or done to a student by a tutor, but is created by the learner as the result of performing actions and experiencing their consequences in the environment. Whatever the kind of tutoring/pedagogy involved, the student must necessarily build his/her own know-how [FL93]. Therefore he/she should be left with as much initiative as possible within the context of the interaction, because in the final analysis, we can never say definitely what suits the student.

In our design of the intelligent tutor for sonar operators, the student can at any time request explanation or help but the system can only deliver advice and explanations in response to the student request. Our aim is to promote the learner participation. We are currently designing an intelligent tutor for air traffic controllers and in this system we are adopting the mixed initiative approach.

3. THE STUDENT MODEL

The student model is composed of static knowledge and dynamic knowledge.

The static knowledge corresponds to the student profile. It is composed of the student tutorials with related assessment and marking, and required level of help. In the intelligent tutor for sonar operators we have defined three levels of help. For each tutorial the instructor can assign a default value for the level of help required. When selecting a tutorial, the student can overwrite the level of help by choosing the one he/she feels is required. The value of the level of help selected by the student is the one stored in the student profile.

The dynamic knowledge correspond to the student actions while solving a problem. Such actions are mapped into the activity model in order to deduce the student's intentions and goals. As pointed out by Boy and Falzon, learning relies very much on a cooperation between the assistant system and the learner [Boy90, Fal90]. Our activity model described above allows the representation of the instructor model and is used to derive student goals based on student actions. The use of our task modelling approach leads to a user-centred design that focuses on the user/student rather than on the system. It describes what the user must learn and so act as a basis for training [AL94].

In our task conceptual model, the notion of action plan offers a structured way to describe the cooperative performance of a task between the learner and the intelligent tutor. It allows direction of user activity in a collaborative way. Providing intelligent assistance to the learner means designing action plans in collaboration with the learner [AL93/1]. The collaboration between the learner and the intelligent tutor is managed using status of tasks. While engaged in a task, the student's actions are mapped onto the activity model in order to derive the goal pursued by the student. The assessment of the student's actions is used to compute task status which then helps to derive context-sensitive tutoring intervention. Such tutoring interventions explain the relationships between tasks and the tasks' internal structure.

4. APPLICATION: AN INTELLIGENT TUTOR FOR SONAR OPERATORS

An intelligent tutor system for sonar operators has been designed based on the above instructional model and student model. This system is described in more detail in {AL94]. The intelligent tutor is composed of the following modules:

(1) The tutorial module which provides a set of tutorials in which the student is asked to identify vessel classes.
(2) The instructor interface which helps the generation of tutorials and allows direct communication between the student and the instructor.

(3) The tutor kernel which is the task engine that manages and controls the task activities.
(4) The student interface which presents the tutorial and provides access to both tutoring knowledge and marking results.
(5) The help module which consists of the implementation of the instructional model. The levels of support and guidance include: provision of guidance regarding possible problem solutions, access to the vessel class database and cross-links to the sonar electronic book.
(6) The marking module which scores the student based on the system assessment of student performance. There are three types of marking: continuous marking, marking at major phases and final marking.
(7) The domain knowledge module which corresponds to the electronic book. It is composed of simulation tools used in the sonar tutorial. It combines hypertext and multi-media techniques to present interactive domain knowledge to the student.

This system supports a rich learning interaction and presents learning as a more rewarding and long lasting activity. It has been built in Toolbook™ running on a PC under Windows™ using M.4™ expert system shell.

5. CONCLUSION

We have presented our instructional model and student model and shown how such models have been used for the design of the intelligent tutor for sonar operators. In this tutoring system the learning process is described in terms of the learner working at the task level, by undertaking tasks while simultaneously at a higher level the system is providing explanation and guidance. Our approach enables a student to manipulate and control his/her progress directly. Such an approach can improve the quality and cost effectiveness of the training and learning process. We have demonstrated that a knowledge-based approach is the answer to the design of effective artificial tutors. The ability to provide context-sensitive assistance to the learner can help promote future computer based training to devices that really teach. This approach is becoming more and more important particularly as we move towards intelligent (advice-giving) system technology.

REFERENCES

[AL94] L. Alem and M. Lee, An Intelligent tutoring System for sonar application in Australia. In proceedings of The second World Congress on Expert Systems, Portugal, January 94.

[AL93/1] L. Alem and M. Lee, An intelligent Assistance in a learning interaction. Workshop on Collaborative problem solving: Theoretical frameworks and innovative systems, AI-ED'93 World Conference on Artificial Intelligence in Education, Edinburgh, August 93.

[AL93/2] L. Alem and M. Lee, TOITS a generic architecture for a task oriented intelligent tutoring system. In Proceeding of the 6th Australian Joint Conference on Artificial Intelligence, AI'93, Melbourne November 1993.

[Boy93] G. Boy. Intelligent Assistant Systems. Academic Press, Harcourt Brace Jovanovich Publisher, 1990.

[BW89] J. Breuker and B. Wielinga, "Models of expertise in knowledge acquisition". Topics in Expert Systems Design: Methodologies and Tools, 1989, North Holland Publishing Co, G. Guida & C. Tasso eds., pp. 1-37.

[Cla87] W. Clancey, Knowledge based tutoring: the GUIDON program. Cambridge, MA: MIT press, 1987.

[CS90] Cumming & Self, Intelligent educational systems: identifying and decoupling the conversational levels. In Instructional Science 19: 1-17, 1990. Kluwer Academic Publishers.

[Fal90] P. Falzon. Cognitive Ergonomics, Understanding, Learning and Design Human-Computer Interaction. Academic Press 1990.

[FL93] M. Futtersack, J.M Labat, Dynamic and interactive pedagogical planning, the quiz example. In AI-ED'93 proceedings, the World Conference on Artificial Intelligence in Education, Edinburgh, August 93.

[Sch93] R.C Schank, Learning via multimedia computers. Communication of the ACM, Vol.36, No 5, May 93.

[TF88] Tueni Li and Fares. AMS: A knowledge-based approach to task representation, manipulation and organisation. COIS-88, Palo Alto.

QUACONTRA: QUALITY CONTROL TRAINING IN THE FOOD INDUSTRY USING AN INTELLIGENT TUTOR

Oscar Castillo
Instituto Tecnologico de Tijuana
P.O. Box 4207, Chula Vista, CA 91909, USA

Patricia Melin
CETYS Tijuana
P.O. Box 4207, Chula Vista CA 91909, USA

ABSTRACT:

We describe a computer program that can be considered an intelligent tutoring system (ITS) for the domain of Quality Control in the Food Industry (QCFI). The ITS has the goal of training new personnel, for the Quality Control lab, with the microbiological and chemical techniques that are needed to perform an evaluation of the quality of the manufacturing process. The goal of the Quality control lab is the application of the minimal number of microbiological and chemical techniques to the samples of food extracted from production lines, so as to have an evaluation of the quality of the production as quickly as possible. The ITS also teaches to the new people in the lab the logic behind a decision on the quality of the production. The ITS simulates a human training expert in this area in the process of teaching QCFI. The computer program contains the knowledge of the human training experts expressed as rules and facts (knowledge base), and uses as a tool of reasoning the inference engine of Prolog. The ITS uses efficiently techniques from Artificial Intelligence (AI) to teach QCFI and also to monitor the actual learning process of the new people in the quality control lab.

INTRODUCTION

We describe a computer program that can be considered an intelligent tutoring system [Mac 94] for the domain of quality control in the food industry [Mel 87]. The process of teaching this domain is difficult since it requires a background on Chemistry and Microbiology [CM 94a, Mel 87]. In this ITS we are assuming that the people have the required prior knowledge mentioned above. When a new person uses the ITS, he or she enters an individualized learning process of the domain of QCFI. The ITS maintains, for each student, all the relevant information about the learning status of the students. Also the ITS monitors the actual learning process of the students. The learning process, considered in this ITS, consists mainly of four steps:

1.- Acquiring the basic concepts of QCFI.
2.- Learning the methods of application of microbiological and chemical techniques to samples of food.
3.- Learning the methods of identification of possible harmful microorganisms or toxic chemicals based on the results of the techniques.
4.- Learning to evaluate the quality of the production based on the identifications of microorganisms or chemicals from the samples of food.

The ITS has the responsibility of giving to the users the best possible information in each of the above steps, so that they can get the best possible understanding of QCFI. Our prototype ITS uses AI techniques [Bra 90] to improve the efficiency of the learning process described above and also to evaluate the level of understanding of the students of this domain. The use of AI techniques enables the right use of the knowledge of the human experts for training in this domain of QCFI.

The reasons why we consider that an ITS is needed for this domain are the following:

1.- There are not enough training experts in quality control for the food industry for all the food manufacturing plants in Mexico.
2.- The domain of quality control for the food industry is a complex one, because of the very different types of microorganisms or toxic chemicals that can be found in food [PR 87], and also because of the many microbiological or chemical methods that can be used, depending on the conditions of the manufacturing process.
3.- There is a need of making more efficient the task of quality control in the laboratory, not only in the time needed to perform the identifications of possible microorganisms or chemicals, but also in the accuracy of the results. The need of increasing this efficiency can be justified with the goals of the company of reducing times and costs of production, and also preventing harm to final consumers.

We believe that an ITS like ours can help solve all of this problems. First of all, the ITS contains the necessary knowledge to simulate training experts in the area of QCFI. Second of all, the ITS has the knowledge to teach to the user/student how to select the microbiological or chemical techniques necessary to make the identifications and at the end to make the final evaluation on the quality of the production. Also, the ITS can help in making more efficient the quality control by training the people in the lab with the techniques and also by teaching them how to make the evaluation on the quality of the production.

Of course, the above features of the system are not sufficient to justify it as an intelligent tutor [Mac 94, Mca 94]. The computer program also has the ability to generate problems for the user/student to solve. The system then compares the solution of the user with the real solution of the problem, to find errors and to evaluate what kind of teaching action needs to do next. For this reasons we believe that our system uses efficiently AI techniques for QCFI education.

DESCRIPTION OF THE INTELLIGENT TUTORING SYSTEM

We describe bellow the general architecture of the ITS, and then we give a detailed description for each of the modules, except for the Inference Engine. The description of the inference engine is based mainly on the Prolog Programming Language [CNV 87] and is not given in this paper.

Architecture of the Intelligent Tutoring System

The architecture of an ITS is very similar to that of an expert system (ES) for a specific domain of application. However, there are also some main differences between an ITS and an ES. One of the main differences is in the Knowledge Base (KB), because for an ITS the KB needs to contain Expert Knowledge about teaching (training) methodologies and not only Expert Knowledge about the domain of application. In figure 1 we show the architecture we used in developing our ITS. We can see in this figure how the KB consists of three parts: Expert Module, Teaching/Learning Module and Student Module. For an ITS the last two modules are the ones that enable the training of new students for the domain and also enable monitoring of their learning process. We will explain in the next section how this is

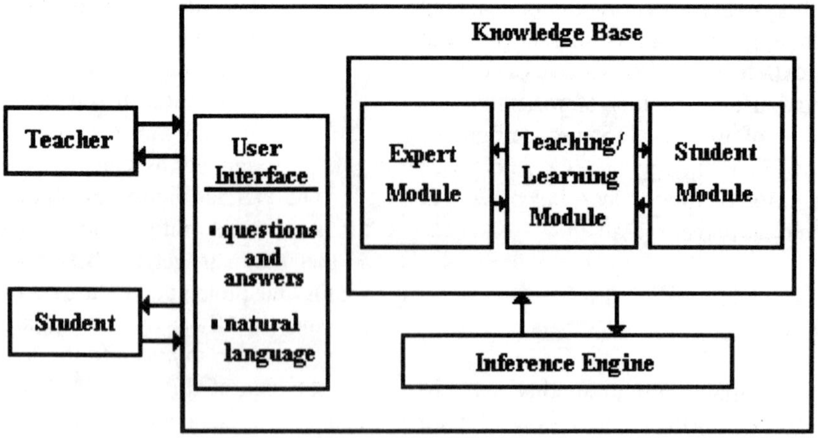

FIGURE 1 General Architecture of the ITS.

accomplish in our ITS for the domain of QCFI. In figure 2 we show how the user interface is divided in its components to give an idea of its structure. The user interface has to process information by the user in the form of questions and answers and also by natural language. In a later section we explain how Natural language Processing (NLP) is handle in this ITS.

Description of the Knowledge Base

The knowledge base of the ITS consists of three modules: the Expert Module, the Teaching/Learning Module and the Student Module. The Expert Module contains the knowledge about the domain of QCFI, i.e., the knowledge about the microbiological or chemical techniques that can be applied to the samples of food, to obtain the identification of microorganisms and chemicals, and as a result of this information, the final evaluation of the quality of the production. The Expert Module also has the knowledge to generate problems for the user/student to solve. The Teaching/Learning Module contains the knowledge about teaching/training methodologies and the knowledge for diagnosis of student's attributes. The Student Module is a data base of the student's performance history and a catalog of student's attributes.

Description of the Expert Module

The expert module consists of four parts: Domain knowledge, Data Base of Basic concepts and problems, Problem generator, and Machine Learning Module. The "Domain knowledge" part consists of rules and cases of QCFI [Mel 87], and additionally heuristics to select the best microbiological or chemical techniques applicable to a given quality control problem. The "Data Base of Basic Concepts and Problems" consists of a set of known examples of identification that can be shown initially to the students. The "problem generator" consists of an algorithm to generate new problems for the user to solve, using the "Domain Knowledge" part and the information about the student's evaluation that the "Teaching/Learning Module" performs on the solution to the problems by the user. Finally, the "Machine Learning Module" gives to the ITS the ability to update its knowledge base [Bra 90] with additional theory or Methods of QCFI, using the information from students, trainers or experts. For details about the microbiological or chemical methods for identification we refer the reader to a prior system developed by the authors [CM 94a].

To give an idea of the way the "Domain Knowledge" is structured, we list below some of the microbiological methods that can be used to make identifications of microorganisms:

1.- Enumeration by direct epifluorescent filter technique
2.- Coliform counting in plate
3.- Baird-Parker Method
4.- Enrichment Method
5.- Pseudomona Counting
6.- Blood-agar Method
7.- Mannitol Method
8.- Coagulase Method
9.- Staphylococcus Method 110
10.- MacConkey agar Method
11.- EMB agar Method

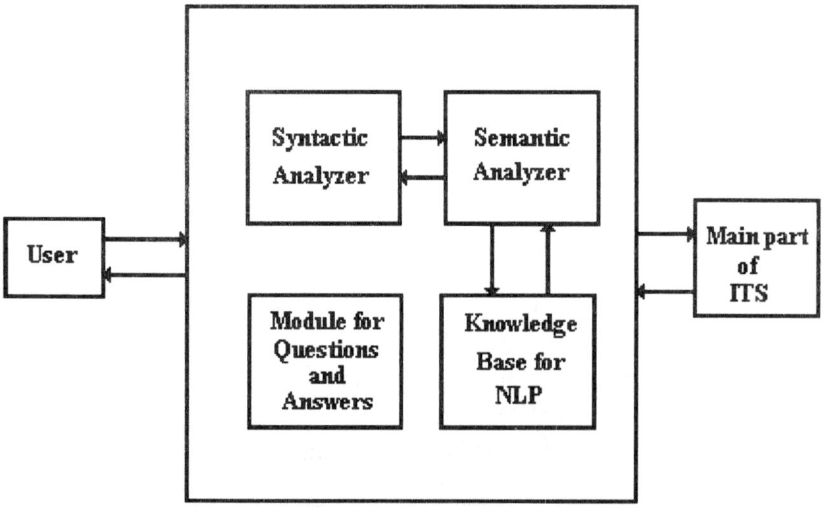

FIGURE 2 Architecture of the User Interface

The types of microorganisms that the intelligent system can identify are the following:

1.- Coliforms
2.- Staphylococcus
3.- Streptococcus
4.- Pseudomonas
5.- Salmonella
6.- Shigella
7.- Other types of Enterobacteriaceae

We show in figure 3 some sample rules of the part of the Expert Module that gives advice of which microbiological methods can be used in the lab.

In figure 4 we also give some sample rules of the part of the Expert Module that makes the identification of microorganisms based on the data from the lab.

Of course, the above rules are only for the identification of one type of microorganism, but there are many more rules in the actual knowledge base for the other types of microorganisms that the ITS can handle at the moment. Also, the Expert Module contains similar rules for the chemical methods of identification that the ITS can suggest to the people in the lab.

The Expert Module also Contains rules to perform the final evaluation of the quality of the manufacturing

Rule 1:
IF Morphology = coccus_in_bunch
AND Gram_Tintion = gram_positive
THEN New_Method = baird_parker_method

Rule 2:
IF Baird_Parker = black
THEN New_Method = blood_agar -method

Rule 3:
IF Blood_Agar = alpha_hemolytic
THEN New_Method = mannitol -method

Rule 4:
IF Mannitol = yellow
THEN New_Method = staphylococcus_method_110

Rule 5:
IF Method_110 = gold
THEN New_Method = coagulase_method

Rule 6:
IF Morphology = bacillus
AND Gram_Tintion = gram_negative
AND Motility = positive
THEN New_Method = mac_conkey_method

Rule 7:
IF Mac_Conkey = not_colored
AND Lactose = negative
THEN New_Method = emb_method

Rule 8:
IF EMB = not_colored
THEN New_Method = s_s_agar_method

Rule 9:
IF S_S = not_colored
AND Indole = negative
AND Gas_production = negative
AND Gelatin = negative
AND KCN = negative
THEN New_Method = tsi_method

FIGURE 3 Sample rules for giving advice of microbiological methods that can be used in the lab

Rule 1:

IF	Morphology = coccus_in_bunch
AND	Gram_Tintion = gram_positive
AND	Baird_Parker = black
AND	Blood_Agar = alpha_hemolytic
AND	Mannitol = yellow
AND	Method_110 = gold
AND	Coagulase = positive
THEN	Identification = staphylococcus_aureus

Rule 2:

IF	Morphology = bacillus
AND	Gram_Tintion = gram_negative
AND	Motility = positive
AND	Mac_Conkey = not_colored
AND	Lactose = negative
AND	EMB = not_colored
AND	S_S = not_colored
AND	Indole = negative
AND	Gas_production = negative
AND	Gelatin = negative
AND	KCN = negative
AND	TSI = h2s_positive
THEN	Identification = salmonella_typhi

FIGURE 4 Sample rules to make the identification of microorganisms in the samples of food

Rule 1:

IF	Microbiological_Conclusion = dangerous_species_can_cause_intoxication
AND	Chemical_Conclusion = toxic_chemicals
THEN	QUALITY = bad_reject_production

Rule 2:

IF	Microbiological_Conclusion = dangerous_species_can_cause_infection
AND	Chemical_Conclusion = toxic_chemicals
THEN	QUALITY = bad_reject_production

Rule 3:

IF	Microbiological_Conclusion = not_dangerous_species
AND	Chemical_Conclusion = high_percentage_salts
THEN	QUALITY = regular_more_tests_to_decide

Rule 4:

IF	Microbiological_Conclusion = not_dangerous_species
AND	Chemical_Conclusion = not_toxic_chemicals
THEN	QUALITY = good_accept_production

Rule 5:

IF	Microbiological_Conclusion = no_species_at_all
AND	Chemical_Conclusion = no_chemicals_at_all
THEN	QUALITY = excellent_accept_production

FIGURE 5 Rules to make final evaluation of quality control in food

process for the food product. This rules decide on the quality of the production, using as information the microorganisms or chemicals found in the samples of food. In figure 5 we show the rules to perform the final evaluation of the quality of the manufacturing process for the food product.

The goal of having a learning Module for our ITS, is to have a computer program that can be updated with new microbiological or Chemical methods and also with new possible microorganisms or toxic chemicals for identification. Of course, the job of updating the System can only be done by the human experts or by the knowledge engineers, and not by other users.

When the learning module is activated the ITS creates new rules for the knowledge base using the data given by the human experts. The method for learning rules that the ITS uses is known as "learning by example" described by Bratko [Bra 90].

Description of the Teaching/Learning Module

The Teaching/Learning Module consists of two parts: The Instructional Knowledge Base and the Controller Module. The "Instructional Knowledge Base" consists of teaching/training methodologies and a rule_base for diagnosis of student's attributes. The "Controller Module" updates the current student's attribute values using the student's history and the current performance. Of course, this Controller uses the Inference Engine as a way to infer the new student's attributes.

To give an idea of the way this Module is structured, we will show some sample rules of the part of the "Instructional Knowledge Base" that performs the evaluation of the learning process. We assume that at the end of a training session the student is given a "test" by the ITS. For example, for stage one of the Learning process (acquiring the basic concepts of QCFI) the evaluation is done using three types of questions: direct questions, thinking questions and explanation questions.

In figure 6 we show the rules to evaluate the level of learning of stage one.

In figure 7 we show the rules needed to make a decision on the acceptance of the student to stage two of the learning process.

Of course, the above rules are only for stage one of the learning process, but there many more rules in the actual "Instructional Knowledge base" for the other stages of the learning process. Also, this module contains other rules for updating the student's attribute values using the student's history and the current performance.

Description of the Student Module

The Student Module consists of two parts: The Student's Attribute Catalog and the Student's Performance History. The Student's Catalog contains information like Cognitive Levels, Thinking Skills, Learning Strategies, Problem Solving Abilities and Domain Knowledge States. The "Student's Performance History" contains information about the past evaluations for each of the students.

In our ITS we classified student's attributes in the following manner. "Problem solving abilities" are classified as: low, intermediate and high-critical thinking. Acquisition of "Domain Knowledge" is considered as: null, shallow, intermediate, deep and expert. The "Learning needs" are classified as: individual, hands-on, experimental, learning cycle. The "thinking skills" are evaluated as one of the following: know, understand, apply, transfer and analyze. The "Cognitive levels" are considered as: concrete, transitional and formal. This classification is very similar to the one described in the papers by Dori [DD 94, DAD 94], however there are some differences reflecting some specific problems in our domain of application.

The student module contains rules to make an evaluation of the student's attributes mentioned above, using as information the results of the tests given to the students in each stage of the learning process. Also, the student module has rules to build the "student's performance history" based on the student's attributes obtained above. Of course, this is also the main information that is used in the "teaching learning module" for the generation of problems and questions for the students during the learning process.

Description of the User Interface

The user interface of the ITS consists of Questions and Answers and Natural Language Processing [Bra 90, CNV 87]. In some cases the input/output has to be in the form of questions and answers by the user, but in other cases a more natural way of interaction is needed (for example with Natural Language). We include in this ITS a module for Natural Language Processing of input by the user, using the advantages that Prolog offers to this end (like the use of DCG's [Bra 90]). An example of the use of this interface is shown in the next section.

In figure 2 was shown that the User Interface consists mainly of four parts: Syntactic Analyzer, Semantic Analyzer, Knowledge base for NLP and Module for Questions and Answers. We will describe briefly each of this modules and how they are implemented in PROLOG. The Syntactic Analyzer consists mainly of a set of rules defining the subset of the English Language that the ITS can handle at the moment. This rules are in Definite Clause Grammar (DCG) form which is very similar to the well known Backus Naur Form (BNF). In PROLOG a computer program in DCG is automatically a syntactic analyzer for the corresponding language. The Semantic Analyzer

Rule 1:

IF Percentage_of_direct_questions ≥ 90
AND Percentage_of_thinking_questions ≥ 80
AND Percentage_of_explanation_questions ≥ 75
THEN Evaluation_of_stage_one = excellent

Rule 2:

IF 80 ≤ Percentage_of_direct_questions < 90
AND 70 ≤ Percentage_of_thinking_questions < 80
AND 60 ≤ Percentage_of_explanation_questions < 75
THEN Evaluation_of_stage_one = good

Rule 3:

IF 70 ≤ Percentage_of_direct_questions < 80
AND 60 ≤ Percentage_of_thinking_questions < 70
AND 50 ≤ Percentage_of_explanation_questions < 60
THEN Evaluation_of_stage_one = regular

Rule 4:

IF Percentage_of_direct_questions < 70
AND Percentage_of_thinking_questions < 60
AND Percentage_of_explanation_questions < 50
THEN Evaluation_of_stage_one = bad

FIGURE 6 Rules to evaluate the level of learning of knowledge in stage one.

Rule 5:

IF Evaluation_of_stage_one = excellent
THEN Decision_learning_process = stage_two

Rule 6:

IF Evaluation_of_stage_one = good
THEN Decision_learning_process = stage_one_revision

Rule 7:

IF Evaluation_of_stage_one = regular
THEN Decision_learning_process = stage_one_strong_revision

Rule 8:

IF Evaluation_of_stage_one = bad
THEN Decision_learning_process = stage_one_almost_all

FIGURE 7 Rules of evaluate possible acceptance of student to stage two of the learning process.

consist of a set of rules in PROLOG that obtain the meaning of a Natural Language Sentence given by the user. Of course, this is done using the information given by the Syntactic Analyzer. The knowledge base for NLP consists of a set of rules in PROLOG containing the knowledge to react to the user needs and also to make the real interface to the ITS. On the other hand, the Module of Questions and Answers is used when the ITS only needs information in a simple form, i.e., in some cases the ITS only needs an answer by the user in the form of yes, no or other simple answer. This module then is only needed in those cases and will work independently of the modules for NLP.

USE AND VALIDATION OF THE INTELLIGENT TUTORING SYSTEM

The intelligent tutoring system has been tested in the quality control lab of a fish processing plant in Mexico, for several months with encouraging results. With the computer program the people in the lab were able to learn how to decide in less time than before (with conventional training methods) which microbiological or chemical methods are appropriate to analyze samples of food. Also, after the microbiological or chemical methods were performed, the computer program provided an efficient method to teach the students how to identify the microorganisms or chemicals that are present in the food. At the end the ITS also provided guidance to the student on how to learn to decide about the final quality of the production, based on the identifications of microorganisms and chemicals made before. For all of this reasons the ITS can be considered an efficient educational (training) tool for the domain of QCFI. The importance of making more efficient this task of training in the domain of QCFI can be justified at least in three ways:

1.-Decreasing the average time needed for training new people in the quality control lab can reduce the costs of production for the company and also the final costs of the food product for the consumers.

2.- Training new people in this domain can help prevent harm to the final consumers of the food in the way of intoxication and possibly death, with obvious losses for the people and for the company.

3.- Training new people in this domain can also help in reducing the time needed to perform the quality control of the production, which in turn results in decreasing the costs of producing the final food product.

After using the ITS as a training tool in the food processing plant, the users suggested several possible improvements to make it more easy to use and to extend the applicability of the computer program. Several of this possible improvements suggested by them and also some that we think can be useful, can be find in the next section.

Finally, we show below in figure 8 a sample session with the ITS to give an idea of its performance.

CONCLUSIONS

We have developed an intelligent tutoring system for QCFI, that can be used to train new people in the Quality Control lab with the microbiological and chemical techniques necessary to make an evaluation of the quality of the production. The ITS monitors the actual training process for each of students until they achieve the required level of understanding of QCFI.

The importance of having an ITS for training new students in this domain can be seen from several points of view. First of all, from the economical point of view the impact of this ITS is in reducing costs of production for the company, because of the more efficient training method resulting from the application of AI and educational techniques to this domain. Second of all, from the social point of view the impact of this ITS is in preventing harm to the final consumers of the food product, which is very important for obvious reasons. Third of all, from the educational point of view the impact of this ITS is in the contribution to the advance of the research in applications and methodologies of AI in Education.

The ITS can be improved in the following ways:

1.- Build a better user interface so that it can be used more easily.

2.- Provide a large domain of types of microorganisms or chemicals that the system can identify.

3.- Use of Multimedia technology to improve even more the efficiency of the learning process.

We are planning to work along the above lines of research in the near future. And also, we are considering adding to this ITS another new method of identification of microorganisms based on the notion of fractal dimension which was developed by the authors [CM 94b]. This alternative method of identification has also been shown very promising in its own right and can also be very interesting to teach it to the students.

QUACONTRA: QUALITY CONTROL TRAINING IN THE FOOD INDUSTRY USING AN INTELLIGENT TUTOR

What is the name of the user?

Jose P.

HI JOSE:YOU HAVE ALREADY PASS THE PHASE OF "LEARNING THE MICROBIOLOGICAL AND CHEMICAL TECHNIQUES", YOU ARE NOW IN THE PHASE OF "LEARNING THE METHODS OF IDENTIFICATION OF MICROORGANISMS AND CHEMICALS".

THE LAST SESSION WE WERE STUDYING METHODS OF IDENTIFICATION OF MICROORGANISMS, AND WE WERE INVESTIGATING THE IDENTIFICATION OF THE STAPHYLOCOCCUS SPECIES

What is the morphology of the microorganisms?

(a) bacillus
(b) coccus in chain
(c) coccus in bunch

>c

What is the gram tintion?

(p) gram positive
(n) gram negative

>p

What is the result of the Baird_Parker method?

(b) black colony
(a) any other color

>b

What is the result of the staphylococcus method 110?

(w) white colony
(g) gold colony

>g

What is the result in the blood_agar method?

(a) α - hemolytic
(b) β - hemolytic
(c) γ - hemolytic

>a

What is the result of the mannitol method?

(r) red colony (does not ferment mannitol)
(y) yellow colony (ferments mannitol)

>y

What is the result of the coagulase method?

(p) positive
(n) negative

>p

Can you make an identification based on the experimental results?

(y) yes
(n) no

>y

What is your identification?

> my identification is Staphylococcus aureus.

Can you explain why?

>the experimental results given above imply this fact.

CONGRATULATIONS JOSE: YOUR IDENTIFICATION OF THE MICROORGANISMS IS ACCURATE, HOWEVER YOU NEED TO IMPROVE ON YOUR EXPLANATION WHY.

Can you tell if the microorganisms found is harmful?

>yes

CONGRATULATIONS JOSE: STAPHYLOCOCCUS AUREUS IS A DANGEROUS SPECIES THAT CAN CAUSE INTOXICATION

Do you want to continue with the session?

>no.

FIGURE 8 Sample session with the ITS

REFERENCES

[Bra90] I. Bratko, "*Prolog Programming for Artificial Intelligence*", Addison Wesley, 1990.

[CM94a] O. Castillo and P. Melin, "*An Intelligent System for the Identification of Microorganisms for Quality Control in the Food Industry*", Proceedings of the International Conference on Data and Knowledge Systems for Manufacturing and Engineering, Vol. 2, p.p. 618-623, Chinese University of Hong Kong, 1994.

[CM94b] O. Castillo and P. Melin, "*Developing a New Method for the Identification of Microorganisms for the Food Industry Using the Fractal Dimension*", Journal of Fractals, Vol.2, Number 3, p.p. 457-460, 1994.

[CNV87] M.A. Covington, D. Nute and A. Vellino, "*Prolog programming in Depth*", Scott Foresman and Co. Computer Books, 1987.

[DAD94] D. Dori, M. Alon and Y.J. Dori, "*Team Training Shell: a Groupware, Multimedia-Supported Application Generator*", Proceedings of ED-Media 94, P.P. 166-171, Vancouver, Canada, 1994

[DD94] Y.J. Dori and D. Dori, "Object-Process Analysis of Intelligent Computer Assisted Instruction Shell: The Polymer Courseware-a case in point", Proceedings of ED-Media 94, p.p. 172-177, Vancouver, Canada, 1994.

[Mac94] S. Mac Donald, "*A Knowledge-base Learning and Testing Systems for Medical Education*", Proceedings of ED-Media 94, p.p. 355-360, Vancouver, Canada, 1994.

[Mca94] G. McCalla, "*Artificial Intelligence and Educational Technology: A natural synergy*", Proceedings of ED-MEDIA 94, p.p. 47-49, Vancouver, Canada, 1994.

[Mel 87] P. Melin, "*Investigacion y Desarrollo de Tecnicas Microbiologicas para el Camaron*", Thesis, Instituto Tecnologico de Tijuana, 1987.

[PR87] G.L. Pettipher and M.V. Rodriguez, "*Rapid Enumeration of Microorganisms in Food by the Direct Epifluorescent Filter Technique*", Applied and Environmental Microbiology, p.p. 115-117, 1987.

A COGNITIVE LOAD APPLICATION IN EXPLANATION

Akihiro Kashihara, Tsukasa Hirashima and Jun'ichi Toyoda
I.S.I.R, Osaka University
8-1, Mihogaoka, Ibaraki, Osaka 567 JAPAN
E-mail: kasihara@ai.sanken.osaka-u.ac.jp

ABSTRACT

Explanation requires a concern about a cognitive load which a hearer will bear in understanding an explanation. Current explanation generation technology has focussed on reducing the load to facilitate the hearer's understanding. However this load reduction approach often fails to reinforce his/her appreciation and retention of the explanation. Another approach to this reinforcement is to apply a load to him/her purposefully. In this paper, we present a framework for the load application, and describe a demonstration system based on the framework. Since the load application is needed especially in educational context, the framework is provided for tutoring. The important point towards the load application is to provide a hearer with an optimal cognitive load that does not go beyond his/her capacity for understanding. Such load control requires an estimate of how much load an explanation imposes on a hearer's understanding process. This paper especially discusses an explanation planner that generates a load-oriented explanation with this estimate.

INTRODUCTION

Explanation generation can be viewed as a task of making a hearer understand an explanation produced by an explainer. In particular, it is important not only to facilitate, but also to reinforce the hearer's appreciation and retention of the explanation. This requires the explainer to take into consideration a cognitive load which the hearer will bear in understanding the explanation [KSM+94].

Previous work on explanation generation has focussed on reducing a hearer's cognitive load to facilitate his/her understanding [Caw91,MS89]. For example, Paris [Par87] has claimed the necessity to tailor explanations according to users' knowledge levels. Suthers et al. [SWC92] have also sought how to generate explanations so that users can understand them as readily as possible. However we believe that this load reduction approach is not always adaptable to reinforcing the hearer's understanding.

Another approach to this reinforcement is to apply a load to a hearer [KSM+94]. Such a load application induces the hearer to pay more attention to the understanding process because he/she finds it more challenging to try a heavier load. In addition, the hearer may reach impasses more often than not. However, supporting explanations that give a way out of the impasses enable him/her to gain a deeper understanding [KHT93,Van88]. Consequently, these effects of the load application are significant for the human-computer interaction, especially in educational context.

In this paper, we present a load application framework for tutoring. In the framework, understanding an explanation is regarded as the processes in which a hearer (a student) builds up a knowledge structure by relating the explanation to his/her already acquired knowledge. Cognitive load is represented as the amount of such knowledge-structuring processes.

This framework also applies a load to a student in the following manner. The knowledge structure, which a student will finally build up, is first set up. An explanation is second presented to him/her. When the student tries to understand this explanation, the two types of cognitive load are applied. These loads are the load of recollecting his/her related knowledge and the load of complementing the knowledge structure by combining the given explanation with his/her knowledge. We have already ascertained that the recollection load and complement load contribute to retaining the knowledge structure built up from the explanation.

This paper also demonstrates an explanation system based on the framework. Most important point towards the computational load application is to provide a student with an optimal cognitive load, which should not be too small, too big, and beyond his/her capacity for understanding. This load control requires an estimate of how much load an explanation imposes on a student's understanding process. This paper shows how this system controls the cognitive load in explanation. In particular, we discuss an explanation planner, called Load-oriented Explanation Planner (LEP).

DESIGN RATIONALE

In order to apply a load to a student, an explainer realized as a computer program requires accomplishing the following points.

First, it is indispensable to deal with the load. This requires representing the load explicitly (**Load representation**). The load representation is determined by a model of how a student understands an explanation. Second, explanations enable the explainer to apply a load. However a student may be overloaded, which may lead to losing his/her motivation to understand. Therefore it is important to estimate how much load the explanations will apply to a student (**Load estimate**). This estimate should be done according to the student's capacity for understanding because the cognitive load applied by the same explanation depends on his/her capacity (**Student modeling**). Third, the important point over the load application is to generate explanations with the load estimate (**Explanation planning**). When, for example, a student is expected to bear a trivial load, the explainer should not provide much information to apply more load. The explainer conversely needs to provide more instructional information when an overload is applied.

Fourth, a student may not bear the load as estimated in his/her understanding processes. Hence it is necessary to set up the educational situation so that the student bears the estimated load (**Situation setting**). This situation setting allows the student to devote his/her attention to overcoming the load [MS90]. Lastly, a student will reach an impasse due to the applied load more often than not. Therefore the supporting explanation is necessary for the student to find a way out of the impasse (**Supporting explanations**). VanLehn [Van88] reported that such supporting instructions promote the student's learning. The load application can be viewed as a technology to produce this impasse-driven learning.

To fulfill the above requirements, we have designed a load-oriented explanation system. We below discuss the framework of our system.

LOAD APPLICATION FRAMEWORK

Explanation Understanding

We deal with the explanation which is described in text form. In understanding an explanation, a student is expected not only to acquire new information from the explanation, but also to recognize what the acquired information means. This recognition requires the student to relate and assimilate the acquired information to his/her own knowledge. Once these processes are completed, he/she will build up a knowledge structure [CLL+93]. In this paper, this knowledge-structuring process is regarded as understanding the explanation.

Many kinds of knowledge structures can be generally constructed from an explanation. However it is difficult for computer to trace the student's knowledge-structuring processes when they are spread out in many directions. Hence we currently focus on a certain process in which a student constructs the knowledge structure to be represented as IS-A hierarchical structure. Moreover, the explanation alone cannot always promote the student's knowledge-structuring. Therefore, we set a knowledge structure as the understanding goal beforehand, and then make the student explain the process of building up the knowledge structure. This self-explanation is useful to make the student aware of the knowledge-structuring [CV89].

These restrictions on explanation understanding allow us to represent the explanation understanding and cognitive load as follows.

Knowledge structure. The goal knowledge structure is represented as IS-A hierarchical network. This network is called goal network. A node in the network indicates an object. The node is also represented by a simplified frame [Min75], in which its name, and attributes with values are described. A link shows the relationship between nodes. There are three pre-defined relationships as follows: 'is a subclass of' and its converse 'has a subclass,' and 'difference'. Each link possesses its own attributes between nodes. The attribute inheritance is not considered since the network is used to help a student structure his/her own knowledge.

Explanation. An explanation is formally represented as explanation unit sequence. Each explanation unit gives a description of what attributes and values a network component (node or link) includes. Hence the unit can be formally identified with the component.

Explanation understanding. Knowledge-structuring process corresponds to composing a goal network from an explanation unit sequence. This is divided into three processes: explanation unit extraction, related knowledge retrieval, and network complement. In the first process, the given text is divided into some explanation units. In the second process, the knowledge, which is related to the units, is retrieved from the already acquired knowledge structure. The retrieved knowledge may facilitate the explanation unit extraction. In the third process, some network components are complemented by combining the retrieved knowledge with the extracted units.

Cognitive Load. A student will bear the cognitive load in each process of composing a goal network. Currently we

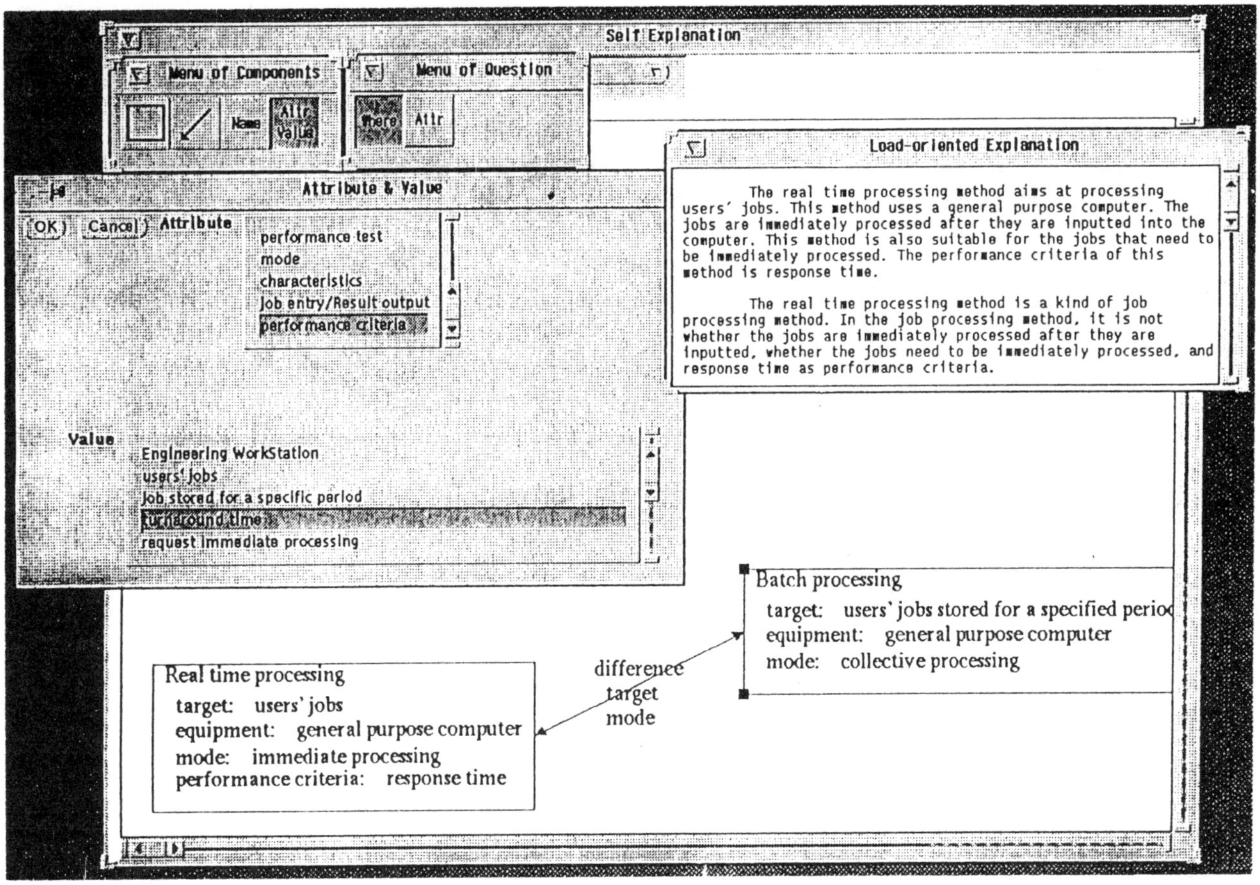

Figure 1. Interface of SEA

focus on the cognitive load which the student will bear in recollecting and complementing some network components which are not included in a given explanation. We assume that the load heaviness can be represented as the number of the components. This number **Nload** is represented as follows: **Nload = Ntotal - Nex-units**. **Ntotal** is the total number of the components in the goal network, and **Nex-units** is the number of the components described in the given explanation. **Nload**, i.e., the load heaviness is controllable by **Nex-units**.

Load Control

Our explanation system consists of LEP and SEA which stands for Self-Explanation Assistant. LEP is charged with the load estimate, explanation planning, and supporting explanation generation; SEA with the situation setting and student modeling.

First, SEA provides a goal network. Second, LEP generates a load-oriented explanation. Displaying this explanation, SEA asks a student to compose the goal network with a self-explanation interface. Figure 1 shows the interface of SEA, which has been implemented in OpenWindows on SUN SPARCstation2. In this interface, the student composes the nodes and links by mouse-selecting the necessary attributes and values from a prepared menu. However, this explanation does not give the sufficient description of the goal network. Therefore the student needs to recollect and complement the deficiencies. SEA acquires a sequence of the student's network composing operations from the interface. Third, LEP uses the acquired information to generate supporting explanations when the student reaches impasses in his/her self-explanation.

In generating a load-oriented explanation, LEP varies **Nex-units** to control **Nload**. For example, LEP reduces the explanation units to apply a heavier load. However, the problem is how to reduce **Nex-units**. This requires LEP to

System> The real time processing method aims at processing user's jobs. This method uses a general purpose computer. The jobs are immediately processed after they are inputted in the computer. This method is also suitable for the jobs that need to be immediately processed. (**Node-c1**) The real time processing method is a kind of job processing method. In the job processing method, it is not considered immediate processing and he jobs do not need to be immediately processed. (**Link-i1**) [1]

System> Explain the vocabularies of batch processing, real time processing, and job processing methods. Moreover, explain the relationships among three vocabularies. [2]

Student>.. ***Student's self-explanation in SEA***

Student> What is the difference between batch processing and real time processing methods? [3]

System> The batch processing method aims at processing the user's jobs stored for a specified period, and collectively processes the jobs. Therefore this processing method is suited for the jobs that need not be immediately processed. On the other hand, the real time processing method aims at processing user's jobs, and immediately processes the jobs after they are inputted into the computer. Therefore this processing method is suitable for the jobs that need to be immediately processed.(**Link-d**) [4]

Figure 2. Sample Dialogue

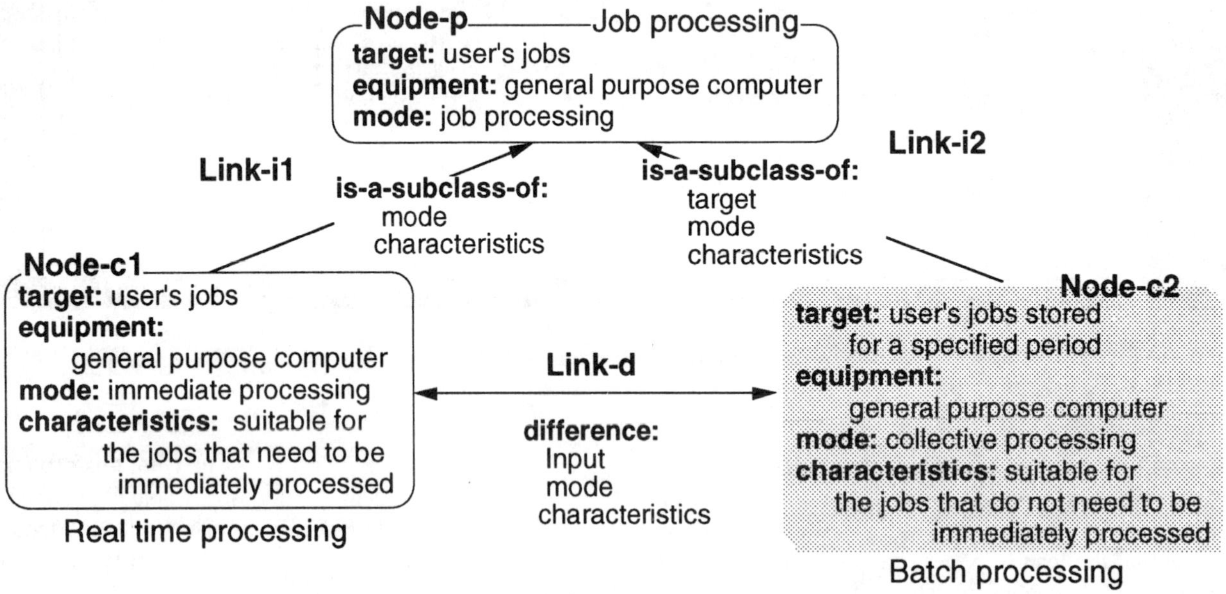

Figure 3. A Goal Network

estimate an explanation effect of how an explanation can help a student recollect and complement network components, and of how much cognitive load is applied. To provide LEP with this estimate, we have built an Explanation Effect Model, EEM for short.

Sample Dialogue

Here we provide a sample dialogue in Figure 2. In this dialogue, the goal network shown in Figure 3 is selected. This network represents a knowledge structure about computer vocabularies, which are job processing, real time processing, and batch processing. This dialogue assumes that a student acquires knowledge about batch processing indicated by Node-c2. It is also assumed that the student knows each of attributes and values in the goal network. Second, the load-oriented explanation (line [1]) is presented. This explanation has the explanation unit sequence as follows: Node-c1 and Link-i1. However, it does not give the description other than these components. Third, SEA requires him/her to compose the goal network (line [2]). In the network composing process, the student needs to recollect Node-c2. He/she also needs to complement Node-p, Link-d and Link-i2. Since he/she reaches an impasse in complementing Link-d (line [3]), LEP generates the supporting explanation that describes Link-d (line [4]).

We now discuss how LEP generates these explanations

Table 1. Explanation Effect Descriptions

Related Knowledge(RK-Scope)	Explanation Unit(DE-Scope)	IE-Scope	Effect Type
Some brother nodes	a 'is a' link	Parent node & Some other 'is a' links	Generalization & Categorizing
A node	Child node	'is a' link	Categorizing
A node & Child node	'has a' link	'is a' link	Categorizing
A node	'has a' link	Child node	Specialization
A node	Parent node	'has a' link	Instantiation
A node & Parent node	'is a' link	'has a' link	Instantiation
A node	'difference' link	Brother node	Differentiation
Some brother nodes	Parent node	'difference' link	Discrimination
A node	Brother node	'difference' link	Discrimination

'is a': 'is a subclass of' 'has a': 'has a subclass'

based on EEM in more detail. The detail of SEA is omitted in this paper [KSM+94].

EXPLANATION EFFECT MODEL

Knowledge-Structuring Scope

Though the same explanation unit is given, the knowledge-structuring scope will change according to what a student has already known. Therefore the scope is schematized as follows: "**Related_knowledge + Explanation_unit → Knowledge-structuring_scope**."

This schema expects that a network component corresponding to **Explanation_unit** is directly formed. This formation is called direct effect of the unit. It is also expected that some components are complemented by means of the unit and **Related_knowledge**. This complement is called indirect effect. The scope of the direct effect (DE) is called DE-Scope; the scope of the indirect effect (IE), IE-Scope. IE-Scope can be viewed as an extension of DE-Scope. **Knowledge-structuring_scope** is composed of DE-Scope, IE-Scope, and the scope of Related_knowledge (RK), called RK-Scope.

The width of IE-Scope to be expected influences the heaviness of the load that a student will bear. EEM currently limits IE-Scope to the adjacency of DE-Scope and RK-Scope. Following this limitation, we have provided several concrete descriptions of the knowledge-structuring schema as shown in Table 1. Let us pick up one example from this table. The bottom of the table describes the following effect. When a student knows a node, the explanation of a brother node enables him/her to form the brother node and complement their 'difference' link.

Indirect effect is divided into three types according to the effect extension from DE-Scope to IE-Scope. These are called bottom-up effect, top-down effect and level-sideways effect. Each effect is moreover divided into two types according to node complement and link complement in IE-Scope. The bottom-up effect is divided into generalization effect and categorizing effect. The generalization effect means that the closest upper node of DE-Scope is complemented. The categorizing effect means that a 'is a subclass of' link between DE-Scope and the upper node is complemented. Similarly, the top-down effect is divided into specialization one and instantiation one; the level-sideways effect, differentiation one and discrimination one.

Figure 4 shows the DE-Scope and IE-Scope of the load-oriented explanation in Figure 2. It is estimated that the explanation unit describing Node-c1 will help the student to form Node-c1 (in DE-Scope[Node-c1]) and to complement Link-d (in IE-Scope[Node-c1]). In this knowledge-structuring, the RK-Scope includes only Node-c2. This effect is an instance of the description in the bottom of Table 1. It is also estimated that the explanation effect of the unit describing Link-i1 appears as DE-Scope[Link-i1] and IE-Scope[Link-i1]. In this estimate, the RK-Scope includes Node-c2 and Node-c1. This effect is an instance of the first of Table 1, in which a generalization effect appears as

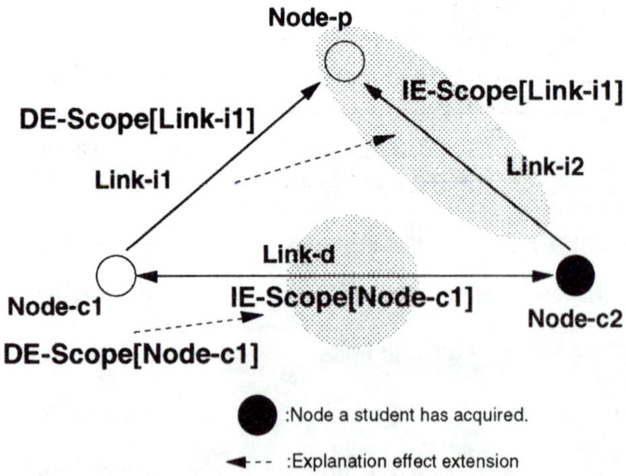

Figure 4. DE-Scope and IE-Scope.

Node-p and a categorizing effect does as Link-i2.

Cognitive Load Heaviness

EEM also has the following assumptions about cognitive load heaviness. First, DE-Scope can be formed with no load. Second, the load of forming RK-Scope and IE-Scope is in proportion to the number of the network components to be recollected and complemented. Forming one component also requires its own load in proportion to the attributes in the component. In addition, the recollection load and complement load follow the assumptions as below [KSM+94].

Recollection load. The load of recollecting a network component depends on the capability for a student to bring back his/her memory. In other words, the higher the capability is, the lighter the load is.

Complement load. The load of complementing a network component depends on the component type and the direction in which the component is complemented from DE-Scope. First, complementing a node requires generalizing (specializing or differentiating) attributes (or values) in a related node, although link complement requires comparing the attributes in a node with the ones in the other node. The generalization (specialization or differentiation) is considered harder than the comparison. Therefore the node complement requires a heavier load than the link complement. Second, the bottom-up complement is harder than the top-down complement or level-sideways complement.

EXPERIMENT

We have experimented with the load application based on EEM. The main purpose is to ascertain if the self-explanation with a load is more effective for retaining the knowledge structure to be constructed from a given explanation than that with no load.

In this experiment, a goal network was composed of two nodes (called Node-1 and Node-2) and a link (called D-Link) between these nodes. Each node included 6 attributes. The link included 2 different attributes between the nodes. Subjects were 32 graduate and undergraduate students. The load was applied to each subject when he/she tried to explain (write) D-Link with his/her knowledge about Node-1 and the explanation of Node-2. This application follows the explanation effect in the bottom of Table 1. We set four conditions on his/her self-explanation of D-Link. These conditions were no load application, the recollection load application, complement load application and both load applications. In order to make each subject do the self-explanation on each condition (within-subject design), we prepared four similar goal networks. The subject was required to explain D-Link of each goal network as follows.

The subject was first required to read the explanation of Node-1 carefully. The aim of this careful reading is to make him/her acquire the knowledge about Node-1. He/she was next required to explain D-Link with looking at the explanation of Node-2. At that time, we provided (did not provide) the subject with the load of recollecting Node-1 by covering (uncovering) the explanation of Node-1. In addition, we provided (did not provide) the subject with the load of complementing D-Link by covering (uncovering) the explanation of D-Link. These treatments enable him/her to explain D-Link on each condition. When the subject reached impasses or made mistakes, he/she was provided correct attributes with values.

After two hours following the subjects' self-explanations, a retention test was conducted. In this test, each subject was requested to reconstruct each goal network [BA93]. The perfect score was 16. Table 2 shows the average score and standard deviation of the self-explanation on each condition. The average scores of the self-explanation with recollection load, complement load, or both loads were higher than the score of the self-explanation with no

Table 2. Scores in Self-Explanation

	No Load	Recollection Load	Complement Load	Both Loads
	Cond-A	Cond-B	Cond-C	Cond-D
N	32	32	32	32
Mean	5.53	6.84	6.00	7.75
SD	3.06	3.30	3.24	3.30

Table 3. Summary of Analysis of Variance on Retention Test

Source	Sum of Squares	df	Means Square	F-ratio		
S	527.88	31.00	17.03			
Sa (Recollection Load)	75.03	1.00	75.03	14.01	0.001	**
S × Sa	165.97	31.00	5.35			
Sb (Complement Load)	15.13	1.00	15.13	1.24	0.274	
S × Sb	378.88	31.00	12.22			
Sa × Sb	1.53	1.00	1.53	0.18	0.672	
S × Sa × Sb	259.47	31.00	8.37			
	1423.88	127.00	11.21	**p<0.01		

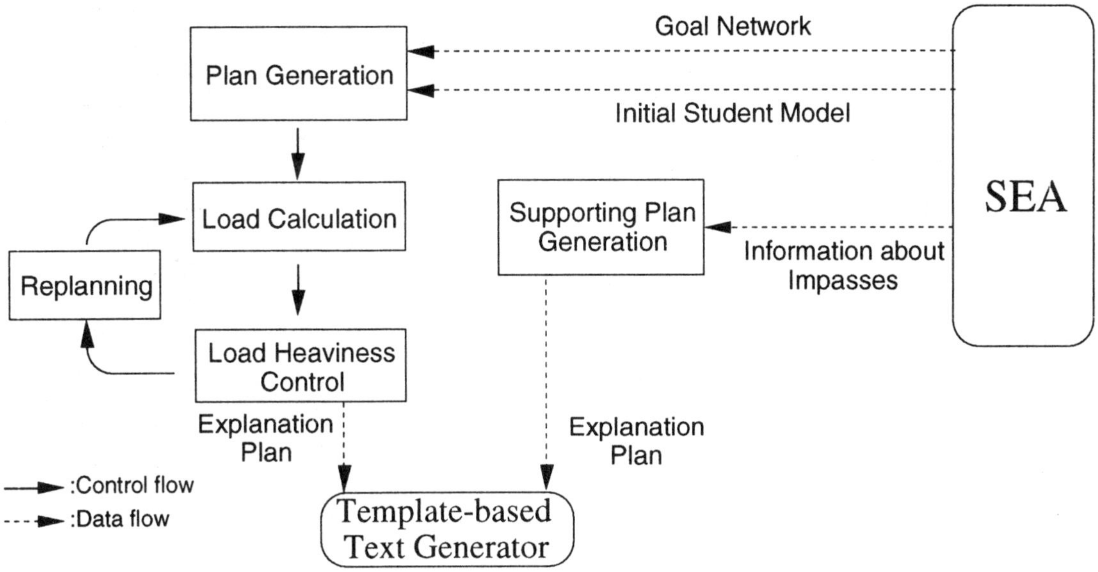

Figure 5. Planning Process in LEP

load. In addition, we performed an analysis of variance on the retention test. Table 3 contains a summary of the analysis. The results of this analysis indicate that there is a significant difference between the average scores of the self-explanation with recollection load (Cond-B and Cond-D) and the others (Cond-A and Cond-C) at the .01 level of significance. From this result, it was ascertained that the load application based on the bottom of Table 1 reinforced the retention of the goal network. Especially in this experiment, the recollection load greatly contributed to the reinforcement of the retention. It is also expected that the load applications based on the others of Table 1 are effective.

We also had another experiment with the other students. After carefully reading the explanations of Node-1 and Node-2, they were required to explain D-Link without looking at the explanations of Node-1, Node-2, and D-Link. In this case, their scores tended to be lower than the scores of the previous self-explanation with both loads. It seems that they bore more load because they needed to recollect Node-1 and besides Node-2. This finding implies that a heavy load has a bad influence on the retention of the goal network. Therefore the load application requires controlling the load.

LOAD-ORIENTED EXPLANATION PLANNER

Figure 5 shows the explanation planning process in LEP. Before planning, SEA makes a student compose a goal network with only all node names included in the network. (The sample dialogue in Figure 2 omits the student modeling.) SEA uses overlay modeling technique [Wen87] to identify which of the components the student has acquired (self-explained), and generates an initial student model. The student model is represented as the network overlaid with the goal network.

Plan Generation

Expecting the knowledge-structuring scope to be formed by each explanation unit, LEP generates an explanation unit sequence. This process is accompanied by the simulation of knowledge-structuring from the initial student model to the goal network. The explanation effect provided by EEM is described in the planning operators. The operators are divided into two types: basic operators and explanation strategies.

The basic operators reflect the schema of the explanation effect. Each operator contains a **Goal**, which represents an explanation effect type provided by the operator; a **RK_Scope** list, which represents a context that the operator is applicable; **ExUnit**, which is an explanation unit placed by the operator; **DE_Scope** and **IE_Scope** lists, which denote the effects of the **ExUnit**. Figure 6 shows a basic operator that provides the discrimination effect in the bottom of Table 1.

The explanation strategies determine the order of searching for basic operators to extend explanation effect. The effect extension starts from the node, called initial node, which a student has acquired. To generate the load-oriented explanation efficiently, LEP first generates a plan

```
(Basic-op9
  (Goal: discrimination)
   (RK_Scope: (node Node))
   (ExUnit: (node (brother Node)))
   (DE_Scope: (brother Node))
   (IE_Scope: (diff Node  (brother Node)))  )
```

Figure 6. A Basic Operator

that applies as heavy load as possible, and subsequently decreases the load to a threshold. In this planning manner, LEP follows the two search preferences: node complement preference and direction preference. These preferences are supported by the assumptions about load heaviness in EEM. First, the basic operators that provide node complement as indirect effect have preference to the ones that provide link complement. Second, the directions of the explanation effect extension are selected in priority order here: bottom-up, top-down, and level-sideways.

Let us account for the planning process with an example shown in Figure 7. This example shows the load-oriented explanation planning for the sample dialogue in Figure 2. First, following node complement preference, LEP decides the direction of explanation effect extension. If the extension is not possible in any directions, link complement is considered. In this example, the bottom-up extension from the initial node(Node-c2) is not possible because the initial student model does not satisfy the **RK_Scope** of the basic operators for bottom-up effect. (The top-down extension is not considered.) Therefore LEP chooses the level-sideways extension from Node-c2 to Node-c1, and searches the basic operator for differentiation effect. As a result, the explanation unit indicated by ExUnit-1 is placed in Plan-A. The plan also includes the DE-Scope and IE-Scope of ExUnit-

```
Plan-A:  ( ((ExUnit  (difference RealTimeProcessing BatchProcessing))        ..... ExUnit-1
            (DE_Scope  (diff RealTimeProcessing BatchProcessing))
            (IE_Scope   RealTimeProcessing) )
           ((ExUnit (is-a RealTimeProcessing JobProcessing))     ..... ExUnit-2
            (DE_Scope  (is-a RealTimeProcessing JobProcessing))
            (IE-Scope   (JobProcessing  (is-a BatchProcessing JobProcessing))) )  )

Plan-B:  ( ((ExUnit RealTimeProcessing)                                       .... ExUnit-1'
            (DE_Scope RealTimeProcessing)
            (IE_Scope  (is-a RealTimeProcessing BatchProcessing)) )
           ((ExUnit (is-a RealTimeProcessing JobProcessing))      ..... ExUnit-2
            (DE_Scope  (is-a RealTimeProcessing JobProcessing))
            (IE_Scope  (JobProcessing  (is-a BatchProcessing JobProcessing))) )  )
```

Figure 7. An Example of Load-oriented Explanation Planning

1. Second, it is assumed that the above differentiation effect is obtained, i.e., the student can acquire Link-d and Node-c1. LEP recursively searches the next basic operator on this assumption. Regarding Node-c1 as the initial node, LEP chooses the operator for the bottom-up extension to Node-p in this example. This operator places ExUnit-2 in Plan-A.

Load Calculation

Next, LEP calculates the load heaviness applied by the generated plan. The heaviness applied by an explanation unit is defined as follows.

$$H(unit) = \frac{\{(Nin + w \times Nil) + (1 - r) \times Nr\}}{N}$$
$$where \quad 0 \le w \le 1, \quad 0 \le r \le 1, \quad Ni = Nin + Nil.$$

H(unit) represents the heaviness. The variable ***N*** represents the total number of attributes in the network components to be formed by the unit. The variables ***Nd***, ***Ni***, and ***Nr*** respectively represent the number of attributes in the DE-Scope, the one in the IE-Scope, and the one in the RK-Scope. These variables are related through $N = Nr + Nd + Ni$. The variables ***Nin*** and ***Nil*** also represent the number of attributes in the nodes and the one in the links.

The above definition follows the EEM's assumptions about load heaviness. The variable ***w*** represents the proportion of link complement load to node complement load. The variable ***r*** represents recollection capability.

LEP defines the load heaviness by the generated plan as follows: ***H(plan)*** = max $\{H(unit_i)\}$. $\{H(unit_i)\}$ is a set of which each element is $H(unit_i)$ by a $unit_i$ in the plan.

Load Heaviness Control & Replanning

In the process of the load heaviness control, ***H(plan)*** is controlled by a threshold, referred to as ***Hth***. This threshold represents a student's capacity for knowledge-structuring. If the heaviness exceeds the threshold, LEP modifies the plan to reduce the load heaviness (**Replanning**); if not, the plan is translated into natural language by the template-based text generator. The threshold enables LEP to generate explanations that apply an appropriate load.

To reduce ***H(plan)***, LEP replaces the basic operator that provides maximum heaviness among $\{H(unit_i)\}$ with a new operator. The replacement follows the node complement preference. If the replaced operator provides a node complement as indirect effect, LEP replaces it with the operator to provide a link complement. If the replaced operator provides a link complement, LEP explains the link directly.

Student Modeling

The variables ***w***, ***r***, and ***Hth*** used in the explanation generation can be viewed as a student's capacity for knowledge-structuring. LEP needs to set these variables properly according to his/her capacity. However, it is difficult to estimate these variables precisely. LEP updates these variables with the information about the student's network composing operations which SEA acquires. The updated variables are used for the next load-oriented explanation planning. Through several self-explanations by a student, LEP refines these variables to individualize the load estimate and load heaviness control. This refinement is as follows.

If a student can compose a goal network with fewer impasses, LEP increases the variable ***Hth***. It means that he/she is expected to sustain a heavier load. If he/she reaches more impasses in complementing links in IE-Scope, LEP increases the variable ***w***. If the student can recollect the network components in RK-Scope with fewer impasses, LEP increases the variable ***r***.

Supporting Explanation Generation

In the supporting explanation planning, LEP generates only the unit corresponding to the network component, in which the student reached an impasse, without the load calculation and load heaviness control.

Example

We return to the example of explanation planning shown in Figure 7. In this example, we assume the variables, $w = 0.5$ and $r = 0.8$. The load heaviness by Plan-A is as follows:

$$H(ExUnit-1) = \frac{\{(4 + 0.5 \times 0) + (1 - 0.8) \times 4\}}{11} = 0.44$$
$$where \quad Nin = N_{Node-c1} = 4, Nil = 0,$$
$$Nr = N_{Node-c2} = 4, and$$
$$N = N_{Node-c2} + N_{Link-d} + N_{Node-c1} = 11$$

$$H(ExUnit-2) = \frac{\{(3 + 0.5 \times 3) + 0.2 \times 8\}}{16} = 0.38$$
$$where \quad Nin = N_{Node-p} = 3, Nil = N_{Link-i2} = 3,$$
$$Nr = N_{Node-c1} + N_{Node-c2} = 8, and$$
$$N = N_{Node-c1} + N_{Node-c2} + N_{Link-i1} + N_{Node-p} + N_{Link-i2} = 16.$$

($N_{component}$ represents the number of the attributes in the component.)

Therefore ***H(Plan-A)*** = max {0.44, 0.38}=0.44. The number 0.44 means that the student needs to recollect and complement about 40 percent of the goal network. If ***Hth***=0.4, LEP needs to modify Plan-A. In this case, the first operator is replaced. Since this operator provides the complement of Node-c1, LEP replaces it with the new operator shown in Figure 6. As a result, ExUnit-1 is replaced with ExUnit-1' in Plan-B shown in Figure 7. The load heaviness by ExUnit-1' is as follows:

$$H(ExUnit-1') = \frac{\{(0+0.5\times 3)+0.2\times 4\}}{11} = 0.21$$

$$\textit{where} \quad Nin = 0, Nil = N_{Link-d} = 3,$$

$$Nr = N_{Node-c2} = 4, \textit{and}$$

$$N = N_{Node-c2} + N_{Node-c1} + N_{Link-d} = 11.$$

Therefore the heaviness is as follows: ***H(Plan-B)*** < 0.4. The load-oriented explanation in Figure 2 is generated from the Plan-B.

CONCLUSION

Towards the cognitive load application in explanation, we have described EEM and LEP. LEP generates the load-oriented explanation by estimating how much cognitive load an explanation applies to a student. EEM provides LEP with the foundation of the load estimate. EEM and LEP are designed to be applicable to any tutoring domain, of which the knowledge structure constructed by an explanation is represented as the IS-A hierarchical network.

We have evaluated such a load-oriented explanation through the experiment of whether subjects can retain a given explanation. As a result of this experiment, the load application has proved to contribute to reinforcing their understanding.

The current system is limited in various ways. For example, EEM is not always a cognitive model adequate to represent human knowledge-structuring process. In addition, our system has not considered the student's misunderstanding. Finally, we will need to refine EEM and to evaluate the cognitive load control in LEP by testing our system.

ACKNOWLEDGMENTS

The authors would like to thank Koichi Matsumura and Akihiro Sugano for their valuable comments. This research is supported in part by Grant-in-Aid for Scientific Research No. 06780164 from the Ministry of Education, Science and Culture of Japan, and in part by The Mazda Foundation's Research Grant.

REFERENCES

[BA93] A.E.Barron, and D.Atkins.Audio Instruction in Multimedia Education: Is Textual Redundancy Important? In *Proceedings of ED-MEDIA '93*, pp.39-46, Orlando, June 1993.

[Caw91] A.Cawsey.Generating Interactive Explanations. In *Proceedings of the Ninth National Conference on Artificial Intelligence*, pp.86-91, AAAI Press, Anaheim, July 1991.

[CLL+93]T.W. Chan, C.C.Lin, S.J.Lin, and H.C.Kuo. *OCTR: A Model of Learning Stages*. In *Proceedings of Artificial Intelligence in Education '93*, pp.257-264, AACE, Edinburgh, August 1993.

[CV89] M.T.H.Chi and K.VanLehn. Self-explanations: How students study use examples in learning to solve problems. Journal of Cognitive Science, 13: pp.145-182, 1989.

[KHT93] A.Kashihara, T.Hirashima, and J.Toyoda. Impasse-oriented Explanation Planning Based on Explanation Effect Model. In *Abridged Proceedings of HCI International '93*, pp.150, Orlando, August 1993.

[KSM+94] A.Kashihara, A.Sugano, K.Matsumura, T.Hirashima, and J.Toyoda. A Cognitive Load Application Approach to Tutoring. In *Proceedings of Fourth International Conference on User Modeling*, Cape Code, pp. 163-168, Cape Code, August 1994.

[Min75] M.Minsky. A framework for representing knowledge. In P.Winston, editor, *The Psychology of Computer Vision*, McGraw-Hill, New York, 1975.

[MS89] J.D.Moore, and W.R.Swartout. Reactive Approach to Explanation. In *Proceedings of 11th Int. Joint Conference on Artificial Intelligence*, pp.1504-1510, Detroit, 1989.

[MS90] J.D.Moore, and W.R.Swartout. Pointing: A Way Toward Explanation Dialogue. In *Proceedings of the Eighth National Conference on Artificial Intelligence*, AAAI Press, pp.457-464, Boston, July - August 1990.

[Par87] C.L.Paris. Combining Discourse Strategies to Generate Descriptions to Users Along a Naive/Expert Spectrum, In *Proceedings of 10th Int. Joint Conference on Artificial Intelligence*, Milan, pp.626-632, August 1987.

[SWC92] D.Suthers, B.Woolf, and M.Cornell. Steps from Explanation Planning to Model Construction Dialogue. In *Proceedings of the Tenth National Conference on Artificial Intelligence*, pp.24-30, AAAI Press, San Jose, Calif., 1992.

[Van88] K.VanLehn. Toward a Theory of Impasse-Driven Learning. In H.Mandl, and H.Lesgold, editors. *Learning Issues for Intelligent Tutoring Systems*, Springer-Verlag, New York, 1988.

[Wen87] E.Wenger. *Artificial Intelligence and Tutoring systems*. Morgan Kaufmann Publishers, Los Altos, Calif, 1987.

INDEX OF AUTHORS